Grundlehren der mathematischen Wissenschaften 338

A Series of Comprehensive Studies in Mathematics

Cédric Villani

Optimal Transport

Old and New

Springer

Cédric Villani
Unité de Mathématiques Pures et Appliquées (UMPA)
École Normale Supérieure de Lyon
46, allée d'Italie
69364 Lyon CX 07
France
cvillani@umpa.ens-lyon.fr

ISSN 0072-7830
ISBN 978-3-540-71049-3 e-ISBN 978-3-540-71050-9

Library of Congress Control Number: 2008932183

Mathematics Subject Classification Numbers (2000): 49-xx, 53-xx, 60-xx

Cover design: WMX Design GmbH, Heidelberg

Printed on acid-free paper

9 8 7 6 5 4 3 2 1

springer.com

Do mo chuisle mo chroí, Aëlle

Preface

When I was first approached for the 2005 edition of the Saint-Flour Probability Summer School, I was intrigued, flattered and scared.[1] Apart from the challenge posed by the teaching of a rather analytical subject to a probabilistic audience, there was the danger of producing a remake of my recent book *Topics in Optimal Transportation.*

However, I gradually realized that I was being offered a unique opportunity to rewrite the whole theory from a different perspective, with alternative proofs and a different focus, and a more probabilistic presentation; plus the incorporation of recent progress. Among the most striking of these recent advances, there was the rising awareness that John Mather's minimal measures had a lot to do with optimal transport, and that both theories could actually be embedded within a single framework. There was also the discovery that optimal transport could provide a robust synthetic approach to Ricci curvature bounds. These links with dynamical systems on one hand, differential geometry on the other hand, were only briefly alluded to in my first book; here on the contrary they will be at the basis of the presentation. To summarize: more probability, more geometry, and more dynamical systems. Of course there cannot be more of everything, so in some sense there is less analysis and less physics, and also there are fewer digressions.

So the present course is by no means a reduction or an expansion of my previous book, but should be regarded as a complementary reading. Both sources can be read independently, or together, and hopefully the complementarity of points of view will have pedagogical value.

[1] Fans of Tom Waits may have identified this quotation.

Throughout the book I have tried to optimize the results and the presentation, to provide complete and self-contained proofs of the most important results, and comprehensive bibliographical notes — a dauntingly difficult task in view of the rapid expansion of the literature. Many statements and theorems have been written specifically for this course, and many results appear in rather sharp form for the first time. I also added several appendices, either to present some domains of mathematics to non-experts, or to provide proofs of important auxiliary results. All this has resulted in a rapid growth of the document, which in the end is about six times (!) the size that I had planned initially. So **the non-expert reader is advised to skip long proofs at first reading**, and concentrate on explanations, statements, examples and sketches of proofs when they are available.

About terminology: For some reason I decided to switch from "transportation" to "transport", but this really is a matter of taste.

For people who are already familiar with the theory of optimal transport, here are some more serious changes.

Part I is devoted to a qualitative description of optimal transport. The dynamical point of view is given a prominent role from the beginning, with Robert McCann's concept of displacement interpolation. This notion is discussed before any theorem about the solvability of the Monge problem, in an abstract setting of "Lagrangian action" which generalizes the notion of length space. This provides a unified picture of recent developments dealing with various classes of cost functions, in a smooth or nonsmooth context.

I also wrote down in detail some important estimates by John Mather, well-known in certain circles, and made extensive use of them, in particular to prove the Lipschitz regularity of "intermediate" transport maps (starting from some intermediate time, rather than from initial time). Then the absolute continuity of displacement interpolants comes for free, and this gives a more unified picture of the Mather and Monge–Kantorovich theories. I rewrote in this way the classical theorems of solvability of the Monge problem for quadratic cost in Euclidean space. Finally, this approach allows one to treat change of variables formulas associated with optimal transport by means of changes of variables that are Lipschitz, and not just with bounded variation.

Part II discusses optimal transport in Riemannian geometry, a line of research which started around 2000; I have rewritten all these applications in terms of Ricci curvature, or more precisely curvature-dimension bounds. This part opens with an introduction to Ricci curvature, hopefully readable without any prior knowledge of this notion.

Part III presents a synthetic treatment of Ricci curvature bounds in metric-measure spaces. It starts with a presentation of the theory of Gromov–Hausdorff convergence; all the rest is based on recent research papers mainly due to John Lott, Karl-Theodor Sturm and myself.

In all three parts, noncompact situations will be systematically treated, either by limiting processes, or by restriction arguments (the restriction of an optimal transport is still optimal; this is a simple but powerful principle). The notion of approximate differentiability, introduced in the field by Luigi Ambrosio, appears to be particularly handy in the study of optimal transport in noncompact Riemannian manifolds.

Several parts of the subject are not developed as much as they would deserve. Numerical simulation is not addressed at all, except for a few comments in the concluding part. The regularity theory of optimal transport is described in Chapter 12 (including the remarkable recent works of Xu-Jia Wang, Neil Trudinger and Grégoire Loeper), but without the core proofs and latest developments; this is not only because of the technicality of the subject, but also because smoothness is not needed in the rest of the book. Still another poorly developed subject is the Monge–Mather–Mañé problem arising in dynamical systems, and including as a variant the optimal transport problem when the cost function is a distance. This topic is discussed in several treatises, such as Albert Fathi's monograph, *Weak KAM theorem in Lagrangian dynamics*; but now it would be desirable to rewrite everything in a framework that also encompasses the optimal transport problem. An important step in this direction was recently performed by Patrick Bernard and Boris Buffoni. In Chapter 8 I shall provide an introduction to Mather's theory, but there would be much more to say.

The treatment of Chapter 22 (concentration of measure) is strongly influenced by Michel Ledoux's book, *The Concentration of Measure Phenomenon*; while the results of Chapters 23 to 25 owe a lot to the monograph by Luigi Ambrosio, Nicola Gigli and Giuseppe Savaré, *Gradient flows in metric spaces and in the space of probability measures*. Both references are warmly recommended complementary reading. One can also consult the two-volume treatise by Svetlozar Rachev and Ludger Rüschendorf, *Mass Transportation Problems*, for many applications of optimal transport to various fields of probability theory.

While writing this text I asked for help from a number of friends and collaborators. Among them, Luigi Ambrosio and John Lott are the ones whom I requested most to contribute; this book owes a lot to their detailed comments and suggestions. Most of Part III, but also significant portions of Parts I and II, are made up with ideas taken from my collaborations with John, which started in 2004 as I was enjoying the hospitality of the Miller Institute in Berkeley. Frequent discussions with Patrick Bernard and Albert Fathi allowed me to get the links between optimal transport and John Mather's theory, which were a key to the presentation in Part I; John himself gave precious hints about the history of the subject. Neil Trudinger and Xu-Jia Wang spent vast amounts of time teaching me the regularity theory of Monge–Ampère equations. Alessio Figalli took up the dreadful challenge to check the entire set of notes from first to last page. Apart from these people, I got valuable help from Stefano Bianchini, François Bolley, Yann Brenier, Xavier Cabré, Vincent Calvez, José Antonio Carrillo, Dario Cordero-Erausquin, Denis Feyel, Sylvain Gallot, Wilfrid Gangbo, Diogo Aguiar Gomes, Nathaël Gozlan, Arnaud Guillin, Nicolas Juillet, Kazuhiro Kuwae, Michel Ledoux, Grégoire Loeper, Francesco Maggi, Robert McCann, Shin-ichi Ohta, Vladimir Oliker, Yann Ollivier, Felix Otto, Ludger Rüschendorf, Giuseppe Savaré, Walter Schachermayer, Benedikt Schulte, Theo Sturm, Josef Teichmann, Anthon Thalmaier, Hermann Thorisson, Süleyman Üstünel, Anatoly Vershik, and others.

Short versions of this course were tried on mixed audiences in the Universities of Bonn, Dortmund, Grenoble and Orléans, as well as the Borel seminar in Leysin and the IHES in Bures-sur-Yvette. Part of the writing was done during stays at the marvelous MFO Institute in Oberwolfach, the CIRM in Luminy, and the Australian National University in Canberra. All these institutions are warmly thanked.

It is a pleasure to thank Jean Picard for all his organization work on the 2005 Saint-Flour summer school; and the participants for their questions, comments and bug-tracking, in particular Sylvain Arlot (great bug-tracker!), Fabrice Baudoin, Jérôme Demange, Steve Evans (whom I also thank for his beautiful lectures), Christophe Leuridan, Jan Obłój, Erwan Saint Loubert Bié, and others. I extend these thanks to the joyful group of young PhD students and maîtres de conférences with whom I spent such a good time on excursions, restaurants, quantum ping-pong and other activities, making my stay in Saint-Flour truly wonderful (with special thanks to my personal driver, Stéphane Loisel, and my table tennis sparring-partner and adversary, François Simenhaus). I will cherish my visit there in memory as long as I live!

Typing these notes was mostly performed on my (now defunct) faithful laptop Torsten, a gift of the Miller Institute. Support by the Agence Nationale de la Recherche and Institut Universitaire de France is acknowledged. My eternal gratitude goes to those who made fine typesetting accessible to every mathematician, most notably Donald Knuth for TEX, and the developers of LATEX, BIBTEX and `XFig`. Final thanks to Catriona Byrne and her team for a great editing process.

As usual, I encourage all readers to report mistakes and misprints. **I will maintain a list of errata, accessible from my Web page.**

Lyon, June 2008 *Cédric Villani*

Contents

Conventions

Axioms

I use the classical axioms of set theory; not the full version of the axiom of choice (only the classical axiom of "countable dependent choice").

Sets and structures

Id is the identity mapping, whatever the space. If A is a set then the function 1_A is the indicator function of A: $1_A(x) = 1$ if $x \in A$, and 0 otherwise. If F is a formula, then 1_F is the indicator function of the set defined by the formula F.

If f and g are two functions, then (f, g) is the function $x \longmapsto (f(x), g(x))$. The composition $f \circ g$ will often be denoted by $f(g)$.

$\mathbb{N}$ is the set of *positive* integers: $\mathbb{N} = \{1, 2, 3, \ldots\}$. A sequence is written $(x_k)_{k \in \mathbb{N}}$, or simply, when no confusion seems possible, (x_k).

$\mathbb{R}$ is the set of real numbers. When I write $\mathbb{R}^n$ it is implicitly assumed that n is a positive integer. The Euclidean scalar product between two vectors a and b in $\mathbb{R}^n$ is denoted interchangeably by $a \cdot b$ or $\langle a, b \rangle$. The Euclidean norm will be denoted simply by $|\cdot|$, independently of the dimension n.

$M_n(\mathbb{R})$ is the space of real $n \times n$ matrices, and I_n the $n \times n$ identity matrix. The trace of a matrix M will be denoted by $\operatorname{tr} M$, its determinant by $\det M$, its adjoint by M^*, and its Hilbert–Schmidt norm $\sqrt{\operatorname{tr}(M^*M)}$ by $\|M\|_{\mathrm{HS}}$ (or just $\|M\|$).

Unless otherwise stated, Riemannian manifolds appearing in the text are finite-dimensional, smooth and complete. If a Riemannian manifold M is given, I shall usually denote by n its dimension, by d the geodesic distance on M, and by vol the volume (= n-dimensional

Hausdorff) measure on M. The tangent space at x will be denoted by T_xM, and the tangent bundle by TM. The norm on T_xM will most of the time be denoted by $|\cdot|$, as in $\mathbb{R}^n$, without explicit mention of the point x. (The symbol $\|\cdot\|$ will be reserved for special norms or functional norms.) If S is a set without smooth structure, the notation T_xS will instead denote the tangent cone to S at x (Definition 10.46).

If Q is a quadratic form defined on $\mathbb{R}^n$, or on the tangent bundle of a manifold, its value on a (tangent) vector v will be denoted by $\langle Q\cdot v, v\rangle$, or simply $Q(v)$.

The open ball of radius r and center x in a metric space $\mathcal{X}$ is denoted interchangeably by $B(x,r)$ or $B_r(x)$. If $\mathcal{X}$ is a Riemannian manifold, the distance is of course the geodesic distance. The closed ball will be denoted interchangeably by $B[x,r]$ or $B_{r]}(x)$. The diameter of a metric space $\mathcal{X}$ will be denoted by $\operatorname{diam}(\mathcal{X})$.

The closure of a set A in a metric space will be denoted by $\overline{A}$ (this is also the set of all limits of sequences with values in A).

A metric space $\mathcal{X}$ is said to be *locally compact* if every point $x\in\mathcal{X}$ admits a compact neighborhood; and *boundedly compact* if every closed and bounded subset of $\mathcal{X}$ is compact.

A map f between metric spaces $(\mathcal{X},d)$ and $(\mathcal{X}',d')$ is said to be C-Lipschitz if $d'(f(x),f(y)) \le C\,d(x,y)$ for all x, y in $\mathcal{X}$. The best admissible constant C is then denoted by $\|f\|_{\mathrm{Lip}}$.

A map is said to be locally Lipschitz if it is Lipschitz on bounded sets, *not necessarily compact* (so it makes sense to speak of a locally Lipschitz map defined almost everywhere).

A curve in a space $\mathcal{X}$ is a continuous map defined on an interval of $\mathbb{R}$, valued in $\mathcal{X}$. For me the words "curve" and "path" are synonymous. The time-t evaluation map e_t is defined by $e_t(\gamma)=\gamma_t=\gamma(t)$.

If γ is a curve defined from an interval of $\mathbb{R}$ into a metric space, its length will be denoted by $\mathcal{L}(\gamma)$, and its speed by $|\dot\gamma|$; definitions are recalled on p. 119.

Usually geodesics will be *minimizing, constant-speed* geodesic curves. If $\mathcal{X}$ is a metric space, $\Gamma(\mathcal{X})$ stands for the space of all geodesics $\gamma:[0,1]\to\mathcal{X}$.

Being given x_0 and x_1 in a metric space, I denote by $[x_0,x_1]_t$ the set of all t-barycenters of x_0 and x_1, as defined on p. 393. If A_0 and A_1 are two sets, then $[A_0,A_1]_t$ stands for the set of all $[x_0,x_1]_t$ with $(x_0,x_1)\in A_0\times A_1$.

Function spaces

$C(\mathcal{X})$ is the space of continuous functions $\mathcal{X} \to \mathbb{R}$, $C_b(\mathcal{X})$ the space of bounded continuous functions $\mathcal{X} \to \mathbb{R}$; and $C_0(\mathcal{X})$ the space of continuous functions $\mathcal{X} \to \mathbb{R}$ converging to 0 at infinity; all of them are equipped with the norm of uniform convergence $\|\varphi\|_\infty = \sup|\varphi|$. Then $C_b^k(\mathcal{X})$ is the space of k-times continuously differentiable functions $u : \mathcal{X} \to \mathbb{R}$, such that all the partial derivatives of u up to order k are bounded; it is equipped with the norm given by the supremum of all norms $\|\partial u\|_{C_b}$, where ∂u is a partial derivative of order at most k; $C_c^k(\mathcal{X})$ is the space of k-times continuously differentiable functions with compact support; etc. When the target space is not $\mathbb{R}$ but some other space $\mathcal{Y}$, the notation is transformed in an obvious way: $C(\mathcal{X};\mathcal{Y})$, etc.

L^p is the Lebesgue space of exponent p; the space and the measure will often be implicit, but clear from the context.

Calculus

The derivative of a function $u = u(t)$, defined on an interval of $\mathbb{R}$ and valued in $\mathbb{R}^n$ or in a smooth manifold, will be denoted by u', or more often by $\dot{u}$. The notation d^+u/dt stands for the upper right-derivative of a real-valued function u: $d^+u/dt = \limsup_{s\downarrow 0}[u(t+s) - u(t)]/s$.

If u is a function of several variables, the partial derivative with respect to the variable t will be denoted by $\partial_t u$, or $\partial u/\partial t$. *The notation u_t does not stand for $\partial_t u$, but for $u(t)$.*

The gradient operator will be denoted by grad or simply ∇; the divergence operator by div or $\nabla\cdot$; the Laplace operator by Δ; the Hessian operator by Hess or ∇^2 (so ∇^2 *does not* stand for the Laplace operator). The notation is the same in $\mathbb{R}^n$ or in a Riemannian manifold. Δ is the divergence of the gradient, so it is typically a nonpositive operator. The value of the gradient of f at point x will be denoted either by $\nabla_x f$ or $\nabla f(x)$. The notation $\widetilde{\nabla}$ stands for the approximate gradient, introduced in Definition 10.2.

If T is a map $\mathbb{R}^n \to \mathbb{R}^n$, ∇T stands for the Jacobian matrix of T, that is the matrix of all partial derivatives $(\partial T_i/\partial x_j)$ $(1 \le i, j \le n)$.

All these differential operators will be applied to (smooth) functions but also to measures, by duality. For instance, the Laplacian of a measure μ is defined via the identity $\int \zeta\, d(\Delta\mu) = \int (\Delta\zeta)\, d\mu$ $(\zeta \in C_c^2)$. The notation is consistent in the sense that $\Delta(f \mathrm{vol}) = (\Delta f)\,\mathrm{vol}$. Similarly, I shall take the divergence of a vector-valued measure, etc.

$f = o(g)$ means $f/g \longrightarrow 0$ (in an asymptotic regime that should be clear from the context), while $f = O(g)$ means that f/g is bounded.

log stands for the natural logarithm with base e.

The positive and negative parts of $x \in \mathbb{R}$ are defined respectively by $x_+ = \max(x, 0)$ and $x_- = \max(-x, 0)$; both are nonnegative, and $|x| = x_+ + x_-$. The notation $a \wedge b$ will sometimes be used for $\min(a, b)$. All these notions are extended in the usual way to functions and also to signed measures.

Probability measures

δ_x is the Dirac mass at point x.

All measures considered in the text are Borel measures on **Polish spaces**, which are complete, separable metric spaces, equipped with their Borel σ-algebra. I shall usually not use the completed σ-algebra, except on some rare occasions (emphasized in the text) in Chapter 5.

A measure is said to be finite if it has finite mass, and locally finite if it attributes finite mass to compact sets.

The space of Borel probability measures on $\mathcal{X}$ is denoted by $P(\mathcal{X})$, the space of finite Borel measures by $M_+(\mathcal{X})$, the space of signed finite Borel measures by $M(\mathcal{X})$. The total variation of μ is denoted by $\|\mu\|_{\mathrm{TV}}$.

The integral of a function f with respect to a probability measure μ will be denoted interchangeably by $\int f(x)\, d\mu(x)$ or $\int f(x)\, \mu(dx)$ or $\int f\, d\mu$.

If μ is a Borel measure on a topological space $\mathcal{X}$, a set N is said to be μ-negligible if N is included in a Borel set of zero μ-measure. Then μ is said to be concentrated on a set C if $\mathcal{X} \setminus C$ is negligible. (If C itself is Borel measurable, this is of course equivalent to $\mu[\mathcal{X} \setminus C] = 0$.) By abuse of language, I may say that $\mathcal{X}$ has full μ-measure if μ is concentrated on $\mathcal{X}$.

If μ is a Borel measure, its support $\operatorname{Spt} \mu$ is the smallest *closed* set on which it is concentrated. The same notation Spt will be used for the support of a continuous function.

If μ is a Borel measure on $\mathcal{X}$, and T is a Borel map $\mathcal{X} \to \mathcal{Y}$, then $T_\# \mu$ stands for the image measure[2] (or push-forward) of μ by T: It is a Borel measure on $\mathcal{Y}$, defined by $(T_\#\mu)[A] = \mu[T^{-1}(A)]$.

The law of a random variable X defined on a probability space $(\Omega, \mathbb{P})$ is denoted by $\operatorname{law}(X)$; this is the same as $X_\#\mathbb{P}$.

The weak topology on $P(\mathcal{X})$ (or topology of weak convergence, or narrow topology) is induced by convergence against $C_b(\mathcal{X})$, i.e. *bounded*

[2] Depending on the authors, the measure $T_\#\mu$ is often denoted by $T\#\mu$, $T_*\mu$, $T(\mu)$, $T\mu$, $\int \delta_{T(a)}\, \mu(da)$, $\mu \circ T^{-1}$, μT^{-1}, or $\mu[T \in \cdot\,]$.

continuous test functions. If $\mathcal{X}$ is Polish, then the space $P(\mathcal{X})$ itself is Polish. Unless explicitly stated, I do not use the weak-$*$ topology of measures (induced by $C_0(\mathcal{X})$ or $C_c(\mathcal{X})$).

When a probability measure is clearly specified by the context, it will sometimes be denoted just by $\mathbb{P}$, and the associated integral, or expectation, will be denoted by $\mathbb{E}$.

If $\pi(dx\,dy)$ is a probability measure in two variables $x \in \mathcal{X}$ and $y \in \mathcal{Y}$, its marginal (or projection) on $\mathcal{X}$ (resp. $\mathcal{Y}$) is the measure $X_{\#}\pi$ (resp. $Y_{\#}\pi$), where $X(x,y) = x$, $Y(x,y) = y$. If (x,y) is random with law $(x,y) = \pi$, then the conditional law of x given y is denoted by $\pi(dx|y)$; this is a measurable function $\mathcal{Y} \to P(\mathcal{X})$, obtained by disintegrating π along its y-marginal. The conditional law of y given x will be denoted by $\pi(dy|x)$.

A measure μ is said to be absolutely continuous with respect to a measure ν if there exists a measurable function f such that $\mu = f\,\nu$.

Notation specific to optimal transport and related fields

If $\mu \in P(\mathcal{X})$ and $\nu \in P(\mathcal{Y})$ are given, then $\Pi(\mu,\nu)$ is the set of all joint probability measures on $\mathcal{X} \times \mathcal{Y}$ whose marginals are μ and ν.

$C(\mu,\nu)$ is the optimal (total) cost between μ and ν, see p. 80. It implicitly depends on the choice of a cost function $c(x,y)$.

For any $p \in [1,+\infty)$, W_p is the Wasserstein distance of order p, see Definition 6.1; and $P_p(\mathcal{X})$ is the Wasserstein space of order p, i.e. the set of probability measures with finite moments of order p, equipped with the distance W_p, see Definition 6.4.

$P_c(\mathcal{X})$ is the set of probability measures on $\mathcal{X}$ with compact support.

If a reference measure ν on $\mathcal{X}$ is specified, then $P^{\mathrm{ac}}(\mathcal{X})$ (resp. $P_p^{\mathrm{ac}}(\mathcal{X})$, $P_c^{\mathrm{ac}}(\mathcal{X})$) stands for those elements of $P(\mathcal{X})$ (resp. $P_p(\mathcal{X})$, $P_c(\mathcal{X})$) which are absolutely continuous with respect to ν.

$\mathcal{DC}_N$ is the displacement convexity class of order N (N plays the role of a dimension); this is a family of convex functions, defined on p. 443 and in Definition 17.1.

U_ν is a functional defined on $P(\mathcal{X})$; it depends on a convex function U and a reference measure ν on $\mathcal{X}$. This functional will be defined at various levels of generality, first in equation (15.2), then in Definition 29.1 and Theorem 30.4.

$U^{\beta}_{\pi,\nu}$ is another functional on $P(\mathcal{X})$, which involves not only a convex function U and a reference measure ν, but also a coupling π and a distortion coefficient β, which is a nonnegative function on $\mathcal{X} \times \mathcal{X}$: See again Definition 29.1 and Theorem 30.4.

The Γ and Γ_2 operators are quadratic differential operators associated with a diffusion operator; they are defined in (14.47) and (14.48).

$\beta_t^{(K,N)}$ is the notation for the distortion coefficients that will play a prominent role in these notes; they are defined in (14.61).

CD(K,N) means "curvature-dimension condition (K,N)", which morally means that the Ricci curvature is bounded below by Kg (K a real number, g the Riemannian metric) and the dimension is bounded above by N (a real number which is not less than 1).

If $c(x,y)$ is a cost function then $\check{c}(y,x) = c(x,y)$. Similarly, if $\pi(dx\,dy)$ is a coupling, then $\check{\pi}$ is the coupling obtained by swapping variables, that is $\check{\pi}(dy\,dx) = \pi(dx\,dy)$, or more rigorously, $\check{\pi} = S_{\#}\pi$, where $S(x,y) = (y,x)$.

Assumptions **(Super)**, **(Twist)**, **(Lip)**, **(SC)**, **(locLip)**, **(locSC)**, **(H∞)** are defined on p. 234, **(STwist)** on p. 299, **(Cut^{n-1})** on p. 303.

Introduction

To start, I shall recall in Chapter 1 some basic facts about couplings and changes of variables, including definitions, a short list of famous couplings (Knothe–Rosenblatt coupling, Moser coupling, optimal coupling, etc.); and some important basic formulas about change of variables, conservation of mass, and linear diffusion equations.

In Chapter 2 I shall present, without detailed proofs, three applications of optimal coupling techniques, providing a flavor of the kind of applications that will be considered later.

Finally, Chapter 3 is a short historical perspective about the foundations and development of optimal coupling theory.

1

Couplings and changes of variables

Couplings are very well-known in all branches of probability theory, but since they will occur again and again in this course, it might be a good idea to start with some basic reminders and a few more technical issues.

Definition 1.1 (Coupling). *Let* $(\mathcal{X}, \mu)$ *and* $(\mathcal{Y}, \nu)$ *be two probability spaces. Coupling* μ *and* ν *means constructing two random variables* X *and* Y *on some probability space* $(\Omega, \mathbb{P})$*, such that* $\mathrm{law}\,(X) = \mu$, $\mathrm{law}\,(Y) = \nu$*. The couple* (X, Y) *is called a coupling of* (μ, ν)*. By abuse of language, the law of* (X, Y) *is also called a coupling of* (μ, ν)*.*

If μ and ν are the only laws in the problem, then without loss of generality one may choose $\Omega = \mathcal{X} \times \mathcal{Y}$. In a more measure-theoretical formulation, coupling μ and ν means constructing a measure π on $\mathcal{X} \times \mathcal{Y}$ such that π admits μ and ν as **marginals** on $\mathcal{X}$ and $\mathcal{Y}$ respectively. The following three statements are equivalent ways to rephrase that marginal condition:

- $(\mathrm{proj}_{\mathcal{X}})_{\#}\pi = \mu$, $(\mathrm{proj}_{\mathcal{Y}})_{\#}\pi = \nu$, where $\mathrm{proj}_{\mathcal{X}}$ and $\mathrm{proj}_{\mathcal{Y}}$ respectively stand for the projection maps $(x, y) \longmapsto x$ and $(x, y) \longmapsto y$;
- For all measurable sets $A \subset \mathcal{X}$, $B \subset \mathcal{Y}$, one has $\pi[A \times \mathcal{Y}] = \mu[A]$, $\pi[\mathcal{X} \times B] = \nu[B]$;
- For all integrable (resp. nonnegative) measurable functions φ, ψ on $\mathcal{X}, \mathcal{Y}$,

$$\int_{\mathcal{X} \times \mathcal{Y}} (\varphi(x) + \psi(y))\, d\pi(x, y) = \int_{\mathcal{X}} \varphi\, d\mu + \int_{\mathcal{Y}} \psi\, d\nu.$$

C. Villani, *Optimal Transport*, Grundlehren der mathematischen Wissenschaften 338,

A first remark about couplings is that they always exist: at least there is the **trivial coupling**, in which the variables X and Y are **independent** (so their joint law is the tensor product $\mu \otimes \nu$). This can hardly be called a coupling, since the value of X does not give any information about the value of Y. Another extreme is when all the information about the value of Y is contained in the value of X, in other words Y is just a function of X. This motivates the following definition (in which X and Y do not play symmetric roles).

Definition 1.2 (Deterministic coupling). *With the notation of Definition 1.1, a coupling (X, Y) is said to be deterministic if there exists a measurable function $T : \mathcal{X} \to \mathcal{Y}$ such that $Y = T(X)$.*

To say that (X, Y) is a deterministic coupling of μ and ν is strictly equivalent to any one of the four statements below:

- (X, Y) is a coupling of μ and ν whose law π is concentrated on the *graph* of a measurable function $T : \mathcal{X} \to \mathcal{Y}$;
- X has law μ and $Y = T(X)$, where $T_{\#}\mu = \nu$;
- X has law μ and $Y = T(X)$, where T is a **change of variables** from μ to ν: for all ν-integrable (resp. nonnegative measurable) functions φ,

$$\int_{\mathcal{Y}} \varphi(y)\, d\nu(y) = \int_{\mathcal{X}} \varphi(T(x))\, d\mu(x); \tag{1.1}$$

- $\pi = (\mathrm{Id}\,, T)_{\#}\mu$.

The map T appearing in all these statements is the same and is uniquely defined μ-almost surely (when the joint law of (X, Y) has been fixed). The converse is true: If T and $\widetilde{T}$ coincide μ-almost surely, then $T_{\#}\mu = \widetilde{T}_{\#}\mu$. It is common to call T the **transport map**: Informally, one can say that T transports the mass represented by the measure μ, to the mass represented by the measure ν.

Unlike couplings, deterministic couplings do not always exist: Just think of the case when μ is a Dirac mass and ν is not. But there may also be infinitely many deterministic couplings between two given probability measures.

Some famous couplings

Here below are some of the most famous couplings used in mathematics — of course the list is far from complete, since everybody has his or her own preferred coupling technique. Each of these couplings comes with its own natural setting; this variety of assumptions reflects the variety of constructions. (This is a good reason to state each of them with some generality.)

1. The **measurable isomorphism.** Let $(\mathcal{X}, \mu)$ and $(\mathcal{Y}, \nu)$ be two Polish (i.e. complete, separable, metric) probability spaces without atom (i.e. no single point carries a positive mass). Then there exists a (nonunique) measurable bijection $T : \mathcal{X} \to \mathcal{Y}$ such that $T_{\#}\mu = \nu$, $(T^{-1})_{\#}\nu = \mu$. In that sense, all atomless Polish probability spaces are isomorphic, and, say, isomorphic to the space $\mathcal{Y} = [0, 1]$ equipped with the Lebesgue measure. Powerful as that theorem may seem, in practice the map T is very singular; as a good exercise, the reader might try to construct it "explicitly", in terms of cumulative distribution functions, when $\mathcal{X} = \mathbb{R}$ and $\mathcal{Y} = [0, 1]$ (issues do arise when the density of μ vanishes at some places). Experience shows that it is quite easy to fall into logical traps when working with the measurable isomorphism, and my advice is to never use it.

2. The **Moser mapping.** Let $\mathcal{X}$ be a smooth compact Riemannian manifold with volume vol, and let f, g be Lipschitz continuous positive probability densities on $\mathcal{X}$; then there exists a deterministic coupling of $\mu = f\,\mathrm{vol}$ and $\nu = g\,\mathrm{vol}$, constructed by resolution of an elliptic equation. On the positive side, there is a somewhat explicit representation of the transport map T, and it is as smooth as can be: if f, g are $C^{k,\alpha}$ then T is $C^{k+1,\alpha}$. The formula is given in the Appendix at the end of this chapter. The same construction works in $\mathbb{R}^n$ provided that f and g decay fast enough at infinity; and it is robust enough to accommodate for variants.

3. The **increasing rearrangement** on $\mathbb{R}$. Let μ, ν be two probability measures on $\mathbb{R}$; define their cumulative distribution functions by

$$F(x) = \int_{-\infty}^{x} d\mu, \qquad G(y) = \int_{-\infty}^{y} d\nu.$$

   Further define their right-continuous inverses by

$$F^{-1}(t) = \inf \Big\{ x \in \mathbb{R};\ F(x) > t \Big\};$$
$$G^{-1}(t) = \inf \Big\{ y \in \mathbb{R};\ G(y) > t \Big\};$$

and set

$$T = G^{-1} \circ F.$$

If μ does not have atoms, then $T_{\#}\mu = \nu$. This rearrangement is quite simple, explicit, as smooth as can be, and enjoys good geometric properties.

4. The **Knothe–Rosenblatt rearrangement** in $\mathbb{R}^n$. Let μ and ν be two probability measures on $\mathbb{R}^n$, such that μ is absolutely continuous with respect to Lebesgue measure. Then define a coupling of μ and ν as follows.

   Step 1: Take the marginal on the first variable: this gives probability measures $\mu_1(dx_1)$, $\nu_1(dy_1)$ on $\mathbb{R}$, with μ_1 being atomless. Then define $y_1 = T_1(x_1)$ by the formula for the increasing rearrangement of μ_1 into ν_1.

   Step 2: Now take the marginal on the first two variables and disintegrate it with respect to the first variable. This gives probability measures $\mu_2(dx_1\,dx_2) = \mu_1(dx_1)\,\mu_2(dx_2|x_1)$, $\nu_2(dy_1\,dy_2) = \nu_1(dy_1)\,\nu_2(dy_2|y_1)$. Then, for each given $y_1 \in \mathbb{R}$, set $y_1 = T_1(x_1)$, and define $y_2 = T_2(x_2; x_1)$ by the formula for the increasing rearrangement of $\mu_2(dx_2|x_1)$ into $\nu_2(dy_2|y_1)$. (See Figure 1.1.)

   Then repeat the construction, adding variables one after the other and defining $y_3 = T_3(x_3; x_1, x_2)$; etc. After n steps, this produces a map $y = T(x)$ which transports μ to ν, and in practical situations might be computed explicitly with little effort. Moreover, the Jacobian matrix of the change of variables T is (by construction) upper triangular with positive entries on the diagonal; this makes it suitable for various geometric applications. On the negative side, this mapping does not satisfy many interesting intrinsic properties; it is not invariant under isometries of $\mathbb{R}^n$, not even under relabeling of coordinates.

5. The **Holley coupling** on a lattice. Let μ and ν be two discrete probabilities on a finite lattice Λ, say $\{0,1\}^N$, equipped with the natural partial ordering ($x \le y$ if $x_n \le y_n$ for all n). Assume that

$$\forall x, y \in \Lambda, \qquad \mu[\inf(x,y)]\,\nu[\sup(x,y)] \ge \mu[x]\,\nu[y]. \qquad (1.2)$$

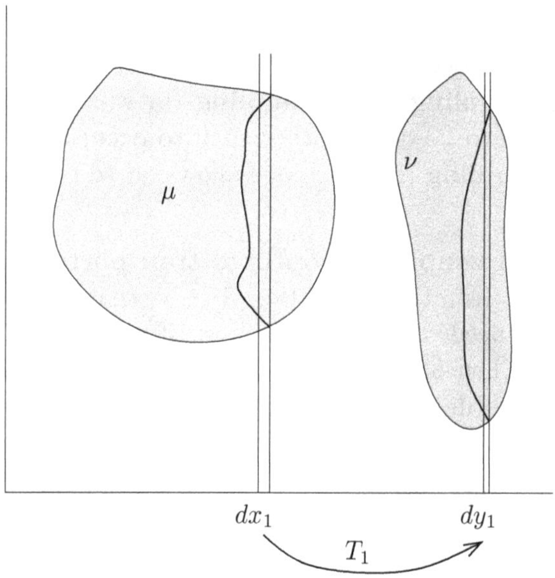

Fig. 1.1. Second step in the construction of the Knothe–Rosenblatt map: After the correspondance $x_1 \to y_1$ has been determined, the conditional probability of x_2 (seen as a one-dimensional probability on a small "slice" of width dx_1) can be transported to the conditional probability of y_2 (seen as a one-dimensional probability on a slice of width dy_1).

Then there exists a coupling (X, Y) of (μ, ν) with $X \leq Y$. The situation above appears in a number of problems in statistical mechanics, in connection with the so-called FKG (Fortuin–Kasteleyn–Ginibre) inequalities. Inequality (1.2) intuitively says that ν puts more mass on large values than μ.

6. **Probabilistic representation formulas** for solutions of partial differential equations. There are hundreds of them (if not thousands), representing solutions of diffusion, transport or jump processes as the laws of various deterministic or stochastic processes. Some of them are recalled later in this chapter.

7. The **exact coupling** of two stochastic processes, or Markov chains. Two realizations of a stochastic process are started at initial time, and when they happen to be in the same state at some time, they are merged: from that time on, they follow the same path and accordingly, have the same law. For two Markov chains which are started independently, this is called the **classical coupling**. There

are many variants with important differences which are all intended to make two trajectories close to each other after some time: the **Ornstein coupling**, the **ε-coupling** (in which one requires the two variables to be close, rather than to occupy the same state), the **shift-coupling** (in which one allows an additional time-shift), etc.

8. The **optimal coupling** or **optimal transport**. Here one introduces a **cost function** $c(x,y)$ on $\mathcal{X} \times \mathcal{Y}$, that can be interpreted as the work needed to move one unit of mass from location x to location y. Then one considers the **Monge–Kantorovich minimization problem**

$$\inf\ \mathbb{E}\ c(X,Y),$$

where the pair (X,Y) runs over all possible couplings of (μ,ν); or equivalently, in terms of measures,

$$\inf \int_{\mathcal{X}\times\mathcal{Y}} c(x,y)\, d\pi(x,y),$$

where the infimum runs over all joint probability measures π on $\mathcal{X}\times\mathcal{Y}$ with marginals μ and ν. Such joint measures are called **transference plans** (or transport plans, or transportation plans); those achieving the infimum are called **optimal transference plans**.

Of course, the solution of the Monge–Kantorovich problem depends on the cost function c. The cost function and the probability spaces here can be very general, and some nontrivial results can be obtained as soon as, say, c is lower semicontinuous and $\mathcal{X},\mathcal{Y}$ are Polish spaces. Even the apparently trivial choice $c(x,y) = 1_{x\neq y}$ appears in the probabilistic interpretation of total variation:

$$\|\mu-\nu\|_{TV} = 2\ \inf\Big\{\mathbb{E}\,1_{X\neq Y};\quad \mathrm{law}\,(X)=\mu,\ \mathrm{law}\,(Y)=\nu\Big\}.$$

Cost functions valued in $\{0,1\}$ also occur naturally in Strassen's duality theorem.

Under certain assumptions one can guarantee that the optimal coupling really is deterministic; the search of deterministic optimal couplings (or Monge couplings) is called the **Monge problem**. A solution of the Monge problem yields a plan to transport the mass at minimal cost with a recipe that associates to each point x a single point y. ("*No mass shall be split.*") To guarantee the existence of solutions to the

Monge problem, two kinds of assumptions are natural: First, c should "vary enough" in some sense (think that the constant cost function will allow for arbitrary minimizers), and secondly, μ should enjoy some regularity property (at least Dirac masses should be ruled out!). Here is a typical result: If $c(x,y) = |x-y|^2$ in the Euclidean space, μ is absolutely continuous with respect to Lebesgue measure, and μ, ν have finite moments of order 2, then there is a unique optimal Monge coupling between μ and ν. More general statements will be established in Chapter 10.

Optimal couplings enjoy several nice properties:

(i) They naturally arise in many problems coming from economics, physics, partial differential equations or geometry (by the way, the increasing rearrangement and the Holley coupling can be seen as particular cases of optimal transport);

(ii) They are quite stable with respect to perturbations;

(iii) They encode good geometric information, if the cost function c is defined in terms of the underlying geometry;

(iv) They exist in smooth as well as nonsmooth settings;

(v) They come with a rich structure: an **optimal cost** functional (the value of the infimum defining the Monge–Kantorovich problem); a **dual variational problem**; and, under adequate structure conditions, a continuous **interpolation**.

On the negative side, it is important to be warned that optimal transport is in general not so smooth. There are known counterexamples which put limits on the regularity that one can expect from it, even for very nice cost functions.

All these issues will be discussed again and again in the sequel. The rest of this chapter is devoted to some basic technical tools.

Gluing

If Z is a function of Y and Y is a function of X, then of course Z is a function of X. Something of this still remains true in the setting of nondeterministic couplings, under quite general assumptions.

Gluing lemma. *Let $(\mathcal{X}_i, \mu_i)$, $i = 1,2,3$, be Polish probability spaces. If (X_1, X_2) is a coupling of (μ_1, μ_2) and (Y_2, Y_3) is a coupling of (μ_2, μ_3),* *then one can construct a triple of random variables* (Z_1, Z_2, Z_3) *such that* (Z_1, Z_2) *has the same law as* (X_1, X_2) *and* (Z_2, Z_3) *has the same law as* (Y_2, Y_3).

It is simple to understand why this is called "gluing lemma": if π_{12} stands for the law of (X_1, X_2) on $\mathcal{X}_1 \times \mathcal{X}_2$ and π_{23} stands for the law of (X_2, X_3) on $\mathcal{X}_2 \times \mathcal{X}_3$, then to construct the joint law π_{123} of (Z_1, Z_2, Z_3) one just has to *glue* π_{12} and π_{23} along their common marginal μ_2. Expressed in a slightly informal way: Disintegrate π_{12} and π_{23} as

$$\pi_{12}(dx_1\, dx_2) = \pi_{12}(dx_1|x_2)\, \mu_2(dx_2),$$
$$\pi_{23}(dx_2\, dx_3) = \pi_{23}(dx_3|x_2)\, \mu_2(dx_2),$$

and then reconstruct π_{123} as

$$\pi_{123}(dx_1\, dx_2\, dx_3) = \pi_{12}(dx_1|x_2)\, \mu_2(dx_2)\, \pi_{23}(dx_3|x_2).$$

Change of variables formula

When one writes the formula for change of variables, say in $\mathbb{R}^n$ or on a Riemannian manifold, a Jacobian term appears, and one has to be careful about two things: the change of variables should be *injective* (otherwise, reduce to a subset where it is injective, or take the multiplicity into account); and it should be somewhat smooth. It is classical to write these formulas when the change of variables is continuously differentiable, or at least Lipschitz:

Change of variables formula. *Let* M *be an* n*-dimensional Riemannian manifold with a* C^1 *metric, let* μ_0, μ_1 *be two probability measures on* M, *and let* $T : M \to M$ *be a measurable function such that* $T_\#\mu_0 = \mu_1$. *Let* ν *be a reference measure, of the form* $\nu(dx) = e^{-V(x)}\,\mathrm{vol}(dx)$, *where* V *is continuous and* vol *is the volume (or* n*-dimensional Hausdorff) measure. Further assume that*

(i) $\mu_0(dx) = \rho_0(x)\, \nu(dx)$ *and* $\mu_1(dy) = \rho_1(y)\, \nu(dy)$;

(ii) T *is injective;*

(iii) T *is locally Lipschitz.*

Then, μ_0*-almost surely,*

$$\rho_0(x) = \rho_1(T(x))\,\mathcal{J}_T(x), \tag{1.3}$$

where $\mathcal{J}_T(x)$ is the Jacobian determinant of T at x, defined by

$$\mathcal{J}_T(x) := \lim_{\varepsilon \downarrow 0} \frac{\nu[T(B_\varepsilon(x))]}{\nu[B_\varepsilon(x)]}. \tag{1.4}$$

The same holds true if T is only defined on the complement of a μ_0-negligible set, and satisfies properties (ii) and (iii) on its domain of definition.

Remark 1.3. When ν is just the volume measure, $\mathcal{J}_T$ coincides with the usual Jacobian determinant, which in the case $M = \mathbb{R}^n$ is the absolute value of the determinant of the Jacobian matrix ∇T. Since V is continuous, it is almost immediate to deduce the statement with an arbitrary V from the statement with $V = 0$ (this amounts to multiplying $\rho_0(x)$ by $e^{V(x)}$, $\rho_1(y)$ by $e^{V(y)}$, $\mathcal{J}_T(x)$ by $e^{V(x)-V(T(x))}$).

Remark 1.4. There is a more general framework beyond differentiability, namely the property of **approximate differentiability**. A function T on an n-dimensional Riemannian manifold is said to be approximately differentiable at x if there exists a function $\widetilde{T}$, differentiable at x, such that the set $\{\widetilde{T} \neq T\}$ has zero density at x, i.e.

$$\lim_{r\to 0} \frac{\operatorname{vol}\left[\{x \in B_r(x);\ T(x) \neq \widetilde{T}(x)\}\right]}{\operatorname{vol}\left[B_r(x)\right]} = 0.$$

It turns out that, roughly speaking, an approximately differentiable map can be replaced, up to neglecting a small set, by a Lipschitz map (this is a kind of differentiable version of Lusin's theorem). So one can prove the Jacobian formula for an approximately differentiable map by approximating it with a sequence of Lipschitz maps.

Approximate differentiability is obviously a local property; it holds true if the distributional derivative of T is a locally integrable function, or even a locally finite measure. So it is useful to know that the change of variables formula still holds true if Assumption (iii) above is replaced by

(iii') T is approximately differentiable.

Conservation of mass formula

The single most important theorem of change of variables arising in continuum physics might be the one resulting from the **conservation of mass** formula,

$$\frac{\partial \rho}{\partial t} + \nabla \cdot (\rho\, \xi) = 0. \tag{1.5}$$

Here $\rho = \rho(t,x)$ stands for the density of a system of particles at time t and position x; $\xi = \xi(t,x)$ for the velocity field at time t and position x; and $\nabla\cdot$ stands for the divergence operator. Once again, the natural setting for this equation is a Riemannian manifold M.

It will be useful to work with particle densities $\mu_t(dx)$ (that are not necessarily absolutely continuous) and rewrite (1.5) as

$$\frac{\partial \mu}{\partial t} + \nabla \cdot (\mu\, \xi) = 0,$$

where the time-derivative is taken in the weak sense, and the divergence operator is defined by duality against continuously differentiable functions with compact support:

$$\int_M \varphi\, \nabla \cdot (\mu\, \xi) = -\int_M (\xi \cdot \nabla \varphi)\, d\mu.$$

The formula of conservation of mass is an **Eulerian** description of the physical world, which means that the unknowns are fields. The next theorem links it with the **Lagrangian** description, in which everything is expressed in terms of particle trajectories, that are integral curves of the velocity field:

$$\xi\big(t, T_t(x)\big) = \frac{d}{dt}\, T_t(x). \tag{1.6}$$

If ξ is (locally) Lipschitz continuous, then the Cauchy–Lipschitz theorem guarantees the existence of a flow T_t locally defined on a maximal time interval, and itself locally Lipschitz in both arguments t and x. Then, for each t the map T_t is a local diffeomorphism onto its image. But the formula of conservation of mass also holds true without any regularity assumption on ξ; one should only keep in mind that if ξ is not Lipschitz, then a solution of (1.6) is not uniquely determined by its value at time 0, so $x \longmapsto T_t(x)$ is not necessarily uniquely defined. Still it makes sense to consider *random* solutions of (1.6).

Mass conservation formula. *Let M be a C^1 manifold, $T \in (0, +\infty]$ and let $\xi(t,x)$ be a (measurable) velocity field on $[0,T) \times M$. Let*

$(\mu_t)_{0\le t<T}$ be a time-dependent family of probability measures on M (continuous in time for the weak topology), such that

$$\int_0^T \int_M |\xi(t,x)|\, \mu_t(dx)\, dt < +\infty.$$

Then, the following two statements are equivalent:

(i) $\mu = \mu_t(dx)$ is a weak solution of the linear (transport) partial differential equation

$$\partial_t \mu + \nabla_x \cdot (\mu\, \xi) = 0$$

on $[0,T) \times M$;

(ii) μ_t is the law at time t of a random solution $T_t(x)$ of (1.6).

If moreover ξ is locally Lipschitz, then $(T_t)_{0\le t<T}$ defines a deterministic flow, and statement (ii) can be rewritten

(ii') $\mu_t = (T_t)_\# \mu_0$.

Diffusion formula

The final reminder in this chapter is very well-known and related to Itô's formula; it was discovered independently (in the Euclidean context) by Bachelier, Einstein and Smoluchowski at the beginning of the twentieth century. It requires a bit more regularity than the Conservation of mass Formula. The natural assumptions on the phase space are in terms of *Ricci curvature*, a concept which will play an important role in these notes. For the reader who has no idea what Ricci curvature means, it is sufficient to know that the theorem below applies when M is either $\mathbb{R}^n$, or a compact manifold with a C^2 metric. By convention, B_t denotes the "standard" Brownian motion on M with identity covariance matrix.

Diffusion theorem. *Let M be a Riemannian manifold with a C^2 metric, such that the Ricci curvature tensor of M is uniformly bounded below, and let $\sigma(t,x) : T_xM \to T_xM$ be a twice differentiable linear mapping on each tangent space. Let X_t stand for the solution of the stochastic differential equation*

$$dX_t = \sqrt{2}\, \sigma(t, X_t)\, dB_t \qquad (0 \le t < T). \tag{1.7}$$

Then the following two statements are equivalent:

(i) $\mu = \mu_t(dx)$ is a weak solution of the linear (diffusion) partial differential equation

$$\partial_t \mu = \nabla_x \cdot \Big((\sigma\sigma^*) \nabla_x \mu \Big)$$

on $M \times [0,T)$, where σ^* stands for the transpose of σ;

(ii) $\mu_t = \mathrm{law}\,(X_t)$ for all $t \in [0,T)$, where X_t solves (1.7).

Example 1.5. In $\mathbb{R}^n$, the solution of the heat equation with initial datum δ_0 is the law of $X_t = \sqrt{2}\, B_t$ (Brownian motion sped up by a factor $\sqrt{2}$).

Remark 1.6. Actually, there is a finer criterion for the diffusion equation to hold true: it is sufficient that the Ricci curvature at point x be bounded below by $-Cd(x_0,x)^2 g_x$ as $x \to \infty$, where g_x is the metric at point x and x_0 is an arbitrary reference point. The exponent 2 here is sharp.

Exercise 1.7. Let M be a smooth compact manifold, equipped with its standard reference volume, and let ρ_0 be a smooth positive probability density on M. Let $(\rho_t)_{t\ge 0}$ be the solution of the heat equation

$$\partial_t \rho = \Delta \rho.$$

Use (ρ_t) to construct a *deterministic* coupling of ρ_0 and ρ_1.

Hint: Rewrite the heat equation in the form of an equation of conservation of mass.

Appendix: Moser's coupling

In this Appendix I shall promote Moser's technique for coupling smooth positive probability measures; it is simple, elegant and powerful, and plays a prominent role in geometry. It is not limited to compact manifolds, but does require assumptions about the behavior at infinity.

Let M be a smooth n-dimensional Riemannian manifold, equipped with a reference probability measure $\nu(dx) = e^{-V(x)}\,\mathrm{vol}(dx)$, where $V \in C^1(M)$. Let $\mu_0 = \rho_0\,\nu$, $\mu_1 = \rho_1\,\nu$ be two probability measures on

M; assume for simplicity that ρ_0, ρ_1 are bounded below by a constant $K > 0$. Further assume that ρ_0 and ρ_1 are locally Lipschitz, and that the equation

$$(\Delta - \nabla V \cdot \nabla)\, u = \rho_0 - \rho_1$$

can be solved for some $u \in C^{1,1}_{\rm loc}(M)$ (that is, ∇u is locally Lipschitz). Then, define a locally Lipschitz vector field

$$\xi(t,x) = \frac{\nabla u(x)}{(1-t)\,\rho_0(x) + t\,\rho_1(x)},$$

with associated flow $(T_t(x))_{0\le t\le 1}$, and a family $(\mu_t)_{0<t<1}$ of probability measures by

$$\mu_t = (1-t)\,\mu_0 + t\,\mu_1.$$

It is easy to check that

$$\partial_t \mu = (\rho_1 - \rho_0)\,\nu,$$

$$\nabla\cdot\big(\mu_t\,\xi(t,\cdot)\big) = \nabla\cdot\big(\nabla u\, e^{-V}\,\mathrm{vol}\big) = e^{-V}\big(\Delta u - \nabla V\cdot\nabla u\big)\,\mathrm{vol} = (\rho_0 - \rho_1)\,\nu.$$

So μ_t satisfies the formula of conservation of mass, therefore $\mu_t = (T_t)_\#\mu_0$. In particular, T_1 pushes μ_0 forward to μ_1.

In the case when M is compact and $V = 0$, the above construction works if ρ_0 and ρ_1 are Lipschitz continuous and positive. Indeed, the solution u of $\Delta u = \rho_0 - \rho_1$ will be of class $C^{2,\alpha}$ for all $\alpha \in (0,1)$, and in particular ∇u will be of class C^1 (in fact $C^{1,\alpha}$). In more general situations, things might depend on the regularity of V, and its behavior at infinity.

Bibliographical notes

An excellent general reference book for the "classical theory" of couplings is the monograph by Thorisson [781]. There one can find an exhaustive treatment of classical couplings of Markov chains or stochastic processes, such as ε-coupling, shift-coupling, Ornstein coupling. The classical theory of optimal couplings is addressed in the two volumes by Rachev and Rüschendorf [696]. This includes in particular the theory of optimal coupling on the real line with a convex cost function, which can be treated in a simple and direct manner [696, Section 3.1].

(In [814], for the sake of consistency of the presentation I treated optimal coupling on $\mathbb{R}$ as a particular case of optimal coupling on $\mathbb{R}^n$, however this has the drawback to involve subtle arguments.)

The Knothe–Rosenblatt coupling was introduced in 1952 by Rosenblatt [709], who suggested that it might be useful to "normalize" statistical data before applying a statistical test. In 1957, Knothe [523] rediscovered it for applications to the theory of convex bodies. It is quite likely that other people have discovered this coupling independently. An infinite-dimensional generalization was studied by Bogachev, Kolesnikov and Medvedev [134, 135].

FKG inequalities were introduced in [375], and have since then played a crucial role in statistical mechanics. Holley's proof by coupling appears in [477]. Recently, Caffarelli [188] has revisited the subject in connection with optimal transport.

It was in 1965 that Moser proved his coupling theorem, for smooth compact manifolds without boundaries [640]; noncompact manifolds were later considered by Greene and Shiohama [432]. Moser himself also worked with Dacorogna on the more delicate case where the domain is an open set with boundary, and the transport is required to fix the boundary [270].

Strassen's duality theorem is discussed e.g. in [814, Section 1.4].

The gluing lemma is due to several authors, starting with Vorob'ev in 1962 for finite sets. The modern formulation seems to have emerged around 1980, independently by Berkes and Philipp [101], Kallenberg, Thorisson, and maybe others. Refinements were discussed e.g. by de Acosta [273, Theorem A.1] (for marginals indexed by an arbitrary set) or Thorisson [781, Theorem 5.1]; see also the bibliographic comments in [317, p. 20]. For a proof of the statement in these notes, it is sufficient to consult Dudley [317, Theorem 1.1.10], or [814, Lemma 7.6]. A comment about terminology: I like the word "gluing" which gives a good indication of the construction, but many authors just talk about "composition" of plans.

The formula of change of variables for C^1 or Lipschitz change of variables can be found in many textbooks, see e.g. Evans and Gariepy [331, Chapter 3]. The generalization to approximately differentiable maps is explained in Ambrosio, Gigli and Savaré [30, Section 5.5]. Such a generality is interesting in the context of optimal transportation, where changes of variables are often very rough (say BV, which means of bounded variation). In that context however, there is more structure:

For instance, changes of variables will typically be given by the gradient of a convex function in $\mathbb{R}^n$, and on such a map one knows slightly more than on a general BV function, because convex functions are twice differentiable almost everywhere (Theorem 14.25 later in these notes). McCann [614] used this property to prove, by slightly more elementary means, the change of variables formula for a gradient of convex function; the proof is reproduced in [814, Theorem 4.8]. It was later generalized by Cordero-Erausquin, McCann and Schmuckenschläger to Riemannian manifolds [246], a case which again can be treated either as part of the general theory of BV changes of variables, or with the help of almost everywhere second derivatives of semiconcave functions.

The formula of conservation of mass is also called the method of characteristics for linear transport equations, and is described in a number of textbooks in partial differential equations, at least when the driving vector field is Lipschitz, see for instance Evans [327, Section 3.2]. An essentially equivalent statement is proven in [814, Theorem 5.34]. Treating vector fields that are only assumed to be locally Lipschitz is not so easy: see Ambrosio, Gigli and Savaré [30, Section 8.1].

The Lipschitz condition can be relaxed into a Sobolev or even a BV condition, but then the flow is determined only almost everywhere, and this becomes an extremely subtle problem, which has been studied by many authors since the pioneering work of DiPerna and Lions [304] at the beginning of the nineties. See Ambrosio [21] for recent progress and references. The version which is stated in these notes, with no regularity assumption, is due to Ambrosio and carefully proved in [30, Section 8.1]. In spite of its appealing and relatively natural character (especially in a probabilistic perspective), this is a very recent research result. Note that, if $T_t(x)$ is not uniquely determined by x, then the solution to the conservation equation starting with a given probability measure might admit several solutions.

A recent work by Lisini [565] addresses a generalization of the formula of conservation of mass in the setting of general Polish spaces. Of course, without any regularity assumption on the space it is impossible to speak of vector fields and partial differential equations; but it is still possible to consider paths in the space of probability measures, and random curves. Lisini's results are most naturally expressed in the language of optimal transport distances; see the bibliographical notes for Chapter 7.

The diffusion formula can be obtained as a simple consequence of the Itô formula, which in the Euclidean setting can be found in any textbook on stochastic differential equations, e.g. [658]. It was recently the hundredth anniversary of the discovery of the diffusion formula by Einstein [322]; or rather rediscovery, since Bachelier already had obtained the main results at the turn of the twentieth century [251, 739]. (Some information about Bachelier's life can be found online at `sjepg.univ-fcomte.fr/sjepgbis/libre/bachelier/page01/page01.htm`.) Fascinating tales about the Brownian motion can be read in Nelson's unconventional book [648], especially Chapters 1–4. For the much more subtle Riemannian setting, one may consult Stroock [759], Hsu [483] and the references therein.

The Brownian motion on a smooth Riemannian manifold is always well-defined, even if the manifold has a wild behavior at infinity (the construction of the Brownian motion is purely local); but in the absence of a good control on the Ricci curvature, there might be several heat kernels, and the heat equation might not be uniquely solvable for a given initial datum. This corresponds to the possibility of a blow-up of the Brownian motion (i.e. the Brownian motion escapes to infinity) in finite time. All this was explained to me by Thalmaier. The sharp criterion $\mathrm{Ric}_x \geq -C\,(1+d(x_0,x)^2)\,g_x$ for avoiding blow-up of the heat equation is based on comparison theorems for Laplace operators. In the version stated here it is due to Ichihara [486]; see also the book by Hackenbroch and Thalmaier [454, p. 544]. Nonexplosion criteria based on curvature have been studied by Gaffney, Yau, Hsu, Karp and Li, Davies, Takeda, Sturm, and Grigor'yan; for a detailed exposition, and many explanations, the reader can consult the survey by Grigor'yan [434, Section 9].

2

Three examples of coupling techniques

In this chapter I shall present three applications of coupling methods. The first one is classical and quite simple, the other two are more original but well-representative of the topics that will be considered later in these notes. The proofs are extremely variable in difficulty and will only be sketched here; see the references in the bibliographical notes for details.

Convergence of the Langevin process

Consider a particle subject to the force induced by a potential $V \in C^1(\mathbb{R}^n)$, a friction and a random white noise agitation. If X_t stands for the position of the particle at time t, m for its mass, λ for the friction coefficient, k for the Boltzmann constant and T for the temperature of the heat bath, then Newton's equation of motion can be written

$$m\,\frac{d^2X_t}{dt^2} = -\nabla V(X_t) - \lambda\, m\,\frac{dX_t}{dt} + \sqrt{kT}\,\frac{dB_t}{dt}, \tag{2.1}$$

where $(B_t)_{t\geq 0}$ is a standard Brownian motion. This is a second-order (stochastic) differential equation, so it should come with initial conditions for both the position X and the velocity $\dot{X}$.

Now consider a large cloud of particles evolving independently, according to (2.1); the question is whether the distribution of particles will converge to a definite limit as $t \to \infty$. In other words: Consider the stochastic differential equation (2.1) starting from some initial distribution $\mu_0(dx\,dv) = \text{law}\,(X_0, \dot{X}_0)$; is it true that law (X_t), or law $(X_t, \dot{X}_t)$, will converge to some given limit law as $t \to \infty$?

C. Villani, *Optimal Transport*, Grundlehren der mathematischen Wissenschaften 338,

Obviously, to solve this problem one has to make some assumptions on the potential V, which should prevent the particles from all escaping at infinity; for instance, we can make the very strong assumption that V is uniformly convex, i.e. there exists $K > 0$ such that the Hessian $\nabla^2 V$ satisfies $\nabla^2 V \geq K I_n$. Some assumptions on the initial distribution might also be needed; for instance, it is natural to assume that the Hamiltonian has finite expectation at initial time:

$$\mathbb{E}\left(V(X_0) + \frac{|\dot{X}_0|^2}{2}\right) < +\infty$$

Under these assumptions, it is true that there is exponential convergence to equilibrium, at least if V does not grow too wildly at infinity (for instance if the Hessian of V is also bounded above). However, I do not know of any simple method to prove this.

On the other hand, consider the limit where the friction coefficient is quite strong, and the motion of the particle is so slow that the acceleration term may be neglected in front of the others: then, up to resetting units, equation (2.1) becomes

$$\frac{dX_t}{dt} = -\nabla V(X_t) + \sqrt{2}\,\frac{dB_t}{dt}, \tag{2.2}$$

which is often called a Langevin process. Now, to study the convergence of equilibrium for (2.2) there is an extremely simple solution by coupling. Consider another random position $(Y_t)_{t\geq 0}$ obeying the same equation as (2.2):

$$\frac{dY_t}{dt} = -\nabla V(Y_t) + \sqrt{2}\,\frac{dB_t}{dt}, \tag{2.3}$$

where the random realization of the Brownian motion is *the same* as in (2.2) (this is the coupling). The initial positions X_0 and Y_0 may be coupled in an arbitrary way, but it is possible to assume that they are independent. In any case, since they are driven by the same Brownian motion, X_t and Y_t will be correlated for $t > 0$.

Since B_t is not differentiable as a function of time, neither X_t nor Y_t is differentiable (equations (2.2) and (2.3) hold only in the sense of solutions of stochastic differential equations); but it is easily checked that $\alpha_t := X_t - Y_t$ is a continuously differentiable function of time, and

$$\frac{d\alpha_t}{dt} = -\big(\nabla V(X_t) - \nabla V(Y_t)\big),$$

so in particular

$$\frac{d}{dt}\frac{|\alpha_t|^2}{2} = -\Big\langle \nabla V(X_t) - \nabla V(Y_t),\ X_t - Y_t \Big\rangle \le -K\,\big|X_t - Y_t\big|^2 = -K\,|\alpha_t|^2.$$

It follows by Gronwall's lemma that

$$|\alpha_t|^2 \le e^{-2Kt}\,|\alpha_0|^2.$$

Assume for simplicity that $\mathbb{E}\,|X_0|^2$ and $\mathbb{E}\,|Y_0|^2$ are finite. Then

$$\mathbb{E}\,|X_t - Y_t|^2 \le e^{-2Kt}\,\mathbb{E}\,|X_0 - Y_0|^2 \le 2\left(\mathbb{E}\,|X_0|^2 + \mathbb{E}\,|Y_0|^2\right) e^{-2Kt}. \quad (2.4)$$

In particular, $X_t - Y_t$ converges to 0 almost surely, and this is independent of the distribution of Y_0.

This in itself would be essentially sufficient to guarantee the existence of a stationary distribution; but in any case, it is easy to check, by applying the diffusion formula, that

$$\nu(dy) = \frac{e^{-V(y)}\,dy}{Z}$$

(where $Z = \int e^{-V}$ is a normalization constant) is stationary: If $\mathrm{law}\,(Y_0) = \nu$, then also $\mathrm{law}\,(Y_t) = \nu$. Then (2.4) easily implies that $\mu_t := \mathrm{law}\,(X_t)$ converges weakly to ν; in addition, the convergence is exponentially fast.

Euclidean isoperimetry

Among all subsets of $\mathbb{R}^n$ with given surface, which one has the largest volume? To simplify the problem, let us assume that we are looking for a bounded open set $\Omega \subset \mathbb{R}^n$ with, say, Lipschitz boundary $\partial\Omega$, and that the measure of $|\partial\Omega|$ is given; then the problem is to maximize the measure of $|\Omega|$. To measure $\partial\Omega$ one should use the $(n-1)$-dimensional Hausdorff measure, and to measure Ω the n-dimensional Hausdorff measure, which of course is the same as the Lebesgue measure in $\mathbb{R}^n$.

It has been known, at least since ancient times, that the solution to this "isoperimetric problem" is the ball. A simple scaling argument shows that this statement is equivalent to the Euclidean **isoperimetric inequality**:

$$\frac{|\partial\Omega|}{|\Omega|^{\frac{n}{n-1}}} \geq \frac{|\partial B|}{|B|^{\frac{n}{n-1}}},$$

where B is any ball.

There are very many proofs of the isoperimetric inequality, and many refinements as well. It is less known that there is a proof by coupling.

Here is a sketch of the argument, forgetting about regularity issues. Let B be a ball such that $|\partial B| = |\partial\Omega|$. Consider a random point X distributed uniformly in Ω, and a random point Y distributed uniformly in B. Introduce the Knothe–Rosenblatt coupling of X and Y: This is a deterministic coupling of the form $Y = T(X)$, such that, at each $x \in \Omega$, the Jacobian matrix $\nabla T(x)$ is triangular with nonnegative diagonal entries. Since the law of X (resp. Y) has uniform density $1/|\Omega|$ (resp. $1/|B|$), the change of variables formula yields

$$\forall x \in \Omega \qquad \frac{1}{|\Omega|} = \big(\det \nabla T(x)\big)\,\frac{1}{|B|}. \tag{2.5}$$

Since ∇T is triangular, the Jacobian determinant of T is $\det(\nabla T) = \prod \lambda_i$, and its divergence $\nabla \cdot T = \sum \lambda_i$, where the nonnegative numbers $(\lambda_i)_{1\leq i\leq n}$ are the eigenvalues of ∇T. Then the arithmetic–geometric inequality $(\prod \lambda_i)^{1/n} \leq (\sum \lambda_i)/n$ becomes

$$\big(\det \nabla T(x)\big)^{1/n} \leq \frac{\nabla \cdot T(x)}{n}.$$

Combining this with (2.5) results in

$$\frac{1}{|\Omega|^{1/n}} \leq \frac{(\nabla \cdot T)(x)}{n\,|B|^{1/n}}.$$

Integrate this over Ω and then apply the divergence theorem:

$$|\Omega|^{1-\frac{1}{n}} \leq \frac{1}{n\,|B|^{\frac{1}{n}}} \int_\Omega (\nabla \cdot T)(x)\,dx = \frac{1}{n\,|B|^{\frac{1}{n}}} \int_{\partial\Omega} (T\cdot\sigma)\,d\mathcal{H}^{n-1}, \tag{2.6}$$

where $\sigma : \partial\Omega \to \mathbb{R}^n$ is the unit outer normal to Ω and $\mathcal{H}^{n-1}$ is the $(n-1)$-dimensional Hausdorff measure (restricted to $\partial\Omega$). But T is valued in B, so $|T\cdot\sigma| \leq 1$, and (2.6) implies

$$|\Omega|^{1-\frac{1}{n}} \leq \frac{|\partial\Omega|}{n\,|B|^{\frac{1}{n}}}.$$

Since $|\partial\Omega| = |\partial B| = n|B|$, the right-hand side is actually $|B|^{1-\frac{1}{n}}$, so the volume of Ω is indeed bounded by the volume of B. This concludes the proof.

The above argument suggests the following problem:

Open Problem 2.1. *Can one devise an optimal coupling between sets (in the sense of a coupling between the uniform probability measures on these sets) in such a way that the total cost of the coupling decreases under some evolution converging to balls, such as mean curvature motion?*

Caffarelli's log-concave perturbation theorem

The previous example was about transporting a set to another, now the present one is in some sense about transporting a whole space to another.

It is classical in geometry to compare a space $\mathcal{X}$ with a "model space" $\mathcal{M}$ that has nice properties and is, e.g., less curved than $\mathcal{X}$. A general principle is that certain inequalities which hold true on the model space can automatically be "transported" to $\mathcal{X}$. The theorem discussed in this section is a striking illustration of this idea.

Let F, G, H, J, L be nonnegative continuous functions on $\mathbb{R}$, with H and J nondecreasing, and let $\ell \in \mathbb{R}$. For a given measure μ on $\mathbb{R}^n$, let $\lambda[\mu]$ be the largest $\lambda \geq 0$ such that, for all Lipschitz functions $h : \mathbb{R}^n \to \mathbb{R}$,

$$\int_{\mathbb{R}^n} L(h)\,d\mu = \ell \quad\Longrightarrow\quad F\left(\int_{\mathbb{R}^n} G(h)\,d\mu\right) \leq \frac{1}{\lambda}\, H\left(\int_{\mathbb{R}^n} J(|\nabla h|)\,d\mu\right). \tag{2.7}$$

Functional inequalities of the form (2.7) are variants of Sobolev inequalities; many of them are well-known and useful. Caffarelli's theorem states that *they can only be improved by log-concave perturbation of the Gaussian distribution.* More precisely, if γ is the standard Gaussian measure and $\mu = e^{-v}\gamma$ is another probability measure, with v convex, then

$$\lambda[\mu] \geq \lambda[\gamma].$$

His proof is a simple consequence of the following remarkable fact, which I shall call **Caffarelli's log-concave perturbation theorem:** *If $d\mu/d\gamma$ is log-concave, then there exists a 1-Lipschitz change* *of variables from the measure* γ *to the measure* μ. In other words, there is a deterministic coupling $(X,\ Y = \mathcal{C}(X))$ of (γ, μ), such that $|\mathcal{C}(x) - \mathcal{C}(y)| \le |x - y|$, or equivalently $|\nabla\mathcal{C}| \le 1$ (almost everywhere). It follows in particular that

$$\big|\nabla(h \circ \mathcal{C})\big| \le |(\nabla h) \circ \mathcal{C}|, \tag{2.8}$$

whatever the function h.

Now it is easy to understand why the existence of the map $\mathcal{C}$ implies (2.7): On the one hand, the definition of change of variables implies

$$\int G(h)\, d\mu = \int G(h \circ \mathcal{C})\, d\gamma, \qquad \int L(h)\, d\mu = \int L(h \circ \mathcal{C})\, d\gamma;$$

on the other hand, by the definition of change of variables again, inequality (2.8) and the nondecreasing property of J,

$$\int J(|\nabla h|)\, d\mu = \int J\big(|\nabla h \circ \mathcal{C}|\big)\, d\gamma \ge \int J\big(|\nabla(h \circ \mathcal{C})|\big)\, d\gamma.$$

Thus, inequality (2.7) is indeed "transported" from the space $(\mathbb{R}^n, \gamma)$ to the space $(\mathbb{R}^n, \mu)$.

Bibliographical notes

It is very classical to use coupling arguments to prove convergence to equilibrium for stochastic differential equations and Markov chains; many examples are described by Rachev and Rüschendorf [696] and Thorisson [781]. Actually, the standard argument found in textbooks to prove the convergence to equilibrium for a positive aperiodic ergodic Markov chain is a coupling argument (but the null case can also be treated in a similar way, as I learnt from Thorisson). Optimal couplings are often well adapted to such situations, but definitely not the only ones to apply.

The coupling method is not limited to systems of independent particles, and sometimes works in presence of correlations, for instance if the law satisfies a nonlinear diffusion equation. This is exemplified in works by Tanaka [777] on the spatially homogeneous Boltzmann equation with Maxwell molecules (the core of Tanaka's argument is reproduced in my book [814, Section 7.5]), or some recent papers [138, 214, 379, 590].

Cattiaux and Guillin [221] found a simple and elegant coupling argument to prove the exponential convergence for the law of the stochastic process

$$dX_t = \sqrt{2}\, dB_t - \widetilde{\mathbb{E}}\, \nabla V(X_t - \widetilde{X}_t)\, dt,$$

where $\widetilde{X}_t$ is an independent copy of X_t, the $\widetilde{\mathbb{E}}$ expectation only bears on $\widetilde{X}_t$, and V is assumed to be a uniformly convex C^1 potential on $\mathbb{R}^n$ satisfying $V(-x) = V(x)$.

It is also classical to couple a system of particles with an auxiliary artificial system to study the limit when the number of particles becomes large. For the Vlasov equation in kinetic theory this was done by Dobrushin [309] and Neunzert [653] several decades ago. (The proof is reproduced in Spohn [757, Chapter 5], and also suggested as an exercise in my book [814, Problem 14].) Later Sznitman used this strategy in a systematic way for the propagation of chaos, and made it very popular, see e.g. his work on the Boltzmann equation [767] or his Saint-Flour lecture notes [768] and the many references included.

In all these works, the "philosophy" is always the same: Introduce some nice coupling and see how it evolves in a certain asymptotic regime (say, either the time, or the number of particles, or both, go to infinity).

It is possible to treat the convergence to equilibrium for the complete system (2.1) by methods that are either analytic [301, 472, 816, 818] or probabilistic [55, 559, 606, 701], but all methods known to me are much more delicate than the simple coupling argument which works for (2.2). It is certainly a nice open problem to find an elementary coupling argument which applies to (2.1). (The arguments in the above-mentioned probabilistic proofs ultimately rely on coupling methods via theorems of convergence for Markov chains, but in a quite indirect way.)

Coupling techniques have also been used recently for proving rather spectacular uniqueness theorems for invariant measures in infinite dimension, see e.g. [321, 456, 457].

Classical references for the isoperimetric inequality and related topics are the books by Burago and Zalgaller [176], and Schneider [741]; and the survey by Osserman [664]. Knothe [523] had the idea to use a "coupling" method to prove geometric inequalities, and Gromov [635, Appendix] applied this method to prove the Euclidean isopetrimetric inequality. Trudinger [787] gave a closely related treatment of the same inequality and some of its generalizations, by means of a clever use of the Monge–Ampère equation (which more or less amounts to the construction of an optimal coupling with quadratic cost function, as will be seen in Chapter 11). Cabré [182] found a surprising simplification of Trudinger's method, based on the solution of just a linear elliptic equation. The "proof" which I gave in this chapter is a variation on Gromov's argument; although it is not rigorous, there is no real difficulty in turning it into a full proof, as was done by Figalli, Maggi and Pratelli [369]. These authors actually prove much more, since they use this strategy to establish a sharp quantitative *stability* of the isoperimetric inequality (if the shape of a set departs from the optimal shape, then its isoperimetric ratio departs from the optimal ratio in a quantifiable way). In the same work one can find a very interesting comparison of the respective performances of the couplings obtained by the Knothe method and by the optimal transport method (the comparison turns very much to the advantage of optimal transport).

Other links between coupling and isoperimetric-type inequalities are presented in Chapter 6 of my book [814], the research paper [587], the review paper [586] and the bibliographical notes at the end of Chapters 18 and 21.

The construction of Caffarelli's map C is easy, at least conceptually: The optimal coupling of the Gaussian measure γ with the measure $\mu = e^{-v}\gamma$, when the cost function is the square of the Euclidean distance, will do the job. But proving that C is indeed 1-Lipschitz is much more of a sport, and involves some techniques from nonlinear partial differential equations [188]. An idea of the core of the proof is explained in [814, Problem 13]. It would be nice to find a softer argument.

Üstünel pointed out to me that, if v is convex and symmetric ($v(-x) = v(x)$), then the Moser transport T from γ to $e^{-v}\gamma$ is contracting, in the sense that $|T(x)| \le |x|$; it is not clear however that T would be 1-Lipschitz.

Caffarelli's theorem has many analytic and probabilistic applications, see e.g. [242, 413, 465]. There is an infinite-dimensional version by Feyel and Üstünel [361], where the Gaussian measure is replaced by the Wiener measure. Another variant was recently studied by Valdimarsson [801].

Like the present chapter, the lecture notes [813], written for a CIME Summer School in 2001, present some applications of optimal transport in various fields, with a slightly impressionistic style.

3

The founding fathers of optimal transport

Like many other research subjects in mathematics, the field of optimal transport was born several times. The first of these births occurred at the end of the eighteenth century, by way of the French geometer Gaspard Monge.

Monge was born in 1746 under the French Ancient Régime. Because of his outstanding skills, he was admitted in a military training school from which he should have been excluded because of his modest origin. He invented descriptive geometry on his own, and the power of the method was so apparent that he was appointed professor at the age of 22, with the understanding that his theory would remain a military secret, for exclusive use of higher officers. He later was one of the most ardent warrior scientists of the French Revolution, served as a professor under several regimes, escaped a death sentence pronounced during the Terror, and became one of Napoleon's closest friends. He taught at École Normale Supérieure and École Polytechnique in Paris. Most of his work was devoted to geometry.

In 1781 he published one of his famous works, *Mémoire sur la théorie des déblais et des remblais* (a "déblai" is an amount of material that is extracted from the earth or a mine; a "remblai" is a material that is input into a new construction). The problem considered by Monge is as follows: Assume you have a certain amount of soil to extract from the ground and transport to places where it should be incorporated in a construction (see Figure 3.1). The places where the material should be extracted, and the ones where it should be transported to, are all known. But the assignment has to be determined: To which destination should one send the material that has been extracted at a certain place? The answer does matter because transport is costly, and

C. Villani, *Optimal Transport*, Grundlehren der mathematischen Wissenschaften 338,

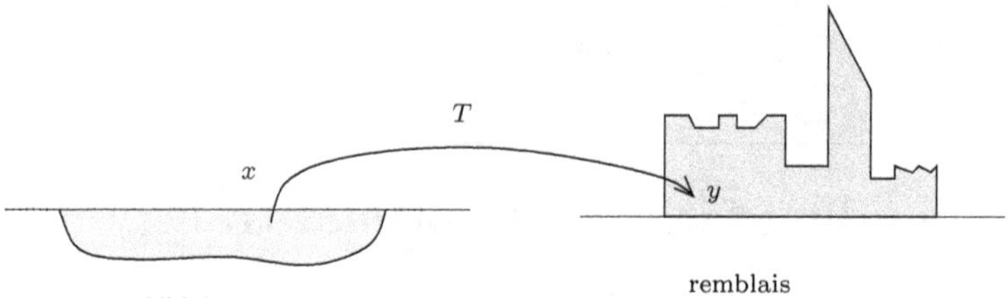

Fig. 3.1. Monge's problem of déblais and remblais

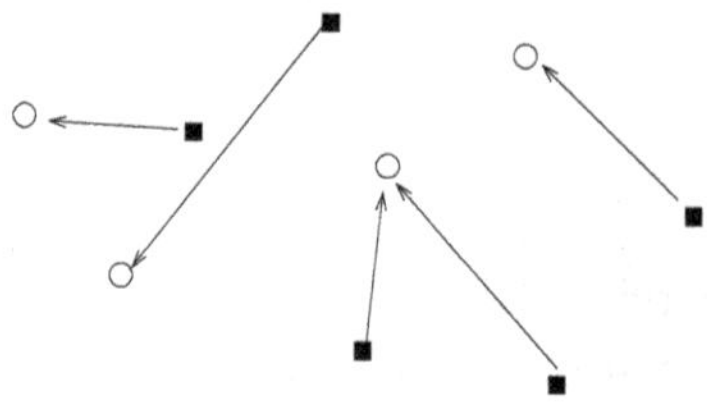

Fig. 3.2. Economic illustration of Monge's problem: squares stand for production units, circles for consumption places.

you want to minimize the total cost. Monge assumed that the transport cost of one unit of mass along a certain distance was given by the product of the mass by the distance.

Nowadays there is a Monge street in Paris, and therein one can find an excellent bakery called *Le Boulanger de Monge.* To acknowledge this, and to illustrate how Monge's problem can be recast in an economic perspective, I shall express the problem as follows. Consider a large number of bakeries, producing loaves, that should be transported each morning to cafés where consumers will eat them. The amount of bread that can be produced at each bakery, and the amount that will be consumed at each café are known in advance, and can be modeled as probability measures (there is a "density of production" and a "density of consumption") on a certain space, which in our case would be Paris (equipped with the natural metric such that the distance between two points is the length of the shortest path joining them). The problem is to find in practice where each unit of bread should go (see Figure 3.2), in such a way as to minimize the total transport cost. So Monge's problem really is the search of an optimal coupling; and to be more precise, he was looking for a *deterministic* optimal coupling.

Monge studied the problem in three dimensions for a continuous distribution of mass. Guided by his beautiful geometric intuition, he made the important observation that transport should go along straight lines that would be orthogonal to a family of surfaces. This study led him to the discovery of *lines of curvature*, a concept that by itself was a great contribution to the geometry of surfaces. His ideas were developed by Charles Dupin and later by Paul Appell. By current mathematical standards, all these arguments were flawed, yet it certainly would be worth looking up all these problems with modern tools.

Much later Monge's problem was rediscovered by the Russian mathematician Leonid Vitaliyevich Kantorovich. Born in 1912, Kantorovich was a very gifted mathematician who made his reputation as a first-class researcher at the age of 18, and earned a position of professor at just the same age as Monge had. He worked in many areas of mathematics, with a strong taste for applications in economics, and later theoretical computer science. In 1938 a laboratory consulted him for the solution of a certain optimization problem, which he found out was representative of a whole class of linear problems arising in various areas of economics. Motivated by this discovery, he developed the tools of linear programming, that later became prominent in economics. The publication of some of his most important works was delayed because of the great care with which Soviet authorities of the time handled the divulgence of scientific research related to economics. In fact (and this is another common point with Monge) for many years it was strictly forbidden for Kantorovich to publicly discuss some of his main discoveries. In the end his work became well-known and in 1975 was awarded the Nobel Prize for economics, jointly with Tjalling Koopmans, "for their contributions to the theory of optimum allocation of resources".

In the case that is of direct interest for us, namely the problem of optimal coupling, Kantorovich stated and proved, by means of functional analytical tools, a duality theorem that would play a crucial role later. He also devised a convenient notion of distance between probability measures: the distance between two measures should be the optimal transport cost from one to the other, if the cost is chosen as the distance function. This distance between probability measures is nowadays called the Kantorovich–Rubinstein distance, and has proven to be particularly flexible and useful.

It was only several years after his main results that Kantorovich made the connection with Monge's work. The problem of optimal coupling has since then been called the **Monge–Kantorovich problem**.

Throughout the second half of the twentieth century, optimal coupling techniques and variants of the Kantorovich–Rubinstein distance (nowadays often called Wasserstein distances, or other denominations) were used by statisticians and probabilists. The "basis" space could be finite-dimensional, or infinite-dimensional: For instance, optimal couplings give interesting notions of distance between probability measures on path spaces. Noticeable contributions from the seventies are due to Roland Dobrushin, who used such distances in the study of particle systems; and to Hiroshi Tanaka, who applied them to study the time-behavior of a simple variant of the Boltzmann equation. By the mid-eighties, specialists of the subject, like Svetlozar Rachev or Ludger Rüschendorf, were in possession of a large library of ideas, tools, techniques and applications related to optimal transport.

During that time, *reparametrization* techniques (yet another word for change of variables) were used by many researchers working on inequalities involving volumes or integrals. Only later would it be understood that optimal transport often provides useful reparametrizations.

At the end of the eighties, three directions of research emerged independently and almost simultaneously, which completely reshaped the whole picture of optimal transport.

One of them was John Mather's work on Lagrangian dynamical systems. Action-minimizing curves are basic important objects in the theory of dynamical systems, and the construction of closed action-minimizing curves satisfying certain qualitative properties is a classical problem. By the end of the eighties, Mather found it convenient to study not only action-minimizing curves, but action-minimizing stationary *measures* in phase space. Mather's measures are a generalization of action-minimizing curves, and they solve a variational problem which in effect is a Monge–Kantorovich problem. Under some conditions on the Lagrangian, Mather proved a celebrated result according to which (roughly speaking) certain action-minimizing measures are automatically concentrated on Lipschitz graphs. As we shall understand in Chapter 8, this problem is intimately related to the construction of a *deterministic* optimal coupling.

The second direction of research came from the work of Yann Brenier. While studying problems in incompressible fluid mechanics, Brenier needed to construct an operator that would act like the projection on the set of measure-preserving mappings in an open set (in probabilistic language, measure-preserving mappings are deterministic couplings of the Lebesgue measure with itself). He understood that he could do so by introducing an optimal coupling: If u is the map for which one wants to compute the projection, introduce a coupling of the Lebesgue measure $\mathcal{L}$ with $u_{\#}\mathcal{L}$. This study revealed an unexpected link between optimal transport and fluid mechanics; at the same time, by pointing out the relation with the theory of Monge–Ampère equations, Brenier attracted the attention of the community working on partial differential equations.

The third direction of research, certainly the most surprising, came from outside mathematics. Mike Cullen was part of a group of meteorologists with a well-developed mathematical taste, working on *semi-geostrophic equations*, used in meteorology for the modeling of atmospheric fronts. Cullen and his collaborators showed that a certain famous change of unknown due to Brian Hoskins could be re-interpreted in terms of an optimal coupling problem, and they identified the minimization property as a *stability* condition. A striking outcome of this work was that optimal transport could arise naturally in partial differential equations which seemed to have nothing to do with it.

All three contributions emphasized (in their respective domain) that *important information can be gained by a qualitative description of optimal transport.* These new directions of research attracted various mathematicians (among the first, Luis Caffarelli, Craig Evans, Wilfrid Gangbo, Robert McCann, and others), who worked on a better description of the structure of optimal transport and found other applications.

An important conceptual step was accomplished by Felix Otto, who discovered an appealing formalism introducing a *differential* point of view in optimal transport theory. This opened the way to a more geometric description of the space of probability measures, and connected optimal transport to the theory of diffusion equations, thus leading to a rich interplay of geometry, functional analysis and partial differential equations.

Nowadays optimal transport has become a thriving industry, involving many researchers and many trends. Apart from meteorology, fluid mechanics and diffusion equations, it has also been applied to such diverse topics as the collapse of sandpiles, the matching of images, and the design of networks or reflector antennas. My book, *Topics in Optimal Transportation*, written between 2000 and 2003, was the first attempt to present a synthetic view of the modern theory. Since then the field has grown much faster than I expected, and it was never so active as it is now.

Bibliographical notes

Before the twentieth century, the main references for the problem of "déblais et remblais" are the memoirs by Monge [636], Dupin [319] and Appell [42]. Besides achieving important mathematical results, Monge and Dupin were strongly committed to the development of society and it is interesting to browse some of their writings about economics and industry (a list can be found online at `gallica.bnf.fr`). A lively account of Monge's life and political commitments can be found in Bell's delightful treatise, *Men of Mathematics* [80, Chapter 12]. It seems however that Bell did dramatize the story a bit, at the expense of accuracy and neutrality. A more cold-blooded biography of Monge was written by de Launay [277]. Considered as one the greatest geologists of his time, not particularly sympathetic to the French Revolution, de Launay documented himself with remarkable rigor, going back to original sources whenever possible. Other biographies have been written since then by Taton [778, 779] and Aubry [50].

Monge originally formulated his transport problem in Euclidean space for the cost function $c(x, y) = |x - y|$; he probably had no idea of the extreme difficulty of a rigorous treatment. It was only in 1979 that Sudakov [765] claimed a proof of the existence of a Monge transport for general probability densities with this particular cost function. But his proof was not completely correct, and was amended much later by Ambrosio [20]. In the meantime, alternative rigorous proofs had been devised first by Evans and Gangbo [330] (under rather strong assumptions on the data), then by Trudinger and Wang [791], and Caffarelli, Feldman and McCann [190].

Kantorovich defined linear programming in [499], introduced his minimization problem and duality theorem in [500], and in [501] applied his theory to the problem of optimal transport; this note can be considered as the act of birth of the modern formulation of optimal transport. Later he made the link with Monge's problem in [502]. His major work in economics is the book [503], including a reproduction of [499]. Another important contribution is a study of numerical schemes based on linear programming, joint with his student Gavurin [505].

Kantorovich wrote a short autobiography for his Nobel Prize [504]. Online at `www.math.nsc.ru/LBRT/g2/english/ssk/legacy.html` are some comments by Kutateladze, who edited his mathematical works. A recent special issue of the *Journal of Mathematical Sciences*, edited by Vershik, was devoted to Kantorovich [810]; this reference contains translations of [501] and [502], as well as much valuable information about the personality of Kantorovich, and the genesis and impact of his ideas in mathematics, economy and computer science. In another historical note [808] Vershik recollects memories of Kantorovich and tells some tragicomical stories illustrating the incredible ideological pressure put on him and other scientists by Soviet authorities at the time.

The "classical" probabilistic theory of optimal transport is exhaustively reviewed by Rachev and Rüschendorf [696, 721]; most notable applications include limit theorems for various random processes. Relations with game theory, economics, statistics, and hypotheses testing are also common (among many references see e.g. [323, 391]).

Mather introduced minimizing measures in [600], and proved his Lipschitz graph theorem in [601]. The explicit connection with the Monge–Kantorovich problem came only recently [105]: see Chapter 8.

Tanaka's contributions to kinetic theory go back to the mid-seventies [644, 776, 777]. His line of research was later taken up by Toscani and collaborators [133, 692]; these papers constituted my first contact with the optimal transport problem. More recent developments in the kinetic theory of granular media appear for instance in [138].

Brenier announced his main results in a short note [154], then published detailed proofs in [156]. Chapter 3 in [814] is entirely devoted to Brenier's **polar factorization theorem** (which includes the existence of the projection operator), its interpretation and consequences. For the sources of inspiration of Brenier, and various links between optimal transport and hydrodynamics, one may consult [155, 158, 159, 160, 163, 170]. Recent papers by Ambrosio and Figalli [24, 25] provide a complete and thorough rewriting of Brenier's theory of generalized incompressible flows.

The semi-geostrophic system was introduced by Eliassen [325] and Hoskins [480, 481, 482]; it is *very* briefly described in [814, Problem 9, pp. 323–326]. Cullen and collaborators wrote many papers on the subject, see in particular [269]; see also the review article [263], the works by Cullen and Gangbo [266], Cullen and Feldman [265] or the recent book by Cullen [262].

Further links between optimal transport and other fields of mathematics (or physics) can be found in my book [814], or in the treatise by Rachev and Rüschendorf [696]. An important source of inspiration was the relation with the qualitative behavior of certain diffusive equations arising from gas dynamics; this link was discovered by Jordan, Kinderlehrer and Otto at the end of the nineties [493], and then explored by several authors [208, 209, 210, 211, 212, 213, 214, 216, 669, 671].

Below is a nonexhaustive list of some other unexpected applications. Relations with the modeling of sandpiles are reviewed by Evans [328], as well as compression molding problems; see also Feldman [353] (this is for the cost function $c(x,y) = |x-y|$). Applications of optimal transport to image processing and shape recognition are discussed by Gangbo and McCann [400], Ahmad [6], Angenent, Haker, Tannenbaum, and Zhu [462, 463], Chazal, Cohen-Steiner and Mérigot [224], and many other contributors from the engineering community (see e.g. [700, 713]). X.-J. Wang [834], and independently Glimm and Oliker [419] (around 2000 and 2002 respectively) discovered that the theoretical problem of designing reflector antennas could be recast in terms of optimal transport for the cost function $c(x,y) = -\log(1 - x \cdot y)$ on S^2; see [402, 419, 660] for further work in the area, and [420] for another version of this problem involving two reflectors.[1] Rubinstein and Wolansky adapted the strategy in [420] to study the optimal design of lenses [712]; and Gutiérrez and Huang to treat a refraction problem [453]. In his PhD Thesis, Bernot [108] made the link between optimal transport, irrigation and the design of networks. Such topics were also considered by Santambrogio with various collaborators [152, 207, 731, 732, 733, 734]; in particular it is shown in [732] that optimal transport theory gives a rigorous basis to some variational constructions used by physicists and hydrologists to study river basin morphology [65, 706]. Buttazzo and collaborators [178, 179, 180] explored city planning via optimal transport. Brenier found a connection to the electrodynamic equations of Maxwell and related models in string theory [161, 162, 163, 164, 165, 166]. Frisch and collaborators linked optimal transport to the problem of reconstruction of the "conditions of the initial Universe" [168, 382, 755]. (The publication of [382] in the prestigious generalist scientific journal *Nature* is a good indication of the current visibility of optimal transport outside mathematics.)

Relations of optimal transport with geometry, in particular Ricci curvature, will be explored in detail in Parts II and III of these notes.

Many generalizations and variants have been studied in the literature, such as the optimal matching [323], the optimal transshipment

[1] According to Oliker, the connection between the two-reflector problem (as formulated in [661]) and optimal transport is in fact much older, since it was first formulated in a 1993 conference in which he and Caffarelli were participating.

(see [696] for a discussion and list of references), the optimal transport of a fraction of the mass [192, 365], or the optimal coupling with more than two prescribed marginals [403, 525, 718, 723, 725]; I learnt from Strulovici that the latter problem has applications in contract theory.

In spite of this avalanche of works, one certainly should not regard optimal transport as a kind of miraculous tool, for "there are no miracles in mathematics". In my opinion this abundance only reflects the fact that optimal transport is a simple, meaningful, natural and therefore universal concept.

Part I

Qualitative description of optimal transport

The first part of this course is devoted to the description and characterization of optimal transport under certain regularity assumptions on the measures and the cost function.

As a start, some general theorems about optimal transport plans are established in Chapters 4 and 5, in particular the Kantorovich duality theorem. The emphasis is on c-cyclically monotone maps, both in the statements and in the proofs. The assumptions on the cost function and the spaces will be very general.

From the Monge–Kantorovich problem one can derive natural distance functions on spaces of probability measures, by choosing the cost function as a power of the distance. The main properties of these distances are established in Chapter 6.

In Chapter 7 a time-dependent version of the Monge–Kantorovich problem is investigated, which leads to an interpolation procedure between probability measures, called displacement interpolation. The natural assumption is that the cost function derives from a Lagrangian action, in the sense of classical mechanics; still (almost) no smoothness is required at that level. In Chapter 8 I shall make further assumptions of smoothness and convexity, and recover some regularity properties of the displacement interpolant by a strategy due to Mather.

Then in Chapters 9 and 10 it is shown how to establish the existence of deterministic optimal couplings, and characterize the associated transport maps, again under adequate regularity and convexity assumptions. The Change of variables Formula is considered in Chapter 11. Finally, in Chapter 12 I shall discuss the regularity of the transport map, which in general is not smooth.

The main results of this part are synthetized and summarized in Chapter 13. A good understanding of this chapter is sufficient to go through Part II of this course.

4

Basic properties

Existence

The first good thing about optimal couplings is that they exist:

Theorem 4.1 (Existence of an optimal coupling). *Let* $(\mathcal{X}, \mu)$ *and* $(\mathcal{Y}, \nu)$ *be two Polish probability spaces; let* $a : \mathcal{X} \to \mathbb{R} \cup \{-\infty\}$ *and* $b : \mathcal{Y} \to \mathbb{R} \cup \{-\infty\}$ *be two upper semicontinuous functions such that* $a \in L^1(\mu)$, $b \in L^1(\nu)$. *Let* $c : \mathcal{X} \times \mathcal{Y} \to \mathbb{R} \cup \{+\infty\}$ *be a lower semicontinuous cost function, such that* $c(x, y) \geq a(x) + b(y)$ *for all* x, y. *Then there is a coupling of* (μ, ν) *which minimizes the total cost* $\mathbb{E}\, c(X, Y)$ *among all possible couplings* (X, Y).

Remark 4.2. The lower bound assumption on c guarantees that the expected cost $\mathbb{E}\, c(X, Y)$ is well-defined in $\mathbb{R} \cup \{+\infty\}$. In most cases of applications — but not all — one may choose $a = 0$, $b = 0$.

The proof relies on basic variational arguments involving the topology of weak convergence (i.e. imposed by bounded continuous test functions). There are two key properties to check: (a) lower semicontinuity, (b) compactness. These issues are taken care of respectively in Lemmas 4.3 and 4.4 below, which will be used again in the sequel. Before going on, I recall **Prokhorov's theorem**: *If* $\mathcal{X}$ *is a Polish space, then a set* $\mathcal{P} \subset P(\mathcal{X})$ *is precompact for the weak topology if and only if it is* tight, *i.e. for any* $\varepsilon > 0$ *there is a compact set* K_ε *such that* $\mu[\mathcal{X} \setminus K_\varepsilon] \leq \varepsilon$ *for all* $\mu \in \mathcal{P}$.

Lemma 4.3 (Lower semicontinuity of the cost functional). *Let* $\mathcal{X}$ *and* $\mathcal{Y}$ *be two Polish spaces, and* $c : \mathcal{X} \times \mathcal{Y} \to \mathbb{R} \cup \{+\infty\}$ *a lower* *semicontinuous cost function. Let* $h : \mathcal{X} \times \mathcal{Y} \to \mathbb{R} \cup \{-\infty\}$ *be an upper semicontinuous function such that* $c \geq h$. *Let* $(\pi_k)_{k\in\mathbb{N}}$ *be a sequence of probability measures on* $\mathcal{X}\times\mathcal{Y}$, *converging weakly to some* $\pi \in P(\mathcal{X}\times\mathcal{Y})$, *in such a way that* $h \in L^1(\pi_k)$, $h \in L^1(\pi)$, *and*

$$\int_{\mathcal{X}\times\mathcal{Y}} h\,d\pi_k \xrightarrow[k\to\infty]{} \int_{\mathcal{X}\times\mathcal{Y}} h\,d\pi.$$

Then

$$\int_{\mathcal{X}\times\mathcal{Y}} c\,d\pi \leq \liminf_{k\to\infty} \int_{\mathcal{X}\times\mathcal{Y}} c\,d\pi_k.$$

In particular, if c *is nonnegative, then* $F : \pi \to \int c\,d\pi$ *is lower semicontinuous on* $P(\mathcal{X} \times \mathcal{Y})$, *equipped with the topology of weak convergence.*

Lemma 4.4 (Tightness of transference plans). *Let* $\mathcal{X}$ *and* $\mathcal{Y}$ *be two Polish spaces. Let* $\mathcal{P} \subset P(\mathcal{X})$ *and* $\mathcal{Q} \subset P(\mathcal{Y})$ *be tight subsets of* $P(\mathcal{X})$ *and* $P(\mathcal{Y})$ *respectively. Then the set* $\Pi(\mathcal{P}, \mathcal{Q})$ *of all transference plans whose marginals lie in* $\mathcal{P}$ *and* $\mathcal{Q}$ *respectively, is itself tight in* $P(\mathcal{X} \times \mathcal{Y})$.

Proof of Lemma 4.3. Replacing c by $c - h$, we may assume that c is a nonnegative lower semicontinuous function. Then c can be written as the pointwise limit of a nondecreasing family $(c_\ell)_{\ell\in\mathbb{N}}$ of continuous real-valued functions. By monotone convergence,

$$\int c\,d\pi = \lim_{\ell\to\infty} \int c_\ell\,d\pi = \lim_{\ell\to\infty}\ \lim_{k\to\infty} \int c_\ell\,d\pi_k \leq \liminf_{k\to\infty} \int c\,d\pi_k.$$

□

Proof of Lemma 4.4. Let $\mu \in \mathcal{P}$, $\nu \in \mathcal{Q}$, and $\pi \in \Pi(\mu,\nu)$. By assumption, for any $\varepsilon > 0$ there is a compact set $K_\varepsilon \subset \mathcal{X}$, independent of the choice of μ in $\mathcal{P}$, such that $\mu[\mathcal{X} \setminus K_\varepsilon] \leq \varepsilon$; and similarly there is a compact set $L_\varepsilon \subset \mathcal{Y}$, independent of the choice of ν in $\mathcal{Q}$, such that $\nu[\mathcal{Y} \setminus L_\varepsilon] \leq \varepsilon$. Then for any coupling (X,Y) of (μ,ν),

$$\mathbb{P}\left[(X,Y) \notin K_\varepsilon \times L_\varepsilon\right] \leq \mathbb{P}\left[X \notin K_\varepsilon\right] + \mathbb{P}\left[Y \notin L_\varepsilon\right] \leq 2\varepsilon.$$

The desired result follows since this bound is independent of the coupling, and $K_\varepsilon \times L_\varepsilon$ is compact in $\mathcal{X} \times \mathcal{Y}$. □

Proof of Theorem 4.1. Since $\mathcal{X}$ is Polish, $\{\mu\}$ is tight in $P(\mathcal{X})$; similarly, $\{\nu\}$ is tight in $P(\mathcal{Y})$. By Lemma 4.4, $\Pi(\mu,\nu)$ is tight in $P(\mathcal{X} \times \mathcal{Y})$, and by Prokhorov's theorem this set has a compact closure. By passing to the limit in the equation for marginals, we see that $\Pi(\mu,\nu)$ is closed, so it is in fact *compact*.

Then let $(\pi_k)_{k\in\mathbb{N}}$ be a sequence of probability measures on $\mathcal{X}\times\mathcal{Y}$, such that $\int c\,d\pi_k$ converges to the infimum transport cost. Extracting a subsequence if necessary, we may assume that π_k converges to some $\pi\in\Pi(\mu,\nu)$. The function $h:(x,y)\longmapsto a(x)+b(y)$ lies in $L^1(\pi_k)$ and in $L^1(\pi)$, and $c\geq h$ by assumption; moreover, $\int h\,d\pi_k=\int h\,d\pi=\int a\,d\mu+\int b\,d\nu$; so Lemma 4.3 implies

$$\int c\,d\pi\leq\liminf_{k\to\infty}\int c\,d\pi_k.$$

Thus π is minimizing. □

Remark 4.5. This existence theorem does not imply that the optimal cost is finite. It might be that *all* transport plans lead to an infinite total cost, i.e. $\int c\,d\pi=+\infty$ for all $\pi\in\Pi(\mu,\nu)$. A simple condition to rule out this annoying possibility is

$$\int c(x,y)\,d\mu(x)\,d\nu(y)<+\infty,$$

which guarantees that at least the independent coupling has finite total cost. In the sequel, I shall sometimes make the stronger assumption

$$c(x,y)\leq c_{\mathcal{X}}(x)+c_{\mathcal{Y}}(y),\qquad (c_{\mathcal{X}},c_{\mathcal{Y}})\in L^1(\mu)\times L^1(\nu),$$

which implies that *any* coupling has finite total cost, and has other nice consequences (see e.g. Theorem 5.10).

Restriction property

The second good thing about optimal couplings is that *any sub-coupling is still optimal.* In words: If you have an optimal transport plan, then any induced sub-plan (transferring part of the initial mass to part of the final mass) has to be optimal too — otherwise you would be able to lower the cost of the sub-plan, and as a consequence the cost of the whole plan. This is the content of the next theorem.

Theorem 4.6 (Optimality is inherited by restriction). *Let* $(\mathcal{X}, \mu)$ *and* $(\mathcal{Y}, \nu)$ *be two Polish spaces,* $a \in L^1(\mu)$, $b \in L^1(\nu)$, *let* $c : \mathcal{X} \times \mathcal{Y} \to \mathbb{R} \cup \{+\infty\}$ *be a measurable cost function such that* $c(x, y) \geq a(x) + b(y)$ *for all* x, y; *and let* $C(\mu, \nu)$ *be the optimal transport cost from* μ *to* ν. *Assume that* $C(\mu, \nu) < +\infty$ *and let* $\pi \in \Pi(\mu, \nu)$ *be an optimal transport plan. Let* $\widetilde{\pi}$ *be a nonnegative measure on* $\mathcal{X} \times \mathcal{Y}$, *such that* $\widetilde{\pi} \leq \pi$ *and* $\widetilde{\pi}[\mathcal{X} \times \mathcal{Y}] > 0$. *Then the probability measure*

$$\pi' := \frac{\widetilde{\pi}}{\widetilde{\pi}[\mathcal{X} \times \mathcal{Y}]}$$

is an optimal transference plan between its marginals μ' *and* ν'.

Moreover, if π *is the unique optimal transference plan between* μ *and* ν, *then also* π' *is the unique optimal transference plan between* μ' *and* ν'.

Example 4.7. If (X, Y) is an optimal coupling of (μ, ν), and $\mathcal{Z} \subset \mathcal{X} \times \mathcal{Y}$ is such that $\mathbb{P}\left[(X, Y) \in \mathcal{Z}\right] > 0$, then the pair (X, Y), conditioned to lie in $\mathcal{Z}$, is an optimal coupling of (μ', ν'), where μ' is the law of X conditioned by the event "$(X, Y) \in \mathcal{Z}$", and ν' is the law of Y conditioned by the same event.

Proof of Theorem 4.6. Assume that π' is not optimal; then there exists π'' such that

$$(\mathrm{proj}_{\mathcal{X}})_{\#}\pi'' = (\mathrm{proj}_{\mathcal{X}})_{\#}\pi' = \mu', \qquad (\mathrm{proj}_{\mathcal{Y}})_{\#}\pi'' = (\mathrm{proj}_{\mathcal{Y}})_{\#}\pi' = \nu', \tag{4.1}$$

yet

$$\int c(x, y)\, d\pi''(x, y) < \int c(x, y)\, d\pi'(x, y). \tag{4.2}$$

Then consider

$$\widehat{\pi} := (\pi - \widetilde{\pi}) + \widetilde{Z}\pi'', \tag{4.3}$$

where $\widetilde{Z} = \widetilde{\pi}[\mathcal{X} \times \mathcal{Y}] > 0$. Clearly, $\widehat{\pi}$ is a nonnegative measure. On the other hand, it can be written as

$$\widehat{\pi} = \pi + \widetilde{Z}(\pi'' - \pi');$$

then (4.1) shows that $\widehat{\pi}$ has the same marginals as π, while (4.2) implies that it has a lower transport cost than π. (Here I use the fact that the total cost is finite.) This contradicts the optimality of π. The conclusion is that π' is in fact optimal.

It remains to prove the last statement of Theorem 4.6. Assume that π is the unique optimal transference plan between μ and ν; and let π'' be any optimal transference plan between μ' and ν'. Define again $\widehat{\pi}$ by (4.3). Then $\widehat{\pi}$ has the same cost as π, so $\widehat{\pi} = \pi$, which implies that $\widetilde{\pi} = \widetilde{Z}\pi''$, i.e. $\pi'' = \pi'$. $\square$

Convexity properties

The following estimates are of constant use:

Theorem 4.8 (Convexity of the optimal cost). *Let $\mathcal{X}$ and $\mathcal{Y}$ be two Polish spaces, let $c : \mathcal{X} \times \mathcal{Y} \to \mathbb{R} \cup \{+\infty\}$ be a lower semicontinuous function, and let C be the associated optimal transport cost functional on $P(\mathcal{X}) \times P(\mathcal{Y})$. Let (Θ, λ) be a probability space, and let μ_θ, ν_θ be two measurable functions defined on Θ, with values in $P(\mathcal{X})$ and $P(\mathcal{Y})$ respectively. Assume that $c(x,y) \geq a(x) + b(y)$, where $a \in L^1(d\mu_\theta\, d\lambda(\theta))$, $b \in L^1(d\nu_\theta\, d\lambda(\theta))$. Then*

$$C\left(\int_\Theta \mu_\theta\, \lambda(d\theta), \int_\Theta \nu_\theta\, \lambda(d\theta)\right) \leq \left(\int_\Theta C(\mu_\theta, \nu_\theta)\, \lambda(d\theta)\right).$$

Proof of Theorem 4.8. First notice that $a \in L^1(\mu_\theta)$, $b \in L^1(\nu_\theta)$ for λ-almost all values of θ. For each such θ, Theorem 4.1 guarantees the existence of an optimal transport plan $\pi_\theta \in \Pi(\mu_\theta, \nu_\theta)$, for the cost c. Then $\pi := \int \pi_\theta\, \lambda(d\theta)$ has marginals $\mu := \int \mu_\theta\, \lambda(d\theta)$ and $\nu := \int \nu_\theta\, \lambda(d\theta)$. Admitting temporarily Corollary 5.22, we may assume that π_θ is a measurable function of θ. So

$$\begin{aligned}
C(\mu,\nu) &\leq \int_{\mathcal{X}\times\mathcal{Y}} c(x,y)\, \pi(dx\, dy) \\
&= \int_{\mathcal{X}\times\mathcal{Y}} c(x,y) \left(\int_\Theta \pi_\theta\, \lambda(d\theta)\right) (dx\, dy) \\
&= \int_\Theta \left(\int_{\mathcal{X}\times\mathcal{Y}} c(x,y)\, \pi_\theta(dx\, dy)\right) \lambda(d\theta) \\
&= \int_\Theta C(\mu_\theta, \nu_\theta)\, \lambda(d\theta),
\end{aligned}$$

and the conclusion follows. $\square$

Description of optimal plans

Obtaining more precise information about minimizers will be much more of a sport. Here is a short list of questions that one might ask:

- Is the optimal coupling unique? smooth in some sense?
- Is there a *Monge coupling*, i.e. a deterministic optimal coupling?
- Is there a geometrical way to characterize optimal couplings? Can one check in practice that a certain coupling is optimal?

About the second question: Why don't we try to apply the same reasoning as in the proof of Theorem 4.1? The problem is that the set of deterministic couplings is in general *not* compact; in fact, this set is often dense in the larger space of all couplings! So we may expect that the *value* of the infimum in the Monge problem coincides with the value of the minimum in the Kantorovich problem; but there is no a priori reason to expect the existence of a Monge minimizer.

Example 4.9. Let $\mathcal{X} = \mathcal{Y} = \mathbb{R}^2$, let $c(x,y) = |x-y|^2$, let μ be $\mathcal{H}^1$ restricted to $\{0\} \times [-1,1]$, and let ν be $(1/2)\,\mathcal{H}^1$ restricted to $\{-1,1\} \times [-1,1]$, where $\mathcal{H}^1$ is the one-dimensional Hausdorff measure. Then there is a unique optimal transport, which for each point $(0,a)$ sends one half of the mass at $(0,a)$ to $(-1,a)$, and the other half to $(1,a)$. This is not a Monge transport, but it is easy to approximate it by (nonoptimal) deterministic transports (see Figure 4.1).

Fig. 4.1. The optimal plan, represented in the left image, consists in splitting the mass in the center into two halves and transporting mass horizontally. On the right the filled regions represent the lines of transport for a deterministic (without splitting of mass) approximation of the optimum.

Bibliographical notes

Theorem 4.1 has probably been known from time immemorial; it is usually stated for nonnegative cost functions.

Prokhorov's theorem is a most classical result that can be found e.g. in [120, Theorems 6.1 and 6.2], or in my own course on integration [819, Section VII-5].

Theorems of the form "infimum cost in the Monge problem = minimum cost in the Kantorovich problem" have been established by Gangbo [396, Appendix A], Ambrosio [20, Theorem 2.1], and Pratelli [687, Theorem B]. The most general results to this date are those which appear in Pratelli's work: Equality holds true if the source space $(\mathcal{X}, \mu)$ is Polish without atoms, and the cost is continuous $\mathcal{X} \times \mathcal{Y} \to \mathbb{R} \cup \{+\infty\}$, with the value $+\infty$ allowed. (In [687] the cost c is bounded below, but it is sufficient that $c(x, y) \geq a(x) + b(y)$, where $a \in L^1(\mu)$ and $b \in L^1(\nu)$ are continuous.)

5

Cyclical monotonicity and Kantorovich duality

To go on, we should become acquainted with two basic concepts in the theory of optimal transport. The first one is a geometric property called cyclical monotonicity; the second one is the Kantorovich dual problem, which is another face of the original Monge–Kantorovich problem. The main result in this chapter is Theorem 5.10.

Definitions and heuristics

I shall start by explaining the concepts of cyclical monotonicity and Kantorovich duality in an informal way, sticking to the bakery analogy of Chapter 3. Assume you have been hired by a large consortium of bakeries and cafés, to be in charge of the distribution of bread from production units (bakeries) to consumption units (cafés). The locations of the bakeries and cafés, their respective production and consumption rates, are all determined in advance. You have written a transference plan, which says, for each bakery (located at) x_i and each café y_j, how much bread should go each morning from x_i to y_j.

As there are complaints that the transport cost associated with your plan is actually too high, you try to reduce it. For that purpose you choose a bakery x_1 that sends part of its production to a distant café y_1, and decide that one basket of bread will be rerouted to another café y_2, that is closer to x_1; thus you will gain $c(x_1, y_2) - c(x_1, y_1)$. Of course, now this results in an excess of bread in y_2, so one basket of bread arriving to y_2 (say, from bakery x_2) should in turn be rerouted to yet another café, say y_3. The process goes on and on until finally you

C. Villani, *Optimal Transport*, Grundlehren der mathematischen Wissenschaften 338,

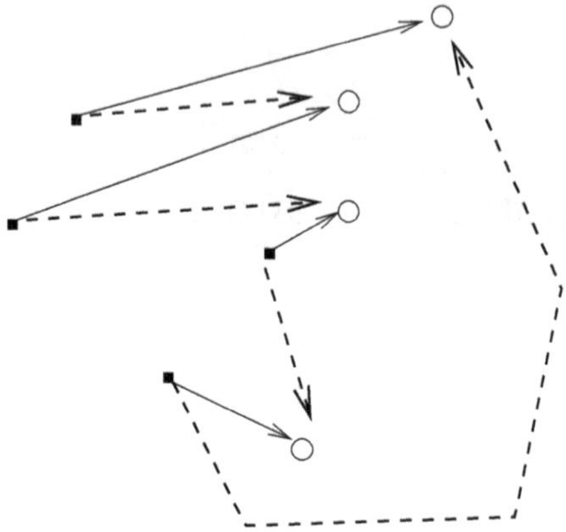

Fig. 5.1. An attempt to improve the cost by a cycle; solid arrows indicate the mass transport in the original plan, dashed arrows the paths along which a bit of mass is rerouted.

redirect a basket from some bakery x_N to y_1, at which point you can stop since you have a new admissible transference plan (see Figure 5.1).

The new plan is (strictly) better than the older one if and only if

$$c(x_1,y_2)+c(x_2,y_3)+\ldots+c(x_N,y_1)<c(x_1,y_1)+c(x_2,y_2)+\ldots+c(x_N,y_N).$$

Thus, if you can find such *cycles* $(x_1,y_1),\ldots,(x_N,y_N)$ in your transference plan, certainly the latter is not optimal. Conversely, if you do *not* find them, then your plan cannot be improved (at least by the procedure described above) and it is likely to be optimal. This motivates the following definitions.

Definition 5.1 (Cyclical monotonicity). *Let $\mathcal{X},\mathcal{Y}$ be arbitrary sets, and $c:\mathcal{X}\times\mathcal{Y}\to(-\infty,+\infty]$ be a function. A subset $\Gamma\subset\mathcal{X}\times\mathcal{Y}$ is said to be c-cyclically monotone if, for any $N\in\mathbb{N}$, and any family $(x_1,y_1),\ldots,(x_N,y_N)$ of points in Γ, holds the inequality*

$$\sum_{i=1}^{N} c(x_i,y_i)\le\sum_{i=1}^{N} c(x_i,y_{i+1}) \tag{5.1}$$

(with the convention $y_{N+1}=y_1$). A transference plan is said to be c-cyclically monotone if it is concentrated on a c-cyclically monotone set.

Informally, a c-cyclically monotone plan is a plan that *cannot be improved*: it is impossible to perturb it (in the sense considered before, by rerouting mass along some cycle) and get something more economical. One can think of it as a kind of local minimizer. It is intuitively obvious that an optimal plan should be c-cyclically monotone; the converse property is much less obvious (maybe it is possible to get something better by radically changing the plan), but we shall soon see that it holds true under mild conditions.

The next key concept is the dual Kantorovich problem. While the central notion in the original Monge–Kantorovich problem is *cost*, in the dual problem it is *price*. Imagine that a company offers to take care of all your transportation problem, buying bread at the bakeries and selling them to the cafés; what happens in between is not your problem (and maybe they have tricks to do the transport at a lower price than you). Let $\psi(x)$ be the price at which a basket of bread is bought at bakery x, and $\phi(y)$ the price at which it is sold at café y. On the whole, the price which the consortium bakery + café pays for the transport is $\phi(y) - \psi(x)$, instead of the original cost $c(x, y)$. This of course is for each unit of bread: if there is a mass $\mu(dx)$ at x, then the total price of the bread shipment from there will be $\psi(x)\,\mu(dx)$.

So as to be competitive, the company needs to set up prices in such a way that

$$\forall (x, y), \qquad \phi(y) - \psi(x) \le c(x, y). \tag{5.2}$$

When you were handling the transportation yourself, your problem was to *minimize the cost*. Now that the company takes up the transportation charge, their problem is to *maximize the profits*. This naturally leads to the **dual Kantorovich problem**:

$$\sup \left\{ \int_{\mathcal{Y}} \phi(y)\, d\nu(y) - \int_{\mathcal{X}} \psi(x)\, d\mu(x); \quad \phi(y) - \psi(x) \le c(x, y) \right\}. \tag{5.3}$$

From a mathematical point of view, it will be imposed that the functions ψ and ϕ appearing in (5.3) be integrable: $\psi \in L^1(\mathcal{X}, \mu)$; $\phi \in L^1(\mathcal{Y}, \nu)$.

With the intervention of the company, the shipment of each unit of bread does not cost more than it used to when you were handling it yourself; so it is obvious that the supremum in (5.3) is no more than the optimal transport cost:

$$\sup_{\phi-\psi\le c}\left\{\int_{\mathcal{Y}}\phi(y)\,d\nu(y)-\int_{\mathcal{X}}\psi(x)\,d\mu(x)\right\}$$
$$\le \inf_{\pi\in\Pi(\mu,\nu)}\left\{\int_{\mathcal{X}\times\mathcal{Y}} c(x,y)\,d\pi(x,y)\right\}. \quad (5.4)$$

Clearly, if we can find a pair (ψ,ϕ) and a transference plan π for which there is equality, then (ψ,ϕ) is optimal in the left-hand side and π is also optimal in the right-hand side.

A pair of price functions (ψ,ϕ) will informally be said to be *competitive* if it satisfies (5.2). For a given y, it is of course in the interest of the company to set the highest possible competitive price $\phi(y)$, i.e. the highest lower bound for (i.e. the infimum of) $\psi(x)+c(x,y)$, among all bakeries x. Similarly, for a given x, the price $\psi(x)$ should be the supremum of all $\phi(y)-c(x,y)$. Thus it makes sense to describe a pair of prices (ψ,ϕ) as *tight* if

$$\phi(y)=\inf_x\Big(\psi(x)+c(x,y)\Big),\qquad \psi(x)=\sup_y\Big(\phi(y)-c(x,y)\Big). \quad (5.5)$$

In words, prices are tight if it is impossible for the company to raise the selling price, or lower the buying price, without losing its competitivity.

Consider an arbitrary pair of competitive prices (ψ,ϕ). We can always improve ϕ by replacing it by $\phi_1(y)=\inf_x\big(\psi(x)+c(x,y)\big)$. Then we can also improve ψ by replacing it by $\psi_1(x)=\sup_y\big(\phi_1(y)-c(x,y)\big)$; then replacing ϕ_1 by $\phi_2(y)=\inf_x\big(\psi_1(x)+c(x,y)\big)$, and so on. It turns out that this process is stationary: as an easy exercise, the reader can check that $\phi_2=\phi_1$, $\psi_2=\psi_1$, which means that after just one iteration one obtains a pair of tight prices. Thus, when we consider the dual Kantorovich problem (5.3), it makes sense to restrict our attention to tight pairs, in the sense of equation (5.5). From that equation we can reconstruct ϕ in terms of ψ, so we can just take ψ as the only unknown in our problem.

That unknown cannot be just any function: if you take a general function ψ, and compute ϕ by the first formula in (5.5), there is no chance that the second formula will be satisfied. In fact this second formula will hold true if and only if ψ is *c-convex*, in the sense of the next definition (illustrated by Figure 5.2).

Definition 5.2 (c-convexity). *Let $\mathcal{X},\mathcal{Y}$ be sets, and $c:\mathcal{X}\times\mathcal{Y}\to(-\infty,+\infty]$. A function $\psi:\mathcal{X}\to\mathbb{R}\cup\{+\infty\}$ is said to be c-convex if it is not identically $+\infty$, and there exists $\zeta:\mathcal{Y}\to\mathbb{R}\cup\{\pm\infty\}$ such that*

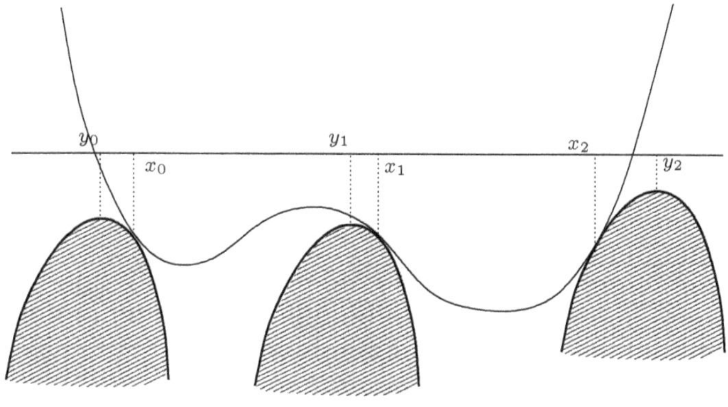

Fig. 5.2. A c-convex function is a function whose graph you can entirely *caress* from below with a tool whose shape is the negative of the cost function (this shape might vary with the point y). In the picture $y_i \in \partial_c\psi(x_i)$.

$$\forall x \in \mathcal{X} \qquad \psi(x) = \sup_{y\in\mathcal{Y}} \Big(\zeta(y) - c(x,y)\Big). \tag{5.6}$$

Then its c-transform is the function ψ^c *defined by*

$$\forall y \in \mathcal{Y} \qquad \psi^c(y) = \inf_{x\in\mathcal{X}} \Big(\psi(x) + c(x,y)\Big), \tag{5.7}$$

and its c-subdifferential is the c-cyclically monotone set defined by

$$\partial_c\psi := \Big\{(x,y) \in \mathcal{X}\times\mathcal{Y}; \quad \psi^c(y) - \psi(x) = c(x,y)\Big\}.$$

The functions ψ *and* ψ^c *are said to be c-conjugate.*

Moreover, the c-subdifferential of ψ *at point* x *is*

$$\partial_c\psi(x) = \Big\{y \in \mathcal{Y}; \quad (x,y) \in \partial_c\psi\Big\},$$

or equivalently

$$\forall z \in \mathcal{X}, \qquad \psi(x) + c(x,y) \le \psi(z) + c(z,y). \tag{5.8}$$

Particular Case 5.3. If $c(x,y) = -x\cdot y$ on $\mathbb{R}^n \times \mathbb{R}^n$, then the c-transform coincides with the usual Legendre transform, and c-convexity is just plain convexity on $\mathbb{R}^n$. (Actually, this is a slight oversimplification: c-convexity is equivalent to plain convexity plus lower semicontinuity! A convex function is automatically continuous on the largest open set Ω where it is finite, but lower semicontinuity might fail at the boundary of Ω.) One can think of the cost function $c(x,y) = -x \cdot y$ as basically the same as $c(x,y) = |x-y|^2/2$, since the "interaction" between the positions x and y is the same for both costs.

Particular Case 5.4. If $c = d$ is a *distance* on some metric space $\mathcal{X}$, then a c-convex function is just a 1-Lipschitz function, and it is its own c-transform. Indeed, if ψ is c-convex it is obviously 1-Lipschitz; conversely, if ψ is 1-Lipschitz, then $\psi(x) \le \psi(y) + d(x,y)$, so $\psi(x) = \inf_y[\psi(y) + d(x,y)] = \psi^c(x)$. As an even more particular case, if $c(x,y) = 1_{x\neq y}$, then ψ is c-convex if and only if $\sup\psi - \inf\psi \le 1$, and then again $\psi^c = \psi$. (More generally, if c satisfies the triangle inequality $c(x,z) \le c(x,y) + c(y,z)$, then ψ is c-convex if and only if $\psi(y) - \psi(x) \le c(x,y)$ for all x, y; and then $\psi = \psi^c$.)

Remark 5.5. There is no measure theory in Definition 5.2, so no assumption of measurability is made, and the supremum in (5.6) is a true supremum, not just an essential supremum; the same for the infimum in (5.7). If c is continuous, then a c-convex function is automatically lower semicontinuous, and its subdifferential is closed; but if c is not continuous the measurability of ψ and $\partial_c\psi$ is not a priori guaranteed.

Remark 5.6. I excluded the case when $\psi \equiv +\infty$ so as to avoid trivial situations; what I called a c-convex function might more properly (!) be called a *proper* c-convex function. This automatically implies that ζ in (5.6) does not take the value $+\infty$ at all if c is real-valued. If c does achieve infinite values, then the correct convention in (5.6) is $(+\infty) - (+\infty) = -\infty$.

If ψ is a function on $\mathcal{X}$, then its c-transform is a function on $\mathcal{Y}$. Conversely, given a function on $\mathcal{Y}$, one may define its c-transform as a function on $\mathcal{X}$. It will be convenient in the sequel to define the latter concept by an *infimum* rather than a supremum. This convention has the drawback of breaking the symmetry between the roles of $\mathcal{X}$ and $\mathcal{Y}$, but has other advantages that will be apparent later on.

Definition 5.7 (c-concavity). *With the same notation as in Definition 5.2, a function $\phi : \mathcal{Y} \to \mathbb{R} \cup \{-\infty\}$ is said to be c-concave if it is not identically $-\infty$, and there exists $\psi : \mathcal{X} \to \mathbb{R} \cup \{\pm\infty\}$ such that $\phi = \psi^c$. Then its c-transform is the function ϕ^c defined by*

$$\forall x \in \mathcal{X} \qquad \phi^c(x) = \sup_{y\in\mathcal{Y}} \Big(\phi(y) - c(x,y)\Big);$$

and its c-superdifferential is the c-cyclically monotone set defined by

$$\partial^c\phi := \Big\{(x,y) \subset \mathcal{X}\times\mathcal{Y};\quad \phi(y) - \phi^c(x) = c(x,y)\Big\}.$$

In spite of its short and elementary proof, the next crucial result is one of the main justifications of the concept of c-convexity.

Proposition 5.8 (Alternative characterization of c-convexity). *For any function $\psi : \mathcal{X} \to \mathbb{R}\cup\{+\infty\}$, let its c-convexification be defined by $\psi^{cc} = (\psi^c)^c$. More explicitly,*

$$\psi^{cc}(x) = \sup_{y\in\mathcal{Y}} \inf_{\widetilde{x}\in\mathcal{X}} \Big(\psi(\widetilde{x}) + c(\widetilde{x},y) - c(x,y)\Big).$$

Then ψ is c-convex if and only if $\psi^{cc} = \psi$.

Proof of Proposition 5.8. As a general fact, for any function $\phi : \mathcal{Y} \to \mathbb{R}\cup\{-\infty\}$ (not necessarily c-convex), one has the identity $\phi^{ccc} = \phi^c$. Indeed,

$$\phi^{ccc}(x) = \sup_{y} \inf_{\widetilde{x}} \sup_{\widetilde{y}} \Big[\phi(\widetilde{y}) - c(\widetilde{x},\widetilde{y}) + c(\widetilde{x},y) - c(x,y)\Big];$$

then the choice $\widetilde{x} = x$ shows that $\phi^{ccc}(x) \le \phi^c(x)$; while the choice $\widetilde{y} = y$ shows that $\phi^{ccc}(x) \ge \phi^c(x)$.

If ψ is c-convex, then there is ζ such that $\psi = \zeta^c$, so $\psi^{cc} = \zeta^{ccc} = \zeta^c = \psi$.

The converse is obvious: If $\psi^{cc} = \psi$, then ψ is c-convex, as the c-transform of ψ^c. □

Remark 5.9. Proposition 5.8 is a generalized version of the Legendre duality in convex analysis (to recover the usual Legendre duality, take $c(x,y) = -x\cdot y$ in $\mathbb{R}^n\times\mathbb{R}^n$).

Kantorovich duality

We are now ready to state and prove the main result in this chapter.

Theorem 5.10 (Kantorovich duality). *Let $(\mathcal{X},\mu)$ and $(\mathcal{Y},\nu)$ be two Polish probability spaces and let $c : \mathcal{X}\times\mathcal{Y} \to \mathbb{R}\cup\{+\infty\}$ be a lower semicontinuous cost function, such that*

$$\forall (x,y)\in\mathcal{X}\times\mathcal{Y},\qquad c(x,y) \ge a(x) + b(y)$$

for some real-valued upper semicontinuous functions $a \in L^1(\mu)$ and $b \in L^1(\nu)$. Then

(i) There is duality:

$$\begin{aligned}
\min_{\pi\in\Pi(\mu,\nu)} \int_{\mathcal{X}\times\mathcal{Y}} c(x,y)\, d\pi(x,y)
&= \sup_{(\psi,\phi)\in C_b(\mathcal{X})\times C_b(\mathcal{Y});\ \phi-\psi\le c} \left(\int_{\mathcal{Y}} \phi(y)\, d\nu(y) - \int_{\mathcal{X}} \psi(x)\, d\mu(x)\right)\\
&= \sup_{(\psi,\phi)\in L^1(\mu)\times L^1(\nu);\ \phi-\psi\le c} \left(\int_{\mathcal{Y}} \phi(y)\, d\nu(y) - \int_{\mathcal{X}} \psi(x)\, d\mu(x)\right)\\
&= \sup_{\psi\in L^1(\mu)} \left(\int_{\mathcal{Y}} \psi^c(y)\, d\nu(y) - \int_{\mathcal{X}} \psi(x)\, d\mu(x)\right)\\
&= \sup_{\phi\in L^1(\nu)} \left(\int_{\mathcal{Y}} \phi(y)\, d\nu(y) - \int_{\mathcal{X}} \phi^c(x)\, d\mu(x)\right),
\end{aligned}$$

and in the above suprema one might as well impose that ψ be c-convex and ϕ c-concave.

(ii) If c is real-valued and the optimal cost $C(\mu,\nu) = \inf_{\pi\in\Pi(\mu,\nu)} \int c\, d\pi$ is finite, then there is a measurable c-cyclically monotone set $\Gamma \subset \mathcal{X}\times\mathcal{Y}$ (closed if a, b, c are continuous) such that for any $\pi \in \Pi(\mu,\nu)$ the following five statements are equivalent:

- *(a) π is optimal;*
- *(b) π is c-cyclically monotone;*
- *(c) There is a c-convex ψ such that, π-almost surely, $\psi^c(y) - \psi(x) = c(x,y)$;*
- *(d) There exist $\psi : \mathcal{X} \to \mathbb{R}\cup\{+\infty\}$ and $\phi : \mathcal{Y} \to \mathbb{R}\cup\{-\infty\}$, such that $\phi(y) - \psi(x) \le c(x,y)$ for all (x,y), with equality π-almost surely;*
- *(e) π is concentrated on Γ.*

(iii) If c is real-valued, $C(\mu,\nu) < +\infty$, and one has the pointwise upper bound

$$c(x,y) \le c_{\mathcal{X}}(x) + c_{\mathcal{Y}}(y), \qquad (c_{\mathcal{X}}, c_{\mathcal{Y}}) \in L^1(\mu)\times L^1(\nu), \tag{5.9}$$

then both the primal and dual Kantorovich problems have solutions, so

$$\begin{aligned}
\min_{\pi\in\Pi(\mu,\nu)} \int_{\mathcal{X}\times\mathcal{Y}} c(x,y)\, d\pi(x,y)
&= \max_{(\psi,\phi)\in L^1(\mu)\times L^1(\nu);\ \phi-\psi\le c} \left(\int_{\mathcal{Y}} \phi(y)\, d\nu(y) - \int_{\mathcal{X}} \psi(x)\, d\mu(x)\right)\\
&= \max_{\psi\in L^1(\mu)} \left(\int_{\mathcal{Y}} \psi^c(y)\, d\nu(y) - \int_{\mathcal{X}} \psi(x)\, d\mu(x)\right),
\end{aligned}$$

and in the latter expressions one might as well impose that ψ *be c-convex and* $\phi = \psi^c$. *If in addition a, b and c are continuous, then there is a closed c-cyclically monotone set* $\Gamma \subset \mathcal{X} \times \mathcal{Y}$, *such that for any* $\pi \in \Pi(\mu,\nu)$ *and for any c-convex* $\psi \in L^1(\mu)$,

$$\begin{cases} \pi \text{ is optimal in the Kantorovich problem if and only if } \pi[\Gamma] = 1; \\ \psi \text{ is optimal in the dual Kantorovich problem if and only if } \Gamma \subset \partial_c\psi. \end{cases}$$

Remark 5.11. When the cost c is continuous, then the support of π is c-cyclically monotone; but for a discontinuous cost function it might a priori be that π is concentrated on a (nonclosed) c-cyclically monotone set, while the support of π is not c-cyclically monotone. So, in the sequel, the words "concentrated on" are not exchangeable with "supported in". There is another subtlety for discontinuous cost functions: It is not clear that the functions ϕ and ψ^c appearing in statements (ii) and (iii) are Borel measurable; it will only be proven that they coincide with measurable functions outside of a ν-negligible set.

Remark 5.12. Note the difference between statements (b) and (e): The set Γ appearing in (ii)(e) is the same for all optimal π's, it only depends on μ and ν. This set is in general not unique. If c is continuous and Γ is imposed to be closed, then one can define a smallest Γ, which is the closure of the union of all the supports of the optimal π's. There is also a largest Γ, which is the intersection of all the c-subdifferentials $\partial_c\psi$, where ψ is such that there exists an optimal π supported in $\partial_c\psi$. (Since the cost function is assumed to be continuous, the c-subdifferentials are closed, and so is their intersection.)

Remark 5.13. Here is a useful practical consequence of Theorem 5.10: Given a transference plan π, if you can cook up a pair of competitive prices (ψ,ϕ) such that $\phi(y) - \psi(x) = c(x,y)$ throughout the support of π, then you know that π is optimal. This theorem also shows that optimal transference plans satisfy very special conditions: if you fix an optimal pair (ψ,ϕ), then mass arriving at y can come from x only if $c(x,y) = \phi(y) - \psi(x) = \psi^c(y) - \psi(x)$, which means that

$$x \in \operatorname*{Arg\,min}_{x' \in \mathcal{X}} \Big(\psi(x') + c(x',y)\Big).$$

In terms of my bakery analogy this can be restated as follows: A café accepts bread from a bakery only if the combined cost of buying the bread there and transporting it here is lowest among all possible bakeries. Similarly, given a pair of competitive prices (ψ, ϕ), if you can cook up a transference plan π such that $\phi(y) - \psi(x) = c(x, y)$ throughout the support of π, then you know that (ψ, ϕ) is a solution to the dual Kantorovich problem.

Remark 5.14. The assumption $c \le c_{\mathcal{X}} + c_{\mathcal{Y}}$ in (iii) can be weakened into

$$\int_{\mathcal{X}\times\mathcal{Y}} c(x,y)\, d\mu(x)\, d\nu(y) < +\infty,$$

or even

$$\begin{cases} \mu\left[\left\{x;\ \int_{\mathcal{Y}} c(x,y)\, d\nu(y) < +\infty\right\}\right] > 0; \\ \\ \nu\left[\left\{y;\ \int_{\mathcal{X}} c(x,y)\, d\mu(x) < +\infty\right\}\right] > 0. \end{cases} \tag{5.10}$$

Remark 5.15. If the variables x and y are swapped, then (μ, ν) should be replaced by (ν, μ) and (ψ, ϕ) by $(-\phi, -\psi)$.

Particular Case 5.16. Particular Case 5.4 leads to the following variant of Theorem 5.10. When $c(x, y) = d(x, y)$ is a distance on a Polish space $\mathcal{X}$, and μ, ν belong to $P_1(\mathcal{X})$, then

$$\inf \mathbb{E}\, d(X,Y) = \sup \mathbb{E}\left[\psi(X) - \psi(Y)\right] = \sup \left\{\int_{\mathcal{X}} \psi\, d\mu - \int_{\mathcal{Y}} \psi\, d\nu\right\}. \tag{5.11}$$

where the infimum on the left is over all couplings (X, Y) of (μ, ν), and the supremum on the right is over all 1-Lipschitz functions ψ. This is the **Kantorovich–Rubinstein formula**; it holds true as soon as the supremum in the left-hand side is finite, and it is very useful.

Particular Case 5.17. Now consider $c(x, y) = -x \cdot y$ in $\mathbb{R}^n \times \mathbb{R}^n$. This cost is not nonnegative, but we have the lower bound $c(x, y) \ge -(|x|^2 + |y|^2)/2$. So if $x \to |x|^2 \in L^1(\mu)$ and $y \to |y|^2 \in L^1(\nu)$, then one can invoke the Particular Case 5.3 to deduce from Theorem 5.10 that

$$\sup \mathbb{E}\,(X \cdot Y) = \inf \mathbb{E}\left[\varphi(X) + \varphi^*(Y)\right] = \inf \left\{\int_{\mathcal{X}} \varphi\, d\mu + \int_{\mathcal{Y}} \varphi^*\, d\nu\right\}, \tag{5.12}$$

where the supremum on the left is over all couplings (X, Y) of (μ, ν), the infimum on the right is over all (lower semicontinuous) convex functions on $\mathbb{R}^n$, and φ^* stands for the usual Legendre transform of φ. In formula (5.12), the signs have been changed with respect to the statement of Theorem 5.10, so the problem is to **maximize the correlation** of the random variables X and Y.

Before proving Theorem 5.10, I shall first informally explain the construction. At first reading, one might be content with these informal explanations and skip the rigorous proof.

Idea of proof of Theorem 5.10. Take an optimal π (which exists from Theorem 4.1), and let (ψ, ϕ) be two competitive prices. Of course, as in (5.4),

$$\int c(x,y)\,d\pi(x,y) \geq \int \phi\,d\nu - \int \psi\,d\mu = \int [\phi(y) - \psi(x)]\,d\pi(x,y).$$

So if both quantities are equal, then $\int [c - \phi + \psi]\,d\pi = 0$, and since the integrand is nonnegative, necessarily

$$\phi(y) - \psi(x) = c(x,y) \qquad \pi(dx\,dy) - \text{almost surely}.$$

Intuitively speaking, whenever there is some transfer of goods from x to y, the prices should be adjusted exactly to the transport cost.

Now let $(x_i)_{0 \leq i \leq m}$ and $(y_i)_{0 \leq i \leq m}$ be such that (x_i, y_i) belongs to the support of π, so there is indeed some transfer from x_i to y_i. Then we hope that

$$\begin{cases} \phi(y_0) - \psi(x_0) = c(x_0, y_0) \\ \phi(y_1) - \psi(x_1) = c(x_1, y_1) \\ \ldots \\ \phi(y_m) - \psi(x_m) = c(x_m, y_m). \end{cases}$$

On the other hand, if x is an arbitrary point,

$$\begin{cases} \phi(y_0) - \psi(x_1) \leq c(x_1, y_0) \\ \phi(y_1) - \psi(x_2) \leq c(x_2, y_1) \\ \ldots \\ \phi(y_m) - \psi(x) \leq c(x, y_m). \end{cases}$$

By subtracting these inequalities from the previous equalities and adding up everything, one obtains

$$\psi(x) \geq \psi(x_0) + \big[c(x_0, y_0) - c(x_1, y_0)\big] + \ldots + \big[c(x_m, y_m) - c(x, y_m)\big].$$

Of course, one can add an arbitrary constant to ψ, provided that one subtracts the same constant from ϕ; so it is possible to decide that $\psi(x_0) = 0$, where (x_0, y_0) is arbitrarily chosen in the support of π. Then

$$\psi(x) \geq \big[c(x_0, y_0) - c(x_1, y_0)\big] + \ldots + \big[c(x_m, y_m) - c(x, y_m)\big], \qquad (5.13)$$

and this should be true for all choices of (x_i, y_i) $(1 \leq i \leq m)$ in the support of π, and for all $m \geq 1$. So it becomes natural to *define* ψ as the supremum of all the functions (of the variable x) appearing in the right-hand side of (5.13). It will turn out that this ψ satisfies the equation

$$\psi^c(y) - \psi(x) = c(x, y) \qquad \pi(dx\, dy)\text{-almost surely.}$$

Then, if ψ and ψ^c are integrable, one can write

$$\int c\, d\pi = \int \psi^c\, d\pi - \int \psi\, d\pi = \int \psi^c\, d\nu - \int \psi\, d\mu.$$

This shows at the same time that π is optimal in the Kantorovich problem, and that the pair (ψ, ψ^c) is optimal in the dual Kantorovich problem. □

Rigorous proof of Theorem 5.10, Part (i). First I claim that *it is sufficient to treat the case when c is nonnegative.* Indeed, let

$$\widetilde{c}(x, y) := c(x, y) - a(x) - b(y) \geq 0, \qquad \Lambda := \int a\, d\mu + \int b\, d\nu \quad \in \mathbb{R}.$$

Whenever $\psi : \mathcal{X} \to \mathbb{R} \cup \{+\infty\}$ and $\phi : \mathcal{Y} \to \mathbb{R} \cup \{-\infty\}$ are two functions, define

$$\widetilde{\psi}(x) := \psi(x) + a(x), \qquad \widetilde{\phi}(y) := \phi(y) - b(y).$$

Then the following properties are readily checked:

$$c \text{ real-valued} \quad \Longrightarrow \quad \widetilde{c} \text{ real-valued}$$

$$c \text{ lower semicontinuous} \quad \Longrightarrow \quad \widetilde{c} \text{ lower semicontinuous}$$

$$\widetilde{\psi} \in L^1(\mu) \Longleftrightarrow \psi \in L^1(\mu); \qquad \widetilde{\phi} \in L^1(\nu) \Longleftrightarrow \phi \in L^1(\nu);$$

$$\forall \pi \in \Pi(\mu, \nu), \qquad \int \widetilde{c}\, d\pi = \int c\, d\pi - \Lambda;$$

$$\forall(\psi,\phi)\in L^1(\mu)\times L^1(\nu),\qquad \int \widetilde{\phi}\,d\nu-\int\widetilde{\psi}\,d\mu=\int\phi\,d\nu-\int\psi\,d\nu\ -\Lambda;$$

$$\psi \text{ is } c\text{-convex} \Longleftrightarrow \widetilde{\psi} \text{ is } \widetilde{c}\text{-convex};$$

$$\phi \text{ is } c\text{-concave} \Longleftrightarrow \widetilde{\phi} \text{ is } \widetilde{c}\text{-concave};$$

$$(\phi,\psi) \text{ are } c\text{-conjugate} \Longleftrightarrow (\widetilde{\phi},\widetilde{\psi}) \text{ are } \widetilde{c}\text{-conjugate};$$

$$\Gamma \text{ is } c\text{-cyclically monotone} \Longleftrightarrow \Gamma \text{ is } \widetilde{c}\text{-cyclically monotone}.$$

Thanks to these formulas, it is equivalent to establish Theorem 5.10 for the cost c or for the nonnegative cost $\widetilde{c}$. So in the sequel, I shall assume, without further comment, that c is nonnegative.

The rest of the proof is divided into five steps.

Step 1: *If $\mu=(1/n)\sum_{i=1}^n\delta_{x_i}$, $\nu=(1/n)\sum_{j=1}^n\delta_{y_j}$, where the costs $c(x_i,y_j)$ are finite, then there is at least one cyclically monotone transference plan.*

Indeed, in that particular case, a transference plan between μ and ν can be identified with a bistochastic $n\times n$ array of real numbers $a_{ij}\in[0,1]$: each a_{ij} tells what proportion of the $1/n$ mass carried by point x_i will go to destination y_j. So the Monge–Kantorovich problem becomes

$$\inf_{(a_{ij})}\sum_{ij}a_{ij}\,c(x_i,y_i)$$

where the infimum is over all arrays (a_{ij}) satisfying

$$\sum_i a_{ij}=1,\qquad \sum_j a_{ij}=1. \tag{5.14}$$

Here we are minimizing a linear function on the compact set $[0,1]^{n\times n}$, so obviously there exists a minimizer; the corresponding transference plan π can be written as

$$\pi=\frac{1}{n}\sum_{ij}a_{ij}\,\delta_{(x_i,y_j)},$$

and its support S is the set of all couples (x_i,y_j) such that $a_{ij}>0$.

Assume that S is not cyclically monotone: Then there exist $N\in\mathbb{N}$ and $(x_{i_1},y_{j_1}),\dots,(x_{i_N},y_{j_N})$ in S such that

$$c(x_{i_1},y_{j_2})+c(x_{i_2},y_{j_3})+\ldots+c(x_{i_N},y_{j_1})<c(x_{i_1},y_{j_1})+\ldots+c(x_{i_N},y_{j_N}). \tag{5.15}$$

Let $a := \min(a_{i_1,j_1},\ldots,a_{i_N,j_N}) > 0$. Define a new transference plan $\widetilde{\pi}$ by the formula

$$\widetilde{\pi} = \pi + \frac{a}{n}\sum_{\ell=1}^{N}\big(\delta_{(x_{i_\ell},y_{j_{\ell+1}})} - \delta_{(x_{i_\ell},y_{j_\ell})}\big).$$

It is easy to check that this has the correct marginals, and by (5.15) the cost associated with $\widetilde{\pi}$ is strictly less than the cost associated with π. This is a contradiction, so S is indeed c-cyclically monotone!

Step 2: *If c is continuous, then there is a cyclically monotone transference plan.*

To prove this, consider sequences of independent random variables $x_i \in \mathcal{X}$, $y_j \in \mathcal{Y}$, with respective law μ, ν. According to the law of large numbers for empirical measures (sometimes called fundamental theorem of statistics, or Varadarajan's theorem), one has, with probability 1,

$$\mu_n := \frac{1}{n}\sum_{i=1}^{n}\delta_{x_i} \longrightarrow \mu, \qquad \nu_n := \frac{1}{n}\sum_{j=1}^{n}\delta_{y_j} \longrightarrow \nu \tag{5.16}$$

as $n \to \infty$, in the sense of weak convergence of measures. In particular, by Prokhorov's theorem, (μ_n) and (ν_n) are tight sequences.

For each n, let π_n be a cyclically monotone transference plan between μ_n and ν_n. By Lemma 4.4, $\{\pi_n\}_{n\in\mathbb{N}}$ is tight. By Prokhorov's theorem, there is a subsequence, still denoted (π_n), which converges weakly to some probability measure π, i.e.

$$\int h(x,y)\,d\pi_n(x,y) \longrightarrow \int h(x,y)\,d\pi(x,y)$$

for all bounded continuous functions h on $\mathcal{X}\times\mathcal{Y}$. By applying the previous identity with $h(x,y) = f(x)$ and $h(x,y) = g(y)$, we see that π has marginals μ and ν, so this is an admissible transference plan between μ and ν.

For each n, the cyclical monotonicity of π_n implies that for all N and $\pi_n^{\otimes N}$-almost all $(x_1,y_1),\ldots,(x_N,y_N)$, the inequality (5.1) is satisfied; in other words, $\pi_n^{\otimes N}$ is concentrated on the set $\mathcal{C}(N)$ of all $((x_1,y_1),\ldots,(x_N,y_N)) \in (\mathcal{X}\times\mathcal{Y})^N$ satisfying (5.1). Since c is continuous, $\mathcal{C}(N)$ is a *closed* set, so the weak limit $\pi^{\otimes N}$ of $\pi_n^{\otimes N}$ is also concentrated on $\mathcal{C}(N)$. Let $\Gamma = \operatorname{Spt}\pi$ (Spt stands for "support"), then

$$\Gamma^N = (\operatorname{Spt}\pi)^N = \operatorname{Spt}(\pi^{\otimes N}) \subset \mathcal{C}(N),$$

and since this holds true for all N, Γ is c-cyclically monotone.

Step 3: *If c is continuous real-valued and π is c-cyclically monotone, then there is a c-convex ψ such that $\partial_c\psi$ contains the support of π.*

Indeed, let Γ again denote the support of π (this is a closed set). Pick any $(x_0, y_0) \in \Gamma$, and define

$$\psi(x) := \sup_{m\in\mathbb{N}} \sup \Big\{ \big[c(x_0,y_0) - c(x_1,y_0)\big] + \big[c(x_1,y_1) - c(x_2,y_1)\big] + \cdots + \big[c(x_m,y_m) - c(x,y_m)\big]; \quad (x_1,y_1),\ldots,(x_m,y_m) \in \Gamma \Big\}. \tag{5.17}$$

By applying the definition with $m = 1$ and $(x_1, y_1) = (x_0, y_0)$, one immediately sees that $\psi(x_0) \geq 0$. On the other hand, $\psi(x_0)$ is the supremum of all the quantities $[c(x_0, y_0) - c(x_1, y_0)] + \ldots + [c(x_m, y_m) - c(x_0, y_m)]$ which by cyclical monotonicity are all nonpositive. So actually $\psi(x_0) = 0$. (In fact this is the only place in this step where c-cyclical monotonicity will be used!)

By renaming y_m as y, obviously

$$\psi(x) = \sup_{y\in\mathcal{Y}} \sup_{m\in\mathbb{N}} \sup_{(x_1,y_1),\ldots,(x_{m-1},y_{m-1}),x_m} \Big\{ \big[c(x_0,y_0) - c(x_1,y_0)\big] + \big[c(x_1,y_1) - c(x_2,y_1)\big] + \cdots + \big[c(x_m,y) - c(x,y)\big]; \quad (x_1,y_1),\ldots,(x_m,y) \in \Gamma \Big\}. \tag{5.18}$$

So $\psi(x) = \sup_y[\zeta(y) - c(x, y)]$, if ζ is defined by

$$\zeta(y) = \sup\Big\{ \big[c(x_0,y_0) - c(x_1,y_0)\big] + \big[c(x_1,y_1) - c(x_2,y_1)\big] + \cdots + c(x_m,y); \quad m\in\mathbb{N}, (x_1,y_1),\ldots,(x_m,y) \in \Gamma \Big\} \tag{5.19}$$

(with the convention that $\zeta = -\infty$ out of $\operatorname{proj}_{\mathcal{Y}}(\Gamma)$). Thus ψ is a c-convex function.

Now let $(\overline{x}, \overline{y}) \in \Gamma$. By choosing $x_m = \overline{x}$, $y_m = \overline{y}$ in the definition of ψ,

$$\psi(x) \geq \sup_m \Big\{ \Big(\sup_{(x_1,y_1),\ldots,(x_{m-1},y_{m-1})} \big[c(x_0,y_0) - c(x_1,y_0)\big] + \cdots + \big[c(x_{m-1},y_{m-1}) - c(\overline{x},y_{m-1})\big] \Big) + \big[c(\overline{x},\overline{y}) - c(x,\overline{y})\big] \Big\}.$$

In the definition of ψ, it does not matter whether one takes the supremum over $m-1$ or over m variables, since one also takes the supremum over m. So the previous inequality can be recast as

$$\psi(x) \geq \psi(\overline{x}) + c(\overline{x}, \overline{y}) - c(x, \overline{y}).$$

In particular, $\psi(x) + c(x, \overline{y}) \geq \psi(\overline{x}) + c(\overline{x}, \overline{y})$. Taking the infimum over $x \in \mathcal{X}$ in the left-hand side, we deduce that

$$\psi^c(\overline{y}) \geq \psi(\overline{x}) + c(\overline{x}, \overline{y}).$$

Since the reverse inequality is always satisfied, actually

$$\psi^c(\overline{y}) = \psi(\overline{x}) + c(\overline{x}, \overline{y}),$$

and this means precisely that $(\overline{x}, \overline{y}) \in \partial_c\psi$. So Γ does lie in the c-subdifferential of ψ.

Step 4: *If c is continuous and bounded, then there is duality.*

Let $\|c\| := \sup c(x,y)$. By Steps 2 and 3, there exists a transference plan π whose support is included in $\partial_c\psi$ for some c-convex ψ, and which was constructed "explicitly" in Step 3. Let $\phi = \psi^c$.

From (5.17), $\psi = \sup \psi_m$, where each ψ_m is a supremum of continuous functions, and therefore lower semicontinuous. In particular, ψ is measurable.[1] The same is true of ϕ.

Next we check that ψ, ϕ are bounded. Let $(x_0, y_0) \in \partial_c\psi$ be such that $\psi(x_0) < +\infty$; then necessarily $\phi(y_0) > -\infty$. So, for any $x \in \mathcal{X}$,

$$\psi(x) = \sup_y \, [\phi(y) - c(x,y)] \;\geq\; \phi(y_0) - c(x, y_0) \;\geq\; \phi(y_0) - \|c\|;$$

$$\phi(y) = \inf_x \, [\psi(x) + c(x,y)] \;\leq\; \psi(x_0) + c(x_0, y) \;\leq\; \psi(x_0) + \|c\|.$$

Re-injecting these bounds into the identities $\psi = \phi^c$, $\phi = \psi^c$, we get

$$\psi(x) \;\leq\; \sup_y \phi(y) \leq\; \psi(x_0) + \|c\|;$$

$$\phi(y) \;\geq\; \inf_x \psi(x) \geq\; \phi(y_0) - \|c\|.$$

So both ψ and ϕ are bounded from above and below.

[1] A lower semicontinuous function on a Polish space is always measurable, even if it is obtained as a supremum of uncountably many continuous functions; in fact it can always be written as a supremum of countably many continuous functions!

Thus we can integrate ϕ, ψ against μ, ν respectively, and, by the marginal condition,

$$\int \phi(y)\,d\nu(y) - \int \psi(x)\,d\mu(x) = \int \big[\phi(y) - \psi(x)\big]\,d\pi(x,y).$$

Since $\phi(y) - \psi(x) = c(x,y)$ on the support of π, the latter quantity equals $\int c(x,y)\,d\pi(x,y)$. It follows that (5.4) is actually an equality, which proves the duality.

Step 5: *If c is lower semicontinuous, then there is duality.*

Since c is nonnegative lower semicontinuous, we can write

$$c(x,y) = \lim_{k\to\infty} c_k(x,y),$$

where $(c_k)_{k\in\mathbb{N}}$ is a nondecreasing sequence of bounded, uniformly continuous functions. To see this, just choose

$$c_k(x,y) = \inf_{(x',y')} \Big\{ \min\Big(c(x',y'),k\Big) + k\Big[d(x,x') + d(y,y')\Big] \Big\};$$

note that c_k is k-Lipschitz, nondecreasing in k, and further satisfies $0 \le c_k(x,y) \le \min(c(x,y),k)$.[2]

By Step 4, for each k we can find π_k, ϕ_k, ψ_k such that ψ_k is bounded and c-convex, $\phi_k = (\psi_k)^c$, and

$$\int c_k(x,y)\,d\pi_k(x,y) = \int \phi_k(y)\,d\nu(y) - \int \psi_k(x)\,d\mu(x).$$

Since c_k is no greater than c, the constraint $\phi_k(y) - \psi_k(x) \le c_k(x,y)$ implies $\phi_k(y) - \psi_k(x) \le c(x,y)$; so all (ϕ_k, ψ_k) are admissible in the dual problem with cost c. Moreover, for each k the functions ϕ_k and ψ_k are uniformly continuous because c itself is uniformly continuous.

By Lemma 4.4, $\Pi(\mu,\nu)$ is weakly sequentially compact. Thus, up to extraction of a subsequence, we can assume that π_k converges to some $\widetilde{\pi} \in \Pi(\mu,\nu)$. For all indices $\ell \le k$, we have $c_\ell \le c_k$, so

$$\begin{aligned}\int c_\ell\,d\widetilde{\pi} &= \lim_{k\to\infty} \int c_\ell\,d\pi_k \\ &\le \limsup_{k\to\infty} \int c_k\,d\pi_k \\ &= \limsup_{k\to\infty} \left(\int \phi_k(y)\,d\nu(y) - \int \psi_k(x)\,d\mu(x) \right).\end{aligned}$$

[2] It is instructive to understand exactly where the lower semicontinuity assumption is used to show $c = \lim c_k$.

On the other hand, by monotone convergence,

$$\int c\,d\widetilde{\pi} = \lim_{\ell\to\infty} \int c_\ell\,d\widetilde{\pi}.$$

So

$$\inf_{\Pi(\mu,\nu)} \int c\,d\pi \le \int c\,d\widetilde{\pi} \le \limsup_{k\to\infty} \left(\int \phi_k(y)\,d\nu(y) - \int \psi_k(x)\,d\mu(x)\right)$$
$$\le \inf_{\Pi(\mu,\nu)} \int c\,d\pi.$$

Moreover,

$$\int \phi_k(y)\,d\nu(y) - \int \psi_k(x)\,d\mu(x) \xrightarrow[k\to\infty]{} \inf_{\Pi(\mu,\nu)} \int c\,d\pi. \qquad (5.20)$$

Since each pair (ψ_k, ϕ_k) lies in $C_b(\mathcal{X}) \times C_b(\mathcal{Y})$, the duality also holds with bounded continuous (and even Lipschitz) test functions, as claimed in Theorem 5.10(i). □

Proof of Theorem 5.10, Part (ii). From now on, I shall assume that the optimal transport cost $C(\mu,\nu)$ is finite, and that c is real-valued. As in the proof of Part (i) I shall assume that c is nonnegative, since the general case can always be reduced to that particular case. Part (ii) will be established in the following way: (a) ⇒ (b) ⇒ (c) ⇒ (d) ⇒ (a) ⇒ (e) ⇒ (b). There seems to be some redundancy in this chain of implications, but this is because the implication (a) ⇒ (c) will be used to construct the set Γ appearing in (e).

(a) ⇒ (b): Let π be an optimal plan, and let $(\phi_k, \psi_k)_{k\in\mathbb{N}}$ be as in Step 5 of the proof of Part (i). Since the optimal transport cost is finite by assumption, the cost function c belongs to $L^1(\pi)$. From (5.20) and the marginal property of π,

$$\int \Big[c(x,y) - \phi_k(y) + \psi_k(x)\Big]\,d\pi(x,y) \xrightarrow[k\to\infty]{} 0,$$

so $c(x,y) - \phi_k(y) + \psi_k(x)$ converges to 0 in $L^1(\pi)$ as $k \to \infty$. Up to choosing a subsequence, we can assume that the convergence is almost sure; then $\phi_k(y_i) - \psi_k(x_i)$ converges to $c(x_i, y_i)$, $\pi(dx_i\,dy_i)$-almost surely, as $k \to \infty$. By passing to the limit in the inequality

$$\sum_{i=1}^N c(x_i, y_{i+1}) \ge \sum_{i=1}^N [\phi_k(y_{i+1}) - \psi_k(x_i)] = \sum_{i=1}^N [\phi_k(y_i) - \psi_k(x_i)]$$

(where by convention $y_{N+1} = y_1$) we see that, $\pi^{\otimes N}$-almost surely,

$$\sum_{i=1}^{N} c(x_i, y_{i+1}) \geq \sum_{i=1}^{N} c(x_i, y_i). \tag{5.21}$$

At this point we know that $\pi^{\otimes N}$ is concentrated on some set $\Gamma_N \subset (\mathcal{X} \times \mathcal{Y})^N$, such that Γ_N consists of N-tuples $((x_1, y_1), \dots, (x_N, y_N))$ satisfying (5.21). Let $\mathrm{proj}_k((x_i, y_i)_{1 \leq i \leq N}) := (x_k, y_k)$ be the projection on the kth factor of $(\mathcal{X} \times \mathcal{Y})^N$. It is not difficult to check that $\Gamma := \cap_{1 \leq k \leq N} \mathrm{proj}_k(\Gamma_N)$ is a c-cyclically monotone set which has full π-measure; so π is indeed c-cyclically monotone.

(b) $\Rightarrow$ (c): Let π be a cyclically monotone transference plan. The function ψ can be constructed just as in Step 3 of the proof of Part (i), only with some differences. First, Γ is not necessarily closed; it is just a Borel set such that $\pi[\Gamma] = 1$. (If Γ is not Borel, make it Borel by modifying it on a negligible set.) With this in mind, define, as in Step 3 of Part (i),

$$\psi(x) := \sup_{m \in \mathbb{N}} \sup \Big\{ \big[c(x_0, y_0) - c(x_1, y_0)\big] + \big[c(x_1, y_1) - c(x_2, y_1)\big] \\ + \dots + \big[c(x_m, y_m) - c(x, y_m)\big]; \quad (x_1, y_1), \dots, (x_m, y_m) \in \Gamma \Big\}. \tag{5.22}$$

From its definition, for any $x \in \mathcal{X}$,

$$\psi(x) \geq c(x_0, y_0) - c(x, y_0) > -\infty.$$

(Here the assumption of c being real-valued is useful.) Then there is no difficulty in proving, as in Step 3, that $\psi(x_0) = 0$, that ψ is c-convex, and that π is concentrated on $\partial_c \psi$.

The rest of this step will be devoted to the *measurability* of ψ, ψ^c and $\partial_c \psi$. These are surprisingly subtle issues, which do not arise if c is continuous; so the reader who only cares for a continuous cost function might go directly to the next step.

First, the measurability of ψ is not clear at all from formula (5.22): This is typically an uncountable supremum of *upper* semicontinuous functions, and there is no a priori reason for this to be Borel measurable.

Since c is nonnegative lower semicontinuous, there is a nondecreasing sequence $(c_\ell)_{\ell \in \mathbb{N}}$ of continuous nonnegative functions, such that $c_\ell(x, y)$ converges to $c(x, y)$ as $\ell \to \infty$, for all (x, y). By Egorov's theorem, for each $k \in \mathbb{N}$ there is a Borel set E_k with $\pi[E_k] \le 1/k$, such that the convergence of c_ℓ to c is uniform on $\Gamma \setminus E_k$. Since π (just as any probability measure on a Polish space) is regular, we can find a *compact* set $\Gamma_k \subset \Gamma \setminus E_k$, such that $\pi[\Gamma_k] \ge 1 - 2/k$. There is no loss of generality in assuming that the sets Γ_k are increasing in k.

On each Γ_k, the sequence (c_ℓ) converges uniformly and monotonically to c; in particular c is continuous on Γ_k. Furthermore, since π is obviously concentrated on the union of all Γ_k, there is no loss of generality in assuming that $\Gamma = \cup \Gamma_k$. We may also assume that $(x_0, y_0) \in \Gamma_1$.

Now, let x be given in $\mathcal{X}$, and for each k, ℓ, m, let

$$F_{m,k,\ell}(x_0, y_0, \ldots, x_m, y_m) := \big[c(x_0, y_0) - c_\ell(x_1, y_0)\big] \\ + \big[c(x_1, y_1) - c_\ell(x_2, y_1)\big] + \cdots + \big[c(x_m, y_m) - c_\ell(x, y_m)\big],$$

for $(x_0, y_0, \ldots, x_m, y_m) \in \Gamma_k^m$. It is clear that $F_{m,k,\ell}$ is a continuous function (because c_ℓ is continuous on $\mathcal{X} \times \mathcal{X}$, and c is continuous on Γ_k). It is defined on the compact set Γ_k^m, and it is nonincreasing as a function of ℓ, with

$$\lim_{\ell \to \infty} F_{m,k,\ell} = F_{m,k},$$

where

$$F_{m,k}(x_0, y_0, \ldots, x_m, y_m) := \big[c(x_0, y_0) - c(x_1, y_0)\big] \\ + \big[c(x_1, y_1) - c(x_2, y_1)\big] + \cdots + \big[c(x_m, y_m) - c(x, y_m)\big].$$

Now I claim that

$$\lim_{\ell \to \infty} \sup_{\Gamma_k^m} F_{m,k,\ell} = \sup_{\Gamma_k^m} F_{m,k}. \tag{5.23}$$

Indeed, by compactness, for each $\ell \in \mathbb{N}$ there is $X_\ell \in \Gamma_k^m$ such that

$$\sup_{\Gamma_k^m} F_{m,k,\ell} = F_{m,k,\ell}(X_\ell);$$

and up to extraction of a subsequence, one may assume that X_ℓ converges to some X. Then by monotonicity, for any $\ell' \le \ell$,

$$\sup_{\Gamma_k^m} F_{m,k,\ell} = F_{m,k,\ell}(X_\ell) \le F_{m,k,\ell'}(X_\ell);$$

and if one lets $\ell \to \infty$, with ℓ' fixed, one obtains

$$\limsup_{\ell\to\infty} \sup_{\Gamma_k^m} F_{m,k,\ell} \le F_{m,k,\ell'}(X).$$

Now let $\ell' \to \infty$, to get

$$\limsup_{\ell\to\infty} \sup_{\Gamma_k^m} F_{m,k,\ell} \le F_{m,k}(X) \le \sup_{\Gamma_k^m} F_{m,k}.$$

The converse inequality

$$\sup_{\Gamma_k^m} F_{m,k} \le \liminf_{\ell\to\infty} \sup_{\Gamma_k^m} F_{m,k,\ell}$$

is obvious because $F_{m,k} \le F_{m,k,\ell}$; so (5.23) is proven.

To summarize: If we let

$$\psi_{m,k,\ell}(x) := \sup\Big\{\big[c(x_0,y_0) - c_\ell(x_1,y_0)\big] + \big[c(x_1,y_1) - c_\ell(x_2,y_1)\big] \\ + \cdots + \big[c(x_m,y_m) - c_\ell(x,y_m)\big];\quad (x_1,y_1),\ldots,(x_m,y_m) \in \Gamma_k\Big\},$$

then we have

$$\lim_{\ell\to\infty} \psi_{m,k,\ell}(x) = \sup\Big\{\big[c(x_0,y_0) - c(x_1,y_0)\big] + \big[c(x_1,y_1) - c(x_2,y_1)\big] \\ + \cdots + \big[c(x_m,y_m) - c(x,y_m)\big];\quad (x_1,y_1),\ldots,(x_m,y_m) \in \Gamma_k\Big\}.$$

It follows easily that, for each x,

$$\psi(x) = \sup_{m\in\mathbb{N}} \sup_{k\in\mathbb{N}} \lim_{\ell\to\infty} \psi_{m,k,\ell}(x).$$

Since $\psi_{m,k,\ell}(x)$ is lower semicontinuous in x (as a supremum of continuous functions of x), ψ itself is measurable.

The measurability of $\phi := \psi^c$ is subtle also, and at the present level of generality it is not clear that this function is really Borel measurable. However, it can be modified on a ν-negligible set so as to become measurable. Indeed, $\phi(y) - \psi(x) = c(x,y)$, $\pi(dx\,dy)$-almost surely, so if one disintegrates $\pi(dx\,dy)$ as $\pi(dx|y)\,\nu(dy)$, then $\phi(y)$ will coincide, $\nu(dy)$-almost surely, with the Borel function $\widetilde{\phi}(y) := \int_{\mathcal{X}}[\psi(x) + c(x,y)]\,\pi(dx|y)$.

Then let Z be a Borel set of zero ν-measure such that $\widetilde{\phi} = \phi$ outside of Z. The subdifferential $\partial_c\psi$ coincides, out of the π-negligible set $\mathcal{X}\times Z$, with the measurable set $\{(x,y) \in \mathcal{X} \times \mathcal{Y};\ \widetilde{\phi}(y) - \psi(x) = c(x,y)\}$. The conclusion is that $\partial_c\psi$ can be modified on a π-negligible set so as to be Borel measurable.

(c) $\Rightarrow$ (d): Just let $\phi = \psi^c$.

(d) $\Rightarrow$ (a): Let (ψ, ϕ) be a pair of admissible functions, and let π be a transference plan such that $\phi - \psi = c$, π-almost surely. The goal is to show that π is optimal. The main difficulty lies in the fact that ψ and ϕ need not be *separately* integrable. This problem will be circumvented by a careful truncation procedure. For $n \in \mathbb{N}$, $w \in \mathbb{R} \cup \{\pm\infty\}$, define

$$T_n(w) = \begin{cases} w & \text{if } |w| \le n \\ n & \text{if } w > n \\ -n & \text{if } w < -n, \end{cases}$$

and

$$\xi(x,y) := \phi(y) - \psi(x); \qquad \xi_n(x,y) := T_n(\phi(y)) - T_n(\psi(x)).$$

In particular, $\xi_0 = 0$. It is easily checked that ξ_n converges *monotonically* to ξ; more precisely,

- $\xi_n(x,y)$ remains equal to 0 if $\xi(x,y) = 0$;
- $\xi_n(x,y)$ increases to $\xi(x,y)$ if the latter quantity is positive;
- $\xi_n(x,y)$ decreases to $\xi(x,y)$ if the latter quantity is negative.

As a consequence, $\xi_n \le (\xi_n)_+ \le \xi_+ \le c$. So $(T_n\phi, T_n\psi)$ is an admissible pair in the dual Kantorovich problem, and

$$\int \xi_n \, d\pi = \int (T_n\phi)\, d\nu - \int (T_n\psi)\, d\mu \le \sup_{\phi'-\psi'\le c} \left(\int \phi' \, d\mu - \int \psi' \, d\nu \right). \tag{5.24}$$

On the other hand, by monotone convergence and since ξ coincides with c outside of a π-negligible set,

$$\int_{\xi\ge 0} \xi_n \, d\pi \xrightarrow[n\to\infty]{} \int_{\xi \ge 0} \xi \, d\pi = \int c \, d\pi;$$

$$\int_{\xi<0} \xi_n \, d\pi \xrightarrow[n\to\infty]{} \int_{\xi<0} \xi \, d\pi = 0.$$

This and (5.24) imply that

$$\int c\, d\pi \le \sup_{\phi'-\psi'\le c} \left(\int \phi' \, d\mu - \int \psi' \, d\nu \right);$$

so π is optimal.

Before completing the chain of equivalences, we should first construct the set Γ. By Theorem 4.1, there is at least one optimal transference plan, say $\widetilde{\pi}$. From the implication (a) $\Rightarrow$ (c), there is some $\widetilde{\psi}$ such that $\widetilde{\pi}$ is concentrated on $\partial_c\widetilde{\psi}$; just choose $\Gamma := \partial_c\widetilde{\psi}$.

(a) $\Rightarrow$ (e): Let $\widetilde{\pi}$ be the optimal plan used to construct Γ, and let $\psi = \widetilde{\psi}$ be the associated c-convex function; let $\phi = \psi^c$. Then let π be another optimal plan. Since π and $\widetilde{\pi}$ have the same cost and same marginals,

$$\begin{aligned}\int c\,d\pi = \int c\,d\widetilde{\pi} &= \lim_{n\to\infty} \int (T_n\phi - T_n\psi)\,d\widetilde{\pi} \\ &= \lim_{n\to\infty} \int (T_n\phi - T_n\psi)\,d\pi,\end{aligned}$$

where T_n is the truncation operator that was already used in the proof of (d) $\Rightarrow$ (a). So

$$\int \big[c(x,y) - T_n\phi(y) + T_n\psi(x)\big]\,d\pi(x,y) \xrightarrow[n\to\infty]{} 0. \tag{5.25}$$

As before, define $\xi(x,y) := \phi(y) - \psi(x)$; then by monotone convergence,

$$\int_{\xi\geq 0} \big[c - T_n\phi + T_n\psi\big]\,d\pi \xrightarrow[n\to\infty]{} \int_{\xi\geq 0} (c-\xi)\,d\pi;$$

$$\int_{\xi< 0} \big[c - T_n\phi + T_n\psi\big]\,d\pi \xrightarrow[n\to\infty]{} \int_{\xi< 0} (c-\xi)\,d\pi.$$

Since $\xi \leq c$, the integrands here are nonnegative and both integrals make sense in $[0,+\infty]$. So by adding the two limits and using (5.25) we get

$$\int (c-\xi)\,d\pi = \lim_{n\to\infty} \int \big[c - T_n\phi + T_n\psi\big] = 0.$$

Since $\xi \leq c$, this proves that c coincides π-almost surely with ξ, which was the desired conclusion.

(e) $\Rightarrow$ (b): This is obvious since Γ is cyclically monotone by assumption. □

Proof of Theorem 5.10, Part (iii). As in the proof of Part (i) we may assume that $c \geq 0$. Let π be optimal, and let ψ be a c-convex function such that π is concentrated on $\partial_c\psi$. Let $\phi := \psi^c$. The goal is to show that under the assumption $c \le c_{\mathcal{X}} + c_{\mathcal{Y}}$, (ψ, ϕ) solves the dual Kantorovich problem.

The point is to prove that ψ and ϕ are integrable. For this we repeat the estimates of Step 4 in the proof of Part (i), with some variants: After securing (x_0, y_0) such that $\phi(y_0)$, $\psi(x_0)$, $c_{\mathcal{X}}(x_0)$ and $c_{\mathcal{Y}}(y_0)$ are finite, we write

$$\begin{aligned}\psi(x) + c_{\mathcal{X}}(x) = \sup_y \big[\phi(y) - c(x,y) + c_{\mathcal{X}}(x)\big] &\ge \sup_y\, [\phi(y) - c_{\mathcal{Y}}(y)]\\ &\ge \phi(y_0) - c_{\mathcal{Y}}(y_0);\end{aligned}$$

$$\begin{aligned}\phi(y) - c_{\mathcal{Y}}(y) = \inf_x \big[\psi(x) + c(x,y) - c_{\mathcal{Y}}(y)\big] &\le \inf_x\, [\psi(x) + c_{\mathcal{X}}(x)]\\ &\le \psi(x_0) + c_{\mathcal{X}}(x_0).\end{aligned}$$

So ψ is bounded below by the μ-integrable function $\phi(y_0) - c_{\mathcal{Y}}(y_0) - c_{\mathcal{X}}$, and ϕ is bounded above by the ν-integrable function $\psi(x_0) + c_{\mathcal{X}}(x_0) + c_{\mathcal{Y}}$; thus both $-\int \psi\, d\mu$ and $\int \phi\, d\nu$ make sense in $\mathbb{R} \cup \{-\infty\}$. Since their sum is $\int (\phi - \psi)\, d\pi = \int c\, d\pi > -\infty$, both integrals are finite. So

$$\int c\, d\pi = \int \phi\, d\nu - \int \psi\, d\mu,$$

and it results from Part (i) of the theorem that both π and (ψ, ϕ) are optimal, respectively in the original and the dual Kantorovich problems.

To prove the last part of (iii), assume that c is continuous; then the subdifferential of any c-convex function is a closed c-cyclically monotone set.

Let π be an arbitrary optimal transference plan, and (ψ, ϕ) an optimal pair of prices. We know that (ψ, ψ^c) is optimal in the dual Kantorovich problem, so

$$\int c(x,y)\, d\pi(x,y) = \int \psi^c\, d\nu - \int \psi\, d\mu.$$

Thanks to the marginal condition, this be rewritten as

$$\int \big[\psi^c(y) - \psi(x) - c(x,y)\big]\, d\pi(x,y) = 0.$$

Since the integrand is nonnegative, it follows that π is concentrated on the set of pairs (x, y) such that $\psi^c(y) - \psi(x) - c(x, y) = 0$, which is precisely the subdifferential of ψ. Thus any optimal transference plan is concentrated on the subdifferential of any optimal ψ. So if Γ is defined as the intersection of all subdifferentials of optimal functions ψ, then Γ also contains the supports of all optimal plans.

Conversely, if $\widetilde{\pi} \in \Pi(\mu,\nu)$ is a transference plan concentrated on Γ, then $\int c\,d\widetilde{\pi} = \int [\psi^c - \psi]\,d\widetilde{\pi} = \int \psi^c\,d\nu - \int \psi\,d\mu$, so $\widetilde{\pi}$ is optimal. Similarly, if $\widetilde{\psi}$ is a c-convex function such that $\partial_c\widetilde{\psi}$ contains Γ, then by the previous estimates $\widetilde{\psi}$ and $\widetilde{\psi}^c$ are integrable against μ and ν respectively, and $\int c\,d\pi = \int [\widetilde{\psi}^c - \widetilde{\psi}]\,d\pi = \int \widetilde{\psi}^c\,d\nu - \int \widetilde{\psi}\,d\mu$, so $(\widetilde{\psi}, \widetilde{\psi}^c)$ is optimal. This concludes the proof. $\square$

Restriction property

The dual side of the Kantorovich problem also behaves well with respect to restriction, as shown by the following results.

Lemma 5.18 (Restriction of c-convexity). *Let $\mathcal{X},\mathcal{Y}$ be two sets and $c : \mathcal{X}\times\mathcal{Y} \to \mathbb{R}\cup\{+\infty\}$. Let $\mathcal{X}' \subset \mathcal{X}$, $\mathcal{Y}' \subset \mathcal{Y}$ and let c' be the restriction of c to $\mathcal{X}'\times\mathcal{Y}'$. Let $\psi : \mathcal{X} \to \mathbb{R}\cup\{+\infty\}$ be a c-convex function. Then there is a c'-convex function $\psi' : \mathcal{X}' \to \mathbb{R}\cup\{+\infty\}$ such that $\psi' \le \psi$ on $\mathcal{X}'$, ψ' coincides with ψ on $\mathrm{proj}_{\mathcal{X}}((\partial_c\psi)\cap(\mathcal{X}'\times\mathcal{Y}'))$ and $\partial_c\psi\cap(\mathcal{X}'\times\mathcal{Y}') \subset \partial_{c'}\psi'$.*

Theorem 5.19 (Restriction for the Kantorovich duality theorem). *Let $(\mathcal{X},\mu)$ and $(\mathcal{Y},\nu)$ be two Polish probability spaces, let $a \in L^1(\mu)$ and $b \in L^1(\nu)$ be real-valued upper semicontinuous functions, and let $c : \mathcal{X}\times\mathcal{Y} \to \mathbb{R}$ be a lower semicontinuous cost function such that $c(x,y) \ge a(x) + b(y)$ for all x, y. Assume that the optimal transport cost $C(\mu,\nu)$ between μ and ν is finite. Let π be an optimal transference plan, and let ψ be a c-convex function such that π is concentrated on $\partial_c\psi$. Let $\widetilde{\pi}$ be a measure on $\mathcal{X}\times\mathcal{Y}$ satisfying $\widetilde{\pi} \le \pi$, and $Z = \widetilde{\pi}[\mathcal{X}\times\mathcal{Y}] > 0$; let $\pi' := \widetilde{\pi}/Z$, and let μ', ν' be the marginals of π'. Let $\mathcal{X}' \subset \mathcal{X}$ be a closed set containing $\mathrm{Spt}\,\mu'$ and $\mathcal{Y}' \subset \mathcal{Y}$ a closed set containing $\mathrm{Spt}\,\nu'$. Let c' be the restriction of c to $\mathcal{X}'\times\mathcal{Y}'$. Then there is a c'-convex function $\psi' : \mathcal{X}' \to \mathbb{R}\cup\{+\infty\}$ such that*

(a) ψ' coincides with ψ on $\mathrm{proj}_{\mathcal{X}}((\partial_c\psi)\cap(\mathcal{X}'\times\mathcal{Y}'))$, which has full μ'-measure;

(b) π' *is concentrated on* $\partial_{c'}\psi'$;

(c) ψ' *solves the dual Kantorovich problem between* $(\mathcal{X}', \mu')$ *and* $(\mathcal{Y}', \nu')$ *with cost* c'.

In spite of its superficially complicated appearance, Theorem 5.19 is very simple. If the reader feels that it is obvious, or alternatively that it is obscure, he or she may just skip the proofs and retain the loose statement that "it is always possible to restrict the cost function".

Proof of Lemma 5.18. Let $\phi = \psi^c$. For $y \in \mathcal{Y}'$, define

$$\phi'(y) = \begin{cases} \phi(y) & \text{if } \exists x' \in \mathcal{X}'; \quad (x', y) \in \partial_c\psi; \\ -\infty & \text{otherwise.} \end{cases}$$

For $x \in \mathcal{X}'$ let then

$$\psi'(x) = \sup_{y \in \mathcal{Y}'} \big[\phi'(y) - c(x,y)\big] = \sup_{y \in \mathcal{Y}'} \big[\phi'(y) - c'(x,y)\big].$$

By construction, ψ' is c'-convex. Since $\phi' \le \phi$ and $\mathcal{Y}' \subset \mathcal{Y}$ it is obvious that

$$\forall x \in \mathcal{X}', \qquad \psi'(x) \le \sup_{y \in \mathcal{Y}} \big[\phi(y) - c(x,y)\big] = \psi(x).$$

Let $x \in \mathrm{proj}_{\mathcal{X}}((\partial_c\psi) \cap (\mathcal{X}' \cap \mathcal{Y}'))$; this means that there is $y \in \mathcal{Y}'$ such that $(x,y) \in \partial_c\psi$. Then $\phi'(y) = \phi(y)$, so

$$\psi'(x) \ge \phi'(y) - c(x,y) = \phi(y) - c(x,y) = \psi(x).$$

Thus ψ' does coincide with ψ on $\mathrm{proj}_{\mathcal{X}}((\partial_c\psi) \cap (\mathcal{X}' \times \mathcal{Y}'))$.

Finally, let $(x,y) \in \partial_c\psi \cap (\mathcal{X}' \times \mathcal{Y}')$, then $\phi'(y) = \phi(y)$, $\psi'(x) = \psi(x)$; so for all $z \in \mathcal{X}'$,

$$\psi'(x) + c'(x,y) = \psi(x) + c(x,y) = \phi(y) = \phi'(y) \le \psi'(z) + c'(z,y),$$

which means that $(x,y) \in \partial_{c'}\psi'$. □

Proof of Theorem 5.19. Let ψ' be defined by Lemma 5.18. To prove (a), it suffices to note that π' is concentrated on $(\partial_c\psi) \cap (\mathcal{X}' \times \mathcal{Y}')$, so μ' is concentrated on $\mathrm{proj}_{\mathcal{X}}((\partial_c\psi) \cap (\mathcal{X}' \times \mathcal{Y}'))$. Then π is concentrated on $\partial_c\psi$, so $\widetilde{\pi}$ is concentrated on $\partial_c\psi \cap (\mathcal{X}' \times \mathcal{Y}')$, which by Lemma 5.18 is included in $\partial_{c'}\psi'$; this proves (b). Finally, (c) follows from Theorem 5.10(ii). □

The rest of this chapter is devoted to various applications of Theorem 5.10.

Application: Stability

An important consequence of Theorem 5.10 is the stability of optimal transference plans. For simplicity I shall prove it under the assumption that c is bounded below.

Theorem 5.20 (Stability of optimal transport). *Let $\mathcal{X}$ and $\mathcal{Y}$ be Polish spaces, and let $c \in C(\mathcal{X} \times \mathcal{Y})$ be a real-valued continuous cost function, $\inf c > -\infty$. Let $(c_k)_{k\in\mathbb{N}}$ be a sequence of continuous cost functions converging uniformly to c on $\mathcal{X}\times\mathcal{Y}$. Let $(\mu_k)_{k\in\mathbb{N}}$ and $(\nu_k)_{k\in\mathbb{N}}$ be sequences of probability measures on $\mathcal{X}$ and $\mathcal{Y}$ respectively. Assume that μ_k converges to μ (resp. ν_k converges to ν) weakly. For each k, let π_k be an optimal transference plan between μ_k and ν_k. If*

$$\forall k \in \mathbb{N}, \qquad \int c_k \, d\pi_k < +\infty,$$

then, up to extraction of a subsequence, π_k converges weakly to some c-cyclically monotone transference plan $\pi \in \Pi(\mu,\nu)$.

If moreover

$$\liminf_{k\in\mathbb{N}} \int c_k \, d\pi_k < +\infty,$$

then the optimal total transport cost $C(\mu,\nu)$ between μ and ν is finite, and π is an optimal transference plan.

Corollary 5.21 (Compactness of the set of optimal plans). *Let $\mathcal{X}$ and $\mathcal{Y}$ be Polish spaces, and let $c(x,y)$ be a real-valued continuous cost function, $\inf c > -\infty$. Let $\mathcal{K}$ and $\mathcal{L}$ be two compact subsets of $P(\mathcal{X})$ and $P(\mathcal{Y})$ respectively. Then the set of optimal transference plans π whose marginals respectively belong to $\mathcal{K}$ and $\mathcal{L}$ is itself compact in $P(\mathcal{X}\times\mathcal{Y})$.*

Proof of Theorem 5.20. Since μ_k and ν_k are convergent sequences, by Prokhorov's theorem they constitute tight sets, and then by Lemma 4.4 the measures π_k all lie in a tight set of $\mathcal{X}\times\mathcal{Y}$; so we can extract a further subsequence, still denoted (π_k) for simplicity, which converges weakly to some $\pi \in \Pi(\mu,\nu)$.

To prove that π is c-monotone, the argument is essentially the same as in Step 2 of the proof of Theorem 5.10(i). Indeed, by Theorem 5.10, each π_k is concentrated on a c_k-cyclically monotone set; so $\pi_k^{\otimes N}$ is concentrated on the set $\mathcal{C}_k(N)$ of $((x_1,y_1),\ldots,(x_N,y_N))$ such that

$$\sum_{1\le i\le N} c_k(x_i,y_i) \le \sum_{1\le i\le N} c_k(x_i,y_{i+1}),$$

where as usual $y_{N+1} = y_1$. So if $\varepsilon > 0$ and N are given, for k large enough $\pi_k^{\otimes N}$ is concentrated on the set $\mathcal{C}_\varepsilon(N)$ defined by

$$\sum_{1\le i\le N} c(x_i,y_i) \le \sum_{1\le i\le N} c(x_i,y_{i+1}) + \varepsilon.$$

Since this is a closed set, the same is true for $\pi^{\otimes N}$, and then by letting $\varepsilon \to 0$ we see that $\pi^{\otimes N}$ is concentrated on the set $\mathcal{C}(N)$ defined by

$$\sum_{1\le i\le N} c(x_i,y_i) \le \sum_{1\le i\le N} c(x_i,y_{i+1}).$$

So the support of π is c-cyclically monotone, as desired.

Now assume that $\liminf_{k\to\infty} \int c_k\, d\pi_k < +\infty$. Then by the same argument as in the proof of Theorem 4.1,

$$\int c\, d\pi \le \liminf_{k\to\infty} \int c_k\, d\pi_k < +\infty.$$

In particular, $C(\mu,\nu) < +\infty$; so Theorem 5.10(ii) guarantees the optimality of π. □

An immediate consequence of Theorem 5.20 is the possibility to select optimal transport plans in a measurable way.

Corollary 5.22 (Measurable selection of optimal plans). *Let $\mathcal{X}$, $\mathcal{Y}$ be Polish spaces and let $c : \mathcal{X}\times\mathcal{Y} \to \mathbb{R}$ be a continuous cost function, $\inf c > -\infty$. Let Ω be a measurable space and let $\omega \longmapsto (\mu_\omega, \nu_\omega)$ be a measurable function $\Omega \to P(\mathcal{X}) \times P(\mathcal{Y})$. Then there is a measurable choice $\omega \longmapsto \pi_\omega$ such that for each ω, π_ω is an optimal transference plan between μ_ω and ν_ω.*

Proof of Corollary 5.22. Let $\mathcal{O}$ be the set of all optimal transference plans, equipped with the weak topology on $P(\mathcal{X}\times\mathcal{Y})$, and let $\Phi : \mathcal{O} \to P(\mathcal{X}) \times P(\mathcal{Y})$ be the map which to π associates its marginals (μ,ν). Obviously Φ is continuous. By Theorem 5.20, $\mathcal{O}$ is closed; in particular it is a Polish space. By Theorem 4.1, Φ is onto. By Corollary 5.21 all pre-images $\Phi^{-1}(\mu,\nu)$ are compact. So the conclusion follows from general theorems of measurable selection (see the bibliographical notes for more information). □

Theorem 5.20 also admits the following corollary about the stability of *transport maps.*

Corollary 5.23 (Stability of the transport map). *Let $\mathcal{X}$ be a locally compact Polish space, and let $(\mathcal{Y}, d)$ be another Polish space. Let $c : \mathcal{X} \times \mathcal{Y} \to \mathbb{R}$ be a lower semicontinuous function with* $\inf c > -\infty$, *and for each $k \in \mathbb{N}$ let $c_k : \mathcal{X} \times \mathcal{Y} \to \mathbb{R}$ be lower semicontinuous, such that c_k converges uniformly to c. Let $\mu \in P(\mathcal{X})$ and let $(\nu_k)_{k\in\mathbb{N}}$ be a sequence of probability measures on $\mathcal{Y}$, converging weakly to $\nu \in P(\mathcal{Y})$. Assume the existence of measurable maps $T_k : \mathcal{X} \to \mathcal{Y}$ such that each $\pi_k := (\mathrm{Id}\,, T_k)_{\#}\mu$ is an optimal transference plan between μ and ν_k for the cost c_k, having finite total transport cost. Further, assume the existence of a measurable map $T : \mathcal{X} \to \mathcal{Y}$ such that $\pi := (\mathrm{Id}\,, T)_{\#}\mu$ is the unique optimal transference plan between μ and ν, for the cost c, and has finite total transport cost. Then T_k converges to T in probability:*

$$\forall \varepsilon > 0 \quad \mu\Big[\Big\{x \in X;\ d\big(T_k(x), T(x)\big) \geq \varepsilon\Big\}\Big] \xrightarrow[k\to\infty]{} 0. \tag{5.26}$$

Remark 5.24. An important assumption in the above statement is the uniqueness of the optimal transport map T.

Remark 5.25. If the measure μ is replaced by a sequence $(\mu_k)_{k\in\mathbb{N}}$ converging weakly to μ, then the maps T_k and T may be far away from each other, even μ_k-almost surely: take for instance $\mathcal{X} = \mathcal{Y} = \mathbb{R}$, $\mu_k = \delta_{1/k}$, $\mu = \delta_0$, $\nu_k = \nu = \delta_0$, $T_k(x) = 0$, $T(x) = 1_{x\neq 0}$.

Proof of Corollary 5.23. By Theorem 5.20 and uniqueness of the optimal coupling between μ and ν, we know that $\pi_k = (\mathrm{Id}\,, T_k)_{\#}\mu_k$ converges weakly to $\pi = (\mathrm{Id}\,, T)_{\#}\mu$.

Let $\varepsilon > 0$ and $\delta > 0$ be given. By Lusin's theorem (in the abstract version recalled in the end of the bibliographical notes) there is a closed set $K \subset \mathcal{X}$ such that $\mu[\mathcal{X} \setminus K] \leq \delta$ and the restriction of T to K is continuous. So the set

$$A_\varepsilon = \Big\{(x, y) \in K \times \mathcal{Y};\ d(T(x), y) \geq \varepsilon\Big\}.$$

is closed in $K \times \mathcal{Y}$, and therefore closed in $\mathcal{X} \times \mathcal{Y}$. Also $\pi[A_\varepsilon] = 0$ since π is concentrated on the graph of T. So, by weak convergence of π_k and closedness of A_ε,

$$\begin{aligned}
0 = \pi[A_\varepsilon] &\geq \limsup_{k\to\infty} \ \pi_k[A_\varepsilon] \\
&= \limsup_{k\to\infty} \ \pi_k\big[\{(x,y) \in K \times \mathcal{Y};\ d(T(x),y) \geq \varepsilon\}\big] \\
&= \limsup_{k\to\infty} \ \mu\big[\{x \in K;\ d(T(x),T_k(x)) \geq \varepsilon\}\big] \\
&\geq \limsup_{k\to\infty} \ \mu\big[\{x \in \mathcal{X};\ d(T(x),T_k(x)) \geq \varepsilon\}\big] - \delta.
\end{aligned}$$

Letting $\delta \to 0$ we conclude that $\limsup \mu[d(T(x),T_k(x)) \geq \varepsilon] = 0$, which was the goal. □

Application: Dual formulation of transport inequalities

Let c be a given cost function, and let

$$C(\mu,\nu) = \inf_{\pi\in\Pi(\mu,\nu)} \int c\, d\pi \tag{5.27}$$

stand for the value of the optimal transport cost of transport between μ and ν.

If ν is a given reference measure, inequalities of the form

$$\forall \mu \in P(\mathcal{X}), \qquad C(\mu,\nu) \leq F(\mu)$$

arise in several branches of mathematics; some of them will be studied in Chapter 22. It is useful to know that if F is a convex function of μ, then there is a nice dual reformulation of these inequalities in terms of the Legendre transform of F. This is the content of the following theorem.

Theorem 5.26 (Dual transport inequalities). *Let $\mathcal{X}$, $\mathcal{Y}$ be two Polish spaces, and ν a given probability measure on $\mathcal{Y}$. Let $F : P(\mathcal{X}) \to \mathbb{R}\cup\{+\infty\}$ be a convex lower semicontinuous function on $P(\mathcal{X})$, and Λ its Legendre transform on $C_b(\mathcal{X})$; more explicitly, it is assumed that*

$$\begin{cases}
\forall \mu \in P(\mathcal{X}), \quad F(\mu) = \displaystyle\sup_{\varphi\in C_b(\mathcal{X})} \left(\int_{\mathcal{X}} \varphi\, d\mu - \Lambda(\varphi)\right) \\[2ex]
\forall \varphi \in C_b(\mathcal{X}), \quad \Lambda(\varphi) = \displaystyle\sup_{\mu\in P(\mathcal{X})} \left(\int_{\mathcal{X}} \varphi\, d\mu - F(\mu)\right).
\end{cases} \tag{5.28}$$

Further, let $c : \mathcal{X} \times \mathcal{Y} \to \mathbb{R} \cup \{+\infty\}$ *be a lower semicontinuous cost function,* $\inf c > -\infty$. *Then, the following two statements are equivalent:*

(i) $\forall \mu \in P(\mathcal{X}), \qquad C(\mu, \nu) \le F(\mu)$;

(ii) $\forall \phi \in C_b(\mathcal{Y}), \qquad \Lambda\left(\int_{\mathcal{Y}} \phi\, d\nu - \phi^c\right) \le 0$, *where* $\phi^c(x) := \sup_y [\phi(y) - c(x, y)]$.

Moreover, if $\Phi : \mathbb{R} \to \mathbb{R}$ *is a nondecreasing convex function with* $\Phi(0) = 0$, *then the following two statements are equivalent:*

(i') $\forall \mu \in P(\mathcal{X}), \qquad \Phi(C(\mu, \nu)) \le F(\mu)$;

(ii') $\forall \phi \in C_b(\mathcal{Y}),\ \forall t \ge 0, \qquad \Lambda\left(t \int_{\mathcal{Y}} \phi\, d\nu - t\, \phi^c - \Phi^*(t)\right) \le 0,$
where Φ^* *stands for the Legendre transform of* Φ.

Remark 5.27. The writing in (ii) or (ii') is not very rigorous since Λ is a priori defined on the set of bounded continuous functions, and ϕ^c might not belong to that set. (It is clear that ϕ^c is bounded from above, but this is all that can be said.) However, from (5.28) $\Lambda(\varphi)$ is a nondecreasing function of φ, so there is in practice no problem to extend Λ to a more general class of measurable functions. In any case, the correct way to interpret the left-hand side in (ii) is

$$\Lambda\left(\int_{\mathcal{Y}} \phi\, d\nu - \phi^c\right) = \sup_{\psi \ge \phi^c} \Lambda\left(\int_{\mathcal{Y}} \phi\, d\nu - \psi\right),$$

where ψ in the supremum is assumed to be bounded continuous.

Remark 5.28. One may simplify (ii') by taking the supremum over t; since Λ is nonincreasing, the result is

$$\Lambda\left(\Phi\left(\int_{\mathcal{Y}} \phi\, d\nu - \phi^c\right)\right) \le 0. \tag{5.29}$$

(This shows in particular that the equivalence (i) $\Leftrightarrow$ (ii) is a particular case of the equivalence (i') $\Leftrightarrow$ (ii').) However, in certain situations it is better to use the inequality (ii') rather than (5.29); see for instance Proposition 22.5.

Example 5.29. The most famous example of inequality of the type of (i) occurs when $\mathcal{X} = \mathcal{Y}$ and $F(\mu)$ is the Kullback information of μ with respect to ν, that is $F(\mu) = H_\nu(\mu) = \int \rho \log \rho\, d\nu$, where $\rho = d\mu/d\nu$;

and by convention $F(\mu) = +\infty$ if μ is not absolutely continuous with respect to ν. Then one has the explicit formula

$$\Lambda(\varphi) = \log\left(\int e^{\varphi}\, d\nu\right).$$

So the two functional inequalities

$$\forall \mu \in P(\mathcal{X}), \quad C(\mu,\nu) \le H_\nu(\mu)$$

and

$$\forall \phi \in C_b(\mathcal{X}), \quad e^{\int \phi\, d\nu} \le \int e^{\phi^c}\, d\nu$$

are equivalent.

Proof of Theorem 5.26. First assume that (i) is satisfied. Then for all $\psi \ge \phi^c$,

$$\begin{aligned} \Lambda\left(\int_{\mathcal{Y}} \phi\, d\nu \; - \; \psi\right) &= \sup_{\mu \in P(\mathcal{X})} \left\{\int_{\mathcal{X}} \left(\int_{\mathcal{Y}} \phi\, d\nu \; - \; \psi\right) d\mu - F(\mu)\right\} \\ &= \sup_{\mu \in P(\mathcal{X})} \left\{\int_{\mathcal{Y}} \phi\, d\nu - \int_{\mathcal{X}} \psi\, d\mu - F(\mu)\right\} \\ &\le \sup_{\mu \in P(\mathcal{X})} \Big[C(\mu,\nu) - F(\mu)\Big] \le 0, \end{aligned}$$

where the easiest part of Theorem 5.10 (that is, inequality (5.4)) was used to go from the next-to-last line to the last one. Then (ii) follows upon taking the supremum over ψ.

Conversely, assume that (ii) is satisfied. Then, for any pair $(\psi,\phi) \in C_b(\mathcal{X}) \times C_b(\mathcal{Y})$ one has, by (5.28),

$$\int_{\mathcal{Y}} \phi\, d\nu - \int_{\mathcal{X}} \psi\, d\mu = \int_{\mathcal{X}} \left(\int_{\mathcal{Y}} \phi\, d\nu - \psi\right) d\mu \le \Lambda\left(\int_{\mathcal{Y}} \phi\, d\nu - \psi\right) + F(\mu).$$

Taking the supremum over all $\psi \ge \phi^c$ yields

$$\int_{\mathcal{Y}} \phi\, d\nu - \int_{\mathcal{X}} \phi^c\, d\mu \le \Lambda\left(\int_{\mathcal{Y}} \phi\, d\nu - \phi^c\right) \; + \; F(\mu).$$

By assumption, the first term in the right-hand side is always nonpositive; so in fact

$$\int_{\mathcal{Y}} \phi\, d\nu - \int_{\mathcal{X}} \phi^c\, d\mu \le F(\mu).$$

Then (i) follows upon taking the supremum over $\phi \in C_b(\mathcal{Y})$ and applying Theorem 5.10 (i).

Now let us consider the equivalence between (i') and (ii'). By assumption, $\Phi(r) \leq 0$ for $r \leq 0$, so $\Phi^*(t) = \sup_r [r\,t - \Phi(r)] = +\infty$ if $t < 0$. Then the Legendre inversion formula says that

$$\forall r \in \mathbb{R}, \quad \Phi(r) = \sup_{t \in \mathbb{R}_+} \big[r\,t - \Phi^*(t)\big].$$

(The important thing is that the supremum is over $\mathbb{R}_+$ and not $\mathbb{R}$.)

If (i') is satisfied, then for all $\phi \in C_b(\mathcal{X})$, for all $\psi \geq \phi^c$ and for all $t \in \mathbb{R}_+$,

$$\begin{aligned}
&\Lambda\left(t\int_{\mathcal{Y}} \phi\,d\nu - t\,\psi - \Phi^*(t)\right)\\
&\qquad = \sup_{\mu \in P(\mathcal{X})} \left[\int_{\mathcal{X}}\left(t\int_{\mathcal{Y}} \phi\,d\nu - t\,\psi - \Phi^*(t)\right) d\mu - F(\mu)\right]\\
&\qquad = \sup_{\mu \in P(\mathcal{X})} \left[t\left(\int_{\mathcal{Y}} \phi\,d\nu - \int_{\mathcal{X}} \psi\,d\mu\right) - \Phi^*(t) - F(\mu)\right]\\
&\qquad \leq \sup_{\mu \in P(\mathcal{X})} \Big[t\,C(\mu,\nu) - \Phi^*(t) - F(\mu)\Big]\\
&\qquad \leq \sup_{\mu \in P(\mathcal{X})} \Big[\Phi(C(\mu,\nu)) - F(\mu)\Big] \leq 0,
\end{aligned}$$

where the inequality $t\,r \leq \Phi(r) + \Phi^*(t)$ was used.

Conversely, if (ii') is satisfied, then for all $(\phi,\psi) \in C_b(\mathcal{X}) \times C_b(\mathcal{Y})$ and $t \geq 0$,

$$\begin{aligned}
t\int_{\mathcal{Y}} \phi\,d\nu - t\int_{\mathcal{X}} \psi\,d\mu - \Phi^*(t) &= \int_{\mathcal{X}}\left(\int_{\mathcal{Y}} t\phi\,d\nu - t\,\psi - \Phi^*(t)\right) d\mu\\
&\leq \Lambda\left(t\int_{\mathcal{Y}} \phi\,d\nu - t\,\psi - \Phi^*(t)\right) + F(\mu);
\end{aligned}$$

then by taking the supremum over $\psi \geq \phi^c$ one obtains

$$\begin{aligned}
t\,C(\mu,\nu) - \Phi^*(t) &\leq \Lambda\left(t\int_{\mathcal{Y}} \phi\,d\nu - t\,\phi^c - \Phi^*(t)\right) + F(\mu)\\
&\leq F(\mu);
\end{aligned}$$

and (i') follows by taking the supremum over $t \geq 0$. □

Application: Solvability of the Monge problem

As a last application of Theorem 5.10, I shall now present the criterion which is used in the large majority of proofs of existence of a deterministic optimal coupling (or Monge transport).

Theorem 5.30 (Criterion for solvability of the Monge problem). *Let $(\mathcal{X},\mu)$ and $(\mathcal{Y},\nu)$ be two Polish probability spaces, and let $a \in L^1(\mu)$, $b \in L^1(\nu)$ be two real-valued upper semicontinuous functions. Let $c : \mathcal{X}\times\mathcal{Y} \to \mathbb{R}$ be a lower semicontinuous cost function such that $c(x,y) \geq a(x) + b(y)$ for all x,y. Let $C(\mu,\nu)$ be the optimal total transport cost between μ and ν. If*

(i) $C(\mu,\nu) < +\infty$;

(ii) for any c-convex function $\psi : \mathcal{X} \to \mathbb{R}\cup\{+\infty\}$, the set of $x \in \mathcal{X}$ such that $\partial_c\psi(x)$ contains more than one element is μ-negligible; then there is a unique (in law) optimal coupling (X,Y) of (μ,ν), and it is deterministic. It is characterized (among all possible couplings) by the existence of a c-convex function ψ such that, almost surely, Y belongs to $\partial_c\psi(X)$. In particular, the Monge problem with initial measure μ and final measure ν admits a unique solution.

Proof of Theorem 5.30. The argument is almost obvious. By Theorem 5.10(ii), there is a c-convex function ψ, and a measurable set $\Gamma \subset \partial_c\psi$ such that any optimal plan π is concentrated on Γ. By assumption there is a Borel set Z such that $\mu[Z] = 0$ and $\partial_c\psi(x)$ contains at most one element if $x \notin Z$. So for any $x \in \mathrm{proj}_{\mathcal{X}}(\Gamma) \setminus Z$, there is exactly one $y \in \mathcal{Y}$ such that $(x,y) \in \Gamma$, and we can then define $T(x) = y$.

Let now π be any optimal coupling. Because it has to be concentrated on Γ, and $Z \times \mathcal{Y}$ is π-negligible, π is also concentrated on $\Gamma \setminus (Z\times\mathcal{Y})$, which is precisely the set of all pairs $(x,T(x))$, i.e. the graph of T. It follows that π is the Monge transport associated with the map T.

The argument above is in fact a bit sloppy, since I did not check the measurability of T. I shall show below how to slightly modify the construction of T to make sure that it is measurable. The reader who does not want to bother about measurability issues can skip the rest of the proof.

Let $(K_\ell)_{\ell\in\mathbb{N}}$ be a nondecreasing sequence of compact sets, all of them included in $\Gamma \setminus (Z \times \mathcal{Y})$, such that $\pi[\cup K_\ell] = 1$. (The family (K_ℓ) exists because π, just as any finite Borel measure on a Polish space, is regular.)

If ℓ is given, then for any x lying in the compact set $J_\ell := \mathrm{proj}_{\mathcal{X}}(K_\ell)$ we can define $T_\ell(x)$ as the unique y such that $(x,y) \in K_\ell$. Then we can define T on $\cup J_\ell$ by stipulating that for each ℓ, T restricts to T_ℓ on J_ℓ. The map T_ℓ is continuous on J_ℓ: Indeed, if $x_m \in T_\ell$ and $x_m \to x$, then the sequence $(x_m, T_\ell(x_m))_{m\in\mathbb{N}}$ is valued in the compact set K_ℓ, so up to extraction it converges to some $(x,y) \in K_\ell$, and necessarily $y = T_\ell(x)$. So T is a Borel map. Even if it is not defined on the whole of $\Gamma \setminus (Z \times \mathcal{Y})$, it is still defined on a set of full μ-measure, so the proof can be concluded just as before. □

Bibliographical notes

There are many ways to state the Kantorovich duality, and even more ways to prove it. There are also several economic interpretations, that belong to folklore. The one which I formulated in this chapter is a variant of one that I learnt from Caffarelli. Related economic interpretations underlie some algorithms, such as the fascinating "auction algorithm" developed by Bertsekas (see [115, Chapter 7], or the various surveys written by Bertsekas on the subject). But also many more classical algorithms are based on the Kantorovich duality [743].

A common convention consists in taking the pair $(-\psi, \phi)$ as the unknown.[3] This has the advantage of making some formulas more symmetric: The c-transform becomes $\varphi^c(y) = \inf_x[c(x,y) - \varphi(x)]$, and then $\psi^c(x) = \inf_y[c(x,y) - \psi(y)]$, so this is the same formula going back and forth between functions of x and functions of y, upon exchange of x and y. Since $\mathcal{X}$ and $\mathcal{Y}$ may have nothing in common, in general this symmetry is essentially cosmetic. The conventions used in this chapter lead to a somewhat natural "economic" interpretation, and will also lend themselves better to a time-dependent treatment. Moreover, they also agree with the conventions used in weak KAM theory, and more generally in the theory of dynamical systems [105, 106, 347]. It might be good to make the link more explicit. In weak KAM theory, $\mathcal{X} = \mathcal{Y}$ is a Riemannian manifold M; a Lagrangian cost function is given on the tangent bundle TM; and $c = c(x,y)$ is a continuous cost function defined from the dynamics, as the minimum action that one

---

[3] The latter pair was denoted (φ, ψ) in [814, Chapter 1], which will probably upset the reader.

should take to go from x to y (as later in Chapter 7). Since in general $c(x,x) \neq 0$, it is not absurd to consider the optimal transport cost $C(\mu,\mu)$ between a measure μ and itself. If M is compact, it is easy to show that there exists a $\overline{\mu}$ that minimizes $C(\mu,\mu)$. To the optimal transport problem between $\overline{\mu}$ and $\overline{\mu}$, Theorem 5.10 associates two distinguished closed c-cyclically monotone sets: a minimal one and a maximal one, respectively $\Gamma_{\min}$ and $\Gamma_{\max} \subset M \times M$. These sets can be identified with subsets of TM via the embedding (initial position, final position) $\longmapsto$ (initial position, initial velocity). Under that identification, $\Gamma_{\min}$ and $\Gamma_{\max}$ are called respectively the Mather and Aubry sets; they carry valuable information about the underlying dynamics. For mnemonic purposes, to recall which is which, the reader might use the resemblance of the name "Mather" to the word "measure". (The Mather set is the one cooked up from the supports of the probability measures.)

In the particular case when $c(x,y) = |x-y|^2/2$ in Euclidean space, it is customary to expand $c(x,y)$ as $|x|^2/2 - x\cdot y + |y|^2/2$, and change unknowns by including $|x|^2/2$ and $|y|^2/2$ into ψ and ϕ respectively, then change signs and reduce to the cost function $x\cdot y$, which is the one appearing naturally in the Legendre duality of convex functions. This is explained carefully in [814, Chapter 2], where reminders and references about the theory of convex functions in $\mathbb{R}^n$ are also provided.

The Kantorovich duality theorem was proven by Kantorovich himself on a compact space in his famous note [501] (even before he realized the connection with Monge's transportation problem). As Kantorovich noted later in [502], the duality for the cost $c(x,y) = |x-y|$ in $\mathbb{R}^n$ implies that transport pathlines are orthogonal to the surfaces $\{\psi = \text{constant}\}$, where ψ is the Kantorovich potential, i.e. the solution of the dual problem; in this way he recovered Monge's celebrated original observation.

In 1958, Kantorovich and Rubinstein [506] made the duality more explicit in the special case when $c(x,y) = d(x,y)$. Much later the statement was generalized by Dudley [316, Lecture 20] [318, Section 11.8], with an alternative argument (partly based on ideas by Neveu) which does not need completeness (this sometimes is useful to handle complete nonseparable spaces); the proof in the first reference contains a gap which was filled by de Acosta [273, Appendix B].[4] Rüschendorf [390,

[4] De Acosta used an idea suggested by Dudley in Saint-Flour, 25 years before my own course!

715], Fernique [356], Szulga [769], Kellerer [512], Feyel [357], and probably others, contributed to the problem.

Modern treatments most often use variants of the Hahn–Banach theorem, see for instance [550, 696, 814]. The proof presented in [814, Theorem 1] first proves the duality when $\mathcal{X}, \mathcal{Y}$ are compact, then treats the general case by an approximation argument; this is somewhat tricky but has the advantage of avoiding the general version of the axiom of choice, since it uses the Hahn–Banach theorem only in the separable space $C(K)$, where K is compact.

Mikami [629] recovered the duality theorem in $\mathbb{R}^n$ using stochastic control, and together with Thieullen [631] extended it to certain classes of stochastic processes.

Ramachandran and Rüschendorf [698, 699] investigated the Kantorovich duality out of the setting of Polish spaces, and found out a necessary and sufficient condition for its validity (the spaces should be "perfect").

In the case when the cost function is a distance, the optimal transport problem coincides with the Kantorovich transshipment problem, for which more duality theorems are available, and a vast literature has been written; see [696, Chapter 6] for results and references. This topic is closely related to the subject of "Kantorovich norms": see [464], [695, Chapters 5 and 6], [450, Chapter 4], [149] and the many references therein.

Around the mid-eighties, it was understood that the study of the dual problem, and in particular the existence of a maximizer, could lead to precious qualitative information about the solution of the Monge–Kantorovich problem. This point of view was emphasized by many authors such as Knott and Smith [524], Cuesta-Albertos, Matrán and Tuero-Díaz [254, 255, 259], Brenier [154, 156], Rachev and Rüschendorf [696, 722], Abdellaoui and Heinich [1, 2], Gangbo [395], Gangbo and McCann [398, 399], McCann [616] and probably others. Then Ambrosio and Pratelli proved the existence of a maximizing pair under the conditions (5.10); see [32, Theorem 3.2]. Under adequate assumptions, one can also prove the existence of a maximizer for the dual problem by direct arguments which do not use the original problem (see for instance [814, Chapter 2]).

The classical theory of convexity and its links with the property of cyclical monotonicity are exposed in the well-known treatise by Rockafellar [705]. The more general notions of c-convexity and

c-cyclical monotonicity were studied by several researchers, in particular Rüschendorf [722]. Some authors prefer to use c-concavity; I personally advocate working with c-convexity, because signs get better in many situations. However, the conventions used in the present set of notes have the disadvantage that the cost function $c(\cdot, y)$ is not c-convex. A possibility to remedy this would be to call $(-c)$-convexity the notion which I defined. This convention (suggested to me by Trudinger) is appealing, but would have forced me to write $(-c)$-convex hundreds of times throughout this book.

The notation $\partial_c\psi(x)$ and the terminology of c-subdifferential is derived from the usual notation $\partial\psi(x)$ in convex analysis. Let me stress that in my notation $\partial_c\psi(x)$ is a set of points, not a set of tangent vectors or differential forms. Some authors prefer to call $\partial_c\psi(x)$ the **contact set** of ψ at x (any y in the contact set is a contact point) and to use the notation $\partial_c\psi(x)$ for a set of tangent vectors (which under suitable assumptions can be identified with the contact set, and which I shall denote by $-\nabla_x c(x, \partial_c\psi_c(x))$, or $\nabla_c^-\psi(x)$, in Chapters 10 and 12).

In [712] the authors argue that c-convex functions should be constructible in practice when the cost function c is convex (in the usual sense), in the sense that such c-convex profiles can be "engineered" with a convex tool.

For the proof of Theorem 5.10, I borrowed from McCann [613] the idea of recovering c-cyclical monotonicity from approximation by combinations of Dirac masses; from Rüschendorf [719] the method used to reconstruct ψ from Γ; from Schachermayer and Teichmann [738] the clever truncation procedure used in the proof of Part (ii). Apart from that the general scheme of proof is more or less the one which was used by Ambrosio and Pratelli [32], and Ambrosio, Gigli and Savaré [30]. On the whole, the proof avoids the use of the axiom of choice, does not need any painful approximation procedure, and leads to the best known results. In my opinion these advantages do compensate for its being somewhat tricky.

About the proof of the Kantorovich duality, it is interesting to notice that "duality for somebody implies duality for everybody" (a rule which is true in other branches of analysis): In the present case, constructing one particular cyclically monotone transference plan allows one to prove the duality, which leads to the conclusion that *all* transference plans should be cyclically monotone. By the way, the latter statement could also be proven directly with the help of a bit of measure theoretical abstract nonsense, see e.g. [399, Theorem 2.3] or [1, 2].

The use of the law of large numbers for empirical measures might be natural for a probabilistic audience, but one should not forget that this is a subtle result: For any bounded continuous test function, the usual law of large numbers yields convergence out of a negligible set, but then one has to find a negligible set that works for *all* bounded continuous test functions. In a compact space $\mathcal{X}$ this is easy, because $C_b(\mathcal{X})$ is separable, but if $\mathcal{X}$ is not compact one should be careful. Dudley [318, Theorem 11.4.1] proves the law of large numbers for empirical measures on general separable metric spaces, giving credit to Varadarajan for this theorem. In the dynamical systems community, these results are known as part of the so-called Krylov–Bogoljubov theory, in relation with ergodic theorems; see e.g. Oxtoby [674] for a compact space.

The equivalence between the properties of optimality (of a transference plan) and cyclical monotonicity, for quite general cost functions and probability measures, was a widely open problem until recently; it was explicitly listed as Open Problem 2.25 in [814] for a quadratic cost function in $\mathbb{R}^n$. The current state of the art is as follows:

- the equivalence is *false* for a general lower semicontinuous cost function with possibly infinite values, as shown by a clever counterexample of Ambrosio and Pratelli [32];
- the equivalence is true for a *continuous* cost function with possibly infinite values, as shown by Pratelli [688];
- the equivalence is true for a *real-valued* lower semicontinuous cost function, as shown by Schachermayer and Teichmann [738]; this is the result that I chose to present in this course. Actually it is sufficient for the cost to be lower semicontinuous and real-valued almost everywhere (with respect to the product measure $\mu \otimes \nu$).
- more generally, the equivalence is true as soon as c is measurable (not even lower semicontinuous!) and $\{c = \infty\}$ is the union of a closed set and a $(\mu \otimes \nu)$-negligible Borel set; this result is due to Beiglböck, Goldstern, Maresch and Schachermayer [79].

Schachermayer and Teichmann gave a nice interpretation of the Ambrosio–Pratelli counterexample and suggested that the correct notion in the whole business is not cyclical monotonicity, but a variant which they named "strong cyclical monotonicity condition" [738].

As I am completing these notes, it seems that the final resolution of this equivalence issue might soon be available, but at the price of a journey into the very guts of measure theory. The following construction was explained to me by Bianchini. Let c be an arbitrary lower semicontinuous cost function with possibly infinite values, and let π be a c-cyclically monotone plan. Let Γ be a c-cyclically monotone set with $\pi[\Gamma] = 1$. Define an equivalence relation $\mathcal{R}$ on Γ as follows: $(x, y) \sim (x', y')$ if there is a finite number of pairs (x_k, y_k), $0 \le k \le N$, such that: $(x, y) = (x_0, y_0)$; either $c(x_0, y_1) < +\infty$ or $c(x_1, y_0) < +\infty$; $(x_1, y_1) \in \Gamma$; either $c(x_1, y_2) < +\infty$ or $c(x_2, y_1) < +\infty$; etc. until $(x_N, y_N) = (x', y')$. The relation $\mathcal{R}$ divides Γ into equivalence classes $(\Gamma_\alpha)_{\alpha \in \Gamma/\mathcal{R}}$. Let p be the map which to a point x associates its equivalence class $\overline{x}$. The set $\Gamma/\mathcal{R}$ in general has no topological or measurable structure, but we can equip it with the largest σ-algebra making p measurable. On $\Gamma \times (\Gamma/\mathcal{R})$ introduce the product σ-algebra. If now the *graph* of p is measurable for this σ-algebra, then π should be optimal in the Monge–Kantorovich problem.

Related to the above discussion is the "transitive c-monotonicity" considered by Beiglböck, Goldstern, Maresch and Schachermayer [79], who also study in depth the links of this notion with c-monotonicity, optimality, strong c-monotonicity in the sense of [738], and a new interesting concept of "robust optimality". The results in [79] unify those of [688] and [738]. A striking theorem is that robust optimality is always equivalent to strong c-monotonicity.

An alternative approach to optimality criteria, based on extensions of the classical saddle-point theory, was developed by Léonard [550].

In most applications, the cost function is continuous, and often rather simple. However, it is sometimes useful to consider cost functions that achieve the value $+\infty$, as in the "secondary variational problem" considered by Ambrosio and Pratelli [32] and also by Bernard and Buffoni [104]. Such is also the case for the optimal transport in Wiener space considered by Feyel and Üstünel [358, 359, 360, 361], for which the cost function $c(x, y)$ is the square norm of $x - y$ in the Cameron–Martin space (so it is $+\infty$ if $x - y$ does not belong to that space). In this setting, optimizers in the dual problem can be constructed via finite-dimensional approximations, but it is not known whether there is a more direct construction by c-monotonicity.

If one uses the cost function $|x - y|^p$ in $\mathbb{R}^n$ and lets $p \to \infty$, then the c-cyclical monotonicity condition becomes $\sup |x_i - y_i| \le \sup |x_i - y_{i+1}|$. Much remains of the analysis of the Kantorovich duality, but there are also noticeable changes [222].

When condition (5.9) (or its weakened version (5.10)) is relaxed, it is not clear in general that the dual Kantorovich problem admits a maximizing pair. Yet this is true for instance in the case of optimal transport in Wiener space; this is an indication that condition (5.10) might not be the "correct" one, although at present no better general condition is known.

Lemma 5.18 and Theorem 5.19 were inspired by a recent work of Fathi and Figalli [348], in which a restriction procedure is used to solve the Monge problem for certain cost functions arising from Lagrangian dynamics in unbounded space; see Theorem 10.28 for more information.

The core argument in the proof of Theorem 5.20 has probably been discovered several times; I learnt it in [30, Proposition 7.13]. At the present level of generality this statement is due to Schachermayer and Teichmann [738].

I learnt Corollary 5.23 from Ambrosio and Pratelli [32], for measures on Euclidean space. The general statement presented here, and its simple proof, were suggested to me by Schulte, together with the counterexample in Remark 5.25. In some situations where more structure is available and optimal transport is smooth, one can refine the convergence in probability into true pointwise convergence [822]. Even when that information is not available, under some circumstances one can obtain the uniform convergence of $c(x, T_k(x))$: Theorem 28.9(v) is an illustration (this principle also appears in the tedious proof of Theorem 23.14).

Measurable selection theorems in the style of Corollary 5.22 go back at least to Berbee [98]. Recently, Fontbona, Guérin and Méléard [374] studied the measurability of the map which to (μ, ν) associates the union of the supports of all optimal transports between μ and ν (measurability in the sense of set-valued functions). This question was motivated by applications in particle systems and coupling of stochastic differential equations.

Theorem 5.26 appears in a more or less explicit form in various works, especially for the particular case described in Example 5.29; see in particular [128, Section 3]. About Legendre duality for convex functions in $\mathbb{R}$, one may consult [44, Chapter 14]. The classical reference textbook about Legendre duality for plainly convex functions in $\mathbb{R}^n$ is [705]. An excellent introduction to the Legendre duality in Banach spaces can be found in [172, Section I.3].

Finally, I shall say a few words about some basic measure-theoretical tools used in this chapter.

The regularity of Borel measures on Polish spaces is proven in [318, p. 225].

Measurable selection theorems provide conditions under which one may select elements satisfying certain conditions, in a measurable way. The theorem which I used to prove Corollary 5.22 is one of the most classical results of this kind: *A surjective Borel map f between Polish spaces, such that the fibers $f^{-1}(y)$ are all compact, admits a Borel right-inverse.* See Dellacherie [288] for a modern proof. There are more advanced selection theorems due to Aumann, Castaing, Kuratowski, Michael, Novikov, Ryll-Nardzewski, Sainte-Beuve, von Neumann, and others, whose precise statements are remarkably opaque for nonexperts; a simplified and readable account can be found in [18, Chapter 18].

Lusin's theorem [352, Theorem 2.3.5], which was used in the proof of Corollary 5.23, states the following: If $\mathcal{X}$ is a locally compact metric space, $\mathcal{Y}$ is a separable metric space, μ is a Borel measure on $\mathcal{X}$, f is a measurable map $\mathcal{X} \to \mathcal{Y}$, and $A \subset \mathcal{X}$ is a measurable set with finite measure, then for each $\delta > 0$ there is a closed set $K \subset A$ such that $\mu[A \setminus K] < \delta$ and the restriction of f to K is continuous.

6

The Wasserstein distances

Assume, as before, that you are in charge of the transport of goods between producers and consumers, whose respective spatial distributions are modeled by probability measures. The farther producers and consumers are from each other, the more difficult will be your job, and you would like to summarize the degree of difficulty with just one quantity. For that purpose it is natural to consider, as in (5.27), the **optimal transport cost** between the two measures:

$$C(\mu,\nu) = \inf_{\pi\in\Pi(\mu,\nu)} \int c(x,y)\, d\pi(x,y), \tag{6.1}$$

where $c(x,y)$ is the cost for transporting one unit of mass from x to y. Here we do not care about the shape of the optimizer but only the *value* of this optimal cost.

One can think of (6.1) as a kind of distance between μ and ν, but in general it does not, strictly speaking, satisfy the axioms of a distance function. However, when the cost is defined in terms of a distance, it is easy to cook up a distance from (6.1):

Definition 6.1 (Wasserstein distances). *Let $(\mathcal{X},d)$ be a Polish metric space, and let $p\in[1,\infty)$. For any two probability measures μ,ν on $\mathcal{X}$, the Wasserstein distance of order p between μ and ν is defined by the formula*

$$\begin{aligned} W_p(\mu,\nu) &= \left(\inf_{\pi\in\Pi(\mu,\nu)} \int_{\mathcal{X}} d(x,y)^p\, d\pi(x,y)\right)^{1/p} \qquad (6.2)\\ &= \inf\left\{\Big[\mathbb{E}\, d(X,Y)^p\Big]^{\frac{1}{p}}, \quad \mathrm{law}\,(X)=\mu, \quad \mathrm{law}\,(Y)=\nu\right\}. \end{aligned}$$

C. Villani, *Optimal Transport*, Grundlehren der mathematischen Wissenschaften 338,

Particular Case 6.2 (Kantorovich–Rubinstein distance). The distance W_1 is also commonly called the Kantorovich–Rubinstein distance (although it would be more proper to save the the terminology Kantorovich–Rubinstein for the *norm* which extends W_1; see the bibliographical notes).

Example 6.3. $W_p(\delta_x, \delta_y) = d(x, y)$. In this example, the distance does not depend on p; but this is not the rule.

At the present level of generality, W_p is still not a distance in the strict sense, because it might take the value $+\infty$; but otherwise it does satisfy the axioms of a distance, as will be checked right now.

Proof that W_p satisfies the axioms of a distance. First, it is clear that $W_p(\mu, \nu) = W_p(\nu, \mu)$.

Next, let μ_1, μ_2 and μ_3 be three probability measures on $\mathcal{X}$, and let (X_1, X_2) be an optimal coupling of (μ_1, μ_2) and (Z_2, Z_3) an optimal coupling of (μ_2, μ_3) (for the cost function $c = d^p$). By the Gluing Lemma (recalled in Chapter 1), there exist random variables (X_1', X_2', X_3') with $\mathrm{law}\,(X_1', X_2') = \mathrm{law}\,(X_1, X_2)$ and $\mathrm{law}\,(X_2', X_3') = \mathrm{law}\,(Z_2, Z_3)$. In particular, (X_1', X_3') is a coupling of (μ_1, μ_3), so

$$\begin{aligned} W_p(\mu_1, \mu_3) \le \big(\mathbb{E}\, d(X_1', X_3')^p\big)^{\frac{1}{p}} &\le \Big(\mathbb{E}\,\big(d(X_1', X_2') + d(X_2', X_3')\big)^p\Big)^{\frac{1}{p}} \\ &\le \Big(\mathbb{E}\, d(X_1', X_2')^p\Big)^{\frac{1}{p}} + \Big(\mathbb{E}\, d(X_2', X_3')^p\Big)^{\frac{1}{p}} \\ &= W_p(\mu_1, \mu_2) + W_p(\mu_2, \mu_3), \end{aligned}$$

where the inequality leading to the second line is an application of the Minkowski inequality in $L^p(\mathbb{P})$, and the last equality follows from the fact that (X_1', X_2') and (X_2', X_3') are optimal couplings. So W_p satisfies the triangle inequality.

Finally, assume that $W_p(\mu, \nu) = 0$; then there exists a transference plan which is entirely concentrated on the diagonal $(y = x)$ in $\mathcal{X} \times \mathcal{X}$. So $\nu = \mathrm{Id}_{\#}\mu = \mu$. □

To complete the construction it is natural to restrict W_p to a subset of $P(\mathcal{X}) \times P(\mathcal{X})$ on which it takes finite values.

Definition 6.4 (Wasserstein space). *With the same conventions as in Definition 6.1, the Wasserstein space of order p is defined as*

$$P_p(\mathcal{X}) := \left\{ \mu \in P(\mathcal{X}); \qquad \int_{\mathcal{X}} d(x_0, x)^p \, \mu(dx) < +\infty \right\},$$

where $x_0 \in \mathcal{X}$ is arbitrary. This space does not depend on the choice of the point x_0. Then W_p defines a (finite) distance on $P_p(\mathcal{X})$.

In words, the Wasserstein space is the space of probability measures which have a *finite moment of order p*. In this course, it will always be equipped with the distance W_p.

Proof that W_p is finite on P_p. Let π be a transference plan between two elements μ and ν in $P_p(\mathcal{X})$. Then the inequality

$$d(x,y)^p \leq 2^{p-1} \left[d(x, x_0)^p + d(x_0, y)^p \right]$$

shows that $d(x,y)^p$ is $\pi(dx\,dy)$-integrable as soon as $d(\,\cdot\,, x_0)^p$ is μ-integrable and $d(x_0,\,\cdot\,)^p$ is ν-integrable. □

Remark 6.5. Theorem 5.10(i) and Particular Case 5.4 together lead to the useful **duality formula for the Kantorovich–Rubinstein distance**: For any μ, ν in $P_1(\mathcal{X})$,

$$W_1(\mu, \nu) = \sup_{\|\psi\|_{\mathrm{Lip}} \leq 1} \left\{ \int_{\mathcal{X}} \psi \, d\mu - \int_{\mathcal{X}} \psi \, d\nu \right\}. \tag{6.3}$$

Among many applications of this formula I shall just mention the following covariance inequality: if f is a probability density with respect to μ then

$$\left(\int f \, d\mu \right) \left(\int g \, d\mu \right) - \int (fg) \, d\mu \leq \|g\|_{\mathrm{Lip}} \, W_1(f\,\mu, \mu).$$

Remark 6.6. A simple application of Hölder's inequality shows that

$$p \leq q \Longrightarrow W_p \leq W_q. \tag{6.4}$$

In particular, the Wasserstein distance of order 1, W_1, is the weakest of all. The most useful exponents in the Wasserstein distances are $p = 1$ and $p = 2$. As a general rule, the W_1 distance is more flexible and easier to bound, while the W_2 distance better reflects geometric features (at least for problems with a Riemannian flavor), and is better adapted when there is more structure; it also scales better with the dimension. Results in W_2 distance are usually stronger, and more difficult to establish, than results in W_1 distance.

Remark 6.7. On the other hand, under adequate regularity assumptions on the cost function and the probability measures, it is possible to control W_p in terms of W_q even for $q < p$; these reverse inequalities express a certain rigidity property of optimal transport maps which comes from c-cyclical monotonicity. See the bibliographical notes for more details.

Convergence in Wasserstein sense

Now we shall study a characterization of convergence in the Wasserstein space. The notation $\mu_k \longrightarrow \mu$ means that μ_k converges weakly to μ, i.e. $\int \varphi \, d\mu_k \to \int \varphi \, d\mu$ for any bounded continuous φ.

Definition 6.8 (Weak convergence in P_p). *Let $(\mathcal{X}, d)$ be a Polish space, and $p \in [1, \infty)$. Let $(\mu_k)_{k\in\mathbb{N}}$ be a sequence of probability measures in $P_p(X)$ and let μ be another element of $P_p(\mathcal{X})$. Then (μ_k) is said to converge weakly in $P_p(\mathcal{X})$ if any one of the following equivalent properties is satisfied for some (and then any) $x_0 \in \mathcal{X}$:*

(i) $\mu_k \longrightarrow \mu$ *and* $\displaystyle\int d(x_0, x)^p \, d\mu_k(x) \longrightarrow \int d(x_0, x)^p \, d\mu(x)$;

(ii) $\mu_k \longrightarrow \mu$ *and* $\displaystyle\limsup_{k\to\infty} \int d(x_0, x)^p \, d\mu_k(x) \leq \int d(x_0, x)^p \, d\mu(x)$;

(iii) $\mu_k \longrightarrow \mu$ *and* $\displaystyle\lim_{R\to\infty} \limsup_{k\to\infty} \int_{d(x_0,x)\geq R} d(x_0, x)^p \, d\mu_k(x) = 0$;

(iv) For all continuous functions φ with $|\varphi(x)| \leq C\big(1 + d(x_0, x)^p\big)$, $C \in \mathbb{R}$, one has

$$\int \varphi(x) \, d\mu_k(x) \longrightarrow \int \varphi(x) \, d\mu(x).$$

Theorem 6.9 (W_p metrizes P_p). *Let $(\mathcal{X}, d)$ be a Polish space, and $p \in [1, \infty)$; then the Wasserstein distance W_p metrizes the weak convergence in $P_p(\mathcal{X})$. In other words, if $(\mu_k)_{k\in\mathbb{N}}$ is a sequence of measures in $P_p(\mathcal{X})$ and μ is another measure in $P(\mathcal{X})$, then the statements*

$$\mu_k \text{ converges weakly in } P_p(\mathcal{X}) \text{ to } \mu$$

and

$$W_p(\mu_k, \mu) \longrightarrow 0$$

are equivalent.

Remark 6.10. As a consequence of Theorem 6.9, convergence in the p-Wasserstein space implies convergence of the moments of order p. There is a stronger statement that the map $\mu \longmapsto (\int d(x_0,x)^p\,\mu(dx))^{1/p}$ is 1-Lipschitz with respect to W_p; in the case of a locally compact length space, this will be proven in Proposition 7.29.

Below are two immediate corollaries of Theorem 6.9 (the first one results from the triangle inequality):

Corollary 6.11 (Continuity of W_p). *If $(\mathcal{X},d)$ is a Polish space, and $p \in [1,\infty)$, then W_p is continuous on $P_p(\mathcal{X})$. More explicitly, if μ_k (resp. ν_k) converges to μ (resp. ν) weakly in $P_p(\mathcal{X})$ as $k \to \infty$, then*

$$W_p(\mu_k,\nu_k) \longrightarrow W_p(\mu,\nu).$$

Remark 6.12. On the contrary, if these convergences are only usual weak convergences, then one can only conclude that $W_p(\mu,\nu) \le \liminf W_p(\mu_k,\nu_k)$: the Wasserstein distance is lower semicontinuous on $P(\mathcal{X})$ (just like the optimal transport cost C, for any lower semicontinuous nonnegative cost function c; recall the proof of Theorem 4.1).

Corollary 6.13 (Metrizability of the weak topology). *Let $(\mathcal{X},d)$ be a Polish space. If $\widetilde{d}$ is a bounded distance inducing the same topology as d (such as $\widetilde{d} = d/(1+d)$), then the convergence in Wasserstein sense for the distance $\widetilde{d}$ is equivalent to the usual weak convergence of probability measures in $P(\mathcal{X})$.*

Before starting the proof of Theorem 6.9, it will be good to make some more comments. The short version of that theorem is that *Wasserstein distances metrize weak convergence.* This sounds good, but after all, there are many ways to metrize weak convergence. Here is a list of some of the most popular ones, defined either in terms of measures μ, ν, or in terms of random variables X, Y with $\mathrm{law}\,(X) = \mu$, $\mathrm{law}\,(Y) = \nu$:

- the **Lévy–Prokhorov distance** (or just Prokhorov distance):

$$d_P(\mu,\nu) = \inf\Big\{\varepsilon > 0;\ \exists X,Y;\ \inf \mathbb{P}\big[d(X,Y) > \varepsilon\big] \le \varepsilon\Big\}; \qquad (6.5)$$

- the **bounded Lipschitz distance** (also called Fortet–Mourier distance):

$$d_{bL}(\mu,\nu) = \sup\left\{\int \varphi\, d\mu - \int \varphi\, d\nu;\quad \|\varphi\|_\infty + \|\varphi\|_{\mathrm{Lip}} \le 1\right\}; \quad (6.6)$$

- the **weak-∗ distance** (on a locally compact metric space):

$$d_{w*}(\mu,\nu) = \sum_{k\in\mathbb{N}} 2^{-k} \left| \int \varphi_k \, d\mu - \int \varphi_k \, d\nu \right|, \tag{6.7}$$

where $(\varphi_k)_{k\in\mathbb{N}}$ is a dense sequence in $C_0(\mathcal{X})$;
- the **Toscani distance** (on $P_2(\mathbb{R}^n)$):

$$d_T(\mu,\nu) = \sup_{\xi\in\mathbb{R}^n\setminus\{0\}} \left(\frac{\left| \int e^{-ix\cdot\xi} \, d\mu(x) - \int e^{-ix\cdot\xi} \, d\nu(x) \right|}{|\xi|^2} \right). \tag{6.8}$$

(Here I implicitly assume that μ, ν have the same mean, otherwise $d_T(\mu,\nu)$ would be infinite; one can also introduce variants of d_T by changing the exponent 2 in the denominator.)

So why bother with Wasserstein distances? There are several answers to that question:

1. Wasserstein distances are rather strong, especially in the way they take care of large distances in $\mathcal{X}$; this is a definite advantage over, for instance, the weak-∗ distance (which in practice is so weak that I advise the reader to never use it). It is not so difficult to combine information on convergence in Wasserstein distance with some smoothness bound, in order to get convergence in stronger distances.
2. The definition of Wasserstein distances makes them convenient to use in problems where optimal transport is naturally involved, such as many problems coming from partial differential equations.
3. The Wasserstein distances have a rich duality; this is especially useful for $p = 1$, in view of (6.3) (compare with the definition of the bounded Lipschitz distance). Passing back and forth from the original to the dual definition is often technically convenient.
4. Being defined by an infimum, Wasserstein distances are often relatively easy to bound from above: The construction of *any* coupling between μ and ν yields a bound on the distance between μ and ν. In the same line of ideas, any C-Lipschitz mapping $f : \mathcal{X} \to \mathcal{X}'$ induces a C-Lipschitz mapping $P_1(\mathcal{X}) \to P_1(\mathcal{X}')$ defined by $\mu \longmapsto f_\#\mu$ (the proof is obvious).

5. Wasserstein distances incorporate a lot of the geometry of the space. For instance, the mapping $x \longmapsto \delta_x$ is an *isometric* embedding of $\mathcal{X}$ into $P_p(\mathcal{X})$; but there are much deeper links. This partly explains why $P_p(\mathcal{X})$ is often very well adapted to statements that combine weak convergence and geometry.

To prove Theorem 6.9 I shall use the following lemma, which has interest on its own and will be useful again later.

Lemma 6.14 (Cauchy sequences in W_p are tight). *Let $\mathcal{X}$ be a Polish space, let $p \geq 1$ and let $(\mu_k)_{k\in\mathbb{N}}$ be a Cauchy sequence in $(P_p(\mathcal{X}), W_p)$. Then (μ_k) is tight.*

The proof is not so obvious and one might skip it at first reading.

Proof of Lemma 6.14. Let $(\mu_k)_{k\in\mathbb{N}}$ be a Cauchy sequence in $P_p(\mathcal{X})$: This means that

$$W_p(\mu_k, \mu_\ell) \longrightarrow 0 \qquad \text{as } k, \ell \to \infty.$$

In particular,

$$\int d(x_0, x)^p \, d\mu_k(x) = W_p\big(\delta_{x_0}, \mu_k\big)^p \leq \Big[W_p(\delta_{x_0}, \mu_1) + W_p(\mu_1, \mu_k)\Big]^p$$

remains bounded as $k \to \infty$.

Since $W_p \geq W_1$, the sequence (μ_k) is also Cauchy in the W_1 sense. Let $\varepsilon > 0$ be given, and let $N \in \mathbb{N}$ be such that

$$k \geq N \Longrightarrow W_1(\mu_N, \mu_k) < \varepsilon^2. \tag{6.9}$$

Then for any $k \in \mathbb{N}$, there is $j \in \{1, \ldots, N\}$ such that $W_1(\mu_j, \mu_k) < \varepsilon^2$. (If $k \geq N$, this is (6.9); if $k < N$, just choose $j = k$.)

Since the finite set $\{\mu_1, \ldots, \mu_N\}$ is tight, there is a compact set K such that $\mu_j[\mathcal{X} \setminus K] < \varepsilon$ for all $j \in \{1, \ldots, N\}$. By compactness, K can be covered by a finite number of small balls: $K \subset B(x_1, \varepsilon) \cup \ldots \cup B(x_m, \varepsilon)$.

Now write

$$U := B(x_1, \varepsilon) \bigcup \cdots \bigcup B(x_m, \varepsilon);$$

$$U_\varepsilon := \Big\{x \in \mathcal{X};\ d(x, U) < \varepsilon\Big\} \ \subset\ B(x_1, 2\varepsilon) \bigcup \cdots \bigcup B(x_m, 2\varepsilon);$$

$$\phi(x) := \left(1 - \frac{d(x, U)}{\varepsilon}\right)_+.$$

Note that $1_U \le \phi \le 1_{U_\varepsilon}$ and ϕ is $(1/\varepsilon)$-Lipschitz. By using these bounds and the Kantorovich–Rubinstein duality (6.3), we find that for $j \le N$ and k arbitrary,

$$\begin{aligned}\mu_k[U_\varepsilon] &\ge \int \phi\, d\mu_k\\ &= \int \phi\, d\mu_j + \left(\int \phi\, d\mu_k - \int \phi\, d\mu_j\right)\\ &\ge \int \phi\, d\mu_j - \frac{W_1(\mu_k,\mu_j)}{\varepsilon}\\ &\ge \mu_j[U] - \frac{W_1(\mu_k,\mu_j)}{\varepsilon}.\end{aligned}$$

On the one hand, $\mu_j[U] \ge \mu_j[K] \ge 1-\varepsilon$ if $j \le N$; on the other hand, for each k we can find $j = j(k)$ such that $W_1(\mu_k,\mu_j) \le \varepsilon^2$. So in fact

$$\mu_k[U_\varepsilon] \ge 1 - \varepsilon - \frac{\varepsilon^2}{\varepsilon} = 1 - 2\varepsilon.$$

At this point we have shown the following: For each $\varepsilon > 0$ there is a finite family $(x_i)_{1\le i\le m}$ such that all measures μ_k give mass at least $1-2\varepsilon$ to the set $Z := \cup B(x_i, 2\varepsilon)$. The point is that Z might not be compact. There is a classical remedy: Repeat the reasoning with ε replaced by $2^{-(\ell+1)}\varepsilon$, $\ell \in \mathbb{N}$; so there will be $(x_i)_{1\le i\le m(\ell)}$ such that

$$\mu_k\left[\mathcal{X} \setminus \bigcup_{1\le i\le m(\ell)} B(x_i,\, 2^{-\ell}\varepsilon)\right] \le 2^{-\ell}\varepsilon.$$

Thus

$$\mu_k[\mathcal{X} \setminus S] \le \varepsilon,$$

where

$$S := \bigcap_{1\le p\le\infty}\ \bigcup_{1\le i\le m(p)} \overline{B(x_i, \varepsilon 2^{-p})}.$$

By construction, S can be covered by finitely many balls of radius δ, where δ is arbitrarily small (just choose ℓ large enough that $2^{-\ell}\varepsilon < \delta$, and then $\overline{B(x_i, 2^{-\ell}\varepsilon)}$ will be included in $B(x_i,\delta)$). Thus S is *totally bounded*, i.e. it can be covered by finitely many balls of arbitrarily small radius. It is also closed, as an intersection of finite unions of closed sets. Since $\mathcal{X}$ is a complete metric space, it follows from a classical result in topology that S is compact. This concludes the proof of Lemma 6.14. □

Proof of Theorem 6.9. Let $(\mu_k)_{k\in\mathbb{N}}$ be such that $\mu_k \to \mu$ in distance W_p; the goal is to show that μ_k converges to μ in $P_p(\mathcal{X})$. First, by Lemma 6.14, the sequence $(\mu_k)_{k\in\mathbb{N}}$ is tight, so there is a subsequence $(\mu_{k'})$ such that $\mu_{k'}$ converges weakly to some probability measure $\widetilde{\mu}$. Then by Lemma 4.3,

$$W_p(\widetilde{\mu},\mu) \le \liminf_{k'\to\infty} W_p(\mu_{k'},\mu) = 0.$$

So $\widetilde{\mu} = \mu$, and the whole sequence (μ_k) has to converge to μ. This only shows the weak convergence in the usual sense, not yet the convergence in $P_p(\mathcal{X})$.

For any $\varepsilon > 0$ there is a constant $C_\varepsilon > 0$ such that for all nonnegative real numbers a, b,

$$(a+b)^p \le (1+\varepsilon)\, a^p + C_\varepsilon\, b^p.$$

Combining this inequality with the usual triangle inequality, we see that whenever x_0, x and y are three points in X, one has

$$d(x_0,x)^p \le (1+\varepsilon)\, d(x_0,y)^p + C_\varepsilon\, d(x,y)^p. \tag{6.10}$$

Now let (μ_k) be a sequence in $P_p(\mathcal{X})$ such that $W_p(\mu_k,\mu) \longrightarrow 0$, and for each k, let π_k be an optimal transport plan between μ_k and μ. Integrating inequality (6.10) against π_k and using the marginal property, we find that

$$\int d(x_0,x)^p\, d\mu_k(x) \le (1+\varepsilon) \int d(x_0,y)^p\, d\mu(y) + C_\varepsilon \int d(x,y)^p\, d\pi_k(x,y).$$

But of course,

$$\int d(x,y)^p\, d\pi_k(x,y) = W_p(\mu_k,\mu)^p \xrightarrow[k\to\infty]{} 0;$$

therefore,

$$\limsup_{k\to\infty} \int d(x_0,x)^p\, d\mu_k(x) \le (1+\varepsilon) \int d(x_0,x)^p\, d\mu(x).$$

Letting $\varepsilon \to 0$, we see that Property (ii) of Definition 6.8 holds true; so μ_k does converge weakly in $P_p(\mathcal{X})$ to μ.

Conversely, assume that μ_k converges weakly in $P_p(\mathcal{X})$ to μ; and again, for each k, introduce an optimal transport plan π_k between μ_k and μ, so that

$$\int d(x,y)^p \, d\pi_k(x,y) \longrightarrow 0.$$

By Prokhorov's theorem, (μ_k) forms a tight sequence; also $\{\mu\}$ is tight. By Lemma 4.4, the sequence (π_k) is itself tight in $P(\mathcal{X} \times \mathcal{X})$. So, up to extraction of a subsequence, still denoted by (π_k), one may assume that

$$\pi_k \longrightarrow \pi \qquad \text{weakly in } P(\mathcal{X} \times \mathcal{X}).$$

Since each π_k is optimal, Theorem 5.20 guarantees that π is an optimal coupling of μ and μ, so this is the (completely trivial) coupling $\pi = (\mathrm{Id}\,, \mathrm{Id}\,)_{\#}\mu$ (in terms of random variables, $Y = X$). Since this is independent of the extracted subsequence, actually π is the limit of the whole sequence (π_k).

Now let $x_0 \in \mathcal{X}$ and $R > 0$. If $d(x,y) > R$, then the largest of the two numbers $d(x,x_0)$ and $d(x_0,y)$ has to be greater than $R/2$, and no less than $d(x,y)/2$. In a slightly pedantic form,

$$\begin{aligned}
&1_{d(x,y)\geq R} \\
&\leq 1_{[d(x,x_0)\geq R/2 \text{ and } d(x,x_0)\geq d(x,y)/2]} + 1_{[d(x_0,y)\geq R/2 \text{ and } d(x_0,y)\geq d(x,y)/2]}.
\end{aligned}$$

So, obviously

$$\begin{aligned}
\big[d(x,y)^p - R^p\big]_+ \leq\ & d(x,y)^p \, 1_{[d(x,x_0)\geq R/2 \text{ and } d(x,x_0)\geq d(x,y)/2]} \\
&+ d(x,y)^p \, 1_{[d(x_0,y)\geq R/2 \text{ and } d(x_0,y)\geq d(x,y)/2]} \\
\leq\ & 2^p d(x,x_0)^p \, 1_{d(x,x_0)\geq R/2} + 2^p d(x_0,y)^p \, 1_{d(x_0,y)\geq R/2}.
\end{aligned}$$

It follows that

$$\begin{aligned}
W_p(\mu_k,\mu)^p &= \int d(x,y)^p \, d\pi_k(x,y) \\
&= \int \big[d(x,y)\wedge R\big]^p \, d\pi_k(x,y) + \int \big[d(x,y)^p - R^p\big]_+ \, d\pi_k(x,y) \\
&\leq \int \big[d(x,y)\wedge R\big]^p \, d\pi_k(x,y) + 2^p \int_{d(x,x_0)\geq R/2} d(x,x_0)^p \, d\pi_k(x,y) \\
&\qquad\qquad + 2^p \int_{d(x_0,y)>R/2} d(x_0,y)^p \, d\pi_k(x,y) \\
&= \int \big[d(x,y)\wedge R\big]^p \, d\pi_k(x,y) + 2^p \int_{d(x,x_0)\geq R/2} d(x,x_0)^p \, d\mu_k(x) \\
&\qquad\qquad + 2^p \int_{d(x_0,y)\geq R/2} d(x_0,y)^p \, d\mu(y).
\end{aligned}$$

Since π_k converges weakly to π, the first term goes to 0 as $k \to \infty$. So

$$\begin{aligned}\limsup_{k\to\infty} W_p(\mu_k,\mu)^p \leq & \lim_{R\to\infty} 2^p \limsup_{k\to\infty} \int_{d(x,x_0)\geq R/2} d(x,x_0)^p \, d\mu_k(x) \\ & + \lim_{R\to\infty} 2^p \limsup_{k\to\infty} \int_{d(x_0,y)\geq R/2} d(x_0,y)^p \, d\mu(y) \\ = & \, 0.\end{aligned}$$

This concludes the argument. □

Control by total variation

The total variation is a classical notion of distance between probability measures. There is, by the way, a classical probabilistic representation formula of the total variation:

$$\|\mu - \nu\|_{TV} \;=\; 2 \inf \mathbb{P}\,[X \neq Y], \tag{6.11}$$

where the infimum is over all couplings (X,Y) of (μ,ν); this identity can be seen as a very particular case of Kantorovich duality for the cost function $1_{x\neq y}$.

It seems natural that a control in Wasserstein distance should be weaker than a control in total variation. This is not completely true, because total variation does not take into account large distances. But one can control W_p by *weighted* total variation:

Theorem 6.15 (Wasserstein distance is controlled by weighted total variation). *Let μ and ν be two probability measures on a Polish space $(\mathcal{X},d)$. Let $p \in [1,\infty)$ and $x_0 \in \mathcal{X}$. Then*

$$W_p(\mu,\nu) \leq 2^{\frac{1}{p'}} \left(\int d(x_0,x)^p \, d|\mu-\nu|(x) \right)^{\frac{1}{p}}, \qquad \frac{1}{p} + \frac{1}{p'} = 1. \tag{6.12}$$

Particular Case 6.16. In the case $p = 1$, if the diameter of $\mathcal{X}$ is bounded by D, this bound implies $W_1(\mu,\nu) \leq D\,\|\mu-\nu\|_{TV}$.

Remark 6.17. The integral in the right-hand side of (6.12) can be interpreted as the Wasserstein distance W_1 for the particular cost function $[d(x_0,x)+d(x_0,y)]1_{x\neq y}$.

Proof of Theorem 6.15. Let π be the transference plan obtained by keeping fixed all the mass shared by μ and ν, and distributing the rest uniformly: this is

$$\pi = (\mathrm{Id}\,, \mathrm{Id}\,)_\#(\mu \wedge \nu) + \frac{1}{a}\,(\mu - \nu)_+ \otimes (\mu - \nu)_-,$$

where $\mu \wedge \nu = \mu - (\mu - \nu)_+$ and $a = (\mu - \nu)_-[X] = (\mu - \nu)_+[X]$. A more sloppy but probably more readable way to write π is

$$\pi(dx\,dy) = (\mu \wedge \nu)(dx)\,\delta_{y=x} + \frac{1}{a}\,(\mu - \nu)_+(dx)\,(\mu - \nu)_-(dy).$$

By using the definition of W_p, the definition of π, the triangle inequality for d, the elementary inequality $(A + B)^p \le 2^{p-1}(A^p + B^p)$, and the definition of a, we find that

$$\begin{aligned}
W_p(\mu,\nu)^p &\le \int d(x,y)^p\,d\pi(x,y)\\
&= \frac{1}{a}\int d(x,y)^p\,d(\mu-\nu)_+(x)\,d(\mu-\nu)_-(y)\\
&\le \frac{2^{p-1}}{a}\int \big[d(x,x_0)^p + d(x_0,y)^p\big]\,d(\mu-\nu)_+(x)\,d(\mu-\nu)_-(y)\\
&\le 2^{p-1}\left[\int d(x,x_0)^p\,d(\mu-\nu)_+(x) + \int d(x_0,y)^p\,d(\mu-\nu)_-(y)\right]\\
&= 2^{p-1}\int d(x,x_0)^p\,d\big[(\mu-\nu)_+ + (\mu-\nu)_-\big](x)\\
&= 2^{p-1}\int d(x,x_0)^p\,d|\mu-\nu|(x).
\end{aligned}$$

$\square$

Topological properties of the Wasserstein space

The Wasserstein space $P_p(\mathcal{X})$ inherits several properties of the base space $\mathcal{X}$. Here is a first illustration:

Theorem 6.18 (Topology of the Wasserstein space). *Let $\mathcal{X}$ be a complete separable metric space and $p \in [1,\infty)$. Then the Wasserstein space $P_p(\mathcal{X})$, metrized by the Wasserstein distance W_p, is also a complete separable metric space. In short: The Wasserstein space over a*

Polish space is itself a Polish space. Moreover, any probability measure can be approximated by a sequence of probability measures with finite support.

Remark 6.19. If $\mathcal{X}$ is compact, then $P_p(\mathcal{X})$ is also compact; but if $\mathcal{X}$ is only locally compact, then $P_p(\mathcal{X})$ is *not* locally compact.

Proof of Theorem 6.18. The fact that $P_p(\mathcal{X})$ is a metric space was already explained, so let us turn to the proof of **separability**. Let $\mathcal{D}$ be a dense sequence in $\mathcal{X}$, and let $\mathcal{P}$ be the space of probability measures that can be written $\sum a_j \delta_{x_j}$, where the a_j are rational coefficients, and the x_j are finitely many elements in $\mathcal{D}$. It will turn out that $\mathcal{P}$ is dense in $P_p(\mathcal{X})$.

To prove this, let $\varepsilon > 0$ be given, and let x_0 be an arbitrary element of $\mathcal{D}$. If μ lies in $P_p(\mathcal{X})$, then there exists a compact set $K \subset \mathcal{X}$ such that

$$\int_{\mathcal{X}\setminus K} d(x_0, x)^p \, d\mu(x) \le \varepsilon^p.$$

Cover K by a finite family of balls $B(x_k, \varepsilon/2)$, $1 \le k \le N$, with centers $x_k \in \mathcal{D}$, and define

$$B'_k = B(x_k, \varepsilon) \setminus \bigcup_{j<k} B(x_j, \varepsilon).$$

Then all B'_k are disjoint and still cover K.

Define f on $\mathcal{X}$ by

$$f(B'_k \cap K) = \{x_k\}, \qquad f(\mathcal{X} \setminus K) = \{x_0\}.$$

Then, for any $x \in K$, $d(x, f(x)) \le \varepsilon$. So

$$\begin{aligned}\int d(x, f(x))^p \, d\mu(x) &\le \varepsilon^p \int_K d\mu(x) \; + \; \int_{\mathcal{X}\setminus K} d(x, x_0)^p \, d\mu(x) \\ &\le \varepsilon^p + \varepsilon^p = 2\,\varepsilon^p.\end{aligned}$$

Since $(\mathrm{Id}\,, f)$ is a coupling of μ and $f_\#\mu$, $W_p(\mu, f_\#\mu) \le 2\varepsilon^p$.

Of course, $f_\#\mu$ can be written as $\sum a_j \delta_{x_j}$, $0 \le j \le N$. This shows that μ might be approximated, with arbitrary precision, by a finite combination of Dirac masses. To conclude, it is sufficient to show that the coefficients a_j might be replaced by rational coefficients, up to a very small error in Wasserstein distance. By Theorem 6.15,

$$W_p\left(\sum_{j\le N} a_j\delta_{x_j},\ \sum_{j\le N} b_j\delta_{x_j}\right) \le 2^{\frac{1}{p'}}\left[\max_{k,\ell} d(x_k,x_\ell)\right]\sum_{j\le N}|a_j-b_j|^{\frac{1}{p}},$$

and obviously the latter quantity can be made as small as possible for some well-chosen rational coefficients b_j.

Finally, let us prove the **completeness**. Again let $(\mu_k)_{k\in\mathbb{N}}$ be a Cauchy sequence in $P_p(\mathcal{X})$. By Lemma 6.14, it admits a subsequence $(\mu_{k'})$ which converges weakly (in the usual sense) to some measure μ. Then,

$$\int d(x_0,x)^p\,d\mu(x) \le \liminf_{k'\to\infty}\int d(x_0,x)^p\,d\mu_{k'}(x) < +\infty,$$

so μ belongs to $P_p(\mathcal{X})$. Moreover, by lower semicontinuity of W_p,

$$W_p(\mu,\mu_{\ell'}) \le \liminf_{k'\to\infty} W_p(\mu_{k'},\mu_{\ell'}),$$

so in particular

$$\limsup_{\ell'\to\infty} W_p(\mu,\mu_{\ell'}) \le \limsup_{k',\ell'\to\infty} W_p(\mu_{k'},\mu_{\ell'}) = 0,$$

which means that $\mu_{\ell'}$ converges to μ in the W_p sense (and not just in the sense of weak convergence). Since (μ_k) is a Cauchy sequence with a converging subsequence, it follows by a classical argument that the whole sequence is converging. □

Bibliographical notes

The terminology of Wasserstein distance (apparently introduced by Dobrushin) is very questionable, since (a) these distances were discovered and rediscovered by several authors throughout the twentieth century, including (in chronological order) Gini [417, 418], Kantorovich [501], Wasserstein [803], Mallows [589] and Tanaka [776] (other contributors being Salvemini, Dall'Aglio, Hoeffding, Fréchet, Rubinstein, Ornstein, and maybe others); (b) the explicit definition of this distance is not so easy to find in Wasserstein's work; and (c) Wasserstein was only interested in the case $p=1$. By the way, also the spelling of Wasserstein is doubtful: the original spelling was Vasershtein. (Similarly, Rubinstein was spelled Rubinshtein.) These issues are discussed in a historical note by Rüschendorf [720], who advocates the denomination of "minimal L^p-metric" instead of "Wasserstein distance". Also Vershik [808] tells about the discovery of the metric by Kantorovich and stands up in favor of the terminology "Kantorovich distance".

However, the terminology "Wasserstein distance" (or "Wasserstein metric") has been extremely successful: at the time of writing, about 30,000 occurrences can be found on the Internet. Nearly all recent papers relating optimal transport to partial differential equations, functional inequalities or Riemannian geometry (including my own works) use this convention. I will therefore stick to this by-now well-established terminology. After all, even if this convention is a bit unfair since it does not give credit to all contributors, not even to the most important of them (Kantorovich), at least it does give credit to somebody.

As I learnt from Bernot, terminological confusion was enhanced in the mid-nineties, when a group of researchers in image processing introduced the denomination of "Earth Mover's distance" [713] for the Wasserstein (Kantorovich–Rubinstein) W_1 distance. This terminology was very successful and rapidly spread by the high rate of growth of the engineering literature; it is already starting to compete with "Wasserstein distance", scoring more than 15,000 occurrences on Internet.

Gini considered the special case where the random variables are discrete and lie on the real line; like Mallows later, he was interested by applications in statistics (the "Gini distance" is often used to roughly quantify the inequalities of wealth or income distribution in a given population). Tanaka discovered applications to partial differential equations. Both Mallows and Tanaka worked with the case $p = 2$, while Gini was interested both in $p = 1$ and $p = 2$, and Hoeffding and Fréchet worked with general p (see for instance [381]). A useful source on the point of view of Kantorovich and Rubinstein is Vershik's review [809].

Kantorovich and Rubinstein [506] made the important discovery that the original Kantorovich distance (W_1 in my notation) can be extended into a *norm* on the set $M(\mathcal{X})$ of signed measures over a Polish space $\mathcal{X}$. It is common to call this extension the **Kantorovich–Rubinstein norm**, and by abuse of language I also used the denomination Kantorovich–Rubinstein metric for W_1. (It would be more proper to call it just the Kantorovich metric, but more or less everything in this subject should be called after Kantorovich.) This norm property is a particular feature of the exponent $p = 1$, and should be taken seriously because it has strong implications in functional analysis. For one thing, the Kantorovich–Rubinstein norm provides an explicit isometric embedding of an arbitrary Polish space in a Banach space.

As pointed out to me by Vershik, the Kantorovich–Rubinstein norm on a metric space $(\mathcal{X}, d)$ can be intrinsically characterized as the maximal norm $\|\cdot\|$ on $M(\mathcal{X})$ which is "compatible" with the distance, in the sense that $\|\delta_x - \delta_y\| = d(x, y)$ for all $x, y \in \mathcal{X}$; this maximality property is a consequence of the duality formula.

Here are a few words about the other probability metrics mentioned in this chapter. The Toscani metric is useful in the theory of the Boltzmann equation, see [812, Section 4.2] and references quoted therein. Together with its variants, it is also handy for studying rates of convergence in the central limit theorem, or certain stable limit theorems [424]. The Lévy–Prokhorov metric appears in a number of textbooks, such as Dudley [318, p. 394]. For the taxonomy of probability metrics and their history, the unavoidable reference is the monograph by Rachev [695], which lists dozens and dozens of metrics together with their main properties and applications. (Many of them are variants, particular cases or extensions of the Wasserstein and Lévy–Prokhorov metrics.) The more recent set of notes by Carrillo and Toscani [216] also presents applications of various probability metrics to problems arising in partial differential equations (in particular the inelastic Boltzmann equation).

Here as in all the rest of this course, I only considered complete separable metric spaces. However, Wasserstein distances also make sense in noncomplete separable metric spaces: The case $p = 1$ was treated by Dudley [318, Lemma 11.8.3], while the general case was recently considered by Clement and Desch [237]. In this reference the triangular inequality is proven by approximation by countable spaces.

The equivalence between the four statements in Definition 6.8 is proven in [814, Theorem 7.12]. I borrowed the proof of Lemma 6.14 from Bolley [136]; and the scheme of proof of Theorem 6.9 from Ambrosio, Gigli and Savaré [30]. There are alternative proofs of Theorem 6.9 in the literature, for instance in [814, Chapter 7]. Similar convergence results had been obtained earlier by various authors, at least in particular cases, see e.g. [260, 468].

In dimension 1, Theorem 6.9 can be proven by simpler methods, and interpreted as a quantitative refinement of Skorokhod's representation theorem, as noticed in [795] or [814, Section 7.3].

The ∞-Wasserstein distance, $W_\infty = \lim_{p\to\infty} W_p$, does not fit in the setting considered in this chapter, in particular because the induced topology is quite stronger than the weak topology of measures. This distance however is useful in a surprising number of problems [208, 212, 222, 466, 617].

The representation formula (6.11) for the total variation distance is a particular case of Strassen's duality theorem, see for instance [814, Section 1.4]. Remark 6.17 is extracted from [427, comments following Remark VI.5].

Theorem 6.15 is a copy–paste from [814, Proposition 7.10], which itself was a slight adaptation of [696, Lemma 10.2.3]. Other upper bounds for the Wasserstein distances are available in the literature; see for instance [527] for the case of the Hamming distance on discrete product spaces.

Results of lower bounds for the Wasserstein distance (in terms of moments for instance) are not so common. One example is Proposition 7.29 in the next chapter. In the particular case of the 2-Wasserstein distance on a Hilbert space, there are lower bounds expressed in terms of moments and covariance matrices [258, 407].

In relation with the ∞-Wasserstein distance, Bouchitté, Jimenez and Rajesh [151] prove the following estimate: If Ω is a bounded Lipschitz open subset of $\mathbb{R}^n$, equipped with the usual Euclidean distance, $\mu(dx) = f(x)\,dx$ and $\nu(dy)$ are probability measures on $\overline{\Omega}$, and the density f is uniformly bounded below, then for any $p > 1$,

$$W_\infty(\mu,\nu)^{p+n} \le \left(\frac{C}{\inf f}\right) W_p(\mu,\nu)^p,$$

where $C = C(p,n,\Omega)$. As mentioned in Remark 6.7, this "converse" estimate is related to the fact that the optimal transport map for the cost function $|x-y|^p$ enjoys some monotonicity properties which make it very rigid, as we shall see again in Chapter 10. (As an analogy: the Sobolev norms $W^{1,p}$ are all topologically equivalent when applied to C-Lipschitz convex functions on a bounded domain.)

Theorem 6.18 belongs to folklore and has probably been proven many times; see for instance [310, Section 14]. Other arguments are due to Rachev [695, Section 6.3], and Ambrosio, Gigli and Savaré [30]. In the latter reference the proof is very simple but makes use of the deep Kolmogorov extension theorem. Here I followed a much more elementary argument due to Bolley [136].

The statement in Remark 6.19 is proven in [30, Remark 7.1.9].

In a Euclidean or Riemannian context, the Wasserstein distance W_2 between two very close measures, say $(1+h_1)\,\nu$ and $(1+h_2)\,\nu$ with h_1, h_2 very small, is approximately equal to the $H^{-1}(\nu)$-norm of $h_1 - h_2$; see [671, Section 7], [814, Section 7.6] or Exercise 22.20. (One may also take a look at [567, 569].) There is in fact a natural one-parameter family of distances interpolating between $H^{-1}(\nu)$ and W_2, defined by a variation on the Benamou–Brenier formula (7.34) (insert a factor $(d\mu_t/d\nu)^{1-\alpha}$, $0 \le \alpha \le 1$ in the integrand of (7.33); this construction is due to Dolbeault, Nazaret and Savaré [312]).

Applications of the Wasserstein distances are too numerous to be listed here; some of them will be encountered again in the sequel. In [150] Wasserstein distances are used to study the best approximation of a measure by a finite number of points. Various authors [700, 713] use them to compare color distributions in different images. These distances are classically used in statistics, limit theorems, and all kinds of problems involving approximation of probability measures [254, 256, 257, 282, 694, 696, 716]. Rio [704] derives sharp quantitative bounds in Wasserstein distance for the central limit theorem on the real line, and surveys the previous literature on this problem. Wasserstein distances are well adapted to study rates of fluctuations of empirical measures, see [695, Theorem 11.1.6], [696, Theorem 10.2.1], [498, Section 4.9], and the research papers [8, 307, 314, 315, 479, 771, 845]. (The most precise results are those in [307]: there it is shown that the average W_1 distance between two independent copies of the empirical measure behaves like $(\int \rho^{1-1/d})/N^{1-1/d}$, where N is the size of the samples, ρ the density of the common law of the random variables, and $d \ge 3$; the proofs are partly based on subadditivity, as in [150].) Quantitative Sanov-type theorems have been considered in [139, 742]. Wasserstein distances are also commonly used in statistical mechanics, most notably in the theory of propagation of chaos, or more generally the mean behavior of large particle systems [768] [757, Chapter 5]; the original idea seems to go back to Dobrushin [308, 309] and has been applied in a large number of problems, see for instance [81, 82, 221, 590, 624]. The original version of the Dobrushin–Shlosman uniqueness criterion [308, 311] in spin systems was expressed in terms of optimal transport distance, although this formulation was lost in most subsequent developments (I learnt this from Ollivier).

Wasserstein distances are also useful in the study of mixing and convergence for Markov chains; the original idea, based on a contraction property, seems to be due to Dobrushin [310], and has been rediscovered since then by various authors [231, 662, 679]. Tanaka proved that the W_2 distance is contracting along solutions of a certain class of Boltzmann equations [776, 777]; these results are reviewed in [814, Section 7.5] and have been generalized in [138, 214, 379, 590].

Wasserstein distances behave well with increasing dimension, and therefore have been successfully used in large or infinite dimension; for instance for the large-time behavior of stochastic partial differential equations [455, 458, 533, 605], or hydrodynamic limits of systems of particles [444].

In a Riemannian context, the W_2 distance is well-adapted to the study of Ricci curvature, in relation with diffusion equations; these themes will be considered again in Part II.

Here is a short list of some more surprising applications. Werner [836] suggested that the W_1 distance is well adapted to quantify some variants of the uncertainty principle in quantum physics. In a recent note, Melleray, Petrov and Vershik [625] use the properties of the Kantorovich–Rubinstein norm to study spaces which are "linearly rigid", in the sense that, roughly speaking, there is only one way to embed them in a Banach space. The beautiful text [809] by Vershik reviews further applications of the Kantorovich–Rubinstein distance to several original topics (towers of measures, Bernoulli automorphisms, classification of metric spaces); see also [808] and the older contribution [807] by the same author.

7

Displacement interpolation

I shall now discuss a **time-dependent** version of optimal transport leading to a *continuous* displacement of measures. There are two main motivations for that extension:

- a time-dependent model gives a more complete description of the transport;
- the richer mathematical structure will be useful later on.

As in the previous chapter I shall assume that the initial and final probability measures are defined on the same Polish space $(\mathcal{X}, d)$. The main additional structure assumption is that the cost is associated with an **action**, which is a way to measure the cost of displacement along a continuous curve, defined on a given time-interval, say $[0, 1]$. So the cost function between an initial point x and a final point y is obtained by minimizing the action among paths that go from x to y:

$$c(x,y) = \inf\Big\{\mathcal{A}(\gamma); \quad \gamma_0 = x, \ \gamma_1 = y; \quad \gamma \in \mathcal{C}\Big\}. \tag{7.1}$$

Here $\mathcal{C}$ is a certain class of continuous curves, to be specified in each particular case of interest, on which the action functional $\mathcal{A}$ is defined.

Of course, Assumption (7.1) is meaningless unless one requires some specific structure on the action functional (otherwise, just choose $\mathcal{A}(\gamma) = c(\gamma_0, \gamma_1)$...). A good notion of action should provide a recipe for choosing optimal paths, and in particular a recipe to interpolate between points in $\mathcal{X}$. It will turn out that under soft assumptions, this interpolation recipe between points can be "lifted" to an interpolation recipe between *probability measures.* This will provide a time-dependent notion of optimal

C. Villani, *Optimal Transport*, Grundlehren der mathematischen Wissenschaften 338,

transport, that will be called **displacement interpolation** (by opposition to the standard *linear* interpolation between probability measures).

This is a key chapter in this course, and I have worked hard to attain a high level of generality, at the price of somewhat lengthy arguments. So the reader should not hesitate to skip proofs at first reading, concentrating on statements and explanations. The main result in this chapter is Theorem 7.21.

Deterministic interpolation via action-minimizing curves

To better understand what an action functional should be, let us start with some examples and informal discussions. Consider a model where the unknown is the position of a given physical system in some position space, say a Riemannnian manifold M. (See the Appendix for reminders about Riemannian geometry if needed.) We learn from classical physics that in the absence of a potential, the action is the integral over time of the (instantaneous) kinetic energy:

$$\mathcal{A}(\gamma) = \int_0^1 \frac{|\dot{\gamma}_t|^2}{2}\, dt,$$

where $\dot{\gamma}_t$ stands for the velocity (or time-derivative) of the curve γ at time t. More generally, an action is classically given by the time-integral of a **Lagrangian** along the path:

$$\mathcal{A}(\gamma) = \int_0^1 L(\gamma_t, \dot{\gamma}_t, t)\, dt. \tag{7.2}$$

Here L is defined on $TM \times [0,1]$, where the smooth manifold M is the position space and the tangent bundle TM is the phase space, which is the space of all possible positions and velocities. It is natural to work in TM because one often deals with second-order differential equations on M (such as Newton's equations), which transform themselves into first-order equations on TM. Typically L would take the form

$$L(x,v,t) = \frac{|v|^2}{2} - V(x) \tag{7.3}$$

where V is a potential; but much more complicated forms are admissible. When V is continuously differentiable, it is a simple particular case of the formula of first variation (recalled in the Appendix) that minimizers of (7.3), with given endpoints, satisfy Newton's equation

$$\frac{d^2x}{dt^2} = -\nabla V(x). \tag{7.4}$$

To make sure that $\mathcal{A}(\gamma)$ is well-defined, it is natural to assume that the path γ is continuously differentiable, or piecewise continuously differentiable, or at least almost everywhere differentiable as a function of t. A classical and general setting is that of **absolutely continuous curves**: By definition, if $(\mathcal{X}, d)$ is a metric space, a continuous curve $\gamma : [0,1] \to \mathcal{X}$ is said to be absolutely continuous if there exists a function $\ell \in L^1([0,1]; dt)$ such that for all intermediate times $t_0 < t_1$ in $[0,1]$,

$$d(\gamma_{t_0}, \gamma_{t_1}) \le \int_{t_0}^{t_1} \ell(t)\, dt. \tag{7.5}$$

More generally, it is said to be absolutely continuous of order p if formula (7.5) holds with some $\ell \in L^p([0,1]; dt)$.

If γ is absolutely continuous, then the function $t \longmapsto d(\gamma_{t_0}, \gamma_t)$ is differentiable almost everywhere, and its derivative is integrable. But the converse is false: for instance, if γ is the "Devil's staircase", encountered in measure theory textbooks (a nonconstant function whose distributional derivative is concentrated on the Cantor set in $[0,1]$), then γ is differentiable almost everywhere, and $\dot\gamma(t) = 0$ for almost every t, even though γ is not constant! This motivates the "integral" definition of absolute continuity based on formula (7.5).

If $\mathcal{X}$ is $\mathbb{R}^n$, or a smooth differentiable manifold, then absolutely continuous paths are differentiable for Lebesgue-almost all $t \in [0,1]$; in physical words, the velocity is well-defined for almost all times.

Before going further, here are some simple and important examples. For all of them, the class of admissible curves is the space of absolutely continuous curves.

Example 7.1. In $\mathcal{X} = \mathbb{R}^n$, choose $L(x,v,t) = |v|$ (Euclidean norm of the velocity). Then the action is just the length functional, while the cost $c(x,y) = |x-y|$ is the Euclidean distance. Minimizing curves are straight lines, with arbitrary parametrization: $\gamma_t = \gamma_0 + s(t)(\gamma_1 - \gamma_0)$, where $s : [0,1] \to [0,1]$ is nondecreasing and absolutely continuous.

Example 7.2. In $\mathcal{X} = \mathbb{R}^n$ again, choose $L(x,v,t) = c(v)$, where c is strictly convex. By Jensen's inequality,

$$c(\gamma_1 - \gamma_0) = c\left(\int_0^1 \dot\gamma_t\,dt\right) \le \int_0^1 c(\dot\gamma_t)\,dt,$$

and this is an equality if and only if $\dot\gamma_t$ is constant. Therefore action-minimizers are straight lines with *constant velocity*: $\gamma_t = \gamma_0 + t\,(\gamma_1 - \gamma_0)$. Then, of course,

$$c(x,y) = c(y-x).$$

Remark 7.3. This example shows that very different Lagrangians can have the same minimizing curves.

Example 7.4. Let $\mathcal{X} = M$ be a smooth Riemannian manifold, TM its tangent bundle, and $L(x,v,t) = |v|^p$, $p \ge 1$. Then the cost function is $d(x,y)^p$, where d is the geodesic distance on M. There are two quite different cases:

- If $p > 1$, minimizing curves are defined by the equation $\ddot\gamma_t = 0$ (zero acceleration), to be understood as $(d/dt)\dot\gamma_t = 0$, where (d/dt) stands for the covariant derivative along the path γ (once again, see the reminders in the Appendix if necessary). Such curves have constant speed ($(d/dt)|\dot\gamma_t| = 0$), and are called **minimizing, constant-speed geodesics**, or simply geodesics.
- If $p = 1$, minimizing curves are geodesic curves parametrized in an arbitrary way.

Example 7.5. Again let $\mathcal{X} = M$ be a smooth Riemannian manifold, and now consider a general Lagrangian $L(x,v,t)$, assumed to be *strictly convex* in the velocity variable v. The characterization and study of extremal curves for such Lagrangians, under various regularity assumptions, is one of the most classical topics in the calculus of variations. Here are some of the basic — which does not mean trivial — results in the field. Throughout the sequel, the Lagrangian L is a C^1 function defined on $TM \times [0,1]$.

- By the first variation formula (a proof of which is sketched in the Appendix), minimizing curves satisfy the Euler–Lagrange equation

$$\frac{d}{dt}\Big[(\nabla_v L)(\gamma_t, \dot\gamma_t, t)\Big] = (\nabla_x L)(\gamma_t, \dot\gamma_t, t), \tag{7.6}$$

which generalizes (7.4). At least this equation should be satisfied for minimizing curves that are sufficiently smooth, say piecewise C^1.

- If there exists $K, C > 0$ such that

$$L(x,v,t) \geq K|v| - C,$$

then the action of a curve γ is bounded below by $K\,\mathcal{L}(\gamma) - C$, where $\mathcal{L}$ is the length; this implies that all action-minimizing curves starting from a given compact K_0 and ending in a given compact K_1 stay within a bounded region.

- If minimizing curves depend smoothly on their position and velocity at some time, then there is also a bound on the velocities along minimizers that join K_0 to K_1. Indeed, there is a bound on $\int_0^1 L(x,v,t)\,dt$; so there is a bound on $L(x,v,t)$ for some t; so there is a bound on the velocity at some time, and then this bound is propagated in time.

- Assume that L is strictly convex and superlinear in the velocity variable, in the following sense:

$$\forall (x,t) \qquad \begin{cases} v \longmapsto L(x,v,t) \quad \text{is convex,} \\ \\ \dfrac{L(x,v,t)}{|v|} \xrightarrow[|v|\to\infty]{} +\infty. \end{cases} \tag{7.7}$$

Then $v \longmapsto \nabla_v L$ is invertible, and (7.6) can be rewritten as a differential equation on the new unknown $\nabla_v L(\gamma, \dot{\gamma}, t)$.

- If in addition L is C^2 and the strict inequality $\nabla_v^2 L > 0$ holds (more rigorously, $\nabla_v^2 L(x,\,\cdot\,,t) \geq K(x)\, g_x$ for all x, where g is the metric and $K(x) > 0$), then the new equation (where x and $p = \nabla_v L(x,v,t)$ are the unknowns) has locally Lipschitz coefficients, and the Cauchy–Lipschitz theorem can be applied to guarantee the unique local existence of Lipschitz continuous solutions to (7.6). Under the same assumptions on L, at least if L does not depend on t, one can show directly that minimizers are of class at least C^1, and therefore satisfy (7.6). Conversely, solutions of (7.6) are locally (in time) minimizers of the action.

- Finally, the convexity of L makes it possible to define its Legendre transform (again, with respect to the velocity variable):

$$H(x,p,t) := \sup_{v \in T_x M} \Big(p \cdot v - L(x,v,t) \Big),$$

which is called the **Hamiltonian**; then one can recast (7.6) in terms of a Hamiltonian system, and access to the rich mathematical world of Hamiltonian dynamics. As soon as L is strictly convex superlinear, the Legendre transform $(x,v) \longmapsto (x, \nabla_v L(x,v,t))$ is a homeomorphism, so assumptions about (x,v) can be re-expressed in terms of the new variables $(x, p = \nabla_v L(x,v,t))$.

- If L does not depend on t, then $H(x, \nabla_v L(x,v))$ is constant along minimizing curves $(x,v) = (\gamma_t, \dot{\gamma}_t)$; if L does depend on t, then $(d/dt) H(x, \nabla_v L(x,v)) = (\partial_t H)(x, \nabla_v L(x,v))$.

Some of the above-mentioned assumptions will come back often in the sequel, so I shall summarize the most interesting ones in the following definition:

Definition 7.6 (Classical conditions on a Lagrangian function). *Let M be a smooth, complete Riemannian manifold, and $L(x,v,t)$ a Lagrangian on $TM \times [0,1]$. In this course, it is said that L satisfies the classical conditions if*

(a) L is C^1 in all variables;

(b) L is a strictly convex superlinear function of v, in the sense of (7.7);

(c) There are constants $K, C > 0$ such that for all $(x,v,t) \in TM \times [0,1]$, $L(x,v,t) \geq K|v| - C$;

(d) There is a well-defined locally Lipschitz flow associated to the Euler–Lagrange equation, or more rigorously to the minimization problem; that is, there is a locally Lipschitz map $(x_0, v_0, t_0; t) \to \phi_t(x_0, v_0, t_0)$ on $TM \times [0,1] \times [0,1]$, with values in TM, such that each action-minimizing curve $\gamma : [0,1] \to M$ belongs to $C^1([0,1]; M)$ and satisfies $(\gamma(t), \dot{\gamma}(t)) = \phi_t(\gamma(t_0), \dot{\gamma}(t_0), t_0)$.

Remark 7.7. Assumption (d) above is automatically satisfied if L is of class C^2, $\nabla_v^2 L > 0$ everywhere and L does not depend on t.

This looks general enough, however there are interesting cases where $\mathcal{X}$ does not have enough differentiable structure for the velocity vector to be well-defined (tangent spaces might not exist, for lack of smoothness). In such a case, it is still possible to define the *speed* along the curve:

$$|\dot{\gamma}_t| := \limsup_{\varepsilon \to 0} \frac{d(\gamma_t, \gamma_{t+\varepsilon})}{|\varepsilon|}. \tag{7.8}$$

This generalizes the natural notion of speed, which is the norm of the velocity vector. Thus it makes perfect sense to write a Lagrangian of the form $L(x,|v|,t)$ in a general metric space $\mathcal{X}$; here L might be essentially any measurable function on $\mathcal{X} \times \mathbb{R}_+ \times [0,1]$. (To ensure that $\int_0^1 L\,dt$ makes sense in $\mathbb{R} \cup \{+\infty\}$, it is natural to assume that L is bounded below.)

Example 7.8. Let $(\mathcal{X}, d)$ be a metric space. Define the **length** of an absolutely continuous curve by the formula

$$\mathcal{L}(\gamma) = \int_0^1 |\dot{\gamma}_t|\,dt. \tag{7.9}$$

Then minimizing curves are called geodesics. They may have variable speed, but, just as on a Riemannian manifold, one can always *reparametrize* them (that is, replace γ by $\widetilde{\gamma}$ where $\widetilde{\gamma}_t = \gamma_{s(t)}$, with s continuous increasing) in such a way that they have constant speed. In that case $d(\gamma_s, \gamma_t) = |t-s|\,\mathcal{L}(\gamma)$ for all $s,t \in [0,1]$.

Example 7.9. Again let $(\mathcal{X}, d)$ be a metric space, but now consider the action

$$\mathcal{A}(\gamma) = \int_0^1 c(|\dot{\gamma}_t|)\,dt,$$

where c is strictly convex and strictly increasing (say $c(|v|) = |v|^p$, $p > 1$). Then,

$$c\big(d(\gamma_0,\gamma_1)\big) \le c(\mathcal{L}(\gamma)) = c\left(\int_0^1 |\dot{\gamma}_t|\,dt\right) \le \int_0^1 c(|\dot{\gamma}_t|)\,dt,$$

with equality in both inequalities if and only if γ is a constant-speed, minimizing geodesic. Thus $c(x,y) = c\big(d(x,y)\big)$ and minimizing curves are also geodesics, but with constant speed. Note that the distance can be recovered from the cost function, just by inverting c. As an illustration, if $p > 1$, and $c(|v|) = |v|^p$, then

$$d(x,y) = \inf\left\{\int_0^1 |\dot{\gamma}_t|^p\,dt; \quad \gamma_0 = x, \quad \gamma_1 = y\right\}^{\frac{1}{p}}.$$

In a given metric space, geodesics might not always exist, and it can even be the case that nonconstant continuous curves do not exist (think of a discrete space). So to continue the discussion we shall have to impose appropriate assumptions on our metric space and our cost function.

Here comes an important observation. When one wants to compute "in real life" the length of a curve, one does not use formula (7.9), but rather subdivides the curve into very small pieces, and approximates the length of each small piece by the distance between its endpoints. The finer the subdivision, the greater the measured approximate length (this is a consequence of the triangle inequality). So by taking finer and finer subdivisions we get an increasing family of measurements, whose upper bound may be taken as the measured length. This is actually an alternative definition of the length, which agrees with (7.9) for absolutely continuous curves, but does not require any further regularity assumption than plain continuity:

$$\mathcal{L}(\gamma) = \sup_{N\in\mathbb{N}} \sup_{0=t_0<t_1<\ldots<t_N=1} \Big[d(\gamma_{t_0}, \gamma_{t_1}) + \cdots + d(\gamma_{t_{N-1}}, \gamma_{t_N}) \Big]. \qquad (7.10)$$

Then one can define a **length space** as a metric space $(\mathcal{X}, d)$ in which, for any two $x, y \in \mathcal{X}$,

$$d(x,y) = \inf_{\gamma\in C([0,1];\mathcal{X})} \Big\{ \mathcal{L}(\gamma); \quad \gamma_0 = x, \ \gamma_1 = y \Big\}. \qquad (7.11)$$

If in addition $\mathcal{X}$ is complete and locally compact, then the infimum is a minimum, in which case the space is said to be a *strictly intrinsic* length space, or **geodesic space**. (By abuse of language, one often says just "length space" for "strictly intrinsic length space".) Such spaces play an important role in modern nonsmooth geometry.

Formulas (7.10) and (7.11) show an intimate link between the length and the distance: The length determines the distance by minimization, but conversely the distance determines the length by subdivision and approximation. The idea behind it is that the length of an "infinitesimal curve" is determined solely by the endpoints. A similar relation holds true for an action which is defined by a general Lagrangian of the form (7.2): indeed, if γ is differentiable at t, then

$$\int_t^{t+\varepsilon} L\big(\gamma_\tau, \dot{\gamma}_\tau, \tau\big)\, d\tau \simeq \varepsilon\, L(\gamma_t, \dot{\gamma}_t, t),$$

and the vector $\dot{\gamma}_t$ is uniquely determined, up to an error $o(\varepsilon)$, by γ_t and $\gamma_{t+\varepsilon}$. Such would not be the case if the function L would also depend on, say, the acceleration of the curve.

This **reconstruction property** plays an important role and it is natural to enforce it in an abstract generalization. To do so, it will be useful to consider an action as a *family* of functionals parametrized by the initial and the final times: So $\mathcal{A}^{s,t}$ is a functional on the set of paths $[s,t] \to \mathcal{X}$. Then we let

$$c^{s,t}(x,y) = \inf \Big\{ \mathcal{A}^{s,t}(\gamma); \quad \gamma_s = x, \ \gamma_t = y; \quad \gamma^{s,t} \in C([s,t];\mathcal{X}) \Big\}. \tag{7.12}$$

In words, $c^{s,t}(x,y)$ is the minimal work needed to go from point x at initial time s, to point y at final time t.

Example 7.10. Consider the Lagrangian $L(x,|v|,t) = |v|^p$. Then

$$c^{s,t}(x,y) = \frac{d(x,y)^p}{(t-s)^{p-1}}.$$

Note a characteristic property of these "power law" Lagrangians: The cost function depends on s,t only through multiplication by a constant. In particular, minimizing curves will be independent of s and t, up to reparametrization.

Abstract Lagrangian action

After all these preparations, the following definition should appear somewhat natural.

Definition 7.11 (Lagrangian action). *Let $(\mathcal{X},d)$ be a Polish space, and let $t_i, t_f \in \mathbb{R}$. A Lagrangian action $(\mathcal{A})^{t_i,t_f}$ on $\mathcal{X}$ is a family of lower semicontinuous functionals $\mathcal{A}^{s,t}$ on $C([s,t];\mathcal{X})$ ($t_i \le s < t \le t_f$), and cost functions $c^{s,t}$ on $\mathcal{X} \times \mathcal{X}$, such that:*

(i) $t_i \le t_1 < t_2 < t_3 \le t_f \Longrightarrow \quad \mathcal{A}^{t_1,t_2} + \mathcal{A}^{t_2,t_3} = \mathcal{A}^{t_1,t_3}$;

(ii) $\forall x,y \in \mathcal{X}$

$$c^{s,t}(x,y) = \inf \Big\{ \mathcal{A}^{s,t}(\gamma); \quad \gamma \in C([s,t];\mathcal{X}); \ \gamma_s = x, \ \gamma_t = y \Big\};$$

(iii) For any curve $(\gamma_t)_{t_i \le t \le t_f}$,

$$\mathcal{A}^{t_i,t_f}(\gamma) = \sup_{N \in \mathbb{N}} \ \sup_{t_i = t_0 \le t_1 \le \ldots \le t_N = t_f} \Big[c^{t_0,t_1}(\gamma_{t_0},\gamma_{t_1}) + c^{t_1,t_2}(\gamma_{t_1},\gamma_{t_2}) + \cdots + c^{t_{N-1},t_N}(\gamma_{t_{N-1}},\gamma_{t_N}) \Big].$$

The functional $\mathcal{A} = \mathcal{A}^{t_i,t_f}$ will just be called the action, and the cost function $c = c^{t_i,t_f}$ the cost associated with the action. A curve $\gamma : [t_i, t_f] \to \mathcal{X}$ is said to be action-minimizing if it minimizes $\mathcal{A}$ among all curves having the same endpoints.

Examples 7.12. (i) To recover (7.2) as a particular case of Definition 7.11, just set

$$\mathcal{A}^{s,t}(\gamma) = \int_s^t L(\gamma_\tau, \dot{\gamma}_\tau, \tau)\, d\tau. \tag{7.13}$$

(ii) A length space is a space in which $\mathcal{A}^{s,t}(\gamma) = \mathcal{L}(\gamma)$ (here $\mathcal{L}$ is the length) defines a Lagrangian action.

If $[t'_i, t'_f] \subset [t_i, t_f]$ then it is clear that $(\mathcal{A})^{t_i,t_f}$ induces an action $(\mathcal{A})^{t'_i,t'_f}$ on the time-interval $[t'_i, t'_f]$, just by restriction.

In the rest of this section I shall take $(t_i, t_f) = (0, 1)$, just for simplicity; of course one can always reduce to this case by reparametrization.

It will now be useful to introduce further assumptions about existence and compactness of minimizing curves.

Definition 7.13 (Coercive action). *Let $(\mathcal{A})^{0,1}$ be a Lagrangian action on a Polish space $\mathcal{X}$, with associated cost functions $(c^{s,t})_{0\le s<t\le 1}$. For any two times s, t $(0 \le s < t \le 1)$, and any two compact sets $K_s, K_t \subset \mathcal{X}$, let $\Gamma^{s,t}_{K_s\to K_t}$ be the set of minimizing paths starting in K_s at time s, and ending in K_t at time t. The action will be called coercive if:*

(i) It is bounded below, in the sense that

$$\inf_{s<t} \inf_{\gamma} \mathcal{A}^{s,t}(\gamma) > -\infty;$$

(ii) If $s < t$ are any two intermediate times, and K_s, K_t are any two nonempty compact sets such that $c^{s,t}(x, y) < +\infty$ for all $x \in K_s$, $y \in K_t$, then the set $\Gamma^{s,t}_{K_s\to K_t}$ is compact and nonempty.

In particular, minimizing curves between any two fixed points x, y with $c(x, y) < +\infty$ should always exist and form a compact set.

Remark 7.14. If each $\mathcal{A}^{s,t}$ has compact sub-level sets (more explicitly, if $\{\gamma;\ \mathcal{A}^{s,t}(\gamma) \le A\}$ is compact in $C([s, t]; \mathcal{X})$ for any $A \in \mathbb{R}$), then the lower semicontinuity of $\mathcal{A}^{s,t}$, together with a standard compactness argument (just as in Theorem 4.1) imply the existence of at least one action-minimizing curve among the set of curves that have prescribed final and initial points. In that case the requirement of nonemptiness in (ii) is fulfilled.

Examples 7.15. (i) If $\mathcal{X}$ is a smooth complete Riemannian manifold and $L(x,v,t)$ is a Lagrangian satisfying the classical conditions of Definition 7.6, then the action defined by (7.13) is coercive.

(ii) If $\mathcal{X}$ is a geodesic length space, then the action defined by $\mathcal{A}^{s,t}(\gamma) = \mathcal{L}(\gamma)^2/(t-s)$ is coercive; in fact minimizers are constant-speed minimizing geodesic curves. On the other hand the action defined by $\mathcal{A}^{s,t}(\gamma) = \mathcal{L}(\gamma)$ is not coercive, since the possibility of reparametrization prevents the compactness of the set of minimizing curves.

Proposition 7.16 (Properties of Lagrangian actions). *Let $(\mathcal{X}, d)$ be a Polish space and $(\mathcal{A})^{0,1}$ a coercive Lagrangian action on $\mathcal{X}$. Then:*

(i) For all intermediate times $s < t$, $c^{s,t}$ is lower semicontinuous on $\mathcal{X} \times \mathcal{X}$, with values in $\mathbb{R} \cup \{+\infty\}$.

(ii) If a curve γ on $[s,t] \subset [0,1]$ is a minimizer of $\mathcal{A}^{s,t}$, then its restriction to $[s',t'] \subset [s,t]$ is also a minimizer for $\mathcal{A}^{s',t'}$.

(iii) For all times $t_1 < t_2 < t_3$ in $[0,1]$, and x_1, x_3 in $\mathcal{X}$,

$$c^{t_1,t_3}(x_1,x_3) = \inf_{x_2 \in \mathcal{X}} \Big(c^{t_1,t_2}(x_1,x_2) + c^{t_2,t_3}(x_2,x_3) \Big); \tag{7.14}$$

and if the infimum is achieved at some point x_2, then there is a minimizing curve which goes from x_1 at time t_1 to x_3 at time t_3, and passes through x_2 at time t_2.

(iv) A curve γ is a minimizer of $\mathcal{A}$ if and only if, for all intermediate times $t_1 < t_2 < t_3$ in $[0,1]$,

$$c^{t_1,t_3}(\gamma_{t_1},\gamma_{t_3}) = c^{t_1,t_2}(\gamma_{t_1},\gamma_{t_2}) + c^{t_2,t_3}(\gamma_{t_2},\gamma_{t_3}). \tag{7.15}$$

(v) If the cost functions $c^{s,t}$ are continuous, then the set Γ of all action-minimizing curves is closed in the topology of uniform convergence;

(vi) For all times $s < t$, there exists a Borel map $S_{s\to t} : \mathcal{X} \times \mathcal{X} \to C([s,t];\mathcal{X})$, such that for all $x, y \in \mathcal{X}$, $S(x,y)$ belongs to $\Gamma^{s,t}_{x\to y}$. In words, there is a measurable recipe to join any two endpoints x and y by a minimizing curve $\gamma : [s,t] \to \mathcal{X}$.

Remark 7.17. The statement in (iv) is a powerful formulation of the minimizing property. It is often quite convenient from the technical point of view, even in a smooth setting, because it does not involve any time-derivative.

Remark 7.18. The continuity assumption in (v) is satisfied in most cases of interest. For instance, if $\mathcal{A}^{s,t}(\gamma) = \mathcal{L}(\gamma)^2/(t-s)$, then $c^{s,t}(x,y) = d(x,y)^2/(t-s)$, which is obviously continuous. Continuity also holds true in the other model example where $\mathcal{X}$ is a Riemannian manifold and the cost is obtained from a Lagrangian function $L(x,v,t)$ on $TM \times [0,1]$ satisfying the classical assumptions; a proof is sketched in the Appendix.

Proof of Proposition 7.16. Let us prove (i). By definition of the coercivity, $c(x,y)$ is never $-\infty$. Let $(x_k)_{k\in\mathbb{N}}$ and $(y_k)_{k\in\mathbb{N}}$ be sequences converging to x and y respectively. Then the family $(x_k) \cup \{x\}$ forms a compact set K_s, and the family $(y_k) \cup \{y\}$ also forms a compact set K_t. By assumption, for each k we can find a minimizing curve $\gamma_k : [s,t] \to \mathcal{X}$ joining x_k to y_k, so γ_k belongs to $\Gamma^{s,t}_{K_s\to K_t}$ which is compact. From $(\gamma_k)_{k\in\mathbb{N}}$ we can extract a subsequence which converges uniformly to some minimizing curve γ. The uniform convergence implies that $x_k = \gamma_k(s) \to \gamma(s)$, $y_k = \gamma_k(t) \to \gamma(t)$, so γ joins x to y. The lower semicontinuity of $\mathcal{A}^{s,t}$ implies that $\mathcal{A}^{s,t}(\gamma) \le \liminf \mathcal{A}^{s,t}(\gamma_k)$; thus

$$c^{s,t}(x,y) \le \mathcal{A}^{s,t}(\gamma) \le \liminf \mathcal{A}^{s,t}(\gamma_k) = \liminf c^{s,t}(x_k,y_k).$$

This establishes the lower semicontinuity of the cost $c^{s,t}$.

Property (ii) is obvious: if the restriction of γ to $[s',t']$ is not optimal, introduce $\widetilde{\gamma}$ on $[s',t']$ such that $\mathcal{A}^{s',t'}(\widetilde{\gamma}) < \mathcal{A}^{s',t'}(\gamma)$. Then the path obtained by concatenating γ on $[s,s']$, $\widetilde{\gamma}$ on $[s',t']$ and γ again on $[t',t]$, has a strictly lower action $\mathcal{A}^{s,t}$ than γ, which is impossible. (Obviously, this is the same line of reasoning as in the proof of the "restriction property" of Theorem 4.6.)

Now, to prove (iii), introduce minimizing curves $\gamma_{1\to 2}$ joining x_1 at time t_1, to x_2 at time t_2, and $\gamma_{2\to 3}$ joining x_2 at time t_2, to x_3 at time t_3. Then define γ on $[t_1,t_3]$ by concatenation of $\gamma_{1\to 2}$ and $\gamma_{2\to 3}$. From the axioms of Definition 7.11,

$$\begin{aligned} c^{t_1,t_3}(x_1,x_3) \le \mathcal{A}^{t_1,t_3}(\gamma) &= \mathcal{A}^{t_1,t_2}(\gamma_{1\to 2}) + \mathcal{A}^{t_2,t_3}(\gamma_{2\to 3}) \\ &= c^{t_1,t_2}(x_1,x_2) + c^{t_2,t_3}(x_2,x_3). \end{aligned}$$

The inequality in (iii) follows by taking the infimum over x_2. Moreover, if there is equality, that is,

$$c^{t_1,t_2}(x_1,x_2) + c^{t_2,t_3}(x_2,x_3) = c^{t_1,t_3}(x_1,x_3),$$

then equality holds everywhere in the above chain of inequalities, so the curve γ achieves the optimal cost $c^{t_1,t_3}(x_1,x_3)$, while passing through x_2 at time t_2.

It is a consequence of (iii) that any minimizer should satisfy (7.15), since the restrictions of γ to $[t_1, t_2]$ and to $[t_2, t_3]$ should both be minimizing. Conversely, let γ be a curve satisfying (7.15) for all (t_1, t_2, t_3) with $t_1 < t_2 < t_3$. By induction, this implies that for each subdivision $0 = t_0 < t_1 \leq \ldots < t_N = 1$,

$$c^{0,1}(\gamma_0, \gamma_1) = \sum_j c^{t_j, t_{j+1}}(\gamma_{t_j}, \gamma_{t_{j+1}}).$$

By point (iii) in Definition 7.11, it follows that $c^{0,1}(\gamma_0, \gamma_1) = \mathcal{A}^{0,1}(\gamma)$, which proves (iv).

If $0 \leq t_1 < t_2 < t_3 \leq 1$, now let $\Gamma(t_1, t_2, t_3)$ stand for the set of all curves satisfying (7.15). If all functions $c^{s,t}$ are continuous, then $\Gamma(t_1, t_2, t_3)$ is closed for the topology of uniform convergence. Then Γ is the intersection of all $\Gamma(t_1, t_2, t_3)$, so it is closed also; this proves statement (v). (Now there is a similarity with the proof of Theorem 5.20.)

For given times $s < t$, let $\Gamma^{s,t}$ be the set of all action-minimizing curves defined on $[s, t]$, and let $E_{s,t}$ be the "endpoints" mapping, defined on $\Gamma^{s,t}$ by $\gamma \longmapsto (\gamma_s, \gamma_t)$. By assumption, any two points are joined by at least one minimizing curve, so $E_{s,t}$ is onto $\mathcal{X} \times \mathcal{X}$. It is clear that $E_{s,t}$ is a continuous map between Polish spaces, and by assumption $E_{s,t}^{-1}(x, y)$ is compact for all x, y. It follows by general theorems of measurable selection (see the bibliographical notes in case of need) that $E_{s,t}$ admits a measurable right-inverse $S_{s \to t}$, i.e. $E_{s,t} \circ S_{s \to t} = \mathrm{Id}$. This proves statement (vi). □

Interpolation of random variables

Action-minimizing curves provide a fairly general framework to interpolate between points, which can be seen as deterministic random variables. What happens when we want to interpolate between genuinely random variables, in a way that is most economic? Since a deterministic point can be identified with a Dirac mass, this new problem contains both the classical action-minimizing problem and the Monge–Kantorovich problem.

Here is a natural recipe. Let c be the cost associated with the Lagrangian action, and let μ_0, μ_1 be two given laws. Introduce an optimal coupling (X_0, X_1) of (μ_0, μ_1), and a random action-minimizing path $(X_t)_{0\le t\le 1}$ joining X_0 to X_1. (We shall see later that such a thing always exists.) Then the random variable X_t is an interpolation of X_0 and X_1; or equivalently the law μ_t is an interpolation of μ_0 and μ_1. This procedure is called **displacement interpolation**, by opposition to the linear interpolation $\mu_t = (1-t)\,\mu_0 + t\,\mu_1$. Note that there is a priori no uniqueness of the displacement interpolation.

Some of the concepts which we just introduced deserve careful attention. In the sequel, e_t will stand for the evaluation at time t: $e_t(\gamma) = \gamma(t)$.

Definition 7.19 (Dynamical coupling). *Let $(\mathcal{X}, d)$ be a Polish space. A dynamical transference plan Π is a probability measure on the space $C([0,1];\mathcal{X})$. A dynamical coupling of two probability measures $\mu_0, \mu_1 \in P(\mathcal{X})$ is a random curve $\gamma : [0,1] \to \mathcal{X}$ such that* law $(\gamma_0) = \mu_0$, law $(\gamma_1) = \mu_1$.

Definition 7.20 (Dynamical optimal coupling). *Let $(\mathcal{X}, d)$ be a Polish space, $(\mathcal{A})^{0,1}$ a Lagrangian action on $\mathcal{X}$, c the associated cost, and Γ the set of action-minimizing curves. A dynamical optimal transference plan is a probability measure Π on Γ such that*

$$\pi_{0,1} := (e_0, e_1)_{\#}\Pi$$

is an optimal transference plan between μ_0 and μ_1. Equivalently, Π is the law of a random action-minimizing curve whose endpoints constitute an optimal coupling of μ_0 and μ_1. Such a random curve is called a dynamical optimal coupling of (μ_0, μ_1). By abuse of language, Π itself is often called a dynamical optimal coupling.

The next theorem is the main result of this chapter. It shows that the law at time t of a dynamical optimal coupling can be seen as a minimizing path in the space of probability measures. In the important case when the cost is a power of a geodesic distance, the corollary stated right after the theorem shows that displacement interpolation can be thought of as a geodesic path in the space of probability measures. ("*A geodesic in the space of laws is the law of a geodesic.*") The theorem also shows that such interpolations can be constructed under quite weak assumptions.

Theorem 7.21 (Displacement interpolation). *Let $(\mathcal{X}, d)$ be a Polish space, and $(\mathcal{A})^{0,1}$ a coercive Lagrangian action on $\mathcal{X}$, with continuous cost functions $c^{s,t}$. Whenever $0 \le s < t \le 1$, denote by $C^{s,t}(\mu, \nu)$ the optimal transport cost between the probability measures μ and ν for the cost $c^{s,t}$; write $c = c^{0,1}$ and $C = C^{0,1}$. Let μ_0 and μ_1 be any two probability measures on $\mathcal{X}$, such that the optimal transport cost $C(\mu_0, \mu_1)$ is finite. Then, given a continuous path $(\mu_t)_{0 \le t \le 1}$, the following properties are equivalent:*

(i) For each $t \in [0,1]$, μ_t is the law of γ_t, where $(\gamma_t)_{0 \le t \le 1}$ is a dynamical optimal coupling of (μ_0, μ_1);

(ii) For any three intermediate times $t_1 < t_2 < t_3$ in $[0,1]$,

$$C^{t_1,t_2}(\mu_{t_1}, \mu_{t_2}) + C^{t_2,t_3}(\mu_{t_2}, \mu_{t_3}) = C^{t_1,t_3}(\mu_{t_1}, \mu_{t_3});$$

(iii) The path $(\mu_t)_{0 \le t \le 1}$ is a minimizing curve for the coercive action functional defined on $P(\mathcal{X})$ by

$$\mathbb{A}^{s,t}(\mu) = \sup_{N \in \mathbb{N}} \sup_{s=t_0<t_1<\ldots<t_N=t} \sum_{i=0}^{N-1} C^{t_i,t_{i+1}}(\mu_{t_i}, \mu_{t_{i+1}}) \tag{7.16}$$

$$= \inf_{\gamma} \mathbb{E}\, \mathcal{A}^{s,t}(\gamma), \tag{7.17}$$

where the last infimum is over all random curves $\gamma : [s,t] \to \mathcal{X}$ such that law $(\gamma_\tau) = \mu_\tau$ *$(s \le \tau \le t)$.*

In that case $(\mu_t)_{0 \le t \le 1}$ is said to be a displacement interpolation between μ_0 and μ_1. There always exists at least one such curve.

Finally, if $\mathcal{K}_0$ and $\mathcal{K}_1$ are two compact subsets of $P(\mathcal{X})$, such that $C^{0,1}(\mu_0, \mu_1) < +\infty$ for all $\mu_0 \in \mathcal{K}_0$, $\mu_1 \in \mathcal{K}_1$, then the set of dynamical optimal transference plans Π with $(e_0)_\# \Pi \in \mathcal{K}_0$ and $(e_1)_\# \Pi \in \mathcal{K}_1$ is compact.

Theorem 7.21 admits two important corollaries:

Corollary 7.22 (Displacement interpolation as geodesics). *Let $(\mathcal{X}, d)$ be a complete separable, locally compact length space. Let $p > 1$ and let $P_p(\mathcal{X})$ be the space of probability measures on $\mathcal{X}$ with finite moment of order p, metrized by the Wasserstein distance W_p. Then, given any two $\mu_0, \mu_1 \in P_p(\mathcal{X})$, and a continuous curve $(\mu_t)_{0 \le t \le 1}$, valued in $P(\mathcal{X})$, the following properties are equivalent:*

(i) μ_t is the law of γ_t, where γ is a random (minimizing, constant-speed) geodesic such that (γ_0, γ_1) is an optimal coupling;

(ii) $(\mu_t)_{0 \le t \le 1}$ is a geodesic curve in the space $P_p(\mathcal{X})$.

Moreover, if μ_0 and μ_1 are given, there exists at least one such curve. More generally, if $\mathcal{K}_0 \subset P_p(\mathcal{X})$ and $\mathcal{K}_1 \subset P_p(\mathcal{X})$ are compact subsets of $P(\mathcal{X})$, then the set of geodesic curves $(\mu_t)_{0 \le t \le 1}$ such that $\mu_0 \in \mathcal{K}_0$ and $\mu_1 \in \mathcal{K}_1$ is compact and nonempty; and also the set of dynamical optimal transference plans Π with $(e_0)_\# \Pi \in \mathcal{K}_0$, $(e_1)_\# \Pi \in \mathcal{K}_1$ is compact and nonempty.

Corollary 7.23 (Uniqueness of displacement interpolation). *With the same assumptions as in Theorem 7.21, if:*

(a) there is a unique optimal transport plan π between μ_0 and μ_1;

(b) $\pi(dx_0\, dx_1)$-almost surely, x_0 and x_1 are joined by a unique minimizing curve;

then there is a unique displacement interpolation $(\mu_t)_{0 \le t \le 1}$ joining μ_0 to μ_1.

Remark 7.24. In Corollary 7.22, $\mathcal{A}^{s,t}(\gamma) = \int_s^t |\dot{\gamma}_\tau|^p \, d\tau$. Then action-minimizing curves in $\mathcal{X}$ are the same, whatever the value of $p > 1$. Yet geodesics in $P_p(\mathcal{X})$ are not the same for different values of p, because a coupling of (μ_0, μ_1) which is optimal for a certain value of p, might well not be for another value.

Remark 7.25. Theorem 7.21 applies to Lagrangian functions $L(x, v, t)$ on a Riemannian manifold *TM*, as soon as L is C^2 and satisfies the classical conditions of Definition 7.6. Then μ_t is the law at time t of a random solution of the Euler–Lagrange equation (7.6).

Remark 7.26. In Theorem 7.21, the minimizing property of the path (μ_t) is expressed in a weak formulation, which makes sense with a lot of generality. But this theorem leaves open certain natural questions:

- Is there a differential equation for geodesic curves, or more generally optimal paths $(\mu_t)_{0 \le t \le 1}$? Of course, the answer is related to the possibility of defining a tangent space in the space of measures.
- Is there a more explicit formula for the action on the space of probability measures, say for a simple enough action on $\mathcal{X}$? Can it be written as $\int_0^1 \mathbb{L}(\mu_t, \dot{\mu}_t, t)\, dt$? (Of course, in Corollary 7.22 this is the case with $\mathbb{L} = |\dot{\mu}|^p$, but this expression is not very "explicit".)
- Are geodesic paths nonbranching? (Does the velocity at initial time uniquely determine the final measure μ_1?)

- Can one identify simple conditions for the existence of a *unique* geodesic path between two given probability measures?

All these questions will be answered affirmatively in the sequel of this course, under suitable regularity assumptions on the space, the action or the probability measures.

Remark 7.27. The assumption of local compactness in Corollary 7.22 is not superficial: it is used to guarantee the coercivity of the action. For spaces that are not locally compact, there might be an analogous theory, but it is certainly more tricky. First of all, selection theorems are not immediately available if one does not assume compactness of the set of geodesics joining two given endpoints. More importantly, the convergence scheme used below to construct a random geodesic curve from a time-dependent law might fail to work. Here we are encountering a general principle in probability theory: Analytic characterizations of stochastic processes (like those based on semigroups, generators, etc.) are essentially available only in locally compact spaces. In spite of all that, there are some representation theorems for Wasserstein geodesics that do not need local compactness; see the bibliographical notes for details.

The proof of Theorem 7.21 is not so difficult, but a bit cumbersome because of measurability issues. For training purposes, the reader might rewrite it in the simpler case where any pair of points is joined by a *unique* geodesic (as in the case of $\mathbb{R}^n$). To help understanding, I shall first sketch the main idea.

Main idea in the proof of Theorem 7.21. The delicate part consists in showing that if (μ_t) is a given action-minimizing curve, then there exists a random minimizer γ such that $\mu_t = \mathrm{law}(\gamma_t)$. This γ will be constructed by dyadic approximation, as follows. First let $(\gamma_0^{(0)}, \gamma_1^{(0)})$ be an optimal coupling of (μ_0, μ_1). (Here the notation $\gamma_0^{(0)}$ could be replaced by just x_0, it does not mean that there is some curve $\gamma^{(0)}$ behind.) Then let $(\gamma_0^{(1)}, \gamma_{1/2}^{(1)})$ be an optimal coupling of $(\mu_0, \mu_{1/2})$, and $((\gamma')_{1/2}^{(1)}, \gamma_1^{(1)})$ be an optimal coupling of $(\mu_{1/2}, \mu_1)$. By gluing these couplings together, I can actually assume that $(\gamma')_{1/2}^{(1)} = \gamma_{1/2}^{(1)}$, so that I have a triple $(\gamma_0^{(1)}, \gamma_{1/2}^{(1)}, \gamma_1^{(1)})$ in which the first two components on the one hand, and the last two components on the other hand, constitute optimal couplings.

Now the key observation is that if $(\gamma_{t_1}, \gamma_{t_2})$ and $(\gamma_{t_2}, \gamma_{t_3})$ are optimal couplings of (μ_{t_1}, μ_{t_2}) and (μ_{t_2}, μ_{t_3}) respectively, and the μ_{t_k} satisfy the equality appearing in (ii), then also $(\gamma_{t_1}, \gamma_{t_3})$ should be optimal. Indeed, by taking expectation in the inequality

$$c^{t_1,t_3}(\gamma_{t_1}, \gamma_{t_3}) \le c^{t_1,t_2}(\gamma_{t_1}, \gamma_{t_2}) + c^{t_2,t_3}(\gamma_{t_2}, \gamma_{t_3})$$

and using the optimality assumption, one obtains

$$\mathbb{E}\, c^{t_1,t_3}(\gamma_{t_1}, \gamma_{t_3}) \le C^{t_1,t_2}(\mu_{t_1}, \mu_{t_2}) + C^{t_2,t_3}(\mu_{t_2}, \mu_{t_3}).$$

Now the fact that (μ_t) is action-minimizing imposes

$$C^{t_1,t_2}(\mu_{t_1}, \mu_{t_2}) + C^{t_2,t_3}(\mu_{t_2}, \mu_{t_3}) = C^{t_1,t_3}(\mu_{t_1}, \mu_{t_3});$$

so actually

$$\mathbb{E}\, c^{t_1,t_3}(\gamma_{t_1}, \gamma_{t_3}) \le C^{t_1,t_3}(\mu_{t_1}, \mu_{t_3}),$$

which means that indeed $(\gamma_{t_1}, \gamma_{t_3})$ is an optimal coupling of (μ_{t_1}, μ_{t_3}) for the cost c^{t_1,t_3}.

So $(\gamma_0^{(1)}, \gamma_1^{(1)})$ is an optimal coupling of (μ_0, μ_1). Now we can proceed in the same manner and define, for each k, a random discrete path $(\gamma^{(k)}_{j\,2^{-k}})$ such that $(\gamma_s^{(k)}, \gamma_t^{(k)})$ is an optimal coupling for all times s, t of the form $j/2^k$. These are only discrete paths, but it is possible to extend them into paths $(\gamma_t^{(k)})_{0\le t\le 1}$ that are minimizers of the action. Of course, if t is not of the form $j/2^k$, there is no reason why law $(\gamma_t^{(k)})$ would coincide with μ_t. But hopefully we can pass to the limit as $k \to \infty$, for each dyadic time, and conclude by a density argument. □

Complete proof of Theorem 7.21. First, if $\mathcal{A}^{s,t}(\gamma)$ is bounded below by a constant $-C$, independently of s, t and γ, then the same is true of the cost functions $c^{s,t}$ and of the total costs $C^{s,t}$. So all the quantities appearing in the proof will be well-defined, the value $+\infty$ being possibly attained. Moreover, the action $\mathbb{A}^{s,t}$ defined by the formula in (iii) will also be bounded below by the same constant $-C$, so Property (i) of Definition 7.13 will be satisfied.

Now let μ_0 and μ_1 be given. According to Theorem 4.1, there exists at least one optimal transference plan π between μ_0 and μ_1, for the cost $c = c^{0,1}$. Let $S_{0\to 1}$ be the mapping appearing in Proposition 7.16(vi), and let

$$\Pi := (S_{0\to 1})_{\#}\pi.$$

Then Π defines the law of a random geodesic γ, and the identity $E_{0,1} \circ S_{0\to 1} = \mathrm{Id}$ implies that the endpoints of γ are distributed according to π. This proves the existence of a path satisfying (i). Now the main part of the proof consists in checking the equivalence of properties (i) and (ii). This will be performed in four steps.

Step 1. Let $(\mu_t)_{0\le t\le 1}$ be any continuous curve in the space of probability measures, and let t_1, t_2, t_3 be three intermediate times. Let $\pi_{t_1\to t_2}$ be an optimal transference plan between μ_{t_1} and μ_{t_2} for the transport cost c^{t_1,t_2}, and similarly let $\pi_{t_2\to t_3}$ be an optimal transference plan between μ_{t_2} and μ_{t_3} for the transport cost c^{t_2,t_3}. By the Gluing Lemma of Chapter 1 one can construct random variables $(\gamma_{t_1}, \gamma_{t_2}, \gamma_{t_3})$ such that $\mathrm{law}\,(\gamma_{t_1}, \gamma_{t_2}) = \pi_{t_1\to t_2}$ and $\mathrm{law}\,(\gamma_{t_2}, \gamma_{t_3}) = \pi_{t_2\to t_3}$ (in particular, $\mathrm{law}\,(\gamma_{t_i}) = \mu_{t_i}$ for $i = 1, 2, 3$). Then, by (7.14),

$$C^{t_1,t_3}(\mu_{t_1}, \mu_{t_3}) \le \mathbb{E}\, c^{t_1,t_3}(\gamma_{t_1}, \gamma_{t_3}) \le \mathbb{E}\, c^{t_1,t_2}(\gamma_{t_1}, \gamma_{t_2}) + \mathbb{E}\, c^{t_2,t_3}(\gamma_{t_2}, \gamma_{t_3}) \\ = C^{t_1,t_2}(\mu_{t_1}, \mu_{t_2}) + C^{t_2,t_3}(\mu_{t_2}, \mu_{t_3}).$$

This inequality holds for any path, optimal or not.

Step 2. Assume that (μ_t) satisfies (i), so there is a dynamical optimal transference plan Π such that $\mu_t = (e_t)_\#\Pi$. Let γ be a random minimizing curve with law Π, and consider the obvious coupling $(\gamma_{t_1}, \gamma_{t_2})$ (resp. $(\gamma_{t_2}, \gamma_{t_3})$) of (μ_{t_1}, μ_{t_2}) (resp. (μ_{t_2}, μ_{t_3})). Then from the definition of the optimal cost and the minimizing property of γ,

$$C^{t_1,t_2}(\mu_{t_1}, \mu_{t_2}) + C^{t_2,t_3}(\mu_{t_2}, \mu_{t_3}) \le \mathbb{E}\, c^{t_1,t_2}(\gamma_{t_1}, \gamma_{t_2}) + \mathbb{E}\, c^{t_2,t_3}(\gamma_{t_2}, \gamma_{t_3}) \\ = \mathbb{E}\, \mathcal{A}^{t_1,t_2}(\gamma) + \mathbb{E}\, \mathcal{A}^{t_2,t_3}(\gamma) = \mathbb{E}\, \mathcal{A}^{t_1,t_3}(\gamma) = \mathbb{E}\, c^{t_1,t_3}(\gamma_{t_1}, \gamma_{t_3}). \quad (7.18)$$

Now choose $t_1 = 0$, $t_2 = t$, $t_3 = 1$. Since by assumption (γ_0, γ_1) is an optimal coupling of (μ_0, μ_1), the above computation implies

$$C^{0,t}(\mu_0, \mu_t) + C^{t,1}(\mu_t, \mu_1) \le C^{0,1}(\mu_0, \mu_1),$$

and since the reverse inequality holds as a consequence of Step 1, actually

$$C^{0,t}(\mu_0, \mu_t) + C^{t,1}(\mu_t, \mu_1) = C^{0,1}(\mu_0, \mu_1).$$

Moreover, equality has to hold in (7.18) (for that particular choice of intermediate times); since $C^{0,1}(\mu_0, \mu_1) < +\infty$ this implies $C^{0,t}(\mu_0, \mu_t) = \mathbb{E}\, c^{0,t}(\gamma_0, \gamma_t)$, which means that (γ_0, γ_t) should actually be an optimal coupling of (μ_0, μ_t). Similarly, (γ_t, γ_1) should be an optimal coupling of (μ_t, μ_1).

Next choose $t_1 = 0$, $t_2 = s$, $t_3 = t$, and apply the previous deduction to discover that (γ_s, γ_t) is an optimal coupling of (μ_s, μ_t). After inserting this information in (7.18) with $s = t_2$ and $t = t_3$, we recover

$$C^{t_1,t_2}(\mu_{t_1}, \mu_{t_2}) + C^{t_2,t_3}(\mu_{t_2}, \mu_{t_3}) \le C^{t_1,t_3}(\mu_{t_1}, \mu_{t_3}).$$

This together with Step 1 proves that (μ_t) satisfies Property (ii). So far we have proven (i) $\Rightarrow$ (ii).

Step 3. Assume that (μ_t) satisfies Property (ii); then we can perform again the same computation as in Step 1, but now all the inequalities have to be equalities. This implies that the random variables $(\gamma_{t_1}, \gamma_{t_2}, \gamma_{t_3})$ satisfy:

(a) $(\gamma_{t_1}, \gamma_{t_3})$ is an optimal coupling of (μ_{t_1}, μ_{t_3}) for the cost c^{t_1,t_3};

(b) $c^{t_1,t_3}(\gamma_{t_1}, \gamma_{t_3}) = c^{t_1,t_2}(\gamma_{t_1}, \gamma_{t_2}) + c^{t_2,t_3}(\gamma_{t_2}, \gamma_{t_3})$ almost surely;

(c) $c^{s,t}(\gamma_s, \gamma_t) < +\infty$ almost surely.

Armed with that information, we proceed as follows. We start from an optimal coupling (γ_0, γ_1) of (μ_0, μ_1), with joint law $\pi_{0\to 1}$. Then as in Step 1 we construct a triple $(\gamma_0^{(1)}, \gamma_{\frac{1}{2}}^{(1)}, \gamma_1^{(1)})$ with $\mathrm{law}\,(\gamma_0^{(1)}) = \mu_0$, $\mathrm{law}\,(\gamma_{\frac{1}{2}}^{(1)}) = \mu_{\frac{1}{2}}$, $\mathrm{law}\,(\gamma_1^{(1)}) = \mu_1$, such that $(\gamma_0^{(1)}, \gamma_{\frac{1}{2}}^{(1)})$ is an optimal coupling of $(\mu_0, \mu_{\frac{1}{2}})$ for the cost $c^{0,\frac{1}{2}}$ and $(\gamma_{\frac{1}{2}}^{(1)}, \gamma_1^{(1)})$ is an optimal coupling of $(\mu_{\frac{1}{2}}, \mu_1)$ for the cost $c^{\frac{1}{2},1}$. From (a) and (b) above we know that $(\gamma_0^{(1)}, \gamma_1^{(1)})$ is an optimal coupling of (μ_0, μ_1) (but $\mathrm{law}\,(\gamma_0^{(1)}, \gamma_1^{(1)})$ might be different from $\mathrm{law}\,(\gamma_0, \gamma_1)$), and moreover $c^{0,1}(\gamma_0^{(1)}, \gamma_1^{(1)}) = c^{0,\frac{1}{2}}(\gamma_0^{(1)}, \gamma_{\frac{1}{2}}^{(1)}) + c^{\frac{1}{2},1}(\gamma_{\frac{1}{2}}^{(1)}, \gamma_1^{(1)})$ almost surely.

Next it is possible to iterate the construction, introducing more and more midpoints. By a reasoning similar to the one above and an induction argument, one can construct, for each integer $k \ge 1$, random variables $(\gamma_0^{(k)}, \gamma_{\frac{1}{2^k}}^{(k)}, \gamma_{\frac{2}{2^k}}^{(k)}, \gamma_{\frac{3}{2^k}}^{(k)}, \ldots \gamma_1^{(k)})$ in such a way that

(a) for any two $i, j \le 2^k$, $(\gamma_{\frac{i}{2^k}}^{(k)}, \gamma_{\frac{j}{2^k}}^{(k)})$ constitutes an optimal coupling of $(\mu_{\frac{i}{2^k}}, \mu_{\frac{j}{2^k}})$,

(b) for any three indices $i_1, i_2, i_3 \le 2^k$, one has

$$c^{\frac{i_1}{2^k},\frac{i_3}{2^k}}\big(\gamma_{\frac{i_1}{2^k}}^{(k)}, \gamma_{\frac{i_3}{2^k}}^{(k)}\big) = c^{\frac{i_1}{2^k},\frac{i_2}{2^k}}\big(\gamma_{\frac{i_1}{2^k}}^{(k)}, \gamma_{\frac{i_2}{2^k}}^{(k)}\big) + c^{\frac{i_2}{2^k},\frac{i_3}{2^k}}\big(\gamma_{\frac{i_2}{2^k}}^{(k)}, \gamma_{\frac{i_3}{2^k}}^{(k)}\big).$$

At this stage it is convenient to extend the random variables $\gamma^{(k)}$, which are only defined for times $j/2^k$, into (random) continuous curves $(\gamma_t^{(k)})_{0\le t\le 1}$. For that we use Proposition 7.16(vi) again, and for any $t \in (i/2^k, (i+1)/2^k)$ we define

$$\gamma_t := e_t\Big(S_{\frac{i}{2^k},\frac{i+1}{2^k}}(\gamma_{\frac{i}{2^k}}, \gamma_{\frac{i+1}{2^k}})\Big).$$

(Recall that e_t is just the evaluation at time t.) Then the law $\Pi^{(k)}$ of $(\gamma_t)_{0\le t\le 1}$ is a probability measure on the set of continuous curves in $\mathcal{X}$.

I claim that $\Pi^{(k)}$ is actually concentrated on minimizing curves. (Skip at first reading and go directly to Step 4.) To prove this, it is sufficient to check the criterion in Proposition 7.16(iv), involving three intermediate times t_1, t_2, t_3. By construction, the criterion holds true if all these times belong to the same time-interval $[i/2^k, (i+1)/2^k]$, and also if they are all of the form $j/2^k$; the problem consists in "crossing subintervals". Let us show that

$$(i-1)\,2^{-k} < s < i\,2^{-k} \le j\,2^{-k} < t < (j+1)\,2^{-k} \quad\Longrightarrow$$
$$\begin{cases} c^{\frac{i-1}{2^k},\frac{j+1}{2^k}}(\gamma_{\frac{i-1}{2^k}}, \gamma_{\frac{j+1}{2^k}}) = c^{\frac{i-1}{2^k},s}(\gamma_{\frac{i-1}{2^k}}, \gamma_s) + c^{s,t}(\gamma_s,\gamma_t) + c^{t,\frac{j+1}{2^k}}(\gamma_t, \gamma_{\frac{j+1}{2^k}}) \\[2ex] c^{s,t}(\gamma_s,\gamma_t) = c^{s,\frac{i}{2^k}}(\gamma_s, \gamma_{\frac{i}{2^k}}) + c^{\frac{i}{2^k},\frac{j}{2^k}}(\gamma_{\frac{i}{2^k}}, \gamma_{\frac{j}{2^k}}) + c^{\frac{j}{2^k},t}(\gamma_{\frac{j}{2^k}}, \gamma_t) \end{cases} \tag{7.19}$$

To prove this, we start with

$$\begin{aligned} c^{\frac{i-1}{2^k},\frac{j+1}{2^k}}(\gamma_{\frac{i-1}{2^k}}, \gamma_{\frac{j+1}{2^k}}) &\le c^{\frac{i-1}{2^k},s}(\gamma_{\frac{i-1}{2^k}}, \gamma_s) + c^{s,t}(\gamma_s,\gamma_t) + c^{t,\frac{j+1}{2^k}}(\gamma_t, \gamma_{\frac{j+1}{2^k}}) \\ \le c^{\frac{i-1}{2^k},s}(\gamma_{\frac{i-1}{2^k}}, \gamma_s) \;+\; c^{s,\frac{i}{2^k}}(\gamma_s, \gamma_{\frac{i}{2^k}}) &\;+\; c^{\frac{i}{2^k},\frac{i+1}{2^k}}(\gamma_{\frac{i}{2^k}}, \gamma_{\frac{i+1}{2^k}}) \\ &+ \;\dots\; + c^{\frac{j}{2^k},t}(\gamma_{\frac{j}{2^k}}, \gamma_t) + c^{t,\frac{j+1}{2^k}}(\gamma_t, \gamma_{\frac{j+1}{2^k}}). \end{aligned} \tag{7.20}$$

Since we have used minimizing curves to interpolate on each dyadic subinterval,

$$c^{\frac{i-1}{2^k},s}(\gamma_{\frac{i-1}{2^k}}, \gamma_s) + c^{s,\frac{i}{2^k}}(\gamma_s, \gamma_{\frac{i}{2^k}}) = c^{\frac{i-1}{2^k},\frac{i}{2^k}}(\gamma_{\frac{i-1}{2^k}}, \gamma_{\frac{i}{2^k}}),$$

etc. So the right-hand side of (7.20) coincides with

$$c^{\frac{i-1}{2^k},\frac{i}{2^k}}(\gamma_{\frac{i-1}{2^k}},\gamma_{\frac{i}{2^k}}) + \ldots + c^{\frac{j}{2^k},\frac{j+1}{2^k}}(\gamma_{\frac{j}{2^k}},\gamma_{\frac{j+1}{2^k}}),$$

and by construction of $\Pi^{(k)}$ this is just $c^{\frac{i-1}{2^k},\frac{j+1}{2^k}}(\gamma_{\frac{i-1}{2^k}},\gamma_{\frac{j+1}{2^k}})$. So there has to be equality everywhere in (7.20), which leads to (7.19). (Here I use the fact that $c^{s,t}(\gamma_s,\gamma_t) < +\infty$.) After that it is an easy game to conclude the proof of the minimizing property for arbitrary times t_1, t_2, t_3.

Step 4. To recapitulate: Starting from a curve $(\mu_t)_{0\le t\le 1}$, we have constructed a family of probability measures $\Pi^{(k)}$ which are all concentrated on the set Γ of minimizing curves, and satisfy $(e_t)_\# \Pi^{(k)} = \mu_t$ for all $t = j/2^k$. It remains to pass to the limit as $k \to \infty$. For that we shall check the tightness of the sequence $(\Pi^{(k)})_{k\in\mathbb{N}}$. Let $\varepsilon > 0$ be arbitrary. Since μ_0, μ_1 are tight, there are compact sets K_0, K_1 such that $\mu_0[\mathcal{X} \setminus K_0] \le \varepsilon$, $\mu_1[\mathcal{X} \setminus K_1] \le \varepsilon$. From the coercivity of the action, the set $\Gamma^{0,1}_{K_0\to K_1}$ of action-minimizing curves joining K_0 to K_1 is compact, and $\Pi\big[\Gamma \setminus \Gamma^{0,1}_{K_0\to K_1}\big]$ is (with obvious notation)

$$\begin{aligned}\mathbb{P}\big[(\gamma_0,\gamma_1) \notin K_0 \times K_1\big] &\le \mathbb{P}\left[\gamma_0 \notin K_0\right] + \mathbb{P}\left[\gamma_1 \notin K_1\right] \\ &= \mu_0[\mathcal{X}\setminus K_0] + \mu_1[\mathcal{X}\setminus K_1] \le 2\varepsilon.\end{aligned}$$

This proves the tightness of the family $(\Pi^{(k)})$. So one can extract a subsequence thereof, still denoted $\Pi^{(k)}$, that converges weakly to some probability measure Π.

By Proposition 7.16(v), Γ is closed; so Π is still supported in Γ. Moreover, for all dyadic time $t = i/2^\ell$ in $[0,1]$, we have, if k is larger than ℓ, $(e_t)_\#\Pi^{(k)} = \mu_t$, and by passing to the limit we find that $(e_t)_\#\Pi = \mu_t$ also.

By assumption, μ_t depends continuously on t. So, to conclude that $(e_t)_\#\Pi = \mu_t$ for all times $t \in [0,1]$ it now suffices to check the continuity of $(e_t)_\#\Pi$ as a function of t. In other words, if φ is an arbitrary bounded continuous function on $\mathcal{X}$, one has to show that

$$\psi(t) = \mathbb{E}\,\varphi(\gamma_t)$$

is a continuous function of t if γ is a random geodesic with law Π. But this is a simple consequence of the continuity of $t \longmapsto \gamma_t$ (for all γ), and Lebesgue's dominated convergence theorem. This concludes Step 4, and the proof of (ii) $\Rightarrow$ (i).

Next, let us check that the two expressions for $\mathcal{A}^{s,t}$ in (iii) do coincide. This is about the same computation as in Step 1 above. Let $s < t$ be given, let $(\mu_\tau)_{s\leq\tau\leq t}$ be a continuous path, and let (t_i) be a subdivision of $[s,t]$. Further, let γ be such that law $(\gamma_\tau) = \mu_\tau$, and let (X_s, X_t) be an optimal coupling of (μ_s, μ_t), for the cost function $c^{s,t}$. Further, let $(\gamma_\tau)_{s\leq\tau\leq t}$ be a random continuous path, such that law $(\gamma_\tau) = \mu_\tau$ for all $\tau \in [s,t]$. Then

$$\sum_i C^{t_i,t_{i+1}}(\mu_{t_i}, \mu_{t_{i+1}}) \leq C^{s,t}(\mu_s, \mu_t) = \mathbb{E}\, c^{s,t}(X_s, X_t)$$
$$\leq \mathbb{E}\, c^{s,t}(\gamma_s, \gamma_t) \leq \mathbb{E}\, \mathcal{A}^{s,t}(\gamma),$$

where the next-to-last inequality follows from the fact that (γ_s, γ_t) is a coupling of (μ_s, μ_t), and the last inequality is a consequence of the definition of $c^{s,t}$. This shows that

$$\sum_i C^{t_i,t_{i+1}}(\mu_{t_i}, \mu_{t_{i+1}}) \leq \mathbb{E}\, \mathcal{A}^{s,t}(\gamma).$$

On the other hand, there is equality in the whole chain of inequalities if $t_0 = s$, $t_1 = t$, $X_s = \gamma_s$, $X_t = \gamma_t$, and γ_τ is a (random) action-minimizing curve. So the two expressions in (iii) do coincide.

Now let us address the equivalence between (ii) and (iii). First, it is clear that $\mathbb{A}^{s,t}$ is lower semicontinuous, since it is defined as a supremum of lower semicontinuous functionals. The inequality $\mathbb{A}^{t_1,t_3} \geq \mathbb{A}^{t_1,t_2} + \mathbb{A}^{t_2,t_3}$ holds true for all intermediate times $t_1 < t_2 < t_3$ (this is a simple consequence of the definitions), and the converse inequality is a consequence of the general inequality

$$s < t_2 < t \Longrightarrow \quad C^{s,t_2}(\mu_s, \mu_{t_2}) \leq C^{s,t_2}(\mu_s, \mu_{t_2}) + C^{t_2,t}(\mu_{t_2}, \mu_t),$$

which we proved in Step 1 above. So Property (i) in Definition 7.11 is satisfied. To check Property (ii) of that definition, take any two probability measures μ_s, μ_t and introduce a displacement interpolation $(\mu_\tau)_{s\leq\tau\leq t}$ for the Lagrangian action restricted to $[s,t]$. Then Property (ii) of Theorem 7.21 implies $\mathbb{A}^{s,t}(\mu) = C^{s,t}(\mu_s, \mu_t)$. Finally, Property (iii) in Definition 7.11 is also satisfied by construction. In the end, $(\mathbb{A})$ does define a Lagrangian action, with induced cost functionals $C^{s,t}$.

To conclude the proof of Theorem 7.21 it only remains to check the coercivity of the action; then the equivalence of (i), (ii) and (iii) will follow from Proposition 7.16(iv). Let $s < t$ be two given times in $[0,1]$, and let $\mathcal{K}_s, \mathcal{K}_t$ be compact sets of probability measures such that $C^{s,t}(\mu_s, \mu_t) < +\infty$ for all $\mu_s \in \mathcal{K}_s$, $\mu_t \in \mathcal{K}_t$. Action-minimizing curves for $\mathbb{A}^{s,t}$ can be written as law $(\gamma_\tau)_{s \le \tau \le t}$, where γ is a random action-minimizing curve $[s,t] \to \mathcal{X}$ such that law $(\gamma_s) \in \mathcal{K}_s$, law $(\gamma_t) \in \mathcal{K}_t$. One can use an argument similar to the one in Step 4 above to prove that the laws Π of such minimizing curves form a tight, closed set; so we have a compact set of dynamical transference plans $\Pi^{s,t}$, that are probability measures on $C([s,t];\mathcal{X})$. The problem is to show that the paths $(e_\tau)_\# \Pi^{s,t}$ constitute a compact set in $C([s,t];P(\mathcal{X}))$. Since the continuous image of a compact set is compact, it suffices to check that the map

$$\Pi^{s,t} \longmapsto \big((e_\tau)_\# \Pi^{s,t}\big)_{s \le \tau \le t}$$

is continuous from $P(C([s,t];\mathcal{X}))$ to $C([s,t];P(\mathcal{X}))$. To do so, it will be convenient to metrize $P(\mathcal{X})$ with the Wasserstein distance W_1, replacing if necessary d by a bounded, topologically equivalent distance. (Recall Corollary 6.13.) Then the uniform distance on $C([s,t];\mathcal{X})$ is also bounded and there is an associated Wasserstein distance $\mathcal{W}_1$ on $P(C([s,t];\mathcal{X}))$. Let Π and $\widetilde{\Pi}$ be two dynamical optimal transference plans, and let $((\gamma_\tau), (\widetilde{\gamma}_\tau))$ be an optimal coupling of Π and $\widetilde{\Pi}$; let also μ_τ, $\widetilde{\mu}_\tau$ be the associated displacement interpolations; then the required continuity follows from the chain of inequalities

$$\sup_{t \in [0,1]} W_1(\mu_t, \widetilde{\mu}_t) \le \sup_{t \in [0,1]} \mathbb{E}\, d(\gamma_t, \widetilde{\gamma}_t) \le \mathbb{E} \sup_{t \in [0,1]} d(\gamma_t, \widetilde{\gamma}_t) = \mathcal{W}_1(\Pi, \widetilde{\Pi}).$$

This proves that displacement interpolations with endpoints lying in given compact sets themselves form a compact set, and concludes the proof of the coercivity of the action ($\mathbb{A}$). □

Remark 7.28. In the proof of the implication (ii) ⇒ (i), instead of defining $\Pi^{(k)}$ on the space of continuous curves, one could instead work with $\Pi^{(k)}$ defined on discrete times, construct by compactness a consistent system of marginals on $\mathcal{X}^{2^\ell+1}$, for all ℓ, and then invoke Kolmogorov's existence theorem to get a Π which is defined on a set of curves. Things however are not that simple, since Kolmogorov's theorem constructs a random measurable curve which is not a priori continuous. Here one has the same conceptual catch as in the construction of Brownian motion as a probability measure on continuous paths.

Proof of Corollary 7.22. Introduce the family of actions

$$\mathcal{A}^{s,t}(\gamma) = \int_s^t |\dot{\gamma}_\tau|^p \, d\tau.$$

Then

$$c^{s,t}(x,y) = \frac{d(x,y)^p}{(t-s)^{p-1}},$$

and all our assumptions hold true for this action and cost. (The assumption of local compactness is used to prove that the action is coercive, see the Appendix.) The important point now is that

$$C^{s,t}(\mu,\nu) = \frac{W_p(\mu,\nu)^p}{(t-s)^{p-1}}.$$

So, according to the remarks in Example 7.9, Property (ii) in Theorem 7.21 means that (μ_t) is in turn a minimizer of the action associated with the Lagrangian $|\dot{\mu}|^p$, i.e. a geodesic in $P_p(\mathcal{X})$. Note that if μ_t is the law of a random optimal geodesic γ_t at time t, then

$$W_p(\mu_t,\mu_s)^p \le \mathbb{E}\, d(\gamma_s,\gamma_t)^p = \mathbb{E}\,(t-s)^p\, d(\gamma_0,\gamma_1)^p = (t-s)^p\, W_p(\mu_0,\mu_1)^p,$$

so the path μ_t is indeed continuous (and actually 1-Lipschitz) for the distance W_p. □

Proof of Corollary 7.23. By Theorem 7.21, any displacement interpolation has to take the form $(e_t)_\# \Pi$, where Π is a probability measure on action-minimizing curves such that $\pi := (e_0,e_1)_\# \Pi$ is an optimal transference plan. By assumption, there is exactly one such π. Let Z be the set of pairs (x_0,x_1) such that there is more than one minimizing curve joining x_0 to x_1; by assumption $\pi[Z]=0$. For $(x_0,x_1)\notin Z$, there is a unique geodesic $\gamma = S(x_0,x_1)$ joining x_0 to x_1. So Π has to coincide with $S_\#\pi$. □

To conclude this section with a simple application, I shall use Corollary 7.22 to derive the Lipschitz continuity of moments, alluded to in Remark 6.10.

Proposition 7.29 (W_p-Lipschitz continuity of p-moments). *Let $(\mathcal{X},d)$ be a locally compact Polish length space, let $p \ge 1$ and $\mu,\nu \in P_p(\mathcal{X})$. Then for any $\varphi \in \mathrm{Lip}(\mathcal{X};\mathbb{R}_+)$,*

$$\left| \left(\int \varphi(x)^p \, \mu(dx) \right)^{\frac{1}{p}} - \left(\int \varphi(y)^p \, \nu(dy) \right)^{\frac{1}{p}} \right| \le \|\varphi\|_{\mathrm{Lip}}\, W_p(\mu,\nu).$$

Proof of Proposition 7.29. Without loss of generality, $\|\varphi\|_{\mathrm{Lip}} = 1$. The case $p = 1$ is obvious from (6.3) and holds true for any Polish space, so I shall assume $p > 1$. Let $(\mu_t)_{0\le t\le 1}$ be a displacement interpolation between $\mu_0 = \mu$ and $\mu_1 = \nu$, for the cost function $c(x,y) = d(x,y)^p$. By Corollary 7.22, there is a probability measure Π on Γ, the set of geodesics in $\mathcal{X}$, such that $\mu_t = (e_t)_\#\Pi$.

Then let $\Psi(t) = \int_{\mathcal{X}} \varphi(x)^p\,\mu_t(dx) = \int_\Gamma \varphi(\gamma_t)^p\,\Pi(d\gamma)$. By Fatou's lemma and Hölder's inequality,

$$\begin{aligned}
\frac{d^+\Psi}{dt} &\le \int_\Gamma \frac{d^+\varphi(\gamma_t)^p}{dt}\,\Pi(d\gamma)\\
&\le p\int \varphi(\gamma_t)^{p-1}\,|\dot\gamma_t|\,\Pi(d\gamma)\\
&\le p\left(\int \varphi(\gamma_t)^p\,\Pi(d\gamma)\right)^{1-\frac1p}\left(\int |\dot\gamma_t|^p\,\Pi(d\gamma)\right)^{\frac1p}\\
&= p\,\Psi(t)^{1-\frac1p}\left(\int d(\gamma_0,\gamma_1)^p\,\Pi(d\gamma)\right)^{\frac1p}\\
&= p\,\Psi(t)^{1-\frac1p}\,W_p(\mu,\nu).
\end{aligned}$$

So $(d^+/dt)[\Psi(t)^{1/p}] \le W_p(\mu,\nu)$, thus $|\Psi(1)^{1/p} - \Psi(0)^{1/p}| \le W_p(\mu,\nu)$, which is the desired result. □

Displacement interpolation between intermediate times and restriction

Again let μ_0 and μ_1 be any two probability measures, $(\mu_t)_{0\le t\le 1}$ a displacement interpolation associated with a dynamical optimal transference plan Π, and $(\gamma_t)_{0\le t\le 1}$ a random action-minimizing curve with law $(\gamma) = \Pi$. In particular (γ_0,γ_1) is an optimal coupling of (μ_0,μ_1). With the help of the previous arguments, it is almost obvious that $(\gamma_{t_0},\gamma_{t_1})$ is also an optimal coupling of (μ_{t_0},μ_{t_1}). What may look at first sight more surprising is that if $(t_0,t_1) \ne (0,1)$, this is the *only* optimal coupling, at least if action-minimizing curves "cannot branch". Furthermore, there is a time-dependent version of Theorem 4.6.

Theorem 7.30 (Interpolation from intermediate times and restriction). *Let $\mathcal{X}$ be a Polish space equipped with a coercive action $(\mathcal{A})$ on $C([0,1];\mathcal{X})$. Let $\Pi \in P(C([0,1];\mathcal{X}))$ be a dynamical optimal* *transport plan associated with a finite total cost. For any* t_0, t_1 *in* $[0,1]$ *with* $0 \le t_0 < t_1 \le 1$*, define the time-restriction of* Π *to* $[t_0, t_1]$ *as* $\Pi^{t_0,t_1} := (r_{t_0,t_1})_\# \Pi$*, where* $r_{t_0,t_1}(\gamma)$ *is the restriction of* γ *to the interval* $[t_0, t_1]$*. Then:*

(i) Π^{t_0,t_1} *is a dynamical optimal coupling for the action* $(\mathcal{A})^{t_0,t_1}$.

(ii) If $\widetilde{\Pi}$ *is a measure on* $C([t_0,t_1];\mathcal{X})$*, such that* $\widetilde{\Pi} \le \Pi^{t_0,t_1}$ *and* $\widetilde{\Pi}[C([t_0,t_1];\mathcal{X})] > 0$*, let*

$$\Pi' := \frac{\widetilde{\Pi}}{\widetilde{\Pi}\big[C([t_0,t_1];\mathcal{X})\big]}, \qquad \mu'_t = (e_t)_\# \Pi'.$$

Then Π' *is a dynamical optimal coupling between* μ'_{t_0} *and* μ'_{t_1}*; and* $(\mu'_t)_{t_0 \le t \le t_1}$ *is a displacement interpolation.*

(iii) Further, assume that action-minimizing curves are uniquely and measurably determined by their restriction to a nontrivial time-interval, and $(t_0,t_1) \ne (0,1)$*. Then,* Π' *in (ii) is the unique dynamical optimal coupling between* μ'_{t_0} *and* μ'_{t_1}*. In particular,* $(\pi')^{t_0,t_1} := (e_{t_0}, e_{t_1})_\# \Pi'$ *is the unique optimal transference plan between* μ'_{t_0} *and* μ'_{t_1}*; and* $\mu'_t := (e_t)_\# \Pi'$ $(t_0 \le t \le t_1)$ *is the unique displacement interpolation between* μ'_{t_0} *and* μ'_{t_1}.

(iv) Under the same assumptions as in (iii), for any $t \in (0,1)$*,* $(\Pi \otimes \Pi)(d\gamma\, d\widetilde{\gamma})$*-almost surely,*

$$[\gamma_t = \widetilde{\gamma}_t] \Longrightarrow [\gamma = \widetilde{\gamma}].$$

In other words, the curves seen by Π *cannot cross at intermediate times. If the costs* $c^{s,t}$ *are continuous, this conclusion extends to all curves* $\gamma, \widetilde{\gamma} \in \operatorname{Spt} \Pi$.

(v) Under the same assumptions as in (iii), there is a measurable map $F_t : \mathcal{X} \to \Gamma(\mathcal{X})$ *such that,* $\Pi(d\gamma)$*-almost surely,* $F_t(\gamma_t) = \gamma$.

Remark 7.31. In Chapter 8 we shall work in the setting of smooth Riemannian manifolds and derive a quantitative variant of the "no-crossing" property expressed in (iv).

Corollary 7.32 (Nonbranching is inherited by the Wasserstein space). *Let* $(\mathcal{X}, d)$ *be a complete separable, locally compact length space and let* $p \in (1,\infty)$*. Assume that* $\mathcal{X}$ *is nonbranching, in the sense that a geodesic* $\gamma : [0,1] \to \mathcal{X}$ *is uniquely determined by its restriction to a nontrivial time-interval. Then also the Wasserstein space* $P_p(\mathcal{X})$ *is nonbranching. Conversely, if* $P_p(\mathcal{X})$ *is nonbranching, then* $\mathcal{X}$ *is nonbranching.*

Proof of Theorem 7.30. Let γ be a random geodesic with law Π. Let γ^{0,t_0}, γ^{t_0,t_1} and $\gamma^{t_1,1}$ stand for the restrictions of γ to the time intervals $[0,t_0]$, $[t_0,t_1]$ and $[t_1,1]$, respectively. Then

$$\begin{aligned} C^{0,t_0}(\mu_0,\mu_{t_0}) + C^{t_0,t_1}(\mu_{t_0},\mu_{t_1}) + C^{t_1,1}(\mu_{t_1},\mu_1) \\ \le \mathbb{E}\, c^{0,t_0}(\gamma^{0,t_0}) + \mathbb{E}\, c^{t_0,t_1}(\gamma^{t_0,t_1}) + \mathbb{E}\, c^{t_1,1}(\gamma^{t_1,1}) \\ = \mathbb{E}\, c^{0,1}(\gamma) = C^{0,1}(\mu_0,\mu_1) \\ \le C^{0,t_0}(\mu_0,\mu_{t_0}) + C^{t_0,t_1}(\mu_{t_0},\mu_{t_1}) + C^{t_1,1}(\mu_{t_1},\mu_1). \end{aligned}$$

So there has to be equality in all the inequalities, and it follows that

$$\mathbb{E}\, c^{t_0,t_1}(\gamma^{t_0},\gamma^{t_1}) = C^{t_0,t_1}(\mu_{t_0},\mu_{t_1}).$$

So γ^{t_0,t_1} is optimal, and Π^{t_0,t_1} is a dynamical optimal transference plan. Statement (i) is proven.

As a corollary of (i), $\pi^{t_0,t_1} = (e_{t_0},e_{t_1})_{\#}\Pi^{t_0,t_1}$ is an optimal transference plan between μ_{t_0} and μ_{t_1}. Let $\widetilde{\pi} := (e_{t_0},e_{t_1})_{\#}\widetilde{\Pi}$. The inequality $\widetilde{\Pi} \le \Pi^{t_0,t_1}$ is preserved by push-forward, so $\widetilde{\pi} \le \pi^{t_0,t_1}$. Also $\widetilde{\pi}[\mathcal{X}\times\mathcal{X}] = \widetilde{\Pi}[C([t_0,t_1];\mathcal{X})] > 0$. By Theorem 4.6, $\pi' := \widetilde{\pi}/\widetilde{\pi}[\mathcal{X}\times\mathcal{X}]$ is an optimal transference plan between its marginals. But π' coincides with $(e_{t_0},e_{t_1})_{\#}\Pi'$, and since Π' is concentrated (just as Π) on action-minimizing curves, Theorem 7.21 guarantees that Π' is a dynamical optimal transference plan between its marginals. This proves (ii). (The continuity of μ'_t in t is shown by the same argument as in Step 4 of the proof of Theorem 7.21.)

To prove (iii), assume, without loss of generality, that $t_0 > 0$; then an action-minimizing curve γ is uniquely and measurably determined by its restriction γ^{0,t_0} to $[0,t_0]$. In other words, there is a measurable function $F^{0,t_0} : \Gamma^{0,t_0} \to \Gamma$, defined on the set of all γ^{0,t_0}, such that any action-minimizing curve $\gamma : [0,1] \to \mathcal{X}$ can be written as $F^{0,t_0}(\gamma^{0,t_0})$. Similarly, there is also a measurable function F^{t_0,t_1} such that $F^{t_0,t_1}(\gamma^{t_0,t_1}) = \gamma$.

By construction, $\widetilde{\Pi}$ is concentrated on the curves γ^{t_0,t_1}, which are the restrictions to $[t_0,t_1]$ of the action-minimizing curves γ. Then let $\overline{\Pi} := (F^{t_0,t_1})_{\#}\widetilde{\Pi}$; this is a probability measure on $C([0,1];\mathcal{X})$. Of course $\overline{\Pi} \le (F^{t_0,t_1})_{\#}\Pi^{t_0,t_1} = \Pi$; so by (ii), $\overline{\Pi}/\overline{\Pi}[C([0,1];\mathcal{X})]$ is optimal. (In words, $\overline{\Pi}$ is obtained from Π by extending to the time-interval $[0,1]$ those curves which appear in the sub-plan Π'.) Then it is easily seen that $\Pi' = (r_{t_0,t_1})_{\#}\overline{\Pi}$, and $\overline{\Pi}[C([0,1];\mathcal{X})] = \Pi'[C([t_0,t_1];\mathcal{X})]$. So it suffices to prove Theorem 7.30(iii) in the case when $\widetilde{\Pi} = \Pi^{t_0,t_1}$, and this will be assumed in the sequel.

Now let γ be a random geodesic with law Π, and $\Pi^{0,t_0} = \mathrm{law}\,(\gamma^{0,t_0})$, $\Pi^{t_0,t_1} = \mathrm{law}\,(\gamma^{t_0,t_1})$, $\Pi^{t_1,1} = \mathrm{law}\,(\gamma^{t_1,1})$. By (i), Π^{t_0,t_1} is a dynamical optimal transference plan between μ_{t_0} and μ_{t_1}; let $\widetilde{\Pi}^{t_0,t_1}$ be another such plan. The goal is to show that $\widetilde{\Pi}^{t_0,t_1} = \Pi^{t_0,t_1}$.

Disintegrate Π^{0,t_0} and $\widetilde{\Pi}^{t_0,t_1}$ along their common marginal μ_{t_0} and glue them together. This gives a probability measure on $C([0,t_0);\mathcal{X}) \times \mathcal{X} \times C((t_0,t_1];\mathcal{X})$, supported on triples $(\gamma, g, \widetilde{\gamma})$ such that $\gamma(t) \to g$ as $t \to t_0^-$, $\widetilde{\gamma}(t) \to g$ as $t \to t_0^+$. Such triples can be identified with continuous functions on $[0,t_1]$, so what we have is in fact a probability measure on $C([0,t_1];\mathcal{X})$. Repeat the operation by gluing this with $\Pi^{t_1,1}$, so as to get a probability measure $\overline{\Pi}$ on $C([0,1];\mathcal{X})$.

Then let $\overline{\gamma}$ be a random variable with law $\overline{\Pi}$: By construction, $\mathrm{law}\,(\overline{\gamma}^{0,t_0}) = \mathrm{law}\,(\gamma^{0,t_0})$, and $\mathrm{law}\,(\overline{\gamma}^{t_1,1}) = \mathrm{law}\,(\gamma^{t_1,1})$, so

$$\begin{aligned}
\mathbb{E}\, c^{0,1}(\overline{\gamma}) &\le \mathbb{E}\, c^{0,t_0}(\overline{\gamma}^{0,t_0}) + \mathbb{E}\, c^{t_0,t_1}(\overline{\gamma}^{t_0,t_1}) + \mathbb{E}\, c^{t_1,1}(\overline{\gamma}^{t_1,1}) \\
&= \mathbb{E}\, c^{0,t_0}(\gamma^{0,t_0}) + \mathbb{E}\, c^{t_0,t_1}(\overline{\gamma}^{t_0,t_1}) + \mathbb{E}\, c^{t_1,1}(\gamma^{t_1,1}) \\
&= C^{0,t_0}(\mu_0,\mu_{t_0}) + C^{t_0,t_1}(\mu_{t_0},\mu_{t_1}) + C^{t_1,1}(\mu_{t_1},\mu_1) \\
&= C^{0,1}(\mu_0,\mu_1).
\end{aligned}$$

Thus $\overline{\Pi}$ is a dynamical optimal transference plan between μ_0 and μ_1. It follows from Theorem 7.21 that there is a random action-minimizing curve $\widehat{\gamma}$ with $\mathrm{law}\,(\widehat{\gamma}) = \overline{\Pi}$. In particular,

$$\mathrm{law}\,(\widehat{\gamma}^{0,t_0}) = \mathrm{law}\,(\gamma^{0,t_0}); \qquad \mathrm{law}\,(\widehat{\gamma}^{t_0,t_1}) = \mathrm{law}\,(\widetilde{\gamma}^{t_0,t_1}).$$

By assumption there is a measurable function F ($F = r_{t_0,t_1} \circ F^{0,t_0}$) such that $g^{t_0,t_1} = F(g^{0,t_0})$, for any action-minimizing curve g. So

$$\mathrm{law}\,(\widetilde{\gamma}^{t_0,t_1}) = \mathrm{law}\,(\widehat{\gamma}^{t_0,t_1}) = \mathrm{law}\,(F(\widehat{\gamma}^{0,t_0})) = \mathrm{law}\,(F(\gamma^{0,t_0})) = \mathrm{law}\,(\gamma^{t_0,t_1}).$$

This proves the uniqueness of the dynamical optimal transference plan joining μ'_{t_0} to μ'_{t_1}. The remaining part of (iii) is obvious since any optimal plan or displacement interpolation has to come from a dynamical optimal transference plan, according to Theorem 7.21.

Now let us turn to the proof of (iv). Since the plan $\pi = (e_0, e_1)_\# \Pi$ is $c^{0,1}$-cyclically monotone (Theorem 5.10(ii)), we have, $\Pi \otimes \Pi(d\gamma\, d\widetilde{\gamma})$-almost surely,

$$c^{0,1}(\gamma_0,\gamma_1) + c^{0,1}(\widetilde{\gamma}_0,\widetilde{\gamma}_1) \le c^{0,1}(\gamma_0,\widetilde{\gamma}_1) + c^{0,1}(\widetilde{\gamma}_0,\gamma_1), \tag{7.21}$$

and all these quantities are finite (almost surely).

If γ and $\widetilde{\gamma}$ are two such paths, assume that $\gamma_t = \widetilde{\gamma}_t = X$ for some $t \in (0,1)$. Then

$$c^{0,1}(\gamma_0,\widetilde{\gamma}_1) \le c^{0,t}(\gamma_0,X) + c^{t,1}(X,\widetilde{\gamma}_1), \tag{7.22}$$

and similarly

$$c^{0,1}(\widetilde{\gamma}_0,\gamma_1) \le c^{0,t}(\widetilde{\gamma}_0,X) + c^{t,1}(X,\gamma_1). \tag{7.23}$$

By adding up (7.22) and (7.23), we get

$$\begin{aligned} c^{0,1}(\gamma_0,\widetilde{\gamma}_1) &+ c^{0,1}(\widetilde{\gamma}_0,\gamma_1) \\ &\le \big[c^{0,t}(\gamma_0,X) + c^{t,1}(X,\gamma_1)\big] + \big[c^{0,t}(\widetilde{\gamma}_0,X) + c^{t,1}(X,\widetilde{\gamma}_1)\big] \\ &= c^{0,1}(\gamma_0,\gamma_1) + c^{0,1}(\widetilde{\gamma}_0,\widetilde{\gamma}_1). \end{aligned}$$

Since the reverse inequality holds true by (7.21), equality has to hold in all intermediate inequalities, for instance in (7.22). Then it is easy to see that the path $\overline{\gamma}$ defined by $\overline{\gamma}(s) = \gamma(s)$ for $0 \le s \le t$, and $\overline{\gamma}(s) = \widetilde{\gamma}(s)$ for $s \ge t$, is a minimizing curve. Since it coincides with γ on a nontrivial time-interval, it has to coincide with γ everywhere, and similarly it has to coincide with $\widetilde{\gamma}$ everywhere. So $\gamma = \widetilde{\gamma}$.

If the costs $c^{s,t}$ are continuous, the previous conclusion holds true not only $\Pi \otimes \Pi$-almost surely, but actually for any two minimizing curves γ, $\widetilde{\gamma}$ lying in the support of Π. Indeed, inequality (7.21) defines a closed set $\mathcal{C}$ in $\Gamma \times \Gamma$, where Γ stands for the set of minimizing curves; so $\operatorname{Spt}\Pi \times \operatorname{Spt}\Pi = \operatorname{Spt}(\Pi \otimes \Pi) \subset \mathcal{C}$.

It remains to prove (v). Let $\Gamma^{0,1}$ be a c-cyclically monotone subset of $\mathcal{X} \times \mathcal{X}$ on which π is concentrated, and let Γ be the set of minimizing curves $\gamma : [0,1] \to \mathcal{X}$ such that $(\gamma_0,\gamma_1) \in \Gamma^{0,1}$. Let $(K_\ell)_{\ell\in\mathbb{N}}$ be a nondecreasing sequence of compact sets contained in Γ, such that $\Pi[\cup K_\ell] = \Pi[\Gamma] = 1$. For each ℓ, we define F_ℓ on $e_t(K_\ell)$ by $F_\ell(\gamma_t) = \gamma$. This map is continuous: Indeed, if $x_k \in e_t(K_\ell)$ converges to x, then for each k we have $x_k = (\gamma_k)_t$ for some $\gamma_k \in K_\ell$, and up to extraction γ_k converges to $\gamma \in K_\ell$, and in particular $(\gamma_k)_t$ converges to γ_t; but then $F_\ell(\gamma_t) = \gamma$. Then we can define F on $\cup K_\ell$ as a map which coincides with F_ℓ on each K_ℓ. (Obviously, this is the same line of reasoning as in Theorem 5.30.) □

Proof of Corollary 7.32. Assume that $\mathcal{X}$ is nonbranching. Then there exists some function F, defined on the set of all curves γ^{0,t_0}, where γ^{0,t_0} is the restriction to $[0,t_0]$ of the geodesic $\gamma : [0,1] \to \mathcal{X}$, such that $\gamma = F(\gamma^{0,t_0})$.

I claim that F is automatically continuous. Indeed, let $(\gamma_n)_{n\in\mathbb{N}}$ be such that γ_n^{0,t_0} converges uniformly on $[0,t_0]$ to some $g : [0,t_0] \to \mathcal{X}$. Since the functions γ_n^{0,t_0} are uniformly bounded, the speeds of all the geodesics γ_n are uniformly bounded too, and the images $\gamma_n([0,1])$ are all included in a compact subset of $\mathcal{X}$. It follows from Ascoli's theorem that the sequence (γ_n) converges uniformly, up to extraction of a subsequence. But then its limit γ has to be a geodesic, and its restriction to $[0,t_0]$ coincides with g. There is at most one such geodesic, so γ is uniquely determined, and the whole sequence γ_n converges. This implies the continuity of F.

Then Theorem 7.30(iii) applies: If $(\mu_t)_{0\le t\le 1}$ is a geodesic in $P_p(\mathcal{X})$, there is only one geodesic between μ_{t_0} and μ_1. So $(\mu_t)_{0\le t\le 1}$ is uniquely determined by its restriction to $[0,t_0]$. The same reasoning could be done for any nontrivial time-interval instead of $[0,t_0]$; so $P_p(\mathcal{X})$ is indeed nonbranching.

The converse implication is obvious, since any geodesic γ in $\mathcal{X}$ induces a geodesic in $P_p(\mathcal{X})$, namely $(\delta_{\gamma(t)})_{0\le t\le 1}$. □

Interpolation of prices

When the path μ_t varies in time, what becomes of the pair of "prices" (ψ,ϕ) in the Kantorovich duality? The short answer is that these functions will also evolve continuously in time, according to **Hamilton–Jacobi equations**.

Definition 7.33 (Hamilton-Jacobi-Hopf-Lax-Oleinik evolution semigroup). *Let $\mathcal{X}$ be a metric space and $(\mathcal{A})^{0,1}$ a coercive Lagrangian action on $\mathcal{X}$, with cost functions $(c^{s,t})_{0\le s<t\le 1}$. For any two functions $\psi : \mathcal{X} \to \mathbb{R}\cup\{+\infty\}$, $\phi : \mathcal{X} \to \mathbb{R}\cup\{-\infty\}$, and any two times $0 \le s < t \le 1$, define*

$$\begin{cases} H_+^{s,t}\psi\,(y) = \inf_{x\in\mathcal{X}} \Big(\psi(x) + c^{s,t}(x,y)\Big); \\[2ex] H_-^{t,s}\phi\,(x) = \sup_{y\in\mathcal{X}} \Big(\phi(y) - c^{s,t}(x,y)\Big). \end{cases}$$

The family of operators $(H_+^{s,t})_{t>s}$ *(resp.* $(H_-^{s,t})_{s<t}$*) is called the forward (resp. backward) Hamilton–Jacobi (or Hopf–Lax, or Lax–Oleinik) semigroup.*

Roughly speaking, $H_+^{s,t}$ gives the values of ψ at time t, from its values at time s; while $H_-^{s,t}$ does the reverse. So the semigroups H_- and H_+ are in some sense inverses of each other. Yet it is not true in general that $H_-^{t,s}H_+^{s,t} = \mathrm{Id}$. Proposition 7.34 below summarizes some of the main properties of these semigroups; the denomination of "semigroup" itself is justified by Property (ii).

Proposition 7.34 (Elementary properties of Hamilton–Jacobi semigroups). *With the notation of Definition 7.33,*

(i) $H_+^{s,t}$ *and* $H_-^{s,t}$ *are order-preserving:* $\psi \le \overline{\psi} \Longrightarrow H_\pm^{s,t}\psi \le H_\pm^{s,t}\overline{\psi}$.

(ii) Whenever $t_1 < t_2 < t_3$ *are three intermediate times in* $[0,1]$,

$$\begin{cases} H_+^{t_2,t_3} H_+^{t_1,t_2} = H_+^{t_1,t_3} \\[2ex] H_-^{t_2,t_1} H_-^{t_3,t_2} = H_-^{t_3,t_1}. \end{cases}$$

(iii) Whenever $s < t$ *are two times in* $[0,1]$,

$$H_-^{t,s} H_+^{s,t} \le \mathrm{Id}\,; \qquad H_+^{s,t} H_-^{t,s} \ge \mathrm{Id}\,.$$

Proof of Proposition 7.34. Properties (i) and (ii) are immediate consequences of the definitions and Proposition 7.16(iii). To check Property (iii), e.g. the first half of it, write

$$H_-^{t,s}(H_+^{s,t}\psi)(x) = \sup_y \inf_{x'} \Big(\psi(x') + c^{s,t}(x',y) - c^{s,t}(x,y)\Big).$$

The choice $x' = x$ shows that the infimum above is bounded above by $\psi(x)$, independently of y; so $H_-^{t,s}(H_+^{s,t}\psi)(x) \le \psi(x)$, as desired. □

The Hamilton–Jacobi semigroup is well-known and useful in geometry and dynamical systems theory. On a smooth Riemannian manifold, when the action is given by a Lagrangian $L(x,v,t)$, strictly convex and superlinear in the velocity variable, then $S_+(t,\,\cdot\,) := H_+^{0,t}\psi_0$ solves the differential equation

$$\frac{\partial S_+}{\partial t}(x,t) + H\big(x, \nabla S_+(x,t), t\big) = 0, \qquad (7.24)$$

where $H = L^*$ is obtained from L by Legendre transform in the v variable, and is called the **Hamiltonian** of the system. This equation provides a bridge between a Lagrangian description of action-minimizing curves, and an Eulerian description: From $S_+(x,t)$ one can reconstruct a velocity field $v(x,t) = \nabla_p H\big(x, \nabla S_+(x,t), t\big)$, in such a way that integral curves of the equation $\dot{x} = v(x,t)$ are minimizing curves. Well, rigorously speaking, that would be the case if S_+ were differentiable! But things are not so simple because S_+ is not in general differentiable everywhere, so the equation has to be interpreted in a suitable sense (called viscosity sense). It is important to note that if one uses the backward semigroup and defines $S_-(x,t) := H_-^{t,1}\psi_t$, then S_- formally satisfies the same equation as S_+, but the equation has to be interpreted with a different convention (backward viscosity). This will be illustrated by the next example.

Example 7.35. On a Riemannian manifold M, consider the simple Lagrangian cost $L(x,v,t) = |v|^2/2$; then the associated Hamiltonian is just $H(x,p,t) = |p|^2/2$. If S is a C^1 solution of $\partial S/\partial t + |\nabla S|^2/2 = 0$, then the gradient of S can be interpreted as the velocity field of a family of minimizing geodesics. But if S_0 is a given Lipschitz function and $S_+(t,x)$ is defined by the forward Hamilton–Jacobi semigroup starting from initial datum S_0, one only has (for all t, x)

$$\frac{\partial S_+}{\partial t} + \frac{|\nabla^- S_+|^2}{2} = 0,$$

where

$$|\nabla^- f|(x) := \limsup_{y \to x} \frac{[f(y) - f(x)]_-}{d(x,y)}, \qquad z_- = \max(-z, 0).$$

Conversely, if one uses the backward Hamilton–Jacobi semigroup to define a function $S_-(x,t)$, then

$$\frac{\partial S_-}{\partial t} + \frac{|\nabla^+ S_-|^2}{2} = 0, \qquad |\nabla^+ f|(x) := \limsup_{y \to x} \frac{[f(y) - f(x)]_+}{d(x,y)},$$

where now $z_+ = \max(z, 0)$. When the Lagrangian is more complicated, things may become much more intricate. The standard convention is to use the forward Hamilton–Jacobi semigroup by default.

We shall now see that the Hamilton–Jacobi semigroup provides a simple answer to the problem of interpolation in dual variables. In the next statement, $\mathcal{X}$ is again a Polish space, $(\mathcal{A})^{0,1}$ a coercive Lagrangian action on $\mathcal{X}$, with associated cost functions $c^{s,t}$; and $C^{s,t}$ stands for the optimal total cost in the transport problem with cost $c^{s,t}$.

Theorem 7.36 (Interpolation of prices). *With the same assumptions and notation as in Definition 7.33, let μ_0, μ_1 be two probability measures on $\mathcal{X}$, such that $C^{0,1}(\mu_0,\mu_1) < +\infty$, and let (ψ_0,ϕ_1) be a pair of $c^{0,1}$-conjugate functions such that any optimal plan $\pi_{0,1}$ between μ_0 and μ_1 has its support included in $\partial_{c^{0,1}}\psi_0$. (Recall Theorem 5.10; under adequate integrability conditions, the pair (ψ_0,ϕ_1) is just a solution of the dual Kantorovich problem.) Further, let $(\mu_t)_{0\le t\le 1}$ be a displacement interpolation between μ_0 and μ_1. Whenever $s < t$ are two intermediate times in $[0,1]$, define*

$$\psi_s := H_+^{0,s}\psi_0, \qquad \phi_t := H_-^{1,t}\phi_1.$$

Then (ψ_s,ϕ_t) is optimal in the dual Kantorovich problem associated to (μ_s,μ_t) and $c^{s,t}$. In particular,

$$C^{s,t}(\mu_s,\mu_t) = \int \phi_t\,d\mu_t - \int \psi_s\,d\mu_s,$$

and

$$\phi_t(y) - \psi_s(x) \le c^{s,t}(x,y),$$

with equality $\pi_{s,t}(dx\,dy)$-almost surely, where $\pi_{s,t}$ is any optimal transference plan between μ_s and μ_t.

Proof of Theorem 7.36. From the definitions,

$$\begin{aligned}\phi_t(y) - \psi_s(x) - c^{s,t}(x,y)&\\ = \sup_{y',x'}\Big[\phi_1(y') - c^{t,1}(y',y) &- \psi_0(x') - c^{0,s}(x',x) - c^{s,t}(x,y)\Big].\end{aligned}$$

Since $c^{0,s}(x',x) + c^{s,t}(x,y) + c^{t,1}(y,y') \ge c^{0,1}(x',y')$, it follows that

$$\phi_t(y) - \psi_s(x) - c^{s,t}(x,y) \le \sup_{y',x'}\Big[\phi_1(y') - \psi_0(x') - c^{0,1}(x',y')\Big] \le 0.$$

So $\phi_t(y) - \psi_s(x) \le c^{s,t}(x,y)$. This inequality does not depend on the fact that (ψ_0,ϕ_1) is a tight pair of prices, in the sense of (5.5), but only on the inequality $\phi_1 - \psi_0 \le c^{0,1}$.

Next, introduce a random action-minimizing curve γ such that $\mu_t = \text{law}\,(\gamma_t)$. Since (ψ_0, ϕ_1) is an optimal pair, we know from Theorem 5.10(ii) that, almost surely,

$$\phi_1(\gamma_1) - \psi_0(\gamma_0) = c^{0,1}(\gamma_0, \gamma_1).$$

From the identity $c^{0,1}(\gamma_0,\gamma_1) = c^{0,s}(\gamma_0,\gamma_s) + c^{s,t}(\gamma_s,\gamma_t) + c^{t,1}(\gamma_t,\gamma_1)$ and the definition of ψ_s and ϕ_t,

$$\begin{aligned} c^{s,t}(\gamma_s,\gamma_t) = \big[\phi_1(\gamma_1) - c^{t,1}(\gamma_t,\gamma_1)\big] - \big[\psi_0(\gamma_0) + c^{0,s}(\gamma_0,\gamma_s)\big] \\ \leq \phi_t(\gamma_t) - \psi_s(\gamma_s). \end{aligned}$$

This shows that actually $c^{s,t}(\gamma_s,\gamma_t) = \phi_t(\gamma_t) - \psi_s(\gamma_s)$ almost surely, so (ψ_s, ϕ_t) has to be optimal in the dual Kantorovich problem between $\mu_s = \text{law}\,(\gamma_s)$ and $\mu_t = \text{law}\,(\gamma_t)$. □

Remark 7.37. In the limit case $s \to t$, the above results become

$$\begin{cases} \phi_t \leq \psi_t \\ \phi_t = \psi_t \quad \mu_t\text{-almost surely} \end{cases}$$

... but it is not true in general that $\phi_t = \psi_t$ everywhere in $\mathcal{X}$.

Remark 7.38. However, the identity $\psi_1 = \phi_1$ holds true everywhere as a consequence of the definitions.

Exercise 7.39. After reading the rest of Part I, the reader can come back to this exercise and check his or her understanding by proving that, for a quadratic Lagrangian:

(i) The displacement interpolation between two balls in Euclidean space is always a ball, whose radius increases linearly in time (here I am identifying a set with the uniform probability measure on this set).

(ii) More generally, the displacement interpolation between two ellipsoids is always an ellipsoid.

(iii) But the displacement interpolation between two sets is in general not a set.

(iv) The displacement interpolation between two spherical caps on the sphere is in general not a spherical cap.

(v) The displacement interpolation between two antipodal spherical caps on the sphere is unique, while the displacement interpolation between two antipodal points can be realized in an infinite number of ways.

Appendix: Paths in metric structures

This Appendix is a kind of crash basic course in Riemannian geometry, and nonsmooth generalizations thereof. Much more detail can be obtained from the references cited in the bibliographical notes.

A (finite-dimensional, smooth) Riemannian manifold is a manifold M equipped with a **Riemannian metric** g: this means that g defines a scalar product on each tangent space T_xM, varying smoothly with x. So if v and w at tangent vectors at x, the notation $v \cdot w$ really means $g_x(v,w)$, where g_x is the metric at x. The degree of smoothness of g depends on the context, but it is customary to consider C^3 manifolds with a C^2 metric. For the purpose of this course, the reader might assume C^∞ smoothness.

Let $\gamma : [0,1] \to M$ be a smooth path,[1] denoted by $(\gamma_t)_{0\le t\le 1}$. For each $t \in (0,1)$, the time-derivative at time t is — by the very definition of tangent space — the tangent vector $v = \dot{\gamma}_t$ in $T_{\gamma_t}M$. The scalar product g gives a way to measure the norm of that vector: $|v|_{T_xM} = \sqrt{v \cdot v}$. Then one can define the **length** of γ by the formula

$$\mathcal{L}(\gamma) = \int_0^1 |\dot{\gamma}_t|\, dt, \tag{7.25}$$

and the **distance**, or geodesic distance, between two points x and y by the formula

$$d(x,y) = \inf\Big\{\mathcal{L}(\gamma); \quad \gamma_0 = x,\ \gamma_1 = y\Big\}. \tag{7.26}$$

After that it is easy to extend the length formula to absolutely continuous curves. Note that any one of the three objects (metric, length, distance) determines the other two; indeed, the metric can be recovered from the distance via the formula

$$|\dot{\gamma}_0| = \lim_{t\downarrow 0} \frac{d(\gamma_0,\gamma_t)}{t}, \tag{7.27}$$

and the usual polarization identity

$$g(v,w) = \frac{1}{4}\Big[g(v+w,v+w) - g(v-w,v-w)\Big].$$

Let TM stand for the collection of all T_xM, $x \in M$, equipped with a manifold structure which is locally product. A point in TM is a pair

[1] For me the words "path" and "curve" are synonymous.

(x, v) with $v \in T_xM$. The map $\pi : (x, v) \longmapsto x$ is the projection of TM onto M; it is obviously smooth and surjective. A function $M \to TM$ is called a vector field: It is given by a tangent vector at each point. So a vector field really is a map $f : x \to (x, v)$, but by abuse of notation one often writes $f(x) = v$. If $\gamma : [0, 1] \to M$ is an injective path, one defines a *vector field along* γ as a path $\xi : [0, 1] \to TM$ such that $\pi \circ \xi = \gamma$.

If $p = p(x)$ is a linear form varying smoothly on T_xM, then it can be identified, thanks to the metric g, to a vector field ξ, via the formula

$$p(x) \cdot v = \xi(x) \cdot v,$$

where $v \in T_xM$, and the dot in the left-hand side just means "$p(x)$ applied to v", while the dot in the right-hand side stands for the scalar product defined by g. As a particular case, if p is the differential of a function f, the corresponding vector field ξ is the *gradient* of f, denoted by ∇f or $\nabla_x f$.

If $f = f(x, v)$ is a function on TM, then one can differentiate it with respect to x or with respect to v. Since $T_{(x,v)}T_xM \simeq T_xM$, both $d_x f$ and $d_v f$ can be seen as linear forms on T_xM; this allows us to define $\nabla_x f$ and $\nabla_v f$, the "gradient with respect to the position variable" and the "gradient with respect to the velocity variable".

Differentiating functions does not pose any particular conceptual problem, but differentiating vector fields is quite a different story. If ξ is a vector field on M, then $\xi(x)$ and $\xi(y)$ live in different vector spaces, so it does not a priori make sense to compare them, unless one *identifies* in some sense T_xM and T_yM. (Of course, one could say that ξ is a map $M \to TM$ and define its differential as a map $TM \to T(TM)$ but this is of little use, because $T(TM)$ is "too large"; it is much better if we can come up with a reasonable notion of derivation which produces a map $TM \to TM$.)

There is in general no canonical way to identify T_xM and T_yM if $x \neq y$; but there is a canonical way to identify $T_{\gamma_0}M$ and $T_{\gamma_t}M$ as t varies continuously. This operation is called **parallel transport**, or Levi-Civita transport. A vector field which is transported in a parallel way along the curve γ will "look constant" to an observer who lives in M and travels along γ. If M is a surface embedded in $\mathbb{R}^3$, parallel transport can be described as follows: Start from the tangent plane at $\gamma(0)$, and then press your plane onto M along γ, in such a way that there is no friction (no slip, no rotation) between the plane and the surface.

With this notion it becomes possible to compute the derivative of a vector field *along a path*: If γ is a path and ξ is a vector field along γ, then the derivative of ξ is another vector field along γ, say $t \to \dot{\xi}(t)$, defined by

$$\dot{\xi}(t_0) = \frac{d}{dt}\bigg|_{t=t_0} \theta_{t\to t_0}(\xi(\gamma_t)),$$

where $\theta_{t\to t_0}$ is the parallel transport from $T_{\gamma_t}M$ to $T_{\gamma_{t_0}}M$ along γ. This makes sense because $\theta_{t\to t_0}\xi(\gamma_t)$ is an element of the fixed vector space $T_{\gamma_{t_0}}M$. The path $t \to \dot{\xi}(t)$ is a vector field along γ, called the **covariant derivative** of ξ along γ, and denoted by $\nabla_{\dot\gamma}\xi$, or, if there is no possible confusion about the choice of γ, $D\xi/Dt$ (or simply $d\xi/dt$). If $M = \mathbb{R}^n$, then $\nabla_{\dot\gamma}\xi$ coincides with $(\dot\gamma \cdot \nabla)\xi$.

It turns out that the value of $\dot{\xi}(t_0)$ only depends on γ_{t_0}, on the values of ξ in a neighborhood of γ_{t_0}, and on the velocity $\dot\gamma_{t_0}$ (not on the whole path γ_t). Thus if ξ is a vector field defined in the neighborhood of a point x, and v is a tangent vector at x, it makes sense to define $\nabla_v\xi$ by the formula

$$\nabla_v\xi(x) = \frac{D\xi}{Dt}(0), \qquad \gamma_0 = x, \quad \dot\gamma_0 = v.$$

The quantity $\nabla_v\xi(x)$ is "the covariant derivative of the vector field ξ at x in direction v." Of course, if ξ and v are two vector fields, one can define a vector field $\nabla_v\xi$ by the formula $(\nabla_v\xi)(x) = (\nabla_{v(x)}\xi)(x)$. The linear operator $v \longmapsto \nabla_v\xi(x)$ is the **covariant gradient** of ξ at x, denoted by $\nabla\xi(x)$ or $\nabla_x\xi$; it is a linear operation $T_xM \to T_xM$.

It is worth noticing explicitly that the notion of covariant derivation coincides with the *convective* derivation used in fluid mechanics (for instance in Euler's equation for an incompressible fluid). I shall sometimes adopt the notation classically used in fluid mechanics: $(\nabla\xi)v = v \cdot \nabla\xi$. (On the contrary, the notation $(\nabla\xi) \cdot v$ should rather be reserved for $(\nabla\xi)^*v$, where $(\nabla\xi)^*$ is the adjoint of $\nabla\xi$; then $\langle v \cdot \nabla\xi, w\rangle = \langle v, \nabla\xi \cdot w\rangle$ and we are back to the classical conventions of $\mathbb{R}^n$.)

The procedure of parallel transport allows one to define the covariant derivation; conversely, the equations of parallel transport along γ can be written as $D\xi/Dt = 0$, where D/Dt is the covariant derivative along γ. So it is equivalent to define the notion of covariant derivation, or to define the rules of parallel transport.

There are (at least) three points of view about the covariant derivation. The first one is the *extrinsic* point of view: Let us think of M as an embedded surface in $\mathbb{R}^N$; that is, M is a subset of $\mathbb{R}^N$, it is equipped with the topology induced by $\mathbb{R}^N$, and the quadratic form g_x is just the usual Euclidean scalar product on $\mathbb{R}^N$, restricted to T_xM. Then the covariant derivative is defined by

$$\dot{\xi}(t) = \Pi_{T_{\gamma_t}M}\left(\frac{d(\xi(\gamma_t))}{dt}\right),$$

where Π_{T_xM} stands for the orthogonal projection (in $\mathbb{R}^N$) onto T_xM. In short, the covariant derivative is the projection of the usual derivative onto the tangent space.

While this definition is very simple, it does not reveal the fact that the covariant derivation and parallel transport are *intrinsic* notions, which are invariant under isometry and do not depend on the embedding of M into $\mathbb{R}^N$, but just on g. An intrinsic way to define covariant derivation is as follows: It is uniquely characterized by the two natural rules

$$\frac{d}{dt}\langle\xi,\zeta\rangle = \langle\dot{\xi},\zeta\rangle + \langle\xi,\dot{\zeta}\rangle; \qquad \frac{D}{Dt}(f\xi) = \dot{f}\,\xi + f\,\dot{\xi}, \tag{7.28}$$

where the dependence of all the expressions on t is implicit; and by the not so natural rule

$$\nabla_\zeta\xi - \nabla_\xi\zeta = [\xi,\zeta].$$

Here $[\xi,\zeta]$ is the Lie bracket of ξ and ζ, which is defined as the unique vector field such that for any function F,

$$(dF)\cdot[\xi,\zeta] = d(dF\cdot\xi)\cdot\zeta - d(dF\cdot\zeta)\cdot\xi.$$

Further, note that in the second formula of (7.28) the symbol $\dot{f}$ stands for the usual derivative of $t \to f(\gamma_t)$; while the symbols $\dot{\xi}$ and $\dot{\zeta}$ stand for the covariant derivatives of the vector fields ξ and ζ along γ.

A third approach to covariant derivation is based on **coordinates**. Let $x \in M$, then there is a neighborhood O of x which is diffeomorphic to some open subset $U \subset \mathbb{R}^n$. Let Φ be a diffeomorphism $U \to O$, and let $(e_1,\ldots,e_n)$ be the usual basis of $\mathbb{R}^n$. A point m in O is said to have *coordinates* $(y^1,\ldots,y^n)$ if $m = \Phi(y^1,\ldots,y^n)$; and a vector $v \in T_mM$ is said to have *components* $v^1,\ldots,v^k$ if $d_{(y^1,\ldots,y^n)}\Phi\cdot(v_1,\ldots,v_k) = v$. Then the *coefficients* of the metric g are the functions g_{ij} defined by $g(v,v) = \sum g_{ij}v^iv^j$.

The coordinate point of view reduces everything to "explicit" computations and formulas in $\mathbb{R}^n$; for instance the derivation of a function

f along the ith direction is defined as $\partial_i f := (\partial/\partial y^i)(f \circ \Phi)$. This is conceptually simple, but rapidly leads to cumbersome expressions. A central role in these formulas is played by the **Christoffel symbols**, which are defined by

$$\Gamma_{ij}^m := \frac{1}{2} \sum_{k=1}^n \Big(\partial_i g_{jk} + \partial_j g_{ki} - \partial_k g_{ij} \Big) g^{km},$$

where (g^{ij}) is by convention the inverse of (g_{ij}). Then the covariant derivation along γ is given by the formula

$$\left(\frac{D\xi}{Dt}\right)^k = \frac{d\xi^k}{dt} + \sum_{ij} \Gamma_{ij}^k \, \dot{\gamma}^i \, \xi^j.$$

Be it in the extrinsic or the intrinsic or the coordinate point of view, the notion of covariant derivative is one of the cornerstones on which differential Riemannian geometry has been constructed.

Another important concept is that of **Riemannian volume**, which I shall denote by vol. It can be defined intrinsically as the n-dimensional Hausdorff measure associated with the geodesic distance (where n is the dimension of the manifold). In coordinates, $\mathrm{vol}(dx) = \sqrt{\det(g)}\, dx$. The Riemannian volume plays the same role as the Lebesgue measure in $\mathbb{R}^n$.

After these reminders about Riemannian calculus, we can go back to the study of action minimization. Let $L(x,v,t)$ be a smooth Lagrangian on $TM \times [0,1]$. To find an equation satisfied by the curves which minimize the action, we can compute the differential of the action. So let γ be a curve, and h a small variation of that curve. (This amounts to considering a family $\gamma_{s,t}$ in such a way that $\gamma_{0,t} = \gamma_t$ and $(d/ds)|_{s=0}\gamma_{s,t} = h(t)$.) Then the infinitesimal variation of the action $\mathcal{A}$ at γ, along the variation h, is

$$d\mathcal{A}(\gamma) \cdot h = \int_0^1 \Big(\nabla_x L(\gamma_t, \dot{\gamma}_t, t) \cdot h(t) + \nabla_v L(\gamma_t, \dot{\gamma}_t, t) \cdot \dot{h}(t) \Big)\, dt.$$

Thanks to (7.28) we can perform an integration by parts with respect to the time variable, and get

$$\begin{aligned} d\mathcal{A}(\gamma) \cdot h = \int_0^1 \left(\nabla_x L - \frac{d}{dt}(\nabla_v L) \right) (\gamma_t, \dot{\gamma}_t, t) \cdot h(t)\, dt \\ + (\nabla_v L)(\gamma_1, \dot{\gamma}_1, 1) \cdot h(1) - (\nabla_v L)(\gamma_0, \dot{\gamma}_0, 0) \cdot h(0). \end{aligned} \tag{7.29}$$

This is the **first variation formula**.

When the endpoints x, y of γ are fixed, the tangent curve h vanishes at $t = 0$ and $t = 1$. Since h is otherwise arbitrary, it is easy to deduce the equation for minimizers:

$$\frac{d}{dt}\nabla_v L = \nabla_x L. \tag{7.30}$$

More explicitly, if a differentiable curve $(\gamma_t)_{0\le t\le 1}$ is minimizing, then

$$\frac{d}{dt}\Big(\nabla_v L(\gamma_t, \dot\gamma_t, t)\Big) = \nabla_x L(\gamma_t, \dot\gamma_t, t), \qquad 0 < t < 1.$$

This is the **Euler–Lagrange equation** associated with the Lagrangian L; to memorize it, it is convenient to write it as

$$\frac{d}{dt}\frac{\partial L}{\partial \dot x} = \frac{\partial L}{\partial x}, \tag{7.31}$$

so that the two time-derivatives in the left-hand side formally "cancel out". Note carefully that the left-hand side of the Euler–Lagrange equation involves the time-derivative of a curve which is valued in TM; so (d/dt) in (7.31) is in fact a covariant derivative along the minimizing curve γ, the same operation as we denoted before by $\nabla_{\dot\gamma}$, or D/Dt.

The most basic example is when $L(x, v, t) = |v|^2/2$. Then $\nabla_v L = v$ and the equation reduces to $dv/dt = 0$, or $\nabla_{\dot\gamma}\dot\gamma = 0$, which is the usual equation of vanishing acceleration. Curves with zero acceleration are called **geodesics**; their equation, in coordinates, is

$$\ddot\gamma^k + \sum_{ij} \Gamma^k_{ij}\,\dot\gamma^i\,\dot\gamma^j = 0.$$

(Note: $\ddot\gamma^k$ is the derivative of $t \to \dot\gamma^k(t)$, *not* the kth component of $\ddot\gamma$.) The speed of such a curve γ is constant, and to stress this fact one can say that these are constant-speed geodesics, by opposition with general geodesics that can be reparametrized in an arbitrary way. Often I shall just say "geodesics" for constant-speed geodesics. It is equivalent to say that a geodesic γ has constant speed, or that its length between any two times $s < t$ is proportional to $t - s$.

An important concept related to geodesics is that of the **exponential map**. If $x \in M$ and $v \in T_xM$ are given, then $\exp_x v$ is defined as $\gamma(1)$, where $\gamma : [0,1] \to M$ is the unique constant-speed geodesic starting from $\gamma(0) = x$ with velocity $\dot\gamma(0) = v$. The exponential map is a convenient notation to handle "all" geodesics of a Riemannian manifold at the same time.

We have seen that minimizing curves have zero acceleration, and the converse is also true *locally*, that is if γ_1 is very close to γ_0. A curve which minimizes the action between its endpoints is called a **minimizing geodesic**, or minimal geodesic, or simply a geodesic. The Hopf–Rinow theorem guarantees that if the manifold M (seen as a metric space) is complete, then any two points in M are joined by at least one minimal geodesic. There might be several minimal geodesics joining two points x and y (to see this, consider two antipodal points on the sphere), but geodesics are:

- *nonbranching*: Two geodesics that are defined on a time interval $[0, t]$ and coincide on $[0, t']$ for some $t' > 0$ have to coincide on the whole of $[0, t]$. Actually, a stronger statement holds true: The velocity of the geodesic at time $t = 0$ uniquely determines the final position at time $t = 1$ (this is a consequence of the uniqueness statement in the Cauchy–Lipschitz theorem).
- *locally unique*: For any given x, there is $r_x > 0$ such that any y in the ball $B_{r_x}(x)$ can be connected to x by a single geodesic $\gamma = \gamma^{x \to y}$, and then the map $y \longmapsto \dot{\gamma}(0)$ is a diffeomorphism (this corresponds to parametrize the endpoint by the initial velocity).
- *almost everywhere unique*: For any x, the set of points y that can be connected to x by several (minimizing!) geodesics is of zero measure. A way to see this is to note that the square distance function $d^2(x, \cdot)$ is locally semiconcave, and therefore differentiable almost everywhere. (See Chapter 10 for background about semiconcavity.)

The set $\Gamma_{x,y}$ of (minimizing, constant speed) geodesics joining x and y might not be single-valued, but in any case it is *compact* in $C([0, 1], M)$, even if M is not compact. To see this, note that (i) the image of any element of $\Gamma_{x,y}$ lies entirely in the ball $B\big(x, d(x, y)\big)$, so $\Gamma_{x,y}$ is uniformly bounded, (ii) elements in $\Gamma_{x,y}$ are $d(x, y)$-Lipschitz, so they constitute an equi-Lipschitz family; (iii) $\Gamma_{x,y}$ is closed because it is defined by the equations $\gamma(0) = x$, $\gamma(1) = y$, $\mathcal{L}(\gamma) \le d(\gamma_0, \gamma_1)$ (the length functional $\mathcal{L}$ is not continuous with respect to uniform convergence, but it is lower semicontinuous, so an upper bound on the length defines a closed set); (iv) M is locally compact, so Ascoli's compactness theorem applies to functions with values in M.

A similar argument shows that for any two given compact sets K_s and K_t, the set of geodesics γ such that $\gamma_s \in K_s$ and $\gamma_t \in K_t$ is compact in $C([s,t];M)$. So the Lagrangian action defined by $\mathcal{A}^{s,t}(\gamma) = \mathcal{L}(\gamma)^2/(t-s)$ is coercive in the sense of Definition 7.13.

Most of these statements can be generalized to the action coming from a Lagrangian function $L(x,v,t)$ on $TM \times [0,1]$, if L is C^2 and satisfies the classical conditions of Definition 7.6. In particular the associated cost functions will be continuous. Here is a sketch of the proof: Let x and y be two given points, and let $x_k \to x$ and $y_k \to y$ be converging sequences. For any $\varepsilon > 0$, small enough,

$$c^{s,t}(x_k, y_k) \le c^{s,s+\varepsilon}(x_k, x) + c^{s+\varepsilon,t-\varepsilon}(x,y) + c^{t-\varepsilon,t}(y, y_k). \qquad (7.32)$$

It is easy to show that there is a uniform bound K on the speeds of all minimizing curves which achieve the costs appearing above. Then the Lagrangian is uniformly bounded on these curves, so $c^{s,s+\varepsilon}(x_k,x) = O(\varepsilon)$, $c^{t-\varepsilon,t}(y,y_k) = O(\varepsilon)$. Also it does not affect much the Lagrangian (evaluated on candidate minimizers) to reparametrize $[s+\varepsilon, t-\varepsilon]$ into $[s,t]$ by a linear change of variables, so $c^{s+\varepsilon,t-\varepsilon}(x,y)$ converges to $c^{s,t}(x,y)$ as $s \to t$. This proves the upper semicontinuity, and therefore the continuity, of $c^{s,t}$.

In fact there is a finer statement: $c^{s,t}$ is *superdifferentiable*. This notion will be explained and developed later in Chapter 10.

Besides the Euclidean space, Riemannian manifolds constitute in some sense the most regular metric structure used by mathematicians. A Riemannian structure comes with many nice features (calculus, length, distance, geodesic equations); it also has a well-defined dimension n (the dimension of the manifold) and carries a natural volume.

Finsler structures constitute a generalization of the Riemannian structure: one has a differentiable manifold, with a norm on each tangent space T_xM, but that norm does not necessarily come from a scalar product. One can then define lengths of curves, the induced distance as for a Riemannian manifold, and prove the existence of geodesics, but the geodesic equations are more complicated.

Another generalization is the notion of **length space** (or intrinsic length space), in which one does not necessarily have tangent spaces, yet one assumes the existence of a length $\mathcal{L}$ and a distance d which are compatible, in the sense that

$$\begin{cases} \mathcal{L}(\gamma) = \displaystyle\int_0^1 |\dot{\gamma}_t|\,dt, \qquad |\dot{\gamma}_t| := \limsup_{\varepsilon\to 0} \frac{d(\gamma_t, \gamma_{t+\varepsilon})}{|\varepsilon|}, \\[2ex] d(x,y) = \inf\Big\{\mathcal{L}(\gamma);\quad \gamma_0 = x,\ \gamma_1 = y\Big\}. \end{cases}$$

In practice the following criterion is sometimes useful: A *complete* metric space $(\mathcal{X}, d)$ is a length space if and only if for any two points in $\mathcal{X}$, and any $\varepsilon > 0$ one can find an ε-midpoint of (x, y), i.e. a point m_ε such that

$$\left|\frac{d(x,y)}{2} - d(x, m_\varepsilon)\right| \leq \varepsilon, \qquad \left|\frac{d(x,y)}{2} - d(y, m_\varepsilon)\right| \leq \varepsilon.$$

Minimizing paths are fundamental objects in geometry. A length space in which any two points can be joined by a minimizing path, or **geodesic**, is called a **geodesic space**, or strictly intrinsic length space, or just (by abuse of language) length space. There is a criterion in terms of midpoints: A complete metric space $(\mathcal{X}, d)$ is a geodesic space if and only if for any two points in $\mathcal{X}$ there is a midpoint, which is of course some $m \in X$ such that

$$d(x, m) = d(m, y) = \frac{d(x,y)}{2}.$$

There is another useful criterion: If the metric space $(\mathcal{X}, d)$ is a complete, *locally compact* length space, then it is geodesic. This is a generalization of the Hopf–Rinow theorem in Riemannian geometry. One can also reparametrize geodesic curves γ in such a way that their speed $|\dot{\gamma}|$ is constant, or equivalently that for all intermediate times s and t, their length between times s and t coincides with the distance between their positions at times s and t.

The same proof that I sketched for Riemannian manifolds applies in geodesic spaces, to show that the set $\Gamma_{x,y}$ of (minimizing, constant speed) geodesics joining x to y is compact; more generally, the set $\Gamma_{K_0 \to K_1}$ of geodesics γ with $\gamma_0 \in K_0$ and $\gamma_1 \in K_1$ is compact, as soon as K_0 and K_1 are compact. So there are important common points between the structure of a length space and the structure of a Riemannian manifold. From the practical point of view, some main differences are that (i) there is no available equation for geodesic curves, (ii) geodesics may "branch", (iii) there is no guarantee that geodesics between x and y are unique for y very close to x, (iv) there is neither a unique notion of dimension, nor a canonical reference measure, (v) there is no guarantee that geodesics will be almost everywhere unique. Still there is a theory of differential analysis on nonsmooth geodesic spaces (first variation formula, norms of Jacobi fields, etc.) mainly in the case where there are lower bounds on the sectional curvature (in the sense of Alexandrov, as will be described in Chapter 26).

Bibliographical notes

There are plenty of classical textbooks on Riemannian geometry, with variable degree of pedagogy, among which the reader may consult [223], [306], [394]. For an introduction to the classical calculus of variations in dimension 1, see for instance [347, Chapters 2–3], [177], or [235]. For an introduction to the Hamiltonian formalism in classical mechanics, one may use the very pedagogical treatise by Arnold [44], or the more complex one by Thirring [780]. For an introduction to analysis in metric spaces, see Ambrosio and Tilli [37]. A wonderful introduction to the theory of length spaces can be found in Burago, Burago and Ivanov [174]. In the latter reference, a Riemannian manifold is defined as a length space which is locally isometric to $\mathbb{R}^n$ equipped with a quadratic form g_x depending smoothly on the point x. This definition is not standard, but it is equivalent to the classical definition, and in some sense more satisfactory if one wishes to emphasize the metric point of view. Advanced elements of differential analysis on nonsmooth metric spaces can be found also in the literature on Alexandrov spaces, see the bibliographical notes of Chapter 26.

I may have been overcautious in the formulation of the classical conditions in Definition 7.6, but in the time-dependent case there are crazy counterexamples showing that nice C^1 Lagrangian functions do not necessarily have C^1 minimizing curves (equivalently, these minimizing curves won't solve the Euler–Lagrange equation); see for instance the constructions by Ball and Mizel [63, 64]. (A backwards search in MathSciNet will give access to many papers concerned with multidimensional analogs of this problem, in relation with the so-called Lavrentiev phenomenon.) I owe these remarks to Mather, who also constructed such counterexamples and noticed that many authors have been fooled by these issues. Mather [601, Section 2 and Appendices], Clarke and Vinter [235, 236, 823] discuss in great detail sufficient conditions under which everything works fine. In particular, if L is C^1, strictly convex superlinear in v *and time-independent*, then minimizing curves are automatically C^1 and satisfy the Euler–Lagrange equation. (The difficult point is to show that minimizers are Lipschitz; after that it is easier to see that they are at least as smooth as the Lagrangian.)

I introduced the abstract concept of "coercive Lagrangian action" for the purpose of this course, but this concept looks so natural to me that I would not be surprised if it had been previously discussed in the literature, maybe in disguised form.

Probability measures on action-minimizing curves might look a bit scary when encountered for the first time, but they were actually rediscovered several times by various researchers, so they are arguably natural objects: See in particular the works by Bernot, Caselles and Morel [109, 110] on irrigation problems; by Bangert [67] and Hohloch [476] on problems inspired by geometry and dynamical systems; by Ambrosio on transport equations with little or no regularity [21, 30]. In fact, in the context of partial differential equations, this approach already appears in the much earlier works of Brenier [155, 157, 158, 159] on the incompressible Euler equation and related systems. One technical difference is that Brenier considers probability measures on the huge (nonmetrizable) space of measurable paths, while the other above-mentioned authors only consider much smaller spaces consisting of continuous, or Lipschitz-continuous functions. There are important subtleties with probability measures on nonmetrizable spaces, and I advise the reader to stay away from them.

Also in relation to the irrigation problem, various models of traffic plans and "dynamic cost function" are studied in [108, 113]; while paths in the space of probability measures are considered in [152].

The Hamilton–Jacobi equation with a quadratic cost function (i.e. $L(x,v,t) = |v|^2$) will be considered in more detail in Chapter 22; see in particular Proposition 22.16. For further information about Hamilton–Jacobi equations, there is an excellent book by Cannarsa and Sinestrari [199]; one may also consult [68, 327, 558] and the references therein. Of course Hamilton–Jacobi equations are closely related to the concept of c-convexity: for instance, it is equivalent to say that ψ is c-convex, or that it is a solution at time 0 of the backward Hamilton–Jacobi semigroup starting at time 1 (with some arbitrary initial datum).

At the end of the proof of Proposition 7.16 I used once again the basic measurable selection theorem which was already used in the proof of Corollary 5.22, see the bibliographical notes on p. 92.

Interpolation arguments involving changes of variables have a long history. The concept and denomination of displacement interpolation was introduced by McCann [614] in the particular case of the quadratic cost in Euclidean space. Soon after, it was understood by Brenier that this procedure could formally be recast as an action minimization problem in the space of measures, which would reduce to the classical geodesic problem when the probability measures are Dirac masses.

In Brenier's approach, the action is defined, at least formally, by the formula

$$\mathbb{A}(\mu) = \inf_{v(t,x)} \left\{ \int_0^1 \int |v(t,x)|^2 \, d\mu_t(x) \, dt; \quad \frac{\partial \mu}{\partial t} + \nabla \cdot (v\mu) = 0 \right\}, \tag{7.33}$$

and then one has the Benamou–Brenier formula

$$W_2(\mu_0, \mu_1)^2 = \inf \mathbb{A}(\mu), \tag{7.34}$$

where the infimum is taken among all paths $(\mu_t)_{0 \le t \le 1}$ satisfying certain regularity conditions. Brenier himself gave two sketches of the proof for this formula [88, 164], and another formal argument was suggested by Otto and myself [671, Section 3]. Rigorous proofs were later provided by several authors under various assumptions [814, Theorem 8.1] [451] [30, Chapter 8] (the latter reference contains the most precise results). The adaptation to Riemannian manifolds has been considered in [278, 431, 491]. We shall come back to these formulas later on, after a more precise qualitative picture of optimal transport has emerged. One of the motivations of Benamou and Brenier was to devise new numerical methods [88, 89, 90, 91]. Wolansky [838] considered a more general situation in the presence of sources and sinks.

There was a rather amazing precursor to the idea of displacement interpolation, in the form of Nelson's theory of "stochastic mechanics". Nelson tried to build up a formalism in which quantum effects would be explained by stochastic fluctuations. For this purpose he considered an action minimization problem which was also studied by Guerra and Morato:

$$\inf \mathbb{E} \int_0^1 |\dot{X}_t|^2 \, dt,$$

where the infimum is over all random paths $(X_t)_{0 \le t \le 1}$ such that $\text{law}\,(X_0) = \mu_0$, $\text{law}\,(X_1) = \mu_1$, and in addition (X_t) solves the stochastic differential equation

$$\frac{dX_t}{dt} = \sigma \frac{dB_t}{dt} + \xi(t, X_t),$$

where $\sigma > 0$ is some coefficient, B_t is a standard Brownian motion, and ξ is a drift, which is an unknown in the problem. (So the minimization is over all possible couplings (X_0, X_1) but also over all drifts!) This formulation is very similar to the Benamou–Brenier formula just alluded to, only there is the additional Brownian noise in it, thus it is more complex in some sense. Moreover, the expected value of the action is always infinite, so one has to renormalize it to make sense of Nelson's problem. Nelson made the incredible discovery that after a change of variables, minimizers of the action produced solutions of the free Schrödinger equation in $\mathbb{R}^n$. He developed this approach for some time, and finally gave up because it was introducing unpleasant non-local features. I shall give references at the end of the bibliographical notes for Chapter 23.

It was Otto [669] who first explicitly reformulated the Benamou–Brenier formula (7.34) as the equation for a geodesic distance on a Riemannian setting, from a formal point of view. Then Ambrosio, Gigli and Savaré pointed out that if one is not interested in the equations of motion, but just in the geodesic property, it is simpler to use the metric notion of geodesic in a length space [30]. Those issues were also developed by other authors working with slightly different formalisms [203, 214].

All the above-mentioned works were mainly concerned with displacement interpolation in $\mathbb{R}^n$. Agueh [4] also considered the case of cost $c(x,y) = |x-y|^p$ $(p > 1)$ in Euclidean space. Then displacement interpolation on Riemannian manifolds was studied, from a heuristic point of view, by Otto and myself [671]. Some useful technical tools were introduced in the field by Cordero-Erausquin, McCann and Schmuckenschläger [246] for Riemannian manifolds; Cordero-Erausquin adapted them to the case of rather general strictly convex cost functions in $\mathbb{R}^n$ [243].

The displacement interpolation for more general cost functions, arising from a smooth Lagrangian, was constructed by Bernard and Buffoni [105], who first introduced in this context Property (ii) in Theorem 7.21. At the same time, they made the explicit link with the Mather minimization problem, which will appear in subsequent chapters. This connection was also studied independently by De Pascale, Gelli and Granieri [278].

In all these works, displacement interpolation took place in a smooth structure, resulting in particular in the uniqueness (almost everywhere) of minimizing curves used in the interpolation, at least if the Lagrangian is nice enough. Displacement interpolation in length spaces, as presented in this chapter, via the notion of dynamical transference plan, was developed more recently by Lott and myself [577]. Theorem 7.21 in this course is new; it was essentially obtained by rewriting the proof in [577] with enough generality added to include the setting of Bernard and Buffoni.

The most natural examples of Lagrangian functions are those taking the form $L(t, x, v) = |v|^2/2 - U(t, x)$, where $U(t, x)$ is a potential energy. In relation with incompressible fluid mechanics, Ambrosio and Figalli [24, 25] studied the case when U is the pressure field (a priori nonsmooth). Another case of great interest is when $U(t, x)$ is the scalar curvature of a manifold evolving in time according to the Ricci flow; then, up to a correct time rescaling, the associated minimal action is known in geometry as Perelman's $\mathcal{L}$-distance. First Topping [782], and then Lott [576] discussed the Lagrangian action induced by the $\mathcal{L}$-distance (and some of its variants) at the level of the space of probability measures. They used this formalism to recover some key results in the theory of Ricci flow.

Displacement interpolation in the case $p = 1$ is quite subtle because of the possibility of reparametrization; it was carefully discussed in the Euclidean space by Ambrosio [20]. Recently, Bernard and Buffoni [104] shed some new light on that issue by making explicit the link with the Mather–Mañé problem. Very roughly, the distance cost function is a typical representative of cost functions that arise from Lagrangians, if one also allows *minimization over the choice of the time-interval* $[0, T] \subset \mathbb{R}$ (rather than fixing, say, $T = 1$). This extra freedom accounts for the degeneracy of the problem.

Lagrangian cost functions of the form $V(\gamma)\,|\dot{\gamma}|$, where V is a "Lyapunov functional", have been used by Hairer and Mattingly [458] in relation to convergence to equilibrium, as a way to force the system to visit a compact set. Such cost functions also appear formally in the modeling of irrigation [108], but in fact this is a nonlinear problem since $V(\gamma)$ is determined by the total mass of particles passing at a given point.

The observation in Remark 7.27 came from a discussion with S. Evans, who pointed out to me that it was difficult, if not impossible, to get characterizations of random processes expressed in terms of the measures when working in state spaces that are not locally compact (such as the space of real trees). In spite of that remark, recently Lisini [565] was able to obtain representation theorems for general absolutely continuous paths $(\mu_t)_{0\le t\le 1}$ in the Wasserstein space $P_p(\mathcal{X})$ $(p > 1)$, as soon as $\int \|\dot{\mu}_t\|_{P_p}^p\,dt < \infty$, where $\mathcal{X}$ is just a Polish space and $\|\dot{\mu}_t\|_{P_p}$ is the metric speed in $P_p(\mathcal{X})$. He showed that such a curve may be written as $(e_t)_\#\Pi$, where Π is the law of a random absolutely continuous curve γ; as a consequence, he could generalize Corollary 7.22 by removing the assumption of local compactness. Lisini also established a metric replacement for the relation of conservation of mass: For almost all t,

$$\mathbb{E}\,|\dot{\gamma}_t|^p \leq \|\dot{\mu}\|_{P_p}^p.$$

He further applied his results to various problems about transport in infinite-dimensional Banach spaces.

Proposition 7.29 is a generalization of a result appearing in [444]. Gigli communicated to me an alternative proof, which is more elementary and needs neither local compactness, nor length space property.

8

The Monge–Mather shortening principle

Monge himself made the following important observation. Consider the transport cost $c(x,y) = |x-y|$ in the Euclidean plane, and two pairs (x_1, y_1), (x_2, y_2), such that an optimal transport maps x_1 to y_1 and x_2 to y_2. (In our language, (x_1, y_1) and (x_2, y_2) belong to the support of an optimal coupling π.) Then either all four points lie on a single line, or the two line segments $[x_1, y_1]$, $[x_2, y_2]$ do not cross, except maybe at their endpoints. The reason is easy to grasp: If the two lines would cross at a point which is not an endpoint of both lines, then, by triangle inequality we would have

$$|x_1 - y_2| + |x_2 - y_1| < |x_1 - y_1| + |x_2 - y_2|,$$

and this would contradict the fact that the support of π is c-cyclically monotone. Stated otherwise: Given two crossing line segments, we can *shorten* the total length of the paths by replacing these lines by the new transport lines $[x_1, y_2]$ and $[x_2, y_1]$ (see Figure 8.1).

Quadratic cost function

For cost functions that do not satisfy a triangle inequality, Monge's argument does not apply, and pathlines can cross. However, it is often the case that the crossing of the *curves* (with the time variable explicitly taken into account) is forbidden. Here is the most basic example: Consider the quadratic cost function in Euclidean space ($c(x,y) = |x-y|^2$), and let (x_1, y_1) and (x_2, y_2) belong to the support of some optimal coupling. By cyclical monotonicity,

C. Villani, *Optimal Transport*, Grundlehren der mathematischen Wissenschaften 338,

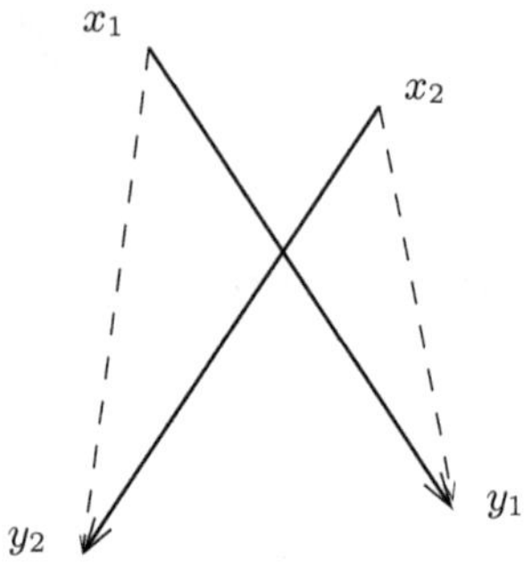

Fig. 8.1. Monge's observation. The cost is Euclidean distance; if x_1 is sent to y_1 and x_2 to y_2, then it is cheaper to send x_1 to y_2 and x_2 to y_1.

$$|x_1 - y_1|^2 + |x_2 - y_2|^2 \le |x_1 - y_2|^2 + |x_2 - y_1|^2. \qquad (8.1)$$

Then let

$$\gamma_1(t) = (1-t)\,x_1 + t\,y_1, \qquad \gamma_2(t) = (1-t)\,x_2 + t\,y_2$$

be the two line segments respectively joining x_1 to y_1, and x_2 to y_2. It may happen that $\gamma_1(s) = \gamma_2(t)$ for some $s, t \in [0,1]$. But if there is a $t_0 \in (0,1)$ such that $\gamma_1(t_0) = \gamma_2(t_0) =: X$, then

$$\begin{aligned}
&|x_1 - y_2|^2 + |x_2 - y_1|^2 \\
&\qquad = |x_1 - X|^2 + |X - y_2|^2 - 2\,\langle X - x_1, X - y_2\rangle \\
&\qquad\qquad + |x_2 - X|^2 + |X - y_1|^2 - 2\,\langle X - x_2, X - y_1\rangle \\
&\qquad = [t_0^2 + (1-t_0)^2]\,\big(|x_1 - y_1|^2 + |x_2 - y_2|^2\big) \\
&\qquad\qquad + 4\,t_0(1-t_0)\,\langle x_1 - y_1, x_2 - y_2\rangle \\
&\qquad \le \big[t_0^2 + (1-t_0)^2 + 2\,t_0(1-t_0)\big]\,\big(|x_1 - y_1|^2 + |x_2 - y_2|^2\big) \\
&\qquad = |x_1 - y_1|^2 + |x_2 - y_2|^2,
\end{aligned}$$

and the inequality is strict unless $x_1 - y_1 = x_2 - y_2$, in which case $\gamma_1(t) = \gamma_2(t)$ for all $t \in [0,1]$. But strict inequality contradicts (8.1). The conclusion is that *two distinct interpolation trajectories cannot meet at intermediate times.*

It is natural to ask whether this conclusion can be reinforced in a quantitative statement. The answer is yes; in fact there is a beautiful identity:

$$\Big|\big((1-t)x_1+ty_1\big)-\big((1-t)x_2+ty_2\big)\Big|^2 = (1-t)^2|x_1-x_2|^2+t^2|y_1-y_2|^2$$
$$+\,t(1-t)\Big(|x_1-y_2|^2+|x_2-y_1|^2-|x_1-y_1|^2-|x_2-y_2|^2\Big). \quad (8.2)$$

To appreciate the consequences of (8.2), let

$$\gamma_1(t)=(1-t)\,x_1+t\,y_1, \qquad \gamma_2(t)=(1-t)\,x_2+t\,y_2.$$

Then (8.1) and (8.2) imply

$$\max\big(|x_1-x_2|,\,|y_1-y_2|\big)\le \max\left(\frac{1}{t},\,\frac{1}{1-t}\right)\,|\gamma_1(t)-\gamma_2(t)|.$$

Since $|\gamma_1(t)-\gamma_2(t)|\le \max(|x_1-x_2|,\,|y_1-y_2|)$ for all $t\in[0,1]$, one can conclude that for any $t_0\in(0,1)$,

$$\sup_{0\le t\le 1}|\gamma_1(t)-\gamma_2(t)|\le \max\left(\frac{1}{t_0},\,\frac{1}{1-t_0}\right)\,|\gamma_1(t_0)-\gamma_2(t_0)|. \quad (8.3)$$

(By the way, this inequality is easily seen to be optimal.) So the uniform distance between the *whole paths* γ_1 and γ_2 can be controlled by their distance at *some* time $t_0\in(0,1)$.

General statement and applications to optimal transport

For the purpose of a seemingly different problem, Mather (not aware of Monge's work, neither of optimal transport) established an estimate which relies on the same idea as Monge's shortening argument — only much more sophisticated — for general cost functions on Lagrangian manifolds. He obtained a quantitative version of these estimates, in a form quite similar to (8.3).

Mather's proof uses three kinds of assumption: (i) the existence of a second-order differential equation for minimizing curves; (ii) an assumption of regularity of the Lagrangian, and (iii) an assumption of strict convexity of the Lagrangian. To quantify the strict convexity, I shall use the following concept: A continuous function L on $\mathbb{R}^n$ will be said to be $(2+\kappa)$-*convex* if it satisfies a (strict) convexity inequality of the form

$$\frac{L(v)+L(w)}{2}-L\left(\frac{v+w}{2}\right)\ge K\,|v-w|^{2+\kappa}$$

for some constant $K>0$.

The next statement is a slight generalization of Mather's estimate; if the reader finds it too dense, he or she can go directly to Corollary 8.2 which is simpler, and sufficient for the rest of this course.

Theorem 8.1 (Mather's shortening lemma). *Let M be a smooth Riemannian manifold, equipped with its geodesic distance d, and let $c(x,y)$ be a cost function on $M \times M$, defined by a Lagrangian $L(x,v,t)$ on $TM \times [0,1]$. Let x_1, x_2, y_1, y_2 be four points on M such that*

$$c(x_1,y_1) + c(x_2,y_2) \le c(x_1,y_2) + c(x_2,y_1).$$

Further, let γ_1 and γ_2 be two action-minimizing curves respectively joining x_1 to y_1 and x_2 to y_2. Let V be a bounded neighborhood of the graphs of γ_1 and γ_2 in $M \times [0,1]$, and S a strict upper bound on the maximal speed along these curves. Define

$$\mathcal{V} := \bigcup_{(x,t)\in V} \Big(x, B_S(0), t\Big) \subset TM \times [0,1].$$

In words, $\mathcal{V}$ is a neighborhood of γ_1 and γ_2, convex in the velocity variable.

Assume that:

(i) minimizing curves for L are solutions of a Lipschitz flow, in the sense of Definition 7.6 (d);

(ii) L is of class $C^{1,\alpha}$ in $\mathcal{V}$ with respect to the variables x and v, for some $\alpha \in (0,1]$ (so $\nabla_x L$ and $\nabla_v L$ are Hölder-α; Hölder-1 meaning Lipschitz);

(iii) L is $(2+\kappa)$-convex in $\mathcal{V}$, with respect to the v variable.

Then, for any $t_0 \in (0,1)$, there is a constant $C_{t_0} = C(L,\mathcal{V},t_0)$, and a positive exponent $\beta = \beta(\alpha,\kappa)$ such that

$$\sup_{0\le t\le 1} d\big(\gamma_1(t),\gamma_2(t)\big) \le \; C_{t_0}\, d\big(\gamma_1(t_0),\gamma_2(t_0)\big)^{\beta}. \tag{8.4}$$

Furthermore, if $\alpha = 1$ and $\kappa = 0$, then one can choose $\beta = 1$ and $C_{t_0} = C(L,\mathcal{V})/\min(t_0, 1-t_0)$.

If L is of class C^2 and $\nabla_v^2 L > 0$, then Assumption (iii) will be true for $\kappa = 0$, so we have the next corollary:

Corollary 8.2 (Mather's shortening lemma again). *Let M be a smooth Riemannian manifold and let $L = L(x, v, t)$ be a C^2 Lagrangian on $TM \times [0, 1]$, satisfying the classical assumptions of Definition 7.6, together with $\nabla_v^2 L > 0$. Let c be the cost function associated to L, and let d be the geodesic distance on M. Then, for any compact $K \subset M$ there is a constant C_K such that, whenever x_1, y_1, x_2, y_2 are four points in K with*

$$c(x_1, y_1) + c(x_2, y_2) \leq c(x_1, y_2) + c(x_2, y_1),$$

and γ_1, γ_2 are action-minimizing curves joining respectively x_1 to y_1 and x_2 to y_2, then for any $t_0 \in (0, 1)$,

$$\sup_{0 \leq t \leq 1} d\big(\gamma_1(t), \gamma_2(t)\big) \leq \frac{C_K}{\min(t_0,\, 1 - t_0)}\, d\big(\gamma_1(t_0), \gamma_2(t_0)\big). \tag{8.5}$$

The short version of the conclusion is that the distance between γ_1 and γ_2 is controlled, *uniformly in t*, by the distance at *any* time $t_0 \in (0, 1)$. In particular, the initial and final distance between these curves is controlled by their distance at any intermediate time. (But the final distance is *not* controlled by the initial distance!) Once again, inequalities (8.4) or (8.5) are quantitative versions of the qualitative statement that the two curves, if distinct, cannot cross except at initial or final time.

Example 8.3. The cost function $c(x, y) = d(x, y)^2$ corresponds to the Lagrangian function $L(x, v, t) = |v|^2$, which obviously satisfies the assumptions of Corollary 8.2. In that case the exponent $\beta = 1$ is admissible. Moreover, it is natural to expect that the constant C_K can be controlled in terms of just a *lower bound on the sectional curvature* of M. I shall come back to this issue later in this chapter (see Open Problem 8.21).

Example 8.4. The cost function $c(x, y) = d(x, y)^{1+\alpha}$ does not satisfy the assumptions of Corollary 8.2 for $0 < \alpha < 1$. Even if the associated Lagrangian $L(x, v, t) = |v|^{1+\alpha}$ is not smooth, the equation for minimizing curves is just the geodesic equation, so Assumption (i) in Theorem 8.1 is still satisfied. Then, by tracking exponents in the proof of Theorem 8.1, one can find that (8.4) holds true with $\beta = (1+\alpha)/(3-\alpha)$. But this is far from optimal: By taking advantage of the homogeneity of the power function, one can prove that the exponent $\beta = 1$ is also admissible, for all $\alpha \in (0,1)$. (It is the constant, rather than the exponent, which deteriorates as $\alpha \downarrow 0$.) I shall explain this argument in the Appendix, in the Euclidean case, and leave the Riemannian case as a delicate exercise. This example suggests that Theorem 8.1 still leaves room for improvement.

The proof of Theorem 8.1 is a bit involved and before presenting it I prefer to discuss some applications in terms of optimal couplings. In the sequel, if K is a compact subset of M, I say that a dynamical optimal transport Π is supported in K if it is supported on geodesics whose image lies entirely inside K.

Theorem 8.5 (The transport from intermediate times is locally Lipschitz). *On a Riemannian manifold M, let c be a cost function satisfying the assumptions of Corollary 8.2, let K be a compact subset of M, and let Π be a dynamical optimal transport supported in K. Then Π is supported on a set of geodesics S such that for any two $\gamma, \widetilde{\gamma} \in S$,*

$$\sup_{0 \le t \le 1} d\big(\gamma(t), \widetilde{\gamma}(t)\big) \le C_K(t_0)\, d\big(\gamma(t_0), \widetilde{\gamma}(t_0)\big). \tag{8.6}$$

In particular, if $(\mu_t)_{0 \le t \le 1}$ is a displacement interpolation between any two compactly supported probability measures on M, and $t_0 \in (0,1)$ is given, then for any $t \in [0,1]$ the map

$$T_{t_0 \to t} : \gamma(t_0) \longmapsto \gamma(t)$$

is well-defined μ_{t_0}-almost surely and Lipschitz continuous on its domain; and it is in fact the unique solution of the Monge problem between μ_{t_0} and μ_t. In other words, the coupling $(\gamma(t_0), \gamma(t))$ is an optimal deterministic coupling.

Example 8.6. On $\mathbb{R}^n$ with $c(x,y) = |x-y|^2$, let $\mu_0 = \delta_0$ and let $\mu = \text{law}\,(X)$ be arbitrary. Then it is easy to check that $\mu_t = \text{law}\,(tX)$, and in fact the random geodesic $\gamma(t)$ is just $t\gamma(1)$. So $\gamma(t) = t\gamma(t_0)/t_0$, which obviously provides a deterministic coupling. This example is easily adapted to more general geometries (see Figure 8.2).

Proof of Theorem 8.5. The proof consists only in formalizing things that by now may look essentially obvious to the reader. First note

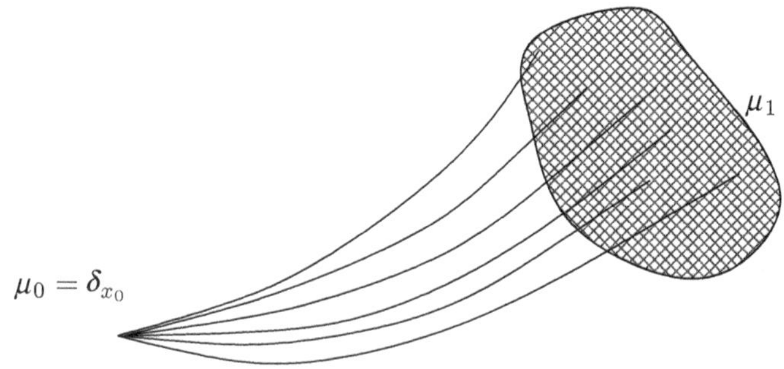

Fig. 8.2. In this example the map $\gamma(0) \to \gamma(1/2)$ is not well-defined, but the map $\gamma(1/2) \to \gamma(0)$ is well-defined and Lipschitz, just as the map $\gamma(1/2) \to \gamma(1)$. Also μ_0 is singular, but μ_t is absolutely continuous as soon as $t > 0$.

that $(e_0, e_1, e_0, e_1)_\#(\Pi \otimes \Pi) = \pi \otimes \pi$, where π is an optimal coupling between μ_0 and μ_1. So if a certain property holds true $\pi \otimes \pi$-almost surely for quadruples, it also holds true $\Pi \otimes \Pi$-almost surely for the endpoints of pairs of curves.

Since π is optimal, it is c-cyclically monotone (Theorem 5.10 (ii)), so, $\pi \otimes \pi(dx\, dy\, d\widetilde{x}\, d\widetilde{y})$-almost surely,

$$c(x, y) + c(\widetilde{x}, \widetilde{y}) \leq c(x, \widetilde{y}) + c(\widetilde{x}, y).$$

Thus, $\Pi \otimes \Pi(d\gamma\, d\widetilde{\gamma})$-almost surely,

$$c(\gamma(0), \gamma(1)) + c(\widetilde{\gamma}(0), \widetilde{\gamma}(1)) \leq c(\gamma(0), \widetilde{\gamma}(1)) + c(\widetilde{\gamma}(0), \gamma(1)).$$

Then (8.6) follows from Corollary 8.2.

Let S be the support of Π; by assumption this is a compact set. Since the inequality (8.6) defines a closed set of pairs of geodesics, actually it has to hold true for *all* pairs $(\gamma, \widetilde{\gamma}) \in S \times S$.

Now define the map $T_{t_0 \to t}$ on the compact set $e_{t_0}(S)$ (that is, the union of all $\gamma(t_0)$, when γ varies over the compact set S), by the formula $T_{t_0 \to t}(\gamma(t_0)) = \gamma(t)$. This map is well-defined, for if two geodesics γ and $\widetilde{\gamma}$ in the support of Π are such that $\gamma(t_0) = \widetilde{\gamma}(t_0)$, then inequality (8.6) imposes $\gamma = \widetilde{\gamma}$. The same inequality shows that $T_{t_0 \to t}$ is actually Lipschitz-continuous, with Lipschitz constant $C_K / \min(t_0, 1 - t_0)$.

All this shows that $(\gamma(t_0), T_{t_0 \to t}(\gamma(t_0)))$ is indeed a Monge coupling of (μ_{t_0}, μ_t), with a Lipschitz map. To complete the proof of the theorem, it only remains to check the uniqueness of the optimal coupling; but this follows from Theorem 7.30(iii). □

The second application is a result of "preservation of absolute continuity".

Theorem 8.7 (Absolute continuity of displacement interpolation). *Let M be a Riemannian manifold, and let $L(x, v, t)$ be a C^2 Lagrangian on $TM \times [0, 1]$, satisfying the classical conditions of Definition 7.6, with $\nabla_v^2 L > 0$; let c be the associated cost function. Let μ_0 and μ_1 be two probability measures on M such that the optimal cost $C(\mu_0, \mu_1)$ is finite, and let $(\mu_t)_{0 \le t \le 1}$ be a displacement interpolation between μ_0 and μ_1. If either μ_0 or μ_1 is absolutely continuous with respect to the volume on M, then also μ_t is absolutely continuous for all $t \in (0, 1)$.*

Proof of Theorem 8.7. Let us assume for instance that μ_1 is absolutely continuous, and prove that μ_{t_0} is absolutely continuous ($0 < t_0 < 1$).

First consider the case when μ_0 and μ_1 are compactly supported. Then the whole displacement interpolation is compactly supported, and Theorem 8.5 applies, so there is a Lipschitz map T solving the Monge problem between μ_{t_0} and μ_{t_1}.

Now if N is a set of zero volume, the inclusion $N \subset T^{-1}(T(N))$ implies

$$\mu_{t_0}[N] \le \mu_{t_0}\big[T^{-1}(T(N))\big] = (T_{\#}\mu_{t_0})[T(N)] = \mu_1[T(N)], \qquad (8.7)$$

and the latter quantity is 0 since $\mathrm{vol}\,[T(N)] \le \|T\|_{\mathrm{Lip}}^n \,\mathrm{vol}\,[N] = 0$ and μ_1 is absolutely continuous. So (8.7) shows that $\mu_{t_0}[N] = 0$ for any Borel set N of zero volume, and this means precisely that μ_{t_0} is absolutely continuous.

Actually, the previous computation is not completely rigorous because $T(N)$ is not necessarily Borel measurable; but this is not serious since $T(N)$ can still be included in a negligible Borel set, and then the proof can be repaired in an obvious way.

Now let us turn to the general case where μ_0 and μ_1 are not assumed to be compactly supported. This situation will be handled by a restriction argument. Assume by contradiction that μ_{t_0} is not absolutely continuous. Then there exists a set Z_{t_0} with zero volume, such that $\mu_{t_0}[Z_{t_0}] > 0$. Let $\mathcal{Z} := \{\gamma \in \Gamma(M);\ \gamma_{t_0} \in Z_{t_0}\}$. Then

$$\Pi[\mathcal{Z}] = \mathbb{P}\left[\gamma_{t_0} \in Z_{t_0}\right] = \mu_{t_0}[Z_{t_0}] > 0.$$

By regularity, there exists a compact set $\mathcal{K} \subset \mathcal{Z}$, such that $\Pi[\mathcal{K}] > 0$. Let

$$\Pi' := \frac{1_{\mathcal{K}}\,\Pi}{\Pi[\mathcal{K}]},$$

and let $\pi' := (e_0, e_1)_\#\Pi'$ be the associated transference plan, and $\mu'_t = (e_t)_\#\Pi'$ the marginal of Π' at time t. In particular,

$$\mu'_1 \le \frac{(e_1)_\#\Pi}{\Pi[\mathcal{K}]} = \frac{\mu_1}{\Pi[\mathcal{K}]},$$

so μ'_1 is still absolutely continuous.

By Theorem 7.30(ii), (μ'_t) is a displacement interpolation. Now, μ'_{t_0} is concentrated on $e_{t_0}(\mathcal{K}) \subset e_{t_0}(\mathcal{Z}) \subset Z_{t_0}$, so μ'_{t_0} is singular. But the first part of the proof rules out this possibility, because μ'_0 and μ'_1 are respectively supported in $e_0(\mathcal{K})$ and $e_1(\mathcal{K})$, which are compact, and μ'_1 is absolutely continuous. □

Proof of Mather's estimates

Now let us turn to the proof of Theorem 8.1. It is certainly more important to grasp the idea of the proof (Figure 8.3) than to follow the calculations, so the reader might be content with the informal explanations below and skip the rigorous proof at first reading.

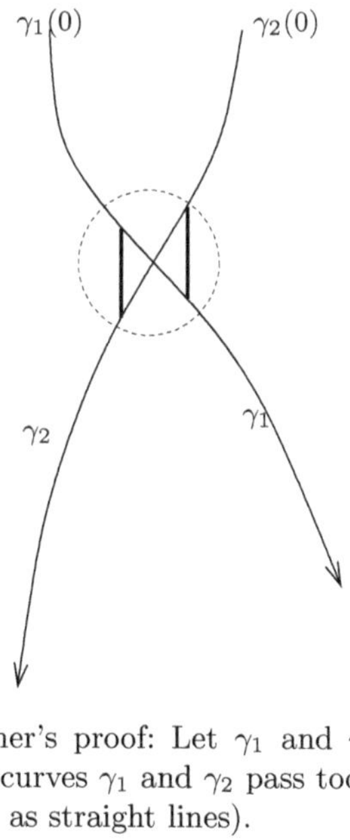

Fig. 8.3. Principle of Mather's proof: Let γ_1 and γ_2 be two action-minimizing curves. If at time t_0 the two curves γ_1 and γ_2 pass too close to each other, one can devise shortcuts (here drawn as straight lines).

Idea of the proof of Theorem 8.1. Assume, to fix the ideas, that γ_1 and γ_2 cross each other at a point m_0 and at time t_0. Close to m_0, these two curves look like two straight lines crossing each other, with respective velocities v_1 and v_2. Now cut these curves on the time interval $[t_0 - \tau, t_0 + \tau]$ and on that interval introduce "deviations" (like a plumber installing a new piece of pipe to short-cut a damaged region of a channel) that join the first curve to the second, and vice versa.

This amounts to replacing (on a short interval of time) two curves with approximate velocity v_1 and v_2, by two curves with approximate velocities $(v_1+v_2)/2$. Since the time-interval where the modification occurs is short, everything is concentrated in the neighborhood of (m_0, t_0), so the modification in the Lagrangian action of the two curves is approximately

$$(2\tau)\left(2\,L\left(m_0, \frac{v_1+v_2}{2}, t_0\right) - \Big[L(m_0, v_1, t_0) + L(m_0, v_2, t_0)\Big]\right).$$

Since $L(m_0, \,\cdot\,, t_0)$ is strictly convex, this quantity is negative if $v_1 \neq v_2$, which means that the total action has been *strictly improved* by the modification. But then $c(x_1, y_2) + c(x_2, y_1) < c(x_1, y_1) + c(x_2, y_2)$, in contradiction to our assumptions. The only possibility available is that $v_1 = v_2$, i.e. at the crossing point the curves have the same position and the same velocity; but then, since they are solutions of a second-order differential inequality, they have to coincide at all times. □

It only remains to make this argument quantitative: If the two curves pass close to each other at time t_0, then their velocities at that time will also be close to each other, and so the trajectories have to be close to each other for all times in $[0, 1]$. Unfortunately this will not be so easy.

Rigorous proof of Theorem 8.1. **Step 1: Localization.** The goal of this step is to show that the problem reduces to a local computation that can be performed as if we were in Euclidean space, and that it is sufficient to control the difference of velocities at time t_0 (as in the above sketchy explanation). If the reader is ready to believe in these two statements, then he or she can go directly to Step 2.

For brevity, let $\gamma_1 \cup \gamma_2$ stand for the union of the images of the minimizing paths γ_1 and γ_2. For any point x in $\mathrm{proj}_M(V)$, there is a small ball $B_{r_x}(x)$ which is diffeomorphic to an open set in $\mathbb{R}^n$, and by compactness one can cover a neighborhood of $\gamma_1 \cup \gamma_2$ by a finite number of such balls B_j, each of them having radius no less than $\delta > 0$. Without loss of generality, all these balls are included in $\mathrm{proj}_M(V)$, and it can be assumed that whenever two points X_1 and X_2 in $\gamma_1 \cup \gamma_2$ are separated by a distance less than $\delta/4$, then one of the balls B_j contains $B_{\delta/4}(X_1) \cup B_{\delta/4}(X_2)$.

If $\gamma_1(t_0)$ and $\gamma_2(t_0)$ are separated by a distance at least $\delta/4$, then the conclusion is obvious. Otherwise, choose τ small enough that $\tau S \le \delta/4$ (recall that S is a strict bound on the maximum speed along the curves); then on the time-interval $[t_0 - \tau, t_0 + \tau]$ the curves never leave the balls $B_{\delta/4}(X_1) \cup B_{\delta/4}(X_2)$, and therefore the whole trajectories of γ_1 and γ_2 on that time-interval have to stay within a single ball B_j. If one takes into account positions, velocities and time, the system is confined within $B_j \times B_S(0) \times [0,1] \subset \mathcal{V}$.

On any of these balls B_j, one can introduce a Euclidean system of coordinates, and perform all computations in that system (write L in those coordinates, etc.). The distance induced on B_j by that system of coordinates will not be the same as the original Riemannian distance, but it can be bounded from above and below by multiples thereof. So we can pretend that we are really working with a Euclidean metric, and all conclusions that are obtained, involving only what happens inside the ball B_j, will remain true up to changing the bounds. Then, for the sake of all computations, we can freely add points as if we were working in Euclidean space.

If it can be shown, in that system of coordinates, that

$$\left|\dot\gamma_1(t_0) - \dot\gamma_2(t_0)\right| \le C\left|\gamma_1(t_0) - \gamma_2(t_0)\right|^\beta, \tag{8.8}$$

then this means that $(\gamma_1(t_0), \dot\gamma_1(t_0))$ and $(\gamma_2(t_0), \dot\gamma_2(t_0))$ are very close to each other in TM; more precisely they are separated by a distance which is $O\big(d(\gamma_1(t_0), \gamma_2(t_0))^\beta\big)$. Then by Assumption (i) and Cauchy–Lipschitz theory this bound will be propagated backward and forward in time, so the distance between $(\gamma_1(t), \dot\gamma_1(t))$ and $(\gamma_2(t), \dot\gamma_2(t))$ will remain bounded by $O\big(d(\gamma_1(t_0), \gamma_2(t_0))^\beta\big)$. Thus to conclude the argument it is sufficient to prove (8.8).

Step 2: Construction of shortcuts. First some notation: Let us write $x_1(t) = \gamma_1(t)$, $x_2(t) = \gamma_2(t)$, $v_1(t) = \dot\gamma_1(t)$, $v_2(t) = \dot\gamma_2(t)$, and also $X_1 = x_1(t_0)$, $V_1 = v_1(t_0)$, $X_2 = x_2(t_0)$, $V_2 = v_2(t_0)$. The goal is to control $|V_1 - V_2|$ by $|X_1 - X_2|$.

Let $x_{12}(t)$ be defined by

$$x_{12}(t) = \begin{cases} x_1(t) & \text{for } t \in [0, t_0 - \tau]; \\[2ex] \frac{x_1(t)+x_2(t)}{2} + \left(\frac{\tau+t-t_0}{2\tau}\right)\left(\frac{x_2(t_0+\tau)-x_1(t_0+\tau)}{2}\right) \\ \qquad\qquad + \left(\frac{\tau-t+t_0}{2\tau}\right)\left(\frac{x_1(t_0-\tau)-x_2(t_0-\tau)}{2}\right) \\ \qquad\qquad\qquad \text{for } t \in [t_0 - \tau, t_0 + \tau]; \\ x_2(t) & \text{for } t \in [t_0 + \tau, 1]. \end{cases}$$

Note that x_{12} is a continuous function of t; it is a path that starts along γ_1, then switches to γ_2. Let $v_{12}(t)$ stand for its time-derivative:

$$v_{12}(t) = \begin{cases} v_1(t) & \text{for } t \in [0, t_0 - \tau]; \\[2ex] \frac{v_1(t)+v_2(t)}{2} + \frac{1}{2\tau}\left(\left[\frac{x_2(t_0-\tau)+x_2(t_0+\tau)}{2}\right] - \left[\frac{x_1(t_0-\tau)+x_1(t_0+\tau)}{2}\right]\right) \\ \qquad\qquad \text{for } t \in [t_0 - \tau, t_0 + \tau]; \\ v_2(t) & \text{for } t \in [t_0 + \tau, 1]. \end{cases}$$

Then the path $x_{21}(t)$ and its time-derivative $v_{21}(t)$ are defined symetrically (see Figure 8.4). These definitions are rather natural: First we try

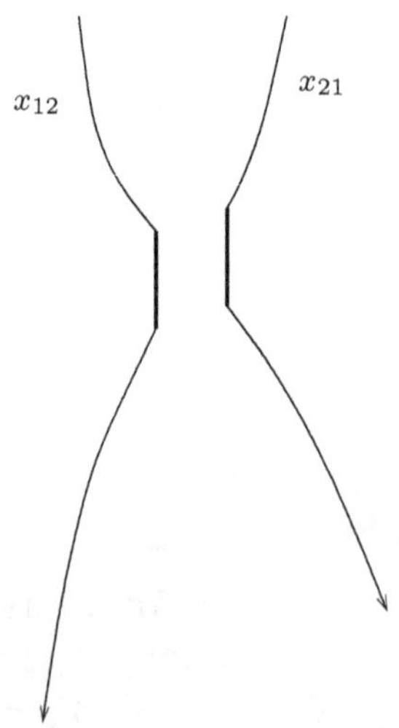

Fig. 8.4. The paths $x_{12}(t)$ and $x_{21}(t)$ obtained by using the shortcuts to switch from one original path to the other.

to construct paths on $[t_0 - \tau, t_0 + \tau]$ whose velocity is about the half of the velocities of γ_1 and γ_2; then we correct these paths by adding simple functions (linear in time) to make them match the correct endpoints.

I shall conclude this step with some basic estimates about the paths x_{12} and x_{21} on the time-interval $[t_0 - \tau, t_0 + \tau]$. For a start, note that

$$\begin{cases} x_{12} - \dfrac{x_1 + x_2}{2} = -\Big(x_{21} - \dfrac{x_1 + x_2}{2}\Big), \\[2ex] v_{12} - \dfrac{v_1 + v_2}{2} = -\Big(v_{21} - \dfrac{v_1 + v_2}{2}\Big). \end{cases} \tag{8.9}$$

In the sequel, the symbol $O(m)$ will stand for any expression which is bounded by Cm, where C only depends on $\mathcal{V}$ and on the regularity bounds on the Lagrangian L on $\mathcal{V}$. From Cauchy–Lipschitz theory and Assumption (i),

$$|v_1 - v_2|(t) + |x_1 - x_2|(t) = O\Big(|X_1 - X_2| + |V_1 - V_2|\Big), \tag{8.10}$$

and then by plugging this back into the equation for minimizing curves we obtain

$$|\dot{v}_1 - \dot{v}_2|(t) = O\Big(|X_1 - X_2| + |V_1 - V_2|\Big).$$

Upon integration in times, these bounds imply

$$x_1(t) - x_2(t) = (X_1 - X_2) + O\big(\tau(|X_1 - X_2| + |V_1 - V_2|)\big); \tag{8.11}$$

$$v_1(t) - v_2(t) = (V_1 - V_2) + O\big(\tau(|X_1 - X_2| + |V_1 - V_2|)\big); \tag{8.12}$$

and therefore also

$$x_1(t)-x_2(t) = (X_1-X_2)+(t-t_0)\,(V_1-V_2)+O\big(\tau^2(|X_1-X_2|+|V_1-V_2|)\big). \tag{8.13}$$

As a consequence of (8.12), if τ is small enough (depending only on the Lagrangian L),

$$|v_1 - v_2|(t) \geq \frac{|V_1 - V_2|}{2} - O\big(\tau|X_1 - X_2|\big). \tag{8.14}$$

Next, from Cauchy–Lipschitz again,

$$x_2(t_0+\tau)-x_1(t_0+\tau) = X_2-X_1+\tau(V_2-V_1)+O\big(\tau^2(|X_1-X_2|+|V_1-V_2|)\big);$$

and since a similar expression holds true with τ replaced by $-\tau$, one has

$$\left[\frac{x_2(t_0+\tau)-x_1(t_0+\tau)}{2}\right]-\left[\frac{x_1(t_0-\tau)-x_2(t_0-\tau)}{2}\right]$$
$$= (X_2-X_1)+O\Big(\tau^2(|X_1-X_2|+|V_1-V_2|)\Big), \quad (8.15)$$

and also

$$\left[\frac{x_2(t_0+\tau)-x_1(t_0+\tau)}{2}\right]+\left[\frac{x_1(t_0-\tau)-x_2(t_0-\tau)}{2}\right]$$
$$= \tau\,(V_2-V_1)+O\Big(\tau^2\,(|X_1-X_2|+|V_1-V_2|)\Big). \quad (8.16)$$

It follows from (8.15) that

$$v_{12}(t)-\frac{v_1(t)+v_2(t)}{2}=O\Big(\frac{|X_1-X_2|}{\tau}+\tau|V_1-V_2|\Big). \quad (8.17)$$

After integration in time and use of (8.16), one obtains

$$\begin{aligned} x_{12}(t)-\frac{x_1(t)+x_2(t)}{2} &= \left[x_{12}(t_0)-\frac{x_1(t_0)+x_2(t_0)}{2}\right]+O\Big(|X_1-X_2|+\tau^2|V_1-V_2|\Big) \\ &= O\Big(|X_1-X_2|+\tau|V_1-V_2|\Big) \quad (8.18) \end{aligned}$$

In particular,

$$|x_{12}-x_{21}|(t)=O\Big(|X_1-X_2|+\tau|V_1-V_2|\Big). \quad (8.19)$$

Step 3: Taylor formulas and regularity of L. Now I shall evaluate the behavior of L along the old and the new paths, using the regularity assumption (ii). From that point on, I shall drop the time variable for simplicity (but it is implicit in all the computations). First,

$$\begin{aligned} L(x_1,v_1)-L\left(\frac{x_1+x_2}{2},v_1\right) \\ = \nabla_x L\left(\frac{x_1+x_2}{2},v_1\right)\cdot\left(\frac{x_1-x_2}{2}\right)+O(|x_1-x_2|^{1+\alpha}); \end{aligned}$$

similarly

$$L(x_2, v_2) - L\left(\frac{x_1 + x_2}{2}, v_2\right)$$
$$= \nabla_x L\left(\frac{x_1 + x_2}{2}, v_2\right) \cdot \left(\frac{x_2 - x_1}{2}\right) + O\big(|x_1 - x_2|^{1+\alpha}\big).$$

Moreover,

$$\nabla_x L\left(\frac{x_1 + x_2}{2}, v_1\right) - \nabla_x L\left(\frac{x_1 + x_2}{2}, v_2\right) = O(|v_1 - v_2|^{\alpha}).$$

The combination of these three identities, together with estimates (8.11) and (8.12), yields

$$\Big(L(x_1, v_1) + L(x_2, v_2)\Big) - \Big(L\Big(\frac{x_1 + x_2}{2}, v_1\Big) + L\Big(\frac{x_1 + x_2}{2}, v_2\Big)\Big)$$
$$= O\Big(|x_1 - x_2|^{1+\alpha} + |x_1 - x_2|\,|v_1 - v_2|^{\alpha}\Big)$$
$$= O\Big(|X_1 - X_2|^{1+\alpha} + \tau|V_1 - V_2|^{1+\alpha} + |X_1 - X_2|\,|V_1 - V_2|^{\alpha}$$
$$+ \tau^{1+\alpha}\,|V_1 - V_2|\,|X_1 - X_2|^{\alpha}\Big).$$

Next, in an analogous way,

$$L(x_{12}, v_{12}) - L\Big(x_{12}, \frac{v_1 + v_2}{2}\Big)$$
$$= \nabla_v L\Big(x_{12}, \frac{v_1 + v_2}{2}\Big) \cdot \Big(v_{12} - \frac{v_1 + v_2}{2}\Big) + O\left(\Big|v_{12} - \frac{v_1 + v_2}{2}\Big|^{1+\alpha}\right),$$

$$L(x_{21}, v_{21}) - L\Big(x_{21}, \frac{v_1 + v_2}{2}\Big)$$
$$= \nabla_v L\Big(x_{21}, \frac{v_1 + v_2}{2}\Big) \cdot \Big(v_{21} - \frac{v_1 + v_2}{2}\Big) + O\left(\Big|v_{21} - \frac{v_1 + v_2}{2}\Big|^{1+\alpha}\right),$$

$$\nabla_v L\left(x_{12}, \frac{v_1 + v_2}{2}\right) - \nabla_v L\left(x_{21}, \frac{v_1 + v_2}{2}\right) = O\big(|x_{12} - x_{21}|^{\alpha}\big).$$

Combining this with (8.9), (8.17) and (8.19), one finds

$$\Big(L(x_{12}, v_{12}) + L(x_{21}, v_{21})\Big) - \Big(L\big(x_{12}, \frac{v_1 + v_2}{2}\big) + L\big(x_{21}, \frac{v_1 + v_2}{2}\big)\Big)$$
$$= O\Big(\Big|v_{12} - \frac{v_1 + v_2}{2}\Big|^{1+\alpha} + \Big|v_{12} - \frac{v_1 + v_2}{2}\Big|\,|x_{12} - x_{21}|^{\alpha}\Big)$$
$$= O\left(\frac{|X_1 - X_2|^{1+\alpha}}{\tau^{1+\alpha}} + \tau^{1+\alpha}|V_1 - V_2|^{1+\alpha}\right).$$

After that,

$$L\Big(x_{12}, \frac{v_1+v_2}{2}\Big) - L\Big(\frac{x_1+x_2}{2}, \frac{v_1+v_2}{2}\Big)$$
$$= \nabla_x L\Big(\frac{x_1+x_2}{2}, \frac{v_1+v_2}{2}\Big)\cdot\Big(x_{12} - \frac{x_1+x_2}{2}\Big) + O\left(\Big|x_{12} - \frac{x_1+x_2}{2}\Big|^{1+\alpha}\right),$$

$$L\Big(x_{21}, \frac{v_1+v_2}{2}\Big) - L\Big(\frac{x_1+x_2}{2}, \frac{v_1+v_2}{2}\Big)$$
$$= \nabla_x L\Big(\frac{x_1+x_2}{2}, \frac{v_1+v_2}{2}\Big)\cdot\Big(x_{21} - \frac{x_1+x_2}{2}\Big) + O\left(\Big|x_{21} - \frac{x_1+x_2}{2}\Big|^{1+\alpha}\right),$$

and now by (8.9) the terms in ∇_x cancel each other exactly upon sommation, so the bound (8.18) leads to

$$\begin{aligned}\Big(L\big(x_{12}, \frac{v_1+v_2}{2}\big) + L\big(x_{21}, \frac{v_1+v_2}{2}\big)\Big) - 2L\Big(\frac{x_1+x_2}{2}, \frac{v_1+v_2}{2}\Big) &= O\left(\Big|x_{21} - \frac{x_1+x_2}{2}\Big|^{1+\alpha}\right)\\ &= O\Big(|X_1 - X_2|^{1+\alpha} + \tau^{1+\alpha}|V_1 - V_2|^{1+\alpha}\Big).\end{aligned}$$

Step 4: Comparison of actions and strict convexity. From our minimization assumption,

$$\mathcal{A}(x_1) + \mathcal{A}(x_2) \le \mathcal{A}(x_{12}) + \mathcal{A}(x_{21}),$$

which of course can be rewritten

$$\int_{t_0-\tau}^{t_0+\tau} \Big(L(x_1(t), v_1(t), t) + L(x_2(t), v_2(t), t) - L(x_{12}(t), v_{12}(t), t) - L(x_{21}(t), v_{21}(t), t)\Big)\, dt \le 0. \tag{8.20}$$

From Step 3, we can replace in the integrand all positions by $(x_1+x_2)/2$, and v_{12}, v_{21} by $(v_1+v_2)/2$, up to a small error. Collecting the various error terms, and taking into account the smallness of τ, one obtains (dropping the t variable again)

$$\frac{1}{2\tau}\int_{t_0-\tau}^{t_0+\tau} \Big\{L\Big(\frac{x_1+x_2}{2}, v_1\Big) + L\Big(\frac{x_1+x_2}{2}, v_2\Big) - 2L\Big(\frac{x_1+x_2}{2}, \frac{v_1+v_2}{2}\Big)\Big\}\, dt \le C\Big(\frac{|X_1 - X_2|^{1+\alpha}}{\tau^{1+\alpha}} + \tau|V_1 - V_2|^{1+\alpha}\Big). \tag{8.21}$$

On the other hand, from the convexity condition (iii) and (8.14), the left-hand side of (8.21) is bounded below by

$$K\,\frac{1}{2\tau}\int_{t_0-\tau}^{t_0+\tau}|v_1-v_2|^{2+\kappa}\,dt \geq K'\Big(|V_1-V_2|-A\tau|X_1-X_2|\Big)^{2+\kappa}. \tag{8.22}$$

If $|V_1-V_2| \leq 2A\tau\,|X_1-X_2|$, then the proof is finished. If this is not the case, this means that $|V_1-V_2|-A\tau|X_1-X_2| \geq |V_1-V_2|/2$, and then the comparison of the upper bound (8.21) and the lower bound (8.22) yields

$$|V_1-V_2|^{2+\kappa} \leq C\Big(\frac{|X_1-X_2|^{1+\alpha}}{\tau^{1+\alpha}}+\tau|V_1-V_2|^{1+\alpha}\Big). \tag{8.23}$$

If $|V_1-V_2|=0$, then the proof is finished. Otherwise, the conclusion follows by choosing τ small enough that $C\tau|V_1-V_2|^{1+\alpha} \leq (1/2)|V_1-V_2|^{2+\kappa}$; then $\tau = O(|V_1-V_2|^{1+\kappa-\alpha})$ and (8.23) implies

$$|V_1-V_2| = O(|X_1-X_2|^{\beta}), \qquad \beta = \frac{1+\alpha}{(1+\alpha)(1+\kappa-\alpha)+2+\kappa}. \tag{8.24}$$

In the particular case when $\kappa=0$ and $\alpha=1$, one has

$$|V_1-V_2|^2 \leq C\left(\frac{|X_1-X_2|^2}{\tau^2}+\tau|V_1-V_2|^2\right),$$

and if τ is small enough this implies just

$$|V_1-V_2| \leq C\,\frac{|X_1-X_2|}{\tau}. \tag{8.25}$$

The upper bound on τ depends on the regularity and strict convexity of τ in $\mathcal{V}$, but also on t_0 since τ cannot be greater than $\min(t_0, 1-t_0)$. This is actually the only way in which t_0 explicitly enters the estimates. So inequality (8.25) concludes the argument. □

Complement: Ruling out focalization by shortening

This section is about the application of the shortening technique to a classical problem in Riemannian geometry; it may be skipped at first reading.

Let M be a smooth Riemannian manifold and let $L = L(x, v, t)$ be a C^2 Lagrangian on $TM \times [0, 1]$, satisfying the classical conditions of Definition 7.6, together with $\nabla_v^2 L > 0$. Let $X_t(x_0, v_0) = X_t(x_0, v_0, 0)$ be the solution at time t of the flow associated with the Lagrangian L, starting from the initial position x_0 at time 0.

It is said that there is **focalization** on another point $x' = X_{t'}(x_0, v_0)$, $t' > 0$, if the differential map $d_{v_0} X_{t'}(x_0, \cdot)$ is singular (not invertible). In words, this means that starting from x_0 it is very difficult to make the curve explore a whole neighborhood of x' by varying its initial velocity; instead, trajectories have a tendency to "concentrate" at time t' along certain preferred directions around x'.

The reader can test his or her understanding of the method presented in the previous section by working out the details of the following problem.

Problem 8.8 (Focalization is impossible before the cut locus). *With the same notation as before, let $\gamma : [0, 1] \to M$ be a minimizing curve starting from some initial point x_0. By using the same strategy of proof as for Mather's estimates, show that, starting from x_0, focalization is impossible at $\gamma(t_*)$ if $0 < t_* < 1$.*

Hint: *Here is a possible reasoning:*

(a) Notice that the restriction of γ to $[0, t_*]$ is the unique minimizing curve on the time-interval $[0, t_*]$, joining x_0 to $x_* = \gamma(t_*)$.

(b) Take y close to x_* and introduce a minimizing curve $\widetilde{\gamma}$ on $[0, t_*]$, joining x_0 to y; show that the initial velocity $\widetilde{v}_0$ of $\widetilde{\gamma}$ is close to the initial velocity v_0 of γ if y is close enough to x_*.

(c) Bound the difference between the action of γ and the action of $\widetilde{\gamma}$ by $O(d(x_*, y))$ (recall that the speeds along γ and $\widetilde{\gamma}$ are bounded by a uniform constant, depending only of the behavior of L in some compact set around γ).

(d) Construct a path $x_0 \to \gamma(1)$ by first going along $\widetilde{\gamma}$ up to time $t = t_* - \tau$ (τ small enough), then using a shortcut from $\widetilde{\gamma}(t_* - \tau)$ to $\gamma(t_* + \tau)$, finally going along γ up to time 1. Show that the gain of action is at least of the order of $\tau |V - \widetilde{V}|^2 - O(d(x_*, y)^2/\tau)$, where $V = \dot{\gamma}(t_*)$ and $\widetilde{V} = \dot{\widetilde{\gamma}}(t_*)$. Deduce that $|V - \widetilde{V}| = O(d(x_*, y)/\tau)$.

(e) Conclude that $|v_0 - \widetilde{v_0}| = O(d(x_*, y)/\tau)$. Use a contradiction argument to deduce that the differential map $d_{v_0} X_t(x_0, \cdot)$ is invertible, and more precisely that its inverse is of size $O((1 - t_*)^{-1})$ as a function of t_*.

In the important case when $L(x,v,t) = |v|^2$, what we have proven is a well-known result in Riemannian geometry; to explain it I shall first recall the notions of **cut locus** and **focal points**.

Let γ be a minimizing geodesic, and let t_c be the largest time such that for all $t < t_c$, γ is minimizing between γ_0 and γ_t. Roughly speaking, $\gamma(t_c)$ is the first point at which the geodesic ceases to be minimizing; γ may or may not be minimizing between $\gamma(0)$ and $\gamma(t_c)$, but it is certainly not minimizing between $\gamma(0)$ and $\gamma(t_c + \varepsilon)$, for any $\varepsilon > 0$. Then the point $\gamma(t_c)$ is said to be a cut point of γ_0 along γ. When the initial position x_0 of the geodesic is fixed and the geodesic varies, the set of all cut points constitutes the *cut locus* of x_0.

Next, two points x_0 and x' are said to be *focal* (or conjugate) if x' can be written as $\exp_{x_0}(t'v_0)$, where the differential $d_{v_0}\exp_{x_0}(t' \cdot\,)$ is *not invertible*. As before, this means that x' can be obtained from x_0 by a geodesic γ with $\dot{\gamma}(0) = v_0$, such that it is difficult to explore a whole neighborhood of x' by slightly changing the initial velocity v_0.

With these notions, the main result of Problem 8.8 can be summarized as follows: *Focalization never occurs before the cut locus.* It can occur either at the cut locus, or after.

Example 8.9. On the sphere S^2, the north pole N has only one cut point, which is also its only focal point, namely the south pole S. Fix a geodesic γ going from $\gamma(0) = N$ to $\gamma(1) = S$, and deform your sphere out of a neighborhood of $\gamma[0,1]$, so as to dig a shortcut that allows you to go from N to $\gamma(1/2)$ in a more efficient way than using γ. This will create a new cut point along γ, and S will not be a cut point along γ any longer (it might still be a cut point along some other geodesic). On the other hand, S will still be the only focal point along γ.

Remark 8.10. If x and y are not conjugate, and joined by a unique minimizing geodesic γ, then it is easy to show that there is a neighborhood U of y such that any z in U is also joined to x by a unique minimizing geodesic. Indeed, any minimizing geodesic has to be close to γ, therefore its initial velocity should be close to $\dot{\gamma}_0$; and by the local inversion theorem, there are neighborhoods W_0 of $\dot{\gamma}_0$ and U of y such that there is a unique correspondence between the initial velocity $\dot{\gamma} \in W_0$ of a minimizing curve starting from x, and the final point $\gamma(1) \in U$. Thus the cut locus of a point x can be separated into two sets:

(a) those points y for which there are at least two distinct minimizing geodesics going from x to y;

(b) those points y for which there is a unique minimizing geodesic, but which are focal points of x.

Introduction to Mather's theory

In this section I shall present an application of Theorem 8.1 to the theory of Lagrangian dynamical systems. This is mainly to give the reader an idea of Mather's motivations, and to let him or her better understand the link between optimal transport and Mather's theory. These results will not play any role in the sequel of the notes.

Theorem 8.11 (Lipschitz graph theorem). *Let M be a compact Riemannian manifold, let $L = L(x, v, t)$ be a Lagrangian function on $TM \times \mathbb{R}$, and $T > 0$, such that*

(a) L is T-periodic in the t variable, i.e. $L(x, v, t+T) = L(x, v, t)$;

(b) L is of class C^2 in all variables;

(c) $\nabla_v^2 L$ is (strictly) positive everywhere, and L is superlinear in v.

Define as usual the action by $\mathcal{A}^{s,t}(\gamma) = \int_s^t L(\gamma_\tau, \dot{\gamma}_\tau, \tau)\, d\tau$. Let $c^{s,t}$ be the associated cost function on $M \times M$, and $C^{s,t}$ the corresponding optimal cost functional on $P(M) \times P(M)$.

Let $\overline{\mu}$ be a probability measure solving the minimization problem

$$\inf_{\mu \in P(\mathcal{X})} C^{0,T}(\mu, \mu), \tag{8.26}$$

and let $(\mu_t)_{0 \le t \le T}$ be a displacement interpolation between $\mu_0 = \overline{\mu}$ and $\mu_T = \overline{\mu}$. Extend (μ_t) into a T-periodic curve $\mathbb{R} \to P(M)$ defined for all times. Then

(i) For all $t_0 < t_1$, the curve $(\mu_t)_{t_0 \le t \le t_1}$ still defines a displacement interpolation;

(ii) The optimal transport cost $C^{t,t+T}(\mu_t, \mu_t)$ is independent of t;

(iii) For any $t_0 \in \mathbb{R}$, and for any $k \in \mathbb{N}$, μ_{t_0} is a minimizer for $C^{t_0,t_0+kT}(\mu, \mu)$.

Moreover, there is a random curve $(\gamma_t)_{t \in \mathbb{R}}$, such that

(iv) For all $t \in \mathbb{R}$, $\mathrm{law}\,(\gamma_t) = \mu_t$;

(v) For any $t_0 < t_1$, the curve $(\gamma_t)_{t_0 \le t \le t_1}$ is action-minimizing;

(vi) The map $\gamma_0 \to \dot{\gamma}_0$ is well-defined and Lipschitz.

Remark 8.12. Since $c^{0,T}$ is not assumed to be nonnegative, the optimal transport problem (8.26) is not trivial.

Remark 8.13. If L does not depend on t, then one can apply the previous result for any $T = 2^{-\ell}$, and then use a compactness argument to construct a constant curve $(\mu_t)_{t\in\mathbb{R}}$ satisfying Properties (i)–(vi) above. In particular μ_0 is a *stationary measure* for the Lagrangian system.

Before giving its proof, let me explain briefly why Theorem 8.11 is interesting from the point of view of the dynamics. A trajectory of the dynamical system defined by the Lagrangian L is a curve γ which is locally action-minimizing; that is, one can cover the time-interval by small subintervals on which the curve is action-minimizing. It is a classical problem in mechanics to construct and study periodic trajectories having certain given properties. Theorem 8.11 does not construct a periodic trajectory, but at least it constructs a random trajectory γ (or equivalently a probability measure Π on the set of trajectories) which is periodic *on average*: The law μ_t of γ_t satisfies $\mu_{t+T} = \mu_t$. This can also be thought of as a probability measure Π on the set of all possible trajectories of the system.

Of course this in itself is not too striking, since there may be a great deal of invariant measures for a dynamical system, and some of them are often easy to construct. The important point in the conclusion of Theorem 8.11 is that the curve γ is not "too random", in the sense that the random variable $(\gamma(0), \dot{\gamma}(0))$ takes values in a Lipschitz *graph*. (If $(\gamma(0), \dot{\gamma}(0))$ were a deterministic element in TM, this would mean that Π just sees a single periodic curve. Here we may have an infinite collection of curves, but still it is not "too large".)

Another remarkable property of the curves γ is the fact that the minimization property holds along *any* time-interval in $\mathbb{R}$, not necessarily small.

Example 8.14. Let M be a compact Riemannian manifold, and let $L(x,v,t) = |v|^2/2 - V(x)$, where V has a unique maximum x_0. Then Mather's procedure selects the probability measure δ_{x_0}, and the stationary curve $\gamma \equiv x_0$ (which is an unstable equilibrium).

It is natural to try to construct more "interesting" measures and curves by Mather's procedure. One way to do so is to change the Lagrangian, for instance replace $L(x,v,t)$ by $L_\omega := L(x,v,t) + \omega(x)\cdot v$, where ω is a vector field on M. Indeed,

- If ω is closed (as a differential form), that is if $\nabla\omega$ is a symmetric operator, then L_ω and L have the same Euler–Lagrange equations, so the associated dynamical system is the same;
- If ω is exact, that is if $\omega = \nabla f$ for some function $f : M \to \mathbb{R}$, then L_ω and L have the same minimizing curves.

As a consequence, one may explore various parts of the dynamics by letting ω vary over the finite-dimensional group obtained by taking the quotient of the closed forms by the exact forms. In particular, one can make sure that the expected mean "rotation number" $\mathbb{E}\left(\frac{1}{T}\int_0^T \dot\gamma\,dt\right)$ takes nontrivial values as ω varies.

Proof of Theorem 8.11. I shall repeatedly use Proposition 7.16 and Theorem 7.21. First, $C^{0,T}(\mu,\mu)$ is a lower semicontinuous function of μ, bounded below by $T(\inf L) > -\infty$, so the minimization problem (8.26) does admit a solution.

Define $\mu_0 = \mu_T = \overline{\mu}$, then define μ_t by displacement interpolation for $0 < t < T$, and extend the result by periodicity.

Let $k \in \mathbb{N}$ be given and let $\widetilde{\mu}$ be a minimizer for the variational problem

$$\inf_{\mu\in P(M)} C^{0,kT}(\mu,\mu).$$

We shall see later that actually $\overline{\mu}$ is a solution of this problem. For the moment, let $(\widetilde{\mu}_t)_{t\in\mathbb{R}}$ be obtained first by taking a displacement interpolation between $\widetilde{\mu}_0 = \widetilde{\mu}$ and $\widetilde{\mu}_{kT} = \widetilde{\mu}$; and then by extending the result by kT-periodicity.

On the one hand,

$$C^{0,kT}(\widetilde{\mu},\widetilde{\mu}) \le C^{0,kT}(\mu_0,\mu_{kT}) \le \sum_{j=0}^{k-1} C^{jT,\,(j+1)T}(\mu_{jT},\mu_{(j+1)T}) = k\,C^{0,1}(\overline{\mu},\overline{\mu}). \tag{8.27}$$

On the other hand, by definition of $\overline{\mu}$,

$$C^{0,T}(\overline{\mu},\overline{\mu}) \le C^{0,T}\Big(\frac{1}{k}\sum_{j=0}^{k-1}\widetilde{\mu}_{jT},\ \frac{1}{k}\sum_{j=0}^{k-1}\widetilde{\mu}_{jT}\Big) \tag{8.28}$$

$$= C^{0,T}\Big(\frac{1}{k}\sum_{j=0}^{k-1}\widetilde{\mu}_{jT},\ \frac{1}{k}\sum_{j=0}^{k-1}\widetilde{\mu}_{(j+1)T}\Big). \tag{8.29}$$

Since $C^{0,T}(\mu,\nu)$ is a convex function of (μ,ν) (Theorem 4.8),

$$C^{0,T}\Big(\frac{1}{k}\sum_{j=0}^{k-1}\widetilde{\mu}_{jT},\,\frac{1}{k}\sum_{j=0}^{k-1}\widetilde{\mu}_{(j+1)T}\Big) \le \frac{1}{k}\sum_{j=0}^{k-1} C^{jT,\,(j+1)T}(\widetilde{\mu}_{jT},\widetilde{\mu}_{(j+1)T})$$
$$= \frac{1}{k}\,C^{0,kT}(\widetilde{\mu}_0,\widetilde{\mu}_{kT}), \tag{8.30}$$

where the last equality is a consequence of Property (ii) in Theorem 7.21. Inequalities (8.29) and (8.30) together imply

$$C^{0,1}(\overline{\mu},\overline{\mu}) \le \frac{1}{k}\,C^{0,kT}(\widetilde{\mu}_0,\widetilde{\mu}_{kT}) = \frac{1}{k}\,C^{0,kT}(\widetilde{\mu},\widetilde{\mu}).$$

Since the reverse inequality holds true by (8.27), in fact all the inequalities in (8.27), (8.29) and (8.30) have to be equalities. In particular,

$$C^{0,kT}(\mu_0,\mu_{kT}) = \sum_{j=0}^{k-1} C^{jT,\,(j+1)T}(\mu_{jT},\mu_{(j+1)T}). \tag{8.31}$$

Let us now check that the identity

$$C^{t_1,t_2}(\mu_{t_1},\mu_{t_2}) + C^{t_2,t_3}(\mu_{t_2},\mu_{t_3}) = C^{t_1,t_3}(\mu_{t_1},\mu_{t_3}) \tag{8.32}$$

holds true for any three intermediate times $t_1 < t_2 < t_3$. By periodicity, it suffices to do this for $t_1 \ge 0$. If $0 \le t_1 < t_2 < t_3 \le T$, then (8.32) is true by the property of displacement interpolation (Theorem 7.21 again). If $jT \le t_1 < t_2 < t_3 \le (j+1)T$, this is also true because of the T-periodicity. In the remaining cases, we may choose k large enough that $t_3 \le kT$. Then

$$\begin{aligned}
&C^{0,kT}(\mu_0,\mu_{kT}) \le C^{0,t_1}(\mu_0,\mu_{t_1}) + C^{t_1,t_3}(\mu_{t_1},\mu_{t_3}) + C^{t_3,kT}(\mu_{t_3},\mu_{kT})\\
&\le C^{0,t_1}(\mu_0,\mu_{t_1}) + C^{t_1,t_2}(\mu_{t_1},\mu_{t_2}) + C^{t_2,t_3}(\mu_{t_2},\mu_{t_3}) + C^{t_3,kT}(\mu_{t_3},\mu_{kT})\\
&\le \sum C^{s_j,s_{j+1}}(\mu_{s_j},\mu_{s_{j+1}}),
\end{aligned} \tag{8.33}$$

where the times s_j are obtained by ordering of $\{0,T,2T,\ldots,kT\} \cup \{t_1,t_2,t_3\}$. On each time-interval $[\ell T,(\ell+1)T]$ we know that (μ_t) is a displacement interpolation, so we can apply Theorem 7.21(ii), and as a result bound the right-hand side of (8.33) by

$$\sum_{\ell} C^{\ell T,(\ell+1)T}(\mu_{\ell T},\mu_{(\ell+1)T}). \tag{8.34}$$

(Consider for instance the particular case when $0 < t_1 < t_2 < T < t_3 < 2T$; then one can write $C^{0,t_1} + C^{t_1,t_2} + C^{t_2,T} = C^{0,T}$, and also $C^{T,t_3} + C^{t_3,2T} = C^{T,2T}$. So $C^{0,t_1} + C^{t_1,t_2} + C^{t_2,T} + C^{T,t_3} + C^{t_3,2T} = C^{0,T} + C^{T,2T}$.)

But (8.34) is just $C^{0,kT}(\mu_0, \mu_{kT})$, as shown in (8.31). So there is in fact equality in all these inequalities, and (8.32) follows. Then by Theorem 7.21, (μ_t) defines a displacement interpolation between any two of its intermediate values. This proves (i). At this stage we have also proven (iii) in the case when $t_0 = 0$.

Now for any $t \in \mathbb{R}$, one has, by (8.32) and the T-periodicity,

$$\begin{aligned} C^{0,T}(\mu_0, \mu_T) &= C^{0,t}(\mu_0, \mu_t) + C^{t,T}(\mu_t, \mu_T) \\ &= C^{t,T}(\mu_t, \mu_T) + C^{T,t+T}(\mu_T, \mu_{t+T}) \\ &= C^{t,t+T}(\mu_t, \mu_{t+T}) \\ &= C^{t,t+T}(\mu_t, \mu_t), \end{aligned}$$

which proves (ii).

Next, let t_0 be given, and repeat the same whole procedure with the initial time 0 replaced by t_0: That is, introduce a minimizer $\widetilde{\mu}$ for $C^{t_0,t_0+T}(\mu, \mu)$, etc. This gives a curve $(\widetilde{\mu}_t)_{t \in \mathbb{R}}$ with the property that $C^{t,t+T}(\widetilde{\mu}_t, \widetilde{\mu}_t) = C^{0,T}(\widetilde{\mu}_0, \widetilde{\mu}_0)$. It follows that

$$\begin{aligned} C^{t_0,t_0+T}(\mu_{t_0}, \mu_{t_0}) &= C^{0,T}(\overline{\mu}, \overline{\mu}) \leq C^{0,T}(\widetilde{\mu}_0, \widetilde{\mu}_0) \\ &= C^{t_0,t_0+T}(\widetilde{\mu}_{t_0}, \widetilde{\mu}_{t_0}) = C^{t_0,t_0+T}(\widetilde{\mu}, \widetilde{\mu}) \leq C^{t_0,t_0+T}(\mu_{t_0}, \mu_{t_0}). \end{aligned}$$

So there is equality everywhere, and μ_{t_0} is indeed a minimizer for $C^{t_0,t_0+T}(\mu, \mu)$. This proves the remaining part of (iii).

Next, let $(\gamma_t)_{0 \leq t \leq T}$ be a random minimizing curve on $[0, T]$ with $\mathrm{law}(\gamma_t) = \mu_t$, as in Theorem 7.21. For each k, define $(\gamma_t^k)_{kT \leq t \leq (k+1)T}$ as a copy of $(\gamma_t)_{0 \leq t \leq T}$. Since μ_t is T-periodic, $\mathrm{law}(\gamma_{kT}^k) = \mathrm{law}(\gamma_{(k+1)T}^k) = \mu_0$, for all k. So we can *glue together* these random curves, just as in the proof of Theorem 7.30, and get random curves $(\gamma_t)_{t \in \mathbb{R}}$ such that $\mathrm{law}(\gamma_t) = \mu_t$ for all $t \in \mathbb{R}$, and each curve $(\gamma_t)_{kT \leq t \leq (k+1)T}$ is action-minimizing. Property (iv) is then satisfied by construction.

Property (v) can be established by a principle which was already used in the proof of Theorem 7.21. Let us check for instance that γ is minimizing on $[0, 2T]$. For this one has to show that (almost surely)

$$c^{t_1,t_2}(\gamma_{t_1}, \gamma_{t_2}) + c^{t_2,t_3}(\gamma_{t_2}, \gamma_{t_3}) = c^{t_1,t_3}(\gamma_{t_1}, \gamma_{t_3}), \tag{8.35}$$

for any choice of intermediate times $t_1 < t_2 < t_3$ in $[0, 2T]$. Assume, without real loss of generality, that $0 < t_1 < t_2 < T < t_3 < 2T$. Then

$$\begin{aligned}
C^{t_1,t_3}(\mu_{t_1},\mu_{t_3}) &\leq \mathbb{E}\, c^{t_1,t_3}(\gamma_{t_1},\gamma_{t_3}) \\
&\leq \mathbb{E}\left[c^{t_1,t_2}(\gamma_{t_1},\gamma_{t_2}) + c^{t_2,t_3}(\gamma_{t_2},\gamma_{t_3})\right] \\
&\leq \mathbb{E}\, c^{t_1,t_2}(\gamma_{t_1},\gamma_{t_2}) + \mathbb{E}\, c^{t_2,T}(\gamma_{t_2},\gamma_T) + \mathbb{E}\, c^{T,t_3}(\gamma_T,\gamma_{t_3}) \\
&= C^{t_1,t_2}(\mu_{t_1},\mu_{t_2}) + C^{t_2,T}(\mu_{t_2},\mu_T) + C^{T,t_3}(\mu_T,\mu_{t_3}) \\
&= C^{t_1,t_3}(\mu_{t_1},\mu_{t_3}),
\end{aligned}$$

where the property of optimality of the path $(\mu_t)_{t\in\mathbb{R}}$ was used in the last step. So all these inequalities are equalities, and in particular

$$\mathbb{E}\left[c^{t_1,t_3}(\gamma_{t_1},\gamma_{t_3}) - c^{t_1,t_2}(\gamma_{t_1},\gamma_{t_2}) - c^{t_2,t_3}(\gamma_{t_2},\gamma_{t_3})\right] = 0.$$

Since the integrand is nonpositive, it has to vanish almost surely. So (8.35) is satisfied almost surely, for given t_1, t_2, t_3. Then the same inequality holds true almost surely for all choices of rational times t_1, t_2, t_3; and by continuity of γ it holds true almost surely at all times. This concludes the proof of (v).

From general principles of Lagrangian mechanics, there is a uniform bound on the speeds of all the curves $(\gamma_t)_{-T\leq t\leq T}$ (this is because γ_{-T} and γ_T lie in a compact set). So for any given $\varepsilon > 0$ we can find δ such that $0 \leq t \leq \delta$ implies $d(\gamma_0, \gamma_t) \leq \varepsilon$. Then if ε is small enough the map $(\gamma_0, \gamma_t) \to (\gamma_0, \dot{\gamma}_0)$ is Lipschitz. (This is another well-known fact in Lagrangian mechanics; it can be seen as a consequence of Remark 8.10.) But from Theorem 8.5, applied with the intermediate time $t_0 = 0$ on the time-interval $[-T, T]$, we know that $\gamma_0 \longmapsto \gamma_t$ is well-defined (almost surely) and Lipschitz continuous. It follows that $\gamma_0 \to \dot{\gamma}_0$ is also Lipschitz continuous. This concludes the proof of Theorem 8.11. □

The story does not end here. First, there is a powerful *dual point of view* to Mather's theory, based on solutions to the dual Kantorovich problem; this is a maximization problem defined by

$$\sup \int (\phi - \psi)\, d\mu, \tag{8.36}$$

where the supremum is over all probability measures μ on M, and all pairs of Lipschitz functions (ψ, ϕ) such that

$$\forall (x, y) \in M \times M, \qquad \phi(y) - \psi(x) \leq c^{0,T}(x, y).$$

Next, Theorem 8.11 suggests that some objects related to optimal transport might be interesting to describe a Lagrangian system. This is indeed the case, and the notions defined below are useful and well-known in the theory of dynamical systems:

Definition 8.15 (Useful transport quantities describing a Lagrangian system). *For each displacement interpolation $(\mu_t)_{t\geq 0}$ as in Theorem 8.11, define*

(i) the Mather critical value as the opposite of the mean optimal transport cost:

$$-M = \overline{c} := \frac{1}{T}\, C^{0,T}(\mu,\mu) = \frac{1}{kT}\, C^{0,kT}(\mu,\mu); \tag{8.37}$$

(ii) the Mather set as the closure of the union of the supports of all measures $V_\#\mu_0$, where $(\mu_t)_{t\geq 0}$ is a displacement interpolation as in Theorem 8.11 and V is the Lipschitz map $\gamma_0 \to (\gamma_0, \dot\gamma_0)$;

(iii) the Aubry set as the set of all $(\gamma_0, \dot\gamma_0)$ such that there is a solution (ϕ, ψ) of the dual problem (8.36) satisfying $H_+^{0,T}\psi(\gamma_1) - \psi(\gamma_0) = c^{0,T}(\gamma_0, \gamma_1)$.

Up to the change of variables $(\gamma_0, \dot\gamma_0) \to (\gamma_0, \gamma_1)$, the Mather and Aubry sets are just the same as $\Gamma_{\min}$ and $\Gamma_{\max}$ appearing in the bibliographical notes of Chapter 5.

Example 8.16. Take a one-dimensional pendulum. For small values of the total energy, the pendulum is confined in a periodic motion, making just small oscillations, going back and forth around its equilibrium position and describing an arc of circle in physical space (see Figure 8.5). For large values, it also has a periodic motion but now it goes always in the same direction, and describes a complete circle ("revolution") in physical space. But if the system is given just the right amount of energy, it will describe a trajectory that is intermediate between these two regimes, and consists in going from the vertical upward position (at time $-\infty$) to the vertical upward position again (at time $+\infty$) after exploring all intermediate angles. There are two such trajectories (one clockwise, and one counterclockwise), which can be called revolutions of infinite period; and they are globally action-minimizing. When $\xi = 0$, the solution of the Mather problem is just the Dirac mass on the unstable equilibrium x_0, and the Mather and Aubry sets Γ are reduced to $\{(x_0, x_0)\}$. When ξ varies in $\mathbb{R}$, this remains the same until ξ reaches

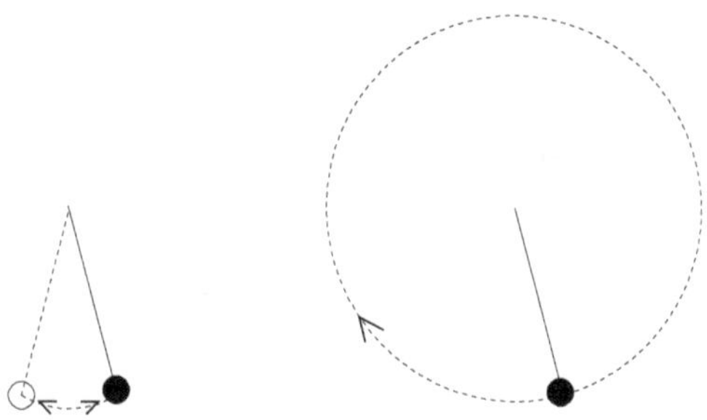

Fig. 8.5. In the left figure, the pendulum oscillates with little energy between two extreme positions; its trajectory is an arc of circle which is described clockwise, then counterclockwise, then clockwise again, etc. On the right figure, the pendulum has much more energy and draws complete circles again and again, either clockwise or counterclockwise.

a certain critical value; above that value, the Mather measures are supported by revolutions. At the critical value, the Mather and Aubry sets differ: the Aubry set (viewed in the variables (x, v)) is the union of the two revolutions of infinite period.

The dual point of view in Mather's theory, and the notion of the Aubry set, are intimately related to the so-called **weak KAM theory**, in which stationary solutions of Hamilton–Jacobi equations play a central role. The next theorem partly explains the link between the two theories.

Theorem 8.17 (Mather critical value and stationary Hamilton–Jacobi equation). *With the same notation as in Theorem 8.11, assume that the Lagrangian L does not depend on t, and let ψ be a Lipschitz function on M, such that $H_+^{0,t}\psi=\psi+c\,t$ for all times $t \geq 0$; that is, ψ is left invariant by the forward Hamilton–Jacobi semigroup, except for the addition of a constant which varies linearly in time. Then, necessarily $c=\overline{c}$, and the pair $(\psi, H_+^{0,T}\psi)=(\psi,\psi+\overline{c}\,T)$ is optimal in the dual Kantorovich problem with cost function $c^{0,T}$, and initial and final measures equal to $\overline{\mu}$.*

Remark 8.18. The equation $H_+^{0,1}\psi = \psi + c\,t$ is a way to reformulate the stationary Hamilton–Jacobi equation $H(x, \nabla\psi(x)) + c = 0$. Yet another reformulation would be obtained by changing the forward Hamilton–Jacobi semigroup for the backward one. Theorem 8.17 does not guarantee the existence of such stationary solutions, it just states that *if* such solutions exist, then the value of the constant c is uniquely determined and can be related to a Monge–Kantorovich problem. In weak KAM theory, one then establishes the existence of these solutions by independent means; see the references suggested in the bibliographical notes for much more information.

Remark 8.19. The constant $-\overline{c}$ (which coincides with Mather's critical value) is often called the **effective Hamiltonian** of the system.

Proof of Theorem 8.17. To fix the ideas, let us impose $T = 1$. Let ψ be such that $H_+^{0,1}\psi = \psi + c$, and let μ be any probability measure on M; then

$$\int (H_+^{0,1}\psi)\, d\mu - \int \psi\, d\mu = \int c\, d\mu = c.$$

By the easy part of the Kantorovich duality, $C^{0,1}(\mu,\mu) \geq c$. By taking the infimum over all $\mu \in P(M)$, we conclude that $\overline{c} \geq c$.

To prove the reverse inequality, it suffices to construct a particular probability measure μ such that $C^{0,1}(\mu,\mu) \leq c$. The idea is to look for μ as a limit of probability measures distributed uniformly over some well-chosen long minimizing trajectories. Before starting this construction, we first remark that since M is compact, there is a uniform bound C on $L(\gamma(t), \dot{\gamma}(t))$, for all action-minimizing curves $\gamma : [0,1] \to M$; and since L is time-independent, this statement trivially extends to all action-minimizing curves defined on time-intervals $[t_0, t_1]$ with $|t_0 - t_1| \geq 1$. Also ψ is uniformly bounded on M.

Now let x be an arbitrary point in M; for any $T > 0$ we have, by definition of the forward Hamilton–Jacobi semigroup,

$$(H_+^{-T,0}\psi)(x) = \inf\left\{\psi(\gamma(-T)) + \int_{-T}^{0} L(\gamma(s), \dot{\gamma}(s))\, ds; \quad \gamma(0) = x\right\},$$

where the infimum is over all action-minimizing curves $\gamma : [-T, 0] \to M$ ending at x. (The advantage in working with negative times is that one can fix one of the endpoints; in the present context where M is compact this is nonessential, but it would become important if M were noncompact.) By compactness, there is a minimizing curve $\gamma = \gamma^{(T)}$; then, by the definition of $\gamma^{(T)}$ and the stationarity of ψ,

$$\frac{1}{T}\int_{-T}^{0} L\big(\gamma^{(T)}(s),\dot\gamma^{(T)}(s)\big)\,ds = \frac{1}{T}\Big[(H_+^{-T,0}\psi)(x) - \psi(\gamma^{(T)}(-T))\Big]$$
$$= \frac{1}{T}\Big(\psi(x) + cT - \psi(\gamma^{(T)}(-T))\Big)$$
$$= c + O\left(\frac{1}{T}\right).$$

In the sequel, I shall write just γ for $\gamma^{(T+1)}$. Of course the estimate above remains unchanged upon replacement of T by $T+1$, so

$$\frac{1}{T}\int_{-(T+1)}^{0} L(\gamma(s),\dot\gamma(s))\,ds = c + O\left(\frac{1}{T}\right).$$

Then define

$$\mu_T := \frac{1}{T}\int_{-(T+1)}^{-1} \delta_{\gamma(s)}\,ds; \qquad \nu_T := \frac{1}{T}\int_{-T}^{0} \delta_{\gamma(s)}\,ds;$$

and $\theta : \gamma(s) \longmapsto \gamma(s+1)$. It is clear that $\theta_{\#}\mu_T = \nu_T$; moreover,

$$c^{0,1}\big(\gamma(s),\theta(\gamma(s))\big) = c^{0,1}\big(\gamma(s),\gamma(s+1)\big) = \int_s^{s+1} L(\gamma(u),\dot\gamma(u))\,du.$$

Thus by Theorem 4.8,

$$C^{0,1}(\mu_T,\nu_T) \le \frac{1}{T}\int_{-(T+1)}^{-1} c^{0,1}\big(\gamma(s),\theta(\gamma(s))\big)\,ds$$
$$= \frac{1}{T}\int_{-(T+1)}^{-1}\left(\int_s^{s+1} L(\gamma(u),\dot\gamma(u))\,du\right)ds$$
$$= \frac{1}{T}\int_{-(T+1)}^{0} L(\gamma(u),\dot\gamma(u))\,a(u)\,du, \tag{8.38}$$

where $a : [-(T+1),0] \to [0,1]$ is defined by

$$a(u) = \int 1_{s\le u\le s+1}\,ds = \begin{cases} 1 & \text{if } -T \le u \le -1;\\ -u & \text{if } -1 \le u \le 0;\\ u+T+1 & \text{if } -(T+1) \le u \le -T.\end{cases}$$

Replacing a by 1 in the integrand of (8.38) involves modifying the integral on a set of measure at most 2; so

$$C^{0,1}(\mu_T,\nu_T) \le \frac{1}{T}\int_{-(T+1)}^{0} L(\gamma(u),\dot\gamma(u))\,du + O\left(\frac{1}{T}\right) = c + O\left(\frac{1}{T}\right). \tag{8.39}$$

Since $P(M)$ is compact, the family $(\mu_T)_{T\in\mathbb{N}}$ converges, up to extraction of a subsequence, to some probability measure μ. Then (up to extraction of the same subsequence) ν_T also converges to μ, since

$$\|\mu_T - \nu_T\|_{TV} = \frac{1}{T}\left\|\int_{-(T+1)}^{-T}\delta_{\gamma(s)}\,ds + \int_{-1}^{0}\delta_{\gamma(s)}\,ds\right\|_{TV} \le \frac{2}{T}.$$

Then from (8.39) and the lower semicontinuity of the optimal transport cost,

$$C^{0,1}(\mu,\mu) \le \liminf_{T\to\infty}\, C^{0,1}(\mu_T,\nu_T)\ \le c.$$

This concludes the proof. □

The next exercise may be an occasion to manipulate the concepts introduced in this section.

Exercise 8.20. With the same assumptions as in Theorem 8.11, assume that L is symmetric in v; that is, $L(x,-v,t) = L(x,v,t)$. Show that $c^{0,T}(x,y) = c^{0,T}(y,x)$. Take an optimal measure $\overline{\mu}$ for the minimization problem (8.26), and let π be an associated optimal transference plan. By gluing together π and $\check{\pi}$ (obtained by exchanging the variables x and y), construct an optimal transference plan for the problem (8.26) with T replaced by $2T$, such that each point x stays in place. Deduce that the curves γ are $2T$-periodic. Show that $c^{0,2T}(x,x) = C^{0,2T}(\overline{\mu},\overline{\mu})$, and deduce that $c^{0,T}(x,y)$ is π-almost surely constant. Construct ψ such that $H_+^{0,2T}\psi = \psi + 2\overline{c}T$, $\overline{\mu}$-almost surely. Next assume that L does not depend on t, and use a compactness argument to construct a ψ and a stationary measure μ, such that $H_+^{0,t}\psi = \psi + \overline{c}t$, for all $t \ge 0$, $\overline{\mu}$-almost surely. *Note that* this is far from proving the existence of a stationary solution of the Hamilton–Jacobi equation, as appearing in Theorem 8.17, for two reasons: First the symmetry of L is a huge simplification; secondly the equation $H_+^{0,t}\psi = \psi + \overline{c}t$ should hold everywhere in M, not just $\overline{\mu}$-almost surely.

Possible extensions of Mather's estimates

As noticed in Example 8.4, it would be desirable to have a sharper version of Theorem 8.1 which would contain as a special case the correct exponents for the Lagrangian function $L(x,v,t) = |v|^{1+\alpha}$, $0 < \alpha < 1$.

But even for a "uniformly convex" Lagrangian there are several extensions of Theorem 8.1 which would be of interest, such as (a) getting rid of the compactness assumption, or at least control the dependence of constants at infinity; and (b) getting rid of the smoothness assumptions. I shall discuss both problems in the most typical case $L(x,v,t) = |v|^2$, i.e. $c(x,y) = d(x,y)^2$.

Intuitively, Mather's estimates are related to the local behavior of geodesics (they should not diverge too fast), and to the convexity properties of the square distance function $d^2(x_0,\,\cdot\,)$. Both features are well captured by lower bounds on the **sectional curvature** of the manifold. There is by chance a generalized notion of sectional curvature bounds, due to Alexandrov, which makes sense in a general metric space, without any smoothness; metric spaces which satisfy these bounds are called **Alexandrov spaces**. (This notion will be explained in more detail in Chapter 26.) In such spaces, one could hope to solve problems (a) and (b) at the same time. Although the proofs in the present chapter strongly rely on smoothness, I would be ready to believe in the following statement (which might be not so difficult to prove):

Open Problem 8.21. *Let $(\mathcal{X}, d)$ be an Alexandrov space with curvature bounded below by $K \in \mathbb{R}$, and let x_1, x_2, y_1, y_2 be four points in $\mathcal{X}$ such that*

$$d(x_1,y_1)^2 + d(x_2,y_2)^2 \le d(x_1,y_2)^2 + d(x_2,y_1)^2.$$

Further, let γ_1 and γ_2 be two constant-speed geodesics respectively joining x_1 to y_1 and x_2 to y_2. Then, for any $t_0 \in (0,1)$, there is a constant C_{t_0}, depending only on K, t_0, and on an upper bound on all the distances involved, such that

$$\sup_{0 \le t \le 1} d\big(\gamma_1(t), \gamma_2(t)\big) \le C_{t_0}\, d\big(\gamma_1(t_0),\, \gamma_2(t_0)\big).$$

To conclude this discussion, I shall mention a much rougher "shortening lemma", which has the advantage of holding true in general metric spaces, even without curvature bounds. In such a situation, in general there may be branching geodesics, so a bound on the distance at one intermediate time is clearly not enough to control the distance between the positions along the whole geodesic curves. One cannot hope either to control the distance between the velocities of these curves, since the velocities might not be well-defined. On the other hand, we may take advantage of the property of *preservation of speed* along the minimizing curves, since this remains true even in a nonsmooth context. The next theorem exploits this to show that if geodesics in a displacement interpolation pass near each other at some intermediate time, then their *lengths* have to be approximately equal.

Theorem 8.22 (A rough nonsmooth shortening lemma). *Let $(\mathcal{X}, d)$ be a metric space, and let γ_1, γ_2 be two constant-speed, minimizing geodesics such that*

$$d\big(\gamma_1(0),\gamma_1(1)\big)^2+d\big(\gamma_2(0),\gamma_2(1)\big)^2 \le d\big(\gamma_1(0),\gamma_2(1)\big)^2+d\big(\gamma_2(0),\gamma_1(1)\big)^2.$$

Let L_1 and L_2 stand for the respective lengths of γ_1 and γ_2, and let D be a bound on the diameter of $(\gamma_1 \cup \gamma_2)([0,1])$. Then

$$|L_1 - L_2| \le \frac{C\sqrt{D}}{\sqrt{t_0(1-t_0)}}\, \sqrt{d\big(\gamma_1(t_0),\gamma_2(t_0)\big)},$$

for some numeric constant C.

Proof of Theorem 8.22. Write $d_{12} = d(x_1, y_2)$, $d_{21} = d(x_2, y_1)$, $X_1 = \gamma_1(t_0)$, $X_2 = \gamma_2(t_0)$. From the minimizing assumption, the triangle inequality and explicit calculations,

$$\begin{aligned}
0 &\le d_{12}^2 + d_{21}^2 - L_1^2 - L_2^2 \\
&\le \Big(d(x_1,X_1) + d(X_1,X_2) + d(X_2,y_2)\Big)^2 \\
&\qquad\qquad + \Big(d(x_2,X_2) + d(X_2,X_1) + d(X_1,y_1)\Big)^2 \\
&= \Big(t_0\, L_1 + d(X_1,X_2) + (1-t_0)\, L_2\Big)^2 \\
&\qquad\qquad + \Big(t_0\, L_2 + d(X_1,X_2) + (1-t_0)\, L_1\Big)^2 - L_1^2 - L_2^2 \\
&= 2\, d(X_1,X_2)\Big(L_1 + L_2 + d(X_1,X_2)\Big) - 2\, t_0(1-t_0)\,(L_1 - L_2)^2.
\end{aligned}$$

As a consequence,

$$|L_1 - L_2| \le \sqrt{\frac{L_1 + L_2 + d(X_1,X_2)}{t_0(1-t_0)}}\, \sqrt{d(X_1,X_2)},$$

and the proof is complete. □

Appendix: Lipschitz estimates for power cost functions

The goal of this Appendix is to prove the following shortening lemma for the cost function $c(x,y) = |x-y|^{1+\alpha}$ in Euclidean space.

Theorem 8.23 (Shortening lemma for power cost functions). *Let $\alpha \in (0,1)$, and let x_1, y_1, x_2, y_2 be four points in $\mathbb{R}^n$, such that*

$$|x_1 - y_1|^{1+\alpha} + |x_2 - y_2|^{1+\alpha} \leq |x_1 - y_2|^{1+\alpha} + |x_2 - y_1|^{1+\alpha}. \tag{8.40}$$

Further, let

$$\gamma_1(t) = (1-t)\,x_1 + t\,y_1, \qquad \gamma_2(t) = (1-t)\,x_2 + t\,y_2.$$

Then, for any $t_0 \in (0,1)$ there is a constant $K = K(\alpha, t_0) > 0$ such that

$$|\gamma_1(t_0) - \gamma_2(t_0)| \geq K \sup_{0 \leq t \leq 1} |\gamma_1(t) - \gamma_2(t)|.$$

Remark 8.24. The proof below is not constructive, so I won't have any quantitative information on the best constant $K(\alpha,t)$. It is natural to think that for each fixed t, the constant $K(\alpha,t)$ (which only depends on α) will go to 0 as $\alpha \downarrow 0$. When $\alpha = 0$, the conclusion of the theorem is false: Just think of the case when x_1, y_1, x_2, y_2 are aligned. But this is the only case in which the conclusion fails, so it might be that a modified statement still holds true.

Proof of Theorem 8.23. First note that it suffices to work in the affine space generated by x_1, y_1, x_2, y_2, which is of dimension at most 3; hence all the constants will be independent of the dimension n. For notational simplicity, I shall assume that $t_0 = 1/2$, which has no important influence on the computations. Let $X_1 := \gamma_1(1/2)$, $X_2 := \gamma_2(1/2)$. It is sufficient to show that

$$|x_1 - x_2| + |y_1 - y_2| \leq C\,|X_1 - X_2|$$

for some constant C, independent of x_1, x_2, y_1, y_2.

Step 1: Reduction to a compact problem by invariance. Exchanging the roles of x and y if necessary, we might assume that $|x_2 - y_2| \leq |x_1 - y_1|$, and then by translation invariance that $x_1 = 0$, by homogeneity that $|x_1 - y_1| = 1$ (treat separately the trivial case $x_1 = y_1$), and by rotation invariance that $y_1 = e$ is a fixed unit vector.

Let $R := |x_2|$, then $|y_2 - x_2| \leq 1$ implies $|x_2 - X_2| \leq 1/2$, so $|X_2| \geq R - 1/2$, and since $|X_1| \leq 1/2$, it follows that $|X_1 - X_2| \geq R - 1$. On the other hand, $|x_1 - x_2| = R$ and $|y_1 - y_2| \leq R+1$. So the conclusion is obvious if $R \geq 2$. Otherwise, $|x_2|$ and $|y_2|$ lie in the ball $B_3(0)$.

Step 2: Reduction to a perturbation problem by compactness. For any positive integer k, let $(x_2^{(k)}, y_2^{(k)})$ be such that $(|x_1 - x_2| + |y_1 - y_2|)/|X_1 - X_2|$ is minimized by $(x_1, y_1, x_2^{(k)}, y_2^{(k)})$ under the constraint $|X_1 - X_2| \geq k^{-1}$.

By compactness, such a configuration does exist, and the value I_k of the infimum goes down with k, and converges to

$$I := \inf \left(\frac{|x_1 - x_2| + |y_1 - y_2|}{|X_1 - X_2|} \right), \tag{8.41}$$

where the infimum is taken over all configurations such that $X_1 \neq X_2$. The strict convexity of $x \to |x|^{1+\alpha}$ together with inequality (8.40) prevent $X_1 = X_2$, unless $(x_1, y_1) = (x_2, y_2)$, in which case there is nothing to prove. So it is sufficient to show that $I > 0$.

Since the sequence $(x_2^{(k)}, y_2^{(k)})$ takes values in a compact set, there is a subsequence thereof (still denoted $(x_2^{(k)}, y_2^{(k)})$) which converges to some $(x_2^{(\infty)}, y_2^{(\infty)})$. By continuity, condition (8.40) holds true with $(x_2, y_2) = (x_2^{(\infty)}, y_2^{(\infty)})$. If (with obvious notation) $|X_1 - X_2^{(\infty)}| > 0$, then the configuration $(x_1, y_1, x_2^{(\infty)}, y_2^{(\infty)})$ achieves the minimum I in (8.41), and that minimum is positive. So the only case there remains to treat is when $X_2^{(\infty)} = X_1$. Then, by strict convexity, condition (8.40) imposes $x_2^{(\infty)} = x_1$, $y_2^{(\infty)} = y_1$. Equivalently, $x_2^{(k)}$ converges to x_1, and $y_2^{(k)}$ to y_1. All this shows that it suffices to treat the case when x_2 is very close to x_1 and y_2 is very close to y_1.

Step 3: Expansions. Now let

$$x_2 = x_1 + \delta x, \qquad y_2 = y_1 + \delta y, \tag{8.42}$$

where δx and δy are vectors of small norm (recall that $x_1 - y_1$ has unit norm). Of course

$$X_1 - X_2 = \frac{\delta x + \delta y}{2}, \qquad x_2 - x_1 = \delta x, \qquad y_2 - y_1 = \delta y;$$

so to conclude the proof it is sufficient to show that

$$\left|\frac{\delta x + \delta y}{2}\right| \geq K(|\delta x| + |\delta y|), \tag{8.43}$$

as soon as $|\delta x|$ and $|\delta y|$ are small enough, and (8.40) is satisfied.

By using the formulas $|a+b|^2 = |a|^2 + 2\langle a,b\rangle + |b|^2$ and

$$(1+\varepsilon)^{\frac{1+\alpha}{2}} = 1 + \frac{(1+\alpha)}{2}\,\varepsilon \;-\; \frac{(1+\alpha)(1-\alpha)}{8}\,\varepsilon^2 + O(\varepsilon^3),$$

one easily deduces from (8.40) that

$$|\delta x - \delta y|^2 - |\delta x|^2 - |\delta y|^2 \leq (1-\alpha)\Big[\langle \delta x - \delta y, e\rangle^2 - \langle \delta x, e\rangle^2 - \langle \delta y, e\rangle^2\Big] + O\Big(|\delta x|^3 + |\delta y|^3\Big).$$

This can be rewritten

$$\langle \delta x, \delta y\rangle - (1-\alpha)\,\langle \delta x, e\rangle\,\langle \delta y, e\rangle \geq O(|\delta x|^3 + |\delta y|^3).$$

Consider the new scalar product

$$\langle\!\langle v, w\rangle\!\rangle := \langle v, w\rangle - (1-\alpha)\,\langle v, e\rangle\,\langle w, e\rangle$$

(which is indeed a scalar product because $\alpha > 0$), and denote the associated norm by $\|v\|$. Then the above conclusion can be summarized into

$$\langle\!\langle \delta x, \delta y\rangle\!\rangle \geq O\big(\|\delta x\|^3 + \|\delta y\|^3\big). \tag{8.44}$$

It follows that

$$\begin{aligned}\left\|\frac{\delta x + \delta y}{2}\right\|^2 &= \frac{1}{4}\Big(\|\delta x\|^2 + \|\delta y\|^2 + 2\langle\!\langle \delta x, \delta y\rangle\!\rangle\Big)\\ &\geq \frac{1}{4}(\|\delta x\|^2 + \|\delta y\|^2) + O(\|\delta x\|^3 + \|\delta y\|^3).\end{aligned}$$

So inequality (8.43) is indeed satisfied if $|\delta x| + |\delta y|$ is small enough. □

Exercise 8.25. Extend this result to the cost function $d(x,y)^{1+\alpha}$ on a Riemannian manifold, when γ and $\widetilde{\gamma}$ stay within a compact set.

Hints: This tricky exercise is only for a reader who feels very comfortable. One can use a reasoning similar to that in Step 2 of the above proof, introducing a sequence $(\gamma^{(k)}, \widetilde{\gamma}^{(k)})$ which is asymptotically the "worst possible", and converges, up to extraction of a subsequence, to $(\gamma^{(\infty)}, \widetilde{\gamma}^{(\infty)})$. There are three cases: (i) $\gamma^{(\infty)}$ and $\widetilde{\gamma}^{(\infty)}$ are distinct geodesic curves which cross; this is ruled out by Theorem 8.1. (ii) $\gamma^{(k)}$ and $\widetilde{\gamma}^{(k)}$ converge to a point; then everything becomes local and one can use the result in $\mathbb{R}^n$, Theorem 8.23. (iii) $\gamma^{(k)}$ and $\widetilde{\gamma}^{(k)}$ converge to a nontrivial geodesic $\gamma^{(\infty)}$; then these curves can be approximated by infinitesimal perturbations of $\gamma^{(\infty)}$, which are described by differential equations (Jacobi equations).

Remark 8.26. Of course it would be much better to avoid the compactness arguments and derive the bounds directly, but I don't see how to proceed.

Bibliographical notes

Monge's observation about the impossibility of crossing appears in his seminal 1781 memoir [636]. The argument is likely to apply whenever the cost function satisfies a triangle inequality, which is always the case in what Bernard and Buffoni have called the Monge–Mañé problem [104]. I don't know of a quantitative version of it.

A very simple argument, due to Brenier, shows how to construct, without any calculations, configurations of points that lead to line-crossing for a quadratic cost [814, Chapter 10, Problem 1].

There are several possible computations to obtain inequalities of the style of (8.3). The use of the identity (8.2) was inspired by a result by Figalli, which is described below.

It is an old observation in Riemannian geometry that two minimizing curves cannot intersect twice and remain minimizing; the way to prove this is the shortcut method already known to Monge. This simple principle has important geometrical consequences, see for instance the works by Morse [637, Theorem 3] and Hedlund [467, p. 722]. (These references, as well as a large part of the historical remarks below, were pointed out to me by Mather.)

At the end of the seventies, Aubry discovered a noncrossing lemma which is similar in spirit, although in a different setting. Together with Le Daeron, he demonstrated the power of this principle in studying the so-called Frenkel–Kantorova model from solid-state physics; see the "Fundamental Lemma" in [52]. In particular, the method of Aubry and Le Daeron provided an alternative proof of results by Mather [598] about the existence of quasiperiodic orbits for certain dynamical systems.[1] The relations between the methods of proof of Aubry and Mather are discussed in [599, 604] and constitute the core of what is usually called the Aubry–Mather theory. Bangert [66, Lemma 3.1] gave a general version of the Aubry–Le Daeron lemma, and illustrated its use in various aspects of the theory of twist diffeomorphisms. (Bangert's paper can be consulted as an entry point for this subject.) He also made the connection with the earlier works of Morse and Hedlund in geometry. There is a related independent study by Bialy and Polterovich [117].

Then Moser [641] showed that the theory of twist diffeomorphisms (at least in certain particular cases) could be embedded in the theory of strictly convex Lagrangian systems, and Denzler [297] adapted the noncrossing arguments of Aubry, Le Daeron and Bangert to this new setting (see for instance Theorem 2.3 there), which in some sense goes back to older geometric works.

Around 1990, Mather came up with two contributions which for our purposes are crucial. The first consists in introducing minimizing *measures* rather than minimizing curves [600]; the second consists in a *quantitative* version of the noncrossing argument, for a general class of strictly convex Lagrangian functions [601, p. 186]. This estimate, which in these notes I called Mather's shortening lemma, was the key technical ingredient in the proof of his fundamental "Lipschitz graph theorem" [601, Theorem 2].

Although the statement in [601] is not really the same as the one which appears in this chapter, the proof really is similar. The idea to use this approach in optimal transport theory came to me when Bernard mentioned Mather's lemma in a seminar where he was presenting his results with Buffoni about the optimal transport problem for rather general Lagrangian functions [105].

In the meantime, an appropriate version of the noncrossing lemma had already been rediscovered (but not in a quantitative version) by researchers in optimal transport. Indeed, the noncrossing property of optimal trajectories, and the resulting estimates about absolute continuity of the displacement interpolant, were some of the key technical tools used by McCann [614] to establish convexity properties of certain functionals along displacement interpolation in $\mathbb{R}^n$ for a quadratic cost;

[1] According to Mather, the chronology is blurred, because Aubry knew similar results somewhat earlier, at least for certain classes of systems, but had never published them; in particular, the discoveries of Aubry and Mather were independent. Further, see the review paper [51].

these statements were generalized by Cordero-Erausquin, McCann and Schmuckenschläger [246] for Riemannian manifolds, and for rather general convex cost functions in $\mathbb{R}^n$ by Cordero-Erausquin [243].

Results similar to Theorems 8.5 and 8.7 are also proven by Bernard and Buffoni [105] via the study of Hamilton–Jacobi equations, in the style of weak KAM theory. This is a bit less elementary but powerful as well. The basic idea is to exploit the fact that solutions of Hamilton–Jacobi equations are automatically semiconcave for positive times; I learnt from Otto the usefulness of this regularization property in the context of optimal transport (see [814, p. 181]). Fathi and Figalli [348] generalized this strategy to noncompact situations. Bernard [102] also used the same idea to recover an important result about the existence of C^1 subsolutions of certain Hamilton–Jacobi equations.

Figalli and Juillet [366] obtained a result similar to Theorem 8.7 when the cost is the squared distance on a degenerate Riemannian structure such as the Heisenberg group or an Alexandrov space with curvature bounded below. Their approach is completely different since it uses the uniqueness of Wasserstein geodesics and the so-called measure contraction property (which is traditionally associated with Ricci curvature bounds but nevertheless holds in the Heisenberg group [496]). Figalli and Juillet note that concentration phenomena arise in the Heisenberg group which are not seen in Riemannian manifolds; and that the Monge–Mather shortening lemma does not hold in this setting.

Theorem 8.11 is a variant of Mather's Lipschitz graph theorem, appearing (up to minor modifications) in Bernard and Buffoni [105, Theorem C]. The core of the proof is also taken from that work.

The acronym "KAM" stands for Kolmogorov, Arnold and Moser; the "classical KAM theory" deals with the stability (with high probability) of perturbed integrable Hamiltonian systems. An account of this theory can be found in, e.g., Thirring [780, Section 3.6]. With respect to weak KAM theory, some important differences are that: (a) classical KAM theory only applies to slight perturbations of integrable systems; (b) it only deals with very smooth objects; (c) it controls the behavior of a large portion of the phase space (the whole of it, asymptotically when the size of the perturbation goes to 0).

The weak KAM theory is much more recent than the classical one; it was developed by several authors, in particular Fathi [344, 345]. A theorem of the existence of a stationary solution of the Hamilton–Jacobi equation can be found in [347, Theorem 4.4.6]. Precursors are

Mather [602, 603] and Mañé [592, 593]. The reader can also consult the book by Fathi [347], and (with a complementary point of view) the one by Contreras and Iturriaga [238]. Also available are some technical notes by Ishii [488], and the review works [604, 752].

The proof of Theorem 8.17, as I wrote it, is a minor variation of an argument shown to me by Fathi. Related considerations appear in a recent work by Bernard and Buffoni [106], who analyze the weak KAM theory in light of the abstract Kantorovich duality. One may also consult [278].

From its very beginning, the weak KAM theory has been associated with the theory of viscosity solutions of Hamilton–Jacobi equations. An early work on the subject (anterior to Mather's papers) is an unpublished preprint by P.-L. Lions, Papanicolaou and Varadhan [564]. Recently, the weak KAM theory has been related to the large-time behavior of Hamilton–Jacobi equations [69, 107, 346, 349, 383, 384, 385, 487, 645, 707]. Aubry sets are also related with the C^1 regularity of Hamilton–Jacobi equations, which has important applications in the theory of dynamical systems [102, 103, 350]. See also Evans and Gomes [332, 333, 334, 423] and the references therein for an alternative point of view.

In this chapter I presented Mather's problem in terms of trajectories and transport cost. There is an alternative presentation in terms of invariant measures, following an idea by Mañé. In Mañé's version of the problem, the unknown is a probability measure $\mu(dx\,dv)$ on the tangent bundle TM; it is stationary in the sense that $\nabla_x \cdot (v\,\mu) = 0$ (this is a stationary kinetic transport equation), and it should minimize the action $\int L(x,v)\,\mu(dx\,dv)$. Then one can show that μ is actually invariant under the Lagrangian flow defined by L. As Gomes pointed out to me, this approach has the drawback that the invariance of μ is not built in from the definition; but it has several nice advantages:

- It makes the graph property trivial if L is strictly convex: Indeed, one can always collapse the measure μ, at each $x \in M$, onto the barycenter $\xi(x) = \int v\,\mu(dv|x)$; this operation preserves the invariance of the measure, and decreases the cost unless μ was already supported on a graph. (Note: This does not give the Lipschitz regularity of the graph!)
- This is a linear programming problem, so it admits a dual problem which is $\inf_\varphi \sup_x H(\nabla_x \varphi, x)$; the value of this infimum is

but another way to characterize the effective Hamiltonian $\overline{H}$, see e.g. [238, 239].

- This is a good starting point for some generalizations, see for instance [422].

I shall conclude with some more technical remarks.

The use of a restriction property to prove the absolute continuity of the displacement interpolant without any compactness assumption was inspired by a discussion with Sturm on a related subject. It was also Sturm who asked me whether Mather's estimates could be generalized to Alexandrov spaces with curvature bounded below.

The theorem according to which a Lipschitz map T dilates the n-dimensional Hausdorff measure by a factor at most $\|T\|_{\mathrm{Lip}}^n$ is an almost immediate consequence of the definitions of Hausdorff measure, see e.g. [174, Proposition 1.7.8].

Alexandrov spaces are discussed at length in the very pedagogical monograph by Burago, Burago and Ivanov [174]. Several characterizations of Alexandrov spaces are given there, and their equivalence is established. For instance, an Alexandrov space has curvature bounded below by K if the square distance function $d(z,\,\cdot\,)^2$ is "no more convex" than the square distance function in the model space having constant sectional curvature K. Also geodesics in an Alexandrov space cannot diverge faster than geodesics in the model space, in some sense. These properties explain why such spaces may be a natural generalized setting for optimal transport. Upper bounds on the sectional curvature, on the other hand, do not seem to be of any help.

Figalli recently solved the Open Problem 8.21 in the special case $K=0$ (nonnegative curvature), with a very simple and sharp argument: He showed that if γ_1 and γ_2 are any two minimizing, constant-speed geodesics in an Alexandrov space $(\mathcal{X},d)$ with nonnegative curvature, and $\gamma_1(0)=x_1$, $\gamma_2(0)=x_2$, $\gamma_1(1)=y_1$, $\gamma_2(1)=y_2$, then

$$\begin{aligned} d(\gamma_1(t),\,\gamma_2(t)) &\geq (1-t)^2\, d(x_1,x_2)^2 \,+\, t^2\, d(y_1,y_2)^2 \\ &+ t(1-t)\,\Big[d(x_1,y_2)^2+d(x_2,y_1)^2-d(x_1,y_1)^2-d(x_2,y_2)^2\Big]. \end{aligned} \tag{8.45}$$

(So in this case there is no need for an upper bound on the distances between x_1,x_2,y_1,y_2.) The general case where K might be negative seems to be quite more tricky. As a consequence of (8.45), Theorem 8.7 holds when the cost is the squared distance on an Alexandrov space with nonnegative curvature; but this can also be proven by the method of Figalli and Juillet [366].

Theorem 8.22 takes inspiration from the no-crossing proof in [246, Lemma 5.3]. I don't know whether the Hölder-1/2 regularity is optimal, and I don't know either whether it is possible/useful to obtain similar estimates for more general cost functions.

9

Solution of the Monge problem I: Global approach

In the present chapter and the next one I shall investigate the solvability of the Monge problem for a Lagrangian cost function. Recall from Theorem 5.30 that it is sufficient to identify conditions under which the initial measure μ does not see the set of points where the c-subdifferential of a c-convex function ψ is multivalued.

Consider a Riemannnian manifold M, and a cost function $c(x, y)$ on $M \times M$, deriving from a Lagrangian function $L(x, v, t)$ on $TM \times [0, 1]$ satisfying the classical conditions of Definition 7.6. Let μ_0 and μ_1 be two given probability measures, and let $(\mu_t)_{0 \leq t \leq 1}$ be a displacement interpolation, written as the law of a random minimizing curve γ at time t.

If the Lagrangian satisfies adequate regularity and convexity properties, Theorem 8.5 shows that the coupling $(\gamma(s), \gamma(t))$ is always deterministic as soon as $0 < s < 1$, however singular μ_0 and μ_1 might be. The question whether one can construct a deterministic coupling of (μ_0, μ_1) is much more subtle, and cannot be answered without regularity assumptions on μ_0. In this chapter, a simple approach to this problem will be attempted, but only with partial success, since eventually it will work out only for a particular class of cost functions, including at least the quadratic cost in Euclidean space (arguably the most important case).

Our main assumption on the cost function c will be:

Assumption (C): *For any c-convex function ψ and any $x \in M$, the c-subdifferential $\partial_c \psi(x)$ is pathwise connected.*

Example 9.1. Consider the cost function $c(x, y) = -x \cdot y$ in $\mathbb{R}^n$. Let y_0 and y_1 belong to $\partial_c \psi(x)$; then, for any $z \in \mathbb{R}^n$ one has

C. Villani, *Optimal Transport*, Grundlehren der mathematischen Wissenschaften 338,

$$\psi(x) + y_0 \cdot (z - x) \leq \psi(z); \qquad \psi(x) + y_1 \cdot (z - x) \leq \psi(z).$$

It follows that $\psi(x)+y_t\cdot(z-x) \leq \psi(z)$, where $y_t := (1-t)\,y_0+t\,y_1$. Thus the line segment $(y_t)_{0\leq t\leq 1}$ is entirely contained in the subdifferential of ψ at x. The same computation applies to $c(x,y) = |x-y|^2/2$, or to any cost function of the form $a(x) - x \cdot y + b(y)$.

Actually, there are not so many examples where Assumption (C) is known to be satisfied. Before commenting more on this issue, let me illustrate the interest of this assumption by showing how it can be used.

Theorem 9.2 (Conditions for single-valued subdifferentials). *Let M be a smooth n-dimensional Riemannian manifold, and c a real-valued cost function, bounded below, deriving from a Lagrangian $L(x,v,t)$ on $TM \times [0,1]$, satisfying the classical conditions of Definition 7.6 and such that:*

(i) Assumption (C) is satisfied.

(ii) The conclusion of Theorem 8.1 (Mather's shortening lemma), in the form of inequality (8.4), holds true for $t_0 = 1/2$ with an exponent $\beta > 1 - (1/n)$, and a uniform constant. More explicitly: Whenever x_1, x_2, y_1, y_2 are four points in M satisfying $c(x_1,y_1) + c(x_2,y_2) \leq c(x_1,y_2) + c(x_2,y_1)$, and γ_1, γ_2 are two action-minimizing curves with $\gamma_1(0) = x_1$, $\gamma_1(1) = y_1$, $\gamma_2(0) = x_2$, $\gamma_2(1) = y_2$, then

$$\sup_{0\leq t\leq 1} d\big(\gamma_1(t), \gamma_2(t)\big) \leq C\, d\big(\gamma_1(1/2), \gamma_2(1/2)\big)^{\beta}. \tag{9.1}$$

Then, for any c-convex function ψ, there is a set $Z \subset M$ of Hausdorff dimension at most $(n-1)/\beta < n$ (and therefore of zero n-dimensional measure), such that the c-subdifferential $\partial_c\psi(x)$ contains at most one element if $x \notin Z$.

Proof of Theorem 9.2. Let Z be the set of points x for which $\psi(x) < +\infty$ but $\partial_c\psi(x)$ is not single-valued; the problem is to show that Z is of dimension at most $(n-1)/\beta$.

Let $x \in M$ with $\psi(x) < +\infty$, and let $y \in \partial_c\psi(x)$. Introduce an action-minimizing curve $\gamma = \gamma^{x,y}$ joining $x = \gamma(0)$ to $y = \gamma(1)$. I claim that the map

$$F : \gamma\left(\frac{1}{2}\right) \longmapsto x$$

is well-defined on its domain of definition, which is the union of all $\gamma^{x,y}(1/2)$. (I mean, $m = \gamma(1/2)$ determines x unambiguously; there cannot be two different points x for which $\gamma(1/2)$ is the same.) Indeed, assume $y \in \partial_c\psi(x)$ and $y' \in \partial_c\psi(x')$, with $\psi(x) < +\infty$, $\psi(x') < +\infty$, and let γ and γ' be minimizing geodesics between x and y on the one hand, x' and y' on the other hand. It follows from the definitions of subdifferential that

$$\begin{cases} \psi(x) \ + c(x,y) \ \leq \psi(x') \ + c(x',y) \\ \psi(x') + c(x',y') \leq \psi(x) + c(x,y'). \end{cases}$$

Thus

$$c(x,y) + c(x',y') \leq c(x,y') + c(x',y).$$

Then by (9.1),

$$d(x,x') \leq C\, d\Big(\gamma\big(\tfrac{1}{2}\big), \gamma'\big(\tfrac{1}{2}\big)\Big)^{\beta}.$$

This implies that $m = \gamma(1/2)$ determines $x = F(m)$ unambiguously, and even that F is Hölder-β. (Obviously, this is the same reasoning as in the proof of Theorem 8.5.)

Now, cover M by a countable number of open sets in which M is diffeomorphic to a subset U of $\mathbb{R}^n$, via some diffeomorphism φ_U. In each of these open sets U, consider the union H_U of all hyperplanes passing through a point of rational coordinates, orthogonal to a unit vector with rational coordinates. Transport this set back to M thanks to the local diffeomorphism, and take the union over all the sets U. This gives a set $D \subset M$ with the following properties: (i) It is of dimension $n-1$; (ii) It meets every nontrivial continuous curve drawn on M (to see this, write the curve locally in terms of φ_U and note that, by continuity, at least one of the coordinates of the curve has to become rational at some time).

Next, let $x \in Z$, and let y_0, y_1 be two distinct elements of $\partial_c\psi(x)$. By assumption there is a continuous curve $(y_t)_{0\leq t\leq 1}$ lying entirely in $\partial_c\psi(x)$. For each t, introduce an action-minimizing curve $(\gamma_t(s))_{0\leq s\leq 1}$ between x and y_t (s here is the time parameter along the curve). Define $m_t := \gamma_t(1/2)$. This is a continuous path, nontrivial (otherwise $\gamma_0(1/2) = \gamma_1(1/2)$, but two minimizing trajectories starting from x cannot cross in their middle, or they have to coincide at all times by (9.1)). So there has to be some t such that $y_t \in D$. Moreover, the map F constructed above satisfies $F(y_t) = x$ for all t. It follows that $x \in F(D)$. (See Figure 9.1.)

As a conclusion, $Z \subset F(D)$. Since D is of Hausdorff dimension $n-1$ and F is β-Hölder, the dimension of $F(D)$ is at most $(n-1)/\beta$. □

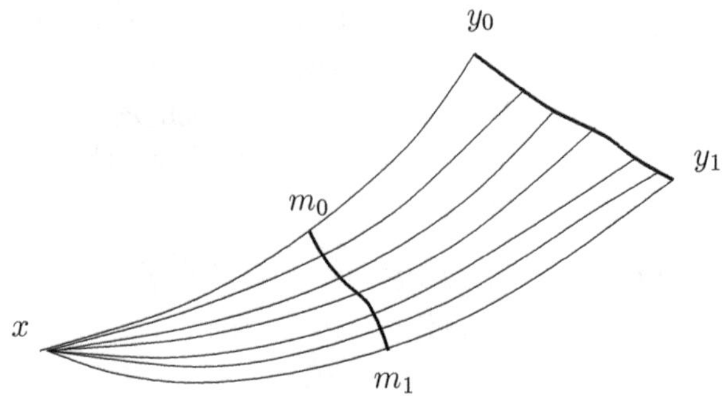

Fig. 9.1. Scheme of proof for Theorem 9.2. Here there is a curve $(y_t)_{0\le t\le 1}$ lying entirely in $\partial_c\psi(x)$, and there is a nontrivial path $(m_t)_{0\le t\le 1}$ obtained by taking the midpoint between x and y_t. This path has to meet D; but its image by $\gamma(1/2)\mapsto\gamma(0)$ is $\{x\}$, so $x\in F(D)$.

Now come the consequences in terms of Monge transport.

Corollary 9.3 (Solution of the Monge problem, I). *Let M be a Riemannian manifold, let c be a cost function on $M\times M$, with associated cost functional C, and let μ, ν be two probability measures on M. Assume that:*

(i) $C(\mu,\nu)<+\infty$;

(ii) the assumptions of Theorem 9.2 are satisfied;

(iii) μ gives zero probability to sets of dimension at most $(n-1)/\beta$.

Then there is a unique (in law) optimal coupling (x,y) of μ and ν; it is deterministic, and characterized (among all couplings of (μ,ν)) by the existence of a c-convex function ψ such that

$$y\in\partial_c\psi(x)\qquad \textit{almost surely.} \tag{9.2}$$

Equivalently, there is a unique optimal transport plan π; it is deterministic, and characterized by the existence of a c-convex ψ such that (9.2) holds true π-almost surely.

Proof of Corollary 9.3. The conclusion is obtained by just putting together Theorems 9.2 and 5.30. □

We have now solved the Monge problem in an absolutely painless way; but under what assumptions? At least we can treat the important cost function $c(x,y)=-x\cdot y$. Indeed the notion of c-convexity reduces to plain convexity (plus lower semicontinuity), and the c-subdifferential of a convex function ψ is just its usual subdifferential, which I shall denote by $\partial\psi$. Moreover, under an assumption of finite second moments, for the Monge problem this cost is just as good as the usual squared Euclidean distance, since $|x-y|^2 = |x|^2 - 2\,x\cdot y + |y|^2$, and $\int(|x|^2+|y|^2)\,d\pi(x,y)$ is independent of the choice of $\pi \in \Pi(\mu,\nu)$. Particular as it may seem, this case is one of the most important for applications, so I shall state the result as a separate theorem.

Theorem 9.4 (Monge problem for quadratic cost, first result). *Let $c(x,y) = |x-y|^2$ in $\mathbb{R}^n$. Let μ, ν be two probability measures on $\mathbb{R}^n$ such that*

$$\int |x|^2\,d\mu(x) + \int |y|^2\,d\nu(y) < +\infty \tag{9.3}$$

and μ does not give mass to sets of dimension at most $n-1$. (This is true in particular if μ is absolutely continuous with respect to the Lebesgue measure.) Then there is a unique (in law) optimal coupling (x,y) of μ and ν; it is deterministic, and characterized, among all couplings of (μ,ν), by the existence of a lower semicontinuous convex function ψ such that

$$y \in \partial\psi(x) \qquad \textit{almost surely.} \tag{9.4}$$

In other words, there is a unique optimal transference π; it is a Monge transport plan, and it is characterized by the existence of a lower semicontinuous convex function ψ whose subdifferential contains $\operatorname{Spt}\pi$.

Remark 9.5. The assumption that μ does not give mass to sets of dimension at most $n-1$ is optimal for the existence of a Monge coupling, as can be seen by choosing $\mu = \mathcal{H}^1|_{[0,1]\times\{0\}}$ (the one-dimensional Hausdorff measure concentrated on the segment $[0,1]\times\{0\}$ in $\mathbb{R}^2$), and then $\nu = (1/2)\,\mathcal{H}^1|_{[0,1]\times\{-1\}\cup[0,1]\times\{+1\}}$. (See Figure 9.2.) It is also optimal for the uniqueness, as can be seen by taking $\mu = (1/2)\,\mathcal{H}^1_{\{0\}\times[-1,1]}$ and $\nu = (1/2)\,\mathcal{H}^1_{[-1,1]\times\{0\}}$. In fact, whenever μ, $\nu \in P_2(\mathbb{R}^n)$ are supported on orthogonal subspaces of $\mathbb{R}^n$, then *any* transference plan is optimal! To see this, define a function ψ by $\psi = 0$ on $\operatorname{Conv}(\operatorname{Spt}\mu)$, $\psi = +\infty$ elsewhere; then ψ is convex lower semicontinuous, $\psi^* = 0$ on $\operatorname{Conv}(\operatorname{Spt}\nu)$, so $\partial\psi$ contains $\operatorname{Spt}\mu \times \operatorname{Spt}\nu$, and any transference plan is supported in $\partial\psi$.

In the next chapter, we shall see that Theorem 9.4 can be improved in at least two ways: Equation (9.4) can be rewritten $y = \nabla\psi(x)$;

Fig. 9.2. The source measure is drawn as a thick line, the target measure as a thin line; the cost function is quadratic. On the left, there is a unique optimal coupling but no optimal Monge coupling. On the right, there are many optimal couplings, in fact any transference plan is optimal.

and the assumption (9.3) can be replaced by the weaker assumption $C(\mu,\nu) < +\infty$ (finite optimal transport cost).

Now if one wants to apply Theorem 9.2 to nonquadratic cost functions, the question arises of how to identify those cost functions $c(x,y)$ which satisfy Assumption (C). Obviously, there might be some geometric obstructions imposed by the domains $\mathcal{X}$ and $\mathcal{Y}$: For instance, if $\mathcal{Y}$ is a *nonconvex* subset of $\mathbb{R}^n$, then Assumption (C) is violated even by the quadratic cost function. But even in the whole of, say, $\mathbb{R}^n$, Assumption (C) is *not* a generic condition, and so far there is only a short list of known examples. These include the cost functions $c(x,y) = \sqrt{1+|x-y|^2}$ on $\mathbb{R}^n \times \mathbb{R}^n$, or more generally $c(x,y) = (1+|x-y|^2)^{p/2}$ $(1 < p < 2)$ on $B_R(0) \times B_R(0) \subset \mathbb{R}^n \times \mathbb{R}^n$, where $R = 1/\sqrt{p-1}$; and $c(x,y) = d(x,y)^2$ on $S^{n-1} \times S^{n-1}$, where d is the geodesic distance on the sphere. For such cost functions, the Monge problem can be solved by combining Theorems 8.1, 9.2 and 5.30, exactly as in the proof of Theorem 9.4.

This approach suffers, however, from two main drawbacks: First it seems to be limited to a small number of examples; secondly, the verification of Assumption (C) is subtle. In the next chapter we shall investigate a more pedestrian approach, which will apply in much greater generality.

I shall end this chapter with a simple example of a cost function which *does not* satisfy Assumption (C).

Proposition 9.6 (Non-connectedness of the c-subdifferential). *Let $p > 2$ and let $c(x,y) = |x-y|^p$ on $\mathbb{R}^2 \times \mathbb{R}^2$. Then there is a c-convex function $\psi : \mathbb{R}^2 \to \mathbb{R}$ such that $\partial_c\psi(0)$ is not connected.*

Proof of Proposition 9.6. For $t \in [-1,1]$ define $y_t = (0,t) \in \mathbb{R}^2$, and

$$\eta_t(x) = -c(x,y_t) + c(0,y_t) + \beta(t) = -\big(x_1^2 + (x_2-t)^2\big)^{p/2} + |t|^p + \beta(t),$$

where β is a smooth even function, $\beta(0) = 0$, $\beta'(t) > 0$ for $|t| > 0$. Further, let $r > 0$ and $X_\pm = (\pm r, 0)$. (The fact that the segments $[X_-, X_+]$ and $[y_{-1}, y_1]$ are orthogonal is not accidental.) Then $\eta_t(0) = \beta(t)$ is an increasing function of $|t|$; while $\eta_t(X_\pm) = -(r^2+t^2)^{p/2} + |t|^p + \beta(t)$ is a decreasing function of $|t|$ if $0 < \beta'(t) < pt\,[(r^2+t^2)^{p/2-1} - t^{p-2}]$, which we shall assume. Now define $\psi(x) = \sup\{\eta_t(x);\ t \in [-1,1]\}$. By construction this is a c-convex function, and $\psi(0) = \beta(1) > 0$, while $\psi(X_\pm) = \eta_0(X_\pm) = -r^p$.

We shall check that $\partial_c\psi(0)$ is not connected. First, $\partial_c\psi(0)$ is not empty: this can be shown by elementary means or as a consequence of Example 10.20 and Theorem 10.24 in the next chapter. Secondly, $\partial_c\psi(0) \subset \{(y_1,y_2) \in \mathbb{R}^2;\ y_1 = 0\}$: This comes from the fact that all functions η_t are decreasing as a function of $|x_1|$. (So ψ is also nonincreasing in $|x_1|$, and if $(y_1,y_2) \in \partial_c\psi(0,0)$, then $(y_1^2+y_2^2)^{p/2} + \psi(0,0) \le |y_2|^p + \psi(y_1,0) \le |y_2|^p + \psi(0,0)$, which imposes $y_1 = 0$.) Obviously, $\partial_c\psi(0)$ is a symmetric subset of the line $\{y_1 = 0\}$. But if $0 \in \partial_c\psi(0)$, then $0 < \psi(0) \le |X_\pm|^p + \psi(X_\pm) = 0$, which is a contradiction. So $\partial_c\psi(0)$ does not contain 0, therefore it is not connected.

(What is happening is the following. When replacing η_0 by ψ, we have surelevated the origin, but we have kept the points $(X_\pm, \eta_0(X_\pm))$ in place, which forbids us to touch the graph of ψ from below at the origin with a translation of η_0.) □

Bibliographical notes

It is classical that the image of a set of Hausdorff dimension d by a Lipschitz map is contained in a set of Hausdorff dimension at most d: See for instance [331, p. 75]. There is no difficulty in modifying the proof to show that the image of a set of Hausdorff dimension d by a Hölder-β map is contained in a set of dimension at most d/β.

The proof of Theorem 9.2 is adapted from a classical argument according to which a real-valued convex function ψ on $\mathbb{R}^n$ has a single-valued subdifferential everywhere out of a set of dimension at most $n-1$; see [11, Theorem 2.2]. The key estimate for the proof of the latter theorem is that $(\mathrm{Id} + \partial\psi)^{-1}$ exists and is Lipschitz; but this can be seen as a very particular case of the Mather shortening lemma. In the next chapter another line of argumentation for that differentiability theorem, more local, will be provided.

The paternity of Theorem 9.4 is shared by Brenier [154, 156] with Rachev and Rüschendorf [722]; it builds upon earlier work by Knott and Smith [524], who already knew that an optimal coupling lying entirely in the subdifferential of a convex function would be optimal. Brenier rewrote the result as a beautiful **polar factorization theorem**, which is presented in detail in [814, Chapter 3].

The nonuniqueness statement in Remark 9.5 was formulated by McCann [613]. Related problems (existence and uniqueness of optimal couplings between measures supported on polygons) are discussed by Gangbo and McCann [400], in relation to problems of shape recognition.

Other forms of Theorem 9.4 appear in Rachev and Rüschendorf [696], in particular an extension to infinite-dimensional separable Hilbert spaces; the proof is reproduced in [814, Second Proof of Theorem 2.9]. (This problem was also considered in [2, 254].) All these arguments are based on duality; then more direct proofs, which do not use the Kantorovich duality explicitly, were found by Gangbo [395], and also Caffarelli [187] (who gives credit to Varadhan for this approach).

A probabilistic approach of Theorem 9.4 was studied by Mikami and Thieullen [628, 630]. The idea is to consider a minimization problem over paths which are not geodesics, but geodesics perturbed by some noise; then let the noise vanish. This is reminiscent of Nelson's approach to quantum mechanics, see the bibliographical notes of Chapters 7 and 23.

McCann [613] extended Theorem 9.4 by removing the assumption of bounded second moment and even the weaker assumption of finite transport cost: *Whenever μ does not charge sets of dimension $n-1$, there exists a unique coupling of (μ, ν) which takes the form $y = \nabla\Psi(x)$, where Ψ is a lower semicontinuous convex function.* The tricky part in this statement is the uniqueness. This theorem will be proven in the next chapter (see Theorem 10.42, Corollary 10.44 and Particular Case 10.45).

Ma, Trudinger and X.-J. Wang [585, Section 7.5] were the first to seriously study Assumption (C); they had the intuition that it was connected to a certain fourth-order differential condition on the cost function which plays a key role in the smoothness of optimal transport.

Later Trudinger and Wang [793], and Loeper [570] showed that the above-mentioned differential condition is essentially, under adequate geometric and regularity assumptions, equivalent to Assumption (C). These issues will be discussed in more detail in Chapter 12. (See in particular Proposition 12.15(iii).)

The counterexample in Proposition 9.6 is extracted from [585]. The fact that $c(x, y) = (1 + |x - y|^2)^{p/2}$ satisfies Assumption (C) on the ball of radius $1/\sqrt{p-1}$ is also taken from [585, 793]. It is Loeper [571] who discovered that the squared geodesic distance on S^{n-1} satisfies Assumption (C); then a simplified argument was devised by von Nessi [824].

As mentioned in the end of the chapter, by combining Loeper's result with Theorems 8.1, 9.2 and 5.30, one can mimick the proof of Theorem 9.4 and get the unique solvability of the Monge problem for the quadratic distance on the sphere, as soon as μ does not see sets of dimension at most $n - 1$. Such a theorem was first obtained for general compact Riemannian manifolds by McCann [616], with a completely different argument.

Other examples of cost functions satisfying Assumption (C) will be listed in Chapter 12 (for instance $|x - y|^{-2}$, or $-|x - y|^p/p$ for $-2 \leq p \leq 1$, or $|x - y|^2 + |f(x) - g(y)|^2$, where f and g are convex and 1-Lipschitz). But these other examples do not come from a smooth convex Lagrangian, so it is not clear whether they satisfy Assumption (ii) in Theorem 9.2.

In the particular case when ν has finite support, one can prove the unique solvability of the Monge problem under much more general assumptions, namely that the cost function is continuous, and μ does not charge sets of the form $\{x;\ c(x, a) - c(x, b) = k\}$ (where a, b, k are arbitrary), see [261]. This condition was recently used again by Gozlan [429].

10

Solution of the Monge problem II: Local approach

In the previous chapter, we tried to establish the almost sure single-valuedness of the c-subdifferential by an argument involving "global" topological properties, such as connectedness. Since this strategy worked out only in certain particular cases, we shall now explore a different method, based on *local* properties of c-convex functions. The idea is that the global question *"Is the c-subdifferential of ψ at x single-valued or not?"* might be much more subtle to attack than the local question *"Is the function ψ differentiable at x or not?"* For a large class of cost functions, these questions are in fact equivalent; but these different formulations suggest different strategies. So in this chapter, the emphasis will be on tangent vectors and gradients, rather than points in the c-subdifferential.

This approach takes its source from the works by Brenier, Rachev and Rüschendorf on the quadratic cost in $\mathbb{R}^n$, around the end of the eighties. It has since then been improved by many authors, a key step being the extension to Riemannian manifolds, first addressed by McCann in 2000.

The main results in this chapter are Theorems 10.28, 10.38 and (to a lesser extent) 10.42, which solve the Monge problem with increasing generality. For Parts II and III of this course, only the particular case considered in Theorem 10.41 is needed.

A heuristic argument

Let ψ be a c-convex function on a Riemannian manifold M, and $\phi = \psi^c$. Assume that $y \in \partial_c\psi(x)$; then, from the definition of c-subdifferential,

C. Villani, *Optimal Transport*, Grundlehren der mathematischen Wissenschaften 338,

one has, for all $\widetilde{x} \in M$,

$$\begin{cases} \phi(y) - \psi(x) = c(x,y) \\ \phi(y) - \psi(\widetilde{x}) \leq c(\widetilde{x},y). \end{cases} \tag{10.1}$$

It follows that

$$\psi(x) - \psi(\widetilde{x}) \leq c(\widetilde{x},y) - c(x,y). \tag{10.2}$$

Now the idea is to see what happens when $\widetilde{x} \to x$, along a given direction. Let w be a tangent vector at x, and consider a path $\varepsilon \to \widetilde{x}(\varepsilon)$, defined for $\varepsilon \in [0, \varepsilon_0)$, with initial position x and initial velocity w. (For instance, $\widetilde{x}(\varepsilon) = \exp_x(\varepsilon w)$; or in $\mathbb{R}^n$, just consider $\widetilde{x}(\varepsilon) = x + \varepsilon w$.) Assume that ψ and $c(\,\cdot\,, y)$ are differentiable at x, divide both sides of (10.2) by $\varepsilon > 0$ and pass to the limit:

$$-\nabla\psi(x) \cdot w \leq \nabla_x c(x,y) \cdot w. \tag{10.3}$$

If then one changes w to $-w$, the inequality will be reversed. So necessarily

$$\nabla\psi(x) + \nabla_x c(x,y) = 0. \tag{10.4}$$

If x is given, this is an equation for y. Since our goal is to show that y is determined by x, then it will surely help if (10.4) admits at most one solution, and this will obviously be the case if $\nabla_x c(x,\,\cdot\,)$ is *injective*. This property (injectivity of $\nabla_x c(x,\,\cdot\,)$) is in fact a classical condition in the theory of dynamical system, where it is sometimes referred to as a **twist condition**.

Three objections might immediately be raised. First, ψ is an unknown of the problem, defined by an infimum, so why would it be differentiable? Second, the injectivity of $\nabla_x c$ as a function of y seems quite hard to check on concrete examples. Third, even if c is given in the problem and a priori quite nice, why should it be differentiable at (x,y)? As a very simple example, consider the square distance function $d(x,y)^2$ on the 1-dimensional circle $S^1 = \mathbb{R}/(2\pi\mathbb{Z})$, identified with $[0, 2\pi)$:

$$d(x,y) = \min\big(|x-y|,\, 2\pi - |x-y|\big).$$

Then $d(x,y)$ is *not* differentiable as a function of x when $|y - x| = \pi$, and of course $d(x,y)^2$ is not differentiable either (see Figure 10.1).

Similar problems would occur on, say, a compact Riemannian manifold, as soon as there is no uniqueness of the geodesic joining x to y.

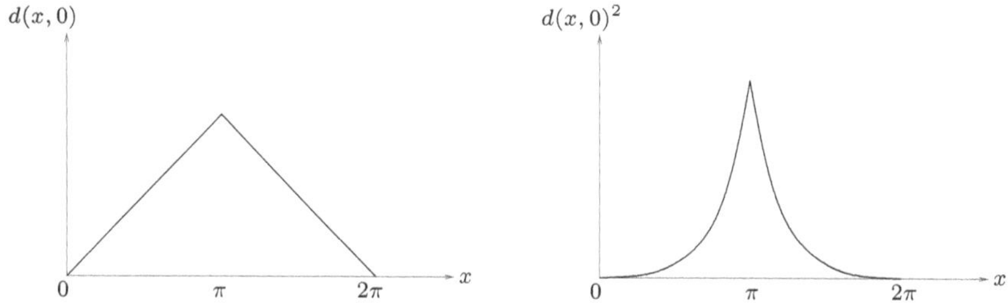

Fig. 10.1. The distance function $d(\cdot, y)$ on S^1, and its square. The upper-pointing singularity is typical. The square distance is not differentiable when $|x-y| = \pi$; still it is superdifferentiable, in a sense that is explained later.

For instance, if N and S respectively stand for the north and south poles on S^2, then $d(x, S)$ fails to be differentiable at $x = N$.

Of course, for any x this happens only for a negligible set of y's; and the cost function is differentiable everywhere else, so we might think that this is not a serious problem. But who can tell us that the optimal transport will not try to take each x (or a lot of them) to a place y such that $c(x, y)$ is not differentiable?

To solve these problems, it will be useful to use some concepts from nonsmooth analysis: subdifferentiability, superdifferentiability, approximate differentiability. The short answers to the above problems are that (a) under adequate assumptions on the cost function, ψ will be differentiable out of a very small set (of codimension at least 1); (b) c will be superdifferentiable because it derives from a Lagrangian, and subdifferentiable wherever ψ itself is differentiable; (c) where it exists, $\nabla_x c$ will be injective because c derives from a *strictly convex* Lagrangian.

The next three sections will be devoted to some basic reminders about differentiability and regularity in a nonsmooth context. For the convenience of the non-expert reader, I shall provide complete proofs of the most basic results about these issues. Conversely, readers who feel very comfortable with these notions can skip these sections.

Differentiability and approximate differentiability

Let us start with the classical definition of differentiability:

Definition 10.1 (Differentiability). *Let $U \subset \mathbb{R}^n$ be an open set. A function $f : U \to \mathbb{R}$ is said to be differentiable at $x \in U$ if there exists* *a vector* $p \in \mathbb{R}^n$ *such that*

$$f(z) = f(x) + \langle p, z - x\rangle + o(|z - x|) \qquad \text{as } z \to x.$$

Then the vector p is uniquely determined; it is called the gradient of f at x, and is denoted by $\nabla f(x)$*; the map* $w \to \langle p, w\rangle$ *is the differential of f at x.*

If U is an open set of a smooth Riemannian manifold M, $f : U \to \mathbb{R}$ *is said to be differentiable at x if it is so when expressed in a local chart around x; or equivalently if there is a tangent vector* $p \in T_x M$ *such that*

$$f(\exp_w x) = f(x) + \langle p, w\rangle + o(|w|) \qquad \text{as } w \to 0.$$

The vector p is again denoted by $\nabla f(x)$*.*

Differentiability is a pointwise concept, which is not invariant under, say, change of Lebesgue equivalence class: If f is differentiable or even C^∞ everywhere, by changing it on a dense countable set we may obtain a function which is discontinuous everywhere, and a fortiori not differentiable. The next notion is more flexible in this respect, since it allows for modification on a negligible set. It relies on the useful concept of **density**. Recall that a measurable set A is said to have density ρ at x if

$$\lim_{r\to 0} \frac{\operatorname{vol}[A \cap B_r(x)]}{\operatorname{vol}[B_r(x)]} = \rho.$$

It is a basic result of measure theory that *a measurable set in* $\mathbb{R}^n$*, or in a Riemannian manifold, has density 1 at almost all of its points.*

Definition 10.2 (Approximate differentiability). *Let U be an open set of a Riemannian manifold M, and let* $f : U \to \mathbb{R} \cup \{\pm\infty\}$ *be a measurable function. Then f is said to be approximately differentiable at* $x \in U$ *if there is a measurable function* $\widetilde{f} : U \to \mathbb{R}$*, differentiable at x, such that the set* $\{\widetilde{f} = f\}$ *has density 1 at x; in other words,*

$$\lim_{r\to 0} \frac{\operatorname{vol}\left[\left\{z \in B_r(x);\ f(z) = \widetilde{f}(z)\right\}\right]}{\operatorname{vol}[B_r(x)]} = 1.$$

Then one defines the approximate gradient of f at x by the formula

$$\widetilde{\nabla} f(x) = \nabla \widetilde{f}(x).$$

Proof that $\widetilde{\nabla} f(x)$ *is well-defined.* Since this concept is local and invariant by diffeomorphism, it is sufficient to treat the case when U is a subset of $\mathbb{R}^n$.

Let $\widetilde{f}_1$ and $\widetilde{f}_2$ be two measurable functions on U which are both differentiable at x and coincide with f on a set of density 1. The problem is to show that $\nabla\widetilde{f}_1(x) = \nabla\widetilde{f}_2(x)$.

For each $r > 0$, let Z_r be the set of points in $B_r(x)$ where either $f(x) \neq \widetilde{f}_1(x)$ or $f(x) \neq \widetilde{f}_2(x)$. It is clear that $\mathrm{vol}\,[Z_r] = o(\mathrm{vol}\,[B_r(x)])$.

Since $\widetilde{f}_1$ and $\widetilde{f}_2$ are continuous at x, one can write

$$\begin{aligned}\widetilde{f}_1(x) &= \lim_{r\to 0} \frac{1}{\mathrm{vol}\,[B_r(x)]}\int \widetilde{f}_1(z)\,dz \\ &= \lim_{r\to 0} \frac{1}{\mathrm{vol}\,[B_r(x)\setminus Z_r]}\int \widetilde{f}_1(z)\,dz = \lim_{r\to 0} \frac{1}{\mathrm{vol}\,[B_r(x)\setminus Z_r]}\int \widetilde{f}_2(z)\,dz \\ &= \lim_{r\to 0} \frac{1}{\mathrm{vol}\,[B_r(x)]}\int \widetilde{f}_2(z)\,dz = \widetilde{f}_2(x).\end{aligned}$$

So let $\widetilde{f}(x)$ be the common value of $\widetilde{f}_1$ and $\widetilde{f}_2$ at x.

Next, for any $z \in B_r(x)\setminus Z_r$, one has

$$\begin{cases}\widetilde{f}_1(z) = \widetilde{f}(x) + \langle \nabla\widetilde{f}_1(x),\, z-x\rangle + o(r),\\[2ex] \widetilde{f}_2(z) = \widetilde{f}(x) + \langle \nabla\widetilde{f}_2(x),\, z-x\rangle + o(r),\end{cases}$$

so

$$\langle \nabla\widetilde{f}_1(x) - \nabla\widetilde{f}_2(x),\, z-x\rangle = o(r).$$

Let $w := \nabla\widetilde{f}_1(x) - \nabla\widetilde{f}_2(x)$; the previous estimate reads

$$x \notin Z_r \Longrightarrow \quad \langle w, z-x\rangle = o(r). \tag{10.5}$$

If $w \neq 0$, then the set of $z \in B_r(x)$ such that $\langle w, z-x\rangle \geq r/2$ has measure at least $K\,\mathrm{vol}\,[B_r(x)]$, for some $K > 0$. If r is small enough, then $\mathrm{vol}\,[Z_r] \leq (K/4)\,\mathrm{vol}\,[B_r(x)] \leq (K/2)\,\mathrm{vol}\,[B_r(x)\setminus Z_r]$, so

$$\mathrm{vol}\left[\left\{z \in B_r(x)\setminus Z_r;\ \langle w, z-x\rangle \geq \frac{r}{2}\right\}\right] \geq \frac{K}{2}\,\mathrm{vol}\,[B_r(x)\setminus Z_r].$$

Then (still for r small enough),

$$\frac{\displaystyle\int_{B_r(x)\setminus Z_r} \big|\langle w, z-x\rangle\big|\,dy}{\mathrm{vol}\,[B_r(x)\setminus Z_r]} \geq \frac{Kr}{4},$$

in contradiction with (10.5). The conclusion is that $w = 0$, which was the goal. □

Regularity in a nonsmooth world

Regularity is a loose concept about the control of "how fast" a function varies. In the present section I shall review some notions of regularity which apply in a nonsmooth context, and act as a replacement for, say, C^1 or C^2 regularity bounds.

Definition 10.3 (Lipschitz continuity). *Let $U \subset \mathbb{R}^n$ be open, and let $f : U \to \mathbb{R}$ be given. Then:*

(i) f is said to be Lipschitz if there exists $L < \infty$ such that

$$\forall x, z \in U, \qquad |f(z) - f(x)| \le L|z - x|.$$

(ii) f is said to be locally Lipschitz if, for any $x_0 \in U$, there is a neighborhood O of x_0 in which f is Lipschitz.

If U is an open subset of a Riemannian manifold M, then $f : U \to \mathbb{R}$ is said to be locally Lipschitz if it is so when expressed in local charts; or equivalently if f is Lipschitz on any compact subset of U, equipped with the geodesic distance on M.

Example 10.4. Obviously, a C^1 function is locally Lipschitz, but the converse is not true (think of $f(x) = |x|$).

Definition 10.5 (Subdifferentiability, superdifferentiability). *Let U be an open set of $\mathbb{R}^n$, and $f : U \to \mathbb{R}$ a function. Then:*

(i) f is said to be subdifferentiable at x, with subgradient p, if

$$f(z) \ge f(x) + \langle p, z - x\rangle + o(|z - x|).$$

The convex set of all subgradients p at x will be denoted by $\nabla^- f(x)$.

(ii) f is said to be uniformly subdifferentiable in U if there is a continuous function $\omega : \mathbb{R}_+ \to \mathbb{R}_+$, such that $\omega(r) = o(r)$ as $r \to 0$, and

$$\forall x \in U \qquad \exists p \in \mathbb{R}^n; \qquad f(z) \ge f(x) + \langle p, z - x\rangle - \omega(|z - x|).$$

(iii) f is said to be locally subdifferentiable (or locally uniformly subdifferentiable) in U if each $x_0 \in U$ admits a neighborhood on which f is uniformly subdifferentiable.

If U is an open set of a smooth manifold M, a function $f : U \to \mathbb{R}$ is said to be subdifferentiable at some point x (resp. locally subdifferentiable in U) if it is so when expressed in local charts.

Corresponding notions of superdifferentiability and supergradients are obtained in an obvious way by just reversing the signs of the inequalities. The convex set of supergradients for f at x is denoted by $\nabla^+ f(x)$.

Examples 10.6. If f has a minimum at $x_0 \in U$, then 0 is a subgradient of f at x_0, whatever the regularity of f. If f has a subgradient p at x and g is smooth, then $f+g$ has a subgradient $p+\nabla g(x)$ at x. If f is convex in U, then it is (uniformly) subdifferentiable at every point in U, by the well-known inequality

$$f(z) \geq f(x) + \langle p, z - x \rangle,$$

which holds true as soon as $p \in \partial f(x)$ and $[x, y] \subset U$. If f is the sum of a convex function and a smooth function, then it is also uniformly subdifferentiable.

It is obvious that differentiability implies both subdifferentiability and superdifferentiability. The converse is true, as shown by the next statement.

Proposition 10.7 (Sub- and superdifferentiability imply differentiability). *Let U be an open set of a smooth Riemannian manifold M, and let $f : U \to \mathbb{R}$ be a function. Then f is differentiable at x if and only if it is both subdifferentiable and superdifferentiable there; and then*

$$\nabla^- f(x) = \nabla^+ f(x) = \{\nabla f(x)\}.$$

Proof of Proposition 10.7. The only nontrivial implication is that if f is both subdifferentiable and superdifferentiable, then it is differentiable. Since this statement is local and invariant by diffeomorphism, let us pretend that $U \subset \mathbb{R}^n$. So let $p \in \nabla^- f(x)$ and $q \in \nabla^+ f(x)$; then

$$f(z) - f(x) \geq \langle p, z - x \rangle - o(|z - x|);$$

$$f(z) - f(x) \leq \langle q, z - x \rangle + o(|z - x|).$$

This implies $\langle p - q, z - x \rangle \leq o(|z - x|)$, which means

$$\lim_{z \to x;\ z \neq x} \left\langle p - q, \frac{z - x}{|z - x|} \right\rangle = 0.$$

Since the unit vector $(z - x)/|z - x|$ can take arbitrary fixed values in the unit sphere as $z \to x$, it follows that $p = q$. Then

$$f(z) - f(x) = \langle p, z - x \rangle + o(|z - x|),$$

which means that f is indeed differentiable at x. This also shows that $p = q = \nabla f(x)$, and the proof is complete. □

The next proposition summarizes some of the most important results about the links between regularity and differentiability:

Theorem 10.8 (Regularity and differentiability almost everywhere). *Let U be an open subset of a smooth Riemannian manifold M, and let $f : U \to \mathbb{R}$ be a function. Let n be the dimension of M. Then:*

(i) If f is continuous, then it is subdifferentiable on a dense subset of U, and also superdifferentiable on a dense subset of U.

(ii) If f is locally Lipschitz, then it is differentiable almost everywhere (with respect to the volume measure).

(iii) If f is locally subdifferentiable (resp. locally superdifferentiable), then it is locally Lipschitz and differentiable out of a countably $(n-1)$-rectifiable set. Moreover, the set of differentiability points coincides with the set of points where there is a unique subgradient (resp. supergradient). Finally, ∇f is continuous on its domain of definition.

Remark 10.9. Statement (ii) is known as **Rademacher's theorem.** The conclusion in statement (iii) is stronger than differentiability almost everywhere, since an $(n-1)$-rectifiable set has dimension $n-1$, and is therefore negligible. In fact, as we shall see very soon, the local subdifferentiability property is stronger than the local Lipschitz property. Reminders about the notion of countable rectifiability are provided in the Appendix.

Proof of Theorem 10.8. First we can cover U by a countable collection of small open sets U_k, each of which is diffeomorphic to an open subset O_k of $\mathbb{R}^n$. Then, since all the concepts involved are local and invariant under diffeomorphism, we may work in O_k. So in the sequel, I shall pretend that U is a subset of $\mathbb{R}^n$.

Let us start with the proof of (i). Let f be continuous on U, and let V be an open subset of U; the problem is to show that f admits at least one point of subdifferentiability in V. So let $x_0 \in V$, and let $r > 0$ be so small that $\overline{B(x_0, r)} \subset V$. Let $B = B(x_0, r)$, let $\varepsilon > 0$ and let g be defined on $\overline{B}$ by $g(x) := f(x) + |x - x_0|^2/\varepsilon$. Since f is continuous, g attains its minimum on $\overline{B}$. But g on ∂B is bounded below by $r^2/\varepsilon - M$, where M is an upper bound for $|f|$ on $\overline{B}$. If $\varepsilon < r^2/(2M)$, then $g(x_0) = f(x_0) \le M < r^2/\varepsilon - M \le \inf_{\partial B} g$; so g cannot achieve its minimum on ∂B, and has to achieve it at some point $x_1 \in B$. Then g is subdifferentiable at x_1, and therefore f also. This establishes (i).

The other two statements are more tricky. Let $f : U \to \mathbb{R}$ be a Lipschitz function. For $v \in \mathbb{R}^n$ and $x \in U$, define

$$D_v f(x) := \lim_{t\to 0} \left[\frac{f(x+tv) - f(x)}{t}\right], \tag{10.6}$$

provided that this limit exists. The problem is to show that for almost any x, there is a vector $p(x)$ such that $D_v f(x) = \langle p(x), v\rangle$ and the limit in (10.6) is uniform in, say, $v \in S^{n-1}$. Since the functions $[f(x+tv) - f(x)]/t$ are uniformly Lipschitz in v, it is enough to prove the pointwise convergence (that is, the mere existence of $D_v f(x)$), and then the limit will automatically be uniform by Ascoli's theorem. So the goal is to show that for almost any x, the limit $D_v f(x)$ exists for any v, and is linear in v.

It is easily checked that:

(a) $D_v f(x)$ is homogeneous in v: $D_{tv} f(x) = t\, D_v f(x)$;

(b) $D_v f(x)$ is a Lipschitz function of v on its domain: in fact, $|D_v f(x) - D_w f(x)| \le L\,|v - w|$, where $L = \|f\|_{\mathrm{Lip}}$;

(c) If $D_w f(x) \to \ell$ as $w \to v$, then $D_v f(x) = \ell$; this comes from the estimate

$$\sup_t \left| \left(\frac{f(x+tv) - f(x)}{t}\right) - \left(\frac{f(x+tv_k) - f(x)}{t}\right)\right| \le \|f\|_{\mathrm{Lip}}\,|v - v_k|.$$

For each $v \in \mathbb{R}^n$, let A_v be the set of $x \in \mathbb{R}^n$ such that $D_v f(x)$ does not exist. The first claim is that *each A_v has zero Lebesgue measure.* This is obvious if $v = 0$. Otherwise, let $H = v^{\perp}$ be the hyperplane orthogonal to v, passing through the origin. For each $x_0 \in H$, let $L_{x_0} = x_0 + \mathbb{R}v$ be the line parallel to v, passing through x_0. The nonexistence of $D_v f(x)$ at $x = x_0 + t_0 v$ is equivalent to the nondifferentiability of $t \longmapsto f(x+tv)$ at $t = t_0$. Since $t \longmapsto f(x+tv)$ is Lipschitz $\mathbb{R} \to \mathbb{R}$, it follows from a well-known result of real analysis that it is differentiable for λ_1-almost all $t \in \mathbb{R}$, where λ_1 stands for the one-dimensional Lebesgue measure. So $\lambda_1[A_v \cap L_{x_0}] = 0$. Then by Fubini's theorem, $\lambda_n[A_v] = \int_H \lambda_1[A_v \cap L_{x_0}]\,dx_0 = 0$, where λ_n is the n-dimensional Lebesgue measure, and this proves the claim.

The problem consists in extending the function $D_v f$ into a *linear* (not just homogeneous) function of v. Let $v \in \mathbb{R}^n$, and let ζ be a smooth compactly supported function. Then, by the dominated convergence theorem,

$$\begin{aligned}(\zeta * D_v f)(x) &= \int \zeta(x-y) \lim_{t\to 0}\left[\frac{f(y+tv)-f(y)}{t}\right] dy \\ &= \lim_{t\to 0} \frac{1}{t}\int \zeta(x-y)\,[f(y+tv)-f(y)]\,dy \\ &= \lim_{t\to 0} \frac{1}{t}\int \big[\zeta(x-y+tv)-\zeta(x-y)\big]\,f(y)\,dy \\ &= \int \langle \nabla\zeta(x-y), v\rangle\, f(y)\,dy.\end{aligned}$$

(Note that $\zeta * D_v f$ is well-defined for any x, even if $D_v f$ is defined only for almost all x.) So $\zeta * D_v f$ depends linearly on v. In particular, if v and w are any two vectors in $\mathbb{R}^n$, then

$$\zeta * [D_{v+w} f - D_v f - D_w f] = 0.$$

Since ζ is arbitrary, it follows that

$$D_v f(x) + D_w f(x) = D_{v+w} f(x) \tag{10.7}$$

for almost all $x \in \mathbb{R}^n \setminus (A_v \cap A_w \cap A_{v+w})$, that is, for almost all $x \in \mathbb{R}^n$.

Now it is easy to conclude. Let $B_{v,w}$ be the set of all $x \in \mathbb{R}^n$ such that $D_v f(x)$, $D_w f(x)$ or $D_{v+w} f(x)$ is not well-defined, or (10.7) does not hold true. Let $(v_k)_{k\in\mathbb{N}}$ be a dense sequence in $\mathbb{R}^n$, and let $B := \bigcup_{j,k\in\mathbb{N}} B_{v_j,v_k}$. Then B is still Lebesgue-negligible, and for each $x \notin B$ we have

$$D_{v_j+v_k} f(x) = D_{v_j} f(x) + D_{v_k} f(x). \tag{10.8}$$

Since $D_v f(x)$ is a Lipschitz continuous function of v, it can be extended uniquely into a Lipschitz continuous function, defined for all $x \notin B$ and $v \in \mathbb{R}^n$, which turns out to be $D_v f(x)$ in view of Property (c). By passing to the limit in (10.8), we see that $D_v f(x)$ is an additive function of v. We already know that it is a homogeneous function of v, so it is in fact linear. This concludes the proof of (ii).

Next let us turn to the proof of (iii). Before going on, I shall first explain in an informal way the **main idea of the proof of statement (iii)**. Suppose for simplicity that we are dealing with a convex function in $\mathbb{R}^n$. If p lies in the subdifferential $\partial\psi(x)$ of ψ at x, then for all $z \in \mathbb{R}^n$,

$$\psi(z) \geq \psi(x) + \langle p,\, z - x \rangle.$$

In particular, if $p \in \partial\psi(x)$ and $p' \in \partial\psi(x')$, then

$$\langle p - p',\, x - x' \rangle \geq 0.$$

If ψ is not differentiable at x, this means that the convex set $\partial\psi(x)$ is not a single point, so it should contain a line segment $[p, p'] \subset \mathbb{R}^n$. For these heuristic explanations, let us *fix* p and p', and let Σ be the set of all $x \in \mathbb{R}^n$ such that $[p, p'] \subset \partial\psi(x)$. Then $\langle p - p', x - x' \rangle \geq 0$ for all $x, x' \in \Sigma$. By exchanging the roles of p and p', we see that actually $\langle p - p', x - x' \rangle = 0$. This implies that Σ is included in a single hyperplane, orthogonal to $p - p'$; in particular its dimension is at most $n - 1$.

The rigorous argument can be decomposed into six steps. In the sequel, ψ will stand for a locally subdifferentiable function.

Step 1: *ψ is locally semiconvex.* Without loss of generality, we may assume that $\omega(r)/r$ is nondecreasing continuous (otherwise replace $\omega(r)$ by $\overline{\omega}(r) = r \sup\{\omega(s)/s;\ s \leq r\}$ which is a nondecreasing continuous function of r); then $\omega(tr) \leq t\,\omega(r)$.

Let $x_0 \in U$, let V be a convex neighborhood of x_0 in U. Let $x, y \in V$, $t \in [0,1]$ and $p \in \nabla^-\psi((1-t)\,x + t\,y)$. Then

$$\psi(x) \geq \psi\big((1-t)\,x + t\,y\big) + \langle t\,(x-y), p \rangle - t\,\omega(|x-y|); \qquad (10.9)$$

$$\psi(y) \geq \psi\big((1-t)\,x + t\,y\big) + \langle (1-t)\,(y-x), p \rangle - (1-t)\,\omega(|x-y|). \quad (10.10)$$

Take the linear combination of (10.9) and (10.10) with respective coefficients $(1-t)$ and t: the result is

$$\psi\big((1-t)\,x + t\,y\big) \leq (1-t)\,\psi(x) + t\,\psi(y) + 2\,t(1-t)\,\omega(|x-y|). \quad (10.11)$$

Step 2: *ψ is locally bounded above.* Let $x_0 \in U$, let $x_1, \ldots, x_N \in U$ be such that the convex hull C of $(x_1, \ldots, x_N)$ is a neighborhood of x_0 ($N = 2n$ will do). Any point of C can be written as $\sum \alpha_i x_i$ where $0 \leq \alpha_i \leq i$, $\sum \alpha_i = 1$. By (10.11) and finite induction,

$$\psi\Big(\sum \alpha_i x_i\Big) \leq \sum \alpha_i\,\psi(x_i) + 2^N \max_{ij}\, \omega(|x_i - x_j|);$$

so ψ is bounded above on C, and therefore in a neighborhood of x_0.

Step 3: *ψ is locally bounded below.* Let $x_0 \in U$, let V be a neighborhood of x_0 on which ψ is bounded above, and let $B = B_r(x_0)$, where r

is such that $B_r(x_0) \subset V$. For any $x \in B$, let $y = 2\,x_0 - x$; then $|x_0 - y| = |x_0 - x| < r$, so $y \in B$, and

$$\psi(x_0) = \psi\left(\frac{x+y}{2}\right) \le \frac{1}{2}\,[\psi(x) + \psi(y)] + \frac{\omega(|x-y|)}{2}.$$

Since $\psi(x_0)$ is fixed and $\psi(y)$ is bounded above, it follows that $\psi(x)$ is bounded below for $x \in B$.

Step 4: *ψ is locally Lipschitz.* Let $x_0 \in U$, let V be a neighborhood of x_0 on which $|\psi| \le M < +\infty$, and let $r > 0$ be such that $B_r(x_0) \subset V$. For any $y, y' \in B_{r/2}(x_0)$, we can write $y' = (1-t)\,y + t\,z$, where $t = |y - y'|/r$, so $z = (y - y')/t + y \in B_r(x_0)$ and $|y - z| = r$. Then $\psi(y') \le (1-t)\,\psi(y) + t\,\psi(z) + 2\,t(1-t)\,\omega(|y-z|)$, so

$$\frac{\psi(y') - \psi(y)}{|y - y'|} = \frac{\psi(y') - \psi(y)}{t\,|y - z|} \le \frac{\psi(y) - \psi(z)}{|y - z|} + \frac{2\,\omega(|y-z|)}{|y-z|}$$
$$\le \frac{2M}{r} + \frac{2\,\omega(r)}{r}.$$

Thus the ratio $[\psi(y') - \psi(y)]/|y' - y|$ is uniformly bounded above in $B_{r/2}(x_0)$. By symmetry (exchange y and y'), it is also uniformly bounded below, and ψ is Lipschitz on $B_{r/2}(x_0)$.

Step 5: *$\nabla^-\psi$ is continuous.* This means that if $p_k \in \nabla^-\psi(x_k)$ and $(x_k, p_k) \to (x, p)$ then $p \in \nabla^-\psi(x)$. To prove this, it suffices to pass to the limit in the inequality

$$\psi(z) \ge \psi(x_k) + \langle p_k, z - x_k\rangle - \omega(|z - x_k|).$$

Step 6: *ψ is differentiable out of an $(n-1)$-dimensional set.* Indeed, let Σ be the set of points x such that $\nabla^-\psi(x)$ is not reduced to a single element. Since $\nabla^-\psi(x)$ is a convex set, for each $x \in \Sigma$ there is a nontrivial segment $[p, p'] \subset \nabla^-\psi(x)$. So

$$\Sigma = \bigcup_{\ell\in\mathbb{N}} \Sigma^{(\ell)},$$

where $\Sigma^{(\ell)}$ is the set of points x such that $\nabla^-\psi(x)$ contains a segment $[p, p']$ of length $1/\ell$ and $|p| \le \ell$. To conclude, it is sufficient to show that each $\Sigma^{(\ell)}$ is countably $(n-1)$-rectifiable, and for that it is sufficient to show that for each $x \in \Sigma^{(\ell)}$ the dimension of the tangent cone $T_x\Sigma^{(\ell)}$ is at most $n-1$ (Theorem 10.48(i) in the First Appendix).

So let $x \in \Sigma^{(\ell)}$, and let $q \in T_x\Sigma^{(\ell)}$, $q \neq 0$. By assumption, there is a sequence $x_k \in \Sigma^{(\ell)}$ such that

$$\frac{x_k - x}{t_k} \longrightarrow q.$$

In particular $|x - x_k|/t_k$ converges to the finite, nonzero limit $|q|$.

For any $k \in \mathbb{N}$, there is a segment $[p_k, p'_k]$, of length ℓ^{-1}, that is contained in $\nabla^-\psi(x_k)$; and $|p_k| \leq \ell$, $|p'_k| \leq \ell + \ell^{-1}$. By compactness, up to extraction of a subsequence one has $x_k \to x$, $p_k \to p$, $p'_k \to p'$, $|p - p'| = \ell^{-1}$. By continuity of $\nabla^-\psi$, both p and p' belong to $\nabla^-\psi(x)$. Then the two inequalities

$$\begin{cases} \psi(x) \geq \psi(x_k) + \langle p'_k,\, x - x_k\rangle - \omega(|x - x_k|) \\ \\ \psi(x_k) \geq \psi(x) + \langle p,\, x_k - x\rangle - \omega(|x - x_k|) \end{cases}$$

combine to yield

$$\langle p - p'_k,\, x - x_k\rangle \geq -2\,\omega(|x - x_k|).$$

So

$$\left\langle p - p'_k,\, \frac{x - x_k}{t_k}\right\rangle \geq -2\, \frac{\omega(|x - x_k|)}{|x - x_k|}\, \frac{|x - x_k|}{t_k}.$$

Passing to the limit, we find

$$\langle p - p', q\rangle \geq 0.$$

But the roles of p and p' can be exchanged, so actually

$$\langle p - p', q\rangle = 0.$$

Since $p - p'$ is nonzero, this means that q belongs to the hyperplane $(p - p')^\perp$. So for each $x \in \Sigma^{(\ell)}$, the tangent cone $T_x\Sigma^{(\ell)}$ is included in a hyperplane.

To conclude the proof of (iii), it only remains to prove the equivalence between differentiability and unique subdifferentiability. If x is a differentiability point of ψ, then we know from Proposition 10.7 that there is a unique subgradient of f at x. Conversely, assume that x is such that $\nabla^-\psi(x) = \{p\}$. Let $(x_k)_{k\in\mathbb{N}}$ be a dense sequence in a neighborhood of x; for each $k \in \mathbb{N}$, let $p_k \in \nabla^-\psi(x_k)$. By definition of subdifferentiability,

$$\begin{aligned}\psi(x) &\geq \psi(x_k) + \langle p_k, x - x_k \rangle - \omega(|x - x_k|) \\ &= \psi(x_k) + \langle p, x - x_k \rangle + \langle p_k - p, x - x_k \rangle - \omega(|x - x_k|).\end{aligned}$$

The continuity of $\nabla^- \psi$ imposes $p_k \to p$ as $x_k \to x$; it follows that $\psi(x) \geq \psi(x_k) + \langle p, x - x_k \rangle - o(|x - x_k|)$. By density,

$$\psi(x) \geq \psi(y) + \langle p, x - y \rangle - o(|x - y|) \qquad \text{as } y \to x.$$

This shows that $p \in \nabla^+ \psi(x)$; then by Proposition 10.7, $p = \nabla \psi(x)$ and the proof is finished. □

Semiconvexity and semiconcavity

Convexity can be expressed without any reference to smoothness, yet it implies a lower bound on the Hessian. In nonsmooth analysis, convexity-type estimates are often used as a replacement for second-order derivative bounds. In this respect the notion of semiconvexity is extremely convenient.

Definition 10.10 (Semiconvexity). *Let U be an open set of a smooth Riemanian manifold and let $\omega : \mathbb{R}_+ \to \mathbb{R}_+$ be continuous, such that $\omega(r) = o(r)$ as $r \to 0$. A function $f : U \to \mathbb{R} \cup \{+\infty\}$ is said to be semiconvex with modulus ω if, for any constant-speed geodesic path $(\gamma_t)_{0 \leq t \leq 1}$, whose image is included in U,*

$$f(\gamma_t) \leq (1-t) f(\gamma_0) + t f(\gamma_1) + t(1-t)\, \omega\big(d(\gamma_0, \gamma_1)\big). \tag{10.12}$$

It is said to be locally semiconvex if for each $x_0 \in U$ there is a neighborhood V of x_0 in U such that (10.12) holds true as soon as $\gamma_0, \gamma_1 \in V$; or equivalently if (10.12) holds true for some fixed modulus ω_K as long as γ stays in a compact subset K of U.

Similar definitions for semiconcavity and local semiconcavity are obtained in an obvious way by reversing the sign of the inequality in (10.12).

Example 10.11. In $\mathbb{R}^n$, semiconvexity with modulus ω means that for any x, y in $\mathbb{R}^n$ and for any $t \in [0, 1]$,

$$f\big((1-t)x + ty\big) \leq (1-t)\, f(x) + t\, f(y) + t(1-t)\, \omega(|x - y|).$$

When $\omega = 0$ this is the usual notion of convexity. In the case $\omega(r) = Cr^2/2$, there is a differential characterization of semiconvexity in terms of Hessian matrices: $f : \mathbb{R}^n \to \mathbb{R}$ is semiconvex with modulus $\omega(r) = Cr^2/2$ if and only if $\nabla^2 f \geq -CI_n$. (If f is not twice differentiable, then $\nabla^2 f$ should be interpreted as the distributional gradient.)

A well-known theorem states that a convex function is subdifferentiable everywhere in the interior of its domain. The next result generalizes this property to semiconvex functions.

Proposition 10.12 (Local equivalence of semiconvexity and subdifferentiability). *Let M be a smooth complete Riemannian manifold. Then:*

(i) If $\psi : M \to \mathbb{R} \cup \{+\infty\}$ is locally semiconvex, then it is locally subdifferentiable in the interior of its domain $D := \psi^{-1}(\mathbb{R})$; and ∂D is countably $(n-1)$-rectifiable;

(ii) Conversely, if U is an open subset of M, and $\psi : U \to \mathbb{R}$ is locally subdifferentiable, then it is also locally semiconvex.

Similar statements hold true with "subdifferentiable" replaced by "superdifferentiable" and "semiconvex" replaced by "semiconcave".

Remark 10.13. This proposition implies that *local* semiconvexity and local subdifferentiability are basically the same. But there is also a *global* version of semiconvexity.

Remark 10.14. We already proved part of Proposition 10.12 in the proof of Theorem 10.8 when $M = \mathbb{R}^n$. Since the concept of semiconvexity is not invariant by diffeomorphism (unless this diffeomorphism is an isometry), we'll have to redo the proof.

Proof of Proposition 10.12. For each $x_0 \in M$, let O_{x_0} be an open neighborhood of x_0. There is an open neighborhood V_{x_0} of x_0 and a continuous function $\omega = \omega_{x_0}$ such that (10.12) holds true for any geodesic whose image is included in V_{x_0}. The open sets $O_{x_0} \cap V_{x_0}$ cover M which is a countable union of compact sets; so we may extract from this family a countable covering of M. If the property is proven in each $O_{x_0} \cap V_{x_0}$, then the conclusion will follow. So it is sufficient to prove (i) in any arbitrarily small neighborhood of any given point x_0. In the sequel, U will stand for such a neighborhood.

Let $D = \psi^{-1}(\mathbb{R})$. If $x_0, x_1 \in D \cap U$, and γ is a geodesic joining x_0 to x_1, then $\psi(\gamma_t) \leq (1-t)\,\psi(\gamma_0) + t\,\psi(\gamma_1) + t\,(1-t)\,\omega(d(\gamma_0, \gamma_1))$ is finite for all $t \in [0,1]$. So D is geodesically convex.

If U is small enough, then (a) any two points in U are joined by a unique geodesic; (b) U is isometric to a small open subset V of $\mathbb{R}^n$ equipped with some Riemannian distance d. Since the property of (geodesic) semiconvexity is invariant by isometry, we may work in V equipped with the distance d. If x and y are two given points in V, let $m(x, y)$ stand for the midpoint (with respect to the geodesic distance d) of x and y. Because d is Riemannian, one has

$$m(x,y) = \frac{x+y}{2} + o(|x-y|).$$

Then let $x \in V$ and let $T_x D$ be the tangent cone of D at x. If p, p' are any two points in $T_x D$, there are sequences $x_k \to x$, $t_k \to 0$, $x'_k \to x$, $t'_k \to 0$ such that

$$\frac{x_k - x}{t_k} \xrightarrow[k\to\infty]{} p; \qquad \frac{x'_k - x}{t'_k} \xrightarrow[k\to\infty]{} p'.$$

Then $m(x_k, x'_k) \in D$ and $m(x_k, x'_k) = (x_k + x'_k)/2 + o(|x_k - x'_k|) = x + t_k\,(p_k + p'_k)/2 + o(t_k)$, so

$$\frac{p+p'}{2} = \lim_{k\to\infty} \frac{m(x_k, x'_k) - x}{t_k} \in T_x D.$$

Thus $T_x D$ is a *convex* cone. This leaves two possibilities: either $T_x D$ is included in a half-space, or it is the whole of $\mathbb{R}^n$.

Assume that $T_x D = \mathbb{R}^n$. If C is a small (Euclidean) cube of side $2r$, centered at x_0, for r small enough any point in a neighborhood of x_0 can be written as a combination of barycenters of the vertices $x_1, \ldots, x_N$ of C, and all these barycenters will lie within a ball of radius $2r$ centered at x_0. (Indeed, let C_0 stand for the set of the vertices of the cube C, then $C_0 = \{x^{(\varepsilon)};\ \varepsilon \in \{\pm 1\}^n\}$, where $x^{(\varepsilon)} = x_0 + r \sum \varepsilon_j e_j$, $(e_j)_{1\le j\le n}$ being the canonical basis of $\mathbb{R}^n$. For each $\varepsilon \in \{\pm 1\}^{n-1}$, let $C_1^{(\varepsilon)}$ be the union of geodesic segments $[x^{(\varepsilon,-1)}, x^{(\varepsilon,1)}]$. Then for $\varepsilon \in \{\pm 1\}^{n-2}$ let $C_2^{(\varepsilon)}$ be the union of geodesic segments between an element of $C_1^{(\varepsilon,-1)}$ and an element of $C_1^{(\varepsilon,1)}$; etc. After n operations we have a simply connected set C_n which asymptotically coincides with the whole (solid) cube as $r \to 0$; so it is a neighborhood of x_0 for r small enough.) Then in the interior of C, ψ is bounded above by $\max\big(\psi(x_1), \ldots, \psi(x_N)\big) + \sup\{\omega(s);\ s \le 4r\}$. This shows at the same time that x_0 lies in the interior of D, and that ψ is bounded above around x_0.

In particular, if $x \in \partial D$, then $T_x D$ cannot be $\mathbb{R}^n$, so it is included in a half space. By Theorem 10.48(ii) in the Appendix, this implies that ∂D is countably $(n-1)$-rectifiable.

In the sequel, x_0 will be an interior point of D and we shall show that ψ is locally subdifferentiable around x_0. We have just seen that ψ is bounded above in a neighborhood of x_0; we shall now see that it is also bounded below. Let $B = B_r(x_0)$; if $r > 0$ is sufficiently small then for any $y \in B$ there is $y' \in B$ such that the midpoint of y and y' is x_0. (Indeed, take the geodesic γ starting from y and going to x_0, say $\gamma(t) = \exp_{x_0}(tv)$, $0 \le t \le 1$, and extend it up to time 2, set $y' = \exp_{x_0}(2v) \in B$. If B is small enough the geodesic is automatically minimizing up to time 2, and $x_0 = m(y, y')$.) Then

$$\psi(x_0) \le \frac{1}{2}\big[\psi(y) + \psi(y')\big] + \sup_{s \le 2r} \omega(s).$$

Since x_0 is fixed and $\psi(y')$ is bounded above, this shows that $\psi(y)$ is bounded below as y varies in B.

Next, let us show that ψ is locally Lipschitz. Let V be a neighborhood of x_0 in which $|\psi|$ is bounded by M. If $r > 0$ is small enough, then for any $y, y' \in B_r(x_0)$ there is $z = z(y, y') \in V$ such that $y' = [y, z]_\lambda$, $\lambda = d(y, y')/4r \in [0, 1/2]$. (Indeed, choose r so small that all geodesics in $B_{5r}(x_0)$ are minimizing, and $B_{5r}(x_0) \subset V$. Given y and y', take the geodesic going from y to y', say $\exp_y(tv)$, $0 \le t \le 1$; extend it up to time $t(\lambda) = 1/(1-\lambda)$, write $z = \exp_y(t(\lambda)\, v)$. Then $d(x_0, z) \le d(x_0, y) + t(\lambda)\, d(y, y') \le d(x_0, y) + 2\, d(y, y') < 5r$.) So

$$\psi(y') \le (1-\lambda)\, \psi(y) + \lambda\, \psi(z) + \lambda(1-\lambda)\, \omega(d(y, z)),$$

whence

$$\frac{\psi(y') - \psi(y)}{d(y, y')} = \frac{\psi(y') - \psi(y)}{\lambda\, d(y, z)} \le \frac{\psi(z) - \psi(y)}{d(y, z)} + \frac{\omega(d(y, z))}{d(y, z)}. \qquad (10.13)$$

Since $d(y, z) = d(y, y')/\lambda = 4r$, (10.13) implies

$$\frac{\psi(y') - \psi(y)}{d(y, y')} \le \frac{2M}{r} + \frac{\omega(r)}{r}.$$

So the ratio $[\psi(y') - \psi(y)]/d(y, y')$ is bounded above in $B_r(x_0)$; by symmetry (exchange y and y') it is also bounded below, and ψ is Lipschitz in $B_r(x_0)$.

The next step consists in showing that there is a uniform modulus of subdifferentiability (at points of subdifferentiability!). More precisely, *if* ψ is subdifferentiable at x, and $p \in \nabla^-\psi(x)$, then for any $w \neq 0$, $|w|$ small enough,

$$\psi(\exp_x w) \geq \psi(x) + \langle p, w \rangle - \omega(|w|). \tag{10.14}$$

Indeed, let $\gamma(t) = \exp_x(tw)$, $y = \exp_x w$, then for any $t \in [0,1]$,

$$\psi(\gamma(t)) \leq (1-t)\,\psi(x) + t\,\psi(y) + t(1-t)\,\omega(|w|),$$

so

$$\frac{\psi(\gamma(t)) - \psi(x)}{t|w|} \leq \frac{\psi(y) - \psi(x)}{|w|} + (1-t)\,\frac{\omega(|w|)}{|w|}.$$

On the other hand, by subdifferentiability,

$$\frac{\psi(\gamma(t)) - \psi(x)}{t|w|} \geq \frac{\langle p, tw \rangle}{t|w|} - \frac{o(t|w|)}{t|w|} = \langle p, \frac{w}{|w|} \rangle - \frac{o(t|w|)}{t|w|}.$$

The combination of the two previous inequalities implies

$$\left\langle p, \frac{w}{|w|} \right\rangle - \frac{o(t|w|)}{t|w|} \leq \frac{\psi(y) - \psi(x)}{|w|} + (1-t)\,\frac{\omega(|w|)}{|w|}.$$

The limit $t \to 0$ gives (10.14). At the same time, it shows that for $|w| \leq r$,

$$\left\langle p, \frac{w}{|w|} \right\rangle \leq \|\psi\|_{\mathrm{Lip}}(B_{2r}(x_0)) + \frac{\omega(r)}{r}.$$

By choosing $w = rp$, we conclude that $|p|$ is bounded above, independently of x. So $\nabla^-\psi$ is *locally bounded* in the sense that there is a uniform bound on the norms of all elements of $\nabla^-\psi(x)$ when x varies in a compact subset of the domain of ψ.

At last we can conclude. Let x be interior to D. By Theorem 10.8(i), there is a sequence $x_k \to x$ such that $\nabla^-\psi(x_k) \neq \emptyset$. For each $k \in \mathbb{N}$, let $p_k \in \nabla^-\psi(x_k)$. As $k \to \infty$, there is a uniform bound on $|p_k|$, so we may extract a subsequence such that $p_k \to p \in \mathbb{R}^n$. For each x_k, for each $w \in \mathbb{R}^n$ small enough,

$$\psi(\exp_{x_k} w) \geq \psi(x_k) + \langle p_k, w \rangle - \omega(|w|).$$

Since ψ is continuous, we may pass to the limit as $k \to \infty$ and recover (for $|w|$ small enough)

$$\psi(\exp_x w) \geq \psi(x) + \langle p, w \rangle - \omega(|w|).$$

So ψ is uniformly subdifferentiable around x, and the proof of (i) is complete.

Statement (ii) is much easier to prove and completely similar to an argument already used in the proof of Theorem 10.8(iii). Let $x \in U$, and let V be a small neighborhood of x, such that f is uniformly subdifferentiable in V with modulus ω. Without loss of generality, assume that $\omega(r)/r$ is a nondecreasing function of r; otherwise replace $\omega(r)$ by $\overline{\omega}(r) = r \sup\{\omega(s)/s;\ 0 < s \leq r\}$. Let $W \subset V$ be a neighborhood of x, small enough that any two points y, y' in W can be joined by a unique geodesic $\gamma^{y,y'}$, whose image is contained in V; by abuse of notation I shall write $y' - y$ for the initial velocity of $\gamma^{y,y'}$.

Then let γ be a geodesic such that $\gamma_0, \gamma_1 \in V$; let $t \in [0,1]$, and let $p \in \nabla^- f(\gamma_t)$. It follows from the subdifferentiability that

$$f(\gamma_1) \leq f(\gamma_t) + \langle p, \gamma_1 - \gamma_t \rangle + \omega\big(d(\gamma_t, \gamma_1)\big).$$

Then since $d(\gamma_t, \gamma_1) = (1-t)\, d(\gamma_0, \gamma_1)$ and $\omega(r)/r$ is nonincreasing,

$$f(\gamma_1) \leq f(\gamma_t) + \langle p, \gamma_1 - \gamma_t \rangle + (1-t)\, \omega\big(d(\gamma_0, \gamma_1)\big). \tag{10.15}$$

Similarly,

$$f(\gamma_0) \leq f(\gamma_t) + \langle p, \gamma_0 - \gamma_t \rangle + t\, \omega\big(d(\gamma_0, \gamma_1)\big). \tag{10.16}$$

Now take the linear combination of (10.15) and (10.16) with coefficients t and $1-t$: Since $t(\gamma_1 - \gamma_t) + (1-t)(\gamma_0 - \gamma_t) = 0$ (in $T_{\gamma_t}M$), we recover

$$(1-t)\, f(\gamma_0) + t\, f(\gamma_1) - f(\gamma_t) \leq 2t\,(1-t)\, \omega(d(\gamma_0, \gamma_1)).$$

This proves that f is semiconvex in W. □

Assumptions on the cost function

Let M be a smooth complete connected Riemannian manifold, let $\mathcal{X}$ be a closed subset of M, let $\mathcal{Y}$ be an arbitrary Polish space, and let $c : M \times \mathcal{Y} \to \mathbb{R}$ be a continuous cost function. (Most of the time, we shall have $\mathcal{X} = M = \mathcal{Y}$.) I shall impose certain assumptions on the behavior of c as a function of x, when x varies in the interior (in M) of $\mathcal{X}$. They will be chosen from the following list (the notation $T_x S$ stands for tangent cone to S at x, see Definition 10.46):

(Super) *$c(x,y)$ is everywhere superdifferentiable as a function of x, for all y.*

(Twist) *On its domain of definition, $\nabla_x c(x,\,\cdot\,)$ is injective: if x,y,y' are such that $\nabla_x c(x,y) = \nabla_x c(x,y')$, then $y=y'$.*

(Lip) *$c(x,y)$ is locally Lipschitz as a function of x, uniformly in y.*

(SC) *$c(x,y)$ is locally semiconcave as a function of x, uniformly in y.*

(locLip) *$c(x,y)$ is locally Lipschitz as a function of x, locally in y.*

(locSC) *$c(x,y)$ is locally semiconcave as a function of x, locally in y.*

$(\mathbf{H}\infty)_1$ *For any x and for any measurable set S which does not "lie on one side of x" (in the sense that $T_x S$ is not contained in a half-space) there is a finite collection of elements $z_1,\ldots,z_k \in S$, and a small open ball B containing x, such that for any y outside of a compact set,*

$$\inf_{w\in B} c(w,y) \ge \inf_{1\le j\le k} c(z_j,y).$$

$(\mathbf{H}\infty)_2$ *For any x and any neighborhood U of x there is a small ball B containing x such that*

$$\lim_{y\to\infty} \sup_{w\in B} \inf_{z\in U} \big[c(z,y) - c(w,y)\big] = -\infty.$$

Our theorems of solvability of the Monge problem will be expressed in terms of these assumptions. (I will write $(\mathbf{H}\infty)$ for the combination of $(\mathbf{H}\infty)_1$ and $(\mathbf{H}\infty)_2$.) There are some obvious implications between them, in particular **(locSC)** implies **(Super)**. Before going any further, I shall give some informal explanations about $(\mathbf{H}\infty)_1$ and $(\mathbf{H}\infty)_2$, which probably look obscure to the reader. Both of them are assumptions about the behavior of $c(x,y)$ as $y\to\infty$, therefore they are void if y varies in a compact set. They are essentially quantitative versions of the following statement: For any y it is possible to lower the cost to go from x to y, by starting from a well-chosen point z close to x. For instance, if c is a radially symmetric cost on $\mathbb{R}^n \times \mathbb{R}^n$, then I would choose z very close to x, "opposite to y".

In the rest of this section, I shall discuss some simple sufficient conditions for all these assumptions to hold true. The first result is that Conditions **(Super)**, **(Twist)**, **(locLip)** and **(locSC)** are satisfied by many Lagrangian cost functions.

Proposition 10.15 (Properties of Lagrangian cost functions). *On a smooth Riemannian manifold M, let $c(x,y)$ be a cost function associated with a C^1 Lagrangian $L(x,v,t)$. Assume that any $x,y \in M$ can be joined by at least one C^1 minimizing curve. Then:*

(i) For any $(x,y) \in M \times M$, and any C^1 minimizing curve γ connecting x to y, the tangent vector $-\nabla_v L(x,\dot{\gamma}_0,0) \in T_x M$ is a supergradient for $c(\,\cdot\,,y)$ at x; in particular, c is superdifferentiable at (x,y) as a function of x.

(ii) If L is strictly convex as a function of v, and minimizing curves are uniquely determined by their initial position and velocity, then c satisfies a twist condition: If c is differentiable at (x,y) as a function of x, then y is uniquely determined by x and $\nabla_x c(x,y)$. Moreover,

$$\nabla_x c(x,y) + \nabla_v L\big(x,\dot{\gamma}(0),0\big) = 0,$$

where γ is the unique minimizing curve joining x to y.

(iii) If L has the property that for any two compact sets K_0 and K_1, the velocities of minimizing curves starting in K_0 and ending in K_1 are uniformly bounded, then c is locally Lipschitz and locally semiconcave as a function of x, locally in y.

Example 10.16. Consider the case $L(x,v,t) = |v|^2$. Then $\nabla_v L = 2v$; and (i) says that $-2v_0$ is a supergradient of $d(\,\cdot\,,y)^2$ at x, where v_0 is the velocity used to go from x to y. This is a generalization of the usual formula in Euclidean space:

$$\nabla_x(|x-y|^2) = 2(x-y) = -2(y-x).$$

Also (ii) says that if $d(x,y)^2$ is differentiable at (x,y) as a function of x, then x and y are connected by a unique minimizing geodesic.[1]

[1] As pointed out to me by Fathi, this implies (by Theorem 10.8(iii)) that d^2 is also differentiable at (x,y) as a function of y!

Remark 10.17. The requirements in (ii) and (iii) are fulfilled if the Lagrangian L is time-independent, C^2, strictly convex superlinear as a function of v (recall Example 7.5). But they also hold true for other interesting cases such as $L(x,v,t) = |v|^{1+\alpha}$, $0 < \alpha < 1$.

Remark 10.18. Part (i) of Proposition 10.15 means that the behavior of the (squared) distance function is typical: if one plots $c(x,y)$ as a function of x, for fixed y, one will always see *upward-pointing* crests as in Figure 10.1, never downward-pointing ones.

Proof of Proposition 10.15. The proof is based on the formula of first variation. Let (x,y) be given, and let $\gamma(t)_{0\le t\le 1}$ be a minimizing curve, C^1 as a function of t, joining x to y. Let $\widetilde{\gamma}$ be another curve, *not necessarily minimizing*, joining $\widetilde{x}$ to $\widetilde{y}$. Assume that $\widetilde{x}$ is very close to x, so that there is a unique geodesic joining x to $\widetilde{x}$; by abuse of notation, I shall write $\widetilde{x} - x$ for the initial velocity of this geodesic. Similarly, let us assume that $\widetilde{y}$ is very close to y. Then, by the formula of first variation,

$$\begin{aligned} \mathcal{A}(\widetilde{\gamma}) = \int_0^1 L(\gamma_t, \dot{\gamma}_t, t)\, dt + \Big[\nabla_v L(\gamma_1, \dot{\gamma}_1, 1)\cdot(\widetilde{y}-y) \\ - \nabla_v L(\gamma_0, \dot{\gamma}_0, 0)\cdot(\widetilde{x}-x)\Big] + \omega\Big(\sup_{0\le t\le 1} d(\gamma_t, \widetilde{\gamma}_t)\Big), \end{aligned} \tag{10.17}$$

where $\omega(r)/r \to 0$, and ω only depends on the behavior of the manifold in a neighborhood of γ, and on a modulus of continuity for the derivatives of L in a neighborhood of $\{(\gamma_t, \dot{\gamma}_t, t)_{0\le t\le 1}\}$. Without loss of generality, we may assume that $\omega(r)/r$ is nonincreasing as $r \downarrow 0$.

Then let $\widetilde{x}$ be arbitrarily close to y. By working in smooth charts, it is easy to construct a curve $\widetilde{\gamma}$ joining $\widetilde{\gamma}_0 = \widetilde{x}$ to $\widetilde{\gamma}_1 = y$, in such a way that $d(\gamma_t, \widetilde{\gamma}_t) \le d(x, \widetilde{x})$. Then by (10.17),

$$c(\widetilde{x}, y) \le \mathcal{A}(\widetilde{\gamma}) \le c(x,y) - \big\langle \nabla_v L(x,v,0), \widetilde{x} - x\big\rangle + \omega(|\widetilde{x} - x|),$$

which proves (i).

Now for the proof of (ii): If $c(\,\cdot\,, y)$ is not only superdifferentiable but plainly differentiable, then by Proposition 10.7 there is just one supergradient, which is the gradient, so $-\nabla_v L(x,v,0) = \nabla_x c(x,y)$. Since L is strictly convex in the v variable, this equation determines v uniquely. By assumption, this in turn determines the whole geodesic γ, and in particular y.

Finally, let us consider (iii). When x and y vary in small balls, the velocity v along the minimizing curves will be bounded by assumption; so the function ω will also be uniform. Then $c(x, y)$ is locally superdifferentiable as a function of x, and the conclusion comes from Proposition 10.12. □

Proposition 10.15 is basically all that is needed to treat quite general cost functions on a compact Riemannian manifold. But for noncompact manifolds, it might be difficult to check Assumptions **(Lip)**, **(SC)** or **(H∞)**. Here are a few examples where this can be done.

Example 10.19. Gangbo and McCann have considered cost functions of the form $c(x, y) = c(x - y)$ on $\mathbb{R}^n \times \mathbb{R}^n$, satisfying the following assumption: *For any given $r > 0$ and $\theta \in (0, \pi)$, if $|y|$ is large enough then there is a cone $K_{r,\theta}(y, e)$, with apex y, direction e, height r and angle θ, such that c takes its maximum on $K_{r,\theta}(y, e)$ at y.* Let us check briefly that this assumption implies **(H∞)$_1$**. (The reader who feels that both assumptions are equally obscure may very well skip this and jump directly to Example 10.20.) Let x and S be given such that $T_x S$ is included in no half-space. So for each direction $e \in S^{n-1}$ there are points z_+ and z_- in S, each of which lies on one side of the hyperplane passing through z and orthogonal to e. By a compactness argument, one can find a finite collection of points $z_1, \ldots, z_k$ in S, an angle $\theta < \pi$ and a positive number $r > 0$ such that for all $e \in S^{n-1}$ and for any w close enough to x, the truncated cone $K_{r,\theta}(w, e)$ contains at least one of the z_j. Equivalently, $K_{r,\theta}(w - y, e)$ contains $z_j - y$. But by assumption, for $|w - y|$ large enough there is a cone $K_{r,\theta}(w - y, e)$ such that $c(z - y) \leq c(w - y)$ for all $z \in K_{r,\theta}(w - y, e)$. This inequality applies to $z = z_j$ (for some j), and then $c(z_j - y) \leq c(w - y)$.

Example 10.20. As a particular case of the previous example, **(H∞)$_1$** holds true if $c = c(x - y)$ is radially symmetric and strictly increasing as a function of $|x - y|$.

Example 10.21. Gangbo and McCann considered cost functions that also satisfy $c(x, y) = c(x - y)$ with c convex and superlinear. This assumption implies **(H∞)$_2$**. Indeed, if x in $\mathbb{R}^n$ and $\varepsilon > 0$ are given, let $z = x - \varepsilon(x - y)/|x - y|$; it suffices to show that

$$c(z - y) - c(x - y) \xrightarrow[y \to \infty]{} -\infty,$$

or equivalently, with $h = x - y$,

$$c(h) - c\Big(h\big(1 - \frac{\varepsilon}{|h|}\big)\Big) \xrightarrow[h\to\infty]{} +\infty.$$

But the inequality

$$c(0) \geq c(p) + \nabla c(p) \cdot (-p)$$

and the superlinearity of c imply $\nabla c(p) \cdot (p/|p|) \to +\infty$ as $p \to \infty$; then, with the notation $h_\varepsilon = h(1 - \varepsilon/|h|)$,

$$c(h) - c(h_\varepsilon) \geq \nabla c(h_\varepsilon) \cdot \frac{\varepsilon h}{|h|} = \varepsilon\, \nabla c(h_\varepsilon) \cdot \frac{h_\varepsilon}{|h_\varepsilon|} \xrightarrow[|h|\to\infty]{} +\infty,$$

as desired.

Example 10.22. If (M, g) is a Riemannian manifold with *nonnegative sectional curvature*, then (as recalled in the Third Appendix) $\nabla_x^2(d(x_0, x)^2/2) \leq g_x$, and it follows that $c(x, y) = d(x, y)^2$ is semiconcave with a modulus $\omega(r) = r^2$. This condition of nonnegative curvature is quite restrictive, but there does not seem to be any good alternative geometric condition implying the semiconcavity of $d(x, y)^2$, uniformly in x and y.

I conclude this section with an open problem:

Open Problem 10.23. *Find simple sufficient conditions on a rather general Lagrangian on an unbounded Riemannian manifold, so that it will satisfy* **(H∞)**.

Differentiability of c-convex functions

We shall now return to optimal transport, and arrive at the core of the analysis of the Monge problem: the study of the regularity of c-convex functions. This includes c-subdifferentiability, subdifferentiability, and plain differentiability.

In Theorems 10.24 to 10.26, M is a complete connected Riemannian manifold of dimension n, $\mathcal{X}$ is a closed subset of M such that the frontier $\partial\mathcal{X}$ (in M) is of dimension at most $n-1$ (for instance it is locally a graph), and $\mathcal{Y}$ is an arbitrary Polish space. The cost function $c : \mathcal{X} \times \mathcal{Y} \to \mathbb{R}$ is assumed to be continuous. The statements will be expressed in terms of the assumptions appearing in the previous section; these assumptions will be made for *interior* points, which are points lying in the interior of $\mathcal{X}$ (viewed as a subset of M).

Theorem 10.24 (c-subdifferentiability of c-convex functions). *Let Assumption* **(H∞)** *be satisfied. Let $\psi : \mathcal{X} \to \mathbb{R} \cup \{+\infty\}$ be a c-convex function, and let Ω be the interior (in M) of its domain $\psi^{-1}(\mathbb{R})$. Then, $\psi^{-1}(\mathbb{R}) \setminus \Omega$ is a set of dimension at most $n-1$. Moreover, ψ is locally bounded and c-subdifferentiable everywhere in Ω. Finally, if $K \subset \Omega$ is compact, then $\partial_c\psi(K)$ is itself compact.*

Theorem 10.25 (Subdifferentiability of c-convex functions). *Assume that* **(Super)** *is satisfied. Let ψ be a c-convex function, and let x be an interior point of $\mathcal{X}$ (in M) such that $\partial_c\psi(x) \neq \emptyset$. Then ψ is subdifferentiable at x. In short:*

$$\partial_c\psi(x) \neq \emptyset \implies \nabla^-\psi(x) \neq \emptyset.$$

More precisely, for any $y \in \partial_c\psi(x)$, one has $-\nabla_x^+ c(x,y) \subset \nabla^-\psi(x)$.

Theorem 10.26 (Differentiability of c-convex functions). *Assume that* **(Super)** *and* **(Twist)** *are satisfied, and let ψ be a c-convex function. Then:*

(i) If **(Lip)** *is satisfied, then ψ is locally Lipschitz and differentiable in $\mathcal{X}$, apart from a set of zero volume; the same is true if* **(locLip)** *and* **(H∞)** *are satisfied.*

(ii) If **(SC)** *is satisfied, then ψ is locally semiconvex and differentiable in the interior (in M) of its domain, apart from a set of dimension at most $n-1$; and the boundary of the domain of ψ is also of dimension at most $n-1$. The same is true if* **(locSC)** *and* **(H∞)** *are satisfied.*

Proof of Theorem 10.24. Let $S = \psi^{-1}(\mathbb{R}) \setminus \partial\mathcal{X}$. (Here $\partial\mathcal{X}$ is the boundary of $\mathcal{X}$ in M, which by assumption is of dimension at most $n-1$.) I shall use the notion of tangent cone defined later in Definition 10.46, and show that if $x \in S$ is such that T_xS is not included in a half-space, then ψ is bounded on a small ball around x. It will follow that x is in fact in the interior of Ω. So for each $x \in S \setminus \Omega$, T_xS will be included in a half-space, and by Theorem 10.48(ii) $S \setminus \Omega$ will be of dimension at most $n-1$. Moreover, this will show that ψ is locally bounded in Ω.

So let x be such that $\psi(x) < +\infty$, and T_xS is not included in a half-space. By assumption, there are points $z_1, \ldots, z_k$ in S, a small ball B around x, and a compact set $K \subset \mathcal{Y}$ such that for any $y \in \mathcal{Y} \setminus K$,

$$\inf_{w \in B} c(w,y) \geq \inf_{1 \leq j \leq k} c(z_j, y).$$

Let ϕ be the c-transform of ψ. For any $y \in \mathcal{Y} \setminus K$,

$$\phi(y) - \inf_{w\in B} c(w,y) \le \phi(y) - \inf_{1\le j\le k} c(z_j,y) \le \sup_{1\le j\le k} \psi(z_j).$$

So

$$\forall w \in B, \quad \forall y \in \mathcal{Y} \setminus K, \qquad \phi(y) - c(w,y) \le \sup_{1\le j\le k} \psi(z_j).$$

When $y \in K$, the trivial bound $\phi(y) - c(w,y) \le \psi(x) + c(x,y) - c(w,y)$ implies

$$\forall w \in B, \quad \psi(w) = \sup_{y\in\mathcal{Y}} [\phi(y) - c(w,y)]$$
$$\le \max\left(\sup_{1\le j\le k} \psi(z_j), \ \sup_{y\in K} \big[c(x,y) + \psi(x) - c(w,y)\big] \right).$$

This shows that ψ is indeed bounded above on B. Since it is lower semicontinuous with values in $\mathbb{R} \cup \{+\infty\}$, it is also bounded below on a neighborhood of x. All in all, ψ is bounded in a neighborhood of x.

Next, let $x \in \Omega$; the goal is to show that $\partial_c\psi(x) \neq \emptyset$. Let U be a small neighborhood of x, on which $|\psi|$ is bounded by M. By assumption there is a compact set K, and a small ball B' in U, such that for all y outside K,

$$\forall z \in B', \quad c(z,y) - c(x,y) \le -(2M+1).$$

So if $y \notin K$, there is a z such that $\psi(z) + c(z,y) \le c(x,y) - (M+1) \le \psi(x) + c(x,y) - 1$, and

$$\phi(y) - c(x,y) \le \inf_{z\in B'} \big[\psi(z) + c(z,y) - c(x,y)\big]$$
$$\le \psi(x) - 1 = \sup_{y'\in\mathcal{Y}} [\phi(y') - c(x,y')] - 1.$$

Then the supremum of $\phi(y) - c(x,y)$ over all $\mathcal{Y}$ is the same as the supremum over only K. But this is a maximization problem for an upper semicontinuous function on a compact set, so it admits a solution, which belongs to $\partial_c\psi(x)$.

The same reasoning can be made with x replaced by w in a small neighborhood B of x, then the conclusion is that $\partial_c\psi(w)$ is nonempty and contained in the compact set K, uniformly for $z \in B$. If $K' \subset \Omega$

is a compact set, we can cover it by a finite number of small open balls B_j such that $\partial_c\psi(B_j)$ is contained in a compact set K_j, so that $\partial_c\psi(K') \subset \cup K_j$. Since on the other hand $\partial_c\psi(K')$ is closed by the continuity of c, it follows that $\partial_c\psi(K')$ is compact. This concludes the proof of Theorem 10.24. □

Proof of Theorem 10.25. Let x be a point of c-subdifferentiability of ψ, and let $y \in \partial_c\psi(x)$. Further, let

$$\phi(y) := \inf \big[\psi(x) + c(x, y)\big]$$

be the c-transform of ψ. By definition of c-subdifferentiability,

$$\psi(x) = \phi(y) - c(x, y). \tag{10.18}$$

Let $w \in T_xM$ and let x_ε be obtained from x by a variation of size $O(\varepsilon)$ in the direction w, say $x_\varepsilon = \exp_x(\varepsilon w)$. From the definition of ϕ, one has of course

$$\psi(x_\varepsilon) \geq \phi(y) - c(x_\varepsilon, y). \tag{10.19}$$

Further, let $p \in \nabla_x^+ c(x, y)$. By (10.18), (10.19) and the superdifferentiability of c,

$$\begin{aligned}\psi(x_\varepsilon) &\geq \phi(y) - c(x_\varepsilon, y)\\ &\geq \phi(y) - c(x, y) - \varepsilon\langle p, w\rangle + o(\varepsilon)\\ &= \psi(x) - \varepsilon\langle p, w\rangle + o(\varepsilon).\end{aligned}$$

This shows that ψ is indeed subdifferentiable at x, with p as a subgradient, and $-p \in \partial_c\psi(x)$. □

Proof of Theorem 10.26. (i) If $\|c(\cdot\,, y)\|_{\mathrm{Lip}} \leq L$, then also $\psi(x) = \sup_y[\phi(y) - c(x, y)]$ satisfies $\|\psi\|_{\mathrm{Lip}} \leq L$. Then by Theorem 10.8(ii) ψ is differentiable everywhere on the interior of $\mathcal{X}$, apart from a set of zero volume.

If c is only locally Lipschitz in x and y, but condition **(H∞)** is ensured, then for each compact set K in $\mathcal{X}$ there is a compact set $K' \subset \mathcal{Y}$ such that

$$\forall x \in K, \qquad \psi(x) = \sup_{y\in\partial_c\psi(x)} [\phi(y) - c(x, y)] = \sup_{y\in K'} [\phi(y) - c(x, y)].$$

The functions inside the supremum are uniformly Lipschitz when x stays in K and y stays in K', so the result of the supremum is again a locally Lipschitz function.

(ii) Assume that $c(x,y)$ is semiconcave, locally in x and uniformly in y. Let K be a compact subset of M, and let γ be a constant-speed geodesic whose image is included in K; then the inequality

$$c(\gamma_t, y) \geq (1-t)\, c(\gamma_0, y) + t\, c(\gamma_1, y) - t(1-t)\,\omega\big(d(\gamma_0,\gamma_1)\big)$$

leads to

$$\begin{aligned}
\psi(\gamma_t) &= \sup_y \big[\phi(y) - c(\gamma_t, y)\big] \\
&\leq \sup_y \Big[\phi(y) - (1-t)\, c(\gamma_0, y) - t\, c(\gamma_1, y)\Big] + t(1-t)\,\omega\big(d(\gamma_0,\gamma_1)\big) \\
&= \sup_y \Big[(1-t)\,\phi(y) - (1-t)\, c(\gamma_0, y) + t\,\phi(y) - t\, c(\gamma_1, y)\Big] \\
&\qquad\qquad + t(1-t)\,\omega\big(d(\gamma_0,\gamma_1)\big) \\
&\leq (1-t)\, \sup_y \big[\phi(y) - c(\gamma_0, y)\big] + t\, \sup_y \big[\phi(y) - c(\gamma_1, y)\big] \\
&\qquad\qquad + t(1-t)\,\omega\big(d(\gamma_0,\gamma_1)\big) \\
&= (1-t)\,\psi(\gamma_0) + t\,\psi(\gamma_1) + t(1-t)\,\omega\big(d(\gamma_0,\gamma_1)\big).
\end{aligned}$$

So ψ inherits the semiconcavity modulus of c as semiconvexity modulus. Then the conclusion follows from Proposition 10.12 and Theorem 10.8(iii). If c is semiconcave in x and y, one can use a localization argument as in the proof of (i). □

Remark 10.27. Theorems 10.24 to 10.26, and (in the Lagrangian case) Proposition 10.15 provide a good picture of differentiability points of c-convex functions: Let c satisfy **(Twist)**, **(Super)** and **(H∞)**, and let x be in the interior of the domain of a c-convex function ψ. If ψ is differentiable at x then $\partial_c\psi(x)$ consists of just one point y, and $\nabla\psi(x) = -\nabla_x c(x,y)$, which in the Lagrangian case also coincides with $\nabla_v L(x,v,0)$, where v is the initial velocity of the unique action-minimizing curve joining x to y. If ψ is not differentiable at x, the picture is not so precise; however, under **(locSC)**, we can use the local semiconvexity of ψ to show that $\nabla^-\psi(x)$ is included in the closed convex hull of $-\nabla_x^+ c(x, \partial_c\psi(x))$, which in the Lagrangian case is also the closed convex hull of $\{\nabla_v L(0,x,\dot\gamma(0))\}$, where γ varies among action-minimizing curves joining x to $\partial_c\psi(x)$. (Use Remark 10.51 in the Second Appendix, the stability of c-subdifferential, the semiconcavity of c, the continuity of $\nabla_x^+ c$, the stability of minimizing curves and the continuity of $\nabla_v L$.) There is in general no reason why $-\nabla_x^+ c(x,\partial_c\psi(x))$ would be convex; we shall come back to this issue in Chapter 12.

Applications to the Monge problem

The next theorem shows how to incorporate the previous information into the optimal transport problem.

Theorem 10.28 (Solution of the Monge problem II). *Let M be a Riemannian manifold, $\mathcal{X}$ a closed subset of M, with* $\dim(\partial\mathcal{X}) \le n-1$, *and $\mathcal{Y}$ an arbitrary Polish space. Let $c : \mathcal{X} \times \mathcal{Y} \to \mathbb{R}$ be a continuous cost function, bounded below, and let $\mu \in P(\mathcal{X})$, $\nu \in P(\mathcal{Y})$, such that the optimal cost $C(\mu, \nu)$ is finite. Assume that:*

(i) c is superdifferentiable everywhere (Assumption **(Super)***);*

(ii) $\nabla_x c(x, \cdot)$ is injective where defined (Assumption **(Twist)***);*

(iii) any c-convex function is differentiable μ-almost surely on its domain of c-subdifferentiability.

Then there exists a unique (in law) optimal coupling (x, y) of (μ, ν); it is deterministic, and there is a c-convex function ψ such that

$$\nabla\psi(x) + \nabla_x c(x, y) = 0 \qquad \textit{almost surely.} \tag{10.20}$$

In other words, there is a unique transport map T solving the Monge problem, and $\nabla\psi(x) + \nabla_x c(x, T(x)) = 0$, $\mu(dx)$-almost surely.

If moreover **(H∞)** *is satisfied, then:*

(a) Equation (10.20) *characterizes the optimal coupling;*

(b) Let Z be the set of points where ψ is differentiable; then one can define a continuous map $x \to T(x)$ on Z by the equation $T(x) \in \partial_c\psi(x)$, and

$$\operatorname{Spt}\nu = \overline{T(\operatorname{Spt}\mu)}. \tag{10.21}$$

Remark 10.29. As a corollary of this theorem, $\nabla\psi$ is uniquely determined μ-almost surely, since the random variable $\nabla\psi(x)$ has to coincide (in law) with $-\nabla_x c(x, y)$.

Remark 10.30. The uniqueness of the optimal transport does not in general imply the uniqueness of the function ψ, even up to an additive constant. However, this becomes true if $\mathcal{X}$ is the closure of a connected open set Ω, such that the density of μ with respect to the volume measure is positive almost everywhere in Ω; indeed, $\nabla\psi$ will then be uniquely defined almost surely in Ω. (This uniqueness theorem does not require μ to be absolutely continuous.)

Remark 10.31. In general $\partial_c\psi(\operatorname{Spt}\mu)$ is larger than $\operatorname{Spt}\nu$. Take for instance $\mathcal{X} = \mathcal{Y} = \mathbb{R}^2$, $c(x,y) = |x-y|^2$, let $B = B_1(0) \subset \mathbb{R}^2$; split B in two halves along the first coordinate, translate the right (resp. left) half ball by a unit vector $(1,0)$ (resp. $(-1,0)$), call B' the resulting set, and define μ (resp. ν) as the normalized measure on B (resp. B'). Then $\partial_c\psi(\operatorname{Spt}\mu)$ will be the whole convex hull of B'. Theorem 12.49 will provide sufficient conditions to control $\partial_c\psi(\operatorname{Spt}\mu)$.

Remark 10.32. If in Theorem 10.28 the cost c derives from a C^1 Lagrangian $L(x,v,t)$, strictly convex in v, such that minimizing curves are uniquely determined by their initial velocity, then Proposition 10.15(ii) implies the following important property of the optimal coupling (x,y): *Almost surely, x is joined to y by a unique minimizing curve.* For instance, if $c(x,y) = d(x,y)^2$, the optimal transference plan π will be concentrated on the set of points (x,y) in $M \times M$ such that x and y are joined by a unique geodesic.

Remark 10.33. Assumption (iii) can be realized in a number of ways, depending on which part of Theorem 10.26 one wishes to use: For instance, it is true if c is Lipschitz on $\mathcal{X}\times\mathcal{Y}$ and μ is absolutely continuous; or if c is locally Lipschitz and μ, ν are compactly supported and μ is absolutely continuous; or if c is locally semiconcave and satisfies **(H∞)** and μ does not charge sets of dimension $n-1$; etc. It is important to note that Assumption (iii) implicitly contains some restrictions about the behavior at infinity, of either the measure μ, or the manifold M, or the cost function c.

Example 10.34. All the assumptions of Theorem 10.28 are satisfied if $\mathcal{X} = M = \mathcal{Y}$ is compact and the Lagrangian L is C^2 and satisfies the classical conditions of Definition 7.6.

Example 10.35. All the assumptions of Theorem 10.28 are satisfied if $\mathcal{X} = M = \mathcal{Y} = \mathbb{R}^n$, c is a C^1 strictly convex function with a bounded Hessian and μ does not charge sets of dimension $n-1$. Indeed, $\nabla_x c$ will be injective by strict convexity of c; and c will be uniformly semiconcave with a modulus Cr^2, so Theorem 10.26 guarantees that c-convex functions are differentiable everywhere apart from a set of dimension at most $n-1$.

Example 10.36. All the assumptions of Theorem 10.28 are satisfied if $\mathcal{X} = M = \mathcal{Y}$, $c(x,y) = d(x,y)^2$, and M has nonnegative sectional curvature; recall indeed Example 10.22. The same is true if M is compact.

Proof of Theorem 10.28. Let π be an optimal transference plan. From Theorem 5.10, there exists a pair of c-conjugate functions (ψ, ϕ) such that $\phi(y) - \psi(x) \leq c(x,y)$ everywhere, with equality π-almost surely. Write again (10.2), at a point x of differentiability of ψ (x should be interior to $\mathcal{X}$, viewed as a subset of M), and choose $\widetilde{x} = \widetilde{x}(\varepsilon) = \gamma(\varepsilon)$, where $\dot{\gamma}(0) = w$; divide by $\varepsilon > 0$ and pass to the lim inf:

$$-\nabla\psi(x) \cdot w \leq \liminf_{\varepsilon \to 0} \frac{c(\widetilde{x}(\varepsilon), y) - c(x,y)}{\varepsilon}. \tag{10.22}$$

It follows that $-\nabla\psi(x)$ is a subgradient of $c(\,\cdot\,, y)$ at x. But by assumption, there exists at least one supergradient of $c(\,\cdot\,, y)$ at x, say G. By Proposition 10.7, $c(\,\cdot\,, y)$ really is differentiable at x, with gradient $-\nabla\psi(x)$.

So (10.20) holds true, and then Assumption (iii) implies $y = T(x) = (\nabla_x c)^{-1}(x, -\nabla\psi(x))$, where $(\nabla_x c)^{-1}$ is the inverse of $x \longmapsto \nabla_x c(x,y)$, viewed as a function of y and defined on the set of x for which $\nabla_x c(x,y)$ exists. (The usual measurable selection theorem implies the measurability of this inverse.)

Thus π is concentrated on the graph of T; or equivalently, $\pi = (\mathrm{Id}\,, T)_{\#}\mu$. Since this conclusion does not depend on the choice of π, but only on the choice of ψ, the optimal coupling is unique.

It remains to prove the last part of Theorem 10.28. From now on I shall assume that **(H∞)** is satisfied. Let $\pi \in \Pi(\mu, \nu)$ be a transference plan, and let ψ be a c-convex function such that (10.20) holds true.

Let Z be the set of differentiability points of ψ, and let $x \in Z$; in particular, x should be interior to $\mathcal{X}$ (in M), and should belong to the interior of the domain of ψ. By Theorem 10.24, there is some $y \in \partial_c\psi(x)$. Let G be a supergradient of $c(\,\cdot\,, y)$ at x; by Theorem 10.25, $-G \in \{\nabla^-\psi(x)\} = \{\nabla\psi(x)\}$, so $-\nabla\psi(x)$ is the only supergradient of $c(\,\cdot\,, y)$ at x (as in the beginning of the proof of Theorem 10.28); $c(\,\cdot\,, y)$ really is differentiable at x and $\nabla_x c(x,y) + \nabla\psi(x) = 0$. By injectivity of $\nabla_x c(x, \cdot\,)$, this equation determines $y = T(x)$ as a function of $x \in Z$. This proves the first part of (b), and also shows that $\partial_c\psi(x) = \{T(x)\}$ for any $x \in Z$.

Now, since π is concentrated on $Z \times \mathcal{Y}$, equation (10.20) implies that that π really is concentrated on the graph of T. A fortiori $\pi[\partial_c\psi] = 1$, so π is c-cyclically monotone, and therefore optimal by Theorem 5.10. This proves (a).

Next, let us prove that T is continuous on Z. Let $(x_k)_{k\in\mathbb{N}}$ be a sequence in Z, converging to $x \in Z$, and let $y_k = T(x_k)$. Assumption

(H∞) and Theorem 10.24 imply that $\partial_c\psi$ transforms compact sets into compact sets; so the sequence $(y_k)_{k\in\mathbb{N}}$ takes values in a compact set, and up to extraction of a subsequence it converges to some $y' \in \mathcal{Y}$. By passing to the limit in the inequality defining the c-subdifferential, we obtain $y' \in \partial_c\psi(x)$. Since $x \in Z$, this determines $y' = T(x)$ uniquely, so the whole sequence $(T(x_k))_{k\in\mathbb{N}}$ converges to $T(x)$, and T is indeed continuous.

Equation (10.21) follows from the continuity of T. Indeed, the inclusion $\operatorname{Spt}\mu \subset T^{-1}(T(\operatorname{Spt}\mu))$ implies

$$\nu[T(\operatorname{Spt}\mu)] = \mu\big[T^{-1}(T(\operatorname{Spt}\mu))\big] \geq \mu[\operatorname{Spt}\mu] = 1;$$

so by definition of support, $\operatorname{Spt}\nu \subset \overline{T(\operatorname{Spt}\mu)}$. On the other hand, if $x \in \operatorname{Spt}\mu \cap Z$, let $y = T(x)$, and let $\varepsilon > 0$; by continuity of T there is $\delta > 0$ such that $T(B_\delta(x)) \subset B_\varepsilon(y)$, and then

$$\nu[B_\varepsilon(y)] = \mu\big[T^{-1}(B_\varepsilon(y))\big] \geq \mu\big[T^{-1}\big(T(B_\delta(x))\big)\big] \geq \mu[B_\delta(x)] > 0;$$

so $y \in \operatorname{Spt}\nu$. This shows that $T(\operatorname{Spt}\mu) \subset \operatorname{Spt}\nu$, and therefore $\overline{T(\operatorname{Spt}\mu)} \subset \operatorname{Spt}\nu$, as desired. This concludes the proof of (b). □

Remark 10.37. Uniqueness in Theorem 10.28 was very easy because we were careful to note in Theorem 5.10 that any optimal π is supported in the c-subdifferential of *any* optimal function ψ for the dual problem. If we only knew that any optimal π is supported in the c-subdifferential of *some* ψ, we still could get uniqueness fairly easily, either by working out a variant of the statement in Theorem 5.10, or by noting that if any optimal measure is concentrated on a graph, then a strict convex combination of such measures has to be concentrated on a graph itself, and this is possible only if the two graphs coincide almost surely.

Removing the conditions at infinity

This section and the next one deal with extensions of Theorem 10.28. Here we shall learn how to cover situations in which no control at infinity is assumed, and in particular Assumption (iii) of Theorem 10.28 might not be satisfied. The short answer is that it is sufficient to replace the *gradient* in (10.20) by an *approximate gradient*. (Actually a little bit more will be lost, see Remarks 10.39 and 10.40 below.)

Theorem 10.38 (Solution of the Monge problem without conditions at infinity). *Let M be a Riemannian manifold and $\mathcal{Y}$ an arbitrary Polish space. Let $c : M \times \mathcal{Y} \to \mathbb{R}$ be a continuous cost function, bounded below, and let $\mu \in P(M)$, $\nu \in P(\mathcal{Y})$, such that the optimal cost $C(\mu, \nu)$ is finite. Assume that:*

(i) c is superdifferentiable everywhere (Assumption **(Super)***);*

(ii) $\nabla_x c(x, \cdot)$ is injective (Assumption **(Twist)***);*

(iii) for any closed ball $B = B_{r]}(x_0)$ and any compact set $K \subset \mathcal{Y}$, the function c' defined on $B \times K$ by restriction of c is such that any c'-convex function on $B \times K$ is differentiable μ-almost surely;

(iv) μ is absolutely continuous with respect to the volume measure.

Then there exists a unique (in law) optimal coupling (x, y) of (μ, ν); it is deterministic, and satisfies the equation

$$\widetilde{\nabla}\psi(x) + \nabla_x c(x, y) = 0 \qquad \textit{almost surely.} \tag{10.23}$$

In other words, there is a unique optimal transport T and it satisfies the equation $\widetilde{\nabla}\psi(x) + \nabla_x c(x, T(x)) = 0$ almost surely.

Remark 10.39. I don't know if (10.23) is a *characterization* of the optimal transport.

Remark 10.40. If Assumption (iv) is weakened into *(iv') μ gives zero mass to sets of dimension at most $n-1$*, then there is still uniqueness of the optimal coupling, and there is a c-convex ψ such that $y \in \partial_c \psi(x)$ almost surely; but it is not clear that equation (10.23) still holds. This uniqueness result is a bit more tricky than the previous one, and I shall postpone its proof to the next section (see Theorem 10.42).

Proof of Theorem 10.38. Let ψ be a c-convex function as in Theorem 5.10. Let π be an optimal transport; according to Theorem 5.10(ii), $\pi[\partial_c \psi] = 1$.

Let x_0 be any point in M. For any $\ell \in \mathbb{N}$, let B_ℓ stand for the closed ball $B_{\ell]}(x_0)$. Let also $(K_\ell)_{\ell \in \mathbb{N}}$ be an increasing sequence of compact sets in $\mathcal{Y}$, such that $\nu[\cup K_\ell] = 1$. The sets $B_\ell \times K_\ell$ fill up the whole of $M \times \mathcal{Y}$, up to a π-negligible set. Let c_ℓ be the restriction of c to $B_\ell \times K_\ell$.

If ℓ is large enough, then $\pi[B_\ell \times K_\ell] > 0$, so we can define

$$\pi_\ell := \frac{1_{B_\ell \times K_\ell}\, \pi}{\pi[B_\ell \times K_\ell]},$$

and then introduce the marginals μ_ℓ and ν_ℓ of π_ℓ. By Theorem 4.6, π_ℓ is optimal in the transport problem from (B_ℓ, μ_ℓ) to (K_ℓ, ν_ℓ), with cost c_ℓ. By Theorem 5.19 we can find a c_ℓ-convex function ψ_ℓ which coincides with ψ μ_ℓ-almost surely, and actually on the whole of $S_\ell := \mathrm{proj}_M((\partial_c\psi) \cap (B_\ell \times K_\ell))$.

The union of all sets S_ℓ covers $\mathrm{proj}_M(\partial_c\psi)$, and therefore also $\mathrm{proj}_M(\mathrm{Spt}\,\pi)$, apart from a μ-negligible set. Let $\widetilde{S}_\ell$ be the set of points in S_ℓ at which S_ℓ has density 1; we know that $S_\ell \setminus \widetilde{S}_\ell$ has zero volume. So the union of all sets $\widetilde{S}_\ell$ still covers M, apart from a μ-negligible set. (Here I have used the absolute continuity of μ.)

By Assumption (iii), each function ψ_ℓ is differentiable apart from a μ-negligible set Z_ℓ. Moreover, by Theorem 10.28, the equation

$$\nabla_x c(x,y) + \nabla\psi_\ell(x) = 0 \tag{10.24}$$

determines the unique optimal coupling between μ_ℓ and ν_ℓ, for the cost c_ℓ. (Note that $\nabla_x c_\ell$ coincides with $\nabla_x c$ when x is in the interior of B_ℓ, and $\mu_\ell[\partial B_\ell] = 0$, so equation (10.24) does hold true π_ℓ-almost surely.)

Now we can define our Monge coupling. For each $\ell \in \mathbb{N}$, and each $x \in \widetilde{S}_\ell \setminus Z_\ell$, ψ_ℓ coincides with ψ on a set which has density 1 at x, so $\nabla\psi_\ell(x) = \widetilde{\nabla}\psi(x)$, and (10.24) becomes

$$\nabla_x c(x,y) + \widetilde{\nabla}\psi(x) = 0. \tag{10.25}$$

This equation is independent of ℓ, and holds π_ℓ-almost surely since $\widetilde{S}_\ell \setminus Z_\ell$ has full μ_ℓ-measure. By letting $\ell \to \infty$ we deduce that π is concentrated on the set of (x,y) satisfying (10.25). By assumption this equation determines $y = \nabla_x c^{-1}(x, -\widetilde{\nabla}\psi(x))$ uniquely as a measurable function of x. Uniqueness follows obviously since π was an arbitrary optimal plan. □

As an illustration of the use of Theorems 10.28 and 10.38, let us see how we can solve the Monge problem for the square distance on a Riemannian manifold. In the next theorem, I shall say that M has *asymptotically nonnegative curvature* if all sectional curvatures σ_x at point x satisfy

$$\sigma_x \geq -\frac{C}{d(x_0,x)^2} \tag{10.26}$$

for some positive constant C and some $x_0 \in M$.

Theorem 10.41 (Solution of the Monge problem for the square distance). *Let M be a Riemannian manifold, and $c(x,y) = d(x,y)^2$. Let μ, ν be two probability measures on M, such that the optimal cost between μ and ν is finite. If μ is absolutely continuous, then there is a unique solution of the Monge problem between μ and ν, and it can be written as*

$$y = T(x) = \exp_x\big(\widetilde{\nabla}\psi(x)\big), \tag{10.27}$$

where ψ is some $d^2/2$-convex function. The approximate gradient can be replaced by a true gradient if any one of the following conditions is satisfied:

(a) μ and ν are compactly supported;

(b) M has nonnegative sectional curvature;

(c) ν is compactly supported and M has asymptotically nonnegative curvature.

Proof. The general theorem is just a particular case of Theorem 10.38.

In case (a), we can apply Theorem 10.28 with $\mathcal{X} = B_{R]}(x_0) = \mathcal{Y}$, where R is large enough that $B_R(x_0)$ contains all geodesics that go from $\operatorname{Spt}\mu$ to $\operatorname{Spt}\nu$. Then the conclusion holds with some c'-convex function ψ, where c' is the restriction of c to $\mathcal{X} \times \mathcal{Y}$:

$$\psi(x) = \sup_{y \in B_{r]}(x_0)} \left[\phi(y) - \frac{d(x,y)^2}{2}\right].$$

To recover a true $d^2/2$-function, it suffices to set $\phi(y) = -\infty$ on $M \setminus \mathcal{Y}$, and let $\psi(x) = \sup_{y\in M}[\phi(y) - d(x,y)^2/2]$ (as in the proof of Lemma 5.18).

In case (b), all functions $d(\,\cdot\,,y)^2/2$ are uniformly semiconcave (as recalled in the Third Appendix), so Theorem 10.28 applies.

In case (c), all functions $d(\,\cdot\,,y)^2/2$, where y varies in the support of ν, are uniformly semiconcave (as recalled in the Third Appendix), so we can choose $\mathcal{Y}$ to be a large closed ball containing the support of ν, and apply Theorem 10.28 again. □

Removing the assumption of finite cost

In this last section, I shall investigate situations where the total transport cost might be infinite. Unless the reader is specifically interested in such a situation, he or she is advised to skip this section, which is quite tricky.

If $C(\mu,\nu) = +\infty$, there is no point in searching for an optimal transference plan. However, it does make sense to look for c-cyclically monotone plans, which will be called **generalized optimal transference plans.**

Theorem 10.42 (Solution of the Monge problem with possibly infinite total cost). *Let $\mathcal{X}$ be a closed subset of a Riemannian manifold M such that $\dim(\partial\mathcal{X}) \le n-1$, and let $\mathcal{Y}$ be an arbitrary Polish space. Let $c : M \times \mathcal{Y} \to \mathbb{R}$ be a continuous cost function, bounded below, and let $\mu \in P(M)$, $\nu \in P(\mathcal{Y})$. Assume that:*

(i) c is locally semiconcave (Assumption **(locSC)***);*

(ii) $\nabla_x c(x,\,\cdot\,)$ is injective (Assumption **(Twist)***);*

(iii) μ does not give mass to sets of dimension at most $n-1$.

Then there exists a unique (in law) coupling (x,y) of (μ,ν) such that $\pi = \mathrm{law}\,(x,y)$ is c-cyclically monotone; moreover this coupling is deterministic. The measure π is called the generalized optimal transference plan between μ and ν. Furthermore, there is a c-convex function $\psi : M \to \mathbb{R} \cup \{+\infty\}$ such that $\pi[\partial_c\psi] = 1$.

- *If Assumption (iii) is reinforced into*

  (iii') μ is absolutely continuous with respect to the volume measure,

then

$$\widetilde{\nabla}\psi(x) + \nabla_x c(x,y) = 0 \qquad \pi(dx\,dy)\text{-almost surely.} \tag{10.28}$$

- *If Assumption (iii) is left as it is, but one adds*

  (iv) the cost function satisfies **(H∞)** *or* **(SC)***,*

then

$$\nabla\psi(x) + \nabla_x c(x,y) = 0 \qquad \pi(dx\,dy)\text{-almost surely,} \tag{10.29}$$

and this characterizes the generalized optimal transport. Moreover, one can define a continuous map $x \to T(x)$ on the set of differentiability points of ψ by the equation $T(x) \in \partial_c\psi(x)$, and then $\mathrm{Spt}\,\nu = \overline{T(\mathrm{Spt}\,\mu)}$.

Remark 10.43. Remark 10.39 applies also in this case.

Proof of Theorem 10.42. Let us first consider the existence problem. Let $(\mu_k)_{k\in\mathbb{N}}$ be a sequence of compactly supported probability measures converging weakly to μ; and similarly let $(\nu_k)_{k\in\mathbb{N}}$ be a sequence of compactly supported probability measures converging weakly to ν. For each k, the total transport cost $C(\mu_k,\nu_k)$ is finite. Then let π_k be an optimal transference plan between μ_k and ν_k; by Theorem 5.10(ii), π_k is c-cyclically monotone. By Lemma 4.4, the sequence $(\pi_k)_{k\in\mathbb{N}}$ converges, up to extraction, to some transference plan $\pi \in \Pi(\mu,\nu)$. By Theorem 5.20, π is c-cyclically monotone. By Step 3 of the proof of Theorem 5.10(i) (Rüschendorf's theorem), there is a c-convex ψ such that $\mathrm{Spt}(\pi) \subset \partial_c\psi$, in particular $\pi[\partial_c\psi] = 1$.

If μ is absolutely continuous, then we can proceed as in the proof of Theorem 10.38 to show that the coupling is deterministic and that (10.28) holds true π-almost surely.

In the case when **(H∞)** or **(SC)** is assumed, we know that ψ is c-subdifferentiable everywhere in the interior of its domain; then we can proceed as in Theorem 10.28 to show that the coupling is deterministic, that (10.29) holds true, and that this equation implies $y \in \partial_c\psi(x)$. So if we prove the uniqueness of the generalized optimal transference plan this will show that (10.28) characterizes it.

Thus it all boils down to proving that *under Assumptions (i)–(iii), the generalized optimal transport is unique.* This will be much more technical, and the reader is advised to skip all the rest of the proof at first reading. The two main ingredients will be the Besicovich density theorem and the implicit function theorem.

Let π be a generalized optimal coupling of μ and ν, and let ψ be a c-convex function such that $\mathrm{Spt}(\pi) \subset \partial_c\psi$. Let $z_0 \in \mathcal{X}$, let $B_\ell = B[z_0,\ell] \cup \mathcal{X}$, and let $(K_\ell)_{\ell\in\mathbb{N}}$ be an increasing sequence of compact subsets of $\mathcal{Y}$, such that $\nu[\cup K_\ell] = 1$. Let $Z_\ell := \pi_\ell[B_\ell \times K_\ell]$, $c_\ell := c|_{B_\ell\times K_\ell}$, $\pi_\ell := 1_{B_\ell\times K_\ell}\pi/Z_\ell$, $S_\ell := \mathrm{proj}_M(\mathrm{Spt}\,\pi_\ell)$; let also μ_ℓ and ν_ℓ be the two marginals of π_ℓ. It is easy to see that S_ℓ is still an increasing family of compact subsets of M, and that $\mu[\cup S_\ell] = 1$.

According to Theorem 5.19, there is a c_ℓ-convex function $\psi_\ell : B_\ell \to \mathbb{R}\cup\{+\infty\}$ which coincides with ψ on S_ℓ. Since c is locally semiconcave, the cost c_ℓ is uniformly semiconcave, and ψ_ℓ is differentiable on S_ℓ apart from a set of dimension $n-1$.

By **Besicovich's density theorem**, the set S_ℓ has μ-density 1 at μ-almost all $x \in S_\ell$; that is

$$\frac{\mu[S_\ell \cap B_r(x)]}{\mu[B_r(x)]} \xrightarrow[r\to 0]{} 1.$$

(The proof of this uses the fact that we are working on a Riemannian manifold; see the bibliographical notes for more information.)

Moreover, the transport plan π_ℓ induced by π on S_ℓ coincides with the deterministic transport associated with the map

$$T_\ell : x \longmapsto (\nabla_x c_\ell)^{-1}(x, -\nabla\psi_\ell(x)).$$

Since π is the nondecreasing limit of the measures $Z_\ell\,\pi_\ell$, it follows that π itself is deterministic, and associated with the transport map T that sends x to $T_\ell(x)$ if $x \in S_\ell$. (This map is well-defined μ-almost surely.)

Then let

$$C_\ell := \Big\{x \in S_\ell;\ x \text{ is interior to } \mathcal{X};\quad S_\ell \text{ has } \mu\text{-density 1 at } x;$$
$$\forall k \geq \ell,\ \psi_k \text{ is differentiable at } x;\quad \nabla_x c(x, T(x)) + \nabla\psi_\ell(x) = 0\Big\}.$$

(Note: There is no reason for $\nabla\psi_\ell(x)$ to be an approximate gradient of ψ at x, because ψ_ℓ is assumed to coincide with ψ only on a set of μ-density 1 at x, not on a set of vol-density 1 at x...)

The sets C_ℓ form a nondecreasing family of bounded Borel sets. Moreover, C_ℓ has been obtained from S_ℓ by deletion of a set of zero volume, and therefore of zero μ-measure. In particular, $\mu[\cup C_\ell] = 1$.

Now let $\widetilde{\pi}$ be another generalized optimal transference plan, and let $\widetilde{\psi}$ be a c-convex function with $\mathrm{Spt}(\widetilde{\pi}) \subset \partial_c\widetilde{\psi}$. We repeat the same construction as above with $\widetilde{\pi}$ instead of π, and get sequences $(\widetilde{Z}_\ell)_{\ell\in\mathbb{N}}$, $(\widetilde{\pi}_\ell)_{\ell\in\mathbb{N}}$, $(\widetilde{c}_\ell)_{\ell\in\mathbb{N}}$, $(\widetilde{\psi}_\ell)_{\ell\in\mathbb{N}}$, $(\widetilde{C}_\ell)_{\ell\in\mathbb{N}}$, such that the sets $\widetilde{C}_\ell$ form a nondecreasing family of bounded Borel sets with $\mu[\cup\widetilde{C}_\ell] = 1$, $\widetilde{\psi}_\ell$ coincides with $\widetilde{\psi}$ on $\widetilde{C}_\ell$. Also we find that $\widetilde{\pi}$ is deterministic and determined by the transport map $\widetilde{T}$, where $\widetilde{T}$ coincides with $\widetilde{T}_\ell$ on S_ℓ.

Next, the sets $C_\ell \cap \widetilde{C}_\ell$ also form a nondecreasing family of Borel sets, and $\mu[\cup(C_\ell \cap \widetilde{C}_\ell)] = \mu[(\cup C_\ell) \cap (\cup\widetilde{C}_\ell)] = 1$ (the nondecreasing property was used in the first equality). Also $C_\ell \cap \widetilde{C}_\ell$ has μ-density 1 at each of its points.

Assume that $T \neq \widetilde{T}$ on a set of positive μ-measure; then there is some $\ell \in \mathbb{N}$ such that $\{T \neq \widetilde{T}\} \cap (C_\ell \cap \widetilde{C}_\ell)$ has positive μ-measure. This implies that $\{T_\ell \neq \widetilde{T}_\ell\} \cap (C_\ell \cap \widetilde{C}_\ell)$ has positive μ-measure, and then

$$\mu\big[\{\nabla\widetilde{\psi}_\ell \neq \nabla\psi\} \cap (C_\ell \cap \widetilde{C}_\ell)\big] > 0.$$

In the sequel, I shall fix such an ℓ.

Let x be a μ-Besicovich point of $E_\ell := (C_\ell \cap \widetilde{C}_\ell) \cap \{\nabla\widetilde{\psi}_\ell \neq \nabla\psi_\ell\}$, i.e. a point at which E_ℓ has μ-density 1. (Such a point exists since E_ℓ has positive μ-measure.) By adding a suitable constant to ψ, we may assume that $\widetilde{\psi}(x) = \psi(x)$. Since ψ_ℓ and $\widetilde{\psi}_\ell$ are semiconvex, we can apply the **implicit function theorem** to deduce that there is a small neighborhood of x in which the set $\{\psi_\ell = \widetilde{\psi}_\ell\}$ has dimension $n-1$. (See Corollary 10.52 in the Second Appendix of this chapter.) Then, for r small enough, Assumption (iii) implies

$$\mu\Big[\{\psi_\ell \neq \widetilde{\psi}_\ell\} \cap B_r(x)\Big] = 0.$$

So at least one of the sets $\{\psi_\ell < \widetilde{\psi}_\ell\} \cap B_r(x)$ and $\{\psi_\ell > \widetilde{\psi}_\ell\} \cap B_r(x)$ has μ-measure at least $\mu[B_r(x)]/2$. Without loss of generality, I shall assume that this is the set $\{\psi_\ell > \widetilde{\psi}_\ell\}$; so

$$\mu\big[\{\psi_\ell > \widetilde{\psi}_\ell\} \cap B_r(x)\big] \geq \frac{\mu[B_r(x)]}{2}. \tag{10.30}$$

Next, ψ_ℓ coincides with ψ on the set S_ℓ, which has μ-density 1 at x, and similarly $\widetilde{\psi}_\ell$ coincides with $\widetilde{\psi}$ on a set of μ-density 1 at x. It follows that

$$\mu\big[\{\psi > \widetilde{\psi}\} \cap \{\psi_\ell > \widetilde{\psi}_\ell\} \cap B_r(x)\big] \geq \mu[B_r(x)]\,\Big(\frac{1}{2} - o(1)\Big) \qquad \text{as } r \to 0. \tag{10.31}$$

Then since x is a Besicovich point of $\{\nabla\psi_\ell \neq \nabla\widetilde{\psi}_\ell\} \cap C_\ell \cap \widetilde{C}_\ell$,

$$\begin{aligned}
&\mu\big[\{\psi > \widetilde{\psi}\} \cap \{\psi_\ell > \widetilde{\psi}_\ell\} \cap \{\nabla\psi_\ell \neq \nabla\widetilde{\psi}_\ell\} \cap (C_\ell \cap \widetilde{C}_\ell) \cap B_r(x)\big] \\
&\quad \geq \mu\big[\{\psi > \widetilde{\psi}\} \cap \{\psi_\ell > \widetilde{\psi}_\ell\} \cap B_r(x)\big] - \mu[B_r(x) \setminus (C_\ell \cap \widetilde{C}_\ell)] \\
&\qquad\qquad \geq \mu[B_r(x)]\,\Big(\frac{1}{2} - o(1) - o(1)\Big).
\end{aligned}$$

As a conclusion,

$$\forall r > 0 \quad \mu\Big[\{\psi > \widetilde{\psi}\} \cap \{\psi_\ell > \widetilde{\psi}_\ell\} \cap \{\nabla\psi_\ell \neq \nabla\widetilde{\psi}_\ell\} \cap (C_\ell \cap \widetilde{C}_\ell) \cap B_r(x)\Big] > 0. \tag{10.32}$$

Now let

$$A := \{\psi > \widetilde{\psi}\}.$$

The proof will result from the next two claims:

Claim 1: $\widetilde{T}^{-1}(T(A)) \subset A$;

Claim 2: The set $\{\psi_\ell > \widetilde{\psi}_\ell\} \cap (C_\ell \cap \widetilde{C}_\ell) \cap \{\nabla\psi_\ell \neq \nabla\widetilde{\psi}_\ell\} \cap \widetilde{T}^{-1}(T(A))$ lies a positive distance away from x.

Let us postpone the proofs of these claims for a while, and see why they imply the theorem. Let $S \subset A$ be defined by

$$S := \{\psi > \widetilde{\psi}\} \cap \{\psi_\ell > \widetilde{\psi}_\ell\} \cap \{\nabla\psi_\ell \neq \nabla\widetilde{\psi}_\ell\} \cap (C_\ell \cap \widetilde{C}_\ell),$$

and let

$$r := d\big(x, S \cap \widetilde{T}^{-1}(T(A))\big)/2.$$

On the one hand, since $S \cap \widetilde{T}^{-1}(T(A)) = \emptyset$ by definition, $\mu[S \cap B(x,r) \cap \widetilde{T}^{-1}(T(A))] = \mu[\emptyset] = 0$. On the other hand, r is positive by Claim 2, so $\mu[S \cap B(x,r)] > 0$ by (10.32). Then

$$\mu\big[A \setminus \widetilde{T}^{-1}(T(A))\big] \geq \mu\big[S \cap B(x,r) \setminus \widetilde{T}^{-1}(T(A))\big] = \mu[S \cap B(x,r)] > 0.$$

Since $\widetilde{T}^{-1}(T(A)) \subset A$ by Claim 1, this implies

$$\mu[\widetilde{T}^{-1}(T(A))] < \mu[A]. \tag{10.33}$$

But then, we can write

$$\mu[A] \leq \mu[T^{-1}(T(A))] = \nu[T(A)] = \nu[\widetilde{T}(A)] = \mu[\widetilde{T}^{-1}(T(A))],$$

which contradicts (10.33). So it all boils down to establishing Claims 1 and 2 above.

Proof of Claim 1: Let $x \in \widetilde{T}^{-1}(T(A))$. Then there exists $y \in A$ such that $T(y) = \widetilde{T}(x)$. Recall that $T(y) \in \partial_c\psi(y)$ and $\widetilde{T}(x) \in \partial_c\widetilde{\psi}(x)$. By using the definition of the c-subdifferential and the assumptions, we can write the following chain of identities and inequalities:

$$\begin{aligned}
\widetilde{\psi}(x) + c(x, \widetilde{T}(x)) &\leq \widetilde{\psi}(y) + c(y, \widetilde{T}(x)) \\
&= \widetilde{\psi}(y) + c(y, T(y)) \\
&< \psi(y) + c(y, T(y)) \\
&\leq \psi(x) + c(x, T(y)) \\
&= \psi(x) + c(x, \widetilde{T}(x)).
\end{aligned}$$

This implies that $\widetilde{\psi}(x) < \psi(x)$, i.e. $x \in A$, and proves Claim 1.

Proof of Claim 2: Assume that this claim is false; then there is a sequence $(x_k)_{k\in\mathbb{N}}$, valued in $\{\psi_\ell > \widetilde{\psi}_\ell\} \cap (C_\ell \cap \widetilde{C}_\ell) \cap \widetilde{T}^{-1}(T(A))$, such that $x_k \to x$. For each k, there is $y_k \in A$ such that $\widetilde{T}(x_k) = T(y_k)$. On $C_\ell \cap \widetilde{C}_\ell$, the transport T coincides with T_ℓ and the transport $\widetilde{T}$ with $\widetilde{T}_\ell$, so $\widetilde{T}(x_k) \in \partial_c \widetilde{\psi}_\ell(x_k)$ and $T(y_k) \in \partial_c \psi_\ell(y_k)$; then we can write, for any $z \in M$,

$$\begin{aligned}\psi_\ell(z) &\geq \psi_\ell(y_k) + c(y_k, T(y_k)) - c(z, T(y_k)) \\ &= \psi_\ell(y_k) + c(y_k, \widetilde{T}(x_k)) - c(z, \widetilde{T}(x_k)) \\ &> \widetilde{\psi}_\ell(y_k) + c(y_k, \widetilde{T}(x_k)) - c(z, \widetilde{T}(x_k)) \\ &\geq \widetilde{\psi}_\ell(x_k) + c(x_k, \widetilde{T}(x_k)) - c(z, \widetilde{T}(x_k))\end{aligned}$$

Since ψ_ℓ is differentiable at x and since c is locally semiconcave by assumption, we can expand the right-hand side and obtain

$$\begin{aligned}\psi_\ell(z) &\geq \widetilde{\psi}_\ell(x_k) + c(x_k, \widetilde{T}(x_k)) - c(z, \widetilde{T}(x_k)) \\ &= \widetilde{\psi}_\ell(x) + \nabla\widetilde{\psi}_\ell(x) \cdot (x_k - x) + o(|x_k - x|) \\ &\qquad - \nabla_x c(x_k, \widetilde{T}(x_k)) \cdot (z - x_k) + o(|z - x_k|). \qquad (10.34)\end{aligned}$$

where $o(|z - x_k|)$ in the last line is uniform in k. (Here I have cheated by pretending to work in $\mathbb{R}^n$ rather than on a Riemannian manifold, but all this is purely local, and invariant under diffeomorphism; so it is really no problem to make sense of these formulas when z is close enough to x_k.) Recall that $\nabla_x c(x_k, \widetilde{T}(x_k)) + \nabla\widetilde{\psi}_\ell(x_k) = 0$; so (10.34) can be rewritten as

$$\begin{aligned}\psi_\ell(z) \geq \nabla\widetilde{\psi}_\ell(x) &+ \nabla\widetilde{\psi}_\ell(x) \cdot (x_k - x) + o(|x_k - x|) \\ &\qquad + \nabla\widetilde{\psi}_\ell(x_k) \cdot (z - x_k) + o(|z - x_k|).\end{aligned}$$

Then we can pass to the limit as $k \to \infty$, remembering that $\nabla\widetilde{\psi}_\ell$ is continuous (because $\widetilde{\psi}_\ell$ is semiconvex, recall the proof of Proposition 10.12(i)), and get

$$\begin{aligned}\psi_\ell(z) &\geq \widetilde{\psi}_\ell(x) + \nabla\widetilde{\psi}_\ell(x) \cdot (z - x) + o(|z - x|) \\ &= \psi_\ell(x) + \nabla\widetilde{\psi}_\ell(x) \cdot (z - x) + o(|z - x|). \qquad (10.35)\end{aligned}$$

(Recall that x is such that $\psi_\ell(x) = \psi(x) = \widetilde{\psi}(x) = \widetilde{\psi}_\ell(x)$.) On the other hand, since ψ_ℓ is differentiable at x, we have

$$\psi_\ell(z) = \psi_\ell(x) + \nabla\psi_\ell(x) \cdot (z - x) + o(|z - x|).$$

Combining this with (10.35), we see that

$$(\nabla\widetilde{\psi}_\ell(x) - \nabla\psi_\ell(x)) \cdot (z - x) \leq o(|z - x|),$$

which is possible only if $\nabla\widetilde{\psi}_\ell(x) - \nabla\psi_\ell(x) = 0$. But this contradicts the definition of x. So Claim 2 holds true, and this concludes the proof of Theorem 10.42. □

The next corollary of Theorem 10.42 is the same as Theorem 10.41 except that the classical Monge problem (search of a transport of minimum cost) has been replaced by the generalized Monge problem (search of a c-monotone transport).

Corollary 10.44 (Generalized Monge problem for the square distance). *Let M be a smooth Riemannian manifold, and let $c(x, y) = d(x, y)^2$. Let μ, ν be two probability measures on M.*

- *If μ gives zero mass to sets of dimension at most $n - 1$, then there is a unique transport map T solving the generalized Monge problem between μ and ν.*
- *If μ is absolutely continuous, then this solution can be written*

$$y = T(x) = \exp_x(\widetilde{\nabla}\psi(x)), \tag{10.36}$$

  where ψ is some $d^2/2$-convex function.
- *If M has nonnegative sectional curvature, or ν is compactly supported and M satisfies (10.26), then equation (10.36) still holds, but in addition the approximate gradient can be replaced by a true gradient.*

Particular Case 10.45. If $M = \mathbb{R}^n$, formula (10.36) becomes

$$y = x + \nabla\psi(x) = \nabla\left(\frac{|\cdot|^2}{2} + \psi\right)(x),$$

where ψ is $|\cdot|^2/2$-convex, or equivalently $|\cdot|^2/2 + \psi$ is convex lower semicontinuous. So (10.36) can be written $y = \nabla\Psi(x)$, where Ψ is convex lower semicontinuous, and we are back to Theorem 9.4.

First Appendix: A little bit of geometric measure theory

The geometric counterpart of differentiability is of course the approximation of a set S by a tangent plane, or hyperplane, or more generally by a tangent d-dimensional space, if d is the dimension of S.

If S is smooth, then there is no ambiguity on its dimension (a curve has dimension 1, a surface has dimension 2, etc.) and the tangent space always exists. But if S is not smooth, this might not be the case, at least not in the usual sense. The notion of **tangent cone** (sometimes called contingent cone) often remedies this problem; it is naturally associated with the notion of **countable d-rectifiability**, which acts as a replacement for the notion of "being of dimension d". I shall recall below some of the basic results about these concepts.

Definition 10.46 (Tangent cone). *If S is an arbitrary subset of $\mathbb{R}^n$, and $x \in \overline{S}$, then the tangent cone T_xS to S at x is defined as*

$$T_xS := \left\{ \lim_{k\to\infty} \frac{x_k - x}{t_k}; \quad x_k \in S,\ x_k \to x,\ t_k > 0,\ t_k \to 0 \right\}.$$

The dimension of this cone is defined as the dimension of the vector space that it generates.

Definition 10.47 (Countable rectifiability). *Let S be a subset of $\mathbb{R}^n$, and let $d \in [0,n]$ be an integer. Then S is said to be countably d-rectifiable if $S \subset \bigcup_{k\in\mathbb{N}} f_k(D_k)$, where each f_k is Lipschitz on a measurable subset D_k of $\mathbb{R}^d$. In particular, S has Hausdorff dimension at most d.*

The next theorem summarizes two results which were useful in the present chapter:

Theorem 10.48 (Sufficient conditions for countable rectifiability).

(i) Let S be a measurable set in $\mathbb{R}^n$, such that T_xS has dimension at most d for all $x \in S$. Then S is countably d-rectifiable.

(ii) Let S be a measurable set in $\mathbb{R}^n$, such that T_xS is included in a half-space, for each $x \in \partial S$. Then ∂S is countably $(n-1)$-rectifiable.

Proof of Theorem 10.48. For each $x \in S$, let π_x stand for the orthogonal projection on T_xS, and let $\pi_x^\perp = \mathrm{Id} - \pi_x$ stand for the orthogonal projection on $(T_xS)^\perp$. I claim that

$$\forall x \in S, \quad \exists r > 0; \quad \forall y \in S, \quad |x-y| \le r \Longrightarrow |\pi_x^\perp(x-y)| \le |\pi_x(x-y)|. \tag{10.37}$$

Indeed, assume that (10.37) is false. Then there is $x \in S$, and there is a sequence $(y_k)_{k\in\mathbb{N}}$ such that $|x - y_k| \le 1/k$ and yet $|\pi_x^\perp(x-y)| > |\pi_x(x-y)|$, or equivalently

$$\left|\pi_x^\perp\left(\frac{x-y_k}{|x-y_k|}\right)\right| > \left|\pi_x\left(\frac{x-y_k}{|x-y_k|}\right)\right|. \tag{10.38}$$

Up to extraction of a subsequence, $w_k := (x-y_k)/|x-y_k|$ converges to $\theta \in T_xS$ with $|\theta| = 1$. Then $|\pi_x w_k| \to 1$ and $|\pi_x^\perp w_k| \to 0$, which contradicts (10.38). So (10.37) is true.

Next, for each $k \in \mathbb{N}$, let

$$S_k := \Big\{x \in S; \quad \text{property (10.37) holds true for } |x-y| \le 1/k\Big\}.$$

It is clear that the sets S_k cover S, so it is sufficient to establish the d-rectifiability of S_k for a given k.

Let $\delta > 0$ be small enough ($\delta < 1/2$ will do). Let Π_d be the set of all orthogonal projections on d-dimensional linear spaces. Since Π_d is compact, we can find a finite family $(\pi_1, \dots, \pi_N)$ of such orthogonal projections, such that for any $\pi \in \Pi_d$ there is $j \in \{1,\dots,N\}$ with $\|\pi - \pi_j\| \le \delta$, where $\|\cdot\|$ stands for the operator norm. So the set S_k is covered by the sets

$$S_{k\ell} := \Big\{x \in S_k; \quad \|\pi_x - \pi_\ell\| \le \delta\Big\}.$$

To prove (i), it suffices to prove that $S_{k\ell}$ is locally rectifiable. We shall show that for any two $x, x' \in S_{k\ell}$,

$$|x-x'| \le \frac{1}{k} \Longrightarrow |\pi_\ell^\perp(x-x')| \le L\,|\pi_\ell(x-x')|, \qquad L = \frac{1+2\delta}{1-2\delta}; \tag{10.39}$$

this will imply that the intersection of $S_{k\ell}$ with a ball of diameter $1/k$ is contained in an L-Lipschitz graph over $\pi_\ell(\mathbb{R}^n)$, and the conclusion will follow immediately.

To prove (10.39), note that, if π, π' are any two orthogonal projections, then $|(\pi^\perp - (\pi')^\perp)(z)| = |(\mathrm{Id} - \pi)(z) - (\mathrm{Id} - \pi')(z)| = |(\pi - \pi')(z)|$, therefore $\|\pi^\perp - (\pi')^\perp\| = \|\pi - \pi'\|$, and

$$\begin{aligned}
|\pi_\ell^\perp(x-x')| &\le |(\pi_\ell^\perp - \pi_x^\perp)(x-x')| + |\pi_x^\perp(x-x')| \\
&\le |(\pi_\ell - \pi_x)(x-x')| + |\pi_x(x-x')| \\
&\le |(\pi_\ell - \pi_x)(x-x')| + |\pi_\ell(x-x')| + |(\pi_\ell - \pi_x)(x-x')| \\
&\le |\pi_\ell(x-x')| + 2\delta|x-x'| \\
&\le (1+2\delta)|\pi_\ell(x-x')| + 2\delta|\pi_\ell^\perp(x-x')|.
\end{aligned}$$

This establishes (10.39), and Theorem 10.48(i).

Now let us turn to part (ii) of the theorem. Let F be a finite set in S^{n-1} such that the balls $(B_{1/8}(\nu))_{\nu\in F}$ cover S^{n-1}. I claim that

$$\forall x \in \partial S, \quad \exists r > 0, \quad \exists \nu \in F, \quad \forall y \in \partial S \cap B_r(x), \quad \langle y-x, \nu\rangle \le \frac{|y-x|}{2}. \tag{10.40}$$

Indeed, otherwise there is $x \in \partial S$ such that for all $k \in \mathbb{N}$ and for all $\nu \in F$ there is $y_k \in \partial S$ such that $|y_k - x| \le 1/k$ and $\langle y_k - x, \nu\rangle > |y_k - x|/2$. By assumption there is $\xi \in S^{n-1}$ such that

$$\forall \zeta \in T_x S, \quad \langle \xi, \zeta\rangle \le 0.$$

Let $\nu \in F$ be such that $|\xi - \nu| < 1/8$ and let $(y_k)_{k\in\mathbb{N}}$ be a sequence as above. Since $y_k \in \partial S$ and $y_k \ne x$, there is $y'_k \in S$ such that $|y_k - y'_k| < |y_k - x|/8$. Then

$$\langle y'_k - x, \xi\rangle \ge \langle y_k - x, \nu\rangle - |y_k - x|\,|\xi - \nu| - |y - y'_k| \ge \frac{|y_k - x|}{4} \ge \frac{|x - y'_k|}{8}.$$

So

$$\left\langle \frac{y'_k - x}{|y'_k - x|}, \xi \right\rangle \ge \frac{1}{8}. \tag{10.41}$$

Up to extraction of a subsequence, $(y'_k - x)/|y'_k - x|$ converges to some $\zeta \in T_x S$, and then by passing to the limit in (10.41) we have $\langle \zeta, \xi\rangle \ge 1/8$. But by definition, ξ is such that $\langle \zeta, \xi\rangle \le 0$ for all $\zeta \in T_x S$. This contradiction establishes (10.40).

As a consequence, ∂S is included in the union of all sets $A_{1/k,\nu}$, where $k \in \mathbb{N}$, $\nu \in F$, and

$$A_{r,\nu} := \left\{ x \in \partial S; \quad \forall y \in \partial S \cap B_r(x), \quad \langle y - x, \nu\rangle \le \frac{|y-x|}{2} \right\}.$$

To conclude the proof of the theorem it is sufficient to show that each $A_{r,\nu}$ is locally the image of a Lipschitz function defined on a subset of an $(n-1)$-dimensional space.

So let $r > 0$ and $\nu \in F$ be given, let $x_0 \in A_{r,\nu}$, and let π be the orthogonal projection of $\mathbb{R}^n$ to $\nu^\perp$. (Explicitly, $\pi(x) = x - \langle x, \nu\rangle\nu$.) We shall show that on $D := A_{r,\nu} \cap B_{r/2}(x_0)$, π is injective and its inverse (on $\pi(D)$) is Lipschitz. To see this, first note that for any two $x, x' \in D$, one has $x' \in B_r(x)$, so, by definition of $A_{r,\nu}$, $\langle x' - x, \nu\rangle \le |x' - x|/2$. By symmetry, also $\langle x - x', \nu\rangle \le |x - x'|/2$, so in fact

$$\big|\langle x - x', \nu\rangle\big| \le \frac{|x - x'|}{2}.$$

Then if $z = \pi(x)$ and $z' = \pi(x')$,

$$|x - x'| \le |z - z'| + \big|\langle x, \nu\rangle - \langle x', \nu\rangle\big| \le |z - z'| + \frac{|x - x'|}{2},$$

so $|x - x'| \le 2\,|z - z'|$. This concludes the proof. □

Second Appendix: Nonsmooth implicit function theorem

Let M be an n-dimensional smooth Riemannian manifold, and $x_0 \in M$. I shall say that a set $M' \subset M$ is a k-dimensional C^r graph (resp. k-dimensional Lipschitz graph) in a neighborhood of x_0 if there are

(i) a smooth system of coordinates around x_0, say

$$x = \zeta(x', y),$$

where ζ is a smooth diffeomorphism from an open subset of $\mathbb{R}^k \times \mathbb{R}^{n-k}$, into a neighborhood O of x_0;

(ii) a C^r (resp. Lipschitz) function $\varphi : O' \to \mathbb{R}^{n-k}$, where O' is an open subset of $\mathbb{R}^k$;

such that for all $x \in O$,

$$x \in M' \iff y = \varphi(x').$$

This definition is illustrated by Figure 10.2.

The following statement is a consequence of the classical implicit function theorem: *If $f : M \to \mathbb{R}$ is of class C^r $(r \ge 1)$, $f(x_0) = 0$ and $\nabla f(x_0) \neq 0$, then the set $\{f = 0\} = f^{-1}(0)$ is an $(n-1)$-dimensional C^r graph in a neighborhood of x_0.*

In this Appendix I shall consider a *nonsmooth* version of this theorem. The following notion will be useful.

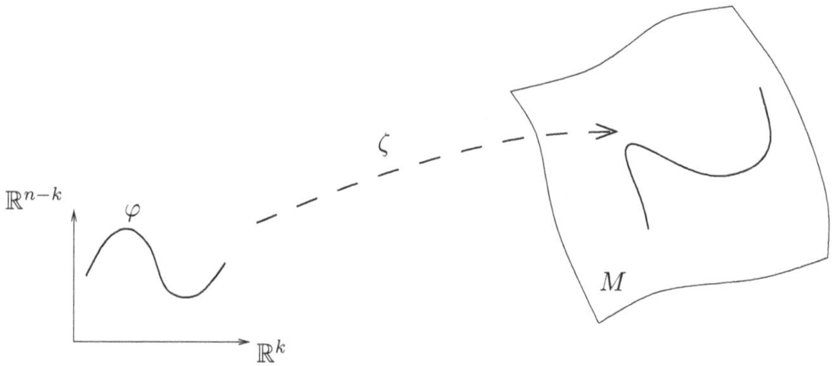

Fig. 10.2. k-dimensional graph

Definition 10.49 (Clarke subdifferential). *Let f be a continuous real-valued function defined on an open subset U of a Riemannian manifold. For each $x \in U$, define $\partial f(x)$ as the convex hull of all limits of sequences $\nabla f(x_k)$, where all x_k are differentiability points of f and $x_k \to x$. In short:*

$$\partial f(x) = \overline{\mathrm{Conv}} \left\{ \lim_{x_k \to x} \nabla f(x_k) \right\}.$$

Here comes the main result of this Appendix. If $(A_i)_{1 \le i \le m}$ are subsets of a vector space, I shall write $\sum A_i = \{\sum_i a_i;\ a_i \in A_i\}$.

Theorem 10.50 (Nonsmooth implicit function theorem). *Let $(f_i)_{1 \le i \le m}$ be real-valued Lipschitz functions defined in an open set U of an n-dimensional Riemannian manifold, and let $x_0 \in U$ be such that:*

(a) $\sum f_i(x_0) = 0$;

(b) $0 \notin \sum \partial f_i(x_0)$.

Then $\{\sum f_i = 0\}$ is an $(n-1)$-dimensional Lipschitz graph around x_0.

Remark 10.51. Let ψ be a convex continuous function defined around some point $x_0 \in \mathbb{R}^n$, and let $p \in \mathbb{R}^n$ such that p does not belong to the Clarke differential of ψ at x_0; then 0 does not belong to the Clarke differential of $\widetilde{\psi} : x \longmapsto \psi(x) - \psi(x_0) - p \cdot (x - x_0) + |x - x_0|^2$ at x_0, and Theorem 10.50 obviously implies the existence of $x \neq x_0$ such that $\widetilde{\psi}(x) = 0$, in particular $\psi(x) < \psi(x_0) + p \cdot (x - x_0)$. So p does not belong to the subdifferential of ψ at x_0. In other words, the subdifferential is included in the Clarke differential. The other inclusion is obvious, so both notions coincide. This justifies a posteriori the notation $\partial\psi$ used in Definition 10.49. An easy localization argument shows that similarly, for any locally semiconvex function ψ defined in a neighborhood of x, $\partial\psi(x) = \nabla^-\psi(x)$.

Corollary 10.52 (Implicit function theorem for two subdifferentiable functions). *Let ψ and $\widetilde{\psi}$ be two locally subdifferentiable functions defined in an open set U of an n-dimensional Riemannian manifold M, and let $x_0 \in U$ be such that ψ, $\widetilde{\psi}$ are differentiable at x_0, and*

$$\psi(x_0) = \widetilde{\psi}(x_0); \qquad \nabla\psi(x_0) \neq \nabla\widetilde{\psi}(x_0).$$

Then there is a neighborhood V of x_0 such that $\{\psi = \widetilde{\psi}\} \cap V$ is an $(n-1)$-dimensional Lipschitz graph; in particular, it has Hausdorff dimension exactly $n-1$.

Proof of Corollary 10.52. Let $f_1 = \psi$, $f_2 = -\widetilde{\psi}$. Since f_1 is locally subdifferentiable and f_2 is locally superdifferentiable, both functions are Lipschitz in a neighborhood of x_0 (Theorem 10.8(iii)). Moreover, ∇f_1 and ∇f_2 are continuous on their respective domain of definition; so $\partial f_i(x_0) = \{\nabla f_i(x_0)\}$ $(i = 1, 2)$. Then by assumption $\sum \partial f_i(x_0) = \{\sum \nabla f_i(x_0)\}$ does not contain 0. The conclusion follows from Theorem 10.50. □

Proof of Theorem 10.50. The statement is purely local and invariant under C^1 diffeomorphism, so we might pretend to be working in $\mathbb{R}^n$. For each i, $\partial f_i(x_0) \subset B(0, \|f_i\|_{\mathrm{Lip}}) \subset \mathbb{R}^n$, so $\partial f_i(x_0)$ is a compact convex subset of $\mathbb{R}^n$; then also $\sum \partial f_i(x_0)$ is compact and convex, and by assumption does not contain 0. By the Hahn–Banach theorem, there are $v \in \mathbb{R}^n$ and $\alpha > 0$ such that

$$\langle p, v\rangle \geq \alpha \qquad \text{for all } p \in \sum \partial f_i(x_0). \tag{10.42}$$

So there is a neighborhood V of x_0 such that $\sum\langle \nabla f_i(x), v\rangle \geq \alpha/2$ at all points $x \in V$ where all functions f_i are differentiable. (Otherwise there would be a sequence $(x_k)_{k\in\mathbb{N}}$, converging to x_0, such that $\sum\langle \nabla f_i(x_k), v\rangle < \alpha/2$, but then up to extraction of a subsequence we would have $\nabla f_i(x_k) \to p_i \in \partial f_i(x_0)$, so $\langle \sum p_i, v\rangle \leq \alpha/2 < \alpha$, which would contradict (10.42).)

Without loss of generality, we may assume that $x_0 = 0$, $v = (e_1, 0, \ldots, 0)$, $V = (-\beta, \beta) \times B(0, r_0)$, where the latter ball is a subset of $\mathbb{R}^{n-1}$ and $r_0 \leq (\alpha\beta)/(4 \sum \|f_i\|_{\mathrm{Lip}})$. Further, let

$$Z' := \Big\{y \in B(0,r_0) \subset \mathbb{R}^{n-1};\\ \lambda_1\big[\{t \in (-\beta,\beta);\ \exists i;\ \nabla f_i(t,y) \text{ does not exist}\}\big] > 0\Big\},$$

and

$$Z = (-\beta,\beta) \times Z'; \qquad D = V \setminus Z.$$

I claim that $\lambda_n[Z] = 0$. To prove this it is sufficient to check that $\lambda_{n-1}[Z'] = 0$. But Z' is the nonincreasing limit of $(Z'_\ell)_{\ell \in \mathbb{N}}$, where

$$Z'_\ell = \Big\{y \in B(0,r_0);\\ \lambda_1\big[\{t \in (-\beta,\beta);\ \exists i;\ \nabla f_i(t,y) \text{ does not exist}\}\big] \geq 1/\ell\Big\}.$$

By Fubini's theorem,

$$\lambda_n\Big[\{x \in O;\ \nabla f_i(x) \text{ does not exist for some } i\}\Big] \geq (\lambda_{n-1}[Z'_\ell]) \times (1/\ell);$$

and the left-hand side is equal to 0 since all f_i are differentiable almost everywhere. It follows that $\lambda_{n-1}[Z'_\ell] = 0$, and by taking the limit $\ell \to \infty$ we obtain $\lambda_{n-1}[Z'] = 0$.

Let $f = \sum f_i$, and let $\partial_1 f = \langle \nabla f, v\rangle$ stand for its partial derivative with respect to the first coordinate. The first step of the proof has shown that $\partial_1 f(x) \geq \alpha/2$ at each point x where all functions f_i are differentiable. So, for each $y \in B(0,r_0) \setminus Z'$, the function $t \to f(t,y)$ is Lipschitz and differentiable λ_1-almost everywhere on $(-\beta,\beta)$, and it satisfies $f'(t,y) \geq \alpha/2$. Thus for all $t, t' \in (-\beta,\beta)$,

$$t < t' \implies \quad f(t',y) - f(t,y) \geq (\alpha/2)\,(t'-t). \tag{10.43}$$

This holds true for all $((t,y),(t',y))$ in $D \times D$. Since $Z = V \setminus D$ has zero Lebesgue measure, D is dense in V, so (10.43) extends to all $((t,y),(t',y)) \in V$.

For all $y \in B(0,r_0)$, inequality (10.43), combined with the estimate

$$|f(0,y)| = |f(0,y) - f(0,0)| \leq \|f\|_{\mathrm{Lip}}\,|y| \leq \frac{\alpha\beta}{4},$$

guarantees that the equation $f(t,y) = 0$ admits exactly one solution $t = \varphi(y)$ in $(-\beta,\beta)$.

It only remains to check that φ is Lipschitz on $B(0,r_0)$. Let $y, z \in B(0,r_0)$, then $f(\varphi(y),y) = f(\varphi(z),z) = 0$, so

$$\left|f(\varphi(y),y) - f(\varphi(z),y)\right| = \left|f(\varphi(z),z) - f(\varphi(z),y)\right|. \tag{10.44}$$

Since the first partial derivative of f is no less than $\alpha/2$, the left-hand side of (10.44) is bounded below by $(\alpha/2)|\varphi(y)-\varphi(z)|$, while the right-hand side is bounded above by $\|f\|_{\mathrm{Lip}}\,|z-y|$. The conclusion is that

$$|\varphi(y) - \varphi(z)| \leq \frac{2\,\|f\|_{\mathrm{Lip}}}{\alpha}\,|z-y|,$$

so φ is indeed Lipschitz. □

Third Appendix: Curvature and the Hessian of the squared distance

The practical verification of the uniform semiconcavity of a given cost function $c(x,y)$ might be a very complicated task in general. In the particular case when $c(x,y) = d(x,y)^2$, this problem can be related to the **sectional curvature** of the Riemannian manifold. In this Appendix I shall recall some results about these links, some of them well-known, other ones more confidential. The reader who does not know about sectional curvature can skip this Appendix, or take a look at Chapter 14 first.

If $M = \mathbb{R}^n$ is the Euclidean space, then $d(x,y) = |x-y|$ and there is the very simple formula

$$\nabla_x^2\left(\frac{|x-y|^2}{2}\right) = I_n,$$

where the right-hand side is just the identity operator on $T_x\mathbb{R}^n = \mathbb{R}^n$.

If M is an arbitrary Riemannian manifold, there is no simple formula for the Hessian $\nabla_x^2 d(x,y)^2/2$, and this operator will in general not be defined in the sense that it can take eigenvalues $-\infty$ if x and y are conjugate points. However, as we shall now see, there is still a recipe to estimate $\nabla_x^2 d(x,y)^2/2$ *from above*, and thus derive estimates of semiconcavity for $d^2/2$.

So let x and y be any two points in M, and let γ be a minimizing geodesic joining y to x, parametrized by arc length; so $\gamma(0) = y$, $\gamma(d(x,y)) = x$. Let $H(t)$ stand for the Hessian operator of $x \to d(x,y)^2/2$ at $x = \gamma(t)$. (The definition of this operator is recalled and discussed in Chapter 14.) On $[0, d(x,y))$ the operator $H(t)$ is well-defined (since the geodesic is minimizing, it is only at $t = d(x,y)$ that eigenvalues $-\infty$ may appear). It starts at $H(0) = \mathrm{Id}$ and then its eigenvectors and eigenvalues vary smoothly as t varies in $(0, d(x,y))$.

The unit vector $\dot{\gamma}(t)$ is an eigenvector of $H(t)$, associated with the eigenvalue $+1$. The problem is to bound the eigenvalues in the orthogonal subspace $S(t) = (\dot{\gamma})^\perp \subset T_{\gamma(t)}M$. So let $(e_2, \ldots, e_n)$ be an orthonormal basis of $S(0)$, and let $(e_2(t), \ldots, e_n(t))$ be obtained by parallel transport of $(e_2, \ldots, e_n)$ along γ; for any t this remains an orthonormal basis of $S(t)$. To achieve our goal, it is sufficient to bound above the quantities $h(t) = \langle H(t) \cdot e_i(t), e_i(t)\rangle_{\gamma(t)}$, where i is arbitrary in $\{2, \ldots, n\}$.

Since $H(0)$ is the identity, we have $h(0) = 1$. To get a differential equation on $h(t)$, we can use a classical computation of Riemannian geometry, about the Hessian of the *distance* (not squared!): If $k(t) = \langle \nabla_x^2 d(y,x) \cdot e_i(t), e_i(t)\rangle_{\gamma(t)}$, then

$$\dot{k}(t) + k(t)^2 + \sigma(t) \leq 0, \tag{10.45}$$

where $\sigma(t)$ is the sectional curvature of the plane generated by $\dot{\gamma}(t)$ and $e_i(t)$ inside $T_{\gamma(t)}M$. To relate $k(t)$ and $h(t)$, we note that

$$\nabla_x \left(\frac{d(y,x)^2}{2}\right) = d(y,x)\, \nabla_x d(y,x);$$

$$\nabla_x^2 \left(\frac{d(y,x)^2}{2}\right) = d(y,x)\, \nabla_x^2 d(y,x) + \nabla_x d(x,y) \otimes \nabla_x d(x,y).$$

By applying this to the tangent vector $e_i(t)$ and using the fact that $\nabla_x d(x,y)$ at $x = \gamma(t)$ is just $\dot{\gamma}(t)$, we get

$$h(t) = d(y, \gamma(t))\, k(t) + \langle \dot{\gamma}(t), e_i(t)\rangle^2 = t\, k(t).$$

Plugging this into (10.45) results in

$$t\, \dot{h}(t) - h(t) + h(t)^2 \leq -t^2\, \sigma(t). \tag{10.46}$$

From (10.46) follow the two comparison results which were used in Theorem 10.41 and Corollary 10.44:

(a) Assume that **the sectional curvatures of M are all nonnegative**. Then (10.46) forces $\dot{h} \leq 0$, so h remains bounded above by 1 for all times. In short:

$$\text{nonnegative sectional curvature} \Longrightarrow \qquad \nabla_x^2 \left(\frac{d(x,y)^2}{2} \right) \le \mathrm{Id}_{T_x M}. \tag{10.47}$$

(If we think of the Hessian as a bilinear form, this is the same as $\nabla_x^2 (d(x,y)^2/2) \le g$, where g is the Riemannian metric.) Inequality (10.47) is rigorous if $d(x,y)^2/2$ is twice differentiable at x; otherwise the conclusion should be reinterpreted as

$$x \longrightarrow \frac{d(x,y)^2}{2} \quad \text{is semiconcave with a modulus } \omega(r) = \frac{r^2}{2}.$$

(b) Assume now that **the sectional curvatures at point x are bounded below by $-C/d(x_0,x)^2$**, where x_0 is an arbitrary point. In this case I shall say that M has *asymptotically nonnegative curvature.* Then if y varies in a compact subset, we have a lower bound like $\sigma(t) \ge -C'/d(y,x)^2 = -C'/t^2$, where C' is some positive constant. So (10.46) implies the differential inequality

$$t\,\dot{h}(t) \le C' + h(t) - h(t)^2.$$

If $h(t)$ ever becomes greater than $\overline{C} := (1 + \sqrt{1 + 4C'^2})/2$, then the right-hand side becomes negative; so h can never go above $\overline{C}$. The conclusion is that

$$M \text{ has asymptotically nonnegative curvature} \Longrightarrow$$
$$\forall y \in K, \quad \nabla_x^2 \left(\frac{d(x,y)^2}{2} \right) \le C(K)\, \mathrm{Id}_{T_x M},$$

where K is any compact subset of M. Again, at points where $d(x,y)^2/2$ is not twice differentiable, the conclusion should be reinterpreted as

$$x \longrightarrow \frac{d(x,y)^2}{2} \quad \text{is semiconcave with a modulus } \omega(r) = C(K)\,\frac{r^2}{2}.$$

Examples 10.53. The previous result applies to any compact manifold, or any manifold which has been obtained from $\mathbb{R}^n$ by modification on a compact set. But it does not apply to the hyperbolic space $\mathbb{H}^n$; in fact, if y is any given point in $\mathbb{H}^n$, then the function $x \to d(y,x)^2$ is not uniformly semiconcave as $x \to \infty$. (Take for instance the unit disk in $\mathbb{R}^2$, with polar coordinates (r,θ) as a model of $\mathbb{H}^2$, then the distance from the origin is $d(r,\theta) = \log((1+r)/(1-r))$; an explicit computation shows that the first (and only nonzero) coefficient of the matrix of the Hessian of $d^2/2$ is $1 + r\,d(r)$, which diverges logarithmically as $r \to 1$.)

Remark 10.54. The exponent 2 appearing in the definition of "asymptotic nonnegative curvature" above is optimal in the sense that for any $p < 2$ it is possible to construct manifolds satisfying $\sigma_x \geq -C/d(x_0, x)^p$ and on which $d(x_0, \cdot\,)^2$ is not uniformly semiconcave.

Bibliographical notes

The key ideas in this chapter were first used in the case of the quadratic cost function in Euclidean space [154, 156, 722].

The existence of solutions to the Monge problem and the differentiability of c-convex functions, for strictly superlinear convex cost functions in $\mathbb{R}^n$ (other than quadratic) was investigated by several authors, including in particular Rüschendorf [717] (formula (10.4) seems to appear there for the first time), Smith and Knott [754], Gangbo and McCann [398, 399]. In the latter reference, the authors get rid of all moment assumptions by avoiding the explicit use of Kantorovich duality. These results are reviewed in [814, Chapter 2]. Gangbo and McCann impose some assumptions of growth and superlinearity, such as the one described in Example 10.19. Ekeland [323] applies similar tools to the optimal matching problem (in which both measures are transported to another topological space).

It is interesting to notice that the structure of the optimal map can be guessed heuristically from the old-fashioned method of Lagrange multipliers; see [328, Section 2.2] where this is done for the case of the quadratic cost function.

The terminology of the twist condition comes from dynamical systems, in particular the study of certain classes of diffeomorphisms in dimension 2 called twist diffeomorphisms [66]. Moser [641] showed that under certain conditions, these diffeomorphisms can be represented as the time-1 map of a strictly convex Lagrangian system. In this setting, the twist condition can be informally recast by saying that the dynamics "twists" vertical lines (different velocities, common position) into oblique curves (different positions).

There are cases of interest where the twist condition is satisfied even though the cost function does not have a Lagrangian structure. Examples are the so-called symmetrized Bregman cost function $c(x, y) = \langle \nabla\phi(x) - \nabla\phi(y), x - y \rangle$ where ϕ is strictly convex [206], or the cost $c(x,y) = |x-y|^2 + |f(x)-g(y)|^2$, where f and g are convex and k-Lipschitz, $k < 1$ [794]. For applications in meteorology, Cullen and Maroofi [268] considered cost functions of the form $c(x,y) = [(x_1-y_1)^2+(x_2-y_2)^2+\varphi(x_3)]/y_3$ in a bounded region of $\mathbb{R}^3$.

Conversely, a good example of an interesting smooth cost function which does not satisfy the twist condition is provided by the restriction of the square Euclidean distance to the sphere, or to a product of convex boundaries [6, 400].

Gangbo and McCann [399] considered not only strictly convex, but also strictly concave cost functions in $\mathbb{R}^n$ (more precisely, strictly concave functions of the distance), which are probably more realistic from an economic perspective, as explained in the introduction of their paper. The main results from [399] are briefly reviewed in [814, Section 2.4]. Further numerical and theoretical analysis for nonconvex cost functions in dimension 1 have been considered by McCann [615], Rüschendorf and Uckelmann [724, 796], and Plakhov [683]. Hsu and Sturm [484] worked on a very nice application of an optimal transport problem with a concave cost to a problem of maximal coupling of Brownian paths.

McCann [616] proved Theorem 10.41 when M is a compact Riemannian manifold and μ is absolutely continuous. This was the first optimal transport theorem on a Riemannian manifold (save for the very particular case of the n-dimensional torus, which was treated before by Cordero-Erausquin [240]). In his paper McCann also mentioned the possibility of covering more general cost functions expressed in terms of the distance.

Later Bernard and Buffoni [105] extended McCann's results to more general Lagrangian cost functions, and imported tools and techniques from the theory of Lagrangian systems (related in particular to Mather's minimization problem). The proof of the basic result of this chapter (Theorem 10.28) is in some sense an extension of the Bernard–Buffoni theorem to its "natural generality". It is clear from the proof that the Riemannian structure plays hardly any role, so it extends for instance to Finsler geometries, as was done in the work of Ohta [657].

Before the explicit link realized by Bernard and Buffoni, several researchers, in particular Evans, Fathi and Gangbo, had become gradually aware of the strong similarities between Monge's theory on the one hand, and Mather's theory on the other. De Pascale, Gelli and Granieri [278] contributed to this story; see also [839].

Fang and Shao rewrote McCann's theorem in the formalism of Lie groups [339]. They used this reformulation as a starting point to derive theorems of unique existence of the optimal transport on the path space over a Lie group. Shao's PhD Thesis [748] contains a synthetic view on these issues, and reminders about differential calculus in Lie groups.

Feyel and Üstünel [358, 359, 360, 362] derived theorems of unique solvability of the Monge problem in the Wiener space, when the cost is the square of the Cameron–Martin distance (or rather pseudo-distance, since it takes the value $+\infty$). Their tricky analysis goes via finite-dimensional approximations.

Ambrosio and Rigot [33] adapted the proof of Theorem 10.41 to cover degenerate (subriemannian) situations such as the Heisenberg group, equipped with either the squared Carnot–Carathéodory metric or the squared Korányi norm. The proofs required a delicate analysis of minimizing geodesics, differentiability properties of the squared distance, and fine properties of BV functions on the Heisenberg group. Then Rigot [702] generalized these results to certain classes of groups. Further work in this area (including the absolute continuity of Wasserstein geodesics at intermediate times, the differentiability of the optimal maps, and the derivation of an equation of Monge–Ampère type) was achieved by Agrachev and Lee [3], Figalli and Juillet [366], and Figalli and Rifford [370].

Another nonsmooth generalization of Theorem 10.41 was obtained by Bertrand [114], who adapted McCann's argument to the case of an Alexandrov space with finite (Hausdorff) dimension and (sectional) curvature bounded below. His analysis makes crucial use of fine regularity results on the structure of Alexandrov spaces, derived by Perelman, Otsu and Shioya [174, 175, 665].

Remark 10.30 about the uniqueness of the potential ψ is due to Loeper [570, Appendix]; it still holds if μ is any singular measure, provided that $d\mu/d\mathrm{vol} > 0$ almost everywhere (even though the transport map might not be unique then).

The use of approximate differentials as in Theorem 10.38 was initiated by Ambrosio and collaborators [30, Chapter 6], for strictly convex cost functions in $\mathbb{R}^n$. The adaptation to Riemannian manifolds is due to Fathi and Figalli [348], with a slightly more complicated (but slightly more general) approach than the one used in this chapter.

The tricky proof of Theorem 10.42 takes its roots in Alexandrov's uniqueness theorem for graphs of prescribed Gauss curvature [16]. (The method can be found in [53, Theorem 10.2].) McCann [613] understood that Alexandrov's strategy could be revisited to yield the uniqueness of a cyclically monotone transport in $\mathbb{R}^n$ without the assumption of finite total cost (Corollary 10.44 in the case when $M = \mathbb{R}^n$). The tricky extension to more general cost functions on Riemannian manifolds was performed later by Figalli [363]. The current proof of Theorem 10.42 is so complicated that the reader might prefer to have a look at [814, Section 2.3.3], where the core of McCann's proof is explained in simpler terms in the particular case $c(x,y) = |x-y|^2$.

The case when the cost function is the distance ($c(x,y) = d(x,y)$) is not covered by Theorem 10.28, nor by any of the theorems appearing in the present chapter. This case is quite more tricky, be it in Euclidean space or on a manifold. The interested reader can consult [814, Section 2.4.6] for a brief review, as well as the research papers [20, 31, 32, 104, 190, 279, 280, 281, 354, 364, 380, 686, 765, 791]. The treatment by Bernard and Buffoni [104] is rather appealing, for its simplicity and links to dynamical system tools. An extreme case (maybe purely academic) is when the cost is the Cameron–Martin distance on the Wiener space; then usual strategies seem to fail, in the first place because of (non)measurability issues [23].

The optimal transport problem with a distance cost function is also related to the *irrigation problem* studied recently by various authors [109, 110, 111, 112, 152], the Bouchitté–Buttazzo variational problem [147, 148], and other problems as well. In this connection, see also Pratelli [689].

The partial optimal transport problem, where only a fixed fraction of the mass is transferred, was studied in [192, 365]. Under adequate assumptions on the cost function, one has the following results: whenever the transferred mass is at least equal to the shared mass between the measures μ and ν, then (a) there is uniqueness of the partial transport map; (b) all the shared mass is at the same time both source and target; (c) the "active" region depends monotonically on the mass transferred, and is the union of the intersection of the supports and a semiconvex set.

To conclude, here are some remarks about the technical ingredients used in this chapter.

Rademacher [697] proved his theorem of almost everywhere differentiability in 1918, for Lipschitz functions of two variables; this was later generalized to an arbitrary number of variables. The simple argument presented in this section seems to be due to Christensen [233]; it can also be found, up to minor variants, in modern textbooks about real analysis such as the one by Evans and Gariepy [331, pp. 81–84]. Ambrosio showed me another simple argument which uses Lebesgue's density theorem and the identification of a Lipschitz function with a function whose distributional derivative is essentially bounded.

The book by Cannarsa and Sinestrari [199] is an excellent reference for semiconvexity and subdifferentiability in $\mathbb{R}^n$, as well as the links with the theory of Hamilton–Jacobi equations. It is centered on semiconcavity rather than semiconvexity (and superdifferentiability rather than subdifferentiability), but this is just a question of convention. Many regularity results in this chapter have been adapted from that source (see in particular Theorem 2.1.7 and Corollary 4.1.13 there). Also the proof of Theorem 10.48(i) is adapted from [199, Theorem 4.1.6 and Corollary 4.1.9]. The core results in this circle of ideas and tools can be traced back to a pioneering paper by Alberti, Ambrosio and Cannarsa [12]. Following Ambrosio's advice, I used the same methods to establish Theorem 10.48(ii) in the present notes.

One often says that $S \subset \mathbb{R}^n$ is d-rectifiable if it can be written as a countable union of C^1 manifolds (submanifolds of $\mathbb{R}^n$), apart from a set of zero $\mathcal{H}^d$ measure. This property seems stronger, but is actually equivalent to Definition 10.47 (see [753, Lemma 11.1]). Stronger notions are obtained by changing C^1 into C^r for some $r \geq 2$. For instance Alberti [9] shows that the $(n-1)$-rectifiability of the nondifferentiability set of a convex function is achieved with C^2 manifolds, which is optimal.

Apart from plain subdifferentiability and Clarke subdifferentiability, other notions of differentiability for nonsmooth functions are discussed in [199], such as Dini derivatives or reachable gradients.

The theory of approximate differentiability (in Euclidean space) is developed in Federer [352, Section 3.1.8]; see also Ambrosio, Gigli and Savaré [30, Section 5.5]. A central result is the fact that any approximately differentiable function coincides, up to a set of arbitrarily small measure, with a Lipschitz function.

The proof of Besicovich's density theorem [331, p. 43] is based on Besicovich's covering lemma. This theorem is an alternative to the more classical Lebesgue density theorem (based on Vitali's covering lemma), which requires the doubling property. The price to pay for Besicovich's theorem is that it only works in $\mathbb{R}^n$ (or a Riemannian manifold, by localization) rather than on a general metric space.

The nonsmooth implicit function theorem in the second Appendix (Theorem 10.50) seems to be folklore in nonsmooth real analysis; the core of its proof was explained to me by Fathi. Corollary 10.52 was discovered or rediscovered by McCann [613, Appendix], in the case where ψ and $\widetilde{\psi}$ are convex functions in $\mathbb{R}^n$.

Everything in the Third Appendix, in particular the key differential inequality (10.46), was explained to me by Gallot. The lower bound assumption on the sectional curvatures $\sigma_x \geq -C/d(x_0, x)^2$ is sufficient to get upper bounds on $\nabla_x^2 d(x,y)^2$ as y stays in a compact set, but it is *not sufficient* to get upper bounds that are uniform in both x and y. A counterexample is developed in [393, pp. 213–214].

The exact computation about the hyperbolic space in Example 10.53 is the extremal situation for a comparison theorem about the Hessian of the squared distance [246, Lemma 3.12]: If M is a Riemannian manifold with sectional curvature bounded below by $\kappa < 0$, then

$$\nabla_x^2 \left(\frac{d(x,y)^2}{2} \right) \leq \frac{\sqrt{|\kappa|}\, d(x,y)}{\tanh(\sqrt{|\kappa|}\, d(x,y))}.$$

As pointed out to me by Ghys, the problem of finding a sufficient condition for the Hessian of $d(x,y)^2$ to be bounded above is related to the problem of whether large spheres $S_r(y)$ centered at y look flat at infinity, in the sense that their second fundamental form is bounded like $O(1/r)$.

11

The Jacobian equation

Transport is but a change of variables, and in many problems involving changes of variables, it is useful to write the Jacobian equation

$$f(x) = g(T(x))\,\mathcal{J}_T(x),$$

where f and g are the respective densities of the probability measures μ and ν with respect to the volume measure (in $\mathbb{R}^n$, the Lebesgue measure), and $\mathcal{J}_T(x)$ is the absolute value of the Jacobian determinant associated with T:

$$\mathcal{J}_T(x) = |\det(\nabla T(x))| = \lim_{r\to 0} \frac{\mathrm{vol}\,[T(B_r(x))]}{\mathrm{vol}\,[B_r(x)]}.$$

There are two important things that one should check before writing the Jacobian equation: First, T should be *injective* on its domain of definition; secondly, it should possess some minimal *regularity*.

So how smooth should T be for the Jacobian equation to hold true? We learn in elementary school that it is sufficient for T to be continuously differentiable, and a bit later that it is actually enough to have T Lipschitz continuous. But that degree of regularity is not always available in optimal transport! As we shall see in Chapter 12, the transport map T might fail to be even continuous.

There are (at least) three ways out of this situation:

(i) Only use the Jacobian equation in situations where the optimal map is smooth. Such situations are rare; this will be discussed in Chapter 12.

(ii) Only use the Jacobian equation for the optimal map between μ_{t_0} and μ_t, where $(\mu_t)_{0\le t\le 1}$ is a compactly supported displacement

C. Villani, *Optimal Transport*, Grundlehren der mathematischen Wissenschaften 338,

interpolation, and t_0 is fixed in $(0,1)$. Then, according to Theorem 8.5, the transport map is essentially Lipschitz. This is the strategy that I shall use in most of this course.

(iii) Apply a more sophisticated theorem of change of variables, covering for instance changes of variables with bounded variation (possibly discontinuous). It is in fact sufficient that the map T be differentiable almost everywhere, or even just approximately differentiable almost everywhere, in the sense of Definition 10.2. Such a theorem is stated below without proof; I shall use it in Chapter 23. The volume measure on M will be denoted by vol.

Theorem 11.1 (Jacobian equation). *Let M be a Riemannian manifold, let $f \in L^1(M)$ be a nonnegative integrable function on M, and let $T : M \to M$ be a Borel map. Define $\mu(dx) = f(x)\,\mathrm{vol}(dx)$ and $\nu := T_\#\mu$. Assume that:*

(i) There exists a measurable set $\Sigma \subset M$, such that $f = 0$ almost everywhere outside of Σ, and T is injective on Σ;

(ii) T is approximately differentiable almost everywhere on Σ.

Let $\widetilde{\nabla} T$ be the approximate gradient of T, and let $\mathcal{J}_T$ be defined almost everywhere on Σ by the equation $\mathcal{J}_T(x) := |\det(\widetilde{\nabla} T(x))|$. Then ν is absolutely continuous with respect to the volume measure if and only if $\mathcal{J}_T > 0$ almost everywhere. In that case ν is concentrated on $T(\Sigma)$, and its density is determined by the equation

$$f(x) = g(T(x))\,\mathcal{J}_T(x). \tag{11.1}$$

In an informal writing:

$$\frac{d(T^{-1})_\#(g\,\mathrm{vol})}{d\mathrm{vol}} = \mathcal{J}_T\,(g \circ T)\,\mathrm{vol}.$$

Theorem 11.1 establishes the Jacobian equation as soon as, say, the optimal transport has locally bounded variation. Indeed, in this case the map T is almost everywhere differentiable, and its gradient coincides with the absolutely continuous part of the distributional gradient $\nabla_{\mathcal{D}'} T$. The property of bounded variation is obviously satisfied for the quadratic cost in Euclidean space, since the second derivative of a convex function is a nonnegative measure.

Example 11.2. Consider two probability measures μ_0 and μ_1 on $\mathbb{R}^n$, with finite second moments; assume that μ_0 and μ_1 are absolutely continuous with respect to the Lebesgue measure, with respective densities

f_0 and f_1. Under these assumptions there exists a unique optimal transport map between μ_0 and μ_1, and it takes the form $T(x) = \nabla\Psi(x)$ for some lower semicontinuous convex function Ψ. There is a unique displacement interpolation $(\mu_t)_{0\le t\le 1}$, and it is defined by

$$\mu_t = (T_t)_\#\mu_0, \qquad T_t(x) = (1-t)\,x + t\,T(x) = (1-t)\,x + t\,\nabla\Psi(x).$$

By Theorem 8.7, each μ_t is absolutely continuous, so let f_t be its density. The map ∇T is of locally bounded variation, and it is differentiable almost everywhere, with Jacobian matrix $\nabla T = \nabla^2\Psi$, where $\nabla^2\Psi$ is the Alexandrov Hessian of Ψ (see Theorem 14.25 later in this course). Then, it follows from Theorem 11.1 that, $\mu_0(dx)$-almost surely,

$$f_0(x) = f_1(\nabla\Psi(x))\,\det(\nabla^2\Psi(x)).$$

Also, for any $t \in [0,1]$,

$$\begin{aligned} f_0(x) &= f_t(T_t(x))\,\det(\nabla T_t(x)) \\ &= f_t\big((1-t)x + t\nabla\Psi(x)\big)\,\det\big((1-t)I_n + t\nabla^2\Psi(x)\big). \end{aligned}$$

If $T_{t_0\to t} = T_t \circ T_{t_0}^{-1}$ stands for the transport map between μ_{t_0} and μ_t, then the equation

$$f_{t_0}(x) = f_t(T_{t_0\to t}(x))\,\det(\nabla T_{t_0\to t}(x))$$

also holds true for $t_0 \in (0,1)$; but now this is just the theorem of change of variables for Lipschitz maps.

In the sequel of this course, with the noticeable expression of Chapter 23, it will be sufficient to use the following theorem of change of variables.

Theorem 11.3 (Change of variables). *Let M be a Riemannian manifold, and $c(x,y)$ a cost function deriving from a C^2 Lagrangian $L(x,v,t)$ on $TM \times [0,1]$, where L satisfies the classical conditions of Definition 7.6, together with $\nabla_v^2 L > 0$. Let $(\mu_t)_{0\le t\le 1}$ be a displacement interpolation, such that each μ_t is absolutely continuous and has density f_t. Let $t_0 \in (0,1)$, and $t \in [0,1]$; further, let $T_{t_0\to t}$ be the (μ_{t_0}-almost surely) unique optimal transport from μ_{t_0} to μ_t, and let $\mathcal{J}_{t_0\to t}$ be the associated Jacobian determinant. Let F be a nonnegative measurable function on $M \times \mathbb{R}_+$ such that*

$$[f_t(y)=0] \implies F(y,f_t(y))=0.$$

Then,

$$\int_M F(y,f_t(y))\,\mathrm{vol}(dy) = \int_M F\left(T_{t_0\to t}(x),\,\frac{f_{t_0}(x)}{\mathcal{J}_{t_0\to t}(x)}\right)\mathcal{J}_{t_0\to t}(x)\,\mathrm{vol}(dx).$$

Furthermore, $\mu_{t_0}(dx)$*-almost surely,* $\mathcal{J}_{t_0\to t}(x)>0$ *for all* $t\in[0,1]$.

Proof of Theorem 11.3. For brevity I shall abbreviate $\mathrm{vol}(dx)$ into just dx. Let us first consider the case when $(\mu_t)_{0\le t\le 1}$ is compactly supported. Let Π be a probability measure on the set of minimizing curves, such that $\mu_t=(e_t)_\#\Pi$. Let $K_t=e_t(\mathrm{Spt}\,\Pi)$ and $K_{t_0}=e_{t_0}(\mathrm{Spt}\,\Pi)$. By Theorem 8.5, the map $\gamma_{t_0}\to\gamma_t$ is well-defined and Lipschitz for all $\gamma\in\mathrm{Spt}\,\Pi$. So $T_{t_0\to t}(\gamma_{t_0})=\gamma_t$ is a Lipschitz map $K_{t_0}\to K_t$. By assumption μ_t is absolutely continuous, so Theorem 10.28 (applied with the cost function $c^{t_0,t}(x,y)$, or maybe $c^{t,t_0}(x,y)$ if $t<t_0$) guarantees that the coupling (γ_t,γ_{t_0}) is deterministic, which amounts to saying that $\gamma_{t_0}\to\gamma_t$ is injective apart from a set of zero probability.

Then we can use the change of variables formula with $g=1_{K_t}$, $T=T_{t_0\to t}$, and we find $f(x)=\mathcal{J}_{t_0\to t}(x)$. Therefore, for any nonnegative measurable function G on M,

$$\begin{aligned}\int_{K_t} G(y)\,dy &= \int_{K_t} G(y)\,d((T_{t_0\to t})_\#\mu)(y)\\ &= \int_{K_{t_0}} (G\circ T_{t_0\to t}(x))\,f(x)\,dx\\ &= \int_{K_{t_0}} G(T_{t_0\to t}(x))\,\mathcal{J}_{t_0\to t}(x)\,dx.\end{aligned}$$

We can apply this to $G(y)=F(y,f_t(y))$ and replace $f_t(T_{t_0\to t}(x))$ by $f_{t_0}(x)/\mathcal{J}_{t_0\to t}(x)$; this is allowed since in the right-hand side the contribution of those x with $f_t(T_{t_0\to t}(x))=0$ is negligible, and $\mathcal{J}_{t_0\to t}(x)=0$ implies (almost surely) $f_{t_0}(x)=0$. So in the end

$$\int_{K_t} F(y,f_t(y))\,dy = \int_{K_{t_0}} F\left(T_{t_0\to t}(x),\,\frac{f_{t_0}(x)}{\mathcal{J}_{t_0\to t}(x)}\right)\mathcal{J}_{t_0\to t}(x)\,dx.$$

Since $f_t(y)=0$ almost surely outside of K_t and $f_{t_0}(x)=0$ almost surely outside of K_{t_0}, these two integrals can be extended to the whole of M.

Now it remains to generalize this to the case when Π is not compactly supported. (Skip this bit at first reading.) Let $(K_\ell)_{\ell\in\mathbb{N}}$ be a nondecreasing sequence of compact sets, such that $\Pi[\cup K_\ell] = 1$. For ℓ large enough, $\Pi[K_\ell] > 0$, so we can consider the restriction Π_ℓ of Π to K_ℓ. Then let $K_{t,\ell}$ and $K_{t_0,\ell}$ be the images of K_ℓ by e_t and e_{t_0}, and of course $\mu_{t,\ell} = (e_t)_\#\Pi_\ell$, $\mu_{t_0,\ell} = (e_{t_0})_\#\Pi_\ell$. Since μ_t and μ_{t_0} are absolutely continuous, so are $\mu_{t,\ell}$ and $\mu_{t_0,\ell}$; let $f_{t,\ell}$ and $f_{t_0,\ell}$ be their respective densities. The optimal map $T_{t_0\to t,\ell}$ for the transport problem between $\mu_{t_0,\ell}$ and $\mu_{t,\ell}$ is obtained as before by the map $\gamma_{t_0} \to \gamma_t$, so this is actually the restriction of $T_{t_0\to t}$ to $K_{t_0,\ell}$. Thus we have the Jacobian equation

$$f_{t_0,\ell}(x) = f_{t,\ell}(T_{t_0\to t}(x))\,\mathcal{J}_{t_0\to t}(x), \tag{11.2}$$

where the Jacobian determinant does not depend on ℓ. This equation holds true almost surely for $x \in K_{\ell'}$, as soon as $\ell' \le \ell$, so we may pass to the limit as $\ell \to \infty$ to get

$$f_{t_0}(x) = f_t(T_{t_0\to t}(x))\,\mathcal{J}_{t_0\to t}(x). \tag{11.3}$$

Since this is true almost surely on $K_{\ell'}$, for each ℓ', it is also true almost surely.

Next, for any nonnegative measurable function G, by monotone convergence and the first part of the proof one has

$$\begin{aligned}
\int_{UK_{t,\ell}} G(y)\,dy &= \lim_{\ell\to\infty}\int_{K_{t,\ell}} G(y)\,dy \\
&= \lim_{\ell\to\infty}\int_{K_{t_0,\ell}} G(T_{t_0\to t}(x))\,\mathcal{J}_{t_0\to t}(x)\,dx \\
&= \int_{UK_{\ell,t_0}} G(T_{t_0\to t}(x))\,\mathcal{J}_{t_0\to t}(x)\,dx.
\end{aligned}$$

The conclusion follows as before by choosing $G(y) = F(y, f_t(x))$ and using the Jacobian equation (11.3), then extending the integrals to the whole of M.

It remains to prove the assertion about $\mathcal{J}_{t_0\to t}(x)$ being positive *for all values of* $t \in [0,1]$, and not just for $t = 1$, or for almost all values of t. The transport map $T_{t_0\to t}$ can be written $\gamma(t_0) \to \gamma(t)$, where γ is a minimizing curve determined uniquely by $\gamma(t_0)$. Since γ is minimizing, we know (recall Problem 8.8) that the map $(\gamma_0, \dot\gamma_0) \to (\gamma_0, \gamma_{t_0})$ is locally invertible. So $T_{t_0\to t}$ can be written as the composition of the maps $F_1 : \gamma(t_0) \to (\gamma(0), \gamma(t_0))$, $F_2 : (\gamma(0), \gamma(t_0)) \to (\gamma(0), \dot\gamma(0))$ and $F_3 : (\gamma(0), \dot\gamma(0)) \to \gamma(t)$. Both F_2 and F_3 have a positive Jacobian determinant, at least if $t < 1$; so if x is chosen in such a way that F_1 has a positive Jacobian determinant at x, then also $T_{t_0 \to t} = F_3 \circ F_2 \circ F_1$ will have a positive Jacobian determinant at x for $t \in [0, 1)$. □

Bibliographical notes

Theorem 11.1 can be obtained (in $\mathbb{R}^n$) by combining Lemma 5.5.3 in [30] with Theorem 3.83 in [26].

In the context of optimal transport, the change of variables formula (11.1) was proven by McCann [614]. His argument is based on Lebesgue's density theory, and takes advantage of Alexandrov's theorem, alluded to in this chapter and proven later as Theorem 14.25: *A convex function admits a Taylor expansion at order 2 at almost each x in its domain of definition.* Since the gradient of a convex function has locally bounded variation, Alexandrov's theorem can be seen essentially as a particular case of the theorem of approximate differentiability of functions with bounded variation. McCann's argument is reproduced in [814, Theorem 4.8].

Along with Cordero-Erausquin and Schmuckenschläger, McCann later generalized his result to the case of Riemannian manifolds [246]. Modulo certain complications, the proof basically follows the same pattern as in $\mathbb{R}^n$. Then Cordero-Erausquin [243] treated the case of strictly convex cost functions in $\mathbb{R}^n$ in a similar way.

Ambrosio pointed out that those results could be retrieved within the general framework of push-forward by approximately differentiable mappings. This point of view has the disadvantage of involving more subtle arguments, but the advantage of showing that it is not a special feature of optimal transport. It also applies to nonsmooth cost functions such as $|x - y|^p$. In fact it covers general strictly convex costs of the form $c(x-y)$ as soon as c has superlinear growth, is C^1 everywhere and C^2 out of the origin. A more precise discussion of these subtle issues can be found in [30, Section 6.2.1].

It is a general feature of optimal transport with strictly convex cost in $\mathbb{R}^n$ that if T stands for the optimal transport map, then the Jacobian matrix ∇T, even if not necessarily nonnegative symmetric, is diagonalizable with nonnegative eigenvalues; see Cordero-Erausquin [243] and

Ambrosio, Gigli and Savaré [30, Section 6.2]. From an Eulerian perspective, that diagonalizability property was already noticed by Otto [666, Proposition A.4]. I don't know if there is an analog on Riemannian manifolds.

Changes of variables of the form $y = \exp_x(\nabla\psi(x))$ (where ψ is not necessarily $d^2/2$-convex) have been used in a remarkable paper by Cabré [181] to investigate qualitative properties of nondivergent elliptic equations (Liouville theorem, Alexandrov–Bakelman–Pucci estimates, Krylov–Safonov–Harnack inequality) on Riemannian manifolds with nonnegative sectional curvature. (See for instance [189, 416, 786] for classical proofs in $\mathbb{R}^n$.) It is mentioned in [181] that the methods extend to sectional curvature bounded below. For the Harnack inequality, Cabré's method was extended to nonnegative *Ricci* curvature by S. Kim [516].

12

Smoothness

The smoothness of the optimal transport map may give information about its qualitative behavior, as well as simplify computations. So it is natural to investigate the regularity of this map.

What characterizes the optimal transport map T is the existence of a c-convex ψ such that (10.20) (or (10.23)) holds true; so it is natural to search for a closed equation on ψ.

To guess the equation, let us work formally without being too demanding about regularity issues. We shall assume that x and y vary in $\mathbb{R}^n$, or in nice subsets of smooth n-dimensional Riemannian manifolds. Let $\mu(dx) = f(x)\,\mathrm{vol}(dx)$ and $\nu(dy) = g(y)\,\mathrm{vol}(dy)$ be two absolutely continuous probability measures, let $c(x,y)$ be a smooth cost function, and let T be a Monge transport. The differentiation of (10.20) with respect to x (once again) leads to

$$\nabla^2\psi(x) + \nabla^2_{xx}c(x,T(x)) + \nabla^2_{xy}c(x,T(x))\cdot\nabla T(x) = 0,$$

which can be rewritten

$$\nabla^2\psi(x) + \nabla^2_{xx}c(x,T(x)) = -\nabla^2_{xy}c(x,T(x))\cdot\nabla T(x). \tag{12.1}$$

The expression on the left-hand side is the Hessian of the function $c(x',T(x)) + \psi(x')$, considered as a function of x' and then evaluated at x. Since this function has a minimum at $x' = x$, its Hessian is nonnegative, so the left-hand side of (12.1) is a nonnegative symmetric operator; in particular its determinant is nonnegative. Take absolute values of determinants on both sides of (12.1):

$$\det\Big(\nabla^2\psi(x) + \nabla^2_{xx}c(x,T(x))\Big) = \Big|\det\nabla^2_{xy}c(x,T(x))\Big|\,\big|\det(\nabla T(x))\big|.$$

C. Villani, *Optimal Transport*, Grundlehren der mathematischen Wissenschaften 338,

Then the Jacobian determinant in the right-hand side can be replaced by $f(x)/g(T(x))$, and we arrive at the basic **partial differential equation of optimal transport:**

$$\det\Big(\nabla^2\psi(x) + \nabla^2_{xx}c(x,T(x))\Big) = \big|\det \nabla^2_{xy}c(x,T(x))\big| \,\frac{f(x)}{g(T(x))}. \qquad (12.2)$$

This becomes a closed equation on ψ in terms of f and g, if one recalls from (10.20) that

$$T(x) = (\nabla_x c)^{-1}\big(x, -\nabla\psi(x)\big), \qquad (12.3)$$

where the inverse is with respect to the y variable.

Remark 12.1. In a genuinely Riemannian context, equation (12.2) at first sight does not need to make sense: change the metric close to $y = T(x)$ and not close to x, then the left-hand side is invariant but the right-hand side seems to change! This is a subtle illusion: indeed, if the metric is changed close to y, then the volume measure also, and the density g has to be modified too. All in all, the right-hand side in (12.2) is invariant.

Now what can be said of (12.2)? Unfortunately not much simplification can be expected, except in special cases. The most important of them is the quadratic cost function, or equivalently $c(x,y) = -x \cdot y$ in $\mathbb{R}^n$. Then (12.2)–(12.3) reduces to

$$\det \nabla^2\psi(x) = \frac{f(x)}{g(\nabla\psi(x))}. \qquad (12.4)$$

This is an instance of the **Monge–Ampère equation**, well-known in the theory of partial differential equations. By extension, the system (12.2)–(12.3) is also called a generalized Monge–Ampère equation.

At this point we may hope that the theory of partial differential equations will help our task quite a bit by providing regularity results for the optimal map in the Monge–Kantorovich problem, at least if we rule out cases where the map is trivially discontinuous (for instance if the support of the initial measure μ is connected, while the support of the final measure ν is not).

However, things are not so simple. As the next few sections will demonstrate, regularity can hold only under certain stringent assumptions on the geometry and the cost function. The identification of these conditions will introduce us to a beautiful chapter of the theory of fully nonlinear partial differential equations; but one sad conclusion will be that **optimal transport is in general not smooth** — even worse, smoothness requires nonlocal conditions which are probably impossible to check effectively, say on a generic Riemannian manifolds. So if we still want to use optimal transport in rather general situations, we'd better find ways to do without smoothness. It is actually a striking feature of optimal transport that this theory can be pushed very far with so little regularity available.

In this chapter I shall first study various counterexamples to identify obstructions to the regularity of the optimal transport, and then discuss some positive results. The following elementary lemma will be useful:

Lemma 12.2. *Let* $(\mathcal{X},\mu)$ *and* $(\mathcal{Y},\nu)$ *be any two Polish probability spaces, let* T *be a continuous map* $\mathcal{X} \to \mathcal{Y}$*, and let* $\pi = (\mathrm{Id}\,,T)_{\#}\mu$ *be the associated transport map. Then, for each* $x \in \operatorname{Spt}\mu$*, the pair* $(x,T(x))$ *belongs to the support of* π*.*

Proof of Lemma 12.2. Let x and $\varepsilon > 0$ be given. By continuity of T, there is $\delta > 0$ such that $T(B_\delta(x)) \subset B_\varepsilon(T(x))$. Without loss of generality, $\delta \le \varepsilon$. Then

$$\begin{aligned}\pi\big[B_\varepsilon(x) \times B_\varepsilon(T(x))\big] &= \mu\big[\{z \in \mathcal{X};\ z \in B_\varepsilon(x) \text{ and } T(z) \in B_\varepsilon(T(x))\}\big]\\ &\ge \mu[B_\varepsilon(x) \cap B_\delta(x)] = \mu[B_\delta(x)] > 0.\end{aligned}$$

Since ε is arbitrarily small, this shows that π attributes positive measure to any neighborhood of $(x,T(x))$, which proves the claim. □

Caffarelli's counterexample

Caffarelli understood that regularity results for (12.2) in $\mathbb{R}^n$ cannot be obtained unless one adds an assumption of *convexity of the support of* ν. Without such an assumption, the optimal transport may very well be discontinuous, as the next counterexample shows.

Theorem 12.3 (An example of discontinuous optimal transport). *There are smooth compactly supported probability densities* f *and* g *on* $\mathbb{R}^n$*, such that the supports of* f *and* g *are smooth and connected,* f *and* g *are (strictly) positive in the interior of their respective supports, and yet the optimal transport between* $\mu(dx) = f(x)\,dx$ *and* $\nu(dy) = g(y)\,dy$*, for the cost* $c(x,y) = |x-y|^2$*, is discontinuous.*

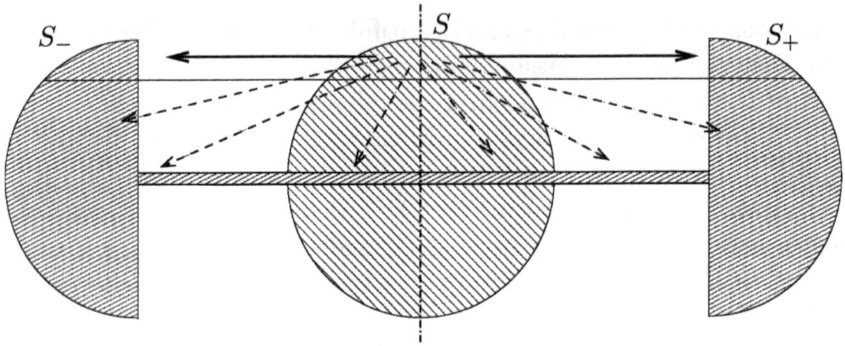

Fig. 12.1. Principle behind Caffarelli's counterexample. The optimal transport from the ball to the "dumb-bells" has to be discontinuous, and in effect splits the upper region S into the upper left and upper right regions S_- and S_+. Otherwise, there should be some transport along the dashed lines, but for some lines this would contradict monotonicity.

Proof of Theorem 12.3. Let f be the indicator function of the unit ball B in $\mathbb{R}^2$ (normalized to be a probability measure), and let $g = g_\varepsilon$ be the (normalized) indicator function of a set C_ε obtained by first separating the ball into two halves B_1 and B_2 (say with distance 2), then building a thin bridge between those two halves, of width $O(\varepsilon)$. (See Figure 12.1.) Let also g be the normalized indicator function of $B_1 \cup B_2$: this is the limit of g_ε as $\varepsilon \downarrow 0$. It is not difficult to see that g_ε (identified with a probability measure) can be obtained from f by a continuous deterministic transport (after all, one can deform B continuously into C_ε; just think that you are playing with clay, then it is possible to massage the ball into C_ε, without tearing off). However, we shall see here that for ε small enough, the optimal transport *cannot be continuous.*

The proof will rest on the stability of optimal transport: If T is the unique optimal transport between μ and ν, and T_ε is an optimal transport betwen μ and ν_ε, then T_ε converges to T in μ-probability as $\varepsilon \downarrow 0$ (Corollary 5.23).

In the present case, choosing $\mu(dx) = f(x)\,dx$, $\nu(dy) = g(y)\,dy$, $c(x,y) = |x-y|^2$, it is easy to figure out that the unique optimal transport T is the one that sends (x,y) to $(x-1,y)$ if $x<0$, and to $(x+1,y)$ if $x>0$.

Let now S, S_- and S_+ be the upper regions in the ball, the left half-ball and the right half-ball, respectively, as in Figure 12.1. As a consequence of the convergence in probability, for ε small enough, a large fraction (say 0.99) of the mass in S has to go to S_- (if it lies on the left) or to S_+ (if it lies on the right). Since the continuous image of a connected set is itself connected, there have to be some points in $T_\varepsilon(S)$ that form a path going from S_- to S_+; and so there is some $x \in S$ such that $T_\varepsilon(x)$ is close to the left-end of the tube joining the half-balls, in particular $T_\varepsilon(x) - x$ has a large downward component. From the convergence in probability again, many of the neighbors $\widetilde{x}$ of x have to be transported to, say, S_-, with nearly horizontal displacements $T(\widetilde{x}) - \widetilde{x}$. If such an $\widetilde{x}$ is picked below x, we shall have $\langle x - \widetilde{x}, T(x) - T(\widetilde{x}) \rangle < 0$; or equivalently, $|x - T(x)|^2 + |\widetilde{x} - T(\widetilde{x})|^2 > |x - T(\widetilde{x})|^2 + |\widetilde{x} - T(x)|^2$. If T is continuous, in view of Lemma 12.2 this contradicts the c-cyclical monotonicity of the optimal coupling. The conclusion is that when ε is small enough, the optimal map T_ε is discontinuous.

The maps f and g in this example are extremely smooth (in fact constant!) in the interior of their support, but they are not smooth as maps defined on $\mathbb{R}^n$. To produce a similar construction with functions that are smooth on $\mathbb{R}^n$, one just needs to regularize f and g_ε a tiny bit, letting the regularization parameter vanish as $\varepsilon \to 0$. □

Loeper's counterexample

Loeper discovered that in a genuine Riemannian setting the smoothness of optimal transport can be prevented by some local geometric obstructions. The next counterexample will illustrate this phenomenon.

Theorem 12.4 (A further example of discontinuous optimal transport). *There is a smooth compact Riemannian surface S, and there are smooth positive probability densities f and g on S, such that the optimal transport between $\mu(dx) = f(x)\,\mathrm{vol}(dx)$ and $\nu(dy) = g(y)\,\mathrm{vol}(dy)$, with a cost function equal to the square of the geodesic distance on S, is discontinuous.*

Remark 12.5. The obstruction has nothing to do with the lack of smoothness of the squared distance. Counterexamples of the same type exist for very smooth cost functions.

Remark 12.6. As we shall see in Theorem 12.44, the surface S in Theorem 12.4 could be replaced by any compact Riemannian manifold admitting a negative sectional curvature at some point. In that sense there is no hope for general regularity results outside the world of nonnegative sectional curvature.

Proof of Theorem 12.4. Let S be a compact surface in $\mathbb{R}^3$ with the following properties: (a) S is invariant under the symmetries $x \to -x$, $y \to -y$; (b) S crosses the axis $(x, y) = (0, 0)$ at exactly two points, namely $O = (0, 0, 0)$ and O'; (c) S coincides in an open ball $B(O, r)$ with the "horse saddle" $(z = x^2 - y^2)$. (Think of S as a small piece of the horse saddle which has been completed into a closed surface.)

Let $x_0, y_0 > 0$ to be determined later. Then let $A_+ = (x_0, 0, x_0^2)$, $A_- = (-x_0, 0, x_0^2)$, and similarly $B_+ = (0, y_0, -y_0^2)$, $B_- = (0, -y_0, -y_0^2)$; see Figure 12.2. In the sequel the symbol $A_\pm$ will stand for "either A_+ or A_-", etc.

If x_0 and y_0 are small enough then the four points A_+, A_-, B_+, B_- belong to a neighborhood of O where S has (strictly) negative curvature, and the unique geodesic joining O to $A_\pm$ (resp. $B_\pm$) satisfies the

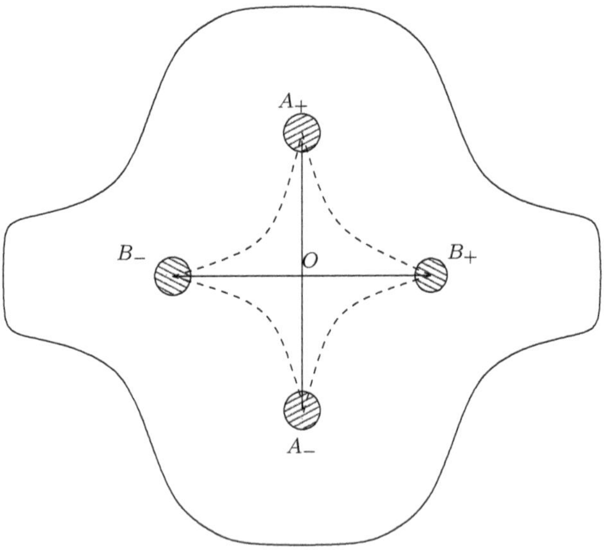

Fig. 12.2. Principle behind Loeper's counterexample. This is the surface S, immersed in $\mathbb{R}^3$, "viewed from above". By symmetry, O has to stay in place. Because most of the initial mass is close to A_+ and A_-, and most of the final mass is close to B_+ and B_-, at least some mass has to move from one of the A-balls to one of the B-balls. But then, because of the modified (negative curvature) Pythagoras inequality, it is more efficient to replace the transport scheme $(A \to B, O \to O)$, by $(A \to O, O \to B)$.

equation $(y = 0)$ (resp. $x = 0$); then the lines $(O, A_\pm)$ and $(O, B_\pm)$ are orthogonal at O. Since we are on a negatively curved surface, Pythagoras's identity in a triangle with a square angle is modified in favor of the diagonal, so

$$d(O, A_\pm)^2 + d(O, B_\pm)^2 < d(A_\pm, B_\pm)^2.$$

By continuity, there is $\varepsilon_0 > 0$ small enough that the balls $B(A_+, \varepsilon_0)$, $B(A_-, \varepsilon_0)$, $B(B_+, \varepsilon_0)$ and $B(B_-, \varepsilon_0)$ are all disjoint and satisfy

$$\begin{aligned}\big[x \in B(A_+, \varepsilon_0) \cup B(A_-, \varepsilon_0), \quad & y \in B(B_+, \varepsilon_0) \cup B(B_-, \varepsilon_0)\big] \\ \Longrightarrow \quad & d(O, x)^2 + d(O, y)^2 < d(x, y)^2. \end{aligned} \tag{12.5}$$

Next let f and g be smooth probability densities on M, even in x and y, such that

$$\int_{B(A_+,\varepsilon_0)\cup B(A_-,\varepsilon_0)} f(x)\, dx > \frac{1}{2}; \qquad \int_{B(B_+,\varepsilon_0)\cup B(B_-,\varepsilon_0)} g(y)\, dy > \frac{1}{2}. \tag{12.6}$$

Let $\mu(dx) = f(x)\, dx$, $\nu(dy) = g(y)\, dy$, let T be the unique optimal transport between the measures μ and ν (for the cost function $c(x, y) = d(x, y)^2$), and let $\widetilde{T}$ be the optimal transport between ν and μ. (T and $\widetilde{T}$ are inverses of each other, at least in a measure-theoretical sense.) I claim that either T or $\widetilde{T}$ is discontinuous.

Indeed, suppose to the contrary that both T and $\widetilde{T}$ are continuous. We shall first see that necessarily $T(O) = O$. Since the problem is symmetric with respect to $x \to -x$ and $y \to -y$, and since there is uniqueness of the optimal transport, T maps O into a point that is invariant under these two transforms, that is, either O or O'. Similarly, $T(O') \in \{O, O'\}$. So we have two cases to dismiss:

Case 1: $T(O) = O'$, $T(O') = O$. Then by Lemma 12.2 the two pairs (O, O') and (O', O) belong to the support of the optimal plan associated to T, which trivially contradicts the cyclical monotonicity since $d(O, O')^2 + d(O', O)^2 > d(O, O)^2 + d(O', O')^2 = 0$.

Case 2: $T(O) = O'$, $T(O') = O'$. Then both (O', O) and (O', O') belong to the support of the optimal plan. Let U and U' be two disjoint neighborhoods of O and O' respectively. By swapping variables x and y we see that (O', O) and (O', O') belong to the support of the optimal plan $(\mathrm{Id}\,, \widetilde{T})_\# \nu$; so for any $\varepsilon > 0$ the map $\widetilde{T}$ has to send a set of positive measure in $B_\varepsilon(O')$ into U, and also a set of positive measure in $B_\varepsilon(O')$ into U'. This contradicts the continuity of $\widetilde{T}$.

So we may assume that $T(O) = O$. By Lemma 12.2 again, (O, O) belongs to the support of the optimal plan π. Then (12.6) implies that there is some transfer of mass from either $B(A_+, \varepsilon_0) \cup B(A_-, \varepsilon_0)$ to $B(B_+, \varepsilon_0) \cup B(B_-, \varepsilon_0)$; in other words, we can find, in the support of the optimal transport, some (x, y) with $x \in B(A_+, \varepsilon_0) \cup B(A_-, \varepsilon_0)$ and $y \in B(B_+, \varepsilon_0) \cup B(B_-, \varepsilon_0)$. From the previous step we know that (O, O) also lies in that support; then by c-cyclical monotonicity,

$$d(x,y)^2 + d(O,O)^2 \leq d(x,O)^2 + d(y,O)^2;$$

but this contradicts (12.5). The proof is complete. □

Smoothness and Assumption (C)

Here as in Chapter 9, I shall say that a cost function c on $\mathcal{X} \times \mathcal{Y}$ satisfies Assumption (C) if for any c-convex function ψ and for any $x \in \mathcal{X}$, the c-subdifferential (or contact set) of ψ at x, $\partial_c\psi(x)$, is connected. Some interesting consequences of this assumption were discussed in Chapter 9, and it was shown that even the simple cost function $c(x, y) = |x - y|^p$ on $\mathbb{R}^n \times \mathbb{R}^n$, $p > 2$, does not satisfy it. Now we shall see that this assumption is more or less *necessary* for the regularity of optimal transport. For simplicity I shall work in a compact setting; it would be easy to extend the proof to more general situations by imposing suitable conditions at infinity.

Theorem 12.7 (Smoothness needs Assumption (C)). *Let $\mathcal{X}$ (resp. $\mathcal{Y}$) be the closure of a bounded open set in a smooth Riemannian manifold M (resp. N), equipped with its volume measure. Let $c : \mathcal{X} \times \mathcal{Y} \to \mathbb{R}$ be a continuous cost function. Assume that there are a c-convex function $\psi : \mathcal{X} \to \mathbb{R}$, and a point $\overline{x} \in \mathcal{X}$, such that $\partial_c\psi(\overline{x})$ is disconnected. Then there exist C^∞ positive probability densities f on $\mathcal{X}$ and g on $\mathcal{Y}$ such that any optimal transport map from $f\,\mathrm{vol}$ to $g\,\mathrm{vol}$ is discontinuous.*

Remark 12.8. Assumption (C) involves both the geometry of $\mathcal{Y}$ and the local properties of c; in some sense the obstruction described in this theorem underlies both Caffarelli's and Loeper's counterexamples.

Remark 12.9. The volume measure in Theorem 12.7 can be replaced by any finite measure giving positive mass to all open sets. (In that sense the Riemannian structure plays no real role.) Moreover, with simple modifications of the proof one can weaken the assumption "$\partial_c\psi(\overline{x})$ is disconnected" into "$\partial_c\psi(\overline{x})$ is not simply connected".

Proof of Theorem 12.7. Note first that the continuity of the cost function and the compactness of $\mathcal{X} \times \mathcal{Y}$ implies the continuity of ψ. In the proof, the notation d will be used interchangeably for the distances on $\mathcal{X}$ and $\mathcal{Y}$.

Let C_1 and C_2 be two disjoint connected components of $\partial_c\psi(\overline{x})$. Since $\partial_c\psi(\overline{x})$ is closed, C_1 and C_2 lie a positive distance apart. Let $r = d(C_1, C_2)/5$, and let $C = \{y \in \mathcal{Y};\ d(y, \partial_c\psi(\overline{x})) \geq 2\,r\}$. Further, let $\overline{y}_1 \in C_1$, $\overline{y}_2 \in C_2$, $B_1 = B_r(\overline{y}_1)$, $B_2 = B_r(\overline{y}_2)$. Obviously, C is compact, and any path going from B_1 to B_2 has to go through C.

Then let $K = \{z \in \mathcal{X};\ \exists y \in C;\ y \in \partial_c\psi(z)\}$. It is clear that K is compact: Indeed, if $z_k \in K$ converges to $z \in \mathcal{X}$, and $y_k \in \partial_c\psi(z_k)$, then without loss of generality y_k converges to some $y \in C$ and one can pass to the limit in the inequality $\psi(z_k) + c(z_k, y_k) \leq \psi(t) + c(t, y_k)$, where $t \in \mathcal{X}$ is arbitrary. Also K is not empty since $\mathcal{X}$ and $\mathcal{Y}$ are compact.

Next I claim that for any $y \in C$, for any x such that $y \in \partial_c\psi(x)$, and for any $i \in \{1, 2\}$,

$$c(\overline{x}, \overline{y}_i) + c(x, y) < c(\overline{x}, y) + c(x, \overline{y}_i). \tag{12.7}$$

Indeed, $y \notin \partial_c\psi(\overline{x})$, so

$$\psi(x) + c(x, y) = \inf_{\widetilde{x} \in \mathcal{X}} \big[\psi(\widetilde{x}) + c(\widetilde{x}, y)\big] < \psi(\overline{x}) + c(\overline{x}, y). \tag{12.8}$$

On the other hand, $y_i \in \partial_c\psi(\overline{x})$, so

$$\psi(\overline{x}) + c(\overline{x}, \overline{y}_i) \leq \psi(x) + c(x, \overline{y}_i). \tag{12.9}$$

The combination of (12.8) and (12.9) implies (12.7).

Next, one can reinforce (12.7) into

$$c(\overline{x}, \overline{y}_i) + c(x, y) \leq c(\overline{x}, y) + c(x, \overline{y}_i) - \varepsilon \tag{12.10}$$

for some $\varepsilon > 0$. This follows easily from a contradiction argument based on the compactness of C and K, and once again the continuity of $\partial_c\psi$.

Then let $\delta \in (0, r)$ be small enough that for any $(x, y) \in K \times C$, for any $i \in \{1, 2\}$, the inequalities

$$d(x,x') \le 2\delta,\ d(\overline{x},\overline{x}') \le 2\delta,\ d(\overline{y}_i,\overline{y}_i') \le 2\delta,\ d(y,y') \le 2\delta$$

imply

$$\begin{aligned} \left|c(\overline{x}',y) - c(\overline{x},y)\right| &\le \frac{\varepsilon}{10}, & \left|c(x',y) - c(x,y)\right| &\le \frac{\varepsilon}{10}, \\ \left|c(\overline{x}',\overline{y}_i') - c(\overline{x},\overline{y}_i)\right| &\le \frac{\varepsilon}{10}, & \left|c(x',\overline{y}_i') - c(x,\overline{y}_i)\right| &\le \frac{\varepsilon}{10}. \end{aligned} \tag{12.11}$$

Let $K^\delta = \{x \in \mathcal{X};\ d(x,K) \le \delta\}$. From the assumptions on $\mathcal{X}$ and $\mathcal{Y}$, $B_\delta(\overline{x})$ has positive volume, so we can fix a smooth positive probability density f on $\mathcal{X}$ such that the measure $\mu = f\,\mathrm{vol}$ satisfies

$$\mu[B_\delta(\overline{x})] \ge \frac{3}{4}; \qquad f \ge \varepsilon_0 > 0 \ \text{ on } K^\delta. \tag{12.12}$$

Also we can construct a sequence of smooth positive probability densities $(g_k)_{k\in\mathbb{N}}$ on $\mathcal{Y}$ such that the measures $\nu_k = g_k\,\mathrm{vol}$ satisfy

$$\nu_k \xrightarrow[k\to\infty]{} \frac{1}{2}\left(\delta_{\overline{y}_1} + \delta_{\overline{y}_2}\right) \quad \text{weakly.} \tag{12.13}$$

Let us assume the existence of a continuous optimal transport T_k sending μ_k to ν_k, for any k. We shall reach a contradiction, and this will prove the theorem.

From (12.13), $\nu_k[B_\delta(\overline{y}_1)] \ge 1/3$ for k large enough. Then by (12.12) the transport T_k has to send some mass from $B_\delta(\overline{x})$ to $B_\delta(\overline{y}_1)$, and similarly from $B_\delta(\overline{x})$ to $B_\delta(\overline{y}_2)$. Since $T_k(B_\delta(\overline{x}))$ is connected, it has to meet C. So there are $y_k \in C$ and $\overline{x}_k' \in B_\delta(\overline{x})$ such that $T_k(\overline{x}_k') = y_k$. Let $x_k \in K$ be such that $y_k \in \partial_c\psi(x_k)$. Without loss of generality we may assume that $x_k \to x_\infty \in K$ as $k \to \infty$. By the second part of (12.12), $m := \mu[B_\delta(x_\infty)] \ge \varepsilon_0\,\mathrm{vol}\,[B_\delta(x_\infty)] > 0$. When k is large enough, $\nu_k[B_\delta(\overline{y}_1) \cup B_\delta(\overline{y}_2)] > 1 - m$ (by (12.13) again), so T_k has to send some mass from $B_\delta(x_\infty)$ to either $B_\delta(\overline{y}_1)$ or $B_\delta(\overline{y}_2)$; say $B_\delta(\overline{y}_1)$. In other words, there is some $x_k' \in B_\delta(x_\infty)$ such that $T(x_k') \in B_\delta(\overline{y}_1)$.

Let us recapitulate: for k large enough,

$$\begin{aligned} &d(\overline{x},\overline{x}_k') \le \delta; \quad T_k(\overline{x}_k') = y_k \in C; \quad y_k \in \partial_c\psi(x_k); \\ &d(x_k,x_k') \le 2\delta; \quad d(T_k(x_k'),\overline{y}_1) \le \delta. \end{aligned} \tag{12.14}$$

By c-cyclical monotonicity of optimal transport and Lemma 12.2,

$$c(\overline{x}_k',T_k(\overline{x}_k')) + c(x_k',T_k(x_k')) \le c(\overline{x}_k',T_k(x_k')) + c(x_k',T_k(\overline{x}_k')).$$

From the inequalities in (12.14) and (12.11), we deduce

$$c(\overline{x}, y_k) + c(x_k, \overline{y}_1) \le c(\overline{x}, \overline{y}_1) + c(x_k, y_k) + \frac{4\,\varepsilon}{10}.$$

Since $y_k \in C \cap \partial_c\psi(x_k)$ and $x_k \in K$, this contradicts (12.10). The proof is complete. □

Regular cost functions

In the previous section we dealt with plainly continuous cost functions; now we shall come back to the differentiable setting used in Chapter 10 for the solution of the Monge problem. Throughout this section, $\mathcal{X}$ will be a closed subset of a Riemannian manifold M and $\mathrm{Dom}\,(\nabla_x c)$ will stand for the set of points $(x, y) \in \mathcal{X} \times \mathcal{Y}$ such that $\nabla_x c(x, y)$ is well-defined. (A priori x should belong to the interior of $\mathcal{X}$ in M.) It will be assumed that $\nabla_x c(x, \cdot\,)$ is one-to-one on its domain of definition (Assumption **(Twist)** in Chapter 10).

Definition 12.10 (c-segment). *A continuous curve $(y_t)_{t\in[0,1]}$ in $\mathcal{Y}$ is said to be a c-segment with base $\overline{x}$ if (a) $(\overline{x}, y_t) \in \mathrm{Dom}\,(\nabla_x c)$ for all t; (b) there are $p_0, p_1 \in T_{\overline{x}}M$ such that $\nabla_x c(\overline{x}, y_t) + p_t = 0$, where $p_t = (1-t)\,p_0 + t\,p_1$.*

In other words a c-segment is the image of a usual segment by a map $(\nabla_x c(\overline{x}, \cdot\,))^{-1}$. Since $(\overline{x}, y_t)$ in the definition always lies in $\mathrm{Dom}\,(\nabla_x c)$, a c-segment is uniquely determined by its base point $\overline{x}$ and its endpoints y_0, y_1; I shall denote it by $[y_0, y_1]_{\overline{x}}$.

Definition 12.11 (c-convexity). *A set $C \subset \mathcal{Y}$ is said to be c-convex with respect to $\overline{x} \in \mathcal{X}$ if for any two points $y_0, y_1 \in C$ there is a c-segment $(y_t)_{0\le t\le 1} = [y_0, y_1]_{\overline{x}}$ (necessarily unique) which is entirely contained in C.*

More generally, a set $C \subset \mathcal{Y}$ is said to be c-convex with respect to a subset $\widetilde{\mathcal{X}}$ of $\mathcal{X}$ if C is c-convex with respect to any $x \in \widetilde{\mathcal{X}}$.

A set $D \subset \mathrm{Dom}\,(\nabla_x c)$ is said to be totally c-convex if for any two points $(\overline{x}, y_0)$ and $(\overline{x}, y_1)$ in D, there is a c-segment $(y_t)_{0\le t\le 1}$ with base $\overline{x}$, such that $(\overline{x}, y_t) \in D$ for all t.

Similar notions of strict c-convexity are defined by imposing that y_t belong to the interior of C if $y_0 \neq y_1$ and $t \in (0,1)$.

Example 12.12. When $\mathcal{X} = \mathcal{Y} = \mathbb{R}^n$ and $c(x,y) = -x \cdot y$ (or $x \cdot y$), c-convexity is just plain convexity in $\mathbb{R}^n$. If $\mathcal{X} = \mathcal{Y} = S^{n-1}$ and $c(x,y) = d(x,y)^2/2$ then $\mathrm{Dom}\,(\nabla_x c(\overline{x},\,\cdot\,)) = S^{n-1} \setminus \{-\overline{x}\}$ (the cut locus is the antipodal point), and $\nabla_x c(\overline{x}, S^{n-1} \setminus \{-\overline{x}\}) = B(0,\pi) \subset T_{\overline{x}}M$ is a convex set. So for any point $\overline{x}$, S^{n-1} minus the cut locus of $\overline{x}$ is c-convex with respect to $\overline{x}$. An equivalent statement is that $S^{n-1} \times S^{n-1} \setminus \{(x,-x)\}$ (which is the whole of $\mathrm{Dom}\,(\nabla_x c)$) is totally $d^2/2$-convex. By abuse of language, one may say that S^{n-1} is $d^2/2$-convex with respect to itself. The same is true of a flat torus with arbitrary sidelengths.

Example 12.13. To construct a Riemannian manifold which is not $(d^2/2)$-convex with respect to itself, start with the Euclidean plane (embedded in $\mathbb{R}^3$), and from the origin draw three lines L_-, L_0 and L_+ directed respectively by the vectors $(1,-1)$, $(1,0)$ and $(1,1)$. Put a high enough mountain on the pathway of L_0, without affecting L_+ or L_-, so that L_0 is minimizing only for a finite time, while $L_\pm$ are still minimizing at all times. The resulting surface is not $d^2/2$-convex with respect to the origin.

Before stating the main definition of this section, I shall now introduce some more notation. If $\mathcal{X}$ is a closed subset of a Riemannian manifold M and $c : \mathcal{X} \to \mathcal{Y} \to \mathbb{R}$ is a continuous cost function, for any x in the interior of $\mathcal{X}$ I shall denote by $\mathrm{Dom}'(\nabla_x c(x,\,\cdot\,))$ the *interior* of $\mathrm{Dom}\,(\nabla_x c(x,\,\cdot\,))$. Moreover I shall write $\mathrm{Dom}'(\nabla_x c)$ for the union of all sets $\{x\} \times \mathrm{Dom}'(\nabla_x c(x,\,\cdot\,))$, where x varies in the interior of $\mathcal{X}$. For instance, if $\mathcal{X} = \mathcal{Y} = M$ is a complete Riemannian manifold and $c(x,y) = d(x,y)^2$ is the square of the geodesic distance, then $\mathrm{Dom}'(\nabla_x c)$ is obtained from $M \times M$ by removing the cut locus, while $\mathrm{Dom}\,(\nabla_x c)$ might be slightly bigger (these facts are recalled in the Appendix).

Definition 12.14 (regular cost function). *A cost $c : \mathcal{X} \times \mathcal{Y} \to \mathbb{R}$ is said to be regular if for any $\overline{x}$ in the interior of $\mathcal{X}$ and for any c-convex function $\psi : \mathcal{X} \to \mathbb{R}$, the set $\partial_c \psi(\overline{x}) \cap \mathrm{Dom}'(\nabla_x c(\overline{x},\,\cdot\,))$ is c-convex with respect to $\overline{x}$.*

The cost c is said to be strictly regular if moreover, for any nontrivial c-segment $(y_t)_{0 \le t \le 1} = [y_0, y_1]_{\overline{x}}$ in $\partial_c \psi(\overline{x})$ and for any $t \in (0,1)$, $\overline{x}$ is the only contact point of ψ^c at y_t, i.e. the only $x \in \mathcal{X}$ such that $y_t \in \partial_c \psi(x)$.

More generally, if D is a subset of $\mathrm{Dom}'(\nabla_x c)$, I shall say that c is regular in D if for any $\overline{x}$ in $\mathrm{proj}_{\mathcal{X}} D$ and any c-convex function $\psi : \mathcal{X} \to \mathbb{R}$, the set $\partial_c\psi(\overline{x}) \cap \{y;\ (\overline{x}, y) \in D\}$ is c-convex. Equivalently, the intersection of D and the graph of $\partial_c\psi$ should be totally c-convex. The notion of strict regularity in D is obtained by modifying this definition in an obvious way.

What does regularity mean? Let ψ be a c-convex function and let $-c(\,\cdot\,, y_0) + a_0$ and $-c(\,\cdot\,, y_1) + a_1$ touch the graph of ψ from below at $\overline{x}$, take any $y_t \in [y_0, y_1]_{\overline{x}}$, $0 < t < 1$, and increase a from $-\infty$ until the function $-c(\,\cdot\,, y_t) + a$ touches the graph of ψ: the regularity property means that $\overline{x}$ should be a contact point (and the only one if the cost is strictly regular).

Before going further I shall discuss several convenient reformulations of the regularity property, in terms of (i) elementary c-convex functions; (ii) subgradients; (iii) connectedness of c-subdifferentials. Assumptions **(Twist)** (twist condition), **(locSC)** (local semiconcavity) and **(H∞)** (adequate behavior at infinity) will be the same as in Chapter 10 (cf. p. 234).

Proposition 12.15 (Reformulation of regularity). *Let $\mathcal{X}$ be a closed subset of a Riemannian manifold M and let $\mathcal{Y}$ be a Polish space. Let $c : \mathcal{X} \times \mathcal{Y} \to \mathbb{R}$ be a continuous cost function satisfying* **(Twist)**. *Then:*

(i) c is regular if and only if for any $(\overline{x}, y_0), (\overline{x}, y_1) \in \mathrm{Dom}'(\nabla_x c)$, the c-segment $(y_t)_{0 \le t \le 1} = [y_0, y_1]_{\overline{x}}$ is well-defined, and for any $t \in [0, 1]$,

$$-c(x, y_t) + c(\overline{x}, y_t) \le \max\Big(-c(x, y_0) + c(\overline{x}, y_0),\ -c(x, y_1) + c(\overline{x}, y_1)\Big) \tag{12.15}$$

(with strict inequality if c is strictly regular, $y_0 \ne y_1$, $t \in (0, 1)$ and $\overline{x} \ne x$).

(ii) If c is a regular cost function satisfying **(locSC)** *and* **(H∞)**, *then for any c-convex $\psi : \mathcal{X} \to \mathbb{R}$ and any $\overline{x}$ in the interior of $\mathcal{X}$ such that $\partial_c\psi(\overline{x}) \subset \mathrm{Dom}'(\nabla_x c(\overline{x}, \cdot))$, one has*

$$\nabla_c^- \psi(\overline{x}) = \nabla^- \psi(\overline{x}),$$

where the c-subgradient $\nabla_c^- \psi(\overline{x})$ is defined by

$$\nabla_c^- \psi(\overline{x}) := -\nabla_x c\big(\overline{x}, \partial_c\psi(\overline{x})\big).$$

(iii) If c satisfies **(locSC)** *and* $\mathrm{Dom}'(\nabla_x c)$ *is totally c-convex, then c is regular if and only if for any c-convex* $\psi : \mathcal{X} \to \mathbb{R}$ *and any* $\overline{x}$ *in the interior of* $\mathcal{X}$,

$$\partial_c\psi(\overline{x}) \cap \mathrm{Dom}'\big(\nabla_x c(\overline{x}, \cdot)\big) \text{ is connected.} \tag{12.16}$$

Remark 12.16. Statement (i) in Proposition 12.15 means that it is sufficient to test Definition 12.14 on the particular functions

$$\psi_{\overline{x},y_0,y_1}(x) := \max\big(-c(x,y_0) + c(\overline{x},y_0),\ -c(x,y_1) + c(\overline{x},y_1)\big), \tag{12.17}$$

where $(\overline{x}, y_0)$ and $(\overline{x}, y_1)$ belong to $\mathrm{Dom}'(\nabla_x c)$. (The functions $\psi_{\overline{x},y_0,y_1}$ play in some sense the role of $x \to |x_1|$ in usual convexity theory.) See Figure 12.3 for an illustration of the resulting "recipe".

Example 12.17. Obviously $c(x,y) = -x \cdot y$, or equivalently $|x-y|^2$, is a regular cost function, but it is not strictly regular. The same is

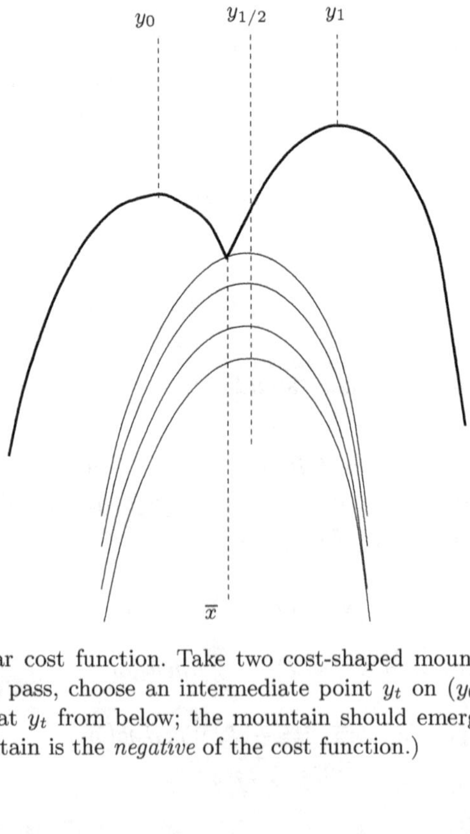

Fig. 12.3. Regular cost function. Take two cost-shaped mountains peaked at y_0 and y_1, let $\overline{x}$ be a pass, choose an intermediate point y_t on $(y_0, y_1)_{\overline{x}}$, and grow a mountain peaked at y_t from below; the mountain should emerge at $\overline{x}$. (Note: the shape of the mountain is the *negative* of the cost function.)

true of $c(x, y) = -|x-y|^2$, although the qualitative properties of optimal transport are quite different in this case. We shall consider other examples later.

Remark 12.18. It follows from the definition of c-subgradient and Theorem 10.24 that for any $\overline{x}$ in the interior of $\mathcal{X}$ and any c-convex $\psi : \mathcal{X} \to \mathbb{R}$,

$$\nabla_c^- \psi(\overline{x}) \subset \nabla^- \psi(\overline{x}).$$

So Proposition 12.15(ii) means that, modulo issues about the domain of differentiability of c, it is equivalent to require that c satisfies the regularity property, or that $\nabla_c^- \psi(\overline{x})$ fills the whole of the convex set $\nabla^- \psi(\overline{x})$.

Remark 12.19. Proposition 12.15(iii) shows that, again modulo issues about the differentiability of c, the regularity property is morally equivalent to Assumption (C) in Chapter 9. See Theorem 12.42 for more.

Proof of Proposition 12.15. Let us start with (i). The necessity of the condition is obvious since y_0, y_1 both belong to $\partial_c \psi_{\overline{x},y_0,y_1}(\overline{x})$. Conversely, if the condition is satisfied, let ψ be any c-convex function $\mathcal{X} \to \mathbb{R}$, let $\overline{x}$ belong to the interior of $\mathcal{X}$ and let $y_0, y_1 \in \partial_c \psi(\overline{x})$. By adding a suitable constant to ψ (which will not change the subdifferential), we may assume that $\psi(\overline{x}) = 0$. Since ψ is c-convex, $\psi \geq \psi_{\overline{x},y_0,y_1}$, so for any $t \in [0, 1]$ and $x \in \mathcal{X}$,

$$\begin{aligned} c(x, y_t) + \psi(x) &\geq c(x, y_t) + \psi_{\overline{x},y_0,y_1}(x) \\ &\geq c(\overline{x}, y_t) + \psi_{\overline{x},y_0,y_1}(\overline{x}) = c(\overline{x}, y_t) + \psi(\overline{x}), \end{aligned}$$

which shows that $y_t \in \partial_c \psi(\overline{x})$, as desired.

Now let c, ψ and $\overline{x}$ be as in Statement (ii). By Theorem 10.25, $-\nabla_x c(\overline{x}, \partial_c \psi(\overline{x}))$ is included in the convex set $\nabla^- \psi(\overline{x})$. Moreover, ψ is locally semiconvex (Theorem 10.26), so $\nabla^- \psi(\overline{x})$ is the convex hull of cluster points of $\nabla \psi(x)$, as $x \to \overline{x}$. (This comes from a localization argument and a similar result for convex functions, recall Remark 10.51.) It follows that any extremal point p of $\nabla^- \psi(\overline{x})$ is the limit of $\nabla \psi(x_k)$ for some sequence $x_k \to \overline{x}$. Then let $y_k \in \partial_c \psi(x_k)$ (these y_k exist by Theorem 10.24 and form a compact set). By Theorem 10.25, $\nabla_x c$ is well-defined at (x_k, y_k) and $-\nabla c(x_k, y_k) = \nabla \psi(x_k)$. Up to extraction of a subsequence, we have $y_k \to y \in \partial_c \psi(\overline{x})$, in particular $(\overline{x}, y)$ lies in the domain of $\nabla_x c$ and $-\nabla c(\overline{x}, y) = p$. The conclusion is that

$$\mathcal{E}(\nabla^-\psi(\overline{x})) \subset \nabla_c^-\psi(\overline{x}) \subset \nabla^-\psi(\overline{x}),$$

where $\mathcal{E}$ stands for "extremal points". In particular, $\nabla_c^-\psi(\overline{x})$ is convex if and only if it coincides with $\nabla^-\psi(\overline{x})$. Part (ii) of the Proposition follows easily.

It remains to prove (iii). Whenever $(\overline{x}, y_0)$ and $(\overline{x}, y_1)$ belong to $\mathrm{Dom}'(\nabla_x c)$, let $\psi_{\overline{x},y_0,y_1}$ be defined by (12.17). Both y_0 and y_1 belong to $\partial_c\psi_{\overline{x},y_0,y_1}(\overline{x})$. If c is regular, then y_0 and y_1 can be connected by a c-segment with base $\overline{x}$, which lies inside $\mathrm{Dom}\,(\nabla_x c) \cap \partial_c\psi(\overline{x})$. This proves (12.16).

Conversely, let us assume that (12.16) holds true, and prove that c is regular. If $\overline{x}, y_0, y_1$ are as above, the c-segment $[y_0, y_1]_{\overline{x}}$ is well-defined by (a), so by part (i) of the Proposition we just have to prove the c-convexity of $\partial_c\psi_{\overline{x},y_0,y_1}$.

Then let $h_0(x) = -c(x, y_0) + c(\overline{x}, y_0)$, $h_1(x) = -c(x, y_1) + c(\overline{x}, y_1)$. By the twist condition, $\nabla h_0(\overline{x}) \neq \nabla h_1(\overline{x})$. Since h_0 and h_1 are semiconvex, we can apply the nonsmooth implicit function theorem (Corollary 10.52) to conclude that the equation $(h_0 = h_1)$ defines an $(n-1)$-dimensional Lipschitz graph G in the neighborhood of $\overline{x}$. This graph is a level set of $\psi = \psi_{\overline{x},y_0,y_1}$, so $\nabla^-\psi(\overline{x})$ is included in the orthogonal of G, which is the line directed by $\nabla h_0(\overline{x}) - \nabla h_1(\overline{x})$. In particular, $\nabla^-\psi(\overline{x})$ is a one-dimensional set.

Let $S = \partial_c\psi(\overline{x}) \cap \mathrm{Dom}\,(\nabla_x c(\overline{x}, \cdot))$. By (b), S is connected, so $S' = \nabla_x c(\overline{x}, S)$ is connected too, and by Theorem 10.25, S' is included in the line $\nabla^-\psi(\overline{x})$. Thus S' is a convex line segment (closed or not) containing $-\nabla_x c(\overline{x}, y_0)$ and $-\nabla_x c(\overline{x}, y_1)$. This finishes the proof of the regularity of c. □

The next result will show that the regularity property of the cost function is a *necessary condition* for a general theory of regularity of optimal transport. In view of Remark 12.19 it is close in spirit to Theorem 12.7.

Theorem 12.20 (Nonregularity implies nondensity of differentiable c-convex functions). *Let $c : \mathcal{X} \times \mathcal{Y} \to \mathbb{R}$ satisfy* **(Twist)**, **(locSC)** *and* **(H∞)**. *Let C be a totally c-convex set contained in $\mathrm{Dom}'(\nabla_x c)$, such that (a) c is not regular in C; and (b) for any $(\overline{x}, y_0), (\overline{x}, y_1)$ in C, $\partial_c\psi_{\overline{x},y_0,y_1}(\overline{x}) \subset \mathrm{Dom}'(\nabla_x c(\overline{x}, \cdot))$. Then for some $(\overline{x}, y_0)$, $(\overline{x}, y_1)$ in C the c-convex function $\psi = \psi_{\overline{x},y_0,y_1}$ cannot be the locally uniform limit of differentiable c-convex functions ψ_k.*

The short version of Theorem 12.20 is that if c is not regular, then differentiable c-convex functions are not dense, for the topology of local uniform convergence, in the set of all c-convex functions. What is happening here can be explained informally as follows. Take a function such as the one pictured in Figure 12.3, and try to tame up the singularity at the "saddle" by letting mountains grow from below and gently surelevate the pass. To do this without creating a new singularity you would like to be able to touch only the pass. Without the regularity assumption, you might not be able to do so.

Before proving Theorem 12.20, let us see how it can be used to prove a nonsmoothness result of the same kind as Theorem 12.7. Let $\overline{x}, y_0, y_1$ be as in the conclusion of Theorem 12.20. We may partition $\mathcal{X}$ in two sets $\mathcal{X}_0$ and $\mathcal{X}_1$ such that $y_i \in \partial_c\psi(x)$ for each $x \in \mathcal{X}_i$ $(i = 0, 1)$. Fix an arbitrary smooth positive probability density f on $\mathcal{X}$, let $a_i = \mu[\mathcal{X}_i]$ and $\nu = a_0\,\delta_{y_0} + a_1\,\delta_{y_1}$. Let T be the transport map defined by $T(\mathcal{X}_i) = y_i$; by construction the associated transport plan has its support included in $\partial_c\psi$, so π is an optimal transport plan and (ψ, ψ^c) is optimal in the dual Kantorovich problem associated with (μ, ν). By Remark 10.30, ψ is the unique optimal map, up to an additive constant, which can be fixed by requiring $\psi(x_0) = 0$ for some $x_0 \in \mathcal{X}$. Then let $(f_k)_{k\in\mathbb{N}}$ be a sequence of smooth probability densities, such that $\nu_k := f_k\,\mathrm{vol}$ converges weakly to ν. For any k let ψ_k be a c-convex function, optimal in the dual Kantorovich problem. Without loss of generality we can assume $\psi_k(x_0) = 0$. Since the functions ψ_k are uniformly continuous, we can extract a subsequence converging to some function $\widetilde{\psi}$. By passing to the limit in both sides of the Kantorovich problem, we deduce that $\widetilde{\psi}$ is optimal in the dual Kantorovich problem, so $\widetilde{\psi} = \psi$. By Theorem 12.20, the ψ_k cannot all be differentiable. So we have proven the following corollary:

Corollary 12.21 (Nonsmoothness of the Kantorovich potential). *With the same assumptions as in Theorem 12.20, if $\mathcal{Y}$ is a closed subset of a Riemannian manifold, then there are smooth positive probability densities f on $\mathcal{X}$ and g on $\mathcal{Y}$, such that the associated Kantorovich potential ψ is not differentiable.*

Now let us prove Theorem 12.20. The following lemma will be useful.

Lemma 12.22. *Let U be an open set of a Riemannian manifold, and let $(\psi_k)_{k\in\mathbb{N}}$ be a sequence of semi-convex functions converging uniformly* *to ψ on U. Let $\overline{x} \in U$ and $p \in \nabla^-\psi(\overline{x})$. Then there exist sequences $x_k \to \overline{x}$, and $p_k \in \nabla^-\psi_k(x_k)$, such that $p_k \to p$.*

Proof of Lemma 12.22. Since this statement is local, we may pretend that U is a small open ball in $\mathbb{R}^n$, centered at $\overline{x}$. By adding a well-chosen quadratic function we may also pretend that ψ is convex. Let $\delta > 0$ and $\widetilde{\psi}_k(x) = \psi_k(x) - \psi(\overline{x}) + \delta|x - \overline{x}|^2/2 - p \cdot (x - \overline{x})$. Clearly $\widetilde{\psi}_k$ converges uniformly to $\widetilde{\psi}(x) = \psi(x) - \psi(\overline{x}) + \delta|x - \overline{x}|^2/2 - p \cdot (x - \overline{x})$. Let $x_k \in \overline{U}$ be a point where $\widetilde{\psi}_k$ achieves its minimum. Since $\widetilde{\psi}$ has a unique minimum at $\overline{x}$, by uniform convergence x_k approaches $\overline{x}$, in particular x_k belongs to U for k large enough. Since $\widetilde{\psi}_k$ has a minimum at x_k, necessarily $0 \in \nabla^-\widetilde{\psi}(x_k)$, or equivalently $p_k := p - \delta(x_k - \overline{x}) \in \nabla^-\psi_k(x_k)$, so $p_k \to p$, which proves the result. □

Proof of Theorem 12.20. Since c is not regular in C, there are $(\overline{x}, y_0)$ and $(\overline{x}, y_1)$ in C such that $\partial_c\psi(\overline{x}) \cap C$ is not c-convex, where $\psi = \psi_{\overline{x},y_0,y_1}$. Assume that ψ is the limit of differentiable c-convex functions ψ_k. Let $p \in \nabla^-\psi(\overline{x})$. By Lemma 12.22 there are sequences $x_k \to \overline{x}$ and $p_k \to p$ such that $p_k \in \nabla^-\psi_k(x_k)$, i.e. $p_k = \nabla\psi_k(x_k)$. Since $\partial_c\psi_k(x_k)$ is nonempty (Proposition 10.24), in fact $-\nabla c(x_k, \partial_c\psi_k(x_k)) = \{p_k\}$, in particular $\partial_c\psi_k(x_k)$ contains a single point, say y_k. By Proposition 10.24 again, y_k stays in a compact set, so we may assume $y_k \to y$. The pointwise convergence of ψ_k to ψ implies $y \in \partial_c\psi(\overline{x})$. Since c is semiconcave, $\nabla_x c$ is continuous on its domain of definition, so $-\nabla_x c(\overline{x}, y) = \lim p_k = p$.

So $-\nabla_x c(\overline{x}, \partial_c\psi(\overline{x}))$ contains the whole of $\nabla^-\psi(\overline{x})$; combining this with Remark 12.18 we conclude that $\partial_c\psi(\overline{x})$ is c-convex, in contradiction with our assumption. □

Now that we are convinced of the importance of the regularity condition, the question naturally arises whether it is possible to translate it in analytical terms, and to check it in practice. The next section will bring a partial answer.

The Ma–Trudinger–Wang condition

Ma, Trudinger and X.-J. Wang discovered a *differential* condition on the cost function, which plays a key role in smoothness estimates, but in the end turns out to be a local version of the regularity property. Before explaining this condition, I shall introduce a new key assumption which will be called the "strong twist" condition. (Recall that the "plain" twist condition is the injectivity of $\nabla_x c(x, \cdot)$ on its domain of definition.)

Let $\mathcal{X}, \mathcal{Y}$ be closed sets in Riemannian manifolds M, N respectively, and as before let $\mathrm{Dom}'(\nabla_x c)$ stand for the set of all $(x, y) \in \mathcal{X} \times \mathcal{Y}$ such that x lies in the interior of $\mathcal{X}$ and y in the interior of $\mathrm{Dom}\,(\nabla_x c(x, \cdot))$. It will be said that c satisfies the **strong twist** condition if

(STwist) $\mathrm{Dom}'(\nabla_x c)$ *is an open set on which c is smooth, $\nabla_x c$ is one-to-one, and the mixed Hessian $\nabla^2_{x,y} c$ is nonsingular.*

Remark 12.23. The invertibility of $\nabla^2_{x,y} c$ implies the local injectivity of $\nabla_x c(x, \cdot)$, by the implicit function theorem; but alone it does not a priori guarantee the global injectivity.

Remark 12.24. One can refine Proposition 10.15 to show that cost functions deriving from a well-behaved Lagrangian do satisfy the strong twist condition. In the Appendix I shall give more details for the important particular case of the squared geodesic distance.

Remark 12.25. One should think of $\nabla^2_{x,y} c$ as a bilinear form on $T_x M \times T_y N$: It takes a pair of tangent vectors $(\xi, \eta) \in T_x M \times T_y N$, and gives back a number, $(\nabla^2 c(x, y)) \cdot (\xi, \eta) = \langle \nabla^2_{x,y} c(x, y) \cdot \xi,\, \eta \rangle$. It will play the role of a Riemannian metric (or rather the negative of a Riemannian metric), except that ξ and η do not necessarily belong to the same tangent space!

The Ma–Trudinger–Wang condition is not so simple and involves fourth-order derivatives of the cost. To write it in an unambiguous way, it will be convenient to use coordinates, with some care. If x and y are given points, let us introduce geodesic coordinates $x_1, \ldots, x_n$ and $y_1, \ldots, y_n$ in the neighborhood of x and y respectively. (This means that one chooses Euclidean coordinates in, say, $T_x M$ and then parameterizes a point $\widetilde{x}$ in the neighborhood of x by the coordinates of the vector ξ such that $\widetilde{x} = \exp_x(\xi)$.) The technical advantage of using geodesic coordinates is that geodesic paths starting from x or y will be straight curves, in particular the Hessian can be computed by just differentiating twice with respect to the coordinate variables. (This advantage is nonessential, as we shall see.)

If u is a function of x, then $u_j = \partial_j u = \partial u / \partial x_j$ will stand for the partial derivative of u with respect to x_j. Indices corresponding to the derivation in y will be written after a comma; so if $a(x, y)$ is a function of x and y, then $a_{i,j}$ stands for $\partial^2 a / \partial x_i\, \partial y_j$. Sometimes I shall use the convention of summation over repeated indices ($a_k\, b^k = \sum_k a_k\, b^k$, etc.), and often the arguments x and y will be implicit.

As noted before, the matrix $(-c_{i,j})$ defined by $c_{i,j}\xi^i\,\eta^j = \langle \nabla^2_{x,y}c\cdot\xi, \eta\rangle$ will play the role of a Riemannian metric; in agreement with classical conventions of Riemannian geometry, I shall denote by $(c^{i,j})$ the coordinates of the inverse of $(c_{i,j})$, and sometimes raise and lower indices according to the rules $-c^{i,j}\xi_i = \xi^j$, $-c_{i,j}\xi^j = \xi_i$, etc. In this case (ξ_i) are the coordinates of a 1-form (an element of $(T_xM)^*$), while (ξ^j) are the coordinates of a tangent vector in T_yM. (I shall try to be clear enough to avoid devastating confusions with the operation of the metric g.)

Definition 12.26 (c-second fundamental form). *Let $c : \mathcal{X} \times \mathcal{Y} \to \mathbb{R}$ satisfy* **(STwist)**. *Let $\Omega \subset \mathcal{Y}$ be open (in the ambient manifold) with C^2 boundary $\partial\Omega$ contained in the interior of $\mathcal{Y}$. Let $(x, y) \in \mathrm{Dom}'(\nabla_x c)$, with $y \in \partial\Omega$, and let n be the outward unit normal vector to $\partial\Omega$, defined close to y. Let (n_i) be defined by $\langle n, \xi\rangle_y = \sum n_i\,\xi^i$ $(\forall \xi \in T_yM)$. Define the quadratic form $\mathrm{I\!I}_c(x, y)$ on $T_y\Omega$ by the formula*

$$\begin{aligned}\mathrm{I\!I}_c(x,y)(\xi) &= \sum_{ijk\ell} (\partial_j n_i - c^{k,\ell} c_{ij,k} n_\ell)\xi^i\xi^j \\ &= \sum_{ijk\ell} c_{i,k}\,\partial_j\big(c^{k,\ell}\,n_\ell\big)\,\xi^i\,\xi^j. \end{aligned} \tag{12.18}$$

Definition 12.27 (Ma–Trudinger–Wang tensor, or c-curvature operator). *Let $c : \mathcal{X} \times \mathcal{Y} \to \mathbb{R}$ satisfy* **(STwist)**. *For any $(x, y) \in \mathrm{Dom}'(\nabla_x c)$, define a quadrilinear form $\mathfrak{S}_c(x, y)$ on the space of bivectors $(\xi, \eta) \in T_xM \times T_yN$ satisfying*

$$\langle \nabla^2_{x,y} c(x, y) \cdot \xi,\ \eta\rangle = 0 \tag{12.19}$$

by the formula

$$\mathfrak{S}_c(x,y)\,(\xi,\eta) = \frac{3}{2} \sum_{ijk\ell rs} \big(c_{ij,r}\, c^{r,s}\, c_{s,k\ell} - c_{ij,k\ell}\big)\,\xi^i\xi^j\eta^k\eta^\ell. \tag{12.20}$$

Remark 12.28. Both for practical and technical purposes, it is often better to use another equivalent definition of $\mathfrak{S}_c$, see formula (12.21) below.

To understand where $\mathfrak{S}_c$ comes from, and to compute it in practice, the key is the change of variables $p = -\nabla_x c(x, y)$. This leads to the following natural definition.

Definition 12.29 (c-exponential). *Let c be a cost function satisfying* **(Twist)**, *define the c-exponential map on the image of $-\nabla_x c$ by the formula $c\text{-}exp_x(p) = (\nabla_x c)^{-1}(x, -p)$.*

In other words, $c\text{-}\exp_x(p)$ is the unique y such that $\nabla_x c(x,y)+p=0$. When $c(x,y) = d(x,y)^2/2$ on $\mathcal{X} = \mathcal{Y} = M$, a complete Riemannian manifold, one recovers the usual exponential of Riemannian geometry, whose domain of definition can be extended to the whole of TM. More generally, if c comes from a time-independent Lagrangian, under suitable assumptions the c-exponential can be defined as the solution at time 1 of the Lagrangian system starting at x with initial velocity v, in such a way that $\nabla_v L(x,v) = p$.

Then, with the notation $p = -\nabla_x c(x,y)$, we have the following reformulation of the c-curvature operator:

$$\mathfrak{S}_c(x,y)(\xi,\eta) = -\frac{3}{2}\frac{d^2}{ds^2}\frac{d^2}{dt^2}\Big(c\big(\exp_x(t\xi),\, (c\text{-}\exp)_x(p+s\eta)\big)\Big), \tag{12.21}$$

where η in the right-hand side is an abuse of notation for the tangent vector at x obtained from η by the operation of $-\nabla^2_{xy}c(x,y)$ (viewed as an operator $T_yM \to T_xM$). In other words, $\mathfrak{S}_c$ is obtained by differentiating the cost function $c(x,y)$ twice with respect to x and twice *with respect to* p, not with respect to y. Getting formula (12.20) from (12.21) is just an exercise, albeit complicated, in classical differential calculus; it involves the differentiation formula for the matrix inverse: $d(M^{-1})\cdot H = -M^{-1}HM^{-1}$.

Particular Case 12.30 (Loeper's identity). If $\mathcal{X} = \mathcal{Y} = M$ is a smooth complete Riemannian manifold, $c(x,y) = d(x,y)^2/2$, and ξ, η are two unit orthogonal vectors in T_xM, then

$$\mathfrak{S}_c(x,x)(\xi,\eta) = \sigma_x(P) \tag{12.22}$$

is the **sectional curvature** of M at x along the plane P generated by ξ and η. (See Chapter 14 for basic reminders about sectional curvature.) To establish (12.22), first note that for any fixed small t, the geodesic curve joining x to $\exp_x(t\xi)$ is orthogonal to $s \to \exp_x(s\eta)$ at $s=0$, so $(d/ds)_{s=0}F(t,s) = 0$. Similarly $(d/dt)_{t=0}F(t,s) = 0$ for any fixed s, so the Taylor expansion of F at $(0,0)$ takes the form

$$\frac{d\big(\exp_x(t\xi), \exp_x(s\eta)\big)^2}{2} = A\,t^2 + B\,s^2 + C\,t^4 + D\,t^2s^2 + E\,s^4 + O(t^6+s^6).$$

Since $F(t,0)$ and $F(0,s)$ are quadratic functions of t and s respectively, necessarily $C = E = 0$, so $\mathfrak{S}_c(x,x) = -6D$ is -6 times the coefficient of t^4 in the expansion of $d(\exp_x(t\xi), \exp_x(t\eta))^2/2$. Then the result follows from formula (14.1) in Chapter 14.

Remark 12.31. Formula (12.21) shows that $\mathfrak{S}_c(x,y)$ is intrinsic, in the sense that it is independent of any choice of geodesic coordinates (this was not obvious from Definition 12.20). However, the geometric interpretation of $\mathfrak{S}_c$ is related to the regularity property, which is independent of the choice of Riemannian structure; so we may suspect that the choice to work in geodesic coordinates is nonessential. It turns out indeed that Definition 12.20 is independent of any choice of coordinates, geodesic or not: We may apply Formula (12.20) by just letting $c_{i,j} = \partial^2 c(x,y)/\partial x_i \, \partial y_j$, $p_i = -c_i$ (partial derivative of $c(x,y)$ with respect to x_i), and replace (12.21) by

$$\mathfrak{S}_c(\overline{x},y)(\xi,\eta) = -\frac{3}{2}\frac{\partial^2}{\partial p_\eta^2}\frac{\partial^2}{\partial x_\xi^2}\, c\left(x, c\text{-}\exp_{\overline{x}}(p)\right)\Big|_{x=\overline{x},\, p=-d_x c(\overline{x},y)}, \qquad (12.23)$$

where the c-exponential is defined in an obvious way in terms of 1-forms (differentials) rather than tangent vectors. This requires some justification since the second differential is not an intrisical concept (except when the first differential vanishes) and might a priori depend on the choice of coordinates. When we differentiate (12.23) twice with respect to x_ξ, there might be an additional term which will be a combination of $\Gamma_{ij}^k(\overline{x})\, \partial_k c(\overline{x}, (c\text{-}\exp_{\overline{x}})(p)) = -\Gamma_{ij}^k(\overline{x})\, p_k$, where the Γ_{ij}^k are the Christoffel symbols. But this additional term is linear in p, so anyway it disappears when we differentiate twice with respect to p_η. (This argument does not need the "orthogonality" condition $\nabla^2 c(x,y)\cdot(\xi,\eta) = 0$.)

Remark 12.32. Even if it is intrinsically defined, from the point of view of Riemannian geometry $\mathfrak{S}_c$ is not a standard curvature-type operator, for at least two reasons. First it involves derivatives of order greater than 2; and secondly it is *nonlocal*, in a strong sense. Take for instance $c(x,y) = d(x,y)^2/2$ on a Riemannian manifold (M,g), fix x and y, compute $\mathfrak{S}_c(x,y)$, then a change of the Riemannian metric g can affect the value of $\mathfrak{S}_c(x,y)$, even if the metric g is left unchanged in a neighborhood of x and y, and even if it is unchanged in a neighborhood of the geodesics joining x to y! Here we are facing the fact that geodesic distance is a highly nonlocal notion.

Remark 12.33. The operator $\mathfrak{S}$ is symmetric under the exchange of x and y, in the sense that if $\check{c}(x,y) = c(y,x)$, then $\mathfrak{S}_{\check{c}}(y,x)(\eta,\xi) = \mathfrak{S}_c(x,y)(\xi,\eta)$. (Here I am assuming implicitly that also $\check{c}$ satisfies the strong twist condition.) This symmetry can be seen from (12.20) by just rearranging indices. To see it directly from (12.21), we may apply Remark 12.31 to change the geodesic coordinate system around $\overline{x}$ and parameterize x by $q = -\nabla_y c(x,\overline{y})$. Then we obtain the nicely symmetric expression for $\mathfrak{S}_c(\overline{x},\overline{y})(\xi,\eta)$:

$$-\frac{3}{2}\frac{\partial^2}{\partial p_\eta^2}\frac{\partial^2}{\partial q_\xi^2}\, c\left(\check{c}\text{-}\exp_{\overline{y}}(q), c\text{-}\exp_{\overline{x}}(p)\right)\Big|_{p=-d_x c(\overline{x},\overline{y}),\, q=-d_y c(\overline{x},\overline{y})}. \qquad (12.24)$$

(Caution: ξ is tangent at x and q at y, so differentiating with respect to q in direction ξ means in fact differentiating in the direction $-\nabla^2_{xy} c\cdot\xi$; recall (12.21). The same for p_η.)

In the sequel, I shall stick to the same conventions as in the beginning of the chapter, so I shall use a fixed Riemannian metric and use Hessians and gradients with respect to this metric, rather than differentials of order 1 and 2.

Theorems 12.35 and 12.36 will provide necessary and sufficient differential conditions for c-convexity and regularity; one should note carefully that these conditions are valid only inside $\mathrm{Dom}'(\nabla_x c)$. I shall use the notation $\check{c}(y,x) = c(x,y)$, $\check{D} = \{(y,x);\ (x,y)\in D\}$. I shall also be led to introduce an additional assumption:

($\mathbf{Cut}^{n-1}$) *For any x in the interior of* $\mathrm{proj}_{\mathcal{X}}(D)$*, the "cut locus"* $\mathrm{cut}_D(x) = \mathrm{proj}_{\mathcal{Y}}(D)\setminus \mathrm{Dom}'(\nabla_x c(x,\cdot))$ *locally has finite $(n-1)$-dimensional Hausdorff measure.*

Example 12.34. Condition ($\mathbf{Cut}^{n-1}$) trivially holds when c satisfies the strong twist condition, D is totally c-convex and *product*. It also holds when c is the squared distance on a Riemannian manifold M, and $D = M\setminus \mathrm{cut}(M)$ is the domain of smoothness of c (see the Appendix).

Now come the main results of this section:

Theorem 12.35 (Differential formulation of c-convexity). *Let $c:\mathcal{X}\times\mathcal{Y}\to\mathbb{R}$ be a cost function satisfying* **(STwist)**. *Let $x\in\mathcal{X}$ and let C be a connected open subset of $\mathcal{Y}$ with C^2 boundary. Let $x\in\mathcal{X}$, such that $\{x\}\times\overline{C}\subset \mathrm{Dom}'(\nabla_x c)$. Then $\overline{C}$ is c-convex with respect to x if and only if $\mathrm{I\!I}_c(x,y)\geq 0$ for all $y\in\partial C$.*

If moreover $\mathrm{I\!I}_c(x,y) > 0$ *for all* $y \in \partial\Omega$ *then* Ω *is strictly c-convex with respect to* x.

Theorem 12.36 (Differential formulation of regularity). *Let* $c : \mathcal{X} \times \mathcal{Y} \to \mathbb{R}$ *be a cost function satisfying* **(STwist)**, *and let* D *be a totally c-convex open subset of* $\mathrm{Dom}'(\nabla_x c)$. *Then:*

(i) If c is regular in D, *then* $\mathfrak{S}_c(x,y) \geq 0$ *for all* $(x,y) \in D$.

(ii) Conversely, if $\mathfrak{S}_c(x,y) \geq 0$ *(resp.* > 0*) for all* $(x,y) \in D$, $\check{c}$ *satisfies* **(STwist)**, $\check{D}$ *is totally* $\check{c}$*-convex, and c satisfies* $(\mathbf{Cut}^{n-1})$ *on* D, *then c is regular (resp. strictly regular) in* D.

Remark 12.37. For Theorem 12.36 to hold true, it is important that no condition be imposed on the sign of $\mathfrak{S}_c(x,y) \cdot (\xi,\eta)$ (or more rigorously, the expression in (12.20)) when ξ and η are not "orthogonal" to each other. For instance, $c(x,y) = \sqrt{1+|x-y|^2}$ is regular, still $\mathfrak{S}_c(x,y) \cdot (\xi,\xi) < 0$ for $\xi = x-y$ and $|x-y|$ large enough.

Remark 12.38. A corollary of Theorem 12.36 and Remark 12.33 is that the regularity of c is more or less equivalent to the regularity of $\check{c}$.

Proof of Theorem 12.35. Let us start with some reminders about classical convexity in Euclidean space. If Ω is an open set in $\mathbb{R}^n$ with C^2 boundary and $x \in \partial\Omega$, let $T_x\Omega$ stand for the tangent space to $\partial\Omega$ at x, and n for the outward normal on $\partial\Omega$ (extended smoothly in a neighborhood of x). The second fundamental form of Ω, evaluated at x, is defined on $T_x\Omega$ by $\mathrm{I\!I}(x)(\xi) = \sum_{ij} \partial_i n_j\, \xi^i\, \xi^j$. A *defining function* for Ω at x is a function Φ defined in a neighborhood of x, such that $\Phi < 0$ in Ω, $\Phi > 0$ outside $\overline{\Omega}$, and $|\nabla\Phi| > 0$ on $\partial\Omega$. Such a function always exists (locally), for instance one can choose $\Phi(x) = \pm d(x,\partial\Omega)$, with $+$ sign when x is outside Ω and $-$ when x is in Ω. (In that case $\nabla\Phi$ is the outward normal on $\partial\Omega$.) If Φ is a defining function, then $n = \nabla\Phi/|\nabla\Phi|$ on $\partial\Omega$, and for all $\xi \perp n$,

$$\partial_i n_j\, \xi^i\, \xi^j = \left(\frac{\Phi_{ij}}{|\nabla\Phi|} - \frac{\Phi_j\, \Phi_{ik}}{|\nabla\Phi|^3} \right) \xi^i\, \xi^j = \frac{\Phi_{ij}}{|\nabla\Phi|}\, \xi^i\, \xi^j.$$

So the condition $\mathrm{I\!I}(x) \geq 0$ on $\partial\Omega$ is equivalent to the condition that $\nabla^2\Phi(x)$ is always nonnegative when evaluated on tangent vectors. In that case, we can always choose a defining function of the form $g(\pm d(x,\Omega))$, with $g(0) = g'(0) = 1$, g strictly convex, so that $\nabla^2\Phi(x)$ is nonnegative in all directions. With a bit more work, one can show that Ω is convex if and only if it is connected and its second fundamental form is nonnegative on the whole of $\partial\Omega$. Moreover, if the second fundamental form is positive then Ω is strictly convex.

Now let C be as in the assumptions of Theorem 12.35 and let $\Omega = -\nabla_x c(x, C) \subset T_x M$. Since $\nabla_x c(x, \cdot)$ is smooth, one-to-one with nonsingular differential, $\partial\Omega$ is smooth and coincides with $-\nabla_x c(x, \partial C)$. Since C is connected, so is Ω, so to prove the convexity of Ω we just have to worry about the sign of its second fundamental form. Let $\overline{y} \in \partial C$ be fixed, and let $\Phi = \Phi(y)$ be a defining function for C in a neighborhood of $\overline{y}$. Then $\Psi = \Phi \circ (-\nabla_x c(x, \cdot))^{-1}$ is a defining function for Ω. By direct computation,

$$\Psi^{ij} = \left(\Phi_{rs} - c_{rs,k}\, c^{k,\ell}\, \Phi_\ell\right) c^{r,i}\, c^{s,j}.$$

(The derivation indices are raised because these are derivatives with respect to the p variables.) Let $\xi = (\xi^j) \in T_y M$ be tangent to C; the bilinear form $\nabla^2 c$ identifies ξ with an element in $(T_x M)^*$ whose coordinates are denoted by $(\xi_i) = (-c_{i,j}\, \xi^j)$. Then

$$\Psi^{ij} \xi_i\, \xi_j = \left(\Phi_{rs} - c_{rs,k}\, c^{k,\ell}\, \Phi_\ell\right) \xi^r\, \xi^s.$$

Since $\xi^s\, n_s = 0$, the nonnegativity of this expression is equivalent to the nonnegativity of $(\partial_r n_s - c^{k,\ell} c_{rs,k} n_\ell)\, \xi^r\, \xi^s$, for all tangent vectors ξ. This establishes the first part of Theorem 12.35, and at the same time justifies formula (12.18). (To prove the equivalence between the two expressions in (12.18), use the temporary notation $n^k = -c^{k,\ell}\, n_\ell$ and note that $\partial_j n_i - c^{k,\ell}\, c_{ij,k}\, n_\ell = -\partial_j(c_{i,k}\, n^k) + c_{ij,k}\, n^k = -c_{i,k}\, \partial_j n^k$.) The statement about strict convexity follows the same lines. □

In the proof of Theorem 12.36 I shall use the following technical lemma:

Lemma 12.39. *If c satisfies* **(STwist)** *and* **(Cut^{n-1})** *then it satisfies the following property of "transversal approximation":*

(TA) *For any $\overline{x}, x$ in the interior of $\mathrm{proj}_{\mathcal{X}}(D)$, any C^2 path $(y_t)_{0\le t\le 1}$ drawn in $\mathrm{Dom}'(\nabla_x c(\overline{x}, \cdot)) \cap \mathrm{proj}_{\mathcal{Y}}(D)$ can be approximated in C^2 topology by a path $(\widehat{y}_t)_{0\le t\le 1}$ such that $\{t \in (0,1);\ \widehat{y}_t \notin \mathrm{Dom}'(\nabla_x c(x, \cdot))\}$ is discrete.*

Lemma 12.39 applies for instance to the squared geodesic distance on a Riemannian manifold. The detailed proof of Lemma 12.39 is a bit tedious, so I shall be slightly sketchy:

Sketch of proof of Lemma 12.39. As in the statement of $(\mathbf{Cut}^{n-1})$, let $\mathrm{cut}_D(x) = \mathrm{proj}_{\mathcal{Y}}(D) \setminus \mathrm{Dom}'(\nabla_x c(x, \cdot))$. Since $\mathrm{cut}_D(x)$ has empty interior, for any fixed $t_0 \in [0,1]$ we can perturb the path y in C^2 topology into a path $\widehat{y}$, in such a way that $\widehat{y}(t_0) \notin \mathrm{cut}_D(x)$. Repeating this operation finitely many times, we can ensure that $\widehat{y}(t_j)$ lies outside $\mathrm{cut}_D(x)$ for each $t_j = j/2^k$, where $k \in \mathbb{N}$ and $j \in \{0, \ldots, 2^k\}$. If k is large enough, then for each j the path $\widehat{y}$ can be written, on the time-interval $[t_j, t_{j+1}]$, in some well-chosen local chart, as a straight line. Moreover, since $\mathrm{cut}_D(x)$ is closed, there will be a small ball $B_j = B(\widehat{y}(t_j), r_j)$, $r_j > 0$, such that on the interval $[t_j - \varepsilon_j, t_j + \varepsilon_j]$ the path $\widehat{y}$ is entirely contained in B_j, and the larger ball $2B_j = B(\widehat{y}(t_j), 2r_j)$ does not meet $\mathrm{cut}_D(x)$. If we prove the approximation property on each interval $[t_{j-1} + \varepsilon_{j-1}, t_j - \varepsilon_j]$, we shall get suitable paths $(\widehat{y}_t)$ on $[t_{j-1} + \varepsilon_{j-1}, t_j - \varepsilon_j]$, approximating (y_t) in C^2; in particular $\widehat{y}_{t_i} \in B_i$, and we can easily "patch together" these pieces of $(\widehat{y}_t)$ in the intervals $[t_j - \varepsilon_j, t_j + \varepsilon_j]$ while staying within $2B_j$.

All this shows that we just have to treat the case when (y_t) takes values in a small open subset U of $\mathbb{R}^n$ and is a straight line. In these coordinates, $\Sigma := \mathrm{cut}_D(x) \cap U$ will have finite $\mathcal{H}^{n-1}$ measure, with $\mathcal{H}^k$ standing for the k-dimensional Hausdorff measure. Without loss of generality, $y_t = t\, e_n$, where $(e_1, \ldots, e_n)$ is an orthonormal basis of $\mathbb{R}^n$ and $-\tau < t < \tau$; and U is the cylinder $B(0,\sigma) \times (-\tau, \tau)$, for some $\sigma > 0$.

For any $z \in B(0,\sigma) \subset \mathbb{R}^{n-1}$, let $y_t^z = (z,t)$. The goal is to show that $\mathcal{H}^{n-1}(dz)$-almost surely, y_t^z intersects Σ in at most finitely many points. To do this one can apply the co-area formula (see the bibliographical notes) in the following form: let $f : (z,t) \longmapsto z$ (defined on U), then

$$\mathcal{H}^{n-1}[\Sigma] \geq \int_{f(\Sigma)} \mathcal{H}^0[\Sigma \cap f^{-1}(z)]\, \mathcal{H}^{n-1}(dz).$$

By assumption the left-hand side is finite, and the right-hand side is exactly $\int \#\{t;\ y_t^z \in \Sigma\}\, \mathcal{H}^{n-1}(dz)$; so the integrand is finite for almost all z, and in particular there is a sequence $z_k \to 0$ such that each $(y_t^{z_k})$ intersects Σ finitely many often. □

Proof of Theorem 12.36. Let us first assume that the cost c is regular on D, and prove the nonnegativity of $\mathfrak{S}_c$.

Let $(\overline{x}, y) \in D$ be given, and let $p \in T_{\overline{x}}M$ be such that $\nabla_x c(\overline{x}, y) + p = 0$. Let ν be a unit vector in $T_{\overline{x}}M$. For $\varepsilon \in [-\varepsilon_0, \varepsilon_0]$ one can define a c-segment by the formula $y(\varepsilon) = (\nabla_{\overline{x}} c)^{-1}(\overline{x}, -p(\varepsilon))$, $p(\varepsilon) = p + \varepsilon\, \nu$;

let $y_0 = y(-\varepsilon_0)$ and $y_1 = y(\varepsilon_0)$. Further, let $h_0(x) = -c(x, y_0) + c(\overline{x}, y_0)$ and $h_1(x) = -c(x, y_1) + c(\overline{x}, y_1)$. By construction, $\nabla_x c(\overline{x}, y_0) - \nabla_x c(\overline{x}, y_1)$ is colinear to ν and nonzero, and $h_0(\overline{x}) = h_1(\overline{x})$.

By the implicit function theorem, the equation $(h_0(x) = h_1(x))$ defines close to $\overline{x}$ an $(n-1)$-dimensional submanifold $\widetilde{M}$, orthogonal to ν at $\overline{x}$. For any $\xi \in T_{\overline{x}}\widetilde{M} = \nu^\perp$ one can define in the neighborhood of $\overline{x}$ a smooth curve $(\gamma(t))_{-\tau \le t \le \tau}$, valued in $\widetilde{M}$, such that $\gamma(0) = \overline{x}$ and $\dot{\gamma}(0) = \xi$.

Let $\psi(x) = \max(h_0(x), h_1(x))$. By construction ψ is c-convex and y_0, y_1 both belong to $\partial_c\psi(\overline{x})$. So

$$\begin{aligned}
\psi(\overline{x}) + c(\overline{x}, y) &\le \psi(\gamma(t)) + c(\gamma(t), y) \\
&= \frac{1}{2}\big(h_0(\gamma(t)) + h_1(\gamma(t))\big) + c(\gamma(t), y) \\
&= \frac{1}{2}\Big[-c(\gamma(t), y_0) + c(\overline{x}, y_0) - c(\gamma(t), y_1) + c(\overline{x}, y_1)\Big] + c(\gamma(t), y).
\end{aligned}$$

Since $\psi(\overline{x}) = 0$, this can be recast as

$$0 \ge \frac{1}{2}\Big(c(\gamma(t), y_0) + c(\gamma(t), y_1)\Big) - \frac{1}{2}\Big(c(\overline{x}, y_0) + c(\overline{x}, y_1)\Big) - c(\gamma(t), y) + c(\overline{x}, y). \tag{12.25}$$

At $t = 0$ the expression on the right-hand side achieves its maximum value 0, so the second t-derivative is nonnegative. In other words,

$$\frac{d^2}{dt^2}\bigg|_{t=0} \left(\frac{1}{2}\big[c(\gamma(t), y_0) + c(\gamma(t), y_1)\big] - c(\gamma(t), y)\right) \le 0.$$

This is equivalent to saying that

$$\big\langle \nabla_x^2 c(\overline{x}, y) \cdot \xi, \xi \big\rangle \ge \frac{1}{2}\Big(\big\langle \nabla_x^2 c(\overline{x}, y_0) \cdot \xi, \xi \big\rangle + \big\langle \nabla_x^2 c(\overline{x}, y_1) \cdot \xi, \xi \big\rangle\Big).$$

(Note: The path $\gamma(t)$ is not geodesic, but this does not matter because the first derivative of the right-hand side in (12.25) vanishes at $t = 0$.)

Since $p = (p_0 + p_1)/2$ and $p_0 - p_1$ was along an arbitrary direction (orthogonal to ξ), this shows precisely that $\langle \nabla_x^2 c(\overline{x}, y) \cdot \xi, \xi \rangle$ is concave as a function of p, after the change of variables $y = y(x, p)$. In particular, the expression in (12.21) is ≥ 0. To see that the expressions in (12.21) and (12.20) are the same, one performs a direct (tedious) computation to check that if η is a tangent vector in the p-space and ξ is a tangent vector in the x-space, then

$$\left(\frac{\partial^4 c(x,y(x,p))}{\partial p_k\,\partial p_\ell\,\partial x_i\,\partial x_j}\right)\xi^i\,\xi^j\,\eta_k\,\eta_\ell = \Big(c_{ij,r}\,c^{r,s}\,c_{s,k\ell} - c_{ij,k\ell}\Big)c^{k,m}\,c^{\ell,n}\,\xi^i\,\xi^j\,\eta_m\,\eta_n$$
$$= \Big(c_{ij,r}\,c^{r,s}\,c_{s,k\ell} - c_{ij,k\ell}\Big)\,\xi^i\,\xi^j\,\eta^k\,\eta^\ell.$$

(Here $\eta^k = -c^{k,m}\,\eta_m$ stand for the coordinates of a tangent vector at y, obtained from η after changing variables $p \to y$, and still denoted η by abuse of notation.) To conclude, one should note that the condition $\xi \perp \eta$, i.e. $\xi^i\,\eta_i = 0$, is equivalent to $c_{i,j}\,\xi^i\,\eta^j = 0$.

Next let us consider the converse implication. Let $(\overline{x}, y_0)$ and $(\overline{x}, y_1)$ belong to D; $p = p_0 = -\nabla_x c(\overline{x}, y_0)$, $p_1 = -\nabla_x c(\overline{x}, y_1)$, $\zeta = p_1 - p_0 \in T_{\overline{x}}M$, $p_t = p + t\zeta$, and $y_t = (\nabla_x c)^{-1}(\overline{x}, -p_t)$. By assumption $(\overline{x}, y_t)$ belongs to D for all $t \in [0,1]$.

For $t \in [0,1]$, let $h(t) = -c(x, y_t) + c(\overline{x}, y_t)$. Let us first assume that $x \in \mathrm{Dom}'(\nabla_y c(\,\cdot\,, y_t))$ for all t; then h is a smooth function of $t \in (0,1)$, and we can compute its first and second derivatives. First,

$$(\dot{y})^i = -c^{i,r}(\overline{x}, y_t)\,\zeta_r =: \zeta^i.$$

Similarly, since (y_t) defines a straight curve in p-space,

$$(\ddot{y})^i = -c^{i,k}\,c_{k,\ell j}\,c^{\ell,r}\,c^{j,s}\,\zeta_r\,\zeta_s = -c^{i,k}\,c_{k,\ell j}\,\zeta^\ell\,\zeta^j.$$

So

$$\dot{h}(t) = -\big[c_{,j}(x, y_t) - c_{,j}(\overline{x}, y_t)\big]\,\zeta^j = c_{i,j}\,\eta^i\,\zeta^j,$$

where $\eta_j = -c_{,j}(x, y_t) + c_{,j}(\overline{x}, y_t)$ and $\eta^i = -c^{j,i}\,\eta_j$.

Next,

$$\begin{aligned}\ddot{h}(t) &= -\big[c_{,ij}(x, y_t) - c_{,ij}(\overline{x}, y_t)\big]\,\zeta^i\,\zeta^j \\ &\qquad\qquad + \big[c_{,j}(x, y_t) - c_{,j}(\overline{x}, y_t)\big]\,c^{j,k}\,c_{k,\ell i}\,\zeta^\ell\,\zeta^i \\ &= -\Big(\big[c_{,ij}(x, y_t) - c_{,ij}(\overline{x}, y_t)\big] - \eta_\ell\,c^{\ell,k}\,c_{k,ij}\Big)\,\zeta^i\,\zeta^j \\ &= -\Big(\big[c_{,ij}(x, y_t) - c_{,ij}(\overline{x}, y_t)\big] - \eta^k\,c_{k,ij}\Big)\,\zeta^i\,\zeta^j.\end{aligned}$$

Now freeze t, y_t, ζ, and let $\Phi(x) = c_{,ij}(x, y_t)\,\zeta^i\,\zeta^j = \langle \nabla_y^2 c(x, y_t)\cdot\zeta, \zeta\rangle$. This can be seen either as a function of x or as a function of $q = -\nabla_y c(x, y_t)$, viewed as a 1-form on $T_{y_t}M$. Computations to go back and forth between the q-space and the x-space are the same as those between the p-space and the y-space. If q and $\overline{q}$ are associated to x and $\overline{x}$ respectively, then $\eta = q - \overline{q}$, and

$$\Big(\big[c_{,ij}(x,y_t) - c_{,ij}(\overline{x},y_t)\big] - \eta^k\, c_{k,ij} \Big)\, \zeta^i\, \zeta^j = \Phi(q) - \Phi(\overline{q}) - d_{\overline{q}}\Phi\cdot(q-\overline{q})$$
$$= \int_0^1 (\nabla_q^2 \Phi)\,\big((1-s)q + s\overline{q}\big)\cdot(q-\overline{q}, q-\overline{q})\,(1-s)\,ds.$$

Computing $\nabla_q^2\Phi$ means differentiating $c(x,y)$ twice with respect to y and then twice with respect to q, with $\nabla_y c(x,y) + q = 0$. According to Remark 12.33, the result is the same as when we first differentiate with respect to x and then with respect to p, so this gives $(-2/3)\mathfrak{S}_c$. Since $q - \overline{q} = -\eta$, we end up with

$$\begin{cases} \dot{h}(t) = \big(\nabla_{x,y}^2 c(x,y_t)\big)\cdot(\eta,\zeta); \\[2ex] \ddot{h}(t) = \dfrac{2}{3}\displaystyle\int_0^1 \mathfrak{S}_c\Big((\nabla_y c)^{-1}\big((1-s)q + s\overline{q}, y_t\big), y_t\Big)\cdot(\eta,\zeta)\,(1-s)\,ds, \end{cases} \tag{12.26}$$

where now $\nabla_y c$ is inverted with respect to the x variable. Here I have slightly abused notation since the vectors η and ζ do not necessarily satisfy $\nabla_{x,y}^2 c\cdot(\eta,\zeta) = 0$; but $\mathfrak{S}_c$ stands for the same analytic expression as in (12.20). Note that the argument of $\mathfrak{S}_c$ in (12.26) is always well-defined because $\check{D}$ was assumed to be $\check{c}$-convex and $x \in \mathrm{Dom}'(\nabla_y c(\,\cdot\,,y_t))$. (So $[\overline{x},x]_{y_t}$ is contained in $\mathrm{Dom}'(\nabla_y c(\,\cdot\,,y_t))$.)

The goal is to show that h achieves its maximum value at $t=0$ or $t=1$. Indeed, the inequality $h(t) \le \max(h(0),h(1))$ is precisely (12.15).

Let us first consider the simpler case when $\mathfrak{S}_c > 0$. If h achieves its maximum at $t_0 \in (0,1)$, then $\dot{h}(t_0) = 0$, so $\nabla_{x,y}^2 c\cdot(\eta,\zeta) = 0$, $\mathfrak{S}_c(\dots)(\eta,\zeta) > 0$, and by (12.26) we have $\ddot{h}(t_0) > 0$ (unless $x = \overline{x}$), which contradicts the fact that t_0 is a maximum. So h has a maximum at $t=0$ or $t=1$, which proves (12.15), and this inequality is strict unless $t=0$ or $t=1$ or $x=\overline{x}$.

To work out the borderline case where $\mathfrak{S}_c$ is only assumed to be nonnegative we shall have to refine the analysis just a bit. By the same density argument as above, we can assume that h is smooth.

Freeze ζ and let η vary in a ball. Since $\mathfrak{S}_c(x,y)\cdot(\eta,\zeta)$ is a quadratic function of η, nonnegative on the hyperplane $\{\zeta_\ell\,\eta^\ell = 0\}$, there is a constant C such that $\mathfrak{S}_c(x,y)\cdot(\eta,\zeta) \ge -C\,|\zeta_\ell\,\eta^\ell| = -C\,|\nabla_{x,y}^2 c\cdot(\zeta,\eta)|$. This constant only depends on an upper bound on the functions of x,y appearing in $\mathfrak{S}_c(x,y)$, and on upper bounds on the norm of $\nabla_{x,y}^2 c$ and its inverse. By homogeneity, $\mathfrak{S}_c(x,y)\cdot(\eta,\zeta) \ge -C\,|\nabla_{x,y}^2 c\cdot(\zeta,\eta)|\,|\zeta|\,|\eta|$, where the constant C is uniform when $x,\overline{x},y_0,y_1$ vary in compact domains. The norms $|\zeta|$ and $|\eta|$ remain bounded as t varies in $[0,1]$, so (12.26) implies

$$\ddot{h}(t) \geq -C\,|\dot{h}(t)|. \tag{12.27}$$

Now let $h_\varepsilon(t) = h(t) + \varepsilon(t-1/2)^k$, where $\varepsilon > 0$ and $k \in 2\mathbb{N}$ will be chosen later. Let t_0 be such that h_ε admits a maximum at t_0. If $t_0 \in (0,1)$, then $\dot{h}_\varepsilon(t_0) = 0$, $\ddot{h}_\varepsilon(t_0) \leq 0$, so

$$\ddot{h}(t_0) = \ddot{h}_\varepsilon(t_0) - \varepsilon\,k(k-1)\,(t_0-1/2)^{k-2} \leq -\varepsilon\,k(k-1)\,(t_0-1/2)^{k-2};$$

$$\dot{h}(t_0) = -\varepsilon\,k\,(t_0-1/2)^{k-1}.$$

Plugging these formulas back in (12.27) we deduce $C\,|t_0-1/2| \geq k-1$, which is impossible if k has been chosen greater than $1+C/2$. Thus h_ε has to reach its maximum either at $t=0$ or $t=1$, i.e.

$$h(t) + \varepsilon\,(t-1/2)^k \leq \max(h(0), h(1)) + \varepsilon\,2^{-k}.$$

Letting $\varepsilon \to 0$, we conclude again to (12.15).

To conclude the proof of the theorem, it only remains to treat the case when x does not belong to $\mathrm{Dom}'(\nabla_y c(\,\cdot\,, y_t))$ for all t, or equivalently when the path (y_t) is not contained in $\mathrm{Dom}'(\nabla_x c(x,\,\cdot\,)$. Let us consider for instance the case $\mathfrak{S}_c > 0$. Thanks to Lemma 12.39, we can approximate (y_t) by a very close path $(\widehat{y}_t)$, in such a way that (y_t) leaves $\mathrm{Dom}'(\nabla_x c(x,\cdot))$ only on a discrete set of times t_j.

Outside of these times, the same computations as before can be repeated with $\widehat{y}_t$ in place of y_t (here I am cheating a bit since $(\widehat{y}_t)$ is not a c-segment any longer, but it is no big deal to handle correction terms). So h cannot achieve a maximum in $(0,1)$ except maybe at some time t_j, and it all amounts to proving that t_j cannot be a maximum of h either. This is obvious if $\dot{h}$ is continuous at t_j and $\dot{h}(t_j) \neq 0$. This is also obvious if $\dot{h}$ is discontinuous at t_j, because by semiconvexity of $t \to -c(x,y_t)$, necessarily $\dot{h}(t_j^+) > \dot{h}(t_j^-)$. Finally, if $\dot{h}$ is continuous at t_j and $\dot{h}(t_j) = 0$, the same computations as before show that $\ddot{h}(t)$ is strictly positive when t is close to (but different from) t_j, then the continuity of $\dot{h}$ implies that h is strictly convex around t_j, so it cannot have a maximum at t_j. □

With Theorems 12.35 and 12.36 at hand it becomes possible to prove or disprove the regularity of certain simple cost functions. Typically, one first tries to exhaust $\mathrm{Dom}'(\nabla_x c(x,\cdot))$ by smooth open sets compactly included in its interior, and one checks the c-convexity of these sets by use of the c-second fundamental form; and similarly for the x variable. Then one checks the sign condition on $\mathfrak{S}_c$. In simple enough situations, this strategy can be worked out successfully.

Example 12.40. If f and g are C^2 convex functions $\mathbb{R}^n \to \mathbb{R}$ with $|\nabla f| < 1$, $|\nabla g| < 1$, then $c(x,y) = |x-y|^2 + |f(x) - g(y)|^2$ satisfies $\mathfrak{S}_c \geq 0$ on $\mathbb{R}^n \times \mathbb{R}^n$, so it is regular on $\mathbb{R}^n \times \mathbb{R}^n$. If $\nabla^2 f$ and $\nabla^2 g$ are positive everywhere, then $\mathfrak{S}_c > 0$, so c is strictly regular on $\mathbb{R}^n \times \mathbb{R}^n$.

Examples 12.41. The cost functions $c(x,y) = \sqrt{1+|x-y|^2}$ in $\mathbb{R}^n \times \mathbb{R}^n$, $\sqrt{1-|x-y|^2}$ in $B(0,1) \times B(0,1) \setminus \{|x-y| \geq 1\}$, $|x-y|^p$ on $\mathbb{R}^n \times \mathbb{R}^n \setminus \{y = x\}$ for $0 < p < 1$, $d(x,y)^2$ on $S^{n-1} \times S^{n-1} \setminus \{y = -x\}$, all satisfy $\mathfrak{S}_c > 0$, and are therefore strictly regular on any totally c-convex subdomain (for instance, $B(0,1/2) \times B(0,1/2)$ for $|x-y|^p$). The same is true of the singular cost functions $|x-y|^p$ $(-2 < p < 0)$, $-\log|x-y|$ on $\mathbb{R}^n \times \mathbb{R}^n \setminus \{y = x\}$, or $-\log|x-y|$ on $S^{n-1} \times S^{n-1} \setminus \{y = \pm x\}$. Also the limit case $c(x,y) = |x-y|^{-2}$ satisfies $\mathfrak{S}_c \geq 0$ on $\mathbb{R}^n \times \mathbb{R}^n \setminus \{x = y\}$.

Theorem 12.42 (Equivalence of regularity conditions). *Let M, N be Riemannian manifolds and let $c : M \times N \to \mathbb{R}$ be a locally semiconcave cost function such that c and $\check{c}$ satisfy* **(STwist)**, *c satisfies* **(Cut^{n-1})**, *and the c-exponential map extends into a continuous map $TM \to N$. Further, assume that* $\mathrm{Dom}'(\nabla_x c)$ *is totally c-convex and* $\mathrm{Dom}'(\nabla_x \check{c})$ *is totally $\check{c}$-convex. Then the following three properties are equivalent:*

(i) c satisfies Assumption (C);

(ii) c is regular;

(iii) c satisfies the Ma–Trudinger–Wang condition $\mathfrak{S}_c \geq 0$.

Remark 12.43. The implications (i) ⇒ (ii) and (ii) ⇒ (iii) remain true without the convexity assumptions. It is a natural open problem whether these assumptions can be completely dispensed with in Theorem 12.42. A bold conjecture would be that (i), (ii) and (iii) are always equivalent and automatically imply the total c-convexity of $\mathrm{Dom}'(\nabla_x c)$.

Proof of Theorem 12.42. Theorem 12.36 ensures the equivalence of (ii) and (iii). We shall now see that (i) and (ii) are also equivalent.

Assume (ii) is satisfied. Let $\psi : M \to \mathbb{R}$ be a c-convex function, and let $\overline{x} \in M$; the goal is to show that $\partial_c \psi(\overline{x})$ is connected. Without loss of generality we may assume $\psi(\overline{x}) = 0$. Let $\psi_{\overline{x},y_0,y_1}$ be defined as in (12.17). The same reasoning as in the proof of Proposition 12.15(i) shows that $\partial_c\psi_{\overline{x},y_0,y_1} \subset \partial_c\psi(\overline{x})$, so it is sufficient to show that $\partial_c\psi_{\overline{x},y_0,y_1}$ is connected. Even if y_0, y_1 do not belong to $\mathrm{Dom}'(\nabla_x c(\overline{x},\cdot))$, the latter set has an empty interior as a consequence of $(\mathbf{Cut}^{n-1})$, so we can find sequences $(y_0^{(k)})_{k\in\mathbb{N}}$ and $(y_1^{(k)})_{k\in\mathbb{N}}$ such that $(\overline{x}, y_i^{(k)}) \in \mathrm{Dom}'(\nabla_x c)$ and $y_i^{(k)} \to y_i$ $(i=0,1)$. In particular, there are $p_i^{(k)} \in T_{\overline{x}}M$, uniquely determined, such that $\nabla_x c(\overline{x}, y_i^{(k)}) + p_i^{(k)} = 0$.

Then let $\psi^{(k)} = \psi_{\overline{x},y_0^{(k)},y_1^{(k)}}$. The c-segment $[y_0^{(k)}, y_1^{(k)}]_{\overline{x}}$ is well-defined and included in $\partial_c\psi^{(k)}(\overline{x})$ (because c is regular). In other words, $c\text{-}\exp_{\overline{x}}((1-t)\,p_0^{(k)} + t\,p_1^{(k)}) \in \partial_c\psi^{(k)}(\overline{x})$. Passing to the limit as $k\to\infty$, after extraction of a subsequence if necessary, we find vectors p_0, p_1 (not necessarily uniquely determined) such that $c\text{-}\exp_{\overline{x}}(p_0) = y_0$, $c\text{-}\exp_{\overline{x}}(p_1) = y_1$, and $c\text{-}\exp_{\overline{x}}((1-t)\,p_0 + t\,p_1) \in \partial_c\psi(\overline{x})$. This proves the desired connectedness property.

Conversely, assume that (i) holds true. Let $\overline{x}, y$ be such that $(\overline{x}, y) \in \mathrm{Dom}'(\nabla_x c)$, and let $p = -\nabla_x c(\overline{x}, y)$. Let ν be a unit vector in T_yM; for $\varepsilon > 0$ small enough, $y_0^{(\varepsilon)} = c\text{-}\exp_{\overline{x}}(p+\varepsilon\nu)$ and $y_1^{(\varepsilon)} = c\text{-}\exp_{\overline{x}}(p-\varepsilon\nu)$ belong to $\mathrm{Dom}'(\nabla_x c(\overline{x},\,\cdot\,))$. Let $\psi^{(\varepsilon)} = \psi_{\overline{x},y_0^{(\varepsilon)},y_1^{(\varepsilon)}}$. As $\varepsilon\to 0$, $\partial_c\psi^{(\varepsilon)}(\overline{x})$ shrinks to $\partial_c\psi_{\overline{x},y,y}(\overline{x}) = \{y\}$ (because $(\overline{x}, y) \in \mathrm{Dom}'(\nabla_x c)$). Since $\mathrm{Dom}'(\nabla_x c)$ is open, for ε small enough the whole set $\partial_c\psi^\varepsilon(\overline{x})$ is included in $\mathrm{Dom}'(\nabla_x c)$. By the same reasoning as in the proof of Proposition 12.15(iii), the connectedness of $\partial_c\psi^\varepsilon(\overline{x})$ implies its c-convexity. Then the proof of Theorem 12.36(i) can be repeated to show that $\mathfrak{S}_c(\overline{x}, y) \geq 0$ along pairs of vectors satisfying the correct orthogonality condition. □

I shall conclude this section with a negative result displaying the power of the differential reformulation of regularity.

Theorem 12.44 (Smoothness of the optimal transport needs nonnegative curvature). *Let M be a compact Riemannian manifold such that the sectional curvature $\sigma_{\overline{x}}(P)$ is negative for some $\overline{x} \in M$ and some plane $P \subset T_{\overline{x}}M$. Then there exist smooth positive probability densities f and g on M such that the optimal transport map T from $f\,\mathrm{vol}$ to $g\,\mathrm{vol}$, with cost $c(x,y) = d(x,y)^2$, is discontinuous.*

The same conclusion holds true under the weaker assumption that $\mathfrak{S}_c(\overline{x},\overline{y})\cdot(\xi,\eta) < 0$ for some $(\overline{x},\overline{y}) \in M\times M$ such that $\overline{y}$ does not belong to the cut locus of $\overline{x}$ and $\nabla^2_{xy}c(\overline{x},\overline{y})\cdot(\xi,\eta) = 0$.

Remark 12.45. A counterexample by Y.-H. Kim shows that the second assumption is strictly weaker than the first one.

Proof of Theorem 12.44. Let ξ, η be orthogonal tangent vectors at x generating P. By Particular Case 12.30, $\mathfrak{S}_c(\overline{x}, \overline{x})(\xi, \eta) < 0$. If we fix a neighborhood V of $\overline{x}$ we can find $r > 0$ such that for any x the set $C_x = (\nabla_x c)^{-1}(x, B_r(0))$ is well-defined. If we take a small enough subdomain of U, containing $\overline{x}$ and define $C = \cap_{x \in U} C_x$, then $U \times C$ is totally c-convex and open. By Theorem 12.36, c is not regular in $U \times C$.

For any y_0, y_1 in C, define $\psi_{\overline{x}, y_0, y_1}$ as in (12.17). If we let $U \times C$ shrink to $\{\overline{x}\} \times \{\overline{x}\}$, $\partial_c \psi_{\overline{x}, y_0, y_1}(\overline{x})$ will converge to $\partial_c \psi_{\overline{x}, \overline{x}, \overline{x}}(\overline{x}) = \{\overline{x}\}$. So if U and C are small enough, $\partial_c \psi_{\overline{x}, y_0, y_1}(\overline{x})$ will be contained in an arbitrarily small ball around $\overline{x}$, a fortiori in $\mathrm{Dom}'(\nabla_x c(\overline{x}, \cdot))$. Then we can apply Theorem 12.20 and Corollary 12.21.

A similar reasoning works for the more general case when $\mathfrak{S}_c(\overline{x}, \overline{y}) \cdot (\xi, \eta) < 0$, since $\partial_c \psi_{\overline{x}, \overline{y}, \overline{y}} = \{\overline{y}\}$ (as long as $\overline{y}$ is not a cut point of $\overline{x}$). $\square$

Differential formulation of c-convexity

Its definition makes the c-convexity property quite difficult to check in general. In contrast, to establish the plain convexity of a smooth function $\mathbb{R}^n \to \mathbb{R}$ it is sufficient to just prove the nonnegativity of its Hessian. If c is an arbitrary cost function, there does not seem to be such a simple differential characterization for c-convexity; but if c is a *regular* cost function there is one, as the next result will demonstrate. The notation is the same as in the previous section.

Theorem 12.46 (Differential criterion for c-convexity). *Let $c : \mathcal{X} \times \mathcal{Y} \to \mathbb{R}$ be a cost function such that c and $\check{c}$ satisfy* **(STwist)**, *and let D be a totally c-convex closed subset of* $\mathrm{Dom}'(\nabla_x c)$ *such that $\check{D}$ is totally $\check{c}$-convex and $\mathfrak{S}_c \geq 0$ on D. Let $\mathcal{X}' = \mathrm{proj}_{\mathcal{X}}(D)$ and let $\psi \in C^2(\mathcal{X}'; \mathbb{R})$ (meaning that ψ is twice continuously differentiable on $\mathcal{X}'$, up to the boundary). If for any $x \in \mathcal{X}'$ there is $y \in \mathcal{Y}$ such that $(x, y) \in D$ and*

$$\begin{cases} \nabla \psi(x) + \nabla_x c(x, y) = 0 \\ \\ \nabla^2 \psi(x) + \nabla_x^2 c(x, y) \geq 0, \end{cases} \tag{12.28}$$

then ψ is c-convex on $\mathcal{X}'$ (or more rigorously, c'-convex, where c' is the restriction of c to D).

Remark 12.47. In view of the discussion at the beginning of this chapter, (12.28) is a *necessary and sufficient* condition for c-convexity, up to issues about the smoothness of ψ and the domain of differentiability of c. Note that the set of y's appearing in (12.28) is not required to be the whole of $\mathrm{proj}_{\mathcal{Y}}(D)$, so in practice one may often enlarge D in the y variable before applying Theorem 12.46.

Remark 12.48. Theorem 12.46 shows that if c is a regular cost, then (up to issues about the domain of definition) c-convexity is a local notion.

Proof of Theorem 12.46. Let ψ satisfy the assumptions of the theorem, and let $(\overline{x},\overline{y}) \in D$ such that $\nabla\psi(\overline{x}) + \nabla_x c(\overline{x},\overline{y}) = 0$. The goal is

$$\forall x \in \mathcal{X}', \qquad \psi(x) + c(x,\overline{y}) \geq \psi(\overline{x}) + c(\overline{x},\overline{y}). \tag{12.29}$$

If this is true then $\psi^c(\overline{y}) = \psi(\overline{x}) + c(\overline{x},\overline{y})$, in particular $\psi^{cc}(\overline{x}) = \sup_y[\psi^c(y) - c(\overline{x},y)] \geq \psi(\overline{x})$, and since this is true for any $\overline{x} \in \mathcal{X}'$ we will have $\psi^{cc} \geq \psi$, therefore $\psi^{cc} = \psi$, so ψ will be c-convex.

The proof of (12.29) is in the same spirit as the proof of the converse implication in Theorem 12.36. Let $x \in \mathcal{X}'$, and let $(x_t)_{0\leq t\leq 1} = [\overline{x},x]_{\overline{y}}$ be the $\check{c}$-segment with base $\overline{y}$ and endpoints $x_0 = \overline{x}$ and $x_1 = x$. Let

$$h(t) = \psi(x_t) + c(x_t,\overline{y}).$$

To prove (12.29) it is sufficient to show that $h(t) \geq h(0)$ for all $t \in [0,1]$.

The $\check{c}$-convexity of $\check{D}$ implies that $(x_t,\overline{y})$ always lies in D. Let $q = -\nabla_y c(x,\overline{y})$, $\overline{q} = -\nabla_y c(\overline{x},\overline{y})$, $\eta = q - \overline{q}$, then as in the proof of Theorem 12.36 we have $(\dot{x})^j = -c^{k,j}\,\eta_k = \eta^j$,

$$\dot{h}(t) = \big[\psi_i(x_t) + c_i(x_t,\overline{y})\big]\,\eta^i = -c_{i,j}(x_t,\overline{y})\,\eta^i\,\zeta^j$$

where $\zeta_i = \psi_i(x_t) + c_i(x_t,\overline{y})$ and $\zeta^j = -c^{j,i}\,\zeta_i$; similarly,

$$\ddot{h}(t) = \Big(\big[\psi_{ij}(x_t) + c_{ij}(x_t,\overline{y})\big] + c_{ij,k}(x_t,\overline{y})\,\zeta^k\Big)\,\eta^i\,\eta^j. \tag{12.30}$$

By assumption there is y_t such that $\nabla\psi(x_t) + \nabla_x c(x_t,y_t) = 0$, in particular $\zeta_i = c_i(x_t,\overline{y}) - c_i(x_t,y_t)$. Then (12.30) can be rewritten as $\Phi(\overline{p}_t) + d_p\Phi(\overline{p}_t)\cdot(p_t - \overline{p}_t)$, where

$$\Phi(y) := \big[\psi_{ij}(x_t) + c_{ij}(x_t,y)\big]\,\eta^i\,\eta^j$$

is seen as a function of $p = -\nabla_x c(x_t,y)$, and of course $p_t = -\nabla_x c(x_t,y_t)$, $\overline{p}_t = -\nabla_x c(x_t,\overline{y})$. (Note that ψ does not contribute to $d_p\Phi$.) After using the c-convexity of D and a Taylor formula, we end up with formulas that are quite similar to (12.26):

$$\begin{cases} \dot h(t) = \nabla^2_{x,y} c(x_t,\overline{y})\cdot(\eta,\zeta) \\ \\ \ddot h(t) = \big[\nabla^2\psi(x_t) + \nabla^2_x c(x_t,y_t)\big]\cdot(\zeta,\zeta) \\ \qquad + \dfrac{2}{3}\displaystyle\int_0^1 \mathfrak{S}_c\Big(x_t,(\nabla_x c)^{-1}\big(x_t,(1-s)p_t + s\overline{p}_t\big)\Big)\cdot(\eta,\zeta)\,(1-s)\,ds. \end{cases} \tag{12.31}$$

By assumption the first term in the right-hand side of $\ddot h$ is nonnegative, so, reasoning as in the proof of Theorem 12.36 we arrive at

$$\ddot h \geq -C\,|\dot h(t)|, \tag{12.32}$$

where C is a positive constant depending on c, ψ, $\overline{x}$, $\overline{y}$, x. We shall see that (12.32), combined with $\dot h(0) = 0$, implies that h is nondecreasing on $[0,1]$, and therefore $h(t) \geq h(0)$, which was our goal.

Assume indeed that $\dot h(t_*) < 0$ for some $t_* \in (0,1]$, and let $t_0 = \sup\{t \leq t_*;\ \dot h(t) = 0\} \geq 0$. For $t \in (t_0,t_*)$ we have $\dot h(t) < 0$ and $(d/dt)\log|\dot h(t)| = \ddot h/\dot h \leq C$, so $\log|\dot h(t)| \geq \log|\dot h(t_*)| - C\,(t_* - t)$, and as $t \to t_0$ we obtain a contradiction since $\log|\dot h(t_0)| = -\infty$. The conclusion is that $\dot h(t) \geq 0$ for all $t \in [0,1]$, and we are done. □

Control of the gradient via c-convexity

The property of c-convexity of the target is the key to get good control of the localization of the gradient of the solution to (12.2). This assertion might seem awkward: After all, we already know that under general assumptions, $\overline{T(\mathrm{Spt}\,\mu)} = \mathrm{Spt}\,\nu$ (recall the end of Theorem 10.28), where the transport T is related to the gradient of ψ by $T(x) = (\nabla_x c)^{-1}(x,-\nabla\psi(x))$; so $\nabla\psi(x)$ always belongs to $-\nabla_x c(\mathrm{Spt}\,\mu,\mathrm{Spt}\,\nu)$ when x varies in $\mathrm{Spt}\,\mu$.

To understand why this is not enough, assume that you are approximating ψ by smooth approximate solutions ψ_k. Then $\nabla\psi_k \longrightarrow \nabla\psi$ at all points of differentiability of ψ, but you have no control of the behavior of $\nabla\psi_k(x)$ if ψ is not differentiable at x! In particular, in the approximation process the point y might very well get beyond the support of ν, putting you in trouble. To guarantee good control of smooth approximations of ψ, you need an information on the whole c-subdifferential $\partial_c\psi(x)$. The next theorem says that such control is available as soon as the target is c-convex.

Theorem 12.49 (Control of c-subdifferential by c-convexity of target). *Let $\mathcal{X}$, $\mathcal{Y}$, $c : \mathcal{X} \times \mathcal{Y} \to \mathbb{R}$, $\mu \in P(\mathcal{X})$, $\nu \in P(\mathcal{Y})$ and $\psi : \mathcal{X} \to \mathbb{R} \cup \{+\infty\}$ satisfy the same assumptions as in Theorem 10.28 (including* **(H∞)***). Let $\Omega \subset \mathcal{X}$ be an open set such that $\operatorname{Spt}\mu = \overline{\Omega}$, and let $C \subset \mathcal{Y}$ be a closed set such that $\operatorname{Spt}\nu \subset C$. Assume that:*

(a) $\Omega \times C \subset \operatorname{Dom}'(\nabla_x c)$;

(b) C is c-convex with respect to Ω.

Then $\partial_c\psi(\Omega) \subset C$.

Proof of Theorem 12.49. We already know from Theorem 10.28 that $T(\overline{\Omega}) \subset C$, where $T(x) = (\nabla_x c)^{-1}(x, -\nabla\psi(x))$ stands for the optimal transport. In particular, $\partial_c\psi(\Omega \cap \operatorname{Dom}(\nabla\psi)) \subset C$. The problem is to control $\partial_c\psi(x)$ when ψ is not differentiable at x.

Let $x \in \Omega$ be such a point, and let $y \in \partial_c\psi(x)$. By Theorem 10.25, $-\nabla_x c(x,y) \in \nabla^-\psi(x)$. Since ψ is locally semiconvex, we can apply Remark 10.51 to deduce that $\nabla^-\psi(x)$ is the convex hull of limits of $\nabla\psi(x_k)$ when $x_k \to x$. Then by Proposition 12.15(ii), there are $L \in \mathbb{N}$ ($L = n+1$ would do), $\alpha_\ell \geq 0$ and $(x_{k,\ell})_{k\in\mathbb{N}}$ ($1 \leq \ell \leq L$) such that $\sum \alpha_\ell = 1$, $x_{k,\ell} \to x$ as $k \to \infty$, and

$$\sum_{\ell=1}^{L} \alpha_\ell \, \nabla\psi(x_{k,\ell}) \xrightarrow[k\to\infty]{} -\nabla_x c(x,y). \tag{12.33}$$

From the observation at the beginning of the proof, $\nabla\psi(x_{k,\ell}) \in -\nabla_x c(x_{k,\ell}, C)$, and as $k \to \infty$ this set converges uniformly (say in the sense of Hausdorff distance) to $-\nabla_x c(x, C)$ which is convex. By passing to the limit in (12.33) we get $-\nabla_x c(x,y) \in -\nabla_x c(x,C)$, so $y \in$C. □

Smoothness results

After a long string of counterexamples (Theorems 12.3, 12.4, 12.7, Corollary 12.21, Theorem 12.44), we can at last turn to positive results about the smoothness of the transport map T. It is indeed absolutely remarkable that a good regularity theory can be developed once all the previously discussed obstructions are avoided, by:

- suitable assumptions of convexity of the domains;
- suitable assumptions of regularity of the cost function.

These results constitute a chapter in the theory of Monge–Ampère-type equations, more precisely for the "second boundary value problem", which means that the boundary condition is not of Dirichlet type; instead, what plays the role of boundary condition is that the image of the source domain by the transport map should be the target domain.

Typically a convexity-type assumption on the target will be needed for local regularity results, while global regularity (up to the boundary) will request convexity of both domains. Throughout this theory the main problem is to get C^2 estimates on the unknown ψ; once these estimates are secured, the equation becomes "uniformly elliptic", and higher regularity follows from the well-developed machinery of uniformly elliptic fully nonlinear second-order partial differential equations combined with the linear theory (Schauder estimates).

It would take much more space than I can afford here to give a fair account of the methods used, so I shall only list some of the main results proven so far, and refer to the bibliographical notes for more information. I shall distinguish three settings which roughly speaking are respectively the quadratic cost function in Euclidean space; regular cost functions; and strictly regular cost functions. The day may come when these results will all be unified in just two categories (regular and strictly regular), but we are not there yet.

In the sequel, I shall denote by $C^{k,\alpha}(\Omega)$ (resp. $C^{k,\alpha}(\overline{\Omega})$) the space of functions whose derivatives up to order k are locally α-Hölder (resp. globally α-Hölder in Ω) for some $\alpha \in (0,1]$, $\alpha = 1$ meaning Lipschitz continuity. I shall say that a C^2-smooth open set $C \subset \mathbb{R}^n$ is uniformly convex if its second fundamental form is uniformly positive on the whole of ∂C. A similar notion of uniform c-convexity can be defined by use of the c-second fundamental form in Definition 12.26. I shall say that a cost function c is uniformly regular if it satisfies $\mathfrak{S}_c(x,y) \geq \lambda\,|\xi|^2\,|\eta|^2$ for some $\lambda > 0$, where $\mathfrak{S}_c$ is defined by (12.20) and $\langle \nabla^2_{xy} c \cdot \xi, \eta \rangle = 0$;

I shall abbreviate this inequality into $\mathfrak{S}_c(x,y) \geq \lambda\,\mathrm{Id}$. When I say that a density is bounded from above and below, this means bounded by positive constants.

Theorem 12.50 (Caffarelli's regularity theory). *Let $c(x,y) = |x-y|^2$ in $\mathbb{R}^n \times \mathbb{R}^n$, and let Ω, Λ be connected bounded open subsets of $\mathbb{R}^n$. Let f, g be probability densities on Ω and Λ respectively, with f and g bounded from above and below. Let $\psi : \Omega \to \mathbb{R}$ be the unique (up to an addive constant) Kantorovich potential associated with the probability measures $\mu(dx) = f(x)\,dx$ and $\nu(dy) = g(y)\,dy$, and the cost c. Then:*

(i) If Λ is convex, then $\psi \in C^{1,\beta}(\Omega)$ for some $\beta \in (0,1)$.

(ii) If Λ is convex, $f \in C^{0,\alpha}(\Omega)$, $g \in C^{0,\alpha}(\Lambda)$ for some $\alpha \in (0,1)$, then $\psi \in C^{2,\alpha}(\Omega)$; moreover, for any $k \in \mathbb{N}$ and $\alpha \in (0,1)$,

$$f \in C^{k,\alpha}(\Omega),\ g \in C^{k,\alpha}(\Lambda) \Longrightarrow \quad \psi \in C^{k+2,\alpha}(\Omega).$$

(iii) If Λ and Ω are C^2 and uniformly convex, $f \in C^{0,\alpha}(\overline{\Omega})$ and $g \in C^{0,\alpha}(\overline{\Lambda})$ for some $\alpha \in (0,1)$, then $\psi \in C^{2,\alpha}(\overline{\Omega})$; more generally, for any $k \in \mathbb{N}$ and $\alpha \in (0,1)$,

$$f \in C^{k,\alpha}(\overline{\Omega}),\ g \in C^{k,\alpha}(\overline{\Lambda}),\ \Omega, \Lambda \in C^{k+2} \Longrightarrow \quad \psi \in C^{k+2,\alpha}(\overline{\Omega}).$$

Theorem 12.51 (Urbas–Trudinger–Wang regularity theory). *Let $\mathcal{X}$ and $\mathcal{Y}$ be the closures of bounded open sets in $\mathbb{R}^n$, and let $c : \mathcal{X} \times \mathcal{Y} \to \mathbb{R}$ be a smooth cost function satisfying* **(STwist)** *and $\mathfrak{S}_c \geq 0$ in the interior of $\mathcal{X} \times \mathcal{Y}$. Let $\Omega \subset \mathcal{X}$ and $\Lambda \subset \mathcal{Y}$ be C^2-smooth connected open sets and let $f \in C(\overline{\Omega})$, $g \in C(\overline{\Lambda})$ be positive probability densities. Let ψ be the unique (up to an additive constant) Kantorovich potential associated with the probability measures $\mu(dx) = f(x)\,dx$ and $\nu(dy) = g(y)\,dy$, and the cost c. If (a) Λ is uniformly c-convex with respect to Ω, and Ω uniformly $\check{c}$-convex with respect to Λ, (b) $f \in C^{1,1}(\overline{\Omega})$, $g \in C^{1,1}(\overline{\Lambda})$, and (c) Λ and Ω are of class $C^{3,1}$, then $\psi \in C^{3,\beta}(\overline{\Omega})$ for all $\beta \in (0,1)$.*

If moreover for some $k \in \mathbb{N}$ and $\alpha \in (0,1)$ we have $f \in C^{k,\alpha}(\overline{\Omega})$, $g \in C^{k,\alpha}(\overline{\Omega})$ and Ω, Λ are of class $C^{k+2,\alpha}$, then $\psi \in C^{k+2,\alpha}(\overline{\Omega})$.

Theorem 12.52 (Loeper–Ma–Trudinger–Wang regularity theory). *Let $\mathcal{X}$ and $\mathcal{Y}$ be the closures of bounded connected open sets in $\mathbb{R}^n$, and let $c : \mathcal{X} \times \mathcal{Y} \to \mathbb{R}$ be a smooth cost function satisfying* **(STwist)** *and $\mathfrak{S}_c \geq \lambda\,\mathrm{Id}$, $\lambda > 0$, in the interior of $\mathcal{X} \times \mathcal{Y}$. Let $\Omega \subset \mathcal{X}$ and $\Lambda \subset \mathcal{Y}$ be two connected open sets, let $\mu \in P(\Omega)$ such that*

$d\mu/dx > 0$ almost everywhere in Ω, and let g be a probability density on Λ, bounded from above and below. Let ψ be the unique (up to an additive constant) Kantorovich potential associated with the probability measures $\mu(dx)$, $\nu(dy) = g(y)\,dy$, and the cost c. Then:

(i) If Λ is c-convex with respect to Ω and if

$$\exists m > n-1,\ \exists C > 0,\ \forall x \in \Omega,\ \forall r > 0,\ \mu[B_r(x)] \le C\,r^m,$$

then $\psi \in C^{1,\beta}(\Omega)$ for some $\beta \in (0,1)$.

(ii) If Λ is uniformly c-convex with respect to Ω, $f \in C^{1,1}(\Omega)$ and $g \in C^{1,1}(\Lambda)$, then $\psi \in C^{3,\beta}(\Omega)$ for all $\beta \in (0,1)$. If moreover for some $k \in \mathbb{N}$ and $\alpha \in (0,1)$ we have $f \in C^{k,\alpha}(\Omega)$, $g \in C^{k,\alpha}(\Lambda)$, then $\psi \in C^{k+2,\alpha}(\Omega)$.

Remark 12.53. Theorem 12.51 shows that the regularity of the cost function is sufficient to build a strong regularity theory. These results are still not optimal and likely to be refined in the near future; in particular one can ask whether $C^\alpha \to C^{2,\alpha}$ estimates are available for plainly regular cost functions (but Caffarelli's methods strongly use the affine invariance properties of the quadratic cost function); or whether interior estimates exist (Theorem 12.52(ii) shows that this is the case for uniformly regular costs).

Remark 12.54. On the other hand, the first part of Theorem 12.52 shows that a uniformly regular cost function behaves better, in certain ways, than the square Euclidean norm! For instance, the condition in Theorem 12.52(i) is automatically satisfied if $\mu(dx) = f(x)\,dx$, $f \in L^p$ for $p > n$; but it also allows μ to be a singular measure. (Such estimates are not even true for the linear Laplace equation!) As observed by specialists, uniform regularity makes the equation much more elliptic.

Remark 12.55. Theorems 12.52 and 12.51 imply a certain converse to Theorem 12.20: Roughly speaking, if the cost function is regular then any c-convex function ψ defined in a uniformly bounded convex domain *can be* approximated uniformly by smooth c-convex functions. In other words, the density of smooth c-convex functions is more or less a necessary and sufficient condition for the regularity property.

All these results are stated only for bounded subsets of $\mathbb{R}^n \times \mathbb{R}^n$, so the question arises whether they can be extended to more general cost functions on Riemannian manifolds. One possibility is to redo the proofs of all these results in curved geometry (with probably additional complications and assumptions). Another possibility is to use a localization argument to reduce the general case to the particular case where the functions are defined in $\mathbb{R}^n$. At the level of the optimal transport problem, such an argument seems to be doomed: If you cut out a small piece $\Omega \subset \mathcal{X}$ and a small piece $\Lambda \subset \mathcal{Y}$, there is in general no hope of being able to choose Ω and Λ in such a way that the optimal transport sends Ω to Λ *and* these domains satisfy adequate c-convexity properties. However, whenever *interior a priori estimates* are available besides the regularity results, this localization strategy is likely to work at the level of the partial differential equation. At least Theorems 12.50 and 12.52 can be complemented with such interior a priori estimates:

Theorem 12.56 (Caffarelli's interior a priori estimates). *Let $\Omega \subset \mathbb{R}^n$ be open and let $\psi : \Omega \to \mathbb{R}$ be a smooth convex function satisfying the Monge–Ampère equation*

$$\det(\nabla^2\psi(x)) = F(x, \nabla\psi(x)) \qquad \text{in } \Omega. \tag{12.34}$$

Let $\kappa_\Omega(\psi)$ stand for the modulus of (strict) convexity of ψ in Ω. Then for any open subdomain Ω' such that $\overline{\Omega'} \subset \Omega$, one has the a priori estimates (for some $\beta \in (0,1)$, for all $\alpha \in (0,1)$, for all $k \in \mathbb{N}$)

$$\|\psi\|_{C^{1,\beta}(\Omega')} \le C\Big(\Omega, \Omega', \|F\|_{L^\infty(\Omega)}, \|\nabla\psi\|_{L^\infty(\Omega)}, \kappa_\Omega(\psi)\Big);$$

$$\|\psi\|_{C^{2,\alpha}(\Omega')} \le C\Big(\alpha, \Omega, \Omega', \|F\|_{C^{0,\alpha}(\Omega)}, \|\nabla\psi\|_{L^\infty(\Omega)}, \kappa_\Omega(\psi)\Big);$$

$$\|\psi\|_{C^{k+2,\alpha}(\Omega')} \le C\Big(k, \alpha, \Omega, \Omega', \|F\|_{C^{k,\alpha}(\Omega)}, \|\nabla\psi\|_{L^\infty(\Omega)}, \kappa_\Omega(\psi)\Big).$$

With Theorem 12.56 and some more work to establish the strict convexity, it is possible to extend Caffarelli's theory to unbounded domains.

Theorem 12.57 (Loeper–Ma–Trudinger–Wang interior a priori estimates). *Let $\mathcal{X}$, $\mathcal{Y}$ be the closures of bounded open sets in $\mathbb{R}^n$, and let $c : \mathcal{X} \times \mathcal{Y} \to \mathbb{R}$ be a smooth cost function satisfying* **(STwist)** *and uniformly regular. Let $\Omega \subset \mathcal{X}$ be a bounded open set and let $\psi : \Omega \to \mathbb{R}$ be a smooth c-convex solution of the Monge–Ampère-type equation*

$$\det\Big(\nabla^2\psi(x) + \nabla_x^2 c\big(x, (\nabla_x c)^{-1}(x, -\nabla\psi(x))\big)\Big) = F(x, \nabla\psi(x)) \qquad \text{in } \Omega. \tag{12.35}$$

Let $\Lambda \subset \mathcal{Y}$ be a strict neighborhood of $\{(\nabla_x c)^{-1}(x, -\nabla\psi(x));\ x \in \Omega\}$, c-convex with respect to Ω. Then for any open subset Ω' such that $\overline{\Omega'} \subset \Omega$, one has the a priori estimates (for some $\beta \in (0,1)$, for all $\alpha \in (0,1)$, for all $k \geq 2$)

$$\|\psi\|_{C^{1,\beta}(\Omega')} \leq C\Big(\Omega, \Omega', c|_{\Omega\times\Lambda}, \|F\|_{L^\infty(\Omega)}, \|\nabla\psi\|_{L^\infty(\Omega)}\Big);$$

$$\|\psi\|_{C^{3,\alpha}(\Omega')} \leq C\Big(\alpha, \Omega, \Omega', c|_{\Omega\times\Lambda}, \|F\|_{C^{1,1}(\Omega)}, \|\nabla\psi\|_{L^\infty(\Omega)}\Big);$$

$$\|\psi\|_{C^{k+2,\alpha}(\Omega')} \leq C\Big(k, \alpha, \Omega, \Omega', c|_{\Omega\times\Lambda}, \|F\|_{C^{k,\alpha}(\Omega)}, \|\nabla\psi\|_{L^\infty(\Omega)}\Big).$$

These a priori estimates can then be extended to more general cost functions. A possible strategy is the following:

(1) Identify a totally c-convex domain $D \subset \mathcal{X} \times \mathcal{Y}$ in which the cost is smooth and uniformly regular. (For instance in the case of $d^2/2$ on S^{n-1} this could be $S^{n-1} \times S^{n-1} \setminus \{d(y, -x) < \delta\}$, i.e. one would remove a strip around the cut locus.)

(2) Prove the continuity of the optimal transport map.

(3) Working on the mass transport condition, prove that $\partial_c\psi$ is entirely included in D. (Still in the case of the sphere, prove that the transport map stays a positive distance away from the cut locus.)

(4) For each small enough domain $\Omega \subset \mathcal{X}$, find a c-convex subset Λ of $\mathcal{Y}$ such that the transport map sends Ω into Λ; deduce from Theorem 12.49 that $\partial_c\psi(\Omega) \subset \Lambda$ and deduce an upper bound on $\nabla\psi$.

(5) Use coordinates to reduce to the case when Ω and Λ are subsets of $\mathbb{R}^n$. (Reduce Ω and Λ if necessary.) Since the tensor $\mathfrak{S}_c$ is intrinsically defined, the uniform regularity property will be preserved by this operation.

(6) Regularize ψ on $\partial\Omega$, solve the associated Monge–Ampère-type equation with regularized boundary data (there is a set of useful techniques for that), and use Theorem 12.57 to obtain interior a priori estimates which are independent of the regularization. Then pass to the limit and get a smooth solution. This last step is well-understood by specialists but requires some skill.

To conclude this chapter, I shall state a regularity result for optimal transport on the sphere, which was obtained by means of the preceding strategy.

Theorem 12.58 (Smoothness of optimal transport on S^{n-1}). *Let S^{n-1} be the unit Euclidean sphere in $\mathbb{R}^n$, equipped with its volume measure, and let d be the geodesic distance on S^{n-1}. Let f and g be $C^{1,1}$ positive probability densities on S^{n-1}. Let ψ be the unique (up to an additive constant) Kantorovich potential associated with the transport of $\mu(dx) = f(x)\,\mathrm{vol}(dx)$ to $\nu(dy) = g(y)\,\mathrm{vol}(dy)$ with cost $c(x,y) = d(x,y)^2$, and let T be the optimal transport map. Then $\psi \in C^{3,\beta}(S^{n-1})$ for all $\beta \in (0,1)$, and in particular $T \in C^{2,\beta}(S^{n-1}, S^{n-1})$. If moreover f, g lie in $C^{k,\alpha}(S^{n-1})$ for some $k \in \mathbb{N}$, $\alpha \in (0,1)$, then $\psi \in C^{k+2,\alpha}(S^{n-1})$ and $T \in C^{k+1,\alpha}(S^{n-1}, S^{n-1})$. (In particular, if f and g are positive and C^∞ then ψ and T are C^∞.)*

Exercise 12.59. Split S^{n-1} into two (geodesically convex) hemispheres S_+ and S_-, according to, say, the sign of the first coordinate. Let $f_\pm$ stand for the uniform probability density on $S_\pm$. Find out the optimal transport between $f_+\,\mathrm{vol}$ and $f_-\,\mathrm{vol}$ (for the cost $c(x,y) = d(x,y)^2$) and explain why it was not a priori expected to be smooth.

Appendix: Strong twist condition for squared distance, and rectifiability of cut locus

In this Appendix I shall explain why the squared geodesic distance satisfies the strong twist condition. The argument relies on a few key results about the cut locus (introduced just after Problem 8.8) which I shall recall without proof. Details can be found in mildly advanced textbooks in Riemannian geometry such as the ones quoted in the bibliographical notes of Chapter 14.

Let (M,g) be a complete smooth Riemannian manifold equipped with its geodesic distance d. Recall that $d(x,\cdot)^2$ is not differentiable at y if and only if there are at least two distinct minimal geodesics going from x to y. It can be shown that the closure of such points y is exactly the cut locus $\mathrm{cut}(x)$ of x. In particular, if $c = d^2$, then $\mathrm{Dom}'(\nabla_x c(x,\cdot)) = M \setminus \mathrm{cut}(x)$. Also $\mathrm{cut}(M) := \bigcup_{x\in M}(\{x\} \times \mathrm{cut}(x))$ is closed, so $\mathrm{Dom}'(\nabla_x c) = (M \times M) \setminus \mathrm{cut}(M)$.

It is easily checked that c is C^∞ outside $\mathrm{cut}(M)$.

Let now x and y be such that $y \notin \mathrm{cut}(x)$; in particular (as we know from Problem 8.8), y is not focal to x, which means that $d\exp_x$ is well-defined and nonsingular at $v_{x,y} = (\exp_x)^{-1}(y)$, which is the initial velocity of the unique geodesic going from x to y. But $\nabla_x c(x,y)$ coincides with $-v_{x\to y}$; so $\nabla^2_{xy} c(x,y) = -\nabla_y((\exp_x)^{-1}) = -(\nabla_v \exp_x)^{-1}$ is nonsingular. This concludes the proof of the strong twist condition.

It is also true that c satisfies $(\mathbf{Cut}^{n-1})$; in fact, for any compact subset K of M and any $x \in M$ one has

$$\mathcal{H}^{n-1}[K \cap \mathrm{cut}(x)] < +\infty. \tag{12.36}$$

This however is a much more delicate result, which will be found in recent research papers.

Bibliographical notes

The denomination Monge–Ampère equation is used for any equation which resembles $\det(\nabla^2\psi) = f$, or more generally

$$\det\big(\nabla^2\psi - A(x,\psi,\nabla\psi)\big) = F\big(x,\psi,\nabla\psi\big). \tag{12.37}$$

Monge himself was probably not aware of the relation between the Monge problem and Monge–Ampère equations; this link was made much later, maybe in the work of Knott and Smith [524]. In any case it is Brenier [156] who made this connection popular among the community of partial differential equations. Accordingly, weak solutions of Monge–Ampère-type equations constructed by means of optimal transport are often called Brenier solutions in the literature. McCann [614] proved that such a solution automatically satisfies the Monge–Ampère equation almost everywhere (see the bibliographical notes for Chapter 11). Caffarelli [185] showed that for a convex target, Brenier's notion of solution is equivalent to the older concepts of Alexandrov solution and viscosity solution. These notions are reviewed in [814, Chapter 4] and a proof of the equivalence between Brenier and Alexandrov solutions is recast there. (The concept of Alexandrov solution is developed in [53, Section 11.2].) Feyel and Üstünel [359, 361, 362] studied the infinite-dimensional Monge–Ampère equation induced by optimal transport with quadratic cost on the Wiener space.

The modern regularity theory of the Monge–Ampère equation was pioneered by Alexandrov [16, 17] and Pogorelov [684, 685]. Since then it has become one of the most prestigious subjects of fully nonlinear partial differential equations, in relation to geometric problems such as the construction of isometric embeddings, or convex hypersurfaces with prescribed (multi-dimensional) Gauss curvature (for instance the Minkowski problem is about the construction of a convex hypersurface whose Gauss curvature is a prescribed function of the normal; the Alexandrov problem is about prescribing the so-called integral curvature). These issues are described in Bakelman's monograph [53]. Recently Oliker [659] has shown that the Alexandrov problem can be recast as an optimal transport problem on the sphere.

A modern account of some parts of the theory of the Monge–Ampère equation can be found in the recent book by Gutiérrez [452]; there is also an unpolished set of notes by Guan [446].

General references for fully nonlinear elliptic partial differential equations are the book by Caffarelli and Cabré [189], the last part of the one by Gilbarg and Trudinger [416], and the user-friendly notes by Krylov [531] or by Trudinger [786]. As a major development, the estimates derived by Evans, Krylov and Safonov [326, 530, 532] allow one to establish the regularity of fully nonlinear second-order equations under an assumption of uniform ellipticity. Eventually these techniques rely on Schauder-type estimates for certain linear equations. Monge–Ampère equations are uniformly elliptic only under a priori estimates on the second derivative of the unknown function; so these a priori estimates must first be established before applying the general theory. An elementary treatment of the basic linear Schauder estimates can be found, e.g., in [835] (together with many references), and a short treatment of the Monge–Ampère equation (based on [833] and on works by other authors) in [490].

Monge–Ampère equations arising in optimal transport have certain distinctive features; one of them is a particular type of boundary condition, sometimes referred to as the second boundary condition. The pioneering papers in this field are due to Delanoë in dimension 2 [284], then Caffarelli [185, 186, 187], Urbas [798, 799], and X.-J. Wang [833] in arbitrary dimension. The first three authors were interested in the case of a quadratic cost function in Euclidean space, while Wang considered the logarithmic cost function on the sphere, which appears in the so-called reflector antenna problem, at least in the particular case of "far field". (At the time of publication of [833] it was not yet understood that the problem treated there really was an optimal transport problem; it was only later that Wang made this remark.)

Wolfson [840] studied the optimal transport problem between two sets of equal area in the plane, with motivations in geometry, and identified obstructions to the existence of smooth solutions in terms of the curvature of the boundary.

Theorem 12.49 is one of the first steps in Caffarelli's regularity theory [185] for the quadratic cost function in Euclidean space; for more general cost functions it is due to Ma, Trudinger and Wang [585].

Theorem 12.50 (together with further refinements) appears in [185, 186, 187]. An informal introduction to the first steps of Caffarelli's techniques can be found in [814, Chapter 4]. The extension to unbounded domains is sketched in [15, Appendix]. Most of the theory was developed with nonconstructive arguments, but Forzani and Maldonado [376] were able to make at least the $C^{1,\beta}$ estimates quantitative. Apart from the $C^{\alpha} \to C^{2,\alpha}$ theory, there is also a difficult $C^0 \to W^{2,p}$ result [184], where p is arbitrarily large and f, g should be positive continuous. (The necessity of both the continuity and the positivity assumptions are demonstrated by counterexamples due to Wang [832]; see also [490].) Caffarelli and McCann [192] studied the regularity of the optimal map in the problem of partial optimal transport, in which only a fraction of the mass is transferred from the source to the target (both measures need not have the same mass); this problem transforms into a (double) obstacle problem for the Monge–Ampère equation. Figalli [365] obtained refined results on this problem by more elementary methods (not relying on the Monge–Ampère equation) and showed that there is in general no regularity if the supports of the source and target measures overlap. Another obstacle problem involving optimal transport, arising from a physical model, has been studied in [737].

Cordero-Erausquin [240] adapted Caffarelli's theory to the case of the torus, and Delanoë [285] and studied the stability of this theory under small perturbations. Roughly speaking he showed the following: Given two smooth probability densities f and g on (say) $\mathbb{T}^n$ and a smooth optimal transport T between $\mu = f\,\mathrm{vol}$ and $\nu = g\,\mathrm{vol}$, it is possible to slightly perturb f, g and the Riemannian metric, in such a way that the resulting optimal transport is still smooth. (Note carefully: How much you are allowed to perturb the metric depends on f and g.)

Urbas [798] considered directly the boundary regularity for uniformly convex domains in $\mathbb{R}^n$, by first establishing a so-called oblique derivative boundary condition, which is a sort of nonlinear version of the Neumann condition. (Actually, the uniform convexity of the domain makes the oblique condition more elliptic in some sense than the Neumann condition.) Fully nonlinear elliptic equations with oblique boundary condition had been studied before in [555, 560, 561], and the connection with the second boundary value problem for the Monge–Ampère equation had been suggested in [562]. Compared to Caffarelli's method, this one only covers the global estimates, and requires higher initial regularity; but it is more elementary.

The generalization of these regularity estimates to nonquadratic cost functions stood as an open problem for some time. Then Ma, Trudinger and Wang [585] discovered that the older interior estimates by Wang [833] could be adapted to general cost functions satisfying the condition $\mathfrak{S}_c > 0$ (this condition was called **(A3)** in their paper and in subsequent works). Theorem 12.52(ii) is extracted from this reference. A subtle caveat in [585] was corrected in [793] (see Theorem 1 there). A key property discovered in this study is that if c is a regular cost function and ψ is c-convex, then any local c-support function for ψ is also a global c-support function (which is nontrivial unless ψ is differentiable); an alternative proof can be derived from the method of Y.-H. Kim and McCann [519, 520].

Trudinger and Wang [794] later adapted the method of Urbas to treat the boundary regularity under the weaker condition $\mathfrak{S}_c \geq 0$ (there called **(A3w)**). The proof of Theorem 12.51 can be found there.

At this point Loeper [570] made three crucial contributions to the theory. First he derived the very strong estimates in Theorem 12.52(i) which showed that the Ma–Trudinger–Wang **(A3)** condition (called **(As)** in Loeper's paper) leads to a theory which is stronger than the Euclidean one in some sense (this was already somehow implicit in [585]). Secondly, he found a geometric interpretation of this condition, namely the regularity property (Definition 12.14), and related it to well-known geometric concepts such as sectional curvature (Particular Case 12.30). Thirdly, he proved that the weak condition **(A3w)** (called **(Aw)** in his work) is mandatory to derive regularity (Theorem 12.21). The psychological impact of this work was important: before that, the Ma–Trudinger–Wang condition could be seen as an obscure ad hoc assumption, while now it became the natural condition.

The proof of Theorem 12.52(i) in [570] was based on approximation and used auxiliary results from [585] and [793] (which also used some of the arguments in [570]... but there is no loophole!)

Loeper [571] further proved that the squared distance on the sphere is a uniformly regular cost, and combined all the above elements to derive Theorem 12.58; the proof is simplified in [572]. In [571], Loeper derived smoothness estimates similar to those in Theorem 12.52 for the far-field reflector antenna problem.

The exponent β in Theorem 12.52(i) is explicit; for instance, in the case when $f = d\mu/dx$ is bounded above and g is bounded below, Loeper obtained $\beta = (4n-1)^{-1}$, n being the dimension. (See [572] for a simplified proof.) However, this is not optimal: Liu [566] improved this into $\beta = (2n-1)^{-1}$, which is sharp.

In a different direction, Caffarelli, Gutiérrez and Huang [191] could get partial regularity for the far-field reflector antenna problem by very elaborate variants of Caffarelli's older techniques. This "direct" approach does not yet yield results as powerful as the a priori estimates by Loeper, Ma, Trudinger and Wang, since only C^1 regularity is obtained in [191], and only when the densities are bounded from above and below; but it gives new insights into the subject.

In dimension 2, the whole theory of Monge–Ampère equations becomes much simpler, and has been the object of numerous studies [744]. Old results by Alexandrov [17] and Heinz [471] imply C^1 regularity of the solution of $\det(\nabla^2\psi) = h$ as soon as h is bounded from above (and strict convexity if it is bounded from below). Loeper noticed that this implied strenghtened results for the solution of optimal transport with quadratic cost in dimension 2, and together with Figalli [368] extended this result to regular cost functions.

Now I shall briefly discuss counterexamples stories. Counterexamples by Pogorelov and Caffarelli (see for instance [814, pp. 128–129]) show that solutions of the usual Monge–Ampère equation are not smooth in general: some strict convexity on the solution is needed, and it has to come from boundary conditions in one way or the other.

The counterexample in Theorem 12.3 is taken from Caffarelli [185], where it is used to prove that the "Hessian measure" (a generalized formulation of the Hessian determinant) cannot be absolutely continuous if the bridge is thin enough; in the present notes I used a slightly different reasoning to directly prove the discontinuity of the optimal transport. The same can be said of Theorem 12.4, which is adapted from Loeper [570]. (In Loeper's paper the contradiction was obtained indirectly as in Theorem 12.44.)

Ma, Trudinger and Wang [585, Section 7.3] generalized Caffarelli's counterexample, showing that for any nonconvex target domain, there are smooth positive densities leading to nonsmooth optimal transport. Their result holds for more general cost functions, up to the replacement of convexity by c-convexity. The method of proof used in [585] is adapted from an older paper by Wang [833] on the reflector antenna problem. A similar strategy was rediscovered by Loeper [570] to construct counterexamples in the style of Theorem 12.44.

Theorem 12.7 was proven for these notes, with the aim of getting an elementary topological (as opposed to differential) version of Loeper's general nonsmoothness result.

An old counterexample by Lewy [744, Section 9.5] shows that the general equation (12.37) needs certain convexity-type assumptions; see [750, Section 3.3] for a rewriting of this construction. In view of Lewy's counterexample, specialists of regularity theory did expect that smoothness would need some assumptions on the cost function, although there was no hint of the geometric insights found by Loeper.

For nonregular cost functions, a natural question consists in describing the possible singularities of solutions of optimal transport problems. Do these singularities occur along reasonably nice curves, or with complicated, fractal-like geometries? No such result seems to be known.

Proposition 12.15, Theorems 12.20 and 12.44, Remarks 12.31 and 12.33, and the direct implication in Theorem 12.36 are all taken from Loeper [570] (up to issues about the domain of definition of $\nabla_x c$). The fact that $\mathfrak{S}_c$ is defined independently of the coordinate system was also noticed independently by Kim and McCann [520]. Theorem 12.35 is due to Trudinger and Wang [793], as well as Examples 12.40 and 12.41, except for the case of S^{n-1} which is due to Loeper [571]. The converse implication in Theorem 12.36 was proven independently by Trudinger and Wang [793] on the one hand, by Y.-H. Kim and McCann [519] on the other. (The result by Trudinger and Wang is slightly more general but the proof is more sophisticated since it involves delicate smoothness theorems, while the argument by Kim and McCann is much more direct.) The proof which I gave in these notes is a simplified version of the Kim–McCann argument, which evolved from discussions with Trudinger.

To go further, a refinement of the Ma–Trudinger–Wang conditions seemed to be necessary. The condition

$$\mathfrak{S}_c(x,y)\cdot(\xi,\eta) \geq K_0\,|\xi|^2\,|\widetilde{\eta}|^2 + C_0\,(\nabla^2_{x,y}c)\cdot(\xi,\eta)$$

is called $\mathrm{MTW}(K_0, C_0)$ in [572]; here $\widetilde{\eta} = -(\nabla^2_{x,y}c) \cdot \eta$. In the particular case $C_0 = +\infty$, one finds again the Ma–Trudinger–Wang condition (strong if $K_0 > 0$, weak if $K_0 = 0$). If c is the squared geodesic Riemannian distance, necessarily $C_0 \geq K_0$. According to [371], the squared distance on the sphere satisfies $\mathrm{MTW}(K_0, K_0)$ for some $K_0 \in (0,1)$ (this improves on [520, 571, 824]); numerical evidence even suggests that this cost satisfies the doubly optimal condition $\mathrm{MTW}(1,1)$.

The proof of Theorem 12.36 was generalized in [572] to the case when the cost is the squared distance and satisfies an $\mathrm{MTW}(K_0, C_0)$ condition. Moreover, one can afford to "slightly bend" the c-segments, and the inequality expressing regularity is also complemented with a remainder term proportional to $K_0\, t(1-t)\, (\inf |\eta|^2)\, |\zeta|^2$ (in the notation of the proof of Theorem 12.36).

Figalli, Kim and McCann [367] have been working on the adaptation of Caffarelli's techniques to cost functions satisfying $\mathrm{MTW}(0,0)$ (which is a reinforcement of the weak regularity property, satisfied by many examples which are not strictly regular).

Remark 12.45 is due to Kim [518], who constructed a smooth perturbation of a very thin cone, for which the squared distance is not a regular cost function, even though the sectional curvature is positive everywhere. (It is not known whether positive sectional curvature combined with some pinching assumption would be sufficient to guarantee the regularity.)

The transversal approximation property **(TA)** in Lemma 12.39 derives from a recent work of Loeper and myself [572], where it is proven for the squared distance on Riemannian manifolds with "nonfocal cut locus", i.e. no minimizing geodesic is focalizing. The denomination "transversal approximation" is because the approximating path $(\widehat{y}_t)$ is constructed in such a way that it goes transversally through the cut locus. Before that, Kim and McCann [519] had introduced a different density condition, namely that $\bigcap_t \mathrm{Dom}'(\nabla_y c(\,\cdot\,, y_t))$ be dense in M. The latter assumption clearly holds on S^n, but is not true for general manifolds. On the contrary, Lemma 12.39 shows that **(TA)** holds with great generality; the principle of the proof of this lemma, based on the co-area formula, was suggested to me by Figalli. The particular co-area formula which I used can be found in [331, p. 109]; it is established in [352, Sections 2.10.25 and 2.10.26].

Property $(\mathbf{Cut}^{n-1})$ for the squared geodesic distance was proven in independent papers by Itoh and Tanaka [489] and Li and Nirenberg [551];

in particular, inequality (12.36) is a particular case of Corollary 1.3 in the latter reference. These results are also established there for quite more general classes of cost functions.

In relation to the conjecture evoked in Remark 12.43, Loeper and I [572] proved that when the cost function is the squared distance on a Riemannian manifold with nonfocal cut locus, the strict form of the Ma–Trudinger–Wang condition automatically implies the uniform c-convexity of all $\mathrm{Dom}'(\nabla_x c(x,\,\cdot\,))$, and a strict version of the regularity property.

It is not so easy to prove that the solution of a Monge–Ampère equation such as (12.1) is the solution of the Monge–Kantorovich problem. There is a standard method based on the strict maximum principle [416] to prove uniqueness of solutions of fully nonlinear elliptic equations, but it requires smoothness (up to the boundary), and cannot be used directly to assess the identity of the smooth solution with the Kantorovich potential, which solves (12.1) only in weak sense. To establish the desired property, the strategy is to prove the c-convexity of the smooth solution; then the result follows from the uniqueness of the Kantorovich potential. This was the initial motivation behind Theorem 12.46, which was first proven by Trudinger and Wang [794, Section 6] under slightly different assumptions (see also the remark in Section 7 of the same work). All in all, Trudinger and Wang suggested three different methods to establish the c-convexity; one of them is a global comparison argument between the strong and the weak solution, in the style of Alexandrov. Trudinger suggested to me that the Kim–McCann strategy from [519] would yield an alternative proof of the c-convexity, and this is the method which I implemented to establish Theorem 12.46. (The proof by Trudinger and Wang is more general in the sense that it does not need the $\check{c}$-convexity; however, this gain of generality might not be so important because the c-convexity of ψ automatically implies the c-convexity of $\partial_c\psi$ and the $\check{c}$-convexity of $\partial^c\psi^c$, up to issues about the domain of differentiability of c.)

In [585] a local uniqueness statement is needed, but this is tricky since c-convexity is a global notion. So the problem arises whether a locally c-convex function (meaning that for each x there is y such that x is a *local* minimizer of $\psi+c(\,\cdot\,,y)$) is automatically c-convex. This local-to-global problem, which is closely related to Theorem 12.46 (and to the possibility of localizing the study of the Monge–Ampère equation (12.35)), is solved affirmatively for uniformly regular cost functions in [793] where a number of variants of c-convexity (local c-convexity, full c-convexity, strict c-convexity, global c-convexity) are carefully discussed. See also [571] for the case of the sphere.

In this chapter I chose to start with geometric (more general) considerations, such as the regularity property, and end up with analytic conditions in terms of the Ma–Trudinger–Wang tensor; but actually the tensor was discovered before the geometric conditions. The role of the Riemannian structure (and geodesic coordinates) in the presentation of this chapter might also seem artificial since it was noticed in Remark 12.31 that the meaningful quantities are actually independent of these choices. As a matter of fact, Kim and McCann [520] develop a framework which avoids any reference to them and identifies the Ma–Trudinger–Wang tensor as the sectional curvature tensor of the mixed second derivative $\partial^2 c/\partial x\,\partial y$, considered as a pseudo-Riemannian metric (with signature (n,n)) on the product manifold. In the same reference they also point out interesting connections with pseudo-Riemannian, Lagrangian and symplectic geometry (related to [840]).

Now some comments about terminology. The terminology of c-curvature was introduced by Loeper [570] after he made the connection between the Ma–Trudinger–Wang tensor and the sectional curvature. It was Trudinger who suggested the term "regular" to describe a matrix-valued function A in (12.37) that would satisfy adequate assumptions of the type $\mathfrak{S}_c \geq 0$. By extension, I used the same denomination for cost functions satisfying the property of Definition 12.14. Kim and McCann [519] call this property **(DASM)** (Double mountain Above Sliding Mountain).

To apply Theorems 12.51 or 12.52 to problems where the cost function is not everywhere differentiable, one first needs to make sure that the c-subdifferential of the unknown function lies within the domain of differentiability. Typically (for the cost function $d(x,y)^2$ on a Riemannian manifold), this means controlling the distance of $T(x)$ to the cut locus of x, where T is the optimal transport. ("Stay away from cut locus!") Until recently, the only manifolds for which this was known to be true independently of the probability densities (say bounded positive) were positively curved manifolds where all geodesics have the same length: the sphere treated in [286] and [571]; and the projective space considered in [287]. In the latter work, it was shown that the "stay-away" property still holds true if the variation of the length of geodesics is small with respect to certain other geometric quantities, and the probability densities satisfy certain size restrictions. Then Loeper and I [572] established smoothness estimates for the optimal transport on C^4 perturbations of the projective space, without any size restriction.

The cut locus is also a major issue in the study of the perturbation of these smoothness results. Because the dependence of the geodesic distance on the Riemannian metric is not smooth near the cut locus, it is not clear whether the Ma–Trudinger–Wang condition is stable under C^k perturbations of the metric, however large k may be. This stability problem, first formulated in [572], is in my opinion extremely interesting; it is solved by Figalli and Rifford [371] near S^2.

Without knowing the stability of the Ma–Trudinger–Wang condition, if pointwise a priori bounds on the probability densities are given, one can afford a C^4 perturbation of the metric and retain the Hölder continuity of optimal transport; or even afford a C^2 perturbation and retain a mesoscopic version of the Hölder continuity [822].

Some of the smoothness estimates discussed in these notes also hold for other more complicated fully nonlinear equations, such as the reflector antenna problem [507] (which in its general formulation does not seem to be equivalent to an optimal transport problem) or the so-called Hessian equations [789, 790, 792, 800], where the dominant term is a symmetric function of the eigenvalues of the Hessian of the unknown. The short survey by Trudinger [788] presents some results of this type, with applications to conformal geometry, and puts this into perspective together with optimal transport. In this reference Trudinger also notes that the problem of the prescribed Schouten tensor resembles an optimal transport problem with logarithmic cost function; this connection had also been made by McCann (see the remarks in [520]) who had long ago noticed the properties of conformal invariance of this cost function.

A topic which I did not address at all is the regularity of certain *sets* solving variational problems involving optimal transport; see [632].

13

Qualitative picture

This chapter is devoted to a recap of the whole picture of optimal transport on a smooth Riemannian manifold M. For simplicity I shall not try to impose the most general assumptions. A good understanding of this chapter is sufficient to attack Part II of this course.

Recap

Let M be a smooth complete connected Riemannian manifold, $L(x, v, t)$ a C^2 Lagrangian function on $TM \times [0, 1]$, satisfying the classical conditions of Definition 7.6, together with $\nabla_v^2 L > 0$. Let $c : M \times M \to \mathbb{R}$ be the induced cost function:

$$c(x, y) = \inf \left\{ \int_0^1 L(\gamma_t, \dot{\gamma}_t, t)\, dt; \quad \gamma_0 = x,\ \gamma_1 = y \right\}.$$

More generally, define

$$c^{s,t}(x, y) = \inf \left\{ \int_s^t L(\gamma_\tau, \dot{\gamma}_\tau, \tau)\, d\tau; \quad \gamma_s = x,\ \gamma_t = y \right\}.$$

So $c^{s,t}(x, y)$ is the optimal cost to go from point x at time s, to point y at time t.

I shall consider three cases: (i) $L(x, v, t)$ arbitrary on a compact manifold; (ii) $L(x, v, t) = |v|^2/2$ on a complete manifold (so the cost is $d^2/2$, where d is the distance); (iii) $L(x, v, t) = |v|^2/2$ in $\mathbb{R}^n$ (so the cost is $|x - y|^2/2$). Throughout the sequel, I denote by μ_0 the initial probability measure, and by μ_1 the final one. When I say "absolutely

C. Villani, *Optimal Transport*, Grundlehren der mathematischen Wissenschaften 338,

continuous" or "singular" this is in reference with the volume measure on the manifold (Lebesgue measure in $\mathbb{R}^n$).

Recall that a *generalized optimal coupling* is a c-cyclically monotone coupling. By analogy, I shall say that a *generalized displacement interpolation* is a path $(\mu_t)_{0\le t\le 1}$ valued in the space of probability measures, such that $\mu_t = \text{law}\,(\gamma_t)$ and γ is a random minimizing curve such that (γ_0, γ_1) is a generalized optimal coupling. These notions are interesting only when the total cost between μ_0 and μ_1 is infinite.

By gathering the results from the previous chapters, we know:

1. There always exists:

- an optimal coupling (or generalized optimal coupling) (x_0, x_1), with law π;
- a displacement interpolation (or generalized displacement interpolation) $(\mu_t)_{0\le t\le 1}$;
- a random minimizing curve γ with law Π;

such that $\text{law}\,(\gamma_t) = \mu_t$, and $\text{law}\,(\gamma_0, \gamma_1) = \pi$. Each curve γ is a solution of the Euler–Lagrange equation

$$\frac{d}{dt}\nabla_v L(\gamma_t, \dot\gamma_t, t) = \nabla_x L(\gamma_t, \dot\gamma_t, t). \tag{13.1}$$

In the case of a quadratic Lagrangian, this equation reduces to

$$\frac{d^2\gamma_t}{dt^2} = 0,$$

so trajectories are just geodesics, or straight lines in $\mathbb{R}^n$. Two trajectories in the support of Π may intersect at time $t = 0$ or $t = 1$, but never at intermediate times.

2. If either μ_0 or μ_1 is absolutely continuous, then so is μ_t, for all $t \in (0,1)$.

3. If μ_0 is absolutely continuous, then the optimal coupling (x_0, x_1) is unique (in law), deterministic ($x_1 = T(x_0)$) and characterized by the equation

$$\nabla\psi(x_0) = -\nabla_x c(x_0, x_1) = \nabla_v L(x_0, \dot\gamma_0, 0), \tag{13.2}$$

where $(\gamma_t)_{0\le t\le 1}$ is the minimizing curve joining $\gamma_0 = x_0$ to $\gamma_1 = x_1$ (it is part of the theorem that this curve is almost surely unique), and ψ is a c-convex function, that is, it can be written as

$$\psi(x) = \sup_{y\in M}\big[\phi(y) - c(x,y)\big]$$

for some nontrivial (i.e. not identically $-\infty$, and never $+\infty$) function ϕ. In case (ii), if nothing is known about the behavior of the distance function at infinity, then the gradient ∇ in the left-hand side of (13.2) should be replaced by an approximate gradient $\widetilde{\nabla}$.

4. Under the same assumptions, the (generalized) displacement interpolation $(\mu_t)_{0\le t\le 1}$ is unique. This follows from the almost sure uniqueness of the minimizing curve joining γ_0 to γ_1, where (γ_0, γ_1) is the optimal coupling. (Corollary 7.23 applies when the total cost is finite; but even if the total cost is infinite, we can apply a reasoning similar to the one in Corollary 7.23. Note that the result does not follow from the $\mathrm{vol}\otimes\mathrm{vol}\,(dx_0\,dx_1)$-uniqueness of the minimizing curve joining x_0 to x_1.)

5. Without loss of generality, one might assume that

$$\phi(y) = \inf_{x\in M} \big[\psi(x) + c(x,y)\big]$$

(these are true supremum and true infimum, not just up to a negligible set). One can also assume without loss of generality that

$$\forall x, y \in M, \qquad \phi(y) - \psi(x) \le c(x,y)$$

and

$$\phi(x_1) - \psi(x_0) = c(x_0, x_1) \qquad \text{almost surely.}$$

6. It is still possible that two minimizing curves meet at time $t = 0$ or $t = 1$, but this event may occur only on a very small set, of dimension at most $n - 1$.

7. All of the above remains true if one replaces μ_0 at time 0 by μ_t at time t, with obvious changes of notation (e.g. replace $c = c^{0,1}$ by $c^{t,1}$); the function ϕ is unchanged, but now ψ should be changed into ψ_t defined by

$$\psi_t(y) = \inf_{x\in M} \big[\psi_0(x) + c^{0,t}(x,y)\big]. \tag{13.3}$$

This ψ_t is a (viscosity) solution of the forward Hamilton–Jacobi equation

$$\partial_t \psi_t + L^*\big(x, \nabla\psi_t(x), t\big) = 0.$$

8. The equation for the optimal transport T_t between μ_0 and μ_t is as follows: $T_t(x)$ is the solution at time t of the Euler–Lagrange equation starting from x with velocity

$$v_0(x) = \big(\nabla_v L(x, \cdot\,, 0)\big)^{-1}(\nabla\psi(x)). \tag{13.4}$$

In particular,

- For the quadratic cost on a Riemannian manifold M, $T_t(x) = \exp_x(t\nabla\psi(x))$: To obtain T_t, flow for time t along a geodesic starting at x with velocity $\nabla\psi(x)$ (or rather $\widetilde{\nabla}\psi(x)$ if nothing is known about the behavior of M at infinity);
- For the quadratic cost in $\mathbb{R}^n$, $T_t(x) = (1-t)\,x + t\,\nabla\Psi(x)$, where $\Psi(x) = |x|^2/2 + \psi(x)$ defines a lower semicontinuous convex function in the usual sense. In particular, the optimal transport from μ_0 to μ_1 is a gradient of convex function, and this property characterizes it uniquely among all admissible transports.

9. Whenever $0 \le t_0 < t_1 \le 1$,

$$\int \psi_{t_1}\, d\mu_{t_1} - \int \psi_{t_0}\, d\mu_{t_0} = C^{t_0,t_1}(\mu_{t_0}, \mu_{t_1})$$
$$= \int_{t_0}^{t_1} \int L\big(x, [(\nabla_v L)(x, \cdot, t)]^{-1}(\nabla\psi_t(x)), t\big)\, d\mu_t(x)\, dt;$$

recall indeed Theorems 7.21 and 7.36, Remarks 7.25 and 7.37, and (13.4).

Simple as they may seem by now, these statements summarize years of research. If the reader has understood them well, then he or she is ready to go on with the rest of this course. The picture is not really complete and some questions remain open, such as the following:

Open Problem 13.1. *If the initial and final densities, ρ_0 and ρ_1, are positive everywhere, does this imply that the intermediate densities ρ_t are also positive? Otherwise, can one identify simple sufficient conditions for the density of the displacement interpolant to be positive everywhere?*

For general Lagrangian actions, the answer to this question seems to be negative, but it is not clear that one can also construct counterexamples for, say, the basic quadratic Lagrangian. My personal guess would be that the answer is about the same as for the smoothness: Positivity of the displacement interpolant is in general false except maybe for some particular manifolds satisfying an adequate structure condition.

Standard approximation procedure

In this last section I have gathered two useful approximation results which can be used in problems where the probability measures are either noncompactly supported, or singular.

In Chapter 10 we have seen how to treat the Monge problem in noncompact situations, without any condition at infinity, thanks to the notion of approximate differentiability. However, in practice, to treat noncompact situations, the simplest solution is often to use again a truncation argument similar to the one used in the proof of approximate differentiability. The next proposition displays the main scheme that one can use to deal with such situations.

Proposition 13.2 (Standard approximation scheme). *Let M be a smooth complete Riemannian manifold, let $c = c(x,y)$ be a cost function associated with a Lagrangian $L(x,v,t)$ on $TM \times [0,1]$, satisfying the classical conditions of Definition 7.6; and let μ_0, μ_1 be two probability measures on M. Let π be an optimal transference plan between μ_0 and μ_1, let $(\mu_t)_{0\le t\le 1}$ be a displacement interpolation and let Π be a dynamical optimal transference plan such that $(e_0,e_1)_\#\Pi = \pi$, $(e_t)_\#\Pi = \mu_t$. Let Γ be the set of all action-minimizing curves, equipped with the topology of uniform convergence; and let $(K_\ell)_{\ell\in\mathbb{N}}$ be a sequence of compact sets in Γ, such that $\Pi[UK_\ell] = 1$. For ℓ large enough, $\Pi[K_\ell] > 0$; then define*

$$Z_\ell := \Pi[K_\ell]; \qquad \Pi_\ell := \frac{1_{K_\ell}\Pi}{Z_\ell};$$

$$\mu_{t,\ell} := (e_t)_\#\Pi_\ell; \qquad \pi_\ell := (e_0,e_1)_\#\Pi_\ell;$$

and let c_ℓ be the restriction of c to $\mathrm{proj}_{M\times M}(K_\ell)$. Then for each ℓ, $(\mu_{t,\ell})_{0\le t\le 1}$ is a displacement interpolation and π_ℓ is an associated optimal transference plan; $\mu_{t,\ell}$ is compactly supported, uniformly in $t\in[0,1]$; and the following monotone convergences hold true:

$$Z_\ell \uparrow 1; \qquad Z_\ell\,\pi_\ell \uparrow \pi; \qquad Z_\ell\,\mu_{t,\ell} \uparrow \mu_t; \qquad Z_\ell\,\Pi_\ell \uparrow \Pi.$$

If moreover μ_0 is absolutely continuous, then there exists a c-convex function $\psi : M \to \mathbb{R}\cup\{+\infty\}$ such that π is concentrated on the graph of the transport $T : x \to (\nabla_x c)^{-1}(x, -\widetilde{\nabla}\psi(x))$. For any ℓ, $\mu_{0,\ell}$ is absolutely continuous, and the optimal transference plan π_ℓ is deterministic. Furthermore, there is a c_ℓ-convex function ψ_ℓ such that ψ_ℓ coincides with

ψ everywhere on $C_\ell := \mathrm{proj}_M(\mathrm{Spt}(\pi_\ell))$; and there is a set Z_ℓ such that $\mathrm{vol}\,[Z_\ell] = 0$ *and for any $x \in C_\ell \setminus Z_\ell$, $\nabla\psi(x) = \nabla\psi_\ell(x)$.*

Still under the assumption that μ_0 is absolutely continuous, the measures $\mu_{t,\ell}$ are also absolutely continuous, and the optimal transport $T_{t_0\to t,\ell}$ between $\mu_{t_0,\ell}$ and $\mu_{t,\ell}$ is deterministic, for any given $t_0 \in [0,1)$ and $t \in [0,1]$. In addition,

$$T_{t_0\to t,\ell} = T_{t_0\to t}, \qquad \mu_{t_0,\ell}\text{-almost surely},$$

where $T_{t_0\to t}$ is the optimal transport from μ_{t_0} to μ_t.

Proof of Proposition 13.2. The proof is quite similar to the argument used in the proof of uniqueness in Theorem 10.42 in a time-independent context. It is no problem to make this into a time-dependent version, since displacement interpolation behaves well under restriction, recall Theorem 7.30. The last part of the theorem follows from the fact that the map $T_{t_0\to t,\ell}$ can be written as $\gamma_{t_0} \mapsto \gamma_t$. □

Remark 13.3. Proposition 13.2 will be used several times throughout this course, for instance in Chapter 17. Its main drawback is that there is absolutely no control of the smoothness of the approximations: Even if the densities ρ_0 and ρ_1 are smooth, the approximate densities $\rho_{0,\ell}$ and $\rho_{1,\ell}$ will in general be discontinuous. In the proof of Theorem 23.14 in Chapter 23, I shall use another approximation scheme which respects the smoothness, but at the price of a loss of control on the approximation of the transport.

Let us now turn to the problem of approximating singular transport problems by smooth ones. If μ_0 and μ_1 are singular, there is a priori no uniqueness of the optimal transference plans, and actually there might be a large number (possibly uncountable) of them. However, the next theorem shows that singular optimal transference plans can always be approximated by nice ones.

Theorem 13.4 (Regularization of singular transport problems). *Let M be a smooth complete Riemannian manifold, and $c : M \times M \to \mathbb{R}$ be a cost function induced by a Lagrangian $L(x,v,t)$ satisfying the classical conditions of Definition 7.6. Further, let μ_0 and μ_1 be two probability measures on M, such that the optimal transport cost between μ_0 and μ_1 is finite, and let π be an optimal transference plan between μ_0 and μ_1. Then there are sequences $(\mu_0^k)_{k\in\mathbb{N}}$, $(\mu_1^k)_{k\in\mathbb{N}}$ and $(\pi^k)_{k\in\mathbb{N}}$ such that*

(i) each π^k *is an optimal transference plan between* μ_0^k *and* μ_1^k*, and any one of the probability measures* μ_0^k*,* μ_1^k *has a smooth, compactly supported density;*

(ii) $\mu_0^k \to \mu_0$*,* $\mu_1^k \to \mu_1$*,* $\pi^k \to \pi$ *in the weak sense as* $k \to \infty$*.*

Proof of Theorem 13.4. By Theorem 7.21, there exists a displacement interpolation $(\mu_t)_{0\le t\le 1}$ between μ_0 and μ_1; let $(\gamma_t)_{0\le t\le 1}$ be such that $\mu_t = \mathrm{law}\,(\gamma_t)$. The assumptions on L imply that action-minimizing curves solve a differential equation with Lipschitz coefficients, and therefore are uniquely determined by their initial position and velocity, a fortiori by their restriction to some time-interval $[0, t_0]$. So for any $t_0 \in (0, 1/2)$, by Theorem 7.30(ii), $(\gamma_{t_0}, \gamma_{1-t_0})$ is the unique optimal coupling between μ_{t_0} and μ_{1-t_0}. Now it is easy to construct a sequence $(\mu_{t_0}^k)_{k\in\mathbb{N}}$ such that $\mu_{t_0}^k$ converges weakly to μ_{t_0} as $k \to \infty$, and each $\mu_{t_0}^k$ is compactly supported with a smooth density. (To construct such a sequence, first truncate to ensure the property of compact support, then localize to charts by a partition of unity, and apply a regularization in each chart.) Similarly, construct a sequence $(\mu_{1-t_0}^k)_{k\in\mathbb{N}}$ such that $\mu_{1-t_0}^k$ converges weakly to μ_{1-t_0}, and each $\mu_{1-t_0}^k$ is compactly supported with a smooth density. Let $\pi_{t_0,1-t_0}^k$ be the unique optimal transference plan between μ_{t_0} and μ_{1-t_0}. By stability of optimal transport (Theorem 5.20), $\pi_{t_0,1-t_0}^k$ converges as $k \to \infty$ to $\pi_{t_0,1-t_0} = \mathrm{law}\,(\gamma_{t_0}, \gamma_{1-t_0})$. Then by continuity of γ, the random variable $(\gamma_{t_0}, \gamma_{1-t_0})$ converges pointwise to (γ_0, γ_1) as $t_0 \to 0$, which implies that $\pi_{t_0,1-t_0}$ converges weakly to π. The conclusion follows by choosing $t_0 = 1/n$, $k = k(n)$ large enough. □

Equations of displacement interpolation

In Chapter 7, we understood that a curve $(\mu_t)_{0\le t\le 1}$ obtained by displacement interpolation solves an action minimization problem in the space of measures, and we wondered whether we could obtain some nice equations for these curves. Here now is a possible answer. For simplicity I shall assume that there is enough control at infinity, that the notion of approximate differentiability can be dispensed with (this is the case for instance if M is compact).

Consider a displacement interpolation $(\mu_t)_{0\le t\le 1}$. By Theorem 7.21, μ_t can be seen as the law of γ_t, where the random path $(\gamma_t)_{0\le t\le 1}$ satisfies the Euler–Lagrange equation (13.1), and so at time t has velocity

$\xi_t(\gamma_t)$, where $\xi_t(x) := \big(\nabla_v L(x,\cdot\,,t)\big)^{-1}(\nabla\psi_t(x))$. By the formula of conservation of mass, μ_t satisfies

$$\frac{\partial \mu_t}{\partial t} + \nabla\cdot(\xi_t\,\mu_t) = 0$$

in the sense of distributions (be careful: ξ_t is not necessarily a gradient, unless L is quadratic). Then we can write down the **equations of displacement interpolation**:

$$\begin{cases} \dfrac{\partial \mu_t}{\partial t} + \nabla\cdot(\xi_t\,\mu_t) = 0; \\ \nabla_v L\big(x,\xi_t(x),t\big) = \nabla\psi_t(x); \\ \psi_0 \text{ is } c\text{-convex}; \\ \partial_t\psi_t + L^*\big(x,\nabla\psi_t(x),t\big) = 0. \end{cases} \tag{13.5}$$

If the cost function is just the square of the distance, then these equations become

$$\begin{cases} \dfrac{\partial \mu_t}{\partial t} + \nabla\cdot(\xi_t\,\mu_t) = 0; \\ \xi_t(x) = \nabla\psi_t(x); \\ \psi_0 \text{ is } d^2/2\text{-convex}; \\ \partial_t\psi_t + \dfrac{|\nabla\psi_t|^2}{2} = 0. \end{cases} \tag{13.6}$$

Finally, for the square of the Euclidean distance, this simplifies into

$$\begin{cases} \dfrac{\partial \mu_t}{\partial t} + \nabla\cdot(\xi_t\,\mu_t) = 0; \\ \xi_t(x) = \nabla\psi_t(x); \\ x \longrightarrow \dfrac{|x|^2}{2} + \psi_0(x) \;\; \text{is lower semicontinuous convex}; \\ \partial_t\psi_t + \dfrac{|\nabla\psi_t|^2}{2} = 0. \end{cases} \tag{13.7}$$

Apart from the special choice of initial datum, the latter system is well-known in physics as the **pressureless Euler equation**, for a potential velocity field.

Quadratic cost function

In a context of Riemannian geometry, it is natural to focus on the quadratic Lagrangian cost function, or equivalently on the cost function $c(x, y) = d(x, y)^2$, and consider the Wasserstein space $P_2(M)$. This will be the core of all the transport proofs in Part II of this course, so a key role will be played by $d^2/2$-convex functions (that is, c-convex functions for $c = d^2/2$). In Part III we shall consider metric structures that are not Riemannian, but still the square of the distance will be the only cost function. So in the remainder of this chapter I shall focus on that particular cost.

The class of $d^2/2$-convex functions might look a bit mysterious, and if they are so important it would be good to have simple characterizations of them. If ψ is $d^2/2$-convex, then $z \to \psi(z) + d(z, y)^2/2$ should have a minimum at x when $y = \exp_x(\nabla\psi(x))$. If in addition ψ is twice differentiable at x, then necessarily

$$\nabla^2\psi(x) \geq -\nabla^2 \left[\frac{d(\,\cdot\,, \exp_x \nabla\psi(x))^2}{2}\right](x). \tag{13.8}$$

However, this is only a necessary condition, and it is not clear that it would imply $d^2/2$-convexity, except for a manifold satisfying the very strong curvature condition $\mathfrak{S}_{d^2/2} \geq 0$ as in Theorem 12.46.

On the other hand, there is a simple and useful general criterion according to which sufficiently small functions are $d^2/2$-convex. This statement will guarantee in particular that any tangent vector $v \in TM$ can be represented as the gradient of a $d^2/2$-convex function.

Theorem 13.5 (C^2-small functions are $d^2/2$-convex). *Let M be a Riemannian manifold, and let K be a compact subset of M. Then, there is $\varepsilon > 0$ such that any function $\psi \in C_c^2(M)$ satisfying*

$$\mathrm{Spt}(\psi) \subset K, \qquad \|\psi\|_{C_b^2} \leq \varepsilon$$

is $d^2/2$-convex.

Example 13.6. Let $M = \mathbb{R}^n$, then ψ is $d^2/2$-convex if $\nabla^2\psi \geq -I_n$. (In this particular case there is no need for compact support, and a one-sided bound on the second derivative is sufficient.)

Proof of Theorem 13.5. Let (M, g) be a Riemannian manifold, and let K be a compact subset of M. Let $K' = \{x \in M;\ d(x, K) \leq 1\}$. For any $y \in M$, the Hessian of $x \to d(x,y)^2/2$ is equal to I_n (or, more rigorously, to the identity on T_xM) at $x = y$; so by compactness one may find $\delta > 0$ such that the Hessian of $x \to d(x,y)^2/2$ remains larger than $I_n/2$ as long as y stays in K' and $d(x,y) < 2\delta$. Without loss of generality, $\delta < 1/2$.

Now let ψ be supported in K, and such that

$$\forall x \in M \qquad |\psi(x)| < \frac{\delta^2}{4}, \quad |\nabla^2\psi(x)| < \frac{1}{4};$$

write

$$f_y(x) = \psi(x) + \frac{d(x,y)^2}{2}$$

and note that $\nabla^2 f_y \geq I_n/4$ in $B_{2\delta}(y)$, so f_y is uniformly convex in that ball.

If $y \in K'$ and $d(x,y) \geq \delta$, then obviously $f_y(x) \geq \delta^2/4 > \psi(y) = f_y(y)$; so the minimum of f_y can be achieved only in $B_\delta(y)$. If there are two distinct such minima, say x_0 and x_1, then we can join them by a geodesic $(\gamma_t)_{0 \leq t \leq 1}$ which stays within $B_{2\delta}(y)$ and then the function $t \to f_y(\gamma_t)$ is uniformly convex (because f_y is uniformly convex in $B_{2\delta}(y)$), and minimum at $t = 0$ and $t = 1$, which is impossible.

If $y \notin K'$, then $\psi(x) \neq 0$ implies $d(x,y) \geq 1$, so $f_y(x) \geq (1/2) - \delta^2/4$, while $f_y(y) = 0$. So the minimum of f_y can only be achieved at x such that $\psi(x) = 0$, and it has to be at $x = y$.

In any case, f_y has exactly one minimum, which lies in $B_\delta(y)$. We shall denote it by $x = T(y)$, and it is characterized as the unique solution of the equation

$$\nabla\psi(x) + \nabla_x \left(\frac{d(x,y)^2}{2} \right) = 0, \tag{13.9}$$

where x is the unknown.

Let x be arbitrary in M, and $y = \exp_x(\nabla\psi(x))$. Then (as a consequence of the first variation formula), $\nabla_x[d(x,y)^2/2] = -\nabla\psi(x)$, so equation (13.9) holds true, and $x = T(y)$. This means that, with the notation $c(x,y) = d(x,y)^2/2$, one has $\psi^c(y) = \psi(x) + c(x,y)$. Then $\psi^{cc}(x) = \sup[\psi^c(y) - c(x,y)] \geq \psi(x)$. Since x is arbitrary, actually we have shown that $\psi^{cc} \geq \psi$; but the converse inequality is always true, so $\psi^{cc} = \psi$, and then ψ is c-convex. □

Remark 13.7. The end of the proof took advantage of a general principle, independent of the particular cost c: If there is a *surjective* map T such that $f_y : x \to \psi(x) + c(x,y)$ is minimum at $T(y)$, then ψ is c-convex.

The structure of $P_2(M)$

A striking discovery made by Otto at the end of the nineties is that the differentiable structure on a Riemannian manifold M induces a kind of differentiable structure in the space $P_2(M)$. This idea takes substance from the following remarks: All of the path $(\mu_t)_{0\le t\le 1}$ is determined from the initial velocity field $\xi_0(x)$, which in turn is determined by $\nabla\psi$ as in (13.4). So it is natural to think of the function $\nabla\psi$ as a kind of "initial velocity" for the path (μ_t). The conceptual shift here is about the same as when we decided that μ_t could be seen either as the law of a random minimizing curve at time t, or as a path in the space of measures: Now we decide that $\nabla\psi$ can be seen either as the field of the initial velocities of our minimizing curves, or as the (abstract) velocity of the path (μ_t) at time $t = 0$.

There is an abstract notion of tangent space $T_x\mathcal{X}$ (at point x) to a metric space $(\mathcal{X}, d)$: in technical language, this is the *pointed Gromov–Hausdorff limit* of the rescaled space. It is a rather natural notion: fix your point x, and zoom onto it, by multiplying all distances by a large factor ε^{-1}, while keeping x fixed. This gives a new metric space $\mathcal{X}_{x,\varepsilon}$, and if one is not too curious about what happens far away from x, then the space $\mathcal{X}_{x,\varepsilon}$ might converge in some nice sense to some limit space, that may not be a vector space, but in any case is a cone. If that limit space exists, it is said to be the tangent space (or tangent cone) to $\mathcal{X}$ at x. (I shall come back to these issues in Part III.)

In terms of that construction, the intuition sketched above is indeed correct: let $P_2(M)$ be the metric space consisting of probability measures on M, equipped with the Wasserstein distance W_2. If μ is absolutely continuous, then the tangent cone $T_\mu P_2(M)$ exists and can be identified isometrically with the closed vector space generated by $d^2/2$-convex functions ψ, equipped with the norm

$$\|\nabla\psi\|_{L^2(\mu;TM)} := \left(\int_M |\nabla\psi|_x^2 \, d\mu(x)\right)^{1/2}.$$

Actually, in view of Theorem 13.5, this is the same as the vector space generated by *all* smooth, compactly supported gradients, completed with respect to that norm.

With what we know about optimal transport, this theorem is not that hard to prove, but this would require a bit too much geometric machinery for now. Instead, I shall spend some time on an important related result by Ambrosio, Gigli and Savaré, according to which any Lipschitz curve in the space $P_2(M)$ admits a *velocity* (which for all t lives in the tangent space at μ_t). Surprisingly, the proof will not require absolute continuity.

Theorem 13.8 (Representation of Lipschitz paths in $P_2(M)$). *Let M be a smooth complete Riemannian manifold, and let $P_2(M)$ be the metric space of all probability measures on M, with a finite second moment, equipped with the metric W_2. Further, let $(\mu_t)_{0\le t\le 1}$ be a Lipschitz-continuous path in $P_2(M)$:*

$$W_2(\mu_s, \mu_t) \le L\,|t-s|.$$

For any $t \in [0,1]$, let H_t be the Hilbert space generated by gradients of continuously differentiable, compactly supported ψ:

$$H_t := \overline{\mathrm{Vect}\big(\{\nabla\psi;\quad \psi \in C^1_c(M)\}\big)}^{L^2(\mu_t;TM)}.$$

Then there exists a measurable vector field $\xi_t(x) \in L^\infty(dt;\, L^2(d\mu_t(x)))$, $\mu_t(dx)\,dt$-almost everywhere unique, such that $\xi_t \in H_t$ for all t (i.e. the velocity field really is tangent along the path), and

$$\partial_t\mu_t + \nabla\cdot(\xi_t\,\mu_t) = 0 \tag{13.10}$$

in the weak sense.

Conversely, if the path $(\mu_t)_{0\le t\le 1}$ satisfies (13.10) *for some measurable vector field $(\xi_t(x))$ whose $L^2(\mu_t)$-norm is bounded by L, almost surely in t, then (μ_t) is a Lipschitz-continuous curve with $\|\dot\mu\| \le L$.*

The proof of Theorem 13.8 requires some analytical tools, and the reader might skip it at first reading.

Proof of Theorem 13.8. Let $\psi : M \to \mathbb{R}$ be a C^1 function, with Lipschitz constant at most 1. For all $s < t$ in $[0,1]$,

$$\left|\int_M \psi\,d\mu_t - \int_M \psi\,d\mu_s\right| \le W_1(\mu_s,\mu_t) \le W_2(\mu_s,\mu_t). \tag{13.11}$$

In particular, $\zeta(t) := \int_M \psi\,d\mu_t$ is a Lipschitz function of t. By Theorem 10.8(ii), the time-derivative of ζ exists for almost all times $t \in [0,1]$. Then let $\pi_{s,t}$ be an optimal transference plan between μ_s and μ_t (for the squared distance cost function). Let

$$\Psi(x,y) := \begin{cases} \dfrac{|\psi(x)-\psi(y)|}{d(x,y)} & \text{if } x \neq y \\[2ex] |\nabla\psi(x)| & \text{if } x = y. \end{cases}$$

Obviously Ψ is bounded by 1, and moreover it is upper semicontinuous.

If t is a differentiability point of ζ, then

$$\begin{aligned}
\left|\frac{d}{dt}\int \psi\, d\mu_t\right| &\leq \liminf_{\varepsilon\downarrow 0} \frac{1}{\varepsilon}\left|\int \psi\, d\mu_t - \int \psi\, d\mu_{t+\varepsilon}\right| \\
&\leq \liminf_{\varepsilon\downarrow 0} \frac{1}{\varepsilon}\int |\psi(y)-\psi(x)|\, d\pi_{t,t+\varepsilon}(x,y) \\
&\leq \left(\liminf_{\varepsilon\downarrow 0}\sqrt{\int \Psi(x,y)^2\, d\pi_{t,t+\varepsilon}(x,y)}\right) \frac{\sqrt{\int d(x,y)^2\, d\pi_{t,t+\varepsilon}(x,y)}}{\varepsilon} \\
&= \left(\liminf_{\varepsilon\downarrow 0}\sqrt{\int \Psi(x,y)^2\, d\pi_{t,t+\varepsilon}(x,y)}\right) \left(\frac{W_2(\mu_t,\mu_{t+\varepsilon})}{\varepsilon}\right) \\
&\leq \left(\liminf_{\varepsilon\downarrow 0}\sqrt{\int \Psi(x,y)^2\, d\pi_{t,t+\varepsilon}(x,y)}\right) L.
\end{aligned}$$

Since Ψ is upper semicontinuous and $\pi_{t,t+\varepsilon}$ converges weakly to $\delta_{x=y}$ (the trivial transport plan where nothing moves) as $\varepsilon \downarrow 0$, it follows that

$$\begin{aligned}
\left|\frac{d}{dt}\int \psi\, d\mu_t\right| &\leq L\sqrt{\int |\Psi(x,x)|^2\, d\mu_t(x)} \\
&= L\sqrt{\int |\nabla\psi(x)|^2\, d\mu_t(x)}.
\end{aligned}$$

Now the key remark is that the time-derivative $(d/dt)\int(\psi + C)\, d\mu_t$ does not depend on the constant C. This shows that $(d/dt)\int \psi\, d\mu_t$ really is a functional of $\nabla\psi$, obviously linear. The above estimate shows that this functional is *continuous* with respect to the norm in $L^2(d\mu_t)$.

Actually, this is not completely rigorous, since this functional is only defined for almost all t, and "almost all" here might depend on ψ. Here is a way to make things rigorous: Let $\mathcal{L}$ be the set of all Lipschitz functions ψ on M with Lipschitz constant at most 1, such that, say, $\psi(x_0) = 0$, where $x_0 \in M$ is arbitrary but fixed once for all, and ψ is supported in a fixed compact $K \subset M$. The set $\mathcal{L}$ is compact in the norm of uniform convergence, and contains a dense sequence $(\psi_k)_{k\in\mathbb{N}}$. By a regularization argument, one can assume that all those functions are actually of class C^1. For each ψ_k, we know that $\int \psi_k \, d\mu_t$ is differentiable for almost all $t \in [0,1]$; and since there are only countably many ζ_k's, we know that for almost every t, each ζ_k is differentiable at time t. The map $(d/dt) \int \alpha \, d\mu_t$ is well-defined at each of these times t, for all α in the vector space H_t generated by all the ψ_k's, and it is continuous if that vector space is equipped with the $L^2(d\mu_t)$ norm. It follows from the Riesz representation theorem that for each differentiability time t there exists a unique vector $\xi_t \in H_t \subset L^2(d\mu_t)$, with norm at most L, such that

$$\frac{d}{dt} \int \psi \, d\mu_t = \int \xi_t \cdot \nabla \psi \, d\mu_t. \tag{13.12}$$

This identity should hold true for any ψ_k, and by density it should also hold true for any $\psi \in C^1(M)$, supported in K.

Let $C^1_K(M)$ be the set of $\psi \in C^1(M)$ that are supported in K. We just showed that there is a negligible set of times, τ_K, such that (13.12) holds true for all $\psi \in C^1_K(M)$ and $t \notin \tau_K$. Now choose an increasing family of compact sets $(K_m)_{m\in\mathbb{N}}$, with $\cup K_m = M$, so that any compact set is included in some K_m. Then (13.12) holds true for all $\psi \in C^1_c(M)$, as soon as t does not belong to the union of τ_{K_m}, which is still a negligible set of times.

But equation (13.12) is really the weak formulation of (13.10). Since ξ_t is uniquely determined in $L^2(d\mu_t)$, for almost all t, actually the vector field $\xi_t(x)$ is $d\mu_t(x)\,dt$-uniquely determined.

To conclude the proof of the theorem, it only remains to prove the converse implication. Let (μ_t) and (ξ_t) solve (13.10). By the equation of conservation of mass, $\mu_t = \operatorname{law}(\gamma_t)$, where γ_t is a (random) solution of

$$\dot{\gamma}_t = \xi_t(\gamma_t).$$

Let $s < t$ be any two times in $[0,1]$. From the formula

$$d(\gamma_s, \gamma_t)^2 = (t-s) \inf \left\{ \int_s^t |\dot{\zeta}_\tau|^2 \, d\tau; \quad \zeta_s = \gamma_s, \ \zeta_t = \gamma_t \right\},$$

we deduce

$$d(\gamma_s, \gamma_t)^2 \le (t-s) \int_s^t |\dot{\gamma}_\tau|^2 \, d\tau \le (t-s) \int_s^t |\xi_t(\gamma_t)|^2 \, d\tau.$$

So

$$\mathbb{E}\, d(\gamma_s, \gamma_t)^2 \le (t-s) \int_s^t |\xi_\tau(x)|^2 \, d\mu_\tau(x)\, d\tau \le (t-s)^2 \|\xi\|_{L^\infty(dt;\, L^2(d\mu_t))}.$$

In particular

$$W_2(\mu_s, \mu_t)^2 \le \mathbb{E}\, d(\gamma_s, \gamma_t)^2 \le L^2 (t-s)^2,$$

where L is an upper bound for the norm of ξ in $L^\infty(L^2)$. This concludes the proof of Theorem 13.8. □

Remark 13.9. With hardly any more work, the preceding theorem can be extended to cover paths that are absolutely continuous of order 2, in the sense defined on p. 115. Then of course the velocity field will not live in $L^\infty(dt;\, L^2(d\mu_t))$, but in $L^2(d\mu_t\, dt)$.

Observe that in a displacement interpolation, the initial measure μ_0 and the initial velocity field $\nabla\psi_0$ uniquely determine the final measure μ_1: this implies that *geodesics in* $P_2(M)$ *are nonbranching*, in the strong sense that their initial position and velocity determine uniquely their final position.

Finally, we can now derive an "explicit" formula for the action functional determining displacement interpolations as minimizing curves. Let $\mu = (\mu_t)_{0\le t\le 1}$ be any Lipschitz (or absolutely continuous) path in $P_2(M)$; let $\xi_t(x) = \nabla\psi_t(x)$ be the associated time-dependent velocity field. By the formula of conservation of mass, μ_t can be interpreted as the law of γ_t, where γ is a random solution of $\dot{\gamma}_t = \xi_t(\gamma_t)$. Define

$$\mathbb{A}(\mu) := \inf \int_0^1 \mathbb{E}_{\mu_t} |\xi_t(\gamma_t)|^2 \, dt, \qquad (13.13)$$

where the infimum is taken over all possible realizations of the random curves γ. By Fubini's theorem,

$$\begin{aligned} \mathbb{A}(\mu) = \inf \mathbb{E} \int_0^1 |\xi_t(\gamma_t)|^2 \, dt &= \inf \mathbb{E} \int_0^1 |\dot{\gamma}_t|^2 \, dt \\ &\ge \mathbb{E} \inf \int_0^1 |\dot{\gamma}_t|^2 \, dt \\ &= \mathbb{E}\, d(\gamma_0, \gamma_1)^2, \end{aligned}$$

and the infimum is achieved if and only if the coupling (γ_0, γ_1) is minimal, and the curves γ are (almost surely) action-minimizing. This shows that displacement interpolations are characterized as the minimizing curves for the action $\mathbb{A}$. Actually $\mathbb{A}$ is the same as the action appearing in Theorem 7.21 (iii), the only improvement is that now we have produced a more explicit form in terms of vector fields.

The expression (13.13) can be made slightly more explicit by noting that the optimal choice of velocity field is the one provided by Theorem 13.8, which is a gradient, so we may restrict the action functional to gradient velocity fields:

$$\mathbb{A}(\mu) := \int_0^1 \mathbb{E}_{\mu_t} |\nabla\psi_t|^2 \, dt; \qquad \frac{\partial \mu_t}{\partial t} + \nabla\cdot(\nabla\psi_t\,\mu_t) = 0. \tag{13.14}$$

Note the formal resemblance to a Riemannian structure: What the formula above says is

$$W_2(\mu_0,\mu_1)^2 = \inf \int_0^1 \|\dot{\mu}_t\|^2_{T_{\mu_t}P_2} \, dt, \tag{13.15}$$

where the norm on the tangent space $T_\mu P_2$ is defined by

$$\begin{aligned} \|\dot{\mu}\|^2_{T_\mu P_2} &= \inf\left\{\int |v|^2\, d\mu; \qquad \dot{\mu} + \nabla\cdot(v\mu) = 0\right\} \\ &= \int |\nabla\psi|^2\, d\mu; \qquad \dot{\mu} + \nabla\cdot(\nabla\psi\,\mu) = 0. \end{aligned}$$

Remark 13.10. There is an appealing physical interpretation, which really is an infinitesimal version of the optimal transport problem. Imagine that you observe the (infinitesimal) evolution of the density of particles moving in a continuum, but don't know the actual velocities of these particles. There might be many velocity fields that are compatible with the observed evolution of density (many solutions of the continuity equation). Among all possible solutions, select the one with minimum kinetic energy. This energy is (up to a factor 1/2) the square norm of your infinitesimal evolution.

Bibliographical notes

Formula (13.8) appears in [246]. It has an interesting consequence which can be described as follows: On a Riemannian manifold, the optimal transport starting from an absolutely continuous probability measure *almost never hits the cut locus*; that is, the set of points x such that the image $T(x)$ belongs to the cut locus of x is of zero probability. Although we already know that almost surely, x and $T(x)$ are joined by a unique geodesic, this alone does not imply that the cut locus is almost never hit, because it is possible that y belongs to the cut locus of x and still x and y are joined by a unique minimizing geodesic. (Recall the discussion after Problem 8.8.) But Cordero-Erausquin, McCann and Schmuckenschläger show that if such is the case, then $d(x,z)^2/2$ fails to be semiconvex at $z=y$. On the other hand, by Alexandrov's second differentiability theorem (Theorem 14.1), ψ is twice differentiable almost everywhere; formula (13.8), suitably interpreted, says that $d(x,\,\cdot\,)^2/2$ is semiconvex at $T(x)$ whenever ψ is twice differentiable at x.

At least in the Euclidean case, and up to regularity issues, the explicit formulas for geodesic curves and action in the space of measures were known to Brenier, around the mid-nineties. Otto [669] took a conceptual step forward by considering formally $P_2(M)$ as an infinite-dimensional Riemannian manifold, in view of formula (13.15). For some time it was used as a purely formal, yet quite useful, heuristic method (as in [671], or later in Chapter 15). It is only recently that rigorous constructions were performed in several research papers, e.g. [30, 203, 214, 577]. The approach developed in this chapter relies heavily on the work of Ambrosio, Gigli and Savaré [30] (in $\mathbb{R}^n$). A more geometric treatment can be found in [577, Appendix A]; see also [30, Section 12.4], and [655, Section 3] (I shall give a few more details in the bibliographical notes of Chapter 26). As I am completing this course, an important contribution to the subject has just been made by Lott [575] who established "explicit" formulas for the Riemannian connection and curvature in $P_2^{\mathrm{ac}}(M)$, or rather in the subset made of smooth positive densities, when M is compact.

The pressureless Euler equations describe the evolution of a gas of particles which interact only when they meet, and then stick together (sticky particles). It is a very degenerate system whose study turns out to be tricky in general [169, 320, 745]. But in applications to optimal transport, it comes in the very particular case of potential flow (the velocity field is a gradient), so the evolution is governed by a simple Hamilton–Jacobi equation.

There are two natural ways to extend a minimizing geodesic $(\mu_t)_{0\le t\le 1}$ in $P_2(M)$ into a geodesic defined (but not necessarily minimizing) at all times. One is to solve the Hamilton–Jacobi equation for all times, in the viscosity sense; then the gradient of the solution will define a velocity field, and one can let (μ_t) evolve by transport, as in (13.6). Another way [30, Example 11.2.9] is to construct a trajectory of the gradient flow of the energy $-W_2(\sigma,\,\cdot\,)^2/2$, where σ is, say, μ_0, and the trajectory starts from an intermediate point, say $\mu_{1/2}$. The existence of this gradient flow follows from [30, Theorems 10.4.12 and 11.2.1], while [30, Theorem 11.2.1] guarantees that it coincides (up to time-reparameterization) with the original minimizing geodesic for short times. (This is close to the construction of quasigeodesics in [678].) It is natural to expect that both approaches give the same result, but as far as I know, this has not been established.

Khesin suggested to me the following nice problem. Let $\mu = (\mu_t)_{t\geq 0}$ be a geodesic in the Wasserstein space $P_2(\mathbb{R}^n)$ (defined with the help of the pressureless Euler equation), and characterize the cut time of μ as the "time of the first shock". If μ_0 is absolutely continuous with positive density, this essentially means the following: let $t_c = \inf\{t_1;\ (\mu_t)_{0\leq t\leq t_1}$ is not a minimizing geodesic$\}$, let $t_s = \sup\{t_0;\ \mu_t$ is absolutely continuous for $t \leq t_0\}$, and show that $t_c = t_s$. Since t_c should be equal to $\sup\{t_2;\ |x|^2/2 + t\,\psi(0,x)$ is convex for $t \leq t_2\}$, the solution of this problem is related to a qualitative study of the way in which convexity degenerates at the first shock. In dimension 1, this problem can be studied very precisely [642, Chapter 1] but in higher dimensions the problem is more tricky. Khesin and Misiołek obtain some results in this direction in [515].

Kloeckner [522] studied the isometries of $P_2(\mathbb{R})$ and found that they are not all induced by isometries of $\mathbb{R}$; roughly speaking, there is one additional "exotic" isometry.

In his PhD thesis, Agueh [4] studied the structure of $P_p(M)$ for $p > 1$ (not necessarily equal to 2). Ambrosio, Gigli and Savaré [30] pushed these investigations further.

Displacement interpolation becomes somewhat tricky in presence of boundaries. In his study of the porous medium equations, Otto [669] considered the case of a bounded open set of $\mathbb{R}^n$ with C^2 boundary.

For many years, the great majority of applications of optimal transport to problems of applied mathematics have taken place in Euclidean setting, but more recently some "genuinely Riemannian" applications have started to pop out. There was an original suggestion to use optimal transport in a three-dimensional Riemannian manifold (actually, a cube equipped with a varying metric) related to image perception and the matching of pictures with different contrasts [289]. In a meteorological context, it is natural to consider the sphere (as a model of the Earth), and in the study of the semi-geostrophic system one is naturally led to optimal transport on the sphere [263, 264]; actually, it is even natural to consider a conformal change of metric which "pinches" the sphere along its equator [263]! For completely different reasons, optimal transport on the sphere was recently used by Otto and Tzavaras [670] in the study of a coupled fluid-polymer model.

Part II

Optimal transport and Riemannian geometry

This second part is devoted to the exploration of Riemannian geometry through optimal transport. It will be shown that the geometry of a manifold influences the qualitative properties of optimal transport; this can be quantified in particular by the effect of *Ricci curvature bounds* on the convexity of certain well-chosen functionals along displacement interpolation. The first hints of this interplay between Ricci curvature and optimal transport appeared around 2000, in works by Otto and myself, and shortly after by Cordero-Erausquin, McCann and Schmuckenschläger.

Throughout, the emphasis will be on the quadratic cost (the transport cost is the square of the geodesic distance), with just a few exceptions. Also, most of the time I shall only handle measures which are absolutely continuous with respect to the Riemannian volume measure.

Chapter 14 is a preliminary chapter devoted to a short and tentatively self-contained exposition of the main properties of Ricci curvature. After going through this chapter, the reader should be able to understand all the rest without having to consult any extra source on Riemannian geometry. The estimates in this chapter will be used in Chapters 15, 16 and 17.

Chapter 15 presents a powerful formal differential calculus on the Wasserstein space, cooked up by Otto.

Chapters 16 and 17 establish the main relations between displacement convexity and Ricci curvature. Not only do Ricci curvature bounds imply certain properties of displacement convexity, but conversely these properties characterize Ricci curvature bounds. These results will play a key role in the rest of the course.

In Chapters 18 to 22 the main theme will be that many classical properties of Riemannian manifolds, that come from Ricci curvature estimates, can be conveniently derived from displacement convexity techniques. This includes in particular estimates about the growth of the volume of balls, Sobolev-type inequalities, concentration inequalities, and Poincaré inequalities.

Then in Chapter 23 it is explained how one can define certain *gradient flows* in the Wasserstein space, and recover in this way well-known equations such as the heat equation. In Chapter 24 some of the functional inequalities from the previous chapters are applied to the qualitative study of some of these gradient flows. Conversely, gradient flows provide alternative proofs to some of these inequalities, as shown in Chapter 25.

The issues discussed in this part are concisely reviewed in the surveys by Cordero-Erausquin [244] and myself [821] (both in French).

Convention: Throughout Part II, unless otherwise stated, a "Riemannian manifold" is a *smooth, complete connected finite-dimensi onal Riemannian manifold distinct from a point, equipped with a smooth metric tensor.*

14

Ricci curvature

Curvature is a generic name to designate a local invariant of a metric space that quantifies the deviation of this space from being Euclidean. (Here "local invariant" means a quantity which is invariant under local isometries.) It is standard to define and study curvature mainly on Riemannian manifolds, for in that setting definitions are rather simple, and the Riemannian structure allows for "explicit" computations. Throughout this chapter, M will stand for a complete connected Riemannian manifold, equipped with its metric g.

The most popular curvatures are: the **sectional curvature** σ (for each point x and each plane $P \subset T_xM$, $\sigma_x(P)$ is a number), the **Ricci curvature** Ric (for each point x, Ric_x is a quadratic form on the tangent space T_xM), and the **scalar curvature** S (for each point x, S_x is a number). All of them can be obtained by reduction of the **Riemann curvature tensor**. The latter is easy to define: If ∇_X stands for the covariant derivation along the vector field X, then

$$\mathrm{Riem}(X,Y) := \nabla_Y \nabla_X - \nabla_X \nabla_Y + \nabla_{[X,Y]};$$

but it is notoriously difficult, even for specialists, to get some understanding of its meaning. The Riemann curvature can be thought of as a tensor with four indices; it can be expressed in coordinates as a nonlinear function of the Christoffel symbols and their partial derivatives.

Of these three notions of curvature (sectional, Ricci, scalar), the sectional one is the most precise; in fact the knowledge of all sectional curvatures is equivalent to the knowledge of the Riemann curvature. The Ricci curvature is obtained by "tracing" the sectional curvature: If e is a given unit vector in T_xM and $(e, e_2, \dots, e_n)$ is an orthonormal basis of T_xM, then $\mathrm{Ric}_x(e,e) = \sum \sigma_x(P_j)$, where P_j $(j = 2, \dots, n)$

C. Villani, *Optimal Transport*, Grundlehren der mathematischen Wissenschaften 338,

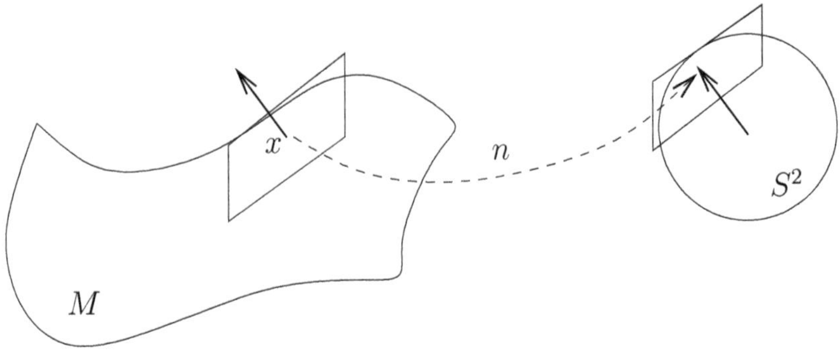

Fig. 14.1. The dashed line gives the recipe for the construction of the Gauss map; its Jacobian determinant is the Gauss curvature.

is the plane generated by $\{e, e_j\}$. Finally, the scalar curvature is the trace of the Ricci curvature. So a control on the sectional curvature is stronger than a control on the Ricci curvature, which in turn is stronger than a control on the scalar curvature.

For a surface (manifold of dimension 2), these three notions reduce to just one, which is the Gauss curvature and whose definition is elementary. Let us first describe it from an extrinsic point of view. Let M be a two-dimensional submanifold of $\mathbb{R}^3$. In the neighborhood of a point x, choose a unit normal vector $n = n(y)$, then this defines locally a smooth map n with values in $S^2 \subset \mathbb{R}^3$ (see Figure 14.1). The tangent spaces T_xM and $T_{n(x)}S^2$ are parallel planes in $\mathbb{R}^3$, which can be identified unambiguously. So the determinant of the differential of n can also be defined without ambiguity, and this determinant is called the curvature. The fact that this quantity is invariant under isometries is one of Gauss's most famous results, a *tour de force* at the time. (To appreciate this theorem, the reader might try to prove it by elementary means.)

As an illustration of this theorem: If you hold a sheet of paper straight, then its equation (as an embedded surface in $\mathbb{R}^3$, and assuming that it is infinite) is just the equation of a plane, so obviously it is not curved. Fine, but now bend the sheet of paper so that it looks like valleys and mountains, write down the horrible resulting equations, give it to a friend and ask him whether it is curved or not. One thing he can do is compute the Gauss curvature from your horrible equations, find that it is identically 0, and deduce that your surface was not curved at all. Well, it looked curved as a surface which was embedded in $\mathbb{R}^3$, but from an *intrinsic* point of view it was not: A tiny creature living on the surface of the sheet, unable to measure the lengths of curves going outside of the surface, would never have noticed that you bent the sheet.

To construct isometries from (M,g) to something else, pick up any diffeomorphism $\varphi : M \to M'$, and equip $M' = \varphi(M)$ with the metric $g' = (\varphi^{-1})^* g$, defined by $g'_x(v) = g_{\varphi^{-1}(x)}(d_x\varphi^{-1}(v))$. Then φ is an isometry between (M,g) and (M',g'). Gauss's theorem says that the curvature computed in (M,g) and the curvature computed in (M',g') are the same, modulo obvious changes (the curvature at point x along a plane P should be compared with the curvature at $\varphi(x)$ along a plane $d_x\varphi(P)$). This is why one often says that the curvature is "invariant under the action of diffeomorphisms".

Curvature is intimately related to the **local behavior of geodesics**. The general rule is that, in the presence of positive curvature, geodesics have a tendency to *converge* (at least for short times), while in the presence of negative curvature they have a tendency to *diverge*. This tendency can usually be felt only at second or third order in time: at first order, the convergence or divergence of geodesics is dictated by the initial conditions. So if, on a space of (strictly) positive curvature, you start two geodesics from the same point with velocities pointing in different directions, the geodesics will start to diverge, but then the tendency to diverge will diminish. Here is a more precise statement, which will show at the same time that the Gauss curvature is an intrinsic notion: From a point $x \in M$, start two constant-speed geodesics with unit speed, and respective velocities v, w. The two curves will spread apart; let $\delta(t)$ be the distance between their respective positions at time t. In a first approximation, $\delta(t) \simeq \sqrt{2(1-\cos\theta)}\, t$, where θ is the angle between v and w (this is the same formula as in Euclidean space). But a more precise study shows that

$$\delta(t) = \sqrt{2(1-\cos\theta)}\, t \left(1 - \frac{\kappa_x \cos^2(\theta/2)}{6}\, t^2 + O(t^4)\right), \tag{14.1}$$

where κ_x is the Gauss curvature at x.

Once the intrinsic nature of the Gauss curvature has been established, it is easy to define the notion of sectional curvature for Riemannian manifolds of any dimension, embedded or not: If $x \in M$ and $P \subset T_xM$, define $\sigma_x(P)$ as the Gauss curvature of the surface which is obtained as the image of P by the exponential map $\exp_x$ (that is, the collection of all geodesics starting from x with a velocity in P). Another equivalent definition is by reduction of the Riemann curvature tensor: If $\{u, v\}$ is an orthonormal basis of P, then $\sigma_x(P) = \langle \mathrm{Riem}\,(u, v)\cdot u,\, v\rangle$.

It is obvious from the first definition of Gauss curvature that the unit two-dimensional sphere S^2 has curvature $+1$, and that the Euclidean plane $\mathbb{R}^2$ has curvature 0. More generally, the sphere $S^n(R)$, of dimension n and radius R, has constant sectional curvature $1/R^2$; while the n-dimensional Euclidean space $\mathbb{R}^n$ has curvature 0. The other classical example is the hyperbolic space, say $\mathbb{H}^n(R) = \{(x, y) \in \mathbb{R}^{n-1} \times (0, +\infty)\}$ equipped with the metric $R^2(dx^2 + dy^2)/y^2$, which has constant sectional curvature $-1/R^2$. These three families (spheres, Euclidean, hyperbolic) constitute the only simply connected Riemannian manifolds with constant sectional curvature, and they play an important role as comparison spaces.

The qualitative properties of optimal transport are also (of course) related to the behavior of geodesics, and so it is natural to believe that curvature has a strong influence on the solution of optimal transport. Conversely, some curvature properties can be read off on the solution of optimal transport. At the time of writing, these links have been best understood in terms of *Ricci curvature*; so this is the point of view that will be developed in the sequel.

This chapter is a tentative crash course on Ricci curvature. Hopefully, a reader who has never heard about that topic before should, by the end of the chapter, know enough about it to understand all the rest of the notes. This is by no means a complete course, since most proofs will only be sketched and many basic results will be taken for granted.

In practice, Ricci curvature usually appears from two points of view: (a) estimates of the **Jacobian determinant of the exponential map**; (b) **Bochner's formula**. These are two complementary points of view on the same phenomenon, and it is useful to know both. Before going on, I shall make some preliminary remarks about Riemannian calculus at second order, for functions which are not necessarily smooth.

Preliminary: second-order differentiation

All curvature calculations involve second-order differentiation of certain expressions. The notion of covariant derivation lends itself well to those computations. A first thing to know is that the exchange of derivatives is still possible. To express this properly, consider a parametrized surface $(s, t) \to \gamma(s, t)$ in M, and write d/dt (resp. d/ds)

for the differentiation along γ, viewed as a function of t with s frozen (resp. as a function of s with t frozen); and D/Dt (resp. D/Ds) for the corresponding covariant differentiation. Then, if $F \in C^2(M)$, one has

$$\frac{D}{Ds}\left(\frac{dF}{dt}\right) = \frac{D}{Dt}\left(\frac{dF}{ds}\right). \tag{14.2}$$

Also a crucial concept is that of the **Hessian operator**. If f is twice differentiable on $\mathbb{R}^n$, its Hessian matrix is just $(\partial^2 f/\partial x_i\, \partial x_j)_{1\le i,j\le n}$, that is, the array of all second-order partial derivatives. Now if f is defined on a Riemannian manifold M, the Hessian operator at x is the linear operator $\nabla^2 f(x) : T_xM \to T_xM$ defined by the identity

$$\nabla^2 f \cdot v = \nabla_v(\nabla f).$$

(Recall that ∇_v stands for the covariant derivation in the direction v.) In short, $\nabla^2 f$ is the covariant gradient of the gradient of f.

A convenient way to compute the Hessian of a function is to differentiate it twice along a geodesic path. Indeed, if $(\gamma_t)_{0\le t\le 1}$ is a geodesic path, then $\nabla_{\dot\gamma}\dot\gamma = 0$, so

$$\begin{aligned}\frac{d^2}{dt^2} f(\gamma_t) = \frac{d}{dt}\langle \nabla f(\gamma_t), \dot\gamma_t\rangle &= \Big\langle \nabla_{\dot\gamma}\nabla f(\gamma_t), \dot\gamma_t\Big\rangle + \Big\langle \nabla f(\gamma_t), \nabla_{\dot\gamma}\dot\gamma_t\Big\rangle \\ &= \big\langle \nabla^2 f(\gamma_t)\cdot\dot\gamma_t, \dot\gamma_t\big\rangle.\end{aligned}$$

In other words, if $\gamma_0 = x$ and $\dot\gamma_0 = v \in T_xM$, then

$$f(\gamma_t) = f(x) + t\,\langle \nabla f(x), v\rangle + \frac{t^2}{2}\big\langle \nabla^2 f(x)\cdot v, v\big\rangle + o(t^2). \tag{14.3}$$

This identity can actually be used to define the Hessian operator.

A similar computation shows that for any two tangent vectors u, v at x,

$$\frac{D}{Ds}\left(\frac{d}{dt} f\big(\exp_x(su+tv)\big)\right) = \big\langle \nabla^2 f(x)\cdot u, v\big\rangle, \tag{14.4}$$

where $\exp_x v$ is the value at time 1 of the constant speed geodesic starting from x with velocity v. Identity (14.4) together with (14.2) shows that if $f \in C^2(M)$, then $\nabla^2 f(x)$ is a symmetric operator: $\langle \nabla^2 f(x)\cdot u, v\rangle_x = \langle \nabla^2 f(x)\cdot v, u\rangle_x$. In that case it will often be convenient to think of $\nabla^2 f(x)$ as a quadratic form on T_xM.

The Hessian is related to another fundamental second-order differential operator, the **Laplacian**, or Laplace–Beltrami operator. The Laplacian can be defined as the trace of the Hessian:

$$\Delta f(x) = \mathrm{tr}\,(\nabla^2 f(x)).$$

Another possible definition is

$$\Delta f = \nabla \cdot (\nabla f),$$

where $\nabla \cdot$ is the **divergence** operator, defined as the negative of the adjoint of the gradient in $L^2(M)$: More explicitly, if ξ is a C^1 vector field on M, its divergence is defined by

$$\forall \zeta \in C_c^\infty(M), \qquad \int_M (\nabla \cdot \xi)\, \zeta \, d\mathrm{vol} = - \int_M \xi \cdot \nabla \zeta \, d\mathrm{vol}.$$

Both definitions are equivalent; in fact, the divergence of any vector field ξ coincides with the trace of the covariant gradient of ξ. When $M = \mathbb{R}^n$, Δf is given by the usual expression $\sum \partial^2_{ii} f$. More generally, in coordinates, the Laplacian reads

$$\Delta f = (\det g)^{-1/2} \sum_{ij} \partial_i \big((\det g)^{1/2} g^{ij}\, \partial_j f\big).$$

In the context of optimal transport, we shall be led to consider Hessian operators for functions f that are not of class C^2, and not even continuously differentiable. However, ∇f and $\nabla^2 f$ will still be well-defined almost everywhere, and this will be sufficient to conduct the proofs. Here I should explain what it means for a function defined almost everywhere to be differentiable. Let ξ be a vector field defined on a domain of a neighborhood U of x; when y is close enough to x, there is a unique velocity $w \in T_x M$ such that $y = \gamma_1$, where γ is the constant-speed geodesic starting from x with initial velocity w; for simplicity I shall write $w = y - x$ (to be understood as $y = \exp_x w$). Then ξ is said to be covariantly differentiable at x in the direction v, if

$$\nabla_v \xi(x) := \lim_{y \to x;\ \frac{y-x}{|y-x|} \to \frac{v}{|v|}} |v| \left(\frac{\theta_{y\to x}\xi(y) - \xi(x)}{|y-x|} \right) \qquad (14.5)$$

exists, where y varies on the domain of definition of ξ, and $\theta_{y \to x}$ is the parallel transport along the geodesic joining y to x. If ξ is defined everywhere in a neighborhood of x, then this is just the usual notion of covariant derivation. Formulas for (14.5) in coordinates are just the same as in the smooth case.

The following theorem is the main result of second differentiability for nonsmooth functions:

Theorem 14.1 (Second differentiability of semiconvex functions). *Let M be a Riemannian manifold equipped with its volume measure, let U be an open subset of M, and let $\psi : U \to \mathbb{R}$ be locally semiconvex with a quadratic modulus of semiconvexity, in the sense of Definition 10.10. Then, for almost every $x \in U$, ψ is differentiable at x and there exists a symmetric operator $A : T_xM \to T_xM$, characterized by one of the following two equivalent properties:*

(i) For any $v \in T_xM$, $\nabla_v(\nabla\psi)(x) = Av$;

(ii) $\psi(\exp_x v) = \psi(x) + \langle \nabla\psi(x), v\rangle + \dfrac{\langle A \cdot v, v\rangle}{2} + o(|v|^2)$ as $v \to 0$.

The operator A is denoted by $\nabla^2\psi(x)$ and called the Hessian of ψ at x. When no confusion is possible, the quadratic form defined by A is also called the Hessian of ψ at x.

The trace of A is denoted by $\Delta\psi(x)$ and called the Laplacian of ψ at x. The function $x \to \Delta\psi(x)$ coincides with the density of the absolutely continuous part of the distributional Laplacian of ψ; while the singular part of this distributional Laplacian is a nonnegative measure.

Remark 14.2. The particular case when ψ is convex $\mathbb{R}^n \to \mathbb{R}$ is known as **Alexandrov's second differentiability theorem**. By extension, I shall use the terminology "Alexandrov's theorem" for the general statement of Theorem 14.1. This theorem is more often stated in terms of Property (ii) than in terms of Property (i); but it is the latter that will be most useful for our purposes.

Remark 14.3. As the proof will show, Property (i) can be replaced by the following more precise statement involving the subdifferential of ψ: If ξ is any vector field valued in $\nabla^-\psi$ (i.e. $\xi(y) \in \nabla^-\psi(y)$ for all y), then $\nabla_v\xi(x) = Av$.

Remark 14.4. For the main part of this course, we shall not need the full strength of Theorem 14.1, but just the particular case when ψ is continuously differentiable and $\nabla\psi$ is Lipschitz; then the proof becomes much simpler, and $\nabla\psi$ is almost everywhere differentiable in the usual sense. Still, on some occasions we shall need the full generality of Theorem 14.1.

Beginning of proof of Theorem 14.1. The notion of local semiconvexity with quadratic modulus is invariant by C^2 diffeomorphism, so it suffices to prove Theorem 14.1 when $M = \mathbb{R}^n$. But a semiconvex function in an open subset U of $\mathbb{R}^n$ is just the sum of a quadratic form and a locally convex function (that is, a function which is convex in any convex subset of U). So it is actually sufficient to consider the special case when ψ is a convex function in a convex subset of $\mathbb{R}^n$. Then if $x \in U$ and B is a closed ball around x, included in U, let ψ_B be the restriction of ψ to B; since ψ is Lipschitz and convex, it can be extended into a Lipschitz convex function on the whole of $\mathbb{R}^n$ (take for instance the supremum of all supporting hyperplanes for ψ_B). In short, to prove Theorem 14.1 it is sufficient to treat the special case of a convex function $\psi : \mathbb{R}^n \to \mathbb{R}$. At this point the argument does not involve any more Riemannian geometry, but only convex analysis; so I shall postpone it to the Appendix (Theorem 14.25). □

The Jacobian determinant of the exponential map

Let M be a Riemannian manifold, and let ξ be a vector field on M (so for each x, $\xi(x)$ lies in T_xM). Recall the definition of the exponential map $T = \exp \xi$: Start from point x a geodesic curve with initial velocity $\xi(x) \in T_xM$, and follow it up to time 1 (it is not required that the geodesic be minimizing all along); the position at time 1 is denoted by $\exp_x \xi(x)$. As a trivial example, in the Euclidean space, $\exp_x \xi(x) = x + \xi(x)$.

The computation of the Jacobian determinant of such a map is a classical exercise in Riemannian geometry, whose solution involves the Ricci curvature. One can take this computation as a theorem about the Ricci curvature (previously defined in terms of sectional or Riemann curvature), or as the mere definition of the Ricci curvature.

So let $x \in M$ be given, and let ξ be a vector field defined in a neighborhood of x, or almost everywhere in a neighborhood of x. Let $(e_1, \ldots, e_n)$ be an orthonormal basis of T_xM, and consider small variations of x in these directions $e_1, \ldots, e_n$, denoted abusively by $x + \delta e_1, \ldots, x + \delta e_n$. (Here $x + \delta e_j$ should be understood as, say, $\exp_x(\delta e_j)$; but it might also be any path $x(\delta)$ with $\dot{x}(0) = e_i$.) As $\delta \to 0$, the infinitesimal parallelepiped P_δ built on $(x + \delta e_1, \ldots, x + \delta e_n)$ has volume $\mathrm{vol}\,[P_\delta] \simeq \delta^n$. (It is easy to make sense of that by using local charts.) The quantity of interest is

$$\mathcal{J}(x) := \lim \frac{\mathrm{vol}\,[T(P_\delta)]}{\mathrm{vol}\,[P_\delta]}.$$

For that purpose, $T(P_\delta)$ can be approximated by the infinitesimal parallelogram built on $T(x+\delta e_1), \ldots, T(x+\delta e_n)$. Explicitly,

$$T(x+\delta e_i) = \exp_{x+\delta e_i}(\xi(x+\delta e_i)).$$

(If ξ is not defined at $x+\delta e_i$ it is always possible to make an infinitesimal perturbation and replace $x+\delta e_i$ by a point which is extremely close and at which ξ is well-defined. Let me skip this nonessential subtlety.)

Assume for a moment that we are in $\mathbb{R}^n$, so $T(x) = x + \xi(x)$; then, by a classical result of real analysis, $\mathcal{J}(x) = |\det(\nabla T(x))| = |\det(I_n + \nabla\xi(x))|$. But in the genuinely Riemannian case, things are much more intricate (unless $\xi(x) = 0$) because the measurement of infinitesimal volumes changes as we move along the geodesic path $\gamma(t,x) = \exp_x(t\,\xi(x))$.

To appreciate this continuous change, let us parallel-transport along the geodesic γ to define a new family $\mathbf{E}(t) = (e_1(t), \ldots, e_n(t))$ in $T_{\gamma(t)}M$. Since $(d/dt)\langle e_i(t), e_j(t)\rangle = \langle \dot{e}_i(t), e_j(t)\rangle + \langle e_i(t), \dot{e}_j(t)\rangle = 0$, the family $\mathbf{E}(t)$ remains an orthonormal basis of $T_{\gamma(t)}M$ for any t. (Here the dot symbol stands for the covariant derivation along γ.) Moreover, $e_1(t) = \dot\gamma(t,x)/|\dot\gamma(t,x)|$. (See Figure 14.2.)

To express the Jacobian of the map $T = \exp\xi$, it will be convenient to consider the whole collection of maps $T_t = \exp(t\,\xi)$. For brevity, let us write

$$T_t(x+\delta\mathbf{E}) = \Big(T_t(x+\delta e_1), \ldots, T_t(x+\delta e_n)\Big);$$

then

$$T_t(x+\delta\mathbf{E}) \simeq T_t(x) + \delta\,\mathbf{J},$$

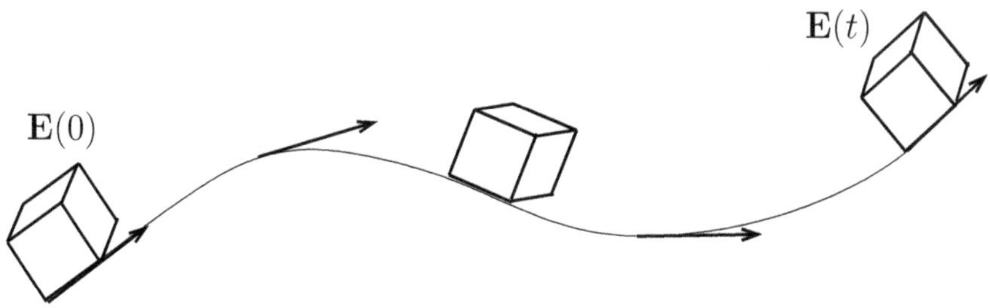

Fig. 14.2. The orthonormal basis $\mathbf{E}$, here represented by a small cube, goes along the geodesic by parallel transport.

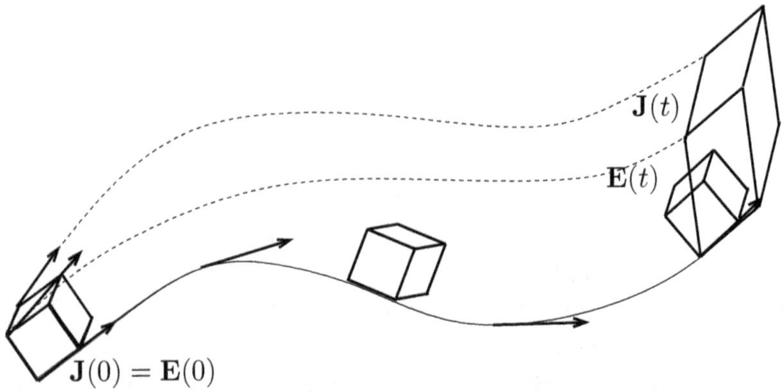

Fig. 14.3. At time $t = 0$, the matrices $\mathbf{J}(t)$ and $\mathbf{E}(t)$ coincide, but at later times they (may) differ, due to geodesic distortion.

where

$$\mathbf{J} = (J_1, \dots, J_n); \qquad J_i(t,x) := \frac{d}{d\delta}\bigg|_{\delta=0} T_t(x + \delta e_i).$$

(See Figure 14.3.)

The vector fields J_i have been obtained by differentiating a family of geodesics depending on a parameter (here δ); such vector fields are called **Jacobi fields** and they satisfy a characteristic linear second-order equation known as the **Jacobi equation**. To write this equation, it will be convenient to express $J_1, \dots, J_n$ in terms of the basis $e_1, \dots, e_n$; so let $J_{ij} = \langle J_i, e_j \rangle$ stand for the jth component of J_i in this basis. The matrix $J = (J_{ij})_{1 \le i,j \le n}$ satisfies the differential equation

$$\ddot{J}(t) + R(t)\, J(t) = 0, \tag{14.6}$$

where $R(t)$ is a matrix which depends on the Riemannian structure at $\gamma(t)$, and can be expressed in terms of the Riemann curvature tensor:

$$R_{ij}(t) = \Big\langle \mathrm{Riem}_{\gamma(t)}(\dot\gamma(t), e_i(t))\, \dot\gamma(t),\ e_j(t) \Big\rangle_{\gamma(t)}. \tag{14.7}$$

(All of these quantities depend implicitly on the starting point x.) The reader who prefers to stay away from the Riemann curvature tensor can take (14.6) as the equation defining the matrix R; the only things that one needs to know about it are (a) $R(t)$ is symmetric; (b) the first row of $R(t)$ vanishes (which is the same, modulo identification, as

$R(t)\,\dot\gamma(t) = 0$); (c) tr $R(t) = \mathrm{Ric}_{\gamma_t}(\dot\gamma_t, \dot\gamma_t)$ (which one can also adopt as a definition of the Ricci tensor); (d) R is invariant under the transform $t \to 1-t$, $E(t) \to -E(1-t)$, $\gamma_t \to \gamma_{1-t}$.

Equation (14.6) is of second order in time, so it should come with initial conditions for both $J(0)$ and $\dot J(0)$. On the one hand, since $T_0(y) = y$,

$$J_i(0) = \frac{d}{d\delta}\Big|_{\delta=0} (x + \delta e_i) = e_i,$$

so $J(0)$ is just the identity matrix. On the other hand,

$$\dot J_i(0) = \frac{D}{Dt}\Big|_{t=0} \frac{d}{d\delta}\Big|_{\delta=0} T_t(x+\delta e_i) = \frac{D}{D\delta}\Big|_{\delta=0} \frac{d}{dt}\Big|_{t=0} T_t(x+\delta e_i)$$
$$= \frac{D}{D\delta}\Big|_{\delta=0} \xi(x+\delta e_i) = (\nabla\xi) e_i,$$

where $\nabla\xi$ is the covariant gradient of ξ. (The exchange of derivatives is justified by the differentiability of ξ at x and the C^∞ regularity of $(t,y,\xi) \to \exp_y(t\xi)$.) So

$$\frac{d}{dt} J_{ij} = \frac{d}{dt}\langle J_i, e_j\rangle = \langle \frac{DJ_i}{Dt}, e_j\rangle = \langle (\nabla\xi) e_i, e_j\rangle.$$

We conclude that the initial conditions are

$$J(0) = I_n, \qquad \dot J(0) = \nabla\xi(x), \tag{14.8}$$

where in the second expression the linear operator $\nabla\xi(x)$ is identified with its matrix in the basis $\mathbf{E}$: $(\nabla\xi)_{ij} = \langle (\nabla\xi) e_i, e_j\rangle = \langle e_i \cdot \nabla\xi, e_j\rangle$. (Be careful, this is the converse of the usual convention $A_{ij} = \langle A e_j, e_i\rangle$; anyway, later we shall work with symmetric operators, so it will not matter.)

From this point on, the problem is about a path $J(t)$ valued in the space $M_n(\mathbb{R})$ of real $n \times n$ matrices, and we can forget about the geometry: Parallel transport has provided a consistent identification of all the tangent spaces $T_{\gamma(t)}M$ with $\mathbb{R}^n$. This path depends on x via the initial conditions (14.8), so in the sequel we shall put that dependence explicitly. It might be very rough as a function of x, but it is very smooth as a function of t. The Jacobian of the map T_t is defined by

$$\mathcal{J}(t,x) = \det J(t,x),$$

and the formula for the differential of the determinant yields

$$\dot{\mathcal{J}}(t,x) = \mathcal{J}(t,x)\ \mathrm{tr}\left(\dot{J}(t,x)\, J(t,x)^{-1}\right), \tag{14.9}$$

at least as long as $J(t,x)$ is invertible (let's forget about that problem for the moment).

So it is natural to set

$$U := \dot{J}\, J^{-1}, \tag{14.10}$$

and look for an equation on U. By differentiating (14.10) and using (14.6), we discover that

$$\dot{U} = \ddot{J}J^{-1} - \dot{J}J^{-1}\dot{J}J^{-1} = -R - U^2$$

(note that J and $\dot{J}$ do not necessarily commute). So the change of variables (14.10) has turned the second-order equation (14.6) into the *first-order* equation

$$\dot{U} + U^2 + R = 0, \tag{14.11}$$

which is of Ricatti type, that is, with a quadratic nonlinearity.

By taking the trace of (14.11), we arrive at

$$\frac{d}{dt}(\mathrm{tr}\ U) + \mathrm{tr}\,(U^2) + \mathrm{tr}\ R = 0.$$

Now the trace of $R(t,x)$ only depends on γ_t and $\dot{\gamma}_t$; in fact, as noticed before, it is precisely the value of the Ricci curvature at $\gamma(t)$, evaluated in the direction $\dot{\gamma}(t)$. So we have arrived at our first important equation involving Ricci curvature:

$$\frac{d}{dt}(\mathrm{tr}\ U) + \mathrm{tr}\,(U^2) + \mathrm{Ric}(\dot{\gamma}) = 0, \tag{14.12}$$

where of course $\mathrm{Ric}(\dot{\gamma})$ is an abbreviation for $\mathrm{Ric}_{\gamma(t)}(\dot{\gamma}(t), \dot{\gamma}(t))$.

Equation (14.12) holds true for any vector field ξ, as long as ξ is covariantly differentiable at x. But in the sequel, I shall only apply it in the particular case when ξ derives from a function: $\xi = \nabla\psi$; and ψ is locally semiconvex with a quadratic modulus of semiconvexity. There are three reasons for this restriction:

(a) In the theory of optimal transport, one only needs to consider such maps;

(b) The semiconvexity of ψ guarantees the almost everywhere differentiability of $\nabla\psi$, by Theorem 14.1;

(c) If $\xi = \nabla\psi$, then $\nabla\xi(x) = \nabla^2\psi(x)$ is *symmetric* and this will imply the symmetry of $U(t,x)$ at all times; this symmetry will allow to derive from (14.12) a closed inequality on $\mathrm{tr}\ U(t,x) = \mathcal{J}(t,x)$.

So from now on, $\xi = \nabla\psi$, where ψ is semiconvex. To prove the symmetry of $U(t,x)$, note that $U(0,x) = I_n$ and $\dot{U}(0,x) = \nabla^2\psi(x)$ (modulo identification) are symmetric, and the time-dependent matrix $R(t,x)$ is also symmetric, so $U(t,x)$ and its transpose $U(t,x)^*$ solve the same differential equation, with the same initial conditions. Then, by the uniqueness statement in the Cauchy–Lipschitz theorem, they have to coincide at all times where they are defined.

Inequality (14.12) cannot be recast as a differential equation involving only the Jacobian determinant (or equivalently $\operatorname{tr} U(t,x)$), since the quantity $\operatorname{tr}(U^2)$ in (14.12) cannot be expressed in terms of $\operatorname{tr} U$. However, the symmetry of U allows us to use the Cauchy–Schwarz inequality, in the form

$$\operatorname{tr}(U^2) \geq \frac{(\operatorname{tr} U)^2}{n};$$

then, by plugging this inequality into (14.12), we obtain an important differential inequality involving Ricci curvature:

$$\frac{d}{dt}(\operatorname{tr} U) + \frac{(\operatorname{tr} U)^2}{n} + \operatorname{Ric}(\dot{\gamma}) \leq 0. \tag{14.13}$$

There are several ways to rewrite this result in terms of the Jacobian determinant $\mathcal{J}(t)$. For instance, by differentiating the formula

$$\operatorname{tr} U = \frac{\dot{\mathcal{J}}}{\mathcal{J}},$$

one obtains easily

$$\frac{d}{dt}(\operatorname{tr} U) + \frac{(\operatorname{tr} U)^2}{n} = \frac{\ddot{\mathcal{J}}}{\mathcal{J}} - \left(1 - \frac{1}{n}\right)\left(\frac{\dot{\mathcal{J}}}{\mathcal{J}}\right)^2.$$

So (14.13) becomes

$$\frac{\ddot{\mathcal{J}}}{\mathcal{J}} - \left(1 - \frac{1}{n}\right)\left(\frac{\dot{\mathcal{J}}}{\mathcal{J}}\right)^2 \leq -\operatorname{Ric}(\dot{\gamma}). \tag{14.14}$$

For later purposes, it will be convenient to define $\mathcal{D}(t) := \mathcal{J}(t)^{1/n}$ (which one can think of as a coefficient of mean distortion); then the left-hand side of (14.14) is exactly $n\ddot{\mathcal{D}}/\mathcal{D}$. So

$$\frac{\ddot{\mathcal{D}}}{\mathcal{D}} \le -\frac{\mathrm{Ric}(\dot\gamma)}{n}. \qquad (14.15)$$

Yet another useful formula is obtained by introducing $\ell(t) := -\log \mathcal{J}(t)$, and then (14.13) becomes

$$\ddot\ell(t) \ge \frac{\dot\ell(t)^2}{n} + \mathrm{Ric}(\dot\gamma). \qquad (14.16)$$

In all of these formulas, we have always taken $t = 0$ as the starting time, but it is clear that we could do just the same with any starting time $t_0 \in [0, 1]$, that is, consider, instead of $T_t(x) = \exp(t\nabla\psi(x))$, the map $T_{t_0\to t}(x) = \exp((t - t_0)\nabla\psi(x))$. Then all the differential inequalities are unchanged; the only difference is that the Jacobian determinant at time $t = 0$ is not necessarily 1.

Taking out the direction of motion

The previous formulas are quite sufficient to derive many useful geometric consequences. However, one can refine them by taking advantage of the fact that *curvature is not felt in the direction of motion.* In other words, if one is traveling along some geodesic γ, one will never be able to detect some curvature by considering variations (in the initial position, or initial velocity) in the direction of γ itself: the path will always be the same, up to reparametrization. This corresponds to the property $R(t)\dot\gamma(t) = 0$, where $R(t)$ is the matrix appearing in (14.6). In short, curvature is felt only in $n - 1$ directions out of n. This loose principle often leads to a refinement of estimates by a factor $(n - 1)/n$.

Here is a recipe to "separate out" the direction of motion from the other directions. As before, assume that the first vector of the orthonormal basis $\mathbf{J}(0)$ is $e_1(0) = \dot\gamma(0)/|\dot\gamma(0)|$. (The case when $\dot\gamma(0) = 0$ can be treated separately.) Set $u_{/\!/} = u_{11}$ (this is the coefficient in U which corresponds to just the direction of motion), and define $U_\perp$ as the $(n - 1) \times (n - 1)$ matrix obtained by removing the first line and first column in U. Of course, $\mathrm{tr}\,(U) = u_{/\!/} + \mathrm{tr}\,(U_\perp)$. Next decompose the Jacobian determinant $\mathcal{J}$ into a parallel and an orthogonal contributions:

$$\mathcal{J} = \mathcal{J}_{/\!/}\mathcal{J}_\perp, \qquad \mathcal{J}_{/\!/}(t) = \exp\left(\int_0^t u_{/\!/}(s)\, ds\right).$$

Further define parallel and orthogonal distortions by

$$\mathcal{D}_{//} = \mathcal{J}_{//}, \qquad \mathcal{D}_\perp = \mathcal{J}_\perp^{\frac{1}{n-1}};$$

and, of course,

$$\ell_{//} = -\log \mathcal{J}_{//}, \qquad \ell_\perp = -\log \mathcal{J}_\perp. \tag{14.17}$$

Since the first row of $R(t)$ vanishes, equation (14.11) implies

$$\dot{u}_{//} = -\sum_j u_{1j}^2 \le -u_{11}^2 = -u_{//}^2.$$

It follows easily that

$$\ddot{\ell}_{//} \ge \dot{\ell}_{//}^2, \tag{14.18}$$

or equivalently

$$\ddot{\mathcal{J}}_{//} \le 0, \tag{14.19}$$

so $\mathcal{D}_{//} = \mathcal{J}_{//}$ is always a concave function of t (this property does not depend on the curvature of M), and the same holds true of course for $\mathcal{D}_{//}$ which coincides with $\mathcal{J}_{//}$.

Now let us take care of the orthogonal part: Putting together (14.9), (14.10), (14.11), (14.18), we find

$$\ddot{\ell}_\perp = -\frac{d}{dt}(\operatorname{tr} U) - \ddot{\ell}_{//} = \operatorname{tr}(U^2) + \operatorname{Ric}(\dot{\gamma}) - \sum u_{1j}^2.$$

Since $\operatorname{tr} U^2 = \operatorname{tr}(U_\perp)^2 + 2\sum u_{1j}^2$, this implies

$$\ddot{\ell}_\perp \ge \operatorname{tr}(U_\perp^2) + \operatorname{Ric}(\dot{\gamma}). \tag{14.20}$$

Then in the same manner as before, one can obtain

$$\ddot{\ell}_\perp \ge \frac{(\dot{\ell}_\perp)^2}{n-1} + \operatorname{Ric}(\dot{\gamma}), \tag{14.21}$$

$$\frac{\ddot{\mathcal{D}}_\perp}{\mathcal{D}_\perp} \le -\frac{\operatorname{Ric}(\dot{\gamma})}{n-1}. \tag{14.22}$$

To summarize: The basic inequalities for $\ell_\perp$ and $\ell_{//}$ are the same as for ℓ, but with the exponent n replaced by $n-1$ in the case of $\ell_\perp$, and 1 in the case of $\ell_{//}$; and the number $\operatorname{Ric}(\dot{\gamma})$ replaced by 0 in the case of $\ell_{//}$. The same for $\mathcal{D}_\perp$ and $\mathcal{D}_{//}$.

Positivity of the Jacobian

Unlike the distance function, the exponential map is always smooth. But this does not prevent the Jacobian determinant $\mathcal{J}(t)$ from vanishing, i.e. the matrix $J(t)$ from becoming singular (not invertible). Then computations such as (14.9) break down. So all the computations performed before are only valid if $\mathcal{J}(t)$ is positive for all $t \in (0,1)$.

In terms of $\ell(t) = -\log \mathcal{J}(t)$, the vanishing of the Jacobian determinant corresponds to a divergence $\ell(t) \to \infty$. Readers familiar with ordinary differential equations will have no trouble believing that these events are not rare: Indeed, ℓ solves the Ricatti-type equation (14.16), and such equations often lead to blow-up in finite time. For instance, consider a function $\ell(t)$ that solves

$$\ddot{\ell} \geq \frac{(\dot{\ell})^2}{n-1} + K,$$

where $K > 0$. Consider a time t_0 where ℓ has a minimum, so $\dot{\ell}(t_0) = 0$. Then, ℓ cannot be defined on a time-interval larger than $[t_0 - T, t_0 + T]$, where $T := \pi\sqrt{(n-1)/K}$. So the Jacobian has to vanish at some time, and we even have a bound on this time. (With a bit more work, this estimate implies the *Bonnet–Myers theorem*, which asserts that the diameter of M cannot be larger than $\pi\sqrt{(n-1)/K}$ if $\mathrm{Ric} \geq Kg$.)

The vanishing of the Jacobian may occur even along geodesics that are minimizing for all times: Consider for instance $\xi(x) = -2x$ in $\mathbb{R}^n$; then the image of $\exp(t\xi)$ is reduced to a single point when $t = 1/2$. However, in the case of optimal transport, the Jacobian cannot vanish at intermediate times, at least for almost all initial points: Recall indeed the last part of Theorem 11.3. This property can be seen as a result of the very special choice of the velocity field ξ, which is the gradient *of a $d^2/2$-convex function*; or as a consequence of the "no-crossing" property explored in Chapter 8. (There is also an interpretation in terms of Jacobi fields, see Proposition 14.31 in the Third Appendix.)

Bochner's formula

So far, we have discussed curvature from a *Lagrangian* point of view, that is, by going along a geodesic path $\gamma(t)$, keeping the memory of the initial position. It is useful to be also familiar with the **Eulerian** point of view, in which the focus is not on the trajectory, but on the velocity field $\xi = \xi(t,x)$. To switch from Lagrangian to Eulerian description, just write

$$\dot{\gamma}(t) = \xi\big(t, \gamma(t)\big). \tag{14.23}$$

In general, this can be a subtle issue because two trajectories might cross, and then there might be no way to define a meaningful velocity field $\xi(t,\,\cdot\,)$ at the crossing point. However, if a smooth vector field $\xi = \xi(0,\,\cdot\,)$ is given, then around any point x_0 the trajectories $\gamma(t,x) = \exp(t\,\xi(x))$ do not cross for $|t|$ small enough, and one can define $\xi(t,x)$ without ambiguity. The covariant differentiation of (14.23) along ξ itself, and the geodesic equation $\ddot{\gamma} = 0$, yield

$$\frac{\partial \xi}{\partial t} + \nabla_\xi \xi = 0, \tag{14.24}$$

which is the **pressureless Euler equation**. From a physical point of view, this equation describes the velocity field of a bunch of particles which travel along geodesic curves without interacting. The derivation of (14.24) will fail when the geodesic paths start to cross, at which point the solution to (14.24) would typically lose smoothness and need reinterpretation. But for the sequel, we only need (14.24) to be satisfied for small values of $|t|$, and locally around x.

All the previous discussion about Ricci curvature can be recast in Eulerian terms. Let $\gamma(t,x) = \exp_x(t\xi(x))$; by the definition of the covariant gradient, we have

$$\dot{\mathbf{J}}(t,x) = \nabla\xi(t,\gamma(t,x))\,\mathbf{J}(t,x)$$

(the same formula that we had before at time $t = 0$). Under the identification of $\mathbb{R}^n$ with $T_{\gamma(t)}M$ provided by the basis $\mathbf{E}(t)$, we can identify $\mathbf{J}$ with the matrix J, and then

$$U(t,x) = \dot{J}(t,x)\,J(t,x)^{-1} = \nabla\xi\big(t,\gamma(t,x)\big), \tag{14.25}$$

where again the linear operator $\nabla\xi$ is identified with its matrix in the basis $\mathbf{E}$.

Then $\operatorname{tr} U(t,x) = \operatorname{tr} \nabla\xi(t,x)$ coincides with the divergence of $\xi(t,\,\cdot\,)$, evaluated at x. By the chain-rule and (14.24),

$$\begin{aligned}\frac{d}{dt}(\operatorname{tr} U)(t,x) &= \frac{d}{dt}(\nabla\cdot\xi)(t,\gamma(t,x))\\ &= \nabla\cdot\left(\frac{\partial\xi}{\partial t}(t,\gamma(t,x))\right) + \dot\gamma(t,x)\cdot\nabla(\nabla\cdot\xi)(t,\gamma(t,x))\\ &= \Big(-\nabla\cdot(\nabla_\xi\xi) + \xi\cdot\nabla(\nabla\cdot\xi)\Big)(t,\gamma(t,x)).\end{aligned}$$

Thus the Lagrangian formula (14.12) can be translated into the Eulerian formula

$$-\nabla\cdot(\nabla_\xi\xi) + \xi\cdot\nabla(\nabla\cdot\xi) + \operatorname{tr}(\nabla\xi)^2 + \operatorname{Ric}(\xi) = 0. \tag{14.26}$$

All functions here are evaluated at $(t,\gamma(t,x))$, and of course we can choose $t=0$, and x arbitrary. So (14.26) is an identity that holds true for any smooth (say C^2) vector field ξ on our manifold M. Of course it can also be established directly by a coordinate computation.[1]

While formula (14.26) holds true for all vector fields ξ, if $\nabla\xi$ is symmetric then two simplifications arise:

(a) $\nabla_\xi\xi = \nabla\xi\cdot\xi = \nabla\frac{|\xi|^2}{2}$;

(b) $\operatorname{tr}(\nabla\xi)^2 = \|\nabla\xi\|_{\mathrm{HS}}^2$, HS standing for Hilbert–Schmidt norm.

So (14.26) becomes

$$-\Delta\frac{|\xi|^2}{2} + \xi\cdot\nabla(\nabla\cdot\xi) + \|\nabla\xi\|_{\mathrm{HS}}^2 + \operatorname{Ric}(\xi) = 0. \tag{14.27}$$

We shall apply it only in the case when ξ is a gradient: $\xi = \nabla\psi$; then $\nabla\xi = \nabla^2\psi$ is indeed symmetric, and the resulting formula is

$$-\Delta\frac{|\nabla\psi|^2}{2} + \nabla\psi\cdot\nabla(\Delta\psi) + \|\nabla^2\psi\|_{\mathrm{HS}}^2 + \operatorname{Ric}(\nabla\psi) = 0. \tag{14.28}$$

The identity (14.26), or its particular case (14.28), is called the **Bochner–Weitzenböck–Lichnerowicz formula**, or just Bochner's formula.[2]

---

[1] With the notation $\nabla_\xi = \xi\cdot\nabla$ (which is classical in fluid mechanics), and $\operatorname{tr}(\nabla\xi)^2 = \nabla\xi\cdot\cdot\nabla\xi$, (14.26) takes the amusing form $-\nabla\cdot\xi\cdot\nabla\xi + \xi\cdot\nabla\nabla\cdot\xi + \nabla\xi\cdot\cdot\nabla\xi + \operatorname{Ric}(\xi) = 0$.

[2] In (14.26) or (14.28) I have written Bochner's formula in purely "metric" terms, which will probably look quite ugly to many geometer readers. An equivalent but more "topological" way to write Bochner's formula is

$$\Delta + \nabla\nabla^* + \operatorname{Ric} = 0,$$

Remark 14.5. With the ansatz $\xi = \nabla\psi$, the pressureless Euler equation (14.24) reduces to the **Hamilton–Jacobi equation**

$$\frac{\partial \psi}{\partial t} + \frac{|\nabla\psi|^2}{2} = 0. \tag{14.29}$$

One can use this equation to obtain (14.28) directly, instead of first deriving (14.26). Here equation (14.29) is to be understood in a viscosity sense (otherwise there are many spurious solutions); in fact the reader might just as well take the identity

$$\psi(t,x) = \inf_{y\in M} \left[\psi(y) + \frac{d(x,y)^2}{2t}\right]$$

as the definition of the solution of (14.29). Then the geodesic curves γ starting with $\gamma(0) = x$, $\dot\gamma(0) = \nabla\psi(x)$ are called **characteristic curves** of equation (14.29).

Remark 14.6. Here I have not tried to derive Bochner's formula for nonsmooth functions. This could be done for semiconvex ψ, with an appropriate "compensated" definition for $-\Delta\frac{|\nabla\psi|^2}{2} + \nabla\psi\cdot\nabla(\Delta\psi)$. In fact, the semiconvexity of $\nabla\psi$ prevents the formation of instantaneous shocks, and will allow the Lagrangian/Eulerian duality for a short time.

Remark 14.7. The operator $U(t,x)$ coincides with $\nabla^2\psi(t,\gamma(t,x))$, which is another way to see that it is symmetric for $t > 0$.

From this point on, we shall only work with (14.28). Of course, by using the Cauchy–Schwarz identity as before, we can bound below $\|\nabla^2\psi\|_{HS}^2$ by $(\Delta\psi)^2/n$; therefore (14.25) implies

$$\Delta\frac{|\nabla\psi|^2}{2} - \nabla\psi\cdot\nabla(\Delta\psi) \geq \frac{(\Delta\psi)^2}{n} + \mathrm{Ric}(\nabla\psi). \tag{14.30}$$

Apart from regularity issues, this inequality is strictly equivalent to (14.13), and therefore to (14.14) or (14.15).

Not so much has been lost when going from (14.28) to (14.30): there is still equality in (14.30) at all points x where $\nabla^2\psi(x)$ is a multiple of the identity.

---

where $\Delta = -(dd^* + d^*d)$ is the Laplace operator on 1-forms, ∇ is the covariant differentiation (under the identification of a 1-form with a vector field) and the adjoints are in $L^2(\mathrm{vol})$. Also I should note that the name "Bochner formula" is attributed to a number of related identities.

One can also take out the direction of motion, $\widehat{\nabla\psi} := (\nabla\psi)/|\nabla\psi|$, from the Bochner identity. The Hamilton–Jacobi equation implies $\partial_t \widehat{\nabla\psi} + \nabla^2\psi \cdot \widehat{\nabla\psi} = 0$, so

$$\partial_t \Big\langle \nabla^2\psi \cdot \widehat{\nabla\psi}, \widehat{\nabla\psi} \Big\rangle \\ = -\big\langle \nabla^2(|\nabla\psi|^2/2) \cdot \widehat{\nabla\psi}, \widehat{\nabla\psi} \big\rangle - 2\, \big\langle \nabla^2\psi \cdot (\nabla^2\psi \cdot \widehat{\nabla\psi}), \widehat{\nabla\psi} \big\rangle,$$

and by symmetry the latter term can be rewritten $-2\,|(\nabla^2\psi)\cdot\widehat{\nabla\psi}|^2$. From this one easily obtains the following refinement of Bochner's formula: Define

$$\Delta_{/\!/} f = \big\langle \nabla^2 f \cdot \widehat{\nabla\psi}, \widehat{\nabla\psi} \big\rangle, \qquad \Delta_\perp = \Delta - \Delta_{/\!/},$$

then

$$\begin{cases} \Delta_{/\!/} \frac{|\nabla\psi|^2}{2} - \nabla\psi \cdot \nabla \Delta_{/\!/}\psi + 2\big|(\nabla^2\psi)\cdot\widehat{\nabla\psi}\big|^2 \geq (\Delta_{/\!/}\psi)^2 \\ \\ \Delta_\perp \frac{|\nabla\psi|^2}{2} - \nabla\psi \cdot \nabla \Delta_\perp\psi - 2\big|(\nabla^2\psi)\cdot\widehat{\nabla\psi}\big|^2 \geq \|\nabla^2_\perp \psi\|^2_{\mathrm{HS}} + \mathrm{Ric}(\nabla\psi). \end{cases} \tag{14.31}$$

This is the "Bochner formula with the direction of motion taken out". I have to confess that I never saw these frightening formulas anywhere, and don't know whether they have any use. But of course, they are equivalent to their Lagrangian counterpart, which will play a crucial role in the sequel.

Analytic and geometric consequences of Ricci curvature bounds

Inequalities (14.13), (14.14), (14.15) and (14.30) are the "working heart" of Ricci curvature analysis. Many gometric and analytic consequences follow from these estimates.

Here is a first example coming from analysis and partial differential equations theory: If the Ricci curvature of M is globally bounded below ($\inf_x \mathrm{Ric}_x > -\infty$), then there exists a unique **heat kernel**, i.e. a measurable function $p_t(x,y)$ ($t > 0$, $x \in M$, $y \in M$), integrable in y, smooth outside of the diagonal $x = y$, such that $f(t,x) := \int p_t(x,y)\, f_0(y)\, d\mathrm{vol}(y)$ solves the heat equation $\partial_t f = \Delta f$ with initial datum f_0.

Here is another example in which some topological information can be recovered from Ricci bounds: If M is a manifold with nonnegative Ricci curvature (for each x, $\mathrm{Ric}_x \geq 0$), and there exists a line in M, that is, a geodesic γ which is minimizing for *all* values of time $t \in \mathbb{R}$, then M is isometric to $\mathbb{R} \times M'$, for some Riemannian manifold M'. This is the **splitting theorem**, in a form proven by Cheeger and Gromoll.

Many *quantitative* statements can be obtained from (i) a lower bound on the Ricci curvature and (ii) an upper bound on the dimension of the manifold. Below is a (grossly nonexhaustive) list of some famous such results. In the statements to come, M is always assumed to be a smooth, complete Riemannian manifold, vol stands for the Riemannian volume on M, Δ for the Laplace operator and d for the Riemannian distance; K is the lower bound on the Ricci curvature, and n is the dimension of M. Also, if A is a measurable set, then A^r will denote its r-neighborhood, which is the set of points that lie at a distance at most r from A. Finally, the "model space" is the simply connected Riemannian manifold with constant sectional curvature which has the same dimension as M, and Ricci curvature constantly equal to K (more rigorously, to Kg, where g is the metric tensor on the model space).

1. Volume growth estimates: The **Bishop–Gromov** inequality (also called Riemannian volume comparison theorem) states that the volume of balls does not increase faster than the volume of balls in the model space. In formulas: for any $x \in M$,

$$\frac{\mathrm{vol}\,[B_r(x)]}{V(r)} \quad \text{is a nonincreasing function of } r,$$

where

$$V(r) = \int_0^r S(r')\,dr',$$

$$S(r) = c_{n,K} \begin{cases} \sin^{n-1}\left(\sqrt{\dfrac{K}{n-1}}\, r\right) & \text{if } K > 0 \\[2ex] r^{n-1} & \text{if } K = 0 \\[2ex] \sinh^{n-1}\left(\sqrt{\dfrac{|K|}{n-1}}\, r\right) & \text{if } K < 0. \end{cases}$$

Here of course $S(r)$ is the surface area of $B_r(0)$ in the model space, that is the $(n-1)$-dimensional volume of $\partial B_r(0)$, and $c_{n,K}$ is a nonessential normalizing constant. (See Theorem 18.8 later in this course.)

2. Diameter estimates: The **Bonnet–Myers theorem** states that, if $K > 0$, then M is compact and more precisely

$$\operatorname{diam}(M) \le \pi\sqrt{\frac{n-1}{K}},$$

with equality for the model sphere.

3. Spectral gap inequalities: If $K > 0$, then the spectral gap λ_1 of the nonnegative operator $-\Delta$ is bounded below:

$$\lambda_1 \ge \frac{n\,K}{n-1},$$

with equality again for the model sphere. (See Theorem 21.20 later in this course.)

4. (Sharp) Sobolev inequalities: If $K > 0$ and $n \ge 3$, let $\mu = \mathrm{vol}/\mathrm{vol}[M]$ be the normalized volume measure on M; then for any smooth function on M,

$$\|f\|^2_{L^{2^\star}(\mu)} \le \|f\|^2_{L^2(\mu)} + \frac{4}{Kn(n-2)}\,\|\nabla f\|^2_{L^2(\mu)}, \qquad 2^\star = \frac{2n}{n-2},$$

and those constants are sharp for the model sphere.

5. Heat kernel bounds: There are many of them, in particular the well-known **Li–Yau estimates**: If $K \ge 0$, then the heat kernel $p_t(x,y)$ satisfies

$$p_t(x,y) \le \frac{C}{\mathrm{vol}\,[B_{\sqrt{t}}(x)]}\,\exp\left(-\frac{d(x,y)^2}{2\,Ct}\right),$$

for some constant C which only depends on n. For $K < 0$, a similar bound holds true, only now C depends on K and there is an additional factor e^{Ct}. There are also pointwise estimates on the derivatives of $\log p_t$, in relation with **Harnack inequalities**.

The list could go on. More recently, Ricci curvature has been at the heart of Perelman's solution of the celebrated Poincaré conjecture, and more generally the topological classification of three-dimensional manifolds. Indeed, Perelman's argument is based on Hamilton's idea to use Ricci curvature in order to define a "heat flow" in the space of metrics, via the partial differential equation

$$\frac{\partial g}{\partial t} = -2\,\mathrm{Ric}(g), \qquad (14.32)$$

where $\mathrm{Ric}(g)$ is the Ricci tensor associated with the metric g, which can be thought of as something like $-\Delta g$. The flow defined by (14.32) is called the Ricci flow. Some time ago, Hamilton had already used its properties to show that a compact simply connected three-dimensional Riemannian manifold with positive Ricci curvature is automatically diffeomorphic to the sphere S^3.

Change of reference measure and effective dimension

For various reasons, one is often led to consider a reference measure ν that is not the volume measure vol, but, say, $\nu(dx) = e^{-V(x)}\,\mathrm{vol}(dx)$, for some function $V : M \to \mathbb{R}$, which in this chapter will always be assumed to be of class C^2. The metric–measure space (M, d, ν), where d stands for the geodesic distance, may be of interest in its own right, or may appear as a limit of Riemannian manifolds, in a sense that will be studied in Part III of these notes.

Of course, such a change of reference measure affects Jacobian determinants; so Ricci curvature estimates will lose their geometric meaning unless one changes the definition of Ricci tensor to take the new reference measure into account. This might perturb the dependence of all estimates on the dimension, so it might also be a good idea to introduce an "effective dimension" N, which may be larger than the "true" dimension n of the manifold.

The most well-known example is certainly the **Gaussian measure** in $\mathbb{R}^n$, which I shall denote by $\gamma^{(n)}$ (do not confuse it with a geodesic!):

$$\gamma^{(n)}(dx) = \frac{e^{-|x|^2}\,dx}{(2\pi)^{n/2}}, \qquad x \in \mathbb{R}^n.$$

It is a matter of experience that most theorems which we encounter about the Gaussian measure can be written just the same in dimension 1 or in dimension n, or even in infinite dimension, when properly interpreted. In fact, the effective dimension of $(\mathbb{R}^n, \gamma^{(n)})$ is infinite, in a certain sense, whatever n. I admit that this perspective may look strange, and might be the result of lack of imagination; but in any case, it will fit very well into the picture (in terms of sharp constants for geometric inequalities, etc.).

So, again let

$$T_t(x) = \gamma(t,x) = \exp_x\big(t\nabla\psi(x)\big);$$

now the Jacobian determinant is

$$\mathcal{J}(t,x) = \lim_{r\downarrow 0} \frac{\nu\big[T_t(B_r(x))\big]}{\nu[B_r(x)]} = \frac{e^{-V(T_t(x))}}{e^{-V(x)}}\, \mathcal{J}_0(t,x),$$

where $\mathcal{J}_0$ is the Jacobian corresponding to $V \equiv 0$ (that is, to $\nu = \mathrm{vol}$).

Then (with dots still standing for derivation with respect to t),

$$(\log \mathcal{J})^{\cdot}(t,x) = (\log \mathcal{J}_0)^{\cdot}(t,x) - \dot\gamma(t,x)\cdot\nabla V(\gamma(t,x)),$$

$$(\log \mathcal{J})^{\cdot\cdot}(t,x) = (\log \mathcal{J}_0)^{\cdot\cdot}(t,x) - \Big\langle \nabla^2 V(\gamma(t,x))\cdot\dot\gamma(t,x),\, \dot\gamma(t,x)\Big\rangle.$$

For later purposes it will be useful to keep track of all error terms in the inequalities. So rewrite (14.12) as

$$(\mathrm{tr}\, U)^{\cdot} + \frac{(\mathrm{tr}\, U)^2}{n} + \mathrm{Ric}(\dot\gamma) = -\left\| U - \left(\frac{\mathrm{tr}\, U}{n}\right) I_n \right\|_{\mathrm{HS}}^2. \qquad (14.33)$$

Then the left-hand side in (14.33) becomes

$$(\log \mathcal{J}_0)^{\cdot\cdot} + \frac{[(\log \mathcal{J}_0)^{\cdot}]^2}{n} + \mathrm{Ric}(\dot\gamma)$$
$$= (\log \mathcal{J})^{\cdot\cdot} + \langle \nabla^2 V(\gamma)\cdot\dot\gamma, \dot\gamma\rangle + \frac{[(\log \mathcal{J})^{\cdot} + \dot\gamma\cdot\nabla V(\gamma)]^2}{n} + \mathrm{Ric}(\dot\gamma).$$

By using the identity

$$\frac{a^2}{n} = \frac{(a+b)^2}{N} - \frac{b^2}{N-n} + \frac{n}{N(N-n)}\left(b - a\,\frac{N-n}{n}\right)^2, \qquad (14.34)$$

we see that

$$\frac{\Big[(\log \mathcal{J})^{\cdot} + \dot\gamma\cdot\nabla V(\gamma)\Big]^2}{n}$$
$$= \frac{\big[(\log \mathcal{J})^{\cdot}\big]^2}{N} - \frac{(\dot\gamma\cdot\nabla V(\gamma))^2}{N-n} + \frac{n}{N(N-n)}\left[\left(\frac{N-n}{n}\right)(\log \mathcal{J})^{\cdot} + \frac{N}{n}\,\dot\gamma\cdot\nabla V(\gamma)\right]^2$$

$$= \frac{\left[(\log \mathcal{J})^{\cdot}\right]^2}{N} - \frac{(\dot\gamma\cdot\nabla V(\gamma))^2}{N-n} + \frac{n}{N(N-n)}\left[\frac{N-n}{n}\,(\log\mathcal{J}_0)^{\cdot} + \dot\gamma\cdot\nabla V(\gamma)\right]^2$$

$$= \frac{\left[(\log \mathcal{J})^{\cdot}\right]^2}{N} - \frac{(\dot\gamma\cdot\nabla V(\gamma))^2}{N-n} + \frac{n}{N(N-n)}\left[\frac{N-n}{n}\,\mathrm{tr}\,U + \dot\gamma\cdot\nabla V(\gamma)\right]^2$$

To summarize these computations it will be useful to introduce some more notation: first, as usual, the negative logarithm of the Jacobian determinant:

$$\ell(t,x) := -\log\mathcal{J}(t,x); \tag{14.35}$$

and then, the **generalized Ricci tensor**:

$$\mathrm{Ric}_{N,\nu} := \mathrm{Ric} + \nabla^2 V - \frac{\nabla V\otimes\nabla V}{N-n}, \tag{14.36}$$

where the tensor product $\nabla V\otimes\nabla V$ is a quadratic form on TM, defined by its action on tangent vectors as

$$\big(\nabla V\otimes\nabla V\big)_x(v) = (\nabla V(x)\cdot v)^2;$$

so

$$\mathrm{Ric}_{N,\nu}(\dot\gamma) = (\mathrm{Ric}+\nabla^2 V)(\dot\gamma) - \frac{(\nabla V\cdot\dot\gamma)^2}{N-n}.$$

It is implicitly assumed in (14.36) that $N\geq n$ (otherwise the correct definition is $\mathrm{Ric}_{N,\nu} = -\infty$); if $N=n$ the convention is $0\times\infty = 0$, so (14.36) still makes sense if $\nabla V = 0$. Note that $\mathrm{Ric}_{\infty,\nu} = \mathrm{Ric}+\nabla^2 V$, while $\mathrm{Ric}_{n,\mathrm{vol}} = \mathrm{Ric}$.

The conclusion of the preceding computations is that

$$\ddot\ell = \frac{\dot\ell^2}{N} + \mathrm{Ric}_{N,\nu}(\dot\gamma) + \left\|U - \left(\frac{\mathrm{tr}\,U}{n}\right)I_n\right\|_{\mathrm{HS}}^2 + \frac{n}{N(N-n)}\left[\left(\frac{N-n}{n}\right)\mathrm{tr}\,U + \dot\gamma\cdot\nabla V(\gamma)\right]^2. \tag{14.37}$$

When $N=\infty$ this takes a simpler form:

$$\ddot\ell = \mathrm{Ric}_{\infty,\nu}(\dot\gamma) + \left\|U - \left(\frac{\mathrm{tr}\,U}{n}\right)I_n\right\|_{\mathrm{HS}}^2 \tag{14.38}$$

When $N<\infty$ one can introduce

$$\mathcal{D}(t) := \mathcal{J}(t)^{\frac{1}{N}},$$

and then formula (14.37) becomes

$$-N\frac{\ddot{\mathcal{D}}}{\mathcal{D}} = \mathrm{Ric}_{N,\nu}(\dot{\gamma}) + \left\| U - \left(\frac{\mathrm{tr}\, U}{n}\right) I_n \right\|_{\mathrm{HS}}^2 + \frac{n}{N(N-n)} \left[\left(\frac{N-n}{n}\right) \mathrm{tr}\, U + \dot{\gamma}\cdot\nabla V(\gamma)\right]^2. \quad (14.39)$$

Of course, it is a trivial corollary of (14.37) and (14.39) that

$$\begin{cases} \ddot{\ell} \geq \dfrac{\dot{\ell}^2}{N} + \mathrm{Ric}_{N,\nu}(\dot{\gamma}) \\ \\ -N\dfrac{\ddot{\mathcal{D}}}{\mathcal{D}} \geq \mathrm{Ric}_{N,\nu}(\dot{\gamma}). \end{cases} \quad (14.40)$$

Finally, if one wishes, one can also take out the direction of motion (skip at first reading and go directly to the next section). Define, with self-explicit notation,

$$\mathcal{J}_\perp(t,x) = \mathcal{J}_{0,\perp}(t,x)\,\frac{e^{-V(T_t(x))}}{e^{-V(x)}},$$

and $\ell_\perp = -\log \mathcal{J}_\perp$, $\mathcal{D}_\perp = \mathcal{J}_\perp^{\frac{1}{N}}$. Now, in place of (14.33), use

$$(\mathrm{tr}\, U_\perp)^{\cdot} + \frac{(\mathrm{tr}\, U_\perp)^2}{n-1} + \mathrm{Ric}(\dot{\gamma}) = -\left\| U_\perp - \left(\frac{\mathrm{tr}\, U_\perp}{n-1}\right) I_{n-1} \right\|_{\mathrm{HS}}^2 - \sum_{j=2}^n u_{1j}^2 \quad (14.41)$$

as a starting point. Computations quite similar to the ones above lead to

$$\ddot{\ell}_\perp = \frac{(\dot{\ell}_\perp)^2}{N-1} + \mathrm{Ric}_{N,\nu}(\dot{\gamma}) + \left\| U_\perp - \left(\frac{\mathrm{tr}\, U_\perp}{n-1}\right) I_{n-1} \right\|_{\mathrm{HS}}^2 + \frac{n-1}{(N-1)(N-n)} \left[\left(\frac{N-n}{n-1}\right) \mathrm{tr}\, U + \dot{\gamma}\cdot\nabla V(\gamma)\right]^2 + \sum_{j=2}^n u_{1j}^2. \quad (14.42)$$

In the case $N = \infty$, this reduces to

$$\ddot{\ell}_\perp = \mathrm{Ric}_{\infty,\nu}(\dot{\gamma}) + \left\| U_\perp - \left(\frac{\mathrm{tr}\, U_\perp}{n-1}\right) I_{n-1} \right\|_{\mathrm{HS}}^2 + \sum_{j=2}^n u_{1j}^2; \quad (14.43)$$

and in the case $N < \infty$, to

$$-N\frac{\ddot{\mathcal{D}}_\perp}{\mathcal{D}_\perp} = \mathrm{Ric}_{N,\nu}(\dot\gamma) + \left\| U_\perp - \left(\frac{\mathrm{tr}\, U_\perp}{n-1}\right) I_{n-1}\right\|_{\mathrm{HS}}^2$$
$$+ \frac{n-1}{(N-1)(N-n)}\left[\left(\frac{N-n}{n-1}\right)\mathrm{tr}\, U + \dot\gamma\cdot\nabla V(\gamma)\right]^2 + \sum_{j=2}^n u_{1j}^2. \quad (14.44)$$

As corollaries,

$$\begin{cases} \ddot\ell_\perp \geq \dfrac{(\dot\ell_\perp)^2}{N-1} + \mathrm{Ric}_{N,\nu}(\dot\gamma) \\ \\ -N\dfrac{\ddot{\mathcal{D}}_\perp}{\mathcal{D}_\perp} \geq \mathrm{Ric}_{N,\nu}(\dot\gamma). \end{cases} \quad (14.45)$$

Generalized Bochner formula and Γ_2 formalism

Of course there is an Eulerian translation of all that. This Eulerian formula can be derived either from the Lagrangian calculation, or from the Bochner formula, by a calculation parallel to the above one; the latter approach is conceptually simpler, while the former is faster. In any case the result is best expressed in terms of the differential operator

$$L = \Delta - \nabla V\cdot\nabla, \quad (14.46)$$

and can be written

$$L\frac{|\nabla\psi|^2}{2} - \nabla\psi\cdot\nabla L\psi$$
$$= \|\nabla^2\psi\|_{\mathrm{HS}}^2 + \big(\mathrm{Ric} + \nabla^2 V\big)(\nabla\psi)$$
$$= \frac{(L\psi)^2}{N} + \mathrm{Ric}_{N,\nu}(\nabla\psi)$$
$$+ \left(\left\|\nabla^2\psi - \left(\frac{\Delta\psi}{n}\right) I_n\right\|_{\mathrm{HS}}^2 + \frac{n}{N(N-n)}\left[\left(\frac{N-n}{n}\right)\Delta\psi + \nabla V\cdot\nabla\psi\right]^2\right).$$

It is convenient to reformulate this formula in terms of the Γ_2 formalism. Given a general linear operator L, one defines the associated Γ operator (or *carré du champ*) by the formula

$$\Gamma(f,g) = \frac{1}{2}\big[L(fg) - fLg - gLf\big]. \tag{14.47}$$

Note that Γ is a bilinear operator, which in some sense encodes the deviation of L from being a derivation operator. In our case, for (14.46),

$$\Gamma(f,g) = \nabla f \cdot \nabla g.$$

Next introduce the Γ_2 operator (or *carré du champ itéré*):

$$\Gamma_2(f,g) = \frac{1}{2}\big[L\Gamma(fg) - \Gamma(f,Lg) - \Gamma(g,Lf)\big]. \tag{14.48}$$

In the case of (14.46), the important formula for later purposes is

$$\Gamma_2(\psi) := \Gamma_2(\psi,\psi) = L\,\frac{|\nabla\psi|^2}{2} - \nabla\psi\cdot\nabla(L\psi). \tag{14.49}$$

Then our previous computations can be rewritten as

$$\begin{aligned}\Gamma_2(\psi) = {} & \frac{(L\psi)^2}{N} + \mathrm{Ric}_{N,\nu}(\nabla\psi)\\ & + \left(\left\|\nabla^2\psi - \left(\frac{\Delta\psi}{n}\right) I_n\right\|_{\mathrm{HS}}^2 + \frac{n}{N(N-n)}\left[\left(\frac{N-n}{n}\right)\Delta\psi + \nabla V\cdot\nabla\psi\right]^2\right).\end{aligned} \tag{14.50}$$

Of course, a trivial corollary is

$$\Gamma_2(\psi) \geq \frac{(L\psi)^2}{N} + \mathrm{Ric}_{N,\nu}(\nabla\psi). \tag{14.51}$$

And as the reader has certainly guessed, one can now take out the direction of motion (this computation is provided for completeness but will not be used): As before, define

$$\widehat{\nabla\psi} = \frac{\nabla\psi}{|\nabla\psi|},$$

then if f is a smooth function, let $\nabla^2_\perp f$ be $\nabla^2 f$ restricted to the space orthogonal to $\nabla\psi$, and $\Delta_\perp f = \mathrm{tr}\,(\nabla^2_\perp f)$, i.e.

$$\Delta_\perp f = \Delta f - \left\langle \nabla^2 f\cdot\widehat{\nabla\psi},\, \widehat{\nabla\psi}\right\rangle,$$

and next,

$$L_\perp f = \Delta_\perp f - \nabla V \cdot \nabla f,$$

$$\Gamma_{2,\perp}(\psi) = L_\perp \frac{|\nabla\psi|^2}{2} - \nabla\psi \cdot \nabla(L_\perp \psi) - 2\big|(\nabla^2\psi)\cdot\widehat{\nabla\psi}\big|^2 - 2|(\nabla^2\psi)\cdot\widehat{\nabla\psi}|^2.$$

Then

$$\Gamma_{2,\perp}(\psi) = \frac{(L_\perp\psi)^2}{N-1} + \mathrm{Ric}_{N,\nu}(\nabla\psi) + \left\| \nabla^2_\perp \psi - \left(\frac{\Delta_\perp\psi}{n-1}\right) I_{n-1} \right\|^2$$
$$+ \frac{n-1}{(N-1)(N-n)} \left[\left(\frac{N-n}{n-1}\right) \Delta_\perp \psi + \nabla V \cdot \nabla\psi \right]^2 + \sum_{j=2}^n (\partial_{1j}\psi)^2.$$

Curvature-dimension bounds

It is convenient to declare that a Riemannian manifold M, equipped with its volume measure, satisfies the **curvature-dimension** estimate $\mathrm{CD}(K,N)$ if its Ricci curvature is bounded below by K and its dimension is bounded above by N: $\mathrm{Ric} \geq K$, $n \leq N$. (As usual, $\mathrm{Ric} \geq K$ is a shorthand for "$\forall x$, $\mathrm{Ric}_x \geq K g_x$.") The number K might be positive or negative. If the reference measure is not the volume, but $\nu = e^{-V}\mathrm{vol}$, then the correct definition is $\mathrm{Ric}_{N,\nu} \geq K$.

Most of the previous discussion is summarized by Theorem 14.8 below, which is all the reader needs to know about Ricci curvature to understand the rest of the proofs in this course. For convenience I shall briefly recall the notation:

- *measures:* vol is the volume on M, $\nu = e^{-V}\mathrm{vol}$ is the reference measure;
- *operators:* Δ is the Laplace(–Beltrami) operator on M, ∇^2 is the Hessian operator, $L = \Delta - \nabla V \cdot \nabla$ is the modified Laplace operator, and $\Gamma_2(\psi) = L(|\nabla\psi|^2/2) - \nabla\psi \cdot \nabla(L\psi)$;
- *tensors:* Ric is the Ricci curvature bilinear form, and $\mathrm{Ric}_{N,\nu}$ is the modified Ricci tensor: $\mathrm{Ric}_{N,\nu} = \mathrm{Ric} + \nabla^2 V - (\nabla V \otimes \nabla V)/(N-n)$, where $\nabla^2 V(x)$ is the Hessian of V at x, identified to a bilinear form;
- *functions:* ψ is an arbitrary function; in formulas involving the Γ_2 formalism it will be assumed to be of class C^3, while in formulas involving Jacobian determinants it will only be assumed to be semi-convex;

- *geodesic paths:* If ψ is a given function on M, $\gamma(t,x) = T_t(x) = \exp_x\big((t-t_0)\nabla\psi(x)\big)$ is the geodesic starting from x with velocity $\dot\gamma(t_0,x) = \nabla\psi(x)$, evaluated at time $t \in [0,1]$; it is assumed that $\mathcal{J}(t,x)$ does not vanish for $t \in (0,1)$; the starting time t_0 may be the origin $t_0 = 0$, or any other time in $[0,1]$;
- *Jacobian determinants:* $\mathcal{J}(t,x)$ is the Jacobian determinant of $T_t(x)$ (with respect to the reference measure ν, not with respect to the standard volume), $\ell = -\log\mathcal{J}$, and $\mathcal{D} = \mathcal{J}^{1/N}$ is the mean distortion associated with (T_t);
- the dot means differentiation with respect to time;
- finally, the subscript $\perp$ in $\mathcal{J}_\perp$, $\mathcal{D}_\perp$, $\Gamma_{2,\perp}$ means that the direction of motion $\dot\gamma = \nabla\psi$ has been taken out (see above for precise definitions).

Theorem 14.8 (CD(K,N) curvature-dimension bound). *Let M be a Riemannian manifold of dimension n, and let $K \in \mathbb{R}$, $N \in [n,\infty]$. Then, the conditions below are all equivalent if they are required to hold true for arbitrary data; when they are fulfilled, M is said to satisfy the* CD(K,N) *curvature-dimension bound:*

(i) $\mathrm{Ric}_{N,\nu} \geq K$;

(ii) $\Gamma_2(\psi) \geq \dfrac{(L\psi)^2}{N} + K|\nabla\psi|^2$;

(iii) $\ddot\ell \geq \dfrac{(\dot\ell)^2}{N} + K|\dot\gamma|^2$.

If $N < \infty$, this is also equivalent to

(iv) $\ddot{\mathcal{D}} + \left(\dfrac{K|\dot\gamma|^2}{N}\right)\mathcal{D} \leq 0$.

Moreover, these inequalities are also equivalent to

(ii') $\Gamma_{2,\perp}(\psi) \geq \dfrac{(L_\perp\psi)^2}{N-1} + K|\nabla\psi|^2$;

(iii') $\ddot\ell_\perp \geq \dfrac{(\dot\ell_\perp)^2}{N-1} + K|\dot\gamma|^2$;

and, in the case $N < \infty$,

(iv') $\ddot{\mathcal{D}}_\perp + \left(\dfrac{K|\dot\gamma|^2}{N-1}\right)\mathcal{D}_\perp \leq 0$.

Remark 14.9. Note carefully that the inequalities (i)–(iv') are required to be true *always*: For instance (ii) should be true for all ψ, all x and all $t \in (0,1)$. The equivalence is that [(i) true for all x] is equivalent to [(ii) true for all ψ, all x and all t], etc.

Examples 14.10 (One-dimensional $\mathrm{CD}(K,N)$ **model spaces).**
(a) Let $K>0$ and $1<N<\infty$, consider

$$M = \left(-\sqrt{\frac{N-1}{K}}\,\frac{\pi}{2},\ \sqrt{\frac{N-1}{K}}\,\frac{\pi}{2}\right) \subset \mathbb{R},$$

equipped with the usual distance on $\mathbb{R}$, and the reference measure

$$\nu(dx) = \cos^{N-1}\left(\sqrt{\frac{K}{N-1}}\,x\right)\,dx;$$

then M satisfies $\mathrm{CD}(K,N)$, although the Hausdorff dimension of M is of course 1. Note that M is not complete, but this is not a serious problem since $\mathrm{CD}(K,N)$ is a local property. (We can also replace M by its closure, but then it is a manifold with boundary.)

(b) For $K<0$, $1\le N<\infty$, the same conclusion holds true if one considers $M=\mathbb{R}$ and

$$\nu(dx) = \cosh^{N-1}\left(\sqrt{\frac{|K|}{N-1}}\,x\right)\,dx.$$

(c) For any $N\in[1,\infty)$, an example of one-dimensional space satisfying $\mathrm{CD}(0,N)$ is provided by $M=(0,+\infty)$ with the reference measure $\nu(dx)=x^{N-1}\,dx$;

(d) For any $K\in\mathbb{R}$, take $M=\mathbb{R}$ and equip it with the reference measure

$$\nu(dx) = e^{-\frac{Kx^2}{2}}\,dx;$$

then M satisfies $\mathrm{CD}(K,\infty)$.

Sketch of proof of Theorem 14.8. It is clear from our discussion in this chapter that (i) implies (ii) and (iii); and (iii) is equivalent to (iv) by elementary manipulations about derivatives. (Moreover, (ii) and (iii) are equivalent modulo smoothness issues, by Eulerian/Lagrangian duality.)

It is less clear why, say, (ii) would imply (i). This comes from formulas (14.37) and (14.50). Indeed, assume (ii) and choose an arbitrary $x_0\in M$, and $v_0\in T_{x_0}M$. Assume, to fix ideas, that $N>n$. Construct a C^3 function ψ such that

$$\begin{cases} \nabla\psi(x_0) = v_0 \\ \nabla^2\psi(x_0) = \lambda_0\, I_n \\ \Delta\psi(x_0) (= n\lambda_0) = -\dfrac{n}{N-n}\big(\nabla V(x_0)\cdot v_0\big). \end{cases}$$

(This is fairly easy by using local coordinates, or distance and exponential functions.) Then all the remainder terms in (14.50) will vanish at x_0, so that

$$K|v_0|^2 = K|\nabla\psi(x_0)|^2 \le \left(\Gamma_2(\psi) - \frac{(L\psi)^2}{N}\right)(x_0)$$
$$= \mathrm{Ric}_{N,\nu}(\nabla\psi(x_0)) = \mathrm{Ric}_{N,\nu}(v_0).$$

So indeed $\mathrm{Ric}_{N,\nu} \ge K$.

The proof goes in the same way for the equivalence between (i) and (ii'), (iii'), (iv'): again the problem is to understand why (ii') implies (i), and the reasoning is almost the same as before; the key point being that the extra error terms in $\partial_{1j}\psi$, $j \ne 2$, all vanish at x_0. □

Many interesting inequalities can be derived from $\mathrm{CD}(K,N)$. It was successfully advocated by Bakry and other authors during the past two decades that $\mathrm{CD}(K,N)$ should be considered as a property of the generalized Laplace operator L. Instead, it will be advocated in this course that $\mathrm{CD}(K,N)$ is a property of the solution of the optimal transport problem, when the cost function is the square of the geodesic distance. Of course, both points of view have their advantages and their drawbacks.

From differential to integral curvature-dimension bounds

There are two ways to characterize the concavity of a function $f(t)$ on a time-interval, say $[0,1]$: the differential inequality $\ddot{f} \le 0$, or the integral bound $f\big((1-\lambda)\,t_0 + \lambda\,t_1\big) \ge (1-\lambda) f(t_0) + \lambda f(t_1)$. If the latter is required to hold true for all $t_0, t_1 \in [0,1]$ and $\lambda \in [0,1]$, then the two formulations are equivalent.

There are two classical generalizations. The first one states that the differential inequality $\ddot{f} + K \le 0$ is equivalent to the integral inequality

$$f\big((1-\lambda)\,t_0 + \lambda\,t_1\big) \ge (1-\lambda)\, f(t_0) + \lambda\, f(t_1) + \frac{Kt(1-t)}{2}\,(t_0 - t_1)^2.$$

Another one is as follows: The differential inequality

$$\ddot{f}(t) + \Lambda f(t) \le 0 \tag{14.52}$$

is equivalent to the integral bound

$$f\Big((1-\lambda)\,t_0+\lambda\,t_1\Big) \geq \tau^{(1-\lambda)}(|t_0-t_1|)\,f(t_0)+\tau^{(\lambda)}(|t_0-t_1|)\,f(t_1), \quad (14.53)$$

where

$$\tau^{(\lambda)}(\theta) = \begin{cases} \dfrac{\sin(\lambda\theta\sqrt{\Lambda})}{\sin(\theta\sqrt{\Lambda})} & \text{if } \Lambda > 0 \\[2ex] \lambda & \text{if } \Lambda = 0 \\[2ex] \dfrac{\sinh(\lambda\theta\sqrt{-\Lambda})}{\sinh(\theta\sqrt{-\Lambda})} & \text{if } \Lambda < 0. \end{cases}$$

A more precise statement, together with a proof, are provided in the Second Appendix of this chapter.

This leads to the following integral characterization of $\mathrm{CD}(K,N)$:

Theorem 14.11 (Integral reformulation of curvature-dimension bounds). *Let M be a Riemannian manifold, equipped with a reference measure $\nu = e^{-V}\mathrm{vol}$, $V \in C^2(M)$, and let d be the geodesic distance on M. Let $K \in \mathbb{R}$ and $N \in [1,\infty]$. Then, with the same notation as in Theorem 14.8, M satisfies $\mathrm{CD}(K,N)$ if and only if the following inequality is always true (for any semiconvex ψ, and almost any x, as soon as $\mathcal{J}(t,x)$ does not vanish for $t \in (0,1)$):*

$$\mathcal{D}(t,x) \geq \tau_{K,N}^{(1-t)}\,\mathcal{D}(0,x) + \tau_{K,N}^{(t)}\,\mathcal{D}(1,x) \qquad (N<\infty) \quad (14.54)$$

$$\ell(t,x) \leq (1-t)\,\ell(0,x) + t\,\ell(1,x) - \frac{Kt(1-t)}{2}\,d(x,y)^2 \qquad (N=\infty), \quad (14.55)$$

where $y = \exp_x(\nabla\psi(x))$ and, in case $N<\infty$,

$$\tau_{K,N}^{(t)} = \begin{cases} \dfrac{\sin(t\alpha)}{\sin\alpha} & \textit{if } K > 0 \\[2ex] t & \textit{if } K = 0 \\[2ex] \dfrac{\sinh(t\alpha)}{\sinh\alpha} & \textit{if } K < 0 \end{cases}$$

where

$$\alpha = \sqrt{\frac{|K|}{N}}\, d(x,y) \qquad (\alpha \in [0,\pi] \text{ if } K > 0).$$

Proof of Theorem 14.11. If $N < \infty$, inequality (14.54) is obtained by transforming the differential bound of (iii) in Theorem 14.8 into an integral bound, after noticing that $|\dot{\gamma}|$ is a constant all along the geodesic γ, and equals $d(\gamma_0, \gamma_1)$. Conversely, to go from (14.54) to Theorem 14.8(iii), we select a geodesic γ, then reparametrize the geodesic $(\gamma_t)_{t_0 \le t \le t_1}$ into a geodesic $[0,1] \to M$, apply (14.54) to the reparametrized path and discover that

$$\mathcal{D}(t,x) \ge \tau_{K,N}^{(1-\lambda)}\, \mathcal{D}(t_0,x) + \tau_{K,N}^{(\lambda)}\, \mathcal{D}(t_1,x) \quad t = (1-\lambda)t_0 + \lambda t_1;$$

where now $\alpha = \sqrt{|K|/N}\, d(\gamma(t_0), \gamma(t_1))$. It follows that $\mathcal{D}(t,x)$ satisfies (14.53) for any choice of t_0, t_1; and this is equivalent to (14.52).

The reasoning is the same for the case $N = \infty$, starting from inequality (ii) in Theorem 14.8. □

The next result states that the the coefficients $\tau_{K,N}^{(t)}$ obtained in Theorem 14.11 can be *automatically improved* if N is finite and $K \neq 0$, by taking out the direction of motion:

Theorem 14.12 (Curvature-dimension bounds with direction of motion taken out). *Let M be a Riemannian manifold, equipped with a reference measure $\nu = e^{-V}\,\mathrm{vol}$, and let d be the geodesic distance on M. Let $K \in \mathbb{R}$ and $N \in [1,\infty)$. Then, with the same notation as in Theorem 14.8, M satisfies $\mathrm{CD}(K,N)$ if and only if the following inequality is always true (for any semiconvex ψ, and almost any x, as soon as $\mathcal{J}(t,x)$ does not vanish for $t \in (0,1)$):*

$$\mathcal{D}(t,x) \ge \tau_{K,N}^{(1-t)}\, \mathcal{D}(0,x) + \tau_{K,N}^{(t)}\, \mathcal{D}(1,x) \tag{14.56}$$

where now

$$\tau_{K,N}^{(t)} = \begin{cases} t^{\frac{1}{N}} \left(\dfrac{\sin(t\alpha)}{\sin\alpha}\right)^{1-\frac{1}{N}} & \textit{if } K > 0 \\ t & \textit{if } K = 0 \\ t^{\frac{1}{N}} \left(\dfrac{\sinh(t\alpha)}{\sinh\alpha}\right)^{1-\frac{1}{N}} & \textit{if } K < 0 \end{cases}$$

and

$$\alpha = \sqrt{\frac{|K|}{N-1}}\, d(x,y) \qquad (\alpha \in [0,\pi] \textit{ if } K > 0).$$

Remark 14.13. When $N < \infty$ and $K > 0$ Theorem 14.12 contains the Bonnet–Myers theorem according to which $d(x,y) \leq \pi\sqrt{(N-1)/K}$. With Theorem 14.11 the bound was only $\pi\sqrt{N/K}$.

Proof of Theorem 14.12. The proof that (14.56) implies $\mathrm{CD}(K,N)$ is done in the same way as for (14.54). (In fact (14.56) is stronger than (14.54).)

As for the other implication: Start from (14.22), and transform it into an integral bound:

$$\mathcal{D}_\perp(t,x) \geq \sigma_{K,N}^{(1-t)}\, \mathcal{D}_\perp(0,x) + \sigma_{K,N}^{(t)}\, \mathcal{D}_\perp(1,x),$$

where $\sigma_{K,N}^{(t)} = \sin(t\alpha)/\sin\alpha$ if $K > 0$; t if $K = 0$; $\sinh(t\alpha)/\sinh\alpha$ if $K < 0$. Next transform (14.19) into the integral bound

$$\mathcal{D}_{//}(t,x) \geq (1-t)\, \mathcal{D}_{//}(0,x) + t\, \mathcal{D}_{//}(1,x).$$

Both estimates can be combined thanks to Hölder's inequality:

$$\begin{aligned} \mathcal{D}(t,x) &= \mathcal{D}_\perp(t,x)^{1-\frac{1}{N}}\, \mathcal{D}_{//}(t,x)^{\frac{1}{N}} \\ &\geq \left(\sigma_{K,N}^{(1-t)}\, \mathcal{D}(0,x) + \sigma_{K,N}^{(t)}\, \mathcal{D}(1,x)\right)^{1-\frac{1}{N}} \\ &\qquad\qquad \left((1-t)\, \mathcal{D}_{//}(0,x) + t\, \mathcal{D}_{//}(1,x)\right)^{\frac{1}{N}} \\ &\geq (\sigma_{K,N}^{(1-t)})^{1-\frac{1}{N}} (1-t)^{\frac{1}{N}}\, \mathcal{D}(0,x) + (\sigma_{K,N}^{(t)})^{\frac{1}{N}} t^{\frac{1}{N}}\, \mathcal{D}_{//}(1,x). \end{aligned}$$

This implies inequality (14.56). □

Estimate (14.56) is sharp in general. The following reformulation yields an appealing interpretation of $\mathrm{CD}(K,N)$ in terms of comparison spaces. In the sequel, I will write Jac_x for the (unoriented) Jacobian determinant evaluated at point x, computed with respect to a given reference measure.

Corollary 14.14 (Curvature-dimension bounds by comparison). *Let M be a Riemannian manifold equipped with a reference measure $\nu = e^{-V}\mathrm{vol}$, $V \in C^2(M)$. Define the $\mathcal{J}$-function of M on $[0,1]\times\mathbb{R}_+\times\mathbb{R}_+$ by the formula*

$$\mathcal{J}_{M,\nu}(t,\delta,J) := \inf\Big\{\mathrm{Jac}_x(\exp(t\xi));\quad |\xi(x)| = \delta;\quad \mathrm{Jac}_x(\exp(\xi)) = J\Big\}, \tag{14.57}$$

where the infimum is over all vector fields ξ defined around x, such that $\nabla\xi(x)$ is symmetric, and $\mathrm{Jac}_x(\exp s\xi) \neq 0$ for $0 \le s < 1$. Then, for any $K \in \mathbb{R}$, $N \in [1,\infty]$ ($K \le 0$ if $N = 1$),

$$(M,\nu) \text{ satisfies } \mathrm{CD}(K,N) \Longleftrightarrow \mathcal{J}_{M,\nu} \ge \mathcal{J}^{(K,N)},$$

where $\mathcal{J}^{(K,N)}$ is the $\mathcal{J}$-function of the model $\mathrm{CD}(K,N)$ space considered in Examples 14.10.

If N is an integer, $\mathcal{J}^{(K,N)}$ is also the $\mathcal{J}$-function of the N-dimensional model space

$$S^{(K,N)} = \begin{cases} S^N\Big(\sqrt{\dfrac{N-1}{K}}\Big) & \text{if } K > 0 \\[2ex] \mathbb{R}^N & \text{if } K = 0 \\[2ex] \mathbb{H}^N\Big(\sqrt{\dfrac{N-1}{|K|}}\Big) & \text{if } K < 0, \end{cases}$$

equipped with its volume measure.

Corollary 14.14 results from Theorem 14.12 by a direct computation of the $\mathcal{J}$-function of the model spaces. In the case of $S^{(K,N)}$, one can also make a direct computation, or note that all the inequalities which were used to obtain (14.56) turn into equalities for suitable choices of parameters.

Remark 14.15. There is a quite similar (and more well-known) formulation of lower *sectional* curvature bounds which goes as follows. Define the $\mathcal{L}$-function of a manifold M by the formula

$$\mathcal{L}_M(t,\delta,L) = \inf\Big\{d(\exp_x(tv),\exp_x(tw));\quad |v| = |w| = \delta;\\ d(\exp_x v, \exp_x w) = L\Big\},$$

where the infimum is over tangent vectors $v, w \in T_x M$. Then M has sectional curvature larger than κ if and only if $\mathcal{L}_M \geq \mathcal{L}^{(\kappa)}$, where $\mathcal{L}^{(\kappa)}$ is the $\mathcal{L}$-function of the reference space $S^{(\kappa)}$, which is $S^2(1/\sqrt{\kappa})$ if $\kappa > 0$, $\mathbb{R}^2$ if $\kappa = 0$, and $\mathbb{H}^2(1/\sqrt{|\kappa|})$ if $\kappa < 0$. By changing the infimum into a supremum and reversing the inequalities, one can also obtain a characterization of upper sectional curvature bounds (under a topological assumption of simple connectivity). The comparison with (14.14) conveys the idea that *sectional curvature bounds measure the rate of separation of geodesics in terms of distances, while Ricci curvature bounds do it in terms of Jacobian determinants.*

Distortion coefficients

Apart from Definition 14.19, the material in this section is not necessary to the understanding of the rest of this course. Still, it is interesting because it will give a new interpretation of Ricci curvature bounds, and motivate the introduction of distortion coefficients, which will play a crucial role in the sequel.

Definition 14.16 (Barycenters). *If A and B are two measurable sets in a Riemannian manifold, and $t \in [0,1]$, a t-barycenter of A and B is a point which can be written γ_t, where γ is a (minimizing, constant-speed) geodesic with $\gamma_0 \in A$ and $\gamma_1 \in B$. The set of all t-barycenters between A and B is denoted by $[A,B]_t$.*

Definition 14.17 (Distortion coefficients). *Let M be a Riemannian manifold, equipped with a reference measure $e^{-V}\mathrm{vol}$, $V \in C(M)$, and let x and y be any two points in M. Then the distortion coefficient $\overline{\beta}_t(x,y)$ between x and y at time $t \in (0,1)$ is defined as follows:*

- *If x and y are joined by a unique geodesic γ, then*

$$\overline{\beta}_t(x,y) = \lim_{r\downarrow 0} \frac{\nu\big[[x,B_r(y)]_t\big]}{\nu[B_{tr}(y)]} = \lim_{r\downarrow 0} \frac{\nu\big[[x,B_r(y)]_t\big]}{t^n\,\nu[B_r(y)]}; \tag{14.58}$$

- *If x and y are joined by several minimizing geodesics, then*

$$\overline{\beta}_t(x,y) = \inf_{\gamma} \limsup_{s\to 1^-} \overline{\beta}_t(x,\gamma_s), \tag{14.59}$$

where the infimum is over all minimizing geodesics joining $\gamma(0) = x$ to $\gamma(1) = y$.

Finally, the values of $\overline{\beta}_t(x,y)$ for $t = 0$ and $t = 1$ are defined by

$$\overline{\beta}_1(x,y) \equiv 1; \qquad \overline{\beta}_0(x,y) := \liminf_{t\to 0^+} \overline{\beta}_t(x,y).$$

The heuristic meaning of distortion coefficients is as follows (see Figure 14.4). Assume you are standing at point x and observing some device located at y. You are trying to estimate the volume of this device, but your appreciation is altered because light rays travel along curved lines (geodesics). If x and y are joined by a unique geodesic, then the coefficient $\overline{\beta}_0(x,y)$ tells by how much you are overestimating; so it is less than 1 in negative curvature, and greater than 1 in positive curvature. If x and y are joined by several geodesics, this is just the same, except that you choose to look in the direction where the device looks smallest.

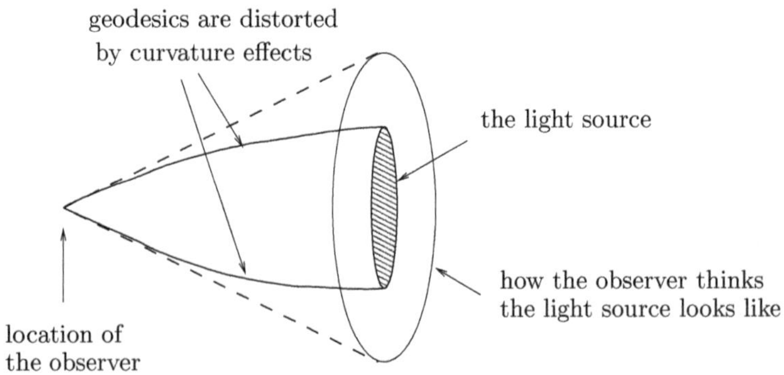

Fig. 14.4. The meaning of distortion coefficients: Because of positive curvature effects, the observer overestimates the surface of the light source; in a negatively curved world this would be the contrary.

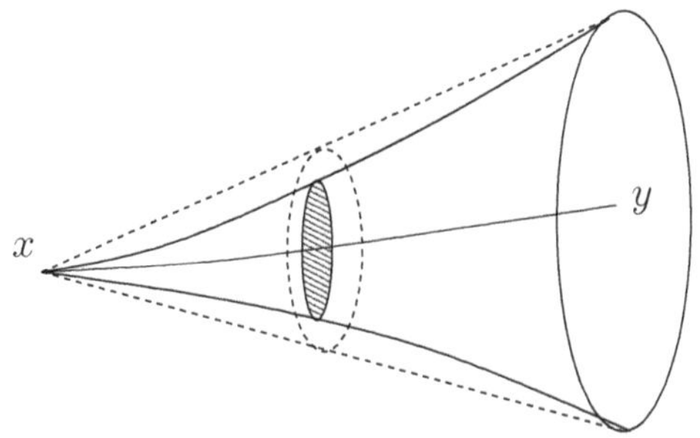

Fig. 14.5. The distortion coefficient is approximately equal to the ratio of the volume filled with lines, to the volume whose contour is in dashed line. Here the space is negatively curved and the distortion coefficient is less than 1.

More generally, $\overline{\beta}_t(x, y)$ compares the volume occupied by the light rays emanating from the light source, when they arrive close to $\gamma(t)$, to the volume that they would occupy in a flat space (see Figure 14.5).

Now let us express distortion coefficients in differential terms, and more precisely *Jacobi fields.* A key concept in doing so will be the notion of *focalization*, which was already discussed in Chapter 8: A point y is said to be focal to another point x if there exists $v \in T_x M$ such that $y = \exp_v x$ and the differential $d_v \exp_x : T_x M \to T_y M$ is not invertible. It is equivalent to say that there is a geodesic γ which visits both x and y, and a Jacobi field J along γ such that $J(x) = 0$, $J(y) = 0$. This concept is obviously symmetric in x and y, and then x, y are said to be conjugate points (along γ).

If x and y are joined by a unique geodesic γ and are not conjugate, then by the local inversion theorem, for r small enough, there is a unique velocity $\xi(z)$ at $z \in B_r(y)$ such that $\exp_z \xi(z) = x$. Then the distortion coefficients can be interpreted as the Jacobian determinant of $\exp \xi$ at time t, renormalized by $(1-t)^n$, which would be the value in Euclidean space. The difference with the computations at the beginning of this chapter is that now the Jacobi field is not defined by its initial value and initial derivative, but rather by its initial value and its *final value*: $\exp_z \xi(z) = x$ independently of z, so the Jacobi field vanishes after a time 1. It will be convenient to reverse time so that $t = 0$ corresponds to x and $t = 1$ to y; thus the conditions are $J(0) = 0$, $J(1) = I_n$. After these preparations it is easy to derive the following:

Proposition 14.18 (Computation of distortion coefficients). *Let M be a Riemannian manifold, let x and y be two points in M.*

Then

$$\overline{\beta}_t(x,y) = \inf_{\gamma} \overline{\beta}_t^{[\gamma]}(x,y),$$

where the infimum is over all minimizing geodesics γ *joining* $\gamma(0) = x$ *to* $\gamma(1) = y$*, and* $\overline{\beta}_t^{[\gamma]}(x,y)$ *is defined as follows:*

- *If* x, y *are not conjugate along* γ*, let* $\mathbf{E}$ *be an orthonormal basis of* $T_y M$ *and define*

$$\overline{\beta}_t^{[\gamma]}(x,y) = \begin{cases} \dfrac{\det \mathbf{J}^{0,1}(t)}{t^n} & \text{if } 0 < t \le 1 \\[2ex] \lim\limits_{s\to 0} \dfrac{\det \mathbf{J}^{0,1}(s)}{s^n} & \text{if } t = 0, \end{cases} \tag{14.60}$$

  where $\mathbf{J}^{0,1}$ *is the unique matrix of Jacobi fields along* γ *satisfying*

$$\mathbf{J}^{0,1}(0) = 0; \qquad \mathbf{J}^{0,1}(1) = \mathbf{E};$$

- *If* x, y *are conjugate along* γ*, define*

$$\overline{\beta}_t^{[\gamma]}(x,y) = \begin{cases} 1 & \text{if } t = 1 \\ +\infty & \text{if } 0 \le t < 1. \end{cases}$$

Distortion coefficients can be explicitly computed for the model $\mathrm{CD}(K,N)$ spaces and depend only on the distance between the two points x and y. These particular coefficients (or rather their expression as a function of the distance) will play a key role in the sequel of these notes.

Definition 14.19 (Reference distortion coefficients). *Given* $K \in \mathbb{R}$*,* $N \in [1,\infty]$ *and* $t \in [0,1]$*, and two points* x, y *in some metric space* $(\mathcal{X}, d)$*, define* $\beta_t^{(K,N)}(x,y)$ *as follows:*

- *If* $0 < t \le 1$ *and* $1 < N < \infty$ *then*

$$\beta_t^{(K,N)}(x,y) = \begin{cases} +\infty & \text{if } K > 0 \text{ and } \alpha > \pi, \\[2ex] \left(\dfrac{\sin(t\alpha)}{t \sin\alpha}\right)^{N-1} & \text{if } K > 0 \text{ and } \alpha \in [0,\pi], \\[2ex] 1 & \text{if } K = 0, \\[2ex] \left(\dfrac{\sinh(t\alpha)}{t \sinh\alpha}\right)^{N-1} & \text{if } K < 0, \end{cases} \tag{14.61}$$

where

$$\alpha = \sqrt{\frac{|K|}{N-1}}\, d(x,y). \tag{14.62}$$

- *In the two limit cases $N \to 1$ and $N \to \infty$, modify the above expressions as follows:*

$$\beta_t^{(K,1)}(x,y) = \begin{cases} +\infty & \text{if } K > 0, \\ 1 & \text{if } K \le 0, \end{cases} \tag{14.63}$$

$$\beta_t^{(K,\infty)}(x,y) = e^{\frac{K}{6}(1-t^2)\, d(x,y)^2}. \tag{14.64}$$

- *For $t = 0$ define $\beta_0^{(K,N)}(x,y) = 1$.*

If $\mathcal{X}$ is the model space for $\mathrm{CD}(K,N)$, as in Examples 14.10, then $\beta^{(K,N)}$ is just the distortion coefficient on $\mathcal{X}$.

If K is positive, then for fixed t, $\beta_t^{(K,N)}$ is an increasing function of α (going to $+\infty$ at $\alpha = \pi$), while for fixed α, it is a decreasing function of t on $[0,1]$. All this is reversed for negative K. On the whole, $\beta_t^{(K,N)}$ is nondecreasing in K and nonincreasing in N. (See Figure 14.6.)

The next two theorems relate distortion coefficients with Ricci curvature lower bounds; they show that (a) distortion coefficients can be interpreted as the "best possible" coefficients in concavity estimates for the Jacobian determinant; and (b) the curvature-dimension bound $\mathrm{CD}(K,N)$ is a particular case of a family of more general estimates characterized by a lower bound on the distortion coefficients.

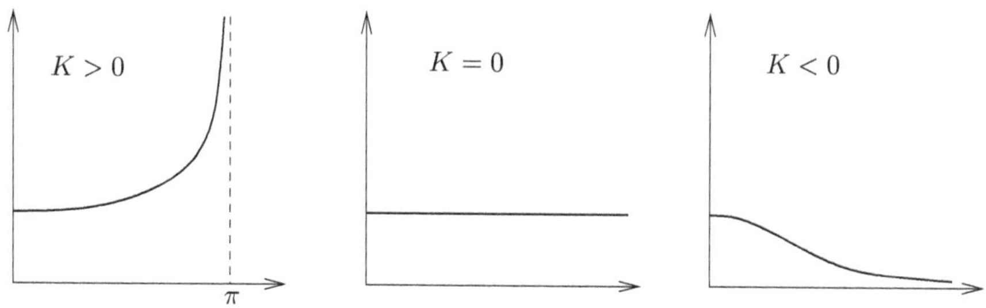

Fig. 14.6. The shape of the curves $\beta_t^{(K,N)}(x,y)$, for fixed $t \in (0,1)$, as a function of $\alpha = \sqrt{|K|/(N-1)}\, d(x,y)$.

Theorem 14.20 (Distortion coefficients and concavity of Jacobian determinant). *Let M be a Riemannian manifold of dimension n, and let x, y be any two points in M. Then if $(\beta_t(x,y))_{0 \le t \le 1}$ and $(\beta_t(y,x))_{0 \le t \le 1}$ are two families of nonnegative coefficients, the following statements are equivalent:*

(a) $\forall t \in [0,1], \quad \beta_t(x,y) \le \overline{\beta}_t(x,y); \quad \beta_t(y,x) \le \overline{\beta}_t(y,x);$

(b) For any $N \ge n$, for any geodesic γ joining x to y, for any $t_0 \in [0,1]$, and for any initial vector field ξ around $x_0 = \gamma(t_0)$, $\nabla\xi(x_0)$ symmetric, let $\mathcal{J}(s)$ stand for the Jacobian determinant of $\exp((s-t_0)\xi)$ at x_0; if $\mathcal{J}(s)$ does not vanish for $0 < s < 1$, then for all $t \in [0,1]$,

$$\begin{cases} \mathcal{J}(t)^{\frac{1}{N}} \ge (1-t)\,\beta_{1-t}(y,x)^{\frac{1}{N}}\,\mathcal{J}(0)^{\frac{1}{N}} + t\,\beta_t(x,y)^{\frac{1}{N}}\,\mathcal{J}(1)^{\frac{1}{N}} & (N < \infty) \\ \\ \log \mathcal{J}(0,x) \ge (1-t)\,\log \mathcal{J}(0) + t\,\log \mathcal{J}(1) & \\ \qquad\qquad + \big[(1-t)\,\log \beta_{1-t}(y,x) + t\,\log \beta_t(x,y)\big] & (N = \infty); \end{cases} \tag{14.65}$$

(c) Property (b) holds true for $N = n$.

Theorem 14.21 (Ricci curvature bounds in terms of distortion coefficients). *Let M be a Riemannian manifold of dimension n, equipped with its volume measure. Then the following two statements are equivalent:*

(a) $\mathrm{Ric} \ge K$;

(b) $\overline{\beta} \ge \beta^{(K,n)}$.

Sketch of proof of Theorem 14.20. To prove the implication (a)⇒(b), it suffices to establish (14.65) for $\beta = \overline{\beta}$. The case $N = \infty$ is obtained from the case $N < \infty$ by passing to the limit, since $\lim_{N\to 0}[N(a^{1/N} - 1)] = \log a$. So all we have to show is that if $n \le N < \infty$, then

$$\mathcal{J}(t)^{\frac{1}{N}} \ge (1-t)\,\overline{\beta}_{1-t}(y,x)^{\frac{1}{N}}\,\mathcal{J}(0)^{\frac{1}{N}} + t\,\overline{\beta}_t(x,y)^{\frac{1}{N}}\,\mathcal{J}(1)^{\frac{1}{N}}.$$

The case when x, y are conjugate can be treated by a limiting argument. (In fact the conclusion is that both $\mathcal{J}(0)$ and $\mathcal{J}(1)$ have to vanish if x and y are conjugate.) So we may assume that x and y are not conjugate, introduce a moving orthonormal basis $\mathbf{E}(t)$, along γ, and define the Jacobi matrices $\mathbf{J}^{1,0}(t)$ and $\mathbf{J}^{0,1}(t)$ by the requirement

$$J^{1,0}(0) = I_n, \quad J^{1,0}(1) = 0; \qquad J^{0,1}(0) = 0, \quad J^{0,1}(1) = I_n.$$

(Here $\mathbf{J}^{1,0}, \mathbf{J}^{0,1}$ are identified with their expressions $J^{1,0}, J^{0,1}$ in the moving basis $\mathbf{E}$.)

As noted after (14.7), the Jacobi equation is invariant under the change $t \to 1-t$, $\mathbf{E} \to -\mathbf{E}$, so $J^{1,0}$ becomes $J^{0,1}$ when one exchanges the roles of x and y, and replaces t by $1-t$. In particular, we have the formula

$$\frac{\det J^{1,0}(t)}{(1-t)^n} = \beta_{1-t}(y,x). \tag{14.66}$$

As in the beginning of this chapter, the issue is to compute the determinant at time t of a Jacobi field $J(t)$. Since the Jacobi fields are solutions of a linear differential equation of the form $\ddot{J} + R\,J = 0$, they form a vector space of dimension $2n$, and they are invariant under right-multiplication by a constant matrix. This implies

$$J(t) = J^{1,0}(t)\,J(0) \;+\; J^{0,1}(t)\,J(1). \tag{14.67}$$

The determinant in dimension n satisfies the following inequality: If X and Y are two $n\times n$ nonnegative symmetric matrices, then

$$\det(X+Y)^{\frac{1}{n}} \geq (\det X)^{\frac{1}{n}} + (\det Y)^{\frac{1}{n}}. \tag{14.68}$$

By combining this with Hölder's inequality, in the form

$$\bigl(a^{\frac{1}{n}} + b^{\frac{1}{n}}\bigr)^{\frac{n}{N}} \geq (1-t)^{\frac{N-n}{N}}\,a^{\frac{1}{N}} + t^{\frac{N-n}{N}}\,b^{\frac{1}{N}},$$

we obtain a generalization of (14.68):

$$\det(X+Y)^{\frac{1}{N}} \geq (1-t)^{\frac{N-n}{N}}(\det X)^{\frac{1}{N}} + t^{\frac{N-n}{N}}(\det Y)^{\frac{1}{N}}. \tag{14.69}$$

If one combines (14.67) and (14.69) one obtains

$$\begin{aligned}
(\det J(t))^{\frac{1}{N}} &\geq (1-t)^{\frac{N-n}{N}}(\det J^{1,0}(t))^{\frac{1}{N}}(\det J(0))^{\frac{1}{N}} \\
&\qquad\qquad + t^{\frac{N-n}{N}}(\det J^{0,1}(t))^{\frac{1}{N}}(\det J(1))^{\frac{1}{N}} \\
&= (1-t)\left[\frac{\det J^{1,0}(t)}{(1-t)^n}\right]^{\frac{1}{N}} \mathcal{J}(0)^{\frac{1}{N}} \;+\; t\left[\frac{\det J^{0,1}(t)}{t^n}\right]^{\frac{1}{N}} \mathcal{J}(1)^{\frac{1}{N}} \\
&= (1-t)\,\overline{\beta}_{1-t}(y,x)^{\frac{1}{N}}\,\mathcal{J}(0)^{\frac{1}{N}} \;+\; t\,\overline{\beta}_t(x,y)^{\frac{1}{N}}\,\mathcal{J}(1)^{\frac{1}{N}},
\end{aligned}$$

where the final equality follows from (14.60) and (14.66).

In this way we have shown that (a) implies (b), but at the price of a gross cheating, since in general the matrices appearing in (14.67) *are not* symmetric! It turns out however that they can be *positively cosymmetrized*: there is a kernel $K(t)$ such that $\det K(t) > 0$, and $K(t)\, J^{1,0}(t)$ and $K(t)\, J^{0,1}(t)\, J(1)$ are both positive symmetric, at least for $t \in (0,1)$. This remarkable property is a consequence of the structure of Jacobi fields; see Propositions 14.30 and 14.31 in the Third Appendix of this chapter.

Once the cosymmetrization property is known, it is obvious how to fix the proof: just write

$$\big(\det(K(t)\, J(t))\big)^{\frac{1}{n}} \geq \big(\det(K(t)\, J^{1,0}(t))\big)^{\frac{1}{n}} + \big(\det(K(t)\, J^{0,1}(t)\, J(1))\big)^{\frac{1}{n}},$$

and then factor out the positive quantity $(\det K(t))^{1/n}$ to get

$$\big(\det J(t)\big)^{\frac{1}{n}} \geq \big(\det(J^{1,0}(t))\big)^{\frac{1}{n}} + \big(\det(J^{0,1}(t)\, J(1))\big)^{\frac{1}{n}}.$$

The end of the argument is as before.

Next, it is obvious that (b) implies (c). To conclude the proof, it suffices to show that (c) $\Rightarrow$ (a). By symmetry and definition of $\overline{\beta}_t$, it is sufficient to show that $\beta_t(x,y) \leq \overline{\beta}_t^{[\gamma]}(x,y)$ for any geodesic γ. If x and y are conjugate along γ then there is nothing to prove. Otherwise, we can introduce $\xi(z)$ in the ball $B_r(y)$ such that for any $z \in B_r(y)$, $\exp_z \xi(z) = x$, and $\exp_z(t\xi(z))$ is the only geodesic joining z to x. Let μ_0 be the uniform probability distribution on $B_r(y)$, and μ_1 be the Dirac mass at x; then $\exp \xi$ is the unique map T such that $T_\#\mu_0 = \mu_1$, so it is the optimal transport map, and therefore can be written as $\exp(\nabla\psi)$ for some $d^2/2$-convex ψ; in particular, $\xi = \nabla\psi$. (Here I have chosen $t_0 = 1$, say.) So we can apply (b) with $N = n$, $\mathcal{D}(1) = 0$, $\mathcal{D}(0) = 1$, $\mathcal{D}(t,x) = \det J^{0,1}(t)$, and obtain

$$\det J^{0,1}(t)^{\frac{1}{n}} \geq t\, \beta_t(x,y)^{\frac{1}{n}}.$$

It follows that $\beta_t(x,y) \leq (\det J^{0,1}(t))/t^n = \overline{\beta}_t^{[\gamma]}(x,y)$, as desired. □

Sketch of proof of Theorem 14.21. To prove (a) $\Rightarrow$ (b), we may apply inequality (14.56) with $n = N$, to conclude that Property (c) in Theorem 14.20 is satisfied with $\beta = \beta^{(K,n)}$; thus $\overline{\beta} \geq \beta^{(K,n)}$. Conversely, if $\overline{\beta} \geq \beta^{(K,n)}$, Theorem 14.20 implies inequality (14.56), which in turn implies $\mathrm{CD}(K,n)$, or equivalently $\mathrm{Ric} \geq K$. □

Remark 14.22. If (M,ν) satisfies $\mathrm{CD}(K,N)$ then (14.65) still holds true with $\beta = \beta^{(K,N)}$ (provided of course that one takes the measure

ν into account when computing Jacobian determinants): this is just a rewriting of Theorem 14.12. However, Theorem 14.20 does not apply in this case, since N is in general larger than the "true dimension" n.

Remark 14.23. Theorems 14.20 and 14.21 suggest a generalization of the CD(K, N) criterion: Given an effective dimension N, define the generalized distortion coefficients $\overline{\beta}_{N,\nu}$ as the best coefficients in (14.65) (the first inequality if $N < \infty$, the second one if $N = \infty$). In this way the condition CD(K, N) might be a particular case of a more general condition CD(β, N), which would be defined by the inequality $\overline{\beta}_{N,\nu} \geq \beta$, where $\beta(x, y)$ would be, say, a given function of the distance between x and y. I shall not develop this idea, because (i) it is not clear at present that it would really add something interesting to the CD(K, N) theory; (ii) the condition CD(β, N) would in general be nonlocal.

Remark 14.24. It is not a priori clear what kind of functions β can occur as distortion coefficients. It is striking to note that, in view of Theorems 14.11 and 14.12, for any given manifold M of dimension n the following two conditions are equivalent, say for $K > 0$:

$$\text{(i) } \forall x, y \in M,\ \forall t \in [0,1],\quad \overline{\beta}_t(x,y) \geq \left(\frac{\sin\left(t\sqrt{\frac{K}{n}}\, d(x,y)\right)}{t\,\sin\left(\sqrt{\frac{K}{n}}\, d(x,y)\right)} \right)^n;$$

$$\text{(ii) } \forall x, y \in M,\ \forall t \in [0,1],\quad \overline{\beta}_t(x,y) \geq \left(\frac{\sin\left(t\sqrt{\frac{K}{n-1}}\, d(x,y)\right)}{t\,\sin\left(\sqrt{\frac{K}{n-1}}\, d(x,y)\right)} \right)^{n-1}.$$

This *self-improvement* property implies restrictions on the possible behavior of $\overline{\beta}$.

First Appendix: Second differentiability of convex functions

In this Appendix I shall provide a proof of Theorem 14.1. As explained right after the statement of that theorem, it suffices to consider the particular case of a convex function $\mathbb{R}^n \to \mathbb{R}$. So here is the statement to be proven:

Theorem 14.25 (Alexandrov's second differentiability theorem). *Let $\varphi : \mathbb{R}^n \to \mathbb{R}$ be a convex function. Then, for Lebesgue-almost every $x \in \mathbb{R}^n$, φ is differentiable at x and there exists a symmetric operator $A : \mathbb{R}^n \to \mathbb{R}^n$, characterized by any one of the following equivalent properties:*

(i) $\nabla\varphi(x+v) = \nabla\varphi(x) + Av + o(|v|)$ *as* $v \to 0$;

(i') $\partial\varphi(x+v) = \nabla\varphi(x) + Av + o(|v|)$ *as* $v \to 0$;

(ii) $\varphi(x+v) = \varphi(x) + \nabla\varphi(x)\cdot v + \dfrac{\langle Av, v\rangle}{2} + o(|v|^2)$ *as* $v \to 0$;

(ii') $\forall v \in \mathbb{R}^n$, $\qquad \varphi(x+tv) = \varphi(x) + t\,\nabla\varphi(x)\cdot v + t^2\,\dfrac{\langle Av, v\rangle}{2} + o(t^2)$ *as* $t \to 0$.

(In (i) the vector v is such that φ is differentiable at $x+v$; in (ii) the notation $o(|v|)$ means a set whose elements are all bounded in norm like $o(|v|)$.)

The operator A is denoted by $\nabla^2\varphi(x)$ and called the Hessian of φ at x. When no confusion is possible, the quadratic form defined by A is also called the Hessian of φ at x. Moreover, the function $x \to \nabla^2\psi(x)$ (resp. $x \to \Delta\psi(x) = \operatorname{tr}(\nabla^2\psi(x))$) is the density of the absolutely continuous part of the distribution $\nabla^2_{\mathcal{D}'}\psi$ (resp. of the distribution $\Delta_{\mathcal{D}'}\psi$).

Before starting the proof, let me recall an elementary lemma about convex functions.

Lemma 14.26. *(i) Let $\varphi : \mathbb{R}^n \to \mathbb{R}$ be a convex function, and let x_0, $x_1, \ldots, x_{n+1} \in \mathbb{R}^n$ such that $B(x_0, 2r)$ is included in the convex hull of $x_1, \ldots, x_{n+1}$. Then,*

$$2\,\varphi(x_0) - \max_{1\le i\le n+1}\varphi(x_i) \le \inf_{B(x_0,2r)}\varphi \le \sup_{B(x_0,2r)}\varphi \le \max_{1\le i\le n+1}\varphi(x_i);$$

$$\|\varphi\|_{\mathrm{Lip}(B(x_0,r))} \le \frac{2\left(\max_{1\le i\le n+1}\varphi(x_i) - \varphi(x_0)\right)}{r}.$$

(ii) If $(\varphi_k)_{k\in\mathbb{N}}$ is a sequence of convex functions which converges pointwise to some function Φ, then the convergence is locally uniform.

Proof of Lemma 14.26. If x lies in $B(x_0, 2r)$ then of course, by convexity, $\varphi(x) \le \max(\varphi(x_1), \ldots, \varphi(x_{n+1}))$. Next, if $z \in B(x_0, 2r)$, then $\widetilde{z} := 2x_0 - z \in B(x_0, 2r)$ and $\varphi(z) \ge 2\,\varphi(x_0) - \varphi(\widetilde{z}) \ge 2\,\varphi(x_0) - \max\varphi(x_i)$.

Next, let $x \in B(x_0, r)$ and let $y \in \partial\varphi(x)$; let $z = x + ry/|y| \in B(x_0, 2r)$. From the subdifferential inequality, $r|y| = \langle y, z-x\rangle \le \varphi(z) - \varphi(x) \le 2\,(\max \varphi(x_i) - \varphi(x_0))$. This proves (i).

Now let $(\varphi_k)_{k\in\mathbb{N}}$ be a sequence of convex functions, let $x_0 \in \mathbb{R}^n$ and let $r > 0$. Let $x_1, \ldots, x_{n+1}$ be such that $B(x_0, 2r)$ is included in the convex hull of $x_1, \ldots, x_{n+1}$. If $\varphi_k(x_j)$ converges for all j, then by (i) there is a uniform bound on $\|\varphi_k\|_{\mathrm{Lip}}$ on $B(x_0, r)$. So if φ_k converges pointwise on $B(x_0, r)$, the convergence has to be uniform. This proves (ii). □

Now we start the proof of Theorem 14.25. To begin with, we should check that the formulations (i), (i'), (ii) and (ii') are equivalent; this will use the convexity of φ.

Proof of the equivalence in Theorem 14.25. It is obvious that (i') ⇒ (i) and (ii) ⇒ (ii'), so we just have to show that (i) ⇒ (ii) and (ii') ⇒ (i').

To prove (i) ⇒ (ii), the idea is to use the mean value theorem; since a priori φ is not smooth, we shall regularize it. Let ζ be a radially symmetric nonnegative smooth function $\mathbb{R}^n \to \mathbb{R}$, with compact support in $B_1(0)$, such that $\int \zeta = 1$. For any $\varepsilon > 0$, let $\zeta_\varepsilon(x) = \varepsilon^{-n}\zeta(x/\varepsilon)$; then let $\varphi_\varepsilon := \varphi * \zeta_\varepsilon$. The resulting function φ_ε is smooth and converges pointwise to φ as $\varepsilon \to 0$; moreover, since φ is locally Lipschitz we have (by dominated convergence) $\nabla\varphi_\varepsilon = (\nabla\varphi) * \zeta_\varepsilon$.

Then we can write

$$\begin{aligned}\varphi(x+v) - \varphi(x) &= \lim_{\varepsilon\to 0}\,[\varphi_\varepsilon(x+v) - \varphi_\varepsilon(x)] \\ &= \lim_{\varepsilon\to 0}\int_0^1 \nabla\varphi_\varepsilon(x+tv)\cdot v\,dt. \qquad (14.70)\end{aligned}$$

Let us assume that $\varepsilon \le |v|$; then, by (i), for all $z \in B_{2\varepsilon}(x)$,

$$\nabla\varphi(z) = \nabla\varphi(x) + A(z-x) + o(|v|).$$

If $y \in B_\varepsilon(x)$, we can integrate this identity against $\zeta_\varepsilon(y-z)\,dz$ (since $\zeta_\varepsilon(y-z) = 0$ if $|y-z| > \varepsilon$); taking into account $\int (z-x)\,\zeta_\varepsilon(z-x)\,dz = 0$, we obtain

$$\nabla\varphi_\varepsilon(y) = \nabla\varphi_\varepsilon(x) + A(y-x) + o(|v|).$$

In particular, $\nabla\varphi_\varepsilon(x+tv) = \nabla\varphi_\varepsilon(x) + tAv + o(|v|)$. By plugging this into the right-hand side of (14.70), we obtain Property (ii).

Now let us prove that (ii') ⇒ (i'). Without loss of generality we may assume that $x = 0$ and $\nabla\varphi(x) = 0$. So the assumption is $\varphi(tw) = t^2$

$\langle Aw, w\rangle/2 + o(t^2)$, for any w. If (i') is false, then there are sequences $x_k \to 0$, $|x_k| \neq 0$, and $y_k \in \partial\varphi(x_k)$ such that

$$\frac{y_k - Ax_k}{|x_k|} \underset{k\to\infty}{\not\longrightarrow} 0. \tag{14.71}$$

Extract an arbitrary sequence from (x_k, y_k) (still denoted (x_k, y_k) for simplicity) and define

$$\varphi_k(w) := \frac{1}{|x_k|^2}\,\varphi(|x_k|w).$$

Assumption (ii) implies that φ_k converges pointwise to Φ defined by

$$\Phi(w) = \frac{\langle Aw, w\rangle}{2}.$$

The functions φ_k are convex, so the convergence is actually locally uniform by Lemma 14.26.

Since $y_k \in \partial\varphi(x_k)$,

$$\forall z \in \mathbb{R}^n, \qquad \varphi(z) \geq \varphi(x_k) + \langle y_k, z - x_k\rangle,$$

or equivalently, with the notation $w_k = x_k/|x_k|$,

$$\forall w \in \mathbb{R}^n, \qquad \varphi_k(w) \geq \varphi_k(w_k) + \left\langle \frac{y_k}{|x_k|}, w - w_k \right\rangle. \tag{14.72}$$

The choice $w = w_k + y_k/|y_k|$ shows that $|y_k|/|x_k| \leq \varphi_k(w) - \varphi_k(w_k)$, so $|y_k|/|x_k|$ is bounded. Up to extraction of a subsequence, we may assume that $w_k = x_k/|x_k| \to \sigma \in S^{n-1}$ and $y_k/|x_k| \to y$. Then we can pass to the limit in (14.72) and recover

$$\forall w \in \mathbb{R}^n, \qquad \Phi(w) \geq \Phi(\sigma) + \langle y, w - \sigma\rangle.$$

It follows that $y \in \partial\Phi(\sigma) = \{A\sigma\}$. So $y_k/|x_k| \to A\sigma$, or equivalently $(y_k - Ax_k)/|x_k| \to 0$. What has been shown is that each subsequence of the original sequence $(y_k - Ax_k)/|x_k|$ has a subsequence which converges to 0; thus the whole sequence converges to 0. This is in contradiction with (14.71), so (i') has to be true. □

Now, before proving Theorem 14.25 in full generality, I shall consider two particular cases which are much simpler.

Proof of Theorem 14.25 in dimension 1. Let $\varphi : \mathbb{R} \to \mathbb{R}$ be a convex function. Then its derivative φ' is nondecreasing, and therefore differentiable almost everywhere. □

Proof of Theorem 14.25 when $\nabla\varphi$ is locally Lipschitz. Let $\varphi : \mathbb{R}^n \to \mathbb{R}$ be a convex function, continuously differentiable and such that $\nabla\varphi$ locally Lipschitz. By Rademacher's theorem, each function $\partial_i\varphi$ is differentiable almost everywhere, where ∂_i stands for the partial derivative with respect to x_i. So the functions $\partial_j(\partial_i\varphi)$ are defined almost everywhere. To conclude the proof, it suffices to show that $\partial_j(\partial_i\varphi) = \partial_i(\partial_j\varphi)$ almost everywhere. To prove this, let ζ be any C^2 compactly supported function; then, by successive use of the dominated convergence theorem and the smoothness of $\varphi * \zeta$,

$$\begin{aligned}(\partial_i\partial_j\varphi) * \zeta = \partial_i(\partial_j\varphi * \zeta) = \partial_i\partial_j(\varphi * \zeta)\\ = \partial_j\partial_i(\varphi * \zeta) = \partial_j(\partial_i\varphi * \zeta) = (\partial_j\partial_i\varphi) * \zeta.\end{aligned}$$

It follows that $(\partial_i\partial_j\varphi - \partial_j\partial_i\varphi) * \zeta = 0$, and since ζ is arbitrary this implies that $\partial_i\partial_j\varphi - \partial_j\partial_i\varphi$ vanishes almost everywhere. This concludes the argument. □

Proof of Theorem 14.25 in the general case. As in the proof of Theorem 10.8(ii), the strategy will be to reduce to the one-dimensional case. For any $v \in \mathbb{R}^n$, $t > 0$, and x such that φ is differentiable at x, define

$$Q_v(t,x) = \frac{\varphi(x+tv) - \varphi(x) - t\,\nabla\varphi(x)\cdot v}{t^2} \geq 0.$$

The goal is to show that for Lebesgue–almost all $x \in \mathbb{R}^n$,

$$q_v(x) := \lim_{t\to 0} Q_v(t,x)$$

exists for all v, and is a quadratic function of v.

Let Dom $q(x)$ be the set of $v \in \mathbb{R}^n$ such that $q_v(x)$ exists. It is clear from the definition that:

(a) $q_v(x)$ is nonnegative and homogeneous of degree 2 in v on Dom $q(x)$;

(b) $q_v(x)$ is a convex function of v on Dom $q(x)$: this is just because it is the limit of the family $Q_v(t,x)$, which is convex in v;

(c) If v is interior to $\overline{\text{Dom}\, q(x)}$ and $q_w(x) \to \ell$ as $w \to v$, $w \in$ Dom $q(x)$, then also $v \in$ Dom $q(x)$ and $q_v(x) = \ell$. Indeed, let $\varepsilon > 0$

and let δ be so small that $|w-v| \leq \delta \Longrightarrow |q_w(x)-\ell| \leq \varepsilon$; then, we can find $v_1, \ldots, v_{n+1}$ in $\mathrm{Dom}\, q(x) \cap B(v,\delta)$ so that v lies in the convex hull of $v_1, \ldots, v_{n+1}$, and then $v_0 \in \mathrm{Dom}\, q(x) \cap B(v,\delta)$, so $v \in B(v_0,\delta)$ and $B(v_0,r)$ is included in the convex hull of $v_1, \ldots, v_{n+1}$. By Lemma 14.26,

$$2\, Q_{v_0}(t,x) - \max Q_{v_i}(t,x) \leq Q_v(t,x) \leq \max Q_{v_i}(t,x).$$

So

$$\ell - 3\varepsilon \leq 2\, q_{v_0}(x) - \max q_{v_i}(x) \leq \liminf_{t\to 0}\; Q_v(t,x) \leq \limsup_{t\to 0}\; Q_v(t,x) \leq \max q_{v_i}(x) \leq \ell + \varepsilon.$$

It follows that $\lim Q_v(t,x) = \ell$, as desired.

Next, we can use the same reasoning as in the proof of Rademacher's theorem (Theorem 10.8(ii)): Let v be given, $v \neq 0$, let us show that $q_v(x)$ exists for almost all x. By Fubini's theorem, it is sufficient to show that $q_v(x)$ exists λ_1-almost everywhere on each line parallel to v. So let $x_0 \in v^\perp$ be given, and let $L_{x_0} = x_0 + \mathbb{R}v$ be the line passing through x_0, parallel to v; the existence of $q_v(x_0+t_0 v)$ is equivalent to the second differentiability of the convex function $\psi : t \to \varphi(x_0+tv)$ at $t = t_0$, and from our study of the one-dimensional case we know that this happens for λ_1-almost all $t_0 \in \mathbb{R}$.

So for each v, the set A_v of $x \in \mathbb{R}^n$ such that $q_v(x)$ does not exist is of zero measure. Let (v_k) be a dense subset of $\mathbb{R}^n$, and let $A = \cup A_{v_k}$: A is of zero measure, and for each $x \in \mathbb{R}^n \setminus A$, $\mathrm{Dom}\, q(x)$ contains all the vectors v_k.

Again, let $x \in \mathbb{R}^n \setminus A$. By Property (b), $q_v(x)$ is a convex function of v, so it is locally Lipschitz and can be extended uniquely into a continuous convex function $r(v)$ on $\mathbb{R}^n$. By Property (c), $r(v) = q_v(x)$, which means that $\mathrm{Dom}\, q(x) = \mathbb{R}^n$.

At this point we know that for almost any x the limit $q_v(x)$ exists for all v, and it is a convex function of v, homogeneous of degree 2. What we do not know is whether $q_v(x)$ is a *quadratic* function of v.

Let us try to solve this problem by a regularization argument. Let ζ be a smooth nonnegative compactly supported function on $\mathbb{R}^n$, with $\int \zeta = 1$. Then $\nabla\varphi * \zeta = \nabla(\varphi * \zeta)$. Moreover, thanks to the nonnegativity of $Q_v(x,t)$ and Fatou's lemma,

$$
\begin{aligned}
(q_v * \zeta)(x) &= \int \lim_{t\downarrow 0} Q_v(y,t)\,\zeta(x-y)\,dy \\
&\le \liminf_{t\downarrow 0} \int Q_v(y,t)\,\zeta(x-y)\,dy \\
&= \liminf_{t\downarrow 0} \frac{1}{t^2}\Big[(\varphi*\zeta)(x+tv) - (\varphi*\zeta)(x) - t\,\nabla(\varphi*\zeta)(x)\cdot v\Big] \\
&= \frac{1}{2}\,\big\langle \nabla^2(\varphi*\zeta)(x)\cdot v,\, v\big\rangle.
\end{aligned}
$$

It is obvious that the right-hand side is a quadratic form in v, but this is only an upper bound on $q_v * \zeta(x)$. In fact, in general $q_v * \zeta$ *does not* coincide with $(1/2)\langle\nabla^2(\varphi*\zeta)v, v\rangle$. The difference is caused by the singular part of the measure $\mu_v := (1/2)\langle\nabla^2\varphi\cdot v, v\rangle$, defined in the distribution sense by

$$
\int \zeta(x)\,\mu_v(dx) = \frac{1}{2}\int \langle \nabla^2\zeta(x)\cdot v, v\rangle\,\varphi(x)\,dx.
$$

This obstacle is the main new difficulty in the proof of Alexandrov's theorem, as compared to the proof of Rademacher's theorem.

To avoid the singular part of the measure μ_v, we shall appeal to Lebesgue's density theory, in the following precise form: Let μ be a locally finite measure on $\mathbb{R}^n$, and let $\rho\,\lambda_n + \mu_s$ be its Lebesgue decomposition into an absolutely continuous part and a singular part. Then, for Lebesgue–almost all $x \in \mathbb{R}^n$,

$$
\frac{1}{\delta^n}\big\|\mu - \rho(x)\lambda_n\big\|_{\mathrm{TV}(B_\delta(x))} \xrightarrow[\delta\to 0]{} 0,
$$

where $\|\cdot\|_{\mathrm{TV}(B_\delta(x))}$ stands for the total variation on the ball $B_\delta(x)$. Such an x will be called a Lebesgue point of μ.

So let ρ_v be the density of μ_v. It is easy to check that μ_v is locally finite, and we also showed that q_v is locally integrable. So, for λ_n-almost all x_0 we have

$$
\begin{cases}
\displaystyle \frac{1}{\delta^n}\int_{B_\delta(x_0)} |q_v(x) - q_v(x_0)|\,dx \xrightarrow[\delta\to 0]{} 0; \\[3ex]
\displaystyle \frac{1}{\delta^n}\big\|\mu_v - \rho_v(x_0)\lambda_n\big\|_{\mathrm{TV}(B_\delta(x_0))} \xrightarrow[\delta\to 0]{} 0.
\end{cases}
$$

The goal is to show that $q_v(x_0) = \rho_v(x_0)$. Then the proof will be complete, since ρ_v is a quadratic form in v (indeed, $\rho_v(x_0)$ is obtained by averaging $\mu_v(dx)$, which itself is quadratic in v). Without loss of generality, we may assume that $x_0 = 0$.

To prove that $q_v(0) = \rho_v(0)$, it suffices to establish

$$\lim_{\delta\to 0} \frac{1}{\delta^n} \int_{B_\delta(0)} |q_v(x) - \rho_v(0)|\, dx = 0, \tag{14.73}$$

To estimate $q_v(x)$, we shall express it as a limit involving points in $B_\delta(x)$, and then use a Taylor formula; since φ is not a priori smooth, we shall regularize it on a scale $\varepsilon \le \delta$. Let ζ be as before, and let $\zeta_\varepsilon(x) = \varepsilon^{-n}\zeta(x/\varepsilon)$; further, let $\varphi_\varepsilon := \varphi * \zeta$.

We can restrict the integral in (14.73) to those x such that $\nabla\varphi(x)$ exists and such that x is a Lebesgue point of $\nabla\varphi$; indeed, such points form a set of full measure. For such an x, $\varphi(x) = \lim_{\varepsilon\to 0} \varphi_\varepsilon(x)$, and $\nabla\varphi(x) = \lim_{\varepsilon\to 0} \nabla\varphi_\varepsilon(x)$. So,

$$\begin{aligned}
&\frac{1}{\delta^n} \int_{B_\delta(0)} |q_v(x) - \rho_v(0)|\, dx \\
&= \frac{1}{\delta^n} \int_{B_\delta(0)} \left| \lim_{t\to 0} \left[\frac{\varphi(x+t\delta v) - \varphi(x) - \nabla\varphi(x)\cdot t\delta v}{t^2\delta^2} \right] - \rho_0(v) \right| dx \\
&= \frac{1}{\delta^n} \int_{B_\delta(0)} \lim_{t\to 0} \lim_{\varepsilon\to 0} \left| \frac{\varphi_\varepsilon(x+t\delta v) - \varphi_\varepsilon(x) - \nabla\varphi_\varepsilon(x)\cdot t\delta v}{t^2\delta^2} - \rho_0(v) \right| dx \\
&= \frac{1}{\delta^n} \int_{B_\delta(0)} \lim_{t\to 0} \lim_{\varepsilon\to 0} \left| \int_0^1 \Big[\langle \nabla^2\varphi_\varepsilon(x+st\delta v)\cdot v, v\rangle - 2\rho_v(0) \Big] (1-s)\, ds \right| dx \\
&\le \liminf_{t\to 0} \liminf_{\varepsilon\to 0} \frac{1}{\delta^n} \int_{B_\delta(0)} \left| \int_0^1 \big[\langle \nabla^2\varphi_\varepsilon(x+st\delta v)\cdot v, v\rangle - 2\rho_v(0) \big] \right. \\
&\qquad\qquad\qquad\qquad\qquad\qquad \left. (1-s)\, ds \right| dx \\
&\le \liminf_{t\to 0} \liminf_{\varepsilon\to 0} \frac{1}{\delta^n} \int_0^1 \int_{B_\delta(st\delta v)} \big| \langle \nabla^2\varphi_\varepsilon(y)\cdot v, v\rangle - \rho_v(0) \big|\, dy\, ds,
\end{aligned}$$

where Fatou's lemma and Fubini's theorem were used successively. Since $B(st\delta v, \delta) \subset B(0, (1+|v|)\delta)$, independently of s and t, we can bound the above expression by

$$\begin{aligned}
&\liminf_{\varepsilon\to 0} \frac{1}{\delta^n} \int_{B(0,(1+|v|)\delta)} \big| \langle \nabla^2\varphi_\varepsilon(y)\cdot v, v\rangle - \rho_v(0) \big|\, dy \\
&\qquad = \liminf_{\varepsilon\to 0} \frac{1}{\delta^n} \int_{B(0,(1+|v|)\delta)} \left| \int \zeta_\varepsilon(y-z) [\mu_v - \rho_v(0)\, \lambda_n](dz) \right| dy \\
&\qquad \le \liminf_{\varepsilon\to 0} \frac{1}{\delta^n} \int_{B(0,(1+|v|)\delta)} \int \zeta_\varepsilon(y-z) |\mu_v - \rho_v(0)\, \lambda_n|(dz)\, dy.
\end{aligned}$$

When y varies in $B(0,(1+|v|)\delta)$, z stays in $B(0,(1+|v|)\delta+\varepsilon)$, which itself is included in $B(0,C\delta)$ with $C=2+|v|$. So, after using Fubini's theorem and integrating out $\zeta_\varepsilon(y-z)\,dy$, we conclude that

$$\frac{1}{\delta^n}\int_{B_\delta(0)} |q_v(x)-\rho_v(0)|\,dx \le \|\mu_v-\rho_v(0)\,\lambda_n\|_{\mathrm{TV}(B(0,C\delta))}.$$

The conclusion is obtained by taking the limit $\delta\to 0$.

Once $\nabla^2\varphi$ has been identified as the density of the distributional Hessian of φ, it follows immediately that $\Delta\varphi := \mathrm{tr}\,(\nabla^2\varphi)$ is the density of the distributional Laplacian of φ. (The trace of a matrix-valued nonnegative measure is singular if and only if the measure itself is singular.) □

Remark 14.27. The notion of a distributional Hessian on a Riemannian manifold is a bit subtle, which is why I did not state anything about it in Theorem 14.1. On the other hand, there is no difficulty in defining the distributional Laplacian.

Second Appendix: Very elementary comparison arguments

There are well-developed theories of comparison estimates for second-order linear differential equations; but the statement to be considered here can be proven by very elementary means.

Theorem 14.28 (One-dimensional comparison for second-order inequalities). *Let $\Lambda\in\mathbb{R}$, and $f\in C([0,1])\cap C^2(0,1)$, $f\ge 0$. Then the following two statements are equivalent:*

(i) $\ddot{f}+\Lambda f\le 0$ in $(0,1)$;

(ii) If $\Lambda<\pi^2$ then for all $t_0,t_1\in[0,1]$,

$$f\big((1-\lambda)t_0+\lambda t_1\big)\ge \tau^{(1-\lambda)}(|t_0-t_1|)\,f(t_0)\;+\;\tau^{(\lambda)}(|t_0-t_1|)\,f(t_1),$$

where

$$\tau^{(\lambda)}(\theta) = \begin{cases} \dfrac{\sin(\lambda\theta\sqrt{\Lambda})}{\sin(\theta\sqrt{\Lambda})} & \text{if } 0 < \Lambda < \pi^2 \\[2ex] \lambda & \text{if } \Lambda = 0 \\[2ex] \dfrac{\sinh(\lambda\theta\sqrt{-\Lambda})}{\sinh(\theta\sqrt{-\Lambda})} & \text{if } \Lambda < 0. \end{cases}$$

If $\Lambda = \pi^2$ *then* $f(t) = c\sin(\pi t)$ *for some* $c \geq 0$*; finally if* $\Lambda > \pi^2$ *then* $f = 0$.

Proof of Theorem 14.28. The easy part is (ii) ⇒ (i). If $\Lambda \geq \pi^2$ this is trivial. If $\Lambda < \pi^2$, take $\lambda = 1/2$, then a Taylor expansion shows that

$$\tau^{(1/2)}(\theta) = \frac{1}{2}\left(1 + \frac{\theta\Lambda^2}{8}\right) + o(\theta^3)$$

and

$$\frac{f(t_0) + f(t_1)}{2} = f\left(\frac{t_0 + t_1}{2}\right) + \frac{(t_0 - t_1)^2}{4}\,\ddot{f}\left(\frac{t_0 + t_1}{2}\right) + o(|t_0 - t_1|^2).$$

So, if we fix $t \in (0,1)$ and let $t_0, t_1 \to t$ in such a way that $t = (t_0+t_1)/2$, we get

$$\tau^{(1/2)}(|t_0 - t_1|)\, f(t_0) + \tau^{(1/2)}(|t_0 - t_1|)\, f(t_1) - f(t) \\ = \frac{(t_0 - t_1)^2}{8}\Big(\ddot{f}(t) + \Lambda f(t) + o(1)\Big).$$

By assumption the left-hand side is nonnegative, so in the limit we recover $\ddot{f} + \Lambda f \leq 0$.

Now consider the reverse implication (ii) ⇒ (i). By abuse of notation, let us write $f(\lambda) = f((1-\lambda)\,t_0 + \lambda\,t_1)$, and denote by a prime the derivation with respect to λ; so $f'' + \Lambda\theta^2 f \leq 0$, $\theta = |t_0 - t_1|$. Let $g(\lambda)$ be defined by the right-hand side of (ii); that is, $\lambda \to g(\lambda)$ is the solution of $g'' + \Lambda\theta^2 g = 0$ with $g(0) = f(0)$, $g(1) = f(1)$. The goal is to show that $f \geq g$ on $[0,1]$.

(a) Case $\Lambda < 0$. Let $a > 0$ be any constant; then $f_a := f + a$ still solves the same differential inequality as f, and $f_a > 0$ (even if we did not assume $f \geq 0$, we could take a sufficiently large for this to be true). Let g_a be defined as the solution of $g_a'' + \Lambda\theta^2 g_a = 0$ with $g_a(0) = f_a(0)$,

$g_a(1) = f_a(1)$. As $a \to 0$, f_a converges to f and g_a converges to g, so it is sufficient to show $f_a \geq g_a$. Therefore, without loss of generality we may assume that f, g are positive, so g/f is continuous.

If g/f attains its maximum at 0 or 1, then we are done. Otherwise, there is $\lambda_0 \in (0,1)$ such that $(g/f)''(\lambda_0) \leq 0$, $(g/f)'(\lambda_0) = 0$, and then the identity

$$\left(\frac{g}{f}\right)'' = \frac{(g'' + \Lambda g)}{f} - \frac{g}{f^2}(f'' + \Lambda f) - 2\frac{f'}{f}\left(\frac{g}{f}\right)' - 2\Lambda\frac{g}{f},$$

evaluated at λ_0, yields $0 > -2\Lambda g/f$, which is impossible.

(b) Case $\Lambda = 0$. This is the basic property of concave functions.

(c) Case $0 < \Lambda < \pi^2$. Let $\theta = |t_0 - t_1| \leq 1$. Since $\theta\sqrt{\Lambda} < \pi$, we can find a function w such that $w'' + \Lambda\theta^2 w \leq 0$ and $w > 0$ on $(0,1)$. (Just take a well-chosen sine or cosine function.) Then $f_a := f + aw$ still satisfies the same differential inequality as f, and it is positive. Let g_a be defined by the equation $g_a'' + \Lambda\theta^2 g_a = 0$ with $g_a(0) = f_a(0)$, $g_a(1) = f_a(1)$. As $a \to 0$, f_a converges to f and g_a to g, so it is sufficient to show that $f_a \geq g_a$. Thus we may assume that f and g are positive, and f/g is continuous.

The rest of the reasoning is parallel to the case $\Lambda < 0$: If f/g attains its minimum at 0 or 1, then we are done. Otherwise, there is $\lambda_0 \in (0,1)$ such that $(f/g)''(\lambda_0) \geq 0$, $(f/g)'(\lambda_0) = 0$, and then the identity

$$\left(\frac{f}{g}\right)'' = \frac{(f'' + \Lambda f)}{g} - \frac{f}{g^2}(g'' + \Lambda g) - 2\frac{g'}{g}\left(\frac{f}{g}\right)' - 2\Lambda\frac{f}{g},$$

evaluated at λ_0, yields $0 < -2\Lambda f/g$, which is impossible.

(d) Case $\Lambda = \pi^2$. Take $t_0 = 0$, $t_1 = 1$. Then let $g(\lambda) = \sin(\pi\lambda)$, and let $h := f/g$. The differential equations $f'' + \Lambda f \leq 0$ and $g'' + \Lambda g = 0$ combine to yield $(h'g^2)' = h''g^2 + 2gh'g' \leq 0$. So $h'g^2$ is nonincreasing. If $h'(\lambda_0) < 0$ for some $t_0 \in (0,1)$, then $h'g^2(\lambda_0) < 0$ for all $\lambda \geq \lambda_0$, so $h'(\lambda) \leq -C/g(\lambda)^2 \leq -C'/(1-\lambda)^2$ as $\lambda \to 1$, where C, C' are positive constants. It follows that $h(\lambda)$ becomes negative for λ close to 1, which is impossible. If on the other hand $h'(\lambda_0) > 0$, then a similar reasoning shows that $h(\lambda)$ becomes negative for λ close to 0. The conclusion is that h' is identically 0, so f/g is a constant.

(e) If $\Lambda > \pi^2$, then for all $t_0, t_1 \in [0,1]$ with $|t_0 - t_1| = \pi/\sqrt{\Lambda}$, the function $f(\lambda) = f(\lambda t_0 + (1-\lambda t_1))$ is proportional to $\sin(\pi\lambda)$, by Case (d). By letting t_0, t_1 vary, it is easy to deduce that f is identically 0. □

Third Appendix: Jacobi fields forever

Let $R : t \longmapsto R(t)$ be a continuous map defined on $[0,1]$, valued in the space of $n \times n$ symmetric matrices, and let $\mathcal{U}$ be the space of functions $u : t \longmapsto u(t) \in \mathbb{R}^n$ solving the second-order linear differential equation

$$\ddot{u}(t) + R(t)\,u(t) = 0. \tag{14.74}$$

By the theory of linear differential equations, $\mathcal{U}$ is a $(2n)$-dimensional vector space and the map $u \longmapsto (u(0), \dot{u}(0))$ is a linear isomorphism $\mathcal{U} \to \mathbb{R}^{2n}$.

As explained in the beginning of this chapter, if $\gamma : [0,1] \to M$ is a geodesic in a Riemannian manifold M, and ξ is a Jacobi field along γ (that is, an infinitesimal variation of geodesic around γ), then the coordinates of $\xi(t)$, written in an orthonormal basis of $T_{\gamma(t)}M$ evolving by parallel transport, satisfy an equation of the form (14.74).

It is often convenient to consider an array $(u_1(t), \ldots, u_n(t))$ of solutions of (14.74); this can be thought of as a time-dependent matrix $t \longmapsto J(t)$ solving the differential (matrix) equation

$$\ddot{J}(t) + R(t)\,J(t) = 0.$$

Such a solution will be called a **Jacobi matrix**. If J is a Jacobi matrix and A is a constant matrix, then JA is still a Jacobi matrix.

Jacobi matrices enjoy remarkable properties, some of which are summarized in the next three statements. In the sequel, the time-interval $[0,1]$ and the symmetric matrix $t \longmapsto R(t)$ are fixed once for all, and the dot stands for time-derivation.

Proposition 14.29 (Jacobi matrices have symmetric logarithmic derivatives). *Let J be a Jacobi matrix such that $J(0)$ is invertible and $\dot{J}(0)\,J(0)^{-1}$ is symmetric. Let $t_* \in [0,1]$ be largest possible such that $J(t)$ is invertible for $t < t_*$. Then $\dot{J}(t)\,J(t)^{-1}$ is symmetric for all $t \in [0,t_*)$.*

Proposition 14.30 (Cosymmetrization of Jacobi matrices). *Let J_0^1 and J_1^0 be Jacobi matrices defined by the initial conditions*

$$J_0^1(0) = I_n, \quad \dot{J}_0^1(0) = 0, \quad J_1^0(0) = 0, \quad \dot{J}_1^0(0) = I_n;$$

so that any Jacobi matrix can be written as

$$J(t) = J_0^1(t)\,J(0) + J_1^0(t)\,\dot{J}(0). \tag{14.75}$$

Assume that $J_1^0(t)$ *is invertible for all* $t \in (0,1]$. *Then:*

(a) $S(t) := J_1^0(t)^{-1} J_0^1(t)$ *is symmetric positive for all* $t \in (0,1]$, *and it is a decreasing function of* t.

(b) There is a unique pair of Jacobi matrices $(J^{1,0}, J^{0,1})$ *such that*

$$J^{1,0}(0) = I_n, \quad J^{1,0}(1) = 0, \quad J^{0,1}(0) = 0, \quad J^{0,1}(1) = I_n;$$

moreover $\dot{J}^{1,0}(0)$ *and* $\dot{J}^{0,1}(1)$ *are symmetric.*

(c) For $t \in [0,1]$ *let*

$$K(t) = t\, J_1^0(t)^{-1},$$

extended by continuity at $t = 0$ *by* $K(0) = I_n$. *If* J *is a Jacobi matrix such that* $J(0)$ *is invertible, and* $J(0)^*\, S(1)\, J(0)^{-1} = S(1)$ *and* $\dot{J}(0)\, J(0)^{-1}$ *are symmetric, then for any* $t \in [0,1]$ *the matrices*

$$K(t)\,\big(J^{1,0}(t)\, J(0)\big)\, J(0)^{-1} \quad \textit{and} \quad K(t)\,\big(J^{0,1}(t)\, J(1)\big)\, J(0)^{-1}$$

are symmetric. Moreover, $\det K(t) > 0$ *for all* $t \in [0,1)$.

Proposition 14.31 (Jacobi matrices with positive determinant). *Let* $S(t)$ *and* $K(t)$ *be the matrices defined in Proposition 14.30. Let* J *be a Jacobi matrix such that* $J(0) = I_n$ *and* $\dot{J}(0)$ *is symmetric. Then the following properties are equivalent:*

(i) $\dot{J}(0) + S(1) > 0$;

(ii) $K(t)\, J^{0,1}(t)\, J(1) > 0$ *for all* $t \in (0,1)$;

(iii) $K(t)\, J(t) > 0$ *for all* $t \in [0,1]$;

(iv) $\det J(t) > 0$ *for all* $t \in [0,1]$.

The equivalence remains true if one replaces the strict inequalities in (i)–(ii) by nonstrict inequalities, and the time-interval $[0,1]$ *in (iii)–(iv) by* $[0,1)$.

Before proving these propositions, it is interesting to discuss their geometric interpretation:

- If $\gamma(t) = \exp_x(t\nabla\psi(x))$ is a minimizing geodesic, then $\dot{\gamma}(t) = \nabla\psi(t,\gamma(t))$, where ψ solves the Hamilton–Jacobi equation

$$\begin{cases} \dfrac{\partial\psi(t,x)}{\partial t} + \dfrac{|\nabla\psi(t,x)|^2}{2} = 0 \\[2ex] \psi(0,\,\cdot\,) = \psi; \end{cases}$$

at least before the first shock of the Hamilton–Jacobi equation. This corresponds to Proposition 14.29 with $J(0) = I_n$, $\dot{J}(0) = \nabla^2\psi(x)$, $\dot{J}(t)\,J(t)^{-1} = \nabla^2\psi(t,\gamma(t))$. (Here $\nabla^2\psi(x)$ and $\nabla^2\psi(t,\gamma(t))$ are identified with the matrix of their respective coordinates, as usual in a varying orthonormal basis).

- The Jacobi matrices $J_0^1(t)$ and $J_1^0(t)$ represent $\partial_x F_t$ and $\partial_v F_t$ respectively, where $F_t(x,v) = (\exp_x(tv), (d/dt)\exp_x(tv))$ is the geodesic flow at time t. So the hypothesis of invertibility of $J_1^0(t)$ in Proposition 14.30 corresponds to an assumption of *nonfocalization* along the geodesic γ.

- The formula
$$\gamma(t) = \exp_x\left(-\nabla_x \frac{d(\,\cdot\,,\gamma(t))^2}{2}\right)$$
yields, after differentiation,
$$0 = J_0^1(t) - J_1^0(t)\,\frac{H(t)}{t},$$
where (modulo identification)
$$H(t) = \nabla_x^2\,\frac{d(\cdot,\gamma(t))^2}{2}.$$
(The extra t in the denominator comes from time-reparameterization.) So $S(t) = J_1^0(t)^{-1}\,J_0^1(t) = H(t)/t$ should be symmetric.

- The assumption
$$J(0) = I_n, \qquad \dot{J}(0) + S(1) \geq 0$$
is the Jacobi field version of the formulas
$$\gamma(t) = \exp_x(t\nabla\psi(x)), \qquad \nabla^2\psi(x) + \nabla_x^2\,\frac{d\big(\,\cdot\,,\exp_x\nabla\psi(x)\big)^2}{2} \geq 0.$$
The latter inequality holds true if ψ is $d^2/2$-convex, so Proposition 14.31 implies the last part of Theorem 11.3, according to which the Jacobian of the optimal transport map remains positive for $0 < t < 1$.

Now we can turn to the proofs of Propositions 14.29 to 14.31.

Proof of Proposition 14.29. The argument was already mentioned in the beginning of this chapter: the matrix $U(t) = \dot{J}(t)\,J(t)^{-1}$ satisfies the Ricatti equation

$$\dot{U}(t) + U(t)^2 + R(t) = 0$$

on $[0, t_*)$; and since R is symmetric the transpose $U(t)^*$ of $U(t)$ satisfies the same equation. By assumption $U(0) = U(0)^*$; so the Cauchy–Lipschitz uniqueness theorem implies $U(t) = U(t)^*$ for all t. □

Proof of Proposition 14.30. First of all, the identity in (14.75) follows immediately from the observation that both sides solve the Jacobi equation with the same initial conditions.

Let now $t \in (0,1]$, $w \in \mathbb{R}^n$ and $\widehat{w} = J_1^0(t)^{-1} J_0^1(t)\,w$; so that $J_1^0(t)\,\widehat{w} = J_0^1(t)\,w$. The function $s \longmapsto u(s) = J_0^1(s)\,w - J_1^0(s)\,\widehat{w}$ belongs to $\mathcal{U}$ and satisfies $u(t) = 0$. Moreover, $w = u(0)$, $\widehat{w} = -\dot{u}(0)$. So the matrix $S(t) = J_1^0(t)^{-1} J_0^1(t)$ can be interpreted as the linear map $u(0) \longmapsto -\dot{u}(0)$, where $s \longmapsto u(s)$ varies in the vector space $\mathcal{U}_t$ of solutions of the Jacobi equation (14.74) satisfying $u(t) = 0$. To prove the symmetry of $S(t)$ it suffices to check that for any two u, v in $\mathcal{U}_t$,

$$\langle u(0), \dot{v}(0)\rangle = \langle \dot{u}(0), v(0)\rangle,$$

where $\langle\,\cdot\,,\,\cdot\,\rangle$ stands for the usual scalar product on $\mathbb{R}^n$. But

$$\begin{aligned}\frac{d}{ds}\Big(\langle u(s), \dot{v}(s)\rangle - \langle \dot{u}(s), v(s)\rangle\Big) &= -\langle u(s), R(s)\,v(s)\rangle + \langle R(s)\,u(s), v(s)\rangle \\ &= 0 \qquad (14.76)\end{aligned}$$

by symmetry of R. This implies the symmetry of S since

$$\langle u(0), \dot{v}(0)\rangle - \langle \dot{u}(0), v(0)\rangle = \langle u(t), \dot{v}(t)\rangle - \langle \dot{u}(t), v(t)\rangle = 0.$$

Remark 14.32. A knowledgeable reader may have noticed that this is a "symplectic" argument (related to the Hamiltonian nature of the geodesic flow): if $\mathbb{R}^n \times \mathbb{R}^n$ is equipped with its natural symplectic form

$$\omega\big((u,\dot{u}),(v,\dot{v})\big) = \langle \dot{u}, v\rangle - \langle \dot{v}, u\rangle,$$

then the flow $(u(s), \dot{u}(s)) \longmapsto (u(t), \dot{u}(t))$, where $u \in \mathcal{U}$, preserves ω. The subspaces $\mathcal{U}_0 = \{u \in \mathcal{U};\ u(0) = 0\}$ and $\dot{\mathcal{U}}_0 = \{u \in \mathcal{U};\ \dot{u}(0) = 0\}$ are *Lagrangian*: this means that their dimension is the half of the dimension of $\mathcal{U}$, and that ω vanishes identically on each of them. Moreover ω is nondegenerate on $\mathcal{U}_0 \times \dot{\mathcal{U}}_0$, providing an identification of these spaces. Then $\mathcal{U}_t$ is also Lagrangian, and if one writes it as a graph in $\mathcal{U}_0 \times \dot{\mathcal{U}}_0$, it is the graph of a symmetric operator.

Back to the proof of Proposition 14.29, let us show that $S(t)$ is a decreasing function of t. To reformulate the above observations we write, for any $w \in \mathbb{R}^n$,

$$\langle S(t)w, w\rangle = -\Big\langle w, \frac{\partial u}{\partial s}(0,t)\Big\rangle,$$

where $u(s,t)$ is defined by

$$\begin{cases} \dfrac{\partial^2 u(s,t)}{\partial s^2} + R(s)\, u(s,t) = 0 \\ u(t,t) = 0 \\ u(0,t) = w. \end{cases} \tag{14.77}$$

So

$$\begin{aligned} \langle \dot{S}(t)w, w\rangle &= -\Big\langle w, \frac{\partial^2 u}{\partial s\, \partial t}(0,t)\Big\rangle \\ &= -\Big\langle w, \frac{\partial}{\partial s}(\partial_t u)(0,t)\Big\rangle = -\Big\langle u(0), \frac{\partial v}{\partial s}(0)\Big\rangle, \end{aligned}$$

where $s \longmapsto v(s) = (\partial_t u)(s,t)$ and $s \longmapsto u(s) = u(s,t)$ are solutions of the Jacobi equation. Moreover, by differentiating the conditions in (14.77) one obtains

$$v(t) + \frac{\partial u}{\partial s}(t) = 0; \qquad v(0) = 0. \tag{14.78}$$

By (14.76) again,

$$-\Big\langle u(0), \frac{\partial v}{\partial s}(0)\Big\rangle = -\Big\langle \frac{\partial u}{\partial s}(0), v(0)\Big\rangle - \Big\langle u(t), \frac{\partial v}{\partial s}(t)\Big\rangle + \Big\langle \frac{\partial u}{\partial s}(t), v(t)\Big\rangle.$$

The first two terms in the right-hand side vanish because $v(0) = 0$ and $u(t) = u(t,t) = 0$. Combining this with the first identity in (14.78) one finds in the end

$$\langle \dot{S}(t)w, w\rangle = -\|v(t)\|^2. \tag{14.79}$$

We already know that $v(0) = 0$; if in addition $v(t) = 0$ then $0 = v(t) = J_1^0(t)\dot{v}(0)$, so (by invertibility of $J_1^0(t)$) $\dot{v}(0) = 0$, and v vanishes identically; then by (14.78) (du/ds) vanishes at $s = t$, and since also $u(t) = 0$ we know that u vanishes identically, which implies $w = 0$. In other words, the right-hand side of (14.79) is strictly negative unless $w = 0$; this means that $S(t)$ is a strictly decreasing function of t. Thus the proof of (a) is finished.

To prove (b) it is sufficient to exhibit the matrices $J^{1,0}(t)$ and $J^{0,1}(t)$ explicitly in terms of J_0^1 and J_1^0:

$$J^{1,0}(t) = J_0^1(t) - J_1^0(t)\, J_1^0(1)^{-1}\, J_0^1(1); \qquad J^{0,1}(t) = J_1^0(t)\, J_1^0(1)^{-1}. \tag{14.80}$$

Moreover, $\dot{J}^{1,0}(0) = -J_1^0(1)^{-1} J_0^1(1)$ and $\dot{J}^{0,1}(t) = \dot{J}_1^0(1)\, J_1^0(1)^{-1}$ are symmetric in view of (a) and Proposition 14.29. Also

$$\begin{aligned} K(t)\, J^{1,0}(t)\, J(0)\, J(0)^{-1} &= t\, J_1^0(t)^{-1} \big(J_0^1(t) - J_1^0(t)\, J_1^0(1)^{-1} J_0^1(1)\big) \\ &= t\, \Big(J_1^0(t)^{-1} J_0^1(t) - J_1^0(1)^{-1}\, J_0^1(1)\Big) \end{aligned}$$

is positive symmetric since by (a) the matrix $S(t) = J_1^0(t)^{-1} J_0^1(t)$ is a *strictly* decreasing function of t. In particular $K(t)$ is invertible for $t > 0$; but since $K(0) = I_n$, it follows by continuity that $\det K(t)$ remains positive on $[0, 1)$.

Finally, if J satisfies the assumptions of (c), then $S(1)\, J(0)^{-1}$ is symmetric (because $S(1)^* = S(1)$). Then

$$\begin{aligned} &\big(K(t)\, J^{0,1}(t)\big)\, J(1)\, J(0)^{-1} \\ &\qquad = \big(t\, J_1^0(t)^{-1} J_1^0(t)\, J_1^0(1)^{-1}\big) \big(J_0^1(1) + J_1^0(1)\, \dot{J}(0)\big)\, J(0)^{-1} \\ &\qquad = t\, \Big(J_1^0(1)^{-1} J_0^1(1)\, J(0)^{-1} + \dot{J}(0)\, J(0)^{-1}\Big) \end{aligned}$$

is also symmetric. □

Proof of Proposition 14.31. Assume (i); then, by the formulas in the end of the proof of Proposition 14.30, with $J(0) = I_n$,

$$K(t)\, J^{1,0}(t) = t\, [S(t) - S(1)]; \qquad K(t)\, J^{0,1}(t)\, J(1) = t\, [S(1) + \dot{J}(0)].$$

As we already noticed, the first matrix is positive for $t \in (0, 1)$; and the second is also positive, by assumption. In particular (ii) holds true.

The implication (ii) ⇒ (iii) is obvious since $J(t) = J^{1,0}(t) + J^{0,1}(t)\, J(1)$ is the sum of two positive matrices for $t \in (0, 1)$. (At $t = 0$ one sees directly $K(0)\, J(0) = I_n$.)

If (iii) is true then $(\det K(t))\,(\det J(t)) > 0$ for all $t \in [0,1)$, and we already know that $\det K(t) > 0$; so $\det J(t) > 0$, which is (iv).

It remains to prove (iv) $\Rightarrow$ (i). Recall that $K(t)\,J(t) = t\,[S(t)+\dot{J}(0)]$; since $\det K(t) > 0$, the assumption (iv) is equivalent to the statement that the symmetric matrix $A(t) = t\,S(t) + t\,\dot{J}(0)$ has positive determinant for all $t \in (0,1]$. The identity $t\,S(t) = K(t)\,J_0^1(t)$ shows that $A(t)$ approaches I_n as $t \to 0$; and since none of its eigenvalues vanishes, $A(t)$ has to remain positive for all t. So $S(t) + \dot{J}(0)$ is positive for all $t \in (0,1]$; but S is a decreasing function of t, so this is equivalent to $S(1) + \dot{J}(0) > 0$, which is condition (i).

The last statement in Proposition 14.31 is obtained by similar arguments and its proof is omitted. □

Bibliographical notes

Recommended textbooks about Riemannian geometry are the ones by do Carmo [306], Gallot, Hulin and Lafontaine [394] and Chavel [223]. All the necessary background about Hessians, Laplace–Beltrami operators, Jacobi fields and Jacobi equations can be found there.

Apart from these sources, a review of comparison methods based on Ricci curvature bounds can be found in [846].

Formula (14.1) does not seem to appear in standard textbooks of Riemannian geometry, but can be derived with the tools found therein, or by comparison with the sphere/hyperbolic space. On the sphere, the computation can be done directly, thanks to a classical formula of spherical trigonometry: If a, b, c are the lengths of the sides of a triangle drawn on the unit sphere S^2, and γ is the angle opposite to c, then $\cos c = \cos a\,\cos b + \sin a\,\sin b\,\cos\gamma$. A more standard computation usually found in textbooks is the asymptotic expansion of the perimeter of a circle centered at x with (geodesic) radius r, as $r \to 0$. Y. H. Kim and McCann [520, Lemma 4.5] recently generalized (14.1) to more general cost functions, and curves of possibly differing lengths.

The differential inequalities relating the Jacobian of the exponential map to the Ricci curvature can be found (with minor variants) in a number of sources, e.g. [223, Section 3.4]. They usually appear in conjunction with volume comparison principles such as the Heintze–Kärcher, Lévy–Gromov, Bishop–Gromov theorems, all of which express the idea that if the Ricci curvature is bounded below by K, and the dimension is less than N, then volumes along geodesic fields grow no faster than volumes in model spaces of constant sectional curvature having dimension N and Ricci curvature identically equal to K. These computations are usually performed in a smooth setting; their adaptation to the nonsmooth context of semiconvex functions has been achieved only recently, first by Cordero-Erausquin, McCann and Schmuckenschläger [246] (in a form that is somewhat different from the one presented here) and more recently by various sets of authors [247, 577, 761].

Bochner's formula appears, e.g., as [394, Proposition 4.15] (for a vector field $\xi = \nabla\psi$) or as [680, Proposition 3.3 (3)] (for a vector field ξ such that $\nabla\xi$ is symmetric, i.e. the 1-form $p \to \xi \cdot p$ is closed). In both cases, it is derived from properties of the Riemannian curvature tensor. Another derivation of Bochner's formula for a gradient vector field is via the properties of the square distance function $d(x_0, x)^2$; this is quite simple, and not far from the presentation that I have followed, since $d(x_0, x)^2/2$ is the solution of the Hamilton–Jacobi equation at time 1, when the initial datum is 0 at x_0 and $+\infty$ everywhere else. But I thought that the explicit use of the Lagrangian/Eulerian duality would make Bochner's formula more intuitive to the readers, especially those who have some experience of fluid mechanics.

There are several other Bochner formulas in the literature; Chapter 7 of Petersen's book [680] is entirely devoted to that subject. In fact "Bochner formula" is a generic name for many identities involving commutators of second-order differential operators and curvature.

The examples (14.10) are by now standard; they have been discussed for instance by Bakry and Qian [61], in relation with spectral gap estimates. When the dimension N is an integer, these reference spaces are obtained by "projection" of the model spaces with constant sectional curvature.

The practical importance of separating out the direction of motion is implicit in Cordero-Erausquin, McCann and Schmuckenschläger [246], but it was Sturm who attracted my attention on this. To implement this idea in the present chapter, I essentially followed the discussion in [763, Section 1]. Also the integral bound (14.56) can be found in this reference.

Many analytic and geometric consequences of Ricci curvature bounds are discussed in Riemannian geometry textbooks such as the one by Gallot, Hulin and Lafontaine [394], and also in hundreds of research papers.

Cordero-Erausquin, McCann and Schmuckenschläger [246, Section 2] express differential inequalities about the Jacobian determinant in terms of volume distortion coefficients; all the discussion about distortion coefficients is inspired from this reference, and most of the material in the Third Appendix is also adapted from that source. This Appendix was born from exchanges with Cordero-Erausquin, who also unwillingly provided its title. It is a pleasure to acknowledge the help of my geometer colleagues (Ghys, Sikorav, Welschinger) in getting the "symplectic argument" for the proof of Proposition 14.29.

Concerning Bakry's approach to curvature-dimension bounds, among many sources one can consult the survey papers [54] and [545].

The almost everywhere second differentiability of convex functions was proven by Alexandrov in 1942 [16]. The proof which I gave in the first Appendix has several points in common with the one that can be found in [331, pp. 241–245], but I have modified the argument to make it look as much as possible like the proof of Rademacher's theorem (Theorem 10.8(ii)). The resulting proof is a bit redundant in some respects, but hopefully it will look rather natural to the reader; also I think it is interesting to have a parallel presentation of the theorems by Rademacher and Alexandrov. Alberti and Ambrosio [11, Theorem 7.10] prove Alexandrov's theorem by a quite different technique, since they *deduce* it from Rademacher's theorem (in the form of the almost everywhere existence of the tangent plane to a Lipschitz graph) together with the area formula. Also they directly establish the differentiability of the gradient, and then deduce the existence of the Hessian; that is, they prove formulation (i) in Theorem 14.1 and then deduce (ii), while in the First Appendix I did it the other way round.

Lebesgue's density theorem can be found for instance in [331, p. 42].

The theorem according to which a nonincreasing function $\mathbb{R} \to \mathbb{R}$ is differentiable almost everywhere is a well-know result, which can be deduced as a corollary of [318, Theorems 7.2.4 and 7.2.7].

15

Otto calculus

Let M be a Riemannian manifold, and let $P_2(M)$ be the associated Wasserstein space of order 2. Recall from Chapter 7 that $P_2(M)$ is a length space and that there is a nice representation formula for the Wasserstein distance W_2:

$$W_2(\mu_0, \mu_1)^2 = \inf \int_0^1 \|\dot{\mu}_t\|_{\mu_t}^2 \, dt, \tag{15.1}$$

where $\|\dot{\mu}\|_\mu$ is the norm of the infinitesimal variation $\dot{\mu}$ of the measure μ, defined by

$$\|\dot{\mu}\|_\mu = \inf \left\{ \int |v|^2 \, d\mu; \qquad \dot{\mu} + \nabla \cdot (v\mu) = 0 \right\}.$$

One of the reasons for the popularity of Riemannian geometry (as opposed to the study of more general metric structures) is that it allows for rather explicit computations. At the end of the nineties, Otto realized that some precious inspiration could be gained by performing computations of a Riemannian nature in the Wasserstein space. His motivations will be described later on; to make a long story short, he needed a good formalism to study certain diffusive partial differential equations which he knew could be considered as gradient flows in the Wasserstein space.

In this chapter, as in Otto's original papers, this problem will be considered from a purely formal point of view, and there will be no attempt at rigorous justification. So the problem is to set up rules for formally differentiating functions (i.e. functionals) on $P_2(M)$. To fix the ideas, and because this is an important example arising in many different contexts, I shall discuss only a certain class of functionals,

C. Villani, *Optimal Transport*, Grundlehren der mathematischen Wissenschaften 338,

that involve (i) a function $V : M \to \mathbb{R}$, used to distort the reference volume measure; and (ii) a function $U : \mathbb{R}_+ \to \mathbb{R}$, twice differentiable (at least on $(0, +\infty)$), which will relate the values of the density of our probability measure and the value of the functional. So let

$$\begin{cases} \nu(dx) := e^{-V(x)} \operatorname{vol}(dx) \\ U_\nu(\mu) := \displaystyle\int_M U(\rho(x))\, d\nu(x), \qquad \mu = \rho\,\nu. \end{cases} \tag{15.2}$$

So far the functional U_ν is only defined on the set of probability measures that are absolutely continuous with respect to ν, or equivalently with respect to the volume measure, and I shall not go beyond that setting before Part III of these notes. If ρ^0 stands for the density of μ with respect to the plain volume, then obviously $\rho^0 = e^{-V}\rho$, so there is the alternative expression

$$U_\nu(\mu) = \int_M U\big(e^V \rho^0\big) e^{-V}\, d\mathrm{vol}, \qquad \mu = \rho^0 \operatorname{vol}.$$

One can think of U as a constitutive law for the **internal energy** of a fluid: this is jargon to say that the energy "contained" in a fluid of density ρ is given by the formula $\int U(\rho)$. The function U should be a property of the fluid itself, and might reflect some microscopic interaction between the particles which constitute it; it is natural to assume $U(0) = 0$.

In the same thermodynamical analogy, one can also introduce the **pressure** law:

$$p(\rho) = \rho\, U'(\rho) - U(\rho). \tag{15.3}$$

The physical interpretation is as follows: if the fluid is enclosed in a domain Ω, then the pressure felt by the boundary $\partial\Omega$ at a point x is normal and proportional to $p(\rho)$ at that point. (Recall that the pressure is defined, up to a sign, as the partial derivative of the internal energy with respect to the volume of the fluid.) So if you consider a homogeneous fluid of total mass 1, in a volume V, then its density is $\rho = 1/V$, the total energy is $V\,U(1/V)$, and the pressure should be $(-d/dV)[V\,U(1/V)] = p(1/V)$; this justifies formula (15.3).

To the pressure p is associated a total pressure $\int p(\rho)\, d\nu$, and one can again consider the influence of small variations of volume on this functional; this leads to the definition of the **iterated pressure**:

$$p_2(\rho) = \rho\, p'(\rho) - p(\rho). \tag{15.4}$$

Both the pressure and the iterated pressure will appear naturally when one differentiates the energy functional: the pressure for first-order derivatives, and the iterated pressure for second-order derivatives.

Example 15.1. Let $m \neq 1$, and

$$U(\rho) = U^{(m)}(\rho) = \frac{\rho^m - \rho}{m-1};$$

then

$$p(\rho) = \rho^m, \qquad p_2(\rho) = (m-1)\,\rho^m.$$

There is an important limit case as $m \to 1$:

$$U^{(1)}(\rho) = \rho \log \rho;$$

then

$$p(\rho) = \rho, \qquad p_2(\rho) = 0.$$

By the way, the linear part $-\rho/(m-1)$ in $U^{(m)}$ does not contribute to the pressure, but has the merit of displaying the link between $U^{(m)}$ and $U^{(1)}$.

Differential operators will also be useful. Let Δ be the Laplace operator on M, then the distortion of the volume element by the function V leads to a natural second-order operator:

$$L = \Delta - \nabla V \cdot \nabla. \tag{15.5}$$

Recall from Chapter 14 the expression of the *carré du champ itéré* associated with L:

$$\Gamma_2(\psi) = L\Big(\frac{|\nabla\psi|^2}{2}\Big) - \nabla\psi \cdot \nabla(L\psi) \tag{15.6}$$

$$= \|\nabla^2\psi\|_{\mathrm{HS}}^2 + \big(\mathrm{Ric} + \nabla^2 V\big)(\nabla\psi); \tag{15.7}$$

the second equality is a consequence of Bochner's formula (14.28), as we shall briefly check. With respect to (14.28), there is an additional term in the left-hand side:

$$\begin{aligned}
&-\nabla V \cdot \nabla \frac{|\nabla\psi|^2}{2} + \nabla\psi \cdot \nabla(\nabla V \cdot \nabla\psi) \\
&\qquad = -\big\langle \nabla^2\psi \cdot \nabla V, \nabla\psi \big\rangle + \big\langle \nabla^2 V \cdot \nabla\psi, \nabla\psi \big\rangle + \big\langle \nabla^2\psi \cdot \nabla V, \nabla\psi \big\rangle \\
&\qquad = \big\langle \nabla^2 V \cdot \nabla\psi, \nabla\psi \big\rangle,
\end{aligned}$$

which is precisely the additional term in the right-hand side.

The next formula is the first important result in this chapter: it gives an "explicit" expression for the gradient of the functional U_ν. For a given measure μ, the gradient of U_ν at μ is a "tangent vector" at μ in the Wasserstein space, so this should be an infinitesimal variation of μ.

Formula 15.2 (Gradient formula in Wasserstein space). *Let μ be absolutely continuous with respect to ν. Then, with the above notation,*

$$\operatorname{grad}_\mu U_\nu = -\nabla\cdot\big(\mu\,\nabla U'(\rho)\big) \tag{15.8}$$
$$= -\nabla\cdot\big(e^{-V}\nabla p(\rho)\big)\,\mathrm{vol} \tag{15.9}$$
$$= -\big(L\,p(\rho)\big)\,\nu. \tag{15.10}$$

Remark 15.3. The expression in the right-hand side of (15.8) is the divergence of a vector-valued measure; recall that $\nabla\cdot m$ is defined in the weak sense by its action on compactly supported smooth functions:

$$\int \phi\, d(\nabla\cdot m) = -\int \nabla\phi\cdot(dm).$$

On the other hand, the divergence in (15.9) is the divergence of a vector field. Note that $\nabla\cdot(\xi\,\mathrm{vol}) = (\nabla\cdot\xi)\,\mathrm{vol}$, so in (15.9) one could put the volume "inside the divergence". All three expressions in Formula (15.2) are interesting, the first one because it writes the "tangent vector" $\operatorname{grad}_\mu U_\nu$ in the normalized form $-\nabla\cdot(\mu\nabla\psi)$, with $\psi = U'(\rho)$; the second one because it gives the result as the divergence of a vector field; the third one because it is stated in terms of the infinitesimal variation of density $\rho = d\mu/d\nu$.

Below are some important examples of application of Formula 15.2.

Example 15.4. Define the H-functional of Boltzmann (opposite of the entropy) by

$$H(\mu) = \int_M \rho\log\rho\, d\mathrm{vol}.$$

Then the second expression in equation (15.8) yields

$$\operatorname{grad}_\mu H = -\Delta\mu,$$

which can be identified with the function $-\Delta\rho$. Thus *the gradient of Boltzmann's entropy is the Laplace operator.* This short statement is one of the first striking conclusions of Otto's formalism.

Example 15.5. Now consider a general $\nu = e^{-V}\mathrm{vol}$, write $\mu = \rho\nu = \rho^0\,\mathrm{vol}$, and define

$$H_\nu(\mu) = \int_M \rho \log \rho \, d\nu = \int_M (\log \rho^0 + V)\, d\mu$$

(this is the H-functional relative to the reference measure ν). Then

$$\mathrm{grad}_\mu H_\nu = -\left(\Delta\rho - \nabla V \cdot \nabla\rho\right)\nu = -\left(L\rho\right)\nu.$$

In short, *the gradient of the relative entropy is the distorted Laplace operator.*

Example 15.6. To generalize Example 15.4 in another direction, consider

$$H^{(m)}(\mu) = \int \frac{\rho^m - \rho}{m-1}\, d\mathrm{vol};$$

then

$$\mathrm{grad}_\mu H^{(m)} = -\Delta(\rho^m).$$

More generally, if ρ is the density with respect to $\nu = e^{-V}\mathrm{vol}$, and

$$H_\nu^{(m)}(\mu) = \int \frac{\rho^m - \rho}{m-1}\, d\nu,$$

then

$$\mathrm{grad}_\mu U_\nu = -\,e^V \nabla \cdot \left(e^{-V}\nabla \rho^m\right)\nu \tag{15.11}$$

$$= -\left(L\rho^m\right)\nu. \tag{15.12}$$

The next formula is about second-order derivatives, or Hessians. Since the Hessian of U_ν at μ is a quadratic form on the tangent space $T_\mu P_2$, I shall write down its expression when evaluated on a tangent vector of the form $-\nabla \cdot (\mu \nabla \psi)$.

Formula 15.7 (Hessian formula in Wasserstein space). *Let μ be absolutely continuous with respect to ν, and let $\dot\mu = -\nabla\cdot(\mu\nabla\psi)$ be a tangent vector at μ. Then, with the above notation,*

$$\mathrm{Hess}_\mu U_\nu(\dot\mu) = \int_M \Gamma_2(\psi)\, p(\rho)\, d\nu + \int_M (L\psi)^2\, p_2(\rho)\, d\nu \tag{15.13}$$

$$= \int_M \Big[\|\nabla^2\psi\|_{\mathrm{HS}}^2 + (\mathrm{Ric} + \nabla^2 V)(\nabla\psi)\Big]\, p(\rho)\, d\nu + \int_M \Big(-\Delta\psi + \nabla V\cdot\nabla\psi\Big)^2 p_2(\rho)\, d\nu. \tag{15.14}$$

Remark 15.8. As expected, this is a quadratic expression in $\nabla\psi$ and its derivatives; and this expression does depend on the measure μ.

Example 15.9. Applying the formula with $U(\rho) = (\rho^m - \rho)/(m-1)$, recalling that $\mu = \rho\,\nu$, one obtains

$$\mathrm{Hess}_\mu\, H_\nu^{(m)}(\dot\mu) =$$
$$\int_M \Big(\|\nabla^2\psi\|_{\mathrm{HS}}^2 + (\mathrm{Ric} + \nabla^2 V)(\nabla\psi) + (m-1)\big(\Delta\psi - \nabla V\cdot\nabla\psi\big)^2\Big)\,\rho^{m-1}\,d\mu.$$

In the limit case $m = 1$, which is $U(\rho) = \rho\log\rho$, this expression simplifies into

$$\mathrm{Hess}_\mu\, H_\nu(\dot\mu) = \int_M \Big(\|\nabla^2\psi\|_{\mathrm{HS}}^2 + (\mathrm{Ric} + \nabla^2 V)(\nabla\psi)\Big)\,d\mu;$$

or equivalently, with the notation of Chapter 14,

$$\mathrm{Hess}_\mu\, H_\nu(\dot\mu) = \int_M \Big(\|\nabla^2\psi\|_{\mathrm{HS}}^2 + \mathrm{Ric}_{\infty,\nu}(\nabla\psi)\Big)\,d\mu.$$

Formulas 15.2 and 15.7 will be justified only at a heuristic level. A rigorous proof would require many more definitions and much more apparatus, as well as regularity and decay assumptions on the measures and the functionals. So here I shall disregard all issues about integrability and regularity, which will be a huge simplification. Still, the proofs will not be completely trivial.

"Proof" of Formula 15.2. When the integration measure is not specified, it will be the volume rather than ν. To understand the proof, it is important to make the distinction between a *gradient* and a *differential*.

Let ζ be such that the tangent vector $\mathrm{grad}_\mu U_\nu$ can be represented as $-\nabla\cdot(\mu\nabla\zeta)$, and let $\partial_t\mu = -\nabla\cdot(\mu\nabla\psi)$ be an arbitrary "tangent vector". The infinitesimal variation of the density $\rho = d\mu/d\nu$ is given by

$$\partial_t\rho = -e^V\,\nabla\cdot\big(\rho\,e^{-V}\nabla\psi\big).$$

By direct computation and integration by parts, the infinitesimal variation of U_ν along that variation is equal to

$$\begin{aligned}\int U'(\rho)\,\partial_t\rho\,d\nu &= -\int U'(\rho)\;\nabla\cdot\big(\rho\,e^{-V}\nabla\psi\big)\\ &= \int \nabla U'(\rho)\cdot\nabla\psi\;\rho\,e^{-V}\\ &= \int \nabla U'(\rho)\cdot\nabla\psi\;d\mu.\end{aligned}$$

By definition of the gradient operator, this should coincide with

$$\big\langle \mathrm{grad}_\mu U_\nu,\, \partial_t \mu \big\rangle = \int \nabla\zeta \cdot \nabla\psi \, d\mu.$$

If this should hold true for all ψ, the only possible choice is that $\nabla U'(\rho) = \nabla\zeta(\rho)$, at least μ-almost everywhere. In any case $\zeta := U'(\rho)$ provides an admissible representation of $\mathrm{grad}_\mu U_\nu$. This proves formula (15.8). The other two formulas are obtained by noting that $p'(\rho) = \rho\, U''(\rho)$, and so

$$\nabla U'(\rho)\rho = \rho\, U''(\rho) \nabla\rho = p'(\rho)\nabla\rho = \nabla p(\rho);$$

therefore

$$\nabla \cdot \big(\mu\, \nabla U'(\rho)\big) = \nabla \cdot \big(e^{-V}\, \rho\, \nabla U'(\rho)\big) = e^{-V}\, L\, p(\rho).$$

□

For the second order (formula (15.7)), things are more intricate. The following identity will be helpful: If ξ is a tangent vector at x on a Riemannian manifold $\mathcal{M}$, and F is a function on $\mathcal{M}$, then

$$\mathrm{Hess}_x F(\xi) = \frac{d^2}{dt^2}\bigg|_{t=0} F(\gamma(t)), \tag{15.15}$$

where $\gamma(t)$ is a geodesic starting from $\gamma(0) = x$ with velocity $\dot\gamma(0) = \xi$. To prove (15.15), it suffices to note that the first derivative of $F(\gamma(t))$ is $\dot\gamma(t) \cdot \nabla F(\gamma(t))$; so the second derivative is $(d/dt)(\dot\gamma(t)) \cdot \nabla F(\gamma(t)) + \big\langle \nabla^2 F(\gamma(t)) \cdot \dot\gamma(t), \dot\gamma(t)\big\rangle$, and the first term vanishes because a geodesic has zero acceleration.

"Proof" of Formula 15.7. The problem consists in differentiating $U_\nu(\mu_t)$ twice along a geodesic path of the form

$$\begin{cases} \partial_t \mu + \nabla \cdot (\mu \nabla\psi) = 0 \\[2ex] \partial_t \psi + \dfrac{|\nabla\psi|^2}{2} = 0. \end{cases}$$

The following integration by parts formula will be useful:

$$\int \nabla f \cdot \nabla g \, d\nu = -\int (Lf) g \, d\nu. \tag{15.16}$$

From the proof of the gradient formula, one has, with the notation $\mu_t = \rho_t \, \nu$,

$$\begin{aligned}\frac{dU_\nu(\mu_t)}{dt} &= \int_M \nabla\psi_t \cdot \nabla U'(\rho_t)\, \rho_t \, d\nu \\ &= \int_M \nabla\psi_t \cdot \nabla p(\rho_t)\, d\nu \\ &= -\int_M (L\psi_t)\, p(\rho_t)\, d\nu.\end{aligned}$$

It remains to differentiate again. To alleviate notation, I shall not write the time variable explicitly. So

$$\frac{d^2U_\nu(\mu)}{dt^2} = -\int (L\partial_t\psi)\, p(\rho)\, d\nu - \int (L\psi) p'(\rho)\partial_t\rho \, d\nu \tag{15.17}$$

$$= \int L\left(\frac{|\nabla\psi|^2}{2}\right) p(\rho)\, d\nu - \int (L\psi)\, p'(\rho)\, \partial_t\mu. \tag{15.18}$$

The last term in (15.18) can be rewritten as

$$\begin{aligned}&\int (L\psi)\, p'(\rho)\, \nabla\cdot(\mu\nabla\psi) \\ &\qquad = -\int \nabla\big((L\psi)p'(\rho)\big)\cdot\nabla\psi\, d\mu \\ &\qquad = -\int \nabla\big((L\psi)p'(\rho)\big)\cdot\nabla\psi\, \rho\, d\nu \\ &= -\int \nabla(L\psi)\cdot\nabla\psi\, p'(\rho)\, \rho\, d\nu - \int (L\psi)\, p''(\rho)\, \rho\, \nabla\rho\cdot\nabla\psi\, d\nu\end{aligned}$$

$$= -\int \nabla(L\psi)\cdot\nabla\psi\, \rho\, p'(\rho)\, d\nu - \int (L\psi)\nabla p_2(\rho)\cdot\nabla\psi\, d\nu. \tag{15.19}$$

The second term in (15.19) needs a bit of reworking: it can be recast as

$$\begin{aligned}&-\int \nabla\big(L\psi\, p_2(\rho)\big)\cdot\nabla\psi\, d\nu - \int (\nabla L\psi)\, p_2(\rho)\cdot\nabla\psi\, d\nu \\ &\qquad = \int (L\psi)^2 p_2(\rho)\, d\nu - \int (\nabla L\psi)\cdot\nabla\psi\, p_2(\rho)\, d\nu,\end{aligned}$$

where (15.16) has been used once more.

By collecting all these calculations,

$$\frac{d^2 U_\nu(\mu)}{dt^2} = \int L\left(\frac{|\nabla\psi|^2}{2}\right) p(\rho)\, d\nu + \int (L\psi)^2 p_2(\rho)\, d\nu \\ + \int (\nabla\psi \cdot \nabla L\psi)\big(p_2(\rho) - \rho\, p'(\rho)\big)\, d\nu.$$

Since $p_2(\rho) - \rho\, p'(\rho) = -p(\rho)$, this transforms into

$$\int \left(L\left(\frac{|\nabla\psi|^2}{2}\right) - \nabla\psi\cdot\nabla L\psi\right) p(\rho)\, d\nu + \int (L\psi)^2 p_2(\rho). \tag{15.20}$$

In view of (15.6)–(15.7), this establishes formula (15.13). □

Exercise 15.10. "Prove" that the gradient of an arbitrary functional $\mathcal{F}$ on $P_2(M)$ can be written

$$\operatorname{grad}_\mu \mathcal{F} = -\nabla\cdot(\mu\nabla\phi), \qquad \phi = \frac{\delta\mathcal{F}}{\delta\mu},$$

where $\delta\mathcal{F}/\delta\mu$ is a function defined by

$$\frac{d}{dt}\mathcal{F}(\mu_t) = \int \left(\frac{\delta\mathcal{F}}{\delta\mu}\right) \partial_t\mu_t.$$

Check that in the particular case

$$\mathcal{F}(\mu) = \int_M F\Big(x, \rho(x), \nabla\rho(x)\Big)\, d\nu(x), \tag{15.21}$$

where $F = F(x,\rho,p)$ is a smooth function of $\rho \in \mathbb{R}_+$, $(x,p) \in TM$, one has

$$\left(\frac{\partial\mathcal{F}}{\partial\mu}\right)(x) = (\partial_\rho F)\big(x, \rho(x), \nabla\rho(x)\big) \\ - (\nabla_x - \nabla V(x))\cdot(\nabla_p F)\big(x,\rho(x),\nabla\rho(x)\big)$$

The following two open problems (loosely formulated) are natural and interesting, and I don't know how difficult they are:

Open Problem 15.11. *Find a nice formula for the Hessian of the functional $\mathcal{F}$ appearing in* (15.21).

Open Problem 15.12. *Find a nice formalism playing the role of the Otto calculus in the space $P_p(M)$, for $p \neq 2$. More generally, are there nice formal rules for taking derivatives along displacement interpolation, for general Lagrangian cost functions?*

To conclude this chapter, I shall come back to the subject of rigorous justification of Otto's formalism. At the time of writing, several theories have been developed, at least in the Euclidean setting (see the bibliographical notes); but they are rather heavy and not yet completely convincing.[1] From the technical point of view, they are based on the natural strategy which consists in truncating and regularizing, then applying the arguments presented in this chapter, then passing to the limit.

A quite different strategy, which I personally recommend, consists in translating all the Eulerian statements in the language of Lagrangian formalism. This is less appealing for intuition and calculations, but somehow easier to justify in the case of optimal transport. For instance, instead of the Hessian operator, one will only speak of the second derivative along geodesics in the Wasserstein space. This point of view will be developed in the next two chapters, and then a rigorous treatment will not be that painful.

Still, in many situations the Eulerian point of view is better for intuition and for understanding, in particular in certain problems involving functional inequalities. The above discussion might be summarized by the slogan *"Think Eulerian, prove Lagrangian"*. This is a rather exceptional situation from the point of view of fluid dynamics, where the standard would rather be "Think Lagrangian, prove Eulerian" (for instance, shocks are delicate to treat in a Lagrangian formalism). Once again, the point is that "there are no shocks" in optimal transport: as discussed in Chapter 8, trajectories do not meet until maybe at final time.

Bibliographical notes

Otto's seminal paper [669] studied the formal Riemannian structure of the Wasserstein space, and gave applications to the study of the porous medium equation; I shall come back to this topic later. With all

[1] I can afford this negative comment since myself I participated in the story.

the preparations of Part I, the computations performed in this chapter may look rather natural, but they were a little conceptual tour de force at the time of Otto's contribution, and had a strong impact on the research community. This work was partly inspired by the desire to understand in depth a previous contribution by Jordan, Kinderlehrer and Otto [493].

Otto's computations were concerned with the case $U(\rho) = \rho^m$ in $\mathbb{R}^n$. Then Otto and I considered $U(\rho) = \rho \log \rho$ on a manifold [671, Section 3]; we computed the Hessian by differentiating twice along geodesics in the Wasserstein space. (To my knowledge, this was the first published work where Ricci curvature appeared in relation to optimal transport.) Functionals of the form $E(\mu) = \int W(x-y)\,\mu(dx)\,\mu(dy)$ in $\mathbb{R}^n$ were later studied by Carrillo, McCann and myself [213]. More recently, Lott and I [577, Appendix E] considered the functionals U_ν presented in this chapter (on a manifold and with a reference measure $e^{-V}\,\mathrm{vol}$).

In my previous book [814, Section 9.1], I already gave formulas for the gradient and Hessian of three basic types of functionals on $P_2(\mathbb{R}^n)$ that I called *internal energy*, *potential energy* and *interaction energy*, and which can be written respectively (with obvious notation) as

$$\int U(\rho(x))\,dx; \qquad \int V\,d\mu; \qquad \frac{1}{2}\int W(x-y)\,d\mu(x)\,d\mu(y). \tag{15.22}$$

A short presentation of the differential calculus in the Wasserstein space can be found in [814, Chapter 8]; other sources dealing with this subject, with some variations in the presentation, are [30, 203, 214, 671, 673].

Apart from computations of gradients and Hessians, little is known about Riemannian calculus in $P_2(M)$. The following issues are natural (I am not sure how important they are, but at least they are natural):

- Is there a Jacobi equation in $P_2(M)$, describing small variations of geodesic fields?
- Can one define Christoffel symbols, at least formally?
- Can one define a Laplace operator?
- Can one define a volume element? A divergence operator?

Recently, Lott [575] partly answered some of these questions by establishing formulas for the Riemannian connection and Riemannian curvature in the subset $P^\infty(M)$ of smooth positive densities, viewed as a subset of $P_2(M)$, when M is compact. In a different direction,

Gigli [415] gave a rigorous construction of a parallel transport along a curve in $P_2(\mathbb{R}^n)$ for which $\int_0^1 \|v_t\|_{\rm Lip}\,dt < +\infty$. (See [29] for improved results.)

The problem whether there exists a natural probability measure ("volume", or "Boltzmann–Gibbs measure") on $P_2(M)$ is, I think, very relevant for applications in geometry or theoretical statistics. Von Renesse and Sturm [827] have managed to construct natural probability measures on $P_2(S^1)$; these measures depend on a parameter β ("inverse temperature") and may be written heuristically as

$$\mathbb{P}_\beta(d\mu) = \frac{e^{-H_\nu(\mu)}\,d\mathrm{vol}(\mu)}{Z_\beta}, \tag{15.23}$$

where ν is the reference measure on S^1, that is, the Lebesgue measure. Their construction strongly uses the one-dimensional assumption, and makes the link with the theory of "Poisson measures" used in nonparametric statistics. A particle approximation was studied in [38].

The point of view that was first advocated by Otto himself, and which I shall adopt in this course, is that the "Otto calculus" should primarily be considered a heuristic tool, and conclusions drawn by its use should then be checked by "direct" means. This might lack elegance, but it is much safer from the point of view of mathematical rigor. Some papers in which this strategy has been used with success are [577, 669, 671, 673, 761]. Recently, Calvez [197] used the Otto formalism to derive complex identities for chemotaxis models of Keller–Segel type, which would have been very difficult to guess otherwise.

In most of these works, rigorous justifications are done in Lagrangian formalism, or by methods which do not use transport at all. The work by Otto and Westdickenberg [673] is an interesting exception: there everything is attacked from an Eulerian perspective (using such tools as regularization of currents on manifolds); see [271] for an elaboration of these ideas, which applies even without smoothness.

All the references quoted above mainly deal with calculus in $P_p(M)$ for $p = 2$. The case $p \neq 2$ is much less well understood; as noticed in [30, p. 10], $P_p(M)$ can be seen as a kind of Finsler structure, and there are also rules to compute derivatives in that space, at least to first order. The most general results to this date are in [30].

A better understood generalization treats the case when geodesics in $P_2(M)$ are replaced by action-minimizing curves, for some Lagrangian action like those considered in Chapter 7; the adaptation of Otto calculus to this situation was worked out by Lott [576], with applications to the study of the Ricci flow.

Let me conclude with some remarks about the functionals considered in this chapter.

Functionals of the form (15.22) appear everywhere in mathematical physics to model all kinds of energies. It would be foolish to try to make a list.

The interpretation of $p(\rho) = \rho U'(\rho) - U(\rho)$ as a pressure associated to the constitutive law U is well-known in thermodynamics, and was explained to me by McCann; the discussion in the present chapter is slightly expanded in [814, Remarks 5.18].

The functional $H_\nu(\mu) = \int \rho \log \rho \, d\nu$ $(\mu = \rho\nu)$ is well-known is statistical physics, where it was introduced by Boltzmann [141]. In Boltzmann's theory of gases, H_ν is identified with the negative of the entropy. It would take a whole book to review the meaning of entropy in thermodynamics and statistical mechanics (see, e.g., [812] for its use in kinetic theory). I should also mention that the H functional coincides with the Kullback information in statistics, and it appears in Shannon's theory of information as an optimal compression rate [747], and in Sanov's theorem as the rate function for large deviations of the empirical measure of independent samples [302, Chapter 3] [296, Theorem 6.2.10].

An interesting example of a functional of the form (15.21) that was considered in relation with optimal transport is the so-called **Fisher information**,

$$I(\mu) = \int \frac{|\nabla \rho|^2}{\rho};$$

see [30, Example 11.1.10] and references there provided. We shall encounter this functional again later.

16

Displacement convexity I

Convexity plays a prominent role in analysis in general. It is most generally used in a vector space $\mathcal{V}$: A function $F : \mathcal{V} \to \mathbb{R} \cup \{+\infty\}$ is said to be convex if for all $x, y \in \mathcal{V}$,

$$\forall t \in [0,1] \qquad F\big((1-t)\,x + t\,y\big) \leq (1-t)\,F(x) + t\,F(y). \tag{16.1}$$

But convexity is also a *metric* notion: In short, convexity in a metric space means convexity along geodesics. Consequently, geodesic spaces are a natural setting in which to define convexity:

Definition 16.1 (Convexity in a geodesic space). *Let $(\mathcal{X}, d)$ be a complete geodesic space. Then a function $F : \mathcal{X} \to \mathbb{R} \cup \{+\infty\}$ is said to be geodesically convex, or just convex, if for any constant-speed geodesic path $(\gamma_t)_{0 \leq t \leq 1}$ valued in $\mathcal{X}$,*

$$\forall t \in [0,1] \qquad F(\gamma_t) \leq (1-t)\,F(\gamma_0) + t\,F(\gamma_1). \tag{16.2}$$

It is said to be weakly convex if for any x_0, x_1 in $\mathcal{X}$ there exists at least one constant-speed geodesic path $(\gamma_t)_{0 \leq t \leq 1}$ with $\gamma_0 = x_0$, $\gamma_1 = x_1$, such that inequality (16.2) *holds true.*

It is a natural problem to identify functionals that are convex on the Wasserstein space. In his 1994 PhD thesis, McCann established and used the convexity of certain functionals on $P_2(\mathbb{R}^n)$ to prove the uniqueness of their minimizers. Since then, his results have been generalized; yet almost all examples which have been treated so far belong to the general class

$$\mathcal{F}(\mu) = \int_{\mathcal{X}^k} I(x_1, \ldots, x_k)\, d\mu(x_1) \ldots d\mu(x_k) \;+\; \int_{\mathcal{X}} U\left(\frac{d\mu}{d\nu}\right) d\nu,$$

C. Villani, *Optimal Transport*, Grundlehren der mathematischen Wissenschaften 338,

where $I(x_1, \ldots, x_k)$ is a certain "k-particle interaction potential", U is a nice function $\mathbb{R}_+ \to \mathbb{R}$, and ν is a reference measure.

In this and the next chapter I shall consider the convexity problem on a general Riemannian manifold M, in the case $I = 0$, so the functionals under study will be the functionals U_ν defined by

$$U_\nu(\mu) = \int_M U(\rho)\, d\nu, \qquad \mu = \rho\,\nu. \tag{16.3}$$

As a start, I shall give some reminders about the notion of convexity and its refinements; then I shall make these notions more explicit in the case of the Wasserstein space $P_2(M)$. In the last section of this chapter I shall use Otto's calculus to guess sufficient conditions under which U_ν satisfies some interesting convexity properties (Guesses 16.6 and 16.7).

Let the reader not be offended if I strongly insist that convexity in the metric space $P_2(M)$ has nothing to do with the convex structure of the space of probability measures. The former concept will be called "convexity along optimal transport" or **displacement convexity**.

Reminders on convexity: differential and integral conditions

The material in this section has nothing to do with optimal transport, and is, for the most part, rather standard.

It is well-known that a function $F : \mathbb{R}^n \to \mathbb{R}$ is convex, in the sense of (16.1), if and only if it satisfies

$$\nabla^2 F \geq 0 \tag{16.4}$$

(nonnegative Hessian) on $\mathbb{R}^n$. The latter inequality should generally be understood in distribution sense, but let me just forget about this subtlety which is not essential here.

The inequality (16.4) is a differential condition, in contrast with the "integral" condition (16.1). There is a more general principle relating a lower bound on the Hessian (differential condition) to a convexity-type inequality (integral condition). It can be stated in terms of the **one-dimensional Green function** (of the Laplace operator with Dirichlet boundary conditions). That Green function is the nonnegative kernel $G(s,t)$ such that for all functions $\varphi \in C([0,1];\mathbb{R}) \cap C^2((0,1);\mathbb{R})$,

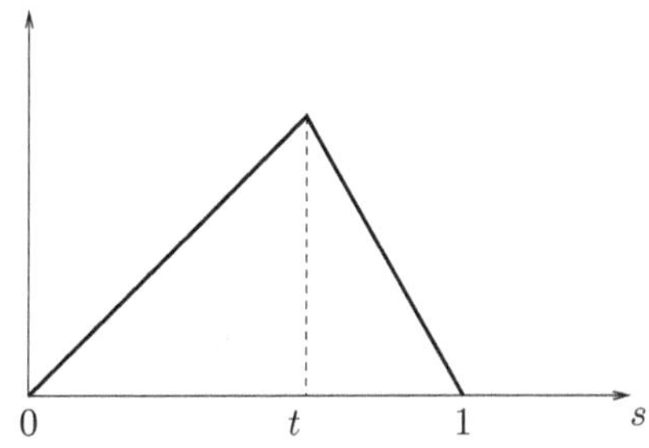

Fig. 16.1. The Green function $G(s,t)$ as a function of s.

$$\varphi(t) = (1-t)\,\varphi(0) + t\,\varphi(1) - \int_0^1 \ddot{\varphi}(s)\,G(s,t)\,ds. \tag{16.5}$$

It is easy to give an explicit expression for G (see Figure 16.1):

$$G(s,t) = \begin{cases} s\,(1-t) & \text{if } s \le t \\ t\,(1-s) & \text{if } s \ge t. \end{cases} \tag{16.6}$$

Then formula (16.5) actually extends to arbitrary continuous functions φ on $[0,1]$, provided that $\ddot{\varphi}$ (taken in a distribution sense) is bounded below by a real number.

The next statement provides the equivalence between several differential and integral convexity conditions in a rather general setting.

Proposition 16.2 (Convexity and lower Hessian bounds). *Let (M,g) be a Riemannian manifold, and let $\Lambda = \Lambda(x,v)$ be a continuous quadratic form on TM; that is, for any x, $\Lambda(x,\,\cdot\,)$ is a quadratic form in v, and it depends continuously on x. Assume that for any constant-speed geodesic $\gamma : [0,1] \to M$,*

$$\lambda[\gamma] := \inf_{0 \le t \le 1} \frac{\Lambda(\gamma_t, \dot{\gamma}_t)}{|\dot{\gamma}_t|^2} > -\infty. \tag{16.7}$$

Then, for any function $F \in C^2(M)$, the following statements are equivalent:

(i) $\nabla^2 F \ge \Lambda$;

(ii) For any constant-speed, minimizing geodesic $\gamma : [0,1] \to M$,

$$F(\gamma_t) \le (1-t)\,F(\gamma_0) + t\,F(\gamma_1) - \int_0^1 \Lambda(\gamma_s, \dot{\gamma}_s)\,G(s,t)\,ds;$$

(iii) For any constant-speed, minimizing geodesic $\gamma : [0,1] \to M$,

$$F(\gamma_1) \geq F(\gamma_0) + \langle \nabla F(\gamma_0), \dot{\gamma}_0 \rangle + \int_0^1 \Lambda(\gamma_t, \dot{\gamma}_t)\,(1-t)\,dt;$$

(iv) For any constant-speed, minimizing geodesic $\gamma : [0,1] \to M$,

$$\langle \nabla F(\gamma_1), \dot{\gamma}_1 \rangle - \langle \nabla F(\gamma_0), \dot{\gamma}_0 \rangle \geq \int_0^1 \Lambda(\gamma_t, \dot{\gamma}_t)\,dt.$$

When these properties are satisfied, F *is said to be* Λ*-convex. The equivalence is still preserved if conditions (ii), (iii) and (iv) are respectively replaced by the following a priori weaker conditions:*

(ii') For any constant-speed, minimizing geodesic $\gamma : [0,1] \to M$,

$$F(\gamma_t) \leq (1-t)\,F(\gamma_0) + t\,F(\gamma_1) - \lambda[\gamma]\,\frac{t(1-t)}{2}\,d(\gamma_0,\gamma_1)^2;$$

(iii') For any constant-speed, minimizing geodesic $\gamma : [0,1] \to M$,

$$F(\gamma_1) \geq F(\gamma_0) + \langle \nabla F(\gamma_0), \dot{\gamma}_0 \rangle + \lambda[\gamma]\,\frac{d(\gamma_0,\gamma_1)^2}{2};$$

(iv') For any constant-speed, minimizing geodesic $\gamma : [0,1] \to M$,

$$\langle \nabla F(\gamma_1), \dot{\gamma}_1 \rangle - \langle \nabla F(\gamma_0), \dot{\gamma}_0 \rangle \geq \lambda[\gamma]\,d(\gamma_0,\gamma_1)^2.$$

Remark 16.3. In the particular case when Λ is equal to λg for some constant $\lambda \in \mathbb{R}$, Property (ii) reduces to Property (ii') with $\lambda[\gamma] = \lambda$. Indeed, since γ has constant speed,

$$\begin{aligned} F(\gamma_t) &\leq (1-t)\,F(\gamma_0) + t\,F(\gamma_1) - \lambda \int_0^1 g(\gamma_s, \dot{\gamma}_s)\,G(s,t)\,ds \\ &= (1-t)\,F(\gamma_0) + t\,F(\gamma_1) - \lambda\,d(\gamma_0,\gamma_1)^2 \int_0^1 G(s,t)\,ds. \end{aligned}$$

By plugging the function $\varphi(t) = t^2$ into (16.5) one sees that $\int_0^1 G(s,t)\,ds = t(1-t)/2$. So (ii) indeed reduces to

$$F(\gamma_t) \leq (1-t)\,F(\gamma_0) + t\,F(\gamma_1) - \frac{\lambda\,t(1-t)}{2}\,d(\gamma_0,\gamma_1)^2. \qquad (16.8)$$

Definition 16.4 (Λ-convexity). *Let M be a Riemannian manifold, and let $\Lambda = \Lambda(x,v)$ be a continuous quadratic form on M, satisfying (16.7). A function $F : M \to \mathbb{R} \cup \{+\infty\}$ is said to be Λ-convex if Property (ii) in Proposition 16.2 is satisfied. In the case when $\Lambda = \lambda g$, $\lambda \in \mathbb{R}$, F will be said to be λ-convex; this means that inequality (16.8) is satisfied. In particular, 0-convexity is just plain convexity.*

Proof of Proposition 16.2. The arguments in this proof will come again several times in the sequel, in various contexts.

Assume that (i) holds true. Consider x_0 and x_1 in M, and introduce a constant-speed minimizing geodesic γ joining $\gamma_0 = x_0$ to $\gamma_1 = x_1$. Then

$$\frac{d^2}{dt^2}F(\gamma_t) = \langle \nabla^2 F(\gamma_t) \cdot \dot{\gamma}_t, \dot{\gamma}_t \rangle \geq \Lambda(\gamma_t, \dot{\gamma}_t).$$

Then Property (ii) follows from identity (16.5) with $\varphi(t) := F(\gamma_t)$.

As for Property (iii), it can be established either by dividing the inequality in (ii) by $t > 0$, and then letting $t \to 0$, or directly from (i) by using the Taylor formula at order 2 with $\varphi(t) = F(\gamma_t)$ again. Indeed, $\dot{\varphi}(0) = \langle \nabla F(\gamma_0), \dot{\gamma}_0 \rangle$, while $\ddot{\varphi}(t) \geq \Lambda(\gamma_t, \dot{\gamma}_t)$.

To go from (iii) to (iv), replace the geodesic γ_t by the geodesic γ_{1-t}, to get

$$F(\gamma_0) \geq F(\gamma_1) - \langle \nabla F(\gamma_1), \dot{\gamma}_1 \rangle + \int_0^1 \Lambda(\gamma_{1-t}, \dot{\gamma}_{1-t})\,(1-t)\,dt.$$

After changing variables in the last integral, this is

$$F(\gamma_0) \geq F(\gamma_1) - \langle \nabla F(\gamma_1), \dot{\gamma}_1 \rangle + \int_0^1 \Lambda(\gamma_t, \dot{\gamma}_t)\,t\,dt,$$

and by adding up (iii), one gets Property (iv).

So far we have seen that (i) $\Rightarrow$ (ii) $\Rightarrow$ (iii) $\Rightarrow$ (iv). To complete the proof of equivalence it is sufficient to check that (iv') implies (i).

So assume (iv'). From the identity

$$\langle \nabla F(\gamma_1), \dot{\gamma}_1 \rangle - \langle \nabla F(\gamma_0), \dot{\gamma}_0 \rangle = \int_0^1 \nabla^2 F(\gamma_t)(\dot{\gamma}_t)\,dt$$

and (iv'), one deduces that for all geodesic paths γ,

$$\lambda[\gamma]\,d(\gamma_0,\gamma_1)^2 \leq \int_0^1 \nabla^2 F(\gamma_t)(\dot{\gamma}_t)\,dt. \tag{16.9}$$

Choose (x_0, v_0) in TM, with $v_0 \neq 0$, and $\gamma(t) = \exp_{x_0}(\varepsilon t v_0)$, where $\varepsilon > 0$; of course γ depends implicitly on ε. Note that $d(\gamma_0, \gamma_1) = \varepsilon |v_0|$. Write

$$\lambda[\gamma] \, \frac{d(\gamma_0, \gamma_1)^2}{\varepsilon^2} \leq \int_0^1 \nabla^2 F(\gamma_t) \left(\frac{\dot{\gamma}_t}{\varepsilon}\right) dt. \tag{16.10}$$

As $\varepsilon \to 0$, $(\gamma_t, \dot{\gamma}_t) \simeq (x_0, \varepsilon v_0)$ in TM, so

$$\lambda[\gamma] = \inf_{0 \leq t \leq 1} \frac{\Lambda(\gamma_t, \dot{\gamma}_t)}{|\dot{\gamma}_t|^2} = \inf_{0 \leq t \leq 1} \frac{\Lambda(\gamma_t, \dot{\gamma}_t/\varepsilon)}{|\dot{\gamma}_t/\varepsilon|^2} \xrightarrow[\varepsilon \to 0]{} \frac{\Lambda(x_0, v_0)}{|v_0|^2}.$$

Thus the left-hand side of (16.10) converges to $\Lambda(x_0, v_0)$. On the other hand, since $\nabla^2 F$ is continuous, the right-hand side obviously converges to $\nabla^2 F(x_0)(v_0)$. Property (i) follows. □

Displacement convexity

I shall now discuss convexity in the context of optimal transport, replacing the manifold M of the previous section by the geodesic space $P_2(M)$. For the moment I shall only consider measures that are absolutely continuous with respect to the volume on M, and denote by $P_2^{ac}(M)$ the space of such measures. It makes sense to study convexity in $P_2^{ac}(M)$ because this is a *geodesically convex* subset of $P_2(M)$: By Theorem 8.7, a displacement interpolation between any two absolutely continuous measures is itself absolutely continuous. (Singular measures will be considered later, together with singular metric spaces, in Part III.)

So let μ_0 and μ_1 be two probability measures on M, absolutely continuous with respect to the volume element, and let $(\mu_t)_{0 \leq t \leq 1}$ be the displacement interpolation between μ_0 and μ_1. Recall from Chapter 13 that this displacement interpolation is uniquely defined, and characterized by the formulas $\mu_t = (T_t)_\# \mu_0$, where

$$T_t(x) = \exp_x(t \, \widetilde{\nabla} \psi(x)), \tag{16.11}$$

and ψ is $d^2/2$-convex. (Forget about the $\widetilde{\ }$ symbol if you don't like it.) Moreover, T_t is injective for $t < 1$; so for any $t < 1$ it makes sense to define the velocity field $v(t, x)$ on $T_t(M)$ by

$$v(t, T_t(x)) = \frac{d}{dt} T_t(x),$$

and one also has

$$v\big(t, T_t(x)\big) = \widetilde{\nabla}\psi_t(T_t(x)),$$

where ψ_t is a solution at time t of the quadratic Hamilton–Jacobi equation with initial datum $\psi_0 = \psi$.

The next definition adapts the notions of convexity, λ-convexity and Λ-convexity to the setting of optimal transport. Below λ is a real number that might nonnegative or nonpositive, while $\Lambda = \Lambda(\mu, v)$ defines for each probability measure μ a quadratic form on vector fields $v : M \to TM$.

Definition 16.5 (Displacement convexity). *With the above notation, a functional $F : P_2^{\mathrm{ac}}(M) \to \mathbb{R} \cup \{+\infty\}$ is said to be:*

- **displacement convex** *if, whenever $(\mu_t)_{0\le t\le 1}$ is a (constant-speed, minimizing) geodesic in $P_2^{\mathrm{ac}}(M)$,*

$$\forall t \in [0,1] \qquad F(\mu_t) \le (1-t)\, F(\mu_0) + t\, F(\mu_1);$$

- **λ-displacement convex**, *if, whenever $(\mu_t)_{0\le t\le 1}$ is a (constant-speed, miminizing) geodesic in $P_2^{\mathrm{ac}}(M)$,*

$$\forall t \in [0,1] \qquad F(\mu_t)\le(1-t)\, F(\mu_0)+t\, F(\mu_1)-\frac{\lambda\, t(1-t)}{2}\, W_2(\mu_0,\mu_1)^2;$$

- **Λ-displacement convex**, *if, whenever $(\mu_t)_{0\le t\le 1}$ is a (constant-speed, minimizing) geodesic in $P_2^{\mathrm{ac}}(M)$, and $(\psi_t)_{0\le t\le 1}$ is an associated solution of the Hamilton–Jacobi equation,*

$$\forall t \in [0,1] \quad F(\mu_t)\le(1-t)F(\mu_0)+tF(\mu_1) - \int_0^1 \Lambda(\mu_s, \widetilde{\nabla}\psi_s) G(s,t)\, ds,$$

where $G(s,t)$ is the one-dimensional Green function of (16.6). *(It is assumed that $\Lambda(\mu_s, \widetilde{\nabla}\psi_s)\, G(s,t)$ is bounded below by an integrable function of $s \in [0,1]$.)*

Of course these definitions are more and more general: Λ-displacement convexity reduces to λ-displacement convexity when $\Lambda(\mu, v) = \lambda\, \|v\|^2_{L^2(\mu)}$; and this in turn reduces to plain displacement convexity when $\lambda = 0$.

Displacement convexity from curvature-dimension bounds

The question is whether the previously defined concepts apply to functionals of the form U_ν, as in (16.3). Of course Proposition 16.2 does not apply, because neither $P_2(M)$ nor $P_2^{\mathrm{ac}}(M)$ are smooth manifolds. However, if one believes in Otto's formalism, then one can hope that displacement convexity, λ-displacement convexity, Λ-displacement convexity of U_ν would be respectively equivalent to

$$\mathrm{Hess}_\mu U_\nu \geq 0, \qquad \mathrm{Hess}_\mu U_\nu \geq \lambda, \qquad \mathrm{Hess}_\mu U_\nu(\dot{\mu}) \geq \Lambda(\mu, \dot{\mu}), \quad (16.12)$$

where $\mathrm{Hess}_\mu U_\nu$ stands for the formal Hessian of U_ν at μ (which was computed in Chapter 15), λ is shorthand for $\lambda \| \cdot \|^2_{L^2(\mu)}$, and $\dot{\mu}$ is identified with $\nabla\psi$ via the usual continuity equation

$$\dot{\mu} + \nabla \cdot (\nabla\psi\, \mu) = 0.$$

Let us try to identify simple sufficient conditions on the manifold M, the reference measure ν and the energy function U, for (16.12) to hold. This quest is, for the moment, just formal; it will be checked later, without any reference to Otto's formalism, that our guess is correct.

To identify conditions for displacement convexity I shall use again the formalism of Chapter 14. Equip the Riemannian manifold M with a reference measure $\nu = e^{-V}\mathrm{vol}$, where V is a smooth function on M, and assume that the resulting space satisfies the curvature-dimension bound $\mathrm{CD}(K, N)$, as in Theorem 14.8, for some $N \in [1, \infty]$ and $K \in \mathbb{R}$. Everywhere in the sequel, ρ will stand for the density of μ with respect to ν.

Consider a continuous function $U : \mathbb{R}_+ \to \mathbb{R}$. I shall assume that U is *convex* and $U(0) = 0$. The latter condition is quite natural from a physical point of view (no matter $\Rightarrow$ no energy). The convexity assumption might seem more artificial, and to justify it I will argue that (i) the convexity of U is necessary for U_ν to be lower semicontinuous with respect to the weak topology induced by the metric W_2; (ii) if one imposes the nonnegativity of the pressure $p(r) = r\,U'(r) - U(r)$, which is natural from the physical point of view, then conditions for displacement convexity will be in the end quite more stringent than just convexity of U; (iii) the convexity of U automatically implies the nonnegativity of the pressure, since $p(r) = r\,U'(r) - U(r) = r\,U'(r) - U(r) + U(0) \geq 0$. For simplicity I shall also impose that U is twice continuously differentiable *everywhere* in $(0,+\infty)$. Finally, I shall assume that ψ in (16.11) is C^2, and I shall avoid the discussion about the domain of definition of U_ν by just considering compactly supported probability measures. Then, from (15.13) and (14.51),

$$\mathrm{Hess}_\mu U_\nu(\dot\mu) = \int_M \Gamma_2(\psi)\, p(\rho)\, d\nu + \int_M (L\psi)^2\, p_2(\rho)\, d\nu \qquad (16.13)$$

$$\geq \int_M \mathrm{Ric}_{N,\nu}(\nabla\psi)\, p(\rho)\, d\nu + \int_M (L\psi)^2 \left[p_2 + \frac{p}{N}\right](\rho)\, d\nu \qquad (16.14)$$

$$\geq K \int_M |\nabla\psi|^2\, p(\rho)\, d\nu + \int_M (L\psi)^2 \left[p_2 + \frac{p}{N}\right](\rho)\, d\nu. \qquad (16.15)$$

To get a bound on this expression, it is natural to assume that

$$p_2 + \frac{p}{N} \geq 0. \qquad (16.16)$$

The set of all functions U for which (16.16) is satisfied will be called the **displacement convexity class of dimension** N and denoted by $\mathcal{DC}_N$. A typical representative of $\mathcal{DC}_N$, for which (16.16) holds as an equality, is $U = U_N$ defined by

$$U_N(\rho) = \begin{cases} -N\left(\rho^{1-\frac{1}{N}} - \rho\right) & (1 < N < \infty) \\ \\ \rho\log\rho & (N = \infty). \end{cases} \qquad (16.17)$$

These functions will come back again and again in the sequel, and the associated functionals will be denoted by $H_{N,\nu}$.

If inequality (16.16) holds true, then

$$\mathrm{Hess}_\mu U_\nu \geq K\Lambda_U,$$

where

$$\Lambda_U(\mu,\dot\mu) = \int_M |\nabla\psi|^2\, p(\rho)\, d\nu. \qquad (16.18)$$

So the conclusion is as follows:

Guess 16.6. *Let M be a Riemannian manifold satisfying a curvature-dimension bound* $\mathrm{CD}(K,N)$ *for some $K \in \mathbb{R}$, $N \in (1,\infty]$, and let U satisfy* (16.16)*; then U_ν is $K\Lambda_U$-displacement convex.*

Actually, there should be an equivalence between the two statements in Guess 16.6. To see this, assume that U_ν is $K\Lambda_U$-displacement convex; pick up an arbitrary point $x_0 \in M$, and a tangent vector $v_0 \in T_{x_0}M$; consider the particular function $U = U_N$, a probability measure μ which is very much concentrated close to x_0, and a function ψ such that $\nabla\psi(x_0) = v_0$ and $\Gamma_2(\psi) + (L\psi)^2/N = \mathrm{Ric}_{N,\nu}(v_0)$ (as in the proof of Theorem 14.8). Then, on the one hand,

$$K\Lambda_U(\mu,\dot\mu) = K\int |\nabla\psi|^2\,\rho^{1-\frac{1}{N}}\,d\nu \simeq K|v_0|^2\int \rho^{1-\frac{1}{N}}\,d\nu; \qquad (16.19)$$

on the other hand, by the choice of U,

$$\mathrm{Hess}_\mu\, U_\nu(\dot\mu) = \int \left[\Gamma_2(\psi) + \frac{(L\psi)^2}{N}\right]\rho^{1-\frac{1}{N}}\,d\nu,$$

but then since μ is concentrated around x_0, this is well approximated by

$$\left[\Gamma_2(\psi) + \frac{(L\psi)^2}{N}\right](x_0)\int \rho^{1-\frac{1}{N}}\,d\nu = \mathrm{Ric}_{N,\nu}(v_0)\int \rho^{1-\frac{1}{N}}\,d\nu.$$

Comparing the latter expression with (16.19) shows that $\mathrm{Ric}_{N,\nu}(v_0) \geq K|v_0|^2$. Since x_0 and v_0 were arbitrary, this implies $\mathrm{Ric}_{N,\nu} \geq K$. Note that this reasoning only used the functional $H_{N,\nu} = (U_N)_\nu$, and probability measures μ that are very concentrated around a given point.

This heuristic discussion is summarized in the following:

Guess 16.7. *If, for each $x_0 \in M$, $H_{N,\nu}$ is $K\Lambda_U$-displacement convex when applied to probability measures that are supported in a small neighborhood of x_0, then M satisfies the* $\mathrm{CD}(K,N)$ *curvature-dimension bound.*

Example 16.8. Condition $\mathrm{CD}(0,\infty)$ with $\nu = \mathrm{vol}$ just means $\mathrm{Ric} \geq 0$, and the statement $U \in \mathcal{DC}_\infty$ just means that the iterated pressure p_2 is nonnegative. The typical example is when $U(\rho) = \rho\log\rho$, and then the corresponding functional is

$$H(\mu) = \int \rho\log\rho\, d\mathrm{vol}, \qquad \mu = \rho\,\mathrm{vol}.$$

Then the above considerations suggest that the following statements are equivalent:

(i) $\mathrm{Ric} \geq 0$;

(ii) If the nonlinearity U is such that the nonnegative iterated pressure p_2 is nonnegative, then the functional U_{vol} is displacement convex;

(iii) H is displacement convex;

(iii') For any $x_0 \in M$, the functional H is displacement convex when applied to probability measures that are supported in a small neighborhood of x_0.

Example 16.9. The above considerations also suggest that the inequality $\mathrm{Ric} \geq Kg$ is equivalent to the K-displacement convexity of the H functional, whatever the value of $K \in \mathbb{R}$.

These guesses will be proven and generalized in the next chapter.

A fluid mechanics feeling for Ricci curvature

Ricci curvature is familiar to physicists because it plays a crucial role in Einstein's theory of general relativity. But what we have been discovering in this chapter is that Ricci curvature can also be given a physical interpretation in terms of *classical fluid mechanics*. To provide the reader with a better feeling of this new point of view, let us imagine how two physicists, the first one used to relativity and light propagation, the second one used to fluid mechanics, would answer the following question: *Describe in an informal way an experiment that can determine whether we live in a nonnegatively Ricci-curved space.*

The light source test: Take a small light source, and try to determine its volume by looking at it from a distant position. If you systematically overestimate the volume of the light source, then you live in a nonnegatively curved space (recall Figure 14.4).

The lazy gas experiment: Take a perfect gas in which particles do not interact, and ask him to move from a certain prescribed density field at time $t = 0$, to another prescribed density field at time $t = 1$. Since the gas is lazy, he will find a way to do so that needs a minimal amount of work (least action path). Measure the entropy of the gas at each time, and check that it always lies *above* the line joining the final and initial entropies. If such is the case, then we know that we live in a nonnegatively curved space (see Figure 16.2).

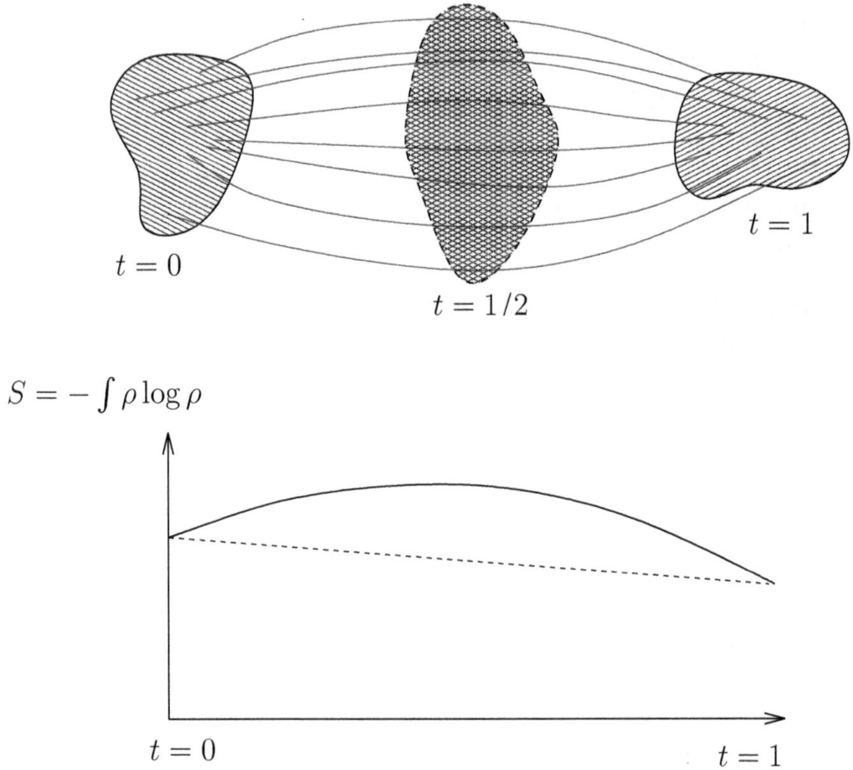

Fig. 16.2. The lazy gas experiment: To go from state 0 to state 1, the lazy gas uses a path of least action. In a nonnegatively curved world, the trajectories of the particles first diverge, then converge, so that at intermediate times the gas can afford to have a lower density (higher entropy).

Bibliographical notes

Convexity has been extensively studied in the Euclidean space [705] and in Banach spaces [172, 324]. I am not aware of textbooks where the study of convexity in more general geodesic spaces is developed, although this notion is now quite frequently used (in the context of optimal transport, see, e.g., [30, p. 50]).

The concept and terminology of displacement convexity were introduced by McCann in the mid-nineties [614]. He identified (16.16) as the basic criterion for convexity in $P_2(\mathbb{R}^n)$, and also discussed other formulations of this condition, which will be studied in the next chapter. Inequality (16.16) was later rediscovered by several authors, in various contexts.

The application of Otto calculus to the study of displacement convexity goes back to [669] and [671]. In the latter reference it was conjectured that nonnegative Ricci curvature would imply displacement convexity of H.

Ricci curvature appears explicitly in Einstein's equations, and will be encountered in any mildly advanced book on general relativity. Fluid mechanics analogies for curvature appear explicitly in the work by Cordero-Erausquin, McCann and Schmuckenschläger [246].

Lott [576] recently pointed out some interesting properties of displacement convexity for functionals explicitly depending on $t \in [0,1]$; for instance the convexity of $t \int \rho_t \log \rho_t \, d\nu + N\, t \log t$ along displacement interpolation is a characterization of $\mathrm{CD}(0,N)$. The Otto formalism is also useful here.

17

Displacement convexity II

In Chapter 16, a conjecture was formulated about the links between displacement convexity and curvature-dimension bounds; its plausibility was justified by some formal computations based on Otto's calculus. In the present chapter I shall provide a rigorous justification of this conjecture. For this I shall use a Lagrangian point of view, in contrast with the Eulerian approach used in the previous chapter. Not only is the Lagrangian formalism easier to justify, but it will also lead to new curvature-dimension criteria based on so-called "distorted displacement convexity".

The main results in this chapter are Theorems 17.15 and 17.37.

Displacement convexity classes

What I shall call displacement convexity class of order N is a family of convex nonlinearities satisfying a certain characteristic differential inequality of second order (recall (16.16)).

Definition 17.1 (Displacement convexity classes). *Let N be a real parameter in $[1,\infty]$. The class $\mathcal{DC}_N$ is defined as the set of continuous convex functions $U:\mathbb{R}_+\to\mathbb{R}$, twice continuously differentiable on $(0,+\infty)$, such that $U(0)=0$, and, with the notation*

$$p(r) = r\,U'(r) - U(r), \qquad p_2(r) = r\,p'(r) - p(r),$$

U satisfies any one of the following equivalent differential conditions:

(i) $p_2 + \dfrac{p}{N} \geq 0$;

C. Villani, *Optimal Transport*, Grundlehren der mathematischen Wissenschaften 338,

(ii) $\dfrac{p(r)}{r^{1-1/N}}$ *is a nondecreasing function of* r;

(iii) $u(\delta) := \left\{\begin{matrix} \delta^N\,U(\delta^{-N}) & (\delta > 0) & \text{if } N < \infty \\ e^\delta\,U(e^{-\delta}) & (\delta \in \mathbb{R}) & \text{if } N = \infty \end{matrix}\right\}$ *is a convex function of* δ.

Remark 17.2. Since U is convex and $U(0) = 0$, the function u appearing in (iii) is automatically nonincreasing.

Remark 17.3. It is clear (from condition (i) for instance) that $\mathcal{DC}_{N'} \subset \mathcal{DC}_N$ if $N' \geq N$. So the smallest class of all is $\mathcal{DC}_\infty$, while $\mathcal{DC}_1$ is the largest (actually, conditions (i)–(iii) are void for $N = 1$).

Remark 17.4. If U belongs to $\mathcal{DC}_N$, then for any $a \geq 0$, $b > 0$, $c \in \mathbb{R}$, the function $r \longmapsto a\,U(br) + cr$ also belongs to $\mathcal{DC}_N$.

Remark 17.5. The requirement for U to be twice differentiable on $(0, +\infty)$ could be removed from many subsequent results involving displacement convexity classes. Still, this regularity assumption will simplify the proofs, without significantly restricting the generality of applications.

Examples 17.6. (i) For any $\alpha \geq 1$, the function $U(r) = r^\alpha$ belongs to all classes $\mathcal{DC}_N$.

(ii) If $\alpha < 1$, then the function $U(r) = -r^\alpha$ belongs to $\mathcal{DC}_N$ if and only if $N \leq (1-\alpha)^{-1}$ (that is, $\alpha \geq 1 - 1/N$). The function $-r^{1-1/N}$ is in some sense the minimal representative of $\mathcal{DC}_N$.

(iii) The function $U_\infty(r) = r \log r$ belongs to $\mathcal{DC}_\infty$. It can be seen as the limit of the functions $U_N(r) = -N(r^{1-1/N} - r)$, which are the same (up to multiplication and addition of a linear function) as the functions appearing in (ii) above.

Proof of the equivalence in Definition 17.1. Assume first $N < \infty$, and write $r(\delta) = \delta^{-N}$. By computation, $u'(\delta) = -Np(r)/r^{1-1/N}$. So u is convex if and only if $p(r)/r^{1-1/N}$ is a nonincreasing function of δ, i.e. a nondecreasing function of r. Thus (ii) and (iii) are equivalent.

Next, by computation again,

$$u''(\delta) = N^2 r^{\frac{2}{N}-1}\left(p_2(r) + \frac{p(r)}{N}\right). \tag{17.1}$$

So u is convex if and only if $p_2 + p/N$ is nonnegative. This shows the equivalence between (i) and (iii).

In the case $N = \infty$, the arguments are similar, with the formulas

$$r(\delta) = e^{-\delta}, \qquad u'(\delta) = -\frac{p(r)}{r}, \qquad u''(\delta) = \frac{p_2(r)}{r}.$$

□

The behavior of functions in $\mathcal{DC}_N$ will play an important role in the sequel of this course. Functions in $\mathcal{DC}_N$ may present singularities at the origin; for example $U_N(r)$ is not differentiable at $r = 0$. It is often possible to get around this problem by replacing $U_N(r)$ by a smooth approximation which still belongs to $\mathcal{DC}_N$, for instance $-N\left(r(r+\varepsilon)^{-1/N} - r\right)$, and later passing to the limit as $\varepsilon \to 0$. The next proposition provides more systematic ways to "regularize" functions in $\mathcal{DC}_N$ near 0 or $+\infty$; at the same time it gives additional information about the behavior of functions in $\mathcal{DC}_N$. The notation $p(r)$ and $p_2(r)$ is the same as in Definition 17.1.

Proposition 17.7 (Behavior of functions in $\mathcal{DC}_N$).

(i) Let $N \in [1, \infty)$, and $\Psi \in C(\mathbb{R}_+; \mathbb{R}_+)$ such that $\Psi(r)/r \to +\infty$ as $r \to \infty$; then there exists $U \in \mathcal{DC}_N$ such that $0 \leq U \leq \Psi$, and $U(r)/r \longrightarrow +\infty$ as $r \to \infty$.

(ii) If $U \in \mathcal{DC}_\infty$, then either U is linear, or there exist constants $a > 0$, $b \in \mathbb{R}$ such that

$$\forall r \geq 0, \qquad U(r) \geq a\, r \log r + b\, r.$$

(iii) Let $N \in [1, \infty]$ and let $U \in \mathcal{DC}_N$. If $r_0 \in (0, +\infty)$ is such that $p(r_0) > 0$, then there is a constant $K > 0$ such that $p'(r) \geq K r^{-1/N}$ for all $r \geq r_0$. If on the contrary $p(r_0) = 0$, then U is linear on $[0, r_0]$. In particular, the set $\{r;\ U''(r) = 0\}$ is either empty, or an interval of the form $[0, r_0]$.

(iv) Let $N \in [1, \infty]$ and let $U \in \mathcal{DC}_N$. Then U is the pointwise nondecreasing limit of a sequence of functions $(U_\ell)_{\ell \in \mathbb{N}}$ in $\mathcal{DC}_N$, such that (a) U_ℓ coincides with U on $[0, r_\ell]$, where r_ℓ is arbitrarily large; (b) for each ℓ there are $a \geq 0$ and $b \in \mathbb{R}$ such that $U_\ell(r) = -a\, r^{1-\frac{1}{N}} + b\, r$ (or $a\, r \log r + b\, r$ if $N = \infty$) for r large enough; (c) $U_\ell'(\infty) \to U'(\infty)$ as $\ell \to \infty$.

(v) Let $N \in [1, \infty]$ and let $U \in \mathcal{DC}_N$. Then U is the pointwise nonincreasing limit of a sequence of functions $(U_\ell)_{\ell \in \mathbb{N}}$ in $\mathcal{DC}_N$, such

that (a) U_ℓ coincides with U on $[r_\ell, +\infty)$, where r_ℓ is an arbitrary real number such that $p'(r_\ell) > 0$; (b) $U_\ell(r)$ is a linear function of r close to the origin; (c) $(U_\ell)'(0) \to U'(0)$ as $\ell \to \infty$.

(vi) In statements (iv) and (v), one can also impose that $U_\ell'' \le C\, U''$, for some constant C independent of ℓ. In statement (v), one can also impose that U_ℓ'' increases nicely from 0, in the following sense: If $[0, r_0]$ is the interval where $U_\ell'' = 0$, then there are $r_1 > r_0$, an increasing function $h : [r_0, r_1] \to \mathbb{R}_+$, and constants K_1, K_2 such that $K_1\, h \le U'' \le K_2\, h$ on $[r_0, r_1]$.

(vii) Let $N \in [1, \infty]$ and let $U \in \mathcal{DC}_N$. Then there is a sequence $(U_\ell)_{\ell \in \mathbb{N}}$ of functions in $\mathcal{DC}_N$ such that $U_\ell \in C^\infty((0, +\infty))$, U_ℓ converges to U monotonically and in $C^2_{\rm loc}((0, +\infty))$; and, with the notation $p_\ell(r) = r\, U_\ell'(r) - U_\ell(r)$,

$$\inf_r \frac{p_\ell(r)}{r^{1-\frac{1}{N}}} \xrightarrow[\ell\to\infty]{} \inf_r \frac{p(r)}{r^{1-\frac{1}{N}}}; \qquad \sup_r \frac{p_\ell(r)}{r^{1-\frac{1}{N}}} \xrightarrow[\ell\to\infty]{} \sup_r \frac{p(r)}{r^{1-\frac{1}{N}}}.$$

Here are some comments about these results. Statements (i) and (ii) show that functions in $\mathcal{DC}_N$ can grow as slowly as desired at infinity if $N < \infty$, but have to grow at least like $r \log r$ if $N = \infty$. Statements (iv) to (vi) make it possible to write any $U \in \mathcal{DC}_N$ as a monotone limit (nonincreasing for small r, nondecreasing for large r) of "very nice" functions $U_\ell \in \mathcal{DC}_N$, which behave linearly close to 0 and like $b\,r - a\,r^{1-1/N}$ (or $a\,r \log r + b\,r$) at infinity (see Figure 17.1). This approximation scheme makes it possible to extend many results which can be proven for very nice nonlinearities, to general nonlinearities in $\mathcal{DC}_N$.

The proof of Proposition 17.7 is more tricky than one might expect, and it is certainly better to skip it at first reading.

Proof of Proposition 17.7. The case $N = 1$ is not difficult to treat separately (recall that $\mathcal{DC}_1$ is the class of all convex continuous functions U with $U(0) = 0$ and $U \in C^2((0, +\infty))$). So in the sequel I shall assume $N > 1$. The strategy will always be the same: First approximate u, then reconstruct U from the approximation, thanks to the formula $U(r) = r\, u(r^{-1/N})$ ($r\, u(\log 1/r)$ if $N = \infty$).

Let us start with the proof of (i). Without loss of generality, we may assume that Ψ is identically 0 on $[0, 1]$ (otherwise, replace Ψ by $\chi \Psi$, where $0 \le \chi \le 1$ and χ is identically 0 on $[0, 1]$, identically 1 on $[2, +\infty)$). Define a function $u : (0, +\infty) \to \mathbb{R}$ by

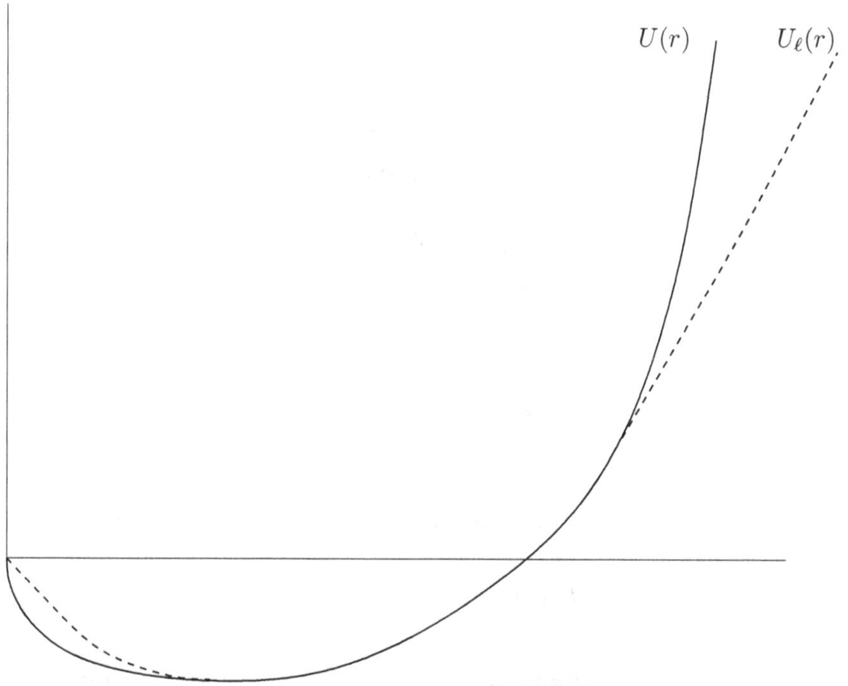

Fig. 17.1. U_ℓ (dashed line) is an approximation of U (solid line); it is linear close to the origin and almost affine at infinity. This regularization can be made without going out of the class $\mathcal{DC}_N$, and without increasing too much the second derivative of U.

$$u(\delta) = \delta^N \Psi(\delta^{-N}).$$

Then $u \equiv 0$ on $[1, +\infty)$, and $\lim_{\delta \to 0^+} u(\delta) = +\infty$.

The problem now is that u is not necessarily convex. So let $\widetilde{u}$ be the lower convex hull of u on $(0, \infty)$, i.e. the supremum of all linear functions bounded above by u. Then $\widetilde{u} \equiv 0$ on $[1, \infty)$ and $\widetilde{u}$ is nonincreasing. Necessarily,

$$\lim_{\delta \to 0^+} \widetilde{u}(\delta) = +\infty. \tag{17.2}$$

Indeed, suppose on the contrary that $\lim_{\delta \to 0^+} \widetilde{u}(\delta) = M < +\infty$, and let $a \in \mathbb{R}$ be defined by $a := \sup_{\delta \geq 0} \frac{M+1-u(\delta)}{\delta}$ (the latter function is nonpositive when δ is small enough, so the supremum is finite). Then $u(\delta) \geq M + 1 - a\delta$, so $\lim_{\delta \to 0^+} \widetilde{u}(\delta) \geq M + 1$, which is a contradiction. Thus (17.2) does hold true.

Then let

$$U(r) := r\, \widetilde{u}(r^{-1/N}).$$

Clearly U is continuous and nonnegative, with $U \equiv 0$ on $[0,1]$. By computation, $U''(r) = N^{-2} r^{-1-1/N} (r^{-1/N} \widetilde{u}''(r^{-1/N}) - (N-1)\, \widetilde{u}'(r^{-1/N}))$. As $\widetilde{u}$ is convex and nonincreasing, it follows that U is convex. Hence $U \in \mathcal{DC}_N$. On the other hand, since $\widetilde{u} \le u$ and $\Psi(r) = r\, u(r^{-1/N})$, it is clear that $U \le \Psi$; and still (17.2) implies that $U(r)/r$ goes to $+\infty$ as $r \to \infty$.

Now consider Property (ii). If $N = \infty$, then the function U can be reconstructed from u by the formula

$$U(r) = r\, u(\log(1/r)), \tag{17.3}$$

As u is convex and nonincreasing, either u is constant (in which case U is linear), or there are constants $a > 0$, $b \in \mathbb{R}$, such that $u(\delta) \ge -a\delta + b$, and then $U(r) \ge -a\, r \log(1/r) + b\, r = a\, r \log r + b\, r$.

Next let us turn to (iii). First assume $N < \infty$. The formula

$$p(r) = r\, U'(r) - U(r) = -\frac{1}{N}\, r^{1-\frac{1}{N}}\, u'(r^{-\frac{1}{N}})$$

shows that $p(r_0) > 0$ if and only if $u'(r_0^{-1/N}) < 0$. Then for any $r \le r_0$, $u'(r^{-1/N}) \le u'(r_0^{-1/N})$, so

$$\begin{aligned} p'(r) = r\, U''(r) &= \frac{r^{-\frac{1}{N}}}{N^2} \Big(r^{-\frac{1}{N}} u''(r^{-\frac{1}{N}}) - (N-1)\, u'(r^{-\frac{1}{N}}) \Big) \\ &\ge - \left(\frac{(N-1)\, u'(r_0^{-\frac{1}{N}})}{N^2} \right) r^{-\frac{1}{N}}. \end{aligned}$$

If on the other hand $u'(r_0^{-1/N}) = 0$, then necessarily $u'(r^{-1/N}) = 0$ for all $r \le r_0$, which means that u is constant on $[r_0^{-1/N}, +\infty)$, so U is linear on $[0, r_0]$.

The reasoning is the same in the case $N = \infty$, with the help of the formulas

$$p(r) = -r\, u'\Big(\log \frac{1}{r}\Big), \qquad U''(r) = \frac{1}{r}\Big(u''\Big(\log\frac{1}{r}\Big) - u'\Big(\log\frac{1}{r}\Big)\Big)$$

and

$$r \ge r_0 \Longrightarrow \qquad p'(r) = r\, U''(r) \ge -u'\Big(\log \frac{1}{r_0}\Big).$$

Now consider statement (iv). The idea is to replace u by an affine function close to the origin, essentially by smoothing of the trivial C^1 approximation by the tangent. First let $N \in [1, \infty)$, let $U \in \mathcal{DC}_N$ and let $u(\delta) = \delta^N U(\delta^{-N})$. We know that u is a nonincreasing, twice differentiable convex function on $(0, +\infty)$. If u is linear close to the origin, there is nothing to prove. Otherwise there is a sequence of positive numbers $(a_\ell)_{\ell \in \mathbb{N}}$ such that $a_{\ell+1} \le a_\ell/4$ and $u'(a_{\ell+1}) < u'(a_\ell/2) < 0$. For each ℓ, construct a C^2 function u_ℓ as follows:

- on $[a_\ell, +\infty)$, u_ℓ coincides with u;
- on $[0, a_\ell]$, $u''_\ell = \chi_\ell\, u''$, where χ_ℓ is a smooth cutoff function such that $0 \le \chi_\ell \le 1$, $\chi_\ell(a_\ell) = 1$, $\chi_\ell(\delta) = 0$ for $\delta \le a_\ell/2$.

Since u_ℓ is convex and $u'_\ell(a_\ell) < 0$, also $u'_\ell < 0$ on $(0, a_\ell]$. By construction, u_ℓ is linear on $[0, a_\ell/2]$. Also $u''_\ell \le u''$, $u'_\ell(a_\ell) = u'(a_\ell)$, $u_\ell(a_\ell) = u(a_\ell)$; by writing the Taylor formula on $[s, a_\ell]$ (with $1/\ell$ as base point), we deduce that $u_\ell(s) \le u(s)$, $u'_\ell(s) \ge u'(s)$ for all $s \le a_\ell$ (and therefore for all s).

For each ℓ, u_ℓ lies above the tangent to u at $1/\ell$; that is

$$u_\ell(s) \ge u(a_\ell) + (s - a_\ell)\, u'(a_\ell) =: T_\ell(s).$$

Since u' is nondecreasing and $u'(a_{\ell+1}) < u'(a_\ell/2)$, the curve T_ℓ lies *strictly* below the curve $T_{\ell+1}$ on $[0, a_\ell/2]$, and therefore on $[0, a_{\ell+1}]$. By choosing χ_ℓ in such a way that $\int_{a_\ell/2}^{a_\ell} \chi_\ell u''$ is very small, we can make sure that u_ℓ is very close to the line $T_\ell(s)$ on $[0, a_\ell]$; and in particular that the whole curve u_ℓ is bounded above by $T_{\ell+1}$ on $[a_{\ell+1}, a_\ell]$. This will ensure that u_ℓ is a nondecreasing function of ℓ.

To recapitulate: $u_\ell \le u$; $u_{\ell+1} \le u_\ell$; $u_\ell = u$ on $[a_\ell, +\infty)$; $0 \le u''_\ell \le u''$; $0 \ge u'_\ell \ge u'$; u_ℓ is affine on $[0, a_\ell/2]$.

Now let

$$U_\ell(r) = r\, u_\ell(r^{-1/N}).$$

By direct computation,

$$U''_\ell(r) = \frac{r^{-1-\frac{1}{N}}}{N^2} \left(r^{-\frac{1}{N}} u''_\ell(r^{-\frac{1}{N}}) - (N-1)\, u'_\ell(r^{-\frac{1}{N}}) \right). \tag{17.4}$$

Since u_ℓ is convex nonincreasing, the above expression is nonnegative; so U_ℓ is convex, and by construction, it lies in $\mathcal{DC}_N$. Moreover U_ℓ satisfies the first requirement in (vi), since $U''_\ell(r)$ is bounded above by $(r^{-1-1/N}/N^2)\,(r^{-1/N} u''(r^{-1/N}) - (N-1)\, u'(r^{-1/N})) = U''(r)$.

In the case $N = \infty$, things are similar, except that now u is defined on the whole of $\mathbb{R}$, the sequence a_ℓ converges to $-\infty$ (say $a_{\ell+1} \le 2a_\ell$), and one should use the formulas

$$U_\ell(r) = r\, u_\ell(\log 1/r); \qquad U_\ell''(r) = \frac{1}{r}\Big(u_\ell''\Big(\log\frac{1}{r}\Big) - u_\ell'\Big(\log\frac{1}{r}\Big)\Big).$$

For (v), the idea is to replace u by a *constant* function for large values of δ. But this cannot be done in a C^1 way, so the smoothing turns out to be more tricky. (Please consider again possibly skipping the rest of this proof.)

I shall distinguish four cases, according to the behavior of u at infinity, and the value of $u'(+\infty) = \lim_{s\to+\infty} u'(s)$. To fix ideas I shall assume that $N < \infty$; but the case $N = \infty$ can be treated similarly. In each case I shall also check the first requirement of (vi), which is $U_\ell'' \le C\, U''$.

Case 1: u is affine at infinity and $u'(+\infty) = 0$. This means that $u(\delta) = c$ for $\delta \ge \delta_0$ large enough, where c is some constant. Then $U(r) = r\, u(r^{-1/N}) = c\, r$ for $r \le \delta_0^{-N}$, and there is nothing to prove.

Case 2: u is affine at infinity and $u'(+\infty) < 0$. Let $a := -u'(+\infty)$, so $u' \le -a$. By assumption there are $\delta_0 > 0$ and $b \in \mathbb{R}$ such that $u(s) = -as+b$ for $s \ge \delta_0$. Let $a_1 \ge \max(1, \delta_0)$. I shall define recursively an increasing sequence $(a_\ell)_{\ell\in\mathbb{N}}$, and C^2 functions u_ℓ such that:

- on $[0, a_\ell]$, u_ℓ coincides with u;
- on $[a_\ell, +\infty)$, $u_\ell''(s) = \chi_\ell(s)/s$, where χ_ℓ is a continuous function with compact support in $(a_\ell, +\infty)$, $0 \le \chi_\ell \le 1$. (So u_ℓ is obtained by integrating this twice, and ensuring the C^1 continuity at $s = \ell$; note that $u''(a_\ell) = 0$, so the result will be C^2.)

Let us choose χ_ℓ to be supported in some interval (a_ℓ, b_ℓ), such that

$$\int_{a_\ell}^{+\infty} \frac{\chi_\ell(s)}{s}\, ds = a.$$

Such a χ_ℓ exists since $\int_{a_\ell}^{+\infty} ds/s = +\infty$. Then we let $a_{\ell+1} \ge b_\ell + 1$.

The function u_ℓ is convex by construction, and affine at infinity; moreover, $u_\ell'(+\infty) = u'(a_{\ell+1}) = u'(a_\ell) + \int_{a_\ell}^{+\infty} \chi_\ell(s)\, ds/s = 0$, so u_ℓ is actually constant at infinity and $u_\ell' \le 0$. Obviously $u_\ell'' \ge u''$, so $u_\ell' \ge u'$ and $u_\ell \ge u$. Also, on $[a_{\ell+1}, +\infty)$, $u_{\ell+1} \le u(a_{\ell+1}) \le u_\ell(a_{\ell+1}) \le u_\ell$, while on $[0, a_{\ell+1}]$, $u_{\ell+1} = u \le u_\ell$; so the sequence (u_ℓ) is nonincreasing in ℓ.

Let $U_\ell(r) = r\,u_\ell(r^{-1/N})$. Formula (17.4) shows again that $U_\ell'' \geq 0$, and it is clear that $U_\ell(0) = 0$, $U_\ell \in C(\mathbb{R}_+) \cap C^2((0,+\infty))$; so $U_\ell \in \mathcal{DC}_N$. It is clear also that $U_\ell \geq U$, U_ℓ coincides with U on $[0, a_\ell^{-N}]$, U_ℓ is linear on $[0, b_\ell^{-N}]$, U_ℓ converges monotonically to U as $\ell \to \infty$, and $U_\ell'(0) = u_\ell(+\infty)$ converges to $u(+\infty) = U'(0) = -\infty$.

It only remains to check the bound $U_\ell'' \leq C U''$. This bound is obvious on $[a_\ell^{-N}, +\infty)$; for $r \leq a_\ell^{-N}$ it results from the formulas

$$\begin{aligned} U_\ell''(r) &= \frac{r^{-1-\frac{1}{N}}}{N^2}\Big(r^{-\frac{1}{N}}\chi_\ell(r^{-\frac{1}{N}})\,r^{\frac{1}{N}} - (N-1)\,u_\ell'(r)\Big) \\ &\leq \frac{r^{-1-\frac{1}{N}}}{N^2}(1 + (N-1)a); \end{aligned}$$

$$U''(r) = \left(\frac{N-1}{N^2}\right) a\, r^{-1-\frac{1}{N}}.$$

So $C = 1 + 1/((N-1)a)$ is admissible.

Case 3: u is not affine at infinity and $u'(+\infty) = 0$. The proof is based again on the same principle, but modified as follows:

- on $[0, a_\ell]$, u_ℓ coincides with u;
- on $[a_\ell, +\infty)$, $u_\ell''(s) = \zeta_\ell(s)\,u''(s)$, where ζ_ℓ is a smooth function identically equal to 1 close to a_ℓ, identically equal to 0 at infinity, with values in $[0, 2]$.

Choose $a_\ell < b_\ell < c_\ell$, and ζ_ℓ supported in $[a_\ell, c_\ell]$, so that $1 \leq \zeta_\ell \leq 2$ on $[a_\ell, b_\ell]$, $0 \leq \zeta_\ell \leq 2$ on $[b_\ell, c_\ell]$, and

$$\int_{a_\ell}^{b_\ell} \zeta_\ell(s)\,u''(s)\,ds > u'(b_\ell) - u'(a_\ell); \qquad \int_{a_\ell}^{c_\ell} \zeta_\ell(s)\,u''(s)\,ds = -u'(a_\ell).$$

This is possible since u' and u'' are continuous and $\int_{a_\ell}^{+\infty} (2u''(s))\,ds = 2(u'(+\infty) - u'(a_\ell)) > u'(+\infty) - u'(a_\ell) > 0$ (otherwise u would be affine on $[a_\ell, +\infty)$). Then choose $a_{\ell+1} > c_\ell + 1$.

The resulting function u_ℓ is convex and it satisfies $u_\ell'(+\infty) = u'(a_\ell) - u'(a_\ell) = 0$, so $u_\ell' \leq 0$ and u_ℓ is constant at infinity.

On $[a_\ell, b_\ell]$, $u_\ell'' \geq u''$, so $u_\ell' \geq u'$ and $u_\ell \geq u$, and these inequalities are strict at b_ℓ. Since u' and u'' are continuous, we can always arrange that b_ℓ is so close to c_ℓ that the inequalities $u_\ell \geq u$ and $u_\ell' \geq u'$ hold true on $[b_\ell, c_\ell]$. Then these inequalities will also hold true on $[c_\ell, +\infty)$ since u_ℓ is constant there, and u is nonincreasing.

Define $U_\ell(r) = r\, u_\ell(r^{-1/N})$. The same reasoning as in the previous case shows that U_ℓ lies in $\mathcal{DC}_N$, $U_\ell \geq U$, U_ℓ is linear on $[0, c_\ell^{-N}]$, U_ℓ converges monotonically to U as $\ell \to \infty$, and $U_\ell'(0) = u_\ell(+\infty)$ converges to $u(+\infty) = U'(0)$. The sequence (U_ℓ) satisfies all the desired properties; in particular the inequalities $u_\ell'' \leq 2u''$ and $u_\ell' \geq u'$ ensure that $U_\ell'' \leq 2U''$.

Case 4: u is not affine at infinity and $u'(+\infty) < 0$. In this case the proof is based on the same principle, and u_ℓ is defined as follows:

- on $[0, a_\ell]$, u_ℓ coincides with u;
- on $[a_\ell, +\infty)$, $u_\ell''(s) = \eta_\ell(s)\, u''(s) + \chi_\ell(s)/s$, where χ_ℓ and η_ℓ are both valued in $[0, 1]$, χ_ℓ is a smooth cutoff function with compact support in $(a_\ell, +\infty)$, and η_ℓ is a smooth function identically equal to 1 close to a_ℓ, and identically equal to 0 close to infinity.

To construct these functions, first choose $b_\ell > a_\ell$ and χ_ℓ supported in $[a_\ell, b_\ell]$ in such a way that

$$\int_{a_\ell}^{b_\ell} \frac{\chi_\ell(s)}{s}\, ds = -\left(\frac{u'(b_\ell) + u'(+\infty)}{2}\right).$$

This is always possible since $\int_{a_\ell}^{+\infty} ds/s = +\infty$, u' is continuous and $-(u'(b_\ell) + u'(+\infty))/2$ approaches the finite limit $-u'(+\infty)$ as $b_\ell \to +\infty$.

Then choose $c_\ell > b_\ell$, and η_ℓ supported in $[a_\ell, c_\ell]$ such that $\eta_\ell = 1$ on $[a_\ell, b_\ell]$ and

$$\int_{b_\ell}^{c_\ell} \eta_\ell\, u'' = \frac{u'(+\infty) - u'(b_\ell)}{2}.$$

This is always possible since $\int_{b_\ell}^{+\infty} u''(s)\, ds = u'(+\infty) - u'(b_\ell) > [u'(+\infty) - u'(b_\ell)]/2 > 0$ (otherwise u would be affine on $[b_\ell, +\infty)$). Finally choose $a_{\ell+1} \geq c_\ell + 1$.

The function u_ℓ so constructed is convex, affine at infinity, and

$$u_\ell'(+\infty) = u'(a_\ell) + \int_{a_\ell}^{b_\ell} u'' + \int_{a_\ell}^{b_\ell} \frac{\chi_\ell(s)}{s}\, ds + \int_{b_\ell}^{c_\ell} \eta_\ell\, u'' = 0.$$

So u_ℓ is actually constant at infinity, and $u_\ell' \leq 0$.

On $[a_\ell, b_\ell]$, $u_\ell'' \geq u''$, $u_\ell'(a_\ell) = u'(a_\ell)$, $u_\ell(a_\ell) = u(a_\ell)$; so $u_\ell' \geq u'$ and $u_\ell \geq u$ on $[a_\ell, b_\ell]$.

On $[b_\ell, +\infty)$, one has $u_\ell' \geq u_\ell'(b_\ell) = (u'(b_\ell) - u'(+\infty))/2 \geq u'(+\infty)$ if $u'(b_\ell) \geq 3u'(+\infty)$. We can always ensure that this inequality holds true by choosing a_1 large enough that $u'(a_1) \geq 3u'(+\infty)$. Then $u_\ell(s) \geq u_\ell(b_\ell) + u'(+\infty)\,(s - b_\ell) \geq u_\ell(b_\ell) + \int_{b_\ell}^s u' = u(s)$; so $u_\ell \geq u$ also on $[b_\ell, +\infty)$.

Define $U_\ell(r) = r\, u_\ell(r^{-1/N})$. All the desired properties of U_ℓ can be shown just as before, except for the bound on U_ℓ'', which we shall now check. On $[a_\ell^{-N}, +\infty)$, $U_\ell'' = U''$. On $[0, a_\ell^{-N})$, with the notation $a = -u'(+\infty)$, we have $u_\ell'(r^{-1/N}) \geq -a$, $u'(r^{-1/N}) \geq -3a$ (recall that we imposed $u'(a_1) \geq -3a$), so

$$U_\ell''(r) \leq \frac{r^{-1-\frac{1}{N}}}{N^2}\Big(r^{-\frac{1}{N}} u''(r^{-\frac{1}{N}}) + 1 + 3(N-1)a\Big),$$

$$U''(r) \geq \frac{r^{-1-\frac{1}{N}}}{N^2}\Big(r^{-\frac{1}{N}} u''(r^{-\frac{1}{N}}) + (N-1)a\Big),$$

and once again $U_\ell'' \leq C U''$ with $C = 3 + 1/((N-1)a)$.

It remains to prove the second part of (vi). This will be done by a further approximation scheme. So let $U \in \mathcal{DC}_N$ be linear close to the origin. (We can always reduce to this case by (v).) If U is linear on the whole of $\mathbb{R}_+$, there is nothing to do. Otherwise, by (iii), the set where U'' vanishes is an interval $[0, r_0]$. The goal is to show that we may approximate U by U_ℓ in such a way that $U_\ell \in \mathcal{DC}_N$, U_ℓ is nonincreasing in ℓ, U_ℓ is linear on some interval $[0, r_0(\ell)]$, and U_ℓ'' increases nicely from 0 on $[r_0(\ell), r_1(\ell))$.

In this case, u is a nonincreasing function, identically equal to a constant on $[s_0, +\infty)$, with $s_0 = r_0^{-1/N}$; and also u' is nonincreasing to 0, so in fact u is strictly decreasing up to s_0. Let $a_1 \in (s_0/2, s_0)$. We can recursively define real numbers a_ℓ and C^2 functions u_ℓ as follows:

- on $(0, a_\ell]$, u_ℓ coincides with u;
- on $[a_\ell, +\infty)$, $(u_\ell)'' = \chi_\ell\, u'' + \eta_\ell(-u')$, where χ_ℓ and η_ℓ are smooth functions valued in $[0, 2]$, $\chi_\ell(r)$ is identically equal to 1 for r close to a_ℓ, and identically equal to 0 for $r \geq b_\ell$; and η_ℓ is compactly supported in $[b_\ell, c_\ell]$ and decreasing to 0 close to c_ℓ; $a_\ell < b_\ell < c_\ell < s_0$.

Let us choose χ_ℓ, η_ℓ, b_ℓ, c_ℓ in such a way that

$$\int_{a_\ell}^{b_\ell} \chi_\ell\, u'' > \int_{a_\ell}^{b_\ell} u''; \qquad \int_{a_\ell}^{b_\ell} \chi_\ell u'' + \int_{b_\ell}^{c_\ell} \eta_\ell(-u') = -u_\ell'(a_\ell);$$

$$\int_{b_\ell}^{c_\ell} \eta_\ell(-u') > 0.$$

This is possible since u', u'' are continuous, $\int_{a_\ell}^{s_0}(2u'') = -2u_\ell'(a_\ell) > -u_\ell'(a_\ell)$, and $(-u')$ is strictly positive on $[a_\ell, s_0]$. It is clear that $u_\ell \geq u$ and $u_\ell' \geq u'$ on $[a_\ell, b_\ell]$, with strict inequalities at b_ℓ; by choosing c_ℓ very close to b_ℓ, we can make sure that these inequalities are preserved on $[b_\ell, c_\ell]$. Then we choose $a_{\ell+1} = (c_\ell + s_0)/2$.

Let us check that $U_\ell(r) := r\, u_\ell(r^{-1/N})$ satisfies all the required properties. To bound U_ℓ'', note that for $r \in [s_0^{-N}, (s_0/2)^{-N}]$,

$$\begin{aligned} U_\ell''(r) &\leq C(N, r_0)\left(u_\ell''(r^{-1/N}) - u_\ell'(r^{-1/N})\right) \\ &\leq 2C(N, r_0)\left(u''(r^{-1/N}) - u'(r^{-1/N})\right) \end{aligned}$$

and

$$U''(r) \geq K(N, r_0)\left(u''(r^{-1/N}) - u'(r^{-1/N})\right),$$

where $C(N, r_0)$, $K(N, r_0)$ are positive constants. Finally, on $[b_\ell, c_\ell]$, $u_\ell'' = \eta_\ell(-u')$ is decreasing close to c_ℓ (indeed, η_ℓ is decreasing close to c_ℓ, and $-u'$ is positive nonincreasing); and of course $-u_\ell'$ is decreasing as well. So $u_\ell''(r^{-1/N})$ and $-u_\ell'(r^{-1/N})$ are increasing functions of r in a small interval $[r_0, r_1]$. This concludes the argument.

To prove (vi), we may first approximate u by a C^∞ convex, nonincreasing function u_ℓ, in such a way that $\|u - u_\ell\|_{C^2((a,b))} \to 0$ for any $a, b > 0$. This can be done in such a way that $u_\ell(s)$ is nondecreasing for small s and nonincreasing for large s; and $u_\ell'(0) \to u'(0)$, $u_\ell'(+\infty) \to u'(+\infty)$. The conclusion follows easily since $p(r)/r^{1-1/N}$ is nondecreasing and equal to $-(1/N)u'(r^{-1/N})$ $(-u'(\log 1/r)$ in the case $N = \infty)$. □

Domain of the functionals U_ν

To each $U \in \mathcal{DC}_N$ corresponds a functional U_ν. However, some conditions might be needed to make sense of $U_\nu(\mu)$. Why is that so? If U is, say, nonnegative, then an integral such as $\int U(\rho)\, d\nu$ always makes sense in $[0, +\infty]$, so U_ν is well-defined on the whole of $P_2^{ac}(M)$. But U might be partially negative, and then one should not exclude the possibility that both the negative and the positive parts of $U(\rho)$ have infinite integrals. The problem comes from infinity and does not arise if M is a compact manifold, or more generally if ν has finite mass.

Theorem 17.8 below solves this issue: It shows that under some integral growth condition on ν, the quantity $U_\nu(\mu)$ is well-defined if μ has finite moments of order p large enough. This suggests that we study U_ν on the set $P_p^{\mathrm{ac}}(M)$ of absolutely continuous measures with finite moment of order p, rather than on the whole space $P_2^{\mathrm{ac}}(M)$.

Since this theorem only uses the metric structure, I shall state it in the context of general Polish spaces rather than Riemannian manifolds.

Theorem 17.8 (Moment conditions make sense of $U_\nu(\mu)$). *Let $(\mathcal{X}, d)$ be a Polish space and let ν be a reference Borel measure on $\mathcal{X}$. Let $N \in [1, \infty]$. Assume that there exists $x_0 \in \mathcal{X}$ and $p \in [2, +\infty)$ such that*

$$\begin{cases} \displaystyle\int_{\mathcal{X}} \frac{d\nu(x)}{[1 + d(x_0, x)]^{p(N-1)}} < +\infty & \text{if } N < \infty, \\[2ex] \displaystyle\exists\, c > 0; \qquad \int_M e^{-c\, d(x_0,x)^p}\, d\nu(x) < +\infty & \text{if } N = \infty. \end{cases} \tag{17.5}$$

Then, for any $U \in \mathcal{DC}_N$, the formula

$$U_\nu(\mu) = \int_{\mathcal{X}} U(\rho)\, d\nu, \qquad \mu = \rho\, \nu$$

unambiguously defines a functional $U_\nu : P_p^{\mathrm{ac}}(\mathcal{X}) \to \mathbb{R} \cup \{+\infty\}$, where $P_p^{\mathrm{ac}}(\mathcal{X})$ is the set of absolutely continuous probability measures on $\mathcal{X}$ with a finite moment of order p.

Even if no such p exists, U_ν is still well-defined on $P_c^{\mathrm{ac}}(\mathcal{X})$, the set of absolutely continuous compactly supported probability measures, provided that ν is finite on compact sets.

Example 17.9. If ν is the Lebesgue measure on $\mathbb{R}^N$, then U_ν is well-defined on $P_2^{\mathrm{ac}}(\mathbb{R}^N)$ for all $U \in \mathcal{DC}_N$, as long as $N \geq 3$. For $N = 2$, Theorem 17.8 allows us to define U_ν on $P_p^{\mathrm{ac}}(\mathbb{R}^N)$, for any $p > 2$. In the case $N = 1$, U_ν is well-defined on $P_c^{\mathrm{ac}}(\mathbb{R}^N)$. All this remains true if $\mathbb{R}^N$ is replaced by an arbitrary N-dimensional Riemannian manifold with nonnegative Ricci curvature. (Indeed, $\mathrm{vol}[B_r(x_0)] = O(r^N)$ for any fixed $x_0 \in M$, so $\int d\nu(x)/[1 + d(x_0, x)]^{p(N-1)} < +\infty$ if $p(N-1) > N$.)

Convention 17.10. In the sequel of this course I shall sometimes write "$p \in [2, +\infty) \cup \{c\}$ satisfying the assumptions of Theorem 17.8" or "$p \in [2, +\infty) \cup \{c\}$ satisfying (17.5)". This means that p is either a real number greater or equal than 2, satisfying (17.5) (the metric space $(\mathcal{X}, d)$ and the reference measure ν should be obvious from the context); or the symbol "c", so that $P_p(\mathcal{X})$ stands for the set $P_c(\mathcal{X})$ of compactly supported probability measures.

Remark 17.11. For any positive constant C, the set of probability measures μ in $P_p(\mathcal{X})$ with $\int d(x_0, x)^p \, d\mu(x) \leq C$ is closed in $P_2(\mathcal{X})$; but in general the whole set $P_p(\mathcal{X})$ is not. Similarly, if K is a given compact subset of $\mathcal{X}$, then the set of probability measures with support in K is compact in $P_2(\mathcal{X})$; but $P_c(\mathcal{X})$ is not closed in general.

Remark 17.12. If $\mathcal{X}$ is a length space (for instance a Riemannian manifold equipped with its geodesic distance), then $P_p(M)$ is a *geodesically convex* subset of $P_q(M)$, for any $q \in (1, +\infty)$. Indeed, let $(\mu_t)_{0 \leq t \leq 1}$ be a geodesic in $P_q(M)$; according to Corollary 7.22, there is a random geodesic γ such that $\mu_t = \text{law}\,(\gamma_t)$; then the bounds $\mathbb{E}\, d(x_0, \gamma_0)^p < +\infty$ and $\mathbb{E}\, d(x_0, \gamma_1)^p < +\infty$ together imply $\mathbb{E}\, d(x_0, \gamma_t)^p < +\infty$, in view of the inequality

$$0 \leq t \leq 1 \Longrightarrow \; d(x_0, \gamma_t)^p \leq 2^{2p-1} \big[d(x_0, \gamma_0)^p + d(x_0, \gamma_1)^p \big].$$

Combining this with Theorem 8.7, we deduce that $P_p^{\mathrm{ac}}(M)$ is geodesically convex in $P_2(M)$, and more precisely

$$\int d(x_0, x)^p \, \mu_t(dx) \leq 2^{2p-1} \left(\int d(x_0, x)^p \, \mu_0(dx) + \int d(x_0, x)^p \mu_1(dx) \right).$$

Thus even if the functional U_ν is a priori only defined on $P_p^{\mathrm{ac}}(M)$, it is not absurd to study its convexity properties along geodesics of $P_2(M)$.

Proof of Theorem 17.8. The problem is to show that under the assumptions of the theorem, $U(\rho)$ is bounded below by a ν-integrable function; then $U_\nu(\mu) = \int U(\rho)\, d\nu$ will be well-defined in $\mathbb{R} \cup \{+\infty\}$.

Suppose first that $N < \infty$. By convexity of u, there is a constant $A > 0$ so that $\delta^N U(\delta^{-N}) \geq -A\delta - A$, which means

$$U(\rho) \geq -A\big(\rho + \rho^{1-\frac{1}{N}}\big). \tag{17.6}$$

Of course, ρ lies in $L^1(\nu)$; so it is sufficient to show that also $\rho^{1-1/N}$ lies in $L^1(\nu)$. But this is a simple consequence of Hölder's inequality:

$$\int_{\mathcal{X}} \rho(x)^{1-\frac{1}{N}}\, d\nu(x)$$
$$= \int_{\mathcal{X}} \big((1+d(x_0,x)^p)\rho(x)\big)^{1-\frac{1}{N}} (1+d(x_0,x)^p)^{-1+\frac{1}{N}}\, d\nu(x)$$
$$\leq \left(\int_{\mathcal{X}} (1+d(x_0,x)^p)\rho(x)\, d\nu(x)\right)^{1-\frac{1}{N}} \left(\int_{\mathcal{X}} (1+d(x_0,x)^p)^{-(N-1)} d\nu(x)\right)^{\frac{1}{N}}.$$

Now suppose that $N=\infty$. By Proposition 17.7(ii), there are positive constants a, b such that

$$U(\rho) \geq a\,\rho \log \rho - b\,\rho. \tag{17.7}$$

So it is sufficient to show that $(\rho\log\rho)_- \in L^1(\nu)$. Write

$$\int_{\mathcal{X}} \rho(x)\, \log \rho(x)\, d\nu(x)$$
$$= \int_{\mathcal{X}} \rho(x)\, e^{c\,d(x_0,x)^p} \log\left(\rho(x)\, e^{c\,d(x_0,x)^p}\right)\, e^{-c\,d(x_0,x)^p}\, d\nu(x)$$
$$- c \int_{\mathcal{X}} d(x_0,x)^p\, \rho(x)\, d\nu(x). \tag{17.8}$$

By Jensen's inequality, applied with the convex function $r \to r\log r$, the probability measure $\frac{e^{-c\,d(x_0,\,\cdot\,)^p}\, d\nu}{\int_{\mathcal{X}} e^{-c\,d(x_0,\,\cdot\,)^p}\, d\nu}$ and the integrable function $\rho\, e^{c\,d(x_0,\,\cdot\,)^p}$, (17.8) can be bounded below by

$$\left(\int_{\mathcal{X}} e^{-cd(x_0,x)^p} d\nu(x)\right) \left(\frac{\int_{\mathcal{X}} \rho d\nu}{\int_{\mathcal{X}} e^{-cd(x_0,x)^p} d\nu(x)}\right) \log\left(\frac{\int_{\mathcal{X}} \rho d\nu}{\int_{\mathcal{X}} e^{-cd(x_0,x)^p} d\nu(x)}\right)$$
$$- c \int_{\mathcal{X}} d(x_0,x)^p \rho(x) d\nu(x).$$

This concludes the argument. □

In the sequel of this chapter, I shall study properties of the functionals U_ν on $P_p^{\mathrm{ac}}(M)$, where M is a Riemannian manifold equipped with its geodesic distance.

Displacement convexity from curvature bounds, revisited

Recall the notation U_N introduced in (16.17) (or in Example 17.6 (iii)). For any $N > 1$, the functional $(U_N)_\nu$ will instead be denoted by $H_{N,\nu}$:

$$H_{N,\nu}(\mu) = \int_M U_N(\rho)\, d\nu, \qquad \mu = \rho\,\nu.$$

I shall often write H_ν instead of $H_{\infty,\nu}$; and I may even write just H if the reference measure is the volume measure. This notation is justified by analogy with **Boltzmann's H functional**: $H(\rho) = \int \rho \log \rho \, d\mathrm{vol}$.

For each $U \in \mathcal{DC}_N$, formula (16.18) defines a functional Λ_U which will later play a role in displacement convexity estimates. It will be convenient to compare this quantity with $\Lambda_N := \Lambda_{U_N}$; explicitly,

$$\Lambda_N(\mu, v) = \int_M |v(x)|^2 \, \rho^{1-\frac{1}{N}}(x)\, d\nu(x), \qquad \mu = \rho\,\nu. \tag{17.9}$$

It is clear that $\Lambda_U \geq K_{N,U}\,\Lambda_N$, where

$$K_{N,U} = \inf_{r>0} \frac{Kp(r)}{r^{1-1/N}} = \begin{cases} K \displaystyle\lim_{r\to 0} \frac{p(r)}{r^{1-1/N}} & \text{if } K > 0 \\ 0 & \text{if } K = 0 \\ K \displaystyle\lim_{r\to\infty} \frac{p(r)}{r^{1-1/N}} & \text{if } K < 0. \end{cases} \tag{17.10}$$

It will also be useful to introduce a *local* version of displacement convexity. In short, a functional U_ν is said to be locally displacement convex if it is displacement convex in the neighborhood of each point.

Definition 17.13 (Local displacement convexity). *Let M be a Riemannian manifold, and let F be defined on a geodesically convex subset of $P_2^{\mathrm{ac}}(M)$, with values in $\mathbb{R} \cup \{+\infty\}$. Then F is said to be locally displacement convex if, for any $x_0 \in M$ there is $r > 0$ such that the convexity inequality*

$$\forall t \in [0,1] \qquad F(\mu_t) \leq (1-t)\, F(\mu_0) + t\, F(\mu_1)$$

holds true as soon as all measures μ_t, $0 \leq t \leq 1$, are supported in the ball $B_r(x_0)$.

The notions of local Λ-displacement convexity and local λ-displacement convexity are defined similarly, by localizing Definition 16.5.

Warning 17.14. When one says that a functional F is locally displacement convex, this *does not* mean that F is displacement convex in a small neighborhood of μ, for any μ. The word "local" refers to the topology of the base space M, not the topology of the Wasserstein space.

The next theorem is a rigorous implementation of Guesses 16.6 and 16.7; it relates curvature-dimension bounds, as appearing in Theorem 14.8, to displacement convexity properties. Recall Convention 17.10.

Theorem 17.15 (**CD bounds read off from displacement convexity**). *Let M be a Riemannian manifold, equipped with its geodesic distance d and a reference measure $\nu = e^{-V}\mathrm{vol}$, $V \in C^2(M)$. Let $K \in \mathbb{R}$ and $N \in (1, \infty]$. Let $p \in [2, +\infty) \cup \{c\}$ satisfy the assumptions of Theorem 17.8. Then the following three conditions are equivalent:*

(i) M satisfies the curvature-dimension criterion $\mathrm{CD}(K, N)$;

(ii) For each $U \in \mathcal{DC}_N$, the functional U_ν is $\Lambda_{N,U}$-displacement convex on $P_p^{\mathrm{ac}}(M)$, where $\Lambda_{N,U} = K_{N,U}\Lambda_N$;

(iii) $H_{N,\nu}$ is locally $K\Lambda_N$-displacement convex;
and then necessarily $N \geq n$, with equality possible only if V is constant.

Remark 17.16. Statement (ii) means, explicitly, that for any displacement interpolation $(\mu_t)_{0 \leq t \leq 1}$ in $P_p^{\mathrm{ac}}(M)$, and for any $t \in [0, 1]$,

$$U_\nu(\mu_t) + K_{N,U} \int_0^1 \int_M \rho_s(x)^{1-\frac{1}{N}} |\widetilde{\nabla}\psi_s(x)|^2 \, d\nu(x) \, G(s,t) \, ds \\ \leq (1-t)\, U_\nu(\mu_0) + t\, U_\nu(\mu_1), \quad (17.11)$$

where ρ_t is the density of μ_t, $\psi_s = H_+^{0,s}\psi$ (Hamilton–Jacobi semigroup), ψ is $d^2/2$-convex, $\exp(\widetilde{\nabla}\psi)$ is the Monge transport $\mu_0 \to \mu_1$, and $K_{N,U}$ is defined by (17.10).

Remark 17.17. If the two quantities in the left-hand side of (17.11) are infinite with opposite signs, the convention is $(+\infty) - (+\infty) = -\infty$, i.e. the inequality is void. This eventuality is ruled out by any one of the following conditions: (a) $K \geq 0$; (b) $N = \infty$; (c) $\mu_0, \mu_1 \in P_q(M)$, where $q > 2N/(N-1)$ is such that $\int (1 + d(x_0, x)^{q(N-1)-2N-\delta})\, \nu(dx) < +\infty$ for some $\delta > 0$. This is a consequence of Proposition 17.24 later in this chapter.

Remark 17.18. The case $N = 1$ is degenerate since U_1 is not defined; but the equivalence (i) $\Leftrightarrow$ (ii) remains true if one defines $K_{N,U}$ to be $+\infty$ if $K > 0$, and 0 if $K \leq 0$. I shall address this case from a slightly different point of view in Theorem 17.41 below. (As stated in that theorem, $N = 1$ is possible only if M is one-dimensional and $\nu = \mathrm{vol}$.)

As a particular case of Theorem 17.15, we now have a rigorous justification of the guess formulated in Example 16.8: The nonnegativity of the Ricci curvature is equivalent to the (local) displacement convexity of Boltzmann's H functional. This is the intersection of two situations where Theorem 17.15 is easier to formulate: (a) the case $N = \infty$; and (b) the case $K = 0$. These cases are important enough to be stated explicitly as corollaries of Theorem 17.15:

Corollary 17.19 ($\mathrm{CD}(K,\infty)$ and $\mathrm{CD}(0,N)$ bounds via optimal transport). *Let M be a Riemannian manifold, $K \in \mathbb{R}$ and $N \in (1,\infty]$; then:*

(a) M satisfies $\mathrm{Ric} \geq Kg$ if and only if Boltzmann's H functional is K-displacement convex on $P_c^{ac}(M)$;

(b) M has nonnegative Ricci curvature and dimension bounded above by N if and only if $H_{N,\mathrm{vol}}$ is displacement convex on $P_c^{ac}(M)$.

Remark 17.20. All these results can be extended to singular measures, so the restriction to absolutely continuous measures is nonessential. I shall come back to these issues in Chapter 30.

Core of the proof of Theorem 17.15. Before giving a complete proof, for pedagogical reasons I shall give the main argument behind the implication (i) $\Rightarrow$ (ii) in Theorem 17.15, in the simple case $K = 0$.

Let $(\mu_t)_{0\leq t\leq 1}$ be a Wasserstein geodesic, with μ_t absolutely continuous, and let ρ_t be the density of μ_t with respect to ν. By change of variables,

$$\int U(\rho_t)\, d\nu = \int U\left(\frac{\rho_0}{\mathcal{J}_t}\right) \mathcal{J}_t \, d\nu,$$

where $\mathcal{J}_t$ is the Jacobian of the optimal transport taking μ_0 to μ_t. The next step consists in rewriting this as a function of the mean distortion. Let $u(\delta) = \delta^N U(\delta^{-N})$, then

$$\int U\left(\frac{\rho_0}{\mathcal{J}_t}\right) \frac{\mathcal{J}_t}{\rho_0}\, \rho_0 \, d\nu = \int u\left(\frac{\mathcal{J}_t^{\frac{1}{N}}}{\rho_0^{\frac{1}{N}}}\right) \rho_0 \, d\nu.$$

The fact that U belongs to $\mathcal{DC}_N$ means precisely that u is convex nonincreasing. The nonnegativity of the (generalized) Ricci curvature means that the argument of u is a concave function of t. Then the convexity of the whole expression follows from the simple fact that the composition of a convex nonincreasing function with a concave function is itself convex. □

Complete proof of Theorem 17.15. Let us start with the implication (i) ⇒ (ii). I shall only treat the case $N < \infty$, since the case $N = \infty$ is very similar. In a **first step**, I shall also assume that μ_0 and μ_1 are compactly supported; this assumption will be relaxed later on.

So let μ_0 and μ_1 be two absolutely continuous, compactly supported probability measures and let $(\mu_t)_{0 \le t \le 1}$ be the unique displacement interpolation between μ_0 and μ_1. It can be written $(T_t)_{\#}\mu_0$, where $T_t(x) = \exp_x(t\nabla\psi(x))$, then let $(\psi_t)_{0 \le t \le 1}$ solve the Hamilton–Jacobi equation with initial datum $\psi_0 = \psi$. The goal is to show that

$$U_\nu(\mu_t) \le (1-t)\, U_\nu(\mu_0) + t\, U_\nu(\mu_1) - K_{N,U} \int_0^1 \int_M \rho_s(x)^{1-\frac{1}{N}}\, |\nabla\psi_s(x)|^2\, d\nu(x)\, G(s,t)\, ds. \qquad (17.12)$$

Note that the last integral is finite since $|\nabla\psi_s(x)|^2$ is almost surely bounded by D^2, where D is the maximum distance between elements of $\mathrm{Spt}(\mu_0)$ and elements of $\mathrm{Spt}(\mu_1)$; and that $\int \rho_s^{1-\frac{1}{N}}\, d\nu \le \nu[\mathrm{Spt}\,\mu_s]^{1/N}$ by Jensen's inequality.

If either $U_\nu(\mu_0) = +\infty$ or $U_\nu(\mu_1) = +\infty$, then there is nothing to prove; so let us assume that these quantities are finite.

Let t_0 be a fixed time in $(0,1)$; on $T_{t_0}(M)$, define, for all $t \in [0,1]$,

$$T_{t_0 \to t}\big(\exp_x(t_0 \nabla\psi(x))\big) = \exp_x(t\nabla\psi(x)).$$

Then $T_{t_0 \to t}$ is the unique optimal transport $\mu_{t_0} \to \mu_t$. Let $\mathcal{J}_{t_0 \to t}$ be the associated Jacobian determinant (well-defined μ_{t_0}-almost surely). Recall from Chapter 11 that μ_t is concentrated on $T_{t_0 \to t}(M)$ and that its density ρ_t is determined by the equation

$$\rho_{t_0}(x) = \rho_t(T_{t_0 \to t}(x))\, \mathcal{J}_{t_0 \to t}(x). \qquad (17.13)$$

Since $U(0) = 0$, we may apply Theorem 11.3 to $F(x) = U(\rho_t(x))$; or more precisely, to the positive part and the negative part of F separately. So

$$\int_M U(\rho_t(x))\,d\nu(x) = \int_M U\big(\rho_t(T_{t_0\to t}(x))\big)\,\mathcal{J}_{t_0\to t}(x)\,d\nu(x).$$

Then formula (17.13) implies

$$\int_M U(\rho_t)\,d\nu = \int_M U\left(\frac{\rho_{t_0}(x)}{\mathcal{J}_{t_0\to t}(x)}\right)\mathcal{J}_{t_0\to t}(x)\,d\nu(x). \tag{17.14}$$

Since the contribution of $\{\rho_{t_0} = 0\}$ does not matter, this can be rewritten

$$\begin{aligned}
U_\nu(\mu_t) &= \int_M U\left(\frac{\rho_{t_0}(x)}{\mathcal{J}_{t_0\to t}(x)}\right)\frac{\mathcal{J}_{t_0\to t}(x)}{\rho_{t_0}(x)}\,\rho_{t_0}(x)\,d\nu(x)\\
&= \int_M U\Big(\delta_{t_0}(t,x)^{-N}\Big)\,\delta_{t_0}(t,x)^N\,d\mu_{t_0}(x)\\
&= \int_M w(t,x)\,d\mu_{t_0}(x),
\end{aligned}$$

where $w(t,x) := U(\delta_{t_0}(t,x)^{-N})\,\delta_{t_0}(t,x)^N$, and

$$\delta_{t_0}(t,x) = \rho_t\big(T_{t_0\to t}(x)\big)^{-\frac{1}{N}} = \left(\frac{\mathcal{J}_{t_0\to t}(x)}{\rho_{t_0}(x)}\right)^{\frac{1}{N}}.$$

Up to a factor which does not depend on t, $\delta_{t_0}(\,\cdot\,,x)$ coincides with $\mathcal{D}(t)$ in the notation of Chapter 14. So, by Theorem 14.8, for μ_{t_0}-almost all x one has

$$\ddot{\delta}_{t_0}(t,x) \le -\frac{K}{N}\,\delta_{t_0}(t,x)\,\big|\nabla\psi_t(T_{t_0\to t}(x))\big|^2.$$

Set $u(\delta) = \delta^N\,U(\delta^{-N})$, so that $w = u\circ\delta$, where δ is shorthand for $\delta_{t_0}(\,\cdot\,,x)$ and x is fixed. Since $u'' \ge 0$ and $u'(\delta) = -Np(r)/r^{1-1/N} \le 0$, one has, with $r = \delta^{-N}$,

$$\begin{aligned}
\frac{\partial^2 w}{\partial t^2} &= \left(\frac{\partial^2 u}{\partial \delta^2}\right)(\dot{\delta}(t))^2 + \left(\frac{\partial u}{\partial \delta}\right)\ddot{\delta}(t)\\
&\ge \left(-N\frac{p(r)}{r^{1-\frac{1}{N}}}\right)\left(-\frac{K}{N}\,\delta(t)\,\big|\nabla\psi_t(T_{t_0\to t}(x))\big|^2\right).
\end{aligned}$$

By combining this with the definition of $K_{N,U}$, one obtains

$$\begin{aligned}
\ddot{w}(t,x) &\ge K_{N,U}\,\delta_{t_0}(t,x)\,\big|\nabla\psi_t(T_{t_0\to t}(x))\big|^2\\
&= K_{N,U}\,\rho_t(T_{t_0\to t}(x))^{-\frac{1}{N}}\big|\nabla\psi_t(T_{t_0\to t}(x))\big|^2.
\end{aligned} \tag{17.15}$$

Since w is a continuous function of t, this implies (recall Proposition 16.2)

$$w(t,x) - (1-t)\,w(0,x) - t\,w(1,x) \\ \leq -K_{N,U} \int_0^1 \rho_s(T_{t_0\to s}(x))^{-\frac{1}{N}} \big|\nabla\psi_s(T_{t_0\to s}(x))\big|^2\, G(s,t)\, ds.$$

Upon integration against μ_{t_0} and use of Fubini's theorem, this inequality becomes

$$\begin{aligned} &U_\nu(\mu_t) - (1-t)\,U_\nu(\mu_0) - t\,U_\nu(\mu_1) \\ &\leq -K_{N,U} \int_M \left(\int_0^1 \rho_s(T_{t_0\to s}(x))^{-\frac{1}{N}} \big|\nabla\psi_s(T_{t_0\to s}(x))\big|^2\, G(s,t)\, ds \right) d\mu_{t_0}(x) \\ &= -K_{N,U} \int_0^1 \int_M \rho_s(T_{t_0\to s}(x))^{-\frac{1}{N}} \big|\nabla\psi_s(T_{t_0\to s}(x))\big|^2\, d\mu_{t_0}(x)\, G(s,t)\, ds \\ &= -K_{N,U} \int_0^1 \int_M \rho_s(y)^{-\frac{1}{N}} |\nabla\psi_s(y)|^2\, d\mu_s(y)\, G(s,t)\, ds \\ &= -K_{N,U} \int_0^1 \int_M \rho_s(y)^{1-\frac{1}{N}} |\nabla\psi_s(y)|^2\, d\nu(y)\, G(s,t)\, ds. \end{aligned}$$

This concludes the proof of Property (ii) when μ_0 and μ_1 have compact support. In a **second step** I shall relax this compactness assumption by a restriction argument. Let $p \in [2,+\infty) \cup \{c\}$ satisfy the assumptions of Theorem 17.8, and let μ_0, μ_1 be two probability measures in $P_p^{\mathrm{ac}}(M)$. Let $(Z_\ell)_{\ell\in\mathbb{N}}$, $(\mu_{t,\ell})_{0\leq t\leq 1,\,\ell\in\mathbb{N}}$ $(\psi_{t,\ell})_{0\leq t\leq 1,\,\ell\in\mathbb{N}}$ be as in Proposition 13.2. Let $\rho_{t,\ell}$ stand for the density of $\mu_{t,\ell}$. By Remark 17.4, the function $U_\ell : r \to U(Z_\ell r)$ belongs to $\mathcal{DC}_N$; and it is easy to check that $K_{N,U_\ell} = Z_\ell^{1-\frac{1}{N}}\, K_{N,U}$. Since the measures $\mu_{t,\ell}$ are compactly supported, we can apply the previous inequality with μ_t replaced by $\mu_{t,\ell}$ and U replaced by U_ℓ:

$$\int U(Z_\ell\,\rho_{t,\ell})\, d\nu \leq (1-t) \int U(Z_\ell\,\rho_{0,\ell})\, d\nu + t \int U(Z_\ell\,\rho_{1,\ell})\, d\nu \\ - Z_\ell^{1-\frac{1}{N}}\, K_{N,U} \int_0^1 \int_M \rho_{s,\ell}(y)^{1-\frac{1}{N}} |\nabla\psi_{s,\ell}(y)|^2\, d\nu(y)\, G(s,t)\, ds. \qquad (17.16)$$

It remains to pass to the limit in (17.16) as $\ell \to \infty$. Recall from Proposition 13.2 that $Z_\ell\,\rho_{t,\ell}$ is a nondecreasing family of functions converging monotonically to ρ_t. Since U_+ is nondecreasing, it follows that

$$U_+(Z_\ell\,\rho_{t,\ell}) \uparrow U_+(\rho_t).$$

On the other hand, the proof of Theorem 17.8 shows that $U_-(r) \leq A(r + r^{1-\frac{1}{N}})$ for some constant $A = A(N, U)$; so

$$U_-(Z_\ell\,\rho_{t,\ell}) \leq A\big(Z_\ell\,\rho_{t,\ell} + Z_\ell^{1-\frac{1}{N}}\,\rho_{t,\ell}^{1-\frac{1}{N}}\big) \leq A\left(\rho_t + \rho_t^{1-\frac{1}{N}}\right). \qquad (17.17)$$

By the proof of Theorem 17.8 and Remark 17.12, the function on the right-hand side of (17.17) is ν-integrable, and then we may pass to the limit by dominated convergence. To summarize:

$$\int U_+(Z_\ell\,\rho_{t,\ell})\,d\nu \xrightarrow[\ell\to\infty]{} \int U_+(\rho_t)\,d\nu \qquad \text{by monotone convergence;}$$

$$\int U_-(Z_\ell\,\rho_{t,\ell})\,d\nu \xrightarrow[\ell\to\infty]{} \int U_-(\rho_t)\,d\nu \qquad \text{by dominated convergence.}$$

So we can pass to the limit in the first three terms appearing in the inequality (17.16). As for the last term, note that $|\nabla\psi_{s,\ell}(y)|^2 = d(y, T_{s\to 1,\ell}(y))^2/(1-s)^2$, at least $\mu_{s,\ell}(dy)$-almost surely; but then according to Proposition 13.2 this coincides with $d(y, T_{s\to 1}(y))^2/(1-s)^2 = |\widetilde{\nabla}\psi_s(y)|^2$. So the last term in (17.16) can be rewritten as

$$K_{N,U}\int_0^1\int_M (Z_\ell\,\rho_{s,\ell}(y))^{1-\frac{1}{N}}\,|\widetilde{\nabla}\psi_s(y)|^2\,d\nu(y)\,G(s,t)\,ds,$$

and by monotone convergence this goes to

$$K_{N,U}\int_0^1\int_M (\rho_s(y))^{1-\frac{1}{N}}\,|\widetilde{\nabla}\psi_s(y)|^2\,d\nu(y)\,G(s,t)\,ds$$

as $\ell \to \infty$. Thus we have passed to the limit in all terms of (17.16), and the proof of (i) $\Rightarrow$ (ii) is complete.

Since the implication (ii) $\Rightarrow$ (iii) is trivial, to conclude the proof of Theorem 17.15 it only suffices to prove (iii) $\Rightarrow$ (i). So let $x_0 \in M$; the goal is to show that $(\mathrm{Ric}_{N,\nu})_{x_0} \geq K g_{x_0}$, where g is the Riemannian metric. Let $r > 0$ be such that $H_{N,\nu}$ is $K\Lambda_N$-displacement convex in $B_r(x_0)$. Let $v_0 \neq 0$ be a tangent vector at x_0. For a start, assume $N > n$. As in the proof of Theorem 14.8, we can construct $\widetilde{\psi} \in C^2(M)$, compactly supported in $B_r(x_0)$, such that $\nabla\widetilde{\psi}(x_0) = v_0$, $\nabla^2\widetilde{\psi}(x_0) = \lambda_0\,I_n$ (I_n is the identity on $T_{x_0}M$) and

$$\left[\Gamma_2(\widetilde{\psi}) - \frac{(L\widetilde{\psi})^2}{N}\right](x_0) = \mathrm{Ric}_{N,\nu}(v_0).$$

Then let $\psi := \theta\widetilde{\psi}$, where θ is a positive real number. If θ is small enough, ψ is $d^2/2$-convex by Theorem 13.5, and $|\nabla\psi| \le r/2$. Let ρ_0 be a smooth probability density, supported in $B_\eta(x_0)$, with $\eta < r/2$; and

$$\mu_0 = \rho_0\,\nu; \qquad \mu_t = \exp(t\nabla\psi)_\#\mu_0.$$

Then $(\mu_t)_{0\le t\le 1}$ is a geodesic in $P_2(M)$, entirely supported in $B_r(x_0)$, so condition (iii) implies

$$H_{N,\nu}(\mu_t) - (1-t)\,H_{N,\nu}(\mu_0) - t\,H_{N,\nu}(\mu_1) \\ \le -K\int_0^1 \left(\int \rho_s(x)^{1-\frac{1}{N}}\,|\nabla\psi_s(x)|^2\,d\nu(x)\right) ds. \tag{17.18}$$

As in the proof of (i) $\Rightarrow$ (ii), let $\mathcal{J}(t,x)$ be the Jacobian determinant of the map $\exp(t\nabla\psi)$ at x, and let $\delta(t,x) = \mathcal{J}(t,x)^{1/N}$. (This amounts to choosing $t_0 = 0$ in the computations above; now this is not a problem since $\exp(t\nabla\psi)$ is for sure Lipschitz.) Further, let $\gamma(t,x) = \exp_x(t\nabla\psi(x))$. Formula (14.39) becomes

$$-N\frac{\ddot\delta(t,x)}{\delta(t,x)} = \mathrm{Ric}_{N,\nu}(\dot\gamma(t,x)) + \left\|U(t,x) - \left(\frac{\mathrm{tr}\;U(t,x)}{n}\right)I_n\right\|_{\mathrm{HS}}^2 \\ + \frac{n}{N(N-n)}\left[\left(\frac{N-n}{n}\right)\mathrm{tr}\;U(t,x) + \dot\gamma(t,x)\cdot\nabla V(\gamma(t,x))\right]^2, \tag{17.19}$$

where $U(0,x) = \nabla^2\psi(x)$, $U(t,x)$ solves the nonlinear differential equation $\dot U + U^2 + R = 0$, and R is defined by (14.7). By using all this information, we shall derive expansions of (17.19) as $\theta \to 0$, $\widetilde{\psi}$ being fixed. First of all, $x = x_0 + O(\theta)$ (this is informal writing to mean that $d(x,x_0) = O(\theta)$); then, by smoothness of the exponential map, $\dot\gamma(t,x) = \theta v_0 + O(\theta^2)$; it follows that $\mathrm{Ric}_{N,\nu}(\dot\gamma(t,x)) = \theta^2\mathrm{Ric}_{N,\nu;x_0}(v_0) + O(\theta^3)$. Next, $U(0) = \theta\nabla^2\widetilde{\psi}(x_0) = \lambda_0\theta I_n$ and $R(t) = O(\theta^2)$; by an elementary comparison argument, $U(t,x) = O(\theta)$, so $\dot U = O(\theta^2)$, and $U(t,x) = \lambda_0\theta I_n + O(\theta^2)$. Also $U - (\mathrm{tr}\;U)I_n/n = O(\theta^2)$, $\mathrm{tr}\;U(t) = \lambda_0\theta n + O(\theta^2)$ and $\dot\gamma(t,x)\cdot\nabla V(\gamma(t,x)) + ((N-n)/n)\,\mathrm{tr}\;U(t,x) = O(\theta^2)$. Plugging all these expansions into (17.19), we get

$$\frac{\ddot\delta(t,x)}{\delta(t,x)} = \frac{1}{N}\Big(-\theta^2\,\mathrm{Ric}_{N,\nu}(v_0) + O(\theta^3)\Big). \tag{17.20}$$

By repeating the proof of (i) $\Rightarrow$ (ii) with $U = U_N$ and using (17.20), one obtains

$$H_{N,\nu}(\mu_t) - (1-t)\,H_{N,\nu}(\mu_0) - t\,H_{N,\nu}(\mu_1)$$
$$\geq -\,\theta^2\big(\mathrm{Ric}_{N,\nu}(v_0) + O(\theta)\big)\int_0^1\int_M \rho_s(y)^{1-\frac{1}{N}}\,d\nu(y)\,G(s,t)\,ds. \quad (17.21)$$

On the other hand, by (17.18),

$$\begin{aligned} H_{N,\nu}(\mu_t) &- (1-t)\,H_{N,\nu}(\mu_0) - t\,H_{N,\nu}(\mu_1) \\ &\leq -K\int_0^1\int_M \rho_s(y)^{1-\frac{1}{N}}\,|\dot\gamma(s,y)|^2\,d\nu(y)\,G(s,t)\,ds \\ &= -K\,\theta^2\big(|v_0|^2 + O(\theta)\big)\int_0^1\int_M \rho_s(y)^{1-\frac{1}{N}}\,d\nu(y)\,G(s,t)\,ds. \end{aligned}$$

Combining this with (17.21) and canceling out multiplicative factors $\theta^2\int_0^1\int \rho_s(y)^{1-\frac{1}{N}}\,d\nu(y)\,G(s,t)\,ds$ on both sides, we obtain $\mathrm{Ric}_{N,\nu}(v_0) \geq K|v_0|^2 + O(\theta)$. The limit $\theta \to 0$ yields

$$\mathrm{Ric}_{N,\nu}(v_0) \geq K|v_0|^2, \quad (17.22)$$

and since v_0 was arbitrary this concludes the proof.

The previous argument was under the assumption $N > n$. If $N = n$ and V is constant, the proof is the same but the modified Ricci tensor $\mathrm{Ric}_{N,\nu}$ is replaced by the usual Ricci tensor.

It remains to rule out the cases when $N = n$ and V is nonconstant, or $N < n$. In these situations (17.19) should be replaced by

$$-N\frac{\ddot\delta(t,x)}{\delta(t,x)} = \big(\mathrm{Ric} + \nabla^2 V\big)(\dot\gamma(t,x)) + \left\|U(t,x) - \left(\frac{\mathrm{tr}\,U(t,x)}{n}\right) I_n\right\|_{\mathrm{HS}}^2$$
$$+\frac{(\mathrm{tr}\,U(t,x))^2}{n} - \frac{\Big(\mathrm{tr}\,U(t,x) - \dot\gamma(t,x)\cdot\nabla V(\gamma(t,x))\Big)^2}{N}. \quad (17.23)$$

(To see this, go back to Chapter 14, start again from (14.33) but this time don't apply (14.34).) Repeat the same argument as before, with ψ satisfying $\nabla\psi(x_0) = v_0$ and $\nabla^2\psi(x_0) = \lambda_0\,I_n$ (now λ_0 is arbitrary). Then instead of (17.22) one retrieves

$$(\mathrm{Ric}+\nabla^2 V)(v_0) + \lambda_0^2\left(\frac{1}{n} - \frac{1}{N}\right) - \frac{(\nabla V(x_0)\cdot v_0)^2}{N} + \frac{2\,\lambda_0(v_0\cdot\nabla V(x_0))}{N} \geq K|v_0|^2. \quad (17.24)$$

If $n = N$, the left-hand side of (17.24) is an affine expression of λ_0, so the inequality is possible only if the slope if zero, i.e. $v_0 \cdot \nabla V(x_0) = 0$. Since v_0 and x_0 are arbitrary, actually $\nabla V = 0$, so V is constant (and $\mathrm{Ric}(v_0) \geq K|v_0|^2$). If $n < N$, what we have on the left of (17.24) is a quadratic expression of λ_0 with negative dominant coefficient, so it cannot be bounded below. This contradiction establishes $N \geq n$. □

Remark 17.21. A completely different (and much more general) proof of the inequality $N \geq n$ will be given later in Corollary 30.13.

Exercise 17.22 (Necessary condition for displacement convexity). This exercise shows that the elements of $\mathcal{DC}_N$ are essentially the only nonlinearities leading to displacement convex functionals. Let N be a positive integer, $M = \mathbb{R}^N$, and let ν be the Lebesgue measure in $\mathbb{R}^N$. Let U be a measurable function $\mathbb{R}_+ \to \mathbb{R}$ such that U_ν is lower semicontinuous and convex on the space $P_c^{\mathrm{ac}}(\mathbb{R}^N)$ (absolutely continuous, compactly supported probability measures), equipped with the distance W_2. Show that (a) U itself is convex lower semicontinuous; (b) $\delta \to \delta^N U(\delta^{-N})$ is convex. *Hint:* To prove (b), consider the geodesic curve $(\mu_\delta)_{\delta>0}$, where μ_δ is the uniform probability measure on $B_\delta(0)$.

Exercise 17.23. Show that if (M,ν) satisfies $\mathrm{CD}(K,N)$ and U belongs to $\mathcal{DC}_N$, then U_ν is $KR^{-1/N}$-displacement convex, when restricted to the geodesically convex set defined by $\rho \leq R$. In short, U_ν is $K\|\rho\|_{L^\infty}^{-1/N}$-displacement convex. *Hint:* Use $U(r) = r^m$ and let $m \to \infty$. (A longer solution is via the proof of Corollary 19.5.)

To conclude this section, I shall establish sufficient conditions for the time-integral appearing in (17.11) to be finite.

Proposition 17.24 (Finiteness of time-integral in displacement convexity inequality). *Let M be a Riemannian manifold equipped with a reference measure $\nu = e^{-V}\mathrm{vol}$, $V \in C(M)$, and let ρ_0, ρ_1 be two probability densities on M. Let ψ be $d^2/2$-convex such that $T = \exp(\widetilde{\nabla}\psi)$ is the optimal Monge transport between $\mu_0 = \rho_0\,\nu$ and $\mu_1 = \rho_1\,\nu$ for the cost function $c(x,y) = d(x,y)^2$. Let $(\mu_t)_{0\leq t\leq 1}$ be the displacement interpolation between μ_0 and μ_1, and let ρ_t be the density of μ_t. Then for any $t \in [0,1)$:*

(i) $\displaystyle\int_M \rho_t\,|\widetilde{\nabla}\psi_t|^2\,d\nu = W_2(\mu_0,\mu_1)^2.$

(ii) Let z be any point in M. If $N \in (1,\infty)$ and $q > 2N/(N-1)$ is such that $\mu_0, \mu_1 \in P_q^{\mathrm{ac}}(M)$ and

$$\exists\,\delta > 0; \qquad \int_M \frac{d\nu(x)}{\big(1+d(z,x)\big)^{q(N-1)-2N-\delta}} < +\infty,$$

then

$$\int_0^1 \left(\int_M \rho_t^{1-\frac{1}{N}} |\widetilde{\nabla}\psi_t|^2 \, d\nu\right) (1-t)\,dt < +\infty.$$

More precisely, there are constants $C = C(N,q,\delta) > 0$ and $\theta = \theta(N,q,\delta) \in (0, 1-1/N)$ such that

$$\begin{aligned}
&(1-t)\int_M \rho_t^{1-\frac{1}{N}} |\widetilde{\nabla}\psi_t|^2\, d\nu \qquad\qquad (17.25)\\
&\le \frac{C}{(1-t)^{1-2\theta}}\, W_2(\mu_0,\mu_1)^{2\theta} \left(\int \frac{\nu(dx)}{(1+d(z,x))^{q(N-1)-2N-\delta}}\right)^{\frac{1}{N}}\\
&\qquad \times \left(1 + \int d(z,x)^q\,\mu_0(dx) + \int d(z,x)^q\,\mu_1(dx)\right)^{(1-\theta)-\frac{1}{N}}.
\end{aligned}$$

Proof of Proposition 17.24. First, $|\widetilde{\nabla}\psi_t(x)| = d(x, T_{t\to 1}(x))/(1-t)$, where $T_{t\to 1}$ is the optimal transport $\mu_t \to \mu_1$. So $\int \rho_t\, |\widetilde{\nabla}\psi_t|^2\, d\nu = W_2(\mu_t,\mu_1)^2/(1-t)^2 = W_2(\mu_0,\mu_1)^2$. This proves (i).

To prove (ii), I shall start from

$$\begin{aligned}
(1-t)\int \rho_t(x)^{1-\frac{1}{N}} |\widetilde{\nabla}\psi_t(x)|^2\, \nu(dx)&\\
&= \frac{1}{1-t}\int \rho_t(x)^{1-\frac{1}{N}}\, d(x, T_{t\to 1}(x))^2\, \nu(dx).
\end{aligned}$$

Let us first see how to bound the integral in the right-hand side, without worrying about the factor $(1-t)^{-1}$ in front. This can be done with the help of Jensen's inequality, in the same spirit as the proof of Theorem 17.8: If $r \ge 0$ is to be chosen later, then

$$\begin{aligned}
&\int \rho_t(x)^{1-\frac{1}{N}}\, d(x, T_{t\to 1}(x))^2\, \nu(dx)\\
&\le \left(\int \rho_t(x)\,\big(1+d(z,x)^r\big)\, d(x,T_{t\to 1}(x))^{2\left(\frac{N}{N-1}\right)}\, \nu(dx)\right)^{\frac{N-1}{N}}\\
&\qquad\qquad \times \left(\int \frac{\nu(dx)}{\big(1+d(z,x)^r\big)^{N-1}}\right)^{\frac{1}{N}}
\end{aligned}$$

$$\leq \overline{C} \left(\int \rho_t(x) \, (1+d(z,x)^r) \, \big(d(z,x)+d(z,T_{t\to 1}(x))\big)^{2\left(\frac{N}{N-1}\right)} \, \nu(dx) \right)^{\frac{N-1}{N}}$$

$$\leq \overline{C} \left(\int \rho_t(x) \left(1+d(z,x)^{r+2\left(\frac{N}{N-1}\right)}+d(z,T_{t\to 1}(x))^{r+2\left(\frac{N}{N-1}\right)}\right) \nu(dx) \right)^{\frac{N-1}{N}}$$

$$= \overline{C} \left(1 + \int \rho_t(x) \, d(z,x)^{r+2\left(\frac{N}{N-1}\right)} + \int \rho_1(y) \, d(z,y)^{r+2\left(\frac{N}{N-1}\right)} \, \nu(dy) \right)^{\frac{N-1}{N}},$$

where

$$\overline{C} = C(r,N,\nu) = C(r,N) \left(\int \frac{\nu(dx)}{\left(1+d(z,x)^r\right)^{N-1}} \right)^{\frac{1}{N}},$$

and $C(r,N)$ stands for some constant depending only on r and N. By Remark 17.12, the previous expression is bounded by

$$C(r,N,\nu) \left(1+ \int d(z,x)^{r+2\left(\frac{N}{N-1}\right)} \mu_0(dx) + \int d(z,x)^{r+2\left(\frac{N}{N-1}\right)} \mu_1(dx) \right)^{\frac{N-1}{N}};$$

and the choice $r = q - 2N/(N-1)$ leads to

$$\begin{aligned} &\int \rho_t(x)^{1-\frac{1}{N}} \, |\widetilde{\nabla}\psi_t(x)|^2 \, \nu(dx) \qquad (17.26) \\ &\leq C(N,q) \left(1 + \int d(z,x)^q \, \mu_0(dx) + \int d(z,x)^q \, \mu_1(dx) \right)^{\frac{N-1}{N}} \\ &\qquad\qquad \times \left(\int \frac{\nu(dx)}{\left(1+d(z,x)\right)^{q(N-1)-2N}} \right)^{\frac{1}{N}}. \end{aligned}$$

This estimate is not enough to establish the convergence of the time-integral, since $\int_0^1 dt/(1-t) = +\infty$; so we need to gain some cancellation as $t \to 1$. The idea is to interpolate with (i), and this is where δ will be useful. Without loss of generality, we may assume $\delta < q(N-1) - 2N$.

Let $\theta \in (0, 1-1/N)$ to be chosen later, and $N' = (1-\theta)N \in (1,\infty)$. By Hölder's inequality,

$$\begin{aligned} &\frac{1}{1-t} \int \rho_t(x)^{1-\frac{1}{N}} \, d(x, T_{t\to 1}(x))^2 \, \nu(dx) \\ &\leq \frac{1}{1-t} \left(\int \rho_t(x) \, d(x, T_{t\to 1}(x))^2 \, \nu(dx) \right)^{\theta} \\ &\qquad\qquad \left(\int \rho_t(x)^{1-\frac{1}{N'}} \, d(x, T_{t\to 1}(x))^2 \, \nu(dx) \right)^{1-\theta} \end{aligned}$$

$$= \frac{1}{1-t}\, W_2(\mu_t,\mu_1)^{2\theta} \left(\int \rho_t(x)^{1-\frac{1}{N'}}\, d(x,T_{t\to 1}(x))^2\, \nu(dx) \right)^{1-\theta}$$
$$= \frac{1}{(1-t)^{1-2\theta}}\, W_2(\mu_0,\mu_1)^{2\theta} \left(\int \rho_t(x)^{1-\frac{1}{N'}}\, d(x,T_{t\to 1}(x))^2\, \nu(dx) \right)^{1-\theta}.$$

Since $N' > 1$ we can apply the preceding computation with N replaced by N':

$$\int \rho_t(x)^{1-\frac{1}{N'}}\, d(x,T_{t\to 1}(x))^2\, \nu(dx)$$
$$\leq C(N',q) \left(1 + \int d(z,x)^q\, \mu_0(dx) + \int d(z,x)^q\, \mu_1(dx) \right)^{1-\frac{1}{N'}} \times \left(\int \frac{\nu(dx)}{(1+d(z,x))^{q(N'-1)-2N'}} \right)^{\frac{1}{N'}}.$$

Then we may choose θ so that $q(N'-1) - 2N' = q(N-1) - 2N - \delta$; that is, $\theta = \delta/((q-2)N) \in (0, 1-1/N)$. The conclusion follows. □

Ricci curvature bounds from distorted displacement convexity

In Theorem 17.15, all the influence of the Ricci curvature bounds lies in the additional term $\int_0^1 (\dots)\, G(s,t)\, ds$. As a consequence, as soon as $K \neq 0$ and $N < \infty$, the formulation involves not only μ_t, μ_0 and μ_1, but the whole geodesic path $(\mu_s)_{0\leq s\leq 1}$. This makes the exploitation of the resulting inequality (in geometric applications, for instance) somewhat delicate, if not impossible.

I shall now present a different formulation, expressed only in terms of μ_t, μ_0 and μ_1. As a price to pay, the functionals $U_\nu(\mu_0)$ and $U_\nu(\mu_1)$ will be replaced by more complicated expressions in which extra distortion coefficients will appear. From the technical point of view, this new formulation relies on the principle that one can "take the direction of motion out", in all reformulations of Ricci curvature bounds that were examined in Chapter 14.

Definition 17.25 (Distorted U_ν functional). *Let $(\mathcal{X}, d)$ be a Polish space equipped with a Borel reference measure ν. Let U be a convex* *function with $U(0) = 0$, let $x \to \pi(dy|x)$ be a family of conditional probability measures on $\mathcal{X}$, indexed by $x \in \mathcal{X}$, and let β be a measurable function $\mathcal{X} \times \mathcal{X} \to (0, +\infty]$. The distorted U_ν functional with distortion coefficient β is defined as follows: For any measure $\mu = \rho\nu$ on $\mathcal{X}$,*

$$U^{\beta}_{\pi,\nu}(\mu) = \int_{\mathcal{X}\times\mathcal{X}} U\left(\frac{\rho(x)}{\beta(x,y)}\right) \beta(x,y)\,\pi(dy|x)\,\nu(dx). \tag{17.27}$$

In particular, if $\pi(dy|x) = \delta_{y=T(x)}$, where $T : \mathcal{X} \to \mathcal{X}$ is measurable, then

$$U^{\beta}_{\pi,\nu}(\mu) = \int_{\mathcal{X}} U\left(\frac{\rho(x)}{\beta(x,T(x))}\right) \beta(x,T(x))\,\nu(dx). \tag{17.28}$$

Remark 17.26. Most of the time, I shall use Definition 17.25 with $\beta = \beta_t^{(K,N)}$, that is, the *reference distortion coefficients* introduced in Definition 14.19.

Remark 17.27. I shall often identify the conditional measures π with the probability measure $\pi(dx\,dy) = \mu(dx)\,\pi(dy|x)$ on $\mathcal{X} \times \mathcal{X}$. Of course the joint measure $\pi(dx\,dy)$ determines the conditional measures $\pi(dy|x)$ only up to a μ-negligible set of x; but this ambiguity has no influence on the value of (17.27) since $U(0) = 0$.

The problems of domain of definition which we encountered for the original U_ν functionals also arise (even more acutely) for the distorted ones. The next theorem almost solves this issue.

Theorem 17.28 (Domain of definition of $U^{\beta}_{\pi,\nu}$). *Let $(\mathcal{X}, d)$ be a Polish space, equipped with a Borel reference measure ν; let $K \in \mathbb{R}$, $N \in [1, \infty]$, and $U \in \mathcal{DC}_N$. Let π be a probability measure on $\mathcal{X} \times \mathcal{X}$, such that the marginal μ of π is absolutely continuous with density ρ. Further, let $\beta : \mathcal{X} \times \mathcal{X} \to (0, +\infty]$ be a measurable function such that*

$$\begin{cases} \beta \text{ is bounded} & (N < \infty) \\[2ex] \displaystyle\int_{\mathcal{X}\times\mathcal{X}} (\log \beta(x,y))_+\,\pi(dx\,dy) < +\infty & (N = \infty). \end{cases} \tag{17.29}$$

If there exists $x_0 \in \mathcal{X}$ and $p \in [2, +\infty)$ such that

$$\begin{cases} \displaystyle\int_{\mathcal{X}} \frac{d\nu(x)}{[1+d(x_0,x)]^{p(N-1)}} < +\infty & (N<\infty) \\ \\ \exists c>0; \qquad \displaystyle\int_{\mathcal{X}} e^{-c\,d(x_0,x)^p}\,d\nu(x) < +\infty & (N=\infty), \end{cases} \tag{17.30}$$

then the integral $U^{\beta}_{\pi,\nu}(\mu)$ appearing in Definition 17.25 makes sense in $\mathbb{R}\cup\{+\infty\}$ as soon as $\mu \in P^{\mathrm{ac}}_p(\mathcal{X})$.

Even if there is no such p, $U^{\beta}_{\pi,\nu}(\mu)$ still makes sense in $\mathbb{R}\cup\{+\infty\}$ if $\mu \in P^{\mathrm{ac}}_c(\mathcal{X})$ and ν is finite on compact sets.

Proof of Theorem 17.28. The argument is similar to the proof of Theorem 17.8. In the case $N<\infty$, β bounded, it suffices to write

$$\beta\, U(\rho/\beta) \geq -a\,\beta\Big(\frac{\rho}{\beta} + \Big(\frac{\rho}{\beta}\Big)^{1-\frac{1}{N}}\Big) = -a\,\rho - b\,\beta^{\frac{1}{N}}\rho^{1-\frac{1}{N}};$$

then the right-hand side is integrable since $\rho^{1-1/N}$ is integrable (as noted in the proof of Theorem 17.8).

In the case $N=\infty$, by Proposition 17.7(ii) there are positive constants a, b such that $U(\rho) \geq a\,\rho\log\rho - b\,\rho$; so

$$\begin{aligned} \beta(x,y)\,U\left(\frac{\rho(x)}{\beta(x,y)}\right) &\geq a\,\beta(x,y)\left(\frac{\rho(x)}{\beta(x,y)}\right)\log\left(\frac{\rho(x)}{\beta(x,y)}\right) - b\,\rho(x) \\ &= a\,\rho(x)\log\rho(x) - a\,\rho(x)\log\beta(x,y) - b\,\rho(x). \end{aligned}$$

We already know by the proof of Theorem 17.8 that $(\rho\log\rho)_-$ and ρ lie in $L^1(\nu)$, or equivalently in $L^1(\pi(dy|x)\,\nu(dx))$. To check the integrability of the negative part of $-a\,\rho\log\beta(x,y)$, it suffices to note that

$$\begin{aligned} \int \rho(x)\,(\log\beta(x,y))_+\,\pi(dy|x)\,\nu(dx) &\leq \int (\log\beta(x,y))_+\pi(dy|x)\,\mu(dx) \\ &= \int (\log\beta(x,y))_+\,\pi(dx\,dy), \end{aligned}$$

which is finite by assumption. This concludes the proof of Theorem 17.28. □

Application 17.29 (Definition of $U^{\beta^{(K,N)}_t}_{\pi,\nu}$). Let $\mathcal{X}=M$ be a Riemannian manifold satisfying a $\mathrm{CD}(K,N)$ curvature-dimension bound, let $t\in[0,1]$ and let $\beta=\beta^{(K,N)}_t$ be the distortion coefficient defined in (14.61)–(14.64).

- If $K \le 0$, then $\beta_t^{(K,N)}$ is bounded;
- If $K > 0$, $N < \infty$ and $\mathrm{diam}\,(M) < D_{K,N} := \pi\sqrt{(N-1)/K}$, then β is bounded;
- If $K > 0$ and $N = \infty$, then $\log \beta_t^{(K,N)}(x,y)$ is bounded above by a constant multiple of $d(x,y)^2$, which is $\pi(dx\,dy)$-integrable whenever π is an optimal coupling arising in some displacement interpolation.

In all three cases, Theorem 17.28 shows that $U^{\beta}_{\pi,\nu}$ is well-defined in $\mathbb{R} \cup \{+\infty\}$, more precisely that the integrand entering the definition is bounded below by an integrable function. The only remaining cases are (a) when $K > 0$, $N < \infty$ and $\mathrm{diam}\,(M)$ coincides with the limit Bonnet–Myers diameter $D_{K,N}$; and (b) when $N = 1$. These two cases are covered by the following definition.

Convention 17.30 (Definition of $U^{\beta}_{\pi,\nu}$ in the limit cases). If either (a) $K > 0$, $N < \infty$ and $\mathrm{diam}\,(M) = \pi\sqrt{(N-1)/K}$ or (b) $N = 1$, I shall define

$$U_{\pi,\nu}^{\beta_t^{(K,N)}}(\mu) = \lim_{N' \downarrow N} U_{\pi,\nu}^{\beta_t^{(K,N')}}(\mu). \tag{17.31}$$

Remark 17.31. The limit in (17.31) is well-defined; indeed, $\beta_t^{(K,N)}$ is increasing as N decreases, and $U(r)/r$ is nondecreasing as a function of r; so $U(\rho(x)/\beta_t^{(K,N)}(x,y))\,\beta_t^{(K,N)}(x,y)$ is a nonincreasing function of N and the limit in (17.31) is monotone. The monotone convergence theorem guarantees that this definition coincides with the original definition (17.27) when it applies, i.e. when the integrand is bounded below by a $\pi(dy|x)\,\nu(dx)$-integrable function.

The combination of Application 17.29 and Convention 17.30 ensures that $U_{\pi,\nu}^{\beta_t^{(K,N)}}(\mu)$ is well-defined as soon as $\int d(x,y)^2\,\pi(dx\,dy) < +\infty$, $\mu \in P_p^{\mathrm{ac}}(M)$, and p satisfies (17.30).

Remark 17.32. In the limit case $\mathrm{diam}\,(M) = D_{K,N}$, it is perfectly possible for $U_{\pi,\nu}^{\beta_t^{(K,N)}}(\mu)$ to take the value $-\infty$. An example is when M is the sphere S^N, $K = N-1$, μ is uniform, $U(r) = -N\,r^{1-1/N}$, and π is the deterministic coupling associated with the map $S : x \to -x$. However, when π is an *optimal* coupling, it is impossible for $U_{\pi,\nu}^{\beta_t^{(K,N)}}(\mu)$ to take the value $-\infty$.

Remark 17.33. If $\operatorname{diam}(M) = D_{K,N}$ then actually M is the sphere $S^N(\sqrt{\frac{N-1}{K}})$ and $\nu = \mathrm{vol}$; but we don't need this information. (The assumption of M being complete without boundary is important for this statement to be true, otherwise the one-dimensional reference spaces of Examples 14.10 provide a counterexample.) See the end of the bibliographical notes for more explanations. In the case $N = 1$, if M is distinct from a point then it is one-dimensional, so it is either the real line or a circle.

Now comes the key notion in this section:

Definition 17.34 (Distorted displacement convexity). *Let M be a Riemannian manifold, equipped with a reference measure ν. Let $(\beta_t(x,y))_{0\le t\le 1}$ be a family of measurable functions $M \times M \to (0,+\infty]$, and let $U : \mathbb{R}_+ \to \mathbb{R}$ be a continuous convex function with $U(0) = 0$. The functional U_ν is said to be displacement convex with distortion (β_t) if, for any geodesic path $(\mu_t)_{0\le t\le 1}$ in the domain of U_ν,*

$$\forall t \in [0,1], \qquad U_\nu(\mu_t) \le (1-t)\, U^{\beta_{1-t}}_{\check{\pi},\nu}(\mu_0) \;+\; t\, U^{\beta_t}_{\pi,\nu}(\mu_1), \qquad (17.32)$$

where π stands for the optimal transference plan between μ_0 and μ_1; and $\check{\pi}$ is obtained from π by switching the variables, that is $\check{\pi} = S_\#\pi$, $S(x_0,x_1) = (x_1,x_0)$.

This notion can be localized as in Definition 17.13.

Remark 17.35. The inequality appearing in (17.32) can be rewritten more explicitly as

$$\begin{aligned}\int U(\rho_t)\, d\nu \\ \le (1-t) \int_{M\times M} U\left(\frac{\rho_0(x_0)}{\beta_{1-t}(x_0,x_1)}\right) \beta_{1-t}(x_0,x_1)\, \pi(dx_0|x_1)\, \nu(dx_0) \\ + t \int_{M\times M} U\left(\frac{\rho_1(x_1)}{\beta_t(x_0,x_1)}\right) \beta_t(x_0,x_1)\, \pi(dx_1|x_0)\, \nu(dx_1).\end{aligned}$$

Remark 17.36. Since $U(r)/r$ is nondecreasing in r, the displacement convexity condition in Definition 17.34 becomes more stringent as β increases.

The next result is an alternative to Theorem 17.15; recall Convention 17.10.

Theorem 17.37 (CD bounds read off from distorted displacement convexity). *Let M be a Riemannian manifold equipped with a reference measure $\nu = e^{-V}\,\mathrm{vol}$, $V \in C^2(M)$. Let $K \in \mathbb{R}$ and $N \in (1, \infty]$; let $\beta_t^{(K,N)}(x, y)$ be defined as in (14.61). Further, let $p \in [2, +\infty) \cup \{c\}$ satisfy (17.30). Then the following three conditions are equivalent:*

(i) M satisfies the curvature-dimension bound $\mathrm{CD}(K, N)$;

(ii) For each $U \in \mathcal{DC}_N$, the functional U_ν is displacement convex on $P_p^{\mathrm{ac}}(M)$ with distortion $(\beta_t^{(K,N)})$;

(iii) $H_{N,\nu}$ is locally displacement convex with distortion $(\beta_t^{(K,N)})$; and then necessarily $N \geq n$, with equality possible only if V is constant.

Before explaining the proof of this theorem, let me state two very natural open problems (I have no idea how difficult they are).

Open Problem 17.38. *Is there a natural "Eulerian" counterpart to Theorem 17.37?*

Open Problem 17.39. *Theorem 17.15 and 17.37 yield two different upper bounds for $U_\nu(\mu_t)$: on the one hand,*

$$\begin{aligned} U_\nu(\mu_t) \leq\ & (1-t)\, U_\nu(\mu_0) + t\, U_\nu(\mu_1) \\ & - K_{N,U} \int_0^1 \left(\int \rho_s(x)^{1-\frac{1}{N}}\, |\widetilde{\nabla}\psi_s|^2\, d\nu \right) G(s,t)\, ds; \qquad (17.33) \end{aligned}$$

on the other hand,

$$\begin{aligned} U_\nu(\mu_t) \leq (1-t) \int_M U\left(\frac{\rho_0(x_0)}{\beta_{1-t}^{(K,N)}(x_0,x_1)} \right) \beta_{1-t}^{(K,N)}(x_0,x_1)\, \pi(dx_1|x_0)\, d\nu(x_0) \\ + t \int_M U\left(\frac{\rho_1(x_1)}{\beta_t^{(K,N)}(x_0,x_1)} \right) \beta_t^{(K,N)}(x_0,x_1)\, \pi(dx_0|x_1)\, d\nu(x_1). \qquad (17.34) \end{aligned}$$

Can one compare those two bounds, and if yes, which one is sharpest?

At least in the case $N = \infty$, inequality (17.34) implies (17.33): see Theorem 30.5 at the end of these notes.

Exercise 17.40. Show, at least formally, that inequalities (17.33) and (17.34) coincide asymptotically when μ_0 and μ_1 approach each other.

Proof of Theorem 17.37. The proof shares many common points with the proof of Theorem 17.15. I shall restrict to the case $N < \infty$, since the case $N = \infty$ is similar.

Let us start with the implication (i) $\Rightarrow$ (ii). In a **first step**, I shall assume that μ_0 and μ_1 are compactly supported, and (if $K > 0$) $\operatorname{diam}(M) < \pi\sqrt{(N-1)/K}$. With the same notation as in the beginning of the proof of Theorem 17.15,

$$\int_M U(\rho_t(x))\,d\nu(x) = \int_M u(\delta_{t_0}(t,x))\,d\mu_{t_0}(x).$$

By applying inequality (14.56) in Theorem 14.12 (up to a factor which only depends on x and t_0, $\mathcal{D}(t)$ coincides with $\delta_{t_0}(t,x)$), and using the decreasing property of u, we get, with the same notation as in Theorem 14.12,

$$\int_M U(\rho_t(x))\,d\nu(x) \le \int_M u\Big(\tau_{K,N}^{(1-t)}\delta_{t_0}(0,x) + \tau_{K,N}^{(t)}\,\delta_{t_0}(1,x)\Big)\,d\mu_{t_0}(x).$$

Next, by the convexity of u, with coefficients t and $1-t$,

$$\begin{aligned}
&\int_M u\Big(\tau_{K,N}^{(1-t)}\,\delta_{t_0}(0,x) + \tau_{K,N}^{(t)}\,\delta_{t_0}(1,x)\Big)\,d\nu(x)\\
&\le (1-t)\int_M u\left(\frac{\tau_{K,N}^{(1-t)}}{1-t}\,\delta_{t_0}(0,x)\right)d\mu_{t_0}(x)\\
&\qquad\qquad + t\int_M u\left(\frac{\tau_{K,N}^{(t)}}{t}\,\delta_{t_0}(1,x)\right)d\mu_{t_0}(x).
\end{aligned}$$

Since $\beta_t^{(K,N)} = (\tau_{K,N}^{(t)}/t)^N$, the right-hand side of the latter inequality can be rewritten as

$$\begin{aligned}
(1-t)&\int_M \frac{\beta_{1-t}^{(K,N)}(x_0,x_1)}{\rho_0(x_0)}\,U\left(\frac{\rho_0(x_0)}{\beta_{1-t}^{(K,N)}(x_0,x_1)}\right)d\pi(x_0,x_1)\\
&+ t\int_M \frac{\beta_t^{(K,N)}(x_0,x_1)}{\rho_1(x_1)}\,U\left(\frac{\rho_1(x_1)}{\beta_t^{(K,N)}(x_0,x_1)}\right)d\pi(x_0,x_1),
\end{aligned}$$

which is the same as the right-hand side of (17.32).

In a **second step**, I shall relax the assumption of compact support by a restriction argument. Let μ_0 and μ_1 be two probability measures in $P_p^{ac}(M)$, and let $(Z_\ell)_{\ell\in\mathbb{N}}$, $(\mu_{t,\ell})_{0\le t\le 1,\,\ell\in\mathbb{N}}$, $(\pi_\ell)_{\ell\in\mathbb{N}}$ be as in

Proposition 13.2. Let $t \in [0,1]$ be fixed. By the first step, applied with the probability measures $\mu_{t,\ell}$ and the nonlinearity $U_\ell : r \to U(Z_\ell\, r)$,

$$(U_\ell)_\nu(\mu_{t,\ell}) \leq (1-t)\,(U_\ell)_{\pi_\ell,\nu}^{\beta_{1-t}^{(K,N)}}(\mu_{0,\ell}) \;+\; t\,(U_\ell)_{\check{\pi}_\ell,\nu}^{\beta_t^{(K,N)}}(\mu_{1,\ell}). \qquad (17.35)$$

It remains to pass to the limit in (17.35) as $\ell \to \infty$. The left-hand side is handled in exactly the same way as in the proof of Theorem 17.15, and the problem is to pass to the limit in the right-hand side. To ease notation, I shall write $\beta_t^{(K,N)} = \beta$. Let us prove for instance

$$(U_\ell)^\beta_{\pi_\ell,\nu}(\mu_{0,\ell}) \xrightarrow[\ell\to\infty]{} U^\beta_{\pi,\nu}(\mu_0). \qquad (17.36)$$

Since μ_0 is absolutely continuous, the optimal transport plan π comes from a deterministic transport T, and similarly the optimal transport π_ℓ comes from a deterministic transport T_ℓ; Proposition 13.2 guarantees that $T_\ell = T$, $\mu_{0,\ell}$-almost surely. So the left-hand side of (17.36) can be rewritten as

$$\int U\left(\frac{Z_\ell\,\rho_{0,\ell}(x_0)}{\beta(x_0,T(x_0))}\right)\beta(x_0,T(x_0))\,\nu(dx_0).$$

Since U_+ is a nondecreasing function and $Z_\ell\,\rho_{0,\ell}$ is a nondecreasing sequence, the contribution of the positive part U_+ is nondecreasing in ℓ. On the other hand, the contribution of the negative part can be controlled as in the proof of Theorem 17.28:

$$\begin{aligned} U_-\left(\frac{Z_\ell\,\rho_{0,\ell}(x_0)}{\beta(x_0,T(x_0))}\right) &\leq A\Big(Z_\ell\,\rho_{0,\ell}(x_0) + \beta(x_0,T(x_0))^{\frac{1}{N}}\,Z_\ell^{1-\frac{1}{N}}\,\rho_{0,\ell}(x_0)^{1-\frac{1}{N}}\Big) \\ &\leq A\big(\rho_0(x_0) + \beta(x_0,T(x_0))^{\frac{1}{N}}\,\rho_0(x_0)^{1-\frac{1}{N}}\big). \end{aligned}$$

Theorem 17.28 (together with Application 17.29) shows that the latter quantity is always integrable. As a conclusion,

$$\begin{aligned} &\int U_+\left(\frac{Z_\ell\,\rho_{0,\ell}(x_0)}{\beta(x_0,T(x_0))}\right)\beta(x_0,T(x_0))\,\nu(dx_0) \\ &\qquad\xrightarrow[\ell\to\infty]{} \int U_+\left(\frac{\rho_0(x_0)}{\beta(x_0,T(x_0))}\right)\beta(x_0,T(x_0))\,\nu(dx_0) \end{aligned}$$

by monotone convergence; while

$$\int U_-\left(\frac{Z_\ell\,\rho_{0,\ell}(x_0)}{\beta(x_0,T(x_0))}\right)\beta(x_0,T(x_0))\,\nu(dx_0) \xrightarrow[\ell\to\infty]{} \int U_-\left(\frac{\rho_0(x_0)}{\beta(x_0,T(x_0))}\right)\beta(x_0,T(x_0))\,\nu(dx_0)$$

by dominated convergence. So (17.36) holds true, and we can pass to the limit in all the terms of (17.35).

In a **third step**, I shall treat the limit case $\mathrm{diam}\,(M) = D_{K,N} = \pi\sqrt{(N-1)/K}$. To do this, note that M satisfies $\mathrm{CD}(K,N')$ for any $N' > N$; then $\mathrm{diam}\,(M) < D_{K,N'}$, so, by the previous step,

$$U_\nu(\mu_t) \le (1-t)\,U_{\pi,\nu}^{\beta_{1-t}^{(K,N')}}(\mu_0) + t\,U_{\check{\pi},\nu}^{\beta_t^{(K,N')}}(\mu_1).$$

The conclusion follows by letting N' decrease to N and recalling Convention 17.30. This concludes the proof of (i) ⇒ (ii).

It is obvious that (ii) ⇒ (iii). So let us now consider the implication (iii) ⇒ (i). For brevity I shall only consider the case $N > n$. (The case $N = n$ and $V =$ constant is similar; the other cases are ruled out in the same way as in the end of the proof of Theorem 17.15.) Let $x_0 \in M$, $v_0 \in T_{x_0}M$; the goal is to show that $\mathrm{Ric}_{N,\nu}(v_0) \ge K$. Construct $\widetilde{\psi}$, ψ and $(\mu_t)_{0\le t\le 1}$ as in the proof of the implication (ii) ⇒ (iii) in Theorem 17.15. Recall (17.21): As $\theta \to 0$,

$$\begin{aligned} &H_{N,\nu}(\mu_t) - (1-t)\,H_{N,\nu}(\mu_0) - t\,H_{N,\nu}(\mu_1) \\ &\ge -\theta^2\big(\mathrm{Ric}_{N,\nu}(v_0) + O(\theta)\big) \int_0^1\int_M \rho_s(y)^{1-\frac{1}{N}}\,d\nu(y)\,G(s,t)\,ds. \qquad (17.37)\end{aligned}$$

The change of variables $x \to T_s(x)$ is smooth and has the Jacobian $\mathcal{J}_{0\to s}(x) = 1 + O(\theta)$. So

$$\begin{aligned}\int \rho_s(x)^{1-\frac{1}{N}}\,\nu(dx) &= \int \rho_s(T_{0\to s}(x))^{1-\frac{1}{N}}\,\mathcal{J}_{0\to s}(x)\,\nu(dx) \\ &= \int \frac{\rho_0(x)^{1-\frac{1}{N}}}{\mathcal{J}_{0\to s}(x)^{1-\frac{1}{N}}}\,\mathcal{J}_{0\to s}(x)\,\nu(dx) \\ &= \int \rho_0(x)^{1-\frac{1}{N}}\,\mathcal{J}_{0\to s}(x)^{\frac{1}{N}}\,\nu(dx) \\ &= (1+O(\theta))\left(\int \rho_0^{1-\frac{1}{N}}\,d\nu\right);\end{aligned}$$

thus (17.37) can be recast as

$$\begin{aligned} H_{N,\nu}(\mu_t) - (1-t)\,H_{N,\nu}(\mu_0) - t\,H_{N,\nu}(\mu_1) \\ \geq -\theta^2\,\mathrm{Ric}_{N,\nu}(v_0)\left(\frac{t(1-t)}{2}\right)\left(\int_M \rho_0^{1-\frac{1}{N}}\,d\nu\right) + O(\theta^3). \quad (17.38)\end{aligned}$$

(Recall that $\int G(s,t)\,ds = t(1-t)/2$.)

On the other hand, with obvious notation, the left-hand side of (17.38) is (by assumption) bounded above by

$$(1-t)\left(H_{N,\pi,\nu}^{\beta_{1-t}^{(K,N)}}(\mu_0) - H_{N,\nu}(\mu_0)\right) + t\left(H_{N,\check{\pi},\nu}^{\beta_t^{(K,N)}}(\mu_1) - H_{N,\nu}(\mu_1)\right). \quad (17.39)$$

Let us see how this expression behaves in the limit $\theta \to 0$; for instance I shall focus on the first term in (17.39). From the definitions,

$$H_{N,\pi,\nu}^{\beta_{1-t}^{(K,N)}}(\mu_0) - H_{N,\nu}(\mu_0) = N\int \rho_0(x)^{1-\frac{1}{N}}\left(1 - \beta_{1-t}^{(K,N)}(x,T(x))^{\frac{1}{N}}\right)d\nu(x), \quad (17.40)$$

where $T = \exp(\nabla\psi)$ is the optimal transport from μ_0 to μ_1. A standard Taylor expansion shows that

$$\beta_{1-t}^{(K,N)}(x,y)^{\frac{1}{N}} = 1 + \frac{K[1-(1-t)^2]}{6N}\,d(x,y)^2 + O(d(x,y)^4);$$

plugging this back into (17.40), we find

$$\begin{aligned} H_{N,\pi,\nu}^{\beta_{1-t}^{(K,N)}}(\mu_0) &- H_{N,\nu}(\mu_0) \\ &= -\frac{K[1-(1-t)^2]}{6}\int \rho_0(x)^{1-\frac{1}{N}}\left(\theta^2|v_0|^2 + O(\theta^3)\right)d\nu(x) \\ &= -\left(\theta^2|v_0|^2 + O(\theta^3)\right)\frac{K[1-(1-t)^2]}{6}\left(\int \rho_0^{1-\frac{1}{N}}\,d\nu\right).\end{aligned}$$

A similar computation can be performed for the second term in (17.39), taking into account $\int \rho_1^{1-\frac{1}{N}}\,d\nu = \int \rho^{1-\frac{1}{N}}\,d\nu + O(\theta)$. Then the whole expression (17.39) is equal to

$$\begin{aligned} -\theta^2 K\left(\frac{(1-t)[1-(1-t)^2] + t[1-t^2]}{6}\right)|v_0|^2\left(\int \rho^{1-\frac{1}{N}}\,d\nu\right) + O(\theta^3) \\ = -\,\theta^2\,\frac{Kt(1-t)}{2}\,|v_0|^2\left(\int \rho^{1-\frac{1}{N}}\,d\nu\right) + O(\theta^3).\end{aligned}$$

Since this is an upper bound for the right-hand side of (17.38), we obtain, after some simplification,

$$\mathrm{Ric}_{N,\nu}(v_0) + O(\theta) \geq K|v_0|^2 + O(\theta),$$

and the conclusion follows upon taking the limit $\theta \to 0$. □

The case $N = 1$ was not addressed in Theorem 17.37, since the functional $H_{1,\nu}$ was not defined, so Property (iii) would not make sense. However, the rest of the theorem holds true:

Theorem 17.41 (One-dimensional CD bounds and displacement convexity). *Let M be an n-dimensional Riemannian manifold, equipped with a reference measure $\nu = e^{-V}\mathrm{vol}$, where $V \in C^2(M)$. Let $K \in \mathbb{R}$ and let $\beta_t^{(K,1)}(x,y)$ be defined as in (14.63). Then the following two conditions are equivalent:*

(i) M satisfies the curvature-dimension bound $\mathrm{CD}(K,1)$;

(ii) For each $U \in \mathcal{DC}_1$, the functional U_ν is displacement convex on $P_c^{\mathrm{ac}}(M)$ with distortion $(\beta_t^{(K,1)})$;

and then necessarily $n = 1$, V is constant and $K \leq 0$.

Proof of Theorem 17.41. When $K > 0$, (i) is obviously false since ν has to be equal to vol (otherwise $\mathrm{Ric}_{1,\nu}$ will take values $-\infty$); but (ii) is obviously false too since $\beta_t^{(K,1)} = +\infty$ for $0 < t < 1$. So we may assume that $K \leq 0$. Then the proof of (i) $\Rightarrow$ (ii) is along the same lines as in Theorem 17.37. As for the implication (ii) $\Rightarrow$ (i), note that $\mathcal{DC}_{N'} \subset \mathcal{DC}_1$ for all $N' < 1$, so M satisfies Condition (ii) in Theorem 17.37 with N replaced by N', and therefore $\mathrm{Ric}_{N',\nu} \geq Kg$. If $N' < 2$, this forces M to be one-dimensional. Moreover, if V is not constant there is x_0 such that $\mathrm{Ric}_{N',\nu} = V'' - (V')^2/(N'-1)$ is $< K$ for N' small enough. So V is constant and $\mathrm{Ric}_{1,\nu} = \mathrm{Ric} = 0$, a fortiori $\mathrm{Ric}_{1,\nu} \geq K$. □

I shall conclude this chapter with an "intrinsic" theorem of displacement convexity, in which the distortion coefficient β only depends on M and not on a priori given parameters K and N. Recall Definition 14.17 and Convention 17.10.

Theorem 17.42 (Intrinsic displacement convexity). *Let M be a Riemannian manifold of dimension $n \geq 2$, and let $\beta_t(x,y)$ be a continuous positive function on $[0,1] \times M \times M$. Let $p \in [2,+\infty) \cup \{c\}$ such* *that* (17.30) *holds true with* $\nu = \mathrm{vol}$ *and* $N = n$. *Then the following three statements are equivalent:*

(i) $\beta \leq \overline{\beta}$;

(ii) For any $U \in \mathcal{DC}_n$, *the functional* U_ν *is displacement convex on* $P_p^{\mathrm{ac}}(M)$ *with distortion coefficients* (β_t);

(iii) The functional H_n *is locally displacement convex with distortion coefficients* (β_t).

The proof of this theorem is along the same lines as the proof of Theorem 17.37, with the help of Theorem 14.20; details are left to the reader.

Bibliographical notes

Historically, the first motivation for the study of displacement convexity was to establish theorems of unique minimization for functionals which are not strictly convex [614]; a recent development can be found in [202]. In the latter reference, the authors note that some independent earlier works by Alberti and Bellettini [10, 13] can be reinterpreted in terms of displacement convexity.

The definition of the displacement convexity classes $\mathcal{DC}_N$ goes back to McCann's PhD thesis [612], in the case $N < \infty$. (McCann required u in Definition 17.1 to be nonincreasing; but this is automatic, as noticed in Remark 17.2.) The definition of $\mathcal{DC}_\infty$ is taken from [577]. Conditions (i), (ii) or (iii) in Definition 17.1 occur in various contexts, in particular in the theory of nonlinear diffusion equations (as we shall see in Chapter 23), so it is normal that these classes of nonlinearities were rediscovered later by several authors. The normalization $U(0) = 0$ is not the only possible one ($U(1) = 0$ would also be convenient in a compact setting), but it has many advantages. In [612] or more recently [577] it is not imposed that U should be twice differentiable on $(0, +\infty)$.

Theorems 17.15 and 17.37 form the core of this chapter. They result from the contributions of many authors and the story is roughly as follows.

McCann [614] established the displacement convexity of U_ν when $M = \mathbb{R}^n$, $n = N$ and ν is the Lebesgue measure. Things were made somewhat simpler by the Euclidean setting (no Jacobi fields, no $d^2/2$-convex functions, etc.) and by the fact that only displacement convexity (as opposed to Λ-displacement convexity) was considered. Apart from that, the strategy was essentially the same as the one used in this chapter, based on a change of variables, except that the reference measure was μ_0 instead of μ_{t_0}. McCann's proof was recast in my book [814, Proof of Theorem 5.15 (i)]; it takes only a few lines, once one has accepted (a) the *concavity of* $\det^{1/n}$ in $\mathbb{R}^n$: that is, if a symmetric matrix $S \leq I_n$ is given, then $t \longmapsto \det(I_n - tS)^{1/n}$ is concave [814, Lemma 5.21]; and (b) the change of variables formula along displacement interpolation.

Later, Cordero-Erausquin, McCann and Schmuckenschläger [246] studied genuinely Riemannian situations, replacing the concavity of $\det^{1/n}$ in $\mathbb{R}^n$ by distortion estimates, and extending the formula of change of variables along displacement interpolation. With these tools they basically proved the displacement convexity of U_ν for $U \in \mathcal{DC}_N$, as soon as M is a Riemannian manifold of dimension $n \leq N$ and nonnegative Ricci curvature, with the reference measure $\nu = \mathrm{vol}$. It is clear from their paper that their argument also yields, for instance, K-displacement convexity of H as soon as $\mathrm{Ric} \geq K$; moreover, they established (i) $\Rightarrow$ (ii) in Theorem 17.42 for compactly supported densities.

Several authors independently felt the need to rewrite more explicitly the connection between Jacobi fields and optimal transport, which was implicit in [246]. This was done simultaneously by Cordero-Erausquin, McCann and Schmuckenschläger [247] again; by Sturm [761]; and by Lott and myself [577]. All those arguments heavily draw on [246], and they are also reminiscent of arguments used in the proof of the Lévy–Gromov isoperimetric inequality. A large part of the proofs was actually devoted to establish the Jacobian estimates on the exponential function, which I recast here as part of Chapter 14.

Modifications needed to replace the volume measure by $\nu = e^{-V}\mathrm{vol}$ were discussed by Sturm [761] for $N = \infty$; and independently by Lott and myself [577] for $N \leq \infty$. For the purpose of this course, all those modifications were included in the section about "change of measure" in Chapter 14.

It was first proven by Sturm and von Renesse [764] that the displacement convexity of H does not only result from, but actually characterizes the nonnegativity of the Ricci curvature. This statement was generalized by Lott and myself [577], and independently Sturm [762].

In a major contribution, Sturm [763] realized the importance and flexibility of the distorted displacement convexity to encode Ricci curvature bounds. He proved Theorem 17.37 in the most important case

$U = U_N$. As we saw, the proof rests on the inequality (14.56) in Theorem 14.12, which is (as far as I know) due to Cordero-Erausquin, McCann and Schmuckenschläger [246] (in the case $n = N$). Then the general formulation with arbitrary $U \in \mathcal{DC}_N$ was worked out shortly after by Lott and myself [578]. All this was for $N < \infty$; but then the case $N = \infty$ works the same, once one has the correct definitions for $\mathcal{DC}_\infty$ and $\beta_t^{(K,\infty)}$.

In a very recent work, Ohta [657] extended these results to Finsler geometries.

Displacement convexity is not the only way to "synthetically" reformulate lower bounds on the Ricci curvature tensor. An alternative approach is via the study of rates of contraction of diffusion processes in Wasserstein distances. For instance, Sturm and von Renesse [764] proved that $\mathrm{Ric} \geq 0$ is equivalent to the property that the heat equation is a contraction in W_p distance, where p is fixed in $[1, \infty)$. Also, Sturm [761] showed that a Riemannian manifold (equipped with the volume measure) satisfies $\mathrm{CD}(0, N)$ if and only if the nonlinear equation $\partial_t \rho = \Delta \rho^m$ is a contraction for $m \geq 1 - 1/N$. (There is a more complicated criterion for $\mathrm{CD}(K, N)$.) As will be explained in Chapter 24, these results are natural in view of the gradient flow structure of these diffusion equations.

Even if one sticks to displacement convexity, there are possible variants in which one allows the functional to explicitly depend on the interpolation time. Lott [576] showed that a measured Riemannian manifold (M, ν) satisfies $\mathrm{CD}(0, N)$ if and only if $t \longmapsto t H_\nu(\mu_t) + N t \log t$ is a convex function of $t \in [0, 1]$ along any displacement interpolation. There is also a more general version of this statement for $\mathrm{CD}(K, N)$ bounds.

Now come some more technical details. The use of Theorem 17.8 to control noncompactly supported probability densities is essentially taken from [577]; the only change with respect to that reference is that I do not try to define U_ν on the whole of P_2^{ac}, and therefore do not require p to be equal to 2.

In this chapter I used restriction arguments to remove the compactness assumption. An alternative strategy consists in using a density argument and stability theorems (as in [577, Appendix E]); these tools will be used later in Chapters 23 and 30. If the manifold has nonnegative sectional curvature, it is also possible to directly apply the argument of change of variables to the family (μ_t), even if it is not compactly supported, thanks to the uniform inequality (8.45).

Another innovation in the proofs of this chapter is the idea of choosing μ_{t_0} as the reference measure with respect to which changes of variables are performed. The advantage of that procedure (which evolved from discussions with Ambrosio) is that the transport map from μ_{t_0} to μ_t is Lipschitz for all times t, as we know from Chapter 8, while the transport map from μ_0 to μ_1 is only of bounded variation. So the proof given in this section only uses the Jacobian formula for *Lipschitz* changes of variables, and not the more subtle formula for BV changes of variables.

Paths $(\mu_t)_{0\le t\le 1}$ defined in terms of transport from a given measure $\widetilde{\mu}$ (not necessarily of the form μ_{t_0}) are studied in [30] in the context of generalized geodesics in $P_2(\mathbb{R}^n)$. The procedure amounts to considering $\mu_t = (T_t)_\#\widetilde{\mu}$ with $T_t(x) = (1-t)\,T_0(x) + t\,T_1(x)$, where T_0 is optimal between $\widetilde{\mu}$ and μ_0, and T_1 is optimal between $\widetilde{\mu}$ and μ_1. Displacement convexity theorems work for these generalized geodesics just as fine as for true geodesics; one reason is that $t \longmapsto \det((1-t)A + tB)^{1/n}$ is concave for any two nonnegative matrices A and B, not just in the case $A = I_n$. These results are useful in error estimates for gradient flows; they have not yet been adapted to a Riemannian context.

The proofs in the present chapter are of Lagrangian nature, but, as I said before, it is also possible to use Eulerian arguments, at the price of further regularization procedures (which are messy but more or less standard), see in particular the original contribution by Otto and Westdickenberg [673]. As pointed out by Otto, the Eulerian point of view, although more technical, has the merit of separating very clearly the input from local smooth differential geometry (Bochner's formula is a purely local statement about the Laplace operator on M, seen as a differential operator on very smooth functions) and the input from global nonsmooth analysis (Wasserstein geodesics involve $d^2/2$-convexity, which is a nonlocal condition; and $d^2/2$-convex functions are in general nonsmooth). Then Daneri and Savaré [271] showed that the Eulerian approach could be applied even without smoothness, roughly speaking by encoding the convexity property into the existence of a suitable gradient flow, which can be defined for nonsmooth data.

Apart from functionals of the form U_ν, most displacement convex functionals presently known are constructed with functionals of the form $\Phi : \mu \longmapsto \int \Phi(x)\,d\mu(x)$, or $\Psi : \mu \longmapsto \int \Psi(x,y)\,d\mu(x)\,d\mu(y)$, where Φ is a given "potential" and Ψ is a given "interaction potential" [84, 213, 214]. Sometimes these functionals appear in disguise [202].

It is easy to show that the displacement convexity of Φ (seen as a function on $P_2(M)$) is implied by the geodesic convexity of Φ, seen as a function on M. Similarly, it is not difficult to show that the displacement convexity of Ψ is implied by the geodesic convexity of Ψ, seen as a function on $M \times M$. These results can be found for instance in my book [814, Theorem 5.15] in the Euclidean setting. (There it is assumed that $\Psi(x,y) = \Psi(x-y)$, with Ψ convex, but it is immediate to generalize the proof to the case where Ψ is convex on $\mathbb{R}^n \times \mathbb{R}^n$.) Under mild assumptions, there is in fact *equivalence* between the displacement convexity of Ψ on $P_2(M)$ and the geodesic convexity of Ψ on $M \times M$. This is the case for instance if $M = \mathbb{R}^n$, $n \geq 2$, and $\Psi(x,y) = \Psi(x-y)$, where $\Psi(z)$ is an even continuous function of z. Contrary to what is stated in [814], this is false in dimension 1; in fact $\mu \longmapsto \int W(x-y)\,\mu(dx)\,\mu(dy)$ is displacement convex on $P_2(\mathbb{R})$ if and only if $z \longmapsto W(z) + W(-z)$ is convex on $\mathbb{R}_+$ (This is because, by monotonicity, $(x,y) \longmapsto (T(x), T(y))$ preserves the set $\{y \geq x\} \subset \mathbb{R}^2$.) As a matter of fact, an interesting example coming from statistical mechanics, where W is not convex on the whole of $\mathbb{R}$, is discussed in [202].

There is no simple displacement convexity statement known for the Coulomb interaction potential; however, Blower [123] proved that

$$E(\mu) = \frac{1}{2} \int_{\mathbb{R}^2} \log \frac{1}{|x-y|}\,\mu(dx)\,\mu(dy)$$

defines a displacement convex functional on $P_2^{\mathrm{ac}}(\mathbb{R})$. Blower further studied what happens when one adds a potential energy to E, and used these tools to establish concentration inequalities for the eigenvalues of some large random matrices. Also Calvez and Carrillo [196, Chapter 7] recently gave a sharp analysis of the defect of displacement convexity for the logarithmic potential in dimension 1 (which arguably should be the worst) with applications to the long-time study of a one-dimensional nonlinear diffusion equation modeling chemotaxis.

Exercise 17.43. Let M be a compact Riemannian manifold of dimension $n \geq 2$, and let Ψ be a continuous function on $M \times M$; show that Ψ defines a displacement functional on $P_2(M)$ if and only if $(x,y) \longmapsto \Psi(x,y) + \Psi(y,x)$ is geodesically convex on $M \times M$. *Hints:* Note that a product of geodesics in M is also a geodesic in $M \times M$. First show that Ψ is locally convex on $M \times M \setminus \Delta$, where $\Delta = \{(x,x)\} \subset M \times M$. Use a density argument to conclude; note that this argument fails if $n = 1$.

I shall conclude with some comments about Remark 17.33. The classical **Cheng–Toponogov theorem** states the following: *If a Riemannian manifold M has dimension N, Ricci curvature bounded below by $K > 0$, and diameter equal to the limit Bonnet–Myers diameter $D_{K,N} = \pi\sqrt{(N-1)/K}$, then it is a sphere.* I shall explain why this result remains true when the reference measure is not the volume, and M is assumed to satisfy $\mathrm{CD}(K,N)$. Cheng's original proof was based on eigenvalue comparisons, but there is now a simpler argument relying on the Bishop–Gromov inequality [846, p. 229]. This proof goes through when the volume measure is replaced by another reference measure ν, and then one sees that $\Psi = -\log(d\nu/d\mathrm{vol})$ should solve a certain differential equation of Ricatti type (replace the inequality in [573, (4.11)] by an equality). Then the second fundamental forms of the spheres in M have to be the same as in S^N, and one gets a formula for the radial derivative of Ψ. After some computations, one finds that M is an n-sphere of diameter $D_{K,N}$; and that the measure ν, in coordinates (r,θ) from the north pole, is $c\,(\sin(kr))^{N-n}$ times the volume, where $c > 0$ and $k = \sqrt{K/(N-1)}$. If $n < N$, the density of $d\nu/d\mathrm{vol}$ vanishes at the north pole, which is not allowed by our assumptions. The only possibility left out is that M has dimension N and ν is a constant multiple of the volume. All this was explained to me by Lott.

18

Volume control

Controlling the volume of balls is a universal problem in geometry. This means of course controlling the volume from above when the radius increases to infinity; but also controlling the volume from below when the radius decreases to 0. The **doubling property** is useful in both situations.

Definition 18.1 (Doubling property). *Let $(\mathcal{X}, d)$ be a metric space, and let μ be a Borel measure on $\mathcal{X}$, not identically 0. The measure μ is said to be doubling if there exists a constant D such that*

$$\forall x \in \mathcal{X}, \quad \forall r > 0, \quad \nu[B_{2r]}(x)] \leq D\,\nu[B_{r]}(x)]. \tag{18.1}$$

The measure μ is said to be locally doubling if for any fixed closed ball $B[z, R] \subset \mathcal{X}$, there is a constant $D = D(z, R)$ such that

$$\forall x \in B[z, R], \quad \forall r \in (0, R), \quad \nu[B_{2r]}(x)] \leq D\,\nu[B_{r]}(x)]. \tag{18.2}$$

Remark 18.2. It is equivalent to say that a measure ν is locally doubling, or that its restriction to any ball $B[z, R]$ (considered as a metric space) is doubling.

Remark 18.3. It does not really matter whether the definition is formulated in terms of open or in terms of closed balls; at worst this changes the value of the constant D.

When the distance d and the reference measure ν are clear from the context, I shall often say that the space $\mathcal{X}$ is doubling (resp. locally doubling), instead of writing that the measure ν is doubling on the metric space $(\mathcal{X}, d)$.

C. Villani, *Optimal Transport*, Grundlehren der mathematischen Wissenschaften 338,

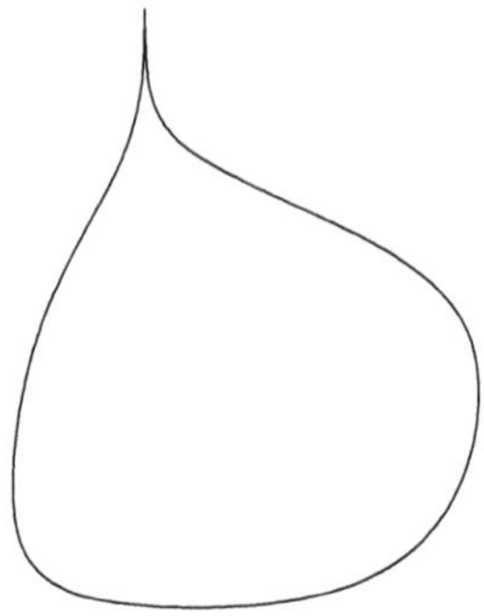

Fig. 18.1. The natural volume measure on this "singular surface" (a balloon with a spine) is not doubling.

It is a standard fact in Riemannian geometry that doubling constants may be estimated, at least locally, in terms of curvature-dimension bounds. These estimates express the fact that the manifold does not contain *sharp spines* (as in Figure 18.1). Of course, it is obvious that a Riemannian manifold has this property, since it is locally diffeomorphic to an open subset of $\mathbb{R}^n$; but curvature-dimension bounds quantify this in terms of the intrinsic geometry, without reference to charts.

Another property which is obvious for a Riemannian manifold, but which doubling makes quantitative, is the fact that the reference measure has full support:

Proposition 18.4 (Doubling measures have full support). *Let* $(\mathcal{X}, d)$ *be a metric space equipped with a locally doubling measure* ν. *Then* $\operatorname{Spt} \nu = \mathcal{X}$.

Proof. Let $x \in \mathcal{X}$, and let $r > 0$. Since ν is nonzero, there is $R > 0$ such that $\nu[B_{R]}(x)] > 0$. Then there is a constant C, possibly depending on x and R, such that ν is C-doubling inside $B_{R]}(x)$. Let $n \in \mathbb{N}$ be large enough that $R \leq 2^n r$; then

$$0 < \nu[B_{R]}(x)] \leq C^n \nu[B_{r]}(x)].$$

So $\nu[B_{r]}(x)] > 0$. Since r is arbitrarily small, x has to lie in the support of ν. □

One of the goals of this chapter is to get doubling constants from curvature-dimension bounds, by means of arguments based on optimal transport. This is not the standard strategy, but it will work just as well as any other, since the results in the end will be optimal. As a preliminary step, I shall establish a "distorted" version of the famous **Brunn–Minkowski inequality**.

Distorted Brunn–Minkowski inequality

The classical Brunn–Minkowski inequality states that whenever A_0 and A_1 are two nonempty compact subsets of $\mathbb{R}^n$, then

$$\left|A_0 + A_1\right|^{\frac{1}{n}} \geq |A_0|^{\frac{1}{n}} + |A_1|^{\frac{1}{n}}, \tag{18.3}$$

where $|\cdot|$ stands for Lebesgue measure, and $A_0 + A_1$ is the set of all vectors of the form $a_0 + a_1$ with $a_0 \in A_0$ and $a_1 \in A_1$. This inequality contains the Euclidean isoperimetric inequality as a limit case (take $A_1 = \varepsilon B(0,1)$ and let $\varepsilon \to 0$).

It is not easy to guess the "correct" generalization of (18.3) to general Riemannian manifolds, and it is only a few years ago that a plausible answer to that problem emerged, in terms of the distortion coefficients (14.61).

In the sequel, I shall use the following notation: if A_0 and A_1 are two nonempty compact subsets of a Riemannian manifold M, then $[A_0, A_1]_t$ stands for the set of all *t-barycenters* of A_0 and A_1, which is the set of all $y \in M$ that can be written as γ_t, where γ is a minimizing, constant-speed geodesic with $\gamma_0 \in A_0$ and $\gamma_1 \in A_1$. Equivalently, $[A_0, A_1]_t$ is the set of all y such that there exists $(x_0, x_1) \in A_0 \times A_1$ with $d(x_0, y)/d(y, x_1) = t/(1-t)$. In $\mathbb{R}^n$, of course $[A_0, A_1]_t = (1-t)\,A_0 + t\,A_1$.

Theorem 18.5 (Distorted Brunn–Minkowski inequality). *Let M be a (complete) Riemannian manifold equipped with a reference measure $\nu = e^{-V}\mathrm{vol}$, $V \in C^2(M)$, satisfying a $\mathrm{CD}(K,N)$ curvature-dimension condition. Let A_0, A_1 be two nonempty compact subsets, and let $t \in (0,1)$. Then:*

- *If $N < \infty$,*

$$\nu\big[[A_0, A_1]_t\big]^{\frac{1}{N}} \geq (1-t)\left[\inf_{(x_0,x_1)\in A_0\times A_1} \beta_{1-t}^{(K,N)}(x_0,x_1)^{\frac{1}{N}}\right] \nu[A_0]^{\frac{1}{N}} + t\left[\inf_{(x_0,x_1)\in A_0\times A_1} \beta_{t}^{(K,N)}(x_0,x_1)^{\frac{1}{N}}\right] \nu[A_1]^{\frac{1}{N}}, \tag{18.4}$$

where $\beta_t^{(K,N)}(x_0,x_1)$ are the distortion coefficients defined in (14.61).

- *If* $N = \infty$,

$$\log \frac{1}{\nu\big[[A_0, A_1]_t\big]} \leq (1-t) \log \frac{1}{\nu[A_0]} + t \log \frac{1}{\nu[A_1]} - \frac{Kt(1-t)}{2} \sup_{x_0 \in A_0,\, x_1 \in A_1} d(x_0, x_1)^2. \quad (18.5)$$

By particularizing Theorem 18.5 to the case when $K = 0$ and $N < \infty$ (so $\beta_t^{(K,N)} = 1$), one can show that nonnegatively curved Riemannian manifolds satisfy a Brunn–Minkowski inequality which is similar to the Brunn–Minkowski inequality in $\mathbb{R}^n$:

Corollary 18.6 (Brunn–Minkowski inequality in nonnegative curvature). *With the same notation as in Theorem 18.5, if M satisfies the curvature-dimension condition* $\mathrm{CD}(0, N)$, $N \in (1, +\infty)$, *then*

$$\nu\big[[A_0, A_1]_t\big]^{\frac{1}{N}} \geq (1-t)\, \nu[A_0]^{\frac{1}{N}} + t\, \nu[A_1]^{\frac{1}{N}}. \quad (18.6)$$

Remark 18.7. When $M = \mathbb{R}^n$, $N = n$, inequality (18.6) reduces to

$$\big|(1-t)A_0 + tA_1\big|^{\frac{1}{n}} \geq (1-t)\, |A_0|^{\frac{1}{n}} + t\, |A_1|^{\frac{1}{n}},$$

where $|\cdot|$ stands for the n-dimensional Lebesgue measure. By homogeneity, this is equivalent to (18.4).

Idea of the proof of Theorem 18.5. Introduce an optimal coupling between a random point γ_0 chosen uniformly in A_0 and a random point γ_1 chosen uniformly in A_1 (as in the proof of isoperimetry in Chapter 2). Then γ_t is a random point (not necessarily uniform) in A_t. If A_t is very small, then the law μ_t of γ_t will be very concentrated, so its density will be very high, but then this will contradict the displacement convexity estimates implied by the curvature assumptions. For instance, consider for simplicity $U(r) = r^m$, $m \geq 1$, $K = 0$: Since $U_\nu(\mu_0)$ and $U_\nu(\mu_1)$ are finite, this implies a bound on $U_\nu(\mu_t)$, and this bound cannot hold if the support of μ_t is too small (in the extreme case where A_t is a single point, μ_t will be a Dirac, so $U_\nu(\mu_t) = +\infty$). Thus, the support of μ_t has to be large enough. It turns out that the optimal estimates are obtained with $U = U_N$, as defined in (16.17). □

Detailed proof of Theorem 18.5. First consider the case $N < \infty$. For brevity I shall write just β_t instead of $\beta_t^{(K,N)}$. By regularity of the measure ν and an easy approximation argument, it is sufficient to treat the case when $\nu[A_0] > 0$ and $\nu[A_1] > 0$. Then one may define $\mu_0 = \rho_0\,\nu$, $\mu_1 = \rho_1\,\nu$, where

$$\rho_0 = \frac{1_{A_0}}{\nu[A_0]}, \qquad \rho_1 = \frac{1_{A_1}}{\nu[A_1]}.$$

In words, μ_{t_0} ($t_0 \in \{0,1\}$) is the law of a random point distributed uniformly in A_{t_0}. Let $(\mu_t)_{0\le t\le 1}$ be the unique displacement interpolation between μ_0 and μ_1, for the cost function $d(x,y)^2$. Since M satisfies the curvature-dimension bound $\mathrm{CD}(K,N)$, Theorem 17.37, applied with $U(r) = U_N(r) = -N\big(r^{1-\frac{1}{N}} - r\big)$, implies

$$\begin{aligned}
&\int_M U_N(\rho_t(x))\,\nu(dx)\\
&\qquad \le (1-t)\int_M U_N\left(\frac{\rho_0(x_0)}{\beta_{1-t}(x_0,x_1)}\right)\beta_{1-t}(x_0,x_1)\,\pi(dx_1|x_0)\,\nu(dx_0)\\
&\qquad\qquad + t\int_M U_N\left(\frac{\rho_1(x_1)}{\beta_t(x_0,x_1)}\right)\beta_t(x_0,x_1)\,\pi(dx_0|x_1)\,\nu(dx_1)\\
&\qquad = (1-t)\int_M U_N\left(\frac{\rho_0(x_0)}{\beta_{1-t}(x_0,x_1)}\right)\frac{\beta_{1-t}(x_0,x_1)}{\rho_0(x_0)}\,\pi(dx_0\,dx_1)\\
&\qquad\qquad + t\int_M U_N\left(\frac{\rho_1(x_1)}{\beta_t(x_0,x_1)}\right)\frac{\beta_t(x_0,x_1)}{\rho_1(x_1)}\,\pi(dx_0\,dx_1),
\end{aligned}$$

where π is the optimal coupling of (μ_0,μ_1), and β_t is a shorthand for $\beta_t^{(K,N)}$; the equality comes from the fact that, say, $\pi(dx_0\,dx_1) = \mu(dx_0)\,\pi(dx_1|x_0) = \rho(x_0)\,\nu(dx_0)\,\pi(dx_1|x_0)$. After replacement of U_N by its explicit expression and simplification, this leads to

$$\begin{aligned}
\int_M \rho_t(x)^{1-\frac{1}{N}}\,\nu(dx) \ge (1-t)&\int_M \rho_0(x)^{-\frac{1}{N}}\beta_{1-t}(x_0,x_1)^{\frac{1}{N}}\,\pi(dx_0\,dx_1)\\
&+ t\int_M \rho_1(x)^{-\frac{1}{N}}\beta_t(x_0,x_1)^{\frac{1}{N}}\,\pi(dx_0\,dx_1). \qquad (18.7)
\end{aligned}$$

Since π is supported in $A_0\times A_1$ and has marginals $\rho_0\,\nu$ and $\rho_1\nu$, one can bound the right-hand side of (18.7) below by

$$(1-t)\,\beta_{1-t}^{\frac{1}{N}}\int_M \rho_0(x_0)^{1-\frac{1}{N}}\,d\nu(x_0) \;+\; t\,\beta_t^{\frac{1}{N}}\int_M \rho_1(x_1)^{1-\frac{1}{N}}\,d\nu(x_1),$$

where β_t stands for the minimum of $\beta_t(x_0,x_1)$ over all pairs $(x_0,x_1)\in A_0\times A_1$. Then, by explicit computation,

$$\int_M \rho_0(x_0)^{1-\frac{1}{N}}\, d\nu(x_0) = \nu[A_0]^{\frac{1}{N}}, \qquad \int_M \rho_1(x_1)^{1-\frac{1}{N}}\, d\nu(x_1) = \nu[A_1]^{\frac{1}{N}}.$$

So to conclude the proof of (18.4) it sufficient to show

$$\int_M \rho_t^{1-\frac{1}{N}}\, d\nu \leq \nu\big[[A_0, A_1]_t\big]^{\frac{1}{N}}.$$

Obviously, μ_t is supported in $A_t = [A_0, A_1]_t$; therefore ρ_t is a probability density on that set. By Jensen's inequality,

$$\begin{aligned}\int_{A_t} \rho_t^{1-\frac{1}{N}}\, d\nu &= \nu[A_t] \int_{A_t} \rho_t^{1-\frac{1}{N}} \frac{d\nu}{\nu[A_t]} \\ &\leq \nu[A_t] \left(\int_{A_t} \rho_t \frac{d\nu}{\nu[A_t]}\right)^{1-\frac{1}{N}} \\ &= \nu[A_t]^{\frac{1}{N}} \left(\int_{A_t} \rho_t\, d\nu\right)^{1-\frac{1}{N}} = \nu[A_t]^{\frac{1}{N}}.\end{aligned}$$

This concludes the proof of (18.4).

The proof in the case $N = \infty$ is along the same lines, except that now it is based on the K-displacement convexity of H_ν and the convexity of $r \longmapsto r \log r$. □

Bishop–Gromov inequality

The Bishop–Gromov inequality states that the volume of balls in a space satisfying $\mathrm{CD}(K,N)$ does not grow faster than the volume of balls in the model space of constant sectional curvature having Ricci curvature equal to K and dimension equal to N. In the case $K = 0$, it takes the following simple form:

$$\frac{\nu[B_r(x)]}{r^N} \qquad \text{is a nonincreasing function of } r.$$

(It does not matter whether one considers the closed or the open ball of radius r.) In the case $K > 0$ (resp. $K < 0$), the quantity on the left-hand side should be replaced by

$$\frac{\nu[B_r(x)]}{\displaystyle\int_0^r \left(\sin\sqrt{\frac{K}{N-1}}\, t\right)^{N-1} dt},$$

$$\text{resp.}\quad \frac{\nu[B_r(x)]}{\displaystyle\int_0^r \left(\sinh\sqrt{\frac{|K|}{N-1}}\,t\right)^{N-1} dt}.$$

Here is a precise statement:

Theorem 18.8 (Bishop–Gromov inequality). *Let M be a Riemannian manifold equipped with a reference measure $\nu = e^{-V}\mathrm{vol}$, satisfying a curvature-dimension condition $\mathrm{CD}(K,N)$ for some $K \in \mathbb{R}$, $1 < N < \infty$. Further, let*

$$s^{(K,N)}(t) = \begin{cases} \left(\sin\sqrt{\dfrac{K}{N-1}}\,t\right)^{N-1} & \text{if } K > 0 \\[2ex] t^{N-1} & \text{if } K = 0 \\[2ex] \left(\sinh\sqrt{\dfrac{|K|}{N-1}}\,t\right)^{N-1} & \text{if } K < 0 \end{cases}$$

Then, for any $x \in M$,

$$\frac{\nu[B_r(x)]}{\displaystyle\int_0^r s^{(K,N)}(t)\,dt}$$

is a nonincreasing function of r.

Proof of Theorem 18.8. Let us start with the case $K = 0$, which is simpler. Let $A_0 = \{x\}$ and $A_1 = B_{r]}(x)$; in particular, $\nu[A_0] = 0$. For any $s \in (0,r)$, one has $[A_0, A_1]_{\frac{s}{r}} \subset B_{s]}(x)$, so by the Brunn–Minkowski inequality (18.6),

$$\nu[B_{s]}(x)]^{\frac{1}{N}} \geq \nu\big[[A_0,A_1]_{\frac{s}{r}}\big]^{\frac{1}{N}} \geq \left(\frac{s}{r}\right)\,\nu[B_{r]}(x)]^{\frac{1}{N}},$$

and the conclusion follows immediately.

Now let us consider the general case. By Lemma 18.9 below, it will be sufficient to check that

$$\frac{\frac{d^+}{dr}\,\nu[B_r]}{s^{(K,N)}(r)} \quad \text{is nonincreasing,} \tag{18.8}$$

where $B_r = B_{r]}(x)$.

Apply Theorem 18.5 with $A_0 = \{x\}$ again, but now $A_1 = B_{r+\varepsilon} \setminus B_r$; then for $t \in (0,1)$ one has $[A_0, A_1]_t \subset B_{t(r+\varepsilon)} \setminus B_t$. Moreover, for $K \geq 0$, one has

$$\beta_t^{(K,N)}(x_0, x_1) \geq \left(\frac{\sin\left(t\sqrt{\frac{K}{N-1}}(r+\varepsilon)\right)}{t\sin\left(\sqrt{\frac{K}{N-1}}(r+\varepsilon)\right)} \right)^{N-1};$$

for $K < 0$ the same formula remains true with sin replaced by sinh, K by $|K|$ and $r+\varepsilon$ by $r-\varepsilon$. In the sequel, I shall only consider $K > 0$, the treatment of $K < 0$ being obviously similar. After applying the above bounds, inequality (18.4) yields

$$\nu\Big[B_{t(r+\varepsilon)} \setminus B_{tr}\Big]^{\frac{1}{N}} \geq t \left(\frac{\sin\left(t\sqrt{\frac{K}{N-1}}\,(r+\varepsilon)\right)}{t\,\sin\left(\sqrt{\frac{K}{N-1}}\,(r+\varepsilon)\right)} \right)^{\frac{N-1}{N}} \nu\Big[B_{r+\varepsilon} \setminus B_r\Big]^{\frac{1}{N}};$$

or, what is the same,

$$\frac{\nu\big[B_{t(r+\varepsilon)} \setminus B_{tr}\big]}{\left(\sin\left(\sqrt{\frac{K}{N-1}}\,t(r+\varepsilon)\right)\right)^{N-1}} \geq t\,\frac{\nu\big[B_{r+\varepsilon} \setminus B_r\big]}{\left(\sin\left(\sqrt{\frac{K}{N-1}}\,(r+\varepsilon)\right)\right)^{N-1}}.$$

If $\phi(r)$ stands for $\nu[B_r]$, then the above inequality can be rewritten as

$$\frac{\phi(tr+t\varepsilon) - \phi(tr)}{\varepsilon\, s^{(K,N)}(t(r+\varepsilon))} \geq \frac{\phi(r+\varepsilon) - \phi(r)}{\varepsilon\, s^{(K,N)}(r+\varepsilon)}.$$

In the limit $\varepsilon \to 0$, this yields

$$\frac{\phi'(tr)}{s^{(K,N)}(tr)} \geq \frac{\phi'(r)}{s^{(K,N)}(r)}.$$

This was for any $t \in [0,1]$, so $\phi'/s^{(K,N)}$ is indeed nonincreasing, and the proof is complete. □

The following lemma was used in the proof of Theorem 18.8. At first sight it seems obvious and the reader may skip its proof.

Lemma 18.9. *Let $a < b$ in $\mathbb{R} \cup \{+\infty\}$, let $g : (a,b) \to \mathbb{R}_+$ be a positive continuous function, integrable at a, and let $G(r) = \int_a^r g(s)\,ds$. Let $F : [a,b) \to \mathbb{R}_+$ be a nondecreasing measurable function satisfying $F(a) = 0$, and let $f(r) = d^+F/dr$ be its upper derivative. If f/g is nonincreasing then also F/G is nonincreasing.*

Proof of Lemma 18.9. Let $h = f/g$; by assumption, h is nonincreasing. In particular, for any $x \geq x_0 > a$, $f(x) \leq g(x)\,h(x_0)$ is locally bounded, so F is locally Lipschitz, and $F(y) - F(x) = \int_x^y f(t)\,dt$ as soon as $y > x > a$. Taking the limit $x \to a$ shows that $F(y) = \int_a^y f(t)\,dt$. So the problem is to show that

$$x \leq y \Longrightarrow \qquad \frac{\int_a^x f(t)\,dt}{\int_a^x g(t)\,dt} \leq \frac{\int_a^y f(t)\,dt}{\int_a^y g(t)\,dt}. \tag{18.9}$$

If $a \leq t \leq x \leq t' \leq y$, then $h(t) \leq h(t')$; so

$$\int_a^x f \int_x^y g = \int_a^x gh \int_x^y g \leq \int_a^x g \int_x^y gh = \int_a^x g \int_x^y f.$$

This implies

$$\frac{\int_a^x f}{\int_a^x g} \leq \frac{\int_x^y f}{\int_x^y g},$$

and (18.9) follows. □

Exercise 18.10. Give an alternative proof of the Bishop–Gromov inequality for $\mathrm{CD}(0, N)$ Riemannian manifolds, using the convexity of $t \longmapsto t\,U_\nu(\mu_t) + N\,t \log t$, for $U \in \mathcal{DC}_N$, when $(\mu_t)_{0 \leq t \leq 1}$ is a displacement interpolation.

Doubling property

From Theorem 18.8 and elementary estimates on the function $s^{(K,N)}$ it is easy to deduce the following corollary:

Corollary 18.11 ($\mathrm{CD}(K,N)$ implies doubling). *Let M be a Riemannian manifold equipped with a reference measure $\nu = e^{-V}\,\mathrm{vol}$, satisfying a curvature-dimension condition $\mathrm{CD}(K, N)$ for some $K \in \mathbb{R}$, $1 < N < \infty$. Then ν is doubling with a constant C that is:*

- *uniform and no more than 2^N if $K \geq 0$;*
- *locally uniform and no more than $2^N\, D(K, N, R)$ if $K < 0$, where*

$$D(K, N, R) = \left[\cosh\left(2\sqrt{\frac{|K|}{N-1}}\,R\right)\right]^{N-1}, \tag{18.10}$$

when restricted to a large ball $B[z, R]$.

The Bishop–Gromov inequality is however more precise than just doubling property: for instance, if $0 < s < r$ then, with the same notation as before,

$$\nu[B_r(x)] \geq \nu[B_s(x)] \geq \left(\frac{V(s)}{V(r)}\right) \nu[B_r(x)],$$

where $V(r)$ is the volume of $B_r(x)$ in the model space. This implies that $\nu[B_r(x)]$ is a continuous function of r. Of course, this property is otherwise obvious, but the Bishop–Gromov inequality provides an explicit modulus of continuity.

Dimension-free bounds

There does not seem to be any "natural" analog of the Bishop–Gromov inequality when $N = \infty$. However, we have the following useful estimates.

Theorem 18.12 (Dimension-free control on the growth of balls). *Let M be a Riemannian manifold equipped with a reference measure $\nu = e^{-V}\mathrm{vol}$, $V \in C^2(M)$, satisfying a $\mathrm{CD}(K,\infty)$ condition for some $K \in \mathbb{R}$. Then, for any $\delta > 0$, there exists a constant $C = C\big(K_-, \delta, \nu[B_\delta(x_0)], \nu[B_{2\delta}(x_0)]\big)$, such that for all $r \geq \delta$,*

$$\nu[B_r(x_0)] \leq e^{Cr}\, e^{(K_-)\frac{r^2}{2}}; \tag{18.11}$$

$$\nu[B_{r+\delta}(x_0) \setminus B_r(x_0)] \leq e^{Cr}\, e^{-K\frac{r^2}{2}} \qquad \text{if } K > 0. \tag{18.12}$$

In particular, if $K' < K$ then

$$\int e^{\frac{K'}{2} d(x_0,x)^2}\, \nu(dx) < +\infty. \tag{18.13}$$

Proof of Theorem 18.12. For brevity I shall write B_r for $B_{r]}(x_0)$. Apply (18.5) with $A_0 = B_\delta$, $A_1 = B_r$, and $t = \delta/(2r) \leq 1/2$. For any minimizing geodesic γ going from A_0 to A_1, one has $d(\gamma_0,\gamma_1) \leq r+\delta$, so

$$d(x_0,\gamma_t) \leq d(x_0,\gamma_0) + d(\gamma_0,\gamma_t) \leq \delta + t(r+\delta) \leq \delta + 2tr \leq 2\delta.$$

So $[A_0, A_1]_t \subset B_{2\delta}$, and by (18.5),

$$\log\frac{1}{\nu[B_{2\delta}]}\leq\Big(1-\frac{\delta}{2r}\Big)\log\frac{1}{\nu[B_\delta]}+\frac{\delta}{2r}\log\frac{1}{\nu[B_r]}+\frac{K_-}{2}\frac{\delta}{2r}\Big(1-\frac{\delta}{2r}\Big)(r+\delta)^2.$$

This implies an estimate of the form

$$\nu[B_r]\leq\exp\Big(a+br+\frac{c}{r}+\frac{K_-r^2}{2}\Big),$$

where a,b,c only depend on δ, $\nu[B_\delta]$ and $\nu[B_{2\delta}]$. Inequality (18.11) follows.

The proof of inequality (18.12) is just the same, but with $A_0=B_\delta$, $A_1=B_{r+\delta}\setminus B_r$, $t=\delta/(3r)$.

To prove (18.13) in the case $K>0$, it suffices to take $\delta=1$ and write

$$\begin{aligned}\int_{\mathcal X}e^{\frac{K'}{2}d(x_0,x)^2}\,\nu(dx)&\leq e^{\frac{K'}{2}}\,\nu[B_1]+\sum_{k\geq 1}e^{\frac{K'}{2}(k+1)^2}\,\nu[B_{k+1}\setminus B_k]\\&\leq e^{\frac{K'}{2}}\,\nu[B_1]+C\sum_{k\geq 1}e^{C(k+1)}e^{\frac{K'}{2}(k+1)^2}e^{-Kk^2}\\&<+\infty.\end{aligned}$$

The case $K\leq 0$ is treated similarly. □

Bibliographical notes

The Brunn–Minkowski inequality in $\mathbb{R}^n$ goes back to the end of the nineteenth century; it was first established by Brunn (for convex sets in dimension 2 or 3), and later generalized by Minkowski (for convex sets in arbitrary dimension) and Lusternik [581] (for arbitrary compact sets). Nowadays, it is still one of the cornerstones of the geometry of convex bodies. Standard references on the Brunn–Minkowski theory are the book by Schneider [741] and the more recent survey paper by Gardner [406]; one can also consult the lecture by Maurey [608], the recent research review by Barthe [71], and his broad-audience introductory text [72].

It is classical to prove the Brunn–Minkowski inequality (in $\mathbb{R}^n$) via changes of variables, usually called *reparametrizations* in this context. McCann [612] noticed that optimal transport does yield a convenient reparametrization; this is a bit more complicated than the reparametrizations classically used in $\mathbb{R}^n$, but it has the advantage of begin defined in more intrinsic terms. McCann's argument is reproduced in [814, Section 6.1]; it is basically the same as the proof of Theorem 18.5, only much simpler because it is in Euclidean space.

At the end of the nineties, it was still not clear what would be the correct extension of that theory to curved spaces. The first hint came when Cordero-Erausquin [241] used the formalism of optimal transport to guess a Prékopa–Leindler inequality on the sphere. In Euclidean space, the Prékopa–Leindler inequality is a well-known functional version of the Brunn–Minkowski inequality (it is discussed for instance in the above-mentioned surveys, and we shall meet it in the next chapter). Cordero-Erausquin, McCann and Schmuckenschläger [246] developed the tools necessary to make this approach rigorous, and also established Prékopa–Leindler inequalities in curved geometry (when the reference measure is the volume). Then Sturm [763] adapted the proof of [246] to get Brunn–Minkowski inequalities for general reference measures. Ohta [657] further generalized these results to Finsler geometries.

The proof of the Bishop–Gromov inequality in the case $K = 0$ is taken from [577]. Apart from that, my presentation in this chapter is strongly inspired by Sturm [763]. In particular, it is from that source that I took the statement of Theorem 18.5 and the proof of the Bishop–Gromov inequality for $K \neq 0$.

Exercice 18.10 was inspired from an exchange with Lott. The convexity property mentioned in the exercise is proven in [576].

More classical proofs of the Bishop–Gromov inequality can be found in reference textbooks, e.g. [394, Theorem 4.19]. The resulting comparison inequality between the volume of balls in a $\mathrm{CD}(K, N)$ Riemannian manifold and in the comparison space is called just the Bishop inequality [394, Theorem 3.101(i)]. Also available is a reversed comparison principle for upper bounds on the *sectional* curvature [394, Theorem 3.101(ii)], due to Gunther.

Lemma 18.9 is a slight variation of [223, Lemma 3.1]; it is apparently due to Gromov [635]. This lemma can also be proven by approximation from its discrete version.

19

Density control and local regularity

The following situation occurs in many problems of local regularity: Knowing a certain estimate on a certain ball $B_r(x_0)$, deduce a better estimate on a smaller ball, say $B_{r/2}(x_0)$. In the fifties, this point of view was put to a high degree of sophistication by De Giorgi in his famous proof of Hölder estimates for elliptic second-order partial differential equations in divergence form; and it also plays a role in the alternative solutions found at the same time by Nash, and later by Moser. When fine analysis on metric spaces started to develop, it became an important issue to understand what were the key ingredients lying at the core of the methods of De Giorgi, Nash and Moser. It is now accepted by many that the two key inequalities are:

- a **doubling inequality** for the reference volume measure;
- a **local Poincaré inequality**, controlling the deviation of a function on a smaller ball by the integral of its gradient on a larger ball. Here is a precise definition:

Definition 19.1 (Local Poincaré inequality). *Let $(\mathcal{X}, d)$ be a Polish metric space and let ν be a Borel measure on $\mathcal{X}$. It is said that ν satisfies a local Poincaré inequality with constant C if, for any Lipschitz function u, any point $x_0 \in \mathcal{X}$ and any radius $r > 0$,*

$$\fint_{B_r(x_0)} \left| u(x) - \langle u \rangle_{B_r(x_0)} \right| d\nu(x) \leq Cr \fint_{B_{2r}(x_0)} |\nabla u(x)| \, d\nu(x), \qquad (19.1)$$

where $\fint_B = (\nu[B])^{-1} \int_B$ stands for the averaged integral over B, and $\langle u \rangle_B = \fint_B u \, d\nu$ for the average of the function u on B.

Let B be a Borel subset of $\mathcal{X}$. It is said that ν satisfies a local Poincaré inequality with constant C in B if inequality (19.1) *holds true under the additional restriction that $B_{2r}(x_0) \subset B$.*

Remark 19.2. The definition of $|\nabla u|$ in a nonsmooth context will be discussed later (see Chapter 20). For the moment the reader does not need to know this notion since this chapter only considers Riemannian manifolds.

Remark 19.3. The word "local" in Definition 19.1 means that the inequality is interested in averages *around some point* x_0. This is in contrast with the "global" Poincaré inequalities that will be considered later in Chapter 21, in which averages are over the whole space.

There are an incredible number of variants of Poincaré inequalities, but I shall stick to the ones appearing in Definition 19.1. Sometimes I shall say that ν satisfies a *uniform* local Poincaré inequality to stress the fact that the constant C is independent of x_0 and r. For most applications this uniformity is not important, all that matters is that inequality (19.1) holds true in the neighborhood of any point x_0; so it is sufficient to prove that ν satisfies a local Poincaré inequality with constant $C = C(R)$ on each ball $B(z, R)$, where z is fixed once for all.

Just as the doubling inequality, the local Poincaré inequality might be ruined by sharp spines, and Ricci curvature bounds will prevent those spines to occur, providing quantitative Poincaré constants (that will be uniform in nonnegative curvature). Again, the goal of this chapter is to prove these facts by using optimal transport. The strategy goes through *pointwise bounds* on the density of the displacement interpolant.

There are at least two ways to prove pointwise bounds on the displacement interpolant. The first one consists in combining the Jacobian equation involving the density of the interpolant (Chapter 11) with the Jacobian determinant estimates derived from the Ricci curvature bounds (Chapter 14). The second way goes via displacement convexity (Chapter 17); it is quite more indirect, but its interest will become apparent later in Chapter 30.

Of course, pointwise bounds do not result directly from displacement convexity, which only yields *integral* bounds on the interpolant; still, it is possible to deduce pointwise bounds from integral bounds by using the stability of optimal transport under restriction (Theorem 4.6).

The idea is simple: a pointwise bound on $\rho_t(x)$ will be achieved by considering integral bounds on a very small ball $B_\delta(x)$, as $\delta \to 0$.

Apart from the local Poincaré inequality, the pointwise control on the density will imply at once the Brunn–Minkowski inequality, and also its functional counterpart, the Prékopa–Leindler inequality. This is not surprising, since a pointwise control is morally stronger than an integral control.

Pointwise estimates on the interpolant density

The next theorem is the key result of this chapter. The notation $[x,y]_t$ stands for the set of all t-barycenters of x and y (as in Theorem 18.5).

Theorem 19.4 ($\mathrm{CD}(K,N)$ **implies pointwise bounds on displacement interpolants).** *Let M be a Riemannian manifold equipped with a reference measure $\nu = e^{-V}\mathrm{vol}$, $V \in C^2(M)$, satisfying a curvature-dimension $\mathrm{CD}(K,N)$ for some $K \in \mathbb{R}$, $N \in (1,\infty]$. Further, let $\mu_0 = \rho_0\,\nu$ and $\mu_1 = \rho_1\,\nu$ be two probability measures in $P_p^{\mathrm{ac}}(M)$, where $p \in [2,+\infty) \cup \{c\}$ satisfies the assumptions of Theorem 17.8. Let $(\mu_t)_{0\le t\le 1}$ be the unique displacement interpolation between μ_0 and μ_1, and let ρ_t stand for the density of μ_t with respect to ν. Then for any $t \in (0,1)$,*

- *If $N < \infty$, one has the pointwise bound*

$$\rho_t(x) \le \sup_{x\in[x_0,x_1]_t} \left((1-t)\left(\frac{\rho_0(x_0)}{\beta_{1-t}^{(K,N)}(x_0,x_1)}\right)^{-\frac{1}{N}} + t\left(\frac{\rho_1(x_1)}{\beta_t^{(K,N)}(x_0,x_1)}\right)^{-\frac{1}{N}} \right)^{-N}, \tag{19.2}$$

  where by convention $\left((1-t)\,a^{-\frac{1}{N}} + t\,b^{-\frac{1}{N}}\right)^{-N} = 0$ if either a or b is 0;
- *If $N = \infty$, one has the pointwise bound*

$$\rho_t(x) \le \sup_{x\in[x_0,x_1]_t} \rho_0(x_0)^{1-t}\,\rho_1(x_1)^t \exp\left(-\frac{Kt(1-t)}{2}\,d(x_0,x_1)^2\right). \tag{19.3}$$

Corollary 19.5 (Preservation of uniform bounds in nonnegative curvature). *With the same notation as in Theorem 19.4, if $K \ge 0$ then*

$$\|\rho_t\|_{L^\infty(\nu)} \le \max\left(\|\rho_0\|_{L^\infty(\nu)}, \|\rho_1\|_{L^\infty(\nu)}\right).$$

As I said before, there are (at least) two possible schemes of proof for Theorem 19.4. The first one is by direct application of the Jacobian estimates from Chapter 14; the second one is based on the displacement convexity estimates from Chapter 17. The first one is formally simpler, while the second one has the advantage of being based on very robust functional inequalities. I shall only sketch the first proof, forgetting about regularity issues, and give a detailed treatment of the second one.

Sketch of proof of Theorem 19.4 by Jacobian estimates. Let $\psi : M \to \mathbb{R} \cup \{+\infty\}$ be a $(d^2/2)$-convex function so that $\mu_t = [\exp(t\widetilde{\nabla}\psi)]_{\#}\mu_0$. Let $\mathcal{J}(t,x)$ stand for the Jacobian determinant of $\exp(t\widetilde{\nabla}\psi)$; then, with the shorthand $x_t = \exp_{x_0}(t\widetilde{\nabla}\psi(x_0))$, the Jacobian equation of change of variables can be written

$$\rho_0(x_0) = \rho_t(x_t)\,\mathcal{J}(t,x_0).$$

Similarly,

$$\rho_0(x_0) = \rho_1(x_1)\,\mathcal{J}(1,x_0).$$

Then the result follows directly from Theorems 14.11 and 14.12: Apply equation (14.56) if $N < \infty$, (14.55) if $N = \infty$ (recall that $\mathcal{D} = \mathcal{J}^{\frac{1}{N}}$, $\ell = -\log \mathcal{J}$). □

Proof of Theorem 19.4 by displacement interpolation. For simplicity I shall only consider the case $N < \infty$, and derive the conclusion from Theorem 17.37. Then the case $N = \infty$ can be treated by adapting the proof of the case $N < \infty$, replacing Theorem 17.37 by Theorem 17.15, and using the function U_∞ defined in (16.17). (Formally, it amounts to taking the limit $N \to \infty$ in (19.2).)

Let $t \in (0,1)$ be given, let $(\mu_s)_{0\le s\le 1}$ be as in the statement of the theorem, and let Π be the law of a random geodesic γ such that $\mathrm{law}(\gamma_s) = \mu_s$. Let y be an arbitrary point in M, and $\delta > 0$; the goal is to estimate from above the probability $\mathbb{P}\left[\gamma_t \in B_\delta(y)\right] = \mu_t[B_\delta(y)]$, so as to recover a bound on $\rho_t(y)$ as $\delta \to 0$.

If $\mathbb{P}\left[\gamma_t \in B_\delta(y)\right] = 0$, then there is nothing to prove. Otherwise we may condition γ by the event "$\gamma_t \in B_\delta(y)$". Explicitly, this means: Introduce γ' such that $\mathrm{law}(\gamma') = \Pi' = (1_{\mathcal{Z}}\Pi)/\Pi[\mathcal{Z}]$, where

$$\mathcal{Z} = \Big\{\gamma \in \Gamma(M);\ \gamma_t \in B_\delta(y)\Big\}.$$

Further, define $\pi' = \mathrm{law}(\gamma'_0, \gamma'_1)$, $\mu'_s = \mathrm{law}(\gamma'_s) = (e_s)_{\#}\Pi'$. Obviously,

$$\Pi' \leq \frac{\Pi}{\Pi[\mathcal{Z}]} = \frac{\Pi}{\mu_t[B_\delta(y)]},$$

so for all $s \in [0,1]$,

$$\mu'_s \leq \frac{\mu_s}{\mu_t[B_\delta(y)]}.$$

In particular, μ'_s is absolutely continuous and its density ρ'_s satisfies (ν-almost surely)

$$\rho'_s \leq \frac{\rho_s}{\mu_t[B_\delta(y)]}. \tag{19.4}$$

When $s = t$, inequality (19.4) can be refined into

$$\rho'_t = \frac{\rho_t \, 1_{B_\delta(y)}}{\mu_t[B_\delta(y)]}, \tag{19.5}$$

since

$$(e_t)_\# \left(\frac{1_{\gamma_t \in B_\delta(y)}}{\mu_t[B_\delta(y)]} \right) = \frac{1_{x \in B_\delta(y)}((e_t)_\# \Pi)}{\mu_t[B_\delta(y)]}.$$

(This is more difficult to write down than to understand!)

From the restriction property (Theorem 4.6), (γ'_0, γ'_1) is an optimal coupling of (μ'_0, μ'_1), and therefore $(\mu'_s)_{0 \leq s \leq 1}$ is a displacement interpolation. By Theorem 17.37 applied with $U(r) = -r^{1-\frac{1}{N}}$,

$$\int_M (\rho'_t)^{1-\frac{1}{N}} \, d\nu \geq (1-t) \int_{M \times M} (\rho'_0(x_0))^{-\frac{1}{N}} \beta_{1-t}(x_0,x_1)^{\frac{1}{N}} \, \pi'(dx_0\, dx_1) \\ + t \int_{M \times M} (\rho'_1(x_1))^{-\frac{1}{N}} \beta_t(x_0,x_1)^{\frac{1}{N}} \, \pi'(dx_0\, dx_1). \tag{19.6}$$

By definition, μ'_t is supported in $B_\delta(y)$, so

$$\int_M (\rho'_t)^{1-\frac{1}{N}} \, d\nu = \int_{B_\delta(y)} (\rho'_t)^{1-\frac{1}{N}} \, d\nu \\ = \nu[B_\delta(y)] \int_{B_\delta(y)} (\rho'_t)^{1-\frac{1}{N}} \, \frac{d\nu}{\nu[B_\delta(y)]}. \tag{19.7}$$

By Jensen's inequality, applied with the concave function $r \to r^{1-\frac{1}{N}}$,

$$\int_{B_\delta(y)} (\rho'_t)^{1-\frac{1}{N}} \frac{d\nu}{\nu[B_\delta(y)]} \leq \left(\int \rho'_t \frac{d\nu}{\nu[B_\delta(y)]}\right)^{1-\frac{1}{N}} = \frac{1}{\nu[B_\delta(y)]^{1-\frac{1}{N}}}.$$

Plugging this into (19.7), we find

$$\int_M (\rho'_t)^{1-\frac{1}{N}} \, d\nu \leq \nu[B_\delta(y)]^{\frac{1}{N}}. \tag{19.8}$$

On the other hand, from (19.4) the right-hand side of (19.6) can be bounded below by

$$\mu_t[B_\delta(y)]^{\frac{1}{N}} \int_{M\times M} \Big[(1-t)\,(\rho_0(x_0))^{-\frac{1}{N}} \beta_{1-t}(x_0,x_1)^{\frac{1}{N}} \\ + \; t\,(\rho_1(x_1))^{-\frac{1}{N}} \beta_t(x_0,x_1)^{\frac{1}{N}}\Big]\, \pi'(dx_0\,dx_1)$$

$$= \mu_t[B_\delta(y)]^{\frac{1}{N}}\; \mathbb{E}\,\Big[(1-t)\,(\rho_0(\gamma'_0))^{-\frac{1}{N}} \beta_{1-t}(\gamma'_0,\gamma'_1)^{\frac{1}{N}} \\ + t\,(\rho_1(\gamma'_1))^{-\frac{1}{N}} \beta_t(\gamma'_0,\gamma'_1)^{\frac{1}{N}}\Big]$$

$$\geq \mu_t[B_\delta(y)]^{\frac{1}{N}}\; \mathbb{E} \inf_{\gamma_t\in[x_0,x_1]_t} \Big[(1-t)\,(\rho_0(\gamma'_0))^{-\frac{1}{N}} \beta_{1-t}(\gamma'_0,\gamma'_1)^{\frac{1}{N}} \\ + t\,(\rho_1(\gamma'_1))^{-\frac{1}{N}} \beta_t(\gamma'_0,\gamma'_1)^{\frac{1}{N}}\Big], \tag{19.9}$$

where the last inequality follows just from the (obvious) remark that $\gamma'_t \in [\gamma'_0,\gamma'_1]_t$. In all of these inequalities, we can restrict π' to the set $\{\rho_0(x_0)>0,\ \rho_1(x_1)>0\}$ which is of full measure.

Let

$$F(x) := \inf_{x\in[x_0,x_1]_t} \Big[(1-t)\,(\rho_0(x_0))^{-\frac{1}{N}} \beta_{1-t}(x_0,x_1)^{\frac{1}{N}} \\ + \; t\,(\rho_1(x_1))^{-\frac{1}{N}} \beta_t(x_0,x_1)^{\frac{1}{N}}\Big];$$

and by convention $F(x)=0$ if either $\rho_0(x_0)$ or $\rho_1(x_1)$ vanishes. (Forget about the measurability of F for the moment.) Then in view of (19.5) the lower bound in (19.9) can be rewritten as

$$\mathbb{E}\,F(\gamma_t') = \int_M F(x)\,d\mu_t'(x) = \frac{\displaystyle\int_{B_\delta(y)} F(x)\,d\mu_t(x)}{\mu_t[B_\delta(y)]}.$$

Combined with the upper bound (19.8), this implies

$$\left(\frac{\mu_t[B_\delta(y)]}{\nu[B_\delta(y)]}\right)^{-\frac{1}{N}} \geq \frac{\displaystyle\int_{B_\delta(y)} F(x)\,d\mu_t(x)}{\mu_t[B_\delta(y)]}. \tag{19.10}$$

Lebesgue's density theorem tells the following: If φ is a locally integrable function, then $\nu(dy)$-almost any y is a Lebesgue point of φ, which means

$$\frac{1}{\nu[B_\delta(y)]}\int_{B_\delta(y)} \varphi(x)\,d\nu(x) \xrightarrow[\delta\downarrow 0]{} \varphi(y).$$

In particular, if y is a Lebesgue point of ρ_t, then

$$\frac{\mu_t[B_\delta(y)]}{\nu[B_\delta(y)]} = \frac{\displaystyle\int_{B_\delta(y)} \rho_t(x)\,d\nu(x)}{\nu[B_\delta(y)]} \xrightarrow[\delta\downarrow 0]{} \rho_t(y).$$

The inequality in (19.10) proves that $F\rho_t$ is locally ν-integrable; therefore also

$$\frac{\displaystyle\int_{B_\delta(y)} F(x)\,d\mu_t(x)}{\nu[B_\delta(y)]} \xrightarrow[\delta\downarrow 0]{} F(y)\,\rho_t(y).$$

If one plugs these two limits in (19.10), one obtains

$$\rho_t(y)^{-\frac{1}{N}} \geq \frac{F(y)\,\rho_t(y)}{\rho_t(y)} = F(y),$$

provided that $\rho_t(y) > 0$; and then $\rho_t(y) \leq F(y)^{-N}$, as desired. In the case $\rho_t(y) = 0$ the conclusion still holds true.

Some final words about measurability. It is not clear (at least to me) that F is measurable; but instead of F one may use the measurable function

$$\widetilde{F}(x) = (1-t)\,\rho_0(\gamma_0)^{-\frac{1}{N}}\,\beta_{1-t}(\gamma_0,\gamma_1)^{\frac{1}{N}} + t\,\rho_1(\gamma_1)^{-\frac{1}{N}}\,\beta_t(\gamma_0,\gamma_1)^{\frac{1}{N}},$$

where $\gamma = F_t(x)$, and F_t is the measurable map defined in Theorem 7.30(v). Then the same argument as before gives $\rho_t(y) \leq \widetilde{F}(y)^{-N}$, and this is obviously bounded above by $F(y)^{-N}$. □

It is useful to consider the particular case when the initial density μ_0 is a Dirac mass and the final mass is the uniform distribution on some set B:

Theorem 19.6 (Jacobian bounds revisited). *Let M be a Riemannian manifold equipped with a reference measure $\nu = e^{-V}\mathrm{vol}$, $V \in C^2(M)$, satisfying a curvature-dimension condition $\mathrm{CD}(K,N)$ for some $K \in \mathbb{R}$, $N \in (1,\infty)$. Let $z_0 \in M$ and let B be a bounded set of positive measure. Further, let $(\mu_t^{z_0})_{0\le t\le 1}$ be the displacement interpolation joining $\mu_0 = \delta_{z_0}$ to $\mu_1 = (1_B\nu)/\nu[B]$. Then the density $\rho_t^{z_0}$ of $\mu_t^{z_0}$ satisfies*

$$\rho_t^{z_0}(x) \le \frac{C(K,N,R)}{t^N\,\nu[B]},$$

where

$$C(K,N,R) = \exp\Big(-\sqrt{(N-1)K_-}\,R\Big), \qquad K_- = \max(-K,0), \tag{19.11}$$

and R is an upper bound on the distances between z_0 and elements of B.

In particular, if $K \ge 0$, then

$$\rho_t^{z_0}(x) \le \frac{1}{t^N\,\nu[B]}.$$

Remark 19.7. Theorem 19.6 is a classical estimate in Riemannian geometry; it is often stated as a bound on the Jacobian of the map $(s,\xi) \longmapsto \exp_x(s\xi)$. It will be a good exercise for the reader to convert Theorem 19.6 into such a Jacobian bound.

Proof of Theorem 19.6. Let z_0 and B be as in the statement of the lemma, and let $\mu_1 = (1_B\,\nu)/\nu[B]$. Consider a displacement interpolation $(\mu_t)_{0\le t\le 1}$ between $\mu_0 = \delta_{z_0}$ and μ_1. Recall from Chapter 13 that μ_t is absolutely continuous for all $t \in (0,1]$. So Theorem 19.4 can be applied to the (reparametrized) displacement interpolation $(\mu'_t)_{0\le t\le 1}$ defined by $\mu'_t = \mu_{t'}$, $t' = t_0 + (1-t_0)t$; this yields

$$\rho_{t'}(x) \le \sup_{x\in[x_0,x_1]_t} \Big[(1-t)\,\beta_{1-t}(x_0,x_1)^{\frac{1}{N}}\,\rho_{t_0}(x_0)^{-\frac{1}{N}} + t\,\beta_t(x_0,x_1)^{\frac{1}{N}}\,\rho_1(x_1)^{-\frac{1}{N}}\Big]^{-N}. \tag{19.12}$$

Clearly, the sum above can be restricted to those pairs (x_0,x_1) such that x_1 lies in the support of μ_1, i.e. $x_1 \in B$; and x_0 lies in the support of μ_{t_0}, which implies $x_0 \in [z_0, B]_{t_0}$. Moreover, since $z \to z^{-N}$ is nonincreasing, one has the obvious bound

$$\rho_{t'}(x) \le \sup_{x\in[x_0,x_1]_t;\ x_0\in[z_0,B]_{t_0};\ x_1\in B} \left[t\,\beta_t(x_0,x_1)^{\frac{1}{N}}\rho_1(x_1)^{-\frac{1}{N}}\right]^{-N}$$
$$= \sup_{x\in[x_0,x_1]_t;\ x_0\in[z_0,B]_{t_0};\ x_1\in B} \frac{\rho_1(x_1)}{t^N\,\beta_t(x_0,x_1)}.$$

Since $\rho_1 = 1_B/\nu[B]$, actually

$$\rho_{t'}(x) \le \frac{S(t_0,z_0,B)}{t^N\nu[B]},$$

where

$$S(t_0,z_0,B) := \sup\Big\{\beta_t(x_0,x_1)^{-\frac{1}{N}};\quad x_0\in[z_0,B]_{t_0},\ x_1\in B\Big\}. \tag{19.13}$$

Now let $t_0 \to 0$ and t go to t', in such a way that t' stays fixed. Since B is bounded, the geodesics linking z_0 to an element of B have a uniformly bounded speed, so the set $[z_0, B]_{t_0}$ is included in a ball $B(z, V t_0)$ for some constant V; this shows that those x_0 appearing in (19.13) converge uniformly to z_0. By continuity of β_t, $S(t_0, z_0, B)$ converges to $S(0, z_0, B)$. Then an elementary estimate of β_t shows that $S(0, z_0, B) \le C(K, N, R)$. This finishes the proof. □

To conclude, I shall state a theorem which holds true with the *intrinsic* distortion coefficients of the manifold, without any reference to a choice of K and N, and without any assumption on the behavior of the manifold at infinity (if the total cost is infinite, we can appeal to the notion of generalized optimal coupling and generalized displacement interpolation, as in Chapter 13). Recall Definition 14.17.

Theorem 19.8 (Intrinsic pointwise bounds on the displacement interpolant). *Let M be an n-dimensional Riemannian manifold equipped with some reference measure $\nu = e^{-V}\mathrm{vol}$, $V \in C^2(M)$, and let $\overline{\beta}$ be the associated distortion coefficients. Let μ_0, μ_1 be two absolutely continuous probability measures on M, let $(\mu_t)_{0\le t\le 1}$ be the unique generalized displacement interpolation between μ_0 and μ_1, and let ρ_t be the density of μ_t with respect to ν. Then one has the pointwise bound*

$$\rho_t(x) \le \sup_{x\in[x_0,x_1]_t} \left((1-t)\left(\frac{\rho_0(x_0)}{\overline{\beta}_{1-t}(x_0,x_1)}\right)^{-\frac{1}{n}} + t\left(\frac{\rho_1(x_1)}{\overline{\beta}_t(x_0,x_1)}\right)^{-\frac{1}{n}}\right)^{-n}, \tag{19.14}$$

where by convention $\left((1-t)\,a^{-\frac{1}{n}} + t\,b^{-\frac{1}{n}}\right)^{-n} = 0$ if either a or b is 0.

Proof of Theorem 19.8. First use the standard approximation procedure of Proposition 13.2 to define probability measures $\mu_{t,\ell}$ with density $\rho_{t,\ell}$, and numbers Z_ℓ such that $Z_\ell \uparrow 1$, $Z_\ell\, \rho_{t,\ell} \uparrow \rho_t$, and $\mu_{t,\ell}$ are compactly supported.

Then we can redo the proof of Theorem 19.4 with $\mu_{0,\ell}$ and $\mu_{1,\ell}$ instead of μ_0 and μ_1, replacing Theorem 17.37 by Theorem 17.42. The result is

$$\rho_{t,\ell}(x) \le \sup_{x\in[x_0,x_1]_t} \left((1-t)\left(\frac{\rho_{0,\ell}(x_0)}{\overline{\beta}_{1-t}(x_0,x_1)}\right)^{-\frac{1}{n}} + t\left(\frac{\rho_{1,\ell}(x_1)}{\overline{\beta}_t(x_0,x_1)}\right)^{-\frac{1}{n}} \right)^{-n}.$$

Since $Z_\ell\, \rho_{t,\ell} \le \rho_t$, it follows that

$$\begin{aligned} Z_\ell\, \rho_{t,\ell}(x) &\le \sup_{x\in[x_0,x_1]_t} \left((1-t)\left(\frac{Z_\ell\, \rho_{0,\ell}(x_0)}{\overline{\beta}_{1-t}(x_0,x_1)}\right)^{-\frac{1}{n}} + t\left(\frac{Z_\ell\, \rho_{1,\ell}(x_1)}{\overline{\beta}_t(x_0,x_1)}\right)^{-\frac{1}{n}} \right)^{-n} \\ &\le \sup_{x\in[x_0,x_1]_t} \left((1-t)\left(\frac{\rho_0(x_0)}{\overline{\beta}_{1-t}(x_0,x_1)}\right)^{-\frac{1}{n}} + t\left(\frac{\rho_1(x_1)}{\overline{\beta}_t(x_0,x_1)}\right)^{-\frac{1}{n}} \right)^{-n}. \end{aligned}$$

The conclusion is obtained by letting $\ell \to \infty$. □

Democratic condition

Local Poincaré inequalities are conditioned, loosely speaking, to the "richness" of the space of geodesics: One should be able to transfer mass between sets by going along geodesics, in such a way that different points use geodesics that do not get "too close to each other". This idea (which is reminiscent of the intuition behind the distorted Brunn–Minkowski inequality) will be more apparent in the following condition. It says that one can use geodesics to redistribute all the mass of a ball in such a way that each point in the ball sends all its mass uniformly over the ball, but no point is visited too often in the process. In the next definition, what I call "uniform distribution on B" is the reference measure ν, conditioned on the ball, that is, $(1_B\nu)/\nu[B]$. The definition is formulated in the setting of a geodesic space (recall the definitions about length spaces in Chapter 7), but in this chapter I shall apply it only in Riemannian manifolds.

Definition 19.9 (Democratic condition). *A Borel measure ν on a geodesic space $(\mathcal{X}, d)$ is said to satisfy the democratic condition* $\mathrm{Dm}(C)$ *for some constant $C > 0$ if the following property holds true: For any closed ball B in $\mathcal{X}$ there is a random geodesic γ such that γ_0 and γ_1 are independent and distributed uniformly in B, and the time-integral of the density of γ_t (with respect ν) never exceeds $C/\nu[B]$.*

The condition is said to hold uniformly if the constant C is independent of the ball $B = B[x, r]$, and locally uniformly if it is independent of B as long as $B[x, 2r]$ remains inside a large fixed ball $B[z, R]$.

A more explicit formulation of the democratic condition is as follows: If μ_t stands for the law of γ_t, then

$$\int_0^1 \mu_t \, dt \le C \, \frac{\nu}{\nu[B]}. \tag{19.15}$$

Theorem 19.10 ($\mathrm{CD}(K, N)$ **implies** Dm). *Let M be a Riemannian manifold equipped with a reference measure $\nu = e^{-V} \mathrm{vol}$, $V \in C^2(M)$, satisfying a curvature-dimension condition* $\mathrm{CD}(K, N)$ *for some $K \in \mathbb{R}$, $N \in (1, \infty)$. Then ν satisfies a locally uniform democratic condition, with an admissible constant $2^N C(K, N, R)$ in a large ball $B[z, R]$, where $C(K, N, R)$ is defined in* (19.11).

In particular, if $K \ge 0$, then ν satisfies the uniform democratic condition $\mathrm{Dm}(2^N)$.

Proof of Theorem 19.10. The proof is largely based on Theorem 19.6.

Let B be a ball of radius r. For any point x_0, let $\mu_t^{x_0}$ be as in the statement of Theorem 19.6; then its density $\rho_t^{x_0}$ (with respect to ν) is bounded above by $C(K, N, R)/(t^N \, \nu[B])$.

On the other hand, $\mu_t^{x_0}$ can be interpreted as the position at time t of a random geodesic γ^{x_0} starting at x_0 and ending at x_1, which is distributed according to μ. By integrating this against $\mu(dx_0)$, we obtain the position at time t of a random geodesic γ such that γ_0 and γ_1 are independent and both distributed according to μ. Explicitly,

$$\mu_t = \mathrm{law}\,(\gamma_t) = \int_M \mu_t^{x_0} \, d\mu(x_0).$$

Obviously, the uniform bound on ρ_t persists upon integration, so

$$\mu_t \le \left[\frac{C(K, N, R)}{t^N \, \nu[B]} \right] \nu. \tag{19.16}$$

Recall that $\mu_t = \text{law}(\gamma_t)$, where γ_0, γ_1 are independent and distributed according to μ. Since geodesics in a Riemannian manifold are almost surely unique, we can throw away a set of zero volume in $B \times B$ such that for each $(x, y) \in (B \times B) \setminus Z$, there is a unique geodesic $(\gamma_t^{x_0,x_1})_{0 \le t \le 1}$ going from x_0 to x_1. Then μ_t is characterized as the law of $\gamma_t^{x_0,x_1}$, where $\text{law}(x_0, x_1) = \mu \otimes \mu$. If we repeat the construction by exchanging the variables x_0 and x_1, and replacing t by $1 - t$, then we get the same path (μ_t), up to reparametrization of time. So

$$\mu_t \le \left[\frac{C(K,N,R)}{(1-t)^N\, \nu[B]}\right] \nu. \tag{19.17}$$

Combining (19.16) and (19.17) and passing to densities, one obtains that, $\nu(dx)$-almost surely,

$$\rho_t(x) \le C(K,N,R) \min\left(\frac{1}{t^N}, \frac{1}{(1-t)^N}\right) \frac{1}{\nu[B]} \le \frac{2^N\, C(K,N,R)}{\nu[B]}, \tag{19.18}$$

which implies Theorem 19.10. □

Remark 19.11. The above bounds (19.18) can be improved as follows. Let $\mu = \rho\nu$ be a probability measure which is absolutely continuous with respect to ν, and otherwise arbitrary. Then there exists a random geodesic γ such that $\text{law}(\gamma_0, \gamma_1) = \mu \otimes \mu$, $\text{law}(\gamma_t)$ admits a density ρ_t with respect to ν, and

$$\|\rho_t\|_{L^p(\nu)} \le C(K,N,R)^{\frac{1}{p'}} \min\left(\frac{1}{t^{N/p'}}, \frac{1}{(1-t)^{N/p'}}\right) \|\rho\|_{L^p(\nu)} \tag{19.19}$$

for all $p \in (1, \infty)$, where $p' = p/(p-1)$ is the conjugate exponent to p.

Local Poincaré inequality

Convention 19.12. If ν satisfies a local Poincaré inequality for some constant C, I shall say that it satisfies a uniform local Poincaré inequality. If ν satisfies a local Poincaré inequality in each ball $B_R(z)$, with a constant that may depend on z and R, I shall just say that ν satisfies a local Poincaré inequality.

Theorem 19.13 (Doubling + democratic imply local Poincaré). *Let $(\mathcal{X}, d)$ be a length space equipped with a reference measure ν satisfying a doubling condition with constant D, and a democratic condition with constant C. Then ν satisfies a local Poincaré inequality with constant $P = 2\,C\,D$.*

If the doubling and democratic conditions hold true inside a ball $B(z,R)$ with constants $C = C(z,R)$ and $D = D(z,R)$ respectively, then ν satisfies a local Poincaré inequality in the ball $B(z,R)$ with constant $P(z,R) = 2\,C(z,R)\,D(z,R)$.

Before giving the proof of Theorem 19.13 I shall state a corollary which follows immediately from this theorem together with Corollary 18.11 and Theorem 19.10:

Corollary 19.14 (CD(K,N) implies local Poincaré). *Let M be a Riemannian manifold equipped with a reference measure $\nu = e^{-V}\mathrm{vol}$, $V \in C^2(M)$, satisfying a curvature-dimension condition* CD(K,N) *for some $K \in \mathbb{R}$, $N \in (1,\infty)$. Then ν satisfies a local Poincaré inequality with constant $P(K,N,R) = 2^{2N+1}\,C(K,N,R)\,D(K,N,R)$, inside any ball $B[z,R]$, where $C(K,N,R)$ and $D(K,N,R)$ are defined by* (19.11) *and* (18.10)*.*

In particular, if $K \geq 0$ then ν satisfies a local Poincaré inequality on the whole of M with constant 2^{2N+1}.

Proof of Theorem 19.13. Let x_0 be a given point in M. Given $r > 0$, write $B = B_{r]}(x_0)$, and $2B = B_{2r]}(x_0)$. As before, let $\mu = (1_B\,\nu)/\nu[B]$. Let $u : B \to \mathbb{R}$ be an arbitrary Lipschitz function. For any $y_0 \in M$, we have

$$u(y_0) - \langle u\rangle_B = \int_M \big(u(y_0) - u(y_1)\big)\,d\mu(y_1).$$

Then

$$\begin{aligned}\fint_B |u - \langle u\rangle_B|\,d\nu &= \int_M \big|u(y_0) - \langle u\rangle_B\big|\,d\mu(y_0)\\ &\leq \int_{B\times B} \big|u(y_0) - u(y_1)\big|\,d\mu(y_0)\,d\mu(y_1). \qquad (19.20)\end{aligned}$$

Next, let us estimate $|u(y_0) - u(y_1)|$ in terms of a constant-speed geodesic path γ joining y_0 to y_1, where $y_0, y_1 \in B$. The length of such a geodesic path is at most $2r$. Then, with the shorthand $g = |\nabla u|$,

$$\big|u(y_0) - u(y_1)\big| \le 2r \int_0^1 g(\gamma(t))\,dt. \tag{19.21}$$

By assumption there is a random geodesic $\gamma : [0,1] \to M$ such that $\mathrm{law}\,(\gamma_0,\gamma_1) = \mu \otimes \mu$ and $\mu_t = \mathrm{law}\,(\gamma_t)$ satisfies (19.15). Integrating (19.21) against the law of γ yields

$$\begin{aligned}\int_{M\times M} |u(y_0) - u(y_1)|\,d\mu(y_0)\,d\mu(y_1) &\le \mathbb{E}\left(2r\int_0^1 g(\gamma(t))\,dt\right) \quad (19.22)\\ &= 2r\int_0^1 \mathbb{E}\,g(\gamma(t))\,dt\\ &= 2r\int_0^1\int_M g\,d\mu_t\,dt.\end{aligned}$$

This, combined with (19.20), implies

$$\fint_B |u - \langle u\rangle_B|\,d\nu \le 2r\int_0^1\int_M g\,d\mu_t\,dt. \tag{19.23}$$

However, a geodesic joining two points in B cannot leave the ball $2B$, so (19.23) and the democratic condition together imply that

$$\fint_B |u - \langle u\rangle_B|\,d\nu \le \frac{2\,C\,r}{\nu[B]}\int_{2B} g\,d\nu. \tag{19.24}$$

By the doubling property, $\frac{1}{\nu[B]} \le \frac{D}{\nu[2B]}$. The conclusion is that

$$\fint_B |u - \langle u\rangle_B|\,d\nu \;\le\; 2\,C\,D\,r \fint_{2B} g\,d\nu. \tag{19.25}$$

This concludes the proof of Theorem 19.13. □

Remark 19.15. With almost the same proof, it is easy to derive the following refinement of the local Poincaré inequality:

$$\int_{B[x,r]} \frac{|u(x) - u(y)|}{d(x,y)}\,d\nu(x)\,d\nu(y) \;\le\; P(K,N,R)\int_{B[x,2r]} |\nabla u|(x)\,d\nu(x).$$

Back to Brunn–Minkowski and Prékopa–Leindler inequalities

To conclude this chapter I shall explain how the Brunn–Minkowski inequality (18.5) comes at once from the pointwise estimates on the interpolant density.

Proof of Theorem 18.5, again. Let μ_0 be the measure ν conditioned on A_0, i.e. $\mu_0 = \rho_0\,\nu$ with $\rho_0 = 1_{A_0}/\nu[A_0]$. Similarly, let $\mu_1 = \rho_1\,\nu$ with $\rho_1 = 1_{A_1}/\nu[A_1]$. Let ρ_t be the density of the displacement interpolant at time t. Then, since ρ_0 vanishes out of A_0, and ρ_1 out of A_1, Theorem 19.4 implies

$$\begin{aligned}\rho_t(x)^{-\frac{1}{N}} \geq (1-t)\left[\inf_{x\in[A_0,A_1]_t} \beta_{1-t}(x_0,x_1)^{\frac{1}{N}}\right]\nu[A_0]^{\frac{1}{N}} \\ + t\left[\inf_{x\in[A_0,A_1]_t} \beta_t(x_0,x_1)^{\frac{1}{N}}\right]\nu[A_1]^{\frac{1}{N}} \\ \geq (1-t)\left[\inf_{(x_0,x_1)\in A_0\times A_1} \beta_{1-t}(x_0,x_1)^{\frac{1}{N}}\right]\nu[A_0]^{\frac{1}{N}} \\ + t\left[\inf_{(x_0,x_1)\in A_0\times A_1} \beta_t(x_0,x_1)^{\frac{1}{N}}\right]\nu[A_1]^{\frac{1}{N}}.\end{aligned}$$

Now integrate this against $\rho_t(x)\,d\nu(x)$: since the right-hand side does not depend on x any longer,

$$\begin{aligned}\int \rho_t(x)^{1-\frac{1}{N}}\,d\nu(x) \geq (1-t)\left[\inf_{x\in[A_0,A_1]_t} \beta_{1-t}(x_0,x_1)^{\frac{1}{N}}\right]\nu[A_0]^{\frac{1}{N}} \\ + t\left[\inf_{x\in[A_0,A_1]_t} \beta_t(x_0,x_1)^{\frac{1}{N}}\right]\nu[A_1]^{\frac{1}{N}}.\end{aligned}$$

On the other hand, ρ_t is concentrated on $[A_0,A_1]_t$, so the same Jensen inequality that was used in the earlier proof of Theorem 18.5 implies

$$\int \rho_t(x)^{1-\frac{1}{N}}\,d\nu(x) \leq \nu\big[[A_0,A_1]_t\big]^{\frac{1}{N}},$$

and inequality (18.4) follows. □

Interestingly enough, Theorem 19.4 also implies the **distorted Prékopa–Leindler inequality**. This is a functional variant of the Brunn–Minkowski inequality, which is sometimes much more convenient to handle. (Here I say that the inequality is “distorted” only because the Prékopa–Leindler inequality is usually stated in $\mathbb{R}^n$, while the Riemannian generalization involves distortion coefficients.) I shall first consider the dimension-free case, which is simpler and does not need distortion coefficients.

Theorem 19.16 (Prékopa–Leindler inequalities). *With the same notation as in Theorem 19.4, assume that (M,ν) satisfies the curvature-dimension condition* $\mathrm{CD}(K,\infty)$. *Let $t \in (0,1)$, and let f, g, h be three nonnegative functions such that the inequality*

$$h(x) \geq \sup_{x\in[x_0,x_1]_t} f(x_0)^{1-t}\, g(x_1)^t \exp\left(-\frac{Kt(1-t)}{2}\, d(x_0,x_1)^2\right) \tag{19.26}$$

is satisfied for all $x \in M$. Then

$$\int h\, d\nu \geq \left(\int f\, d\nu\right)^{1-t} \left(\int g\, d\nu\right)^t.$$

Proof of Theorem 19.16. By an easy homogeneity argument, we may assume $\int f = \int g = 1$. Then write $\rho_0 = f$, $\rho_1 = g$; by Theorem 19.4, the displacement interpolant ρ_t between $\rho_0\,\nu$ and $\rho_1\,\nu$ satisfies (19.3). From (19.26), $h \geq \rho_t$. It follows that $\int h \geq \int \rho_t = 1$, as desired. □

Remark 19.17. Let (M,ν) be a Riemannian manifold satisfying a curvature-dimension bound $\mathrm{CD}(K,\infty)$ with $K > 0$, and let $A \subset M$ be a compact set such that $\nu[A] > 0$. Apply the Prékopa–Leindler inequality with $t = 1/2$, $f = 1_A$, $g = \exp(K\,d(x,A)^2/4)$ and $h = 1$: This shows that

$$\int_M e^{\frac{K\,d(x,A)^2}{4}}\, d\nu(x) < +\infty, \tag{19.27}$$

and one easily deduces that ν admits square-exponential moments (something which we already know from Theorem 18.12).

I shall conclude with the dimension-dependent form of the Prékopa–Leindler inequality, which will require some more notation. For any $a,b \geq 0$, $t \in [0,1]$, $q \in \mathbb{R} \setminus \{0\}$, define

$$\mathcal{M}_t^q(a,b) := \Big[(1-t)\,a^q \;+\; t\,b^q\Big]^{\frac{1}{q}},$$

with the convention that $\mathcal{M}_t^q(a,b) = 0$ if either a or b is 0; and $\mathcal{M}_t^{-\infty}(a,b) = \min(a,b)$.

Theorem 19.18 (Finite-dimension distorted Prékopa–Leindler inequality). *With the same notation as in Theorem 19.4, assume that* (M, ν) *satisfies a curvature-dimension condition* $\mathrm{CD}(K, N)$ *for some* $K \in \mathbb{R}$, $N \in (1, \infty)$. *Let* f, g *and* h *be three nonnegative functions on* M *satisfying*

$$h(x) \geq \sup_{x \in [x_0, x_1]_t} \mathcal{M}_t^q \left(\frac{f(x_0)}{\beta_{1-t}^{(K,N)}(x_0, x_1)}, \frac{g(x_1)}{\beta_t^{(K,N)}(x_0, x_1)} \right), \qquad q \geq -\frac{1}{N}; \tag{19.28}$$

then

$$\int h \, d\nu \geq \mathcal{M}_t^{\frac{q}{1+Nq}} \left(\int f \, d\nu, \int g \, d\nu \right). \tag{19.29}$$

Proof of Theorem 19.18. The proof is quite similar to the proof of Theorem 19.16, except that now N is finite. Let f, g and h satisfy the assumptions of the theorem, define $\rho_0 = f/\|f\|_{L^1}$, $\rho_1 = g/\|g\|_{L^1}$, and let ρ_t be the density of the displacement interpolant at time t between $\rho_0 \, \nu$ and $\rho_1 \, \nu$. Let $\mathcal{M}$ be the right-hand side of (19.29); the problem is to show that $\int (h/\mathcal{M}) \geq 1$, and this is obviously true if $h/\mathcal{M} \geq \rho_t$. In view of Theorem 19.4, it is sufficient to establish

$$\frac{h(x)}{\mathcal{M}} \geq \sup_{x \in [x_0, x_1]_t} \mathcal{M}_t^{\frac{1}{N}} \left(\frac{\beta_{1-t}(x_0, x_1)}{\rho_0(x_0)}, \frac{\beta_t(x_0, x_1)}{\rho_1(x_1)} \right)^{-1}. \tag{19.30}$$

In view of the assumption of h and the form of $\mathcal{M}$, it is sufficient to check that

$$\frac{1}{\mathcal{M}_t^{\frac{1}{N}} \left(\frac{\beta_{1-t}(x_0,x_1)}{\rho_0(x_0)}, \frac{\beta_t(x_0,x_1)}{\rho_1(x_1)} \right)} \leq \frac{\mathcal{M}_t^q \left(\frac{f(x_0)}{\beta_{1-t}(x_0,x_1)}, \frac{g(x_1)}{\beta_t(x_0,x_1)} \right)}{\mathcal{M}_t^{\frac{q}{1+Nq}} (\|f\|_{L^1}, \|g\|_{L^1})}.$$

But this is a consequence of the following computation:

$$\begin{aligned} \frac{1}{\mathcal{M}_t^{-s}(a^{-1}, b^{-1})} &= \mathcal{M}_t^s(a, b) \\ &\leq \mathcal{M}_t^q \left(\frac{a}{c}, \frac{b}{d} \right) \mathcal{M}_t^r(c, d) = \frac{\mathcal{M}_t^q \left(\frac{a}{c}, \frac{b}{d} \right)}{\mathcal{M}_t^{-r}(c, d)}, \qquad (19.31) \\ &\quad \frac{1}{q} + \frac{1}{r} = \frac{1}{s}, \qquad q + r \geq 0, \end{aligned}$$

where the two equalities in (19.31) are obvious by homogeneity, and the central inequality is a consequence of the two-point Hölder inequality (see the bibliographical notes for references). □

Bibliographical notes

The main historical references concerning interior regularity estimates are by De Giorgi [274], Nash [646] and Moser [638, 639]. Their methods were later considerably developed in the theory of elliptic partial differential equations, see e.g. [189, 416]. Moser's Harnack inequality is a handy technical tool to recover the most famous regularity results. The relations of this inequality with Poincaré and Sobolev inequalities, and the influence of Ricci curvature on it, were studied by many authors, including in particular Saloff-Coste [726] and Grigor'yan [433].

Lebesgue's density theorem can be found in most textbooks about measure theory, e.g. Rudin [714, Chapter 7].

Local Poincaré inequalities admit many variants and are known under many names in the literature, in particular "weak Poincaré inequalities", by contrast with "strong" Poincaré inequalities, in which the larger ball is not $B[x, 2r]$ but $B[x, r]$. In spite of that terminology, both inequalities are in some sense equivalent [461]. Sometimes one replaces the ball $B[x, 2r]$ by a smaller ball $B[x, \lambda r]$, $\lambda > 1$. One sometimes says that the inequality (19.1) is of type $(1, 1)$ because there are L^1 norms on both sides. Inequality (19.1) also implies the other main members of the family of local Poincaré inequalities, see for instance Heinonen [469, Chapters 4 and 9]. There are equivalent formulations of these inequalities in terms of modulus and capacity, see e.g. [510, 511] and the many references therein. The study of Poincaré inequalities in metric spaces has turned into a surprisingly large domain of research.

Local Poincaré inequalities are also used to study large-scale geometry, see e.g. [250]. Further, inequality (19.1), applied to the whole space, is equivalent to **Cheeger's isoperimetric inequality**:

$$\nu[\Omega] \leq \frac{1}{2} \Longrightarrow \quad |\partial\Omega|_\nu \geq K\,\nu[\Omega], \tag{19.32}$$

where ν is a reference probability measure on $\mathcal{X}$, Ω is a Borel subset of $\mathcal{X}$, and $|\partial\Omega|_\nu$ is the ν-surface of Ω. Cheeger's inequality in turn implies the usual Poincaré inequality [225, 609, 610]. See [633] and the references therein for more details.

Theorem 19.6 is a classical estimate, usually formulated in terms of Jacobian estimates, see e.g. Saloff-Coste [728, p. 179]; the differences in the formulas are due to the convention that geodesics might be parametrized by arc length rather than defined on $[0, 1]$. The transport-based proof was devised by Lott and myself [578].

The "intrinsic" bounds appearing in Theorem 19.8 go back to [246] (in the compactly supported case) and [363, Section 3] (in the general case). The methods applied in these references are different. The restriction strategy which I used to prove Theorems 19.4 and 19.8 is an amplification of the transport-based proof of Theorem 19.6 from [578]. A nice alternative strategy, also based on restriction, was suggested by Sturm [763, Proof of Proposition IV.2]. Instead of conditioning with respect to the values of γ_t, Sturm conditions with respect to the values of (γ_0, γ_1); this has the technical drawback to modify the values of ρ_t, but one can get around this difficulty by a two-step limit procedure.

The proofs of Theorems 19.10 and 19.13 closely follow [578]. It was Pajot who pointed out to me the usefulness of Jacobian estimates expressed by Theorem 19.6 (recall Remark 19.7) for proving local Poincaré inequalities. The democratic condition $\mathrm{Dm}(C)$ was explicitly introduced in [578], but it is somehow implicit in earlier works, such as Cheeger and Colding [230]. Other proofs of the local Poincaré inequality from optimal transport, based on slightly different but quite close arguments, were found independently by von Renesse [825] and Sturm [763].

The general strategy of proof behind Theorem 19.13 is rather classical, be it in the context of Riemannian manifolds or groups or graphs; see e.g. [249] and the references therein.

The classical Prékopa–Leindler inequality in Euclidean space goes back to [548, 690]; see [406] for references and its role in the Brunn–Minkowski theory. Although in principle equivalent to the Brunn–Minkowski inequality, it is sometimes rather more useful, see e.g. a famous application by Maurey [607] for concentration inequalities. Bobkov and Ledoux [131] have shown how to use this inequality to derive many functional inequalities such as logarithmic Sobolev inequalities, to be considered in Chapter 21.

In the Euclidean case, the stronger version of the Prékopa–Leindler inequality which corresponds to Theorem 19.18 was established by Borell [144], Brascamp and Lieb [153], and others. The proof of Theorem 19.18 from Theorem 19.4 follows the argument given at the very end of [246]. The inequality used in (19.31) appears in [406, Lemma 10.1].

The Prékopa–Leindler inequality on manifolds, Theorem 19.16, shows up in a recent research paper by Cordero-Erausquin, McCann and Schmuckenschläger [247]. In that reference, displacement convexity is established independently of the Prékopa–Leindler inequality, but with similar tools (namely, the Jacobian estimates in Chapter 14). The presentation that I have adopted makes it clear that the Prékopa–Leindler inequality, and even the stronger pointwise bounds in Theorem 19.4, can really be seen as a consequence of displacement convexity inequalities (together with the restriction property). This determination to derive everything from displacement convexity, rather than directly from Jacobian estimates, will find a justification in Part III of this course: In some sense the notion of displacement convexity is softer and more robust.

In $\mathbb{R}^N$, there is also a "stronger" version of Theorem 19.18 in which the exponent q can go down to $-1/(N-1)$ instead of $-1/N$; it reads

$$h\big((1-t)x_0 + tx_1\big) \geq \mathcal{M}_t^q(f(x_0), g(x_1)) \implies$$
$$\int h(z)\,dz \geq \mathcal{M}_t^{\frac{q}{1+q(N-1)}}\big(m_i(f), m_i(g)\big) \cdot \mathcal{M}_t^1\left(\frac{1}{m_i(f)}\int f,\ \frac{1}{m_i(g)}\int g\right), \tag{19.33}$$

where $i \in \{1, \dots, N\}$ is arbitrary and

$$m_i(f) = \sup_{x_i \in \mathbb{R}} \int_{\mathbb{R}^{N-1}} f(x)\,dx_1 \dots dx_{i-1}\,dx_{i+1} \dots dx_N.$$

It was recently shown by Bobkov and Ledoux [132] that this inequality can be used to establish optimal Sobolev inequalities in $\mathbb{R}^N$ (with the usual Prékopa–Leindler inequality one can apparently reach only the logarithmic Sobolev inequality, that is, the dimension-free case [131]). See [132] for the history and derivation of (19.33).

20

Infinitesimal displacement convexity

The goal of the present chapter is to translate displacement convexity inequalities of the form "the graph of a convex function lies below the chord" into inequalities of the form "the graph of a convex function lies above the tangent" — just as in statements (ii) and (iii) of Proposition 16.2. This corresponds to the limit $t \to 0$ in the convexity inequality.

The main results in this chapter are the HWI inequality (Corollary 20.13); and its generalized version, the distorted HWI inequality (Theorem 20.10).

Time-derivative of the energy

As a preliminary step, a useful lower bound will now be given for a derivative of $U_\nu(\mu_t)$, where $(\mu_t)_{0\le t\le 1}$ is a Wasserstein geodesic and U_ν an energy functional with a reference measure ν. This computation hardly needs any regularity on the space, and for later use I shall state it in a more general setting than Riemannian manifolds.

In the next theorem, I consider a locally compact, complete geodesic space $\mathcal{X}$ equipped with a distance d and a locally finite measure ν. Then $U : [0, +\infty) \to \mathbb{R}$ is a continuous convex function, twice differentiable on $(0, +\infty)$. To U is associated the functional

$$U_\nu(\mu) = \int_{\mathcal{X}} U(\rho)\, d\nu \qquad \mu = \rho\,\nu.$$

The statement below will involve norms of gradients. Even though there is no natural notion for the gradient ∇f of a function f defined on a nonsmooth length space, there are still natural definitions for the *norm* of the gradient, $|\nabla f|$. The most common one is

$$|\nabla f|(x) := \limsup_{y \to x} \frac{|f(y) - f(x)|}{d(x,y)}. \tag{20.1}$$

Rigorously speaking, this formula makes sense only if x is not isolated, which will always be the case in the sequel. A slightly finer notion is the following:

$$|\nabla^- f|(x) := \limsup_{y \to x} \frac{[f(y) - f(x)]_-}{d(x,y)}, \tag{20.2}$$

where $a_- = \max(-a, 0)$ stands for the negative part of a (which is a nonnegative number!). It is obvious that $|\nabla^- f| \leq |\nabla f|$, and both notions coincide with the usual one if f is differentiable. Note that $|\nabla^- f|(x)$ is automatically 0 if x is a local minimum of f.

Theorem 20.1 (Differentiating an energy along optimal transport). *Let $(\mathcal{X}, d, \nu)$ and U be as above, and let $(\mu_t)_{0 \leq t \leq 1}$ be a geodesic in $P_2(\mathcal{X})$, such that each μ_t is absolutely continuous with respect to ν, with density ρ_t, and $U(\rho_t)_-$ is ν-integrable for all t. Further assume that ρ_0 is Lipschitz continuous, $U(\rho_0)$ and $\rho_0\, U'(\rho_0)$ are ν-integrable, and U' is Lipschitz continuous on $\rho_0(\mathcal{X})$. Then*

$$\liminf_{t \downarrow 0} \left[\frac{U_\nu(\mu_t) - U_\nu(\mu_0)}{t} \right] \geq \\ - \int_{\mathcal{X}} U''(\rho_0(x_0))\, |\nabla^- \rho_0|(x_0)\, d(x_0, x_1)\, \pi(dx_0\, dx_1), \tag{20.3}$$

where π is an optimal coupling of (μ_0, μ_1) associated with the geodesic path $(\mu_t)_{0 \leq t \leq 1}$.

Remark 20.2. The technical assumption on the negative part of $U(\rho_t)$ being integrable is a standard way to make sure that $U_\nu(\mu_t)$ is well-defined, with values in $\mathbb{R} \cup \{+\infty\}$. As for the assumption about U' being Lipschitz on $\rho_0(\mathcal{X})$, it means in practice that either U is twice (right-)differentiable at the origin, or ρ_0 is bounded away from 0.

Remark 20.3. Here is a more probabilistic reformulation of (20.3) (which will also make more explicit the link between π and μ_t): Let γ be a random geodesic such that $\mu_t = \text{law}\,(\gamma_t)$, then

$$\liminf_{t \downarrow 0} \left[\frac{U_\nu(\mu_t) - U_\nu(\mu_0)}{t} \right] \geq - \mathbb{E} \left[U''(\rho_0(\gamma_0))\, |\nabla^- \rho_0|(\gamma_0)\, d(\gamma_0, \gamma_1) \right].$$

Proof of Theorem 20.1. By convexity,

$$U(\rho_t) - U(\rho_0) \geq U'(\rho_0)\,(\rho_t - \rho_0), \tag{20.4}$$

where $U'(0)$ is the right-derivative of U at 0.

On the one hand, $U(\rho_0)$ and $U(\rho_t)_-$ are ν-integrable by assumption, so the integral in the left-hand side of (20.4) makes sense in $\mathbb{R}\cup\{+\infty\}$ (and the integral of each term is well-defined). On the other hand, $\rho_0\,U'(\rho_0)$ is integrable by assumption, while $\rho_t\,U'(\rho_0)$ is bounded above by $(\max U')\rho_t$, which is integrable; so the integral of the right-hand side makes sense in $\mathbb{R}\cup\{-\infty\}$. All in all, inequality (20.4) can be integrated into

$$\begin{aligned} U_\nu(\mu_t) - U_\nu(\mu_0) &\geq \int U'(\rho_0)\rho_t\,d\nu - \int U'(\rho_0)\rho_0\,d\nu \\ &= \int U'(\rho_0)\,d\mu_t - \int U'(\rho_0)\,d\mu_0. \end{aligned}$$

Now let γ be a random geodesic, such that $\mu_t = \mathrm{law}\,(\gamma_t)$. Then the above inequality can be rewritten

$$\begin{aligned} U_\nu(\mu_t) - U_\nu(\mu_0) &\geq \mathbb{E}\,U'(\rho_0(\gamma_t)) - \mathbb{E}\,U'(\rho_0(\gamma_0)) \\ &= \mathbb{E}\Big[U'(\rho_0(\gamma_t)) - U'(\rho_0(\gamma_0))\Big]. \end{aligned}$$

Since U' is nondecreasing,

$$U'(\rho_0(\gamma_t)) - U'(\rho_0(\gamma_0)) \geq \big[U'(\rho_0(\gamma_t)) - U'(\rho_0(\gamma_0))\big]\,1_{\rho_0(\gamma_0)>\rho_0(\gamma_t)}.$$

Multiplying and dividing by $\rho_0(\gamma_t) - \rho_0(\gamma_0)$, and then by $d(\gamma_0,\gamma_t)$, one arrives at

$$\Big[U_\nu(\mu_t) - U_\nu(\mu_0)\Big] \geq \\ \mathbb{E}\left(\frac{U'(\rho_0(\gamma_t)) - U'(\rho_0(\gamma_0))}{\rho_0(\gamma_t) - \rho_0(\gamma_0)}\right)\left(\frac{\rho_0(\gamma_t) - \rho_0(\gamma_0)}{d(\gamma_0,\gamma_t)}\,1_{\rho_0(\gamma_0)>\rho_0(\gamma_t)}\right) d(\gamma_0,\gamma_t).$$

After division by t and use of the identity $d(\gamma_0,\gamma_t) = t\,d(\gamma_0,\gamma_1)$, one obtains in the end

$$\frac{1}{t}\Big[U_\nu(\mu_t) - U_\nu(\mu_0)\Big] \geq \\ \mathbb{E}\left(\frac{U'(\rho_0(\gamma_t)) - U'(\rho_0(\gamma_0))}{\rho_0(\gamma_t) - \rho_0(\gamma_0)}\right)\left(\frac{\rho_0(\gamma_t) - \rho_0(\gamma_0)}{d(\gamma_0,\gamma_t)}\,1_{\rho_0(\gamma_0)>\rho_0(\gamma_t)}\right) d(\gamma_0,\gamma_1). \tag{20.5}$$

It remains to pass to the limit in the right-hand side of (20.5) as $t \to 0$. Since ρ_0 is continuous, for almost each geodesic γ one has $\rho_0(\gamma_t) \to \rho_0(\gamma_0) > 0$ as $t \to 0$, and in particular,

$$\frac{U'(\rho_0(\gamma_t)) - U'(\rho_0(\gamma_0))}{\rho_0(\gamma_t) - \rho_0(\gamma_0)} \xrightarrow[t\to 0]{} U''(\rho_0(\gamma_0)),$$

Similarly,

$$\liminf_{t\to 0} \left(\frac{\rho_0(\gamma_t) - \rho_0(\gamma_0)}{d(\gamma_0, \gamma_t)} 1_{\rho_0(\gamma_0)>\rho_0(\gamma_t)} \right) \geq -|\nabla^- \rho_0|(\gamma_0).$$

So, if $v_t(\gamma)$ stands for the integrand in the right-hand side of (20.5), one has

$$\liminf_{t\to 0}\ v_t(\gamma) \geq -U''(\rho_0(\gamma_0))\,|\nabla^- \rho_0|(\gamma_0)\, d(\gamma_0, \gamma_1).$$

On the other hand, ρ_0 is Lipschitz by assumption, and also U' is Lipschitz on the range of ρ_0. So $|v_t(\gamma)| \leq C d(\gamma_0, \gamma_1)$, where C is the product of the Lipschitz constants of ρ_0 and U'. This uniform domination makes it possible to apply Fatou's lemma, in the form $\liminf_{t\to 0} \mathbb{E}\, v_t(\gamma) \geq \mathbb{E}\, \liminf v_t(\gamma)$. Thus

$$\liminf_{t\to 0} \frac{1}{t}\Big[U_\nu(\mu_t) - U_\nu(\mu_0)\Big] \geq -\mathbb{E}\, U''(\rho_0(\gamma_0))\,|\nabla^- \rho_0|(\gamma_0)\, d(\gamma_0, \gamma_1),$$

as desired. □

Remark 20.4. This theorem does not assume smoothness of $\mathcal{X}$, and does not either assume structural restrictions on the function U. On the other hand, when $\mathcal{X}$ is a Riemannian manifold of dimension n, $\nu = e^{-V}\mathrm{vol}$, and μ is compactly supported, then there is a more precise result:

$$\lim_{t\to 0} \frac{[U_\nu(\mu_t) - U_\nu(\mu_0)]}{t} = -\int p(\rho_0)(L\psi)\, d\nu, \tag{20.6}$$

where ψ is such that $T = \exp(\nabla\psi)$ is the unique optimal transport from μ_0 to μ_1, and $L\psi = \Delta\psi - \nabla V \cdot \nabla\psi$ (defined almost everywhere). It is not clear a priori how this compares with the result of Theorem 20.1, but then, under slightly more stringent regularity assumptions, one can justify the integration by parts formula

$$-\int p(\rho_0)\, L\psi\, d\nu \geq \int \rho_0\, U''(\rho_0) \nabla\rho_0 \cdot \nabla\psi\, d\nu \tag{20.7}$$

(note indeed that $p'(r) = r\,U''(r)$). Since $\pi = (\rho_0\,\nu) \otimes \delta_{x_1=T(x_0)}$ with $T = \exp\nabla\psi$, the right-hand side can be rewritten

$$\int U''(\rho_0)\nabla\rho_0 \cdot \nabla\psi\, d\pi.$$

As $|\nabla\psi(x_0)| = d(x_0, x_1)$, this integral is obviously an upper bound for the expression in (20.3). In the present chapter, the more precise result (20.6) will not be useful, but later in Chapter 23 we shall have to go through it (see the proof of Theorem 23.14). More comments are in the bibliographical notes.

Exercise 20.5. Use Otto's calculus to guess that $(d/dt)U_\nu(\mu_t)$ should coincide with the right-hand side of (20.7).

HWI inequalities

Recall from Chapters 16 and 17 that $\mathrm{CD}(K,N)$ bounds imply convexity properties of certain functionals U_ν along displacement interpolation. For instance, if a Riemannian manifold M, equipped with a reference measure ν, satisfies $\mathrm{CD}(0,\infty)$, then by Theorem 17.15 the Boltzmann H functional is displacement convex.

If $(\mu_t)_{0\le t\le 1}$ is a geodesic in $P_2^{\mathrm{ac}}(M)$, for $t \in (0,1]$ the convexity inequality

$$H_\nu(\mu_t) \le (1-t)\, H_\nu(\mu_0) + t\, H_\nu(\mu_1)$$

may be rewritten as

$$\frac{H_\nu(\mu_t) - H_\nu(\mu_0)}{t} \le H_\nu(\mu_1) - H_\nu(\mu_0).$$

Under suitable assumptions we may then apply Theorem 20.1 to pass to the limit as $t \to 0$, and get

$$-\int \frac{|\nabla\rho_0(x_0)|}{\rho_0(x_0)}\, d(x_0,x_1)\, \pi(dx_0\, dx_1) \le H_\nu(\mu_1) - H_\nu(\mu_0).$$

This implies, by Cauchy–Schwarz inequality,

$$\begin{aligned} H_\nu(\mu_0) - H_\nu(\mu_1) &\le \sqrt{\int d(x_0,x_1)^2 \pi(dx_0\, dx_1)}\ \sqrt{\int \frac{|\nabla\rho_0(x_0)|}{\rho_0(x_0)^2}\, \pi(dx_0\, dx_1)} \\ &= W_2(\mu_0,\mu_1) \sqrt{\int \frac{|\nabla\rho_0|^2}{\rho_0}\, d\nu}, \end{aligned} \tag{20.8}$$

where I have used the fact that the first marginal of π is $\mu_0 = \rho_0\,\nu$.

Inequality (20.8) is the **HWI inequality**: It is expressed in terms of

- the H-functional $H_\nu(\mu) = \int \rho \log \rho \, d\nu$ (as usual $\rho = d\mu/d\nu$);
- the Wasserstein distance of order 2, W_2;
- the Fisher information I, defined by $I_\nu(\mu) = \int \frac{|\nabla \rho|^2}{\rho}\, d\nu$.

The present section is devoted to establishing such inequalities.

For technical reasons (such as the treatment of small values of ρ_0 in noncompact manifolds, or finite-dimensional generalizations) it will be convenient to recast this discussion in the more general setting of *distorted* HWI inequalities, which involve distortion coefficients. Let $\beta_t = \beta_t^{(K,N)}$ be the reference distortion coefficients defined in (14.61). Note that $\beta_1(x_0, x_1) = 1$, $\beta_0'(x_0, x_1) = 0$, where the prime stands for partial derivation with respect to t. For brevity I shall write

$$\beta(x_0, x_1) = \beta_0(x_0, x_1); \qquad \beta'(x_0, x_1) = \beta_1'(x_0, x_1).$$

By explicit computation,

$$\beta(x_0, x_1) = \begin{cases} \left(\frac{\alpha}{\sin \alpha}\right)^{N-1} > 1 & \text{if } K > 0 \\ 1 & \text{if } K = 0 \\ \left(\frac{\alpha}{\sinh \alpha}\right)^{N-1} < 1 & \text{if } K < 0, \end{cases} \tag{20.9}$$

$$\beta'(x_0, x_1) = \begin{cases} -(N-1)\left(1 - \frac{\alpha}{\tan \alpha}\right) < 0 & \text{if } K > 0 \\ 0 & \text{if } K = 0 \\ (N-1)\left(\frac{\alpha}{\tanh \alpha} - 1\right) > 0 & \text{if } K < 0, \end{cases} \tag{20.10}$$

where

$$\alpha = \sqrt{\frac{|K|}{N-1}}\, d(x_0, x_1).$$

Moreover, a standard Taylor expansion shows that, as $\alpha \to 0$ while K is fixed (which means that either $d(x_0, x_1) \to 0$ or $N \to \infty$), then

$$\beta \simeq 1 - \frac{K}{6}\, d(x_0, x_1)^2, \qquad \beta' \simeq -\frac{K}{3}\, d(x_0, x_1)^2,$$

whatever the sign of K.

The next definition is a generalization of the classical notion of Fisher information:

Definition 20.6 (Generalized Fisher information). *Let U be a continuous convex function $\mathbb{R}_+ \to \mathbb{R}$, twice continuously differentiable on $(0,+\infty)$. Let M be a Riemannian manifold, equipped with a Borel reference measure ν. Let $\mu \in P^{ac}(M)$ be a probability measure on M, whose density ρ is locally Lipschitz. Define*

$$I_{U,\nu}(\mu) = \int \rho\, U''(\rho)^2\, |\nabla\rho|^2\, d\nu = \int \frac{|\nabla p(\rho)|^2}{\rho}\, d\nu = \int \rho\, |\nabla U'(\rho)|^2\, d\nu, \tag{20.11}$$

where $p(r) = r\,U'(r) - U(r)$.

Particular Case 20.7 (Fisher information). When $U(r) = r\log r$, (20.11) becomes

$$I_\nu(\mu) = \int \frac{|\nabla\rho|^2}{\rho}\, d\nu.$$

Remark 20.8. The identity in (20.11) comes from the chain-rule:

$$\nabla p(\rho) = p'(\rho)\, \nabla\rho = \rho\, U''(\rho)\, \nabla\rho = \rho\, \nabla U'(\rho).$$

(Strictly speaking this is true only if $\rho > 0$, but the integral in (20.11) may be restricted to the set $\{\rho > 0\}$.) Also, in Definition 20.6 one can replace $|\nabla\rho|$ by $|\nabla^-\rho|$ and $|\nabla p(\rho)|$ by $|\nabla^- p(\rho)|$ since a locally Lipschitz function is differentiable almost everywhere.

Remark 20.9. If $p(\rho) \in L^1_{\mathrm{loc}}(M)$, then the convexity of $(p,r) \to |p|^2/r$ on $\mathbb{R}^n \times \mathbb{R}_+$ makes it possible to define $\int |\nabla p(\rho)|^2/\rho\, d\nu$ in $[0,+\infty]$ even if ρ is not locally Lipschitz. In particular, $I_\nu(\mu)$ makes sense in $[0,+\infty]$ for all probability measures μ (with the understanding that $I_\nu(\mu) = +\infty$ if μ is singular). I shall not develop this remark, and in the sequel shall only consider densities which are locally (and even globally) Lipschitz.

Theorem 20.10 (Distorted HWI inequality). *Let M be a Riemannian manifold equipped with a reference measure $\nu = e^{-V}\mathrm{vol}$, $V \in C^2(M)$, satisfying a curvature-dimension bound $\mathrm{CD}(K,N)$ for some $K \in \mathbb{R}$, $N \in (1,\infty]$. Let $U \in \mathcal{DC}_N$ and let $p(r) = r\,U'(r) - U(r)$. Let $\mu_0 = \rho_0\,\nu$ and $\mu_1 = \rho_1\,\nu$ be two absolutely continuous probability measures such that:*

(a) $\mu_0, \mu_1 \in P^{ac}_p(M)$, where $p \in [2,+\infty) \cup \{c\}$ satisfies (17.30);

(b) ρ_0 is Lipschitz.

If $N=\infty$, further assume $\rho_0 \log_+ \rho_0$ and $\rho_1 \log_+ \rho_1$ belong to $L^1(\nu)$. If $K>0$, further assume $\rho_0 U'(\rho_0) \in L^1(\nu)$. If $K>0$ and $N=\infty$, further assume $p(\rho_0)^2/\rho_0 \in L^1(\nu)$. Then

$$\begin{aligned}\int_M U(\rho_0)\,d\nu \leq \int_{M\times M} U\left(\frac{\rho_1(x_1)}{\beta(x_0,x_1)}\right)\beta(x_0,x_1)\,\pi(dx_0|x_1)\,\nu(dx_1)\\ + \int_{M\times M} p(\rho_0(x_0))\,\beta'(x_0,x_1)\,\pi(dx_1|x_0)\,\nu(dx_0)\\ + \int_{M\times M} U''(\rho_0(x_0))\,|\nabla\rho_0(x_0)|\,d(x_0,x_1)\,\pi(dx_0\,dx_1), \qquad (20.12)\end{aligned}$$

where π is the unique optimal coupling of (μ_0,μ_1) and the coefficients β, β' are defined in (20.9)–(20.10).

In particular,

(i) If $K=0$ and $U_\nu(\mu_1)<+\infty$, then

$$\begin{aligned}U_\nu(\mu_0)-U_\nu(\mu_1) &\leq \int U''(\rho_0(x_0))\,|\nabla\rho_0(x_0)|\,d(x_0,x_1)\,\pi(dx_0\,dx_1)\\ &\leq W_2(\mu_0,\mu_1)\,\sqrt{I_{U,\nu}(\mu_0)}. \qquad (20.13)\end{aligned}$$

(ii) If $N=\infty$ and $U_\nu(\mu_1)<+\infty$, then

$$\begin{aligned}&U_\nu(\mu_0)-U_\nu(\mu_1)\\ &\quad\leq \int U''(\rho_0(x_0))\,|\nabla\rho_0(x_0)|\,d(x_0,x_1)\,\pi(dx_0\,dx_1) - K_{\infty,U}\,\frac{W_2(\mu_0,\mu_1)^2}{2}\\ &\quad\leq W_2(\mu_0,\mu_1)\,\sqrt{I_{U,\nu}(\mu_0)} - K_{\infty,U}\,\frac{W_2(\mu_0,\mu_1)^2}{2}, \qquad (20.14)\end{aligned}$$

where $K_{\infty,U}$ is defined in (17.10).

(iii) If $N<\infty$, $K\geq 0$ and $U_\nu(\mu_1)<+\infty$ then

$$\begin{aligned}&U_\nu(\mu_0)-U_\nu(\mu_1)\\ &\leq \int U''(\rho_0(x_0))\,|\nabla\rho_0(x_0)|\,d(x_0,x_1)\,\pi(dx_0\,dx_1)\\ &\qquad\qquad - K\lambda_{N,U}\,\max\big(\|\rho_0\|_{L^\infty(\nu)},\|\rho_1\|_{L^\infty(\nu)}\big)^{-\frac{1}{N}}\,\frac{W_2(\mu_0,\mu_1)^2}{2}\\ &\leq W_2(\mu_0,\mu_1)\,\sqrt{I_{U,\nu}(\mu_0)}\\ &\qquad\qquad - K\lambda_{N,U}\,\max\big(\|\rho_0\|_{L^\infty(\nu)},\|\rho_1\|_{L^\infty(\nu)}\big)^{-\frac{1}{N}}\,\frac{W_2(\mu_0,\mu_1)^2}{2}, \qquad (20.15)\end{aligned}$$

where

$$\lambda_{N,U} = \lim_{r \to 0} \frac{p(r)}{r^{1-\frac{1}{N}}}. \qquad (20.16)$$

Exercise 20.11. When U is well-behaved, give a more direct derivation of (20.14), via plain displacement convexity (rather than distorted displacement convexity). The same for (20.15), with the help of Exercise 17.23.

Remark 20.12. As the proof will show, Theorem 20.10(iii) extends to negative curvature modulo the following changes: replace $\lim_{r\to 0}$ in (20.16) by $\lim_{r\to\infty}$; and $(\max(\|\rho_0\|_{L^\infty}, \|\rho_1\|_{L^\infty}))^{-1/N}$ in (20.15) by $\max(\|1/\rho_0\|_{L^\infty}, \|1/\rho_1\|_{L^\infty})^{1/N}$. (This result is not easy to derive by plain displacement convexity.)

Corollary 20.13 (HWI inequalities). *Let M be a Riemannian manifold equipped with a reference measure $\nu = e^{-V}\mathrm{vol}$, $V \in C^2(M)$, satisfying a curvature-dimension bound $\mathrm{CD}(K,\infty)$ for some $K \in \mathbb{R}$. Then:*

(i) Let $p \in [2,+\infty) \cup \{c\}$ satisfy (17.30) for $N = \infty$, and let $\mu_0 = \rho_0\,\nu$, $\mu_1 = \rho_1\,\nu$ be any two probability measures in $P_p^{\mathrm{ac}}(M)$, such that $H_\nu(\mu_1) < +\infty$ and ρ_0 is Lipschitz; then

$$H_\nu(\mu_0) - H_\nu(\mu_1) \le W_2(\mu_0,\mu_1)\sqrt{I_\nu(\mu_0)} - K\frac{W_2(\mu_0,\mu_1)^2}{2}.$$

(ii) If $\nu \in P_2(M)$ then for any $\mu \in P_2(M)$,

$$H_\nu(\mu) \le W_2(\mu,\nu)\sqrt{I_\nu(\mu)} - K\frac{W_2(\mu,\nu)^2}{2}. \qquad (20.17)$$

Remark 20.14. The HWI inequality plays the role of a nonlinear interpolation inequality: it shows that the Kullback information H is controlled by a bit of the Fisher information I (which is stronger, in the sense that it involves smoothness) and the Wasserstein distance W_2 (which is weaker). A related "linear" interpolation inequality is $\|h\|_{L^2} \le \sqrt{\|h\|_{H^{-1}}\,\|h\|_{H^1}}$, where H^1 is the Sobolev space defined by the L^2-norm of the gradient, and H^{-1} is the dual of H^1.

Proof of Corollary 20.13. Statement (i) follows from Theorem 20.10 by choosing $N = \infty$ and $U(r) = r\log r$. Statement (ii) is obtained by approximation: One just needs to find a sequence of probability densities $\rho_{0,k} \to \rho_0$ in such a way that each ρ_0 is Lipschitz and

$H_\nu(\rho_{0,k}\,\nu) \longrightarrow H_\nu(\mu)$, $W_2(\rho_{0,k}\,\nu,\nu) \longrightarrow W_2(\mu,\nu)$, $I_\nu(\rho_{0,k}\,\nu) \longrightarrow I_\nu(\mu)$. I shall not go into this argument and refer to the bibliographical notes for more information. □

Proof of Theorem 20.10. First recall from the proof of Theorem 17.8 that $U_-(\rho_0)$ is integrable; since $\rho_0\,U'(\rho_0) \geq U(\rho_0)$, the integrability of $\rho_0\,U'(\rho_0)$ implies the integrability of $U(\rho_0)$. Moreover, if $N = \infty$ then $U(r) \geq a\,r\log r - b\,r$ for some positive constants a, b (unless U is linear). So

$$\big[\rho_0\,U'(\rho_0) \in L^1\big] \Longrightarrow \big[U(\rho_0) \in L^1\big] \overset{\text{if } N=\infty}{\Longrightarrow} \big[\rho_0 \log_+ \rho_0 \in L^1\big].$$

The proof of (20.12) will be performed in three steps.

Step 1: In this step I shall assume that U and β are nice. More precisely,

- *If $N < \infty$ then U is Lipschitz, U' is Lipschitz and β, β' are bounded;*
- *If $N = \infty$ then $U(r) = O(r\log(2+r))$ and U' is Lipschitz.*

Let $(\mu_t = \rho_t\,\nu)_{0\leq t\leq 1}$ be the unique Wasserstein geodesic joining μ_0 to μ_1. Recall from Theorem 17.37 the displacement convexity inequality

$$\begin{aligned}
&\int_M U(\rho_t)\,d\nu \\
&\leq (1-t)\int_{M\times M} U\left(\frac{\rho_0(x_0)}{\beta_{1-t}(x_0,x_1)}\right)\beta_{1-t}(x_0,x_1)\,\pi(dx_1|x_0)\,\nu(dx_0) \\
&\qquad + t\int_{M\times M} U\left(\frac{\rho_1(x_1)}{\beta_t(x_0,x_1)}\right)\beta_t(x_0,x_1)\,\pi(dx_0|x_1)\,\nu(dx_1),
\end{aligned}$$

and transform this into

$$\begin{aligned}
&\int_{M\times M} U\left(\frac{\rho_0}{\beta_{1-t}}\right)\beta_{1-t}\,\pi\,d\nu \leq \int_{M\times M} U\left(\frac{\rho_1}{\beta_t}\right)\beta_t\,\pi\,d\nu \\
&+ \int_{M\times M}\left[\frac{U\left(\frac{\rho_0}{\beta_{1-t}}\right)\beta_{1-t} - U(\rho_0)}{t}\right]\pi\,d\nu - \frac{1}{t}\int_M \big[U(\rho_t) - U(\rho_0)\big]\,d\nu.
\end{aligned} \tag{20.18}$$

The problem is to pass to the limit as $t \to 0$. Let us consider the four terms in (20.18) one after the other.

First term of (20.18)*:* If $K = 0$ there is nothing to do.

If $K > 0$ then $\beta_t(x_0, x_1)$ is a decreasing function of t; since $U(r)/r$ is a nondecreasing function of r it follows that $U(\rho_0/\beta)\,\beta \le U(\rho_0/\beta_{1-t})\,\beta_{1-t} \uparrow U(\rho_0)$ (as $t \to 0$). By the proof of Theorem 17.28, $U_-(\rho_0/\beta)$ is integrable, so we may apply the monotone convergence theorem to conclude that

$$\int U\left(\frac{\rho_0(x_0)}{\beta_{1-t}(x_0,x_1)}\right) \beta_{1-t}(x_0,x_1)\,\pi(dx_1|x_0)\,\nu(dx_0) \xrightarrow[t\to 0]{} \int U(\rho_0)\,d\nu. \tag{20.19}$$

If $K < 0$ then β_{1-t} is an increasing function of t, $U(\rho_0/\beta)\,\beta \ge U(\rho_0/\beta_{1-t})\,\beta_{1-t} \downarrow U(\rho_0)$, and now we should check the integrability of $U_+(\rho_0/\beta)\,\beta$. In the case $N < \infty$, this is a consequence of the Lipschitz continuity of U. In the case $N = \infty$, this comes from

$$\begin{aligned}
&\int U\left(\frac{\rho_0(x_0)}{\beta(x_0,x_1)}\right) \beta(x_0,x_1)\,\pi(dx_1|x_0)\,\nu(dx_0) \\
&\qquad \le C \int \rho_0(x_0)\,\log\left(2 + \rho_0(x_0) + \frac{1}{\beta(x_0,x_1)}\right) \pi(dx_1|x_0)\,\nu(dx_0) \\
&\qquad \le C \int \rho_0 \log(2+\rho_0)\,d\nu + C \int \log\left(2 + \frac{1}{\beta(x_0,x_1)}\right) \pi(dx_0\,dx_1) \\
&\qquad \le C\left(1 + \int \rho_0 \log \rho_0\,d\nu + W_2(\mu_0,\mu_1)^2\right),
\end{aligned}$$

where C stands for various numeric constants. Then (20.19) also holds true for $K < 0$.

Second term of (20.18)*:* This is the same as for the first term except that the inequalities are reversed. If $K > 0$ then $U(\rho_1) \ge U(\rho_1/\beta_t)\,\beta_t \downarrow U(\rho_1/\beta)\,\beta$, and to pass to the limit it suffices to check the integrability of $U_+(\rho_1)$. If $N < \infty$ this follows from the Lipschitz continuity of U, while if $N = \infty$ this comes from the assumption $\rho_1 \log_+ \rho_1 \in L^1(\nu)$. If $K < 0$ then $U(\rho_1) \le U(\rho_1/\beta_t)\,\beta_t \uparrow U(\rho_1/\beta)\,\beta$, and now we can conclude because $U_-(\rho_1)$ is integrable by Theorem 17.8. In either case,

$$\begin{aligned}
&\int U\left(\frac{\rho_1(x_1)}{\beta_t(x_0,x_1)}\right) \beta_t(x_0,x_1)\,\pi(dx_0|x_1)\,\nu(dx_1) \\
&\qquad \xrightarrow[t\to 0]{} \int U\left(\frac{\rho_1(x_1)}{\beta(x_0,x_1)}\right) \beta(x_0,x_1)\,\pi(dx_0|x_1)\,\nu(dx_1). \qquad (20.20)
\end{aligned}$$

Third term of (20.18)*:* This term exists only if $K \ne 0$. By convexity of U, the function $b \longmapsto U(r/b)$ is convex, with derivative $-p(r/b)$; so

$$U(\rho_0) - U\left(\frac{\rho_0}{\beta_{1-t}}\right)\beta_{1-t} \geq -p\left(\frac{\rho_0}{\beta_{1-t}}\right)(1-\beta_{1-t});$$

or equivalently

$$\frac{U\left(\frac{\rho_0}{\beta_{1-t}}\right)\beta_{1-t} - U(\rho_0)}{t} \leq p\left(\frac{\rho_0}{\beta_{1-t}}\right)\left(\frac{1-\beta_{1-t}}{t}\right). \tag{20.21}$$

Since U is convex, p is nondecreasing. If $K > 0$ then β_{1-t} decreases as t decreases to 0, so $p(\rho_0/\beta_{1-t})$ increases to $p(\rho_0)$, while $(1-\beta_{1-t}(x_0,x_1))/t$ increases to $\beta'(x_0,x_1)$; so the right-hand side of (20.21) increases to $p(\rho)\,\beta'$. The same is true if $K < 0$ (the inequalities are reversed but the product of two decreasing nonnegative functions is nondecreasing). Moreover, for $t = 1$ the left-hand side of (20.21) is integrable. So the monotone convergence theorem implies

$$\limsup_{t\downarrow 0} \int \left[\frac{U\left(\frac{\rho_0(x_0)}{\beta_{1-t}(x_0,x_1)}\right)\beta_{1-t}(x_0,x_1)}{t}\right] \pi(dx_1|x_0)\,\nu(dx_0)$$
$$\leq \int p(\rho_0(x_0))\,\beta'(x_0,x_1)\,\pi(dx_1|x_0)\,\nu(dx_0). \tag{20.22}$$

Fourth term of (20.18)*:* By Theorem 20.1,

$$\limsup_{t\downarrow 0}\left(-\frac{1}{t}\int [U(\rho_t) - U(\rho_0)]\,d\pi\right)$$
$$\leq \int U''(\rho_0(x_0))\,|\nabla^-\rho_0|(x_0)\,d(x_0,x_1)\,\pi(dx_0\,dx_1). \tag{20.23}$$

All in all, (20.12) follows from (20.19), (20.20), (20.22), (20.23).

Step 2: *Relaxation of the assumptions on* U.

By Proposition 17.7 we can find a sequence $(U_\ell)_{\ell\in\mathbb{N}}$ such that U_ℓ coincides with U on $[\ell^{-1},\ell]$, $U_\ell(r)$ is nonincreasing in ℓ for $r \leq 1$ and nondecreasing for $r \geq 1$, U_ℓ is linear close to the origin, U_ℓ' is Lipschitz, $U_\ell'' \leq CU''$, and

- if $N < \infty$, U_ℓ is Lipschitz;
- if $N = \infty$, $U_\ell(r) = O(r\log(2+r))$.

Then by Step 1, with the notation $p_\ell(r) = r\,U_\ell'(r) - U_\ell(r)$,

$$\int_M U_\ell(\rho_0)\,d\nu \le \int_{M\times M} U_\ell\left(\frac{\rho_1(x_1)}{\beta(x_0,x_1)}\right)\beta(x_0,x_1)\,\pi(dx_0|x_1)\,\nu(dx_1)$$
$$+\int_{M\times M} p_\ell(\rho_0(x_0))\,\beta'(x_0,x_1)\,\pi(dx_1|x_0)\,\nu(dx_0)$$
$$+\int_{M\times M} U''_\ell(\rho_0(x_0))\,|\nabla\rho_0(x_0)|\,d(x_0,x_1)\,\pi(dx_0\,dx_1). \tag{20.24}$$

Passing to the limit in $\int U_\ell(\rho_0)$ and $\int U_\ell(\rho_1/\beta)\,\beta$ is performed as in the proof of Theorem 17.37.

Next I claim that

$$\limsup_{\ell\to\infty}\int_{M\times M} p_\ell(\rho_0(x_0))\,\beta'(x_0,x_1)\,\pi(dx_1|x_0)\,\nu(dx_0)$$
$$\le \int_{M\times M} p(\rho_0(x_0))\,\beta'(x_0,x_1)\,\pi(dx_1|x_0)\,\nu(dx_0). \tag{20.25}$$

To prove (20.25), first note that $p(0)=0$ (because $p(r)/r^{1-1/N}$ is nondecreasing), so $p_\ell(r)\to p(r)$ for all r, and the integrand in the left-hand side converges to the integrand in the right-hand side.

Moreover, since $p_\ell(0)=0$ and $p'_\ell(r)=r\,U''_\ell(r)\le C\,r\,U''(r)=C\,p'(r)$, we have $0\le p_\ell(r)\le C\,p(r)$. Then:

- If $K=0$ then $\beta'=0$ and there is nothing to prove.
- If $K<0$ then $\beta'>0$. If $\int p(\rho_0(x_0))\,\beta'(x_0,x_1)\,\pi(dx_1|x_0)\,\nu(dx_0)<+\infty$ then the left-hand side converges to the right-hand side by dominated convergence; otherwise the inequality is obvious.
- If $K>0$ and $N<\infty$ then β' is bounded and we may conclude by dominated convergence as soon as $\int p(\rho_0(x_0))\,d\nu(x_0)<+\infty$. This in turn results from the fact that $\rho_0\,U'(\rho_0),U_-(\rho_0)\in L^1(\nu)$.
- If $K>0$ and $N=\infty$, then $\beta'(x_0,x_1)=-(K/3)\,d(x_0,x_1)^2$, so the same reasoning applies if

$$\int p(\rho_0(x_0))\,d(x_0,x_1)^2\,\pi(dx_1|x_0)\,\nu(dx_0)<+\infty. \tag{20.26}$$

But the left-hand side of (20.26) is bounded by

$$\sqrt{\int\frac{p(\rho_0(x_0))^2}{\rho_0(x_0)}\,\nu(dx_0)}\sqrt{\int d(x_0,x_1)^2\,\pi(dx_0\,dx_1)}$$
$$=\sqrt{\int\frac{p(\rho_0)^2}{\rho_0}\,d\nu}\;W_2(\mu_0,\mu_1),$$

which is finite by assumption.

It remains to take care of the last term in (20.24), i.e. show that

$$\limsup_{\ell\to\infty} \int U_\ell''(\rho_0(x_0))\,|\nabla\rho_0(x_0)|\,d(x_0,x_1)\,\pi(dx_0\,dx_1) \leq \int U''(\rho_0(x_0))\,|\nabla\rho_0(x_0)|\,d(x_0,x_1)\,\pi(dx_0\,dx_1).$$

If the integral on the right-hand side is infinite, the inequality is obvious. Otherwise the left-hand side converges to the right-hand side by dominated convergence, since $U_\ell''(\rho_0(x_0)) \leq C\,U''(\rho_0(x_0))$.

In the end we can pass to the limit in (20.24), and recover

$$\begin{aligned}\int_M U(\rho_0)\,d\nu \leq \int_{M\times M} U\left(\frac{\rho_1(x_1)}{\beta(x_0,x_1)}\right)\beta(x_0,x_1)\,\pi(dx_0|x_1)\,\nu(dx_1)\\ + \int_{M\times M} p(\rho_0(x_0))\,\beta'(x_0,x_1)\,\pi(dx_1|x_0)\,\nu(dx_0)\\ + \int_{M\times M} U''(\rho_0(x_0))\,|\nabla\rho_0(x_0)|\,d(x_0,x_1)\,\pi(dx_0\,dx_1),\end{aligned} \tag{20.27}$$

Step 3: *Relaxation of the assumption on β.*

If $N<\infty$ I have assumed that β,β' are bounded, which is true if $K\leq 0$ or if $\mathrm{diam}\,(M) < D_{K,N} = \pi\sqrt{(N-1)/K}$. The only problem is when $K>0$ and $\mathrm{diam}\,(M) = D_{K,N}$. In this case it suffices to establish (20.27) with N replaced by $N'>N$ and then pass to the limit as $N'\downarrow N$. Explicitly:

$$\begin{aligned}\int_M U(\rho_0)\,d\nu \leq \int_{M\times M} U\left(\frac{\rho_1(x_1)}{\beta_0^{K,N'}(x_0,x_1)}\right)\beta_0^{K,N'}(x_0,x_1)\,\pi(dx_0|x_1)\nu(dx_1)\\ + \int_{M\times M} p(\rho_0(x_0))\,(\beta^{K,N'})_1'(x_0,x_1)\,\pi(dx_1|x_0)\,\nu(dx_0)\\ + \int_{M\times M} U''(\rho_0(x_0))\,|\nabla\rho_0(x_0)|\,d(x_0,x_1)\,\pi(dx_0\,dx_1).\end{aligned}$$

Passing to the limit is allowed because the right-hand side is decreasing as $N'\downarrow N$. Indeed, $\beta^{(K,N')}$ is increasing, so $U(\rho_1/\beta^{K,N'})\,\beta^{(K,N')}$ is decreasing; and $(\beta^{(K,N')})_1'$ is decreasing. This concludes the proof of (20.12).

Next, (20.13) is obtained by considering the particular case $K = 0$ in (20.12) (so $\beta = 1$ and $\beta' = 0$), and then applying the Cauchy–Schwarz inequality:

$$\int U''(\rho_0(x_0))\,|\nabla\rho_0(x_0)|\,d(x_0,x_1)\,\pi(dx_0\,dx_1)$$
$$\leq \sqrt{\int d(x_0,x_1)^2\,\pi(dx_0\,dx_1)}\,\sqrt{\int U''(\rho_0(x_0))^2\,|\nabla\rho_0(x_0)|^2\,\pi(dx_0\,dx_1)}.$$

The case $N = \infty$ requires a bit more work. Let $u(\delta) = U(e^{-\delta})\,e^{\delta}$, then u is convex, and $u'(\delta) = -e^{\delta}\,p(e^{-\delta})$, so

$$U(e^{-\delta_1})\,e^{\delta_1} \geq U(e^{-\delta_2})\,e^{\delta_2} - e^{\delta_2}\,p(e^{-\delta_2})\,(\delta_1 - \delta_2).$$

In particular,

$$\begin{aligned}
&U\left(\frac{\rho_1(x_1)}{\beta(x_0,x_1)}\right)\frac{\beta(x_0,x_1)}{\rho_1(x_1)}\\
&\leq U(\rho_1(x_1))\,\frac{1}{\rho_1(x_1)}\\
&\qquad\qquad + \frac{\beta(x_0,x_1)}{\rho_1(x_1)}\,p\left(\frac{\rho_1(x_1)}{\beta(x_0,x_1)}\right)\left(\log\frac{1}{\rho_1(x_1)} - \log\frac{\beta(x_0,x_1)}{\rho_1(x_1)}\right)\\
&= \frac{U(\rho_1(x_1))}{\rho_1(x_1)} - \frac{\beta(x_0,x_1)}{\rho_1(x_1)}\,p\left(\frac{\rho_1(x_1)}{\beta(x_0,x_1)}\right)\frac{K}{6}\,d(x_0,x_1)^2\\
&\leq \frac{U(\rho_1(x_1))}{\rho_1(x_1)} - \frac{K_{\infty,U}}{6}\,d(x_0,x_1)^2.
\end{aligned}$$

Thus

$$\begin{aligned}
&\int U\left(\frac{\rho_1(x_1)}{\beta(x_0,x_1)}\right)\beta(x_0,x_1)\,\pi(dx_0|x_1)\,\nu(dx_1)\\
&\leq \int U(\rho_1(x_1))\,\nu(dx_1) - \frac{K_{\infty,U}}{6}\int \rho_1(x_1)\,d(x_0,x_1)^2\,\pi(dx_0|x_1)\,\nu(dx_1)\\
&= \int U(\rho_1)\,d\nu - \frac{K_{\infty,U}}{6}\,W_2(\mu_0,\mu_1)^2. \qquad (20.28)
\end{aligned}$$

On the other hand,

$$\begin{aligned}\int p(\rho_0(x_0))\,\beta'(x_0,x_1)\,\pi(dx_1|x_0)\,\nu(dx_0) \\ &\leq -\frac{K_{\infty,U}}{3}\int \rho_0(x_0)\,d(x_0,x_1)^2\,\pi(dx_1|x_0)\,\nu(dx_0) \\ &= -\frac{K_{\infty,U}}{3}\,W_2(\mu_0,\mu_1)^2. \end{aligned} \tag{20.29}$$

Plugging (20.28) and (20.29) into (20.12) finishes the proof of (20.14). The proof of (iii) is along the same lines: I shall establish the identity

$$\begin{aligned}\int U\left(\frac{\rho_1(x_1)}{\beta(x_0,x_1)}\right)\beta(x_0,x_1)\,\pi(dx_0|x_1)\,\nu(dx_1) \\ + \int p(\rho_0(x_0))\,\beta'(x_0,x_1)\,\pi(dx_1|x_0)\,\nu(dx_0) \\ \leq U_\nu(\mu_1) - K\lambda\left(\frac{(\sup\rho_0)^{-\frac{1}{N}}}{3}+\frac{(\sup\rho_1)^{-\frac{1}{N}}}{6}\right)W_2(\mu_0,\mu_1)^2. \end{aligned} \tag{20.30}$$

This combined with Corollary 19.5 will lead from (20.12) to (20.15).

So let us prove (20.30). By convexity of $s \longmapsto s^N\,U(s^{-N})$,

$$\begin{aligned} U\left(\frac{\rho_1(x_1)}{\beta(x_0,x_1)}\right)\frac{\beta(x_0,x_1)}{\rho_1(x_1)} \leq \frac{U(\rho_1(x_1))}{\rho_1(x_1)} \\ +N\left(\frac{\beta(x_0,x_1)}{\rho_1(x_1)}\right)^{1-\frac{1}{N}} p\left(\frac{\rho_1(x_1)}{\beta(x_0,x_1)}\right)\left[\left(\frac{\beta(x_0,x_1)}{\rho_1(x_1)}\right)^{\frac{1}{N}} - \frac{1}{\rho_1(x_1)^{\frac{1}{N}}}\right], \end{aligned}$$

which is the same as

$$\begin{aligned} U\left(\frac{\rho_1(x_1)}{\beta(x_0,x_1)}\right)\beta(x_0,x_1) \leq U(\rho_1(x_1)) \\ + N\,p\left(\frac{\rho_1(x_1)}{\beta(x_0,x_1)}\right)\beta(x_0,x_1)^{1-\frac{1}{N}}\left(\beta(x_0,x_1)^{\frac{1}{N}}-1\right). \end{aligned}$$

As a consequence,

$$\begin{aligned} \int U\left(\frac{\rho_1(x_1)}{\beta(x_0,x_1)}\right)\beta(x_0,x_1)\,\pi(dx_0|x_1)\,\nu(dx_1) \leq U_\nu(\mu_1) \\ +N\int p\left(\frac{\rho_1(x_1)}{\beta(x_0,x_1)}\right)\beta(x_0,x_1)^{1-\frac{1}{N}}\left(\beta(x_0,x_1)^{\frac{1}{N}}-1\right)\pi(dx_0|x_1)\,\nu(dx_1). \end{aligned} \tag{20.31}$$

Since $K \geq 0$, $\beta(x_0, x_1)$ coincides with $(\alpha/\sin\alpha)^{N-1}$, where $\alpha = \sqrt{K/(N-1)}\, d(x_0, x_1)$. By the elementary inequality

$$0 \leq \alpha \leq \pi \Longrightarrow \qquad N\left(\left(\frac{\alpha}{\sin\alpha}\right)^{\frac{N-1}{N}} - 1\right) \geq \left(\frac{N-1}{6}\right)\alpha^2 \tag{20.32}$$

(see the bibliographical notes for details), the right-hand side of (20.31) is bounded above by

$$U_\nu(\mu_1) - \frac{K}{6}\int p\left(\frac{\rho_1(x_1)}{\beta(x_0, x_1)}\right)\beta(x_0, x_1)^{1-\frac{1}{N}}\,\pi(dx_0|x_1)\,\nu(dx_1)$$

$$\leq U_\nu(\mu_1) - \frac{K}{6}\left[\inf_{r>0}\frac{p(r)}{r^{1-\frac{1}{N}}}\right]\int \rho_1(x_1)^{1-\frac{1}{N}}\,d(x_0, x_1)^2\,\pi(dx_0|x_1)\,\nu(dx_1)$$

$$\leq U_\nu(\mu_1)$$
$$\qquad - \frac{K}{6}\left(\lim_{r\to 0}\frac{p(r)}{r^{1-\frac{1}{N}}}\right)(\sup\rho_1)^{-\frac{1}{N}}\int \rho_1(x_1)\,d(x_0, x_1)^2\,\pi(dx_0|x_1)\,\nu(dx_1)$$

$$= U_\nu(\mu_1) - \frac{K}{6}\,\lambda\,(\sup\rho_1)^{-\frac{1}{N}}\,W_2(\mu_0, \mu_1)^2, \tag{20.33}$$

where $\lambda = \lambda_{N,U}$.

On the other hand, since $\beta'(x_0, x_1) = -(N-1)(1 - (\alpha/\tan\alpha)) < 0$, we can use the elementary inequality

$$0 < \alpha \leq \pi \Longrightarrow \qquad (N-1)\left(1 - \frac{\alpha}{\tan\alpha}\right) \geq (N-1)\,\frac{\alpha^2}{3} \tag{20.34}$$

(see the bibliographical notes again) to deduce

$$\int p(\rho_0(x_0))\,\beta'(x_0, x_1)\,\pi(dx_1|x_0)\,\nu(dx_0) \tag{20.35}$$

$$\leq \left(\inf_{r>0}\frac{p(r)}{r^{1-\frac{1}{N}}}\right)\int \rho_0(x_0)^{1-\frac{1}{N}}\,\beta'(x_0, x_1)\,\pi(dx_1|x_0)\,\nu(dx_0)$$

$$\leq \left(\lim_{r\to 0}\frac{p(r)}{r^{1-\frac{1}{N}}}\right)(\sup\rho_0)^{-\frac{1}{N}}\int \rho_0(x_0)\,\beta'(x_0, x_1)\,\pi(dx_1|x_0)\,\nu(dx_0)$$

$$\leq \left(\lim_{r\to 0}\frac{p(r)}{r^{1-\frac{1}{N}}}\right)(\sup\rho_0)^{-\frac{1}{N}}\int \rho_0(x_0)\left(\frac{K d(x_0, x_1)^2}{3}\right)\pi(dx_1|x_0)\,\nu(dx_0)$$

$$= \frac{K\lambda\,(\sup\rho_0)^{-\frac{1}{N}}}{3}\int d(x_0, x_1)^2\,\pi(dx_0\,dx_1)$$

$$= \frac{K\lambda\,(\sup\rho_0)^{-\frac{1}{N}}}{3}\,W_2(\mu_0, \mu_1)^2. \tag{20.36}$$

The combination of (20.33) and (20.36) implies (20.30) and concludes the proof of Theorem 20.10. □

Bibliographical notes

Formula (20.6) appears as Theorem 5.30 in my book [814] when the space is $\mathbb{R}^n$; there were precursors, see for instance [669, 671]. The integration by parts leading from (20.6) to (20.7) is quite tricky, especially in the noncompact case; this will be discussed later in more detail in Chapter 23 (see Theorem 23.14 and the bibliographical notes). Here I preferred to be content with Theorem 20.1, which is much less technical, and still sufficient for most applications known to me. Moreover, it applies to nonsmooth spaces, which will be quite useful in Part III of this course. The argument is taken from my joint work with Lott [577] (where the space $\mathcal{X}$ is assumed to be compact, which simplifies slightly the assumptions).

Fisher introduced the Fisher information as part of his theory of "efficient statistics" [373]. It plays a crucial role in the Cramér–Rao inequality [252, Theorem 12.11.1], determines the asymptotic variance of the maximum likelihood estimate [802, Chapter 4] and the rate function for large deviations of time-averages of solutions of heat-like equations [313]. The Boltzmann–Gibbs–Shannon–Kullback information on the one hand, the Fisher information on the other hand, play the two leading roles in information theory [252, 295]. They also have a leading part in statistical mechanics and kinetic theory (see e.g. [817, 812]).

The HWI inequality was established in my joint work with Otto [671]; it obviously extends to any reasonable K-displacement convex functional. A precursor inequality was studied by Otto [669]. An application to a "concrete" problem of partial differential equations can be found in [213, Section 5]. Recently Gao and Wu [405] used the HWI inequality to derive new criteria of uniqueness for certain spin systems.

It is shown in [671, Appendix] and [814, Proof of Theorem 9.17, Step 1] how to devise approximating sequences of smooth densities in such a way that the H_ν and I_ν functionals pass to the limit. By adapting these arguments one may conclude the proof of Corollary 20.13.

The role of the HWI inequality as an interpolation inequality is briefly discussed in [814, Section 9.4] and turned into application in [213, Proof of Theorem 5.1]: in that reference we study rates of convergence for certain nonlinear partial differential equations, and combine a bound on the Fisher information with a convergence estimate in Wasserstein distance, to establish a convergence estimate in a stronger sense (L^1 norm, for instance).

The HWI inequality is also interesting as an "infinite-dimensional" interpolation inequality; this is applied in [445] to the study of the limit behavior of the entropy in a hydrodynamic limit.

A slightly different derivation of the HWI inequality is due to Cordero-Erausquin [242]; a completely different derivation is due to Bobkov, Gentil and Ledoux [127]. Variations of these inequalities were studied by Agueh, Ghoussoub and Kang [5]; and Cordero-Erausquin, Gangbo and Houdré [245].

There is no well-identified analog of the HWI inequality for nonquadratic cost functions. For nonquadratic costs in $\mathbb{R}^n$, some inequalities in the spirit of HWI are established in [76], where they are used to derive various isoperimetric-type inequalities.

The first somewhat systematic studies of HWI-type inequalities in the case $N < \infty$ are due to Lott and myself [577, 578].

The elementary inequalities (20.32) and (20.34) are proven in [578, Section 5], where they are used to derive the Lichnerowicz spectral gap inequality (Theorem 21.20 in Chapter 21).

21

Isoperimetric-type inequalities

It is a fact of experience that several inequalities with isoperimetric content can be retrieved by considering the above-tangent formulation of displacement convexity. Here is a possible heuristic explanation for this phenomenon. Assume, for the sake of the discussion, that the initial measure is the normalized indicator function of some set A. Think of the functional U_ν as the internal energy of some fluid that is initially confined in A. In a displacement interpolation, some of the mass of the fluid will have to flow out of A, leading to a variation of the energy (typically, more space available means less density and less energy). The decrease of energy at initial time is related to the amount of mass that is able to flow out of A at initial time, and that in turn is related to the surface of A (a small surface leads to a small variation, because not much of the fluid can escape). So by controlling the decrease of energy, one should eventually gain control of the surface of A.

The functional nature of this approach makes it possible to replace the set A by some arbitrary probability measure $\mu = \rho\,\nu$. Then, what plays the role of the "surface" of A is some integral expression involving $\nabla\rho$. Any inequality expressing the domination of an integral expression of ρ by an integral expression of ρ and $\nabla\rho$ will be loosely referred to as a **Sobolev**-type, or **isoperimetric**-type inequality. Of course there are many many variants of such inequalities.

Logarithmic Sobolev inequalities

A probability measure ν on a Riemannian manifold is said to satisfy a logarithmic Sobolev inequality if the functional H_ν is dominated by (a constant multiple of) the functional I_ν. Here is a more precise definition:

C. Villani, *Optimal Transport*, Grundlehren der mathematischen Wissenschaften 338,

Definition 21.1 (Logarithmic Sobolev inequality). *Let M be a Riemannian manifold, and ν a probability measure on M. It is said that ν satisfies a logarithmic Sobolev inequality with constant λ if, for any probability measure $\mu = \rho\,\nu$ with ρ Lipschitz, one has*

$$H_\nu(\mu) \le \frac{1}{2\lambda} I_\nu(\mu). \tag{21.1}$$

Explicitly, inequality (21.1) means

$$\int \rho \log \rho \, d\nu \le \frac{1}{2\lambda} \int \frac{|\nabla \rho|^2}{\rho} \, d\nu. \tag{21.2}$$

Equivalently, for any function u (regular enough) one should have

$$\int u^2 \log(u^2)\, d\nu - \left(\int u^2 \, d\nu\right) \log \left(\int u^2 \, d\nu\right) \le \frac{2}{\lambda} \int |\nabla u|^2 \, d\nu. \tag{21.3}$$

To go from (21.2) to (21.3), just set $\rho = u^2/(\int u^2\, d\nu)$ and notice that $\nabla|u| \le |\nabla u|$.

The Lipschitz regularity of ρ allows one to define $|\nabla \rho|$ pointwise, for instance by means of (20.1). Everywhere in this chapter, $|\nabla \rho|$ may also be replaced by the quantity $|\nabla^- \rho|$ appearing in (20.2); in fact both expressions coincide almost everywhere if u is Lipschitz.

This restriction of Lipschitz continuity is unnecessary, and can be relaxed with a bit of work. For instance, if $\nu = e^{-V}\mathrm{vol}$, $V \in C^2(M)$, then one can use a little bit of distribution theory to show that the quantity $\int |\nabla \rho|^2/\rho \, d\nu$ is well-defined in $[0, +\infty]$, and then (21.1) makes sense. But in the sequel, I shall just stick to Lipschitz functions. The same remark applies to other functional inequalities which will be encountered later: dimension-dependent Sobolev inequalities, Poincaré inequalities, etc.

Logarithmic Sobolev inequalities are *dimension-free* Sobolev-type inequalities: the dimension of the space does not appear explicitly in (21.3). This is one reason why these inequalities are extremely popular in various branches of statistical mechanics, mathematical statistics, quantum field theory, and more generally the study of phenomena in high or infinite dimension. They are also used in geometry and partial differential equations, including Perelman's recent work on the Ricci flow and the Poincaré conjecture.

At this stage of the course, the next theorem, a famous result in Riemannian geometry, may seem almost trivial.

Theorem 21.2 (Bakry–Émery theorem). *Let M be a Riemannian manifold equipped with a reference probability measure $\nu = e^{-V}\mathrm{vol}$, $V \in C^2(M)$, satisfying a curvature assumption* $\mathrm{CD}(K,\infty)$ *for some $K > 0$. Then ν satisfies a logarithmic Sobolev inequality with constant K, i.e.*

$$H_\nu \leq \frac{I_\nu}{2K}. \tag{21.4}$$

Example 21.3. For the Gaussian measure $\gamma(dx) = (2\pi)^{-n/2}e^{-|x|^2/2}$ in $\mathbb{R}^n$, one has

$$H_\gamma \leq \frac{I_\gamma}{2}, \tag{21.5}$$

independently of the dimension. This is the **Stam–Gross logarithmic Sobolev inequality**. By scaling, for any $K > 0$ the measure $\gamma_K(dx) = (2\pi/K)^{-n/2}e^{-K|x|^2/2}\,dx$ satisfies a logarithmic Sobolev inequality with constant K.

Remark 21.4. More generally, if $V \in C^2(\mathbb{R}^n)$ and $\nabla^2 V \geq K I_n$, Theorem 21.2 shows that $\nu(dx) = e^{-V(x)}\,dx$ satisfies a logarithmic Sobolev inequality with constant K. When $V(x) = K|x|^2/2$ the constant K is optimal in (21.4).

Remark 21.5. The curvature assumption $\mathrm{CD}(K,\infty)$ is quite restrictive; however, there are known perturbation theorems which immediately extend the range of application of Theorem 21.2. For instance, if ν satisfies a logarithmic Sobolev inequality, v is a bounded function and $\widetilde{\nu} = e^{-v}\nu/Z$ is another probability measure obtained from ν by multiplication by e^{-v}, then also $\widetilde{\nu}$ satisfies a logarithmic Sobolev inequality (Holley–Stroock perturbation theorem). The same is true if v is unbounded, but satisfies $\int e^{\alpha|\nabla v|^2}\,d\nu < \infty$ for α large enough.

Proof of Theorem 21.2. By Theorem 18.12, ν admits square-exponential moments, in particular it lies in $P_2(M)$. Then from Corollary 20.13(ii) and the inequality $ab \leq Ka^2/2 + b^2/(2K)$,

$$H_\nu(\mu) \leq W_2(\mu,\nu)\,\sqrt{I_\nu(\mu)} - \frac{KW_2(\mu,\nu)^2}{2} \leq \frac{I_\nu(\mu)}{2K}.$$

□

Open Problem 21.6. *For Riemannian manifolds satisfying* $\mathrm{CD}(K,N)$ *with $N < \infty$, the optimal constant in the logarithmic Sobolev inequality is not K but $KN/(N-1)$. Can this be proven by a transport argument?*

In the next section, some finite-dimensional Sobolev inequalities will be addressed, but it is not at all clear that they are strong enough to lead to the solution of Problem 21.6. Before examining these issues, I shall state an easy variation of Theorem 21.2. Recall Definition 20.6.

Theorem 21.7 (Sobolev-L^∞ interpolation inequalities). *Let M be a Riemannian manifold, equipped with a reference probability measure $\nu = e^{-V}\mathrm{vol}$, $V \in C^2(M)$, satisfying a* $\mathrm{CD}(K,N)$ *curvature-dimension bound for some $K > 0$, $N \in (1,\infty]$. Further, let $U \in \mathcal{DC}_N$. Then, for any Lipschitz-continuous probability density ρ, if $\mu = \rho\,\nu$, one has the inequality*

$$0 \le U_\nu(\mu) - U_\nu(\nu) \le \frac{(\sup \rho)^{\frac{1}{N}}}{2K\lambda}\, I_{U,\nu}(\mu), \tag{21.6}$$

where

$$\lambda = \lim_{r\to 0} \left(\frac{p(r)}{r^{1-\frac{1}{N}}} \right).$$

Proof of Theorem 21.7. The proof of the inequality on the right-hand side of (21.6) is the same as for Theorem 21.2, using Theorem 20.10(iii). The inequality on the left-hand side is a consequence of Jensen's inequality: $U_\nu(\mu) = \int U(\rho)\,d\nu \ge U(\int \rho\,d\nu) = U(1) = U_\nu(\nu)$. □

Sobolev inequalities

Sobolev inequalities are one among several classes of functional inequalities with isoperimetric content; they are extremely popular in the theory of partial differential equations. They look like logarithmic Sobolev inequalities, but with powers instead of logarithms, and they take dimension into account explicitly.

The most basic Sobolev inequality is in Euclidean space: If u is a function on $\mathbb{R}^n$ such that $\nabla u \in L^p(\mathbb{R}^n)$ $(1 \le p < n)$ and u vanishes at infinity (in whatever sense, see e.g. Remark 21.13 below), then u automatically lies in $L^{p^\star}(\mathbb{R}^n)$ where $p^\star = (np)/(n-p) > p$. More quantitatively, there is a constant $S = S(n,p)$ such that

$$\|u\|_{L^{p^\star}(\mathbb{R}^n)} \le S\,\|\nabla u\|_{L^p(\mathbb{R}^n)}.$$

There are other versions for $p = n$ (in which case essentially $\exp(c\,u^{n'})$ is integrable, $n' = n/(n-1)$), and $p > n$ (in which case u is Hölder-continuous). There are also very many variants for a function u defined on a set Ω that might be a reasonable open subset of either $\mathbb{R}^n$ or a Riemannian manifold M. For instance (say for $p < n$),

$$\|u\|_{L^{p^\star}(\Omega)} \le A\,\|\nabla u\|_{L^p(\Omega)} + C\,\|u\|_{L^{p^\sharp}(\partial\Omega)}, \qquad p^\sharp = \frac{(n-1)\,p}{n-p},$$

$$\|u\|_{L^{p^\star}(\Omega)} \le A\,\|\nabla u\|_{L^p(\Omega)} + B\,\|u\|_{L^q(\Omega)}, \qquad q \ge 1,$$

etc. One can also quote the **Gagliardo–Nirenberg interpolation inequalities**, which typically take the form

$$\|u\|_{L^{p^\star}} \le G\,\|\nabla u\|_{L^p}^{1-\theta}\,\|u\|_{L^q}^{\theta}, \qquad 1 \le p < n, \quad 1 \le q < p^\star, \quad 0 \le \theta \le 1,$$

with some restrictions on the exponents. I will not say more about Sobolev-type inequalities, but there are entire books devoted to them.

In a Riemannian setting, there is a famous family of Sobolev inequalities obtained from the curvature-dimension bound $\mathrm{CD}(K,N)$ with $K > 0$ and $2 < N < \infty$:

$$1 \le q \le \frac{2N}{N-2} \quad \Longrightarrow$$
$$\frac{c}{q-2}\left[\left(\int |u|^q\,d\nu\right)^{\frac{2}{q}} - \int |u|^2\,d\nu\right] \le \int |\nabla u|^2\,d\nu, \qquad c = \frac{NK}{N-1}. \tag{21.7}$$

When $q \to 2$, (21.7) reduces to Bakry–Émery's logarithmic Sobolev inequality. The other most interesting member of the family is obtained when q coincides with the critical exponent $2^\star = (2N)/(N-2)$, and then (21.7) becomes

$$\|u\|^2_{L^{\frac{2N}{N-2}}(M)} \le \|u\|^2_{L^2(M)} + \left(\frac{4}{N-2}\right)\left(\frac{N-1}{KN}\right)\|\nabla u\|^2_{L^2(M)}. \tag{21.8}$$

There is no loss of generality in assuming $u \ge 0$, since the inequality for general u follows easily from the inequality for nonnegative u. Let us then change unknowns by choosing $\rho = u^{2N/(N-2)}$. By homogeneity, there is also no loss of generality in assuming that $\mu := \rho\,\nu$ is a probability measure. Then inequality (21.8) transforms into

$$H_{N/2,\nu}(\mu) = -\frac{N}{2}\int \left(\rho^{1-\frac{2}{N}} - \rho\right) d\nu$$
$$\le \frac{1}{2K}\int \frac{|\nabla\rho|^2}{\rho}\left(\frac{(N-1)(N-2)}{N^2}\rho^{-\frac{2}{N}}\right) d\nu. \tag{21.9}$$

The way in which I have written inequality (21.9) might look strange, but it has the merit of showing very clearly how the limit $N \to \infty$ leads to the logarithmic Sobolev inequality $H_{\infty,\nu}(\mu) \le (1/2K) \int (|\nabla \rho|^2/\rho)\, d\nu$.

I don't know whether (21.9), or more generally (21.7), can be obtained by transport. Instead, I shall derive related inequalities, whose relation to (21.9) is still unclear.

Remark 21.8. It is possible that (21.8) implies (21.6) if $U = U_N$. This would follow from the inequality

$$H_{N,\nu} \le \left(\frac{N-1}{N-2}\right) (\sup \rho)^{\frac{1}{N}}\, H_{N/2,\nu},$$

which should not be difficult to prove, or disprove.

Theorem 21.9 (Sobolev inequalities from $\mathrm{CD}(K,N)$**).** *Let M be a Riemannian manifold, equipped with a reference measure $\nu = e^{-V}\mathrm{vol}$, $V \in C^2(M)$, satisfying a curvature-dimension inequality $\mathrm{CD}(K,N)$ for some $K > 0$, $1 < N < \infty$. Then, for any probability density ρ, Lipschitz continuous and strictly positive, and $\mu = \rho\,\nu$, one has*

$$H_{N,\nu}(\mu) = -N \int_M (\rho^{1-\frac{1}{N}} - \rho)\, d\nu \le \int_M \Theta^{(N,K)}(\rho, |\nabla \rho|)\, d\nu, \qquad (21.10)$$

where

$$\Theta^{(N,K)}(r,g) = \\ r \sup_{0 \le \alpha \le \pi} \left(\frac{N-1}{N} \frac{g}{r^{1+\frac{1}{N}}} \sqrt{\frac{N-1}{K}}\, \alpha + N \left(1 - \left(\frac{\alpha}{\sin\alpha}\right)^{1-\frac{1}{N}}\right) \right. \\ \left. + \; (N-1)\left(\frac{\alpha}{\tan\alpha} - 1\right) r^{-\frac{1}{N}} \right). \qquad (21.11)$$

As a consequence,

$$H_{N,\nu}(\mu) \le \frac{1}{2K} \int_M \frac{|\nabla \rho|^2}{\rho} \left(\left(\frac{N-1}{N}\right)^2 \frac{\rho^{-\frac{2}{N}}}{\frac{1}{3} + \frac{2}{3}\rho^{-\frac{1}{N}}} \right) d\nu. \qquad (21.12)$$

Remark 21.10. By taking the limit as $N \to \infty$ in (21.12), one recovers again the logarithmic Sobolev inequality of Bakry and Émery, with the sharp constant. For fixed N, the exponents appearing in (21.12) are sharp: For large ρ, the integrand in the right-hand side behaves like $|\nabla\rho|^2\rho^{-(1+2/N)} = c_N|\nabla\rho^{\frac{1}{2^\star}}|^2$, so the critical Sobolev exponent $2^\star$ does govern this inequality. On the other hand, the *constants* appearing in (21.12) are definitely not sharp; for instance it is obvious that they do not imply exponential integrability as $N \to 2$.

Open Problems 21.11. *Is inequality* (21.10) *stronger, weaker, or not comparable to inequality* (21.9)*? Does* (21.12) *follow from* (21.9)*? Can one find a transport argument leading to* (21.9)*?*

Proof of Theorem 21.9. Start from Theorem 20.10 and choose $U(r) = -N(r^{1-\frac{1}{N}} - r)$. After some straightforward calculations, one obtains

$$H_{N,\nu}(\mu) \leq \int_M \theta^{(N,K)}(\rho, |\nabla\rho|, \alpha),$$

where $\alpha = \sqrt{K/(N-1)}\, d(x_0, x_1) \in [0,\pi]$, and $\theta^{(N,K)}$ is an explicit function such that

$$\Theta^{(N,K)}(r,g) = \sup_{\alpha\in[0,\pi]} \theta^{(N,K)}(r,g,\alpha).$$

This is sufficient to prove (21.10).

To go from (21.10) to (21.12), one can use the elementary inequalities (20.32) and (20.34) and compute the supremum explicitly. $\square$

Now I shall consider the case of the Euclidean space $\mathbb{R}^n$, equipped with the Lebesgue measure, and show that sharp Sobolev inequalities can be obtained by a transport approach. The proof will take advantage of the scaling properties in $\mathbb{R}^n$.

Theorem 21.12 (Sobolev inequalities in $\mathbb{R}^n$). *Whenever u is a Lipschitz, compactly supported function on $\mathbb{R}^n$, then*

$$\|u\|_{L^{p^\star}(\mathbb{R}^n)} \leq S_n(p)\, \|\nabla u\|_{L^p(\mathbb{R}^n)}, \qquad 1 \leq p < n, \quad p^\star = \frac{np}{n-p}, \tag{21.13}$$

where the constant $S_n(p)$ is given by

$$S_n(p) = \inf\left\{ \frac{p(n-1)}{n(n-p)} \frac{\left(\int |g|\right)^{\frac{1}{p^\star}} \left(\int |y|^{p'}|g(y)|\,dy\right)^{\frac{1}{p'}}}{\int |g|^{1-\frac{1}{n}}} \right\}, \qquad p' = \frac{p}{p-1},$$

and the infimum is taken over all functions $g \in L^1(\mathbb{R}^n)$, not identically 0.

Remark 21.13. The assumption of Lipschitz continuity for u can be removed, but I shall not do so here. Actually, inequality (21.13) holds true as soon as u is locally integrable and vanishes at infinity, in the sense that the Lebesgue measure of $\{|u| \geq r\}$ is finite for any $r > 0$.

Remark 21.14. The constant $S_n(p)$ in (21.13) is optimal.

Proof of Theorem 21.12. Choose $M = \mathbb{R}^n$, $\nu =$ Lebesgue measure, and apply Theorem 20.10 with $K = 0$, $N = n$, and $\mu_0 = \rho_0\,\nu$, $\mu_1 = \rho_1\,\nu$, both of them compactly supported. By formula (20.14) in Theorem 20.10(i),

$$H_{n,\nu}(\mu_0) - H_{n,\nu}(\mu_1) \\ \leq \left(1 - \frac{1}{n}\right) \int_{\mathbb{R}^n\times\mathbb{R}^n} \rho_0(x_0)^{-(1+\frac{1}{n})}\,|\nabla\rho_0|(x_0)\,d(x_0,x_1)\,\pi(dx_0\,dx_1).$$

Then Hölder's inequality and the marginal property of π imply

$$H_{n,\nu}(\mu_0) - H_{n,\nu}(\mu_1) \leq \left(1-\frac{1}{n}\right)\left(\int_{\mathbb{R}^n} \rho_0^{-p(1+\frac{1}{n})}\,|\nabla\rho_0|^p\,d\mu_0\right)^{\frac{1}{p}} \\ \left(\int_{\mathbb{R}^n\times\mathbb{R}^n} d(x_0,x_1)^{p'}\,\pi(dx_0\,dx_1)\right)^{\frac{1}{p'}},$$

where $p' = p/(p-1)$. This can be rewritten

$$n\int \rho_1^{1-\frac{1}{n}}\,d\nu \leq n\int \rho_0^{1-\frac{1}{n}}\,d\nu \\ + \left(1-\frac{1}{n}\right)\left(\rho_0^{-p(1+\frac{1}{n})}\,|\nabla\rho_0|^p\,d\mu_0\right)^{\frac{1}{p}} W_{p'}(\mu_0,\mu_1). \qquad (21.14)$$

Now I shall use a homogeneity argument. Fix ρ_1 and ρ_0 as above, and define $\rho_0^{(\lambda)}(x) = \lambda^n \rho_0(\lambda x)$. On the one hand,

$$\int \left(\rho_0^{(\lambda)}\right)^{1-\frac{1}{n}}\,d\nu = \lambda^{-1}\int \rho_0^{1-\frac{1}{n}}\,d\nu \xrightarrow[\lambda\to\infty]{} 0;$$

on the other hand,

$$\int_{\mathbb{R}^n} (\rho_0^{(\lambda)})^{-p(1+\frac{1}{n})}\,|\nabla\rho_0^{(\lambda)}|^p\,d\mu_0^{(\lambda)} \qquad \text{does not depend on } \lambda.$$

Moreover, as $\lambda \to \infty$, the probability measure $\mu_0^{(\lambda)} = \rho_0^{(\lambda)}\,\nu$ converges weakly to the Dirac mass δ_0 at the origin; so

$$W_{p'}(\mu_0^{(\lambda)}, \mu_1) \longrightarrow W_{p'}(\delta_0, \mu_1) = \left(\int |y|^{p'}\, d\mu_1(y)\right)^{\frac{1}{p'}}.$$

After writing (21.14) for $\mu_0 = \mu_0^{(\lambda)}$ and then passing to the limit as $\lambda \to \infty$, one obtains

$$n \int \rho_1^{1-\frac{1}{n}}\, d\nu \leq \left(\int_{\mathbb{R}^n} \rho_0^{-p(1+\frac{1}{n})}\, |\nabla \rho_0|^p\, d\mu_0\right)^{\frac{1}{p}} \left(\int |y|^{p'}\, d\mu_1(y)\right)^{\frac{1}{p'}}. \tag{21.15}$$

Let us change unknowns and define $\rho_0 = u^{1/p^\star}$, $\rho_1 = g$; inequality (21.15) then becomes

$$1 \leq \frac{p\,(n-1)}{n\,(n-p)} \left(\frac{\left(\int |y|^{p'}\, g(y)\, dy\right)^{\frac{1}{p'}}}{\int g^{(1-\frac{1}{n})}} \right) \|\nabla u\|_{L^p},$$

where u and g are only required to satisfy $\int u^{p^\star} = 1$, $\int g = 1$. The inequality (21.13) follows by homogeneity again. □

To conclude this section, I shall assume $\mathrm{Ric}_{N,\nu} \geq K < 0$ and derive Sobolev inequalities for compactly supported functions. Since I shall not be concerned here with optimal constants, I shall only discuss the limit case $p = 1$, $p^\star = n/(n-1)$, which implies the general inequality for $p < n$ (via Hölder's inequality), up to a loss in the constants.

Theorem 21.15 (CD (K,N) implies L^1-Sobolev inequalities). *Let M be a Riemannian manifold equipped with a reference measure $\nu = e^{-V}\mathrm{vol}$, satisfying a curvature-dimension bound $\mathrm{CD}(K,N)$ for some $K < 0$, $N \in (1,\infty)$. Then, for any ball $\mathcal{B} = B(z,R)$, $R \geq 1$, there are constants A and B, only depending on a lower bound on K, and upper bounds on N and R, such that for any Lipschitz function u supported in $\mathcal{B}$,*

$$\|u\|_{L^{\frac{N}{N-1}}} \leq A\, \|\nabla u\|_{L^1} + B\, \|u\|_{L^1}. \tag{21.16}$$

Proof of Theorem 21.15. Inequality (21.16) remains unchanged if ν is multiplied by a positive constant. So we might assume, without loss of generality, that $\nu[B(z,R)] = 1$.

Formula (20.12) in Theorem 20.10 implies

$$N - \int \rho_0^{1-\frac{1}{N}}\, d\nu \le N - \int \rho_1(x_1)^{1-\frac{1}{N}}\, \beta(x_0,x_1)^{\frac{1}{N}}\, \pi(dx_0|x_1)\, \nu(dx_1) \\ + \int \rho_0(x_0)^{1-\frac{1}{N}}\, \beta'(x_0,x_1)\, \pi(dx_1|x_0)\, \nu(dx_0) \\ + \frac{1}{N}\int \rho_0(x_0)^{-1-\frac{1}{N}}\, |\nabla\rho_0(x_0)|\, d(x_0,x_1)\, \pi(dx_0\, dx_1). \quad (21.17)$$

Choose $\rho_1 = 1_{B(z,R)}/\nu[B(z,R)]$ (the normalized indicator function of the ball). The arguments of β and β' in (21.17) belong to $B(z,R)$, so the coefficients β and β' remain bounded by some explicit function of N, K and R, while the distance $d(x_0,x_1)$ remains bounded by $2R$. So there are constants $\delta(K,N,R) > 0$ and $\overline{C}(K,N,R)$ such that

$$-\int \rho_0^{1-\frac{1}{N}}\, d\nu \le -\,\delta(K,N,R)\,\nu[\mathcal{B}]^{\frac{1}{N}} \\ + \overline{C}(K,N,R)\left[\int \rho_0^{1-\frac{1}{N}} + \int \rho_0^{-\frac{1}{N}}|\nabla\rho_0|\right]. \quad (21.18)$$

Recall that $\nu[\mathcal{B}] = 1$; then after the change of unknowns $\rho_0 = u^{N/(N-1)}$, inequality (21.18) implies

$$1 \le S(K,N,R)\,\left[\|\nabla u\|_{L^1(M)} + \|u\|_{L^1(M)}\right],$$

for some explicit constant $S = (\overline{C}+1)/\delta$. This holds true under the constraint $1 = \int \rho = \int u^{N/(N-1)}$, and then inequality (21.16) follows by homogeneity. □

Isoperimetric inequalities

Isoperimetric inequalities are sometimes obtained as limits of Sobolev inequalities applied to indicator functions. The most classical example is the equivalence between the Euclidean isoperimetric inequality

$$\frac{|\partial A|}{|A|^{\frac{n-1}{n}}} \ge \frac{|\partial B^n|}{|B^n|^{\frac{n-1}{n}}}$$

already considered in Chapter 2, and the *optimal* Sobolev inequality $\|u\|_{L^{n/(n-1)}(\mathbb{R}^n)} \le S_n(1)\,\|\nabla u\|_{L^1(\mathbb{R}^n)}$.

As seen before, there is a proof of the optimal Sobolev inequality in $\mathbb{R}^n$ based on transport, and of course this leads to a proof of the Euclidean isoperimetry. There is also a more direct path to a transport-based proof of isoperimetry, as explained in Chapter 2.

Besides the Euclidean one, the most famous isoperimetric inequality in differential geometry is certainly the **Lévy–Gromov inequality**, which states that if A is a reasonable set in a manifold (M, g) with dimension n and Ricci curvature bounded below by $K > 0$, then

$$\frac{|\partial A|}{|M|} \geq \frac{|\partial B|}{|S|},$$

where B is a spherical cap in the model sphere S (that is, the sphere with dimension N and Ricci curvature K) such that $|B|/|S| = |A|/|M|$. In other words, isoperimetry in M is at least as strong as isoperimetry in the model sphere.

I don't know if the Lévy–Gromov inequality can be retrieved from optimal transport, and I think this is one of the most exciting open problems in the field. Indeed, there seems to be no "reasonable" proof of the Lévy–Gromov inequality, in the sense that the only known arguments rely on subtle results from geometric measure theory, about the rectifiability of certain extremal sets. A softer argument would be conceptually very satisfactory. I record this in the form of a loosely formulated open problem:

Open Problem 21.16. *Find a transport-based, soft proof of the Lévy–Gromov isoperimetric inequality.*

The same question can be asked for the **Gaussian isoperimetry**, which is the infinite-dimensional version of the Lévy–Gromov inequality. In this case however there are softer approaches based on functional versions; see the bibliographical notes for more details.

Poincaré inequalities

Poincaré inequalities are related to Sobolev inequalities, and often appear as limit cases of them. (I am sorry if the reader begins to be bored by this litany: Logarithmic Sobolev inequalities are limits of Sobolev inequalities, isoperimetric inequalities are limits of Sobolev inequalities, Poincaré inequalities are limits of Sobolev inequalities...) Here in this section I shall only consider *global* Poincaré inequalities, which are rather different from the local inequalities considered in Chapter 19.

Definition 21.17 (Poincaré inequality). *Let M be a Riemannian manifold, and ν a probability measure on M. It is said that ν satisfies a Poincaré inequality with constant λ if, for any $u \in L^2(\mu)$ with u Lipschitz, one has*

$$\left\|u - \langle u \rangle\right\|^2_{L^2(\nu)} \leq \frac{1}{\lambda}\,\|\nabla u\|^2_{L^2(\nu)}, \qquad \langle u \rangle = \int u\, d\nu. \tag{21.19}$$

Remark 21.18. Throughout Part II, I shall always assume that ν is absolutely continuous with respect to the volume measure. This implies that Lipschitz functions are ν-almost everywhere differentiable.

Inequality (21.19) can be reformulated into

$$\left[\int u\, d\nu = 0\right] \Longrightarrow \qquad \|u\|^2_{L^2} \leq \frac{\|\nabla u\|^2_{L^2}}{\lambda}.$$

This writing makes the formal connection with the logarithmic Sobolev inequality very natural. (The Poincaré inequality is obtained as the limit of the logarithmic Sobolev inequality when one sets $\mu = (1+\varepsilon u)\,\nu$ and lets $\varepsilon \to 0$.)

Like Sobolev inequalities, Poincaré inequalities express the domination of a function by its gradient; but unlike Sobolev inequalities, they do not include any gain of integrability. Poincaré inequalities have spectral content, since the best constant λ can be interpreted as the spectral gap for the Laplace operator on M.[1] There is no Poincaré inequality on $\mathbb{R}^n$ equipped with the Lebesgue measure (the usual "flat" Laplace operator does not have a spectral gap), but there is a Poincaré inequality on, say, any compact Riemannian manifold equipped with its volume measure.

Poincaré inequalities are implied by logarithmic Sobolev inequalities, but the converse is false.

Example 21.19 (Exponential measure). Let $\nu(dx) = e^{-|x|}\,dx$ be the exponential measure on $[0, +\infty)$. Then ν satisfies a Poincaré inequality (with constant 1). On the other hand, it does not satisfy any

[1] This is one reason to take λ (universally accepted notation for spectral gap) as the constant defining the Poincaré inequality. Unfortunately this is not consistent with the convention that I used for local Poincaré inequalities; another choice would have been to call λ^{-1} the Poincaré constant.

logarithmic Sobolev inequality. The same conclusions hold true for the double-sided exponential measure $\nu(dx) = e^{-|x|}\,dx/2$ on $\mathbb{R}$. More generally, the measure $\nu_\beta(dx) = e^{-|x|^\beta}\,dx/Z_\beta$ satisfies a Poincaré inequality if and only if $\beta \geq 1$, and a logarithmic Sobolev inequality if and only if $\beta \geq 2$.

Establishing Poincaré inequalities for various measures is an extremely classical problem on which a lot has been written. Here is one of the oldest results in the field:

Theorem 21.20 (Lichnerowicz's spectral gap inequality). *Let M be a Riemannian manifold equipped with a reference Borel measure $\nu = e^{-V}\mathrm{vol}$, $V \in C^2(M)$, satisfying a curvature-dimension condition $\mathrm{CD}(K,N)$ for some $K > 0$, $N \in (1,\infty]$. Then ν satisfies a Poincaré inequality with constant $KN/(N-1)$.*

In other words, if $\mathrm{CD}(K,N)$ holds true, then for any Lipschitz function f on M with $\int f\,d\nu = 0$, one has

$$\left[\int f\,d\nu = 0\right] \Longrightarrow \qquad \int f^2\,d\nu \leq \frac{N-1}{KN}\int |\nabla f|^2\,d\nu. \tag{21.20}$$

Remark 21.21. Let $L = \Delta - \nabla V\cdot\nabla$, then (21.20) means that L admits a spectral gap of size at least $KN/(N-1)$:

$$\lambda_1(-L) \geq \frac{KN}{N-1}.$$

Proof of Theorem 21.20. In the case $N < \infty$, apply (21.12) with $\mu = (1+\varepsilon f)\,\nu$, where ε is a small positive number, f is Lipschitz and $\int f\,d\nu = 0$. Since M has a finite diameter, f is bounded, so μ is a probability measure for ε small enough. Then, by standard Taylor expansion of the logarithm function,

$$H_{N,\nu}(\mu) = \varepsilon\int f\,d\nu + \varepsilon^2\left(\frac{N-1}{N}\int\frac{f^2}{2}\,d\nu\right) + o(\varepsilon^2),$$

and the first term on the right-hand side vanishes by assumption. Similarly,

$$\int\frac{|\nabla\rho|^2}{\rho}\left(\frac{\rho^{-\frac{2}{N}}}{\frac{1}{3}+\frac{2}{3}\rho^{-\frac{1}{N}}}\right) = \varepsilon^2\int|\nabla f|^2\,d\nu + o(\varepsilon^2).$$

So (21.12) implies

$$\frac{N-1}{N}\int \frac{f^2}{2}\,d\nu \leq \frac{1}{2K}\left(\frac{N-1}{N}\right)^2 \int |\nabla f|^2\,d\nu,$$

and then inequality (21.20) follows.

In the case $N = \infty$, start from inequality (21.4) and apply a similar reasoning. (It is in fact a well-known property that a logarithmic Sobolev inequality with constant K implies a Poincaré inequality with constant K.) □

Bibliographical notes

Popular sources dealing with classical isoperimetric inequalities are the book by Burago and Zalgaller [176], and the survey by Osserman [664]. A very general discussion of isoperimetric inequalities can be found in Bobkov and Houdré [129]. The subject is related to Poincaré inequalities, as can be seen through Cheeger's isoperimetric inequality (19.32). As part of his huge work on concentration of measure, Talagrand has put forward the use of isoperimetric inequalities in product spaces [772].

There are entire books devoted to logarithmic Sobolev inequalities; this subject goes back at least to Nelson [649] and Gross [441], in relation to hypercontractivity and quantum field theory; but it also has its roots in earlier works by Stam [758], Federbush [351] and Bonami [142]. A gentle introduction, and references, can be found in [41]. The 1992 survey by Gross [442], the Saint-Flour course by Bakry [54] and the book by Royer [711] are classical references. Applications to concentration of measure and deviation inequalities can also be found in those sources, or in Ledoux's synthesis works [544, 546].

The first and most famous logarithmic Sobolev inequality is the one that holds true for the Gaussian reference measure in $\mathbb{R}^n$ (equation (21.5)). At the end of the fifties, Stam [758] established an inequality which can be recast (after simple changes of functions) as the usual logarithmic Sobolev inequality, found fifteen years later by Gross [441]. Stam's inequality reads $\mathcal{N} I \geq 1$, where I is the Fisher information, and $\mathcal{N}$ is the "power entropy". (In dimension n, this inequality should be replaced by $\mathcal{N} I \geq n$.) The main difference between these inequalities is that Stam's is expressed in terms of the Lebesgue reference measure, while Gross's is expressed in terms of the Gaussian reference measure. Although Stam is famous for his information-theoretical inequalities, it was only at the beginning of the nineties that specialists identified a version of the logarithmic Sobolev inequality in his work. (I learnt it from Carlen.) I personally use the name "Stam–Gross logarithmic Sobolev inequality" for (21.5); but this is of course debatable. Stam's argument was slightly flawed because of regularity issues, see [205, 783] for corrected versions.

At present, there are more than fifteen known proofs of (21.5); see Gross [442] for a partial list.

The Bakry–Émery theorem (Theorem 21.2) was proven in [56] by a semigroup method which will be reinterpreted in Chapter 25 as a gradient flow argument. The proof was rewritten in a language of partial differential equations in [43], with emphasis on the link with the convergence to equilibrium for the heat-like equation $\partial_t \rho = L\rho$.

The proof of Theorem 21.2 given in these notes is essentially the one that appeared in my joint work with Otto [671]. When the manifold M is $\mathbb{R}^n$ (and V is K-convex), there is a slightly simpler variant of that argument, due to Cordero-Erausquin [242]; there are also two quite different proofs, one by Caffarelli [188] (based on his log concave perturbation theorem) and one by Bobkov and Ledoux [131] (based on the Brunn–Minkowski inequality in $\mathbb{R}^n$). It is likely that the distorted Prékopa–Leindler inequality (Theorem 19.16) can be used to derive an alternative proof of the Bakry–Émery theorem in the style of Bobkov–Ledoux.

In the definition of logarithmic Sobolev inequality (Definition 21.1), I imposed a Lipschitz condition on the probability density ρ. Relaxing this condition can be done by tedious approximation arguments in the style of [671, Appendix], [43], and [814, Proof of Theorem 9.17, Step 1]. (In these references this is done only for the Euclidean space.) By the way, the original proof by Bakry and Émery [56] is complete only for compact manifolds. (Only formal computations were established in this paper, and integrations by parts were not taken care of rigorously, neither the proof of ergodicity of the diffusion in the entropy sense. For a compact manifold this is nothing; but on a noncompact manifold, one should be careful about the behavior of the density at infinity.) In his PhD thesis, Demange worked out much more complicated situations in full detail, so there is no doubt that the Bakry–Émery strategy can be made rigorous also on noncompact manifolds, although the proof is probably nowhere to be found.

The Holley–Stroock perturbation theorem for logarithmic Sobolev inequalities, explained in Remark 21.5, was proven in [478]. The other criterion mentioned in Remark 21.5, namely $\int e^{\alpha|\nabla v|^2}\,d\nu < \infty$ for α large enough, is due to Aida [7]. Related results can be found in a paper by F.-Y. Wang [829].

Another theorem by F.-Y. Wang [828] shows that logarithmic Sobolev inequalities follow from curvature-dimension lower bounds together with square-exponential moments. More precisely, if (M,ν) satisfies $\mathrm{CD}(-K,\infty)$ for some $K>0$, and $\int e^{(\frac{K}{2}+\varepsilon)\,d(x,x_0)^2}\,\nu(dx) < +\infty$ for some $\varepsilon>0$ and $x_0\in M$, then ν satisfies a logarithmic Sobolev inequality. Barthe and Kolesnikov [76] derived more general results in the same spirit.

Logarithmic Sobolev inequalities in $\mathbb{R}^n$ for the measure $e^{-V(x)}\,dx$ require a sort of quadratic growth of the potential V, while Poincaré inequalities require a sort of linear growth. (Convexity is sufficient, as shown by Bobkov [126]; there is now an elementary proof of this fact [74], while refinements and generalizations can be found in [633].) It is natural to ask what happens in between, that is, when $V(x)$ behaves like $|x|^\beta$ as $|x|\to\infty$, with $1<\beta<2$. This subject has been studied by Latała and Oleszkiewicz [540], Barthe, Cattiaux and Roberto [75] and Gentil, Guillin and Miclo [410]. The former set of authors chose to focus on functional inequalities which interpolate between Poincaré and log Sobolev, and seem to be due to Beckner [78]; on the contrary, the latter set of authors preferred to focus on modified versions of logarithmic Sobolev inequalities, following the steps of Bobkov and Ledoux [130]. (Modified logarithmic Sobolev inequalities will be studied later, in Chapter 22.)

On the real line, there is a characterization of logarithmic Sobolev inequalities, in terms of weighted Hardy-type inequalities, due to Bobkov and Götze [128]; see also Barthe and Roberto [77]. In the simpler context of Poincaré inequalities, the relation with Hardy inequalities goes back at least to Muckenhoupt [643].

The refinement of the constant in the logarithmic Sobolev inequalities by a dimensional factor of $N/(N-1)$ is somewhat tricky; see for instance Ledoux [541]. As a limit case, on S^1 there is a logarithmic Sobolev inequality with constant 1, although the Ricci curvature vanishes identically.

The normalized volume measure on a compact Riemannian manifold always satisfies a logarithmic Sobolev inequality, as Rothaus [710] showed long ago. But even on a compact manifold, this result does not diminish the interest of the Bakry–Émery theorem, because the constant in Rothaus' argument is not explicit. Saloff-Coste [727] proved a partial converse: If M has finite volume and Ricci curvature tensor bounded below by K, and the normalized volume measure satisfies a logarithmic Sobolev inequality with constant $\lambda > 0$, then M is compact and there is an explicit upper bound on its diameter:

$$\operatorname{diam}(M) \leq C\,\sqrt{\dim(M)}\,\max\left(\frac{1}{\sqrt{\lambda}},\,\frac{K_-}{\lambda}\right), \tag{21.21}$$

where C is numeric. (In view of the Bakry–Émery theorem, this can be thought of as a generalization of the Bonnet–Myers bound.) Simplified proofs of this result were given by Ledoux [544], then by Otto and myself [671, Theorem 4].

Like their logarithmic relatives, Sobolev inequalities also fill up books, but usually the emphasis is more on regularity issues. (In fact, for a long time logarithmic Sobolev inequalities and plain Sobolev inequalities were used and studied by quite different communities.) A standard reference is the book by Maz'ja [611], but there are many alternative sources.

A good synthetic discussion of the family (21.7) is in the course by Ledoux [545], which reviews many results obtained by Bakry and collaborators about Sobolev inequalities and $\mathrm{CD}(K,N)$ curvature-dimension bounds (expressed in terms of Γ and Γ_2 operators). There it is shown how to deduce some geometric information from (21.7), including the Myers diameter bound (following [58]).

Demange recently obtained a derivation of (21.9) which is, from my point of view, very satisfactory, and will be explained later in Chapter 25. By Demange's method one can establish the following generalization of (21.9): Under adequate regularity assumptions, if (M,ν) satisfies the curvature-dimension bound $\mathrm{CD}(K,N)$, $U \in \mathcal{DC}_N$, and A is defined by $A(0) = 0$ and $A(1) = 0$, $A''(r) = r^{-1/N}U''(r)$, then for any probability density ρ,

$$\int_M A(\rho)\,d\nu \leq \frac{1}{2K}\int_M \rho^{1-\frac{1}{N}}\left|\nabla U'(\rho)\right|^2 d\nu.$$

Many variants, some of them rather odd-looking, appear in Demange's work [290, 291, 293]. For instance, he is able to establish seemingly sharp inequalities for nonlinearities U satisfying the following condition:

$$\frac{d}{dr}\left[r\left(\frac{r\,U''(r)}{U'(r)}+\frac{1}{N}\right)\right] \geq \frac{9N}{4(N+2)}\left(\frac{r\,U''(r)}{U'(r)}+\frac{1}{N}\right)^2.$$

Demange also suggested that (21.8) might imply (21.6). It is interesting to note that (21.6) can be proven by a simple transport argument, while no such thing is known for (21.8).

The proof of Theorem 21.9 is taken from a collaboration with Lott [578].

The use of transport methods to study isoperimetric inequalities in $\mathbb{R}^n$ goes back at least to Knothe [523]. Gromov [635, Appendix] revived the interest in Knothe's approach by using it to prove the isoperimetric inequality in $\mathbb{R}^n$. Recently, the method was put to a higher degree of sophistication by Cordero-Erausquin, Nazaret and myself [248]. In this work, we recover general optimal Sobolev inequalities in $\mathbb{R}^n$, together with some families of optimal Gagliardo–Nirenberg inequalities. (The proof of the Sobolev inequalities is reproduced in [814, Theorem 6.21].) The results themselves are not new, since optimal Sobolev inequalities in $\mathbb{R}^n$ were established independently by Aubin, Talenti and Rodemich in the seventies (see [248] for references), while the optimal Gagliardo–Nirenberg inequalities were discovered recently by Del Pino and Dolbeault [283]. However, I think that all in all the transport approach is simpler, especially for the Gagliardo–Nirenberg family. In [248] the optimal Sobolev inequalities came with a "dual" family of inequalities, that can be interpreted as a particular case of so-called Faber–Krahn inequalities; there is still (at least for me) some mystery in this duality.

In the present chapter, I have modified slightly the argument of [248] to avoid the use of Alexandrov's theorem about second derivatives of convex functions (Theorem 14.25). The advantage is to get a more elementary proof; however, the computations are less precise, and some useful "magic" cancellations (such as $x + (\nabla\varphi - x) = \nabla\varphi$) are not available any longer; I used a homogeneity argument to get around this problem. A drawback of this approach is that the discussion about cases of equality is not possible any longer (anyway a clean discussion of equality cases requires much more effort; see [248, Section 4]). The proof presented here should work through if $\mathbb{R}^n$ is replaced by a cone with nonnegative Ricci curvature, although I did not check details.

The homogeneity (under dilations) of the function $-\int \rho^{1-1/N}$ is used in this argument to transform a seemingly nonoptimal inequality into the optimal one; I wonder whether a similar argument could lead from (21.6) to (21.8) on the sphere — but what would play the role of homogeneity? In fact, homogeneity also underlies [248]: if one wishes to interpret the proof in terms of "Otto calculus" it all amounts to the fact that $-\int \rho^{1-1/N}$ is displacement convex *and homogeneous*. The Euclidean analog is the following: if Φ is convex on a Euclidean space and minimal at 0 then $\Phi(X) \leq X \cdot \nabla\Phi(X) + |\nabla\Phi(X)|^2/2$; but if in addition Φ is λ-homogeneous ($\Phi(tx) = t^\lambda \Phi(x)$) then $X \cdot \nabla\Phi(X) = \lambda\,\Phi(X)$ (Euler relation), so

$$\Phi(X) \leq X \cdot \nabla\Phi(X) + \frac{|\nabla\Phi(X)|^2}{2} = \lambda\,\Phi(X) + \frac{|\nabla\Phi(X)|^2}{2},$$

so if $\lambda < 1$ we obtain $\Phi(X) \leq |\nabla\Phi(X)|^2/(2(1-\lambda))$.

Another inequality for which it would be interesting to have a transport proof is the Hardy–Littlewood–Sobolev inequality. Recently Calvez and Carrillo [198] obtained such a proof in dimension 1; further investigation is under way.

After [248], Maggi and I pushed the method even further [587], to recover "very optimal" Sobolev inequalities with trace terms, in $\mathbb{R}^n$. This settled some problems that had been left open in a classical work by Brézis and Lieb [173]. Much more information can be found in [587]; recently we also wrote a sequel [588] in which limit cases (such as inequalities of Moser–Trudinger type) are considered.

As far as all these applications of transport to Sobolev or isoperimetric inequalities in $\mathbb{R}^n$ are concerned, the Knothe coupling works just about as fine as the optimal coupling. (As a matter of fact, many such inequalities are studied in [134, 135] via the Knothe coupling.) But the choice of coupling method becomes important for refinements such as the characterization of minimizers [248], or even more establishing quantitative stability estimates around minimizers. This point is discussed in detail by Figalli, Maggi and Pratelli [369] who use the optimal coupling to refine Gromov's proof of isoperimetry, to the point of obtaining a beautiful (and sharp) quantitative isoperimetric theorem, giving a lower bound on how much the isoperimetric ratio of a set departs from the optimal ratio, in terms of how much the shape of this set departs from the optimal shape.

Interestingly enough, the transport method in all of these works is insensitive to the choice of norm in $\mathbb{R}^n$. (I shall come back to this observation in the concluding chapter.) Lutwak, Yang and Zhang [582] developed this remark and noticed that if a function f is given, then the problem of minimizing $\|\nabla f\|_{L^1}$ over all norms on $\mathbb{R}^n$ can be related to Minkowski's problem of prescribed Gauss curvature. The isoperimetric problem for a non-Euclidean norm, also known as the Wulff problem, is not an academic issue, since it is used in the modeling of surface energy (see the references in [369]).

In a Euclidean context, there are other more classical methods to attack these problems, such as rearrangement or symmetrization. For instance, quantitative symmetrization inequalities were used in [389] to prove the stability of the sharp Euclidean isoperimetric inequality. But even in this case, as noticed by Maggi [586], the most efficient strategy seems to be a combination of quantitative symmetrization inequalities with optimal transport techniques (for an auxiliary one-dimensional problem in that case). Similarly, the stability of optimal Sobolev inequalities is treated in [234, 388] by a combination of quantitative symmetrization (to reduce to radially symmetric functions) and optimal transport (with a quantitative version of [248] on radially symmetric functions).

In any case, if the reader is looking for a transport argument related to some geometric inequality in $\mathbb{R}^n$, I personally advise him or her to try the Knothe coupling first, and if this turns out to be insufficient because of some geometric reason, to go on with the optimal transport.

The Lévy–Gromov inequality was first conjectured by Lévy in the case when the manifold M is the boundary of a uniformly convex set (so the *sectional* curvatures are bounded below by a positive constant). Lévy thought he had a proof, but his argument was faulty and repaired by Gromov [438]. A lecture about Gromov's proof is available in [663]; one may also consult [394, Section 4.H].

There have also been some striking works by Bobkov, Bakry and Ledoux on the infinite-dimensional version of the Lévy–Gromov inequality (often called Gaussian isoperimetry); for this inequality there is an elegant functional formulation [57, 125], and the extremal subsets are half-spaces, rather than balls [145, 766]. On that subject I warmly recommend (as usual) the synthesis works by Ledoux [542, 543]. Further, see [76] where various isoperimetric inequalities are obtained via optimal transport.

In finite dimension, functional versions of the Lévy–Gromov inequality are also available [70, 740], but the picture is not so clear as in the infinite-dimensional case.

The Lichnerowicz spectral gap theorem is usually encountered as a simple application of the Bochner formula; it was obtained almost half a century ago by Lichnerowicz [553, p. 135]; see [100, Section D.I.] or [394, Theorem 4.70] for modern presentations. The above proof of Theorem 21.20 is a variant of the one which appears in my joint work with Lott [578]. Although less simple than the classical proof, it has the advantage, for the purpose of these notes, to be based on optimal transport. This is actually, to my knowledge, the first time that the dimensional refinement in the constants by a factor $N/(N-1)$ in an "infinite-dimensional functional inequality" has been obtained from a transport argument.

22

Concentration inequalities

The theory of concentration of measure is a collection of results, tools and recipes built on the idea that if a set A is given in a metric probability space $(\mathcal{X}, d, \mathbb{P})$, then the enlargement $A^r := \{x;\ d(x, A) \leq r\}$ might acquire a very high probability as r increases. There is an equivalent statement that Lipschitz functions $\mathcal{X} \to \mathbb{R}$ are "almost constant" in the sense that they have a very small probability of deviating from some typical quantity, for instance their mean value. This theory was founded by Lévy and later developed by many authors, in particular V. Milman, Gromov and Talagrand.

To understand the relation between the two sides of concentration (sets and functions), it is most natural to think in terms of **median**, rather than mean value. By definition, a real number m_f is a median of the random variable $f : \mathcal{X} \to \mathbb{R}$ if

$$\mathbb{P}\,[f \geq m_f] \geq \frac{1}{2}; \qquad \mathbb{P}\,[f \leq m_f] \geq \frac{1}{2}.$$

Then the two statements

(a) $\forall A \subset \mathcal{X},\ \ \forall r \geq 0, \quad \mathbb{P}\,[A] \geq 1/2 \Longrightarrow \mathbb{P}\,[A^r] \geq 1 - \psi(r)$

(b) $\forall f \in \mathrm{Lip}(\mathcal{X}),\ \ \forall r \geq 0, \quad \mathbb{P}\,[f > m_f + r] \leq \psi(r/\|f\|_{\mathrm{Lip}})$

are equivalent. Indeed, to pass from (a) to (b), first reduce to the case $\|f\|_{\mathrm{Lip}} = 1$ and let $A = \{f \leq m_f\}$; conversely, to pass from (b) to (a), let $f = d(\,\cdot\,, A)$ and note that 0 is a median of f.

The typical and most emblematic example of concentration of measure occurs in the Gaussian probability space $(\mathbb{R}^n, \gamma)$:

$$\gamma[A] \geq \frac{1}{2} \Longrightarrow \quad \forall r \geq 0, \quad \gamma[A^r] \geq 1 - e^{-\frac{r^2}{2}}. \tag{22.1}$$

Here is the translation in terms of Lipschitz functions: If X is a Gaussian random variable with law γ, then for all Lipschitz functions $f : \mathbb{R}^n \to \mathbb{R}$,

$$\forall r \geq 0, \quad \mathbb{P}\left[f(X) \geq \mathbb{E}\, f(X) + r\right] \leq \exp\left(-\frac{r^2}{2\,\|f\|_{\mathrm{Lip}}^2}\right). \tag{22.2}$$

Another famous example is the unit sphere S^N: if σ^N stands for the normalized volume on S^N, then the formulas above can be replaced by

$$\sigma^N[A] \geq \frac{1}{2} \Longrightarrow \quad \sigma^N[A^r] \geq 1 - e^{-\frac{(N-1)}{2}\,r^2},$$

$$\mathbb{P}\left[f(X) \geq \mathbb{E}\, f(X) + r\right] \leq \exp\left(-\frac{(N-1)\,r^2}{2\,\|f\|_{\mathrm{Lip}}^2}\right).$$

In this example we see that the phenomenon of concentration of measure becomes more and more important as the dimension increases to infinity.

In this chapter I shall review the links between optimal transport and concentration, focusing on certain *transport inequalities*. The main results will be Theorems 22.10 (characterization of Gaussian concentration), 22.14 (concentration via Ricci curvature bounds), 22.17 (concentration via logarithmic Sobolev inequalities), 22.22 (concentration via Talagrand inequalities) and 22.25 (concentration via Poincaré inequalities). The chapter will be concluded by a recap, and a technical appendix about Hamilton–Jacobi equations, of independent interest.

Optimal transport and concentration

As first understood by Marton, there is a simple and robust functional approach to concentration inequalities based on optimal transport. One can encode some information about the concentration of measure with respect to some reference measure ν, by functional inequalities of the form

$$\forall \mu \in P(\mathcal{X}), \quad C(\mu, \nu) \leq \mathcal{E}_\nu(\mu), \tag{22.3}$$

where $C(\mu, \nu)$ is the optimal transport cost between μ and ν, and $\mathcal{E}_\nu$ is some local nonlinear functional ("energy") of μ, involving for instance the integral of a function of the density of μ with respect to ν.

This principle may be heuristically understood as follows. To any measurable set A, associate the conditional measure $\mu_A = (1_A/\nu[A])\,\nu$. If the measure of A is not too small, then the associated energy $\mathcal{E}_\nu(\mu_A)$ will not be too high, and by (22.3) the optimal transport cost $C(\mu_A,\nu)$ will not be too high either. In that sense, the whole space $\mathcal{X}$ can be considered as a "small enlargement" of just A.

Here is a fluid mechanics analogy: imagine μ as the density of a fluid. The term on the right-hand side of (22.3) measures how difficult it is to prepare μ, for instance to confine it within a set A (this has to do with the measure of A); while the term on the left-hand side says how difficult it is for the fluid to invade the whole space, after it has been prepared initially with density μ.

The most important class of functional inequalities of the type (22.3) occurs when the cost function is of the type $c(x,y) = d(x,y)^p$, and the "energy" functional is the square root of Boltzmann's H functional,

$$H_\nu(\mu) = \int \rho \log \rho \, d\nu, \qquad \mu = \rho\,\nu,$$

with the understanding that $H_\nu(\mu) = +\infty$ if μ is not absolutely continuous with respect to ν. Here is a precise definition of these functional inequalities:

Definition 22.1 (T_p inequality). *Let $(\mathcal{X},d)$ be a Polish space and let $p \in [1,\infty)$. Let ν be a reference probability measure in $P_p(\mathcal{X})$, and let $\lambda > 0$. It is said that ν satisfies a T_p inequality with constant λ if*

$$\forall \mu \in P_p(\mathcal{X}), \qquad W_p(\mu,\nu) \le \sqrt{\frac{2\,H_\nu(\mu)}{\lambda}}. \tag{22.4}$$

These inequalities are often called transportation-cost inequalities, or **Talagrand inequalities**, although the latter denomination is sometimes restricted to the case $p = 2$.

Remark 22.2. Since $W_p \le W_q$ for $p \le q$, the T_p inequalities become stronger and stronger as p increases. The inequalities T_1 and T_2 have deserved most attention. It is an experimental fact that T_1 is more handy and flexible, while T_2 has more geometric content, and behaves better in large dimensions (see for instance Corollary 22.6 below).

There are two important facts to know about T_p inequalities when p varies in the range $[1,2]$: they admit a *dual formulation*, and they *tensorize*. These properties are described in the two propositions below.

Proposition 22.3 (Dual formulation of T_p). *Let $(\mathcal{X}, d)$ be a Polish space, $p \in [1,2]$ and $\nu \in P_p(\mathcal{X})$. Then the following two statements are equivalent:*

(a) ν satisfies $T_p(\lambda)$;

(b) For any $\varphi \in C_b(\mathcal{X})$,

$$\begin{cases} \forall t \geq 0 \quad \displaystyle\int e^{\lambda t \inf_{y \in \mathcal{X}} \left[\varphi(y) + \frac{d(x,y)^p}{p}\right]} \nu(dx) \leq e^{\lambda\left(\frac{1}{p}-\frac{1}{2}\right) t^{\frac{2}{2-p}}} \, e^{\lambda t \int \varphi \, d\nu} & (p < 2) \\ \displaystyle\int e^{\lambda \inf_{y \in \mathcal{X}} \left[\varphi(y) + \frac{d(x,y)^2}{2}\right]} \nu(dx) \leq e^{\lambda \int \varphi \, d\nu} & (p = 2). \end{cases} \tag{22.5}$$

Particular Case 22.4 (Dual formulation of T_1). Let $(\mathcal{X}, d)$ be a Polish space and $\nu \in P_1(\mathcal{X})$, then the following two statements are equivalent:

(a) ν satisfies $T_1(\lambda)$;

(b) For any $\varphi \in C_b(\mathcal{X})$,

$$\forall t \geq 0 \quad \int e^{t \inf_{y \in \mathcal{X}} \left[\varphi(y) + d(x,y)\right]} \nu(dx) \leq e^{\frac{t^2}{2\lambda}} \, e^{t \int \varphi \, d\nu}. \tag{22.6}$$

Proposition 22.5 (Tensorization of T_p). *Let $(\mathcal{X}, d)$ be a Polish space, $p \in [1,2]$ and let $\nu \in P_p(\mathcal{X})$ be a reference probability measure satisfying an inequality $T_p(\lambda)$. Then for any $N \in \mathbb{N}$, the measure $\nu^{\otimes N}$ satisfies an inequality $T_p(N^{1-\frac{2}{p}}\lambda)$ on $(\mathcal{X}^N, d_p, \nu^{\otimes N})$, where the product distance d_p is defined by*

$$d_p\big((x_1, \ldots, x_N); (y_1, \ldots, y_N)\big) = \Big(\sum_{i=1}^N d(x_i, y_i)^p\Big)^{\frac{1}{p}}.$$

Corollary 22.6 (T_2 inequalities tensorize exactly). *If ν satisfies $T_2(\lambda)$, then also $\mu^{\otimes N}$ satisfies $T_2(\lambda)$ on $(\mathcal{X}^N, d_2, \nu^{\otimes N})$, for any $N \in \mathbb{N}$.*

Proof of Proposition 22.3. Proposition 22.3 will be obtained as a consequence of Theorem 5.26. Recall the Legendre representation of the H-functional: For any $\lambda > 0$,

$$\begin{cases} \forall \mu \in P(\mathcal{X}), & \dfrac{H_\nu(\mu)}{\lambda} = \sup_{\varphi \in C_b(\mathcal{X})} \left[\displaystyle\int \varphi \, d\mu - \frac{1}{\lambda} \log \left(\int_{\mathcal{X}} e^{\lambda\varphi} \, d\nu \right) \right], \\ \forall \varphi \in C_b(\mathcal{X}), & \dfrac{1}{\lambda} \log \left(\displaystyle\int_{\mathcal{X}} e^{\lambda\varphi} \, d\nu \right) = \sup_{\mu \in P(\mathcal{X})} \left[\int \varphi \, d\mu - \frac{H_\nu(\mu)}{\lambda} \right]. \end{cases} \tag{22.7}$$

(See the bibliographical notes for proofs of these identities.)

Let us first treat the case $p = 2$. Apply Theorem 5.26 with $c(x,y) = d(x,y)^2/2$, $F(\mu) = (1/\lambda) H_\nu(\mu)$, $\Lambda(\varphi) = (1/\lambda)\log\left(\int e^{\lambda\varphi}\, d\nu\right)$. The conclusion is that ν satisfies $T_2(\lambda)$ if and only if

$$\forall \phi \in C_b(\mathcal{X}), \quad \log \int \exp\left(\lambda \int \phi \, d\nu - \lambda \phi^c \right) d\nu \leq 0,$$

i.e.

$$\int e^{-\lambda \phi^c} \, d\nu \leq e^{-\lambda \int \phi \, d\nu},$$

where $\phi^c(x) := \sup_y \left(\phi(y) - d(x,y)^2/2\right)$. Upon changing ϕ for $\varphi = -\phi$, this is the desired result. Note that the Particular Case 22.4 is obtained from (22.5) by choosing $p = 1$ and performing the change of variables $t \to \lambda t$.

The case $p < 2$ is similar, except that now we appeal to the equivalence between (i') and (ii') in Theorem 5.26, and choose

$$c(x,y) = \frac{d(x,y)^p}{p}; \qquad \Phi(r) = \frac{p^{\frac{2}{p}}}{2}\, r^{\frac{2}{p}}\, 1_{r \geq 0}; \qquad \Phi^*(t) = \left(\frac{1}{p} - \frac{1}{2}\right) t^{\frac{2}{2-p}}.$$

□

Proof of Proposition 22.5. First we need to set up a bit of notation. Let $\mu = \mu(dx_1\, dx_2 \ldots dx_N)$ be a probability measure on $\mathcal{X}^N$, and let $(x_1, \ldots, x_N) \in \mathcal{X}^N$ be distributed randomly according to μ. I shall write $\mu_1(dx_1)$ for the law of x_1, $\mu_2(dx_2|x_1)$ for the conditional law of x_2 given x_1, $\mu_3(dx_3|x_1,x_2)$ for the conditional law of x_3 given x_1 and x_2, etc. I shall also use the shorthand $x^i = (x_1, x_2, \ldots, x_i)$ (with the convention that $x^0 = \emptyset$), and write μ^i for the law of x^i.

The proof is reminiscent of the strategy used to construct the Knothe–Rosenblatt coupling. First choose an optimal coupling (for the cost function $c = d^p$) between $\mu_1(dx_1)$ to $\nu(dy_1)$, call it $\pi_1(dx_1\, dy_1)$. Then for each x_1, choose an optimal coupling between $\mu_2(dx_2|x_1)$ and $\nu(dy_2)$, call it $\pi_2(dx_2\, dy_2|x_1)$. Then for each (x_1, x_2), choose an optimal coupling between $\mu_3(dx_3|x_1,x_2)$ and $\nu(dy_3)$, call it $\pi_3(dx_3\,dy_3|x_1,x_2)$; etc. In the end, glue these plans together to get a coupling

$$\begin{aligned}&\pi(dx_1\,dy_1\,dx_2\,dy_2\ldots dx_N\,dy_N)\\&=\pi_1(dx_1\,dy_1)\,\pi_2(dx_2\,dy_2|x_1)\,\pi_3(dx_3\,dy_3|x_1,x_2)\ldots\\&\qquad\qquad\ldots\pi_N(dx_N\,dy_N|x_1,\ldots,x_{N-1}).\end{aligned}$$

In more compact notation,

$$\pi(dx\,dy)=\pi_1(dx_1\,dy_1)\,\pi_2(dx_2\,dy_2|x^1)\ldots\pi_N(dx_N\,dy_N|x^{N-1}).$$

Here something should be said about the measurability, since there is a priori no canonical way to choose $\pi_i(\,\cdot\,|x^{i-1})$ as a function of x^{i-1}. But Corollary 5.22 ensures that this choice can be made in a measurable way.

By the definition of d_p,

$$\begin{aligned}\mathbb{E}_\pi\, d_p(x,y)^p&=\sum_{i=1}^N\mathbb{E}_\pi\, d(x_i,y_i)^p\\&=\sum_{i=1}^N\int\Big[\mathbb{E}_{\pi(\,\cdot\,|x^{i-1})}\,d(x_i,y_i)^p\Big]\,\pi^{i-1}(dx^{i-1}\,dy^{i-1})\\&=\sum_{i=1}^N\int\Big[\mathbb{E}_{\pi(\,\cdot\,|x^{i-1})}\,d(x_i,y_i)^p\Big]\,\mu^{i-1}(dx^{i-1}),\end{aligned}\tag{22.8}$$

where of course

$$\pi^i(dx^i\,dy^i)=\pi_1(dx_1\,dy_1)\,\pi_2(dx_2\,dy_2|x^1)\ldots\pi_i(dx_i\,dy_i|x^{i-1}).$$

For each i and each $x^{i-1}=(x_1,\ldots,x_{i-1})$, the measure $\pi(\,\cdot\,|x^{i-1})$ is an optimal transference plan between its marginals. So the right-hand side of (22.8) can be rewritten as

$$\sum_{i=1}^N\int W_p\big(\mu_i(\,\cdot\,|x^{i-1}),\nu\big)^p\,\mu^{i-1}(dx^{i-1}).$$

Since this cost is achieved for the transference plan π, we obtain the key estimate

$$W_p(\mu,\nu^{\otimes N})^p\le\sum_{i=1}^N\int W_p\big(\mu_i(\,\cdot\,|x^{i-1}),\nu\big)^p\,\mu^{i-1}(dx^{i-1}).\tag{22.9}$$

By assumption, ν satisfies $T_p(\lambda)$, so the right-hand side in (22.9) is bounded above by

$$\sum_i \int \left(\frac{2}{\lambda} H_\nu(\mu_i(\,\cdot\,|x^{i-1}))\right)^{\frac{p}{2}} \mu^{i-1}(dx^{i-1}). \qquad (22.10)$$

Since $p \leq 2$, we can apply Hölder's inequality, in the form $\sum_{i\leq N} a_i^{p/2} \leq N^{1-p/2}(\sum a_i)^{p/2}$, and bound (22.10) by

$$N^{1-\frac{p}{2}} \left(\frac{2}{\lambda}\right)^{\frac{p}{2}} \left[\int \left(\sum_{i=1}^N H_\nu(\mu_i(\,\cdot\,|x^{i-1}))\right) \mu^{i-1}(dx^{i-1})\right]^{\frac{p}{2}}. \qquad (22.11)$$

But the formula of **additivity of entropy** (Lemma 22.8 below) states that

$$\sum_{1\leq i\leq N} \int H_\nu(\mu_i(dx_i|x^{i-1}))\, \mu^{i-1}(dx^{i-1}) = H_{\nu^{\otimes N}}(\mu). \qquad (22.12)$$

Putting all the previous bounds back together, we end up with

$$W_p(\mu, \nu^{\otimes N})^p \leq N^{1-\frac{p}{2}} \left(\frac{2}{\lambda}\right)^{\frac{p}{2}} H_{\nu^{\otimes N}}(\mu)^{\frac{p}{2}},$$

which is equivalent to the desired inequality. □

Remark 22.7. The same proof shows that the inequality

$$\forall \mu \in P(\mathcal{X}), \qquad C(\mu,\nu) \leq H_\nu(\mu)$$

implies

$$\forall \mu \in P(\mathcal{X}^N), \qquad C^N(\mu,\nu) \leq H_{\nu^{\otimes N}}(\mu),$$

where C^N is the optimal transport cost associated with the cost function

$$c^N(x,y) = \sum c(x_i,y_i)$$

on $\mathcal{X}^N$.

The following important lemma was used in the course of the proof of Proposition 22.5.

Lemma 22.8 (Additivity of entropy). *Let $N \in \mathbb{N}$, let $\mathcal{X}_1, \ldots, \mathcal{X}_N$ be Polish spaces, $\nu_i \in P(\mathcal{X}_i)$ $(1 \le i \le N)$, $\mathcal{X} = \prod \mathcal{X}_i$, $\nu = \bigotimes \nu_i$, and $\mu \in P(\mathcal{X})$. Then, with the same notation as in the beginning of the proof of Proposition 22.5,*

$$H_\nu(\mu) = \sum_{1 \le i \le N} \int H_{\nu_i}\big(\mu_i(dx_i|x^{i-1})\big)\, \mu^{i-1}(dx^{i-1}). \tag{22.13}$$

Proof of Lemma 22.8. By induction, it suffices to treat the case $N = 2$. Let $\rho = \rho(x_1, x_2)$ be the density of μ with respect to $\nu_1 \otimes \nu_2$. By an easy approximation argument based on the monotone convergence theorem, it is sufficient to establish (22.13) in the case when ρ is bounded.

The measure $\mu_1(dx_1)$ has density $\int \rho(x_1, x_2)\, \nu_2(dx_2)$, while the conditional measure $\mu_2(dx_2|x_1)$ has density $\rho(x_1, x_2)/(\int \rho(x_1, x_2)\, \nu_2(dx_2))$. From this and the additive property of the logarithm, we deduce

$$\begin{aligned}
&\int H_\nu\big(\mu_2(\,\cdot\,|x_1)\big)\, \mu_1(dx_1) \\
&= \int \left(\int \frac{\rho(x_1, x_2)}{\int \rho(x_1, x_2')\, \nu_2(dx_2')} \log \frac{\rho(x_1, x_2)}{\int \rho(x_1, x_2')\, \nu_2(dx_2')}\, \nu_2(dx_2)\right) \\
&\qquad\qquad\qquad\qquad \left(\int \rho(x_1, x_2')\, \nu_2(dx_2')\right) \nu_1(dx_1) \\
&= \iint \rho(x_1, x_2) \log \rho(x_1, x_2)\, \nu_2(dx_2)\, \nu_1(dx_1) \\
&\qquad - \int \left(\int \rho(x_1, x_2)\, \nu_2(dx_2)\right) \log\left(\int \rho(x_1, x_2)\, \nu_2(dx_2)\right) \nu_1(dx_1) \\
&= H_\nu(\mu) - H_{\nu_1}(\mu_1).
\end{aligned}$$

This concludes the proof. □

Exercise 22.9. Give an alternative proof of Proposition 22.5 based on the dual formulation of T_p inequalities (Proposition 22.3).

Gaussian concentration and T_1 inequality

Gaussian concentration is a loose terminology meaning that some reference measure enjoys properties of concentration which are similar to those of the Gaussian measure. In this section we shall see that a certain form of Gaussian concentration is *equivalent* to a T_1 inequality. Once again, this principle holds in very general metric spaces.

Theorem 22.10 (Gaussian concentration). *Let $(\mathcal{X}, d)$ be a Polish space, equipped with a reference probability measure ν. Then the following properties are all equivalent:*

(i) ν lies in $P_1(\mathcal{X})$ and satisfies a T_1 inequality;

(ii) There is $\lambda > 0$ such that for any $\varphi \in C_b(\mathcal{X})$,

$$\forall t \geq 0 \qquad \int e^{t \inf_{y\in\mathcal{X}} \left[\varphi(y)+d(x,y)\right]}\, \nu(dx) \leq e^{\frac{t^2}{2\lambda}}\, e^{t\int \varphi\, d\nu}.$$

(iii) There is a constant $C > 0$ such that for any Borel set $A \subset \mathcal{X}$,

$$\nu[A] \geq \frac{1}{2} \Longrightarrow \quad \forall r > 0, \quad \nu[A^r] \geq 1 - e^{-C\, r^2}.$$

(iv) There is a constant $C > 0$ such that

$$\forall f \in L^1(\nu) \cap \mathrm{Lip}(\mathcal{X}),\ \forall \varepsilon > 0,$$
$$\nu\Big[\Big\{x \in \mathcal{X};\ f(x) \geq \int f\, d\nu + \varepsilon\Big\}\Big] \leq \exp\left(-C \frac{\varepsilon^2}{\|f\|_{\mathrm{Lip}}^2}\right);$$

(v) There is a constant $C > 0$ such that

$$\forall f \in L^1(\nu) \cap \mathrm{Lip}(\mathcal{X}),\ \forall \varepsilon > 0,\ \forall N \in \mathbb{N},$$
$$\nu^{\otimes N}\Big[\Big\{x \in \mathcal{X}^N;\ \frac{1}{N}\sum_{i=1}^{N} f(x_i) \geq \int f\, d\nu + \varepsilon\Big\}\Big] \leq \exp\left(-C \frac{N\, \varepsilon^2}{\|f\|_{\mathrm{Lip}}^2}\right);$$

(vi) There is a constant $C > 0$ such that

$$\forall f \in \mathrm{Lip}(\mathcal{X}),\ \forall\ m_f = \text{median of } f,\ \ \forall \varepsilon > 0,$$
$$\nu\Big[\Big\{x \in \mathcal{X};\ f(x) \geq m_f + \varepsilon\Big\}\Big] \leq \exp\left(-C \frac{\varepsilon^2}{\|f\|_{\mathrm{Lip}}^2}\right);$$

(vii) For any $x_0 \in \mathcal{X}$ there is a constant $a > 0$ such that

$$\int e^{a\, d(x_0,x)^2}\, \nu(dx) < +\infty;$$

(viii) There exists $a > 0$ such that

$$\int e^{a\, d(x,y)^2}\, \nu(dx)\, \nu(dy) < +\infty;$$

(ix) There exist $x_0 \in \mathcal{X}$ *and* $a > 0$ *such that*

$$\int e^{a\, d(x_0,x)^2}\, \nu(dx) < +\infty.$$

Remark 22.11. One should not overestimate the power of Theorem 22.10. The simple (too simple?) criterion (ix) behaves badly in large dimensions, and in practice might lead to terrible constants at the level of (iii). In particular, this theorem alone is unable to recover dimension-free concentration inequalities such as (22.1) or (22.2). Statement (v) is dimension-independent, but limited to particular observables of the form $(1/N)\sum f(x_i)$. Here we see some limitations of the T_1 inequality.

Proof of Theorem 22.10. We shall establish (i) $\Rightarrow$ (ii) $\Rightarrow$ (iv) $\Rightarrow$ (vii), (i) $\Rightarrow$ (v) $\Rightarrow$ (iv), (i) $\Rightarrow$ (iii) $\Rightarrow$ (vi) $\Rightarrow$ (vii) $\Rightarrow$ (viii) $\Rightarrow$ (ix) $\Rightarrow$ (i), and this will prove the theorem.

(i) $\Rightarrow$ (ii) was already seen in Particular Case 22.4.

To prove (ii) $\Rightarrow$ (iv), it suffices to treat the case $\|f\|_{\mathrm{Lip}} = 1$ (replace ε by $\varepsilon/\|f\|_{\mathrm{Lip}}$ and f by $f/\|f\|_{\mathrm{Lip}}$). Then if f is 1-Lipschitz,

$$\inf_{y\in\mathcal{X}} \big[f(y) + d(x,y)\big] = f(x),$$

so (ii) implies

$$\int e^{t\, f(x)}\, \nu(dx) \le e^{\frac{t^2}{2\lambda}}\, e^{t\int f\, d\nu}.$$

With the shorthand $\langle f\rangle = \int f\, d\nu$, this is the same as

$$\int e^{t\,(f-\langle f\rangle)}\, d\nu \le e^{\frac{t^2}{2\lambda}}.$$

Then by the exponential Chebyshev inequality,

$$\nu\Big[\{f - \langle f\rangle \ge \varepsilon\}\Big] \le e^{-t\varepsilon} \int e^{t\,(f-\langle f\rangle)}\, d\nu \le e^{-t\varepsilon} e^{\frac{t^2}{2\lambda}};$$

and (iv) is obtained by optimizing in t. ($C = \lambda/2$ does the job.)

Now let us prove (iv) $\Rightarrow$ (vii). Let ν satisfy (iv) and let $x_0 \in \mathcal{X}$. First we shall check that $d(\,\cdot\,, x_0) \in L^1(\nu)$. Let $m \in \mathbb{N}$, and let $f_m = d(\,\cdot\,, x_0) \wedge m$; then $f_m \in L^1(\nu) \cap \mathrm{Lip}(\mathcal{X})$, so

$$\nu\Big[f_m \ge s + \int f_m\, d\nu\Big] \le e^{-C\, s^2}.$$

It follows that for any $A \le m$,

$$\begin{aligned}
\int f_m^2\,d\nu &= \int_0^{+\infty} 2s\,\nu[f_m \ge s]\,ds \\
&\le \int_0^A 2s\,\nu[f_m \ge s]\,ds + \int_A^{\int f_m\,d\nu} 2s\,\nu[f_m \ge s]\,ds \\
&\qquad\qquad + \int_{\int f_m\,d\nu}^{+\infty} 2s\,\nu\big[f_m \ge s\big]\,ds \\
&\le A^2 + \nu[f_m \ge A] \int_A^{\int f_m\,d\nu} 2s\,ds \\
&\qquad + \int_0^{+\infty} 2\Big(s + \int f_m\,d\nu\Big)\,\nu\Big[f_m \ge s + \int f_m\,d\nu\Big]\,ds \\
&\le A^2 + \nu[f_m \ge A]\left(\int f_m\,d\nu\right)^2 + \int_0^{+\infty} 2s\,e^{-C\,s^2}\,ds \\
&\qquad + 2\left(\int f_m\,d\nu\right)\int_0^{+\infty} e^{-C\,s^2}\,ds \\
&\le A^2 + \nu[f \ge A]\left(\int f_m\,d\nu\right)^2 + \int_0^{+\infty} 2s\,e^{-C\,s^2}\,ds + \frac{1}{4}\left(\int f_m\,d\nu\right)^2 \\
&\qquad + 8\left(\int_0^{+\infty} e^{-C\,s^2}\,ds\right)^2 \\
&\le A^2 + \left(\int f_m^2\,d\nu\right)\left(\nu[f \ge A] + \frac{1}{4}\right) + \overline{C},
\end{aligned}$$

where $\overline{C} = \int_0^{+\infty} 2s\,e^{-C\,s^2}\,ds + 8\left(\int_0^{+\infty} e^{-C\,s^2}\,ds\right)^2$ is a finite constant. If A is large enough, then $\nu[f \ge A] \le 1/4$, and the above inequality implies $\int f_m^2\,d\nu \le 2(A^2 + \overline{C})$. By taking $m \to \infty$ we deduce that $\int f^2\,d\nu < +\infty$, in particular $f \in L^1(\nu)$. So we can apply directly (iv) to $f = d(\,\cdot\,, x_0)$, and we get that for any $a < C$,

$$\begin{aligned}
\int e^{a\,d(x,x_0)^2}\,\nu(dx) &= \int_0^{+\infty} 2ase^{as^2}\,\nu[f \ge s]\,ds \\
&= \int_0^{\int f\,d\nu} 2ase^{as^2}\,\nu[f \ge s]\,ds \\
&\qquad + \int_0^{+\infty} 2a\Big(s + \int f\,d\nu\Big)\,e^{a(s+\int f\,d\nu)^2}\,\nu\Big[f \ge s + \int f\,d\nu\Big]\,ds \\
&\le e^{a(\int f\,d\nu)^2} + \int_0^{+\infty} 2a\Big(s + \int f\,d\nu\Big)\,e^{a(s+\int f\,d\nu)^2}e^{-\,Cs^2}\,ds \ < +\infty.
\end{aligned}$$

This proves (vii).

The next implication is (i) ⇒ (v). If ν satisfies $T_1(\lambda)$, by Proposition 22.5 $\nu^{\otimes N}$ satisfies $T_1(\lambda/N)$ on $\mathcal{X}^N$ equipped with the distance $d_1(x,y) = \sum d(x_i, y_i)$. Let $F : \mathcal{X}^N \to \mathbb{R}$ be defined by

$$F(x) = \frac{1}{N}\sum_{i=1}^N f(x_i).$$

If f is Lipschitz then $\|F\|_{\mathrm{Lip}} = \|f\|_{\mathrm{Lip}}/N$. Moreover, $\int F\,d\nu^{\otimes N} = \int f\,d\nu$. So if we apply (iv) with $\mathcal{X}$ replaced by $\mathcal{X}^N$ and f replaced by F, we obtain

$$\begin{aligned}
\nu^{\otimes N}\left[\left\{x \in \mathcal{X}^N;\ \frac{1}{N}\sum_{i=1}^N f(x_i) \geq \int f\,d\nu + \varepsilon\right\}\right] \\
&= \nu^{\otimes N}\left[\left\{x \in \mathcal{X}^N;\ F(x) \geq \int F\,d\nu + \varepsilon\right\}\right] \\
&\leq \exp\left(-(C/N)\frac{\varepsilon^2}{(\|f\|_{\mathrm{Lip}}/N)^2}\right) \\
&= \exp\left(-C\frac{N\varepsilon^2}{\|f\|_{\mathrm{Lip}}^2}\right),
\end{aligned}$$

where $C = \lambda/2$ (cf. the remark at the end of the proof of (i) ⇒ (iv)).

The implication (v) ⇒ (iv) is trivial.

Let us now consider the implication (i) ⇒ (iii). Assume that

$$\forall \mu \in P_1(\mathcal{X}), \qquad W_1(\mu,\nu) \leq C\sqrt{H_\nu(\mu)}. \tag{22.14}$$

Choose A with $\nu[A] \geq 1/2$, and $\mu = (1_A\,\nu)/\nu[A]$. If $\nu[A^r] = 1$ there is nothing to prove, otherwise let $\widetilde{\mu} = (1_{\mathcal{X}\setminus A^r}\,\nu)/\nu[\mathcal{X}\setminus A^r]$. By an immediate computation,

$$H_\nu(\mu) = \log\frac{1}{\nu[A]} \leq \log 2, \qquad H_\nu(\widetilde{\mu}) = \log\left(\frac{1}{1-\nu[A^r]}\right).$$

By (22.14) and the triangle inequality for the distance W_1,

$$W_1(\mu,\widetilde{\mu}) \leq W_1(\mu,\nu) + W_1(\widetilde{\mu},\nu) \leq C\sqrt{\log 2} + C\sqrt{\log\left(\frac{1}{1-\nu[A^r]}\right)}. \tag{22.15}$$

On the other hand, it is obvious that $W_1(\mu,\widetilde{\mu}) \geq r$ (all the mass has to go from A to $\mathcal{X}\setminus A^r$, so each unit of mass should travel a distance at least r). So (22.15) implies

$$r \leq C\sqrt{\log 2} + C\sqrt{\log\left(\frac{1}{1-\nu[A^r]}\right)},$$

therefore

$$\nu[A^r] \geq 1 - \exp\left[-\left(\frac{r}{C} - \sqrt{\log 2}\right)^2\right].$$

This establishes a bound of the type $\nu[A^r] \geq 1 - ae^{-C\,r^2}$, and (iii) follows. (To get rid of the constant a, it suffices to note that $\nu[A^r] \geq \max(1/2,\, 1-ae^{-cr^2}) \geq 1 - e^{-c'r^2}$ for c' well-chosen.)

To prove (iii) $\Rightarrow$ (vi), let $A = \{y;\ f(y) \leq m_f\}$. By the very definition of a median, A has probability at least $1/2$, so (iii) implies $\nu[A^r] \geq 1 - e^{-Cr^2}$. On the other hand, $\{f \geq m_f + \varepsilon\}$ is included in $\mathcal{X} \setminus A^r$ for any $r < \varepsilon/\|f\|_{\mathrm{Lip}}$. (Indeed, if $f(x) \geq m_f + \varepsilon$ and $y \in A$ then $f(x) - f(y) \geq \varepsilon$, so $d(x,y) \geq \varepsilon/\|f\|_{\mathrm{Lip}} > r$.) This leads to the bound $\nu[\{f \geq m_f + \varepsilon\}] \leq e^{-C(\varepsilon/\|f\|_{\mathrm{Lip}})^2}$, which is Property (vi).

To show (vi) $\Rightarrow$ (vii), let A be a compact set such that $\nu[A] \geq 1/2$; also, let $x_0 \in A$, and let R be the diameter of A. Further, let $f(x) = d(x, A)$; then f is a 1-Lipschitz function admitting 0 for median. So (vi) implies

$$\nu\big[d(x,x_0) \geq R + r\big] \leq \nu[d(x,A) \geq r] \leq e^{-C\,r^2}.$$

It follows that for any $a < C$,

$$\begin{aligned}
\int e^{a\,d(x,x_0)^2}\,\nu(dx) &= \int_0^{+\infty} \nu\big[d(\,\cdot\,,x_0)^2 \geq s\big]\,2ase^{as^2}\,ds \\
&\leq \int_0^R 2ase^{as^2}\,ds + \int_R^{+\infty} \nu\big[d(\,\cdot\,,x_0)^2 \geq s\big]\,2ase^{as^2}\,ds \\
&\leq e^{aR^2} + \int_R^{\infty} e^{-C(s-R)^2} 2ase^{as^2}\,ds \; < +\infty.
\end{aligned}$$

To prove (vii) $\Rightarrow$ (viii), pick up any $x_0 \in \mathcal{X}$ and write

$$\begin{aligned}
\int e^{a\,d(x,y)^2}\,\nu(dx)\,\nu(dy) &\leq \int e^{2a\,d(x,x_0)^2 + 2a\,d(x_0,y)^2}\,\nu(dx)\,\nu(dy) \\
&= \left(\int e^{2a\,d(x,x_0)^2}\,\nu(dx)\right)^2.
\end{aligned}$$

The implication (viii) $\Rightarrow$ (ix) is obvious.

It only remains to establish (ix) $\Rightarrow$ (i). If ν satisfies (ix), then obviously $\nu \in P_1(\mathcal{X})$. To prove that ν satisfies T_1, we shall establish the **weighted Csiszár–Kullback–Pinsker inequality**

$$\big\|d(x_0,\cdot\,)\,(\mu-\nu)\big\|_{TV} \le \frac{\sqrt{2}}{a}\left(1+\log\int_{\mathcal{X}} e^{a\,d(x_0,x)^2}\,d\nu(x)\right)^{1/2}\sqrt{H_\nu(\mu)}. \tag{22.16}$$

Inequality (22.16) implies a T_1 inequality, since Theorem 6.15 yields

$$W_1(\mu,\nu) \le \big\|d(x_0,\cdot\,)\,(\mu-\nu)\big\|_{TV}.$$

So let us turn to the proof of (22.16). We may assume that μ is absolutely continuous with respect to ν, otherwise (22.16) is trivial. Then let f be the density of μ, and let $u = f-1$, so that

$$\mu = (1+u)\,\nu;$$

note that $u \ge -1$ and $\int u\,d\nu = 0$. We also define

$$h(v) := (1+v)\log(1+v) - v \ge 0, \qquad v \in [-1,+\infty);$$

so that

$$H_\nu(\mu) = \int_{\mathcal{X}} h(u)\,d\nu. \tag{22.17}$$

Finally, let $\varphi(x) = a\,d(x_0,x)$.

Since $h(0) = h'(0) = 0$, Taylor's formula (with integral remainder) yields

$$h(u) = u^2\int_0^1 \frac{1-t}{1+tu}\,dt,$$

so

$$H_\nu(\mu) = \int_{\mathcal{X}}\int_0^1 \frac{u^2(x)\,(1-t)}{1+tu(x)}\,d\nu(x)\,dt.$$

On the other hand, by Cauchy–Schwarz inequality on $(0,1)\times\mathcal{X}$,

$$\left(\int_0^1 (1-t)\,dt\right)^2\left(\int_{\mathcal{X}} \varphi|u|\,d\nu\right)^2$$

$$
\begin{aligned}
&= \left(\int_{(0,1)\times\mathcal{X}} (1-t)\,\varphi(x)|u(x)|\,d\nu(x)\,dt\right)^2 \\
&\le \left(\iint (1-t)\,(1+tu(x))\,\varphi^2(x)\,d\nu(x)\,dt\right) \\
&\qquad\qquad \times \left(\iint \frac{1-t}{1+tu(x)}|u(x)|^2\,d\nu(x)\,dt\right);
\end{aligned}
$$

thus

$$
\left(\int \varphi\,|u|\,d\nu\right)^2 \le C H_\nu(\mu), \tag{22.18}
$$

where

$$
C := \frac{\displaystyle\iint (1-t)\,(1+tu)\,\varphi^2\,d\nu\,dt}{\left(\displaystyle\int_0^1 (1-t)\,dt\right)^2}. \tag{22.19}
$$

The numerator can be rewritten as follows:

$$
\begin{aligned}
&\iint (1-t)\,(1+tu)\,\varphi^2\,d\nu\,dt \\
&\qquad = \int (1-t)t\,dt \int (1+u)\,\varphi^2\,d\nu + \int (1-t)^2\,dt \int \varphi^2\,d\nu \\
&\qquad = \frac{1}{6}\int \varphi^2\,d\mu + \frac{1}{3}\int \varphi^2\,d\nu.
\end{aligned} \tag{22.20}
$$

From the Legendre representation of the H functional,

$$
\int \varphi^2\,d\mu \le H_\nu(\mu) + \log \int e^{\varphi^2}\,d\nu, \tag{22.21}
$$

and Jensen's inequality, in the form

$$
\int \varphi^2\,d\nu \le \log \int e^{\varphi^2}\,d\nu, \tag{22.22}
$$

we deduce that the right-hand side of (22.20) is bounded above by

$$
\frac{1}{6} H_\nu(\mu) + \frac{1}{2}\log \int e^{\varphi^2}\,d\nu.
$$

Plugging this into (22.19) and (22.18), we conclude that

$$
\left(\int \varphi\,|u|\,d\nu\right)^2 \le \left(\frac{2}{3}H + 2L\right) H, \tag{22.23}
$$

where H stands for $H_\nu(\mu)$ and L for $\log \int e^{\varphi^2}\,d\nu$.

The preceding bound is relevant only for "small" values of H. To handle large values, note that

$$\begin{aligned}\left(\int \varphi|u|\,d\nu\right)^2 &\leq \int \varphi^2|u|\,d\nu \int |u|\,d\nu \\ &\leq \left(\int \varphi^2\,d\mu + \int \varphi^2\,d\nu\right)\left(\int d\mu + \int d\nu\right) \\ &\leq (H+2L)\,2 \end{aligned} \tag{22.24}$$

where I have successively used the Cauchy–Schwarz inequality, the inequality $|u| \leq 1+u+1$ on $[-1,+\infty)$ (which results in $|u|\,\nu \leq \mu+\nu$), and finally (22.21) and (22.22).

By (22.23) and (22.24),

$$\left(\int \varphi|u|\,d\nu\right)^2 \leq \min\left((2H)\big(\frac{H}{3}+L\big)\,,\,2(H+2L)\right).$$

Then the elementary inequality

$$\min\left(At^2+Bt,\, t+D\right) \leq Mt, \qquad M = \frac{1}{2}\left\{1+B+\sqrt{(B-1)^2+4AD}\right\}$$

implies

$$\int \varphi|u|\,d\nu \leq m\sqrt{H(\mu|\nu)}$$

where

$$m = \sqrt{1+L+\sqrt{(L-1)^2+\frac{8}{3}L}} \leq \sqrt{2}\,\sqrt{L+1}.$$

This concludes the proof. □

Remark 22.12 (CKP inequality). In the particular case $\varphi = 1$, we can replace the inequality (22.21) by just $\int d\mu = 1$; then instead of (22.23) we obtain

$$\|\mu-\nu\|_{TV} \leq \sqrt{2\,H_\nu(\mu)}. \tag{22.25}$$

This is the classical **Csiszár–Kullback–Pinsker** (CKP) inequality, with the sharp constant $\sqrt{2}$.

Remark 22.13. If ν satisfies $T_2(\lambda)$, then also $\nu^{\otimes N}$ satisfies $T_2(\lambda)$, independently of N; so one might hope to improve the concentration inequality appearing in Theorem 22.10(v). But now the space $\mathcal{X}^N$ should be equipped with the d_2 distance, for which the function $F : x \to (1/N) \sum f(x_i)$ is only $\sqrt{N}$-Lipschitz! In the end, T_2 does not lead to any improvement of Theorem 22.10(v). This is not in contradiction with the fact that T_2 is significantly stronger than T_1 (as we shall see in the sequel); it just shows that we cannot tell the difference when we consider observables of the particular form $(1/N) \sum \varphi(x_i)$. If one is interested in more complicated observables (such as nonlinear functionals, or suprema as in Example 22.36 below) the difference between T_1 and T_2 might become considerable.

Talagrand inequalities from Ricci curvature bounds

In the previous section the focus was on T_1 inequalities; in this and the next two sections we shall consider the stronger T_2 inequalities (Talagrand inequalities).

The most simple criterion for T_2 to hold is expressed in terms of *Ricci curvature bounds*:

Theorem 22.14 ($\mathrm{CD}(K,\infty)$ implies $T_2(K)$). *Let M be a Riemannian manifold, equipped with a reference probability measure $\nu = e^{-V}\mathrm{vol}$, $V \in C^2(M)$, satisfying a $\mathrm{CD}(K,\infty)$ curvature-dimension bound for some $K > 0$. Then ν belongs to $P_2(M)$ and satisfies the Talagrand inequality $T_2(K)$. In particular, ν satisfies Gaussian concentration bounds.*

Proof of Theorem 22.14. It follows by Theorem 18.12 that ν lies in $P_2(M)$; then the inequality $T_2(K)$ comes from Corollary 20.13(i) with $\mu_0 = \nu$ and $\mu_1 = \mu$. Since $T_2(K)$ implies $T_1(K)$, Theorem 22.10 shows that ν satisfies Gaussian concentration bounds. □

Example 22.15. The standard Gaussian γ on $\mathbb{R}^N$ satisfies $\mathrm{CD}(1,\infty)$, and therefore $T_2(1)$ too. This is independent of N.

The links between Talagrand inequalities and dimension free concentration bounds will be considered further in Theorem 22.22.

Relation with log Sobolev and Poincaré inequalities

So far we learnt that logarithmic Sobolev inequalities follow from curvature bounds, and that Talagrand inequalities also result from the same bounds. We also learnt from Chapter 21 that logarithmic Sobolev inequalities imply Poincaré inequalities. Actually, Talagrand inequalities are *intermediate* between these two inequalities: a logarithmic Sobolev inequality implies a Talagrand inequality, which in turn implies a Poincaré inequality. In some sense however, Talagrand is closer to logarithmic Sobolev than to Poincaré: For instance, in nonnegative curvature, the validity of the Talagrand inequality is equivalent to the validity of the logarithmic Sobolev inequality — up to a degradation of the constant by a factor 1/4.

To establish these properties, we shall use, for the first time in this course, a **semigroup argument**. As discovered by Bobkov, Gentil and Ledoux, it is indeed convenient to consider inequality (22.5) from a dynamical point of view, with the help of the (forward) Hamilton–Jacobi semigroup defined as in Chapter 7 by

$$\begin{cases} H_0\,\varphi = \varphi, \\ \\ (H_t\,\varphi)(x) = \inf_{y\in M}\left[\varphi(y) + \dfrac{d(x,y)^2}{2t}\right] \qquad (t>0,\ x\in M). \end{cases} \tag{22.26}$$

The next proposition summarizes some of the nice properties of the semigroup $(H_t)_{t\geq 0}$. Recall the notation $|\nabla^- f|$ from (20.2).

Proposition 22.16 (Some properties of the quadratic Hamilton–Jacobi semigroup). *Let f be a bounded continuous function on a Riemannian manifold M. Then:*

(i) For any $s,t\geq 0$, $H_t H_s f = H_{t+s} f$.

(ii) For any $x\in M$, $\inf f \leq (H_t f)(x) \leq f(x)$; moreover, the infimum over M in (22.26) can be restricted to the ball $B[x,\sqrt{Ct}]$, where $C := 2\,(\sup f - \inf f)$.

(iii) For any $t>0$, $H_t f$ is Lipschitz and locally semiconcave (with a quadratic modulus of semiconcavity) on M.

(iv) For any $x\in M$, $H_t f(x)$ is a nonincreasing function of t, converging monotonically to $f(x)$ as $t\to 0$. In particular, $\lim_{t\to 0} H_t f = f$, locally uniformly.

(v) For any $t\geq 0$, $s>0$, $x\in M$,

$$\frac{|H_{t+s}f(x) - H_t f(x)|}{s} \leq \frac{\|H_t f\|^2_{\mathrm{Lip}(B[x,\sqrt{Cs}])}}{2}.$$

(vi) For any $x \in M$ and $t \geq 0$,

$$\liminf_{s\to 0^+} \frac{(H_{t+s}f)(x) - (H_t f)(x)}{s} \geq -\frac{|\nabla^- H_t f(x)|^2}{2}. \tag{22.27}$$

(vii) For any $x \in M$ and $t > 0$,

$$\lim_{s\to 0^+} \frac{(H_{t+s}f)(x) - (H_t f)(x)}{s} = -\frac{|\nabla^- H_t f(x)|^2}{2}. \tag{22.28}$$

The proof of Proposition 22.16 is postponed to the Appendix, where a more general statement will be provided (Theorem 22.46).

Now we are ready for the main result of this section.

Theorem 22.17 (Logarithmic Sobolev $\Rightarrow$ T_2 $\Rightarrow$ Poincaré). *Let M be a Riemannian manifold equipped with a reference probability measure $\nu \in P_2(M)$. Then:*

(i) If ν satisfies a logarithmic Sobolev inequality with some constant $K > 0$, then it also satisfies a Talagrand inequality with constant K.

(ii) If ν satisfies a Talagrand inequality with some constant $K > 0$, then it also satisfies a Poincaré inequality with constant K.

Remark 22.18. Theorem 22.17 has the important advantage over Theorem 22.14 that logarithmic Sobolev inequalities are somewhat easy to perturb (recall Remark 21.5), while there are few known perturbation criteria for T_2. One of them is as follows: if ν satisfies T_2 and $\widetilde{\nu} = e^{-v}\nu$ with v bounded, then there is a constant C such that

$$\forall \mu \in P_2(M), \quad W_2(\mu,\nu) \leq C\Big(\sqrt{H_\nu(\mu)} + H_\nu(\mu)^{\frac{1}{4}}\Big). \tag{22.29}$$

Remark 22.19. Part (ii) of Theorem 22.17 shows that the T_2 inequality on a Riemannian manifold contains spectral information, and imposes qualitative restrictions on measures satisfying T_2. For instance, the support of such a measure needs to be connected. (Otherwise take $u = a$ on one connected component, $u = b$ on another, $u = 0$ elsewhere, where a and b are two constants chosen in such a way that $\int u\, d\nu = 0$. Then $\int |\nabla u|^2\, d\nu = 0$, while $\int u^2\, d\nu > 0$.) This remark shows that T_2 does not result from just decay estimates, in contrast with T_1.

Proof of Theorem 22.17, part (i). Let ν satisfy a logarithmic Sobolev inequality with constant $K > 0$. By the dual reformulation of $T_2(K)$ (Proposition 22.3 for $p = 2$), it is sufficient to show

$$\forall g \in C_b(M), \qquad \int_M e^{K(Hg)}\, d\nu \leq e^{\int_M g\, d\nu}, \tag{22.30}$$

where

$$(Hg)(x) = \inf_{y\in M}\left[g(y) + \frac{d(x,y)^2}{2}\right].$$

Define

$$\phi(t) = \frac{1}{Kt}\log\left(\int_M e^{KtH_t g}\, d\nu\right). \tag{22.31}$$

Since g is bounded, Proposition 22.16(ii) implies that $H_t g$ is bounded, uniformly in t. Thus

$$e^{KtH_t g} = 1 + Kt\int_M H_t g\, d\nu + O(t^2) \tag{22.32}$$

and

$$\phi(t) = \int_M H_t g\, d\nu + O(t). \tag{22.33}$$

By Proposition 22.16(iv), $H_t g$ converges pointwise to g as $t \to 0^+$; then by the dominated convergence theorem,

$$\lim_{t\to 0^+}\phi(t) = \int_M g\, d\nu. \tag{22.34}$$

So it all amounts to showing that $\phi(1) \leq \lim_{t\to 0^+}\phi(t)$, and this will obviously be true if $\phi(t)$ is nonincreasing in t. To prove this, we shall compute the time-derivative $\phi'(t)$. We shall go slowly, so the hasty reader may go directly to the result, which is formula (22.41) below.

Let $t \in (0,1]$ be given. For $s > 0$, we have

$$\begin{aligned}\frac{\phi(t+s)-\phi(t)}{s} &= \frac{1}{s}\left(\frac{1}{K(t+s)} - \frac{1}{Kt}\right)\log\int_M e^{K(t+s)H_{t+s}g}\, d\nu \\ &+ \frac{1}{Kts}\left(\log\int_M e^{K(t+s)H_{t+s}g}\, d\nu - \log\int_M e^{KtH_t g}\, d\nu\right).\end{aligned} \tag{22.35}$$

As $s \to 0^+$, $e^{K(t+s)H_{t+s}g}$ converges pointwise to $e^{Kt\,H_t g}$, and is uniformly bounded. So the first term in the right-hand side of (22.35) converges, as $s \to 0^+$, to

$$- \frac{1}{Kt^2} \log \left(\int_M e^{Kt\, H_t g} \, d\nu \right). \tag{22.36}$$

On the other hand, the second term in the right-hand side of (22.35) converges to

$$\frac{1}{Kt \displaystyle\int e^{Kt\, H_t g} \, d\nu} \lim_{s \to 0^+} \left[\frac{1}{s} \left(\int_M e^{K(t+s) H_{t+s} g} \, d\nu - \int_M e^{Kt\, H_t g} \, d\nu \right) \right], \tag{22.37}$$

provided that the latter limit exists.

To evaluate the limit in (22.37), we decompose the expression inside square brackets into

$$\int_M \left(\frac{e^{K(t+s) H_{t+s} g} - e^{Kt\, H_{t+s} g}}{s} \right) d\nu + \int_M \left(\frac{e^{Kt\, H_{t+s} g} - e^{Kt\, H_t g}}{s} \right) d\nu. \tag{22.38}$$

The integrand of the first term in the above formula can be rewritten as $(e^{Kt\, H_{t+s} g})(e^{Ks\, H_{t+s} g} - 1)/s$, which is uniformly bounded and converges pointwise to $(e^{Kt\, H_t g}) Kt\, H_t g$ as $s \to 0^+$. So the first integral in (22.38) converges to $\int_M (K\, H_t g) e^{Kt\, H_t g} \, d\nu$.

Let us turn to the second term of (22.38). By Proposition 22.16(vii), for each $x \in M$,

$$H_{t+s} g(x) = H_t g(x) - s \left(\frac{|\nabla^- H_t g(x)|^2}{2} + o(1) \right),$$

and therefore

$$\lim_{s \to 0^+} \frac{e^{Kt H_{t+s} g(x)} - e^{Kt\, H_t g(x)}}{s} = -\, Kt e^{Kt\, H_t g} \, \frac{|\nabla^- H_t g(x)|^2}{2}. \tag{22.39}$$

On the other hand, parts (iii) and (v) of Proposition 22.16 imply that

$$H_{t+s} g = H_t g + O(s).$$

Since $H_t g(x)$ is uniformly bounded in t and x,

$$\frac{e^{Kt H_{t+s} g} - e^{Kt\, H_t g}}{s} = O(1) \qquad \text{as } s \to 0^+. \tag{22.40}$$

By (22.39), (22.40) and the dominated convergence theorem,

$$\lim_{s\to 0^+} \int_M \left(\frac{e^{KtH_{t+s}g} - e^{KtH_t g}}{s} \right) d\nu = -Kt \int_M \frac{|\nabla^- H_t g|^2}{2}\, e^{KtH_t g}\, d\nu.$$

In summary, for any $t > 0$, ϕ is right-differentiable at t and

$$\begin{aligned}
\frac{d^+\phi(t)}{dt} &:= \lim_{s\to 0^+} \left[\frac{\phi(t+s)-\phi(t)}{s}\right] \\
&= \frac{1}{Kt^2 \int_M e^{KtH_t g}\, d\nu} \left[- \left(\int_M e^{KtH_t g}\, d\nu\right) \log \left(\int_M e^{KtH_t g}\, d\nu\right) \right. \\
&\qquad \left. + \int_M (KtH_t g)\, e^{KtH_t g}\, d\nu - \frac{1}{2K} \int_M \big(Kt|\nabla^- H_t g|\big)^2 e^{KtH_t g}\, d\nu \right].
\end{aligned} \tag{22.41}$$

Because ν satisfies a logarithmic Sobolev inequality with constant K, the quantity inside square brackets is nonpositive. So ϕ is nonincreasing and the proof is complete. □

Before attacking the proof of Theorem 22.17(ii), it might be a good idea to think over the next exercise, so as to understand more "concretely" why Talagrand inequalities are related to Poincaré inequalities.

Exercise 22.20. Use Otto's calculus to show that, at least formally,

$$\|h\|_{H^{-1}(\nu)} = \lim_{\varepsilon\to 0} \frac{W_2\big((1+\varepsilon h)\nu,\, \nu\big)}{\varepsilon},$$

where h is smooth and bounded (and compactly supported, if you wish), $\int h\, d\nu = 0$, and the dual Sobolev norm $H^{-1}(\nu)$ is defined by

$$\|h\|_{H^{-1}(\nu)} = \sup_{h\neq 0} \frac{\|h\|_{L^2(\nu)}}{\|\nabla h\|_{L^2(\nu)}} = \big\|\nabla(L^{-1}h)\big\|_{L^2(\nu)},$$

where as before $L = \Delta - \nabla V \cdot \nabla$. Deduce that, at least formally, the Talagrand inequality reduces, in the limit when $\mu = (1+\varepsilon h)\,\nu$ and $\varepsilon \to 0$, to the **dual Poincaré inequality**

$$\left[\int h\, d\nu = 0\right] \Longrightarrow \qquad \|h\|_{H^{-1}(\nu)} \le \frac{\|h\|_{L^2(\nu)}}{\sqrt{K}}.$$

Proof of Theorem 22.17, part (ii). Let $h : M \to \mathbb{R}$ be a bounded Lipschitz function satisfying $\int_M h\,d\nu = 0$. Introduce

$$\psi(t) = \int_M e^{KtH_t h}\,d\nu.$$

From the dual formulation of Talagrand's inequality (Proposition 22.3 for $p = 2$), $\psi(t)$ is bounded above by $\exp(Kt\int_M h\,d\nu) = 1$; hence ψ has a maximum at $t = 0$. Combining this with $\int h\,d\nu = 0$, we find

$$0 \le \limsup_{t\to 0^+} \left(\frac{1-\psi(t)}{Kt^2}\right) = \limsup_{t\to 0^+} \int_M \left(\frac{1 + Kt\,h - e^{K_t\,H_t h}}{Kt^2}\right) d\nu. \tag{22.42}$$

By the boundedness of $H_t h$ and Proposition 22.16(iv),

$$\begin{aligned} e^{KtH_t h} &= 1 + KtH_t h + \frac{K^2t^2}{2}\,(H_t h)^2 + O(t^3) \\ &= 1 + KtH_t h + \frac{K^2t^2}{2}\,h^2 + o(t^2). \end{aligned} \tag{22.43}$$

So the right-hand side of (22.42) equals

$$\limsup_{t\to 0^+} \int_M \left(\frac{h - H_t h}{t}\right) d\nu - \frac{K}{2}\int_M h^2\,d\nu.$$

By Proposition 22.16(v), $(h - H_t h)/t$ is bounded; so we can apply Fatou's lemma, in the form

$$\limsup_{t\to 0^+} \int_M \left(\frac{h - H_t h}{t}\right) d\nu \le \int_M \limsup_{t\to 0^+} \left(\frac{h - H_t h}{t}\right) d\nu.$$

Then Proposition 22.16(vi) implies that

$$\int_M \limsup_{t\to 0^+} \left(\frac{h - H_t h}{t}\right) d\nu \le \int_M \frac{|\nabla^- h|^2}{2}\,d\nu.$$

All in all, the right-hand side of (22.42) can be bounded above by

$$\frac{1}{2}\int_M |\nabla^- h|^2\,d\nu - \frac{K}{2}\int_M h^2\,d\nu. \tag{22.44}$$

So (22.44) is always nonnegative, which concludes the proof of the Poincaré inequality. □

To close this section, I will show that the Talagrand inequality does imply a logarithmic Sobolev inequality under strong enough curvature assumptions.

Theorem 22.21 (T_2 sometimes implies log Sobolev). *Let M be a Riemannian manifold and let $\nu = e^{-V}\mathrm{vol} \in P_2(M)$ be a reference measure on M, $V \in C^2(M)$. Assume that ν satisfies a Talagrand inequality $T_2(\lambda)$, and a curvature-dimension inequality $\mathrm{CD}(K,\infty)$ for some $K > -\lambda$. Then ν also satisfies a logarithmic Sobolev inequality with constant*

$$\widetilde{\lambda} = \max\left[\frac{\lambda}{4}\left(1+\frac{K}{\lambda}\right)^2,\ K\right].$$

Proof of Theorem 22.21. From the assumptions and Corollary 20.13(ii), the nonnegative quantities $H = H_\nu(\mu)$, $W = W_2(\mu,\nu)$ and $I = I_\nu(\mu)$ satisfy the inequalities

$$H \le W\sqrt{I} - \frac{\lambda W^2}{2}, \qquad W \le \sqrt{\frac{2H}{K}}.$$

It follows by an elementary calculation that $H \le I/(2\widetilde{\lambda})$, so ν satisfies a logarithmic Sobolev inequality with constant $\widetilde{\lambda}$. □

Talagrand inequalities and Gaussian concentration

We already saw in Theorem 22.10 that the T_1 inequality implies Gaussian concentration bounds. Now we shall see that the stronger T_2 inequality implies *dimension free* concentration bounds; roughly speaking, this means that $\nu^{\otimes N}$ satisfies concentration inequalities with constants independent of N.

Theorem 22.22 (T_2 and dimension free Gaussian concentration). *Let $(\mathcal{X},d)$ be a Polish space, equipped with a reference probability measure ν. Then the following properties are equivalent:*

(i) ν lies in $P_2(\mathcal{X})$ and satisfies a T_2 inequality.

(ii) There is a constant $C > 0$ such that for any $N \in \mathbb{N}$ and any Borel set $A \subset \mathcal{X}^N$,

$$\nu[A] \geq \frac{1}{2} \implies \forall r > 0, \quad \nu^{\otimes N}[A^r] \geq 1 - e^{-Cr^2};$$

here the enlargement A^r is defined with the d_2 distance,

$$d_2(x,y) = \sqrt{\sum_{i=1}^{N} d(x_i, y_i)^2}.$$

(iii) There is a constant $C > 0$ such that for any $N \in \mathbb{N}$ and any $f \in \mathrm{Lip}(\mathcal{X}^N, d_2)$ (resp. $\mathrm{Lip}(\mathcal{X}^N, d_2) \cap L^1(\nu^{\otimes N})$),

$$\nu^{\otimes N}\Big[\big\{x \in \mathcal{X}^N;\ f(x) \geq m + r\big\}\Big] \leq e^{-C \frac{r^2}{\|f\|_{\mathrm{Lip}}^2}},$$

where m is a median of f (resp. the mean value of f) with respect to the measure $\nu^{\otimes N}$.

Remark 22.23. The dependence of the constants can be made more precise: If ν satisfies $T_2(K)$, then there are $a, r_0 > 0$ such that (with obvious notation) $\nu^{\otimes N}[A^r] \geq 1 - a\, e^{-K(r-r_0)^2/2}$ and $\nu^{\otimes N}[\{f \geq m + r\}] \leq a\, e^{-K(r-r_0)^2/2}$, for all $N \in \mathbb{N}$ and $r \geq r_0$. Conversely, these inequalities imply $T_2(K)$.

Proof of Theorem 22.22. If (i) is satisfied, then by Proposition 22.5, $\nu^{\otimes N}$ satisfies T_2 (and therefore T_1) with a uniform constant. Then we can repeat the proof of (i) ⇒ (iii) in Theorem 22.10, with ν replaced by $\nu^{\otimes N}$. Since the constants obtained in the end are independent of N, this proves (ii).

The implication (ii) ⇒ (iii) follows the same lines as in Theorem 22.10.

The implication (iii) ⇒ (i) is more subtle. For any $x \in \mathcal{X}^N$, define the empirical measure $\widehat{\mu}_x^N$ by

$$\widehat{\mu}_x^N = \frac{1}{N} \sum_{i=1}^{N} \delta_{x_i} \in P(\mathcal{X}),$$

and let

$$f_N(x) = W_2\big(\widehat{\mu}_x^N, \nu\big).$$

By the triangle inequality for W_2 and Theorem 4.8, for any $x, y \in \mathcal{X}^N$ one has

$$|f_N(x) - f_N(y)|^2 \le W_2\Big(\frac{1}{N}\sum_{i=1}^N \delta_{x_i}, \frac{1}{N}\sum_{i=1}^N \delta_{y_i}\Big)^2$$
$$\le \frac{1}{N}\sum_{i=1}^N W_2(\delta_{x_i}, \delta_{y_i})^2 = \frac{1}{N}\sum_{i=1}^N d(x_i, y_i)^2;$$

so f_N is $(1/\sqrt{N})$-Lipschitz in distance d_2. By (iii),

$$\forall r > 0, \quad \nu^{\otimes N}\big[f_N \ge m_N + r\big] \le e^{-C\,N\,r^2}, \tag{22.45}$$

where m_N is either a median, or the mean value of f_N.

Let us check that m_N goes to zero as $N \to \infty$. According to Varadarajan's theorem, $\widehat{\mu}_x^N$ converges weakly to ν for all x outside of a $\nu^{\otimes \mathbb{N}}$-negligible set. (Here I am slightly abusing notation since x stands both for a sequence and for the first N elements of this sequence.) Moreover, if x_0 is any element in $\mathcal{X}$ and $\varphi(x) = d(x_0, x)^2$, then $\int \varphi\, d\widehat{\mu}_x^N = (1/N)\sum_{i=1}^N d(x_0, x_i)^2$ is (under $\nu^{\otimes \mathbb{N}}$) a sum of independent, identically distributed variables with finite moment of order 1; and thus, by the strong law of large numbers, converges to $\int d(x_0, x)^2\, d\nu(x) = \int \varphi\, d\nu$. Combining this information with Theorem 6.9 we see that $W_2(\widehat{\mu}_x^N, \nu) \longrightarrow 0$ as $N \to \infty$, for $\nu^{\otimes \mathbb{N}}$-almost all sequences $(x_i)_{i\in\mathbb{N}}$. By Lebesgue's dominated convergence theorem, for any $t > 0$ we have $\nu^{\otimes \mathbb{N}}[W_2(\widehat{\mu}^N, \nu) \ge t] \to 0$; this implies that any sequence (m_N) of medians of f_N converges to zero.

The convergence of the mean value requires an additional estimate:

$$\int W_2(\widehat{\mu}^N, \nu)^2\, d\nu^{\otimes N} \le \int W_2(\delta_x, \nu)^2\, d\nu(x)$$
$$\le \int W_2(\delta_x, \delta_y)^2\, d\nu(x)\, d\nu(y) = \int d(x,y)^2\, d\nu(x)\, d\nu(y) < +\infty;$$

so f_N is bounded in $L^2(\nu^{\otimes \mathbb{N}})$, in particular it is equi-integrable. Then the almost sure convergence of f_N to 0 implies the convergence also in L^1; a fortiori the mean value of f_N goes to zero.

Since $m_N \to 0$, (22.45) implies

$$\liminf_{N\to\infty} \left(-\frac{1}{N}\log \nu^{\otimes N}\Big[\Big\{x;\ W_2(\widehat{\mu}_x^N, \nu) \ge r\Big\}\Big]\right) \ge C\, r^2. \tag{22.46}$$

The left-hand side is a large deviation estimate for the empirical measure of independent identically distributed samples; since the set

$W_2(\mu,\nu) > r$ is open in the topology of $P_2(\mathcal{X})$, a suitable version of **Sanov's theorem** (see the bibliographical notes) yields

$$\limsup_{N\to\infty}\left(-\frac{1}{N}\log \nu^{\otimes N}\left[\left\{x;\ W_2(\widehat{\mu}_x^N,\nu)\geq r\right\}\right]\right) \leq \inf\left\{H_\nu(\mu);\ \nu\in P_2(\mathcal{X});\ W_2(\mu,\nu)>r\right\}. \quad (22.47)$$

Combining (22.46) and (22.47) gives

$$\inf\left\{H_\nu(\mu);\ \nu\in P_2(\mathcal{X});\ W_2(\mu,\nu)>r\right\}\geq C\,r^2,$$

which is equivalent to $W_2(\mu,\nu)\leq\sqrt{H_\nu(\mu)/C}$, hence (i). □

Poincaré inequalities and quadratic-linear transport cost

So far we have encountered transport inequalities involving the quadratic cost function $c(x,y) = d(x,y)^2$, and the linear cost function $c(x,y) = d(x,y)$. Remarkably, Poincaré inequalities can be recast in terms of transport cost inequalities for a cost function which behaves quadratically for small distances, and linearly for large distances. As discovered by Bobkov and Ledoux, they can also be rewritten as **modified logarithmic Sobolev inequalities**, which are just usual logarithmic Sobolev inequalities, except that there is a Lipschitz constraint on the logarithm of the density of the measure. These two reformulations of Poincaré inequalities will be discussed below.

Definition 22.24 (Quadratic-linear cost). *Let $(\mathcal{X},d)$ be a metric space. The quadratic-linear cost $c_{q\ell}$ on $\mathcal{X}$ is defined by*

$$c_{q\ell}(x,y)=\begin{cases} d(x,y)^2 & \text{if } d(x,y)\leq 1;\\ d(x,y) & \text{if } d(x,y)>1.\end{cases}$$

In a compact notation, $c_{q\ell}(x,y)=\min(d(x,y)^2,d(x,y))$. The optimal total cost associated with $c_{q\ell}$ will be denoted by $C_{q\ell}$.

Theorem 22.25 (Reformulations of Poincaré inequalities). *Let M be a Riemannian manifold equipped with a reference probability measure $\nu=e^{-V}\mathrm{vol}$. Then the following statements are equivalent:*

(i) ν satisfies a Poincaré inequality;

(ii) There are constants $c, K > 0$ such that for any Lipschitz probability density ρ,

$$|\nabla \log \rho| \leq c \implies \qquad H_\nu(\mu) \leq \frac{I_\nu(\mu)}{K}, \qquad \mu = \rho\,\nu; \tag{22.48}$$

(iii) $\nu \in P_1(M)$ and there is a constant $C > 0$ such that

$$\forall \mu \in P_1(M), \qquad C_{q\ell}(\mu, \nu) \leq C\, H_\nu(\mu). \tag{22.49}$$

Remark 22.26. The equivalence between (i) and (ii) remains true when the Riemannian manifold M is replaced by a general metric space. On the other hand, the equivalence with (iii) uses at least a little bit of the Riemannian structure (say, a local Poincaré inequality, a local doubling property and a length property).

Remark 22.27. The equivalence between (i), (ii) and (iii) can be made more precise. As the proof will show, if ν satisfies a Poincaré inequality with constant λ, then for any $c < 2\sqrt{\lambda}$ there is an explicit constant $K = K(c) > 0$ such that (22.48) holds true; and the $K(c)$ converges to λ as $c \to 0$. Conversely, if for each $c > 0$ we call $K(c)$ the best constant in (22.48), then ν satisfies a Poincaré inequality with constant $\lambda = \lim_{c\to 0} K(c)$. Also, in (ii) $\Rightarrow$ (iii) one can choose $C = \max(4/K,\, 2/c)$, while in (iii) $\Rightarrow$ (i) the Poincaré constant can be taken equal to C^{-1}.

Theorem 22.25 will be obtained by two ingredients: The first one is the Hamilton–Jacobi semigroup with a nonquadratic Lagrangian; the second one is a generalization of Theorem 22.17, stated below. The notation L^* will stand for the Legendre transform of the convex function $L : \mathbb{R}_+ \to \mathbb{R}_+$, and L'' for the distributional derivative of L (well-defined on $\mathbb{R}_+$ once L has been extended by 0 on $\mathbb{R}_-$).

Theorem 22.28 (From generalized log Sobolev to transport to generalized Poincaré). *Let M be a Riemannian manifold equipped with its geodesic distance d and with a reference probability measure $\nu = e^{-V}\mathrm{vol} \in P_2(M)$. Let L be a strictly increasing convex function $\mathbb{R}_+ \to \mathbb{R}_+$ such that $L(0) = 0$ and L'' is bounded above; let $c_L(x, y) = L(d(x, y))$ and let C_L be the optimal transport cost associated with the cost function c_L. Further, assume that $L(r) \leq C(1 + r)^p$ for some $p \in [1, 2]$ and some $C > 0$. Then:*

(i) Further, assume that $L^*(ts) \le t^2 L^*(s)$ *for all* $t \in [0,1]$, $s \ge 0$. *If there is* $\lambda \in (0,1]$ *such that* ν *satisfies the generalized logarithmic Sobolev inequality with constant* λ*:*

For any $\mu = \rho\nu \in P(M)$ *such that* $\log \rho \in \mathrm{Lip}(M)$,

$$H_\nu(\mu) \le \frac{1}{\lambda} \int L^*\big(|\nabla \log \rho|\big)\, d\mu; \tag{22.50}$$

then ν *also satisfies the following transport inequality:*

$$\forall \mu \in P_p(M), \qquad C_L(\mu,\nu) \le \frac{H_\nu(\mu)}{\lambda}. \tag{22.51}$$

(ii) If ν *satisfies* (22.51)*, then it also satisfies the generalized Poincaré inequality with constant* λ*:*

$$\forall f \in \mathrm{Lip}(M),\ \|f\|_{\mathrm{Lip}} \le L'(\infty),$$
$$\int f\, d\nu = 0 \Longrightarrow \qquad \int f^2\, d\nu \le \frac{2}{\lambda} \int L^*(|\nabla f|)\, d\nu.$$

Proof of Theorem 22.28. The proof is the same as the proof of Theorem 22.17. After picking up $g \in C_b(M)$, one introduces the function

$$\phi(t) = \frac{1}{\lambda t} \log \int e^{\lambda t\, H_t g}\, d\nu, \qquad H_t g(x) = \inf_{y \in M} \left[g(y) + t\, L\left(\frac{d(x,y)}{t}\right)\right].$$

The properties of $H_t g$ are investigated in Theorem 22.46 in the Appendix.

It is not always true that ϕ is continuous at $t = 0$, but at least the monotonicity of $H_t g$ implies

$$\lim_{t \to 0^+} \phi(t) \le \phi(0).$$

Theorem 22.46 makes it possible to compute the right derivative of ϕ as in the proof of Theorem 22.17(i):

$$\begin{aligned}
\frac{d^+\phi(t)}{dt} &:= \lim_{s \to 0^+} \left[\frac{\phi(t+s) - \phi(t)}{s}\right] \\
&= \frac{1}{\lambda t^2 \displaystyle\int_M e^{\lambda t H_t g}\, d\nu} \left[-\left(\int_M e^{\lambda t H_t g}\, d\nu\right) \log\left(\int_M e^{\lambda t H_t g}\, d\nu\right) \right. \\
&\qquad \left. + \int_M (\lambda t H_t g) e^{\lambda t H_t g}\, d\nu \;-\; \frac{1}{2\lambda} \int_M \big(\lambda^2 t^2 L^*(|\nabla^- H_t g|)\big)\, e^{\lambda t H_t g}\, d\nu \right]
\end{aligned}$$

$$\leq \frac{1}{\lambda t^2 \int_M e^{\lambda t H_t g}\, d\nu} \Bigg[- \left(\int_M e^{\lambda t H_t g}\, d\nu \right) \log \left(\int_M e^{\lambda t H_t g}\, d\nu \right) \\ + \int_M (\lambda t H_t g) e^{\lambda t H_t g}\, d\nu - \frac{1}{2\lambda} \int_M L^* \left(\lambda t |\nabla^- H_t g| \right) e^{\lambda t H_t g}\, d\nu \Bigg], \tag{22.52}$$

where the inequality $L^*(\lambda t s) \leq \lambda^2 t^2 L^*(s)$ was used. By assumption, the quantity inside square brackets is nonpositive, so ϕ is nonincreasing on $(0,1]$, and therefore on $[0,1]$. The inequality $\phi(1) \leq \phi(0)$ can be recast as

$$\frac{1}{\lambda} \log \int_M e^{\lambda \inf_{y \in M} \left[g(y) + L(d(x,y)) \right]}\, \nu(dx) \leq \int_M g\, d\nu,$$

which by Theorem 5.26 is the dual formulation of (22.51).

As for part (ii) of the theorem, it is similar to part (ii) of Theorem 22.17.□

Now we have enough tools at our disposal to carry on the proof of Theorem 22.25.

Proof of Theorem 22.25. We shall start with the proof of (i) $\Rightarrow$ (ii). Let $f = \log \rho - \int (\log \rho)\, d\nu$; so $\int f\, d\nu = 0$ and the assumption in (ii) reads $|\nabla f| \leq c$. Moreover, with $a = \int (\log \rho)\, d\nu$ and $X = \int e^f\, d\nu$,

$$I_\nu(\mu) = e^a \int |\nabla f|^2\, e^f\, d\nu;$$

$$\begin{aligned} H_\nu(\mu) &= \int (f+a) e^{f+a}\, d\nu - \left(\int e^{f+a}\, d\nu \right) \log \left(\int e^{f+a}\, d\nu \right) \\ &= e^a \left(\int f e^f\, d\nu - \int e^f\, d\nu + 1 \right) - e^a (X \log X - X + 1) \\ &\leq e^a \left(\int f e^f\, d\nu - \int e^f\, d\nu + 1 \right). \end{aligned}$$

So it is sufficient to prove

$$|\nabla f| \leq c \Longrightarrow \qquad \int \left(f e^f - e^f + 1 \right) d\nu \leq \frac{1}{K} \int |\nabla f|^2 e^f\, d\nu. \tag{22.53}$$

In the sequel, c is any constant satisfying $0 < c < 2\sqrt{\lambda}$. Inequality (22.53) will be proven by two auxiliary inequalities:

$$\int f^2\,d\nu \le e^{c\sqrt{5/\lambda}} \int f^2 e^{-|f|}\,d\nu; \tag{22.54}$$

$$\int f^2 e^f\,d\nu \le \frac{1}{\lambda}\left(\frac{2\sqrt{\lambda}+c}{2\sqrt{\lambda}-c}\right)^2 \int |\nabla f|^2 e^f\,d\nu. \tag{22.55}$$

Note that the upper bound on $|\nabla f|$ is crucial in both inequalities.

Once (22.54) and (22.55) are established, the result is immediately obtained. Indeed, the right-hand side of (22.54) is obviously bounded by the left-hand side of (22.55), so both expressions are bounded above by a constant multiple of $\int |\nabla f|^2 e^f\,d\nu$. On the other hand, an elementary study shows that

$$\forall f \in \mathbb{R}, \qquad f e^f - e^f + 1 \le \max\,(f^2, f^2 e^f),$$

so (22.53) holds true for some explicit constant $K(c)$.

To obtain (22.54), we proceed as follows. The elementary inequality $2|f|^3 \le \delta f^2 + \delta^{-1} f^4$ ($\delta > 0$) integrates up to

$$\begin{aligned} 2\int |f|^3\,d\nu &\le \delta \int f^2\,d\nu + \delta^{-1}\int f^4\,d\nu \\ &= \delta\int f^2\,d\nu + \delta^{-1}\left(\int f^2\,d\nu\right)^2 + \delta^{-1}\left[\int (f^2)^2\,d\nu - \left(\int f^2\,d\nu\right)^2\right]. \end{aligned} \tag{22.56}$$

By the Poincaré inequality, $\int f^2\,d\nu \le (1/\lambda)\int |\nabla f|^2\,d\nu \le c^2/\lambda$, which implies $(\int f^2\,d\nu)^2 \le (c^2/\lambda)\int f^2\,d\nu$. Also by the Poincaré inequality,

$$\begin{aligned} \int (f^2)^2\,d\nu - \left(\int f^2\,d\nu\right)^2 &\le (1/\lambda)\int |\nabla (f^2)|^2\,d\nu \\ &= (4/\lambda)\int f^2 |\nabla f|^2\,d\nu \le (4c^2/\lambda)\int f^2\,d\nu. \end{aligned}$$

Plugging this information back into (22.56), we obtain

$$2\int |f|^3\,d\nu \le \left(\delta + \frac{5c^2}{\delta\lambda}\right)\int f^2\,d\nu.$$

The choice $\delta = \sqrt{5c^2/\lambda}$ yields

$$\int |f|^3\,d\nu \le c\sqrt{\frac{5}{\lambda}}\int f^2\,d\nu. \tag{22.57}$$

By Jensen's inequality, applied with the convex function $x \to e^{-|x|}$ and the probability measure $\sigma = f^2 \, \nu / (\int f^2 \, d\nu)$,

$$\int f^2 e^{-|f|} \, d\nu = \left(\int e^{-|f|} \, d\sigma \right) \left(\int f^2 \, d\nu \right) \geq e^{-\int |f| \, d\sigma} \left(\int f^2 \, d\nu \right);$$

in other words

$$\int f^2 \, d\nu \leq \exp\left(\frac{\int |f|^3 \, d\nu}{\int f^2 \, d\nu} \right) \int f^2 e^{-|f|} \, d\nu.$$

Combining this inequality with (22.57) finishes the proof of (22.54).

To establish (22.55), we first use the condition $\int f \, d\nu = 0$ and the Poincaré inequality to write

$$\begin{aligned}
&\left(\int f e^{f/2} \, d\nu \right)^2 \\
&= \frac{1}{4} \left(\int [f(x) - f(y)] \, [e^{f(x)/2} - e^{f(y)/2}] \, d\nu(x) \, d\nu(y) \right)^2 \\
&\leq \frac{1}{4} \left(\int |f(x) - f(y)|^2 \, d\nu(x) \, d\nu(y) \right) \left(\int [e^{f(x)/2} - e^{f(y)/2}]^2 \, d\nu(x) \, d\nu(y) \right) \\
&= \left(\int f^2 \, d\nu - \left(\int f \, d\nu \right)^2 \right) \left(\int e^f \, d\nu - \left(\int e^{f/2} \, d\nu \right)^2 \right) \\
&\leq \frac{1}{\lambda^2} \left(\int |\nabla f|^2 \, d\nu \right) \left(\int |\nabla (e^{f/2})|^2 \, d\nu \right) \\
&\leq \frac{c^2}{4\lambda^2} \int |\nabla f|^2 e^f \, d\nu.
\end{aligned} \qquad (22.58)$$

Next, also by the Poincaré inequality and the chain-rule,

$$\begin{aligned}
&\int f^2 e^f \, d\nu - \left(\int f e^{f/2} \, d\nu \right)^2 \\
&\leq \frac{1}{\lambda} \int \left| \nabla (f e^{f/2}) \right|^2 d\nu \\
&= \frac{1}{\lambda} \int |\nabla f|^2 \left(1 + \frac{f}{2} \right)^2 e^f \, d\nu \\
&= \frac{1}{\lambda} \left(\int |\nabla f|^2 e^f \, d\nu + \int |\nabla f|^2 f e^f \, d\nu + \frac{1}{4} \int |\nabla f|^2 f^2 \, e^f \, d\nu \right) \\
&\leq \frac{1}{\lambda} \left(\int |\nabla f|^2 e^f \, d\nu + c \sqrt{\int |\nabla f|^2 e^f \, d\nu} \sqrt{\int f^2 e^f \, d\nu} + \frac{c^2}{4} \int f^2 e^f \, d\nu \right).
\end{aligned} \qquad (22.59)$$

By adding up (22.58) and (22.59), we obtain

$$\int f^2 e^f \, d\nu \le \left(\frac{1}{\lambda} + \frac{c^2}{4\lambda^2}\right) \int |\nabla f|^2 e^f \, d\nu + \frac{c}{\lambda}\sqrt{\int |\nabla f|^2 e^f \, d\nu}\sqrt{\int f^2 e^f \, d\nu} + \frac{c^2}{4\lambda}\int f^2 e^f \, d\nu.$$

This inequality of degree 2 involving the two quantities $\sqrt{\int f^2 e^f \, d\nu}$ and $\sqrt{\int |\nabla f|^2 e^f \, d\nu}$ can be transformed into (22.55). (Here the fact that $c^2/(4\lambda) < 1$ is crucial.) This completes the proof of (i) ⇒ (ii).

Now we shall see that (ii) ⇒ (iii). Let ν satisfy a modified logarithmic Sobolev inequality as in (22.48). Then let $L(s) = cs^2/2$ for $0 \le s \le 1$, $L(s) = c(s - 1/2)$ for $s > 1$. The function L so defined is convex, strictly increasing and $L'' \le c$. Its Legendre transform L^* is quadratic on $[0, c]$ and identically $+\infty$ on $(c, +\infty)$. So (22.48) can be rewritten

$$H_\nu(\mu) \le \frac{2c}{K} \int L^*(|\nabla \log \rho|) \, d\mu.$$

Since $L^*(tr) \le t^2 L^*(r)$ for all $t \in [0, 1]$, $r \ge 0$, we can apply Theorem 22.28(i) to deduce the modified transport inequality

$$C_L(\mu, \nu) \le \max\left(\frac{2c}{K}, 1\right) H_\nu(\mu), \tag{22.60}$$

which implies (iii) since $C_{q\ell} \le (2/c)\, C_L$.

It remains to check (iii) ⇒ (i). If ν satisfies (iii), then it also satisfies $C_L(\mu, \nu) \le C\, H_\nu(\mu)$, where L is as before (with $c = 1$); as a consequence, it satisfies the generalized Poincaré inequality of Theorem 22.28(ii). Pick up any Lipschitz function f and apply this inequality to εf, where ε is small enough that $\varepsilon \|f\|_{\mathrm{Lip}} < 1$; the result is

$$\int f \, d\nu = 0 \Longrightarrow \quad \varepsilon^2 \int f^2 \, d\nu \le (2C) \int L^*(\varepsilon |\nabla^- f|) \, d\nu.$$

Since L^* is quadratic on $[0, 1]$, factors ε^2 cancel out on both sides, and we are back with the usual Poincaré inequality. □

Exercise 22.29. Prove directly the implication (ii) ⇒ (i).

Let us now see the implications of Theorem 22.25 in terms of concentration of measure.

Theorem 22.30 (Measure concentration from Poincaré inequality). *Let M be a Riemannian manifold equipped with its geodesic distance, and with a reference probability measure $\nu = e^{-V}\mathrm{vol}$. Assume that ν satisfies a Poincaré inequality with constant λ. Then there is a constant $C = C(\lambda) > 0$ such that for any Borel set A,*

$$\forall r \geq 0, \qquad \nu[A^r] \geq 1 - \frac{e^{-C\min(r,r^2)}}{\nu[A]}. \tag{22.61}$$

Moreover, for any $f \in \mathrm{Lip}(M)$ (resp. $\mathrm{Lip}(M) \cap L^1(\nu)$),

$$\nu\Big[\big\{x;\ f(x) \geq m + r\big\}\Big] \leq e^{-C\min\left(\frac{r}{\|f\|_{\mathrm{Lip}}},\, \frac{r^2}{\|f\|_{\mathrm{Lip}}^2}\right)}, \tag{22.62}$$

where m is a median (resp. the mean value) of f with respect to ν.

Proof of Theorem 22.30. The proof of (22.61) is similar to the implication (i) $\Rightarrow$ (iii) in Theorem 22.10. Define $B = M \setminus A^r$, and let $\nu_A = (1_A)\nu/\nu[A]$, $\nu_B = (1_B)\nu/\nu[B]$. Obviously, $C_{q\ell}(\nu_A, \nu_B) \geq \min(r, r^2)$. The inequality $\min(a+b, (a+b)^2) \leq 4\,[\min(a, a^2) + \min(b, b^2)]$ makes it possible to adapt the proof of the triangle inequality for W_1 (given in Chapter 6) and get $C_{q\ell}(\nu_A, \nu_B) \leq 4\,[C_{q\ell}(\nu_A, \nu) + C_{q\ell}(\nu_B, \nu)]$. Thus

$$\min(r, r^2) \leq 4[C_{q\ell}(\nu_A, \nu) + C_{q\ell}(\nu_B, \nu)].$$

By Theorem 22.28, ν satisfies (22.49), so there is $C > 0$ such that

$$\begin{aligned}\min(r, r^2) &\leq C\big(H_\nu(\nu_A) + H_\nu(\nu_B)\big)\\ &= C\Big(\log\frac{1}{\nu[A]} + \log\frac{1}{1-\nu[A^r]}\Big),\end{aligned}$$

and (22.61) follows immediately. Then (22.62) is obtained by arguments similar to those used before in the proof of Theorem 22.10. □

Example 22.31. The exponential measure $\nu(dx) = (1/2)e^{-|x|}\,dx$ does not admit Gaussian tails, so it fails to satisfy properties of Gaussian concentration expressed in Theorem 22.10. However, it does satisfy a Poincaré inequality. So (22.61), (22.62) hold true for this measure.

Now, consider the problem of concentration of measure in a *product space*, say $(M^N, \nu^{\otimes N})$, where ν satisfies a Poincaré inequality. We may equip M^N with the metric

$$d_2(x,y) = \sqrt{\sum_i d(x_i,y_i)^2};$$

then $\mu^{\otimes N}$ will satisfy a Poincaré inequality with the same constant as ν, and we may apply Theorem 22.30 to study concentration in $(M^N, d_2, \nu^{\otimes N})$.

However, there is a more interesting approach, due to Talagrand, in which one uses both the distance d_2 and the distance

$$d_1(x,y) = \sum_i d(x_i,y_i).$$

Here is the procedure: Given a Borel set $A \subset M^N$, first enlarge it by r in distance d_2 (that is, consider all points which lie at a distance less than r from A); then enlarge the result by r^2 in distance d_1. This is explained in the next theorem, where $A^{r;d}$ stands for the enlargement of A by r in distance d, and $\|f\|_{\mathrm{Lip}(\mathcal{X},d)}$ stands for the Lipschitz norm of f on $\mathcal{X}$ with respect to the distance d.

Theorem 22.32 (Product measure concentration from Poincaré inequality). *Let M be a Riemannian manifold equipped with its geodesic distance d and a reference probability measure $\nu = e^{-V}\mathrm{vol}$. Assume that ν satisfies a Poincaré inequality with constant λ. Then there is a constant $C = C(\lambda)$ such that for any $N \in \mathbb{N}$, and for any Borel set $A \subset M^N$,*

$$\nu^{\otimes N}[A] \geq \frac{1}{2} \implies \nu^{\otimes N}\Big[(A^{r;d_2})^{r^2;d_1}\Big] \geq 1 - e^{-C\,r^2}. \tag{22.63}$$

Moreover, for any $f \in \mathrm{Lip}(M^N, d_1) \cap \mathrm{Lip}(M^N, d_2)$ (resp. $\mathrm{Lip}(M^N, d_1) \cap \mathrm{Lip}(M^N, d_2) \cap L^1(\nu^{\otimes N})$),

$$\nu^{\otimes N}\Big[\{x;\ f(x) \geq m + r\}\Big] \leq e^{-C\min\left(\frac{r}{\|f\|_{\mathrm{Lip}(M^N,d_1)}},\, \frac{r^2}{\|f\|^2_{\mathrm{Lip}(M^N,d_2)}}\right)}, \tag{22.64}$$

where m is a median of f (resp. the mean value of f) with respect to the measure $\nu^{\otimes N}$.

Conversely, if (22.63) or (22.64) holds true, then ν satisfies a Poincaré inequality.

Proof of Theorem 22.32. Once again, the equivalence (22.63) $\Leftrightarrow$ (22.64) is based on arguments similar to those used in the proof of Theorem 22.10; in the sequel I shall use (22.63).

Let us assume that ν satisfies a Poincaré inequality, and prove (22.63). By Theorem 22.25, ν satisfies a transport-cost inequality of the form

$$\forall \mu \in P_1(M), \quad C_{q\ell}(\mu,\nu) \le C\, H_\nu(\mu).$$

On M^N define the cost

$$c(x,y) = \sum c_{q\ell}(x_i,y_i),$$

and let $\overline{C}$ be the associated optimal cost functional.

By Remark 22.7, $\nu^{\otimes N}$ satisfies an inequality of the form

$$\forall \mu \in P_1(M^N), \quad \overline{C}(\mu,\nu^{\otimes N}) \le C\, H_{\nu^{\otimes N}}(\mu). \tag{22.65}$$

Let A be a Borel set of M^N with $\nu^{\otimes N}[A] \ge 1/2$, and let $r > 0$ be given. Let $B = M^N \setminus (A^{r;d_2})^{r^2;d_1}$. Let ν_B be obtained by conditioning ν on B (that is, $\nu_B = (1_B)\nu/\nu[B]$). Consider the problem of transporting ν_B to ν optimally, with the cost c. At least a portion $\nu^{\otimes N}[A] \ge 1/2$ of the mass has to go to from B to A, so

$$\overline{C}(\nu_B,\nu^{\otimes N}) \ge \frac{1}{2} \inf_{x\in A,\, y\in B} c(x,y) \;=: \frac{1}{2}\, c(A,B).$$

On the other hand, by (22.65),

$$\overline{C}(\mu,\nu^{\otimes N}) \le C\, H_{\nu^{\otimes N}}(\mu) = C\,\log\frac{1}{\nu[B]}.$$

By combining these two inequalities, we get

$$\nu\Big[(A^{r;d_2})^{r^2;d_1}\Big] = 1 - \nu[B] \ge 1 - e^{-1/(2C)\,c(A,B)}.$$

To prove (22.63), it only remains to check that

$$c(A,B) \ge r^2.$$

So let $x = (x_1, \dots, x_N) \in A$, and let $y \in M^N$ such that $c(x,y) < r^2$; the goal is to show that $y \in (A^{r;d_2})^{r^2;d_1}$. For each $i \in \{1, \dots, N\}$, define $z_i = x_i$ if $d(x_i, y_i) > 1$, $z_i = y_i$ otherwise. Then

$$d_2(x,z)^2 = \sum_{d(x_i,y_i)\le 1} d(x_i,y_i)^2 \le \sum_i c_{q\ell}(x_i,y_i) = c(x,y) < r^2;$$

so $z \in A^{r;d_2}$. Similarly,

$$d_1(z,y) = \sum_{d(x_i,y_i)>1} d(x_i,y_i) \le \sum_i c_{q\ell}(x_i,y_i) = c(x,y) < r^2;$$

so y lies at a distance at most r^2 from z, in distance d_1. This concludes the proof.

Let us now assume that ν satisfies (22.63), and prove the modified Talagrand inequality appearing in Theorem 22.25(iii), which will in turn imply the Poincaré inequality. The first step is to note that

$$(A^{r;d_2})^{r^2;d_1} \subset A^{c;4r^2} := \Big\{x \in \mathcal{X}^N;\ \exists a \in A;\ c(a,x) \le 4r^2\Big\}. \tag{22.66}$$

Indeed, separating the four cases (a) $d(a_i,z_i) \le 1,\ d(y_i,z_i) \le 1$; (b) $d(a_i,z_i) \le 1,\ d(y_i,z_i) \ge 1$; (c) $d(a_i,z_i) \ge 1,\ d(a_i,y_i) \ge 1/2$; (d) $d(a_i,z_i) \ge 1,\ d(a_i,y_i) \le 1/2,\ d(y_i,z_i) \ge d(a_i,z_i)/2$, one obtains the elementary inequality

$$c_{q\ell}(a_i,z_i) \le 2\left[d(a_i,y_i)^2 + d(y_i,z_i)\right],$$

and (22.66) follows from the definitions and summation over i.

As in the proof of Theorem 22.22, let $\widehat{\mu}_x^N = (1/N)\sum_i \delta_{x_i}$, let $f_N(x) = C_{q\ell}(\widehat{\mu}_x^N, \nu)$ let m_N be a median of f_N (with respect to the probability measure $\nu^{\otimes N}$) and let $A = \{x;\ f_N(x) \le m_N\}$, so that A has probability at least $1/2$. If $x \in (A^{r;d_2})^{r^2;d_1}$ then by (22.66) there is $a \in A$ such that $c(a,x) \le 4r^2$, and by Theorem 4.8,

$$C_{q\ell}(\widehat{\mu}_x^N, \widehat{\mu}_a^N) \le \frac{4r^2}{N}.$$

On the other hand, as in the proof of Theorem 22.30, one has

$$C_{q\ell}(\widehat{\mu}_x^N, \nu) \le 4\left[C_{q\ell}(\widehat{\mu}_x^N, \widehat{\mu}_a^N) + C_{q\ell}(\widehat{\mu}_a^N, \nu)\right].$$

It follows by means of Property (22.63) that

$$\nu^{\otimes N}\left[C_{q\ell}(\widehat{\mu}_x^N, \nu) > 4\,m_N + 16\,\frac{r^2}{N}\right] \le \nu^{\otimes N}\left[\mathcal{X}^N \setminus A^{c;4r^2}\right]$$
$$\le \nu^{\otimes N}\left[\mathcal{X}^N \setminus (A^{r;d_2})^{r^2;d_1}\right] \le e^{-C\,r^2}.$$

The rest of the argument is as in Theorem 22.22. □

Example 22.33. Let $\nu(dx)$ be the exponential measure $e^{-|x|}dx/2$ on $\mathbb{R}$, then $\nu^{\otimes N}(dx) = (1/2^N)e^{-\sum |x_i|}\prod dx_i$ on $\mathbb{R}^N$. Theorem 22.32 shows that for every Borel set $A \subset \mathbb{R}^N$ with $\nu^{\otimes N}[A] \ge 1/2$ and any $\delta > 0$,

$$\nu^{\otimes N}\left[A + B_r^{d_2} + B_{r^2}^{d_1}\right] \ge 1 - e^{-cr^2} \tag{22.67}$$

where B_r^d stands for the ball of center 0 and radius r in $\mathbb{R}^N$ for the distance d.

Remark 22.34. Strange as this may seem, inequality (22.67) contains (up to numerical constants) the Gaussian concentration of the Gaussian measure! Indeed, let $T : \mathbb{R} \to \mathbb{R}$ be the increasing rearrangement of the exponential measure ν onto the one-dimensional Gaussian measure γ (so $T_\#\nu = \gamma$, $(T^{-1})_\#\gamma = \nu$). An explicit computation shows that

$$|T(x) - T(y)| \le C \min\left(|x-y|, \sqrt{|x-y|}\right) \tag{22.68}$$

for some numeric constant C. Let $T_N(x_1, \ldots, x_N) = (T(x_1), \ldots, T(x_N))$; obviously $(T_N)_\#(\nu^{\otimes N}) = \gamma^{\otimes N}$, $(T_N)_\#^{-1}(\gamma^{\otimes N}) = \nu^{\otimes N}$. Further, let A be any Borel set in $\mathbb{R}^N$, and let $y \in T_N^{-1}(A) + B_r^{d_2} + B_{r^2}^{d_1}$. This means that there are w and x such that $T_N(w) \in A$, $|x - w|_2 \le r$, $|y - x|_1 \le r^2$. Then by (22.68),

$$\begin{aligned}
|T_N(w) - T_N(y)|_2^2 &= \sum |T(w_i) - T(y_i)|^2 \\
&\le C^2 \sum_i \min\left(|w_i - y_i|, |w_i - y_i|^2\right) \\
&\le C^2 \Biggl(\sum_{|w_i - x_i| \ge |x_i - y_i|} 2|w_i - x_i| + \sum_{|w_i - x_i| < |x_i - y_i|} 4|x_i - y_i|^2 \Biggr) \\
&\le 4C^2 \Bigl(\sum |x_i - w_i| + \sum |x_i - y_i|^2\Bigr) \\
&\le 8C^2 r^2;
\end{aligned}$$

so $T_N(y) \in A + B_{\sqrt{8}Cr}^{d_2}$. In summary, if $C' = \sqrt{8}C$, then

$$T_N\left(T_N^{-1}(A) + B_r^{d_2} + B_{r^2}^{d_1}\right) \subset A + B_{C'r}^{d_2}.$$

As a consequence, if $A \subset \mathbb{R}^N$ is any Borel set satisfying $\gamma^{\otimes N}[A] \ge 1/2$, then $\nu^{\otimes N}[T_N^{-1}(A)] = \gamma^{\otimes N}[A] \ge 1/2$, and

$$\begin{aligned}
\gamma^{\otimes N}[A^{C'r}] &\ge \gamma^{\otimes N}\left[T_N\left(T_N^{-1}(A) + B_r^{d_2} + B_{r^2}^{d_1}\right)\right] \\
&= \nu^{\otimes N}\left[T_N^{-1}(A) + B_r^{d_2} + B_{r^2}^{d_1}\right] \\
&\ge 1 - e^{-cr^2}
\end{aligned}$$

for some numeric constant $c > 0$. This is precisely the Gaussian concentration property as it appears in Theorem 22.10(iii) — in a dimension-free form.

Remark 22.35. In certain situations, (22.67) provides sharper concentration properties for the Gaussian measure, than the usual Gaussian concentration bounds. This might look paradoxical, but can be explained by the fact that Gaussian concentration considers *arbitrary* sets A, while in many problems one is led to study the concentration of measure around certain very particular sets, for instance with a "cubic" structure; then inequality (22.67) might be very efficient.

Example 22.36. Let $A = \{x \in \mathbb{R}^N;\ \max |x_i| \le m\}$ be the centered cube of side $2m$, where $m = m(N) \to \infty$ is chosen in such a way that $\gamma^{\otimes N}[A] \ge 1/2$. (It is a classical fact that $m = O(\sqrt{\log N})$ will do, but we don't need that information.) If $r \ge 1$ is small with respect to m, then the enlargement of the cube is dominated by the behavior of T close to $T^{-1}(m)$. Since $T(x)$ behaves approximately like $\sqrt{x}$ for large values of x, $T^{-1}(m)$ is of the order m^2; and close to m^2 the Lipschitz norm of T is $O(1/m)$. Then the computation before can be sharpened into

$$T_N\big(T_N^{-1}(A) + B_r^{d_2} + B_{r^2}^{d_1}\big) \subset A + B_{C'r^2/m}^{d_2};$$

so the concentration of measure can be felt with enlargements by a distance of the order of $r^2/m \ll r$.

Dimension-dependent inequalities

There is no well-identified analog of Talagrand inequalities that would take advantage of the finiteness of the dimension to provide sharper concentration inequalities. In this section I shall suggest some natural possibilities, focusing on positive curvature for simplicity; so M will be compact. This compactness assumption is not a serious restriction: If $(M, e^{-V}\mathrm{vol})$ is any Riemannian manifold satisfying a $\mathrm{CD}(K,N)$ inequality for some $K \in \mathbb{R}$, $N < \infty$, and $e^{-V}\mathrm{vol}$ satisfies a Talagrand inequality, then it can be shown that M is compact.

Theorem 22.37 (Finite-dimensional transport-energy inequalities). *Let M be a Riemannian manifold equipped with a probability* *measure* $\nu = e^{-V}\mathrm{vol}$, $V \in C^2(M)$, *satisfying a curvature-dimension bound* $\mathrm{CD}(K,N)$ *for some* $K > 0$, $N \in (1,\infty)$. *Then, for any* $\mu = \rho\,\nu \in P_2(M)$,

$$\int_M \left[N\left(\frac{\alpha}{\sin\alpha}\right)^{1-\frac{1}{N}} \rho(x_0)^{-\frac{1}{N}} - (N-1)\frac{\alpha}{\tan\alpha}\right] \pi(dx_0\,dx_1) \le 1, \tag{22.69}$$

where $\alpha(x_0,x_1) = \sqrt{K/(N-1)}\,d(x_0,x_1)$, *and* π *is the unique optimal coupling between* μ *and* ν. *Equivalently,*

$$\int_M \left[N\left(\frac{\alpha}{\sin\alpha}\right)^{1-\frac{1}{N}} - (N-1)\frac{\alpha}{\tan\alpha} - 1\right] \pi(dx_0\,dx_1) \\ \le \int \left(\frac{\alpha}{\sin\alpha}\right)^{1-\frac{1}{N}} \left[(N-1)\rho - N\rho^{1-\frac{1}{N}} + 1\right] d\nu. \tag{22.70}$$

Remark 22.38. The function $(N-1)r - N r^{1-\frac{1}{N}} + 1$ is nonnegative, and so is the integrand in the right-hand side of (22.70). If the coefficient $\alpha/\sin\alpha$ above would be replaced by 1, then the right-hand side of (22.70) would be just $\int [(N-1)\rho - N\rho^{1-\frac{1}{N}} + 1]\,d\nu = H_{N,\nu}(\rho)$.

Corollary 22.39 (Further finite-dimensional transport-energy inequalities). *With the same assumptions and notation as in Theorem 22.37, the following inequalities hold true:*

$$\forall p \in (1,\infty) \qquad H_{Np,\nu}(\mu) \ge \\ \int \left[(Np-1) - (N-1)\frac{\alpha}{\tan\alpha} - N(p-1)\left(\frac{\sin\alpha}{\alpha}\right)^{\frac{1}{p-1}\left(1-\frac{1}{N}\right)}\right] d\pi; \tag{22.71}$$

$$2H_{N,\nu}(\mu) - \int \rho^{1-\frac{1}{N}} \log\rho\, d\nu \ge \\ \int \left[(2N-1) - (N-1)\frac{\alpha}{\tan\alpha} - N\exp\left(1 - \left(\frac{\alpha}{\sin\alpha}\right)^{1-\frac{1}{N}}\right)\right] d\pi; \tag{22.72}$$

$$H_{\infty,\nu}(\mu) \ge (N-1)\int \left(1 - \frac{\alpha}{\tan\alpha} + \log\frac{\alpha}{\sin\alpha}\right) d\pi. \tag{22.73}$$

Proof of Theorem 22.37. Apply Theorem 20.10 with $U(r) = -N\,r^{1-1/N}$, $\rho_0 = 1$, $\rho_1 = \rho$: This yields the inequality

$$-N \le -\int N\,\rho^{1-\frac{1}{N}} \left(\frac{\alpha}{\sin\alpha}\right)^{1-\frac{1}{N}} d\pi(\,\cdot\,|\,\cdot\,)\,d\nu$$
$$-\int (N-1)\left(1-\frac{\alpha}{\tan\alpha}\right) d\pi(\,\cdot\,|\,\cdot\,)\,d\nu.$$

Since π has marginals $\rho\,\nu$ and ν, this is the same as (22.69).

To derive (22.70) from (22.69), it is sufficient to check that

$$\int NQ\,d\pi = \int Q\big[(N-1)\rho+1\big]\,d\nu,$$

where $Q = (\alpha/\sin\alpha)^{1-\frac{1}{N}}$. But this is immediate because Q is a symmetric function of x_0 and x_1, and π has marginals $\mu=\rho\,\nu$ and ν, so

$$\int Q(x_0,x_1)\,d\nu(x_0) = \int Q(x_0,x_1)\,d\nu(x_1) = \int Q(x_0,x_1)\,d\pi(x_0,x_1)$$
$$= \int Q(x_0,x_1)\,\rho(x_0)\,d\nu(x_0).$$

□

Proof of Corollary 22.39. Write again $Q=(\alpha/\sin\alpha)^{1-\frac{1}{N}}$. The classical Young inequality can be written $ab \le a^p/p + b^{p'}/p'$, where $p'=p/(p-1)$ is the conjugate exponent to p; so

$$Np\rho^{1-\frac{1}{Np}} = (Np\rho)\left[\rho^{-\frac{1}{N}}Q\right]^{\frac{1}{p}} Q^{-\frac{1}{p}} \le (Np\rho)\left[\frac{\rho^{-\frac{1}{N}}Q}{p} + \frac{Q^{-\frac{p'}{p}}}{p'}\right].$$

By integration of this inequality and (22.69),

$$H_{N,\nu}(\mu) + Np = -Np\int \rho^{1-\frac{1}{Np}}$$
$$\ge -\int N\,\rho^{-\frac{1}{N}}\,Q\,d\pi - N(p-1)\int Q^{-\frac{1}{p-1}}\,d\pi$$
$$\ge -1 - \int (N-1)\,\frac{\alpha}{\tan\alpha}\,d\pi - N(p-1)\int Q^{-\frac{1}{p-1}}\,d\pi,$$

which is the same as (22.71).

Then (22.72) and (22.73) are obtained by taking the limits $p\to 1$ and $p\to\infty$, respectively. Equivalently, one can apply the inequalities $ab \le a\log a - 2a + e^{b+1}$ and $ab \le a\log a - a + e^b$ instead of Young's inequality; more precisely, to get (22.72) from (22.69), one can write

$$N\rho^{1-\frac{1}{N}} \log \rho^{\frac{1}{N}} = (N\rho^{1-\frac{1}{N}} e^{-Q})(e^Q \log \rho^{\frac{1}{N}})$$
$$\leq (N\rho^{1-\frac{1}{N}} e^{-Q})\Big(e^Q Q - 2e^Q + e\rho^{\frac{1}{N}}\Big);$$

and to get (22.73) from (22.69),

$$N\rho \log Q = (N\rho^{1-\frac{1}{N}})\rho^{\frac{1}{N}} \log Q \leq (N\rho^{1-\frac{1}{N}})\Big(\rho^{\frac{1}{N}} \log \rho^{\frac{1}{N}} - \rho^{\frac{1}{N}} + Q\Big).$$

□

All the inequalities appearing in Corollary 22.39 can be seen as refinements of the Talagrand inequality appearing in Theorem 22.14; concentration inequalities derived from them take into account, for instance, the fact that the distance between any two points can never exceed $\pi\sqrt{(N-1)/K}$.

Exercise 22.40. Give a more direct derivation of inequality (22.73), based on the fact that $U(r) = r \log r$ lies in $\mathcal{DC}_N$.

Exercise 22.41. Use the inequalities proven in this section, and the result of Exercise 22.20, to recover, at least formally, the inequality

$$\left[\int h\, d\nu = 0\right] \Longrightarrow \qquad \|h\|^2_{H^{-1}(\nu)} \leq \frac{KN}{N-1}\, \|h\|^2_{L^2(\nu)}$$

under an assumption of curvature-dimension bound $\mathrm{CD}(K,N)$. Now turn this into a rigorous proof, assuming as much smoothness on h and on the density of ν as you wish. (*Hint:* When $\varepsilon \to 0$, the optimal transport between $(1+\varepsilon h)\,\nu$ and ν converges in measure to the identity map; this enables one to pass to the limit in the distortion coefficients.)

Remark 22.42. If one applies the same procedure to (22.71), one recovers a constant $K(Np)/(Np-1)$, which reduces to the correct constant only in the limit $p \to 1$. As for inequality (22.73), it leads to just K (which would be the limit $p \to \infty$).

Remark 22.43. Since the Talagrand inequality implies a Poincaré inequality without any loss in the constants, and the optimal constant in the Poincaré inequality is $KN/(N-1)$, it is natural to ask whether this is also the optimal constant in the Talagrand inequality. The answer is affirmative, in view of Theorem 22.17, since the logarithmic Sobolev inequality also holds true with the same constant. But I don't know of any transport proof of this!

Open Problem 22.44. *Find a direct transport argument to prove that the curvature-dimension* $\mathrm{CD}(K,N)$ *with* $K>0$ *and* $N<\infty$ *implies* $T_2(\widetilde{K})$ *with* $\widetilde{K}=KN/(N-1)$, *rather than just* $T_2(K)$.

Note that inequality (22.73) does not solve this problem, since by Remark 22.42 it only implies the Poincaré inequality with constant K.

I shall conclude with a very loosely formulated open problem, which might be nonsense:

Open Problem 22.45. *In the Euclidean case, is there a particular variant of the Talagrand inequality which takes advantage of the homogeneity under dilations, just as the usual Sobolev inequality in* $\mathbb{R}^n$*? Is it useful?*

Recap

In the end, the main results of this chapter can be summarized by just a few diagrams:

• **Relations between functional inequalities:** By combining Theorems 21.2, 22.17, 22.10 and elementary inequalities, one has

$$\begin{array}{ccccccccc} \mathrm{CD}(K,\infty) & \Longrightarrow & (\mathrm{LS}) & \Longrightarrow & (T_2) & \Longrightarrow & (\mathrm{P}) & \Longrightarrow & (\exp_1) \\ & & & & \Downarrow & & & & \\ & & & & (T_1) & \Longleftrightarrow & (\exp_2) & \Longrightarrow & (\exp_1) \end{array}$$

All these symbols designate properties of the reference measure ν: (LS) stands for logarithmic Sobolev inequality, (P) for Poincaré inequality, $\exp_2$ means that ν has a finite square-exponential moment, and $\exp_1$ that it has a finite exponential moment.

• **Reformulations of Poincaré inequality:** Theorem 22.25 can be visualized as

$$(\mathrm{P}) \Longleftrightarrow (\mathrm{LSLL}) \Longleftrightarrow (T_{q\ell})$$

where (LSLL) means logarithmic Sobolev for log-Lipschitz functions, and $(T_{q\ell})$ designates the transportation-cost inequality involving the quadratic-linear cost.

• **Concentration properties via functional inequalities:** The three main such results proven in this chapter are

$(T_1) \Longleftrightarrow$ *Gaussian concentration* (Theorem 22.10)

$(T_2) \Longleftrightarrow$ *dimension free* Gaussian concentration (Theorem 22.22)

$(P) \Longleftrightarrow$ dimension free *exponential* concentration (Theorem 22.32)

Appendix: Properties of the Hamilton–Jacobi semigroup

This Appendix is devoted to the proof of Theorem 22.46 below, which was used in the proof of Theorem 22.28 (and also in the proof of Theorem 22.17 via Proposition 22.16). It says that if a nice convex Lagrangian $L(|v|)$ is given on a Riemannian manifold, then the solution $f(t,x)$ of the associated Hamilton–Jacobi semigroup satisfies (a) certain regularity properties which go beyond differentiability; (b) the *pointwise* differential equation

$$\frac{\partial f}{\partial t} + L^*\big(|\nabla^- f(x)|\big) = 0,$$

where $|\nabla^- f(x)|$ is defined by (20.2).

Recall the notion of semiconcavity from Definition 10.10; throughout this Appendix I shall say "semiconcave" for "semiconcave with a quadratic modulus". To say that L is locally semiconcave is equivalent to saying that the (distributional) second derivative L'' is locally bounded above on $\mathbb{R}$, once L has been extended by 0 on $\mathbb{R}_-$.

Theorem 22.46 (Some properties of the Hamilton–Jacobi semigroup on a manifold). *Let $L : \mathbb{R}_+ \to \mathbb{R}_+$ be a strictly increasing, locally semiconcave, convex continuous function with $L(0) = 0$. Let M be a Riemannian manifold equipped with its geodesic distance d. For any $f \in C_b(M)$, define the evolution $(H_t f)_{t\geq 0}$ by*

$$\begin{cases} H_0 f = f \\[2ex] (H_t f)(x) = \inf_{y\in M} \left[f(y) + t\, L\left(\dfrac{d(x,y)}{t}\right)\right] \qquad (t>0,\ x\in M). \end{cases} \tag{22.74}$$

Then:

(i) For any $s,t \geq 0$, $H_s\, H_t f = H_{t+s} f$.

(ii) For any $x \in M$, $\inf f \le (H_t f)(x) \le f(x)$; moreover, for any $t > 0$ the infimum over M in (22.74) can be restricted to the closed ball $B[x, R(f,t)]$, where

$$R(f,t) = t\, L^{-1}\left(\frac{\sup f - \inf f}{t}\right).$$

(iii) For any $t > 0$, $H_t f$ is Lipschitz and locally semiconcave on M; moreover $\|H_t f\|_{\mathrm{Lip}} \le L'(\infty)$.

(iv) For any $t > 0$, $H_{t+s} f$ is nonincreasing in s, and converges monotonically and locally uniformly to $H_t f$ as $s \to 0$; this conclusion extends to $t = 0$ if $\|f\|_{\mathrm{Lip}} \le L'(\infty)$.

(v) For any $t \ge 0$, $s > 0$, $x \in M$,

$$\frac{|H_{t+s} f(x) - H_t f(x)|}{s} \le L^*\Big(\|H_t f\|_{\mathrm{Lip}(B[x,R(f,s)])}\Big).$$

(vi) For any $x \in M$ and $t > 0$,

$$\liminf_{s \downarrow 0} \frac{(H_{t+s} f)(x) - (H_t f)(x)}{s} \ge -\, L^*\big(|\nabla^- H_t f(x)|\big);$$

this conclusion extends to $t = 0$ if $\|f\|_{\mathrm{Lip}} \le L'(\infty)$.

(vii) For any $x \in M$ and $t > 0$,

$$\lim_{s \downarrow 0} \frac{(H_{t+s} f)(x) - (H_t f)(x)}{s} = -\, L^*\big(|\nabla^- H_t f|\big);$$

this conclusion extends to $t = 0$ if $\|f\|_{\mathrm{Lip}} \le L'(\infty)$ and f is locally semiconcave.

Remark 22.47. If $L'(\infty) < +\infty$ then in general $H_t f$ is *not* continuous as a function of t at $t = 0$. This can be seen by the fact that $\|H_t f\|_{\mathrm{Lip}} \le L'(\infty)$ for all $t > 0$.

Remark 22.48. There is no measure theory in Theorem 22.46, and conclusions hold for *all* (not just almost all) $x \in M$.

Proof of Theorem 22.46. First, note that the inverse L^{-1} of L is well-defined $\mathbb{R}_+ \to \mathbb{R}_+$ since L is strictly increasing and goes to $+\infty$ at infinity. Also $L'(\infty) = \lim_{r\to\infty}(L(r)/r)$ is well-defined in $(0, +\infty]$. Further, note that

$$L^*(p) = \sup_{r\geq 0} \big[p\,r - L(r)\big]$$

is a convex nondecreasing function of p, satisfying $L^*(0) = 0$.

Let $x, y, z \in M$ and $t, s > 0$. Since L is increasing and convex,

$$\begin{aligned} L\left(\frac{d(x,y)}{t+s}\right) &\leq L\left(\frac{d(x,z)+d(z,y)}{t+s}\right) \\ &\leq \frac{t}{t+s}\,L\left(\frac{d(x,z)}{t}\right) + \frac{s}{t+s}\,L\left(\frac{d(z,y)}{s}\right), \end{aligned}$$

with equality if $d(x,z)/t = d(z,y)/s$, i.e. if z is an $s/(t+s)$-barycenter of x and y. (There always exists such a z.) So

$$(t+s)\,L\left(\frac{d(x+y)}{t+s}\right) = \inf_{z\in M}\left[t\,L\left(\frac{d(x,z)}{t}\right) + s\,L\left(\frac{d(x,y)}{s}\right)\right].$$

This implies (i).

The lower bound in (ii) is obvious since $L \geq 0$, and the upper bound follows from the choice $y = x$ in (22.74). Moreover, if $d(x,y) > R(f,t)$, then

$$\begin{aligned} f(x) + t\,L\left(\frac{d(x,y)}{t}\right) &> (\inf f) + t\,L\left(\frac{R(f,t)}{t}\right) \\ &= (\inf f) + (\sup f - \inf f) = \sup f; \end{aligned}$$

so the infimum in (22.74) may be restricted to those $y \in M$ such that $d(x,y) \leq R(f,t)$. Note that $R(f,t)$ is finite for all $t > 0$.

When y varies in $B[x, R(f,t)]$, the function $t\,L(d(x,y)/t)$ remains C-Lipschitz, where $C = L'(R(f,t)/t) < +\infty$. So $H_t f$ is an infimum of uniformly Lipschitz functions, and is therefore Lipschitz. It is obvious that $C \leq L'(\infty)$.

To prove (iii) it remains to show that $H_t f$ is locally semiconcave for $t > 0$. Let $(\gamma_t)_{0\leq t\leq 1}$ be a minimizing geodesic in M, then for $\lambda \in [0,1]$,

$$\begin{aligned} &H_t f(\gamma_\lambda) - (1-\lambda)\,H_t f(\gamma_0) - \lambda\,H_t f(\gamma_1) \\ &= \inf_{z_\lambda}\,\inf_{z_0}\,\sup_{z_1}\Bigg\{f(z_\lambda) - (1-\lambda)\,f(z_0) - \lambda\,f(z_1) \\ &\qquad + t\left[L\left(\frac{d(z_\lambda,\gamma_\lambda)}{t}\right) - (1-\lambda)\,L\left(\frac{d(z_0,\gamma_0)}{t}\right) - \lambda\,L\left(\frac{d(z_1,\gamma_1)}{t}\right)\right]\Bigg\} \\ &\geq t\,\inf_{z}\left[L\left(\frac{d(z,\gamma_\lambda)}{t}\right) - (1-\lambda)\,L\left(\frac{d(z,\gamma_0)}{t}\right) - \lambda\,L\left(\frac{d(z,\gamma_1)}{t}\right)\right], \end{aligned}$$

where the latter inequality has been obtained by choosing $z = z_0 = z_1 = z_\lambda$. The infimum may be restricted to a large ball containing the balls of radius $R(f,t)$ centered at γ_0, γ_λ and γ_1. When the image of γ is contained in a compact set K we may therefore find a large ball B (depending on K) such that

$$H_t f(\gamma_\lambda) - (1-\lambda)\, H_t f(\gamma_0) - \lambda\, H_t f(\gamma_1)$$
$$\geq t \inf_{z\in B} \left[L\left(\frac{d(z,\gamma_\lambda)}{t}\right) - (1-\lambda)\, L\left(\frac{d(z,\gamma_0)}{t}\right) - \lambda\, L\left(\frac{d(z,\gamma_1)}{t}\right)\right]. \tag{22.75}$$

When z varies in B, the distance function $d(z,\,\cdot\,)$ is uniformly semiconcave (with a quadratic modulus) on the compact set K; recall indeed the computations in the Third Appendix of Chapter 10. Let $F = L(\,\cdot\,/t)$, restricted to a large interval where $d(z,\,\cdot\,)$ takes values; and let $\varphi = d(z,\,\cdot\,)$, restricted to K. Since F is semiconcave increasing Lipschitz and φ is semiconcave Lipschitz, their composition $F \circ \varphi$ is semiconcave, and the modulus of semiconcavity is uniform in z. So there is $C = C(K)$ such that

$$\inf_{z\in B} \left[L\left(\frac{d(z,\gamma_\lambda)}{t}\right) - (1-\lambda)\, L\left(\frac{d(z,\gamma_0)}{t}\right) - \lambda\, L\left(\frac{d(z,\gamma_1)}{t}\right)\right]$$
$$\geq -\, C\lambda(1-\lambda)\, d(\gamma_0,\gamma_1)^2.$$

This and (22.75) show that $H_t f$ is locally semiconcave and conclude the proof of (iii).

To prove (iv), let $g = H_t f$. It is clear that $H_s g$ is a nonincreasing function of s since $s\, L(d(x,y)/s)$ is itself a nonincreasing function of s. I shall now distinguish two cases.

Case 1: $L'(\infty) = +\infty$; then $\lim_{s\to 0} R(g,s) = (\sup g - \inf g)/L'(\infty) = 0$. For any $x \in M$,

$$g(x) \geq H_s g(x) = \inf_{d(x,y)\leq R(g,s)} \left[g(y) + s\, L\left(\frac{d(x,y)}{s}\right)\right]$$
$$\geq \inf_{d(x,y)\leq R(g,s)} g(y),$$

and this converges to $g(x)$ as $s \to 0$, locally uniformly in x.

Case 2: $L'(\infty) < +\infty$; then $\lim_{s\to 0} R(g,s) > 0$ (except if g is constant, a case which I omit since it is trivial). Since $\|g\|_{\mathrm{Lip}} \leq L'(\infty)$,

$$g(y) \geq g(x) - L'(\infty)\, d(x,y),$$

so

$$\begin{aligned} g(x) \geq H_s g(x) &\geq g(x) + \inf_{d(x,y)\leq R(g,s)} \left[s\, L\left(\frac{d(x,y)}{s}\right) - L'(\infty)\, d(x,y) \right] \\ &\geq g(x) + \left[s\, L\left(\frac{R(g,s)}{s}\right) - L'(\infty)\, R(g,s) \right], \qquad (22.76) \end{aligned}$$

where I used the fact that $s\,L(d/s) - L'(\infty)\, d$ is a nonincreasing function of d (to see this, note that $L'(d/s) - L'(\infty) \leq 0$, where $L'(r)$ is the right-derivative of L at r). By definition, $s\,L(R(g,s)/s) = \sup g - \inf g$, so (22.76) becomes

$$H_s g(x) \geq g(x) + \Big[(\sup g - \inf g) - L'(\infty)\, R(g,s) \Big].$$

As $s \to 0$, the expression inside square brackets goes to 0, and $H_s g(x)$ converges uniformly to $g(x)$. So (iv) is established.

To prove (v), again let $g = H_t f$, then

$$\begin{aligned} 0 &\leq g(x) - H_s g(x) \\ &= \sup_{d(x,y)\leq R(g,s)} \left[g(x) - g(y) - s\, L\left(\frac{d(x,y)}{s}\right) \right] \\ &\leq s \left\{ \sup_{d(x,y)\leq R(g,s)} \left(\frac{[g(y)-g(x)]_-}{d(x,y)} \right) \frac{d(x,y)}{s} - L\left(\frac{d(x,y)}{s}\right) \right\} \\ &\leq s\, L^*\left(\sup_{d(x,y)\leq R(g,s)} \frac{[g(y)-g(x)]_-}{d(x,y)} \right), \qquad (22.77) \end{aligned}$$

where I have used the inequality $p\,r \leq L(r) + L^*(p)$. Statement (v) follows at once from (22.77). Moreover, if $L'(\infty) = +\infty$, then L^* is continuous on $\mathbb{R}_+$, so by the definition of $|\nabla^- g|$ and the fact that $R(g,s) \to 0$,

$$\begin{aligned} \limsup_{s\downarrow 0} \frac{g(x) - H_s g(x)}{s} &\leq L^*\left(\lim_{s\downarrow 0} \sup_{d(x,y)\leq R(g,s)} \frac{[g(y)-g(x)]_-}{d(x,y)} \right) \\ &= L^*\big(|\nabla^- g(x)|\big), \end{aligned}$$

which proves (vi) in the case $L'(\infty) = +\infty$.

If $L'(\infty) < +\infty$, things are a bit more intricate. If $\|g\|_{\mathrm{Lip}} \leq L'(\infty)$, then of course $|\nabla^- g(x)| \leq L'(\infty)$. I shall distinguish two situations:

- If $|\nabla^- g(x)| = L'(\infty)$, the same argument as before shows

$$\frac{g(x) - H_s g(x)}{s} \leq L^*(\|g\|_{\mathrm{Lip}}) \leq L^*(L'(\infty)) = L^*(|\nabla^- g(x)|).$$

- If $|\nabla^- g(x)| < L'(\infty)$, I claim that there is a function $\alpha = \alpha(s)$, *depending on* x, such that $\alpha(s) \longrightarrow 0$ as $s \to 0$, and

$$H_s g(x) = \inf_{d(x,y) \leq \alpha(s)} \left[g(y) + s\, L\left(\frac{d(x,y)}{s}\right)\right]. \qquad (22.78)$$

If this is true then the same argument as in the case $L'(\infty) = +\infty$ will work.

So let us prove (22.78). By Lemma 22.49 below, there exists $\delta > 0$ such that for all $y \in B[x, R(g,s)]$,

$$g(x) - g(y) \leq \big(L'(\infty) - \delta\big)\, d(x,y). \qquad (22.79)$$

For any $\alpha_0 > 0$ we can find s_0 such that

$$\alpha \geq \alpha_0,\ s \leq s_0 \Longrightarrow \qquad \frac{s}{\alpha} L\left(\frac{\alpha}{s}\right) \geq L'(\infty) - \frac{\delta}{2}.$$

So we may define a function $\alpha(s) \to 0$ such that

$$\alpha \geq \alpha(s) \Longrightarrow \qquad s\, L\left(\frac{\alpha}{s}\right) \geq \left(L'(\infty) - \frac{\delta}{2}\right)\alpha. \qquad (22.80)$$

If $d(x,y) \geq \alpha(s)$, (22.79) and (22.80) imply

$$\begin{aligned} g(y) + s\, L\left(\frac{d(x,y)}{s}\right) &\geq g(x) + s\, L\left(\frac{d(x,y)}{s}\right) - \big(L'(\infty) - \delta\big)\, d(x,y) \\ &\geq g(x) + \frac{\delta}{2}\, d(x,y) \ > g(x). \end{aligned}$$

So the infimum of $[g(y) + s\, L(d(x,y)/s)]$ may be restricted to those $y \in M$ such that $d(x,y) \leq \alpha(s)$, and (22.78) is true. Thus (vi) holds true in all cases.

It only remains to prove (vii). Let $g = H_t f$; as we already know, $\|g\|_{\mathrm{Lip}} \leq L'(\infty)$ and g is locally semiconcave. The problem is to show

$$\liminf_{s \downarrow 0} \left[\frac{g(x) - H_s g(x)}{s}\right] \geq L^*(|\nabla^- g(x)|).$$

This is obvious if $|\nabla^- g(x)| = 0$, so let us assume $|\nabla^- g(x)| > 0$. (Note that $|\nabla^- g(x)| < +\infty$ since g is Lipschitz.)

By the same computation as before,

$$\frac{g(x) - H_s g(x)}{s} = \frac{1}{s} \sup_{d(x,y) \le R(g,s)} \left[g(x) - g(y) - s\, L\left(\frac{d(x,y)}{s}\right)\right].$$

First, assume $L'(\infty) = +\infty$, so L^* is defined on the whole of $\mathbb{R}_+$. Let $q \in \partial L(|\nabla^- g(x)|)$. As $s \to 0$,

$$\frac{R(g,s)}{s} \longrightarrow L^{-1}(\infty) = +\infty.$$

So for s small enough, $R(g,s) > s\, q$. This implies

$$\begin{aligned} \frac{g(x) - H_s g(x)}{s} &\ge \frac{1}{s} \sup_{d(x,y) = s\, q} \left[g(x) - g(y) - s\, L\left(\frac{d(x,y)}{s}\right)\right] \\ &= \sup_{d(x,y) = s\, q} \left[\left(\frac{g(x) - g(y)}{d(x,y)}\right) q - L(q)\right]. \end{aligned} \tag{22.81}$$

Let

$$\psi(r) = \sup_{d(x,y) = r} \left[\frac{g(x) - g(y)}{d(x,y)}\right].$$

If it can be shown that

$$\psi(r) \xrightarrow[r \to 0]{} |\nabla^- g(x)|, \tag{22.82}$$

then we can pass to the limit in (22.81) and recover

$$\liminf_{s \downarrow 0} \frac{g(x) - H_s g(x)}{s} \ge |\nabla^- g(x)|\, q - L(q) = L^*(|\nabla^- g(x)|).$$

If $L'(\infty) < +\infty$ and $|\nabla^- g(x)| = L'(\infty)$, the above reasoning fails because $\partial L^*(|\nabla^- g(x)|)$ might be empty. However, for any $\theta < |\nabla^- g(x)|$ we may find $q \in \partial L^*(\theta)$, then the previous argument shows that

$$\liminf_{s \downarrow 0} \frac{g(x) - H_s g(x)}{s} \ge L^*(\theta);$$

the conclusion is obtained by letting $\theta \to |\nabla^- g(x)|$ and using the lower semicontinuity of L^*.

So it all boils down to checking (22.82). This is where the semiconcavity of g will be useful. (Indeed (22.82) might fail for an arbitrary Lipschitz function.) The problem can be rewritten

$$\lim_{r\to 0} \psi(r) = \lim_{r\to 0} \sup_{s\le r} \psi(s);$$

so it is enough to show that ψ does have a limit at 0.

Let S_r denote the sphere of center x and radius r. If r is small enough, for any $z \in S_r$ there is a unique geodesic joining x to z, and the exponential map induces a bijection between $S_{r'}$ and S_r, for any $r' \in (0, r]$. Let $\lambda = r'/r \in (0, 1]$; for any $y \in S_{r'}$ we can find a unique geodesic γ such that $\gamma_0 = x$, $\gamma_\lambda = y$, $\gamma_1 \in S_r$. By semiconcavity, there is a constant $C = C(x, r)$ such that

$$g(\gamma_\lambda) - (1-\lambda)\, g(\gamma_0) - \lambda\, g(\gamma_1) \ge -C\,\lambda(1-\lambda)\, d(\gamma_0,\gamma_1)^2.$$

This can be rewritten

$$\frac{g(\gamma_0) - g(\gamma_1)}{d(\gamma_0,\gamma_1)} - \frac{g(\gamma_0) - g(\gamma_\lambda)}{\lambda\, d(\gamma_0,\gamma_1)} \ge -C\,\lambda(1-\lambda)\, d(\gamma_0,\gamma_1);$$

or equivalently,

$$\frac{g(x) - g(\gamma_1)}{d(x,\gamma_1)} - \frac{g(x) - g(y)}{d(x,y)} \ge -C\,(r - r').$$

So

$$d(x,y) = r' \Longrightarrow \qquad \psi(r) - \frac{g(x) - g(y)}{d(x,y)} \ge -C\,(r - r').$$

By taking the supremum over y, we conclude that

$$\psi(r) - \psi(r') \ge -C\,(r - r').$$

In particular $\psi(r) + C\,r$ is a nondecreasing function of r, so ψ has a limit as $r \to 0$. This concludes the proof. □

The following lemma was used in the proof of Theorem 22.46:

Lemma 22.49. *Let M be a Riemannian manifold (or more generally, a geodesic space), and let L, R be positive numbers. If $g : M \to \mathbb{R}$ is L-Lipschitz and $|\nabla^- g(x)| < L$ for some $x \in M$ then there is $\delta > 0$ such that for any $y \in B[x, R]$,*

$$g(x) - g(y) \le (L - \delta)\, d(x,y).$$

Proof of Lemma 22.49. By assumption

$$\limsup_{y\to x} \frac{[g(y)-g(x)]_-}{d(x,y)} < L.$$

So there are $r>0$, $\eta>0$ such that if $d(x,z)\le r$ then

$$g(x) \le g(z) + (L-\eta)\, d(x,z). \tag{22.83}$$

Let $y\in B[z,R]$ and let γ be a geodesic joining $\gamma(0)=x$ to $\gamma(1)=y$; let $z=\gamma(r/R)$. Then $d(x,z)=(r/R)\,d(x,y)\le r$, so (22.83) holds true. As a consequence,

$$\begin{aligned} g(x)-g(y) &= [g(x)-g(z)] + [g(z)-g(y)] \\ &\le (L-\eta)\, d(x,z) + L\, d(z,y) \\ &= L\, d(x,y) - \eta\, d(x,z) \\ &\le \left(L - \eta\,\frac{r}{R}\right) d(x,y), \end{aligned}$$

which proves the lemma. □

Bibliographical notes

Most of the literature described below is reviewed with more detail in the synthesis works of Ledoux [542, 543, 546]. Selected applications of the concentration of measure to various parts of mathematics (Banach space theory, fine study of Brownian motion, combinatorics, percolation, spin glass systems, random matrices, etc.) are briefly developed in [546, Chapters 3 and 8]. The role of T_p inequalities in that theory is discussed in [546, Chapter 6], [41, Chapter 8], and [427]. One may also take a look at Massart's Saint-Flour lecture notes [597].

Lévy is often quoted as the founding father of concentration theory. His work might have been forgotten without the determination of V. Milman to make it known. The modern period of concentration of measure starts with a work by Milman himself on the so-called Dvoretzy theorem [634].

The Lévy–Gromov isoperimetric inequality [435] is a way to get sharp concentration estimates from Ricci curvature bounds. Gromov has further worked on the links between Ricci curvature and concentration, see his very influential book [438], especially Chapter $3\frac{1}{2}$ therein. Also Talagrand made decisive contributions to the theory of concentration of measure, mainly in product spaces, see in particular [772, 773]. Dembo [294] showed how to recover several of Talagrand's results in an elegant way by means of information-theoretical inequalities.

T_p inequalities have been studied for themselves at least since the beginning of the nineties [695]; the Csiszár–Kullback–Pinsker inequality can be considered as their ancestor from the sixties (see below). Sometimes it is useful to consider more general transport inequalities of the form $C(\mu, \nu) \leq H_\nu(\mu)$, or even $C(\mu, \widetilde{\mu}) \leq H_\nu(\mu) + H_\nu(\widetilde{\mu})$ (recall Theorem 22.30 and its proof).

It is easy to show that transport inequalities are stable under weak convergence [305, Lemma 2.2]. They are also stable under push-forward [425].

Proposition 22.3 was studied by Rachev [695] and Bobkov and Götze [128], in the cases $p = 1$ and $p = 2$. These duality formulas were later systematically exploited by Bobkov, Gentil and Ledoux [127, 131]. The Legendre reformulation of the H functional can be found in many sources (for instance [577, Appendix B] when $\mathcal{X}$ is compact).

The tensorization argument used in Proposition 22.5 goes back to Marton [595]. The measurable selection theorem used in the construction of the coupling π can be found e.g. in [288]. As for Lemma 22.8, it is as old as information theory, since Shannon [746] used it to motivate the introduction of entropy in this context. After Marton's work, this tensorization technique has been adapted to various situations, such as weakly dependent Markov chains; see [124, 140, 305, 596, 621, 703, 730, 841]. Relations with the so-called Dobrushin(–Shlosman) mixing condition appear in some of these works, for instance [596, 841]; in fact, the original version of the mixing condition [308, 311] was formulated in terms of optimal transport!

It is also Marton [595] who introduced the simple argument by which T_p inequalities lead to concentration inequalities (implication (i) $\Rightarrow$ (iii) in Theorem 22.10), and which has since then been reproduced in nearly all introductions to the subject. She used it mainly with the so-called *Hamming distance*: $d((x_i)_{1\leq i\leq n}, (y_i)_{1\leq i\leq n}) = \sum 1_{x_i \neq y_i}$.

There are alternative functional approaches to the concentration of measure: via logarithmic Sobolev inequalities [546, Chapter 5] [41,

Chapter 7]; and via Brunn–Minkowski, Prékopa–Leindler, or isoperimetric inequalities [546, Chapter 2]; one may also consult [547] for a short review on these various approaches. For instance, (19.27) immediately implies

$$\nu[A^r] \geq 1 - \frac{e^{-\frac{Kr^2}{4}}}{\nu[A]}.$$

This kind of inequality goes back to Gromov and V. Milman [440], who also studied concentration from Poincaré inequalities (as Borovkov and Utev [146] did independently at about the same time). The relations between Poincaré inequalities and concentration are reviewed by E. Milman [633], who also shows that a weak concentration estimate, together with the $\mathrm{CD}(0,\infty)$ criterion, implies Poincaré (and even the stronger Cheeger isoperimetric inequality). The tight links between all these functional inequalities show that these various strategies are in some sense related.

First introduced by Herbst, the Laplace transform became an important tool in some of these developments, especially in the hands of Ledoux and coworkers — as can be seen in many places in [546].

Theorem 22.10 has been obtained by patching together results due to Bobkov and Götze [128], Djellout, Guillin and Wu [305], and Bolley and myself [140], together with a few arguments from folklore. There is an alternative proof of (ix) $\Rightarrow$ (i) based on the following fact, well-known to specialists: *Let X be a centered real random variable such that $\mathbb{E}\, e^{X^2} < \infty$, then the Laplace transform of X is bounded above by a Gaussian Laplace transform.*

The Csiszár–Kullback–Pinsker (CKP) inequality (22.25) was found independently by Pinsker [682], Kullback [534] and Csiszár [253]. The popular short proof by Pinsker is based on an obscure inequality on the real line [41, ineq. (8.4)]. The approach used in Remark 22.12 is taken from [140]; it takes inspiration from an argument which I heard in a graduate course by Talagrand. In Exercise 29.22 I shall propose yet another line of reasoning which is closer to Csiszár's original argument, and which was the basis to the generalization in the context of quantum physics [473]. (This reference was pointed out to me by Seiringer.)

Weighted CKP inequalities such as (22.16) were introduced in my paper with Bolley [140]; then Gozlan and Léonard [430] studied similar inequalities from the point of view of the theory of large deviations. More information can be found in Gozlan's PhD thesis [427]. Different kinds of generalizations of the CKP inequality appear in [140, 669, 797, 842], together with applications.

Talagrand [774] proved Theorem 22.14 when ν is the Gaussian measure in $\mathbb{R}^n$, using a change of variables in the one-dimensional case, and then a tensorization argument (Corollary 22.6). This strategy was developed by Blower [122] who proved Theorem 22.14 when $M = \mathbb{R}^n$, $\nu(dx) = e^{-V(x)}\,dx$, $\nabla^2 V \geq K > 0$; see also Cordero-Erausquin [242]. Generalizations to nonquadratic costs appear in [245]. Also Kolesnikov [526] made systematic use of this approach in infinite-dimensional situations and for various classes of inequalities. More recently, the same strategy was used by Barthe [73] to recover the modified transport inequalities for the exponential measure on the half-line (a particular case of Theorem 22.25).

Otto and I [671] found an alternative approach to Theorem 22.14, via the HWI inequality (which at the time of [671] had been established only in $\mathbb{R}^n$). The proof which I have used in this chapter is the same as the proof in [671], modulo the extension of the HWI inequality to general Riemannian manifolds.

There are several other schemes of proof for Theorem 22.14. One consists in combining Theorems 21.2 and 22.17(i). When $M = \mathbb{R}^n$, there is an argument based on Caffarelli's log concave perturbation theorem [188] (exercise). Yet another proof has been given by Bobkov and Ledoux [131], based on the Brunn–Minkowski inequality, or its functional counterpart, the Prékopa–Leindler inequality (in this work there are interesting extensions to cases where the convexity assumptions are not the standard ones and the cost might not be quadratic). Bobkov and Ledoux only worked in $\mathbb{R}^n$, but it is quite possible that their strategy can be extended to genuinely Riemannian situations, by means of the "Riemannian" Prékopa–Leindler inequality stated in Theorem 19.16.

Theorem 22.17 (log Sobolev implies T_2 implies Poincaré) was first proven by Otto and myself [671]; the Otto calculus had first been used to get an idea of the result. Our proof relied on a heat semigroup argument, which will be explained later in Chapter 25. The "dual" strategy which I have used in this chapter, based on the Hamilton–Jacobi semigroup, is due to Bobkov, Gentil and Ledoux [127]. In [671] it was assumed that the Ricci curvature of the manifold M is bounded below, and this assumption was removed in [127]. This is because the proof in [671] used a heat semigroup, which has infinite speed of propagation and is influenced by the asymptotic behavior of the manifold, while the argument in [127] was based on the Hopf–Lax semigroup, for which there is only finite speed of propagation (if the initial datum is bounded). The methods in [127] were pushed in [412] to treat nonquadratic cost functions. Infimum convolutions in the style of the Hopf–Lax formula also play a role in [131, 132], in relation with logarithmic or plain Sobolev inequalities. Much later, Gozlan [429] found a third proof of Theorem 22.17, based on Theorem 22.22 (which is itself based on Sanov's theorem).

Various generalizations of the proof in [671] were studied by Cattiaux and Guillin [219]; see the bibliographical notes of Chapter 25 for more details.

The proof of [127] was adapted by Lott and myself [579] to compact length spaces $(\mathcal{X}, d)$ equipped with a reference measure ν that is locally doubling and satisfies a local Poincaré inequality; see Theorem 30.28 in the last chapter of these notes. In fact the proof of Theorem 22.17, as I have written it, is essentially a copy–paste from [579]. (A detailed proof of Proposition 22.16 is also provided there.) Then Gozlan [429] relaxed these assumptions even further.

If M is a compact Riemannian manifold, then the normalized volume measure on M satisfies a Talagrand (T_2) inequality: This results from the existence of a logarithmic Sobolev inequality [710] and Theorem 22.17. Moreover, by [671, Theorem 4], the diameter of M can be bounded in terms of the constant in the Talagrand inequality, the dimension of M and a lower bound on the Ricci curvature, just as in (21.21) (where now λ stands for the constant in the Talagrand inequality). (The same bound certainly holds true even if M is not a priori assumed to be compact, but this was not explicitly checked in [671].) There is an analogous result where Talagrand inequality is replaced by logarithmic Sobolev inequality [544, 727].

The remarkable result according to which dimension free Gaussian concentration bounds are *equivalent* to T_2 inequality (Theorem 22.22) is due to Gozlan [429]; the proof of (iii) $\Rightarrow$ (i) in Theorem 22.22 is extracted from this paper. Gozlan's argument relies on Sanov's theorem in large deviation theory [296, Theorem 6.2.10]; this classical result states that the rate of deviation of the empirical measure of independent, identically distributed samples is the (Kullback) information with respect to their common law; in other words, under adequate conditions,

$$-\frac{1}{N}\log \nu^{\otimes N}[\widehat{\mu}_x^N \in A] \simeq \inf \Big\{ H_\nu(\mu);\ \mu \in A \Big\}.$$

A simple proof of the particular estimate (22.47) is provided in the Appendix of [429]. The observation that Talagrand inequalities and

Sanov's theorem match well goes back at least to [139]; but Gozlan's theorem uses this ingredient with a quite new twist.

Varadarajan's theorem (law of large numbers for empirical measures) was already used in the proof of Theorem 5.10; it is proven for instance in [318, Theorem 11.4.1]. It is anyway implied by Sanov's theorem.

Theorem 22.10 shows that T_1 is quite well understood, but many questions remain open about the more interesting T_2 inequality. One of the most natural is the following: given a probability measure ν satisfying T_2, and a bounded function v, does $e^{-v}\nu/(\int e^{-v}\,d\nu)$ also satisfies a T_2 inequality? For the moment, the only partial result in this direction is (22.29). This formula was first established by Blower [122] and later recovered with simpler methods by Bolley and myself [140].

If one considers probability measures of the form $e^{-V(x)}\,dx$ with $V(x)$ behaving like $|x|^\beta$ for large $|x|$, then the critical exponents for concentration-type inequalities are the same as we already discussed for isoperimetric-type inequalities: If $\beta \geq 2$ there is the T_2 inequality, while for $\beta = 1$ there is the transport inequality with linear-quadratic cost function. What happens for intermediate values of β has been investigated by Gentil, Guillin and Miclo in [410], by means of modified logarithmic Sobolev inequalities in the style of Bobkov and Ledoux [130]. Exponents $\beta > 2$ have also been considered in [131].

It was shown in [671] that (Talagrand) $\Rightarrow$ (log Sobolev) in $\mathbb{R}^n$, if the reference measure ν is log concave (with respect to the Lebesgue measure). It was natural to conjecture that the same argument would work under an assumption of nonnegative curvature (say $\mathrm{CD}(0,\infty)$); Theorem 22.21 shows that such is indeed the case.

It is only recently that Cattiaux and Guillin [219] produced a counterexample on the real line, showing that the T_2 inequality does not necessarily imply a log Sobolev inequality. Their counterexample takes the form $d\nu = e^{-V}\,dx$, where V oscillates rather wildly at infinity, in particular V'' is not bounded below. More precisely, their potential looks like $V(x) = |x|^3 + 3x^2\sin^2 x + |x|^\beta$ as $x \to +\infty$; then ν satisfies a logarithmic Sobolev inequality only if $\beta \geq 5/2$, but a T_2 inequality as soon as $\beta > 2$. Counterexamples with V'' bounded below have still not yet been found.

Even more recently, Gozlan [425, 426, 428] exhibited a characterization of T_2 and other transport inequalities on $\mathbb{R}$, for certain classes of measures. He even identified situations where it is useful to deduce logarithmic Sobolev inequalities from T_2 inequalities. Gentil, Guillin and Miclo [411] considered transport inequalities on $\mathbb{R}$ for log-concave probability measures. This is a rather active area of research. For instance, consider a transport inequality of the form $C(\mu,\nu) \le H_\nu(\mu)$, where the cost function is $c(x,y) = \theta(a\,|x-y|)$, $a > 0$, and $\theta : \mathbb{R}_+ \to \mathbb{R}_+$ is convex with $\theta(r) = r^2$ for $0 \le r \le 1$. If $\nu(dx) = e^{-V}\,dx$ with $V'' = o(V'^2)$ at infinity and $\limsup_{x\to+\infty} \theta'(\lambda\,x)/V'(x) < +\infty$ for some $\lambda > 0$, then there exists $a > 0$ such that the inequality holds true.

Theorem 22.14 admits an almost obvious generalization: if $\mathcal{F}$ is uniformly K-displacement convex and has a minimum at ν, then

$$\frac{K W_2(\mu,\nu)^2}{2} \le \mathcal{F}(\mu) - \mathcal{F}(\nu). \tag{22.84}$$

Such inequalities have been studied in [213, 671] and have proven useful in the study of certain partial differential equations: see e.g. [213] (various generalizations of the inequalities considered there appear in [245, 412]). In Section 5 of [213], (22.84) is combined with the HWI inequality and the convergence of the functional $\mathcal{F}$, to deduce convergence in total variation. By the way, this is one of the rare applications in finite (fixed) dimension that I know where a T_2-type inequality has a real advantage on a T_1-type inequality.

Optimal transport inequalities in *infinite dimension* have started to receive a lot of attention recently, for instance on the Wiener space. A major technical difficulty is that the natural distance in this problem, the so-called Cameron–Martin distance, takes the value $+\infty$ "most of the time". (So it is not a real distance, but rather a pseudo-distance.) Gentil [408, Section 5.8] established the T_2 inequality for the Wiener measure by using the logarithmic Sobolev inequality on the Wiener space, and adapting the proof of Theorem 22.17(i) to that setting. Feyel and Üstünel [359] on the one hand, and Djellout, Guillin and Wu [305, Section 6] on the other, suggested a more direct approach based on Girsanov's formula. Interestingly enough, the T_2 inequality on the Wiener space implies the T_2 inequality on the Gaussian space, just by "projection" under the map $(x_t)_{0\le t\le 1} \to x_1$; this gives another proof of Talagrand's original inequality (with the optimal constant) for the Gaussian measure. There are closely related works by Wu and Zhang [843, 844]. Also F.-Y. Wang [830, 831] studied another kind of Talagrand inequality on the path space over an arbitrary Riemannian manifold.

In his recent PhD thesis, Shao [748] studied T_2 inequalities on the path space and loop space constructed over a compact Lie group G. (The path space is equipped with the Wiener measure over G.) Together with Fang [341], he adapted the strategy based on the Girsanov formula, to get a T_2 inequality on the path space, and also on the path space over the loop space; then by reduction he gets a T_2 inequality on the loop space (equipped with a measure associated with the Brownian motion on loop space). This approach however only seems to give results when the loop space is equipped with the topology of uniform convergence, not with the more natural Cameron–Martin distance. I refer to [341] for more explanations.

Fang and Shao [340] also extended Theorem 22.17 (Logarithmic Sobolev implies Talagrand inequality) to an infinite-dimensional setting, via the study of the Hamilton–Jacobi semigroup in infinite dimension. Thanks to known results about logarithmic Sobolev inequalities on loop spaces (studied by Driver, Lohrentz and others), they recover a T_2 inequality on the loop space, now for the Cameron–Martin distance. The technical core of these results is the analysis of the Hamilton–Jacobi for semi-distances in infinite dimension, performed in [749].

Very recently, Fang and Shao [339] used Talagrand inequalities to obtain results of unique existence of optimal transport in the Wiener space over a Lie group, when the target measure ν is the Wiener measure and the source measure μ satisfies $H_\nu(\mu) < +\infty$. In the standard (Gaussian) Wiener space, Feyel and Üstünel [359] have solved the same problem in more generality, but so far their results have not been extended outside the Gaussian setting.

A quite different direction of generalization is in the field of free probability, where analogs of the Talagrand inequality have been established by various authors [118, 474, 475].

The equivalence between Poincaré inequalities and modified transport inequalities, as expressed in Theorem 22.25, has a long history. Talagrand [770] had identified concentration properties satisfied by the exponential measure, or a product of exponential measures. He established the following precise version of (22.67):

$$\nu^{\otimes N}\big[A + 6\sqrt{r}B_1^{d_2} + 9rB_1^{d_1}\big] \geq 1 - \frac{e^{-r}}{\nu^{\otimes N}[A]}.$$

A proof can be found in [546, Theorem 4.16]. It is also Talagrand who noticed that concentration inequalities for the product exponential measure are in some sense stronger than concentration inequalities for the Gaussian measure (Remark 22.34 and Example 22.36, which I copied from [546]). Then Maurey [607] found a simple approach to concentration inequalities for the product exponential measure. Later, Talagrand [774] made the connection with transport inequalities for the quadratic-linear cost. Bobkov and Ledoux [130] introduced modified logarithmic Sobolev inequalities, and showed their equivalence with Poincaré inequalities. (The proof of (i) $\Rightarrow$ (ii) is copied almost verbatim from [130].) Very recently, alternative methods based on Lyapunov functionals have been developed to handle these inequalities in an elementary way [447].

Bobkov and Ledoux also showed how to recover concentration inequalities directly from their modified logarithmic Sobolev inequalities, showing in some sense that the concentration properties of the exponential measure were shared by all measures satisfying a Poincaré inequality. Finally, Bobkov, Gentil and Ledoux [127] understood how to deduce quadratic-linear transport inequalities from modified logarithmic Sobolev inequalities, thanks to the Hamilton–Jacobi semigroup. The proof of the direct implication in Theorem 22.28 is just an expanded version of the arguments suggested in [127]; the proof of the converse follows Gozlan [429].

In the particular case when $\nu(dx) = e^{-|x|}\,dx$ on $\mathbb{R}_+$, there are simpler proofs of Theorem 22.25, also with improved constants; see for instance the above-mentioned works by Talagrand or Ledoux, or a recent remark by Barthe [73]. One may also note that deviation estimates with bounds like $\exp(-c \min(t,t^2))$ for sums of independent real variables go back to the elementary Bernstein inequalities (see [96] or [775, Theorem A.2.1]).

The treatment of dimension-dependent Talagrand-type inequalities in the last section is inspired from a joint work with Lott [578]. That topic had been addressed before, with different tools, by Gentil [409]; it would be interesting to compare precisely his results with the ones in this chapter. I shall also mention another dimension-dependent inequality in Remark 24.14.

Theorem 22.46 (behavior of solutions of Hamilton–Jacobi equations) has been obtained by generalizing the proof of Proposition 22.16 as it appears in [579]. When $L'(\infty) = +\infty$, the proof is basically the same, while there are a few additional technical difficulties if $L'(\infty) < +\infty$. In fact Proposition 22.16 was established in [579] in a more general context, namely when M is a finite-dimensional Alexandrov spaces with (sectional) curvature locally bounded below. The same extension probably holds for Theorem 22.46, although part (vii) would require a bit more thinking because the inequalities defining Alexandrov spaces are in terms of the squared distance, not the distance.

The study of Hamilton–Jacobi equations is an old topic (see the reference texts [68, 199, 327, 558] and the many references therein); so I do not exclude that Theorem 22.46 might be found somewhere in the literature, maybe in disguised form. Bobkov and Ledoux recently established closely related results [132, Lemma A] for the quadratic Hamilton–Jacobi equation in a finite-dimensional Banach space.

I shall conclude by listing some further applications of T_p inequalities which I did not previously mention.

Relations of T_p inequalities with the so-called slicing problem are discussed in [623].

These inequalities are also useful to study the propagation of chaos or the mean behavior of particle systems [221, 590].

As already noticed before, the functional H_ν appears in Sanov's theorem as the rate function for the deviations of the empirical mean of independent samples; this explains why T_p inequalities are handy tools for a quantitative study of concentration of the empirical measure associated with certain particle systems [139]. The links with large deviation theory were further explored in [355, 429, 430, 448]. If one is interested in the concentration of *time averages*, then one should replace the Kullback information H_ν by the Fisher information I_ν, as was understood by Donsker and Varadhan [313].[1] As a matter of fact, Guillin, Léonard, Wu and Yao [448] have established that the functional inequality

$$\alpha\big(W_1(\mu,\nu)\big) \le I_\nu(\mu),$$

where α is an increasing function, $\alpha(0) = 0$, is equivalent to the concentration inequality

$$\mathbb{P}\left[\frac{1}{t}\int_0^t \varphi(X_s)\,ds > \int \varphi\,d\nu + \varepsilon\right] \le \left\|\frac{d\mu}{d\nu}\right\|_{L^2(\nu)} e^{-t\,\alpha\left(\frac{\varepsilon}{\|\varphi\|_{\mathrm{Lip}}}\right)},$$

[1] A related remark, which I learnt from Ben Arous, is that the logarithmic Sobolev inequality compares the rate functions of two large deviation principles, one for the empirical measure of independent samples and the other one for the empirical time-averages.

where $(X_s)_{s\geq 0}$ is the symmetric diffusion process with invariant measure ν, $\mu = \text{law}\,(X_0)$, and φ is an arbitrary Lipschitz function. (Compare with Theorem 22.10(v).)

23

Gradient flows I

Take a Riemannian manifold M and a function $\Phi : M \to \mathbb{R}$, which for the sake of this exposition will be assumed to be continuously differentiable. The gradient of Φ, denoted by $\nabla\Phi$, is the vector field defined by the equation

$$d_x\Phi \cdot v = \langle \nabla_x\Phi, v\rangle_x,$$

where v is an arbitrary vector in the tangent space T_xM, $d_x\Phi$ stands for the differential of Φ at x, and $\langle \cdot\,, \cdot\rangle_x$ is the scalar product on T_xM. In other words, if $(\gamma_t)_{-\varepsilon<t<\varepsilon}$ is a smooth path in M, with $\gamma_0 = x$, then

$$\left[\frac{d}{dt}\bigg|_{t=0} \gamma_t = v\right] \Longrightarrow \qquad \frac{d}{dt}\bigg|_{t=0} \Phi(\gamma_t) = \langle \nabla_x\Phi, v\rangle_x.$$

If $|v|$ is given, then in order to make the latter derivative as large as possible, the best choice is to take v colinear to $\nabla_x\Phi$. In that sense $\nabla_x\Phi$ indicates *the direction in which Φ increases most rapidly.*

The gradient flow associated to Φ is the flow induced by the differential equation

$$\frac{dX}{dt} = -\operatorname{grad}_X\Phi.$$

One may think of it heuristically as a flow which *makes Φ decrease as fast as possible.* Stated in this way, this idea is of course grossly false: for instance, $\dot{X} = -\lambda \operatorname{grad}_X\Phi$, $\lambda > 1$, will make Φ decrease even faster; but in the last section of this chapter I shall make the statement more precise and hopefully more convincing.

An important consequence of the definition of gradient flow is the following neat formula for the time-derivative of the energy:

C. Villani, *Optimal Transport*, Grundlehren der mathematischen Wissenschaften 338,

$$\frac{d}{dt}\Phi(X(t)) = -\left|\mathrm{grad}_{X(t)}\Phi\right|^2.$$

Gradient flows (as Hamiltonian flows) are everywhere in physics and mathematics. In mechanics, they often describe the behavior of damped Hamiltonian systems, in an asymptotic regime in which dissipative effects play such an important role, that the effects of forcing and dissipation compensate each other. The basic example one should think of is

$$\ddot{X} = -\lambda\,\mathrm{grad}_X\Phi - \lambda\,\dot{X}$$

(acceleration = forcing − friction), in the limit $\lambda \to +\infty$ (which means strong friction).

Gradient flows in Wasserstein space

Around the end of the nineties, Jordan, Kinderlehrer and Otto made the important discovery that *a number of well-known partial differential equations can be reformulated as gradient flows in the Wasserstein space.* The most emblematic example is that of the heat equation,

$$\partial_t \rho = \Delta\rho,$$

say in Euclidean space for simplicity. It is classical that this equation can be seen as a gradient flow, for instance for the quadratic functional $\Phi(\rho) = \int |\nabla\rho|^2\,dx$ in $L^2(\mathbb{R}^n)$. But the Jordan–Kinderlehrer–Otto formulation describes the heat equation as a gradient flow *in the space of probability measures,* and with a natural "information-theoretical" content. In this new approach, the functional Φ is the negative of the entropy: $\Phi(\rho) = \int \rho \log\rho\,dx$.

To better understand this point of view, Otto developed a set of calculation rules which I dubbed "Otto calculus" in Chapter 15. We have already seen several applications of this calculus, at least for heuristic purposes.

In this chapter, I shall describe in which *rigorous* sense one can say that certain equations are gradient flows in the Wasserstein space. Before that, it will be necessary to get a good understanding of gradient flows in abstract metric spaces, a subject which is important in itself.

Reformulations of gradient flows

There are several ways to reformulate gradient flows in a weak sense, so as to obtain definitions that are general (for nonsmooth energies, or nonsmooth spaces), and stable (under some limit process). They usually require a convexity-type assumption on the energy Φ. Here I shall present some of these reformulations and explain why they are equivalent to the classical formulation when used in a smooth setting. Recall the definitions of $|\nabla^-\Phi|$ and $\nabla^-\Phi$ (or $\partial\Phi$):

$$|\nabla^-\Phi(x)| = \limsup_{y\to x} \frac{[\Phi(y)-\Phi(x)]_-}{d(x,y)};$$

$$\nabla^-\Phi(x)=\Big\{v \in T_xM;\ \forall w\in T_xM,\ \Phi\big(\exp_x(\varepsilon w)\big) \geq \Phi(x)+\varepsilon\langle v,w\rangle+o(\varepsilon)\Big\}.$$

(It is not a priori assumed that $o(\varepsilon)$ is uniform in w.) Proposition 16.2 will also be useful.

Proposition 23.1 (Reformulations of gradient flows). *Let M be a Riemannian manifold. Let $\Lambda = \Lambda(x,v)$ be a quadratic form on TM, satisfying (16.7), and let Φ be a differentiable function $M \to \mathbb{R}$, Λ-convex in the sense of Proposition 16.2. Let $X : (t_1,t_2) \to M$ be a continuous path, and let $t \in (t_1,t_2)$ be a time where X is differentiable. Then, the following statements are equivalent:*

(i) $\dot X(t) = -\operatorname{grad}_{X(t)}\Phi$;

(ii) $\dfrac{|\dot X(t)|^2 + |\nabla^-\Phi(X(t))|^2}{2} = -\dfrac{d}{dt}\Phi(X(t))$;

(iii) $-\dot X(t) \in \nabla^-\Phi(X(t))$;

(iv) For any $y\in M$, and any geodesic $(\gamma_s)_{0\leq s\leq 1}$ joining $\gamma_0 = X(t)$ to $\gamma_1 = y$,

$$\frac{d^+}{dt}\left(\frac{d(X(t),y)^2}{2}\right) \leq \frac{d^+}{ds}\bigg|_{s=0} \Phi(\gamma_s) := \limsup_{s\downarrow 0} \frac{\Phi(\gamma_s)-\Phi(\gamma_0)}{s};$$

(v) For any $y\in M$, and any geodesic $(\gamma_s)_{0\leq s\leq 1}$ joining $\gamma_0 = X(t)$ to $\gamma_1 = y$,

$$\frac{d^+}{dt}\left(\frac{d(X(t),y)^2}{2}\right) \leq \Phi(y) - \Phi(X(t)) - \int_0^1 \Lambda(\gamma_s,\dot\gamma_s)\,(1-s)\,ds;$$

(vi) For any $y \in M$, and any geodesic $(\gamma_s)_{0\le s\le 1}$ joining $\gamma_0 = X(t)$ to $\gamma_1 = y$,

$$\frac{d^+}{dt}\left(\frac{d(X(t),y)^2}{2}\right) \le \Phi(y) - \Phi(X(t)) - \lambda[\gamma]\,\frac{d(X(t),y)^2}{2},$$

where $\lambda[\gamma]$ is defined by (16.7).

Remark 23.2. As the proof will show, the equivalence between (iii), (iv), (v) and (vi) does not require the differentiability of Φ; it is sufficient that Φ be valued in $\mathbb{R}\cup\{+\infty\}$ and $\Phi(X(t)) < +\infty$.

Remark 23.3. The most well-known case is when $\Lambda = 0$ (Φ is convex), and then (v) becomes just

$$\frac{d^+}{dt}\left(\frac{d(X(t),y)^2}{2}\right) \le \Phi(y) - \Phi(X(t)).$$

Remark 23.4. Statements (i) to (iii) do not explicitly depend on Λ, so here the assumption of Λ-convexity is not essential. But as soon as one wants to generalize Proposition 23.1 by dropping some smoothness assumptions, it might be important to know that Φ is Λ-convex for some Λ. Note that in formulations (iv) to (vi), one can always replace Λ by $\Lambda' \le \Lambda$, and the equivalence still holds true, independently of the choice of Λ'! In particular, if $\Lambda(x,v) \ge \lambda\,|v|^2$ for some $\lambda \in \mathbb{R}$, i.e. when Φ is λ-convex in the sense of Definition 16.4, one may replace $\Lambda(x,v)$ by $\lambda\,|v|^2$.

Remark 23.5. If one wants to use Proposition 23.1 to characterize a curve $(X(t))$ as a gradient flow, the natural regularity assumption is that X be an absolutely continuous function of t, in the sense of (7.5). This will imply the existence of the derivative $\dot{X}(t)$ for almost all t, and in addition this will guarantee that the values of X are uniquely determined by $X(0)$ and the values of $\dot{X}$.

Before going on with the proof of Proposition 23.1, I shall briefly explain its interest. Property (ii) involves speeds (norms of velocities) rather than velocities; this is interesting also in a nonsmooth setting, where the speed might be well-defined even though the velocity is not. Property (iii) has the advantage of being formulated in terms of subgradients (or subdifferentials), which are often well-defined even if Φ is not differentiable (for instance if Φ is semiconvex), and quite stable.

Finally, Properties (iv) to (vi) are quite handy to study gradient flows in abstract metric spaces. As a matter of fact, in the sequel I shall use (iv) to define gradient flows in the Wasserstein space.

Proof of Proposition 23.1. By the chain-rule, and Cauchy–Schwarz and Young's inequalities,

$$-\frac{d}{dt}\Phi(X(t)) = \Big\langle -\nabla\Phi(X(t)),\, \dot{X}(t)\Big\rangle \\ \leq \big|\nabla\Phi(X(t))\big|\,\big|\dot{X}(t)\big| \leq \frac{|\nabla\Phi(X(t))|^2 + |\dot{X}(t)|^2}{2},$$

with equality if and only if $\nabla\Phi(X(t))$ and $\dot{X}(t)$ have the same norm and opposite directions. So (i) is equivalent to (ii).

Next, if Φ is differentiable then

$$|\nabla^-\Phi(x)| = |\nabla\Phi(x)|, \qquad \nabla^-\Phi(x) = \{\nabla\Phi(x)\};$$

in particular (i) and (iii) are equivalent.

Now, let us check the equivalence of statements (iii) to (vi). Let y be given, and let γ be a geodesic path joining $\gamma(0) = X(t)$ to $\gamma(1) = y$. By the formula of first variation (7.29),

$$\frac{d^+}{dt}\left(\frac{d(X(t),y)^2}{2}\right) \leq -\left\langle \dot{\gamma}(0), \dot{X}(t)\right\rangle_{X(t)} \tag{23.1}$$

(the distance decreases if $\dot{X}$ is in the direction of $\dot{\gamma}(0)$). On the other hand, $\gamma_s = \exp(sw)$, where $w = \dot{\gamma}(0)$, so if (iii) holds true then as $s \to 0$,

$$\frac{\Phi(\gamma_s) - \Phi(\gamma_0)}{s} \geq \langle -\dot{X}(t), \dot{\gamma}_0\rangle + o(1).$$

Consequently,

$$\frac{d^+}{dt}\frac{d(X(t),y)^2}{2} \leq \liminf_{s\downarrow 0} \frac{\Phi(\gamma_s) - \Phi(\gamma_0)}{s},$$

which obviously implies (iv).

Next, since Φ is Λ-convex,

$$\frac{\Phi(\gamma_s) - \Phi(\gamma_0)}{s} \leq \Phi(\gamma_1) - \Phi(\gamma_0) - \int_0^1 \Lambda(\gamma_\tau, \dot{\gamma}_\tau)\,\frac{G(\tau,s)}{s}\,d\tau,$$

where G is the Green function defined in (16.5). We can pass to the limit as $s \to 0$ since $G(\tau, s)/s \le 1 - \tau$ and $\Lambda(\gamma_\tau, \dot\gamma_\tau)$ is bounded below for $\tau \in [0,1]$. So

$$\limsup_{s \downarrow 0} \frac{\Phi(\gamma_s) - \Phi(\gamma_0)}{s} \le \Phi(\gamma_1) - \Phi(\gamma_0) - \int_0^1 \Lambda(\gamma_\tau, \dot\gamma_\tau)\,(1-\tau)\,d\tau,$$

and it becomes clear that (iv) $\Rightarrow$ (v).

The implication (v) $\Rightarrow$ (vi) is trivial since $\int_0^1 \Lambda(\gamma_s, \dot\gamma_s)\,(1-s)\,ds \ge \lambda[\gamma]\,d(\gamma_0,\gamma_1)^2/2$. So it only remains to check (vi) $\Rightarrow$ (iii).

Again, let t be given, $w \in T_{X(t)}M$, $y = \exp_{X(t)}(\varepsilon w)$. If ε is small enough there is a unique geodesic γ joining $\gamma(0) = X(t)$ to $\gamma(1) = y$, namely $\gamma(s) = \exp_{X(t)}(s\varepsilon w)$. Then $|\dot\gamma| = \varepsilon |w|$, and

$$\frac{d}{dt}\left(\frac{d(X(t),y)^2}{2}\right) = -\langle \dot\gamma(0), \dot X(t)\rangle = -\langle \varepsilon w, \dot X(t)\rangle.$$

So (v) implies

$$\begin{aligned} -\langle \varepsilon w, \dot X(t)\rangle &= \frac{d^+}{dt}\left(\frac{d(X(t),y)^2}{2}\right) \\ &\le \Phi(\exp_{X(t)} \varepsilon w) - \Phi(X(t)) + \lambda[\gamma]\,\frac{d(X(t),y)^2}{2} \\ &= \Phi(\exp_{X(t)} \varepsilon w) - \Phi(X(t)) - \lambda[\gamma]\,\frac{\varepsilon^2\,|w|^2}{2}. \end{aligned}$$

As a consequence,

$$\Phi(\exp_{X(t)} \varepsilon w) \ge \Phi(X(t)) + \varepsilon\langle w, -\dot X(t)\rangle + o(\varepsilon),$$

which precisely means that $-\dot X(t) \in \nabla^-\Phi(X(t))$, so (iii) is satisfied. This concludes the proof of Proposition 23.1. □

Gradient flows in metric spaces

Now we are ready to study an abstract definition of gradient flows. In the sequel I shall use the following notation:

Notation 23.6 (Locally absolutely continuous paths). Let $(\mathcal{X}, d)$ be a metric space and $T \in (0, +\infty]$. I shall denote by $\mathrm{AC}_{\mathrm{loc}}((0,T);\mathcal{X})$ the set of paths $\gamma : (0,T) \to \mathcal{X}$ such that there is a measurable function $\ell : (0,T) \to \mathbb{R}_+$ satisfying $d(\gamma_s, \gamma_t) \le \int_s^t \ell(\tau)\, d\tau$ for all $s < t$ in $(0,T)$, and

$$0 < t_1 < t_2 < T \Longrightarrow \qquad \int_{t_1}^{t_2} \ell(\tau)\, d\tau < +\infty.$$

Proposition 23.1 suggests the following definition for gradient flows in possibly nonsmooth length spaces. Recall that a geodesic space is a strictly intrinsic length space, that is, a length space in which any two points are connected by a (constant-speed, minimizing) geodesic.

Definition 23.7 (Gradient flows in a geodesic space). *Let $(\mathcal{X}, d)$ be a geodesic space and let $\Phi : \mathcal{X} \to \mathbb{R} \cup \{+\infty\}$. Let $T \in (0, +\infty]$ and let $X \in C([0,T);\mathcal{X}) \cap \mathrm{AC}_{\mathrm{loc}}((0,T);\mathcal{X})$. Then X is said to be a trajectory of the gradient flow associated with the energy Φ if (a) $\Phi(X(t)) < +\infty$ for all $t > 0$; and (b) for any $y \in \mathcal{X}$ and for almost any $t > 0$, there is a geodesic $(\gamma_s)_{0 \le s \le 1}$ joining $\gamma_0 = X(t)$ to $\gamma_1 = y$, such that*

$$\frac{d^+}{dt}\left(\frac{d(X(t), y)^2}{2}\right) \le \left.\frac{d^+}{ds}\right|_{s=0} \Phi(\gamma_s).$$

Remark 23.8. If Φ is λ-convex, then property (b) in the previous definition implies

$$\frac{d^+}{dt}\left(\frac{d(X(t), y)^2}{2}\right) \le \Phi(y) - \Phi(X(t)) - \lambda\, \frac{d(X(t), y)^2}{2}. \tag{23.2}$$

The proof is the same as for the implication (iv) ⇒ (v) in Proposition 23.1. I could have used Inequality (23.2) to define gradient flows in metric spaces, at least for λ-convex functions; but Definition 23.7 is more general.

Proposition 23.1 guarantees that the concept of abstract gradient flow coincides with the usual one when $\mathcal{X}$ is a Riemannian manifold equipped with its geodesic distance. In the sequel, I shall apply Definition 23.7 in the Wasserstein space $P_2(M)$, where M is a Riemannian manifold (sometimes with additional geometric assumptions). To avoid complications I shall in fact use Definition 23.7 in $P_2^{\mathrm{ac}}(M)$, that is, restricting to absolutely continuous probability measures. This might look a bit dangerous, because

$\mathcal{X} = P_2^{ac}(M)$ is not complete, but after all it is a geodesic space in its own right, as a geodesically convex subset of $P_2(M)$ (recall Theorem 8.7), and I shall not need completeness. Of course, this does not mean that it is not interesting to study gradient flows in the whole of $P_2(M)$.

To go on with this program, I have to

- compute the (upper) derivative of the distance function;
- compute the subdifferential of a given energy functional.

This will be the purpose of the next two sections, more precisely Theorems 23.9 and 23.14. Proofs will be long because I tried to achieve what looked like the "correct" level of generality; so the reader should focus on the results at first reading.

Derivative of the Wasserstein distance

The next theorem will at the same time study the differentiability of the Wasserstein distance, and give simple sufficient conditions for a path in the Wasserstein space to be absolutely continuous, in the sense of (7.5).

Theorem 23.9 (Derivative of the Wasserstein distance). *Let M be a Riemannian manifold, and $[t_1, t_2) \subset \mathbb{R}$. Let (μ_t) and $(\widehat{\mu}_t)$ be two weakly continuous curves $[t_1, t_2) \to P(M)$. Assume that $\mu_t, \widehat{\mu}_t \in P_2^{ac}(M)$ for all $t \in (t_1, t_2)$, and that μ_t, $\widehat{\mu}_t$ solve the continuity equations*

$$\frac{\partial \mu_t}{\partial t} + \nabla \cdot (\xi_t\, \mu_t) = 0, \qquad \frac{\partial \widehat{\mu}_t}{\partial t} + \nabla \cdot (\widehat{\xi}_t\, \widehat{\mu}_t) = 0, \tag{23.3}$$

where $\xi_t = \xi_t(x)$, $\widehat{\xi}_t = \widehat{\xi}_t(x)$ are locally Lipschitz vector fields and

$$\int_{t_1}^{t_2} \left(\int_M |\xi_t|^2\, d\mu_t + \int_M |\widehat{\xi}|^2\, d\widehat{\mu}_t \right) dt < +\infty.$$

Then $t \to \mu_t$ and $t \to \widehat{\mu}_t$ are Hölder-1/2 continuous and absolutely continuous. Moreover, for almost any $t \in (t_1, t_2)$,

$$\frac{d}{dt}\left(\frac{W_2(\mu_t, \widehat{\mu}_t)^2}{2} \right) = -\int_M \langle \widetilde{\nabla}\psi_t, \xi_t \rangle\, d\mu_t \; - \int_M \langle \widetilde{\nabla}\widehat{\psi}_t, \widehat{\xi}_t \rangle\, d\widehat{\mu}_t, \tag{23.4}$$

where ψ_t, $\widehat{\psi}_t$ are $(d^2/2)$-convex functions such that

$$\exp(\widetilde{\nabla}\psi_t)_\# \mu_t = \widehat{\mu}_t, \qquad \exp(\widetilde{\nabla}\widehat{\psi}_t)_\# \widehat{\mu}_t = \mu_t.$$

Remark 23.10. Recall that Theorem 10.41 gives a list of a few conditions under which the approximate gradient $\widetilde{\nabla}$ can be replaced by the usual gradient ∇ in the formulas above.

Exercise 23.11. Guess formula (23.4) by means of Otto calculus.

Remark 23.12. For the purpose of this chapter, the superdifferentiability of the Wasserstein distance would be enough. However, the plain differentiability will be useful later on.

Proof of Theorem 23.9. Without loss of generality I shall assume that $\tau = |t_1 - t_2|$ is finite.

A crucial ingredient in the proof is the flow associated with the velocity fields ξ and $\widehat{\xi}$. If t and s both belong to $[t_1, t_2)$, define the characteristics (or flow, or trajectory map) $T_{t\to s} : M \to M$ associated with ξ by the differential equation

$$\begin{cases} T_{t\to t}(x) = x; \\ \\ \dfrac{d}{ds} T_{t\to s}(x) = \xi_s(T_{t\to s}x). \end{cases} \tag{23.5}$$

(If $\xi_s(x)$ is the velocity field at time s and position x, then $T_{t\to s}(x)$ is the position at time s of a particle which was at time t at x and then followed the flow.)

By the formula of conservation of mass, for all $t, s \in [t_1, t_2)$,

$$\mu_s = (T_{t\to s})_\# \mu_t.$$

The idea is to compose the transport $T_{t\to s}$ with some optimal transport; this will not result in an optimal transport, but at least it will provide bounds on the Wasserstein distance.

In other words, $\mu_t = \operatorname{law}(\gamma_t)$, where γ_t is a random solution of $\dot{\gamma}_t = \xi_t(\gamma_t)$. Restricting the time-interval slightly if necessary, we may assume that these curves are defined on the closed interval $[t_1, t_2]$. Each of these curves is continuous and therefore bounded.

If γ solves $\dot{\gamma}_t = \xi_t(\gamma_t)$, then $d(\gamma_s, \gamma_t) \le \int |\xi_\tau(\gamma_\tau)|\, d\tau$. Since (γ_s, γ_t) is a coupling of (μ_s, μ_t), it follows from the very definition of W_2 that

$$\begin{aligned}
W_2(\mu_s,\mu_t) &\le \sqrt{\mathbb{E}\left(\int_s^t |\xi_\tau(\gamma_\tau)|\,d\tau\right)^2}\\
&\le \sqrt{\mathbb{E}\,(s-t)\left(\int_s^t |\xi_\tau(\gamma_\tau)|^2\,d\tau\right)}\\
&= \sqrt{|s-t|}\sqrt{\int_s^t \mathbb{E}\,|\xi_\tau(\gamma_\tau)|^2\,d\tau}\\
&= \sqrt{|s-t|}\sqrt{\int_s^t \left(\int |\xi_\tau|^2\,d\mu_\tau\right)d\tau}\\
&\le \frac{1}{2}\left(|s-t| + \int_s^t \left(\int |\xi_\tau|^2\,d\mu_\tau\right)d\tau\right).
\end{aligned}$$

This shows at the same time that $t \to \mu_t$ is Hölder-1/2, and that it is absolutely continuous: Indeed, $W_2(\mu_s,\mu_t) \le \int_s^t \ell(\tau)\,d\tau$ with

$$\ell(\tau) = \frac{1}{2}\left(1 + \int |\xi_\tau|^2\,d\mu_\tau\right).$$

The rest of the proof is decomposed into four steps. All intermediate results have their own interest.

Step 1: *$t \to W_2(\mu_t,\sigma)$ is superdifferentiable at each Lebesgue point of $t \to \int |\xi_t|^2\,d\mu_t$.*

In this step, the path $\widehat{\mu}_t$ will be constant and equal to some fixed $\sigma \in P_2^{\mathrm{ac}}(M)$. Let $t \in (t_1,t_2)$ be such that

$$\begin{cases}
\dfrac{1}{s}\displaystyle\int_0^s \left(\int |\xi_{t+\tau}|^2\,d\mu_\tau\right)d\tau \xrightarrow[s\to 0]{} \int |\xi_t|^2\,d\mu_t;\\[2ex]
\dfrac{1}{s}\displaystyle\int_0^s \left(\int |\xi_{t-\tau}|^2\,d\mu_\tau\right)d\tau \xrightarrow[s\to 0]{} \int |\xi_t|^2\,d\mu_t.
\end{cases} \tag{23.6}$$

Let ψ_t be a $d^2/2$-convex function such that $\exp(\widetilde{\nabla}\psi_t)_{\#}\mu_t = \sigma$. We shall see that

$$\frac{W_2(\mu_{t+s},\sigma)^2}{2} \le \frac{W_2(\mu_t,\sigma)^2}{2} - s\int \langle \widetilde{\nabla}\psi_t, \xi_t\rangle\,d\mu_t + o(s). \tag{23.7}$$

Remark 23.13. By Lebesgue's density theorem, (23.6) holds true for almost all t, so this step will already establish the superdifferentiability formula for almost all t, which is all we need in this chapter to identify gradient flows.

Back to the proof of (23.7): By symmetry, it is sufficient to establish it for $s > 0$. Let $T = \exp(\widetilde{\nabla}\widehat{\psi})$ be the optimal (Monge) transport $\sigma \to \mu_t$. Then

$$\frac{W_2(\mu_t,\sigma)^2}{2} = \frac{1}{2}\int d(x,T(x))^2\,d\sigma(x). \tag{23.8}$$

For any $s > 0$ small enough, $(T_{t\to t+s})_{\#}\mu_t = \mu_{t+s}$; so $T_{t\to t+s}\circ T$ is a transport $\sigma \to \mu_{t+s}$. By definition of the Wasserstein distance,

$$\frac{W_2(\mu_{t+s},\sigma)^2}{2} \le \frac{1}{2}\int d\Big(x, T_{t\to t+s}\circ T(x)\Big)^2\,d\sigma(x).$$

This, combined with (23.8), implies

$$\frac{1}{s}\left(\frac{W_2(\mu_{t+s},\sigma)^2}{2} - \frac{W_2(\mu_t,\sigma)^2}{2}\right) \le \int \left(\frac{d\big(x, T_{t\to t+s}\circ T(x)\big)^2 - d(x,T(x))^2}{2s}\right) d\sigma(x). \tag{23.9}$$

The maps $\exp(\widetilde{\nabla}\psi)$ and $\exp(\widetilde{\nabla}\widehat{\psi})$ are inverse to each other in the almost sure sense. So for $\sigma(dx)$-almost all x, there is a minimizing geodesic connecting $T(x)$ to x with initial velocity $\widetilde{\nabla}\psi(T(x))$; then by the formula of first variation,

$$\limsup_{s\downarrow 0}\left[\frac{d\big(x, T_{t\to t+s}\circ T(x)\big)^2 - d(x,T(x))^2}{2s}\right] \le -\Big\langle \xi_t(T(x)), \widetilde{\nabla}\psi(T(x))\Big\rangle.$$

So if we can pass to the lim sup as $s \to 0$ in (23.9), it will follow that

$$\begin{aligned}\frac{d^+}{dt}\left(\frac{W_2(\mu_t,\sigma)^2}{2}\right) &\le -\int_M \big\langle \xi_t(T(x)), \widetilde{\nabla}\psi(T(x))\big\rangle\,d\sigma(x)\\ &= -\int \big\langle \xi_t(y), \widetilde{\nabla}\psi(y)\big\rangle\,d(T_{\#}\sigma)(y)\\ &= -\int \Big\langle \xi_t(y), \widetilde{\nabla}\psi(y)\Big\rangle\,d\mu_t(y),\end{aligned}$$

and this will establish the desired (23.7).

So we should check that we can indeed pass to the lim sup in (23.9). Let $v(s,x)$ be the integrand in the right-hand side of (23.9): If $0 < s \leq 1$ then

$$\begin{aligned}
v(s,x) &= \frac{d\big(x, T_{t\to t+s}\circ T(x)\big)^2 - d(x,T(x))^2}{2s} \\
&\leq \left|\frac{d\big(x, T_{t\to t+s}\circ T(x)\big) - d(x,T(x))}{s}\right| \left(\frac{d\big(x, T_{t\to t+s}\circ T(x)\big) + d(x,T(x))}{2}\right) \\
&\leq \frac{d\big(T(x), T_{t\to t+s}(T(x))\big)}{s}\left(d(x,T(x)) + \frac{d\big(T(x), T_{t\to t+s}(T(x))\big)}{2}\right) \\
&\leq \frac{d\big(T(x), T_{t\to t+s}(T(x))\big)}{s}\left(d(x,T(x)) + \frac{d\big(T(x), T_{t\to t+s}(T(x))\big)}{2}\right) \\
&\leq \frac{d(x,T(x))^2}{2} + \frac{d\big(T(x), T_{t\to t+s}(x)\big)^2}{s^2} \qquad =: w(s,x).
\end{aligned}$$

Note that $x \to d(x,T(x))^2 \in L^1(\sigma)$, since

$$\int d(x,T(x))^2\, d\sigma(x) = W_2(\sigma,\mu_t)^2 \; < +\infty. \tag{23.10}$$

Moreover,

$$\begin{aligned}
\int \frac{d\big(T(x), T_{t\to t+s}(T(x))\big)^2}{s^2}\, d\sigma(x) &= \int \frac{d(y, T_{t\to t+s}(y))^2}{s^2}\, d\mu_t(y) \\
&\leq \int \frac{1}{s^2}\left(\int_0^s \big|\xi_{t+\tau}(T_{t\to t+\tau}(y))\big|\, d\tau\right)^2 d\mu_t(y) \\
&\leq \int \frac{1}{s}\int_0^s \big|\xi_{t+\tau}(T_{t\to t+\tau}(y))\big|^2\, d\tau\, d\mu_t(y) \\
&\leq \frac{1}{s}\int_0^s \int \big|\xi_{t+\tau}(T_{t\to t+\tau}(y))\big|^2\, d\mu_t(y)\, d\tau \\
&= \frac{1}{s}\int_0^s \left(\int |\xi_{t+\tau}(z)|^2\, d\mu_{t+\tau}(z)\right) d\tau.
\end{aligned}$$

By assumption, the latter quantity converges as $s \to 0$ to

$$\int |\xi_t(x)|^2\, d\mu_t(x) = \int |\xi_t(T(x))|^2\, d\sigma(x).$$

Since $d(T(x), T_{t\to t+s}(T(x)))^2/s^2 \longrightarrow |\xi_t(T(x))|^2$ as $s \to 0$, we can combine this with (23.10) to deduce that

$$\limsup_{s\downarrow 0} \int w(s,x)\,d\sigma(x) \le \int \left[\lim_{s\downarrow 0} w(s,x)\right]\,d\sigma(x).$$

By Fatou's lemma, in fact

$$\lim_{s\downarrow 0} \int w(s,x)\,d\sigma(x) = \int \left[\lim_{s\downarrow 0} w(s,x)\right]\,d\sigma(x).$$

So the domination $v(s,x) \le w(s,x)$ is sufficient to apply again Fatou's lemma, in the form

$$\limsup_{s\downarrow 0} \int v(s,x)\,d\sigma(x) \le \int \left[\limsup_{s\downarrow 0} v(s,x)\right]\,d\sigma(x),$$

and conclude the proof of this step.

Step 2: *If ξ grows at most linearly, then differentiability holds for all t.*

In this step I shall assume that there are $z \in M$ and $C > 0$ such that for all $x \in M$ and $t \in (t_1, t_2)$,

$$|\xi_t(x)| \le C\,\big(1 + d(z,x)\big).$$

Under this assumption I shall show, with the same notation as in Step 1, that $t \to W_2(\mu_t,\sigma)^2$ is differentiable on the whole of (t_1,t_2), and

$$\frac{d}{dt}\left(\frac{W_2(\mu_t,\sigma)^2}{2}\right) = -\int_M \langle \widetilde{\nabla}\psi_t, \xi_t\rangle\,d\mu_t.$$

I shall start with some estimates on the flow $T_{t\to s}$. From the assumptions,

$$\left|\frac{d}{ds}\,d(z,T_{t\to s}(x))\right| \le \big|\xi_s(T_{t\to s}(x))\big| \le C\big(1 + d(z,T_{t\to s}(x))\big),$$

which implies $1+d(z,T_{t\to s}(x)) \le e^{C\tau}(1+d(z,x))$. As a consequence, we have $(d/ds)\,d(x,T_{t\to s}(x)) \le C(1+e^{C\tau}(1+d(z,x)))$, so $d(x,T_{t\to s}(x)) \le |s-t|\,C(1+e^{C\tau}(1+d(z,x)))$. To summarize: There is a constant C such that for all $y \in M$, and $t,s \in (t_1,t_2)$.

$$\begin{cases} d(z,T_{t\to s}(x)) \le C(1+d(z,x)); \\ d(x,T_{t\to s}(x)) \le C|s-t|\,(1+d(z,x)). \end{cases} \tag{23.11}$$

In the sequel, the symbol C will stand for other constants that may depend only on τ and the Lipschitz constant of ξ.

Next, let us check that the second moment of μ_t is bounded by a constant independent of t. From the continuity equation (23.3), $(T_{t\to s})_\#\mu_t = \mu_s$. Combining this and (23.11), we deduce that for any fixed time $t_0 \in (t_1, t_2)$,

$$\int d(z,x)^2\,\mu_t(dx) = \int d(z, T_{t_0\to t}(x))^2\,\mu_{t_0}(dx) \\ \leq C\int (1+d(z,x))^2\,\mu_{t_0}(dx) \;<+\infty. \tag{23.12}$$

It is worth noticing that the assumptions also imply the Lipschitz continuity of $t \to \mu_t$. Indeed, by definition of W_2 and (23.11),

$$W_2(\mu_t,\mu_s)^2 \leq \int d(x, T_{t\to s}(x))^2\,\mu_t(dx) \\ \leq C\,|t-s|^2 \int (1+d(z,x))^2\,\mu_t(dx);$$

taking square roots and using (23.12), we deduce

$$W_2(\mu_t,\mu_s) \leq C\,|t-s|.$$

To prove the superdifferentiability of $W_2(\mu_t,\sigma)^2$, by Step 1 it is sufficient to check the continuity of $t \to \int |\xi_t|^2\,d\mu_t$. Let $t \in (t_1,t_2)$ be fixed, then

$$\int |\xi_{t+s}|^2\,d\mu_{t+s} = \int \left|\xi_{t+s}\circ T_{t\to t+s}(x)\right|^2 d\mu_t(x). \tag{23.13}$$

For any x, the integrand $|\xi_{t+s}\circ T_{t\to t+s}(x)|^2$ is a continuous function of s, and by (23.11) it is bounded above by $C\,(1+d(z,T_{t\to t+s}(x)))^2 \leq C'\,(1+d(z,x))^2$, where C, C' are positive constants. Since the latter function is integrable with respect to μ_t, we can apply the dominated convergence theorem to show that (23.13) converges to $\int |\xi_t|^2\,d\mu_t$ as $s\to 0$. This establishes the desired continuity, and the superdifferentiability of $W_2(\mu_t,\sigma)^2$.

The proof of subdifferentiability is a bit more tricky and the reader might wish to skip the rest of this step.

As before, to establish the subdifferentiability, it is sufficient to prove the right-subdifferentiability, more precisely

$$\liminf_{s\downarrow 0} \frac{W_2(\mu_{t+s},\sigma)^2 - W_2(\mu_t,\sigma)^2}{s} \geq -\int \langle \xi_t, \widetilde{\nabla}\psi\rangle\,d\mu_t.$$

For each $s > 0$, let $T^{(s)}$ be the optimal transport between σ and μ_{t+s}. As $s \downarrow 0$ we can extract a subsequence $s_k \to 0$, such that

$$\liminf_{s\downarrow 0} \left(\frac{W_2(\mu_{t+s},\sigma)^2 - W_2(\mu_t,\sigma)^2}{2s} \right) = \lim_{k\to\infty} \left(\frac{W_2(\mu_{t+s_k},\sigma)^2 - W_2(\mu_t,\sigma)^2}{2s_k} \right).$$

Changing signs and reasoning as in Step 1, we obtain

$$\limsup_{s\downarrow 0} \left(\frac{W_2(\mu_t,\sigma)^2 - W_2(\mu_{t+s},\sigma)^2}{2s} \right) \le \limsup_{k\to\infty} \int v_k(x)\,\sigma(dx), \tag{23.14}$$

where

$$v_k(x) = \frac{d\Big(x, T_{t+s_k\to t} \circ T^{(t+s_k)}(x)\Big)^2 - d\big(x, T^{(t+s_k)}(x)\big)^2}{2s_k}.$$

Since $T^{(t)}$ is the unique optimal transport between σ and μ_t, and since $s \to \mu_{t+s}$ is continuous with respect to the weak topology, we know from Corollary 5.23 that $T^{(t+s_k)}$ converges to $T^{(t)}$ in probability, with respect to the measure σ. Extracting a further subsequence if necessary, we may assume that $T^{(t+s_k)}$ converges $\sigma(dx)$-almost surely to $T^{(t)}(x)$.

Next, the square distance d^2 is locally superdifferentiable, so

$$\begin{aligned}\frac{d\big(x, T_{t+s_k\to t}(y)\big)^2}{2} &\le \frac{d(x,y)^2}{2} + s_k \left\langle \xi_t(y), \dot\gamma(1)\right\rangle_y + o\big(d(y, T_{t+s_k\to t}(y)\big) \\ &\le \frac{d(x,y)^2}{2} + s_k \left\langle \xi_t(y), \dot\gamma(1)\right\rangle_y + o(s_k),\end{aligned}$$

where γ is a geodesic joining x to y, and the $o(s_k)$ is uniform in a neighborhood of y. So if $y_k \to y$, then

$$\limsup_{k\to\infty} \frac{d\big(x, T_{t+s_k\to t}(y_k)\big)^2 - d(x,y_k)^2}{2s_k} \le \langle \xi_t(y), \dot\gamma(1)\rangle.$$

Applying this to $y_k = T^{(t+s_k)}(x) \to T^{(t)}(x)$ (σ-almost surely), we deduce that

$$\limsup_{k\to\infty} v_k(x) \le v(x) := \big\langle \xi_t(T^{(t)}(x)), \dot\gamma_x(1)\big\rangle_{T^{(t)}(x)}, \tag{23.15}$$

where γ_x is the geodesic joining x to $T^{(t)}(x)$: in particular, $\sigma(dx)$-almost surely, $\dot{\gamma}_x(1) = -\widetilde{\nabla}\psi(T^{(t)}(x))$. In view of (23.14) and (23.15), the proof will be complete if we show

$$\limsup_{k\to\infty} \int v_k\, d\sigma \leq \int v\, d\sigma.$$

Let us bound the functions v_k. For each x and k, we can use (23.11) to derive

$$\begin{aligned}
& d\Big(x, T_{t+s_k\to t} \circ T^{(t+s_k)}(x)\Big)^2 - d\big(x, T^{(t+s_k)}(x)\big)^2 \\
& \leq \Big(d\big(x, T_{t+s_k\to t}(T^{(t+s_k)}(x))\big) + d\big(x, T^{(t+s_k)}(x)\big)\Big) \\
& \qquad\qquad\qquad \times d\Big(T^{(t+s_k)}(x), T_{t+s_k\to t}(T^{(t+s_k)}(x))\Big) \\
& \leq C\Big(1 + d(z,x) + d\big(z, T^{(t+s_k)}(x)\big)\Big)\, s_k\, d\big(z, T^{(t+s_k)}(x)\big) \\
& \leq C\, s_k\, \Big(1 + d(z,x)^2 + d\big(z, T^{(t+s_k)}(x)\big)^2\Big).
\end{aligned}$$

So

$$v_k(x) \leq C\,\Big(1 + d(z,x)^2 + d\big(z, T^{(t+s_k)}(x)\big)^2\Big).$$

Let $\chi : \mathbb{R}_+ \to [0,1]$ be a continuous cutoff function, $\chi(r) = 1$ for $r \leq 1$, $\chi(r) = 0$ for $r \geq 2$, and for any $R \geq 1$ let $\chi_R(r) = \chi(r/R)$. (This is a continuous approximation of the indicator function $1_{r\leq R}$.) When $\chi_R(1 + d(z,x) + d(z, T^{(t+s_k)}(x))) \neq 0$, $v_k(x)$ stays bounded like $O(R^2)$. So we can invoke Fatou's lemma as in Step 1: for any fixed R,

$$\begin{aligned}
\limsup_{k\to\infty} \int \chi_R\big(1 + d(z,x) + d(z, T^{(t+s_k)}(x))\big)\, v_k(x)\, \sigma(dx) \\
\leq \int \chi_R\big(1 + d(z,x) + d(z, T^{(t)}(x))\big)\, v(x)\, \sigma(dx).
\end{aligned}$$

To conclude the argument it suffices to show that

$$\lim_{R\to\infty} \int (1-\chi_R)\big(1 + d(z,x) + d(z, T^{(t)}(x))\big)\, |v(x)|\, \sigma(dx) = 0; \quad (23.16)$$

$$\lim_{R\to\infty} \limsup_{k\to\infty} \int (1-\chi_R)\big(1+d(z,x)+d(z, T^{(t+s_k)}(x))\big)\, |v_k(x)|\, \sigma(dx) = 0. \tag{23.17}$$

Of course $\chi_R(r) \neq 1$ only if $r > R$. Then, say for $R \geq 3$,

$$\begin{aligned}
&(1-\chi_R)\big(1 + d(z,x) + d(z,T^{(t)}(x))\big)\,|v(x)| \\
&\leq C\,1_{1+d(z,x)+d(z,T^{(t)}(x))\geq R}\Big(1 + d(z,x)^2 + d(z,T^{(t)}(x))^2\Big) \\
&\leq (3C)\Big[1_{d(z,x)\geq R/3}\, d(z,x)^2 + 1_{d(z,T^{(t)}(x))\geq R/3}\, d(z,T^{(t)}(x))^2\Big].
\end{aligned}$$

So the integral in (23.16) is bounded by

$$\begin{aligned}
&(3C)\left[\int_{d(z,x)\geq R/3} d(z,x)^2\,\sigma(dx) + \int_{d(z,T^{(t)}(x))\geq R/3} d(z,T^{(t)}(x))^2\,\sigma(dx)\right] \\
&\qquad \leq (3C)\left[\int_{d(z,x)\geq R/3} d(z,x)^2\,\sigma(dx) + \int_{d(z,y)\geq R/3} d(z,y)^2\,\mu_t(dy)\right],
\end{aligned}$$

which does converge to 0 as $R \to \infty$.

Similarly, the integral in (23.17) is bounded by

$$(3C)\left[\int_{d(z,x)\geq R/3} d(z,x)^2\,\sigma(dx) + \int_{d(z,y)\geq R/3} d(z,y)^2\,\mu_{t+s_k}(dy)\right].$$

Since μ_{t+s_k} converges to μ_t in distance W_2, it follows from Theorem 6.9 and Definition 6.8 that

$$\lim_{R\to\infty}\limsup_{k\to\infty}\int_{d(z,y)\geq R/3} d(z,y)^2\,\mu_{t+s_k}(dy) = 0.$$

So (23.17) holds, and the proof of subdifferentiability is complete.

Step 3: *Doubling of variables.*

Now let ξ_t, $\widehat{\xi}_t$ satisfy the same assumptions as in Step 2:

$$|\xi_t(x)| \leq C\,(1 + d(z,x)), \qquad |\widehat{\xi}_t(x)| \leq C\,(1 + d(z,x)).$$

By Step 2, $s \to W_2(\mu_s,\widehat{\mu}_t)$ and $t \to W_2(\mu_s,\widehat{\mu}_t)$ are differentiable for all s,t. To conclude to the differentiability of $t \to W_2(\mu_t,\widehat{\mu}_t)$, we can use Lemma 23.28 in the Appendix, provided that we check that, say, $s \to W_2(\mu_s,\widehat{\mu}_t)$ is (locally) absolutely continuous in s, uniformly in t. This will result from the triangle inequality:

$$\begin{aligned}
&W_2(\mu_s,\widehat{\mu}_t)^2 - W_2(\mu_{s'},\widehat{\mu}_t)^2 \\
&\qquad = \Big[W_2(\mu_s,\widehat{\mu}_t) + W_2(\mu_{s'},\widehat{\mu}_t)\Big]\,\Big[W_2(\mu_s,\widehat{\mu}_t) - W_2(\mu_{s'},\widehat{\mu}_t)\Big] \\
&\qquad \leq \Big[W_2(\mu_s,\widehat{\mu}_t) + W_2(\mu_{s'},\widehat{\mu}_t)\Big]\,W_2(\mu_s,\mu_{s'}) \\
&\qquad \leq \Big[W_2(\mu_s,\sigma) + W_2(\mu_{s'},\sigma) + 2\,W_2(\widehat{\mu}_t,\sigma)\Big]\,W_2(\mu_s,\mu_{s'}),
\end{aligned}$$

where σ is any arbitrary element of $P_2(M)$. The quantity inside square brackets is bounded (in fact it is a Lipschitz function of s, s' and t), and the path (μ_s) is Lipschitz in W_2 distance; so in fact

$$W_2(\mu_s, \widehat{\mu}_t)^2 - W_2(\mu_{s'}, \widehat{\mu}_t)^2 \leq C\,|s - s'|$$

for some constant C.

This concludes the proof of Theorem 23.9 for vector fields which grow at most linearly at infinity.

Step 4: *Integral reformulation and restriction argument.*

In this last step I shall complete the proof of Theorem 23.9. Let $\xi_t, \widehat{\xi}_t$ satisfy the assumptions of the theorem. Let z be a fixed point in M; consider the increasing sequence of events $A_k = \{\sup_t d(z, \gamma_t) \leq k\}$. For k large enough the event A_k has positive probability and it makes sense to condition γ by it. Then let $\mu_{t,k}$ be the law of this conditioned path, evaluated at time t: explicitly,

$$\mu_{t,k} = (e_t)_\# \Pi_k, \qquad \Pi_k(d\gamma) = \frac{1_{\gamma \in A_k}\, \Pi(d\gamma)}{\Pi[A_k]},$$

where of course e_t is the evaluation at time t. Let $Z_k := \Pi[A_k]$. Then $Z_k \uparrow 1$, $Z_k\, \mu_{t,k} \uparrow \mu_t$ as $k \to \infty$.

For each k, $\mu_{t,k}$ solves the *same* continuity equation as μ_t:

$$\frac{\partial \mu_{t,k}}{\partial t} + \nabla \cdot (\xi_t\, \mu_{t,k}) = 0. \tag{23.18}$$

But by definition $\mu_{t,k}$ is concentrated on the ball $B[z,k]$, so in (23.18) we may replace ξ_t by $\xi_{t,k} = \xi \chi_k$, where χ_k is a smooth cutoff function, $0 \leq \chi_k \leq 1$, $\chi_k = 1$ on $B[z,k]$, $\chi_k = 0$ outside of $B[z, 2k]$.

Let $\widehat{A}_k$, $\widehat{Z}_k$, $\widehat{\mu}_{t,k}$ and $\widehat{\xi}_{t,k}$ be defined similarly in terms of $\widehat{\xi}$ and $\widehat{\mu}_t$.

Since $\xi_{t,k}$ and $\widehat{\xi}_{t,k}$ are compactly supported, we may apply the result of Step 3: for all $t \in (t_1, t_2)$,

$$\frac{d}{dt}\left(\frac{W_2(\mu_{t,k}, \widehat{\mu}_{t,k})^2}{2}\right) = -\int \langle \widetilde{\nabla} \psi_{t,k}, \xi_{t,k} \rangle \, d\mu_{t,k} - \int \langle \widetilde{\nabla} \widehat{\psi}_{t,k}, \widehat{\xi}_{t,k} \rangle \, d\widehat{\mu}_{t,k}, \tag{23.19}$$

where $\exp(\widetilde{\nabla} \psi_{t,k})$ and $\exp(\widetilde{\nabla} \widehat{\psi}_{t,k})$ are the optimal transports $\mu_{t,k} \to \widehat{\mu}_{t,k}$ and $\widehat{\mu}_{t,k} \to \mu_{t,k}$.

Since $t \to \mu_{t,k}$ and $t \to \widehat{\mu}_{t,k}$ are Lipschitz paths, also $W_2(\mu_{t,k}, \widehat{\mu}_{t,k})$ is a Lipschitz function of t, so (23.19) integrates up to

$$\frac{W_2(\mu_{t,k},\widehat{\mu}_{t,k})^2}{2} = \frac{W_2(\mu_{0,k},\widehat{\mu}_{0,k})^2}{2} - \int_0^t \left(\int \langle \widetilde{\nabla}\psi_{s,k}, \xi_{s,k}\rangle \, d\mu_{s,k} + \int \langle \widetilde{\nabla}\widehat{\psi}_{s,k}, \widehat{\xi}_{s,k}\rangle \, d\widehat{\mu}_{s,k} \right) ds. \quad (23.20)$$

Since $t \to \mu_t$ and $t \to \widehat{\mu}_t$ are absolutely continuous paths, a computation similar to the one in Step 3 shows that $W_2(\mu_t,\widehat{\mu}_t)$ is absolutely continuous in t, in particular differentiable at almost all $t \in (t_1,t_2)$. If we can pass to the limit as $k \to \infty$ in (23.20), then we shall have

$$\frac{W_2(\mu_t,\widehat{\mu}_t)^2}{2} = \frac{W_2(\mu_0,\widehat{\mu}_0)^2}{2} - \int_0^t \left(\int \langle \widetilde{\nabla}\psi_s, \xi_s\rangle \, d\mu_s + \int \langle \widetilde{\nabla}\widehat{\psi}_{s,k}, \widehat{\xi}_{s,k}\rangle \, d\widehat{\mu}_{s,k} \right) ds, \quad (23.21)$$

and the desired result will be obtained by differentiating again in t. So it all amounts to passing to the limit in (23.20).

First, since $Z_k \uparrow 1$ and $Z_k\,\mu_{s,k} \uparrow \mu_s$, $\mu_{s,k}$ converges in total variation (and a fortiori weakly) to μ_s. Similarly, $\widehat{\mu}_{s,k}$ converges in total variation to $\widehat{\mu}_s$.

Next, the restriction property of optimal transport implies that $\exp(\widetilde{\nabla}\psi_{s,k})$ coincides with $\exp(\widetilde{\nabla}\psi_s)$, $\mu_{s,k}$-almost surely. If x is outside a set N_0 of zero μ_s-probability, we know from Theorem 10.28 and Remark 10.32 that there is a unique geodesic joining x to $\exp_x(\widetilde{\nabla}\psi(x))$; then the initial velocity is uniquely determined by the final position, so $\widetilde{\nabla}\psi_{s,k}$ coincides with $\widetilde{\nabla}\psi_s$, $\mu_{s,k}$-almost surely.

Now let us check that $\mu_{s,k}$ converges to μ_s in $P_2(M)$. It follows from its definition that $\mu_{s,k}$ coincides with μ_s/Z_k on the set $e_s(A_k)$. Let S stand for the support of Π; then

$$\begin{aligned}
&\int d(z,x)^2 \, d|\mu_{s,k} - \mu_s(x)| \\
&\quad \le (Z_k^{-1}-1)\int d(z,x)^2\,\mu_s(dx) + \frac{1}{Z_k}\int d(z,x)^2\,|Z_k\mu_{s,k}-\mu_s|(dx) \\
&\quad \le (Z_k^{-1}-1)\int d(z,x)^2\,\mu_s(dx) + \frac{1}{Z_k}\int_{e_s(S)\setminus e_s(A_k)} d(z,x)^2\,\mu_s(dx) \\
&\quad \le (Z_k^{-1}-1)\int d(z,x)^2\,\mu_s(dx) + \frac{1}{Z_k}\int_{S\setminus A_k} d(z,\gamma_s)^2\,\Pi(d\gamma).
\end{aligned}$$

By Theorem 6.15, this implies that $W_2(\mu_{s,k},\mu_s) \to 0$ for each s. Similarly, $W_2(\widehat{\mu}_{s,k},\widehat{\mu}_s) \to 0$.

Moreover, if k is large enough that $Z_k \geq 1/2$, we have the uniform bound

$$\int d(z,x)^2\, d|\mu_{s,k} - \mu_s(x)| \leq 2 \int d(z,x)^2\, \mu_s(dx),$$

which is a continuous function of s (because μ_s is continuous in W_2; recall Theorem 6.9). So there is a uniform bound (independent of s) on $W_2(\mu_{s,k}, \mu_s)$. Similarly, there is a uniform bound on $W_2(\widehat{\mu}_{s,k}, \widehat{\mu}_s)$.

Since $W_2(\mu_{t,k}, \mu_t)$ and $W_2(\widehat{\mu}_{t,k}, \widehat{\mu}_t)$ converge to 0 as $k \to \infty$, by Corollary 6.11 the first two terms in (23.20) converge to the first two terms in (23.21). To conclude the argument it suffices to check the convergence of the last two terms. Let us show for instance that

$$\int_0^t \left(\int \langle \widetilde{\nabla} \psi_{s,k}, \xi_s \rangle \, d\mu_{s,k} \right) ds \xrightarrow[k\to\infty]{} \int_0^t \left(\int \langle \widetilde{\nabla} \psi_s, \xi_s \rangle \, d\mu_s \right) ds. \qquad (23.22)$$

First observe that the integrand in (23.22) is dominated by an integrable function of s. Indeed, there is a constant C such that

$$\begin{aligned} \left| \int \langle \widetilde{\nabla} \psi_{s,k}, \xi_s \rangle \, d\mu_{s,k} \right| &\leq \sqrt{\int |\widetilde{\nabla} \psi_{s,k}|^2 \, d\mu_{s,k}} \sqrt{\int |\xi_s|^2 \, d\mu_{s,k}} \\ &\leq W_2(\widehat{\mu}_{s,k}, \mu_{s,k}) \sqrt{\frac{1}{Z_k} \int |\xi_s|^2 \, d\mu_s} \\ &\leq C \sqrt{\int |\xi_s|^2 \, d\mu_s}, \end{aligned}$$

and the latter function lies in $L^2(ds)$. So it is sufficient to prove that for almost all s,

$$\int \langle \widetilde{\nabla} \psi_{s,k}, \xi_s \rangle \, \rho_{s,k} \, d\nu \xrightarrow[k\to\infty]{} \int \langle \widetilde{\nabla} \psi_s, \xi_s \rangle \, \rho_s \, d\nu, \qquad (23.23)$$

where of course $\rho_{s,k}$ is the density of $\mu_{s,k}$ with respect to ν. It is sufficient to prove this for a subsequence of some arbitrary subsequence (in k), so we may assume that $\rho_{s,k}$ converges ν-almost everywhere to ρ_s. Recalling that $\widetilde{\nabla} \psi_{s,k}$ coincides with $\widetilde{\nabla} \psi_s$, $\mu_{s,k}$-almost surely, we just have to show

$$\int \langle \widetilde{\nabla} \psi_s, \xi_s (\rho_{s,k} - \rho_s) \rangle \, d\nu \xrightarrow[k\to\infty]{} 0.$$

This is easily done by dominated convergence: Indeed, $\rho_{s,k} \leq 2\rho_s$ and $|\widetilde{\nabla} \psi_s| \, |\xi_s| \, \rho_s$ is integrable, since

$$\int |\widetilde{\nabla}\psi_s|\, |\xi_s|\, \rho_s\, d\nu \leq \sqrt{\int |\widetilde{\nabla}\psi_s|^2\, \rho_s\, d\nu}\, \sqrt{\int |\xi_s|^2\, \rho_s\, d\nu}$$
$$= W_2(\mu_s, \widehat{\mu}_s)\, \sqrt{\int |\xi_s|^2\, d\mu_s} < +\infty.$$

This concludes the proof of Theorem 23.9. □

Subdifferential of energy functionals

The next problem to be addressed is the differentiation of an energy functional U_ν, along a path in the Wasserstein space $P_2(M)$, or rather in $P_2^{\mathrm{ac}}(M)$. This problem is easy to solve formally by means of Otto calculus, but the rigorous justification is definitely not trivial, especially when M is noncompact. The proof of the next statement will use Alexandrov's second differentiability theorem (Theorem 14.1), some elements of distribution theory, and many technical tricks. I shall denote by $W^{1,1}_{\mathrm{loc}}(M)$ the space of functions f which are locally integrable in M and whose distributional gradient ∇f is defined by a locally integrable function. Recall Convention 17.10.

Theorem 23.14 (Computation of subdifferentials in Wasserstein space). *Let M be a Riemannian manifold, equipped with a reference measure $\nu = e^{-V}\mathrm{vol}$, $V \in C^2(M)$, satisfying a curvature-dimension bound $\mathrm{CD}(K,N)$ for some $K \in \mathbb{R}$, $N \in (1,\infty]$. Let $U \in \mathcal{DC}_N$, $p(r) = r\,U'(r) - U(r)$, let μ and σ belong to $P_2^{\mathrm{ac}}(M)$, and let ρ be the density of μ with respect to ν. Let ψ be a $d^2/2$-convex function such that $T = \exp(\widetilde{\nabla}\psi)$ is the unique Monge transport $\mu \to \sigma$, and for $t \in [0,1]$ let $\mu_t = (\exp(t\widetilde{\nabla}\psi))_\#\mu$. Assume that:*

(i) $p(\rho) \in W^{1,1}_{\mathrm{loc}}$;

(ii) $U_\nu(\mu) < +\infty$;

(iii) $K_{N,U} > -\infty$, *where $K_{N,U}$ is defined in* (17.10).

If M is noncompact, further assume that:

(iv) $I_{U,\nu}(\mu) := \displaystyle\int \frac{|\nabla p(\rho)|^2}{\rho}\, d\nu < +\infty$; *and*

(v) $\mu, \sigma \in P_p^{\mathrm{ac}}(M)$, *where $p \in [2,+\infty) \cup \{c\}$ satisfies* (17.5).

If M is noncompact, $N < \infty$ and $K < 0$, reinforce (v) into:

(v') $\mu, \sigma \in P_q^{\mathrm{ac}}(M)$, where $q \in ((2N)/(N-1), +\infty) \cup \{c\}$ satisfies

$$\exists\, \delta > 0; \qquad \int \frac{\nu(dx)}{(1 + d(x_0, x))^{q(N-1)-2N-\delta}} < +\infty. \tag{23.24}$$

Then

$$\liminf_{t \downarrow 0} \frac{U_\nu(\mu_t) - U_\nu(\mu)}{t} \geq \int \langle \widetilde{\nabla} \psi, \nabla p(\rho) \rangle \, d\nu; \tag{23.25}$$

and

$$U_\nu(\sigma) \geq U_\nu(\mu) + \int \langle \widetilde{\nabla} \psi, \nabla p(\rho) \rangle \, d\nu \\ + K_{N,U} \int_0^1 \left(\int |\widetilde{\nabla} \psi_t(x)|^2 \, \rho_t(x)^{1-\frac{1}{N}} \, \nu(dx) \right) (1-t) \, dt. \tag{23.26}$$

Particular Case 23.15 (Displacement convexity of H: above-tangent formulation). If $N = \infty$ and $U(r) = r \log r$, Formula (23.26) becomes

$$H_\nu(\sigma) \geq H_\nu(\mu) + \int_M \langle \widetilde{\nabla} \psi, \nabla \rho \rangle \, d\nu + K \frac{W_2(\mu, \sigma)^2}{2}. \tag{23.27}$$

Remark 23.16. By the Cauchy–Schwarz inequality, (23.27) implies

$$H_\nu(\sigma) \geq H_\nu(\mu) - \sqrt{\int_M \rho \, |\widetilde{\nabla} \psi|^2 \, d\nu} \sqrt{\int_M \frac{|\nabla \rho|^2}{\rho} \, d\nu} + K \frac{W_2(\mu, \sigma)^2}{2} \\ = H_\nu(\mu) - W_2(\mu, \sigma) \sqrt{I_\nu(\mu)} + K \frac{W_2(\mu, \sigma)^2}{2}.$$

In this sense (23.27) is a precise version of the HWI inequality appearing in Corollary 20.13.

Remark 23.17. If $K_{N,U} = -\infty$ (i.e. $K < 0$ and $p(r)/r^{1-1/N} \to +\infty$ as $r \to \infty$) then (23.26) remains obviously true but I don't know about (23.25).

Remark 23.18. As soon as $\rho \in W^{1,1}_{\mathrm{loc}}(M)$ we can write $\nabla p(\rho) = p'(\rho) \nabla \rho = \rho \, U''(\rho) \nabla \rho = \rho \, \nabla U'(\rho)$, so (23.25) becomes

$$\liminf_{t\downarrow 0} \frac{U_\nu(\mu_t) - U_\nu(\mu)}{t} \geq \int \langle \widetilde{\nabla}\psi, \nabla U'(\rho)\rangle \, d\mu, \tag{23.28}$$

which is the result that one would have formally guessed by using Otto's calculus.

Proof of Theorem 23.14. The complete proof is quite tricky and it is strongly advised to just focus on the compactly supported case at first reading. I have divided the argument into seven steps.

Step 1: Computation of the $\liminf$ **in the compactly supported case.** To begin with, I shall assume that μ and σ are compactly supported, and *compute* the lower derivative:

$$\liminf_{t\downarrow 0} \frac{U_\nu(\mu_t) - U_\nu(\mu_0)}{t} = -\int p(\rho)\,(L\psi)\,d\nu, \tag{23.29}$$

where the function $L\psi$ is obtained from the measure $L\psi$ (understood in the sense of distributions) by keeping only the absolutely continuous part (with respect to the volume measure).

The argument starts as in the proof of Theorem 17.15, with a change of variables:

$$\begin{aligned}U_\nu(\mu_t) = \int U(\rho_t(x))\,d\nu(x) &= \int U\Big(\rho_t(\exp_x t\nabla\psi(x))\Big)\,\mathcal{J}_{0\to t}(x)\,d\nu(x)\\ &= \int U\left(\frac{\rho_0(x)}{\mathcal{J}_{0\to t}(x)}\right)\,\mathcal{J}_{0\to t}(x)\,d\nu(x),\end{aligned}$$

where ρ_0 is the same as ρ, and $\mathcal{J}_{0\to t}$ is the Jacobian determinant associated with the map $\exp(t\nabla\psi)$ (the reference measure being ν). *Note carefully* that here I am using the Jacobian formula for a change of variables which a priori is not Lipschitz; also note that $T = \exp(\nabla\psi)$ (there is no need to use approximate gradients since everything is compactly supported). Upon use of $\mu = \rho_0\,\nu$, it follows that

$$\frac{U_\nu(\mu_t) - U_\nu(\mu)}{t} = \int w(t,x)\,\mu(dx), \tag{23.30}$$

where

$$w(t,x) = \frac{u(t,x) - u(0,x)}{t}, \qquad u(t,x) = U\left(\frac{\rho_0(x)}{\mathcal{J}_{0\to t}(x)}\right)\frac{\mathcal{J}_{0\to t}(x)}{\rho_0(x)}. \tag{23.31}$$

By Theorem 14.1, for μ-almost all x we have the Taylor expansion

$$\det\big(d_x \exp(t\nabla\psi)\big) = 1 + t\,\Delta\psi(x) + o(t) \qquad \text{as } t \to 0,$$

so

$$\begin{aligned}\mathcal{J}_{0\to t}(x) &= e^{-\left[V(\exp t\nabla\psi(x)) - V(x)\right]} \det\big(d_x \exp(t\nabla\psi)\big) \\ &= \big(1 - t\,\nabla V(x)\cdot\nabla\psi(x)\big)\,\big(1 + t\,\Delta\psi(x) + o(t)\big) \\ &= 1 + t\,(L\psi)(x) + o(t). \end{aligned} \tag{23.32}$$

On the other hand, for given r, the derivative of $\delta \to (\delta/r)\,U(r/\delta)$ at $\delta = 1$ is $-\,p(r)/r$. This and (23.32) imply that for almost all x where $\rho_0(x) > 0$,

$$\lim_{t\downarrow 0} w(t,x) \;=\; -\,(L\psi(x))\left(\frac{p(\rho_0(x))}{\rho_0(x)}\right).$$

So to establish (23.29), it suffices to prove

$$\lim_{k\to\infty} \int w(t_k,x)\,\mu(dx) = \int \Big[\lim_{k\to\infty} w(t_k,x)\Big]\,\mu(dx), \tag{23.33}$$

where $(t_k)_{k\in\mathbb{N}}$ is an arbitrary sequence of positive times decreasing to 0.

First consider the case $K \geq 0$, which is simpler. From the estimates in the proof of Theorem 17.15 (recall (17.15)) we know that $u(t,x)$ is a convex function of t; then $w(t,x)$ is nonincreasing as $t \downarrow 0$, and (23.33) follows from the monotone convergence theorem.

Now consider the case when $K < 0$. As in the estimates in the proof of Theorem 17.15 (see (17.15) again),

$$\frac{d^2\,u(t,x)}{dt^2} \geq K_{N,U}\,\rho_t\big(\exp(t\nabla\psi(x))\big)^{-\frac{1}{N}}\big|\nabla\psi_t(\exp(t\nabla\psi(x)))\big|^2,$$

and by assumption $K_{N,U}$ is finite. Note that $|\nabla\psi_t(\exp(t\nabla\psi(x)))| = d(x,T(x))$, which is bounded ($\mu$-almost surely) by the maximum distance between points in the support of μ and points in the support of ν. So there is a positive constant $\overline{C}$ such that

$$\frac{d^2\,u(t,x)}{dt^2} \geq -\,\overline{C}\,\rho_t\big(\exp(t\nabla\psi(x))\big)^{-\frac{1}{N}}. \tag{23.34}$$

Let

$$R(t,x) = \overline{C} \int_0^t \int_0^s \rho_\tau\big(\exp(\tau\nabla\psi(x))\big)^{-\frac{1}{N}}\,d\tau\,ds; \tag{23.35}$$

$$\widetilde{u}(t,x) = u(t,x) + R(t,x); \qquad \widetilde{w}(t,x) = \frac{\widetilde{u}(t,x) - \widetilde{u}(0,x)}{t}.$$

Then $t \to \widetilde{u}(t,x)$ is a convex function, so the previous reasoning applies to show that

$$\lim_{k\to\infty} \int \widetilde{w}(t_k,x)\,\mu(dx) = \int \lim_{k\to\infty} \widetilde{w}(t_k,x)\,\mu(dx).$$

To conclude the proof, it suffices to check that the additional term $R(s,x)$ does not count in the limit, i.e.

$$\lim_{k\to\infty} \left(\frac{R(t_k,x)}{t_k}\right) = 0, \qquad \mu(dx)\text{-almost surely,} \tag{23.36}$$

$$\lim_{k\to\infty} \int \left(\frac{R(t_k,x)}{t_k}\right) \mu(dx) = 0. \tag{23.37}$$

If (23.37) is true, then also (23.36) will be true up to extraction of a subsequence; but since the sequence t_k is already arbitrary, this will imply that $R(t,x) \to 0$ as $t \to 0$, μ-almost surely. So we just have to check (23.37). By (23.35) and Fubini,

$$\begin{aligned}
\int \left(\frac{R(t_k,x)}{t_k}\right) \mu(dx) &= \frac{1}{t}\int_0^t \int_0^s \left(\int \rho_\tau\big(\exp(\tau\nabla\psi(x))\big)^{-\frac{1}{N}}\,\mu(dx)\right) d\tau\,ds \\
&= \frac{1}{t}\int_0^t \int_0^s \left(\int \rho_\tau(y)^{-\frac{1}{N}}\,\mu_\tau(dy)\right) d\tau\,ds \\
&= \frac{1}{t}\int_0^t \int_0^s \left(\int \rho_\tau(y)^{1-\frac{1}{N}}\,\nu(dy)\right) d\tau\,ds.
\end{aligned} \tag{23.38}$$

By Jensen's inequality, $\int \rho_\tau^{1-1/N}\,d\nu \le \nu[S]^{1/N} (\int \rho_\tau\,d\nu)^{1-1/N} = \nu[S]^{1/N}$, where S is the support of μ_τ, and this is bounded independently of $\tau \in [0,1]$. So (23.38) is bounded like $O(t^{-1}\int_0^t\int_0^s d\tau\,ds) = O(t)$. This proves (23.37) and concludes the argument.

Step 2: Extension of $\nabla\psi$. The function ψ might not be finite outside $\operatorname{Spt}\mu$, which might cause problems in the sequel. In this step, we shall see that it is possible to extend ψ into a function which is finite everywhere.

Let π be the optimal transference plan between μ and σ, and let $T = \exp(\nabla\psi)$ be the Monge transport $\mu \to \sigma$. Let $\widetilde{c}$ be the restriction of $c(x,y) = d(x,y)^2$ to $(\operatorname{Spt}\mu) \times (\operatorname{Spt}\sigma)$. By Theorem 5.19, there is a $\widetilde{c}$-convex function $\widetilde{\psi}$ such that $\operatorname{Spt}\pi \subset \partial_{\widetilde{c}}\widetilde{\psi}$, so $\exp(\nabla\widetilde{\psi})$ is a Monge transport between μ and σ. By uniqueness of the Monge transport (Theorem 10.28),

$$\exp_x(\nabla\widetilde{\psi}(x)) = \exp_x(\nabla\psi(x)), \qquad \mu(dx)\text{-almost surely.} \tag{23.39}$$

Also recall from Remark 10.32 that μ-almost surely, x and $T(x) = \exp_x(\nabla\psi(x))$ are joined by a unique geodesic. So (23.39) implies

$$\nabla\widetilde{\psi}(x) = \nabla\psi(x), \qquad \mu(dx)\text{-almost surely.}$$

Now let ϕ be the $\widetilde{c}$-transform of $\widetilde{\psi}$: then

$$\widetilde{\psi}(x) = \sup_{y\in\operatorname{Spt}\sigma}\left(\phi(y) - \frac{d(x,y)^2}{2}\right). \tag{23.40}$$

This formula can be used to define $\widetilde{\psi}$ outside of Spt μ, everywhere on M; the resulting function will still be $d^2/2$-convex.

Since Spt σ is bounded, the function inside brackets in (23.40) is locally Lipschitz in x, uniformly in y, so the extended function $\widetilde{\psi}$ is also locally Lipschitz. Also, since the function $d(x,y)^2/2$ is locally semiconcave in x, uniformly in $y \in \operatorname{Spt}\sigma$, the function $\widetilde{\psi}$ is locally semiconvex (recall Theorem 10.26).

To summarize: We have constructed a locally Lipschitz, locally semiconvex, $d^2/2$-convex function $\widetilde{\psi}$ such that $\nabla\widetilde{\psi}$ coincide μ-almost surely with $\nabla\psi$. In the sequel, I shall work with $\widetilde{\psi}$ and drop the tilde symbol, so ψ will be defined in the whole of M.

Step 3: Integration by parts. In this step I shall show that

$$-\int p(\rho)\, L\psi\, d\nu \geq \int \langle\nabla\psi, \nabla p(\rho)\rangle\, d\nu, \tag{23.41}$$

where $L\psi = \Delta\psi - \nabla V\cdot\nabla\psi$ is understood in the sense of Alexandrov (Theorem 14.1) or equivalently, as the absolutely continuous part of the distribution $L\psi$.

Since ρ is compactly supported, so is $p(\rho)$. By assumption $p(\rho)$ lies in $W^{1,1}(M)$. By regularization (in local charts, or using a C^∞ mollifier kernel with properties similar to those in Definition 29.34), we can construct a sequence $(\zeta_k)_{k\in\mathbb{N}}$ of nonnegative functions in $C_c^\infty(M)$ such that

$$\zeta_k \xrightarrow[L^1]{} p(\rho), \qquad \nabla\zeta_k \xrightarrow[L^1]{} \nabla p(\rho), \tag{23.42}$$

and all functions ζ_k are supported in a fixed compact set W.

By Theorem 14.1, the function $\Delta\psi$ is bounded above by the distribution $\Delta\psi$; on the other hand, the function $\nabla V \cdot \nabla\psi$ coincides with the distribution $\nabla V \cdot \nabla\psi$; so the function $L\psi$ is bounded above by the distribution $L\psi$ (denoted $L_{\mathcal{D}'}\psi$). This implies

$$\begin{aligned}\int \zeta_k (L\psi)\, d\nu &\leq \int \langle \zeta_k, L_{\mathcal{D}'}\psi\rangle\, d\nu \\ &= -\int \langle \nabla\zeta_k, \nabla\psi\rangle\, d\nu. \end{aligned} \tag{23.43}$$

The function ψ is Lipschitz in W, so $\nabla\psi$ is bounded; combining this with (23.42) we get

$$\int \langle \nabla\zeta_k, \nabla\psi\rangle\, d\nu \xrightarrow[k\to\infty]{} \int \langle \nabla p(\rho), \nabla\psi\rangle\, d\nu. \tag{23.44}$$

Next, the function $\Delta\psi$ is bounded below on W because ψ is semi-convex (or because, by (13.8),

$$\begin{aligned}\Delta\psi(x) &\geq -\,\Delta\big|_{z=x} \frac{d(z, \exp_x \nabla\psi(x))^2}{2} \\ &\geq -\sup\left\{\Delta_x \frac{d(x,y)^2}{2};\; y \in (\exp \nabla\psi)(W)\right\},\end{aligned}$$

which is finite, recall the Third Appendix of Chapter 10). So $L\psi$ is bounded below on W, and Fatou's lemma applies to show

$$\int p(\rho)\,(L\psi)\, d\nu = \int \big(\lim_{k\to\infty} \zeta_k\big)\,(L\psi)\, d\nu \leq \liminf_{k\to\infty} \int \zeta_k (L\psi)\, d\nu. \tag{23.45}$$

The combination of (23.43), (23.44) and (23.45) implies (23.41).

Step 4: Integral reformulation. We shall now take advantage of the displacement convexity properties of U_ν to reformulate the differential condition

$$\liminf_{t\downarrow 0} \frac{U_\nu(\mu_t) - U_\nu(\mu_0)}{t} \geq \int \langle \nabla\psi, \nabla p(\rho_0)\rangle\, d\nu \tag{23.46}$$

into the integral (in t) condition

$$\begin{aligned} U_\nu(\mu_1) \geq U_\nu(\mu_0) &+ \int \langle \nabla\psi, \nabla p(\rho_0)\rangle\, d\nu \\ &+ K_{N,U} \int_0^1 \left(\int \rho_t(x)^{1-\frac{1}{N}} |\nabla\psi_t(x)|^2\, \nu(dx)\right)(1-t)\, dt. \end{aligned} \tag{23.47}$$

The advantage of the integral formulation is that it is quite stable under limits. At the same time, this will establish Theorem 23.14 in the case when μ and σ are compactly supported.

The strategy is the same as in the proof of (iv) $\Rightarrow$ (v) in Proposition 23.1. Recall from Theorem 17.15 that for any $t \in (0,1)$,

$$U_\nu(\mu_t) \leq (1-t)\,U_\nu(\mu_0) + t\,U_\nu(\mu_1) \\ - K_{N,U} \int_0^1 \left(\int_M \rho_s(x)^{1-\frac{1}{N}} |\nabla\psi_s(x)|^2\,\nu(dx) \right) G(s,t)\,ds,$$

where $G(s,t)$ is the one-dimensional Green function of the Laplacian. By subtracting $U_\nu(\mu_0)$ on both sides and dividing by t, we get

$$\frac{U_\nu(\mu_t) - U_\nu(\mu_0)}{t} \leq U_\nu(\mu_1) - U_\nu(\mu_0) \\ - K_{N,U} \int_0^1 \left(\int \rho_s(x)^{1-\frac{1}{N}} |\nabla\psi_s(x)|^2\,\nu(dx) \right) \frac{G(s,t)}{t}\,ds. \quad (23.48)$$

Then we can use Steps 1 and 2 to pass to the lim inf in the left-hand side.

As for the right-hand side of (23.48), we note that if B is a large ball containing the supports of all ρ_s $(0 \leq s \leq 1)$, and $D = \mathrm{diam}\,(B)$, then

$$\int \rho_s(x)^{1-\frac{1}{N}} |\nabla\psi_s(x)|^2\,\nu(dx) \leq D^2 \int \rho_s^{1-\frac{1}{N}}\,d\nu \\ \leq D^2\,\nu[B]^{\frac{1}{N}} \; < +\infty.$$

(Here I have used Jensen's inequality again as in Step 1.) So the quantity inside brackets in the right-hand side of (23.48) is uniformly (in t) bounded. Since $G(s,t)/t$ converges to $1-s$ in $L^1(ds)$ as $t \to 0$, we can pass to the limit in the right-hand side of (23.48) too. The final result is

$$\int \langle \nabla\psi, \nabla p(\rho)\rangle\,d\nu \leq U_\nu(\mu_1) - U_\nu(\mu_0) \\ - K_{N,U} \int_0^1 \left(\int \rho_s(x)^{1-\frac{1}{N}} |\nabla\psi_s(x)|^2\,\nu(dx) \right) (1-s)\,ds,$$

which is the same as (23.47) (or (23.26) since $\mu_0 = \mu$, $\mu_1 = \sigma$).

The last three steps consist in a long string of approximations to go from the compact to the noncompact case.

Step 5: Removal of compactness assumption, for nice pressure laws. In this step I shall use an approximation argument to extend the validity of (23.47) (or a relaxation thereof) to the noncompact case. The difficulty is that the standard approximation scheme of Proposition 13.2 does not in general preserve any smoothness of ρ_0, which makes it problematic to pass to the limit in the term $\int \langle \nabla\psi, \nabla p(\rho_0)\rangle$.

So let M, ν, U, $(\mu_t)_{0\le t\le 1}$, $(\rho_t)_{0\le t\le 1}$, $(\psi_t)_{0\le t\le 1}$ satisfy the assumptions of the theorem. Later on I shall make some additional assumptions on the function U, which will be removed in the next step.

I shall distinguish two cases, according to whether the support of μ_1 is compact or not. Somewhat surprisingly, the two arguments will be different.

Case 1: $\mathrm{Spt}(\mu_1)$ **is compact.** In this case one can use a slight modification of the standard approximation scheme. Let $\chi : \mathbb{R}_+ \to \mathbb{R}$ be a smooth nondecreasing function with $0 \le \chi \le 1$, $\chi(r) = 1$ for $r \le 1$, $\chi(r) = 0$ for $r \ge 2$. Let $z \in M$ be arbitrary, and for $k \in \mathbb{N}$ let $\chi_k(r) = \chi(d(z,x)/k)$. Of course $0 \le \chi_k \le 1$, χ_k is identically equal to 1 on $B[z,k]$ and identically equal to 0 outside $B[z,2k]$; moreover, χ_k is C/k-Lipschitz, where $C = \|\chi\|_{\mathrm{Lip}}$. For k large enough, $\int \chi_k\, d\mu_0 \ge 1/2$; then let us set

$$Z_k = \int \chi_k\, d\mu_0; \qquad \rho_{0,k} = \frac{\chi_k\, \rho_0}{Z_k}; \qquad \mu_{0,k} = \rho_{0,k}\, \nu;$$

$$\Pi_k(d\gamma) = \frac{\chi_k(\gamma_0)\, \Pi(d\gamma)}{Z_k}.$$

Of course, $(e_0)_\# \Pi_k = \mu_{0,k}$. For each $t \in [0,1]$ we let $\mu_{t,k} = (e_t)_\# \Pi_k$, and define $\rho_{t,k}$ as the density of $\mu_{t,k}$. Then $Z_k \uparrow 1$, $Z_k\, \rho_{t,k} \uparrow \rho_t$, and for each k, all the measures $\mu_{t,k}$ are supported in a common compact set. (This is because μ_1 is compactly supported!) Moreover, the optimal transport $\exp(\nabla\psi_{t,k})$ between $\mu_{t,k}$ and $\mu_{1,k}$ coincides, $\mu_{t,k}$-almost surely, with $\exp(\widetilde{\nabla}\psi_t)$; and the gradient $\nabla\psi_k$ coincides, $\mu_{0,k}$-almost surely, with the approximate gradient $\widetilde{\nabla}\psi$.

For each k, we can apply the results of Step 4 with μ_t replaced by $\mu_{t,k}$ and U replaced by $U_k = U(Z_k\, \cdot)$. We obtain

$$U_{k,\nu}(\mu_{1,k}) \geq U_{k,\nu}(\mu_{0,k}) + \int \langle \nabla\psi_k, \nabla p_k(\rho_{0,k})\rangle\, d\nu$$
$$+ K_{N,U_k} \int_0^1 \left(\int_M \rho_{t,k}(x)^{1-\frac{1}{N}} \, |\nabla\psi_{t,k}(x)|^2\, \nu(dx) \right) (1-t)\, dt, \quad (23.49)$$

where $p_k(r) = r\, U_k'(r) - U_k(r)$.

We can pass to the limit in the first, second and fourth terms in (23.49) exactly as in the proof of Theorem 17.15. (The key observation is that $\int U_k(\mu_{t,k}) = \int U(Z_k\, \rho_{t,k})\, d\nu$, and the monotone convergence theorem applies to $U_+(Z_k\, \rho_{t,k})$.) It remains to take care of the third term, which is

$$\int \langle \nabla\psi_k, \nabla p_k(\rho_{0,k})\rangle\, d\nu = \int \langle \widetilde{\nabla}\psi, Z_k\, \nabla p(\chi_k\, \rho_0)\rangle\, d\nu$$
$$= \int \left\langle \widetilde{\nabla}\psi, Z_k\, p'(\chi_k\, \rho_0)\, (\rho_0 \nabla\chi_k + \chi_k \nabla\rho_0) \right\rangle\, d\nu. \quad (23.50)$$

Here I used the chain-rule formula $\nabla p(\chi_k\, \rho_0) = p'(\chi_k\, \rho_0)\, \nabla(\chi_k\, \rho_0)$, which requires a bit of justification. (Don't hesitate to skip this paragraph.) The problem is that the assumption $p(\rho) \in W^{1,1}_{\mathrm{loc}}(M)$ does not imply that ρ itself lies in $W^{1,1}_{\mathrm{loc}}(M)$. A possible justification is as follows. Let $[0, r_0]$ be the interval on which $p' = 0$ (recall Proposition 17.7(iii)); then for any $a > r_0$, $b > a$, the map $p(r) \to \min(a, \max(r, b)) =: \varphi_{a,b}(r)$ is Lipschitz. So $\varphi_{a,b}(\rho)$ lies in $W^{1,1}_{\mathrm{loc}}(M)$, and $1_{a\leq\rho\leq b}\nabla p(\rho) = p'(\rho) 1_{a\leq\rho\leq b} \nabla\rho$; this identity defines $\nabla\rho$ almost everywhere as a function on the set $\{a \leq \rho_0 \leq b\}$. Since a, b are arbitrary, this also establishes $1_{\rho>r_0}\nabla p(\rho) = p'(\rho) 1_{\rho>r_0}\nabla\rho$. But for $\rho \leq r_0$, $p(\rho)$ is a constant and it results by a classical theorem (see the bibliographical notes) that $\nabla p(\rho) = 0$ almost everywhere. Also $p'(\rho) = 0$ if $\rho \leq r_0$, so we may decide that $p'(\rho) 1_{\rho\leq r_0}\nabla\rho = 0$ even if $\nabla\rho$ might not be defined as a function for $\rho \leq r_0$. The same reasoning also applies with ρ replaced by $\chi_k\, \rho$, because $\chi_k\, \rho$ is never greater than ρ (so $p(\chi_k\, \rho_0)$ is constant for $\rho \leq r_0$).

Now let us come back to the control of (23.50). Since p' is continuous, $Z_k \to 1$, $\chi_k \to 1$ and $\nabla\chi_k \to 0$, it is clear that the integrand in (23.50) converges pointwise to $\langle \widetilde{\nabla}\psi, p'(\rho_0)\, \nabla\rho_0\rangle$. To conclude, it suffices to check that the dominated convergence theorem applies, i.e. that

$$|\widetilde{\nabla}\psi|\, p'(\chi_k\, \rho_0) \Big(\rho_0\, |\nabla\chi_k| + \chi_k\, |\nabla\rho_0| \Big) \quad (23.51)$$

is bounded by a fixed L^1 function.

The first term in (23.51) is easy to dominate if we assume that p is Lipschitz: Then there is a constant C such that

$$|\widetilde{\nabla}\psi|\, p'(\chi_k\,\rho_0)\rho_0\,|\nabla\chi_k| \le C\,\rho_0\,|\widetilde{\nabla}\psi|,$$

and the latter is integrable since $\int \rho_0\,|\widetilde{\nabla}\psi| \le \sqrt{\int \rho_0|\widetilde{\nabla}\psi|^2} = W_2(\mu_0,\mu_1)$.

To control the second term in (23.51), I shall assume the existence of finite positive constants A, B, b, c such that

$$\rho_1 \le \rho_0 \le A \Longrightarrow \qquad p'(\rho_1) \le C\,p'(\rho_0); \tag{23.52}$$

$$\rho_0 \ge A \Longrightarrow \qquad p'(\rho_0) \ge c\,\rho_0^{-\frac{1}{N}}; \tag{23.53}$$

$$\rho_0 \ge B \Longrightarrow \qquad p'(\rho_0) = b\,\rho_0^{-\frac{1}{N}}. \tag{23.54}$$

(Here as below, C is a notation for various constants which may depend on U.) With these assumptions we can reason as follows:

- If $\rho_0 \le A$, then

$$|\widetilde{\nabla}\psi|\, p'(\chi_k\,\rho_0)\,\chi_k\,|\nabla\rho_0| \le C\,|\widetilde{\nabla}\psi|\,p'(\rho_0)|\nabla\rho_0| = C\,|\widetilde{\nabla}\psi|\,|\nabla p(\rho_0)|.$$

- If $\rho_0 \ge A$ and $\chi_k\,\rho_0 \ge B$, then $\chi_k\,p'(\chi_k\,\rho_0) = b\,\chi_k^{1-1/N}\rho_0^{-1/N}$, so

$$\begin{aligned}
|\widetilde{\nabla}\psi|\,\chi_k\,p'(\chi_k\,\rho_0)\,|\nabla\rho_0| &\le C\,|\widetilde{\nabla}\psi|\,\chi_k^{1-\frac{1}{N}}\,\rho_0^{-\frac{1}{N}}\,|\nabla\rho_0| \\
&\le C\,|\widetilde{\nabla}\psi|\,\rho_0^{-\frac{1}{N}}\,|\nabla\rho_0| \\
&\le C\,|\widetilde{\nabla}\psi|\,p'(\rho_0)\,|\nabla\rho_0|.
\end{aligned}$$

- If $\rho_0 \ge A$ and $\chi_k\,\rho_0 \le B$, then $\chi_k\,p'(\chi_k\,\rho_0) \le C\,\chi_k \le C\,\chi_k^{1/N} \le C\,B^{1/N}\,\rho_0^{-1/N}$, so

$$\begin{aligned}
|\widetilde{\nabla}\psi|\,\chi_k\,p'(\chi_k\,\rho_0)\,|\nabla\rho_0| &\le C\,|\widetilde{\nabla}\psi|\,\rho_0^{-\frac{1}{N}}\,|\nabla\rho_0| \\
&\le C\,|\widetilde{\nabla}\psi|\,p'(\rho_0)\,|\nabla\rho_0|.
\end{aligned}$$

To summarize: In all cases, $|\widetilde{\nabla}\psi|\,\chi_k\,p'(\chi_k\,\rho_0)\,|\nabla\rho_0|$ is bounded by $C\,|\widetilde{\nabla}\psi|\,|\nabla p(\rho_0)|$, and the latter quantity is integrable since

$$\begin{aligned}
\int |\widetilde{\nabla}\psi|\,|\nabla p(\rho_0)|\,d\nu &\le \sqrt{\int \rho_0|\widetilde{\nabla}\psi|^2\,d\nu}\,\sqrt{\int \frac{|\nabla p(\rho_0)|^2}{\rho_0}\,d\nu} \\
&= W_2(\mu_0,\mu_1)\,\sqrt{I_{U,\nu}(\mu_0)}.
\end{aligned} \tag{23.55}$$

So we can pass to the limit in the third term of (23.49), and the proof is complete.

Case 2: $\mathrm{Spt}(\mu_1)$ **is not compact.** In this case we shall definitely not use the standard approximation scheme by restriction, but instead a more classical procedure of smooth truncation.

Again let $\chi : \mathbb{R}_+ \to \mathbb{R}_+$ be a smooth nondecreasing function with $0 \le \chi \le 1$, $\chi(r) = 1$ for $r \le 1$, $\chi(r) = 0$ for $r \ge 2$; now we require in addition that $(\chi')^2/\chi$ is bounded. (It is easy to construct such a cutoff function rather explicitly.) Then we define $\chi_k(x) = \chi(d(z,x)/k)$, where z is an arbitrary point in M. For k large enough, $Z_{1,k} := \int \chi_k \, d\mu_1 \ge 1/2$. Then we choose $\ell = \ell(k)$ large enough that $Z_{0,k} := \int \chi_\ell \, d\mu_0$ is larger than $Z_{1,k}$; this is possible since $Z_{1,k} < 1$ (otherwise μ_1 would be compactly supported). Then we let

$$\mu_{0,k} = \frac{\chi_{\ell(k)} \, \mu_0}{Z_{0,k}}; \qquad \mu_{1,k} = \frac{\chi_k \, \mu_1}{Z_{1,k}}.$$

For each k, these are two compactly supported, absolutely continuous probability measures; let $(\mu_{t,k})_{0 \le t \le 1}$ be the displacement interpolation joining them, and let $\rho_{t,k}$ be the density of $\mu_{t,k}$. Further, let ψ_k be a $d^2/2$-convex function so that $\exp(\nabla\psi_k)$ is the optimal (Monge) transport μ_0 and μ_1, and let $\psi_{t,k}$ be deduced from ψ_k by the Hamilton–Jacobi forward semigroup.

Note carefully: It is obvious from the construction that $Z_{0,k}\, \rho_{0,k} \uparrow \rho_0$, $Z_{1,k}\, \rho_{1,k} \uparrow \rho_1$, but there is a priori no monotone convergence relating $\rho_{t,k}$ to ρ_t! Instead, we have the following information. Since $\mu_{0,k} \to \mu_0$ and $\mu_{1,k} \to \mu_1$, Corollary 7.22 shows that the geodesic curves $(\mu_{t,k})_{0 \le t \le 1}$ converge, up to extraction of a subsequence, to some geodesic curve $(\mu_{t,\infty})_{0 \le t \le 1}$ joining μ_0 to μ_1. (The convergence holds true for all t.) Since (μ_t) is the unique such curve, actually $\mu_{t,\infty} = \mu_t$, which shows that $\mu_{t,k}$ converges weakly to μ_t for all $t \in [0,1]$.

For each k, we can apply the results of Step 4 with U replaced by $U_k = U_k(Z_{1,k} \, \cdot)$ and μ_t replaced by $\mu_{t,k}$:

$$\begin{aligned} U_{k,\nu}(\mu_{1,k}) \ge U_{k,\nu}(\mu_{0,k}) &+ \int \langle \nabla\psi_k, \nabla p_k(\rho_{0,k}) \rangle \, d\nu \\ &+ K_{N,U_k} \int_0^1 \left(\int_M \rho_{t,k}(x)^{1-\frac{1}{N}} \, |\nabla\psi_{t,k}(x)|^2 \, \nu(dx) \right) (1-t) \, dt, \end{aligned} \tag{23.56}$$

where $p_k(r) = r\, U_k'(r) - U_k(r)$. The problem is to pass to the limit as $k \to \infty$. We shall consider all four terms in (23.56) separately, and use a few results which will be proven later on in Part III of this course (in a more general context).

First term of (23.56)*:* Since $U_{k,\nu}(\mu_{1,k}) = \int U(Z_{1,k}\,\rho_{1,k})\,d\nu$ and $Z_{1,k}\,\rho_{1,k} = \chi_k\,\rho_1$ converges monotonically to ρ_1, the same arguments as in the proof of Theorem 17.15 apply to show that

$$U_{k,\nu}(\mu_{1,k}) \xrightarrow[k\to\infty]{} U_\nu(\mu_1). \tag{23.57}$$

Second term of (23.56)*:* Since $Z_{1,k}\,\rho_{0,k}$ converges to μ_0 (in total variation, a fortiori weakly), we can use the lower semicontinuity of the convex functional U_ν (see Theorem 30.6(i) later in these notes), to pass to the lim inf:

$$U_\nu(\mu_0) \le \liminf_{k\to\infty} U_{k,\nu}(\mu_{0,k}). \tag{23.58}$$

(Theorem 30.6 is proven under the assumption that $U(r) \ge -c\,r$ for some $c \in \mathbb{R}$, so let us make this assumption here too.) By the way, notice that the treatment of μ_0 and μ_1 is not symmetric.

Third term of (23.56)*:* First note that $\nabla\psi_k$ converges μ_0-almost surely to $\widetilde{\nabla}\psi$, and therefore also in μ_0-probability. The argument is the same as in the proof of (ii) in Theorem 23.9.

Now our goal is to show that

$$\int \big\langle \nabla\psi_k, \nabla p_k(\rho_{0,k})\big\rangle\,d\nu - \int \langle \widetilde{\nabla}\psi, \nabla p(\rho_0)\rangle\,d\nu \xrightarrow[k\to\infty]{} 0. \tag{23.59}$$

The left-hand side of (23.59) can be decomposed into two terms as follows:

$$\int \big\langle \nabla\psi_k - \widetilde{\nabla}\psi, \nabla p_k(\rho_{0,k})\big\rangle\,d\nu + \int \langle \widetilde{\nabla}\psi, \nabla p_k(\rho_{0,k}) - \nabla p(\rho_0)\rangle\,d\nu. \tag{23.60}$$

Both terms will be treated separately.

First term in (23.60)*:* By Cauchy–Schwarz, this term is bounded by

$$\sqrt{\int \rho_{0,k}\,|\nabla\psi_k - \widetilde{\nabla}\psi|^2\,d\nu}\,\sqrt{\int \frac{|\nabla p_k(\rho_{0,k})|^2}{\rho_{0,k}}\,d\nu}. \tag{23.61}$$

Let us expand the square norm:

$$\begin{aligned}
&\int \rho_{0,k}\,|\nabla\psi_k - \widetilde{\nabla}\psi|^2\,d\nu \\
&= \int \rho_{0,k}\,|\nabla\psi_k|^2\,d\nu + \int \rho_{0,k}\,|\widetilde{\nabla}\psi|^2\,d\nu - 2\int \rho_{0,k}\,\langle\nabla\psi_k, \widetilde{\nabla}\psi\rangle\,d\nu \\
&= \int \rho_{0,k}\,|\nabla\psi_k|^2\,d\nu - \int \rho_{0,k}\,|\widetilde{\nabla}\psi|^2\,d\nu - 2\int \rho_{0,k}\,\langle\nabla\psi_k - \widetilde{\nabla}\psi, \widetilde{\nabla}\psi\rangle\,d\nu.
\end{aligned} \tag{23.62}$$

Observe that:

(a) $\int \rho_{0,k}\,|\nabla\psi_k|^2\,d\nu = W_2(\mu_{0,k},\mu_{1,k})^2$. To prove that this converges to $W_2(\mu_0,\mu_1)^2$, by Corollary 6.11 it suffices to check that

$$W_2(\mu_{0,k},\mu_0) \longrightarrow 0; \qquad W_2(\mu_{1,k},\mu_1) \longrightarrow 0;$$

but this is an immediate consequence of the construction of $\mu_{0,k}$, $\mu_{1,k}$, Theorem 6.9 and Definition 6.8.

(b) $\int \rho_{0,k}\,|\widetilde{\nabla}\psi|^2\,d\nu = (1/Z_{0,k})\int \chi_k\,|\widetilde{\nabla}\psi|^2\,d\mu_0 \longrightarrow \int |\widetilde{\nabla}\psi|^2\,d\mu_0 = W_2(\mu_0,\mu_1)^2$ by monotone convergence.

(c) For any $\varepsilon, M > 0$,

$$\begin{aligned}
&\left|\int \rho_{0,k}\langle\nabla\psi_k - \widetilde{\nabla}\psi, \widetilde{\nabla}\psi\rangle\,d\nu\right| \\
&\le \varepsilon\int \rho_{0,k}\,|\widetilde{\nabla}\psi|\,d\nu + \int_{|\nabla\psi_k - \widetilde{\nabla}\psi|\ge\varepsilon} \rho_{0,k}\,|\nabla\psi_k - \widetilde{\nabla}\psi|\,|\widetilde{\nabla}\psi|\,d\nu \\
&\le \varepsilon\sqrt{\int \rho_{0,k}\,|\widetilde{\nabla}\psi|^2\,d\nu} + \sqrt{\int \rho_{0,k}\,|\nabla\psi_k - \widetilde{\nabla}\psi|^2\,d\nu} \\
&\qquad\qquad \times \sqrt{\int_{|\nabla\psi_k - \widetilde{\nabla}\psi|\ge\varepsilon} \rho_{0,k}\,|\widetilde{\nabla}\psi|^2\,d\nu} \\
&\le \varepsilon\sqrt{\int \rho_0\,|\widetilde{\nabla}\psi|^2\,d\nu} + \sqrt{2}\left(\sqrt{\int \rho_{0,k}\,|\nabla\psi_k|^2\,d\nu} + \sqrt{\int \rho_{0,k}\,|\widetilde{\nabla}\psi|^2\,d\nu}\right) \\
&\qquad\qquad \times \sqrt{M^2\int_{|\nabla\psi_k - \widetilde{\nabla}\psi|\ge\varepsilon} \rho_{0,k}\,d\nu + \int_{|\widetilde{\nabla}\psi|\ge M} \rho_{0,k}\,|\widetilde{\nabla}\psi|^2\,d\nu}
\end{aligned}$$

$$\leq \varepsilon \sqrt{\int \rho_0 \, |\widetilde{\nabla}\psi|^2 \, d\nu} + \sqrt{2} \left(\sqrt{\int \rho_{0,k} \, |\nabla\psi_k|^2 \, d\nu} + \sqrt{2} \sqrt{\int \rho_0 \, |\widetilde{\nabla}\psi|^2 \, d\nu} \right)$$

$$\times \sqrt{M^2 \int_{|\nabla\psi_k - \widetilde{\nabla}\psi| \geq \varepsilon} \rho_{0,k} \, d\nu + \int_{|\widetilde{\nabla}\psi| \geq M} \rho_{0,k} \, |\widetilde{\nabla}\psi|^2 \, d\nu}$$

$$\leq \varepsilon \, W_2(\mu_0, \mu_1) + 2 \Big(W_2(\mu_{0,k}, \mu_{1,k}) + W_2(\mu_0, \mu_1) \Big)$$

$$\times \sqrt{2M^2 \mu_0 \big[\{ |\nabla\psi_k - \widetilde{\nabla}\psi| \geq \varepsilon \} \big] + 2 \int_{|\widetilde{\nabla}\psi| \geq M} \rho_0 \, |\widetilde{\nabla}\psi|^2 \, d\nu}.$$

By letting $\varepsilon \to 0$ and then $M \to \infty$, we conclude that

$$\int \rho_{0,k} \, \langle \nabla\psi_k - \widetilde{\nabla}\psi, \widetilde{\nabla}\psi \rangle \, d\nu \xrightarrow[k\to\infty]{} 0.$$

Plugging back the results of (a), (b) and (c) into (23.62), we obtain

$$\int \rho_{0,k} \, |\nabla\psi_k - \widetilde{\nabla}\psi|^2 \, d\nu \xrightarrow[k\to\infty]{} 0. \tag{23.63}$$

The inequality $Z_{1,k} \leq Z_{0,k}$ implies $Z_{1,k} \, \rho_{0,k} \leq \rho_0 \, \chi_k \leq \rho_0$. This makes it possible to justify the chain-rule as in Step 1: $\nabla p_k(\rho_{0,k}) = p_k'(\rho_{0,k}) \nabla \rho_{0,k}$. Next, since $\rho_{0,k} \in W^{1,1}_{\mathrm{loc}}(M)$ and p is Lipschitz, we can write

$$\int \frac{|\nabla p_k(\rho_{0,k})|^2}{\rho_{0,k}} \, d\nu = \int \frac{\big| p_k'(\rho_{0,k}) \, \nabla \rho_{0,k} \big|^2}{\rho_{0,k}} \, d\nu$$

$$= \frac{Z_{1,k}^2}{Z_{0,k}} \, \frac{p'\left(\frac{Z_{1,k}}{Z_{0,k}} \, \chi_k \, \rho_0 \right)^2 \, \big| \chi_k \, \nabla \rho_0 + \rho_0 \, \nabla \chi_k \big|^2}{\rho_0 \, \chi_k} \, d\nu$$

$$\leq 4 \int p'\left(\frac{Z_{1,k}}{Z_{0,k}} \, \chi_k \, \rho_0 \right)^2 \left(\chi_k \, \frac{|\nabla \rho_0|^2}{\rho_0} + \rho_0 \, \frac{|\nabla \chi_k|^2}{\chi_k} \right) d\nu.$$

Since $Z_{1,k}/Z_{0,k} \leq 1$, we can use conditions (23.52) to (23.54), and a reasoning similar to the one in Step 1, to find a constant C such that

$$p'\left(\frac{Z_{1,k}}{Z_{0,k}} \, \chi_k \, \rho_0 \right)^2 \chi_k \, \frac{|\nabla \rho_0|^2}{\rho_0} \leq C \, p'(\rho_0)^2 \frac{|\nabla \rho_0|^2}{\rho_0} = C \, \frac{|\nabla p(\rho_0)|^2}{\rho_0},$$

which is integrable by assumption. Also, since p' and $(\chi')^2/\chi$ are bounded, there is a constant C such that

$$p'\left(\frac{Z_{1,k}}{Z_{0,k}}\,\chi_k\,\rho_0\right)^2\rho_0\,\frac{|\nabla\chi_k|^2}{\chi_k}\leq C\rho_0,$$

which is of course integrable. The conclusion is that the second integral in (23.61) is bounded. Combining this with (23.63), we conclude that the whole expression in (23.61) converges to 0.

Second term of (23.60)*:* We wish to show that $\int\langle\widetilde{\nabla}\psi,\nabla p_k(\rho_{0,k})\rangle\,d\nu$ converges to $\int\langle\widetilde{\nabla}\psi,\nabla p(\rho_0)\rangle\,d\nu$. As before, we can find a constant C such that

$$|\nabla p_k(\rho_{0,k})|\leq C\,|\nabla p(\rho_0)|.$$

Then the conclusion follows by the dominated convergence theorem, since $\int|\widetilde{\nabla}\psi|\,|\nabla p(\rho_0)|\,d\nu$ is integrable by (23.55). This finishes the treatment of the third term in (23.56).

Fourth term of (23.56)*:* I shall consider three cases:

Case (I): $N=\infty$. Then as in Proposition 17.24(i) the fourth term of (23.56) can be rewritten as $Z_k\,K_{\infty,U}\,W_2(\mu_{0,k},\mu_{1,k})^2/2$, and this converges to $K_{\infty,U}\,W_2(\mu_0,\mu_1)^2/2$ as $k\to\infty$.

Case (II): $N<\infty$ and $K\geq 0$. Then let us just say that the fourth term of (23.56) is nonnegative.

Case (III): $N<\infty$ and $K<0$. This case is much more tricky. By assumption, $\int(1+d(z,x)^q)\,\mu_0(dx)+\int(1+d(z,y)^q)\,\mu_1(dy)<+\infty$, where q satisfies (23.24). Then it follows from the construction of $\mu_{0,k}$ and $\mu_{1,k}$ that

$$\sup_{k\in\mathbb{N}}\int(1+d(z,x)^q)\,\mu_{0,k}(dx)<+\infty;\quad\sup_{k\in\mathbb{N}}\int(1+d(z,y)^q)\,\mu_{1,k}(dy)<+\infty.$$

By Proposition 17.24(ii), there is a function $\eta\in L^1((0,1);dt)$ such that

$$\forall t\in(0,1),\qquad\int\rho_{t,k}(x)^{1-\frac{1}{N}}\,|\nabla\psi_{t,k}(x)|^2\,\nu(dx)\;\leq\eta(t).\qquad(23.64)$$

I claim that

$$\int_0^1\left(\int\rho_t(x)^{1-\frac{1}{N}}|\widetilde{\nabla}\psi_t(x)|^2\,\nu(dx)\right)(1-t)\,dt$$
$$\geq\limsup_{k\to\infty}\int_0^1\left(\int\rho_{t,k}(x)^{1-\frac{1}{N}}|\nabla\psi_{t,k}(x)|^2\,\nu(dx)\right)(1-t)\,dt.\quad(23.65)$$

By (23.64) the integrands in the right-hand side of (23.65) are bounded above by an integrable function, so the claim will follow by Fatou's lemma if we can show that for each $t \in (0,1)$,

$$\int \rho_t(x)^{1-\frac{1}{N}} |\widetilde{\nabla}\psi_t(x)|^2 \,\nu(dx) \ge \limsup_{k\to\infty} \int \rho_{t,k}(x)^{1-\frac{1}{N}} |\nabla\psi_{t,k}(x)|^2 \,\nu(dx). \tag{23.66}$$

In the sequel, t will be fixed in $(0,1)$.

To establish (23.66), we may cut out large values of x, by introducing a cutoff function $\chi_\ell(x)$ ($0 \le \chi_\ell \le 1$, $\chi_\ell = 0$ outside $B[z,2\ell]$, $\chi_\ell = 1$ inside $B[z,\ell]$): this is possible since, by Jensen's inequality as in the beginning of the proof of Proposition 17.24(ii),

$$\begin{aligned}
&\int (1-\chi_\ell(x))\,\rho_t(x)^{1-\frac{1}{N}}\,|\widetilde{\nabla}\psi_t(x)|^2\,\nu(dx)\\
&\le C(t,N,q)\left(1+\int d(z,x)^q\,\mu_0(dx)+\int d(z,x)^q\mu_1(dx)\right)^{1-\frac{1}{N}}\\
&\qquad\qquad\times\left(\int \frac{(1-\chi_\ell(x))^N\,\nu(dx)}{(1+d(z,x))^{q(N-1)-2N}}\right)^{\frac{1}{N}}\\
&\xrightarrow[\ell\to\infty]{} 0,
\end{aligned}$$

and similar bounds hold with ρ_t, $\widetilde{\nabla}\psi_t$ replaced by $\rho_{t,k}$, $\nabla\psi_{t,k}$ (uniformly in k).

Now we shall prove a *locally uniform* bound on $|\nabla\psi_{t,k}|$: for any fixed $t \in (0,1)$,

$$\mu_{t,k}(dx) - \text{almost surely}, \quad x \in B[z,2\ell] \Longrightarrow |\nabla\psi_{t,k}(x)| \le C, \tag{23.67}$$

where C may depend on ℓ and t, but not on k.

The argument will rely on cyclical monotonicity. First we note that, the sequences $(\mu_{0,k})$ and $(\mu_{1,k})$ being tight, there is $\ell \in \mathbb{N}$ such that

$$\forall k \in \mathbb{N}, \qquad \mu_{0,k}\big[B[z,\ell]\big] \ge 3/4, \quad \mu_{1,k}\big[B[z,\ell]\big] \ge 3/4;$$

then a proportion at least $1/2$ of the mass is transferred from $B[z,\ell]$ to $B[z,\ell]$. In particular, if Π_k is a dynamical optimal transference plan associated with $(\mu_{t,k})_{0\le t\le 1}$, and $\Pi_k = \mathrm{law}\,(\gamma)$, then

$$\Pi_k\Big[\Big\{d(z,\gamma_0) \le \ell \text{ and } d(z,\gamma_1) \le \ell\Big\}\Big] \ge \frac{1}{2}. \tag{23.68}$$

Next we recall that $|\nabla\psi_{t,k}(\gamma_t)| = d(\gamma_t,\gamma_1)/(1-t)$, so

$$\begin{aligned}\mu_{t,k}\Big[|\nabla\psi_{t,k}(x)| \geq C \text{ and } d(z,x) \leq 2\ell\Big] \\ &= \Pi_k\left[\frac{d(\gamma_t,\gamma_1)}{1-t} \geq C \quad\text{and}\quad d(z,\gamma_t)\leq 2\ell\right] \\ &= \Pi_k\left[d(\gamma_0,\gamma_1) \geq C\left(\frac{1-t}{t}\right) \quad\text{and}\quad d(z,\gamma_t)\leq 2\ell\right]. \qquad (23.69)\end{aligned}$$

Comnbining (23.68) and (23.69), we conclude that

$$\mu_{t,k}\Big[|\nabla\psi_{t,k}(x)| \geq C \text{ and } d(z,x)\leq 2\ell\Big] \leq 2(\Pi_k\otimes\Pi_k)[E], \qquad (23.70)$$

where E is an event involving two random geodesics γ and $\widetilde{\gamma}$, defined by

$$d(z,\gamma_0)\leq\ell,\quad d(z,\gamma_1)\leq\ell,\quad d(z,\widetilde{\gamma}_t)\leq 2\ell,\quad d(\widetilde{\gamma}_0,\widetilde{\gamma}_1)\geq C\left(\frac{1-t}{t}\right).$$

If these inequalities hold true, then

$$\begin{aligned}& d(\gamma_0,\gamma_1)^2 + d(\widetilde{\gamma}_0,\widetilde{\gamma}_1)^2 - d(\gamma_0,\widetilde{\gamma}_1)^2 - d(\widetilde{\gamma}_0,\widetilde{\gamma}_1)^2 \\ &\geq d(\widetilde{\gamma}_0,\widetilde{\gamma}_1)^2 - \big[d(\gamma_0,z)+d(z,\widetilde{\gamma}_t)+d(\widetilde{\gamma}_t,\widetilde{\gamma}_1)\big]^2 \\ &\qquad\qquad - \big[d(\widetilde{\gamma}_0,\widetilde{\gamma}_t)+d(\widetilde{\gamma}_t,z)+d(z,\widetilde{\gamma}_1)\big]^2 \\ &\geq d(\widetilde{\gamma}_0,\widetilde{\gamma}_1)^2 - 2(1+\delta^{-1})(3\ell)^2 - (1+\delta)\left[d(\widetilde{\gamma}_t,\widetilde{\gamma}_1)^2 + d(\widetilde{\gamma}_0,\widetilde{\gamma}_t)^2\right] \\ &= d(\widetilde{\gamma}_0,\widetilde{\gamma}_1)^2\left[1-(1+\delta)((1-t)^2+t^2)\right] - 18(1+\delta^{-1})\ell^2 \\ &\geq C^2\left(\frac{1-t}{t}\right)^2\left[1-(1+\delta)(1-2t(1-t))\right] - 18(1+\delta^{-1})\ell^2,\end{aligned}$$

and this quantity is positive if δ is chosen small enough and then C large enough. Then the event E has zero $\Pi_k\otimes\Pi_k$ measure since $(e_0,e_1)_\#\Pi_k$ is cyclically monotone (Theorems 5.10 and 7.21). So the left-hand side in (23.70) vanishes, and (23.67) is true. A similar result holds for $\mu_{t,k}$ and $\nabla\psi_{t,k}$ replaced by μ_t and $\widetilde{\nabla}\psi_t$, respectively.

Let $Z_k = \int \chi_\ell\, d\mu_{t,k}$ (which is positive if ℓ is large enough), and let $\widehat{\Pi}_k$ be the probability measure on geodesics defined by

$$\widehat{\Pi}_k(d\gamma) = \frac{\chi_\ell(\gamma_t)\,\widehat{\Pi}_k(d\gamma)}{Z_k}.$$

This is still a dynamical optimal transference plan, so it is associated with a displacement interpolation $(\widehat{\mu}_{s,k})_{0\le s\le 1}$, and a velocity field $\nabla\widehat{\psi}_{s,k}$ (recall Theorem 7.21, or the recap provided in Chapter 13). The definition implies that

$$\widehat{\mu}_{t,k} = \frac{\chi_\ell\,\mu_{t,k}}{Z_k}.$$

Moreover, the "no-crossing property" (in the form of Theorem 7.30(v)) implies that $|\nabla\widehat{\psi}_{t,k}| = |\nabla\psi_{t,k}|$, at least $\widehat{\mu}_{t,k}$-almost surely, or equivalently $\mu_{t,k}$-almost surely on $\operatorname{Spt}\chi_\ell$. We can also make a similar construction for Π and μ_s in place of Π_k and $\mu_{s,k}$, and get (with obvious notation) $\widehat{\Pi}$, $\widehat{\mu}_s$, $\nabla\widehat{\psi}_s$ $(0\le s\le 1)$. It is easily seen that $\widehat{\mu}_{0,k}$ and $\widehat{\mu}_{1,k}$ converge to $\widehat{\mu}_0$ and $\widehat{\mu}_1$, respectively.

The bound (23.67) implies that all plans $\widehat{\Pi}_k$ are supported in a *uniform* compact set. Combining this with the fact that $\widehat{\Pi}$ is the unique dynamical optimal transference plan between its endpoints (Corollary 7.23), it results from Theorem 28.9(v) (proven later in these notes) that the speed fields $|\nabla\psi_{t,k}|$ converge *uniformly*. More precisely, there are nonnegative measurable maps w_k coinciding $\widehat{\mu}_{t,k}$-almost surely with $|\nabla\widehat{\psi}_{t,k}|^2$, and w coinciding $\widehat{\mu}_t$-almost surely with $|\nabla\widehat{\psi}_t|^2$, such that

$$\sup_{x\in B[x,2\ell]} |w_k(x)-w(x)| \xrightarrow[k\to\infty]{} 0. \tag{23.71}$$

On the other hand, $\int \rho_{t,k}^{1-1/N}\,d\nu$ is bounded independently of k (by Theorem 17.8; the moment condition used there is weaker than the one which is presently enforced). This and (23.71) imply

$$\int \chi_\ell(x)\,\rho_{t,k}(x)^{1-\frac{1}{N}}\,|w_k(x)-w(x)|\,\nu(dx) \xrightarrow[k\to\infty]{} 0. \tag{23.72}$$

Since $\mu_{t,k}$ converges weakly to μ_t, the *concavity* of the function $\Phi(r)=r^{1-1/N}$ implies

$$\limsup_{k\to\infty}\int \rho_{t,k}^{1-\frac{1}{N}}\,w\,\chi_\ell\,d\nu \le \int \rho_t^{1-\frac{1}{N}}\,w\,\chi_\ell\,d\nu \tag{23.73}$$

(see Theorem 29.20 later in Chapter 29 and change signs; note that $\chi_\ell\,w\,\nu$ is a compactly supported measure).

Combining (23.72) and (23.73) yields

$$\begin{aligned}
&\limsup_{k\to\infty} \int \rho_{t,k}(x)^{1-\frac{1}{N}}\, \chi_\ell(x)\, |\nabla\psi_{t,k}(x)|^2\, \nu(dx) \\
&= \limsup_{k\to\infty} \int \rho_{t,k}(x)^{1-\frac{1}{N}}\, \chi_\ell(x)\, w_k(x)\, \nu(dx) \\
&= \limsup_{k\to\infty} \int \rho_{t,k}(x)^{1-\frac{1}{N}}\, \chi_\ell(x)\, w(x)\, \nu(dx) \\
&\qquad \le \int \rho_t(x)^{1-\frac{1}{N}}\, \chi_\ell(x)\, w(x)\, \nu(dx) \\
&\qquad = \int \rho_t(x)^{1-\frac{1}{N}}\, \chi_\ell(x)\, |\widetilde{\nabla}\psi_t(x)|^2\, \nu(dx).
\end{aligned}$$

This completes the proof of (23.66). Then we can at last pass to the lim sup in the fourth term of (23.56).
Let us recapitulate: In this step we have shown that

- if $N = \infty$, then

$$U_\nu(\mu_1) \ge U_\nu(\mu_0) + \int \langle \widetilde{\nabla}\psi, \nabla p(\rho_0)\rangle\, d\nu + \frac{K_{\infty,U}\, W_2(\mu_0,\mu_1)^2}{2};$$

- if $N < \infty$ (or $N = \infty$) and $K \ge 0$, then

$$U_\nu(\mu_1) \ge U_\nu(\mu_0) + \int \langle \widetilde{\nabla}\psi, \nabla p(\rho_0)\rangle\, d\nu;$$

- if $N < \infty$ and $K < 0$, then

$$\begin{aligned}
U_\nu(\mu_1) \ge U_\nu(\mu_0) &+ \int \langle \widetilde{\nabla}\psi, \nabla p(\rho_0)\rangle\, d\nu \\
&+ K_{N,U} \int_0^1 \left(\int \rho_t(x)^{1-\frac{1}{N}} |\widetilde{\nabla}\psi_t(x)|^2\, \nu(dx)\right) (1-t)\, dt.
\end{aligned}$$

Step 6: Extension to general pressure laws. The previous step was performed under some regularity assumptions on the nonlinearity U (and therefore on the pressure $p(r) = r\,U'(r) - U(r)$): namely, p was assumed to be Lipschitz, and to satisfy Conditions (23.52) to (23.54).

If p does not satisfy these assumptions, then we can always apply Proposition 17.7 to write U as the monotone limit of U_ℓ, where $U_\ell(r)$ is nondecreasing (resp. nonincreasing) in ℓ for $r \ge R$ (resp. $r \le R$), such that each U_ℓ satisfies:

(a) $p_\ell(r) = a\, r^{1-1/N}$ for r large enough, where $p_\ell(r) = r\, U_\ell'(r) - U_\ell(r)$ (this holds true for $N < \infty$ as well as for $N = \infty$);

(b) $p_\ell(r) = 0$ for r small enough;

(c) If $[0, r_0(\ell)]$ is the interval on which p_ℓ vanishes, then there are $r_1 > r_0$, a function h nondecreasing on $[r_0, r_1]$, and constants K, C such that $Kh \le p_\ell' \le Ch$; in particular, if $r \le r' \le r_1$, then $p_\ell'(r) \le (C/K)\, p_\ell'(r')$;

(d) $p_\ell'(r) \ge K_0\, r^{-1/N}$ for $r \ge r_1$.

This implies that each U_ℓ satisfies all the assumptions for Step 5 to go through. Moreover, Proposition 17.7 guarantees that $U_\ell'' \le C\, U''$ for some constant C which does not depend on ℓ; in particular, $p_\ell' \le C p'$. Admitting for a while that this implies

$$|\nabla p_\ell(\rho_0)| \le C\, |\nabla p(\rho_0)| \tag{23.74}$$

(which is formally obvious, but requires a bit of care since p is not assumed to be Lipschitz), we obtain

$$\int \frac{|\nabla p_\ell(\rho_0)|^2}{\rho_0}\, d\nu \le C^2 \int \frac{|\nabla p(\rho_0)|^2}{\rho_0}\, d\nu < +\infty.$$

So we can write the result of Step 5 with U replaced by U_ℓ and p replaced by p_ℓ, and it remains to pass to the limit as $\ell \to \infty$.

There is no difficulty in showing that $(U_\ell)_\nu(\mu_{t_0})$ converges to $U_\nu(\mu_{t_0})$ for $t_0 \in \{0, 1\}$: This is done by monotone convergence, as in the proof of Theorem 17.15.

If $K \ge 0$, let us just forget about the time-integral. If $K < 0$, then the condition $K_{N,U} > -\infty$ means $p(r) = O(r^{1-1/N})$; then the construction of U_ℓ implies that K_{N,U_ℓ} converges to $K_{N,U}$, so there is no difficulty in passing to the limit in the time-integral. The last subtle point consists in showing that

$$\int \langle \widetilde{\nabla}\psi, \nabla p_\ell(\rho_0)\rangle\, d\nu \xrightarrow[\ell\to\infty]{} \int \langle \widetilde{\nabla}\psi, \nabla p(\rho_0)\rangle\, d\nu. \tag{23.75}$$

First we check that $\nabla p_\ell(\rho_0)$ converges ν-almost everywhere to $\nabla p(\rho_0)$. Let $r_0 \ge 0$ be such that $U'' = 0$ on $(0, r_0]$. For any $a > r_0$, $b > a$, if $a \le \rho_0 \le b$ we can write ρ_0 as a Lipschitz function of $p(\rho_0)$, which by assumption lies in $W^{1,1}_{\mathrm{loc}}$; this implies $1_{a\le\rho_0\le b} \nabla p(\rho_0) = 1_{a\le\rho_0\le b}\, p'(\rho_0)\, \nabla\rho_0$. So

$$1_{a\le\rho_0\le b}\nabla p_\ell(\rho_0) = p'_\ell(\rho_0)\,1_{a\le\rho_0\le b}\,\nabla\rho_0 \xrightarrow[\ell\to\infty]{} 1_{a\le\rho_0\le b}\nabla p(\rho_0).$$

This proves that $\nabla p_\ell(\rho_0)$ converges almost surely to $\nabla p(\rho_0)$ on each set $\{r_0 < a \le \rho_0 \le b\}$, and therefore on the whole of $\{\rho_0 > r_0\}$. On the other hand, if $\rho_0 \le r_0$ then $p(\rho_0) = 0$, so $\nabla p(\rho_0)$ vanishes almost surely on $\{\rho_0 \le r_0\}$ (this is a well-known theorem from distribution theory; see the bibliographical notes in case of need), and also $\nabla p_\ell(\rho_0) = 0$ on that set. This proves the almost everywhere convergence of $\nabla p_\ell(\rho_0)$ to $\nabla p(\rho_0)$. At the same time, this reasoning proves (23.74).

So to pass to the limit in (23.75) it suffices to prove that the integrand is dominated by an integrable function. But

$$\begin{aligned}\big|\langle\widetilde{\nabla}\psi, \nabla p_\ell(\rho_0)\rangle\big| &\le |\widetilde{\nabla}\psi|\,|\nabla p_\ell(\rho_0)|\\ &\le C\,|\widetilde{\nabla}\psi|\,|\nabla p(\rho_0)|\end{aligned}$$

which is integrable, since, again,

$$\begin{aligned}\int |\widetilde{\nabla}\psi|\,|\nabla p(\rho_0)|\,d\nu &\le \sqrt{\int \rho_0\,|\widetilde{\nabla}\psi|^2\,d\nu}\sqrt{\int \frac{|\nabla p(\rho_0)|^2}{\rho_0}\,d\nu}\\ &= W_2(\mu_0,\mu_1)\,\sqrt{I_{U,\nu}(\mu_0)}.\end{aligned}$$

Step 7: Differential reformulation and conclusion. To conclude the proof of Theorem 23.14, I shall distinguish three cases:

Case (I): $N < \infty$ and $K \ge 0$. Then we know

$$U_\nu(\mu_1) \ge U_\nu(\mu_0) + \int \langle\widetilde{\nabla}\psi, \nabla p(\rho_0)\rangle\,d\nu. \tag{23.76}$$

This is a priori weaker than (23.26), but we shall manage to improve this inequality thanks to the displacement convexity properties of U_ν. First, by applying (23.76) with μ_t replaced by μ_1 and ψ replaced by $t\psi$, and then passing to the limit (which is also an infimum) as $t \to 0$, we obtain

$$\liminf_{t\downarrow 0} \frac{U_\nu(\mu_t) - U_\nu(\mu_0)}{t} \ge \int \langle\widetilde{\nabla}\psi, \nabla p(\rho_0)\rangle\,d\nu. \tag{23.77}$$

By Theorem 17.15,

$$U_\nu(\mu_t) + K_{N,U} \int_0^1 \left(\int \rho_s(x)^{1-\frac{1}{N}} \, |\widetilde{\nabla}\psi_s(x)|^2 \, \nu(dx) \right) G(s,t) \, ds$$
$$\leq (1-t) \, U_\nu(\mu_0) + t \, U_\nu(\mu_1),$$

where $G(s,t)$ is the Green function of the one-dimensional Laplace operator; in particular $G(s,t) = t(1-s)$ for $s \in [t,1]$. This combined with (23.77) implies

$$\int \langle \widetilde{\nabla}\psi, \nabla p(\rho_0) \rangle \, d\nu + K_{N,U} \int_t^1 \left(\int \rho_s^{1-\frac{1}{N}} \, |\widetilde{\nabla}\psi_s|^2 \, d\nu \right) (1-s) \, ds$$
$$\leq \frac{U_\nu(\mu_t) - U_\nu(\mu_0)}{t} + K_{N,U} \int_0^1 \left(\int \rho_s^{1-\frac{1}{N}} \, |\widetilde{\nabla}\psi_s|^2 \, d\nu \right) \frac{G(s,t)}{t} \, ds$$
$$\leq U_\nu(\mu_1) - U_\nu(\mu_0),$$

and the result follows by letting $t \to 0$. (This works for $N < \infty$ as well as for $N = \infty$; recall Proposition 17.24(i).)

Case (II): $N = \infty$ and $K < 0$. Then

$$U_\nu(\mu_1) \geq U_\nu(\mu_0) + \int \langle \widetilde{\nabla}\psi, \nabla p(\rho_0) \rangle \, d\nu + \frac{K W_2(\mu_0,\mu_1)^2}{2}. \tag{23.78}$$

This proves (23.26). To establish (23.25), let us write (23.78) with μ_1 replaced by μ_t, and ψ replaced by $t\psi$ $(0 \leq t \leq 1)$; this gives

$$U_\nu(\mu_t) \geq U_\nu(\mu_0) + t \int \langle \widetilde{\nabla}\psi, \nabla p(\rho_0) \rangle \, d\nu + t^2 \, \frac{K W_2(\mu_0,\mu_1)^2}{2}.$$

After dividing by t and passing to the lim inf, we recover (23.77) again.

Case (III): $N < \infty$ and $K < 0$. Then we have

$$U_\nu(\mu_1) \geq U_\nu(\mu_0) + \int \langle \widetilde{\nabla}\psi, \nabla p(\rho_0) \rangle \, d\nu$$
$$+ K_{N,U} \int_0^1 \left(\int \rho_s(x)^{1-\frac{1}{N}} |\widetilde{\nabla}\psi_s(x)|^2 \, \nu(dx) \right) (1-s) \, ds,$$

which is the same as (23.26). If we change μ_1 for μ_t, then ψ should be replaced by $t\psi$, ρ_s by ρ_{st} and ψ_s by $t\psi_{st}$; the result is

$$\begin{aligned} U_\nu(\mu_t) &\geq U_\nu(\mu_0) + t \int \langle \widetilde{\nabla}\psi, \nabla p(\rho_0)\rangle \, d\nu \\ &\qquad + t^2 \, K_{N,U} \int_0^1 \left(\int \rho_{st}(x)^{1-\frac{1}{N}} |\widetilde{\nabla}\psi_{st}(x)|^2 \, \nu(dx) \right) (1-s)\, ds \\ &\geq U_\nu(\mu_0) + t \int \langle \widetilde{\nabla}\psi, \nabla p(\rho_0)\rangle \, d\nu \\ &\qquad + t^2 \, \frac{K_{N,U}}{2} \left(\sup_{0\leq\tau\leq t} \int \rho_\tau(x)^{1-\frac{1}{N}} |\widetilde{\nabla}\psi_\tau(x)|^2 \, \nu(dx) \right). \end{aligned}$$

The first part of the proof of Proposition 17.24(ii) shows that the expression inside brackets is uniformly bounded as soon as, say, $t \leq 1/2$. So

$$U_\nu(\mu_t) \geq U_\nu(\mu_0) + t \int \langle \widetilde{\nabla}\psi, \nabla p(\rho_0)\rangle \, d\nu - O(t^2) \qquad \text{as } t \to 0,$$

and we can conclude as before.

This finishes the proof of Theorem 23.14 in all cases. □

Diffusion equations as gradient flows

Now we are equipped to identify certain nonlinear diffusive equations as gradient flows in the Wasserstein space.

Theorem 23.19 (Diffusion equations as gradient flows in the Wasserstein space). *Let M be a Riemannian manifold equipped with a reference measure $\nu = e^{-V}\mathrm{vol}$, $V \in C^2(M)$, satisfying a $\mathrm{CD}(K,N)$ curvature-dimension bound for some $K \in \mathbb{R}$, $N \in (1,\infty]$. Let $L = \Delta - \nabla V \cdot \nabla$. Let U be a nonlinearity in $\mathcal{DC}_N$, such that $U \in C^3(0,+\infty)$; and let $p(r) = r\,U'(r) - U(r)$. Let $\rho = \rho_t(x)$ be a smooth (C^1 in t, C^2 in x) positive solution of the partial differential equation*

$$\frac{\partial \rho_t}{\partial t} = L\, p(\rho_t), \tag{23.79}$$

and let $\mu_t = \rho_t\, \nu$. Assume that $U_\nu(\mu_t) < +\infty$ for all $t > 0$; and that for all $0 < t_1 < t_2$,

$$\int_{t_1}^{t_2} I_{U,\nu}(\mu_t)\, dt < +\infty.$$

- *If $K < 0$, further assume that $p(r) = O(r^{1-1/N})$ as $r \to \infty$.*
- *If M is noncompact, further assume, with the same notation as in Theorem 23.14, that $\mu_t \in P_p^{\mathrm{ac}}(M)$.*

- *If M is noncompact, $K < 0$ and $N < \infty$, reinforce the latter assumption into $\mu_t \in P_q^{\mathrm{ac}}(M)$, where q is as in Theorem 23.14.*

Then $(\mu_t)_{t\geq 0}$ is a trajectory of the gradient flow associated with the energy functional U_ν in $P_2^{\mathrm{ac}}(M)$.

Remark 23.20. If (ρ_t) is reasonably well-behaved at infinity (a fortiori if M is compact), then $t \to U_\nu(\mu_t)$ is nonincreasing (see Theorem 24.2(ii) in the next chapter). Thus the assumption $U_\nu(\mu_t) < +\infty$ is satisfied as soon as $U_\nu(\mu_0) < +\infty$. However, it is interesting to cover also cases where $U_\nu(\mu_0) = +\infty$.

Remark 23.21. The finiteness of $I_{U,\nu}(\mu_t) = \int \rho_t \, |\nabla U'(\rho_t)|^2 \, d\nu$ is a reinforcement of the condition $p(\rho) \in W^{1,1}_{\mathrm{loc}}(M)$, since $\int |\nabla p(\rho)| \, d\nu \leq \sqrt{I_{U,\nu}(\mu)}$.

Example 23.22. If M is compact, any smooth positive solution of $\partial_t \rho = \Delta \rho$ can be seen as a trajectory of the gradient flow associated with the energy $H(\mu) = \int \rho \log \rho$. Similarly, any smooth positive solution of $\partial_t \rho = \Delta \rho + \nabla \cdot (\rho \, \nabla V)$ can be seen as a trajectory of the gradient flow associated with the energy $F(\mu) = \int \rho \log \rho + \int \rho V$. (With respect to the previous example, this amounts to changing the reference measure vol into e^{-V}vol.) If M has dimension n, any smooth positive solution of $\partial_t \rho = \Delta \rho^m$, $m \geq 1 - 1/n$, can be seen as a trajectory of the gradient flow associated with the energy $E(\mu) = (m-1)^{-1} \int \rho^m$. All these statements can be generalized to noncompact manifolds, under adequate global smoothness and decay assumptions. For instance, any smooth positive solution of $\partial_t \rho = \Delta \rho$ in $\mathbb{R}^n$, with $\int \rho_0(x) |x|^2 \, dx < +\infty$, is a trajectory of the gradient flow associated with the H functional.

Proof of Theorem 23.19. First note that the assumptions of the theorem imply $K_{N,U} > -\infty$.

Because U is C^3 on $(0, +\infty)$, the function $U'(\rho)$ is C^2, so

$$\xi_t(x) := -\nabla U'(\rho_t(x))$$

is a C^1 vector field. Then (23.79) can be rewritten as

$$\frac{\partial \mu_t}{\partial t} + \nabla \cdot (\xi_t \, \mu_t) = 0.$$

Let $\sigma \in P_2^{\mathrm{ac}}(M)$. By Theorem 23.9, the definition of ξ and the identity $\rho \, U''(\rho) = p'(\rho)$, for almost any t,

$$\frac{d}{dt}\left(\frac{W_2(\mu_t,\sigma)^2}{2}\right) = -\int \langle \widetilde{\nabla}\psi_t, \xi_t\rangle\, d\mu_t$$
$$= \int \langle \widetilde{\nabla}\psi_t, \nabla U'(\rho_t)\rangle\, d\mu_t$$
$$= \int \langle \widetilde{\nabla}\psi_t, \nabla p(\rho_t)\rangle\, d\nu, \qquad (23.80)$$

where $\exp(\widetilde{\nabla}\psi_t)$ is the Monge transport $\mu_t \to \sigma$.

Let $(\mu^{(s)})$ be the displacement interpolation joining $\mu^{(0)} = \mu_t$ to $\mu^{(1)} = \sigma$. By Theorem 23.14,

$$\liminf_{s\downarrow 0} \frac{U_\nu(\mu^{(s)}) - U_\nu(\mu^{(0)})}{s} \geq \int \bigl\langle \widetilde{\nabla}\psi_t, \nabla p(\rho_t)\bigr\rangle\, d\nu. \qquad (23.81)$$

The combination of (23.80) and (23.81) implies

$$\frac{d^+}{dt}\left(\frac{W_2(\mu_t,\sigma)^2}{2}\right) \leq \limsup_{s\downarrow 0} \frac{U_\nu(\mu^{(s)}) - U_\nu(\mu^{(0)})}{s},$$

and the conclusion follows from Definition 23.7. □

In Theorem 23.19 I assumed the smoothness of the density; but in many situations there are regularization theorems for such (a priori nonlinear) diffusion equations, so the smoothness assumption can be relaxed in the end. Such is the case for the heat equation. Here is a result for this case, stated without proof (for simplicity I shall only consider compact manifolds; noncompact manifolds would require moment estimates):

Corollary 23.23 (Heat equation as a gradient flow). *Let M be a compact Riemannian manifold curvature, let $V \in C^2(M)$, and let $L = \Delta - \nabla V \cdot \nabla$. Let $\mu_0 \in P_2(M)$, and let $\mu_t = \rho_t \nu$ solve*

$$\frac{\partial \rho_t}{\partial t} = L\rho_t.$$

Then $(\mu_t)_{t>0}$ is a trajectory of the gradient flow associated with the energy functional

$$H_\nu(\mu) = \int \rho \log \rho \, d\nu, \qquad \mu = \rho\nu$$

in the Wasserstein space $P_2^{\mathrm{ac}}(M)$.

In particular, the gradient flow associated with H_{vol} is the standard heat equation

$$\frac{\partial \rho}{\partial t} = \Delta\rho.$$

Remark 23.24. The distinction between $P_2(M)$ and $P_2^{ac}(M)$ is not essential here, but I have to do it since in Theorems 23.9 and 23.14 I have only worked with absolutely continuous measures.

Remark 23.25. As I already said at the beginning of this chapter, the heat equation can be seen as a gradient flow in various ways. For instance, take for simplicity the basic heat equation in $\mathbb{R}^n$, in the form $\partial_t u = \Delta u$, then it can be interpreted as the gradient flow of the functional $E(u) = (1/2) \int |\nabla u|^2$ for the usual Hilbert structure imposed by the L^2 norm; or as the gradient flow of the functional $E(u) = \int u^2$ for the Hilbert structure induced by the H^{-1} norm (say on the subspace $\int u = 0$). But the interesting new feature coming from Theorem 23.19 is that now the heat equation can be seen as the gradient flow of a nice functional which has statistical (or thermodynamical) meaning; and in such a way that it is naturally set in the space of probability measures. There are other reasons why this new interpretation seems "natural"; see the bibliographical notes for more information.

Stability

A good point of our weak formulation of gradient flows is that it comes with stability estimates almost for free. This is illustrated by the next theorem, in which regularity assumptions are far from optimal.

Theorem 23.26 (Stability of gradient flows in the Wasserstein space). *Let $\mu_t, \widehat{\mu}_t$ be two solutions of (23.79), satisfying the assumptions of Theorem 23.19 with either $K \geq 0$ or $N = \infty$. Let $\lambda = K_{\infty,U}$ if $N = \infty$; and $\lambda = 0$ if $N < \infty$ and $K \geq 0$. Then, for all $t \geq 0$,*

$$W_2(\mu_t, \widehat{\mu}_t) \leq e^{-\lambda t}\, W_2(\mu_0, \widehat{\mu}_0).$$

Remark 23.27. I don't know how seriously the restrictions on K and N should be taken.

Proof of Theorem 23.26. By Theorem 23.9, for almost any t,

$$\frac{d}{dt}\left(\frac{W_2(\mu_t, \widehat{\mu}_t)^2}{2}\right) = \int \langle \widetilde{\nabla}\psi_t, \nabla U'(\rho_t)\rangle \, d\mu_t + \int \langle \widetilde{\nabla}\widehat{\psi}_t, \nabla U'(\widehat{\rho}_t)\rangle \, d\widehat{\mu}_t, \tag{23.82}$$

where $\exp(\widetilde{\nabla}\psi_t)$ (resp. $\exp(\widetilde{\nabla}\widehat{\psi}_t)$) is the optimal transport $\mu_t \to \widehat{\mu}_t$ (resp. $\widehat{\mu}_t \to \mu_t$).

By the chain-rule and Theorem 23.14,

$$\int \langle \widetilde{\nabla}\psi_t, \nabla U'(\rho_t)\rangle \, d\mu_t = \int \langle \widetilde{\nabla}\psi_t, \nabla p(\rho_t)\rangle \, d\nu$$
$$\leq U_\nu(\widehat{\mu}_t) - U_\nu(\mu_t) - \lambda \frac{W_2(\mu_t, \widehat{\mu}_t)^2}{2}. \qquad (23.83)$$

Similarly,

$$\int \langle \widetilde{\nabla}\widehat{\psi}_t, \nabla U'(\widehat{\rho}_t)\rangle \, d\widehat{\mu}_t \leq U_\nu(\mu_t) - U_\nu(\widehat{\mu}_t) - \lambda \frac{W_2(\widehat{\mu}_t, \mu_t)^2}{2}. \qquad (23.84)$$

The combination of (23.82), (23.83) and (23.84) implies

$$\frac{d}{dt}\left(\frac{W_2(\mu_t, \widehat{\mu}_t)^2}{2}\right) \leq -2\lambda \left(\frac{W_2(\mu_t, \widehat{\mu}_t)^2}{2}\right).$$

Then the result follows from Gronwall's lemma. □

General theory and time-discretization

This last section evokes some key issues which I shall not develop, although they are closely related to the material in the rest of this chapter.

There is a general theory of gradient flows in metric spaces, based for instance on Definition 23.7, or other variants appearing in Proposition 23.1. Motivations for these developments come from both pure and applied mathematics. This theory was pushed to a high degree of sophistication by many researchers, in particular De Giorgi and his school. A key role is played by discrete-time **approximation schemes**, the simplest of which can be stated as follows:

1. Choose your initial datum X_0;
2. Choose a time step τ, which in the end will decrease to 0;
3. Let $X_1^{(\tau)}$ be a minimizer of $X \longmapsto \Phi(X) + \dfrac{d(X_0, X)^2}{2\tau}$; then define inductively $X_{k+1}^{(\tau)}$ as a minimizer of $X \longmapsto \Phi(X) + \dfrac{d(X_k^{(\tau)}, X)^2}{2\tau}$.
4. Pass to the limit in $X_k^{(\tau)}$ as $\tau \to 0$, $k\tau \to t$, hopefully recover a function $X(t)$ which is the value of the gradient flow at time t.

Such schemes sometimes provide an excellent way to construct the gradient flow, and they may be useful in numerical simulations. They also give a more precise formulation of the statement according to which gradient flows make the energy decrease "as fast as possible".

There are strong results for the convergence of time-discretized gradient flow as $\tau \to 0$; see the bibliographical notes for details.[1]

The time-discretization procedure also suggests a better understanding of the gradient flow in Wasserstein distance, as I shall explain in a slightly informal way. Consider, as in Theorem 23.19, the partial differential equation

$$\frac{\partial \rho}{\partial t} = L\, p(\rho).$$

Suppose you know the density $\rho(t)$ at some time t, and look for the density $\rho(t+dt)$ at a later time, where dt is infinitesimally small. To do this, minimize the quantity

$$U_\nu(\mu_{t+dt}) - U_\nu(\mu_t) + \frac{W_2(\mu_t, \mu_{t+dt})^2}{2\, dt}.$$

By using the interpretation of the Wasserstein distance between infinitesimally close probability measures, this can also be rewritten as

$$\frac{W_2(\mu_t, \mu_{t+dt})^2}{dt} \simeq \inf \left\{ \int |v|^2\, d\mu_t; \quad \frac{\partial \mu}{\partial t} + \nabla \cdot (\mu v) = 0 \right\}.$$

To go from $\mu(t)$ to $\mu(t+dt)$, what you have to do is *find a velocity field v inducing an infinitesimal variation $d\mu = -\nabla \cdot (\mu v)\, dt$, so as to minimize the infinitesimal quantity*

$$dU_\nu + K dt, \tag{23.85}$$

where $U_\nu(\mu) = \int U(\rho)\, d\nu$, and K is the kinetic energy $(1/2) \int |v|^2\, d\mu$ (so $K dt$ is the infinitesimal action). For the heat equation $\frac{\partial \rho}{\partial t} = \Delta \rho$, $\nu = \mathrm{vol}$, $U_\nu(\mu) = \int \rho \log \rho\, d\nu$, we are back to the example discussed at the beginning of this chapter, and (23.85) can be rewritten as an "infinitesimal variation of free energy", say

$$K dt - dS,$$

with S standing for the entropy.

[1] These notes stop before what some readers might consider as the most interesting part, namely try to *construct* solutions by use of the gradient flow interpretation.

There is an important moral here: Behind many *nonequilibrium* equations of statistical mechanics, there is a variational principle involving entropy and energy, or functionals alike — just as in equilibrium statistical mechanics.

Appendix: A lemma about doubling variables

The following important lemma was used in the proof of Theorems 23.9 and 23.26.

Lemma 23.28 (Differentiation through doubling of variables). *Let $F = F(s,t)$ be a function $[0,T] \times [0,T] \to \mathbb{R}$, locally absolutely continuous in s, uniformly in t; and locally absolutely continuous in t, uniformly in s. Then $t \to F(t,t)$ is absolutely continuous, and for almost all t_0,*

$$\frac{d}{dt}\bigg|_{t=t_0} F(t,t) \le \limsup_{t\uparrow t_0} \left(\frac{F(t,t_0) - F(t_0,t_0)}{t-t_0}\right) + \limsup_{t\downarrow t_0} \left(\frac{F(t_0,t) - F(t_0,t_0)}{t-t_0}\right). \tag{23.86}$$

If moreover $F(t_0,\cdot\,)$ and $F(\,\cdot\,,t_0)$ are differentiable at all times, for almost all t_0, then the inequality (23.86) *can be reinforced into the equality*

$$\frac{d}{dt}\bigg|_{t=t_0} F(t,t) = \frac{d}{dt}\bigg|_{t=t_0} F(t,t_0) + \frac{d}{dt}\bigg|_{t=t_0} F(t_0,t). \tag{23.87}$$

Explicitly, to say that F is locally absolutely continuous in s, uniformly in t, means that there is a fixed function $u \in L^1_{\mathrm{loc}}(dt)$ such that

$$\sup_{0\le t\le T} \big|F(s,t) - F(s',t)\big| \le \int_s^{s'} u(\tau)\,d\tau.$$

Remark 23.29. Lemma 23.28 does not allow us to conclude to (23.87) if it is only known that for any t_0, $F(t,t_0)$ and $F(t_0,t)$ are differentiable almost everywhere as functions of t. Indeed, it might be a priori that differentiability fails precisely at $t = t_0$, for all t_0.

Proof of Lemma 23.28. By assumption there are functions $u \in L^1_{\mathrm{loc}}(dt)$ and $v \in L^1_{\mathrm{loc}}(ds)$ such that

$$\begin{cases} \sup_{0\le t\le T} \bigl|F(s,t) - F(s',t)\bigr| \le \int_s^{s'} u(\tau)\,d\tau \\ \\ \sup_{0\le s\le T} \bigl|F(s,t) - F(s,t')\bigr| \le \int_t^{t'} v(\tau)\,d\tau. \end{cases}$$

Without loss of generality we may take $u = v$.

Let $f(t) = F(t,t)$. Then we have $|f(s) - f(t)| \le |F(s,s) - F(s,t)| + |F(s,t) - F(t,t)| \le 2\int_s^t u(\tau)\,d\tau$; so f is locally absolutely continuous.

Let $\dot f$ stand for the derivative of f. Since f is absolutely continuous, $\dot f$ is also (almost everywhere) the distributional derivative of f. The goal is to show that $\dot f(t)$ is bounded above by the right-hand side of (23.86). If this is true, then the rest of the proof follows easily: Indeed, if $F(s,t)$ is differentiable in s and t then

$$\frac{d}{dt}\Big|_{t=t_0} F(t,t) \le \frac{d}{dt}\Big|_{t=t_0} F(t,t_0) + \frac{d}{dt}\Big|_{t=t_0} F(t_0,t),$$

and the reverse inequality will be obtained by changing F for $-F$.

Let ζ be a C^∞ nonnegative function supported in $(0,1)$. For h small enough, $\zeta(\,\cdot + h)$ is also supported in $(0,1)$, and

$$\begin{aligned}\int \dot f\zeta = -\int f\dot\zeta = \lim_{h\downarrow 0}\int_0^1 f(t)\left[\frac{\zeta(t-h)-\zeta(t)}{h}\right] dt \\ = \lim_{h\downarrow 0}\int_0^1 \zeta(t)\left[\frac{f(t+h)-f(t)}{h}\right] dt.\end{aligned}$$

Replacing f by its expression in terms of F, we get

$$\begin{aligned}\int \dot f\zeta &= \lim_{h\downarrow 0}\left\{\int_0^1 \zeta(t)\left[\frac{F(t+h,t+h)-F(t,t+h)}{h}\right] dt \right.\\ &\qquad\left. + \int_0^1 \zeta(t)\left[\frac{F(t,t+h)-F(t,t)}{h}\right] dt\right\} \\ &\le \limsup_{h\downarrow 0}\int_0^1 \zeta(t-h)\left[\frac{F(t,t)-F(t-h,t)}{h}\right] dt \\ &\qquad + \limsup_{h\downarrow 0}\int_0^1 \zeta(t)\left[\frac{F(t,t+h)-F(t,t)}{h}\right] dt. \end{aligned} \tag{23.88}$$

In the first integral on the right-hand side of (23.88), it is possible to replace $\zeta(t-h)$ by $\zeta(t)$; indeed, since ζ and $\zeta(\cdot - h)$ are supported in $(\delta, 1-\delta)$ for some $\delta > 0$, we may write

$$\begin{aligned}
&\left|\int_0^1 [\zeta(t-h) - \zeta(t)] \left(\frac{F(t,t) - F(t-h,t)}{h}\right) dt\right| \\
&\qquad \le \|\zeta\|_{\mathrm{Lip}} \int_\delta^{1-\delta} |F(t,t) - F(t-h,t)|\, dt \\
&\qquad \le \|\zeta\|_{\mathrm{Lip}} \int_0^1 \left(\int_{t-h}^t u(\tau)\, d\tau\right) dt \\
&\qquad = \|\zeta\|_{\mathrm{Lip}} \int_\delta^{1-\delta} \left(\int_\tau^{\min(\tau+h,1)} dt\right) u(\tau)\, d\tau \\
&\qquad \le \|\zeta\|_{\mathrm{Lip}}\, h \int_\delta^{1-\delta} u(\tau)\, d\tau \ = O(h).
\end{aligned}$$

To summarize:

$$\begin{aligned}
\int_0^1 \zeta \dot{f} \le \limsup_{h\downarrow 0} \int_0^1 \zeta(t) \left[\frac{F(t,t) - F(t-h,t)}{h}\right] dt& \\
+ \limsup_{h\downarrow 0} \int_0^1 \zeta(t) \left[\frac{F(t,t+h) - F(t,t)}{h}\right] dt.& \qquad (23.89)
\end{aligned}$$

By assumption,

$$\left|\frac{F(t,t) - F(t-h,t)}{h}\right| \le \frac{1}{h}\int_{t-h}^t u(\tau)\, d\tau;$$

and by Lebesgue's density theorem, the right-hand side converges to u in $L^1_{\mathrm{loc}}(dt)$ as $h \to 0$. This makes it possible to apply Fatou's lemma, in the form

$$\begin{aligned}
\limsup_{h\downarrow 0} \int_0^1 \zeta(t) \left[\frac{F(t,t) - F(t-h,t)}{h}\right] dt& \\
\le \int_0^1 \zeta(t) \limsup_{h\downarrow 0} \left[\frac{F(t,t) - F(t-h,t)}{h}\right] dt.& \qquad (23.90)
\end{aligned}$$

Similarly,

$$\begin{aligned}
\limsup_{h\downarrow 0} \int_0^1 \zeta(t) \left[\frac{F(t,t+h) - F(t,t)}{h}\right] dt& \\
\le \int_0^1 \zeta(t) \limsup_{h\downarrow 0} \left[\frac{F(t,t+h) - F(t,t)}{h}\right] dt.& \qquad (23.91)
\end{aligned}$$

Plugging (23.90) and (23.91) back in (23.89), we find

$$\int_0^1 \zeta\dot{f} \leq \int \zeta(t)\left\{\limsup_{h\downarrow 0}\left(\frac{F(t,t)-F(t-h,t)}{h}\right)\right.$$
$$\left.+\limsup_{h\downarrow 0}\left(\frac{F(t,t+h)-F(t,t)}{h}\right)\right\}dt.$$

Since ζ is arbitrary, $\dot{f}$ is bounded above by the expression in curly brackets, almost everywhere. This concludes the proof. □

Bibliographical notes

Historically, the development of the theory of abstract gradient flows was initiated by De Giorgi and coworkers (see e.g. [19, 275, 276]) on the basis of the time-discretized variational scheme; and by Bénilan [95] on the basis of the variational inequalities involving the square distance, as in Proposition 23.1(iv)–(vi). The latter approach has the advantage of incorporating stability and uniqueness as a built-in feature, while the former is more efficient in establishing existence. Bénilan introduced his method in the setting of Banach spaces, but it applies just as well to abstract metric spaces. Both approaches work in the Wasserstein space. De Giorgi also introduced the formulation in Proposition 23.1(ii), which is an alternative "intrinsic" definition for gradient flows in metric spaces.

Basically all methods used to construct abstract gradient flows rely on some convexity property. But the interplay is in both directions: the existence of an energy-decreasing flow satisfying (23.2) implies the λ-convexity of the energy Φ. When (23.2) is replaced by a suitable time-integrated formulation this becomes a powerful way to study convexity properties in metric spaces; see [271] for a neat application to the study of displacement convexity.

Currently, *the* reference for abstract gradient flows is the recent monograph by Ambrosio, Gigli and Savaré [30]. (There is a short version in Ambrosio's Santander lecture notes [22], and a review by Gangbo [397].) There the reader will find the most precise results known to this day, apart from some very recent refinements. More than half of the book is devoted to gradient flows in the space of probability measures on $\mathbb{R}^n$ (or a separable Hilbert space). Issues about the replacement of $P_2(\mathbb{R}^n)$ by $P_2^{ac}(\mathbb{R}^n)$ are carefully discussed there. Some abstract results from [30] are improved by Gigli [415] (with the help of an interesting quasi-minimization theorem which can be seen as a consequence of the Ekeland variational principle).

The results presented in this chapter extend some of the results in [30] to $P_2(M)$, where M is a Riemannian manifold, sometimes at the price of less precise conclusions. In another direction of generalization, Fang, Shao and Sturm [342] have considered gradient flows in $P_2(\mathcal{W})$, where $\mathcal{W}$ is an abstract Wiener space.

Other treatments of gradient flows in nonsmooth structures, under various curvature assumptions, are due to Perelman and Petrunin [678], Lytchak [584], Ohta [655] and Savaré [735]; the first two references are concerned with Alexandrov spaces, while the latter two deal with so-called 2-uniform spaces. The assumption of 2-uniform smoothness is relevant for optimal transport, since the Wasserstein space over a Riemannian manifold is not an Alexandrov space in general (except for nonnegative curvature). All these works address the construction of gradient flows under various sets of geometric assumptions.

The classical theory of gradient flows in Hilbert spaces, mostly for convex functionals, based on Remark 23.3, is developed in Brézis [171] and other sources; it is also implicitly used in several parts of the popular book by J.-L. Lions [557].

The differentiability of the Wasserstein distance in $P_2^{ac}(\mathbb{R}^n)$, and in fact in $P_p^{ac}(\mathbb{R}^n)$ (for $1 < p < \infty$), is proven in [30, Theorems 10.2.2 and 10.2.6, Corollary 10.2.7]. The assumption of absolute continuity of the probability measures is not crucial for the superdifferentiability (actually in [30, Theorem 10.2.2] there is no such assumption). For the subdifferentiability, this assumption is only used to guarantee the uniqueness of the transference plan. Proofs in [30] slightly differ from the proofs in the present chapter.

There is a more general statement that the Wasserstein distance $W_2(\sigma, \mu_t)$ is almost surely (in t) differentiable along any absolutely continuous curve $(\mu_t)_{0\leq t\leq 1}$, without any assumption of absolute continuity of the measures [30, Theorem 8.4.7]. (In this reference, only the Euclidean space is considered, but this should not be a problem.)

There is a lot to say about the genesis of Theorem 23.14, which can be considered as a refinement of Theorem 20.1. The exact computation of Step 1 appears in [669, 671] for particular functions U, and in [814, Theorem 5.30] for general functions U; all these references only consider $M = \mathbb{R}^n$. The procedure of extension of $\nabla\psi$ (Step 2) appears e.g. in [248, Proof of Theorem 2] (in the particular case of convex functions). The integration by parts of Step 3 appears in many papers; under adequate assumptions, it can be justified in the whole of $\mathbb{R}^n$ (without any assumption of compact support): see [248, Lemma 7], [214, Lemma 5.12], [30, Lemma 10.4.5]. The proof in [214, 248] relies on the possibility to find an exhaustive sequence of cutoff functions with Hessian uniformly bounded, while the proof in [30] uses the fact that in $\mathbb{R}^n$, the distance to a convex set is a convex function. None of these arguments seems to apply in more general noncompact Riemannian manifolds (the second proof would probably work in nonnegative curvature), so I have no idea whether the integration by parts in the proof of Theorem 23.14 could be performed without compactness assumption; this is the reason why I went through the painful[2] approximation procedure used in the end of the proof of Theorem 23.14.

It is interesting to compare the two strategies used in the extension from compact to noncompact situations, in Theorem 17.15 on the one hand, and in Theorem 23.14 on the other. In the former case, I could use the standard approximation scheme of Proposition 13.2, with an excellent control of the displacement interpolation and the optimal transport. But for Theorem 23.14, this seems to be impossible because of the need to control the smoothness of the approximation of ρ_0; as a consequence, passing to the limit is more delicate. Further, note that Theorem 17.15 was used in the proof of Theorem 23.14, since it is the convexity properties of U_ν along displacement interpolation which allows us to go back and forth between the integral and the differential (in the t variable) formulations.

The argument used to prove that the first term of (23.61) converges to 0 is reminiscent of the well-known argument from functional analysis, according to which convergence in weak L^2 combined with convergence of the L^2 norm imply convergence in strong L^2.

At some point I have used the following theorem: *If* $u \in W^{1,1}_{\mathrm{loc}}(M)$, *then for any constant* c, $\nabla u = 0$ *almost everywhere on* $\{u = c\}$. This classical result can be found e.g. in [554, Theorem 6.19].

Another strategy to attack Theorem 23.14 would have been to start from the "curve-above-tangent" formulation of the convexity of $\mathcal{J}_t^{1/N}$, where $\mathcal{J}_t$ is the Jacobian determinant. (Instead I used the

[2] Still an intense experience!

"curve-below-chord" formulation of convexity via Theorem 17.15.) I don't know if technical details can be completed with this approach.

The interpretation of the linear Fokker–Planck equation $\partial_t \rho = \Delta \rho + \nabla \cdot (\rho \nabla V)$ as the limit of a discretized scheme goes back to the pioneering work of Jordan, Kinderlehrer and Otto [493]. In that sense the Fokker–Planck equation can be considered as the abstract gradient flow corresponding to the free energy $\Phi(\rho) = \int \rho \log \rho + \int \rho V$. The proof (slightly rewritten) appears in my book [814, Section 8.4]. It is based on the three main estimates which are more or less at the basis of the whole theory of abstract gradient flows: If τ is the time step, and $X_k^{(\tau)}$ the position at step k of the discretized system, then

$$\begin{cases} \Phi(X_n^{(\tau)}) = O(1); \\ \\ \displaystyle\sum_{j=1}^{\infty} \frac{d(X_j^{(\tau)}, X_{j+1}^{(\tau)})^2}{2\tau} = O(1); \\ \\ \displaystyle\tau \sum_{j=1}^{\infty} \big\|\mathrm{grad}\Phi(X_j^{(\tau)})\big\|^2 = O(1). \end{cases}$$

Here I have assumed that Φ is bounded below (which is the case when Φ is the free energy functional). When $\inf \Phi = -\infty$, there are still estimates of the same type, only quite more complicated [30, Section 3.2]. Ambrosio and Savaré recently found a simplified proof of error estimates and convergence for time-discretized gradient flows [34].

Otto applied the same method to various classes of nonlinear diffusion equations, including porous medium and fast diffusion equations [669], and parabolic p-Laplace type equations [666], but also more exotic models [667, 668] (see also [737]). For background about the theory of porous medium and fast diffusion equations, the reader may consult the review texts by Vázquez [804, 806].

In his work about porous medium equations, Otto also made two important conceptual contributions: First, he introduced the abstract formalism allowing him to interpret these equations as gradient flows, directly at the continuous level (without going through the time-discretization). Secondly, he showed that certain features of the porous medium equations (qualitative behavior, rates of convergence to equilibrium) were best seen via the new gradient flow interpretation.

The psychological impact of this work on specialists of optimal transport was important. Otto's approach was developed by various authors, including Carrillo, McCann and myself [213, 214], Ambrosio, Gigli and Savaré [30], and others. As an example of recent application, Carrillo and Calvez [198] applied the same methodology to a one-dimensional variant of the Keller–Segel chemotaxis model, and showed that its qualitative behavior can be conveyed by diagrams depicting the behavior of a free energy in the Wasserstein space.

The setting adopted in [30, 214, 814] is the following: Let E denote an energy functional of the form

$$E(\mu) = \int_{\mathbb{R}^n} U(\rho(x))\,dx + \int_{\mathbb{R}^n} V(x)\,d\mu(x) + \frac{1}{2}\int_{\mathbb{R}^n\times\mathbb{R}^n} W(x-y)\,d\mu(x)\,d\mu(y),$$

where as usual ρ is the density of μ, and $U(0) = 0$; then under certain regularity assumptions, the associated gradient flow with respect to the 2-Wasserstein distance W_2 is

$$\frac{\partial \rho}{\partial t} = \Delta p(\rho) + \nabla\cdot(\rho\,\nabla V) + \nabla\cdot\big(\rho\,\nabla(\rho * W)\big),$$

where as usual $p(r) = r\,U'(r) - U(r)$. (When $p(r) = r$, the above equation is a special case of McKean–Vlasov equation.) The most general results of this kind can be found in [30]. Such equations arise in a number of physical models; see e.g. [214]. As an interesting particular case, the logarithmic interaction in dimension 2 gives rise to a form of the Keller–Segel model for chemotaxis; variants of this model are studied by means of optimal transport in [121, 198]; (see also [196, Chapter 7]).

Other interesting gradient flows are obtained by choosing for the energy functional:

- the Fisher information

$$I(\mu) = \int \frac{|\nabla\rho|^2}{\rho};$$

  then the resulting fourth-order, nonlinear partial differential equation is a quantum drift-diffusion equation [30, Example 11.1.10], which also appears in the modeling of interfaces in spin systems. The gradient flow interpretation of this equation was recently studied rigorously by Gianazza, Savaré and Toscani [414]. See [497] and the many references there quoted for other recent contributions on this model.

- the squared H^{-1} norm

$$\|\mu\|_{H^{-1}}^2 = \|\nabla \Delta^{-1}\rho\|_{L^2}^2;$$

then the resulting equation appears in the Ginzburg–Landau dynamics. This idea has been in the air for a few years at a purely formal level; recently, Ambrosio and Serfaty [36] have justified this guess rigorously. (The functional treated there is not exactly the squared H^{-1} norm, but it is related; at the same time they showed how this problem gives some insight into the tricky issue of boundary conditions.)

- the negative squared W_2 distance, $\mu \longmapsto -W_2(\sigma,\mu)^2/2$, where σ is a reference measure. (In dimension 2, this is as a kind of dissipative variant of the semi-geostrophic equations.) The resulting gradient flow can be considered as the geodesic flow on $P_2(M)$ (the case $M = \mathbb{R}^n$ is treated in [30, Example 11.2.9 and Theorem 11.2.10]).

Gradient flows with respect to the Wasserstein distances W_p with $p \neq 2$ were considered in [666] and lead to other classes of well-known diffusion equations, such as p-Laplace equations $\partial_t \rho = \nabla \cdot (|\nabla\rho|^{p-2}\nabla\rho)$. A large part of the discussion can be transposed to that case [4, 594], but things become quite more difficult.

Brenier [164] has suggested that certain cost functions with "relativistic" features could be physically relevant, say $c(x,y) = c(x-y)$ with

$$c(v) = 1 - \sqrt{1 - \frac{|v|^2}{c^2}} \qquad \text{or} \qquad c(v) = \sqrt{1 + \frac{|v|^2}{c^2}} - 1.$$

By applying the general formalism of gradient flows with such cost functions, he derived relativistic-like heat equations, such as

$$\frac{\partial \rho}{\partial t} = \nabla \cdot \left(\frac{\rho \nabla \rho}{\sqrt{\rho^2 + \varepsilon^2 |\nabla \rho|^2}} \right).$$

This looked a bit like a formal game, but it was later found out that related equations were common in the physical literature about flux-limited diffusion processes [627], and that in fact Brenier's very equation had already been considered by Rosenau [708]. A rigorous treatment of these equations leads to challenging analytical difficulties, which triggered several recent technical works, see e.g. [39, 40, 618] and the references therein. By the way, the cost function $1-\sqrt{1-|x-y|^2}$ was later found to have applications in the design of lenses [712].

Lemma 23.28 in the Appendix is borrowed from [30, Lemma 4.3.4]. As Ambrosio pointed out to me, the argument is quite reminiscent of Kruzkhov's doubling method for the proof of uniqueness in the theory of scalar conservation laws, see for instance the nice presentation in [327, Sections 10.2 and 11.4]. It is important to note that the almost everywhere differentiability of F in both variables separately is not enough to apply this lemma.

The stability theorem (Theorem 23.26) is a particular case of more abstract general results, see for instance [30, Theorem 4.0.4(iv)].

In their study of gradient flows, Ambrosio, Gigli and Savaré point out that it is useful to construct curves (μ_t) satisfying the convexity-type inequality

$$\begin{aligned} W_2(\sigma,\mu_t)^2 \geq (1-t)\, W_2(\sigma,\mu_0)^2 + t\, W_2(\sigma,\mu_1)^2 \\ - t(1-t) \int d\big(T_0(x),T_1(x)\big)^2\,\sigma(dx), \quad (23.92) \end{aligned}$$

where T_i $(i=0,1)$ is the optimal transport between σ and μ_i. (Their construction is performed only in $\mathbb{R}^n$ [30, Proposition 9.3.12].) At first sight (23.92) seems to contradict the fact that $P_2(M)$ does not have sectional curvature bounded above in the Alexandrov sense; but there is in fact no contradiction since the curve $(\mu_t)_{0\leq t\leq 1}$ will depend on the measure σ.

I shall now make some more speculative comments about the interpretation of the results, and their possible extensions.

For the most part, equilibrium statistical mechanics rests on the idea that the equilibrium measure is obtained by the minimization of a thermodynamical functional such as the free energy. The principle according to which *nonequilibrium* statistical mechanics may also be understood through variational principles is much more original; I first heard it explicitly in a seminar by Kinderlehrer (June 1997 in Paris), about the interpretation of the Fokker–Planck equation by means of Wasserstein distance. Independently of optimal transport theory, the same idea has been making its way in the community of physicists, where it may be attributed to Prigogine. There is ongoing research in that direction, in relation to large deviations and fluctuation of currents, performed by Gabrieli, Landim, Derrida, Lebowitz, Speer, Jona Lasinio and others. It seems to me that both approaches (optimal transport on the one hand, large deviations on the other) have a lot in common, although the formalisms look very different. By the way, some links between optimal transport and large deviations have recently been explored in a book by Feng and Kurtz [355] as well as in [429, 430, 448].

As for the idea to rewrite (possibly linear) diffusion equations as nonlinear transport equations, it may at first sight seem odd, but looks like the right point of view in many problems, and has been used for some time in numerical analysis.

So far I have mainly discussed gradient flows associated with cost functions that are quadratic ($p = 2$), or at least strictly convex. But there are some quite interesting models of gradient flows for, say, the cost function which is equal to the distance ($p = 1$). Such equations have been used for instance in the modeling of sandpiles [46, 48, 329, 336, 691] or compression molding [47]. These issues are briefly reviewed by Evans [328].

Also gradient flows of certain "energy" functionals in the Wasserstein space may be used to *define* certain diffusive semigroups; think for instance of the problem of constructing the heat semigroup on a nonsmooth metric space. Recently Ohta [655] showed that the heat equation could be introduced in this way on Alexandrov spaces of nonnegative sectional curvature. An independent contribution by Savaré [735] addresses the same problem on certain classes of metric spaces including Alexandrov spaces with curvature bounded below. On such spaces, the heat equation can be constructed by other means [537], but it might be that the optimal transport approach will apply to even more singular situations, and in any case this provides a genuinely different approach of the problem.

One can also hope to treat subriemannian situations in the same way; at least from a formal point of view, this is addressed in [514].

It can also be expected that the gradient flow interpretation will provide powerful general tools to study the *stability* of diffusion equations. In a striking recent application of this principle, Ambrosio, Savaré and Zambotti [35] proved the stability of the linear diffusion process associated with the Fokker–Planck equation, in finite or infinite dimension, with respect to the weak convergence of the invariant probability measure, as soon as the latter is log concave. Savaré [735, 736] also used this interpretation to prove the stability of the heat semigroup (on manifolds, or on certain metric-measure spaces) under measured Gromov–Hausdorff convergence.

Very recently, variational problems taking the form of a discretized gradient flow have made their way in mathematical economics or decision theory; in these models the negative of the energy can be thought of as, say, the reward or the benefits obtained from a certain skill or method or decision, while the cost function can be interpreted as the effort or difficulty which one has to spend or endure in order to learn this skill or change one's habits or take the decision. As an entry point to that literature, the reader may take a look at a paper by Attouch and Soubeyran [49]. It is interesting to note that the gradient flows in this kind of literature would be more of the kind $p = 1$ than of the kind $p = 2$.

One may speculate about the possible applications of all these gradient flow interpretations in numerical simulation. In [121] the authors studied a numerical scheme based on the discretization of a gradient flow in Wasserstein space, which has the striking advantage of preserving qualitative features such as finite-time blow-up or long-time extinction. The simulation is however very costly.

This chapter was only concerned with gradient flows. The situation about Hamiltonian flows is anything but clear. In [814, Section 8.3.2] one can find some examples of equations that one would like to intepret as Hamiltonian equations with respect to the distance W_2, and other equations that one would like to interpret as dissipative Hamiltonian equations. There are many other ones.

An important example of a "Hamiltonian equation" is the semi-geostrophic system and its variants [86, 87, 265, 266, 268, 421, 569]. The well-known Euler–Poisson and Vlasov–Poisson models should belong to this class too, but also some strange variants suggested by Brenier and Loeper such as the Euler–Monge–Ampère equation [567] or its kinetic counterpart, the Vlasov–Monge–Ampère equation [170].

Among many examples of "dissipative Hamiltonian equations", I shall mention the rescaled two-dimensional incompressible Navier–Stokes equation in vorticity formulation (for nonnegative vorticity), as studied by Gallay and Wayne [392]. Caglioti, Pulvirenti and Rousset [193, 194] have used this interpretation of Navier–Stokes in terms of optimal transport, to derive related equations attempting to describe "quasi-stationary states", or intermediate time-asymptotics.

About the rigorous justification of the Hamiltonian formalism in the Wasserstein space, there are recent works by Ambrosio and Gangbo [27, 28] covering certain classes of Hamiltonian equations, yet not as wide as one could wish. Cullen, Gangbo and Pisante [267] have studied the approximation of some of these equations by particle systems. There is also a work by Gangbo, Nguyen and Tudorascu on the one-dimensional Euler–Poisson model [401]. Their study provides evidence of striking "pathological" behavior: If one defines variational solutions of Euler–Poisson as the minimizing paths for the natural action, then a solution might very well start absolutely continuous at initial time, collapse on a Dirac mass in finite time, stay in this state for a positive time, and then spread again. Also Wolansky [837] obtained coagulation–fragmentation phenomena through a Hamiltonian formalism (based on an internal energy rather than an interaction energy).

The precise sense of the "Hamiltonian structure" should be taken with some care. It was suggested to me some time ago by Ghys that this really is a *Poisson structure*, in the spirit of Kirillov. This guess was justified and completely clarified (at least formally) by Lott [575], who also made the link with previous work by Weinstein and collaborators (see [575, Section 6] for explanations). Further contributions are due to Khesin and Lee [514], who also studied the sense in which the semigeostrophic system defines a Hamiltonian flow [513].

A particularly interesting "dissipative Hamiltonian equation" that should have an interpretation in terms of optimal transport is the kinetic Fokker–Planck equation, with or without self-interaction. Huang and Jordan [485] studied this model in the setting of gradient flows (which is not natural here since we are rather dealing with a Hamiltonian flow with dissipation). A quite different contribution by Carlen and Gangbo [204] approaches the kinetic Fokker–Planck equation by a splitting scheme alternating free transport and time-discretized gradient flow in the velocity variable.

More recently, Gangbo and Westdickenberg [404] suggested an approximation scheme for the isentropic Euler system based on a kinetic approximation (so one works with probability measures on $\mathbb{R}^n \times \mathbb{R}^n$), with a splitting scheme alternating free transport, time-discretized gradient flow in the position variable, and an optimal transport problem to reconstruct the velocity variable. Their scheme, which has some hidden relation to the Huang–Jordan contribution, can be reinterpreted in terms of minimization of an action that would involve the squared acceleration rather than the squared velocity; and there are heuristic arguments to believe that this system should converge to a physical solution satisfying Dafermos's entropy criterion (the physical energy, which is formally conserved, should decrease as much as possible). Numerical simulations based on this scheme perform surprisingly well, at least in dimension 1.

In the big picture also lies the work of Nelson [201, 647, 648, 650, 651, 652] on the foundations of quantum mechanics. Nelson showed that the usual Schrödinger equation can be derived from a principle of least action over solutions of a stochastic differential equation, where the noise is fixed but the drift is unknown. Other names associated with this approach are Guerra, Morato and Carlen. The reader may consult [343, Chapter 5] for more information. In Chapter 7 of the same reference, I have briefly made the link with the optimal transport problem. Von Renesse [826] explicitly reformulated the Schrödinger equation as a Hamiltonian system in Wasserstein space.

A more or less equivalent way to see Nelson's point of view (explained to me by Carlen) is to study the critical points of the action

$$\mathcal{A}(\rho, m) = \int_0^1 \Big(K(\rho_t, m_t) - F(\rho_t) \Big)\, dt, \tag{23.93}$$

where $\rho = \rho(t,x)$ is a time-dependent density (say on $\mathbb{R}^n$), $m = m(t,x)$ is a time-dependent momentum density, $K(\rho, m) = \int |m|^2/(2\rho)$ is the kinetic energy and $F(\rho) = I(\rho) = \int |\nabla \rho|^2/\rho$ is the Fisher information. The density and momentum should satisfy the equation of mass conservation, namely $\partial_t \rho + \nabla \cdot m = 0$. At least formally, critical points of (23.93), for fixed ρ_0, ρ_1, satisfy the Euler–Lagrange equation

$$\begin{cases} \partial_t \rho + \nabla \cdot (\rho \nabla \varphi) = 0 \\[2ex] \partial_t \varphi + \dfrac{|\nabla \varphi|^2}{2} = 2\, \dfrac{\Delta \sqrt{\rho}}{\sqrt{\rho}}; \end{cases} \tag{23.94}$$

the pressure term $(\Delta\sqrt{\rho})/\sqrt{\rho}$ is sometimes called the "Bohm potential". Then the change of unknown $\psi = \sqrt{\rho}\, e^{i\varphi}$ transforms (23.94) into the usual linear Schrödinger equation.

Variational problems of the form of (23.93) can be used to derive many systems of Hamiltonian type, and some of these actions are interesting in their own right. The choice $F = 0$ gives just the squared 2-Wasserstein distance; this is the Benamou–Brenier formula [814, Theorem 8.1]. The functional $F(\rho) = -\int |\nabla \Delta^{-1}(\rho - 1)|^2$ appears in the so-called reconstruction problem in cosmology [568] and leads to the Euler–Poisson equations (see also [401]). As a final example, $F(\rho) = -(\pi^2/6)\int \rho^3$ appears (in dimension $n = 1$) in the qualitative description of certain random matrix models, and leads to the isentropic Euler equations with *negative* cubic pressure law, as first realized by Matytsin; see [449] for rigorous justification and references. (Some simple remarks about uniqueness and duality for such negative pressure models can be found in [811].) Unexpectedly, the same variational problem appears in a seemingly unrelated problem of stochastic control [563].

24

Gradient flows II: Qualitative properties

Consider a Riemannian manifold M, equipped with a reference measure $\nu = e^{-V}\mathrm{vol}$, and a partial differential equation of the form

$$\frac{\partial \rho}{\partial t} = L\, p(\rho), \tag{24.1}$$

where $p(r) = r\,U'(r) - U(r)$, U is a given nonlinearity, the unknown $\rho = \rho(t,x)$ is a probability density on M and $L = \Delta - \nabla V \cdot \nabla$.

Theorem 23.19 provides an interpretation of (24.1) as a gradient flow in the Wasserstein space $P_2(M)$. What do we gain from that information? A first possible answer is a new physical insight. Another one is a set of recipes and estimates associated with gradient flows; this is what I shall illustrate in this chapter.

As in the previous chapter, I shall use the following conventions:

- M is a complete Riemannian manifold, d its geodesic distance and vol its volume;
- $\nu = e^{-V}\mathrm{vol}$ is a reference measure on M;
- $L = \Delta - \nabla V \cdot \nabla$ is a linear differential operator admitting ν as invariant measure;
- U is a convex nonlinearity with $U(0) = 0$; typically U will belong to some $\mathcal{DC}_N$ class;
- $p(r) = r\,U'(r) - U(r)$ is the pressure function associated to U;
- $\mu_t = \rho_t\,\nu$ is the solution of a certain partial differential equation $\partial_t \rho_t = L\,p(\rho_t)$ (sometimes I shall say that μ is the solution, sometimes that ρ is the solution);
- $U_\nu(\mu) = \displaystyle\int U(\rho)\,d\nu; \quad I_{U,\nu}(\mu) = \int \rho\,|\nabla U'(\rho)|^2\,d\nu = \int \frac{|\nabla p(\rho)|^2}{\rho}\,d\nu.$

C. Villani, *Optimal Transport*, Grundlehren der mathematischen Wissenschaften 338,

Calculation rules

Having put equation (24.1) in gradient flow form, one may use Otto's calculus to shortcut certain formal computations, and quickly get relevant results, without risks of computational errors. When it comes to rigorous justification, things however are not so nice, and regularity issues should — alas! — be addressed.[1] For the most important of these gradient flows, such as the heat, Fokker–Planck or porous medium equations, these regularity issues are nowadays under good control.

Examples 24.1. Consider a power law nonlinearity $U(r) = r^m$, $m > 0$. For $m > 1$ the resulting equation (24.1) is called a porous medium equation, and for $m < 1$ a fast diffusion equation. These equations are usually studied under the restriction $m > 1 - (2/n)$, because for $m \le 1 - (2/n)$ the solution might fail to exist (there is in general loss of mass at infinity in finite time, or even in no time, at least if $M = \mathbb{R}^n$). If M is compact and ρ_0 is positive, then there is a unique C^∞, positive solution. For $m > 1$, if ρ_0 vanishes somewhere, the solution in general fails to have C^∞ regularity at the boundary of the support of ρ. For $m < 1$, adequate decay conditions at infinity are needed.

To avoid inflating the size of this chapter much further, I shall *not* go into these regularity issues, and be content with theorems that will be conditional to the regularity of the solution.

Theorem 24.2 (Computations for gradient flow diffusion equations). *Let $\rho = \rho(t,x)$ be a solution of (24.1) defined and continuous on $[0,T) \times M$. Further, let A be a convex nonlinearity, C^2 on $(0,+\infty)$. Assume that:*

(a) ρ is bounded and positive on $[0,\theta) \times M$, for any $\theta < T$;
(b) ρ is C^3 in the x variable and C^1 in the t variable on $(0,T) \times M$;
(c) U is C^4 on $(0,T) \times M$;
(d) V is C^4 on M;
(e) For any $t > 0$, $\exists \delta > 0$;

$$\sup_{|s-t|<\delta} \frac{1}{|t-s|}\Big(|\rho_t - \rho_s| + |U(\rho_t) - U(\rho_s)| \\ + |L\,U'(\rho_t)\,p(\rho_t) - L\,U'(\rho_s)\,p(\rho_s)|\Big) \in L^1(d\nu);$$

[1] Sometimes the gradient flow structure allows one to dispense with regularity, but I shall not explore this possibility.

(f) ρ, $p(\rho)$, $Lp(\rho)$, $p_2(\rho)$, $\nabla p_2(\rho)$, $U'(\rho)$, $\nabla U'(\rho)$, $LU'(\rho)$, $\nabla LU'(\rho)$, $L|\nabla U'(\rho)|^2$, $L(\nabla U'(\rho)\,\nabla LU'(\rho))$ *and* e^{-V} *satisfy adequate growth/decay conditions at infinity.*

Then the following formulas hold true:

(i) $\forall t > 0,\quad \frac{d}{dt}\int A(\rho_t)\,d\nu = -\int p'(\rho_t)\,A''(\rho_t)|\nabla\rho_t|^2\,d\nu;$

(ii) $\forall t > 0,\quad \frac{d}{dt}U_\nu(\mu_t) = -I_{U,\nu}(\mu_t);$

(iii) $\forall t > 0,$

$$\frac{d}{dt}I_{U,\nu}(\mu_t) = -2\int_M \Big[\|\nabla^2 U'(\rho_t)\|_{\mathrm{HS}}^2 + \big(\mathrm{Ric}+\nabla^2 V\big)\big(\nabla U'(\rho_t)\big)\Big]\,p(\rho_t)\,d\nu + \int_M \big(LU'(\rho_t)\big)^2 p_2(\rho_t)\,d\nu;$$

(iv) $\forall \sigma \in P_2^{\mathrm{ac}}(M),\quad \left|\frac{d}{dt}W_2(\sigma,\mu_t)\right| \le \sqrt{I_{U,\nu}(\mu_t)}$ *for almost all* $t > 0$.

Particular Case 24.3. When $U(r) = r\log r$, Formula (ii) becomes a famous identity: *the Fisher information is the time-derivative of the entropy along the heat semigroup.* (What I call entropy is not H_ν but $-H_\nu$; this agrees with the physicists' convention.)

In the sequel, a *smooth solution* of (24.1) will be a solution satisfying Assumptions *(a)* to *(f)* above.

Remark 24.4. I do not wish to be precise about the conditions at infinity needed in Assumption *(f)*, because there are a large number of possible assumptions. The point is to be able to justify a certain number of integrations by parts, using integrability and moment conditions. If $V = 0$, this works for instance if ρ, $p(\rho)$ and $p_2(\rho)$ have finite moments of all orders and $U'(\rho)$ and all its derivatives up to, say, order 5, have polynomial growth; but there are many alternative sets of assumptions. When V is not zero, there might be issues about the density of $C_c^\infty(M)$ in the weighted Sobolev spaces $H^1(e^{-V})$ and $H^2(e^{-V})$ which are associated with the operator L. These problems might be worsened by the behavior of the manifold M at infinity.

Before giving a rigorous proof, I shall first provide a formal argument for Theorem 24.2 based on Otto's calculus.

Formal proof of Theorem 24.2. By Formula 15.2,

$$\begin{aligned}\frac{d}{dt}\int A(\rho_t)\,d\nu &= -\Big\langle \mathrm{grad}_{\mu_t} A_\nu,\ \mathrm{grad}_{\mu_t} U_\nu\Big\rangle \\ &= -\int \rho_t\,\nabla A'(\rho_t)\cdot\nabla U'(\rho_t)\,d\nu \\ &= -\int \rho_t\,U''(\rho_t)\,A''(\rho_t)\,|\nabla\rho_t|^2\,d\nu \\ &= -\int p'(\rho_t)\,A''(\rho_t)\,|\nabla\rho_t|^2\,d\nu.\end{aligned}$$

This leads to formula (i). The choice $A = U$ gives

$$\begin{aligned}\frac{d}{dt}\int U(\rho_t)\,d\nu &= -\Big\|\mathrm{grad}_{\mu_t} U_\nu\Big\|^2 \\ &= -\int \rho_t\,|\nabla U'(\rho_t)|^2\,d\nu = -I_{U,\nu}(\mu_t),\end{aligned}$$

which is (ii).

Next, we can differentiate the previous expression once again along the gradient flow $\dot{\mu} = -\mathrm{grad}U_\nu(\mu)$:

$$\frac{d}{dt}\Big\|\mathrm{grad}_{\mu_t} U_\nu\Big\|^2 = -2\Big\langle \mathrm{Hess}_{\mu_t}\cdot\mathrm{grad}_{\mu_t} U_\nu,\ \mathrm{grad}_{\mu_t} U_\nu\Big\rangle,$$

and then (iii) follows from Formula 15.7.

As for (iv), this is just a particular case of the general formula

$$\big|(d/dt)d(X_0,\gamma(t))\big| \le |\dot{\gamma}(t)|_{\gamma(t)}.$$

□

Rigorous proof of Theorem 24.2. A crucial observation is that (24.1) can be rewritten $\partial_t\rho = \nabla_\nu\cdot(\rho_t\nabla U'(\rho_t))$, where $\nabla_\nu\cdot$ stands for the negative of the adjoint of the gradient operator in $L^2(\nu)$. (Explicitly: $\nabla_\nu\cdot u = \nabla\cdot u - \nabla V\cdot u$.) Then the proofs of (i) and (ii) are obtained by just repeating the arguments by which Formula 15.2 was established. This is a succession of differentiations under the integral symbol, chain-rules and integrations by parts:

$$
\begin{aligned}
\frac{d}{dt}\int A(\rho_t)\,d\nu &= \int \partial_t[A(\rho_t)]\,d\nu \\
&= \int A'(\rho_t)\,(\partial_t\rho_t)\,d\nu \\
&= \int A'(\rho_t)\,\nabla_\nu\cdot(\rho_t\,\nabla U'(\rho_t))\,d\nu \\
&= -\int \nabla A'(\rho_t)\,\rho_t\,\nabla U'(\rho_t)\,d\nu,
\end{aligned}
$$

and then the rest of the computation is the same as before.

The justification of (iii) is more tricky. First write

$$
\begin{aligned}
-\int \rho\,|\nabla U'(\rho)|^2\,d\nu &= \int U'(\rho)\,\nabla_\nu\cdot(\rho\,\nabla U'(\rho))\,d\nu \\
&= \int U'(\rho)\,Lp(\rho)\,d\nu = \int LU'(\rho)\,p(\rho)\,d\nu,
\end{aligned}
$$

where the self-adjointness of L with respect to the measure ν was used. Then

$$
\begin{aligned}
&\frac{d}{dt}\int LU'(\rho_t)\,p(\rho_t)\,d\nu \\
&= \int \partial_t\big(LU'(\rho_t)\big)\,p(\rho_t)\,d\nu + \int LU'(\rho_t)\,\partial_t\big(p(\rho_t)\big)\,d\nu \\
&= \int L(\partial_t U'(\rho_t))\,p(\rho_t)\,d\nu + \int LU'(\rho_t)\,p'(\rho_t)\,\nabla_\nu\cdot(\rho_t\,\nabla U'(\rho_t))\,d\nu.
\end{aligned}
\tag{24.2}
$$

On the other hand,

$$
\begin{aligned}
\partial_t U'(\rho_t) = U''(\rho_t)\,\partial_t\rho_t &= U''(\rho_t)\,\nabla_\nu\cdot(\rho_t\,\nabla U'(\rho_t)) \\
&= U''(\rho_t)\,\nabla\rho_t\cdot\nabla U'(\rho_t) + \rho_t\,U''(\rho_t)\,LU'(\rho_t) \\
&= |\nabla U'(\rho_t)|^2 + \rho_t\,U''(\rho_t)\,LU'(\rho_t).
\end{aligned}
$$

Plugging this back into (24.2), we obtain

$$
\begin{aligned}
&\frac{d}{dt}\int LU'(\rho_t)\,p(\rho_t)\,d\nu \\
&= \int L|\nabla U'(\rho_t)|^2\,p(\rho_t)\,d\nu + \int L\Big(\rho_t\,U''(\rho_t)\,LU'(\rho_t)\Big)\,p(\rho_t)\,d\nu \\
&\qquad + \int LU'(\rho_t)\,p'(\rho_t)\,\nabla_\nu\cdot(\rho_t\,\nabla U'(\rho_t))\,d\nu.
\end{aligned}
\tag{24.3}
$$

The last two terms in this formula are actually equal: Indeed, if ρ is smooth then

$$\int L\Big(\rho\, U''(\rho)\, LU'(\rho)\Big)\, p(\rho)\, d\nu = \int \rho\, U''(\rho)\, LU'(\rho)\, Lp(\rho)\, d\nu$$
$$= \int p'(\rho)\, LU'(\rho)\, \nabla_\nu \cdot (\rho\, \nabla U'(\rho))\, d\nu.$$

So the expression appearing in (24.3) is exactly *twice* the expression appearing in (15.18), up to the replacement of ψ by $-U'(\rho_t)$. At this point, to arrive at formula (iii), it suffices to repeat the computations leading from (15.18) to (15.20), and to apply Bochner's formula, say in the form (15.6)–(15.7).

The proof of (iv) is simple: By Theorem 23.9, for almost all t,

$$\frac{d}{dt}\left(\frac{W_2(\mu_t,\sigma)^2}{2}\right) = -\int \langle \nabla p(\rho_t), \widetilde{\nabla}\psi\rangle\, d\nu,$$

where $\exp(\widetilde{\nabla}\psi)$ is the optimal transport $\mu_t \to \sigma$. It follows that

$$\frac{d^+}{dt}\left(\frac{W_2(\mu_t,\sigma)^2}{2}\right) \le \sqrt{\int \frac{|\nabla p(\rho_t)|^2}{\rho_t}\, d\nu}\, \sqrt{\int \rho_t\, |\widetilde{\nabla}\psi|^2\, d\nu}$$
$$= \sqrt{I_{U,\nu}(\mu_t)}\, W_2(\mu_t,\sigma);$$

so for almost all $t > 0$,

$$\frac{d^+}{dt} W_2(\mu_t,\sigma) \le \left(\frac{1}{2\, W_2(\mu_t,\sigma)}\right) \frac{d^+}{dt} W_2(\mu_t,\sigma)^2 \le \sqrt{I_{U,\nu}(\mu_t)}. \qquad (24.4)$$

This is the desired estimate. $\square$

As a final remark, Theorem 24.2 automatically implies some integrated (in time) "regularity" a priori estimates for (24.1), as the next corollary shows.

Corollary 24.5 (Integrated regularity for gradient flows). *With the same assumptions as in Theorem 24.2, one has* $U_\nu(\mu_t) \le U_\nu(\mu_0)$ *and*

$$\int_0^{+\infty} \left[\limsup_{s\downarrow 0} \frac{W_2(\mu_t,\mu_{t+s})}{s}\right]^2 dt \le \int_0^{+\infty} I_{U,\nu}(\mu_t) dt \le U_\nu(\mu_0) - (\inf U_\nu).$$

Remark 24.6. If U_ν is bounded below, this corollary yields exactly the regularity which is a priori required in Theorem 23.19. It also shows that $t \to \mu_t$ belongs to $\mathrm{AC}_2((0,+\infty); P_2(M))$ (absolute continuity of order 2) in the sense that there is $\ell \in L^2(dt)$ such that $W_2(\mu_t, \mu_s) \leq \int_s^t \ell(\tau)\,d\tau$. (Take $\ell(t) = \limsup_{s\to 0} W_2(\mu_t, \mu_{t+s})/s$.) Finally, the bound $\int U_{U,\nu}(\mu_t)\,dt < +\infty$ is the assumption of Theorem 23.9.

Large-time behavior

Otto's calculus, described in Chapter 15, was first developed to estimate rates of equilibration for certain nonlinear diffusion equations. The next theorem illustrates this.

Theorem 24.7 (Equilibration in positive curvature). *Let M be a Riemannian manifold equipped with a reference measure $\nu = e^{-V}$, $V \in C^4(M)$, satisfying a curvature-dimension bound $\mathrm{CD}(K,N)$ for some $K > 0$, $N \in (1,\infty]$, and let $U \in \mathcal{DC}_N$. Then:*

(i) **(exponential convergence to equilibrium)** *Any smooth solution $(\mu_t)_{t\geq 0}$ of (24.1) satisfies the following estimates:*

$$\begin{cases} (a) \qquad [U_\nu(\mu_t) - U_\nu(\nu)] \leq e^{-2K\lambda t}\,[U_\nu(\mu_0) - U_\nu(\nu)] \\[2ex] (b) \qquad I_{U,\nu}(\mu_t) \leq e^{-2K\lambda t}\, I_{U,\nu}(\mu_0) \\[2ex] (c) \qquad W_2(\mu_t,\nu) \leq e^{-K\lambda t}\, W_2(\mu_0,\nu), \end{cases} \tag{24.5}$$

where

$$\lambda := \left(\lim_{r\to 0} \frac{p(r)}{r^{1-\frac{1}{N}}}\right) \left(\sup_{x\in M} \rho_0(x)\right)^{-\frac{1}{N}}. \tag{24.6}$$

In particular, λ is independent of ρ_0 if $N = \infty$.

(ii) **(exponential contraction)** *Any two smooth solutions $(\mu_t)_{t\geq 0}$ and $(\widetilde{\mu}_t)_{t\geq 0}$ of (24.1) satisfy*

$$W_2(\mu_t, \widetilde{\mu}_t) \leq e^{-K\lambda t}\, W_2(\mu_0, \widetilde{\mu}_0), \tag{24.7}$$

where

$$\lambda := \left(\lim_{r\to 0} \frac{p(r)}{r^{1-\frac{1}{N}}}\right) \left[\max\left(\sup_{x\in M} \rho_0(x),\ \sup_{x\in M} \rho_1(x)\right)\right]^{-\frac{1}{N}}. \tag{24.8}$$

Example 24.8. Smooth solutions of the Fokker–Planck equation

$$\frac{\partial \rho}{\partial t} = L\rho \tag{24.9}$$

converge to equilibrium at least as fast as $O(e^{-Kt})$, in W_2 distance, in the entropy sense (i.e. in the sense of the convergence of $\sqrt{H_\nu(\mu)}$ to 0), and in the Fisher information sense.

Remark 24.9. At least formally, these properties are in fact *general properties of gradient flows*: Let F be a function defined on a geodesically convex subset of a Riemannian manifold (M, g); $\operatorname{Hess} F \geq \lambda\, g$, $\lambda > 0$; X_∞ is the minimizer of F; and X, $\widetilde{X}$ are two trajectories of the gradient flow associated with F. Then we have three neat estimates: (a) $[F(X(t)) - F(X_\infty)] \leq e^{-\lambda t}\, [F(X(0)) - F(X_\infty)]$; (b) $|\nabla F(X(t))| \leq e^{-\lambda t}\, |\nabla F(X(0))|$; (c) $d(X(t), \widetilde{X}(t)) \leq e^{-\lambda t}\, d(X(0), \widetilde{X}(0))$. The proof of these inequalities will be a good exercise for the reader.

Remark 24.10. The rate of decay $O(e^{-\lambda t})$ is optimal for (24.9) if dimension is not taken into account; but if N is finite, the optimal rate of decay is $O(e^{-\lambda t})$ with $\lambda = KN/(N-1)$. The method presented in this chapter is not clever enough to catch this sharp rate.

Remark 24.11. I believe that the preceding results of convergence are satisfactory as I have stated them, i.e. in terms of convergence of natural, physically meaningful functionals. However, it is also often possible to get similar rates of decay for more classical distances such as the L^1 norm, thanks to the Csiszár–Kullback–Pinsker inequality (22.25) and generalizations thereof.

Remark 24.12. If $N < \infty$, Theorem 24.7 proves convergence to equilibrium with a rate that depends on the initial datum. However, if the solution $(\rho_t)_{t \geq 0}$ satisfies *uniform* smoothness bounds, it is often possible to reinforce the statement $\rho_t \xrightarrow[L^1]{} 1$ into $\rho_t \xrightarrow[L^\infty]{} 1$. Then we can choose ρ_T as new initial datum, and get

$$t \geq T \Longrightarrow \quad U_\nu(\mu_t) \leq e^{-K\lambda_T\,(t-T)}\, U_\nu(\mu_T) \;\leq e^{-K\lambda_T\,(t-T)}\, U_\nu(\mu_0),$$

where $\lambda_T = (\lim p(r)/r^{1-1/N})\,(\sup \rho_T)^{-\frac{1}{N}} \longrightarrow \lambda_\infty = (\lim p(r)/r^{1-1/N})$ as $T \to \infty$. It follows that μ_t converges to ν as $O(e^{-K\widetilde{\lambda} t})$ for any $\widetilde{\lambda} > \lambda$.

Proof of Theorem 24.7. Let $H(t) = U_\nu(\mu_t)$. Theorem 24.2(ii) reads $H'(t) = -I_{U,\nu}(\mu_t)$. Let $\lambda_0 := \lim_{r\to 0} p(r)/r^{1-1/N}$. The (modified) Sobolev inequality of Theorem 21.7 implies

$$U_\nu(\mu_t) \le \frac{(\sup \rho_t)^{\frac{1}{N}}}{2K\lambda_0}\, I_{U,\nu}(\mu_t).$$

Thus,

$$\frac{d}{dt}H(t) \le -2K\lambda_0(\sup \rho_t)^{-1/N} H(t). \tag{24.10}$$

Theorem 24.2(i) with $A(r) = r^p$, $p \ge 2$, gives

$$\frac{d}{dt}\int \rho^p \, d\nu = -p(p-1)\int \rho\, U''(\rho)\rho^{p-2}|\nabla\rho|^2\, d\nu \le 0.$$

So $\|\rho_t\|_{L^p}^p$ is a nonincreasing function of t, and therefore

$$\forall t \ge 0 \qquad \|\rho_t\|_{L^p(\nu)} \le \|\rho_0\|_{L^p(\nu)}.$$

Passing to the limit as $p \to \infty$ yields

$$\forall t \ge 0, \qquad \sup \rho_t \le \sup \rho_0. \tag{24.11}$$

Plugging this back into (24.10), we get

$$\frac{d}{dt}H(t) \le -2K\lambda_0\,(\sup\rho_0)^{-1/N} H(t) = -2K\lambda\, H(t),$$

and then (24.5)(a) follows.

Next, if $U \in \mathcal{DC}_N$, and $\mathrm{CD}(K,N)$ is enforced, we can write, as in (16.13),

$$\begin{aligned}
-\frac{1}{2}\frac{d}{dt}\, & I_{U,\nu}(\mu_t) \\
&= \int_M \Gamma_2(U'(\rho_t))\, p(\rho_t)\, d\nu + \int_M (LU'(\rho_t))^2\, p_2(\rho_t)\, d\nu \\
&\ge \int_M \mathrm{Ric}_{N,\nu}(\nabla U'(\rho_t))\, p(\rho_t)\, d\nu + \int_M (LU'(\rho_t))^2 \left[p_2 + \frac{p}{N}\right](\rho_t)\, d\nu \\
&\ge K\int_M |\nabla U'(\rho_t)|^2\, p(\rho_t)\, d\nu + \int_M (LU'(\rho_t))^2 \left[p_2 + \frac{p}{N}\right](\rho_t)\, d\nu \\
&\ge K\int_M |\nabla U'(\rho_t)|^2\, p(\rho_t)\, d\nu \\
&\ge K\lambda_0(\sup\rho_t)^{-\frac{1}{N}} \int_M |\nabla U'(\rho_t)|^2\, \rho_t\, d\nu \\
&= K\lambda_0(\sup\rho_t)^{-\frac{1}{N}}\, I_{U,\nu}(\mu_t).
\end{aligned}$$

This implies (24.5)(b).

It remains to establish (24.7) (of which (24.5)(c) obviously is a corollary). The strategy is the same as for Theorem 23.26. Let $t > 0$ be given.

First, the assumption $K \geq 0$ implies $\sup \rho_t \leq \sup \rho_0$, $\sup \widetilde{\rho}_t \leq \sup \widetilde{\rho}_0$ (recall (24.11)).

If $(\mu^{(s)} = \rho^{(s)} \nu)_{0 \leq s \leq 1}$ is the displacement interpolation between $\rho_t \nu$ and $\widetilde{\rho}_t \nu$, then since $r \to r^m$ lies in $\mathcal{DC}_N$ and $K \geq 0$, we can use the displacement convexity to write

$$\forall s \in [0,1], \qquad \int (\rho^{(s)})^m \, d\nu \leq \max \left(\int (\rho^{(0)})^m \, d\nu, \int (\rho^{(1)})^m \, d\nu \right);$$

by raising to the power $1/m$ and letting $m \to \infty$, we obtain

$$\sup \rho^{(s)} \leq \max (\sup \rho_t, \sup \widetilde{\rho}_t) \leq \max (\sup \rho_0, \sup \widetilde{\rho}_0), \tag{24.12}$$

where all the suprema are essential suprema.

Now let $T = \exp(\widetilde{\nabla} \psi)$ be the optimal transport from μ_t to $\widetilde{\mu}_t$, where ψ is $d^2/2$-convex. If $(\psi^{(s)})_{0 \leq s \leq 1}$ is obtained from ψ by action of the Hamilton–Jacobi semigroup, then by Proposition 17.24(i),

$$\int_0^1 \left(\int (\rho^{(s)})^{1-\frac{1}{N}} \, |\widetilde{\nabla} \psi^{(s)}|^2 \, d\nu \right) (1-t) \, dt \\ \geq \Big[\max (\sup \rho_0, \sup \widetilde{\rho}_0) \Big]^{-\frac{1}{N}} \frac{W_2(\mu_t, \widetilde{\mu}_t)^2}{2}.$$

So if we apply Theorem 23.14 with σ replaced by $\widetilde{\mu}_t$ and μ replaced by μ_t, we deduce

$$U_\nu(\widetilde{\mu}_t) \geq U_\nu(\mu_t) + \int \langle \widetilde{\nabla} \psi, \nabla p(\rho_t) \rangle \, d\nu + \frac{K \lambda \, W_2(\mu_t, \widetilde{\mu}_t)^2}{2}, \tag{24.13}$$

where λ is defined by (24.8).

Similarly, if $\exp(\widetilde{\nabla} \widetilde{\psi})$ is the optimal transport between $\widetilde{\mu}_t$ and μ_t, then

$$U_\nu(\mu_t) \geq U_\nu(\widetilde{\mu}_t) + \int \langle \widetilde{\nabla} \widetilde{\psi}, \nabla p(\widetilde{\rho}_t) \rangle \, d\nu + \frac{K \lambda \, W_2(\mu_t, \widetilde{\mu}_t)^2}{2}. \tag{24.14}$$

By (24.13) and (24.14),

$$\int \langle \widetilde{\nabla} \psi, \nabla p(\rho_t) \rangle \, d\nu + \int \langle \widetilde{\nabla} \widetilde{\psi}, \nabla p(\widetilde{\rho}_t) \rangle \, d\nu \leq -K \lambda \, W_2(\mu_t, \widetilde{\mu}_t)^2. \tag{24.15}$$

On the other hand, Theorem 23.9 shows that $(d/dt)(W_2(\mu_t,\widetilde{\mu}_t)^2/2)$ is equal to the left-hand side of (24.15), for almost all t. We conclude that

$$\frac{d^+}{dt} W_2(\mu_t,\widetilde{\mu}_t)^2 \le -2K\lambda\, W_2(\mu_t,\widetilde{\mu}_t)^2, \tag{24.16}$$

and the desired result follows. □

Remark 24.13. Here is an alternative scheme of proof for Theorem 24.7(ii). The problem is to estimate

$$\int \langle \nabla U'(\rho_t), \nabla\psi\rangle\, d\mu_t + \int \langle \nabla U'(\widetilde{\rho}_t), \nabla\widetilde{\psi}\rangle\, d\widetilde{\mu}_t.$$

(Let us forget about the approximate gradients.) Introduce the displacement interpolation $(\mu_t^{(s)})_{0\le s\le 1}$ with $\mu_t^{(0)} = \mu_t$, $\mu_t^{(1)} = \widetilde{\mu}_t$, and let $\psi^{(s)}$ be the solution of the Hamilton–Jacobi equation starting from $\psi^{(0)} = \psi$, so that $\psi^{(1)} = -\widetilde{\psi}$ (recall Remarks 5.15 and 7.38). Dropping the tildes for notational simplicity, the problem is to estimate from below

$$\int \langle \nabla U'(\rho^{(0)}), \nabla\psi^{(0)}\rangle\, d\mu^{(0)} - \int \langle \nabla U'(\rho^{(1)}), \nabla\psi^{(1)}\rangle\, d\mu^{(1)} = f(0) - f(1),$$

where $f(s) = \int \langle \nabla U'(\rho^{(s)}), \nabla\psi^{(s)}\rangle\, d\mu^{(s)}$. This can be done by estimating the time-derivative of f, and considering the quantity

$$\frac{f(s+h)-f(s)}{h}$$
$$= \frac{1}{h}\left\{\int \langle \nabla U'(\rho^{(s+h)}), \nabla\psi^{(s+h)}\rangle\, d\mu^{(s+h)} - \int \langle \nabla U'(\rho^{(s)}), \nabla\psi^{(s)}\rangle\, d\mu^{(s)}\right\}.$$

Note that $\mu^{(s+h)} = \exp(\frac{h}{1-s}\nabla\psi^{(s)})_\#\mu^{(s)}$; also if $\mathit{\Pi}_v$ stands for the parallel transport along the curve $\exp(\tau v)$ $(0\le\tau\le 1)$, then

$$\left(\mathit{\Pi}_{(\frac{h}{1-s})\nabla\psi^{(s)}}\right)(\nabla\psi^{(s)}) = \nabla\psi^{(s+h)}\Bigl(\exp\bigl(\frac{h}{1-s}\nabla\psi^{(s)}\bigr)\Bigr).$$

Using these identities and the fact that parallel transport preserves the scalar product, we deduce

$$\int \langle \nabla U'(\rho^{(s+h)}), \nabla\psi^{(s+h)}\rangle\, d\mu^{(s+h)}$$
$$= \int \Bigl\langle \nabla U'(\rho^{(s+h)}), \nabla\psi^{(s+h)}\Bigr\rangle \circ \exp\Bigl(\frac{h}{1-s}\nabla\psi^{(s)}\Bigr)\, d\mu^{(s)}$$
$$= \int \Bigl\langle \mathit{\Pi}^{-1}_{\frac{h}{1-s}\nabla\psi^{(s)}} \nabla U'\Bigl(\rho^{(s+h)}\Bigl(\exp\Bigl(\frac{h}{1-s}\nabla\psi^{(s)}\Bigr)\Bigr)\Bigr), \nabla\psi^{(s)}\Bigr\rangle\, d\mu^{(s)}.$$

Then, at least formally, df/ds coincides with

$$\int \left\langle \lim_{h\downarrow 0} \left(\frac{\Pi^{-1}_{\frac{h}{1-s}\nabla\psi^{(s)}} \nabla U'\Big(\rho^{(s+h)}\big(\exp(\frac{h}{1-s}\nabla\psi^{(s)})\big)\Big) - U'(\rho)}{h} \right), \nabla\psi^{(s)} \right\rangle d\mu^{(s)}.$$

From this point it is clear that one can take the computation to the end, with adequate input from Riemannian geometry. Although this approach is in some sense more direct, I preferred to use the other strategy based on displacement convexity estimates, because all the input from differential geometry was already contained in those estimates.

Remark 24.14. In the particular case $\widetilde{\mu}_t = \nu$, inequality (24.13) and Theorem 23.9 imply the following: *Let M satisfy a* $\mathrm{CD}(K,N)$ *condition with $K \geq 0$, let $U \in \mathcal{DC}_N$ with $U(1) = 0$, let $\mu = \rho\,\nu \in P_2^{\mathrm{ac}}(M)$, and let*

$$\lambda := \left(\lim_{r\to 0} \frac{p(r)}{r^{1-\frac{1}{N}}} \right) \left(\sup_{x\in M} \rho_0(x) \right)^{-\frac{1}{N}};$$

if $(\mu_t)_{0\leq t\leq 1}$ is a smooth solution of the gradient flow $\partial_t\rho_t = L\,p(\rho_t)$ starting from $\mu_0 = \mu$, then for almost all t,

$$-\left.\frac{d^+}{dt}\right|_{t=0} \left(\frac{W_2(\mu_t,\nu)^2}{2} \right) \geq U_\nu(\mu) + \frac{K\lambda\, W_2(\mu,\nu)^2}{2}. \tag{24.17}$$

On the other hand, by using the same arguments as in the proof of Theorem 22.14 and in (24.12), it is easy to establish the Talagrand-type inequality

$$U_\nu(\mu) \geq \frac{K\lambda\, W_2(\mu,\nu)^2}{2}.$$

So (24.17) is a reinforcement of (24.16).

Example 24.15. When M is $\mathbb{R}^n$ equipped with the standard Gaussian measure γ, we have the following inequalities relating the functionals H_γ (Kullback information), I_γ (Fisher information) and W_2 (Wasserstein distance) along the Ornstein–Uhlenbeck semigroup $\partial_t\rho_t(x) = \Delta\rho_t(x) - x\cdot\nabla\rho_t(x)$:

$$\begin{cases} -\dfrac{d}{dt} H_\gamma(\mu_t) = I_\gamma(\mu_t) \geq 2\, H_\gamma(\mu_t); \\[2ex] -\dfrac{d}{dt}\left(\dfrac{W_2(\mu_t,\gamma)^2}{2}\right) \geq H_\gamma(\mu_t) + \dfrac{W_2(\mu_t,\gamma)^2}{2} \geq W_2(\mu_t,\gamma)^2. \end{cases}$$

Short-time behavior

A popular and useful topic in the study of diffusion processes consists in establishing **regularization estimates** in short time. Typically, a certain functional used to quantify the regularity of the solution (for instance, the supremum of the unknown or some Lebesgue or Sobolev norm) is shown to be bounded like $O(t^{-\kappa})$ for some characteristic exponent κ, independent of the initial datum (or depending only on certain weak estimates on the initial datum), when $t > 0$ is small enough. Here I shall present some slightly unconventional estimates of this type.

Theorem 24.16 (Short-time regularization for gradient flows). *Let M be a Riemannian manifold satisfying a curvature-dimension bound $\mathrm{CD}(K,\infty)$, $K \in \mathbb{R}$; let $\nu = e^{-V}\mathrm{vol} \in P_2(M)$, with $V \in C^4(M)$, and let $U \in \mathcal{DC}_\infty$ with $U(1) = 0$. Further, let $(\mu_t)_{t\geq 0}$ be a smooth solution of (24.1). Then:*

(i) If $K \geq 0$ then for any $t \geq 0$,

$$t^2\, I_{U,\nu}(\mu_t) + 2t\, U_\nu(\mu_t) + W_2(\mu_t,\nu)^2 \leq W_2(\mu_0,\nu)^2.$$

In particular,

$$U_\nu(\mu_t) \leq \frac{W_2(\mu_0,\nu)^2}{2t}, \tag{24.18}$$

$$I_{U,\nu}(\mu_t) \leq \frac{W_2(\mu_0,\nu)^2}{t^2}. \tag{24.19}$$

(ii) If $K \geq 0$ and $t \geq s > 0$, then

$$W_2(\mu_s,\mu_t) \leq \min\left(\sqrt{2\,U_\nu(\mu_s)}\,\sqrt{|t-s|},\ \sqrt{I_{U,\nu}(\mu_s)}\,|t-s|\right) \tag{24.20}$$

$$\leq W_2(\mu_0,\nu)\,\min\left(\frac{\sqrt{|t-s|}}{\sqrt{s}},\ \frac{|t-s|}{s}\right). \tag{24.21}$$

(iii) If $K < 0$, the previous conclusions become

$$U_\nu(\mu_t) \leq \frac{e^{2Ct}\,W_2(\mu_0,\nu)^2}{2t}; \qquad I_{U,\nu}(\mu_t) \leq \frac{e^{2Ct}\,W_2(\mu_0,\nu)^2}{t^2};$$

$$W_2(\mu_s,\mu_t) \leq e^{Ct}\min\left(\sqrt{2\,U_\nu(\mu_s)}\,\sqrt{|t-s|},\ \sqrt{I_{U,\nu}(\mu_s)}\,|t-s|\right)$$
$$\leq e^{2Ct}\,W_2(\mu_0,\nu)\,\min\left(\frac{\sqrt{|t-s|}}{\sqrt{s}},\ \frac{|t-s|}{s}\right),$$

with $C = -K$.

Particular Case 24.17. When $U(\rho) = \rho \log \rho$, inequalities (24.18) and (24.19) become

$$H_\nu(\mu_t) \le \frac{W_2(\mu_0,\nu)^2}{2t}, \qquad I_\nu(\mu_t) \le \frac{W_2(\mu_0,\nu)^2}{t^2}. \tag{24.22}$$

Under a $\mathrm{CD}(K,\infty)$ bound ($K < 0$) there is an additional factor e^{-2Kt}.

Remark 24.18. Theorem 24.16 should be thought of as an *a priori estimate.* If life is not unfair, one can then remove the assumption of smoothness by a density argument, and transform (24.18), (24.19) into genuine regularization estimates. This is true at least for the Particular Case 24.17.

Remark 24.19. Inequalities (24.20) and (24.21) establish the following estimates: The curve $(\mu_t)_{t\ge 0}$, viewed as a function of time t, is Hölder-1/2 close to $t = 0$, and Lipschitz away from $t = 0$, if $U_\nu(\mu_0)$ is finite. If $I_{U,\nu}(\mu_0)$ is finite, then the curve is Lipschitz all along.

Remark 24.20. Theorem 24.7 gave upper bounds on $U_\nu(\mu_t) - U_\nu(\nu)$ like $O(e^{-\kappa t})$, with a constant depending on $U_\nu(\mu_0)$. But now we can combine Theorem 24.7 with Theorem 24.16 to get an exponential decay with a constant that does not depend on $U_\nu(\mu_0)$, but only on $W_2(\mu_0,\nu)$. By approximation, this will lead to results of convergence that do not need the finiteness of $U_\nu(\mu_0)$.

Remark 24.21. I would bet that the estimates in (24.22) are optimal in general (although they would deserve more thinking) as far as the dependence on μ_0 and t is concerned. On the other hand, if μ_0 is given, these bounds are terrible estimates for the short-time behavior of the Kullback and Fisher informations as functions of just t. Indeed, the correct scale for the Kullback information $H_\nu(\mu_t)$ is $O(\log(1/t))$, and for the Fisher information it is $O(1/t)$, as can be checked easily in the particular case when $M = \mathbb{R}^n$ and ν is the Gaussian measure.

Proof of Theorem 24.16. First note that $U(1) = 0$ implies $U_\nu(\mu) \ge U_\nu(\nu) = 0$.

Let $t > 0$ be given, and let $\exp(\widetilde{\nabla}\psi)$ be the optimal transport between μ_t and ν, where as usual ψ is $d^2/2$-convex. Since $U_\nu(\nu) = 0$ and $K \ge 0$, Theorem 23.14 implies

$$U_\nu(\mu_t) + \int \langle \widetilde{\nabla}\psi, \nabla p(\rho_t)\rangle \le 0. \tag{24.23}$$

On the other hand, by Theorem 23.9, for almost all t,

$$\frac{d^+}{dt} W_2(\mu_t, \nu)^2 \leq 2 \int \langle \widetilde{\nabla}\psi, \nabla p(\rho_t) \rangle \, d\nu. \tag{24.24}$$

The combination of (24.23) and (24.24) implies

$$\frac{d^+}{dt} W_2(\mu_t, \nu)^2 \leq -2\, U_\nu(\mu_t). \tag{24.25}$$

Now introduce

$$\psi(t) := a(t)\, I_{U,\nu}(\mu_t) + b(t)\, U_\nu(\mu_t) + c(t)\, W_2(\mu_t, \nu)^2,$$

where $a(t)$, $b(t)$ and $c(t)$ will be determined later.

Because of the assumption of nonnegative curvature, the quantity $I_{U,\nu}(\mu_t)$ is nonincreasing with time. (Set $K = 0$ in (24.5)(b).) Combining this with (24.25) and Theorem 24.2(ii), we get

$$\frac{d^+\psi}{dt} \leq [a'(t) - b(t)]\, I_{U,\nu}(\mu_t) \; + [b'(t) - 2c(t)]\, U_\nu(\mu_t) \; + c'(t)\, W_2(\mu_t, \nu)^2.$$

If we choose

$$a(t) \equiv t^2, \qquad b(t) \equiv 2t, \qquad c(t) \equiv 1,$$

then ψ has to be nonincreasing as a function of t, and this implies (i).

Let us now prove (ii). By Theorem 24.2(iv), for almost all $t > s \geq 0$,

$$\frac{d^+}{dt} W_2(\mu_s, \mu_t) \leq \sqrt{I_{U,\nu}(\mu_t)} \leq \sqrt{I_{U,\nu}(\mu_s)},$$

so

$$W_2(\mu_s, \mu_t) \leq \sqrt{I_{U,\nu}(\mu_s)}\, |t - s|. \tag{24.26}$$

On the other hand, by Theorems 23.9 and 23.14 (more precisely (23.26) with $K = 0$, σ replaced by μ_t and μ replaced by μ_s),

$$\frac{d^+}{dt} W_2(\mu_s, \mu_t)^2 \leq 2\,[U_\nu(\mu_s) - U_\nu(\mu_t)] \leq 2\, U_\nu(\mu_s).$$

So

$$W_2(\mu_s, \mu_t)^2 \leq 2\, U_\nu(\mu_s)\, |t - s|. \tag{24.27}$$

Then (ii) results from the combination of (24.26) and (24.27), together with (i).

The proof of (iii) is pretty much the same, with the following modifications:

$$\frac{d\,I_{U,\nu}(\mu_t)}{dt} \le (-2K)\,I_{U,\nu}(\mu_t);$$

$$\frac{d^+}{dt}\,W_2(\mu_t,\nu)^2 \le -2\,U_\nu(\mu_t) + (-2K)\,W_2(\mu_t,\nu)^2;$$

$$\psi(t) := e^{2Kt}\Big(t^2\,I_{U,\nu}(\mu_t) + 2t\,U_\nu(\mu_t) + W_2(\mu_t,\nu)^2\Big).$$

Details are left to the reader. (The estimates in (iii) can be somewhat refined.) □

Exercise 24.22. Assuming $\mathrm{CD}(0,\infty)$, establish the estimate

$$I_{U,\nu}(\mu_t) \le \frac{U_\nu(\mu_0)}{t}.$$

Remark 24.23. There are many known regularization results in short time, for certain of the gradient flows considered in this chapter. The two most famous examples are:

- the **Li–Yau estimates**, which give lower bounds on $\Delta \log \rho_t$, for a solution of the heat equation on a Riemanian manifold, under certain curvature-dimension conditions. For instance, if M satisfies $\mathrm{CD}(0,N)$, then
$$\Delta \log \rho_t \ge -\frac{N}{2t};$$
- the **Aronson–Bénilan estimates**, which give lower bounds on $\Delta \rho_t^{m-1}$ for solutions of the nonlinear diffusion equation $\partial_t \rho = \Delta \rho^m$ in $\mathbb{R}^n$, where $1 - 2/n < m < 1$:
$$\frac{m}{m-1}\,\Delta(\rho_t^{m-1}) \ge -\frac{n}{\lambda t}, \qquad \lambda = 2 - n(1-m).$$

There is an obvious similarity between these two estimates, and both can be interpreted as a lower bound on the rate of divergence of the vector field which drives particles in the gradient flow interpretation of these partial differential equations. I think it would be very interesting to have a unified proof of these inequalities, under certain geometric conditions. For instance one could try to use the gradient flow interpretation of the heat and nonlinear diffusion equations, and maybe some localization by restriction.

Bibliographical notes

In [669], Otto advocated the use of his formalism both for the purpose of finding new schemes of proof, and for giving a new understanding of certain results.

What I call the Fokker–Planck equation is

$$\frac{\partial \mu}{\partial t} = \Delta \mu + \nabla \cdot (\mu_t \nabla V).$$

This is in fact an equation on *measures*. It can be recast as an equation on *functions* (densities):

$$\frac{\partial \rho}{\partial t} = \Delta \rho - \nabla V \cdot \nabla \rho.$$

From the point of view of stochastic processes, the relation between these two formalisms is the following: μ_t can be thought of as law (X_t), where X_t is the stochastic process defined by $dX_t = \sqrt{2}\, dB_t - \nabla V(X_t)\, dt$ ($B_t =$ standard Brownian motion on the manifold), while $\rho_t(x)$ is defined by the equation $\rho_t(x) = \mathbb{E}_x \rho_0(X_t)$ (the subscript x means that the process X_t starts at $X_0 = x$). In the particular case when V is a quadratic potential in $\mathbb{R}^n$, the evolution equation for ρ_t is often called the Ornstein–Uhlenbeck equation.

The observation that the Fisher information I_ν is the time-derivative of the entropy functional $-H_\nu$ along the heat semigroup seems to first appear in a famous paper by Stam [758] at the end of the fifties, in the case $M = \mathbb{R}$ (equipped with the Lebesgue measure). Stam gives credit to de Bruijn for that remark. The generalization appearing in Theorem 24.2(ii) has been discovered and rediscovered by many authors.

Theorem 24.2(iii) goes back to Bakry and Émery [56] for the case $U(r) = r \log r$. After many successive generalizations, the statement as I wrote it was formally derived in [577, Appendix D]. To my knowledge, the argument given in the present chapter is the first rigorous one to be written down in detail (modulo the technical justifications of the integrations by parts), although it is a natural expansion of previous works.

Theorem 24.2(iv) was proven by Otto and myself [671] for $\sigma = \mu_0$. The case $\sigma = \nu$ is also useful and was considered in [219].

Regularity theory for porous medium equations has been the object of many works, see in particular the synthesis works by Vázquez [804, 805, 806]. When one studies nonlinear diffusions by means of optimal transport theory, the regularity theory is the first thing to worry about. In a Riemannian context, Demange [291, 292, 290, 293] presents many approximation arguments based on regularization, truncation, etc. in great detail. Going into these issues would have led me to considerably expand the size of this chapter; but ignoring them completely would have led to incorrect proofs.

It has been known since the mid-seventies that logarithmic Sobolev inequalities yield rates of convergence to equilibrium for heat-like equations, and that these estimates are independent of the dimension. For certain problems of convergence to equilibrium involving entropy, logarithmic Sobolev inequalities are quite more convenient than spectral tools. This is especially true in infinite dimension, although logarithmic Sobolev inequalities are also very useful in finite dimension. For more information see the bibliographical notes of Chapter 21.

As recalled in Remark 24.11, convergence in the entropy sense implies convergence in total variation. In [220] various functional methods leading to convergence in total variation are examined and compared.

Around the mid-nineties, Toscani [784, 785] introduced the logarithmic Sobolev inequality in kinetic theory, where it was immediately recognized as a powerful tool (see e.g. [300]). The links between logarithmic Sobolev inequalities and Fokker–Planck equations were re-investigated by the kinetic theory community, see in particular [43] and the references therein. The emphasis was more on *proving* logarithmic Sobolev inequalities thanks to the study of the convergence to equilibrium for Fokker–Planck equations, than the reverse. So the key was the study of convergence to equilibrium in the Fisher information sense, as in Chapter 25; but the final goal really was convergence in the entropy sense. To my knowledge, it is only in a recent study of certain algorithms based on stochastic integration [549], that convergence in the Fisher information sense in itself has been found useful. (In this work some constructive criteria for exponential convergence in Fisher information are given; for instance this is true for the heat equation $\partial_t \rho = \Delta \rho$, under a $\mathrm{CD}(K,\infty)$ bound ($K < 0$) and a logarithmic Sobolev inequality.)

Around 2000, it was discovered independently by Otto [669], Carrillo and Toscani [215] and Del Pino and Dolbeault [283] that the same "information-theoretical" tools could be used for nonlinear equations of the form

$$\frac{\partial \rho}{\partial t} = \Delta \rho^m \tag{24.28}$$

in $\mathbb{R}^n$. Such equations are called porous medium equations for $m > 1$, and fast diffusion equations for $m < 1$. For these models there is no convergence to equilibrium: the solution disperses at infinity. But there is a well-known scaling, due to Barenblatt, which transforms (24.28) into

$$\frac{\partial \rho}{\partial t} = \Delta \rho^m + \nabla_x \cdot (\rho x). \tag{24.29}$$

Then, up to rescaling space and time, it is equivalent to understand the convergence to equilibrium for (24.29), or to understand the asymptotic behavior for (24.28), that is, how fast it approaches a certain known self-similar profile.

The extra drift term in (24.29) acts like the confinement by a quadratic potential, and this in effect is equivalent to imposing a curvature condition $\mathrm{CD}(K,\infty)$ ($K > 0$). This explains why there is an approach based on generalized logarithmic Sobolev inequalities, quite similar to the proof of Theorem 24.7.

These problems can be attacked without any knowledge of optimal transport. In fact, among the authors quoted before, only Otto did use optimal transport, and this was not at the level of proofs, but only at the level of intuition. Later in [671], Otto and I gave a more direct proof of logarithmic Sobolev inequality based on the HWI inequality. The same strategy was applied again in my joint work with Carrillo and McCann [213], for more general equations involving also a (simple) nonlinear drift.

In [213] the basic equation takes the form

$$\frac{\partial \rho}{\partial t} = \sigma \Delta \rho + \nabla \cdot (\rho \nabla V) + \nabla \cdot \Big(\rho \, \nabla (\rho * \nabla W)\Big), \tag{24.30}$$

where $\sigma \in \mathbb{R}_+$ and $W = W(x-y)$ is some interaction potential on $\mathbb{R}^n$. These equations (a particular instance of McKean–Vlasov equations) appeared in the modeling of granular media [92, 93, 622], either with $\sigma = 0$ or with $\sigma > 0$, in particular in dimension 1. See the review paper [820] for much more information. Similar equations also appear in the theory of self-interacting diffusion processes [83, 84, 85, 521, 535]. (Some of the ingredients of [213] are used again in [521], along with many subtle probabilistic arguments, in the study of the confining effect of self-interaction.)

The study of exponential convergence for (24.30) leads to interesting issues, some of them briefly reviewed in [815, 820]. There are criteria for exponential convergence in terms of the convexity of V and W.

These problems can also be set on a Riemannian manifold M (replace $W(x-y)$ by $W(x,y)$), and then Ricci curvature estimates come into play [761]. In the particular case of linear diffusion in $\mathbb{R}^n$, there are alternative approaches to these convergence results, more directly based on coupling arguments [221, 590, 591]. In the other particular case where (24.30) is set in dimension 1, $\sigma = 0$ and $W(z) = |z|^3/3$, the solution converges to a Dirac mass, and there is a self-similar scaling allowing one to refine the study of the rate of convergence. A somewhat surprising (at least so it was for us) result of Caglioti and myself [195] states that the refinement obtained by this method is necessarily small; the argument is based on a proof of "slow convergence" for a rescaled equation, which uses the 1-Wasserstein distance W_1.

The strategy based on displacement convexity does not apply directly to (24.30) when the potential W is not convex. However, there is an interesting interplay between the diffusion and the effect of the interaction potential. Such an effect was studied in [213] for (24.30) when $\sigma > 0$, $V = 0$ and $W = |z|^3$: even though the interaction potential is degenerately convex, the spreading caused by the diffusion is sufficient to make it "effectively" uniformly convex. Even more striking, Calvez and Carrillo [198] established convergence to equilibrium (in Wasserstein distance) for (24.30) with $V = 0$ in the *nonconvex* case $W(z) = |z|^k/k$ for $-1 < k < 1$ ($k = 0$ corresponds to $\log |z|$). This also works if the linear diffusion $\Delta\rho$ is replaced by $\Delta\rho^m$ with $m + k > 1$.

Demange [290, 291, 292, 293] studied the fast diffusion equation $\partial_t \rho = \Delta \rho^{1-1/N}$ on a Riemannian manifold, under a curvature-dimension condition $\mathrm{CD}(K,N)$. He used the Sobolev inequality, in the form

$$\begin{aligned} H_{N/2}(\mu) &\leq \frac{(N-2)(N-1)}{2K} \int \rho^{-1-\frac{2}{N}} |\nabla \rho|^2 \, d\nu \\ &\leq \frac{(N-2)(N-1)}{2K} (\sup \rho)^{-\frac{1}{N}} \int \rho^{1-\frac{1}{N}} |\nabla \rho|^2 \, d\nu \end{aligned}$$

to obtain a differential inequality such as

$$\frac{dH_{N/2}(\mu_t)}{dt} \leq -\left(\frac{N-2}{N-1}\right) (\sup \rho)^{-\frac{1}{N}} \, \frac{H_{N/2}(\mu_t)}{2K},$$

and deduced an estimate of the form

$$H_{N/2}(\mu_t) = O\big(e^{-(\lambda_N + \varepsilon)\, t}\big),$$

where λ_N is the presumably optimal rate that one would obtain without the $(\sup \rho)$ term, and $\varepsilon > 0$ is arbitrarily small. His estimate is slightly stronger than the one which I derived in Theorem 24.7 and Remark 24.12, but the asymptotic rate is the same.

All the methods described before apply to the study of the time asymptotics of the porous medium equation $\partial_t \rho = \Delta \rho^m$, but only under the restriction $m \geq 1 - 1/N$. In that regime one can use time-rescaling and tools similar to the ones described in this chapter, to prove that the solutions become close to Barenblatt's self-similar solution.

When $m < 1-1/N$, displacement convexity and related tricks do not apply any more. This is why it was rather a sensation when Carrillo and Vázquez [217] applied the Aronson–Bénilan estimates to the problem of asymptotic behavior for fast diffusion equations with exponents m in $(1 - \frac{2}{N}, 1 - \frac{1}{N})$, which is about the best that one can hope for, since Barenblatt profiles do not exist for $m \leq 1 - 2/N$.

Here we see the limits of Otto's formalism: such results as the dimensional refinement of the rate of convergence for diffusive equations (Remark 24.10), or the Carrillo–Vázquez estimates, rely on inequalities of the form

$$\int p(\rho)\, \Gamma_2\big(\nabla U'(\rho)\big)\, d\nu + \int p_2(\rho)\, (LU'(\rho))^2\, d\nu \geq \dots$$

in which ones takes advantage of the fact that *the same* function ρ appears in the terms $p(\rho)$ and $p_2(\rho)$ on the one hand, and in the terms $\nabla U'(\rho)$ and $LU'(\rho)$ on the other. The technical tool might be changes of variables for the Γ_2 (as in [541]), or elementary integration by parts (as in [217]); but I don't see any interpretation of these tricks in terms of the Wasserstein space $P_2(M)$.

The story about the rates of equilibration for fast diffusion equations does not end here. At the same time as Carrillo and Vázquez obtained their main results, Denzler and McCann [298, 299] computed the spectral gap for the linearized fast diffusion equations in the same interval of exponents $(1 - \frac{2}{N}, 1 - \frac{1}{N})$. This study showed that the rate of convergence obtained by Carrillo and Vázquez is off the value suggested by the linearized analysis by a factor 2 (except in the radially symmetric case where they obtain the optimal rate thanks to a comparison method). The connection between the nonlinear and the linearized dynamics is still unclear, although some partial results have been obtained by McCann and Slepčev [619]. More recently, S.J. Kim and McCann [517] have derived optimal rates of convergence for the "fastest" nonlinear diffusion equations, in the range $1-2/N < m \leq 1-2/(N+2)$, by comparison methods involving Newtonian potentials. Another work by Cáceres and Toscani [183] also recovers some of the results of Denzler and McCann by means of completely different methods with their roots in kinetic theory. There is still ongoing research to push the rates of convergence and the range of admissible nonlinearities, in particular by Denzler, Koch, McCann and probably others.

In dimension 2, the limit case $m = 0$ corresponds to a logarithmic diffusion; it is related to geometric problems, such as the evolution of conformal surfaces or the Ricci flow [806, Chapter 8].

More general nonlinear diffusion equations of the form $\partial_t \rho = \Delta p(\rho)$ have been studied by Biler, Dolbeault and Esteban [119], and Carrillo, Di Francesco and Toscani [210, 211] in $\mathbb{R}^n$. In the latter work the rescaling procedure is recast in a more geometric and physical interpretation, in terms of temperature and projections; a sequel by Carrillo and Vázquez [218] shows that the intermediate asymptotics can be complicated for well-chosen nonlinearities. Nonlinear diffusion equations on manifolds were also studied by Demange [291] under a $CD(K, N)$ curvature-dimension condition, $K > 0$.

Theorem 24.7(ii) is related to a long tradition of study of contraction rates in Wasserstein distance for diffusive equations [231, 232, 458, 662]. Sturm and Renesse [764] noted that such contraction rates *characterize* nonnegative Ricci curvature; Sturm [761] went on to give various characterizations of $CD(K, N)$ bounds in terms of contraction rates for possibly nonlinear diffusion equations.

In the one-dimensional case ($M = \mathbb{R}$) there are alternative methods to get contraction rates in W_2 distance, and one can also treat larger classes of models (for instance viscous conservation laws), and even obtain decay in W_p for any p; see for instance [137, 212]. Recently, Brenier found a re-interpretation of these one-dimensional contraction properties in terms of monotone operators [167]. Also the asymptotic behavior of certain conservation laws has been analyzed in this way [208, 209] (with the help of the strong "W_∞ distance"!).

Another model for which contraction in W_2 distance has been established is the Boltzmann equation, in the particular case of a spatially homogeneous gas of Maxwellian molecules. This contraction property was discovered by Tanaka [644, 776, 777]; see [138] for recent work on the subject. Some striking uniqueness results have been obtained by Fournier and Mouhot [377, 379] with a related method (see also [378]).

To conclude this discussion about contraction estimates, I shall briefly discuss some links with Perelman's analysis of the backward version of Hamilton's Ricci flow. A first observation by McCann and Topping [620] is that the evolution by this flow forces the heat equation to be a contraction in Wasserstein distance, even if the Ricci curvature is not everywhere nonnegative. McCann and Topping also established a converse result *characterizing* the Ricci flow in terms of contractive semigroups. Related topics were independently studied by Carfora [200].

These investigations were pushed further by Topping [782] and Lott [576] with the help of the formalism of displacement interpolation for Lagrangian actions (recall Chapter 7). For instance, define

$$\mathcal{L}_0^{t_0,t_1}(x,y) = \inf\left\{\frac{1}{2}\int_{t_0}^{t_1}\Big(\|\dot{\gamma}(t)|^2 + S(\gamma(t),t)\Big)\,dt\right\},$$

where the infimum is taken over all C^1 paths $\gamma : [t_0,t_1] \to M$ such that $\gamma(t_0) = x$ and $\gamma(t_1) = y$, and $S(x,t)$ is the scalar curvature of M (evolving under backward Ricci flow) at point x and time t. As in Chapter 7, this induces a Hamilton–Jacobi equation, and an action in the space of measures. Then it is shown in [576] that $H(\mu_t) - \int \phi_t\,d\mu_t$ is convex in t along the associated displacement interpolation, where (ϕ_t) is a solution of the Hamilton–Jacobi equation. Other theorems in [576, 782] deal with a variant of $\mathcal{L}_0$ in which some time-rescalings have been performed. Not only do these results generalize the contraction property of [620], but they also imply Perelman's estimates of monotonicity of the so-called W-entropy and reduced volume functionals (which were an important tool in the proof of the Poincaré conjecture).

I shall now comment on short-time decay estimates. The short-time behavior of the entropy and Fisher information along the heat flow (Theorem 24.16) was studied by Otto and myself around 1999 as a technical ingredient to get certain a priori estimates in a problem of hydrodynamical limits. This work was not published, and I was quite surprised to discover that Bobkov, Gentil and Ledoux [127, Theorem 4.3] had found similar inequalities and applied them to get a new proof of the HWI inequality. Otto and I published our method [672] as a comment to [127]; this is the same as the proof of Theorem 24.16. It can be considered as an adaptation, in the context of the Wasserstein space, of some classical estimates about gradient flows in Hilbert spaces, that can be found in Brézis [171, Théorème 3.7]. The result of Bobkov, Gentil and Ledoux is actually more general than ours, because these authors seem to have sharp constants under $\mathrm{CD}(K,\infty)$ for all values of $K \in \mathbb{R}$, while it is not clear that our method is sharp for $K \neq 0$. For $K = 0$ both methods yield exactly the same result, which was a bit of a puzzle to me. It would be interesting to clarify all this.

In relation to Remark 24.21, I was asked the following question by Guionnet (and I am unable to answer): Given a solution (μ_t) of the heat equation $\partial_t \rho = L\rho$, is it true that $t\, I_\nu(\mu_t)$ converges to a finite limit as $t \to 0$? If yes, then by De L'Hospital's rule, this is also the limit of $H_\nu(\mu_t)/|\log t|$ as $t \to 0$. In the particular case when $\mu_0 = f\,\nu + \sum_{k=1}^{N} a_k\,\delta_{x_k}$, with f smooth, it is not difficult to show that $t\,I_\nu(\mu_t)$ converges to $\sum a_k$. This question is motivated by some problems in free probability theory.

Inequality (24.25) goes back to [672], under adequate regularity assumptions, for the main case of interest which is $U(r) = r \log r$.

Hölder-1/2 estimates in time are classical for gradient flows; in the context of the Wasserstein space, they appeared in several works, for instance [30].

In [214] and [30] there were some investigations about the possibility to directly use Otto's formalism to perform the proof of Theorem 24.2 and the other theorems in this chapter.

The Li–Yau heat kernel estimates go back to [552]; they were refined by Davies [272], then by Bakry and Qian [60]; the latter paper is closely related to certain issues that will be addressed in the next chapter. In any case, the Bochner formula and various forms of maximum principles are the main ingredients behind these estimates. Recently, Bakry and Ledoux [59] derived improved forms of the Li–Yau estimates, and made the connection with the theory of logarithmic Sobolev inequalities.

There are similarities between the Li–Yau parabolic Harnack inequality and the study of Perelman's $\mathcal{L}$-functional; in view of Topping's work [782], this seems to give one further reason to hope for a direct relation between optimal transport and Li–Yau inequalities.

The Aronson–Bénilan estimates were established in [45]. There is some overlap between the Aronson–Bénilan and Li–Yau bounds; see [580] for a common framework. (Early attempts were performed by Carrillo and myself.)

Recently Demange has obtained short-time regularization estimates like $\sup \rho_t = O(t^{-N})$ for the fast diffusion equation $\partial_t \rho = \Delta \rho^{1-1/N}$ in positive curvature, which are optimal in a certain sense.

In this chapter as in the previous one, I have only been interested in *gradient flows*; but there are probably other questions about the qualitative behavior of *Hamiltonian* flows which make sense in relation to optimal transport. For instance, if one were able to construct "Gibbs measures" of the form (15.23) on the set $P_2(M)$, where M is a symplectic manifold, then they would be natural candidates to be relevant invariant measures for Hamiltonian flows in $P_2(M)$. Take for instance $M = \mathbb{T}^2$, and define the Hamiltonian as $H(\mu) = \int G(x,y)\,\mu(dx)\,\mu(dy)$, where $G(x,y)$ is the fundamental solution of the Laplace operator on $\mathbb{T}^2$; then the associated "Hamiltonian flow" should be the two-dimensional Euler equation. For this equation the problem of constructing invariant measures was considered long ago [94, 693] without real success (see however [14]); it is natural to ask whether the optimal transport approach provides a path to attack this problem.

25

Gradient flows III: Functional inequalities

In the preceding chapter certain functional inequalities were used to provide quantitative information about the behavior of solutions to certain partial differential equations. In the present chapter, conversely, the behavior of solutions to certain partial differential equations will help establish certain functional inequalities.

For the kind of inequalities that will be encountered in this chapter, this principle has been explored in depth since the mid-eighties, starting with Bakry and Émery's heat semigroup proof of Theorem 21.2. Nowadays, one can prove this theorem by more direct means (as I did in Chapter 21); nevertheless, the heat semigroup argument is still of interest, and not only for historical reasons. Indeed it has been the basis for many generalizations, some of which are still out of reach of alternative methods.

Optimal transport appears in this game from two different perspectives. On the one hand, several inequalities involving optimal transport have been proven by diffusion semigroup methods. On the other hand, optimal transport has provided a re-interpretation of these methods, since several diffusion equations can be understood as gradient flows with respect to a structure induced by optimal transport. This interpretation has led to a more synthetic and geometric picture of the field; and Otto's calculus has provided a way to shortcut some intricate computations.

That being said, I have to admit that there are limitations to this point of view. It is true that some of the most important computations in Bakry's Γ_2 calculus can be understood in terms of optimal transport; but some other parts of the formalism, in particular those based on changes of functions, have remained inaccessible so far. Usually such

C. Villani, *Optimal Transport*, Grundlehren der mathematischen Wissenschaften 338,

manipulations are useful to treat functional inequalities involving a natural class of functions whose dimension "does not match" the dimension of the curvature-dimension condition. More explicitly: It is usually okay to interpret in terms of optimal transport a proof involving functions in $\mathcal{DC}_\infty$ under a curvature-dimension assumption $\mathrm{CD}(K,\infty)$. Such is also the case for a proof involving functions in $\mathcal{DC}_N$ under a curvature-dimension assumption $\mathrm{CD}(K,N)$. But to get the correct constants for an inequality involving functions in $\mathcal{DC}_N$ under a condition $\mathrm{CD}(K,N')$, $N' < N$, may be much more of a problem.

In this chapter, I shall discuss three examples which can be worked out nicely. The first one is an alternative proof of Theorem 21.2, following the original argument of Bakry and Émery. The second example is a proof of the optimal Sobolev inequality (21.8) under a $\mathrm{CD}(K,N)$ condition, recently discovered by Demange. The third example is an alternative proof of Theorem 22.17, along the lines of the original proof by Otto and myself.

The proofs in this chapter will be sloppy in the sense that I shall not go into smoothness issues, or rather admit auxiliary regularity results which are not trivial, especially in unbounded manifolds. These regularity issues are certainly the main drawback of the gradient flow approach to functional inequalities — to the point that many authors prefer to just ignore these difficulties!

I shall use the same conventions as in the previous chapters: U will be a nonlinearity belonging to some displacement convexity class, and $p(r) = r\,U'(r) - U(r)$ will be the associated pressure function; $\nu = e^{-V}\mathrm{vol}$ will be a reference measure, and L will be the associated Laplace-type operator admitting ν as invariant measure. Moreover,

$$U_\nu(\mu) = \int U(\rho)\,d\nu, \qquad I_{U,\nu}(\mu) = \int \rho\,|\nabla U'(\rho)|^2\,d\nu = \int \frac{|\nabla p(\rho)|^2}{\rho}\,d\nu,$$

$$H_{N,\nu}(\mu) = -N\int (\rho^{1-\frac{1}{N}} - \rho)\,d\nu, \quad I_{N,\nu}(\mu) = \left(1 - \frac{1}{N}\right)^2 \int \rho^{-1-\frac{2}{N}}|\nabla\rho|^2\,d\nu,$$

$$H_{\infty,\nu}(\mu) = H_\nu(\mu) = \int \rho\log\rho\,d\nu, \qquad I_{\infty,\nu}(\mu) = I_\nu(\mu) = \int \frac{|\nabla\rho|^2}{\rho}\,d\nu,$$

where ρ always stands for the density of μ with respect to ν.

Logarithmic Sobolev inequalities revisited

Theorem 25.1 (Infinite-dimensional Sobolev inequalities from Ricci curvature). *Let M be a Riemannian manifold equipped with a reference measure ν satisfying a curvature-dimension bound* $\mathrm{CD}(K,\infty)$ *for some $K > 0$, and let $U \in \mathcal{DC}_\infty$. Further, let $\lambda := \lim_{r\to 0} p(r)/r$. Then, for all $\mu \in P_2^{\mathrm{ac}}(M)$,*

$$U_\nu(\mu) - U_\nu(\nu) \le \frac{I_{U,\nu}(\mu)}{2K\lambda}.$$

Particular Case 25.2 (Bakry–Émery theorem again). If (M,ν) satisfies $\mathrm{CD}(K,\infty)$ for some $K > 0$, then the following logarithmic Sobolev inequality holds true:

$$\forall \mu \in P^{\mathrm{ac}}(M), \qquad H_\nu(\mu) \le \frac{I_\nu(\mu)}{2K}.$$

Sloppy proof of Theorem 25.1. By using Theorem 17.7(vii) and an approximation argument, we may assume that ρ is smooth, that U is smooth on $(0,+\infty)$, that the solution $(\rho_t)_{t\ge 0}$ of the gradient flow

$$\frac{\partial \rho}{\partial t} = L\, p(\rho_t),$$

starting from $\rho_0 = \rho$ is smooth, that $U_\nu(\mu_0)$ is finite, and that $t \to U_\nu(\mu_t)$ is continuous at $t = 0$.

For notational simplicity, let

$$H(t) := U_\nu(\mu_t), \qquad I(t) := I_{U,\nu}(\mu_t).$$

From Theorems 24.2(ii) and 24.7(i)(b),

$$\frac{dH(t)}{dt} = -I(t), \qquad I(t) \le I(0)\, e^{-2K\lambda t}.$$

By Theorem 24.7(i)(a), $H(t) \to 0$ as $t \to \infty$. So

$$H(0) = \int_0^{+\infty} I(t)\, dt \le I(0) \int_0^{+\infty} e^{-2K\lambda t}\, dt = \frac{I(0)}{2K\lambda},$$

which is the desired result. □

Sobolev inequalities revisited

Theorem 25.3 (Generalized Sobolev inequalities under Ricci curvature bounds). *Let M be a Riemannian manifold equipped with a reference measure $\nu = e^{-V}$, $V \in C^2(M)$, satisfying a curvature-dimension bound* $\mathrm{CD}(K,N)$ *for some $K > 0$, $N \in [1,\infty)$. Let $U \in \mathcal{DC}_N$ with $U'' > 0$ on $(0,+\infty)$, and let $A \in C(\mathbb{R}_+) \cap C^2((0,+\infty))$ be such that $A(0) = A(1) = 0$ and $A''(r) = r^{-\frac{1}{N}} U''(r)$. Then, for any probability density ρ on M,*

$$\int_M A(\rho)\, d\nu \le \frac{1}{2K\lambda} \int_M \rho\, |\nabla U'(\rho)|^2\, d\nu, \tag{25.1}$$

where

$$\lambda = \lim_{r \downarrow 0} \frac{p(r)}{r^{1-\frac{1}{N}}}.$$

Remark 25.4. For a given U, there might not necessarily exist a suitable A. For instance, if $U = U_N$, it is only for $N > 2$ that we can construct A.

Particular Case 25.5 (Sobolev inequalities). Whenever $N > 2$, let

$$U(r) = U_N(r) = -N(r^{1-\frac{1}{N}} - r), \qquad A(r) = -\frac{N(N-1)}{2(N-2)}\,(r^{1-\frac{2}{N}} - r);$$

then (25.1) reads

$$H_{\frac{N}{2},\nu}(\mu) \le \frac{1}{2K} \left(\frac{N-2}{N-1}\right) I_{N,\nu}(\mu),$$

which can also be rewritten in the form of (21.9) or (21.8).

Sloppy proof of Theorem 25.3. By density, we may assume that the density ρ_0 of μ is smooth; we may also assume that A and U are smooth on $(0,+\infty)$ (recall Proposition 17.7(vii)). Let $(\rho_t)_{t\ge 0}$ be the solution of the gradient flow equation

$$\frac{\partial \rho}{\partial t} = \nabla \cdot (\rho \nabla U'(\rho)), \tag{25.2}$$

and as usual $\mu_t = \rho_t\, \nu$. It can be shown that ρ_t is uniformly bounded below by a positive number as $t \to \infty$.

By Theorem 24.2(iii),

$$\frac{d}{dt} I_{U,\nu}(\mu_t) \leq -2K\lambda \int_M \rho_t^{1-\frac{1}{N}} |\nabla U'(\rho_t)|^2 \, d\nu. \tag{25.3}$$

On the other hand, from the assumption $A''(r) = r^{-\frac{1}{N}} U''(r)$,

$$\nabla A'(\rho) = \rho^{-\frac{1}{N}} \, \nabla U'(\rho).$$

So Theorem 24.2(i) implies

$$\begin{aligned} \frac{d}{dt} \int A(\rho_t) \, d\nu &= - \int_M \rho_t \nabla A'(\rho_t) \cdot \nabla U'(\rho_t) \, d\nu \\ &= - \int_M \rho_t^{1-\frac{1}{N}} \left|\nabla U'(\rho_t)\right|^2 d\nu. \end{aligned} \tag{25.4}$$

The combination of (25.3) and (25.4) leads to

$$-\frac{d}{dt} A_\nu(\mu_t) \leq - \left(\frac{1}{2K\lambda}\right) \frac{d}{dt} I_{U,\nu}(\mu_t). \tag{25.5}$$

As $t \to \infty$, $I_{U,\nu}(\mu_t)$ and $U_\nu(\mu_t)$ converge to 0 (Theorem 24.7(i)). Since ρ_t is uniformly bounded below and U'' is uniformly positive on the range of ρ_t, this implies that $\rho_t \to 1$ in $L^1(\nu)$, and also that $A_\nu(\mu_t)$ converges to 0. Then one can integrate both sides of (25.5) from $t = 0$ to $t = \infty$, and recover

$$A_\nu(\mu_0) \leq \left(\frac{1}{2K\lambda}\right) I_{U,\nu}(\mu_0),$$

as desired. □

From log Sobolev to Talagrand, revisited

Theorem 25.6 (From Sobolev-type inequalities to concentration inequalities). *Let M be a Riemannian manifold equipped with a reference probability measure $\nu = e^{-V} \mathrm{vol} \in P_2^{\mathrm{ac}}(M)$, $V \in C^2(M)$. Let $U \in \mathcal{DC}_\infty$. Assume that for any $\mu \in P_2^{\mathrm{ac}}(M)$, holds the inequality*

$$U_\nu(\mu) - U_\nu(\nu) \leq \frac{1}{2K} I_{U,\nu}(\mu). \tag{25.6}$$

Further assume that the Cauchy problem associated with the gradient flow $\partial_t \rho = L\, p(\rho)$ admits smooth solutions for smooth initial data. Then, for any $\mu \in P_2^{ac}(M)$, holds the inequality

$$\frac{W_2(\mu,\nu)^2}{2} \leq \frac{U_\nu(\mu) - U_\nu(\nu)}{K}.$$

Particular Case 25.7 (From Log Sobolev to Talagrand). If the reference measure ν on M satisfies a logarithmic Sobolev inequality with constant K, and a curvature-dimension bound $\mathrm{CD}(K',\infty)$ for some $K' \in \mathbb{R}$, then it also satisfies a Talagrand inequality with constant K:

$$\forall \mu \in P_2^{ac}(M), \qquad W_2(\mu,\nu) \leq \sqrt{\frac{2\, H_\nu(\mu)}{K}}. \tag{25.7}$$

Sloppy proof of Theorem 25.6. By a density argument, we may assume that μ has a smooth density μ_0, and let $(\mu_t)_{t\geq 0}$ evolve according to the gradient flow (25.2). By Theorem 24.2(ii),

$$\frac{d}{dt}\, U_\nu(\mu_t) = -I_{U,\nu}(\mu_t).$$

In particular, $(d/dt)U_\nu(\mu_t) \leq -2KU_\nu(\mu_t)$, so $U_\nu(\mu_t)$ converges to 0 as $t \to \infty$ (exponentially fast).

By Theorem 24.2(iv), for almost all t,

$$\frac{d^+}{dt}\, W_2(\mu_0,\mu_t) \leq \sqrt{I_{U,\nu}(\mu_t)}.$$

On the other hand, by assumption,

$$\sqrt{I_{U,\nu}(\mu_t)} \leq \frac{I_{U,\nu}(\mu_t)}{\sqrt{2KU_\nu(\mu_t)}} = -\frac{d}{dt}\sqrt{\frac{2\,U_\nu(\mu_t)}{K}}. \tag{25.8}$$

From (24.4) and (25.8),

$$\frac{d^+}{dt}\, W_2(\mu_0,\mu_t) \leq -\frac{d}{dt}\sqrt{\frac{2\,U_\nu(\mu_t)}{K}}.$$

Stated otherwise: If

$$\psi(t) := W_2(\mu_0,\mu_t) + \sqrt{\frac{2\,U_\nu(\mu_t)}{K}},$$

then $d^+\psi/dt \leq 0$, i.e. ψ is nonincreasing as a function of t, and so

$$\lim_{t\to\infty} \psi(t) \le \psi(0). \tag{25.9}$$

Let us now check that μ_t converges weakly to ν. Inequality (25.9) implies that $W_2(\mu_0, \mu_t)$ remains bounded as $t \to \infty$; so $\int d(z,x)\,\mu_t(dx) \le \int d(z,x)\,\mu_0(dx) + W_1(\mu_0,\mu_t)$ is also uniformly bounded, and $\{\mu_t\}$ is tight as $t \to \infty$. Up to extraction of a sequence of times, μ_t converges weakly to some measure $\widetilde{\mu}$. On the other hand, the functional inequality (25.6) forces U'' to be positive on $(0,+\infty)$, and then the convergence $U_\nu(\mu_t) \to 0 = U_\nu(\nu)$ is easily seen to imply $\rho_t \xrightarrow[t\to\infty]{} 1$ almost surely. This combined with the weak convergence of μ to $\widetilde{\mu}$ imposes $\widetilde{\mu} = \nu$; so μ_t does converge weakly to ν. As a consequence,

$$\begin{aligned} W_2(\mu_0,\nu) &\le \liminf_{t\to\infty} W_2(\mu_0,\mu_t) \\ &= \liminf_{t\to\infty} \psi(t) \\ &\le \psi(0) = \sqrt{(2\,U_\nu(\mu_0))/K}, \end{aligned}$$

which proves the claim. □

Appendix: Comparison of proofs

The proofs in the present chapter were based on gradient flows of displacement convex functionals, while proofs in Chapters 21 and 22 were more directly based on displacement interpolation. How do these two strategies compare to each other?

From a formal point of view, they are not so different as one may think. Take the case of the heat equation,

$$\frac{\partial \rho}{\partial t} = \Delta \rho,$$

or equivalently

$$\frac{\partial \rho}{\partial t} + \nabla \cdot \big(\rho \nabla(-\log \rho)\big) = 0.$$

The evolution of ρ is determined by the "vector field" $\rho \to (-\log \rho)$, in the space of probability densities. Rescale time and the vector field itself as follows:

$$\varphi_\varepsilon(t,x) = -\varepsilon \log \rho\left(\frac{\varepsilon t}{2}, x\right).$$

Then φ_ε satisfies the equation

$$\frac{\partial \varphi_\varepsilon}{\partial t} + \frac{|\nabla \varphi_\varepsilon|^2}{2} = \frac{\varepsilon}{2} \Delta \varphi_\varepsilon.$$

Passing to the limit as $\varepsilon \to 0$, one gets, at least formally, the Hamilton–Jacobi equation

$$\frac{\partial \varphi}{\partial t} + \frac{|\nabla \varphi|^2}{2} = 0,$$

which is in some sense the equation driving displacement interpolation.

There is a general principle here: After suitable rescaling, the velocity field associated with a gradient flow resembles the velocity field of a geodesic flow. Here might be a possible way to see this. Take an arbitrary smooth function U, and consider the evolution

$$\dot{x}(t) = -\nabla U(x(t)).$$

Turn to Eulerian formalism, consider the associated vector field v defined by

$$\frac{d}{dt} X(t, x_0) = -\nabla U(X(t, x_0)) =: -v\big(t, X(t, x_0)\big),$$

and rescale by

$$v_\varepsilon(t, x_0) = \varepsilon\, v\big(\varepsilon t,\; X(\varepsilon t, x_0)\big).$$

then one can check that, as $\varepsilon \to 0$,

$$\nabla_{x_0} v_\varepsilon(t, x_0) \simeq \varepsilon \nabla^2 U(x_0).$$

It follows by an explicit calculation that

$$\frac{\partial v_\varepsilon}{\partial t} + v_\varepsilon \cdot \nabla v_\varepsilon \simeq 0.$$

So as $\varepsilon \to 0$, $v_\varepsilon(t, x)$ should asymptotically satisfy the equation of a geodesic vector field (pressureless Euler equation).

There is certainly more to say on the subject, but whatever the interpretation, the Hamilton–Jacobi equations can always be squeezed out of the gradient flow equations after some suitable rescaling. Thus we may expect the gradient flow strategy to be more precise than the displacement convexity strategy. This is also what the use of Otto's calculus suggests: Proofs based on gradient flows need a control of $\operatorname{Hess} U_\nu$ only in the direction $\operatorname{grad} U_\nu$, while proofs based on displacement convexity need a control of $\operatorname{Hess} U_\nu$ in all directions. This might explain why there is at present no displacement convexity analogue of Demange's proof of the Sobolev inequality (so far only weaker inequalities with nonsharp constants have been obtained).

On the other hand, proofs based on displacement convexity are usually rather simpler, and more robust than proofs based on gradient flows: no issues about the regularity of the semigroup, no subtle interplay between the Hessian of the functional and the "direction of evolution"...

In the end we can put some of the main functional inequalities discussed in these notes in a nice array. Below, "LSI" stands for "Logarithmic Sobolev inequality"; "T" for "Talagrand inequality"; and "Sob_2" for the Sobolev inequality with exponent 2. So $\mathrm{LSI}(K)$, $\mathrm{T}(K)$, $\mathrm{HWI}(K)$ and $\mathrm{Sob}_2(K,N)$ respectively stand for (21.4), (22.4) (with $p = 2$), (20.17) and (21.8).

Theorem	Gradient flow proof	Displ. convexity proof
$\mathrm{CD}(K,\infty) \Rightarrow \mathrm{LSI}(K)$	Bakry–Émery	Otto–Villani
$\mathrm{LSI}(K) \Rightarrow \mathrm{T}(K)$	Otto–Villani	Bobkov–Gentil–Ledoux
$\mathrm{CD}(K,\infty) \Rightarrow \mathrm{HWI}(K)$	Bobkov–Gentil–Ledoux	Otto–Villani
$\mathrm{CD}(K,N) \Rightarrow \mathrm{Sob}_2(K,N)$	Demange	??

Bibliographical notes

Stam used a heat semigroup argument to prove an inequality which is equivalent to the Gaussian logarithmic Sobolev inequality in dimension 1 (recall the bibliographical notes for Chapter 21). His argument was not completely rigorous because of regularity issues, but can be repaired; see for instance [205, 783].

The proof of Theorem 25.1 in this chapter follows the strategy by Bakry and Émery, who were only interested in the Particular Case 25.2. These authors used a set of calculus rules which has been dubbed the "Γ_2 calculus". They were not very careful about regularity issues, and for that reason the original proof probably cannot be considered as completely rigorous (in particular for noncompact manifolds, in which regularity issues are not so innocent, even if the curvature-dimension condition prevents the blow-up of the heat semigroup). However, recently Demange [291] carried out complete proofs for much more delicate situations, so there is no reason to doubt that the Bakry–Émery argument can be made fully rigorous. Also, when the manifold is $\mathbb{R}^n$ equipped with a reference density e^{-V}, the proof was carefully rewritten by Arnold, Markowich, Toscani and Unterreiter [43], in the language of partial differential equations. This paper was the sequel to a simpler paper by Toscani [785] considering the particular case of the Gaussian measure.

The Bakry–Émery strategy was applied independently by Otto [669] and by Carrillo and Toscani [215] to study the asymptotic behavior of porous medium equations. Since then, many authors have applied it to various classes of nonlinear equations, see e.g. [213, 217].

The interpretation of the Bakry–Émery proof as a gradient flow argument was developed in my paper with Otto [671]. This interpretation was of much help when we considered more complicated nonlinear situations in [213].

In this chapter as in [671] the gradient flow interpretation was used only as a help to understanding; but the gradient flow formalism can also be used more directly, see for instance [655].

Theorem 25.3 is due to Demange [291]. Demange did not only treat the inequality (21.9), but also the whole family (21.7). A disturbing remark is that for many members of this family, several distinct gradient flows can be used to yield the same functional inequality. Demange also discussed other criteria than $U \in \mathcal{DC}_N$, allowing for finer results if, say, $U \in \mathcal{DC}_N$ but the curvature-dimension bound is $\mathrm{CD}(K, N')$ for some $N' < N$; at this point he uses formulas of change of variables for Γ_2 operators. He found a mysterious structure condition on the nonlinearity U, which in many cases leads to finer results than the $\mathcal{DC}_N$ condition:

$$r q'(r) + q(r) \geq \frac{9N}{4(N+2)}\, q(r)^2, \qquad q(r) = \frac{r U''(r)}{U'(r)} + \frac{1}{N}. \tag{25.10}$$

(Note that $q \equiv 0$ for $U = U_N$.)

Demange worked on arbitrary noncompact manifolds by using a careful truncation procedure; he restricted the equation to bounded open subsets and imposed Dirichlet boundary conditions. (Neumann's boundary conditions would be more natural, for instance because they preserve the mass; but the Dirichlet boundary conditions have the major technical advantage of coming with a monotonicity principle.) All of Demange's results still seem to be out of reach of more direct methods based on displacement interpolation.

The proof of Theorem 25.6 was implemented in my joint work with Otto [671]. The proof there is (hopefully!) complete, but we only considered Particular Case 25.7 (certainly the most important). We carefully checked that the curvature bound $\mathrm{CD}(K', \infty)$ prevents the blow-up of the heat equation. Maybe one can still make the proof work without that lower bound assumption, by truncating the logarithmic Sobolev inequality and the Talagrand inequality, and then working in an arbitrarily large bounded open subset of the manifold, imposing Neumann boundary conditions. In any case, to treat noncompact manifolds without lower bounds on the curvature, it is certainly easier to use the proof of Theorem 22.17, based on the Bobkov–Gentil–Ledoux method.

Later, Biane and Voiculescu [118] adapted our argument to free probability theory, deriving a noncommutative analog of the Talagrand inequality; what plays the role of the Gaussian measure is now Wigner's semi-circular law. In their paper, they also discuss many generalizations, some of which seem to have no classical counterpart so far.

F.-Y. Wang [831], and Cattiaux and Guillin [219] have worked out several other variants and applications of our scheme of proof. Cattiaux and Guillin also noticed that one could replace the original argument based on an upper estimate of $(d/dt)W_2(\mu_0, \mu_t)$, by a lower estimate of $(d/dt)W_2(\mu_t, \nu)$.

The observation that the Hamilton–Jacobi equation can be obtained from the heat equation after proper rescaling is quite old, and it is now a classical exercise in the theory of viscosity solutions (see e.g. [335]). Bobkov, Gentil and Ledoux [127] observed that this could constitute a bridge between the two main existing strategies for logarithmic Sobolev inequalities. Links with the theory of large deviations have been investigated in [335, 355].

As for the final array in the Appendix, the corresponding papers are those of Bakry and Émery [56], Otto and Villani [671], Bobkov, Gentil and Ledoux [127], and Demange [290, 291].

Part III

Synthetic treatment of Ricci curvature

The last part of these notes is devoted to a direction of research which was mainly explored by Lott, Sturm and myself from 2004 on.

In Chapter 17 it was proven that lower Ricci curvature bounds influence displacement convexity properties of certain classes of functionals; but also that these properties **characterize** lower Ricci curvature bounds. So we may "transform the theorem into a definition" and express the property "Ricci curvature is bounded below by K" in terms of certain displacement convexity properties. This approach is **synthetic**, in the sense that it does not rely on analytic computations (of the Ricci tensor...), but rather on the properties of certain objects which play an important role in some geometric arguments.

This point of view has the advantage of applying to nonsmooth spaces, just as lower (or upper) sectional curvature bounds can be defined in nonsmooth metric spaces by Alexandrov's method. An important difference however is that the notion of generalized Ricci curvature will be defined not only in terms of distances, but also in terms of reference measures. So the basic object will not be a metric space, but a **metric-measure space**, that is, a metric space equipped with a reference measure.

Chapters 26 and 27 are preparatory. In Chapter 26 I shall try to convey in some detail the meaning of the word "synthetic", with a simple illustration about convex functions; then Chapter 27 will be devoted to some reminders about the convergence of metric-measure spaces.

The next two chapters constitute the core of this part. In Chapter 28 I will consider optimal transport in possibly nonsmooth spaces, and establish various properties of stability of optimal transport under convergence of metric-measure spaces. Then in Chapter 29 I shall present a synthetic definition of the curvature-dimension condition $\mathrm{CD}(K,N)$ in a nonsmooth context, and prove that it too is stable. Here is a geometric consequence of these results that can be stated without any reference to optimal transport: If a Riemannian manifold is the limit of a sequence of $\mathrm{CD}(K,N)$ Riemannian manifolds, then it, too, satisfies $\mathrm{CD}(K,N)$.

The last chapter will present the state of the art concerning the qualitative geometric and analytic properties enjoyed by metric-measure spaces satisfying curvature-dimension conditions, with complete proofs.

The issues discussed in this part are concisely reviewed in my seminar proceedings [821] (in French), or the survey paper by Lott [573], whose presentation is probably more geometer-friendly.

Convention: Throughout Part III, geodesics are *constant-speed, minimizing geodesics.*

26

Analytic and synthetic points of view

The present chapter is devoted to a simple pedagogical illustration of the opposition between the "analytic" and "synthetic" points of view. Consider the following two definitions for convexity on $\mathbb{R}^n$:

(i) A convex function is a function φ which is twice continuously differentiable, and whose Hessian $\nabla_x^2 \varphi$ is nonnegative at each $x \in \mathbb{R}^n$;

(ii) A convex function is a function φ such that for all $x, y \in \mathbb{R}^n$, and $\lambda \in [0,1]$,

$$\varphi\big((1-\lambda)x + \lambda y\big) \leq (1-\lambda)\,\varphi(x) + \lambda\,\varphi(y).$$

How can we compare these two definitions?

1) When applied to C^2 functions, both definitions coincide, but the second one is obviously *more general.* Not only is it expressed without any reference to second differentiability, but there are examples, such as $\varphi(x) = |x|$, which satisfy (ii) but not (i). So Definition (ii) really is an extension of Definition (i).

2) Definition (ii) is *more stable* than Definition (i). Here is what I mean by that: Take a sequence $(\varphi_k)_{k\in\mathbb{N}}$ of convex functions, converging to some other function φ; how do I know that φ is convex? To pass to the limit in Definition (i), I would need the convergence to be very strong, say in $C^2(\mathbb{R}^n)$. (Let's forget here about the notion of distributional convergence, which would solve the problem but is much less elementary.) On the other hand, I can pass to the limit in Definition (ii) assuming only, say, pointwise convergence. So Definition (ii) is much easier to "pass to the limit in" — even if the limit is known to be smooth.

C. Villani, *Optimal Transport*, Grundlehren der mathematischen Wissenschaften 338,

3) Definition (ii) is also a better *starting point* for studying properties of convex functions. In this set of notes, most of the time, when I used some convexity, it was via (ii), not (i).

4) On the other hand, if I give you a particular function (by its explicit analytic expression, say), and ask you whether it is convex, it will probably be a nightmare to check Definition (ii) directly, while Definition (i) might be workable: You just need to *compute* the second derivative and check its sign. Probably this is the method that will work most easily for the huge majority of candidate convex functions that you will meet in your life, if you don't have any extra information on them (like they are the limit of some family of functions...).

5) Definition (i) is naturally *local*, while Definition (ii) is global (and probably this is related to the fact that it is so difficult to check). In particular, Definition (i) involves an object (the second derivative) which can be used to quantify the "strength of convexity" *at each point.* Of course, one may define a convex function as a function satisfying (ii) *locally*, i.e. when x and y stay in the neighborhood of any given point; but then locality does not enter in such a simple way as in (i), and the issue immediately arises whether a function which satisfies (ii) locally, also satisfies (ii) globally.

In the above discussion, Definition (i) can be thought of as **analytic** (it is based on the computation of certain objects), while Definition (ii) is **synthetic** (it is based on certain qualitative properties which are the basis for proofs). Observations 1–5 above are in some sense typical: synthetic definitions ought to be more general and more stable, and they should be usable directly to prove interesting results; on the other hand, they may be difficult to check in practice, and they are usually less precise (and less "local") than analytic definitions.

In classical Euclidean geometry, the analytic approach consists in introducing Cartesian coordinates and making computations with equations of lines and circles, sines and cosines, etc. The synthetic approach, on the other hand, is more or less the one that was used already by ancient Greeks (and which is still taught, or at least should be taught, to our kids, for developing the skill of proof-making): It is not based on computations, but on axioms *à la* Euclid, qualitative properties of lines, angles, circles and triangles, construction of auxiliary points, etc. The analytic approach is conceptually simple, but sometimes leads to very horrible computations; the synthetic approach is often lighter, but requires better intuition and clever elementary arguments. It is also usually (but this is of course a matter of taste) more elegant.

In "Riemannian" geometry, curvature is traditionally defined by a purely analytic approach: From the Riemannian scalar product one can compute several functions which are called sectional curvature, Ricci curvature, scalar curvature, etc. For instance, for any $x \in M$, the sectional curvature at point x is a *function* which associates to each 2-dimensional plane $P \subset T_x M$ a number $\sigma_x(P)$, for which there is an explicit expression in terms of a basis of P, and a certain combination of derivatives of the metric at x. Intuitively, $\sigma_x(P)$ measures the speed of convergence of geodesics that start at x, with velocities spanning the plane P. A lot of geometric information can be retrieved from the bounds on the sectional curvature. Then a space is said to have nonnegative sectional curvature if $\sigma_x(P)$ is nonnegative, for all P and for all x.

However, there is also a famous synthetic point of view on sectional curvature, due to Alexandrov. In Alexandrov's approach one does not try to define what the curvature is, but what it means to have nonnegative curvature: By definition, a geodesic space $(\mathcal{X}, d)$ is said to have Alexandrov curvature bounded below by K, or to be an **Alexandrov space** with curvature bounded below by K, if triangles in $\mathcal{X}$ are no more "skinny" than reference triangles drawn on the model space $\mathbb{R}^2$. More precisely: If xyz is a triangle in $\mathcal{X}$, $x_0 y_0 z_0$ is a triangle drawn on $\mathbb{R}^2$ with $d(x_0, y_0) = d(x, y)$, $d(y_0, z_0) = d(y, z)$, $d(z_0, x_0) = d(z, x)$, x' is a midpoint between y and z, and x'_0 a midpoint between y_0 and z_0, then one should have $d(x_0, x'_0) \leq d(x, x')$. (See Figure 26.1.)

There is an excellent analogy with the previous discussion for convex functions. The Alexandrov definition is equivalent to the analytic one in case it is applied to a smooth Riemannian manifold; but it is more general, since it applies for instance to a cone (say, the two-dimensional cone embedded in $\mathbb{R}^3$, constructed over a circular basis). It is also more stable; in particular, it passes to the limit under **Gromov–Hausdorff convergence**, a notion that will be described in the sequel. It can still be used as the starting point for many properties involving sectional curvature. On the other hand, it is in general difficult to check directly, and there is no associated notion of *curvature* (when one says "Alexandrov space of nonnegative curvature", the words "nonnegative" and "curvature" do not make sense independently of each other).

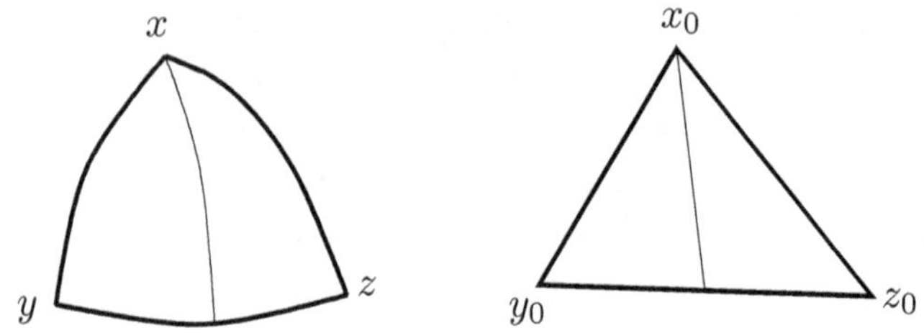

Fig. 26.1. The triangle on the left is drawn in $\mathcal{X}$, the triangle on the right is drawn on the model space $\mathbb{R}^2$; the lengths of their edges are the same. The thin geodesic lines go through the apex to the middle of the basis; the one on the left is longer than the one on the right. In that sense the triangle on the left is **fatter** than the triangle on the right. If all triangles in $\mathcal{X}$ look like this, then $\mathcal{X}$ has nonnegative curvature. (Think of a triangle as the belly of some individual, the belt being the basis, and the neck being the apex; of course the line going from the apex to the middle of the basis is the tie. The fatter the individual, the longer his tie should be.)

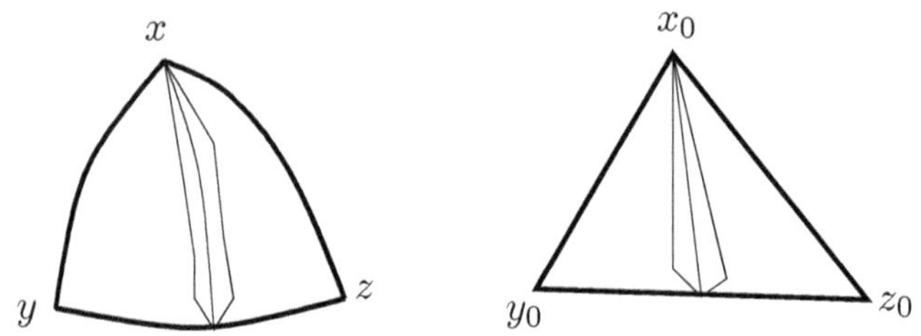

Still there is a generalization of what it means to have curvature bounded below by $K \in \mathbb{R}$, where K is an arbitrary real number, not necessarily 0. It is obtained by replacing the model space $\mathbb{R}^2$ by the model space with constant curvature K, that is:

- the sphere $S^2(1/\sqrt{K})$ with radius $R = 1/\sqrt{K}$, if $K > 0$;
- the plane $\mathbb{R}^2$, if $K = 0$;
- the hyperbolic space $\mathbb{H}(1/\sqrt{|K|})$ with "hyperbolic radius" $R = 1/\sqrt{|K|}$, if $K < 0$; this can be realized as the half-plane $\mathbb{R}\times(0,+\infty)$, equipped with the metric $g_{(x,y)}(dx\,dy) = (dx^2 + dy^2)/(|K|y^2)$.

Geodesic spaces satisfying these inequalities are called Alexandrov spaces with curvature bounded below by K; all the remarks which were made above in the case $K = 0$ apply in this more general case. There is also a symmetric notion of Alexandrov spaces with curvature bounded above, obtained by just reversing the inequalities.

This generalized notion of *sectional* curvature bounds has been explored by many authors, and quite strong results have been obtained concerning the geometric and analytic implications of such bounds. But until recently the synthetic treatment of lower *Ricci* curvature bounds stood as an open problem. The thesis developed in the rest of these notes is that optimal transport provides a solution to this problem (maybe not the only one, but so far the only one which seems to be acceptable).

Bibliographical notes

In close relation to the topics discussed in this chapter, there is an illuminating course by Gromov [437], which I strongly recommend to the reader who wants to learn about the meaning of curvature.

Alexandrov spaces are also called **CAT** spaces, in honor of Cartan, Alexandrov and Toponogov. But the terminology of CAT space is often restricted to Alexandrov spaces with *upper* sectional bounds. So a CAT(K) space typically means an Alexandrov space with "sectional curvature bounded above by K". In the sequel, I shall only consider lower curvature bounds.

There are several good sources for Alexandrov spaces, in particular the book by Burago, Burago and Ivanov [174] and the synthesis paper by Burago, Gromov and Perelman [175]. Analysis on Alexandrov spaces has been an active research topic in the past decade [536, 537, 539, 583, 584, 665, 676, 677, 678, 681, 751].

There is also a notion of "approximate" Alexandrov spaces, called CAT$_\delta$(K) spaces, in which a fixed "resolution error" δ is allowed in the defining inequalities [439]. (In the case of upper curvature bounds, this notion has applications to the theory of hyperbolic groups.) Such spaces are not necessarily geodesic spaces, not even length spaces, they can be discrete; a pair of points (x_0, x_1) will not necessarily admit a midpoint, but there will be a δ-approximate midpoint (that is, m such that $|d(x_0, m) - d(x_0, x_1)/2| \leq \delta$, $|d(x_1, m) - d(x_0, x_1)/2| \leq \delta$).

The open problem of developing a satisfactory synthetic treatment of Ricci curvature bounds was discussed in the above-mentioned book by Gromov [437, pp. 84–85], and more recently in a research paper by Cheeger and Colding [228, Appendix 2]. References about recent developments related to optimal transport will be given in the sequel.

Although this is not really the point of this chapter, I shall take this opportunity to briefly discuss the structure of the Wasserstein space over Alexandrov spaces. In Chapter 8 I already mentioned that Alexandrov spaces with lower curvature bounds might be the natural setting for certain regularity issues associated with optimal transport; recall indeed Open Problem 8.21 and the discussion before it. Another issue is how sectional (or Alexandrov) curvature bounds influence the geometry of the Wasserstein space.

It was shown by Lott and myself [577, Appendix A] that if M is a compact Riemannian manifold then M has nonnegative sectional curvature if and only if $P_2(M)$ is an Alexandrov space with nonnegative curvature. Independently, Sturm [762, Proposition 2.10(iv)] proved the more general result according to which $\mathcal{X}$ is an Alexandrov space with nonnegative curvature if and only if $P_2(\mathcal{X})$ is an Alexandrov space with nonnegative curvature. A study of tangent cones was performed in [577, Appendix A]; it is shown in particular that tangent cones at absolutely continuous measures are Hilbert spaces.

All this suggested that the notion of Alexandrov curvature matched well with optimal transport. However, at the same time, Sturm showed that if $\mathcal{X}$ is not nonnegatively curved, then $P_2(\mathcal{X})$ cannot be an Alexandrov space (morally, the curvature takes all values in $(-\infty, +\infty)$ at Dirac masses); this negative result was recently developed in [575]. To circumvent this obstacle, Ohta [655, Section 3] suggested replacing the Alexandrov property by a weaker condition known as 2-uniform convexity and used in Banach space theory (see e.g. [62]). Savaré [735, 736] came up independently with a similar idea. By definition, a geodesic space $(\mathcal{X}, d)$ is 2-uniform with a constant $S \geq 1$ if, given any three points $x, y, z \in \mathcal{X}$, and a minimizing geoesic γ joining y to z, one has

$$\forall t \in [0,1], \qquad d(x,\gamma_t)^2 \geq (1-t)\,d(x,y)^2 + t\,d(x,z)^2 - S^2\,t(1-t)\,d(y,z)^2.$$

(When $S = 1$ this is exactly the inequality defining nonnegative Alexandrov curvature.) Ohta shows that (a) any Alexandrov space with curvature bounded below is locally 2-uniform; (b) $\mathcal{X}$ is 2-uniformly smooth with constant S if and only if $P_2(\mathcal{X})$ is 2-uniformly smooth with the same constant S. He further uses the 2-uniform smoothness to study the structure of tangent cones in $P_2(\mathcal{X})$. Both Ohta and Savaré use these inequalities to construct gradient flows in the Wasserstein space over Alexandrov spaces.

Even when $\mathcal{X}$ is a smooth manifold with nonnegative curvature, many technical issues remain open about the structure of $P_2(M)$ as an Alexandrov space. For instance, the notion of the tangent cone used in the above-mentioned works is the one involving the space of directions, but does it coincide with the notion derived from rescaling and Gromov–Hausdorff convergence? (See Example 27.16. In finite dimension there are theorems of equivalence of the two definitions, but now we are in a genuinely infinite-dimensional setting.) How should one define a regular point of $P_2(M)$: as an absolutely continuous measure, or an absolutely continuous measure with positive density? And in the latter case, do these measures form a totally convex set? Do singular measures form a very small set in some sense? Can one define and use quasi-geodesics in $P_2(M)$? And so on.

27

Convergence of metric-measure spaces

The central question in this chapter is the following: What does it mean to say that a metric-measure space $(\mathcal{X}, d_{\mathcal{X}}, \nu_{\mathcal{X}})$ is "close" to another metric-measure space $(\mathcal{Y}, d_{\mathcal{Y}}, \nu_{\mathcal{Y}})$? We would like to have an answer that is as "intrinsic" as possible, in the sense that it should depend only on the metric-measure properties of $\mathcal{X}$ and $\mathcal{Y}$.

So as not to inflate this chapter too much, I shall omit many proofs when they can be found in accessible references, and prefer to insist on the main stream of ideas.

Hausdorff topology

There is a well-established notion of distance between compact sets in a given metric space, namely the **Hausdorff distance**. If $\mathcal{X}$ and $\mathcal{Y}$ are two compact subsets of a metric space $(\mathcal{Z}, d)$, their Hausdorff distance is

$$d_H(\mathcal{X}, \mathcal{Y}) = \max\left(\sup_{x \in \mathcal{X}} d(x, \mathcal{Y}),\ \sup_{y \in \mathcal{Y}} d(y, \mathcal{X})\right),$$

where as usual $d(a, B) = \inf\{d(a, b);\ b \in B\}$ is the distance from the point a to the set B. The choice of notation is significant: Think of $\mathcal{X}$ and $\mathcal{Y}$ not just as subsets, but rather as metric subspaces of $\mathcal{Z}$.

The statement "$d_H(\mathcal{X}, \mathcal{Y}) \leq r$" can be recast informally as follows: "If we *inflate* (enlarge) $\mathcal{Y}$ by a distance r, then the resulting set covers $\mathcal{X}$; and conversely if we inflate $\mathcal{X}$ by a distance r, the resulting set covers $\mathcal{Y}$." (See Figure 27.1.)

C. Villani, *Optimal Transport*, Grundlehren der mathematischen Wissenschaften 338,

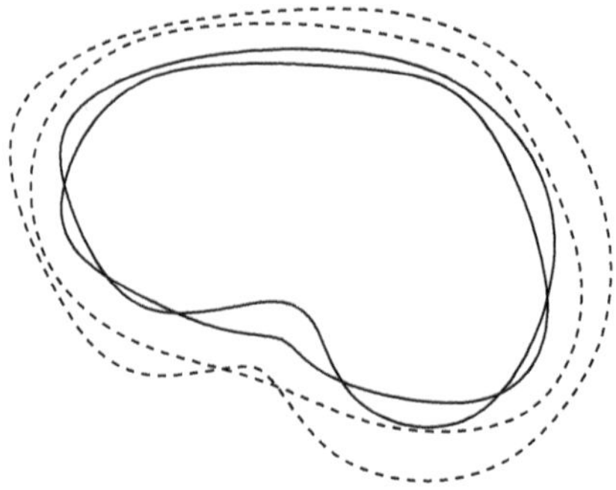

Fig. 27.1. In solid lines, the borders of the two sets $\mathcal{X}$ and $\mathcal{Y}$; in dashed lines, the borders of their enlargements. The width of enlargement is just sufficient that any of the enlarged sets covers both $\mathcal{X}$ and $\mathcal{Y}$.

The Hausdorff distance can be thought of as a set-theoretical analog of the Prokhorov distance between probability measures (of course, historically the former came first). This will become more apparent if I rewrite the Hausdorff distance as

$$d_H(A,B) = \inf\Big\{r > 0; \quad A \subset B^{r]} \ \text{ and } \ B \subset A^{r]}\Big\},$$

and the Prokhorov distance as

$$d_P(\mu,\nu) = \inf\Big\{r > 0; \quad \forall C, \quad \mu[C] \le \nu[C^{r]}]+r \ \text{ and } \nu[C] \le \mu[C^{r]}]+r\Big\},$$

where C stands for an arbitrary closed set, $C^{r]}$ is the set of all points whose distance to C is no more than r, i.e. the union of all *closed* balls $B[x,r]$, $x \in C$.

The analogy between the two notions goes further: While the Prokhorov distance can be defined in terms of couplings, the Hausdorff distance can be defined in terms of **correspondences**. By definition, a correspondence (or relation) between two sets $\mathcal{X}$ and $\mathcal{Y}$ is a subset $\mathcal{R}$ of $\mathcal{X} \times \mathcal{Y}$: if $(x,y) \in \mathcal{R}$, then x and y are said to be in correspondence; it is required that each $x \in \mathcal{X}$ should be in correspondence with at least one y, and each $y \in \mathcal{Y}$ should be in correspondence with at least one x. Then we have the two very similar formulas:

$$\begin{cases} d_P(\mu,\nu) = \inf\Big\{r > 0; \quad \exists \text{ coupling } (X,Y) \text{ of } (\mu,\nu); \\ \qquad\qquad\qquad\qquad\qquad\qquad \mathbb{P}\left[d(X,Y) > r\right] \le r\Big\}; \\ d_H(\mu,\nu) = \inf\Big\{r > 0; \quad \exists \text{ correspondence } \mathcal{R} \text{ in } \mathcal{X} \times \mathcal{Y}; \\ \qquad\qquad\qquad\qquad\qquad \forall (x,y) \in \mathcal{R}, \quad d(x,y) \le r\Big\}. \end{cases}$$

Moreover, it is easy to guess an "optimal" correspondence: Just define

$$(x,y) \in \mathcal{R} \iff \big[d(x,y) = d(x,\mathcal{Y}) \text{ or } d(y,x) = d(y,\mathcal{X}) \big].$$

So each $(x,y) \in \mathcal{R}$ satisfies $d(x,y) \leq d_H(\mathcal{X},\mathcal{Y})$, with equality for at least one pair. (Indeed, the maximum in the definition of the Hausdorff distance is obviously achieved.)

Like their probabilistic counterparts, correspondences can be **glued** together: if $\mathcal{R}_{12}$ is a correspondence between $\mathcal{X}_1$ and $\mathcal{X}_2$, and $\mathcal{R}_{23}$ is a correspondence between $\mathcal{X}_2$ and $\mathcal{X}_3$, one may define a correspondence $\mathcal{R}_{13} = \mathcal{R}_{23} \circ \mathcal{R}_{12}$ between $\mathcal{X}_1$ and $\mathcal{X}_3$ by

$$(x_1,x_3) \in \mathcal{R}_{13} \iff \big[\exists x_2 \in \mathcal{X}_2;\ (x_1,x_2) \in \mathcal{R}_{12} \text{ and } (x_2,x_3) \in \mathcal{R}_{23}\big].$$

The next observation is that the Hausdorff distance really is a distance! Indeed:

(i) It is obviously symmetric ($d_H(\mathcal{X},\mathcal{Y}) = d_H(\mathcal{Y},\mathcal{X})$);

(ii) Because it is defined on compact (hence bounded) sets, the infimum in the definition is a nonnegative *finite* number;

(iii) If $d_H(\mathcal{X},\mathcal{Y}) = 0$, then any $x \in \mathcal{X}$ satisfies $d(x,\mathcal{Y}) \leq \varepsilon$, for any $\varepsilon > 0$; so $d(x,\mathcal{Y}) = 0$, and since $\mathcal{Y}$ is assumed to be compact (hence closed), this implies $\mathcal{X} = \mathcal{Y}$;

(iv) if $\mathcal{X}_1$, $\mathcal{X}_2$ and $\mathcal{X}_3$ are given, introduce optimal correspondences $\mathcal{R}_{12}$ and $\mathcal{R}_{23}$ in the correspondence representation of the Hausdorff measure; then the composed representation $\mathcal{R}_{13} = \mathcal{R}_{23} \circ \mathcal{R}_{12}$ is such that any $(x_1,x_3) \in \mathcal{R}_{13}$ satisfies $d(x_1,x_3) \leq d(x_1,x_2) + d(x_2,x_3) \leq d_H(\mathcal{X}_1,\mathcal{X}_2) + d_H(\mathcal{X}_2,\mathcal{X}_3)$ for some x_2. So d_H satisfies the triangle inequality.

Then one may define the metric space $\mathcal{H}(\mathcal{Z})$ as the space of all compact subsets of $\mathcal{Z}$, equipped with the Hausdorff distance. There is a nice statement that if $\mathcal{Z}$ is compact then $\mathcal{H}(\mathcal{Z})$ is also a compact metric space.

So far everything is quite simple, but soon it will become a bit more complicated, which is a good reason to go slowly.

The Gromov–Hausdorff distance

The Hausdorff distance only compares subsets of a given underlying space. But how can we compare different metric spaces with possibly nothing in common? First one would like to say that two spaces which are *isometric* really are the same. Recall the definition of an isometry: If $(\mathcal{X}, d)$ and $(\mathcal{X}', d')$ are two metric spaces, a map $f : \mathcal{X} \to \mathcal{X}'$ is called an isometry if:

(a) it preserves distances: for all $x, y \in \mathcal{X}$, $d'(f(x), f(y)) = d(x, y)$;

(b) it is surjective: for any $x' \in \mathcal{X}'$ there is $x \in \mathcal{X}$ with $f(x) = x'$.

An isometry is automatically injective, so it has to be a bijection, and its inverse f^{-1} is also an isometry. Two metric spaces are said to be **isometric** if there exists an isometry between them. If two spaces $\mathcal{X}$ and $\mathcal{X}'$ are isometric, then any statement about $\mathcal{X}$ which can be expressed in terms of just the distance, is automatically "transported" to $\mathcal{X}'$ by the isometry.

This motivates the desire to work with *isometry classes*, rather than metric spaces. By definition, an isometry class $\overline{\mathcal{X}}$ is the set of all metric spaces which are isometric to some given space $\mathcal{X}$. Instead of "isometry class", I shall often write "abstract metric space". All the spaces in a given isometry class have the same topological properties, so it makes sense to say of an abstract metric space that it is compact, or complete, etc.

This looks good, but a bit frightening: There are so many metric spaces around that the concept of abstract metric space seems to be ill-posed from the set-theoretical point of view (just like there is no "set of all sets"). However, things becomes much more friendly when one realizes that any compact metric space, being *separable*, is isometric to the completion of $\mathbb{N}$ for a suitable metric. (To see this, introduce a dense sequence (x_k) in your favorite space $\mathcal{X}$, and define $d(k, \ell) = d_{\mathcal{X}}(x_k, x_\ell)$.) Then we might think of an isometry class as a subset of the set of all distances on $\mathbb{N}$; this is still huge, but at least it makes sense from a set-theoretical point of view.

Now the problem is to find a good distance on the set of abstract compact metric spaces. The natural concept here is the **Gromov–Hausdorff distance**, which is obtained by formally *taking the quotient of the Hausdorff distance by isometries*: If $(\mathcal{X}, d_{\mathcal{X}})$ and $(\mathcal{Y}, d_{\mathcal{Y}})$ are two compact metric spaces, define

$$d_{GH}(\mathcal{X}, \mathcal{Y}) = \inf\, d_H(\mathcal{X}', \mathcal{Y}'), \tag{27.1}$$

where the infimum is taken over all *isometric embeddings* $\mathcal{X}'$, $\mathcal{Y}'$ of $\mathcal{X}$ and $\mathcal{Y}$ into a common metric space $\mathcal{Z}$; this means that $\mathcal{X}'$ is isometric to $\mathcal{X}$, $\mathcal{Y}'$ is isometric to $\mathcal{Y}$, and both $\mathcal{X}'$ and $\mathcal{Y}'$ are subspaces of $\mathcal{Z}$.

Of course, there is no loss of generality in choosing $\mathcal{Z} = \mathcal{X}' \cup \mathcal{Y}'$, but let me insist: the metric on $\mathcal{X}'$, $\mathcal{Y}'$ has to be the metric induced by $\mathcal{Z}$! In that situation I shall say that $(\mathcal{X}', \mathcal{Y}')$ constitute a **metric coupling** of the abstract spaces $(\mathcal{X}, \mathcal{Y})$. Two metric couplings $(\mathcal{X}', \mathcal{Y}')$ and $(\mathcal{X}'', \mathcal{Y}'')$ will be said to be isometric if there is an isometry $F : (\mathcal{X}' \cup \mathcal{Y}') \to (\mathcal{X}'' \cup \mathcal{Y}'')$ which restricts to isometries $\mathcal{X}' \to \mathcal{X}''$ and $\mathcal{Y}' \to \mathcal{Y}''$.

Representation by semi-distances

As we know, all the probabilistic information contained in a coupling (X, Y) of two probability spaces $(\mathcal{X}, \nu_{\mathcal{X}})$ and $(\mathcal{Y}, \nu_{\mathcal{Y}})$ is summarized by a *joint probability measure on the product space* $\mathcal{X} \times \mathcal{Y}$. There is an analogous statement for metric couplings: All the geometric information contained in a metric coupling $(\mathcal{X}', \mathcal{Y}')$ of two abstract metric spaces $\mathcal{X}$ and $\mathcal{Y}$ is summarized by a *semi-distance on the disjoint union* $\mathcal{X} \sqcup \mathcal{Y}$. Here are the definitions:

- a semi-distance on a set $\mathcal{Z}$ is a map $d : \mathcal{Z} \times \mathcal{Z} \to [0, +\infty)$ satisfying $d(x, x) = 0$, $d(x, y) = d(y, x)$, $d(x, z) \le d(x, y) + d(y, z)$, but *not necessarily* $[d(x, y) = 0 \Longrightarrow x = y]$;
- the disjoint union $\mathcal{X} \sqcup \mathcal{Y}$ is the union of two disjoint isometric copies of $\mathcal{X}$ and $\mathcal{Y}$. The particular way in which this disjoint union is constructed does not matter; for instance, take any representative of $\mathcal{X}$ (still denoted $\mathcal{X}$ for simplicity), any representative of $\mathcal{Y}$, and set $\mathcal{X} \sqcup \mathcal{Y} = (\{0\} \times \mathcal{X}) \cup (\{1\} \times \mathcal{Y})$. Then $\{0\} \times \mathcal{X}$ is isometric to $\mathcal{X}$ via the map $(0, x) \to x$, etc.

However, not all semi-distances are allowed. In a probabilistic context, the only admissible couplings of two measures $\nu_{\mathcal{X}}$ and $\nu_{\mathcal{Y}}$ are those whose joint law π has marginals $\nu_{\mathcal{X}}$ and $\nu_{\mathcal{Y}}$. There is a similar principle for metric couplings: If $(\mathcal{X}, d_{\mathcal{X}})$ and $(\mathcal{Y}, d_{\mathcal{Y}})$ are two given abstract metric spaces, the only admissible semi-distances on $\mathcal{X} \sqcup \mathcal{Y}$ are those whose restriction to $\mathcal{X} \times \mathcal{X}$ (resp. $\mathcal{Y} \times \mathcal{Y}$) coincides with $d_{\mathcal{X}}$ (resp. $d_{\mathcal{Y}}$). When that condition is satisfied, it will be possible to reconstruct a metric coupling from the semi-distance, by just "taking the quotient" of $\mathcal{X} \sqcup \mathcal{Y}$ by the semi-distance d, in other words deciding that two points a and b with $d(a, b) = 0$ really are the same.

All this is made precise by the following statement:

Proposition 27.1 (Metric couplings as semi-distances). *Let* $(\mathcal{X}, d_{\mathcal{X}})$ *and* $(\mathcal{Y}, d_{\mathcal{Y}})$ *be two disjoint metric spaces, and let* $\mathcal{X} \sqcup \mathcal{Y}$ *be their union. Then:*

(i) Let $(\mathcal{X}', \mathcal{Y}')$ *be a metric coupling of* $\mathcal{X}$ *and* $\mathcal{Y}$*; let* $f : \mathcal{X} \to \mathcal{X}'$ *and* $g : \mathcal{Y} \to \mathcal{Y}'$ *be isometries, and let* $(\mathcal{Z}, d_{\mathcal{Z}})$ *be the ambient metric space containing* $\mathcal{X}'$ *and* $\mathcal{Y}'$*. Whenever* a, b *belong to* $\mathcal{X} \sqcup \mathcal{Y}$*, define*

$$d(a,b) = \begin{cases} d_{\mathcal{X}}(a,b) & \text{if } a,b \in \mathcal{X} \\ d_{\mathcal{Y}}(a,b) & \text{if } a,b \in \mathcal{Y} \\ d_{\mathcal{Z}}(f(a), g(b)) & \text{if } a \in \mathcal{X},\ b \in \mathcal{Y} \\ d_{\mathcal{Z}}(g(a), f(b)) & \text{if } a \in \mathcal{Y},\ b \in \mathcal{X}. \end{cases}$$

Then d *is a semi-distance on* $\mathcal{X} \sqcup \mathcal{Y}$*, whose restriction to* $\mathcal{X} \times \mathcal{X}$ *(resp.* $\mathcal{Y} \times \mathcal{Y}$*) coincides with* $d_{\mathcal{X}}$ *(resp.* $d_{\mathcal{Y}}$*).*

(ii) Conversely, let d *be a semi-distance on* $\mathcal{X} \sqcup \mathcal{Y}$*, whose restriction to* $\mathcal{X} \times \mathcal{X}$ *(resp.* $\mathcal{Y} \times \mathcal{Y}$*) coincides with* $d_{\mathcal{X}}$ *(resp.* $d_{\mathcal{Y}}$*). On* $\mathcal{X} \sqcup \mathcal{Y}$*, define the relation* $\mathcal{R}$ *by the property*

$$x\, \mathcal{R}\, x' \iff d(x, x') = 0.$$

This is an equivalence relation, so one may define

$$\mathcal{Z} = (\mathcal{X} \sqcup \mathcal{Y})/d := (\mathcal{X} \sqcup \mathcal{Y})/\mathcal{R}$$

as the set of classes of equivalence in $\mathcal{X} \sqcup \mathcal{Y}$*. Write* $\overline{x}$ *for the equivalence class of* x*, and define*

$$d_{\mathcal{Z}}(\overline{a}, \overline{b}) = d(a, b)$$

(this does not depend on the choice of representatives a, b*, but only on the equivalence classes* $\overline{a}, \overline{b}$*). Then* $x \to \overline{x}$ *is an isometric embedding of* $\mathcal{X}$ *into* $(\mathcal{Z}, d_{\mathcal{Z}})$*, and similarly* $y \to \overline{y}$ *is an isometric embedding of* $\mathcal{Y}$ *into* $(\mathcal{Z}, d_{\mathcal{Z}})$*.*

The reader should have no difficulty in writing down the proof of Proposition 27.1; just be patient enough and make sure that you consider all cases!

Now the following property should not come as a surprise:

Theorem 27.2 (Metric gluing lemma). *Let* $(\mathcal{X}_1, d_1)$*,* $(\mathcal{X}_2, d_2)$*,* $(\mathcal{X}_3, d_3)$ *be three abstract compact metric spaces. If* $(\mathcal{X}_1', \mathcal{X}_2')$ *is a metric coupling of* $(\mathcal{X}_1, \mathcal{X}_2)$ *and* $(\mathcal{X}_2'', \mathcal{X}_3'')$ *is a metric coupling of* $(\mathcal{X}_2, \mathcal{X}_3)$*, then* *there is a triple of metric spaces* $(\widetilde{\mathcal{X}}_1, \widetilde{\mathcal{X}}_2, \widetilde{\mathcal{X}}_3)$, *all subspaces of a common metric space* $(\mathcal{Z}, d_{\mathcal{Z}})$, *such that* $(\widetilde{\mathcal{X}}_1, \widetilde{\mathcal{X}}_2)$ *is isometric (as a coupling) to* $(\mathcal{X}_1', \mathcal{X}_2')$, *and* $(\widetilde{\mathcal{X}}_2, \widetilde{\mathcal{X}}_3)$ *is isometric to* $(\mathcal{X}_2'', \mathcal{X}_3'')$.

Sketch of proof of Theorem 27.2. By means of Proposition 27.1, the metric coupling $(\mathcal{X}_1', \mathcal{X}_2')$ may be thought of as a semi-distance d_{12} on $\mathcal{X}_1 \sqcup \mathcal{X}_2$; and similarly, $(\mathcal{X}_2'', \mathcal{X}_3'')$ may be thought of as a semi-distance d_{23} on $\mathcal{X}_2 \sqcup \mathcal{X}_3$. Then, for $x_1 \in \mathcal{X}_1$ and $x_3 \in \mathcal{X}_3$, define

$$d_{13}(x_1, x_3) = \inf_{x_2 \in \mathcal{X}_2} \big[d_{12}(x_1, x_2) + d_{23}(x_2, x_3)\big].$$

Next, on $\mathcal{X}_1 \sqcup \mathcal{X}_2 \sqcup \mathcal{X}_3$ introduce the semi-distance

$$d(a,b) = \begin{cases} d_{12}(a,b) & \text{if } a, b \in \mathcal{X}_1 \sqcup \mathcal{X}_2 \\ d_{23}(a,b) & \text{if } a, b \in \mathcal{X}_2 \sqcup \mathcal{X}_3 \\ d_{13}(a,b) & \text{if } a \in \mathcal{X}_1 \text{ and } b \in \mathcal{X}_3 \\ d_{13}(b,a) & \text{if } a \in \mathcal{X}_3 \text{ and } b \in \mathcal{X}_1. \end{cases}$$

This is a semi-distance; so one can define

$$\mathcal{Z} = (\mathcal{X}_1 \sqcup \mathcal{X}_2 \sqcup \mathcal{X}_3)/d,$$

and repeat the same reasoning as in Proposition 27.1. □

Representation by approximate isometries

If a correspondence $\mathcal{R} \subset \mathcal{X} \times \mathcal{Y}$ preserves distances, in the sense that $d(x, x') = d(y, y')$ for all (x, y), (x', y') in $\mathcal{R}$, then it is almost obvious that $\mathcal{R}$ is the graph of an isometry between $\mathcal{X}$ and $\mathcal{Y}$. To measure how far a correspondence is from being an isometry, define its **distortion** by the formula

$$\mathrm{dis}\,(\mathcal{R}) = \sup_{(x,y),(x',y') \in \mathcal{R}} \Big| d_{\mathcal{Y}}(y, y') - d_{\mathcal{X}}(x, x') \Big|.$$

Then it can be shown that

$$d_{GH}(\mathcal{X}, \mathcal{Y}) = \frac{1}{2} \inf \mathrm{dis}\,(\mathcal{R}), \tag{27.2}$$

where the infimum is over all correspondences $\mathcal{R}$ between $\mathcal{X}$ and $\mathcal{Y}$.

There is an even more handy way to evaluate Gromov–Hausdorff distances, in terms of **approximate isometries**. By definition, an ε-isometry between $(\mathcal{X}, d_{\mathcal{X}})$ and $(\mathcal{Y}, d_{\mathcal{Y}})$ is a map $f : \mathcal{X} \to \mathcal{Y}$ that is "almost an isometry", which means:

(a') it almost preserves distances: for all x, x' in $\mathcal{X}$,

$$\Big|d(f(x), f(x')) - d(x, x')\Big| \le \varepsilon;$$

(b') it is almost surjective:

$$\forall y \in \mathcal{Y} \quad \exists x \in \mathcal{X}; \qquad d(f(x), y) \le \varepsilon.$$

In particular, $d_H(f(\mathcal{X}), \mathcal{Y}) \le \varepsilon$.

Remark 27.3. Heuristically, an ε-isometry is a map that you can't distinguish from an isometry if you are short-sighted, that is, if you measure all distances with a possible error of about ε.

It is not clear whether one can reformulate the Gromov–Hausdorff distance in terms of ε-isometries, but at least from the qualitative point of view this works fine: It can be shown that

$$\frac{2}{3}\, d_{GH}(\mathcal{X}, \mathcal{Y}) \;\le\; \inf\Big\{\varepsilon;\ \exists f\ \varepsilon\text{-isometry } \mathcal{X} \to \mathcal{Y}\Big\} \;\le\; 2\, d_{GH}(\mathcal{X}, \mathcal{Y}). \tag{27.3}$$

Indeed, if f is an ε-isometry, define a relation $\mathcal{R}$ by

$$(x, y) \in \mathcal{R} \iff \quad d(f(x), y) \le \varepsilon;$$

then $\operatorname{dis}(\mathcal{R}) \le 3\varepsilon$, and the left inequality in (27.3) follows by formula (27.2). Conversely, if $\mathcal{R}$ is a relation with distortion η, then for any $\varepsilon > \eta$ one can define an ε-isometry f whose graph is included in $\mathcal{R}$: The idea is to define $f(x) = y$, where y is such that $(x, y) \in \mathcal{R}$. (See the comments at the end of the bibliographical notes.)

The symmetry between $\mathcal{X}$ and $\mathcal{Y}$ seems to have been lost in (27.3), but this is not serious, because any approximate isometry admits an **approximate inverse**: If f is an ε-isometry $\mathcal{X} \to \mathcal{Y}$, then there is a (4ε)-isometry $f' : \mathcal{Y} \to \mathcal{X}$ such that for all $x \in \mathcal{X}$, $y \in \mathcal{Y}$,

$$d_{\mathcal{X}}\big(f' \circ f(x),\, x\big) \le 3\varepsilon, \qquad d_{\mathcal{Y}}\big(f \circ f'(y),\, y\big) \le \varepsilon. \tag{27.4}$$

Such a map f' will be called an ε-*inverse* of f.

To construct f', consider the relation $\mathcal{R}$ induced by f, whose distortion is no more than 3ε; then flip the components of $\mathcal{R}$ to get a relation $\mathcal{R}'$ from $\mathcal{Y}$ to $\mathcal{X}$, with (obviously) the same distortion as $\mathcal{R}$, and construct a (4ε)-isometry $f' : \mathcal{Y} \to \mathcal{X}$ whose graph is a subset of $\mathcal{R}$. Then $(f(x), x) \in \mathcal{R}'$ and $(f(x), f'(f(x))) \in \mathcal{R}'$, so $d(f'(f(x)), x) \leq \operatorname{dis}(\mathcal{R}') + d(f(x), f(x)) \leq 3\varepsilon$. Similarly, the identity $(f'(y), y) \in \mathcal{R}$ implies $d(f(f'(y)), y) \leq \varepsilon$.

If there exists an ε-isometry between $\mathcal{X}$ and $\mathcal{Y}$, then I shall say that $\mathcal{X}$ and $\mathcal{Y}$ are ε-isometric. This terminology has the drawback that the order of $\mathcal{X}$ and $\mathcal{Y}$ matters: if $\mathcal{X}$ and $\mathcal{Y}$ are ε-isometric, then $\mathcal{Y}$ and $\mathcal{X}$ are not necessarily ε-isometric; but at least they are (4ε)-isometric, so from the qualitative point of view this is not a problem.

Lemma 27.4 (Approximate isometries converge to isometries). *Let $\mathcal{X}$ and $\mathcal{Y}$ be two compact metric spaces, and for each $k \in \mathbb{N}$ let f_k be an ε_k-isometry, where $\varepsilon_k \to 0$. Then, up to extraction of a subsequence, f_k converges to an isometry.*

Sketch of proof of Lemma 27.4. Introduce a dense subset S of $\mathcal{X}$. For each $x \in \mathcal{X}$, the sequence $(f_k(x))$ is valued in the compact set $\mathcal{Y}$, and so, up to extraction of a subsequence, it converges to some $f(x) \in \mathcal{Y}$. By a diagonal extraction, we may assume that $f_k(x) \to f(x)$ for all $x \in \mathcal{X}$. By passing to the limit in the inequality satisfied by f_k, we see that f is distance-preserving. By uniform continuity, it may be extended into a distance-preserving map $\mathcal{X} \to \mathcal{Y}$.

Similarly, there is a distance-preserving map $\mathcal{Y} \to \mathcal{X}$, obtained as a limit of approximate inverses of f_k, and denoted by g. The composition $g \circ f$ is a distance-preserving map $\mathcal{X} \to \mathcal{X}$, and since $\mathcal{X}$ is compact it follows from a well-known theorem that $g \circ f$ is a bijection. As a consequence, both f and g are bijective, so they are isometries. □

Remark 27.5. The above proof establishes the pointwise convergence of (a subsequence of) f_k to f, but in fact one can impose the uniform convergence; see Theorem 27.10.

The Gromov–Hausdorff space

After all these preparations, we may at last understand why d_{GH} is a honest distance:

(i) It is obviously symmetric.

(ii) Let $\mathcal{X}$ and $\mathcal{Y}$ be two compact metric spaces; define $\mathcal{Z}$ to be the disjoint union $\mathcal{X} \sqcup \mathcal{Y}$, equip $\mathcal{X}$ and $\mathcal{Y}$ with their respective distances, and extend this into a distance d on $\mathcal{X} \sqcup \mathcal{Y}$ by letting $d(x,y) = D > 0$ as soon as $(x,y) \in \mathcal{X} \times \mathcal{Y}$. If D is chosen large enough, this is indeed a distance; so the injections $(x,y) \to x$ and $(x,y) \to y$ realize a metric coupling of $(\mathcal{X},\mathcal{Y})$. Thus the infimum in (27.1) is not $+\infty$.

(iii) Obviously, $d_{GH}(\mathcal{X},\mathcal{X}) = 0$. Conversely, if $\mathcal{X}$ and $\mathcal{Y}$ are two abstract compact metric spaces such that $d_{GH}(\mathcal{X},\mathcal{Y}) = 0$, introduce any two representatives of these isometry classes (still denoted by $\mathcal{X}$ and $\mathcal{Y}$ for simplicity), then for each $k \in \mathbb{N}$ there is an $(1/k)$-isometry $f_k : \mathcal{X} \to \mathcal{Y}$. Up to extraction of a subsequence, f_k converges to a true isometry by Lemma 27.4, so $\mathcal{X}$ and $\mathcal{Y}$ are isometric, and define the same isometry class.

(iv) Finally, the triangle inequality follows easily from the metric gluing lemma — just as the triangle inequality for the Wasserstein distance was a consequence of the probabilistic gluing lemma.

So the set $(\mathcal{GH}, d_{GH})$ of all classes of isometry of compact metric spaces, equipped with the Gromov–Hausdorff distance, is itself a complete separable metric space. An explicit countable dense subset of $\mathcal{G}H$ is provided by the family of all finite subsets of $\mathbb{N}$ with rational-valued distances. Convergence in the Gromov–Hausdorff distance is called **Gromov–Hausdorff convergence**.

It is equivalent to express the Gromov–Hausdorff convergence in terms of embeddings and Hausdorff distance, or in terms of distortions of correspondences, or in terms of approximate isometries. This leads to the following definition.

Definition 27.6 (Gromov–Hausdorff convergence). *Let $(\mathcal{X}_k)_{k\in\mathbb{N}}$ be a sequence of compact metric spaces, and let $\mathcal{X}$ be a compact metric space. Then it is said that $\mathcal{X}_k$ converges to $\mathcal{X}$ in the Gromov–Hausdorff topology if any one of the three equivalent statements is satisfied:*

(i) $d_{GH}(\mathcal{X}_k,\mathcal{X}) \longrightarrow 0$;

(ii) There exist correspondences $\mathcal{R}_k$ between $\mathcal{X}_k$ and $\mathcal{X}$ such that $\operatorname{dis}\mathcal{R}_k \longrightarrow 0$;

(iii) There exist ε_k-isometries $f_k : \mathcal{X}_k \to \mathcal{X}$, for some sequence $\varepsilon_k \to 0$.

This convergence will be denoted by $\mathcal{X}_k \xrightarrow{GH} \mathcal{X}$, or just $\mathcal{X}_k \longrightarrow \mathcal{X}$.

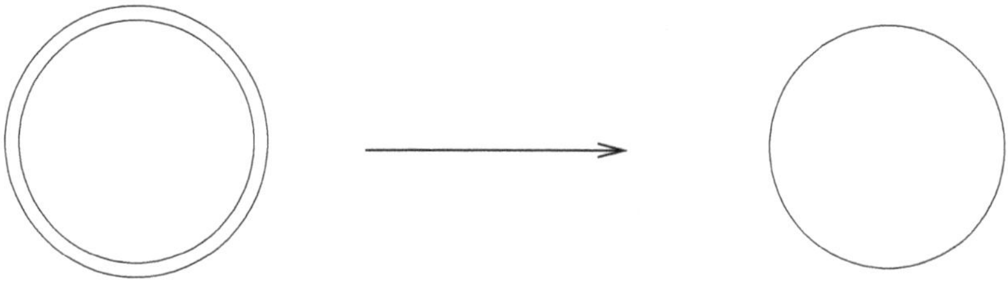

Fig. 27.2. A very thin tire (2-dimensional manifold) is very close, in Gromov–Hausdorff sense, to a circle (1-dimensional manifold)

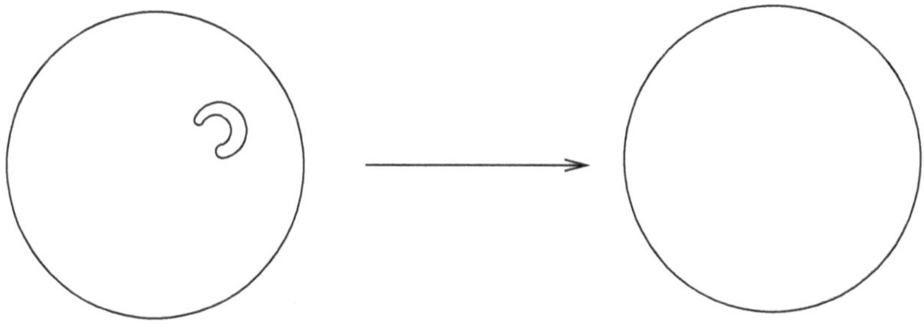

Fig. 27.3. A balloon with a very small handle (not simply connected) is very close to a balloon without handle (simply connected).

Remark 27.7. Keeping Remark 27.3 in mind, two spaces are close in Gromov–Hausdorff topology if they look the same to a short-sighted person (see Figure 27.2). I learnt from Lott the expression *Mr. Magoo topology* to convey this idea.

Remark 27.8. It is important to allow the approximate isometries to be discontinuous. Figure 27.3 below shows a simple example where two spaces $\mathcal{X}$ and $\mathcal{Y}$ are very close to each other in Gromov–Hausdorff topology, although there is no continuous map $\mathcal{X} \to \mathcal{Y}$. (Still there is a celebrated convergence theorem by Gromov showing that such behavior is ruled out by bounds on the curvature.)

The Gromov–Hausdorff topology enjoys the very nice property that *any geometric statement which can be expressed in terms of the distances between a finite number of points automatically passes to the limit.* For example, consider the statement "Any pair (x, y) of points in $\mathcal{X}$ admits a midpoint", which (under a completeness assumption) is characteristic of a geodesic space. This only involves the distance between configurations of three points (x, y and the candidate midpoint), so it passes to the limit. Then geodesics can be reconstructed from successive midpoints. In this way one can easily prove:

Theorem 27.9 (Convergence of geodesic spaces). *Let $(\mathcal{X}_k)_{k\in\mathbb{N}}$ be a sequence of compact geodesic spaces converging to $\mathcal{X}$ in Gromov–Hausdorff topology; then $\mathcal{X}$ is a geodesic space. Moreover, if f_k is an ε_k-isometry $\mathcal{X}_k \to \mathcal{X}$, and γ_k is a geodesic curve in $\mathcal{X}_k$ such that $f_k \circ \gamma_k$ converges to some curve γ in $\mathcal{X}$, then γ is a geodesic.*

Gromov–Hausdorff topology and nets

Given $\varepsilon > 0$, a set $\mathcal{N}$ in a metric space $(\mathcal{X}, d)$ is called an ε-net (in $\mathcal{X}$) if the enlargement $S^{\varepsilon]}$ covers $\mathcal{X}$; in other words, for any $x \in \mathcal{X}$ there is $y \in \mathcal{N}$ such that $d(x,y) \le \varepsilon$.

If $\mathcal{N}$ is an ε-net in $\mathcal{X}$, clearly the distance between $\mathcal{N}$ and $\mathcal{X}$ is at most ε. And if $\mathcal{X}$ is compact, then it admits finite ε-nets for all $\varepsilon > 0$, so it can be approximated in Gromov–Hausdorff topology by a sequence of finite sets.

In fact, it is another nice feature of the Gromov–Hausdorff topology that it ultimately *always* reduces to convergence of finite ε-nets. More precisely, $\mathcal{X}_n \longrightarrow \mathcal{X}$ in the Gromov–Hausdorff topology if and only if for any $\varepsilon > 0$ there exists a finite ε-net $\{x_1, \dots, x_k\}$ in $\mathcal{X}$, and for n large enough there is an ε-net $\{x_1^{(n)}, \dots, x_k^{(n)}\}$ in $\mathcal{X}_n$, and for all $j \le k$, $x_j^{(n)} \longrightarrow x_j$.

This leads to the main compactness criterion in Gromov–Hausdorff topology. Recall that a metric space $\mathcal{X}$ is said to be **totally bounded** if for any $\varepsilon > 0$ it can be covered by a finite number of balls of radius ε. If $N(\varepsilon)$ is the minimal number of such balls, then the function $\varepsilon \longmapsto N(\varepsilon)$ can be thought of as a "modulus of total boundedness". Then the following statement, due to Gromov, is vaguely reminiscent of Ascoli's theorem:

Theorem 27.10 (Compactness criterion in Gromov–Hausdorff topology). *A family $\mathcal{F}$ of compact metric spaces is precompact in the Gromov–Hausdorff topology if and only if it is uniformly totally bounded, in the sense that for any $\varepsilon > 0$ there is $N = N(\varepsilon)$ such that any $\mathcal{X} \in \mathcal{F}$ contains an ε-net of cardinality at most N.*

Noncompact spaces

There is no problem in extending the definition of the Gromov–Hausdorff distance to noncompact spaces, except of course that the resulting "distance" might be infinite. But even when it is finite, this notion is of limited use. A good analogy is the concept of uniform convergence of functions, which usually is too strong a notion for noncompact spaces, and should be replaced by *locally uniform* convergence, i.e. uniform convergence on any compact subset.

At first sight, it does not seem to make much sense to define a notion of local Gromov–Hausdorff convergence. Indeed, if a sequence $(\mathcal{X}_k)_{k\in\mathbb{N}}$ of metric spaces is given, there is a priori no canonical family of compact sets in $\mathcal{X}_k$ to compare to a family of compact sets in $\mathcal{X}$. So we had better impose the existence of these compact sets on each member of the sequence. The idea is to exhaust the space $\mathcal{X}$ by compact sets $K^{(\ell)}$ in such a way that each $K^{(\ell)}$ (equipped with the metric induced by $\mathcal{X}$) is a Gromov–Hausdorff limit of corresponding compact sets $K_k^{(\ell)} \subset \mathcal{X}_k$ (each of them with the induced metric), as $k \to \infty$.

The next definition makes this more precise. (Recall that a Polish space is a complete separable metric space.)

Definition 27.11 (Local Gromov–Hausdorff convergence). *Let $(\mathcal{X}_k)_{k\in\mathbb{N}}$ be a family of Polish spaces, and let $\mathcal{X}$ be another Polish space. It is said that $\mathcal{X}_k$ converges to $\mathcal{X}$ in the local Gromov–Hausdorff topology if there are nondecreasing sequences of compact sets $(K_k^{(\ell)})_{\ell\in\mathbb{N}}$ in each $\mathcal{X}_k$, and $(K^{(\ell)})_{\ell\in\mathbb{N}}$ in $\mathcal{X}$, such that:*

(i) $\bigcup_\ell K^{(\ell)}$ is dense in $\mathcal{X}$;

(ii) for each fixed ℓ, $K_k^{(\ell)}$ converges to $K^{(\ell)}$ in Gromov–Hausdorff sense as $k \to \infty$.

If one works in length spaces, as will be the case in the rest of this course, the above definition does not seem so good because $K^{(\ell)}$ will in general not be a strictly intrinsic (i.e. geodesic) length space: Geodesics joining elements of $K^{(\ell)}$ might very well leave $K^{(\ell)}$ at some intermediate time; so properties involving geodesics might not pass to the limit. This is the reason for requirement (iii) in the following definition.

Definition 27.12 (Geodesic local Gromov–Hausdorff convergence). *Let $(\mathcal{X}_k)_{k\in\mathbb{N}}$ be a family of geodesic Polish spaces, and let $\mathcal{X}$ be a Polish space. It is said that $\mathcal{X}_k$ converges to $\mathcal{X}$ in the geodesic* *local Gromov–Hausdorff topology if there are nondecreasing sequences of compact sets* $(K_k^{(\ell)})_{\ell\in\mathbb{N}}$ *in each* $\mathcal{X}_k$*, and* $(K^{(\ell)})_{\ell\in\mathbb{N}}$ *in* $\mathcal{X}$*, such that (i) and (ii) in Definition 27.11 are satisfied, and in addition:*

(iii) For each $\ell \in \mathbb{N}$*, there exists* ℓ' *such that all geodesics starting and ending in* $K_k^{(\ell)}$ *have their image contained in* $K_k^{(\ell')}$*.*

Then $\mathcal{X}$ *is automatically a geodesic space.*

If $\mathcal{X}$ is boundedly compact (that is, all closed balls are compact), there is a rather natural choice of exhaustive family of compact sets in $\mathcal{X}$: Pick up an arbitrary point $\star \in \mathcal{X}$, and consider the closed balls $B[\star, R_\ell]$, where R_ℓ is any sequence of positive real numbers going to infinity, say $R_\ell = \ell$. One can fix the sequence R_ℓ once for all, and then the notion of convergence only depends on the choice of the "reference point" or "base point" $\star$ (the point from which the convergence is seen). This suggests that the basic objects should not be just metric spaces, but rather **pointed metric spaces**. By definition, a pointed metric space consists of a triple $(\mathcal{X}, d, \star)$, where $(\mathcal{X}, d)$ is a metric space and $\star$ is some point in $\mathcal{X}$. Sometimes I shall just write $(\mathcal{X}, \star)$ or even just $\mathcal{X}$ as a shorthand for the triple $(\mathcal{X}, d, \star)$.

It is equivalent for a geodesic space to be boundedly compact or to be locally compact; so in the sequel the basic regularity assumption, when considering pointed Gromov–Hausdorff convergence, will be **local compactness**.

All the notions that were introduced in the previous section can be generalized in a completely obvious way to pointed metric spaces: A pointed isometry between $(\mathcal{X}, \star_\mathcal{X})$ and $(\mathcal{Y}, \star_\mathcal{Y})$ is an isometry which sends $\star_\mathcal{X}$ to $\star_\mathcal{Y}$; the pointed Gromov–Hausdorff distance d_{pGH} between two pointed spaces $(\mathcal{X}, \star_\mathcal{X})$ and $(\mathcal{Y}, \star_\mathcal{Y})$ is obtained as an infimum of Hausdorff distances over all pointed isometric embeddings; a pointed correspondence is a correspondence such that $\star_\mathcal{X}$ is in correspondence with $\star_\mathcal{Y}$; a pointed ε-isometry is an ε-isometry which sends $\star_\mathcal{X}$ to $\star_\mathcal{Y}$, etc. Then Definition 27.6 can be trivially transformed into a pointed notion of convergence, expressing the fact that *for each* R*, the closed ball* $B[\star_k, R]$ *in* $\mathcal{X}_k$ *converges to the closed ball* $B[\star, R]$ *in* $\mathcal{X}$. By an easy extraction argument, this is equivalent to the following alternative definition.

Definition 27.13 (Pointed Gromov–Hausdorff convergence). *Let* $(\mathcal{X}_k, \star_k)$ *be a sequence of pointed locally compact geodesic Polish spaces, and let* $(\mathcal{X}, \star)$ *be a pointed locally compact Polish space. Then* *it is said that $\mathcal{X}_k$ converges to $\mathcal{X}$ in the pointed Gromov–Hausdorff topology if any one of the three equivalent statements is satisfied:*

(i) There is a sequence $R_k \to \infty$ such that

$$d_{pGH}\Big(B[\star_k, R_k],\, B[\star, R_k]\Big) \longrightarrow 0;$$

(ii) There is a sequence $R_k \to \infty$, and there are pointed correspondences $\mathcal{R}_k$ between $B[\star_k, R_k]$ and $B[\star, R_k]$ such that

$$\mathrm{dis}\,(\mathcal{R}_k) \longrightarrow 0;$$

(iii) There are sequences $R_k \to \infty$ and $\varepsilon_k \to 0$, and pointed ε_k-isometries $f_k : B[\star_k, R_k] \to B[\star, R_k]$ with $\varepsilon_k \to 0$.

Remark 27.14. This notion of convergence implies the geodesic local convergence, as defined in Definition 27.12. Indeed, (i) and (ii) of Definition 27.11 are obviously satisfied, and (iii) follows from the fact that if a geodesic curve has its endpoints in $B[\star, R]$, then its image lies entirely inside $B[\star, R']$ with $R' = 2R$.

Example 27.15 (Blow-up). Let M be a Riemannian manifold of dimension n, and x a point in M. For each k, consider the pointed metric space $\mathcal{X}_k = (M, kd, x)$, where x is the basepoint, and kd is just the original geodesic distance on M, dilated by a factor k. Then $\mathcal{X}_k$ converges in the pointed Gromov–Hausdorff topology to the tangent space T_xM, pointed at 0 and equipped with the metric g_x (it is a Euclidean space). This is true as soon as M is just differentiable at x, in the sense of the existence of the tangent space.

Example 27.16. More generally, if $\mathcal{X}$ is a given metric space, and x is a point in $\mathcal{X}$, one can define the rescaled pointed spaces $\mathcal{X}_k = (\mathcal{X}, kd, x)$; if this sequence converges in the pointed Gromov–Hausdorff topology to some metric space $\mathcal{Y}$, then $\mathcal{Y}$ is said to be the tangent space, or **tangent cone**, to $\mathcal{X}$ at x. In many cases, the tangent cone coincides with the metric cone built on some length space Σ, which itself can be thought of as the space of tangent directions. (By definition, if (B, d) is a length space, the metric cone over B is obtained by considering $B \times [0, \infty)$, gluing together all the points in the fiber $B \times \{0\}$, and equipping the resulting space with the *cone metric*: $d_c((x,t),(y,s)) = \sqrt{t^2 + s^2 - 2ts\cos d(x,y)}$ when $d(x,y) \le \pi$, $d_c((x,t),(y,s)) = t + s$ when $d(x,y) > \pi$.)

Example 27.17. For any $p \in [1, \infty)$, define the ℓ^p norm on $\mathbb{R}^n$ by the usual formula $\|x\|_{\ell^p} = (\sum |x_i|^p)^{1/p}$; and let $\mathcal{X}_p$ be the space $\mathbb{R}^n$, equipped with the ℓ^p norm, pointed at 0. Then, as $p \to \infty$, $\mathcal{X}_p$ converges in the pointed Gromov–Hausdorff topology to $\mathcal{X}_\infty$, which is $\mathbb{R}^n$ equipped with the ℓ^∞ norm, $\|x\|_{\ell^\infty} = \sup |x_i|$. In $\mathcal{X}_p$, geodesics are segments of the form $(1-t)\,a + t\,b$, in particular they are nonbranching (two distinct geodesics cannot coincide on a nontrivial time interval), and unique (any two points are joined by a unique geodesic path). In contrast, geodesics in $\mathcal{X}_\infty$ are branching and definitely nonunique (any two distinct points can be joined by an uncountable set of geodesic paths). We see in this example that neither the nonbranching property, nor the property of uniqueness of geodesics, are preserved under Gromov–Hausdorff convergence. Moreover, the huge majority of geodesics in $\mathcal{X}_\infty$ cannot be realized as limits of geodesics on $\mathcal{X}_p$.

Remark 27.18. Consider pointed geodesic spaces $(\mathcal{X}_k, \star_k)$ and $(\mathcal{X}, \star)$, and let f_k be a pointed ε_k-isometry $B_{R_k]}(\star_k) \to B_{R_k]}(\star)$. Then let $R'_k \leq R_k$. It is clear that the distortion of f_k on $B_{R'_k]}(\star_k)$ is no more than the distortion of f_k on $B_{R_k]}(\star_k)$. Also if x belongs to $B_{R'_k]}(\star)$, and $\mathcal{X}$ is a geodesic space, then there is $x' \in B_{R'_k]}(\star)$ with $d(x, x') = 2\varepsilon_k$ and $d(\star, x') \leq R'_k - 2\varepsilon_k$; then there is $x'_k \in B_{R_k]}(\star_k)$ such that $d(f_k(x'_k), x') \leq \varepsilon_k$, so $d(\star, f_k(x'_k)) \leq R'_k - \varepsilon_k$, and then by the distortion property $d(\star_k, x'_k) \leq R'_k - \varepsilon_k + \varepsilon_k = R'_k$; on the other hand, $d(x'_k, x) \leq 2\varepsilon_k + \varepsilon_k = 3\varepsilon_k$. The conclusion is that the restriction of f_k to $B_{R'_k]}(\star_k)$ defines a $(3\varepsilon_k)$-isometry $B_{R'_k]}(\star_k) \to B_{R'_k]}(\star)$. In other words, it is always possible to *reduce* R_k while keeping the property of approximate isometry, provided that one changes ε_k for $3\varepsilon_k$.

Remark 27.19 (important). In the theory of Gromov–Hausdorff convergence, it is often imposed that $R_k = (\varepsilon_k)^{-1}$ in Definition 27.13. This is consistent with Example 27.15, and also with most tangent cones that are usually encountered. However, I shall not do so.

Functional analysis on Gromov–Hausdorff converging sequences

Many theorems about metric spaces still hold true, after appropriate modification, for *converging sequences* of metric spaces. Such is the case for some of the basic compactness theorems in functional analysis: Ascoli's theorem and Prokhorov's theorem. I shall not need these results outside the setting of compact spaces, so I shall be sketchy about their formulation in the noncompact case; the reader can easily fill in the gaps.

Proposition 27.20 (Ascoli theorem in Gromov–Hausdorff converging sequences). *Let $(\mathcal{X}_k)_{k\in\mathbb{N}}$ be a sequence of compact metric spaces, converging in the Gromov–Hausdorff topology to some compact metric space $\mathcal{X}$, by means of ε_k-approximations $f_k : \mathcal{X}_k \to \mathcal{X}$, admitting approximate inverses f'_k; and let $(\mathcal{Y}_k)_{k\in\mathbb{N}}$ be another sequence of compact metric spaces converging to $\mathcal{Y}$ in the Gromov–Hausdorff topology, by means of ε_k-approximations $g_k : \mathcal{Y}_k \to \mathcal{Y}$. Let $(\alpha_k)_{k\in\mathbb{N}}$ be a sequence of maps $\mathcal{X}_k \to \mathcal{Y}_k$ that are asymptotically equicontinuous, in the sense that for every $\varepsilon > 0$, there are $\delta = \delta(\varepsilon) > 0$ and $N = N(\varepsilon) \in \mathbb{N}$ so that for all $k \geq N$,*

$$d_{\mathcal{X}_k}(x_k, x'_k) \leq \delta \qquad \Longrightarrow \qquad d_{\mathcal{Y}_k}\big(\alpha_k(x_k), \alpha_k(x'_k)\big) \leq \varepsilon. \tag{27.5}$$

Then after passing to a subsequence, the maps $g_k \circ \alpha_k \circ f'_k : \mathcal{X} \to \mathcal{Y}$ converge uniformly to a continuous map $\alpha : \mathcal{X} \to \mathcal{Y}$.

This statement extends to locally compact spaces converging in the pointed Gromov–Hausdorff topology, and locally asymptotically uniformly equicontinuous maps, provided that the conclusion is weakened into locally uniform convergence.

Remark 27.21. In Proposition 27.20, the maps $g_k \circ \alpha_k \circ f'_k$ may be discontinuous, yet they will converge uniformly.

Proposition 27.22 (Prokhorov theorem in Gromov–Hausdorff converging sequences). *Let $(\mathcal{X}_k)_{k\in\mathbb{N}}$ be a sequence of compact metric spaces, converging in the Gromov–Hausdorff topology to some compact metric space $\mathcal{X}$, by means of ε_k-approximations $f_k : \mathcal{X}_k \to \mathcal{X}$. For each k, let μ_k be a probability measure on $\mathcal{X}_k$. Then, after extraction of a subsequence, $(f_k)_\#\mu_k$ converges in the weak topology to a probability measure μ on $\mathcal{X}$ as $k \to \infty$.*

This statement extends to Polish spaces converging by means of local Gromov–Hausdorff approximations, provided that the probability measures μ_k are uniformly tight with respect to the sequences $(K_k^{(\ell)})$ appearing in the definition of local Gromov–Hausdorff approximation.

Remark 27.23. In the previous proposition, the fact that the maps f_k are approximate isometries is useful only in the noncompact case; otherwise it just boils down to the compactness of $P(\mathcal{X})$ when $\mathcal{X}$ is compact.

Now comes another simple compactness criterion for which I shall provide a proof. Recall that a locally finite measure is a measure attributing finite mass to balls.

Proposition 27.24 (Compactness of locally finite measures). *Let $(\mathcal{X}_k, d_k, \star_k)_{k\in\mathbb{N}}$ be a sequence of pointed locally compact Polish spaces converging in the pointed Gromov–Hausdorff topology to some pointed locally compact Polish space $(\mathcal{X}, d, \star)$, by means of pointed ε_k-isometries f_k with $\varepsilon_k \to 0$. For each $k \in \mathbb{N}$, let ν_k be a locally finite Borel measure on $\mathcal{X}_k$. Assume that for each $R > 0$, there is a finite constant $M = M(R)$ such that*

$$\forall k \in \mathbb{N}, \qquad \nu_k[B_{R]}(\star_k)] \leq M.$$

Then, there is a locally finite measure ν such that, up to extraction of a subsequence,

$$(f_k)_\#\nu_k \xrightarrow[k\to\infty]{} \nu$$

in the weak-$*$ topology (that is, convergence against compactly supported continuous test functions).

Proof of Proposition 27.24. For any fixed $R > 0$, $(f_k)_\#\nu_k[B_{R]}(\star)] = \nu_k[(f_k)^{-1}(B_{R]}(\star))] \leq \nu_k[B_{R+\varepsilon_k]}(\star_k)]$ is uniformly bounded by $M(R+1)$ for k large enough. Since on the other hand $B_{R]}(\star)$ is compact, we may extract a subsequence in k such that $(f_k)_\#\nu_k[B_{R]}(\star)]$ converges to some finite measure ν_R in the weak-$*$ topology of $B_{R]}(\star)$. Then the result follows by taking $R = \ell \to \infty$ and applying a diagonal extraction. $\square$

Remark 27.25. There is an easy extension of Proposition 27.24 to local Gromov–Hausdorff convergence.

Adding the measure

Now let us switch from metric spaces to **metric-measure spaces**, which are triples $(\mathcal{X}, d, \nu)$, where d is a distance on $\mathcal{X}$ and ν a Borel measure on $\mathcal{X}$. (For brevity I shall sometimes write just $\mathcal{X}$ instead of $(\mathcal{X}, d, \nu)$.) So the question is to measure how far two metric-measure spaces $\mathcal{X}$ and $\mathcal{Y}$ are from each other.

There is a nontrivial choice to be made:

(a) Either we insist that metric-measure spaces are metric spaces in the first place, so two metric-measure spaces should be declared close only if they are close in terms of both the metric and the measure;

(b) Or we think that only the measure is relevant, and we should disregard sets of zero or small measure when estimating how far two metric-measure spaces are.

In the first case, one should identify two spaces $(\mathcal{X}, d, \nu)$ and $(\mathcal{X}', d', \nu')$ only if they are *isomorphic* as metric-measure spaces, which means that there exists a measurable bijection $f : \mathcal{X} \to \mathcal{X}'$ such that f is an isometry, and f preserves the measure: $f_\# \nu = \nu'$. Such a map is naturally called a **measure-preserving isometry**, and its inverse f^{-1} is automatically a measure-preserving isometry. (Note: It is not enough to impose that $\mathcal{X}$ and $\mathcal{X}'$ are isomorphic as metric spaces, and isomorphic as measure spaces; the same map should do the job for both isomorphisms.)

In the second case, one should identify sets that are isomorphic up to a zero measure set; so a natural thing to do is to declare that $(\mathcal{X}, d, \nu)$ and $(\mathcal{X}', d', \nu')$ are the same if there is a measure-preserving isometry between $\operatorname{Spt} \nu$ and $\operatorname{Spt} \nu'$, seen as subspaces of $\mathcal{X}$ and $\mathcal{X}'$ respectively.

Figure 27.4 is a classical example of a convergence which holds true in the sense of (b), not in the sense of (a).

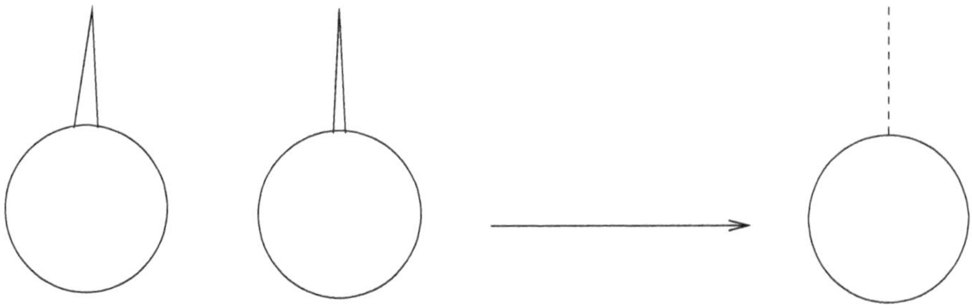

Fig. 27.4. An example of "reduction of support" that can arise in measured Gromov–Hausdorff convergence. This is a balloon with a very thin spike; in the Gromov–Hausdorff sense it is approximated by a balloon to which a one-dimensional spike is attached, that carries no measure.

Now it is easy to cook up notions of distance between metric-measure spaces. For a start, let us restrict to compact probability spaces. Pick up a distance which metrizes the weak topology on the set of probability measures, such as the Prokhorov distance d_P, and introduce the **Gromov–Hausdorff–Prokhorov distance** by the formula

$$d_{GHP}(\mathcal{X},\mathcal{Y}) = \inf \Big\{ d_H(\mathcal{X}',\mathcal{Y}') + d_P(\nu_{\mathcal{X}'},\nu_{\mathcal{Y}'}) \Big\},$$

where the infimum is taken over all measure-preserving isometric embeddings $f : (\mathcal{X},\nu_{\mathcal{X}}) \to (\mathcal{X}',\nu_{\mathcal{X}'})$ and $g : (\mathcal{Y},\nu_{\mathcal{Y}}) \to (\mathcal{Y}',\nu_{\mathcal{Y}'})$ into a *common* metric space $\mathcal{Z}$. That choice would correspond to point of view (a), while in point of view (b) one would rather use the **Gromov–Prokhorov distance**, which is defined, with the same notation, as

$$d_{GP}(\mathcal{X},\mathcal{Y}) = \inf\ d_P(\nu_{\mathcal{X}'},\nu_{\mathcal{Y}'}).$$

(The metric structure of $\mathcal{X}$ and $\mathcal{Y}$ has not disappeared since the infimum is only over isometries.)

Both d_{GHP} and d_{GP} satisfy the triangle inequality, as can be checked by a gluing argument again. (Now one should use both the metric and the probabilistic gluing!) Then there is no difficulty in checking that d_{GHP} is a honest distance on classes of metric-measure isomorphisms, with point of view (a). Similarly, d_{GP} is a distance on classes of metric-measure isomorphisms, with point of view (b), but now it is quite nontrivial to check that $[d_{GP}(\mathcal{X},\mathcal{Y}) = 0] \Longrightarrow [\mathcal{X} = \mathcal{Y}]$. I shall not insist on this issue, for in the sequel I shall focus on point of view (a).

There are several variants of these constructions:

1. *Use other distances on probability metrics.* Essentially everybody agrees on the Hausdorff distance to measure distances between sets, but as we know, there are many natural choices of distances between probability measures. In particular, one can replace the Prokhorov distance by the Wasserstein distance of order p, and thus obtain the **Gromov–Hausdorff–Wasserstein distance of order p**:

$$d_{GHW_p}(\mathcal{X},\mathcal{Y}) = \inf \Big\{ d_H(\mathcal{X}',\mathcal{Y}') + W_p(\nu_{\mathcal{X}'},\nu_{\mathcal{Y}'}) \Big\},$$

where the infimum is over isometric embeddings; and of course the **Gromov–Wasserstein distance of order p**:

$$d_{GW_p}(\mathcal{X},\mathcal{Y}) = \inf\ W_p(\nu_{\mathcal{X}'},\nu_{\mathcal{Y}'}).$$

2. *Measure distances between spaces on which the measure is finite but not necessarily normalized to 1.* This obviously amounts to measuring distances between finite nonnegative measures that are not necessarily normalized. There are two rather natural strategies (and many variants). The first one consists in using the bounded Lipschitz distance, as defined in (6.6), which makes sense for arbitrary signed measures; in this way one can define the "Gromov–Hausdorff–bounded-Lipschitz distance" d_{GHbL} and the "Gromov–bounded-Lipschitz distance" d_{GbL}, the definitions of which should be obvious to the reader. Another possibility consists in comparing the normalized metric spaces, and then adding a penalty which takes into account the discrepancy between the total masses. For instance, if μ and ν are defined on some common space $\mathcal{Z}$, one may let

$$d(\mu,\nu) = d_P\left(\frac{\mu}{\mu[\mathcal{Z}]}, \frac{\nu}{\nu[\mathcal{Z}]}\right) + \Big|\mu[\mathcal{Z}] - \nu[\mathcal{Z}]\Big|.$$

One may also replace d_P by W_p, or whatever; and change the penalty (why not something like $|\log(\mu[\mathcal{Z}]/\nu[\mathcal{Z}])|$?); etc. So there is a tremendous number of "natural" possibilities.

3. *Consider noncompact spaces.* Here also, there are many possible frameworks, and the reader is free to consider this variety as a wealth or as a big mess. A first possibility is to just ignore the fact that spaces are noncompact: this is not reasonable if one sticks to philosophy (a), because the Hausdorff distance between noncompact spaces is too rigid; but it makes perfect sense with philosophy (b), at least for finite measures (say probability measures). Then, one may apply distances d_{GHW_p} to noncompact situations, or variants designed to handle measures that are not probability measures. When the measures are only σ-finite, this simple approach has to be modified. A possibility consists in localizing as in Definition 27.11. Another possibility, which makes sense in a locally compact context, consists in pointing as in Definition 27.13 (and one may also impose the same condition as in Remark 27.19).

Convention: In the sequel of this chapter, when $(\mathcal{X}, d, \nu)$ is a metric space equipped with a measure, I shall always implicitly assume that ν is nonzero, and that it is:

- σ-finite if $(\mathcal{X}, d)$ is only assumed to be Polish;
- locally finite if $(\mathcal{X}, d)$ is assumed in addition to be locally compact.

(Given a locally compact metric space $(\mathcal{X}, d)$ and an arbitrary point $\star \in \mathcal{X}$, $\mathcal{X}$ is the union of the closed balls $B_{k]}(\star)$, so any locally finite measure on $\mathcal{X}$ is automatically σ-finite.)

Convergence and doubling property

The discussion of the previous section showed that one should be cautious about which notion of convergence is used. However, whenever they are available, *doubling estimates*, in the sense of Definition 18.1, basically rule out the discrepancy between approaches (a) and (b) above. The idea is that doubling prevents the formation of sharp spikes as in Figure 27.4. This discussion is not so clearly made in the literature that I know, so in this section I shall provide more careful proofs.

Proposition 27.26 (Doubling lets metric and metric-measure approaches coincide). *Let $(\mathcal{X}, \mu)$ and $(\mathcal{Y}, \nu)$ be two compact Polish probability spaces with diameter at most R. Assume that both μ and ν are doubling with a constant D. Then*

$$d_{GP}(\mathcal{X}, \mathcal{Y}) \leq d_{GHP}(\mathcal{X}, \mathcal{Y}) \leq \Phi_{R,D}\big(d_{GP}(\mathcal{X}, \mathcal{Y})\big), \tag{27.6}$$

where

$$\Phi_{R,D}(\delta) = \max\left(8\,\delta,\ R\,(16\,\delta)^{\frac{1}{\log_2 D}}\right) + \delta$$

is a function that goes to 0 as $\delta \to 0$, at a rate that is controlled in terms of just upper bounds on R and D.

Proof of Proposition 27.26. The inequality on the left of (27.6) is trivial, so let us focus on the right inequality. To start with, let $x \in \mathcal{X}$, $\varepsilon > 0$, then

$$1 = \mu[\mathcal{X}] = \mu[B_{R]}(x)] \leq D^N\, \mu[B_{\varepsilon/4]}(x)],$$

where

$$N = \left\lceil \log_2 \frac{4R}{\varepsilon} \right\rceil \leq \log_2 \frac{R}{\varepsilon} + 3,$$

and after a few manipulations this leads to

$$\mu[B_{\varepsilon/4]}(x)] \geq \frac{1}{8}\left(\frac{\varepsilon}{R}\right)^{\log_2 D}.$$

Now let $(\mathcal{X}', \mu')$ and $(\mathcal{Y}', \nu')$ be two isomorphic copies of $(\mathcal{X}, \mu)$ and $(\mathcal{Y}, \nu)$ in some metric space $\mathcal{Z}$. Let ε be the Hausdorff distance between $\mathcal{X}'$ and $\mathcal{Y}'$, and δ the Prokhorov distance between μ' and ν'; the goal is to control $\varepsilon + \delta$ in terms of δ alone. If $\varepsilon \le 8\delta$, then we are done, so we might assume that $\delta < \varepsilon/8$. Since $\varepsilon > 0$, there is, say, some $x \in \mathcal{X}'$ such that the ball $B_{\varepsilon/2]}(x)$ does not intersect $\mathcal{Y}'$.

The doubling property of $(\mathcal{X}, \mu)$ is of course transferred to $(\mathcal{X}', \mu')$, so by the previous estimate

$$\mu'[B_{\varepsilon/4]}(x)] \ge \frac{1}{8}\left(\frac{\varepsilon}{R}\right)^{\log_2 D}. \tag{27.7}$$

By definition of the Prokhorov distance,

$$\mu'[B_{\varepsilon/4]}(x)] \le \nu'[B_{\varepsilon/4+2\delta]}(x)] + 2\delta. \tag{27.8}$$

From (27.7) and (27.8) it follows that

$$\nu'[B_{\varepsilon/4+2\delta]}(x)] \ge \frac{1}{8}\left(\frac{\varepsilon}{R}\right)^{\log_2 D} - 2\delta. \tag{27.9}$$

Since $\delta < \varepsilon/8$, the ball $B_{\varepsilon/4+2\delta]}$ is included in $B_{\varepsilon/2]}(x)$, which does not intersect $\mathcal{Y}'$; so the left-hand side in (27.9) has to be 0. Thus

$$\varepsilon \le R\,(16\delta)^{\frac{1}{\log_2 D}};$$

and then the conclusion is easily obtained. □

Proposition 27.26 is better appreciated in view of the next exercise:

Exercise 27.27. Let $(\mathcal{X}_k, d_k, \nu_k)$ be a sequence of Polish probability spaces converging in the sense of d_{GP} to $(\mathcal{X}, d, \nu)$. Assume that each ν_k is doubling, with a uniform bound on the doubling constant. Prove that ν is also doubling.

The combination of Proposition 27.26 and Exercise 27.27 yields the following corollary:

Corollary 27.28 (d_{GP} convergence and doubling imply d_{GHP} convergence). *Let $(\mathcal{X}_k, d_k, \nu_k)$ be a family of Polish probability spaces satisfying a uniform doubling condition, uniformly bounded, and converging to $(\mathcal{X}, d, \nu)$ in the Gromov–Prokhorov sense. Then $(\mathcal{X}_k, d_k, \nu_k)$ also converges in the Gromov–Hausdorff–Prokhorov sense to $(\mathcal{X}, d, \nu)$. In particular, $(\mathcal{X}_k, d_k)$ converges to $(\mathcal{X}, d)$ in the Gromov–Hausdorff sense.*

To summarize once again: Qualitatively, the distinction between points of view (a) and (b) is nonessential when one deals with the convergence of probability spaces satisfying a uniform doubling estimate. A more careful discussion would extend this conclusion to metric-measure spaces that are not necessarily probability spaces; and then to the pointed convergence of metric-measure spaces, provided that the doubling constant on a ball of radius R (around the base point of each space) only depends on R.

When doubling estimates are not available, things are not so simple and it does matter whether one adheres to philosophy (a) or (b). Point of view (b) is the one that was mainly developed by Gromov, in relation to the phenomenon of *concentration of measure*. It is also the point of view that was adopted by Sturm in his study of the stability of displacement convexity. Nevertheless, I shall prefer to stick here to point of view (a), partly because this is the approach which Lott and I adopted for the study of the stability of optimal transport, partly because it can be argued that point of view (a) provides a more precise notion of convergence and description of the limit space. For instance, in the example of Figure 27.4, the fact that the limit space has a spike carrying zero measure retains information about the asymptotic shape of the converging sequence. Of course, this will not prevent me from throwing away pieces with zero measure by restricting to the support of the measure, when that is possible.

Doubling has another use in the present context: It leads to uniform total boundedness estimates, and therefore to compactness statements via Theorem 27.10.

Proposition 27.29 (Doubling implies uniform total boundedness). *Let $(\mathcal{X}, d)$ be a Polish space with diameter bounded above by R, equipped with a finite (nonzero) D-doubling measure ν. Then for any $\varepsilon > 0$ there is a number $N = N(\varepsilon)$, only depending on R, D and ε, such that $\mathcal{X}$ can be covered with N balls of radius ε.*

Proof of Proposition 27.29. Without loss of generality, we might assume that $\nu[\mathcal{X}] = 1$. Let $r = \varepsilon/2$, and let n be such that $R \leq 2^n r$. Choose an arbitrary point x_1 in $\mathcal{X}$; then a point x_2 in $\mathcal{X} \setminus (B_{2r]}(x_1))$, a point x_3 in $\mathcal{X} \setminus (B_{2r]}(x_1) \cup B_{2r]}(x_2))$, and so forth. All the balls $B_{r]}(x_j)$ are disjoint, and by the doubling property each of them has measure at least D^{-n}. So $\mathcal{X} \setminus (B_{r]}(x_0) \cup \ldots \cup B_{r]}(x_k))$ has measure at most $1 - kD^{-n}$, and therefore D^n is an upper bound on the number of points x_j that can be chosen.

Now let $x \in \mathcal{X}$. There is at least one index j such that $d(x, x_j) < 2r$; otherwise x would lie in the complement of the union of all the balls $B_{2r}(x_j)$, and could be added to the family $\{x_j\}$. So $\{x_j\}$ constitutes a $2r$-net in $\mathcal{X}$, with cardinality at most $N = D^n$. This concludes the proof. □

Measured Gromov–Hausdorff topology

After all this discussion I can state a precise definition of the notion of convergence that will be used in the sequel for metric-measure spaces: this is the **measured Gromov–Hausdorff topology**. It is associated with the convergence of spaces as metric spaces *and* as measure spaces. This concept can be defined quantitatively in terms of, e.g., the distance d_{GHP} and its variants, but I shall be content with a purely topological (qualitative) definition. As in the case of the plain Gromov–Hausdorff topology, there is a convenient reformulation in terms of approximate isometries.

Definition 27.30 (Measured Gromov–Hausdorff topology). *Let $(\mathcal{X}_k, d_k, \nu_k)_{k\in\mathbb{N}}$ and $(\mathcal{X}, d, \nu)$ be compact metric spaces, equipped with finite nonzero measures. It is said that $\mathcal{X}_k$ converges to $\mathcal{X}$ in the measured Gromov–Hausdorff topology if there are measurable ε_k-isometries $f_k : \mathcal{X}_k \to \mathcal{X}$ such that $\varepsilon_k \to 0$ and*

$$(f_k)_{\#}\nu_k \xrightarrow[k\to\infty]{} \nu$$

in the weak topology of measures.

If $(\mathcal{X}_k, d_k, \nu_k)$ and $(\mathcal{X}, d, \nu)$ are Polish spaces, not necessarily compact, equipped with σ-finite measures, it is said that $\mathcal{X}_k$ converges to $\mathcal{X}$ in the local measured Gromov–Hausdorff topology if there are nondecreasing sequences of compact sets $(K_k^{(\ell)})_{\ell\in\mathbb{N}}$ for each k, and $(K^{(\ell)})_{\ell\in\mathbb{N}}$, such that for each ℓ, the space $K_k^{(\ell)}$, seen as a subspace of $\mathcal{X}_k$, converges in the measured Gromov–Hausdorff topology to $K^{(\ell)}$ as $k \to \infty$; and the union of all $K^{(\ell)}$ is dense in $\mathcal{X}$.

If the spaces $(\mathcal{X}_k, d_k, \nu_k, \star_k)$ and $(\mathcal{X}, d, \nu, \star)$ are locally compact pointed Polish spaces equipped with locally finite measures, it is said that $\mathcal{X}_k$ converges to $\mathcal{X}$ in the pointed measured Gromov–Hausdorff *topology if there are sequences $R_k \to \infty$ and $\varepsilon_k \to 0$, and measurable pointed ε_k-isometries $B[\star_k, R_k] \to B[\star, R_k]$, such that*

$$(f_k)_\# \nu_k \xrightarrow[k\to\infty]{} \nu,$$

where the convergence is now in the weak-$*$ topology (convergence against compactly supported continuous functions).

Remark 27.31. As already remarked for the plain Gromov–Hausdorff topology, one might also require that $R_k = (\varepsilon_k)^{-1}$, but I shall not do so here.

From the material in this chapter it is easy to derive the following compactness criterion:

Theorem 27.32 (Compactness in measured Gromov–Hausdorff topology). *(i) Let $R > 0$, $D > 0$, and $0 < m \le M$ be finite positive constants, and let $\mathcal{F}$ be a family of compact metric-measure spaces, such that (a) for each $(\mathcal{X}, d, \nu) \in \mathcal{F}$ the diameter of $(\mathcal{X}, d)$ is bounded above by $2R$; (b) the measure ν has a doubling constant bounded above by D; and (c) $m \le \nu[\mathcal{X}] \le M$. Then $\mathcal{F}$ is precompact in the measured Gromov–Hausdorff topology. In particular, any weak cluster space $(\mathcal{X}_\infty, d_\infty, \nu_\infty)$ satisfies* $\operatorname{Spt} \nu_\infty = \mathcal{X}_\infty$.

(ii) Let $\mathcal{F}$ be a family of locally compact pointed Polish metric-measure spaces. Assume that for each R, there is a constant $D = D(R)$ such that for each $(\mathcal{X}, d, \nu, \star) \in \mathcal{F}$ the measure ν is D-doubling on the ball $B_{R]}(\star)$. Further, assume the existence of $m, M > 0$ such that $m \le \nu[B_{1]}(\star)] \le M$ for all $(\mathcal{X}, d, \nu) \in \mathcal{F}$. Then $\mathcal{F}$ is precompact in the pointed measured Gromov–Hausdorff topology. In particular, any weak cluster space $(\mathcal{X}_\infty, d_\infty, \nu_\infty)$ satisfies $\operatorname{Spt} \nu_\infty = \mathcal{X}_\infty$.

Remark 27.33. A particular case of Theorem 27.32 is when all measures are normalized to have unit mass.

Proof of Theorem 27.32. Part (i) follows from the combination of Proposition 27.29, Theorem 27.10 and Proposition 27.22. Part (ii) follows in addition from the definition of pointed measured Gromov–Hausdorff convergence and Proposition 27.24. Note that in (ii), the doubling assumption is used to prevent the formation of “spikes”, but also to ensure uniform bounds on the mass of balls of radius R for *any* R, once it is known for *some* R. (Here I chose $R = 1$, but of course any other choice would have done.) □

The following simple but fundamental corollary is obtained by combining Theorem 27.32 with the Bishop–Gromov inequality (Theorem 18.8):

Corollary 27.34 (Gromov's precompactness theorem). *Let $K \in \mathbb{R}$, $N \in (1, \infty]$ and $D \in (0, +\infty)$. Let $\mathcal{M}(K, N, D)$ be the set of Riemannian manifolds (M, g) such that $\dim(M) \leq N$, $\mathrm{Ric}_M \geq Kg$ and $\mathrm{diam}\,(M) \leq D$, equipped with their geodesic distance and their volume measure. Then $\mathcal{M}(K, N, D)$ is precompact in the measured Gromov–Hausdorff topology.*

Bibliographical notes

It is well-known to specialists, but not necessarily obvious to others, that the Prokhorov distance, as defined in (6.5), coincides with the expression given in the beginning of this chapter; see for instance [814, Remark 1.29].

Gromov's influential book [438] is one of the founding texts for the Gromov–Hausdorff topology. Some of the motivations, developments and applications of Gromov's ideas can can be found in the research papers [116, 386, 675] written shortly after the publication of that book.

My presentation of the Gromov–Hausdorff topology mainly followed the very pedagogical book of Burago, Burago and Ivanov [174]. (These authors do not use the terminology "geodesic space" but rather "strictly intrinsic length space".) One can find there the proofs of Theorems 27.9 and 27.10. Besides Gromov's own book, other classical sources about the convergence of metric spaces and metric-measure spaces are the book by Petersen [680, Chapter 10], and the survey by Fukaya [387]. Also a short presentation is available in the Saint-Flour lecture notes by S. Evans.

Information about Mr. Magoo, the famous short-sighted cartoon character, can be found on the website `www.toontracker.com`.

The definition of a metric cone can be found in [174, Section 3.6.2], and the notion of a tangent cone is explained in [174, p. 321]. Recall from Example 27.16 that the tangent cone may be defined in two different ways: either as the metric cone over the space of directions, or as the Gromov–Hausdorff limit of rescaled spaces; read carefully Remarks 9.1.40 to 9.1.42 in [174] to avoid traps (or see [626]). For

Alexandrov spaces with curvature bounded below, both notions coincide, see [174, Section 10.9] and the references therein.

A classical source for the measured Gromov–Hausdorff topology is Gromov's book [438]. The point of view mainly used there consists in forgetting the Gromov–Hausdorff distance and "using the measure to kill infinity"; so the distances that are found there would be of the sort of d_{GW_p} or d_{GP}. An interesting example is Gromov's "box" metric $\underline{\square}_1$, defined as follows [438, pp. 116–117]. If d and d' are two metrics on a given probability space $\mathcal{X}$, define $\underline{\square}_1(d, d')$ as the infimum of $\varepsilon > 0$ such that $|d - d'| \leq \varepsilon$ outside of a set of measure at most ε in $\mathcal{X} \times \mathcal{X}$ (the subscript 1 means that the measure of the discarded set is at most $1 \times \varepsilon$). Take the particular metric space $I = [0, 1]$, equipped with the usual distance and with the Lebesgue measure λ, as reference space. If $(\mathcal{X}, d, \nu)$ and $(\mathcal{X}', d', \nu')$ are two Polish probability spaces, define $\underline{\square}_1(\mathcal{X}, \mathcal{X}')$ as the infimum of $\square_1(d \circ \phi, d' \circ \phi')$ where ϕ (resp. ϕ') varies in the set of all measure-preserving maps $\phi : I \to \mathcal{X}$ (resp. $\phi' : [0, 1] \to \mathcal{X}'$).

Sturm made a detailed study of d_{GW_2} (denoted by $\mathbf{D}$ in [762]) and advocated it as a natural distance between classes of equivalence of probability spaces in the context of optimal transport. He proved that $\mathbf{D}$ satisfies the length property, and compared it with Gromov's box distance as follows:

$$\begin{cases} \mathbf{D}(\mathcal{X}, \mathcal{Y}) \leq \left(\max\big(\operatorname{diam}(\mathcal{X}), \operatorname{diam}(\mathcal{Y})\big) + \dfrac{1}{4}\right) \underline{\square}_1(\mathcal{X}, \mathcal{Y})^{\frac{1}{2}} \\ \\ \mathbf{D}(\mathcal{X}, \mathcal{Y}) \geq (1/2)^{3/2}\, \underline{\square}_1(\mathcal{X}, \mathcal{Y})^{\frac{3}{2}}. \end{cases}$$

The alternative point of view in which one takes care of both the metric and the measure was introduced by Fukaya [386]. This is the one which was used by Lott and myself in our study of displacement convexity in geodesic spaces [577].

The pointed Gromov–Hausdorff topology is presented for instance in [174]; it has become very popular as a way to study tangent spaces in the absence of smoothness. In the context of optimal transport, the pointed Gromov–Hausdorff topology was used independently in [30, Section 12.4] and in [577, Appendix A].

I introduced the definition of *local* Gromov–Hausdorff topology for the purpose of these notes; it looks to me quite natural if one wants to preserve the idea of pointing in a setting that might not necessarily be locally compact. This is not such a serious issue and the reader who does not like this notion can still go on with the rest of these notes. Still, let me recommend it as a natural concept to treat the Gromov–Hausdorff convergence of the Wasserstein space over a noncompact metric space (see Chapter 28).

The statement of completeness of the Gromov–Hausdorff space appears in Gromov's book [438, p. 78]. Its proof can be found, e.g., in Fukaya [387, Theorem 1.5], or in the book by Petersen [680, Chapter 10, Proposition 1.7].

The theorem briefly alluded to in the end of Remark 27.8 states the following: If M is an n-dimensional compact Riemannian manifold, and $(M_k)_{k\in\mathbb{N}}$ is a sequence of n-dimensional compact Riemannian manifolds converging to M, with uniform upper and lower bounds on the sectional curvatures, and a volume which is uniformly bounded below, then M_k is diffeomorphic to M for k large enough. This result is due to Gromov (after precursors by Shikata); see, e.g., [387, Chapter 3] for a proof and references.

The Gromov–Hausdorff topology is not the only one used to compare Riemannian manifolds; for instance, some authors have defined a "spectral distance" based on properties of the heat kernel [97, 492, 508, 509].

I shall conclude with some technical remarks.

The theorem according to which a distance-preserving map $\mathcal{X} \to \mathcal{X}$ is a bijection if $\mathcal{X}$ is compact can be found in [174, Theorem 1.6.14].

Proposition 27.20 appears, in a form which is not exactly the one that I used, but quite close to, in [436, p. 66] and [443, Appendix A]. The reader should have no difficulty in adapting the statements there into Proposition 27.20 (or redo the proof of the Ascoli theorem). Proposition 27.22 is rather easy to prove, and anyway in the next chapter we shall prove some more complicated related theorems.

That a locally compact geodesic space is automatically boundedly compact is part of a generalized version of the Hopf–Rinow theorem [174, Theorem 2.5.28].

Finally, the construction of approximate isometries from correspondences, as performed in [174], uses the full axiom of choice (on p. 258: "For each x, choose $f(x) \in Y$"). So I should sketch a proof which does not use it. Let $\mathcal{R}$ be a correspondence with distortion η; the problem is to construct an ε-isometry $f : \mathcal{X} \to \mathcal{Y}$ for any $\varepsilon > \eta$. Let $\mathcal{D}$ be a countable dense subset in $\mathcal{X}$. Choose δ so small that $2\delta + \eta < \varepsilon$. Cover $\mathcal{X}$ by finitely many disjoint sets A_k, such that each A_k is included in some ball $B(x_k, \delta)$, with $x_k \in \mathcal{D}$. Then for each $x \in \mathcal{D}$ choose $f(x)$ in relation with x. (This only uses the countable version of the axiom of choice.) Finally, for each $x \in A_k$ define $f(x) = f(x_k)$. It is easy to check that $\operatorname{dis}(f) \leq 2\delta + \eta < \varepsilon$.

(This axiomatic issue is also the reason why I work with approximate inverses that are (4ε)-isometries rather than (3ε)-isometries.)

28

Stability of optimal transport

This chapter is devoted to the following theme: Consider a family of geodesic spaces $\mathcal{X}_k$ which converges to some geodesic space $\mathcal{X}$; does this imply that certain basic objects in the theory of optimal transport on $\mathcal{X}_k$ "pass to the limit"? In this chapter I shall show that the answer is affirmative: One of the main results is that the Wasserstein space $P_2(\mathcal{X}_k)$ converges, in (local) Gromov–Hausdorff sense, to the Wasserstein space $P_2(\mathcal{X})$. Then I shall consider the stability of dynamical optimal transference plans, and related objects (displacement interpolation, kinetic energy, etc.). Compact spaces will be considered first, and will be the basis for the subsequent treatment of noncompact spaces.

Optimal transport in a nonsmooth setting

Most of the objects that were introduced and studied in the context of optimal transport on Riemannian manifolds still make sense on a general metric-measure length space $(\mathcal{X}, d, \nu)$, satisfying certain regularity assumptions. I shall assume here that $(\mathcal{X}, d)$ is a **locally compact, complete separable geodesic space** equipped with a σ-finite reference Borel measure ν. From general properties of such spaces, plus the results in Chapters 6 and 7:

- the cost function $c(x, y) = d(x, y)^2$ is associated with the coercive Lagrangian action $\mathcal{A}(\gamma) = \mathcal{L}(\gamma)^2$, and minimizers are constant-speed, minimizing geodesics, the collection of which is denoted by $\Gamma(\mathcal{X})$;

C. Villani, *Optimal Transport*, Grundlehren der mathematischen Wissenschaften 338,

- for any given μ_0, μ_1 in $P_2(\mathcal{X})$, the optimal total cost $C(\mu_0,\mu_1)$ is finite and there exists at least one optimal transference plan $\pi \in P(\mathcal{X}\times\mathcal{X})$ with marginals μ_0 and μ_1;
- the 2-Wasserstein space $P_2(\mathcal{X})$, equipped with the 2-Wasserstein distance, is a complete separable geodesic space;
- a displacement interpolation $(\mu_t)_{0\le t\le 1}$ can be defined either as a geodesic in $P_2(\mathcal{X})$, or as $(e_t)_\#\Pi$, where e_t is the evaluation at time t, and Π is a dynamical optimal transference plan, i.e. the law of a random geodesic whose endpoints form an optimal coupling of (μ_0,μ_1).

One can also introduce the interpolant density $\rho_t = \rho_t(x)$ as the density of μ_t with respect to the reference measure ν.

Many of the statements that were available in the Riemannian setting can be recast in terms of these objects. An important difference however is the absence of any "explicit" description of optimal couplings in terms of $d^2/2$-convex maps ψ. So expressions involving $\nabla\psi$ will not a priori make sense, unless we find an intrinsic reformulation in terms of the above-mentioned objects. For instance,

$$\int \rho_0(x)|\nabla\psi(x)|^2\,d\nu(x) = \int d\Big(x, \exp_x \nabla\psi(x)\Big)^2\,d\mu_0(x) = W_2(\mu_0,\mu_1)^2. \tag{28.1}$$

There is a more precise procedure which allows one to make sense of $|\nabla\psi|$, even if $\nabla\psi$ itself does not. The crucial observation, as in (28.1), is that $|\nabla\psi(x)|$ can be identified with the *length* $\mathcal{L}(\gamma)$ of the geodesic γ joining $\gamma(0)=x$ to $\gamma(1)=y$. In the next paragraph I shall develop this remark. The hasty reader can skip this bit and go directly to the section about the convergence of Wasserstein spaces.

Kinetic energy and speed

Definition 28.1 (Kinetic energy). *Let $\mathcal{X}$ be a locally compact Polish geodesic space, and let $\Pi \in P(\Gamma(\mathcal{X}))$ be a dynamical transference plan. For each $t\in(0,1)$ define the associated kinetic energy $\varepsilon_t(dx)$ by the formula*

$$\varepsilon_t = (e_t)_\#\left(\frac{\mathcal{L}^2}{2}\,\Pi\right).$$

If ε_t *is absolutely continuous with respect to* μ_t*, define the speed field* $|v|(t,x)$ *by the formula*

$$|v|(t,x) = \sqrt{2\,\frac{d\varepsilon_t}{d\mu_t}}.$$

Remark 28.2. If $\mathcal{X}$ is compact then $\varepsilon_t \le C\mu_t$ with $C = (\operatorname{diam} \mathcal{X})^2/2$; so $|v|$ is well-defined (up to modification on a set of zero μ_t-measure) and almost surely bounded by $\sqrt{2C} = \operatorname{diam}(\mathcal{X})$.

Remark 28.3. If γ is a geodesic curve, then $\mathcal{L}(\gamma) = |\dot{\gamma}|(t)$, whatever $t \in (0,1)$. Assume that $\mathcal{X}$ is a Riemannian manifold M, and geodesics in the support of Π do not cross at intermediate times. (As we know from Chapter 8, this is the case if Π is an *optimal* dynamical transference plan.) Then for each $t \in (0,1)$ and $x \in M$ there is at most one geodesic $\gamma = \gamma^{x,t}$ such that $\gamma(t) = x$. So

$$\varepsilon_t(dx) = \left(\frac{|\dot{\gamma}^{x,t}(t)|^2}{2}\right)\,[(e_t)_\#\Pi](dx) = \left(\frac{|\dot{\gamma}^{x,t}(t)|^2}{2}\right)\,\mu_t(dx);$$

and $|v|(t,x)$ really is $|\dot{\gamma}^{x,t}|$, that is, the speed at time t and position x. Thus Definition 28.1 is consistent with the usual notions of kinetic energy and speed field (speed = norm of the velocity).

Particular Case 28.4. Let M be a Riemannian manifold, and let μ_0 and μ_1 be two probability measures in $P_2(M)$, μ_0 being absolutely continuous with respect to the volume measure. Let ψ be a $d^2/2$-convex function such that $\exp(\nabla\psi)$ is the optimal transport from μ_0 to μ_1, and let ψ_t be obtained by solving the forward Hamilton–Jacobi equation $\partial_t\psi_t + |\nabla\psi_t|^2/2 = 0$ starting from the initial datum $\psi_0 = \psi$. Then the speed $|v|(t,x)$ coincides, μ_t-almost surely, with $|\nabla\psi_t(x)|$.

The kinetic energy is a nonnegative measure, while the field speed is a function. Both objects will enjoy good stability properties under Gromov–Hausdorff approximation. But under adequate assumptions, the velocity field will also enjoy compactness properties in the *uniform* topology. This comes from the next statement.

Theorem 28.5 (Regularity of the speed field). *Let* $(\mathcal{X},d)$ *be a compact geodesic space, let* $\Pi \in P(\Gamma(\mathcal{X}))$ *be a dynamical optimal transference plan, let* $(\mu_t)_{0\le t\le 1}$ *be the associated displacement interpolation,* *and* $|v| = |v|(t,x)$ *the associated speed field. Then, for each* $t \in (0,1)$ *one can modify* $|v|(t,\,\cdot\,)$ *on a* μ_t*-negligible set in such a way that for* all $x,y \in \mathcal{X}$,

$$\Big|\, |v|(t,x) \;-\; |v|(t,y) \,\Big| \;\le\; \frac{C\sqrt{\mathrm{diam}\,(\mathcal{X})}}{\sqrt{t(1-t)}}\;\sqrt{d(x,y)}, \tag{28.2}$$

where C *is a numeric constant. In particular,* $|v|(t,\,\cdot\,)$ *is Hölder-1/2.*

Proof of Theorem 28.5. Let t be a fixed time in $(0,1)$. Let γ_1 and γ_2 be two minimizing geodesics in the support of Π, and let $x = \gamma_1(t)$, $y = \gamma_2(t)$. Then by Theorem 8.22,

$$\big|\mathcal{L}(\gamma_1) - \mathcal{L}(\gamma_2)\big| \le \frac{C\sqrt{\mathrm{diam}\,(\mathcal{X})}}{\sqrt{t(1-t)}}\;\sqrt{d(x,y)}. \tag{28.3}$$

Let $\mathcal{X}_t$ be the union of all $\gamma(t)$, for γ in the support of Π. For a given $x \in \mathcal{X}_t$, there might be several geodesics γ passing through x, but (as a special case of (28.3)) they will all have the same length; define $|v|(t,x)$ to be that length. This is a measurable function, since it can be rewritten

$$|v|(t,x) = \int_\Gamma \mathcal{L}(\gamma)\,\Pi\big(d\gamma\,|\,\gamma(t)=x\big),$$

where $\Pi(d\gamma|\gamma(t)=x)$ is of course the disintegration of Π with respect to $\mu_t = \mathrm{law}\,(\gamma_t)$. Then $|v|(t,\,\cdot\,)$ is an admissible density for ε_t, and as a consequence of (28.3) it satisfies (28.2) for all $x,y \in \mathcal{X}_t$.

To extend $|v|(t,x)$ on the whole of $\mathcal{X}$, I shall adapt the proof of a well-known extension theorem for Lipschitz functions. Let $H := C\sqrt{\mathrm{diam}\,\mathcal{X}}/(t(1-t))$, so that $|v|$ is Hölder-1/2 on $\mathcal{X}_t$ with constant H. Define, for $x \in \mathcal{X}$,

$$w(x) := \inf_{y\in\mathcal{X}_t}\Big[H\sqrt{d(x,y)} + |v|(t,y)\Big].$$

It is clear that $w \ge 0$, and the estimate (28.2) easily implies that $w(x) = |v|(t,x)$ for any $x \in \mathcal{X}_t$. Next, whenever x and x' are two points in $\mathcal{X}$, one has

$$\begin{aligned}
w(x)-w(x') &= \inf_{y\in\mathcal{X}_t}\Big[H\sqrt{d(x,y)}+|v|(t,y)\Big]-\inf_{y'\in\mathcal{X}_t}\Big[H\sqrt{d(x',y')}+|v|(t,y')\Big] \\
&= \sup_{y'\in\mathcal{X}_t}\inf_{y\in\mathcal{X}_t}\Big[H\sqrt{d(x,y)}-H\sqrt{d(x',y')}+|v|(t,y)-|v|(t,y')\Big] \\
&\le H\sup_{y'\in\mathcal{X}_t}\inf_{y\in\mathcal{X}_t}\Big[\sqrt{d(x,y)}-\sqrt{d(x',y')}+\sqrt{d(y,y')}\Big] \\
&\le H\sup_{y'\in\mathcal{X}_t}\Big[\sqrt{d(x,y')}-\sqrt{d(x',y')}\Big]. \qquad (28.4)
\end{aligned}$$

But

$$\sqrt{d(x,y')}\le\sqrt{d(x,x')+d(x',y')}\le\sqrt{d(x,x')}+\sqrt{d(x',y')};$$

so (28.4) is bounded above by $H\sqrt{d(x,x')}$.

To summarize: w coincides with $|v|(t,\,\cdot\,)$ on $\mathcal{X}_t$, and it satisfies the same Hölder-1/2 estimate. Since μ_t is concentrated on $\mathcal{X}_t$, w is also an admissible density for ε_t, so we can take it as the new definition of $|v|(t,\,\cdot\,)$, and then (28.2) holds true on the whole of $\mathcal{X}$. □

Convergence of the Wasserstein space

The main goal of this section is the proof of the convergence of the Wasserstein space $P_2(\mathcal{X})$, as expressed in the next statement.

Theorem 28.6 (If $\mathcal{X}_k$ converges then $P_2(\mathcal{X}_k)$ also). *Let $(\mathcal{X}_k)_{k\in\mathbb{N}}$ and $\mathcal{X}$ be compact metric spaces such that*

$$\mathcal{X}_k \xrightarrow{GH} \mathcal{X}.$$

Then

$$P_2(\mathcal{X}_k) \xrightarrow{GH} P_2(\mathcal{X}).$$

Moreover, if $f_k:\mathcal{X}_k\to\mathcal{X}$ are approximate isometries, then the maps $(f_k)_\#:P_2(\mathcal{X}_k)\to P_2(\mathcal{X})$, defined by $(f_k)_\#(\mu)=(f_k)_\#\mu$, are approximate isometries too.

Theorem 28.6 will come as an immediate corollary of the following more precise results:

Proposition 28.7 (If f is an approximate isometry then $f_\#$ also). *Let $f : (\mathcal{X}_1, d_1) \to (\mathcal{X}_2, d_2)$ be an ε-isometry between two Polish spaces. Then the map $f_\#$ is an $\widetilde{\varepsilon}$-isometry between $P_2(\mathcal{X}_1)$ and $P_2(\mathcal{X}_2)$, where*

$$\widetilde{\varepsilon} = 6\,\varepsilon + 2\sqrt{\varepsilon\,(2\,\mathrm{diam}\,(\mathcal{X}_2) + \varepsilon)} \leq 8\big(\varepsilon + \sqrt{\varepsilon\ \mathrm{diam}\,(\mathcal{X}_2)}\big). \qquad (28.5)$$

Remark 28.8. The map $f_\#$ is continuous if and only if f itself is continuous (which in general is not the case).

Proof of Proposition 28.7. Let f be an ε-isometry, and let f' be an ε-inverse for f. Recall that f' is a (4ε)-isometry and satisfies (27.4).

Given μ_1 and μ_1' in $P_2(\mathcal{X}_1)$, let π_1 be an optimal transference plan between μ_1 and μ_1'. Define

$$\pi_2 := (f, f)_\# \pi_1.$$

Obviously, π_2 is a transference plan between $f_\#\mu_1$ and $f_\#\mu_1'$; so

$$\begin{aligned} W_2\big(f_\#\mu_1, f_\#\mu_1'\big)^2 &\leq \int_{\mathcal{X}_2\times\mathcal{X}_2} d_2(x_2,y_2)^2\, d\pi_2(x_2,y_2) \\ &= \int_{\mathcal{X}_1\times\mathcal{X}_1} d_2\big(f(x_1), f(y_1)\big)^2\, d\pi_1(x_1,y_1). \end{aligned} \qquad (28.6)$$

The identity

$$\begin{aligned} &\Big|d_2(f(x_1), f(y_1))^2 - d_1(x_1,y_1)^2\Big| \\ &\quad = \Big|d_2(f(x_1), f(y_1)) - d_1(x_1,y_1)\Big|\,\Big(d_2(f(x_1), f(y_1)) + d_1(x_1,y_1)\Big), \end{aligned}$$

implies

$$\Big|d_2(f(x_1), f(y_1))^2 - d_1(x_1,y_1)^2\Big| \leq \varepsilon\,\big(\mathrm{diam}\,(\mathcal{X}_1) + \mathrm{diam}\,(\mathcal{X}_2)\big).$$

Plugging this bound into (28.6), we deduce that

$$W_2\big(f_\#\mu_1, f_\#\mu_1'\big)^2 \leq W_2(\mu_1,\mu_1')^2 + \varepsilon\,\big(\mathrm{diam}\,(\mathcal{X}_1) + \mathrm{diam}\,(\mathcal{X}_2)\big), \quad (28.7)$$

hence

$$W_2\big(f_\#\mu_1, f_\#\mu_1'\big) \leq W_2(\mu_1,\mu_1') + \sqrt{\varepsilon\,\big(\mathrm{diam}\,(\mathcal{X}_1) + \mathrm{diam}\,(\mathcal{X}_2)\big)}. \quad (28.8)$$

It follows from the definition of an ε-isometry that $\operatorname{diam}(\mathcal{X}_1) \le \operatorname{diam}(\mathcal{X}_2) + \varepsilon$; so (28.8) leads to

$$W_2\big(f_\# \mu_1, f_\# \mu_1'\big) \le W_2(\mu_1, \mu_1') + \sqrt{\varepsilon\,(2\operatorname{diam}(\mathcal{X}_2) + \varepsilon)}, \qquad (28.9)$$

which shows that $f_\#$ does not increase distances much.

Exchanging the roles of $\mathcal{X}_1$ and $\mathcal{X}_2$, and applying (28.8) to the map f' and the measures $f_\#\mu_1$ and $f_\#\mu_1'$, together with $\operatorname{diam}(\mathcal{X}_1) \le \operatorname{diam}(\mathcal{X}_2) + \varepsilon$ again, we obtain

$$W_2\Big((f')_\#(f_\#\mu_1), (f')_\#(f_\#\mu_1')\Big) \le W_2\big(f_\#\mu_1, f_\#\mu_1'\big) + \sqrt{4\varepsilon\,(2\operatorname{diam}(\mathcal{X}_2) + \varepsilon)}. \qquad (28.10)$$

(The factor 4 is because f' is a (4ε)-isometry.) Since $f' \circ f$ is an admissible Monge transport between μ_1 and $(f' \circ f)_\#\mu_1$, or between μ_1' and $(f' \circ f)_\#\mu_1'$, which moves points by a distance at most 3ε, we have

$$W_2\Big((f' \circ f)_\#\mu_1, \mu_1\Big) \le 3\varepsilon; \qquad W_2\Big((f' \circ f)_\#\mu_1', \mu_1'\Big) \le 3\varepsilon.$$

Then by (28.10) and the triangle inequality,

$$\begin{aligned} W_2(\mu_1, \mu_1') &\le W_2\big(\mu_1, (f' \circ f)_\#\mu_1\big) + W_2\big((f' \circ f)_\#\mu_1, (f' \circ f)_\#\mu_1'\big) \\ &\qquad + W_2\big((f' \circ f)_\#\mu_1', \mu_1'\big) \\ &\le 3\varepsilon + W_2\big(f_\#\mu_1, f_\#\mu_1'\big) + \sqrt{4\varepsilon(2\operatorname{diam}(\mathcal{X}_2) + \varepsilon)} + 3\varepsilon. \end{aligned} \qquad (28.11)$$

Equations (28.9) and (28.11) together show that $f_\#$ distorts distances by at most $\widetilde{\varepsilon}$.

It remains to show that $f_\#$ is approximately surjective. To do this, pick up some $\mu_2 \in P_2(\mathcal{X}_2)$, and consider the Monge transport $f \circ f'$ from μ_2 to $(f \circ f')_\#\mu_2$. By (27.4), $f \circ f'$ moves points by a distance at most ε, so $W_2(\mu_2, f_\#(f'_\#\mu_2)) \le \varepsilon$. This concludes the proof that $f_\#$ is an $\widetilde{\varepsilon}$-isometry. □

Compactness of dynamical transference plans and related objects

The issue now is to show that dynamical transference plans enjoy good stability properties in a Gromov–Hausdorff approximation. The main technical difficulty comes from the fact that ε-isometries, being in general discontinuous, will not map geodesic paths into continuous paths. So we shall be led to work on the horribly large space of *measurable* paths $[0,1] \to \mathcal{X}$. I shall daringly embed this space in the even much larger space of probability measures on $[0,1] \times \mathcal{X}$, via the identification

$$\gamma \longmapsto \overline{\gamma} = (\mathrm{Id}\,, \gamma)_{\#}\lambda, \tag{28.12}$$

where λ is the Lebesgue measure on $[0,1]$. In loose notation,

$$\overline{\gamma}(dt\,dx) = \delta_{x=\gamma(t)}\,dt. \tag{28.13}$$

Of course, the first marginal of such a measure is always the Lebesgue measure. (That is, if $\tau : [0,1] \times \mathcal{X} \to [0,1]$ is the projection on the first factor then $\tau_{\#}\overline{\gamma} = \lambda$.) Moreover, the uniqueness of conditional measures shows that if $\overline{\gamma_1} = \overline{\gamma_2}$, then $\gamma_1 = \gamma_2$ λ-almost surely, and therefore actually $\gamma_1 = \gamma_2$ (because γ_1, γ_2 are continuous).

I shall think of the injection $i : \Gamma \to P([0,1]\times\mathcal{X})$ defined by $i(\gamma) = \overline{\gamma}$ as an "inclusion"; any $\Pi \in P(\Gamma)$ can be identified with its push-forward $i_{\#}\Pi \in P(P([0,1]\times\mathcal{X}))$. This point of view is reminiscent of the theory of Young measures; one of its advantages is that the space $P([0,1]\times\mathcal{X})$ is separable, while the space of measurable paths with values in $\mathcal{X}$ is not.

The next theorem expresses the stability of the main objects associated with transport (optimal or not). Recall that e_t stands for the evaluation at time t, and $\mathcal{L}(\gamma)$ for the length of the curve γ.

Theorem 28.9 (Optimal transport is stable under Gromov–Hausdorff convergence). *Let $(\mathcal{X}_k, d_k)_{k\in\mathbb{N}}$ and $(\mathcal{X}, d)$ be compact geodesic spaces such that $\mathcal{X}_k$ converges in the Gromov–Hausdorff topology as $k \to \infty$, by means of approximate isometries $f_k : \mathcal{X}_k \to \mathcal{X}$. For each $k \in \mathbb{N}$, let Π_k be a Borel probability measure on $\Gamma(\mathcal{X}_k)$; further, let $\pi_k = (e_0, e_1)_{\#}\Pi_k$, $\mu_{k,t} = (e_t)_{\#}\Pi_k$, and $\varepsilon_{k,t} = (e_t)_{\#}[(\mathcal{L}^2/2)\,\Pi_k]$. Then, after extraction of a subsequence, still denoted with the index k for simplicity, there is a dynamical transference plan Π on $\mathcal{X}$, with associated transference plan $\pi(dx\,dy)$, measure-valued path $(\mu_t(dx))_{0\le t\le 1}$, and kinetic energy $\varepsilon_t(dx)$, such that:*

(i) $\lim_{k\to\infty} (f_k\circ)_{\#}\Pi_k = \Pi$ *in the weak topology on* $P(P([0,1] \times \mathcal{X}))$*;*

(ii) $\lim_{k\to\infty} (f_k, f_k)_{\#}\pi_k = \pi$ *in the weak topology on* $P(\mathcal{X} \times \mathcal{X})$*;*

(iii) $\lim_{k\to\infty} (f_k)_{\#}\mu_{k,t} = \mu_t$ *in* $P_2(\mathcal{X})$ *uniformly in t; more explicitly,*

$$\lim_{k\to\infty} \sup_{t\in[0,1]} W_2(\mu_{k,t}, \mu_t) = 0;$$

(iv) $\lim_{k\to\infty} (f_k)_\# \varepsilon_{k,t} = \varepsilon_t$, *in the weak topology of measures, for each* $t \in (0,1)$.

Assume further that each Π_k *is an optimal dynamical transference plan, for the square distance cost function. Then:*

(v) For each $t \in (0,1)$, *there is a choice of the speed fields* $|v_k|$ *associated with the plans* Π_k, *such that* $\lim_{k\to\infty} |v_k| \circ f'_k = |v|$, *in the uniform topology;*

(vi) The limit Π *is an optimal dynamical transference plan, so* π *is an optimal transference plan and* $(\mu_t)_{0\le t\le 1}$ *is a displacement interpolation.*

Remark 28.10. In (i) $f_k\circ$ is the map $\gamma \to f_k \circ \gamma$, which maps continuous paths $[0,1] \to \mathcal{X}_k$ into measurable maps $[0,1] \to \mathcal{X}$ (identified to probability measures on $[0,1] \times \mathcal{X}$).

Proof of Theorem 28.9. The proof is quite technical, so the reader might skip it at first reading and go directly to the last section of this chapter. In a first step, I shall establish the compactness of the relevant objects, and in a second step pass to the limit.

It will be convenient to regularize rough paths with the help of some continuous mollifiers. For $\delta \in (0, 1/2)$, define

$$\varphi^\delta(s) = \frac{\delta + s}{\delta^2}\, 1_{-\delta\le s<0} + \frac{\delta - s}{\delta^2}\, 1_{0\le s\le\delta} \tag{28.14}$$

and

$$\varphi^\delta_+(s) = \varphi^\delta(s-\delta), \qquad \varphi^\delta_-(s) = \varphi^\delta(s+\delta). \tag{28.15}$$

Then $\operatorname{supp} \varphi^\delta_+ \subset [0, 2\delta]$ and $\operatorname{supp} \varphi^\delta_- \subset [-2\delta, 0]$. These functions have a graph that looks like a sharp "tent hat"; their integral (on the real line) is equal to 1, and as $\delta \to 0$ they converge in the weak topology to the Dirac mass δ_0 at the origin.

In the sequel, $f_k : \mathcal{X}_k \to \mathcal{X}$ will be an ε_k-isometry, and the sequence ε_k goes to 0 as $k \to \infty$.[1]

[1] In this proof ε_k and $\varepsilon_{k,t}$ stand for completely different objects, but I hope this is not confusing. (As usual the choice of notation is a delicate exercise.)

Step 1: Compactness. First, $[0,1]\times\mathcal{X}$ is a compact metric space, so the same holds true for $P([0,1]\times\mathcal{X})$ and $P(P([0,1]\times\mathcal{X}))$. Hence, after extraction of a subsequence, the sequence $((f_k\circ)_\#\Pi_k)_{k\in\mathbb{N}}$ converges to some $\Pi\in P(P([0,1]\times\mathcal{X}))$. Taking a further subsequence, we can assume that $\lim_{k\to\infty}(f_k,f_k)_\#\pi_k=\pi$ for some $\pi\in P(\mathcal{X}\times\mathcal{X})$.

Next, since $\mathcal{X}$ is bounded and $\mathcal{X}_k\to\mathcal{X}$, there is a uniform bound C on the diameters $\operatorname{diam}(\mathcal{X}_k)$. So the lengths of all geodesics $\gamma_k\in\Gamma(\mathcal{X}_k)$ are all bounded by C, and $d(\gamma_k(s),\gamma_k(t))\le C|s-t|$ for all times $s,t\in[0,1]$. It follows that $W_2(\mu_{k,s},\mu_{k,t})\le C|s-t|$, as $(e_s,e_t)_\#\Pi_k$ is a particular transference plan between $\mu_{k,s}$ and $\mu_{k,t}$. This shows that the paths $(\mu_{k,t})_{k\in\mathbb{N},\,t\in[0,1]}$ are uniformly continuous in t, with a uniform modulus of continuity. On the other hand, by Theorem 28.6, $(P_2(\mathcal{X}_k))_{k\in\mathbb{N}}$ converges in the Gromov–Hausdorff topology to $P_2(\mathcal{X})$. By Proposition 27.20, there is a subsequence (in k) of the family $((f_k)_\#\mu_{k,t})_{t\in[0,1],k\in\mathbb{N}}$ which converges uniformly to a continuous curve $(\mu_t)_{t\in[0,1]}\in C([0,1];P(\mathcal{X}))$.

Next, for each $t\in(0,1)$ and each $k\in\mathbb{N}$, the total mass of the measure $\varepsilon_{k,t}$ is bounded by $\operatorname{diam}(\mathcal{X}_k)^2/2\le C^2/2$; so the same holds true for the measures $(f_k)_\#\varepsilon_{k,t}$, which therefore constitute a precompact family in the space of nonnegative measures on $\mathcal{X}$. So up to extraction, we may assume that $(f_k)_\#\varepsilon_{k,t}$ converges weakly to some measure ε_t.

To conclude the proof of (i)–(iv), we should establish that
(a) Π is actually concentrated on $\Gamma(\mathcal{X})$;
(b) $\pi=(e_0,e_1)_\#\Pi$;
(c) $\mu_t=(e_t)_\#\Pi$;
(d) $\varepsilon_t=(e_t)_\#(\mathcal{L}^2\Pi)/2$.

Step 2: Embedding in probability measures and passing to the limit. I shall first mollify the condition "$\mathcal{L}(\gamma)=d(\gamma(0),\gamma(1))$" (which characterizes geodesics) in such a way that the resulting condition will pass to the limit under weak convergence of probability measures. Given $\delta\in(0,1/2)$, a continuous path $\gamma:[0,1]\to\mathcal{X}$, and times t_0,s_0 with $0\le t_0<s_0\le 1$, define

$$\mathcal{L}^\delta_{t_0\to s_0}(\gamma)=\int_0^1\int_0^1 d(\gamma(t),\gamma(s))\,\varphi^\delta_+(t-t_0)\,\varphi^\delta_-(s-s_0)\,dt\,ds.$$

(This is an approximation of the length of γ between times t_0 and s_0.) The function $\mathcal{L}^\delta_{t_0\to s_0}$ can be extended into a continuous function on $P([0,1]\times\mathcal{X})$, still denoted $\mathcal{L}^\delta_{t_0\to s_0}$ for simplicity:

$$\mathcal{L}^{\delta}_{t_0\to s_0}(\sigma)=\int_{[0,1]\times\mathcal{X}}\int_{[0,1]\times\mathcal{X}} d(x,y)\,\varphi^{\delta}_{+}(t-t_0)\,\varphi^{\delta}_{-}(s-s_0)\,d\sigma(t,x)\,d\sigma(s,y).$$

Since f_k is an ε_k-isometry, for any geodesic $\gamma_k \in \Gamma(\mathcal{X}_k)$,

$$\begin{aligned}
&\mathcal{L}^{\delta}_{t_0\to s_0}(f_k\circ\gamma_k)\\
&\quad=\int_0^1\int_0^1 \Big[d\big(\gamma_k(t),\gamma_k(s)\big)+O(\varepsilon_k)\Big]\,\varphi^{\delta}_{+}(t-t_0)\,\varphi^{\delta}_{-}(s-s_0)\,dt\,ds\\
&\quad=\int_0^1\int_0^1 d\big(\gamma_k(t),\gamma_k(s)\big)\,\varphi^{\delta}_{+}(t-t_0)\,\varphi^{\delta}_{-}(s-s_0)\,dt\,ds+O(\varepsilon_k)\\
&\quad=d\big(\gamma_k(0),\gamma_k(1)\big)\int_0^1\int_0^1 |s-t|\,\varphi^{\delta}_{+}(t-t_0)\,\varphi^{\delta}_{-}(s-s_0)\,dt\,ds+O(\varepsilon_k)\\
&\quad=d\big(\gamma_k(0),\gamma_k(1)\big)\Big(|s_0-t_0|+O(\delta)\Big)\;+O(\varepsilon_k).
\end{aligned}$$

In particular, taking $t_0=0$ and $s_0=1$ gives

$$\mathcal{L}^{\delta}_{0\to 1}(f_k\circ\gamma_k)=d\big(\gamma_k(0),\gamma_k(1)\big)\,(1+O(\delta))\;+O(\varepsilon_k).$$

Since all of the lengths $d(\gamma_k(0),\gamma_k(1))$ are uniformly bounded, we conclude that there is a constant C such that for all t_0 and s_0 as above,

$$\begin{cases}\Big|\mathcal{L}^{\delta}_{t_0\to s_0}(f_k\circ\gamma_k)-|s_0-t_0|\,\mathcal{L}^{\delta}_{0\to 1}(f_k\circ\gamma_k)\Big|\;\le\;C(\delta+\varepsilon_k);\\[2ex]
\mathcal{L}^{\delta}_{0\to 1}(f_k\circ\gamma_k)\le C.\end{cases}$$

Now for $\varepsilon,\delta>0$, define

$$\begin{aligned}\Gamma_{\varepsilon,\delta}(\mathcal{X})=\Big\{\sigma\in P([0,1]\times\mathcal{X});\quad \tau_{\#}\sigma=\lambda;\\
\Big|\mathcal{L}^{\delta}_{t_0\to s_0}(\sigma)-|s_0-t_0|\,\mathcal{L}^{\delta}_{0\to 1}(\sigma)\Big|\le C(\delta+\varepsilon);\quad \mathcal{L}^{\delta}_{0\to 1}(\sigma)\le C\Big\}.\end{aligned}\tag{28.16}$$

It is easy to see that $\Gamma_{\varepsilon,\delta}(\mathcal{X})$ is closed in $P([0,1]\times\mathcal{X})$. Moreover, for k large enough one has $\varepsilon_k\le\varepsilon$ and then $i(f_k\circ\gamma_k)\in\Gamma_{\varepsilon,\delta}(\mathcal{X})$ for any geodesic $\gamma_k\in\Gamma(\mathcal{X}_k)$. It follows that $(f_k\circ)_{\#}\Pi_k\in P(\Gamma_{\varepsilon,\delta}(\mathcal{X}))$ for k large enough; by passing to the limit, also $\Pi\in P(\Gamma_{\varepsilon,\delta}(\mathcal{X}))$. Since ε,δ are arbitrarily small,

$$\Pi\in P\Bigg(\bigcap_{\varepsilon,\delta>0}\Gamma_{\varepsilon,\delta}(\mathcal{X})\Bigg).$$

So to conclude the proof of (a) it suffices to prove

$$\bigcap_{\varepsilon,\delta>0} \Gamma_{\varepsilon,\delta}(\mathcal{X}) = \Gamma(\mathcal{X}).$$

Let $\sigma \in \bigcap_{\varepsilon,\delta>0} \Gamma_{\varepsilon,\delta}(\mathcal{X})$. Taking $\varepsilon \to 0$ in (28.16), we get

$$\left| \mathcal{L}^{\delta}_{t_0\to s_0}(\sigma) - |s_0 - t_0|\, \mathcal{L}^{\delta}_{0\to 1}(\sigma) \right| \leq \delta. \tag{28.17}$$

In particular,

$$\mathcal{L}^{\delta}_{t_0\to s_0}(\sigma) \leq C(|s_0 - t_0| + \delta). \tag{28.18}$$

In Lemma 28.11 below it will be shown that, as a consequence of (28.18), σ can be written as $(\mathrm{Id}\,,\gamma)_{\#}\lambda$ for some Lipschitz-continuous curve $\gamma : [0,1] \to \mathcal{X}$. Once that is known, the end of the proof of (a) is straightforward: Since

$$\mathcal{L}^{\delta}_{t_0\to s_0}(\sigma) = d\big(\gamma(t_0), \gamma(s_0)\big) + O(\delta),$$

the inequality (28.17) becomes, in the limit $\delta \to 0$,

$$d\big(\gamma(t_0), \gamma(s_0)\big) = |s_0 - t_0|\, d\big(\gamma(0), \gamma(1)\big).$$

This implies that $\mathcal{L}(\gamma) = d(\gamma(0), \gamma(1))$, so γ is a geodesic curve. This concludes the proof of (a), and of part (i) of Theorem 28.9 at the same time.

Now I shall use a similar reasoning for the convergence of the marginals of Π. Given $\Phi \in C(\mathcal{X} \times \mathcal{X})$ and $\gamma \in \Gamma(\mathcal{X})$, define

$$\Phi^{\delta}(\gamma) = \int_0^1 \int_0^1 \Phi\big(\gamma(t), \gamma(s)\big)\, \varphi^{\delta}_{+}(t)\, \varphi^{\delta}_{-}(s-1)\, dt\, ds.$$

As before, this extends to a continuous function on $P([0,1] \times \mathcal{X})$ by

$$\Phi^{\delta}(\sigma) = \int_{[0,1]\times\mathcal{X}} \int_{[0,1]\times\mathcal{X}} \Phi(x,y)\, \varphi^{\delta}_{+}(t)\, \varphi^{\delta}_{-}(s-1)\, d\sigma(t,x)\, d\sigma(s,y).$$

By part (i) of the theorem,

$$\int_{\Gamma(\mathcal{X}_k)} \Phi^{\delta}(f_k \circ \gamma_k)\, d\Pi_k(\gamma_k) \longrightarrow \int_{\Gamma(\mathcal{X})} \Phi^{\delta}(\gamma)\, d\Pi(\gamma). \tag{28.19}$$

Let us examine the behavior of both sides of (28.19) as $\delta \to 0$. If γ is a geodesic on $\mathcal{X}$, the continuity of Φ and γ implies that $\Phi^\delta(\gamma) \longrightarrow \Phi(\gamma(0), \gamma(1))$ as $\delta \to 0$. Then by dominated convergence,

$$\int_{\Gamma(\mathcal{X})} \Phi^\delta(\gamma)\, d\Pi(\gamma) \longrightarrow \int_{\Gamma(\mathcal{X})} \Phi\big(\gamma(0), \gamma(1)\big)\, d\Pi(\gamma)$$
$$= \int \Phi\, d(e_0, e_1)_\# \Pi = \int \Phi\, d\pi. \tag{28.20}$$

As for the left-hand side of (28.19), things are not so immediate because $f_k \circ \gamma_k$ may be discontinuous. However, for $0 \le t \le 2\delta$ one has

$$d\big(f_k(\gamma_k(0)), f_k(\gamma_k(t))\big) = d_k\big(\gamma_k(0), \gamma_k(t)\big) + O(\varepsilon_k) = O(\delta + \varepsilon_k),$$

where the implicit constant in the right-hand side is independent of γ_k. Similarly, for $1 - 2\delta \le s \le 1$, one has

$$d\big(f_k(\gamma_k(s)), f_k(\gamma_k(1))\big) = O(\delta + \varepsilon_k).$$

Then it follows from the uniform continuity of Φ that

$$\sup \left| \Phi\Big(f_k(\gamma_k(t)),\, f_k(\gamma_k(s))\Big) - \Phi\Big(f_k(\gamma_k(0)),\, f_k(\gamma_k(1))\Big) \right| \longrightarrow 0$$

as $\delta \to 0$ and $k \to \infty$, where the supremum is over all $\gamma_k \in \Gamma(\mathcal{X}_k)$, $t \in [0, 2\delta]$ and $s \in [1-2\delta, 1]$. So in this limit, the left-hand side of (28.19) is well approximated by

$$\int_{\Gamma(\mathcal{X}_k)} \Phi\big(f_k(\gamma_k(0)), f_k(\gamma_k(1))\big)\, d\Pi_k(\gamma_k)$$
$$= \int_{\mathcal{X}\times\mathcal{X}} \Phi\, d\Big[(f_k, f_k)_\# \big((e_0, e_1)_\# \Pi_k\big)\Big]$$
$$= \int_{\mathcal{X}\times\mathcal{X}} \Phi\, d(f_k, f_k)_\# \pi_k. \tag{28.21}$$

The comparison of (28.19), (28.20) and (28.21) shows that $(f_k, f_k)_\# \pi_k$ converges to π, which concludes the proof of (b).

As for (c) we just have to show that $\lim_{k\to\infty} (f_k)_\# \mu_{k,t_0} = \mu_{t_0}$ for all $t_0 \in [0, 1]$. The argument is quite similar to the proof of (b). Assume for example that $t_0 < 1$. Given $\Phi \in C(\mathcal{X})$, define

$$\Phi^\delta_{t_0}(\gamma) = \int_0^1 \Phi(\gamma(t))\, \varphi^\delta_+(t - t_0)\, dt.$$

This extends to a continuous function on $P([0,1]\times\mathcal{X})$, so by (i),

$$\int_{\Gamma(\mathcal{X}_k)} \Phi^\delta_{t_0}(f_k\circ\gamma_k)\,d\Pi_k(\gamma_k) \longrightarrow \int_{\Gamma(\mathcal{X})} \Phi^\delta_{t_0}(\gamma)\,d\Pi(\gamma).$$

The right-hand side converges to $\int_{\mathcal{X}}\Phi(x)\,d\mu_{t_0}(x)$ as $\delta\to 0$, while the left-hand side is well approximated by $\int_{\mathcal{X}}\Phi(f_k(x))\,d\mu_{k,t_0}(x)$. The conclusion follows.

The proof of (d) is obtained by a similar reasoning.

Let us finally turn to the proof of statements (v) and (vi) in the theorem. In the sequel, it will be assumed that each Π_k is an *optimal* dynamical optimal transference plan. In view of Theorem 28.5, for each $t\in(0,1)$, the speed fields $|v_{k,t}|$ can be chosen in such a way that they satisfy a uniform Hölder-1/2 estimate. Then the precompactness of $|v_{k,t}|$ results from Ascoli's theorem, in the form of Proposition 27.20. So up to extraction, we may assume that $|v_{k,t}|\circ f'_k$ converges uniformly to some function $|v_t|$. It remains to show that $|v_t|^2/2$ is an admissible density for ε, at each time $t\in(0,1)$. For simplicity I shall omit the time variable, so t is implicit and fixed in $(0,1)$. Since there is a uniform bound on the diameter of the spaces $\mathcal{X}_k$ (and therefore on $|v_k|$), the function $|v_k|^2\circ f'_k$ converges uniformly to $|v|^2$. By uniform continuity of $|v_k|^2$, the difference between $|v_k|^2$ and $|v_k|^2\circ(f'_k\circ f_k)$ is bounded by $\eta(k)$, where $\eta(k)\to 0$ as $k\to\infty$. After going back to the definitions of push-forward and weak convergence, it follows that

$$\lim_{k\to\infty}\left(|v_k|^2\circ f'_k\right)(f_k)_\#\mu_k = \lim_{k\to\infty}(f_k)_\#\left(|v_k|^2\,\mu_k\right) = 2\lim_{k\to\infty}(f_k)_\#\varepsilon_k = 2\,\varepsilon. \tag{28.22}$$

Since $|v_k|^2\circ f'_k$ converges *uniformly* to $|v|^2$, and $(f_k)_\#\mu_k$ converges weakly to μ, the left-hand side in (28.22) converges weakly to $|v|^2\,\mu$. This implies that $|v|^2/2$ is an admissible density for the kinetic energy ε, and concludes the proof of (v).

The proof of (vi) is easy now. Since $\pi=\lim(f_k,f_k)_\#\pi_k$ and f_k is an approximate isometry,

$$\begin{aligned}\int d(x_0,x_1)^2\,d\pi(x_0,x_1) &= \lim_{k\to\infty}\int d(f_k(x_0),f_k(x_1))^2\,d\pi_k(x_0,x_1)\\ &= \lim_{k\to\infty}\int d_k(x_0,x_1)^2\,d\pi_k(x_0,x_1).\end{aligned} \tag{28.23}$$

By assumption, π_k is optimal for each k, so

$$\int d_k(x_0,x_1)^2\,d\pi_k(x_0,x_1) = W_2(\mu_{0,k},\mu_{1,k})^2. \tag{28.24}$$

By Theorem 28.6, $(f_k)_\#$ is an approximate isometry $P_2(\mathcal{X}_k) \to P_2(\mathcal{X})$, so

$$\lim_{k\to\infty} W_2(\mu_{0,k},\mu_{1,k})^2 = \lim_{k\to\infty} W_2\big((f_k)_\#\mu_{0,k},(f_k)_\#\mu_{1,k}\big)^2 = W_2(\mu_0,\mu_1)^2, \tag{28.25}$$

where the latter limit follows from the continuity of W_2 under weak convergence (Corollary 6.11). By combining (28.23), (28.24) and (28.25), we deduce that $\int d(x_0,x_1)^2\,d\pi(x_0,x_1) = W_2(\mu_0,\mu_1)^2$, so π is an optimal transference plan. Thus Π is an optimal dynamical transference plan, and the proof of (vi) is complete.

(**Note:** Since $(\mu_{k,t})_{0\le t\le 1}$ is a geodesic path in $P_2(\mathcal{X}_k)$ (recall Corollary7.22), and $(f_k)_\#$ is an approximate isometry $P_2(\mathcal{X}_k) \to P_2(\mathcal{X})$, Theorem 27.9 implies directly that the limit path $\mu_t = (f_k)_\#\mu_{k,t}$ is a geodesic in $P_2(\mathcal{X})$; however, I am not sure that the whole theorem can be proven by this approach.) □

To complete the proof of Theorem 28.9, it only remains to establish the next lemma, which was used in the proof of statement (a).

Lemma 28.11. *Let $(\mathcal{X},d)$ be a compact geodesic space. Let σ be a probability measure on $[0,1]\times\mathcal{X}$ satisfying (28.18). Then there is a Lipschitz curve $\gamma:[0,1]\to\mathcal{X}$ such that $\sigma(dt\,dx) = \overline{\gamma}(dt\,dx) = \delta_{x=\gamma(t)}\,dt$.*

Proof of Lemma 28.11. First disintegrate σ with respect to its first marginal λ: There is a family $(\nu_t)_{0\le t\le 1}$, measurable as a map from $[0,1]$ to $P(\mathcal{X})$ and unique up to modification on a set of zero Lebesgue measure in $[0,1]$, such that

$$\sigma(dt\,dx) = \nu_t(dx)\,dt.$$

The goal is to show that, up to modification of ν_t on a negligible set of times t,

$$\nu_t(dx) = \delta_{x=\gamma(t)},$$

where γ is Lipschitz.

The argument will be divided into three steps. It will be convenient to use W_1, the 1-Wasserstein distance.

Step 1: almost-everywhere Lipschitz continuity. Let β be an arbitrary nonnegative continuous function on $[0,1] \times [0,1]$. Integrating (28.18) with respect to β yields

$$\int_0^1 \int_0^1 \beta(t_0, s_0)\, \mathcal{L}^{\delta}_{t_0 \to s_0}(\sigma)\, dt_0\, ds_0 \leq C \int_0^1 \int_0^1 \beta(t_0, s_0)\, \big(|s_0 - t_0| + \delta\big)\, dt_0\, ds_0. \tag{28.26}$$

The left-hand side of (28.26) can be written as

$$\int \varphi^{\delta}_{+}(t - t_0)\, \varphi^{\delta}_{-}(s - s_0)\, d\nu_t(x)\, dt\, d\nu_s(y)\, ds\, dt_0\, ds_0 = \int_0^1 \int_0^1 F^{\delta}(t, s)\, \Lambda(t, s)\, dt\, ds, \tag{28.27}$$

where

$$\begin{cases} F^{\delta}(t, s) = \displaystyle\int_0^1 \int_0^1 \beta(t_0, s_0)\, \varphi^{\delta}_{+}(t - t_0)\, \varphi^{\delta}_{-}(s - s_0)\, dt_0\, ds_0 \\ \Lambda(t, s) = \displaystyle\int_{\mathcal{X} \times \mathcal{X}} d(x, y)\, d\nu_t(x)\, d\nu_s(y). \end{cases}$$

Since $F^{\delta}(t, s)$ converges to $\beta(t, s)$ in $C([0,1] \times [0,1])$ as $\delta \to 0$, the expression in (28.27) converges to

$$\int_{\mathcal{X} \times [0,1]} \int_{\mathcal{X} \times [0,1]} \beta(t, s)\, d(x, y)\, d\nu_t(x)\, dt\, d\nu_s(y)\, ds.$$

Now plug this back into (28.26) and let $\delta \to 0$ to conclude that

$$\int_{\mathcal{X} \times [0,1]} \int_{\mathcal{X} \times [0,1]} \beta(t, s)\, d(x, y)\, d\nu_t(x)\, dt\, d\nu_s(y)\, ds \leq C \int_0^1 \int_0^1 \beta(t, s)\, |s - t|\, dt\, ds.$$

As β is arbitrary, we actually have

$$\int_{\mathcal{X} \times \mathcal{X}} d(x, y)\, \nu_t(dx)\, \nu_s(dy) \leq C\, |t - s|$$

for $(\lambda \otimes \lambda)$-almost all (t, s) in $[0,1] \times [0,1]$. In particular,

$$W_1(\nu_t, \nu_s) \leq C\, |t - s| \qquad \text{for almost all } (t, s) \in [0,1] \times [0,1]. \tag{28.28}$$

Step 2: true Lipschitz continuity. Now let us show that ν_t can be modified on a negligible set of times t so that (28.28) holds for *all* $(t,s) \in [0,1] \times [0,1]$.

For small $\varepsilon > 0$ and $t \in [\varepsilon, 1-\varepsilon]$, define

$$\nu_t^\varepsilon = \frac{1}{2\varepsilon} \int_{-\varepsilon}^{\varepsilon} \nu_{t+\tau}\, d\tau. \tag{28.29}$$

Then by Theorem 4.8,

$$W_1(\nu_t^\varepsilon, \nu_s^\varepsilon) \le \frac{1}{2\varepsilon} \int_{-\varepsilon}^{\varepsilon} W_1(\nu_{t+\tau}, \nu_{s+\tau})\, d\tau \le C|t-s| + O(\varepsilon). \tag{28.30}$$

Next, let $(\psi_k)_{k\in\mathbb{N}}$ be a countable dense subset of $C(\mathcal{X})$. For any k,

$$\int_{\mathcal{X}} \psi_k\, d\nu_t^\varepsilon = \frac{1}{2\varepsilon} \int_{t-\varepsilon}^{t+\varepsilon} \left(\int_{\mathcal{X}} \psi_k(x)\, d\nu_\tau(x) \right) d\tau. \tag{28.31}$$

Since the expression inside parentheses is a bounded measurable function of τ, Lebesgue's density theorem ensures that as $\varepsilon \to 0$, the right-hand side of (28.31) converges to $\int_{\mathcal{X}} \psi_k\, d\nu_t$ for almost all t. So there is a negligible subset of $[0,1]$, say N_k, such that if $t \notin N_k$ then

$$\lim_{\varepsilon\to 0} \int_{\mathcal{X}} \psi_k\, d\nu_t^\varepsilon = \int_{\mathcal{X}} \psi_k\, d\nu_t. \tag{28.32}$$

Let $N = \bigcup_{k=1}^\infty N_k$; this is a negligible subset of $[0,1]$. For all $t \notin N$, equation (28.32) holds for all k. This proves that $\lim_{\varepsilon\to 0} \nu_t^\varepsilon = \nu_t$ in the weak topology, for almost all t.

Now for arbitrary $t \in (0,1)$, there is a sequence of times $t_j \to t$, such that $\nu_{t_j}^\varepsilon$ converges to ν_{t_j} as $\varepsilon \to 0$. Then for ε and ε' sufficiently small,

$$\begin{aligned} W_1(\nu_t^\varepsilon, \nu_t^{\varepsilon'}) &\le W_1(\nu_{t_j}^\varepsilon, \nu_{t_j}^{\varepsilon'}) + 2C\,|t-t_j| \\ &\le W_1(\nu_{t_j}^\varepsilon, \nu_{t_j}) + W_1(\nu_{t_j}^{\varepsilon'}, \nu_{t_j}) + 2C\,|t-t_j|. \end{aligned} \tag{28.33}$$

It follows that $\lim_{\varepsilon,\varepsilon'\to 0} W_1(\nu_t^\varepsilon, \nu_t^{\varepsilon'}) = 0$. Since $(P(\mathcal{X}), W_1)$ is a complete metric space (Theorem 6.18), in fact $\lim_{\varepsilon\to 0} \nu_t^\varepsilon$ exists for *all* (not just almost all) $t \in (0,1)$. The limit coincides with ν_t for almost all $t \in (0,1)$, so it defines the same measure $\sigma(dt\,dx)$. Redefine ν_t on a negligible set of times if necessary, so that the limit is ν_t for all $t \in (0,1)$. It is possible to pass to the limit in (28.30) as $\varepsilon \to 0$, and recover $W_1(\nu_t, \nu_s) \le C|t-s|$ for all $t, s \in (0,1)$. Of course this extends to $t, s \in [0,1]$ by continuity.

Step 3: Conclusion. From the previous step, $W_1(\nu_t, \nu_{t_0}) \le C\delta$ if $|t - t_0| \le \delta$. It results from the definition of $\mathcal{L}^\delta_{t_0 \to s_0}$ that

$$\begin{aligned}
&\mathcal{L}^\delta_{t_0 \to s_0}(\sigma) \\
&= \int_{\mathcal{X}\times[0,1]} \int_{\mathcal{X}\times[0,1]} d(x,y)\, \varphi^\delta_+(t-t_0)\, \varphi^\delta_-(s-s_0)\, d\nu_{t_0}(x)\, dt\, d\nu_{s_0}(y)\, ds \\
&\qquad\qquad\qquad\qquad\qquad\qquad\qquad\qquad + O(\delta) \\
&= \int_{\mathcal{X}\times\mathcal{X}} d(x,y)\, d\nu_{t_0}(x)\, d\nu_{s_0}(y) \; + O(\delta).
\end{aligned}$$

Plugging this back into (28.18) and taking $\delta \to 0$, we obtain

$$\int_{\mathcal{X}\times\mathcal{X}} d(x,y)\, d\nu_{t_0}(x)\, d\nu_{s_0}(y) \le C\,|s_0 - t_0|.$$

This holds for all t_0 and s_0, so we can choose $s_0 = t_0$ and obtain

$$\int_{\mathcal{X}\times\mathcal{X}} d(x,y)\, d\nu_{t_0}(x)\, d\nu_{t_0}(y) = 0.$$

This is possible only if ν_{t_0} is a Dirac measure. Hence for any $t_0 \in [0,1]$ there is $\gamma(t_0) \in \mathcal{X}$ such that $\nu_{t_0} = \delta_{\gamma(t_0)}$. Then $d(\gamma(t), \gamma(s)) = W_1(\nu_t, \nu_s) \le C|t-s|$, so γ is Lipschitz continuous. This concludes the proof of Lemma 28.11. □

Exercise 28.12 (Gromov–Hausdorff stability of the dual Kantorovich problem). With the same notation as in Theorem 28.9, let ψ_k be an optimal $d_k^2/2$-convex function in the dual Kantorovich problem between $\mu_{k,0}$ and $\mu_{k,1}$. Show that up to extraction of a subsequence, there are constants $c_k \in \mathbb{R}$ such that $\psi_k \circ f'_k + c_k$ converges uniformly to some $d^2/2$-convex function ψ which is optimal in the dual Kantorovich problem between μ_0 and μ_1.

Noncompact spaces

It will be easy to extend the preceding results to noncompact spaces, by localization. These generalizations will not be needed in the sequel, so I shall not be very precise in the proofs.

Theorem 28.13 (Pointed convergence of $\mathcal{X}_k$ implies local convergence of $P_2(\mathcal{X}_k)$). *Let $(\mathcal{X}_k, d_k, \star_k)$ be a sequence of locally compact geodesic Polish spaces converging in the pointed Gromov–Hausdorff topology to some locally compact Polish space $(\mathcal{X}, d, \star)$. Then $P_2(\mathcal{X}_k)$ converges to $P_2(\mathcal{X})$ in the geodesic local Gromov–Hausdorff topology.*

Remark 28.14. If a basepoint $\star$ is given in $\mathcal{X}$, there is a natural choice of basepoint for $P_2(\mathcal{X})$, namely $\delta_\star$. However, $P_2(\mathcal{X})$ is in general *not* locally compact, and it does not make sense to consider the *pointed* convergence of $P_2(\mathcal{X}_k)$ to $P_2(\mathcal{X})$.

Remark 28.15. Theorem 28.13 admits the following extension: If $(\mathcal{X}_k, d_k)$ converges to $(\mathcal{X}, d)$ in the geodesic local Gromov–Hausdorff topology, then also $P_2(\mathcal{X}_k)$ converges to $P_2(\mathcal{X})$ in the geodesic local Gromov–Hausdorff topology. The proof is almost the same and is left to the reader.

Proof of Theorem 28.13. Let $R_\ell \to \infty$ be a given increasing sequence of positive numbers. Define

$$K^{(\ell)} = P_2\big(B_{R_\ell]}(\star)\big) \subset P_2(\mathcal{X}),$$

$$K_k^{(\ell)} = P_2\big(B_{R_\ell]}(\star_k)\big) \subset P_2(\mathcal{X}_k),$$

where the inclusion is understood in an obvious way (a probability measure on a subset of $\mathcal{X}$ can be seen as the restriction of a probability measure on $\mathcal{X}$). Since $B_{R_\ell]}(\star)$ is a compact set, $K^{(\ell)}$ is compact too, and so is $K_k^{(\ell)}$, for each k and each ℓ. Moreover, the union of all K_k is dense in $P_2(\mathcal{X})$, as a corollary of Theorem 6.18.

For each ℓ, there is a sequence $(f_k)_{k\in\mathbb{N}}$ such that each f_k is an ε_k-isometry $B_{R_\ell]}(\star_k) \to B_{R_\ell]}(\star)$, where $\varepsilon_k \to 0$. From Proposition 28.7, $(f_k)_\#$ is an $\widetilde{\varepsilon}_{k,\ell}$-isometry $K_k^{(\ell)} \to K^{(\ell)}$, with

$$\widetilde{\varepsilon}_{k,\ell} \leq 8\,(\varepsilon_k + \sqrt{2R_\ell\varepsilon_k}),$$

which goes to 0 as $k \to \infty$. So all the requirements of Definition 27.11 are satisfied, and $P_2(\mathcal{X}_k)$ does converge to $P_2(\mathcal{X})$ in the local Gromov–Hausdorff topology.

To check condition (iii) appearing in Definition 27.12, it suffices to recall that any geodesic in $P_2(B_{R_\ell]}(\star_k))$ can be written as the law of a random geodesic joining points in $B_{R_\ell]}(\star_k)$; such a geodesic has its image contained in $B_{2R_\ell]}(\star_k)$, so we just need to choose ℓ' large enough that $R_{\ell'} \geq 2R_\ell$. $\square$

Exercise 28.16. Write down an analog of Theorem 28.9 for noncompact metric spaces, replacing Gromov–Hausdorff convergence of $\mathcal{X}_k$ by pointed Gromov–Hausdorff convergence, and using an appropriate "tightness" condition. *Hint:* Recall that if K is a given compact then the set of geodesics whose endpoints lie in K is itself compact in the topology of uniform convergence.

Bibliographical notes

Theorem 28.6 is taken from [577, Section 4], while Theorem 28.13 is an adaptation of [577, Appendix E]. Theorem 28.9 is new. (A part of this theorem was included in a preliminary version of [578], and later removed from that reference.)

The discussion about push-forwarding dynamical transference plans is somewhat subtle. The point of view adopted in this chapter is the following: when an approximate isometry f is given between two spaces, use it to push-forward a dynamical transference plan Π, via $(f\circ)_\#\Pi$. The advantage is that this is the same map that will push-forward the measure and the dynamical plan. The drawback is that the resulting object $(f\circ)_\#\Pi$ is **not** a dynamical transference plan, in fact it may not even be supported on continuous paths. This leads to the kind of technical sport that we've encountered in this chapter, embedding into probability measures on probability measures and so on.

Another option would be as follows: Given two spaces $\mathcal{X}$ and $\mathcal{Y}$, with an approximate isometry $f : \mathcal{X} \to \mathcal{Y}$, and a dynamical transference plan Π on $\Gamma(\mathcal{X})$, define a **true** dynamical transference plan on $\Gamma(\mathcal{Y})$, which is a good approximation of $(f\circ)_\#\Pi$. The point is to construct a recipe which to any geodesic γ in $\mathcal{X}$ associates a geodesic $S(\gamma)$ in $\mathcal{Y}$ that is "close enough" to $f \circ \gamma$. This strategy was successfully implemented in the final version of [578, Appendix]; it is much simpler, and still quite sufficient for some purposes. The example treated in [578] is the stability of the "democratic condition" considered by Lott and myself; but certainly this simplified version will work for many other stability issues. On the other hand, I don't know if it is enough to treat such topics as the stability of general weak Ricci bounds, which will be considered in the next chapter.

The study of the kinetic energy measure and the speed field occurred to me during a parental meeting of the Crèche Le Rêve en Couleurs in Lyon. My motivations for regularity estimates on the speed are explained in the bibliographical notes of Chapter 29, and come from a direction of research which I have more or less left aside for the moment. I also used it in the tedious proof of Theorem 23.14.

The last section of [822] contains a solution of Exercise 28.12, together with a statement of upper semicontinuity of $d^2/2$-subdifferentials under Gromov–Hausdorff approximation.

29

Weak Ricci curvature bounds I: Definition and Stability

In Chapter 14 I discussed several reformulations of the CD(K,N) curvature-dimension bound for a smooth manifold (M,g) equipped with a reference measure ν whose density (with respect to the volume element) is smooth. For instance, here is a possible reformulation of CD(K,N) for $N<\infty$: For any C^2 function $\psi : M \to \mathbb{R}$, let $\mathcal{J}(t,\cdot)$ be the Jacobian determinant of $T_t : x \longmapsto \exp_x(t\nabla\psi(x))$, and let $\mathcal{D}(t,x) = \mathcal{J}(t,x)^{\frac{1}{N}}$; then, with the notation of Theorem 14.11,

$$\mathcal{D}(t,x) \geq \tau_{K,N}^{(1-t)}\, \mathcal{D}(0,x) + \tau_{K,N}^{(t)}\, \mathcal{D}(1,x). \tag{29.1}$$

How to generalize this definition in such a way that it would make sense in a possibly nonsmooth metric-measure space? This is definitely not obvious since (i) there might be no good notion of gradient, and (ii) there might be no good notion of exponential map either.

There are many definitions that one may try, but so far the only approach that yields acceptable results is the one based on displacement convexity. Recall from Chapters 16 and 17 two displacement convexity inequalities that characterize CD(K,N): Let μ_0 and μ_1 be two compactly supported (for simplification) absolutely continuous probability measures, let $\pi = (\mathrm{Id}\,, \exp\nabla\psi)_{\#}\mu_0$ be the optimal coupling of (μ_0,μ_1), let $(\rho_t\,\nu)_{0\leq t\leq 1}$ be the displacement interpolation between $\mu_0 = \rho_0\,\nu$ and $\mu_1 = \rho_1\,\nu$; let $(v_t)_{0\leq t\leq 1}$ be the associated velocity field ($v_t = \widetilde{\nabla}\psi_t$ in the notation of Remark 17.16); then for any $U \in \mathcal{DC}_N$, $t\in[0,1]$,

$$\int U(\rho_t)\,d\nu \leq (1-t)\,U_\nu(\mu_0) + t\,U_\nu(\mu_1) \\ - K_{N,U}\int_0^1 \rho_s(x)^{1-\frac{1}{N}}|v_s(x)|^2\,G(s,t)\,ds, \tag{29.2}$$

and

$$\int U(\rho_t)\,d\nu \;\le\; (1-t)\int_{M\times M} U\left(\frac{\rho_0(x_0)}{\beta_{1-t}^{(K,N)}(x_0,x_1)}\right)\beta_{1-t}^{(K,N)}(x_0,x_1)\,\pi(dx_1|x_0)\,\nu(dx_0)$$
$$+\,t\int_{M\times M} U\left(\frac{\rho_1(x_1)}{\beta_t^{(K,N)}(x_0,x_1)}\right)\beta_t^{(K,N)}(x_0,x_1)\,\pi(dx_0|x_1)\,\nu(dx_1). \tag{29.3}$$

Here $G(s,t)$ is the one-dimensional Green function of (16.6), $K_{N,U}$ is defined by (17.10), and the distortion coefficients $\beta_t^{(K,N)} = \beta_t$ are those appearing in (14.61).

Which of these formulas should we choose for the extension to nonsmooth spaces? When $K = 0$, both inequalities reduce to just

$$\int U(\rho_t)\,d\nu \le (1-t)\int U(\rho_0)\,d\nu + t\int U(\rho_1)\,d\nu. \tag{29.4}$$

In the case $N < \infty$, formula (29.3) is much more convenient to establish functional inequalities; while in the case $N = \infty$ it is formula (29.2) which is easier to use. However, it will turn out that in the case $N = \infty$, (29.2) is an immediate consequence of (29.3). All this concurs to suggest that (29.3) is the correct choice on which we should base the general definition.

Now we would like to adapt these formulas to a nonsmooth context. This looks simpler than working with (29.1), but there are still a few issues to take into account.

(i) **First issue: Nonuniqueness of the displacement interpolation.** There is a priori no reason to expect uniqueness of the displacement interpolation in a nonsmooth context. We may require the distorted displacement convexity (29.3) along every displacement interpolation, i.e. every geodesic in Wasserstein space; but this is not a good idea for stability issues. (Recall Example 27.17: in general the geodesics in the limit space cannot be realized as limits of geodesics.) Instead, we shall only impose a **weak displacement convexity** property: For any μ_0 and μ_1 there should be *some* geodesic $(\mu_t)_{0\le t\le 1}$ along which inequality (29.3) holds true.

To appreciate the difference between "convexity" and "weak convexity", note the following: If F is a function defined on a geodesic space $\mathcal{X}$, then the two statements "F is convex along each geodesic $(\gamma_t)_{0\le t\le 1}$" and "For any x_0 and x_1, there is a geodesic $(\gamma_t)_{0\le t\le 1}$ joining x_0 to x_1, such that $F(\gamma_t) \le (1-t)\,F(\gamma_0) + t\,F(\gamma_1)$" are not equivalent in general. (They become equivalent under some regularity assumption on $\mathcal{X}$, for instance if any two close enough points in $\mathcal{X}$ are joined by a unique geodesic.)

(ii) **Second issue: Treatment of the singular part.** Even if μ_0 and μ_1 are absolutely continuous with respect to ν, there is no guarantee that the Wasserstein interpolant μ_t will also be absolutely continuous. Also for stability issues it will be useful to work with singular measures, since $P_2^{\mathrm{ac}}(\mathcal{X},\nu)$ is not closed under weak convergence.

Then the problem arises to devise a "correct" definition for the integral functionals of the density which appear in the displacement convexity inequalities, namely

$$\begin{cases} U_\nu(\mu) = \displaystyle\int_{\mathcal{X}} U\left(\frac{d\mu}{d\nu}\right) d\nu \\[2ex] U_{\pi,\nu}^{\beta}(\mu) = \displaystyle\int_{\mathcal{X}\times\mathcal{X}} U\left(\frac{1}{\beta(x,y)}\,\frac{d\mu}{d\nu}(x)\right) \beta(x,y)\,\pi(dy|x)\,\nu(dx), \end{cases}$$

when μ is not absolutely continuous with respect to ν. It would be a mistake to keep the same definition and replace $d\mu/d\nu$ by the density of the absolutely continuous part of μ with respect to ν. In fact there is only one natural extension of the functionals U and $U_{\pi,\nu}^{\beta}$; before giving it explicitly, I shall try to motivate it. Think of the singular part of μ as something which "always has infinite density". Assume that the respective contributions of finite and infinite values of the density decouple, so that one would define separately the contributions of the absolutely continuous part μ_{ac} and of the singular part μ_s. Only the asymptotic behavior of $U(r)$ as $r\to\infty$ should count when one defines the contribution of μ_s. Finally, if $U(r)$ were increasing like cr, it is natural to assume that $U_\nu(\mu_s)$ should be $\int_{\mathcal{X}} c\,d\mu_s = c\,\mu_s[\mathcal{X}]$. So it is the *asymptotic slope* of U that should matter. Since U is convex, there is a natural notion of asymptotic slope of U:

$$U'(\infty) := \lim_{r\to\infty} \frac{U(r)}{r} = \lim_{r\to\infty} U'(r) \qquad \in \mathbb{R}\cup\{+\infty\}. \tag{29.5}$$

This suggests the addition of a new term $U'(\infty)\,\mu_s[\mathcal{X}]$ in the definitions of $U_\nu(\mu)$ and $U_{\pi,\nu}^{\beta}(\mu)$.

Integral functionals of singular measures

The discussion in the previous paragraph should make the following definition natural.

Definition 29.1 (Integral functionals for singular measures). *Let $(\mathcal{X}, d, \nu)$ be a locally compact metric-measure space, where ν is locally finite; let $U : \mathbb{R}_+ \to \mathbb{R}$ be a continuous convex function with $U(0) = 0$, and let μ be a measure on $\mathcal{X}$, compactly supported. Let*

$$\mu = \rho\nu + \mu_s$$

be the Lebesgue decomposition of μ into absolutely continuous and singular parts. Then:

(i) define the integral functional U_ν, with nonlinearity U and reference measure ν, by

$$U_\nu(\mu) := \int_{\mathcal{X}} U(\rho(x))\,\nu(dx) \;+\; U'(\infty)\,\mu_s[\mathcal{X}];$$

(ii) if $x \to \pi(dy|x)$ is a family of probability measures on $\mathcal{X}$, indexed by $x \in \mathcal{X}$, and β is a measurable function $\mathcal{X} \times \mathcal{X} \to (0, +\infty]$, define the integral functional $U^\beta_{\pi,\nu}$ with nonlinearity U, reference measure ν, coupling π and distortion coefficient β, by

$$U^\beta_{\pi,\nu}(\mu) := \int_{\mathcal{X}\times\mathcal{X}} U\left(\frac{\rho(x)}{\beta(x,y)}\right) \beta(x,y)\,\pi(dy|x)\,\nu(dx) \;+\; U'(\infty)\,\mu_s[\mathcal{X}]. \tag{29.6}$$

Remark 29.2. It is clear that $U^\beta_{\pi,\nu}$ reduces to U_ν when $\beta \equiv 1$, i.e. when there is no distortion. In the sequel, I shall use Definition 29.1 only with the special coefficients $\beta_t(x,y) = \beta_t^{(K,N)}(x,y)$ defined in (14.61)–(14.64).

Remark 29.3. Remark 17.27 applies here too: I shall often identify π with the probability measure $\pi(dx\,dy) = \mu(dx)\,\pi(dy|x)$ on $\mathcal{X} \times \mathcal{X}$.

Remark 29.4. The new definition of U_ν takes care of the subtleties linked to singularities of the measure μ; there are also subtleties linked to the behavior at infinity, but I shall take them into account only in the next chapter. For the moment I shall only consider compactly supported displacement interpolations.

Remark 29.5. For $U^{\beta}_{\pi,\nu}$ the situation is worse because (29.6) might not be a priori well-defined in $\mathbb{R} \cup \{+\infty\}$ if β is *unbounded* (recall Remark 17.32). In the sequel, this ambiguity will occur in two limit cases: one is when $\mathcal{X}$ satisfies $\mathrm{CD}(K,N)$ and has exactly the limit Bonnet–Myers diameter $D_{K,N} = \pi\sqrt{(N-1)/K}$; the other is when $N = 1$. In both cases I shall use Convention 17.30 to make sense of $U^{\beta}_{\pi,\nu}(\mu)$. It will turn out a posteriori that $U^{\beta}_{\pi,\nu}(\mu)$ is never $-\infty$ when π arises from an optimal transport.

For later use I record here two elementary lemmas about the functionals $U^{\beta}_{\pi,\nu}$. The reader may skip them at first reading and go directly to the next section.

First, there is a handy way to rewrite $U^{\beta}_{\pi,\nu}(\mu)$ when $\mu_s = 0$:

Lemma 29.6 (Rewriting of the distorted U_ν functional). *With the notation of Definition 29.1,*

$$U^{\beta}_{\pi,\nu}(\mu) = \int_{\mathcal{X}\times\mathcal{X}} U\left(\frac{\rho(x)}{\beta(x,y)}\right) \frac{\beta(x,y)}{\rho(x)}\, \pi(dx\, dy) \tag{29.7}$$

$$= \int_{\mathcal{X}\times\mathcal{X}} v\left(\frac{\rho(x)}{\beta(x,y)}\right) \pi(dx\, dy), \tag{29.8}$$

where $v(r)=U(r)/r$, with the conventions $U(0)/0 = U'(0) \in [-\infty,+\infty)$, $U(\infty)/\infty = U'(\infty) \in (-\infty,+\infty]$, and $\rho=0$ on $\operatorname{Spt}\mu_s$.

Proof of Lemma 29.6. The identity is formally obvious if one notes that $\rho(x)\,\pi(dy|x)\,\nu(dx) = \pi(dy|x)\,\mu(dx) = \pi(dy\,dx)$; so all the subtlety lies in the fact that in (29.7) the convention is $U(0)/0 = 0$, while in (29.8) it is $U(0)/0 = U'(0)$. Switching between both conventions is allowed because the set $\{\rho = 0\}$ is anyway of zero π-measure. □

Secondly, the functionals $U^{\beta}_{\pi,\nu}$ (and the functionals U_ν) satisfy a principle of "rescaled subadditivity", which might at first sight seem contradictory with the convexity property, but is not at all.

Lemma 29.7 (Rescaled subadditivity of the distorted U_ν functionals). *Let $(\mathcal{X},d,\nu)$ be a locally compact metric-measure space, where ν is locally finite, and let β be a positive measurable function on $\mathcal{X}\times\mathcal{X}$. Let U be a continuous convex function with $U(0) = 0$. Let $\mu_1,\dots,\mu_k$ be probability measures on $\mathcal{X}$, let $\pi_1,\dots,\pi_k$ be probability measures on $\mathcal{X}\times\mathcal{X}$, and let $Z_1,\dots,Z_k$ be positive numbers with $\sum Z_j = 1$. Then, with the notation $U_a(r) = a^{-1}U(ar)$, one has*

$$U^{\beta}_{\sum_j Z_j\, \pi_j,\nu}\Big(\sum Z_j\, \mu_j\Big) \geq \sum_j Z_j\, (U_{Z_j})^{\beta}_{\pi_j,\nu}(\mu_j),$$

with equality if the measures μ_k are singular with respect to each other.

Proof of Lemma 29.7. By induction, it is sufficient to treat the case $k = 2$. The following inequality will be useful: If X_1, X_2, p_1, p_2 are any four nonnegative numbers then (with the convention $U(0)/0 = U'(0)$)

$$\frac{U(X_1 + X_2)}{X_1 + X_2}(p_1 + p_2) \geq \frac{U(X_1)}{X_1}\, p_1 + \frac{U(X_2)}{X_2}\, p_2. \tag{29.9}$$

This comes indeed from $U(r)/r$ being a nonincreasing function of r.

Now, in view of Lemma 29.6,

$$U^{\beta}_{Z_1\pi_1+Z_2\pi_2,\nu}(Z_1\mu_1 + Z_2\mu_2)$$
$$= \int U\left(\frac{Z_1\rho_1(x)+Z_2\rho_2(x)}{\beta(x,y)}\right) \frac{\beta(x,y)}{Z_1\rho_1(x) + Z_2\rho_2(x)}\, (Z_1\pi_1 + Z_2\pi_2)(dx\, dy);$$

$$U^{\beta}_{Z_1,\pi_1,\nu}(\mu_1) = \int U\left(\frac{Z_1\rho_1(x)}{\beta(x,y)}\right) \frac{\beta(x,y)}{Z_1\rho_1(x)}\, \pi_1(dx\, dy);$$

$$U^{\beta}_{Z_2,\pi_2,\nu}(\mu_2) = \int U\left(\frac{Z_2\rho_2(x)}{\beta(x,y)}\right) \frac{\beta(x,y)}{Z_2\rho_2(x)}\, \pi_2(dx\, dy).$$

So to prove the lemma it is sufficient to show that

$$U\left(\frac{Z_1\rho_1 + Z_2\rho_2}{\beta}\right) \frac{\beta}{Z_1\rho_1 + Z_2\rho_2}\, (Z_1\pi_1 + Z_2\pi_2)$$
$$\geq U\left(\frac{Z_1\rho_1}{\beta}\right) \frac{\beta}{Z_1\rho_1}\, (Z_1\pi_1) + U\left(\frac{Z_2\rho_2}{\beta}\right) \frac{\beta}{Z_2\rho_2}\, (Z_2\pi_2). \tag{29.10}$$

But this is an obvious consequence of (29.9) with $X_1 = Z_1\rho_1(x)/\beta(x,y)$, $X_2 = Z_2\rho_2(x)/\beta(x,y)$, $p_1 = d(Z_1\pi_1)/d(Z_1\pi_1 + Z_2\pi_2)(x,y)$, and $p_2 = d(Z_2\pi_2)/d(Z_1\pi_1 + Z_2\pi_2)(x,y)$. □

Synthetic definition of the curvature-dimension bound

In the next definition I shall say that an optimal transference plan π is associated with a displacement interpolation $(\mu_t)_{0\leq t\leq 1}$ if there is a dynamical optimal transference plan Π such that $\mu_t = (e_t)_{\#}\Pi$,

$\pi = (e_0, e_1)_{\#}\pi$. (Equivalently, there is a random geodesic γ such that $\pi = \text{law}(\gamma_0, \gamma_1)$ and $\mu_t = \text{law}(\gamma_t)$.) Also, if π is a given probability measure on $\mathcal{X}\times\mathcal{X}$, I shall denote by $\check{\pi}$ the probability measure obtained from π by "exchanging x and y"; more rigorously, $\check{\pi} = S_{\#}\pi$, where $S(x,y) = (y,x)$.

Definition 29.8 (Weak curvature-dimension condition). *Let $K \in \mathbb{R}$ and $N \in [1,\infty]$. A locally compact, complete σ-finite metric-measure geodesic space $(\mathcal{X}, d, \nu)$ is said to satisfy a weak* $\mathrm{CD}(K,N)$ *condition, or to be a weak* $\mathrm{CD}(K,N)$ *space, if the following is satisfied: Whenever μ_0 and μ_1 are two compactly supported probability measures with $\operatorname{Spt}\mu_0, \operatorname{Spt}\mu_1 \subset \operatorname{Spt}\nu$, there exist a displacement interpolation $(\mu_t)_{0\le t\le 1}$ and an associated optimal coupling π of (μ_0,μ_1) such that, for all $U \in \mathcal{DC}_N$ and for all $t \in [0,1]$,*

$$U_\nu(\mu_t) \le (1-t)\, U_{\pi,\nu}^{\beta_{1-t}^{(K,N)}}(\mu_0) \;+\; t\, U_{\check{\pi},\nu}^{\beta_t^{(K,N)}}(\mu_1). \tag{29.11}$$

Roughly speaking, the weak $\mathrm{CD}(K,N)$ condition states that the functionals U_ν are "jointly" weakly displacement convex with distortion coefficients $(\beta_t^{(K,N)})$, for all $U \in \mathcal{DC}_N$. This is a property of the triple $(\mathcal{X}, d, \nu)$, but for simplicity I shall often abbreviate the statement "$(\mathcal{X}, d, \nu)$ satisfies a weak $\mathrm{CD}(K,N)$ condition" into "$\mathcal{X}$ satisfies a weak $\mathrm{CD}(K,N)$ condition", with the understanding that the distance and reference measure should be clear from the context.

Before going any further, I shall make explicit the fact that this definition is an extension of the usual one, and connect the synthetic notion of weak $\mathrm{CD}(K,N)$ space with the corresponding analytic notion (considered in Chapter 14, and defined for instance in terms of the modified Ricci curvature tensor (14.36)).

Theorem 29.9 (Smooth weak $\mathrm{CD}(K,N)$ spaces are $\mathrm{CD}(K,N)$ manifolds). *Let (M,g) be a smooth Riemannian manifold, equipped with its geodesic distance d, its volume measure* vol, *and a reference measure $\nu = e^{-V}\mathrm{vol}$, where $V \in C^2(M)$. Then, (M,d,ν) is a weak* $\mathrm{CD}(K,N)$ *space if and only if (M,g,ν) satisfies the* $\mathrm{CD}(K,N)$ *curvature-dimension bound; or equivalently, if the modified Ricci tensor* $\mathrm{Ric}_{N,\nu}$ *satisfies* $\mathrm{Ric}_{N,\nu} \ge Kg$.

Proof of Theorem 29.9. By Theorem 10.41 (unique solvability of the Monge problem), Corollary 7.23 (uniqueness of the displacement interpolation) and Theorem 17.37 (characterization of $\mathrm{CD}(K,N)$ via distorted displacement convexity), (M,g,ν) satisfies the $\mathrm{CD}(K,N)$ bound if and only if (29.11) holds true as soon as μ_0, μ_1 are absolutely continuous with respect to ν. So it only remains to show that if (29.11) holds for absolutely continuous μ_0, μ_1 then it also holds for singular measures μ_0, μ_1. This will follow from Corollary 29.23 later in this chapter.[1] □

The end of this section is devoted to a series of comments about the definition of weak $\mathrm{CD}(K,N)$ spaces.

• In Definition 29.8 I was careful to impose displacement convexity inequalities along *some* Wasserstein geodesic, because such geodesics might not be unique. There are two possible reasons for this nonuniqueness: one is the a priori lack of smoothness of the metric space $\mathcal{X}$; the other is the fact that μ_0, μ_1 might be singular. Even on a Riemannian manifold, it is easy to construct examples of measures μ_0, μ_1 which are joined by more than one displacement interpolation (just take $\mu_0 = \delta_{x_0}$, $\mu_1 = \delta_{x_1}$, where x_0 and x_1 are joined by multiple geodesics). However, it will turn out later that displacement convexity inequalities hold along *all* Wasserstein geodesics if the space $\mathcal{X}$ satisfies some mild regularity assumption, namely if it is nonbranching (see Theorem 30.32).

• An important property of the classical $\mathrm{CD}(K,N)$ condition is that it is more and more stringent as K increases and as N decreases. The next proposition shows that the same is true in a nonsmooth setting.

Proposition 29.10 (Consistency of the $\mathrm{CD}(K,N)$ conditions). *The weak condition* $\mathrm{CD}(K,N)$ *becomes more and more stringent as* K *increases, and as* N *decreases.*

Proof of Proposition 29.10. First, the class $\mathcal{DC}_N$ becomes smaller as N increases, which means fewer conditions to satisfy. Next, recall that $\beta_t^{(K,N)}$ and $\beta_{1-t}^{(K,N)}$ are increasing in K and decreasing in N (as noticed right after Definition 14.19); since $U(r)/r$ is nonincreasing, the quantities $\beta_{1-t}^{(K,N)}\, U(\rho_0/\beta_{1-t}^{(K,N)})$ and $\beta_t^{(K,N)}\, U(\rho_1/\beta_t^{(K,N)})$ are nondecreasing in N and nonincreasing in K. The conclusion follows immediately. □

• In the case $K > 0$ and $N < \infty$, the coefficient $\beta_t^{(K,N)}(x,y)$ takes the value $+\infty$ if $0 < t < 1$ and $d(x,y) \geq D_{K,N} := \pi\sqrt{(N-1)/K}$. In that case the natural convention is $\infty\, U(r/\infty) = U'(0)\, r$, in accordance with Lemma 29.6. With this convention, Definition 29.8 implies that

[1] This is one of the rare instances in this book where a result is used before it is proven; but I think the resulting presentation is more clear and pedagogical.

the diameter of the support of ν is automatically bounded above by $D_{K,N}$. Otherwise, take $x_0, x_1 \in \operatorname{Spt}\nu$ with $d(x_0,x_1) > D_{K,N}$ and choose $r > 0$ small enough that $d(x_0', x_1') > D_{K,N}$ for all $x_0' \in B_r(x_0)$, $x_1' \in B_r(x_1)$. Take $\rho_0 = 1_{B_r(x_0)}/\nu[B_r(x_0)]$ and $\rho_1 = 1_{B_r(x_1)}/\nu[B_r(x_1)]$ in the definition of the weak CD(K,N) bound. Then the coefficients β_t appearing in the right-hand side of (29.11) are identically $+\infty$, and the measures have no singular part; so that inequality becomes just

$$U_\nu(\mu_t) \leq U'(0)\Big((1-t)\int \rho_0\, d\nu \;+\; t\int \rho_1\, d\nu\Big) = U'(0). \qquad (29.12)$$

Now choose $U(r) = -r^{1-1/N}$: Then $U'(0) = -\infty$, so (29.12) implies $U_\nu(\mu) = -\infty$. On the other hand, by Jensen's inequality, $U_\nu(\mu_t) \geq -\nu[S]\left(\int \rho_t\, d\nu/\nu[S]\right)^{1-1/N} \geq -\nu[S]^{1/N}$, where S stands for the support of μ_t; so $U_\nu(\mu_t)$ cannot be $-\infty$. This contradiction proves the claim. Let me record the conclusion in the form of a separate statement:

Proposition 29.11 (Bonnet–Myers diameter bound for weak CD(K,N) spaces). *If $(\mathcal{X}, d, \nu)$ is a weak* CD(K,N) *space with $K > 0$ and $N < \infty$, then*

$$\operatorname{diam}(\operatorname{Spt}\nu) \leq D_{K,N} := \pi\sqrt{\frac{N-1}{K}}.$$

As a corollary, when one uses inequality (29.11) in a weak CD(K,N) space, the distortion coefficients appearing in the right-hand side are in fact always finite, except possibly in the two limit cases $N = 1$ and $\operatorname{diam}(\operatorname{Spt}\nu) = D_{K,N}$. (In both cases, $U^\beta_{\pi,\nu}(\mu)$ is defined as in Convention 17.30.)

- To check Definition 29.8, it is not really necessary to establish inequality (29.11) for the whole class $\mathcal{DC}_N$: It is sufficient to restrict to members of $\mathcal{DC}_N$ that are nonnegative, and Lipschitz (for $N < \infty$), or behaving at infinity like $O(r \log r)$ (for $N = \infty$). This is the content of the next statement.

Proposition 29.12 (Sufficient condition to be a weak CD(K,N) space). *In Definition 29.8, it is equivalent to require that inequality* (29.11) *hold for all $U \in \mathcal{DC}_N$, or just for those $U \in \mathcal{DC}_N$ which are nonnegative and satisfy:*

- *U is Lipschitz, if $N < \infty$;*

- *U is locally Lipschitz and $U(r) = a\,r\log r + b\,r$ for r large enough, if $N = \infty$ ($a \geq 0$, $b \in \mathbb{R}$).*

Proof of Proposition 29.12. Let us assume that $(\mathcal{X}, d, \nu)$ satisfies Definition 29.8, except that (29.11) holds true only for those U satisfying the above conditions; and we shall check that inequality (29.11) holds true for all $U \in \mathcal{DC}_N$. Thanks to Convention 17.30 it is sufficient to prove it under the assumption that $\beta_t^{(K,N)}$ is bounded; then the two limit cases $N = 1$ and $\operatorname{diam}(\mathcal{X}) = D_{K,N}$ are treated by taking another dimension $N' > N$ and letting $N' \downarrow N$ (the same reasoning will be used later to establish Corollary 29.23 and Theorem 29.24).

The proof will be performed in three steps.

Step 1: Relaxing the nonnegativity condition. Let $U \in \mathcal{DC}_N$ be Lipschitz (for $N < \infty$) or locally Lipschitz and behaving like $a\,r\log r + b\,r$ for r large enough (for $N = \infty$), but not necessarily nonnegative. Then we can decompose U as

$$U(r) = \widetilde{U}(r) - A\,r,$$

where $\widetilde{U} \in \mathcal{DC}_N \cap \mathrm{Lip}(\mathbb{R}_+, \mathbb{R}_+)$ and $A \geq 0$ (choose $A = \max(-U'(0), 0)$). By assumption, with the same notation as in Definition 29.8, one has the inequality

$$\widetilde{U}_\nu(\mu_t) \leq (1-t)\, \widetilde{U}_{\pi,\nu}^{\beta_{1-t}^{(K,N)}}(\mu_0) \;+\; t\, \widetilde{U}_{\check{\pi},\nu}^{\beta_t^{(K,N)}}(\mu_1). \tag{29.13}$$

Write $\mu_t = \rho_t\,\nu + (\mu_t)_s$ for the Lebesgue decomposition of μ_t with respect to ν. The replacement of $\widetilde{U}$ by U amounts to adding to the left-hand side

$$A\left(\int \rho_t\, d\nu + (\mu_t)_s[\mathcal{X}]\right) = A\,\mu_t[\mathcal{X}] = A,$$

and to the right-hand side

$$A\,(1-t)\int_{\mathcal{X}\times\mathcal{X}} \frac{\rho_0(x_0)}{\beta_{1-t}^{(K,N)}(x_0,x_1)}\,\beta_{1-t}^{(K,N)}(x_0,x_1)\,\pi(dx_1|x_0)\,\nu(dx_0)$$
$$+\,A\,t\int_{\mathcal{X}\times\mathcal{X}} \frac{\rho_1(x_1)}{\beta_t^{(K,N)}(x_0,x_1)}\,\beta_t^{(K,N)}(x_0,x_1)\,\pi(dx_0|x_1)\,\nu(dx_1) \;=A$$

also. So (29.13) also holds true with $\widetilde{U}$ replaced by U.

Step 2: Behavior at the origin. Let $U \in \mathcal{DC}_N$ be such that U is Lipschitz at infinity ($N < \infty$) or behaves like $a\,r \log r + b\,r$ at infinity ($N = \infty$). By Proposition 17.7(v), there is a nonincreasing sequence $(U_\ell)_{\ell \in \mathbb{N}}$, converging pointwise to U, coinciding with U on $[1, +\infty)$, with $U_\ell'(0) > -\infty$ and $U_\ell'(0) \to U'(0)$. Each U_ℓ is locally Lipschitz, and one has, by Step 1,

$$(U_\ell)_\nu(\mu_t) \leq (1-t)\,(U_\ell)_{\pi,\nu}^{\beta_{1-t}^{(K,N)}}(\mu_0) \;+\; t\,(U_\ell)_{\check{\pi},\nu}^{\beta_t^{(K,N)}}(\mu_1). \tag{29.14}$$

The problem is to pass to the limit as $\ell \to \infty$. In the left-hand side this is obvious, since $U \leq U_\ell$. In the right-hand side, this will follow from the monotone convergence theorem as soon as we have verified that the integrands are bounded above, uniformly in ℓ, by integrable functions. Let us check this for the first term in the right-hand side of (29.14): If $\rho_0(x_0)/\beta_{1-t}(x_0, x_1) \geq 1$, then

$$(U_\ell)\left(\frac{\rho_0(x_0)}{\beta_{1-t}^{(K,N)}(x_0,x_1)}\right)\beta_{1-t}^{(K,N)}(x_0,x_1) = U\left(\frac{\rho_0(x_0)}{\beta_{1-t}^{(K,N)}(x_0,x_1)}\right)\beta_{1-t}^{(K,N)}(x_0,x_1);$$

and otherwise, the same expression is bounded above by

$$(U_\ell)'(1)\,\frac{\rho_0(x_0)}{\beta_{1-t}^{(K,N)}(x_0,x_1)}\;\beta_{1-t}^{(K,N)}(x_0,x_1) \leq U'(1)\,\rho_0(x_0).$$

In both cases, the upper bound belongs to $L^1\big(\pi(dx_0|x_1)\,\nu(dx_1)\big)$, which proves the claim.

Step 3: Behavior at infinity. Finally we consider the case of a general $U \in \mathcal{DC}_N$, and approximate it at infinity so that it has the desired behavior. The reasoning is pretty much the same as for Step 2. Let us assume for instance $N < \infty$. By Proposition 17.7(iv), there is a nondecreasing sequence $(U_\ell)_{\ell \in \mathbb{N}}$, converging pointwise to U, with the desired behavior at infinity, and $U_\ell'(\infty) \to U'(\infty)$. By Step 2,

$$(U_\ell)_\nu(\mu_t) \leq (1-t)\,(U_\ell)_{\pi,\nu}^{\beta_{1-t}^{(K,N)}}(\mu_0) \;+\; t\,(U_\ell)_{\check{\pi},\nu}^{\beta_t^{(K,N)}}(\mu_1),$$

and it remains to pass to the limit as $\ell \to \infty$. In the right-hand side, this is obvious since $U_\ell \leq U$. The left-hand side may be rewritten as

$$\int U_\ell(\rho_t)\,d\nu + U'_\ell(\infty)\,(\mu_t)_s[\mathcal{X}]. \tag{29.15}$$

Then we know that $U'_\ell(\infty) \to U'(\infty)$, so we may pass to the limit in the second term of (29.15). To pass to the limit in the first term by monotone convergence, it suffices to check that $U_\ell(\rho_t)$ is bounded below, uniformly in ℓ, by a ν-integrable function. But this is true since, for instance, U_0 which is bounded below by an affine function of the form $r \to -C(r+1)$, $C \geq 0$; so $U_\ell(\rho_t) \geq -C\rho_t - C\,1_{\rho_t>0}$, and the latter function is integrable since ρ_t has compact support. □

- In Definition 29.8 I imposed μ_0 and μ_1 to be compactly supported. This assumption can actually be relaxed, but the definition which one gets by so doing is not stronger; see Theorem 30.5 in the next chapter for more details. Conversely, one could also have imposed μ_0 and μ_1 to be absolutely continuous, or even to have continuous densities, but the definition would not be weaker (see Corollary 29.23).
- Finally, here are some **examples** of weak $\mathrm{CD}(K,N)$ spaces:

Example 29.13. Let V be a continuous function $\mathbb{R}^n \to \mathbb{R}$ with $\int e^{-V(x)}\,dx < \infty$, let $\nu(dx) = e^{-V(x)}\,dx$, and let d_2 be the Euclidean distance. Then the space $(\mathbb{R}^n, d_2, \nu)$ satisfies the usual $\mathrm{CD}(K,\infty)$ condition if V is C^2 and $\nabla^2 V \geq K\,I_n$ in the classical sense. It satisfies the weak $\mathrm{CD}(K,\infty)$ condition *without any regularity assumption on* V, as soon as $\nabla^2 V \geq K\,I_n$ in the sense of distributions, which means that V is K-convex. For instance, if V is merely convex, then $(\mathbb{R}^n, d_2, \nu)$ satisfies the weak $\mathrm{CD}(0,\infty)$ condition. To see this, note that if $\mu(dx) = \rho(x)\,dx$, then

$$H_\nu(\mu) = \int \rho(x)\log\rho(x)\,dx + \int \rho(x)\,V(x)\,dx = H(\mu) + \int V\,d\mu;$$

then the first term is always displacement convex, and the second is displacement convex if V is convex (simple exercise).

Remark 29.14. If V is not convex, then one can find $x_0, x_1 \in \mathbb{R}^n$ and $t \in [0,1]$ such that

$$V\big((1-t)x_0 + t x_1\big) > (1-t)\,V(x_0) + t\,V(x_1).$$

Then let ρ be a compactly supported probability density, and $\rho^\epsilon = \epsilon^{-n}\rho(\,\cdot\,/\epsilon)$: then $\rho^\epsilon(x)\,dx$ converges weakly to δ_0, so for ϵ small enough,

$$\int V(x)\,\rho^\epsilon(x-(1-t)x_0-tx_1)\,dx \;>\; (1-t)\int V(x)\,\rho^\epsilon(x-x_0)\,dx + t\int V(x)\,\rho^\epsilon(x-x_1)\,dx.$$

On the other hand, $\int \rho^\epsilon(x-v)\log\rho^\epsilon(x-v)\,dx$ is independent of $v\in\mathbb{R}^n$; so

$$H_{e^{-V}dx}\big(\rho^\epsilon(\cdot-(1-t)x_0-tx_1)\big)\,dx \;>\; (1-t)\,H_{e^{-V}dx}\big(\rho^\epsilon(\cdot-x_0)\big) + t\,H_{e^{-V}dx}\big(\rho^\epsilon(\cdot-x_1)\big).$$

Since the path $(\rho^\epsilon(x-(1-s)x_0-sx_1)\,dx)_{0\le s\le 1}$ is a geodesic interpolation (this is the translation at uniform speed, corresponding to $\nabla\psi=$ constant), we see that $(\mathbb{R}^n,d_2,e^{-V(x)}\,dx)$ cannot be a weak $\mathrm{CD}(0,\infty)$ space. So the conclusion of Example 29.13 can be refined as follows: $(\mathbb{R}^n,d_2,e^{-V(x)}\,dx)$ is a weak $\mathrm{CD}(0,\infty)$ space *if and only if* V is convex.

Example 29.15. Let M be a smooth compact n-dimensional Riemannian manifold with nonnegative Ricci curvature, and let G be a compact Lie group acting isometrically on M. (See the bibliographical notes for references on these notions.) Then let $\mathcal{X}=M/G$ and let $q: M\to\mathcal{X}$ be the quotient map. Equip $\mathcal{X}$ with the quotient distance $d(x,y)=\inf\{d_M(x',y');\ q(x')=x, q(y')=y\}$, and with the measure $\nu=q_\#\mathrm{vol}_M$. The resulting space $(\mathcal{X},d,\nu)$ is a weak $\mathrm{CD}(0,n)$ space, that in general will not be a manifold. (There will typically be singularities at fixed points of the group action.)

Example 29.16. It will be shown in the concluding chapter that $(\mathbb{R}^n,\|\cdot\|,\lambda_n)$ is a weak $\mathrm{CD}(K,N)$ space, where $\|\cdot\|$ is *any* norm on $\mathbb{R}^n$, and λ_n is the n-dimensional Lebesgue measure. This example proves that a weak $\mathrm{CD}(K,N)$ space may be "strongly" branching (recall the discussion in Example 27.17).

Example 29.17. Let $\mathcal{X}=\prod_{i=1}^\infty T_i$, where $T_i=\mathbb{R}/(\varepsilon_i\,\mathbb{Z})$ is equipped with the usual distance d_i and the normalized Lebesgue measure λ_i, and $\varepsilon_i=2\,\mathrm{diam}\,(T_i)$ is some positive number. If $\sum\varepsilon_i^2<+\infty$ then the product distance $d=\sqrt{\sum d_i^2}$ turns $\mathcal{X}$ into a compact metric space. Equip $\mathcal{X}$ with the product measure $\nu=\prod\lambda_i$; then $(\mathcal{X},d,\nu)$ is a weak $\mathrm{CD}(0,\infty)$ space. (Indeed, it is the measured Gromov–Hausdorff limit of $\mathcal{X}=\prod_{j=1}^k T_j$ which is $\mathrm{CD}(0,k)$, hence $\mathrm{CD}(0,\infty)$; and it will be shown in Theorem 29.24 that the $\mathrm{CD}(0,\infty)$ property is stable under measured Gromov–Hausdorff limits.)

The remaining part of the present chapter is devoted to a proof of stability for the weak $\mathrm{CD}(K,N)$ property.

Continuity properties of the functionals U_ν and $U^\beta_{\pi,\nu}$

In this section, I shall explain some of the remarkable properties of the integral functionals appearing in Definition 29.1. For the moment it will be sufficient to restrict to the case of a *compact* space $\mathcal{X}$, and it will be convenient to consider that U_ν and $U^\beta_{\pi,\nu}$ are defined on the set of all (nonnegative) finite signed Borel measures, not necessarily *probability* measures. (Actually, in Definition 29.1 it was not assumed that μ is a probability measure.) One may even think of these functionals as defined on the whole vector space $M(\mathcal{X})$ of finite Borel measures on $\mathcal{X}$, with the convention that their value is $+\infty$ if μ is not nonnegative; then U_ν and $U^\beta_{\pi,\nu}$ are true convex functional on $M(\mathcal{X})$.

It will be convenient to study the functionals U_ν by means of their **Legendre representation.** Generally speaking, the Legendre representation of a convex functional Φ defined on a vector space E is an identity of the form

$$\Phi(x) = \sup \Big\{ \langle \Lambda, x\rangle - \Psi(\Lambda)\Big\},$$

where Λ varies over a certain subset of E^*, and Ψ is a convex functional of Λ. Usually, Λ varies over the whole set E^*, and $\Psi(\Lambda) = \sup_{x\in E}[\langle \Lambda, x\rangle - \Phi(x)]$ is the Legendre transform of Φ; but here we don't really want to do so, because nobody knows what the huge space $M(\mathcal{X})^*$ looks like. So it is better to restrict to subspaces of $M(\mathcal{X})^*$. There are several natural possible choices, resulting in various Legendre representations; which one is most convenient depends on the context. Here are the ones that will be useful in the sequel.

Definition 29.18 (Legendre transform of a real-valued convex function). *Let $U : \mathbb{R}_+ \to \mathbb{R}$ be a continuous convex function with $U(0) = 0$; its Legendre transform is defined on $\mathbb{R}$ by*

$$U^*(p) = \sup_{r\in\mathbb{R}_+} \Big[pr - U(r)\Big].$$

It is easy to check that U^* is a convex function, taking the value $-U(0) = 0$ on $(-\infty, U'(0)]$ and $+\infty$ on $(U'(\infty), +\infty)$.

Proposition 29.19 (Legendre representation of U_ν). *Let $U : \mathbb{R}_+ \to \mathbb{R}$ be a continuous convex function with $U(0) = 0$, let $\mathcal{X}$ be a compact metric space, equipped with a finite reference measure ν. Then, whenever μ is a finite measure on $\mathcal{X}$,*

$$(i)\ \ U_\nu(\mu) = \sup\left\{\int_{\mathcal{X}} \varphi\, d\mu - \int_{\mathcal{X}} U^*(\varphi)\, d\nu;\quad \varphi \in L^\infty(\mathcal{X});\quad \varphi \le U'(\infty)\right\}$$

$$(ii)\ \ U_\nu(\mu) = \sup\left\{\int_{\mathcal{X}} \varphi\, d\mu - \int_{\mathcal{X}} U^*(\varphi)\, d\nu;\quad \varphi \in C(\mathcal{X}),\right.$$
$$\left. U'\left(\frac{1}{M}\right) \le \varphi \le U'(M);\quad M \in \mathbb{N}\right\}.$$

The deceiving simplicity of these formulas hides some subtleties: For instance, it is in general impossible to drop the restriction $\varphi \le U'(\infty)$ in (i), so the supremum is not taken over the whole vector space $L^\infty(\mathcal{X})$ but only on a subspace thereof. Proposition 29.19 can be proven by elementary tools of measure theory; see the bibliographical notes for references and comments.

The next statement gathers the three important properties on which the main results of this chapter rest: (i) $U_\nu(\mu)$ is *lower semicontinuous* in (μ, ν); (ii) $U_\nu(\mu)$ is *never increased by push-forward*; (iii) μ can be regularized in such a way that $U^\beta_{\pi,\nu}(\mu)$ is *upper semicontinuous* in (π, μ) along the approximation. In the next statement, $M_+(\mathcal{X})$ will stand for the set of finite (nonnegative) Borel measures on $\mathcal{X}$, and $L^1_+(\nu)$ for the set of nonnegative ν-integrable measurable functions on $\mathcal{X}$.

Theorem 29.20 (Continuity and contraction properties of U_ν and $U^\beta_{\pi,\nu}$). *Let $(\mathcal{X}, d)$ be a compact metric space, equipped with a finite measure ν. Let $U : \mathbb{R}_+ \to \mathbb{R}_+$ be a convex continuous function, with $U(0) = 0$. Further, let $\beta(x, y)$ be a continuous positive function on $\mathcal{X} \times \mathcal{X}$. Then, with the notation of Definition 29.1:*

(i) $U_\nu(\mu)$ is a weakly lower semicontinuous function of both μ and ν in $M_+(\mathcal{X})$. More explicitly, if $\mu_k \to \mu$ and $\nu_k \to \nu$ in the weak topology of convergence against bounded continuous functions, then

$$U_\nu(\mu) \le \liminf_{k\to\infty} U_{\nu_k}(\mu_k).$$

(ii) U_ν satisfies a contraction principle in both μ and ν; that is, if $\mathcal{Y}$ is another compact space, and $f : \mathcal{X} \to \mathcal{Y}$ is any measurable function, then

$$U_{f_\#\nu}(f_\#\mu) \le U_\nu(\mu).$$

(iii) If U "grows at most polynomially", in the sense that

$$\forall r > 0, \qquad r\,U'(r) \le C\,(U(r)_+ + r), \tag{29.16}$$

then for any probability measure $\mu \in P(\mathcal{X})$, with $\operatorname{Spt}\mu \subset \operatorname{Spt}\nu$, there is a sequence $(\mu_k)_{k\in\mathbb{N}}$ of probability measures converging weakly to μ, such that each μ_k has a continuous density, and for any sequence $(\pi_k)_{k\in\mathbb{N}}$ converging weakly to π in $P(\mathcal{X} \times \mathcal{X})$, such that π_k admits μ_k as first marginal and $\operatorname{Spt}\pi_k \subset (\operatorname{Spt}\nu) \times (\operatorname{Spt}\nu)$,

$$\limsup_{k\to\infty}\, U^\beta_{\pi_k,\nu}(\mu_k) \le U^\beta_{\pi,\nu}(\mu). \tag{29.17}$$

Remark 29.21. The assumption of polynomial growth in (iii) is obviously true if U is Lipschitz; or if $U(r)$ behaves at infinity like $a\,r\log r + b\,r$ (or like a polynomial).

Exercise 29.22. Use Property (ii) in the case $U(r) = r\log r$ to recover the Csiszár–Kullback–Pinsker inequality (22.25). *Hint:* Take f to be valued in $\{0,1\}$, apply Csiszár's two-point inequality

$$\forall x, y \in [0,1], \qquad x\log\frac{x}{y} + (1-x)\log\left(\frac{1-x}{1-y}\right) \ge 2\,(x-y)^2$$

and optimize the choice of f.

Proof of Theorem 29.20. To prove (i), note that U^* is continuous on $[U'(1/M), U'(M))$; so if φ is continuous with values in $[U'(1/M), U'(M)]$, then $U^*(\varphi)$ is also continuous. Then Proposition 29.19(ii) can be rewritten as

$$U_\nu(\mu) = \sup_{(\varphi,\psi)\in\mathcal{U}} \left\{ \int \varphi\,d\mu + \int \psi\,d\nu \right\},$$

where $\mathcal{U}$ is a certain subset of $C(\mathcal{X}) \times C(\mathcal{X})$. In particular, $U_\nu(\mu)$ is a supremum of continuous functions of (μ,ν); it follows that U_ν is lower semicontinuous.

To prove (ii), pick up any $\varphi \in L^\infty(\mathcal{X})$ with $\varphi \le U'(\infty)$. Then

$$\int_{\mathcal{X}} (\varphi\circ f)\,d\mu - \int_{\mathcal{X}} U^*(\varphi\circ f)\,d\nu = \int_{\mathcal{Y}} \varphi\,d(f_\#\mu) - \int_{\mathcal{Y}} U^*(\varphi)\,d(f_\#\nu).$$

If $\varphi \le U'(\infty)$, then also $\varphi\circ f \le U'(\infty)$; similarly, if φ is bounded, then also $\varphi\circ f$ is bounded. So

$$\sup_{\psi\in L^\infty;\ \psi\le U'(\infty)} \left\{\int_{\mathcal{X}} \psi\,d\mu - \int_{\mathcal{X}} U^*(\psi)\,d\nu\right\} \le \sup_{\varphi\in L^\infty;\ \varphi\le U'(\infty)} \left\{\int_{\mathcal{Y}} \varphi\,d(f_\#\mu) - \int_{\mathcal{Y}} U^*(\varphi)\,d(f_\#\nu)\right\}.$$

By Proposition 29.19(i), the left-hand side coincides with $U_\nu(\mu)$, and the right-hand side with $U_{f_\#\mu}(f_\#\nu)$. This concludes the proof of the contraction property (ii).

Now let us consider the proof of (iii), which is a bit tricky. For pedagogical reasons I shall first treat a simpler case.

Particular case: $\beta \equiv 1$. (Then there is no need for any restriction on the growth of U.)

Let $\varepsilon = \varepsilon_k$ be a sequence in $(0,1)$, $\varepsilon_k \to 0$, and let $K_\varepsilon(x,y)$ be a sequence of symmetric continuous nonnegative functions on $\mathcal{X}\times\mathcal{X}$ satisfying

$$\begin{cases} \forall x\in \operatorname{Spt}\nu, & \displaystyle\int K_\varepsilon(x,y)\,\nu(dy) = 1;\\ \forall x,y\in\mathcal{X}, & d(x,y)\ge\varepsilon \Longrightarrow K_\varepsilon(x,y)=0. \end{cases}$$

Such kernels induce a *regularization* of probability measures, as recalled in the First Appendix. On $S = \operatorname{Spt}\nu$, define

$$\rho_\varepsilon(x) = \int_{\operatorname{Spt}\nu} K_\varepsilon(x,y)\,\mu(dy);$$

this is a (uniformly) continuous function on a compact set. By the Tietze–Urysohn extension theorem, ρ_ε can be extended into a continuous function $\widetilde{\rho}_\varepsilon$ on the whole of $\mathcal{X}$. Of course, ρ_ε and $\widetilde{\rho}_\varepsilon$ coincide ν-almost everywhere. We shall see that $\mu_\varepsilon = \rho_\varepsilon\,\nu$ (or, more explicitly, $\mu_k = \rho_{\varepsilon_k}\,\nu$) does the job for statement (iii).

Let us first assume that μ is absolutely continuous, and let ρ be its density. Since $\operatorname{Spt}\mu$ and $\operatorname{Spt}\mu_\varepsilon$ are included in $S = \operatorname{Spt}\nu$,

$$U_\nu(\mu) = \int_{\mathcal{X}} U(\rho)\,d\nu = \int_S U(\rho)\,d\nu; \qquad U_\nu(\mu_\varepsilon) = \int_S U(\rho_\varepsilon)\,d\nu.$$

So up to changing $\mathcal{X}$ for S, we might just assume that $\mathrm{Spt}(\nu) = \mathcal{X}$. Then for each $x \in \mathcal{X}$, $K_\varepsilon(x,y)\,\nu(dy)$ is a probability measure on $\mathcal{X}$, and by Jensen's inequality,

$$U(\rho_\varepsilon(x)) = U\left(\int_{\mathcal{X}} K_\varepsilon(x,y)\,\rho(y)\,\nu(dy)\right) \le \int_{\mathcal{X}} K_\varepsilon(x,y)\,U(\rho(y))\,\nu(dy). \tag{29.18}$$

Now integrate both sides of the latter inequality against $\nu(dx)$; this is allowed because $U(r) \ge -C(r+1)$ for some finite constant C, so the left-hand side of (29.18) is bounded below by the integrable function $-C\,(\rho_\varepsilon + 1)$. After integration, one has

$$\int_{\mathcal{X}} U(\rho_\varepsilon)\,d\nu \le \int_{\mathcal{X}\times\mathcal{X}} K_\varepsilon(x,y)\,U(\rho(y))\,\nu(dy)\,\nu(dx).$$

But $K_\varepsilon(x,y)\,\nu(dx)$ is a probability measure for any $y \in \mathcal{X}$, so

$$\int_{\mathcal{X}\times\mathcal{X}} K_\varepsilon(x,y)\,U(\rho(y))\,\nu(dy)\,\nu(dx) = \int_{\mathcal{X}} U(\rho(y))\,\nu(dy) = \int_{\mathcal{X}} U(\rho)\,d\nu.$$

To summarize: $U_\nu(\mu_\varepsilon) \le U_\nu(\mu)$ for all $\varepsilon > 0$, and then the conclusion follows.

If μ is not absolutely continuous, define

$$\rho_{a,\varepsilon}(x) = \int_{\mathrm{Spt}\,\nu} K_\varepsilon(x,y)\,\rho(y)\,\nu(dy); \qquad \rho_{s,\varepsilon}(x) = \int_{\mathrm{Spt}\,\nu} K_\varepsilon(x,y)\,\mu_s(dy),$$

where $\mu = \rho\,\nu + \mu_s$ is the Lebesgue decomposition of μ with respect to ν. Then $\rho_\varepsilon = \rho_{a,\varepsilon} + \rho_{s,\varepsilon}$. By convexity of U, for any $\theta \in (0,1)$,

$$\begin{aligned}
\int U(\rho_\varepsilon)\,d\nu &\le (1-\theta)\int U\left(\frac{\rho_{a,\varepsilon}}{1-\theta}\right) d\nu + \theta \int U\left(\frac{\rho_{s,\varepsilon}}{\theta}\right) d\nu \\
&\le (1-\theta)\int U\left(\frac{\rho_{a,\varepsilon}}{1-\theta}\right) d\nu + U'(\infty)\int \rho_{s,\varepsilon}\,d\nu \\
&= (1-\theta)\int U\left(\frac{\rho_{a,\varepsilon}}{1-\theta}\right) d\nu + U'(\infty)\,\mu_s[\mathcal{X}].
\end{aligned}$$

It is easy to pass to the limit as $\theta \to 0$ (use the monotone convergence theorem for the positive part of U, and the dominated convergence theorem for the negative part). Thus

$$\int U(\rho_\varepsilon)\,d\nu \le \int U(\rho_{a,\varepsilon})\,d\nu + U'(\infty)\,\mu_s[\mathcal{X}].$$

The first part of the proof shows that $\int U(\rho_{a,\varepsilon})\,d\nu \le \int U(\rho)\,d\nu$, so

$$\int U(\rho_\varepsilon)\,d\nu \le \int U(\rho)\,d\nu + U'(\infty)\,\mu_s[\mathcal{X}] \ = U_\nu(\mu),$$

and this again implies the conclusion.

General case with β variable: This is much, much more tricky and I urge the reader to skip this case at first encounter.

Before starting the proof, here are a few remarks. The assumption of polynomial growth implies the following estimate: For each $B > 0$ there is a constant C such that

$$\sup_{B^{-1}\le a\le B} \left(\frac{U_+(ar)}{a}\right) \le C\,(U_+(r) + r). \tag{29.19}$$

Let us check (29.19) briefly. If $U \le 0$, there is nothing to prove. If $U \ge 0$, the polynomial growth assumption amounts to $r\,U'(r) \le C(U(r)+r)$, so $r\,(U'(r)+1) \le 2C\,(U(r)+r)$; then $(d/dt)\log[U(t\,r)+t\,r] \le 2C/t$, so

$$U(ar) + ar \le (U(r) + r)\,a^{2C}, \tag{29.20}$$

whence the conclusion. Finally, if U does not have a constant sign, this means that there is $r_0 > 0$ such that $U(r) \le 0$ for $r \le r_0$, and $U(r) > 0$ for $r > r_0$. Then:

- if $r < B^{-1}\,r_0$, then $U_+(ar) = 0$ for all $a \le B$ and (29.19) is obviously true;
- if $r > B\,r_0$, then $U(ar) > 0$ for all $a \in [B^{-1}, B]$ and one establishes (29.20) as before;
- if $B^{-1}\,r_0 \le r \le B\,r_0$, then $U_+(ar)$ is bounded above for $a \le B$, while r is bounded below, so obviously (29.19) is satisfied for some well-chosen constant C.

Next, we may dismiss the case when the right-hand side of the inequality in (29.17) is $+\infty$ as trivial; so we might assume that

$$\int \beta(x,y)\,U_+\left(\frac{\rho(x)}{\beta(x,y)}\right)\pi(dy|x)\,\nu(dx) < +\infty; \qquad U'(\infty)\,\mu_s[\mathcal{X}] < +\infty. \tag{29.21}$$

Since $B^{-1} \leq \beta(x,y) \leq B$ for some $B > 0$, (29.19) implies the existence of a constant C such that

$$\begin{aligned} C^{-1} \int U_+(\rho(x))\,\nu(dx) &= C^{-1} \int U_+(\rho(x))\,\pi(dy|x)\,\nu(dx) \\ &\leq \int \beta(x,y)\,U_+\left(\frac{\rho(x)}{\beta(x,y)}\right)\pi(dy|x)\,\nu(dx) \\ &\leq C \int U_+(\rho(x))\,\pi(dy|x)\,\nu(dx) = C\int U_+(\rho(x))\,\nu(dx). \end{aligned}$$

So (29.21) implies the integrability of $U_+(\rho)$.

After these preliminaries, we can go on with the proof. In the sequel, the symbol C will stand for constants that may vary from place to place but only depend on the nonlinearity U and the distortion coefficients β. I shall write μ_ε for $\mu_k = \mu_{\varepsilon_k}$, to emphasize the role of ε as a regularization parameter. For consistency, I shall also write π_ε instead of π_k.

Step 1: Reduction to the case $\operatorname{Spt}\nu = \mathcal{X}$**.** This step is essentially trivial. Let $S = \operatorname{Spt}\nu$. By assumption $\pi_\varepsilon[(\mathcal{X}\times\mathcal{X})\setminus(S\times S)] = 0$, so $\pi_\varepsilon[y\notin S|x] = 0$, $\rho_\varepsilon\,\nu(dx)$-almost surely; equivalently, $\pi_\varepsilon[y\notin S|x] = 0$, $\nu(dx)$-almost surely on $\{\rho_\varepsilon > 0\}$. Since $U(0) = 0$, values of x such that $\rho_\varepsilon(x) = 0$ do not affect the integral in the left-hand side of (29.17), so we might restrict this integral to $y \in S$. Then the $\nu(dx)$ integration allows us to further restrict the integral to $x \in S$.

Since each π_ε is concentrated on the closed set $S\times S$, the same is true for the weak limit π; then the same reasoning as above applies for the right-hand side of (29.17), and that integral can also be restricted to $S\times S$.

It only remains to check that the assumption of weak convergence $\pi_\varepsilon \to \pi$ is preserved under restriction to $S\times S$. Let φ be a continuous function on $S\times S$. Since $S\times S$ is compact, φ is uniformly continuous, so by Tietze–Urysohn's theorem it can be extended into a continuous function on the whole of $\mathcal{X}\times\mathcal{X}$, still denoted φ. Then

$$\int_{S\times S}\varphi\,d\pi_\varepsilon = \int_{\mathcal{X}\times\mathcal{X}}\varphi\,d\pi_\varepsilon \xrightarrow[\varepsilon\to 0]{} \int_{\mathcal{X}\times\mathcal{X}}\varphi\,d\pi = \int_{S\times S}\varphi\,d\pi,$$

so π is indeed the weak limit of π_ε, when viewed as a probability measure on $S\times S$.

In the sequel all the discussion will be restricted to S, so I shall assume $\operatorname{Spt}\nu = \mathcal{X}$.

Step 2: Reduction to the case $U'(0) > -\infty$. The problem now is to get rid of possibly very large negative values of U' close to 0. For any $\delta > 0$, define $U'_\delta(r) := \max(U'(\delta), U'(r))$ and

$$U_\delta(r) = \int_0^r U'_\delta(s)\, ds.$$

Since U' is nondecreasing, U'_δ converges monotonically to $U' \in L^1(0,r)$ as $\delta \to 0$. It follows that $U_\delta(r)$ decreases to $U(r)$, for all $r > 0$. Let us check that all the assumptions which we imposed on U, still hold true for U_δ. First, $U_\delta(0) = 0$. Also, since U'_δ is nondecreasing, U_δ is convex. Finally, U_δ has polynomial growth; indeed:

- if $r \le \delta$, then $r\,(U_\delta)'(r) = r\,U'(\delta)$ is bounded above by a constant multiple of r;
- if $r > \delta$, then $r\,(U_\delta)'(r) = r\,U'(r)$, which is bounded (by assumption) by $C(U(r)_+ + r)$, and this obviously is bounded by $C(U_\delta(r)_+ + r)$, for $U \le U_\delta$.

The next claim is that the integral

$$\int \beta(x,y)\, U_\delta\left(\frac{\rho(x)}{\beta(x,y)}\right) \pi(dy|x)\,\nu(dx)$$

makes sense and is not $+\infty$. Indeed, as we have just seen, there is a constant C such that $(U_\delta)_+ \le C(U_+(r) + r)$. Then the contribution of the linear part Cr is finite, since

$$\int \beta(x,y) \left(\frac{\rho(x)}{\beta(x,y)}\right) \pi(dy|x)\,\nu(dx) = \int \rho(x)\,\nu(dx) \le 1;$$

and the contribution of $C\,U_+(r)$ is also finite in view of (29.21). So

$$\int \beta(x,y)\,(U_\delta)_+ \left(\frac{\rho(x)}{\beta(x,y)}\right) \pi(dy|x)\,\nu(dx) \ < +\infty,$$

which proves the claim.

Now assume that Theorem 29.20(iii) has been proved with U_δ in place of U. Then, for any $\delta > 0$,

$$\begin{aligned}
\limsup_{\varepsilon\downarrow 0} \int \beta(x,y)\, U\left(\frac{\rho_\varepsilon(x)}{\beta(x,y)}\right) \pi_\varepsilon(dy|x)\,\nu(dx) \\
&\le \limsup_{\varepsilon\downarrow 0} \int \beta(x,y)\, U_\delta\left(\frac{\rho_\varepsilon(x)}{\beta(x,y)}\right) \pi_\varepsilon(dy|x)\,\nu(dx) \\
&\le \int \beta(x,y)\, U_\delta\left(\frac{\rho(x)}{\beta(x,y)}\right) \pi(dy|x)\,\nu(dx).
\end{aligned}$$

But by monotone convergence,

$$\int \beta(x,y)\, U_\delta\left(\frac{\rho(x)}{\beta(x,y)}\right) \pi(dy|x)\, \nu(dx) = \int \beta(x,y)\, U\left(\frac{\rho(x)}{\beta(x,y)}\right) \pi(dy|x)\, \nu(dx),$$

and inequality (29.17) follows.

To summarize: It is sufficient to establish (29.17) with U replaced by U_δ, and thus we may assume that U is bounded below by a linear function $r \to -Kr$.

Step 3: Reduction to the case when $U \geq 0$. As we have already seen, adding a linear function Kr to U does not alter the assumptions on U; it does not change the conclusion either, because this only adds the constant K to both sides of the inequality (29.17). For the right-hand side, this is a consequence of

$$\int \beta(x,y) K\, \frac{\rho_\varepsilon(x,y)}{\beta(x,y)}\, \pi_\varepsilon(dy|x)\, \nu(dx) = K \int \pi_\varepsilon(dy|x)\, (\rho_\varepsilon\, \nu)(dx) = K \int \pi_\varepsilon(dx\, dy) = K;$$

and for the right-hand side the computation is similar, once one has noticed that the first marginal of π is the weak limit of the first marginal μ_ε of π_ε, i.e. μ (as recalled in the First Appendix).

So in the sequel I shall assume that $U \geq 0$.

Step 4: Treatment of the singular part. To take care of the singular part, the reasoning is similar to the one already used in the particular case $\beta = 1$: Write $\mu = \rho\nu + \mu_s$, and

$$\rho_{a,\varepsilon}(x) = \int_{\mathcal{X}} K_\varepsilon(x,y)\, \rho(y)\, \nu(dy); \qquad \rho_{s,\varepsilon}(x) = \int_{\mathcal{X}} K_\varepsilon(x,y)\, \mu_s(dy).$$

Then by convexity of U, for any $\theta \in (0,1)$,

$$\begin{aligned} U^\beta_{\pi,\nu}(\mu_\varepsilon) &\leq (1-\theta)\, U^\beta_{\pi,\nu}\left(\frac{\rho_{a,\varepsilon}\, \nu}{1-\theta}\right) + \theta\, U^\beta_{\pi,\nu}\left(\frac{\rho_{s,\varepsilon}\, \nu}{\theta}\right) \\ &\leq (1-\theta)\, U^\beta_{\pi,\nu}\left(\frac{\rho_{a,\varepsilon}\, \nu}{1-\theta}\right) + U'(\infty)\, \mu_s[\mathcal{X}], \end{aligned}$$

and the limit $\theta \to 0$ yields

$$U^\beta_{\pi,\nu}(\mu_\varepsilon) \le U^\beta_{\pi,\nu}(\rho_{a,\varepsilon}\,\nu) + U'(\infty)\,\mu_s[\mathcal{X}].$$

In the next two steps I shall focus on the first term $U^\beta_{\pi,\nu}(\rho_{a,\varepsilon}\,\nu)$; I shall write ρ_ε for $\rho_{a,\varepsilon}$.

Step 5: Approximation of β. For any two points x, y in $\mathcal{X}$, define

$$\beta_\varepsilon(x,y) = \int_{\mathcal{X}\times\mathcal{X}} K_\varepsilon(\overline{x},x)\,K_\varepsilon(\overline{y},y)\,\beta(\overline{x},\overline{y})\,\nu(d\overline{x})\,\nu(d\overline{y}).$$

The measure $K_\varepsilon(x,\overline{x})K_\varepsilon(y,\overline{y})\,\nu(d\overline{x})\,\nu(d\overline{y})$ is a probability measure on $\mathcal{X}\times\mathcal{X}$, supported in $\{d(x,\overline{x})\le\varepsilon,\ d(y,\overline{y})\le\varepsilon\}$, so

$$\begin{aligned}|\beta_\varepsilon(x,y)-\beta(x,y)| &= \left|\int K_\varepsilon(x,\overline{x})\,K_\varepsilon(y,\overline{y})[\beta(\overline{x},\overline{y})-\beta(x,y)]\nu(d\overline{x})\,\nu(d\overline{y})\right| \\ &\le \sup\Big\{|\beta(\overline{x},\overline{y})-\beta(x,y)|;\ d(x,\overline{x})\le\varepsilon,\ d(y,\overline{y})\le\varepsilon\Big\}.\end{aligned}$$

The latter quantity goes to 0 uniformly in x and y by uniform continuity of β; so β_ε converges uniformly to β. The goal now is to replace β by β_ε in the left-hand side of the desired inequality (29.17).

The map $w : b \longmapsto b\,U(r/b)$ is continuously differentiable and (since $U\ge 0$),

$$|w'(b)| = p(r/b) \le (r/b)\,U'(r/b) \le C(U(r/b)+r/b).$$

So if, $\beta\le\widetilde{\beta}$ are two positive real numbers, then

$$\left|\widetilde{\beta}\,U\left(\frac{\rho}{\widetilde{\beta}}\right) - \beta\,U\left(\frac{\rho}{\beta}\right)\right| = \int_\beta^{\widetilde{\beta}} p\left(\frac{\rho}{b}\right)\,db \le C\,|\widetilde{\beta}-\beta| \sup_{\frac{\rho}{\widetilde{\beta}}\le r\le\frac{\rho}{\beta}} (U(r)+r).$$

Now assume that β and $\widetilde{\beta}$ are bounded from above and below by positive constants, say $B^{-1}\le\beta,\widetilde{\beta}\le B$, then by (29.19) there is $C>0$, depending only on B, such that

$$\left|\widetilde{\beta}\,U\left(\frac{\rho}{\widetilde{\beta}}\right) - \beta\,U\left(\frac{\rho}{\beta}\right)\right| \le C\,|\widetilde{\beta}-\beta|\,(U(\rho)+\rho).$$

Apply this estimate with $\rho=\rho_\varepsilon(x)$, $\beta=\beta(x,y)$, and $\{\beta,\widetilde{\beta}\} = \{\beta(x,y),\beta_\varepsilon(x,y)\}$; this is allowed since β is bounded from above and below by positive constants, and the same is true of β_ε since it has been obtained by averaging β. So there is a constant C such that for all $x,y\in\mathcal{X}$,

$$\left|\beta_\varepsilon(x,y)\,U\left(\frac{\rho_\varepsilon(x)}{\beta_\varepsilon(x,y)}\right) - \beta(x,y)\,U\left(\frac{\rho_\varepsilon(x)}{\beta(x,y)}\right)\right| \\ \leq C\left|\beta_\varepsilon(x,y) - \beta(x,y)\right|\big(U(\rho_\varepsilon(x)) + \rho_\varepsilon(x)\big). \quad (29.22)$$

Then

$$\left|\int \beta_\varepsilon(x,y)\,U\left(\frac{\rho_\varepsilon(x)}{\beta_\varepsilon(x,y)}\right)\pi_\varepsilon(dy|x)\,\nu(dx) \right. \\ \left. - \int \beta(x,y)\,U\left(\frac{\rho_\varepsilon(x)}{\beta(x,y)}\right)\pi_\varepsilon(dy|x)\,\nu(dx)\right|$$

$$\leq \int \left|\beta_\varepsilon(x,y)\,U\left(\frac{\rho_\varepsilon(x)}{\beta_\varepsilon(x,y)}\right) - \beta(x,y)\,U\left(\frac{\rho_\varepsilon(x)}{\beta(x,y)}\right)\right|\pi_\varepsilon(dy|x)\,\nu(dx)$$

$$\leq C\int \left|\beta_\varepsilon(x,y) - \beta(x,y)\right|\big(U(\rho_\varepsilon(x)) + \rho_\varepsilon(x)\big)\,\pi_\varepsilon(dy|x)\,\nu(dx)$$

$$\leq C\left(\sup_{x,y\in\mathcal{X}}\left|\beta_\varepsilon(x,y) - \beta(x,y)\right|\right)\int \big[U(\rho_\varepsilon(x)) + \rho_\varepsilon(x)\big]\,\nu(dx)$$

$$\leq C\left(\sup_{x,y\in\mathcal{X}}\left|\beta_\varepsilon(x,y) - \beta(x,y)\right|\right)\int \big[U(\rho) + \rho\big]\,d\nu,$$

where the last inequality follows from Jensen's inequality as in the proof of the particular case $\beta \equiv 1$. To summarize:

$$\limsup_{\varepsilon\downarrow 0}\left|\int \beta_\varepsilon(x,y)\,U\left(\frac{\rho_\varepsilon(x)}{\beta_\varepsilon(x,y)}\right)\pi_\varepsilon(dy|x)\,\nu(dx) \right. \\ \left. - \int \beta(x,y)\,U\left(\frac{\rho_\varepsilon(x)}{\beta(x,y)}\right)\pi_\varepsilon(dy|x)\,\nu(dx)\right| = 0.$$

So the desired conclusion is equivalent to

$$\limsup_{\varepsilon\downarrow 0}\int \beta_\varepsilon(x,y)\,U\left(\frac{\rho_\varepsilon(x)}{\beta_\varepsilon(x,y)}\right)\pi_\varepsilon(dy|x)\,\nu(dx) + U'(\infty)\,\mu_s[\mathcal{X}] \\ \leq \int \beta(x,y)\,U\left(\frac{\rho(x)}{\beta(x,y)}\right)\pi(dy|x)\,\nu(dx) + U'(\infty)\,\mu_s[\mathcal{X}].$$

Step 6: Convexity inequality. This is a key step. By Legendre representation,

$$\beta\, U\left(\frac{\rho}{\beta}\right) = \beta \sup_{p\in\mathbb{R}} \left(p\,\frac{\rho}{\beta} - U^*(p)\right)$$
$$= \sup_{p\in\mathbb{R}} \big[p\rho - \beta\, U^*(p)\big],$$

so $\beta\, U(\rho/\beta)$ is a (jointly) convex function of $(\beta,\rho) \in (0,+\infty)\times\mathbb{R}_+$.

On the other hand, β_ε and ρ_ε are averaged values of $\beta(\overline{x},\overline{y})$ and $\rho(\overline{x})$, respectively, over the probability measure $K_\varepsilon(x,\overline{x})K_\varepsilon(y,\overline{y})\,\nu(d\overline{x})\,\nu(d\overline{y})$. So by Jensen's inequality,

$$\beta_\varepsilon(x,y)\, U\left(\frac{\rho_\varepsilon(x)}{\beta_\varepsilon(x,y)}\right)$$
$$\le \int K_\varepsilon(x,\overline{x})\, K_\varepsilon(y,\overline{y})\, \beta(\overline{x},\overline{y})\, U\left(\frac{\rho(\overline{x})}{\beta(\overline{x},\overline{y})}\right) \nu(d\overline{x})\,\nu(d\overline{y}).$$

To conclude the proof of (29.17) it suffices to show

$$\int K_\varepsilon(x,\overline{x})\, K_\varepsilon(y,\overline{y})\, \beta(\overline{x},\overline{y})\, U\left(\frac{\rho(\overline{x})}{\beta(\overline{x},\overline{y})}\right) \nu(d\overline{x})\,\nu(d\overline{y})\,\pi_\varepsilon(dy|x)\,\nu(dx)$$
$$+ U'(\infty)\,\mu_s[\mathcal{X}]$$
$$\xrightarrow[\varepsilon\to 0]{} \int \beta(x,y)\, U\left(\frac{\rho(x)}{\beta(x,y)}\right) \pi(dy|x)\,\nu(dx) + U'(\infty)\,\mu_s[\mathcal{X}]. \qquad (29.23)$$

This will be the object of the final three steps.

Step 7: Approximation by a continuous function. Let us start with some explanations. Forget about the singular part for simplicity, and define

$$f(x,y) = \beta(x,y)\, U\left(\frac{\rho(x)}{\beta(x,y)}\right),$$
$$\omega_\varepsilon(dx\,dy) = \left(\int K_\varepsilon(\overline{x},x)\, K_\varepsilon(\overline{y},y)\,\pi_\varepsilon(d\overline{y}|\overline{x})\,\nu(d\overline{x})\right)\nu(dy)\,\nu(dx),$$
$$\omega(dx\,dy) = \pi(dy|x)\,\nu(dx).$$

With this notation, the goal (29.23) can be rewritten as

$$\int f\, d\omega_\varepsilon \longrightarrow \int f\, d\omega. \qquad (29.24)$$

This is not trivial, in particular because f is a priori not continuous. The obvious solution is to try to replace f by a continuous approximation, but then we run into another problem: the conditional measure

$\pi(dy|x)$ is completely arbitrary if $\rho(x) = 0$. This is not serious as long as we multiply it by $f(x,y)$, since $f(x,y)$ vanishes when $\rho(x)$ does, but this might become annoying when $f(x,y)$ is replaced by a continuous approximation.

Instead, let us rather define

$$g(x,y) = \frac{f(x,y)}{\rho(x)} = \frac{\beta(x,y)}{\rho(x)}\, U\left(\frac{\rho(x)}{\beta(x,y)}\right),$$

with the convention $U(0)/0 = U'(0)$, $U(\infty)/\infty = U'(\infty)$, and we impose that $\rho(x)$ is finite outside of $\operatorname{Spt}\mu_s$ and takes the value $+\infty$ on $\operatorname{Spt}\mu_s$. If $U'(\infty) = +\infty$, then the finiteness of $U^{\beta}_{\pi,\nu}(\mu)$ imposes $\mu_s[\mathcal{X}] = 0$, so $\operatorname{Spt}\mu_s = \emptyset$; in particular, g takes values in $\mathbb{R}$.

We further define

$$\widetilde{\pi}_\varepsilon(dx\,dy) = \left(\int K_\varepsilon(\overline{x},x)\,K_\varepsilon(\overline{y},y)\,\pi_\varepsilon(d\overline{y}|\overline{x})\,\nu(d\overline{x})\right)\nu(dy)\,\mu(dx).$$

Note that the x-marginal of $\widetilde{\pi}_\varepsilon$ is

$$\begin{aligned}
&\left(\int K_\varepsilon(\overline{x},x)\,K_\varepsilon(\overline{y},y)\,\pi_\varepsilon(d\overline{y}|\overline{x})\,\nu(d\overline{x})\,\nu(dy)\right)\mu(dx)\\
&= \left(\int K_\varepsilon(\overline{x},x)\,\pi_\varepsilon(d\overline{y}|\overline{x})\,\nu(d\overline{x})\right)\mu(dx)\\
&= \left(\int K_\varepsilon(\overline{x},x)\,\nu(d\overline{x})\right)\mu(dx) \;= \mu(dx).
\end{aligned}$$

In particular,

$$\int_{\operatorname{Spt}\mu_s} g(x,y)\,\widetilde{\pi}_\varepsilon(dx\,dy) = U'(\infty)\,\mu_s(dx).$$

Then the goal (29.23) becomes

$$\int g(x,y)\,\widetilde{\pi}_\varepsilon(dx\,dy) \xrightarrow[\varepsilon\to 0]{} \int g(x,y)\,\pi(dx\,dy). \tag{29.25}$$

- If $U'(\infty) = +\infty$, then $\operatorname{Spt}\mu_s = \emptyset$ and for each x, $g(x,y)$ is a continuous function of y; moreover, since β is bounded from above and below by positive constants, (29.19) implies

$$\sup_y \left[\beta(x,y)\,U\left(\frac{\rho(x)}{\beta(x,y)}\right)\right] \le C\Big(U(\rho(x)) + \rho(x)\Big).$$

In particular,

$$\int_{\mathcal{X}} \sup_y g(x,y)\,d\mu(x) = \int_{\mathcal{X}} \sup_y \left[\beta(x,y)\,U\left(\frac{\rho(x)}{\beta(x,y)}\right)\right]\,d\nu(x) \;<+\infty;$$

in other words, g belongs to the vector space $L^1\big((\mathcal{X},\mu);\,C(\mathcal{X})\big)$ of μ-integrable functions valued in the normed space $C(\mathcal{X})$ (equipped with the norm of uniform convergence).

- If $U'(\infty) < +\infty$ then $g(x,\,\cdot\,)$ is also continuous for all x (it is identically equal to $U'(\infty)$ if $x \in \operatorname{Spt}\mu_s$), and $\sup_y g(x,y) \leq U'(\infty)$ is obviously $\mu(dx)$-integrable; so the conclusion $g \in L^1\big((\mathcal{X},\mu);\,C(\mathcal{X})\big)$ still holds true.

By Lemma 29.36 in the Second Appendix, there is a sequence $(\Psi_k)_{k\in\mathbb{N}}$ in $C(\mathcal{X}\times\mathcal{X})^{\mathbb{N}}$ such that

$$\int_{\mathcal{X}} \sup_y \big|g(x,y) - \Psi_k(x,y)\big|\,d\mu(x) \xrightarrow[k\to\infty]{} 0,$$

and each function $\Psi_k(x,\,\cdot\,)$ is identically equal to $U'(\infty)$ when x varies in $\operatorname{Spt}\mu_s$.

Since the x-marginal of π is μ,

$$\left|\int \Big(g(x,y) - \Psi_k(x,y)\Big)\,\pi(dx\,dy)\right| \\ \leq \int \sup_y \big|f(x,y) - \Psi_k(x,y)\big|\,\mu(dx) \xrightarrow[k\to\infty]{} 0.$$

A similar computation applies with $\widetilde{\pi}_\varepsilon$ in place of π. So it is also true that

$$\limsup_{\varepsilon\downarrow 0} \left|\int \Big(g(x,y) - \Psi_k(x,y)\Big)\,\widetilde{\pi}_\varepsilon(dx\,dy)\right| \xrightarrow[k\to\infty]{} 0.$$

After these estimates, to conclude the proof it is sufficient to show that for any fixed k,

$$\int \Psi_k(x,y)\,\widetilde{\pi}_\varepsilon(dx\,dy) \xrightarrow[\varepsilon\downarrow 0]{} \int \Psi_k(x,y)\,\pi(dx\,dy). \qquad (29.26)$$

In the sequel I shall drop the index k and write just Ψ for Ψ_k. It will be useful in the sequel to know that $\Psi(x,y) = U'(\infty)$ when $x \in \operatorname{Spt}\mu_s$; apart from that, Ψ might be any continuous function.

Step 8: Variations on a regularization theme. Let $\widetilde{\pi}_\varepsilon^{(a)}$ be the contribution of $\rho\nu$ to $\widetilde{\pi}_\varepsilon$. Explicitly,

$$\widetilde{\pi}_\varepsilon^{(a)}(dx\,dy) = \left(\int K_\varepsilon(\overline{x},x)\,K_\varepsilon(\overline{y},y)\,\pi_\varepsilon(d\overline{y}|\overline{x})\,\rho(x)\,\nu(d\overline{x})\right)\nu(dy)\,\nu(dx).$$

Then let

$$\overline{\pi}_\varepsilon^{(a)}(dx\,dy) = \left(\int K_\varepsilon(\overline{x},x)\,K_\varepsilon(\overline{y},y)\,\rho(\overline{x})\,\nu(d\overline{x})\,\pi_\varepsilon(d\overline{y}|\overline{x})\right)\nu(dy)\,\nu(dx);$$

$$\widehat{\pi}_\varepsilon^{(a)}(dx\,dy) = \left(\int K_\varepsilon(\overline{x},x)\,K_\varepsilon(\overline{y},y)\,\rho_\varepsilon(\overline{x})\,\nu(d\overline{x})\,\pi_\varepsilon(d\overline{y}|\overline{x})\right)\nu(dy)\,\nu(dx).$$

We shall check that $\widetilde{\pi}_\varepsilon^{(a)}$ is well approximated by $\overline{\pi}_\varepsilon^{(a)}$ and $\widehat{\pi}_\varepsilon^{(a)}$. First of all,

$$\begin{aligned}
&\|\overline{\pi}_\varepsilon^{(a)} - \widetilde{\pi}_\varepsilon^{(a)}\|_{TV} \\
&\qquad \le \int K_\varepsilon(\overline{x},x)\,K_\varepsilon(\overline{y},y)\,|\rho(\overline{x})-\rho(x)|\,\pi_\varepsilon(d\overline{y}|\overline{x})\,\nu(d\overline{x})\,\nu(dy)\,\nu(dx) \\
&\qquad = \int K_\varepsilon(\overline{x},x)\,|\rho(\overline{x})-\rho(x)|\,\pi_\varepsilon(d\overline{y}|\overline{x})\,\nu(d\overline{x})\,\nu(dx) \\
&\qquad = \int K_\varepsilon(\overline{x},x)\,|\rho(\overline{x})-\rho(x)|\,\nu(d\overline{x})\,\nu(dx).
\end{aligned} \tag{29.27}$$

Let us show that the integral in (29.27) converges to 0. Since $C(\mathcal{X})$ is dense in $L^1(\mathcal{X},\nu)$, there is a sequence of continuous functions $(\psi_j)_{j\in\mathbb{N}}$ converging to ρ in $L^1(\mathcal{X},\nu)$. Then

$$\int K_\varepsilon(\overline{x},x)\,|\rho(\overline{x})-\psi_j(\overline{x})|\,\nu(d\overline{x})\,\nu(dx) = \int |\rho(\overline{x})-\psi_j(\overline{x})|\,\nu(d\overline{x}) \xrightarrow[j\to\infty]{} 0, \tag{29.28}$$

and the convergence is uniform in ε. Symetrically,

$$\int K_\varepsilon(\overline{x},x)\,|\rho(x)-\psi_j(x)|\,\nu(d\overline{x})\,\nu(dx) \xrightarrow[j\to\infty]{} 0 \qquad \text{uniformly in } \varepsilon. \tag{29.29}$$

On the other hand, for each j, the uniform continuity of ψ_j guarantees that

$$\begin{aligned}
&\int K_\varepsilon(\overline{x},x)\,|\psi_j(x)-\psi_j(\overline{x})|\,\nu(d\overline{x})\,\nu(dx) \\
&\qquad\qquad \le \nu[\mathcal{X}] \sup_{d(x,\overline{x})\le\varepsilon} |\psi_j(x)-\psi_j(\overline{x})| \xrightarrow[\varepsilon\to 0]{} 0.
\end{aligned} \tag{29.30}$$

The combination of (29.27), (29.28), (29.29) and (29.30) shows that

$$\|\overline{\pi}^{(a)}_\varepsilon - \widetilde{\pi}^{(a)}_\varepsilon\|_{TV} \longrightarrow 0. \tag{29.31}$$

Next,

$$\begin{aligned}
&\|\overline{\pi}^{(a)}_\varepsilon - \widehat{\pi}^{(a)}_\varepsilon\|_{TV} \\
&\quad\le \int K_\varepsilon(\overline{x},x)\,K_\varepsilon(\overline{y},y)\,|\rho_\varepsilon(\overline{x})-\rho(\overline{x})|\,\pi_\varepsilon(d\overline{y}|\overline{x})\,\nu(d\overline{x})\,\nu(dy)\,\nu(dx)\\
&\quad= \int K_\varepsilon(\overline{x},x)\,|\rho_\varepsilon(\overline{x})-\rho(\overline{x})|\,\pi_\varepsilon(d\overline{y}|\overline{x})\,\nu(d\overline{x})\,\nu(dx)\\
&\quad= \int K_\varepsilon(\overline{x},x)\,|\rho_\varepsilon(\overline{x})-\rho(\overline{x})|\,\nu(d\overline{x})\,\nu(dx)\\
&\quad= \int |\rho_\varepsilon(\overline{x})-\rho(\overline{x})|\,\nu(d\overline{x})\\
&\quad= \int \left|\int K_\varepsilon(\overline{x},\overline{y})\,\big[\rho(\overline{y})-\rho(\overline{x})\big]\,\nu(d\overline{y})\right|\nu(d\overline{x})\\
&\quad\le \int K_\varepsilon(\overline{x},\overline{y})\,|\rho(\overline{y})-\rho(\overline{x})|\,\nu(d\overline{y})\,\nu(d\overline{x}),
\end{aligned}$$

and this goes to 0 as we already saw before; so

$$\|\widehat{\pi}^{(a)}_\varepsilon - \overline{\pi}^{(a)}_\varepsilon\|_{TV} \longrightarrow 0, \tag{29.32}$$

By (29.31) and (29.32), $\|\widehat{\pi}^{(a)}_\varepsilon - \widetilde{\pi}^{(a)}_\varepsilon\|_{TV} \longrightarrow 0$ as $\varepsilon \to 0$. In particular,

$$\int \Psi\,d\widehat{\pi}^{(a)}_\varepsilon - \int \Psi\,d\widetilde{\pi}^{(a)}_\varepsilon \xrightarrow[\varepsilon\downarrow 0]{} 0. \tag{29.33}$$

Now let

$$\widetilde{\pi}^{(s)}_\varepsilon(dx\,dy) = \left(\int K_\varepsilon(\overline{x},x)\,K_\varepsilon(\overline{y},y)\,\pi_\varepsilon(d\overline{y}|\overline{x})\,\nu(d\overline{x})\right)\,\nu(dy)\,\mu_s(dx);$$

$$\widehat{\pi}^{(s)}_\varepsilon(dx\,dy) = \left(\int K_\varepsilon(\overline{x},x)\,K_\varepsilon(\overline{y},y)\,\mu_{s,\varepsilon}(d\overline{x})\,\pi_\varepsilon(d\overline{y}|\overline{x})\right)\,\nu(dy)\,\nu(dx);$$

so that $\widetilde{\pi}_\varepsilon(dx\,dy) = \widetilde{\pi}^{(a)}_\varepsilon(dx\,dy) + \widetilde{\pi}^{(s)}_\varepsilon(dx\,dy)$. Further, define

$$\begin{aligned}
\widehat{\pi}_\varepsilon(dx\,dy) &= \widehat{\pi}^{(a)}_\varepsilon(dx\,dy) + \widehat{\pi}^{(s)}_\varepsilon(dx\,dy)\\
&= \left(\int K_\varepsilon(\overline{x},x)\,K_\varepsilon(\overline{y},y)\,\pi_\varepsilon(d\overline{x}\,d\overline{y})\right)\,\nu(dy)\,\nu(dx).
\end{aligned}$$

Since $\Psi(x,y) = U'(\infty)$ when $x \in \operatorname{Spt} \mu_s$, we have

$$\int \Psi \, d\widetilde{\pi}_\varepsilon^{(s)} = \int_{\operatorname{Spt} \mu_s} \Psi \, d\widetilde{\pi}_\varepsilon^{(s)} = U'(\infty)\, \mu_s[\mathcal{X}],$$

while

$$\int \Psi \, d\widehat{\pi}_\varepsilon^{(s)} = U'(\infty)\, \mu_{s,\varepsilon}[\mathcal{X}] = U'(\infty)\, \mu_s[\mathcal{X}].$$

Combining this with (29.33), we can conclude that

$$\int \Psi \, d\widehat{\pi}_\varepsilon - \int \Psi \, d\widetilde{\pi}_\varepsilon \xrightarrow[\varepsilon \downarrow 0]{} 0. \tag{29.34}$$

At this point, to finish the proof of the theorem it suffices to establish $\int \Psi \, d\widehat{\pi}_\varepsilon \longrightarrow \int \Psi \, d\pi$; which is true if $\widehat{\pi}_\varepsilon$ converges weakly to π.

Step 9: Duality. Proving the convergence of $\widehat{\pi}_\varepsilon$ to π will be easy because $\widehat{\pi}_\varepsilon$ is a kind of regularization of π_ε, and it will be possible to "transfer the regularization to the test function" by duality. Indeed:

$$\begin{aligned}\int \Psi(x,y)\, \widehat{\pi}_\varepsilon(dx\, dy) &= \int \Psi(x,y)\, K_\varepsilon(\overline{x},x)\, K_\varepsilon(\overline{y},y)\, \pi_\varepsilon(d\overline{y}\, d\overline{x})\, \nu(dy)\, \nu(dx) \\ &= \int \Psi_\varepsilon(\overline{x},\overline{y})\, \pi_\varepsilon(d\overline{y}\, d\overline{x}),\end{aligned}$$

where

$$\Psi_\varepsilon(\overline{x},\overline{y}) = \int \Psi(x,y)\, K_\varepsilon(\overline{x},x)\, K_\varepsilon(\overline{y},y)\, \nu(dy)\, \nu(dx).$$

By the same classical argument that we already used several times, Ψ_ε converges uniformly to Ψ:

$$\begin{aligned}&\left|\Psi_\varepsilon(\overline{x},\overline{y}) - \Psi(\overline{x},\overline{y})\right| \\ &\qquad = \left|\int [\Psi(x,y) - \Psi(\overline{x},\overline{y})]\, K_\varepsilon(\overline{x},x)\, K_\varepsilon(\overline{y},y)\, \nu(dx)\, \nu(dy)\right| \\ &\qquad \leq \sup_{d(\overline{x},x)\leq\varepsilon,\ d(\overline{y},y)\leq\varepsilon} \left|\Psi(x,y) - \Psi(\overline{x},\overline{y})\right| \xrightarrow[\varepsilon \downarrow 0]{} 0;\end{aligned}$$

Since on the other hand π_ε converges weakly to π, we conclude that

$$\int \Psi_\varepsilon \, d\pi_\varepsilon \longrightarrow \int \Psi \, d\pi,$$

and the proof is complete. □

I shall conclude this section with a corollary of Theorem 29.20:

Corollary 29.23 (Another sufficient condition to be a weak CD(K,N) space). *In Definition 29.8 it is equivalent to require inequality* (29.11) *for all probability measures μ_0, μ_1; or only when μ_0, μ_1 are absolutely continuous with continuous densities.*

Proof of Corollary 29.23. Assume that $(\mathcal{X}, d, \nu)$ satisfies the assumptions of Definition 29.8, except that μ_0, μ_1 are required to be absolutely continuous. The goal is to show that the absolute continuity condition can be relaxed.

By Proposition 29.12, we may assume that U has polynomial growth, in the sense of Theorem 29.20(iii). Let us also assume that $\beta_t^{(K,N)}$ is continuous.

Let μ_0, μ_1 be two possibly singular probability measures such that $\operatorname{Spt}\mu_0, \operatorname{Spt}\mu_1 \subset \nu$. By Theorem 29.20(iii), there are sequences of probability measures $\mu_{k,0} \to \mu_0$ and $\mu_{k,1} \to \mu_1$, all absolutely continuous and with continuous densities, such that for any $\pi_k \in \Pi(\mu_{k,0}, \mu_{k,1})$, converging weakly to π,

$$\begin{cases} \limsup_{k\to\infty} U^{\beta_{1-t}^{(K,N)}}_{\pi_k,\nu}(\mu_{k,0}) \le U^{\beta_{1-t}^{(K,N)}}_{\pi,\nu}(\mu_0); \\ \limsup_{k\to\infty} U^{\beta_t^{(K,N)}}_{\check{\pi}_k,\nu}(\mu_{k,1}) \le U^{\beta_t^{(K,N)}}_{\check{\pi},\nu}(\mu_{k,0}). \end{cases} \qquad (29.35)$$

For each $k \in \mathbb{N}$, there is a displacement interpolation $(\mu_{k,t})_{0\le t\le 1}$ and an associated optimal transference plan $\pi_k \in P(\mathcal{X} \times \mathcal{X})$ such that

$$U_\nu(\mu_{k,t}) \le (1-t)\, U^{\beta_{1-t}^{(K,N)}}_{\pi_k,\nu}(\mu_{k,0}) + t\, U^{\beta_t^{(K,N)}}_{\check{\pi}_k,\nu}(\mu_{k,1}). \qquad (29.36)$$

By Theorem 28.9 (in the very simple case when $\mathcal{X}_k = \mathcal{X}$ for all k), we may extract a subsequence such that $\mu_{k,t} \longrightarrow \mu_t$ in $P_2(\mathcal{X})$, for each $t \in [0,1]$, and $\pi_k \longrightarrow \pi$ in $P(\mathcal{X}\times\mathcal{X})$, where $(\mu_t)_{0\le t\le 1}$ is a displacement interpolation and $\pi \in \Pi(\mu_0, \mu_1)$ is an associated optimal transference plan. Then by Theorem 29.20(i),

$$U_\nu(\mu_t) \le \liminf_{k\to\infty} U_\nu(\mu_{k,t}).$$

Combining this with (29.35) and (29.36), we deduce that

$$U_\nu(\mu_t) \le (1-t)\, U^{\beta_{1-t}^{(K,N)}}_{\pi,\nu}(\mu_0) + t\, U^{\beta_t^{(K,N)}}_{\check{\pi},\nu}(\mu_1),$$

as required.

It remains to treat the case when $\beta_t^{(K,N)}$ is not continuous. By Proposition 29.11 this can occur only if $N = 1$ or $\operatorname{diam}(\mathcal{X}) = \pi\sqrt{(N-1)/K}$. In both cases, Proposition 29.10 and the previous proof show that

$$\forall N' > N, \qquad U_\nu(\mu_t) \le (1-t)\, U_{\pi,\nu}^{\beta_{1-t}^{(K,N')}}(\mu_0) + t\, U_{\check{\pi},\nu}^{\beta_t^{(K,N')}}(\mu_1).$$

Then the conclusion is obtained by letting $N' \downarrow N$. □

Stability of Ricci bounds

Now we have all the tools to prove the main result of this chapter: *The weak curvature-dimension bound* $\mathrm{CD}(K,N)$ *passes to the limit.* Once again, the compact case will imply the general statement.

Theorem 29.24 (Stability of weak CD(K,N) under MGH). *Let $(\mathcal{X}_k, d_k, \nu_k)_{k\in\mathbb{N}}$ be a sequence of compact metric-measure geodesic spaces converging in the measured Gromov–Hausdorff topology to a compact metric-measure space $(\mathcal{X}, d, \nu)$. Let $K \in \mathbb{R}$ and $N \in [1,\infty]$. If each $(\mathcal{X}_k, d_k, \nu_k)$ satisfies the weak curvature-dimension condition* $\mathrm{CD}(K,N)$*, then also $(\mathcal{X}, d, \nu)$ satisfies* $\mathrm{CD}(K,N)$.

Theorem 29.25 (Stability of weak CD(K,N) under pMGH). *Let $(\mathcal{X}_k, d_k, \nu_k)_{k\in\mathbb{N}}$ be a sequence of locally compact, complete, separable σ-finite metric-measure geodesic spaces converging in the pointed measured Gromov–Hausdorff topology to a locally compact, complete separable σ-finite metric-measure space $(\mathcal{X}, d, \nu)$. Let $K \in \mathbb{R}$ and $N \in [1,\infty]$. If each $(\mathcal{X}_k, d_k, \nu_k)$ satisfies the weak curvature-dimension condition* $\mathrm{CD}(K,N)$*, then also $(\mathcal{X}, d, \nu)$ satisfies* $\mathrm{CD}(K,N)$.

Remark 29.26. An easy variant of Theorem 29.25 is as follows: If $(\mathcal{X}_k, d_k, \nu_k)$ converges to $(\mathcal{X}, d, \nu)$ in the geodesic local Gromov–Hausdorff topology, and each $(\mathcal{X}_k, d_k, \nu_k)$ satisfies $\mathrm{CD}(K,N)$, then also $(\mathcal{X}, d, \nu)$ satisfies $\mathrm{CD}(K,N)$.

Proof of Theorem 29.24. Let $(\mathcal{X}_k, d_k, \nu_k)_{k\in\mathbb{N}}$ be a sequence of metric-measure spaces satisfying the assumptions of Theorem 29.24. From the characterization of measured Gromov–Hausdorff convergence, we know that there are measurable functions $f_k : \mathcal{X}_k \to \mathcal{X}$ such that:

(i) f_k is an ε_k-isometry $(\mathcal{X}_k, d_k) \to (\mathcal{X}, d)$, with $\varepsilon_k \to 0$;

(ii) $(f_k)_\# \nu_k$ converges weakly to ν.

Let ρ_0, ρ_1 be two probability densities on $(\mathcal{X}, \nu)$; let $\mu_0 = \rho_0\, \nu$, $\mu_1 = \rho_1\, \nu$.

Let $\varepsilon = (\varepsilon_m)_{m \in \mathbb{N}}$ be a sequence going to 0; for each $t_0 \in \{0, 1\}$, let (ρ_{ε,t_0}) be a sequence of continuous probability densities satisfying the conclusion of Theorem 29.20(iii) with $\mu = \mu_{t_0}$, and let $\mu_{\varepsilon,t_0} = \rho_{\varepsilon,t_0}\, \nu$ be the associated measure. In particular, μ_{ε,t_0} converges weakly to μ_{t_0} as $\varepsilon \to 0$.

Still for $t_0 \in \{0, 1\}$, define

$$\mu^k_{\varepsilon,t_0} := \frac{(\rho_{\varepsilon,t_0} \circ f_k)\, \nu_k}{Z^k_{\varepsilon,t_0}}, \qquad Z^k_{\varepsilon,t_0} = \int (\rho_{\varepsilon,t_0} \circ f_k)\, d\nu_k.$$

Since ρ_{ε,t_0} is continuous and $(f_k)_\# \nu_k$ converges weakly to ν, we have

$$Z^k_{\varepsilon,t_0} = \int \rho_{\varepsilon,t_0}\, d((f_k)_\# \nu_k) \xrightarrow[k \to \infty]{} \int \rho_{\varepsilon,t_0}\, d\nu = 1;$$

in particular $Z^k_{\varepsilon,t_0} > 0$ for k large enough, and then μ^k_{ε,t_0} is a probability measure on $\mathcal{X}_k$.

Let $\psi \in C(\mathcal{X})$, then

$$\begin{aligned} \int \psi\, d((f_k)_\# \mu^k_{\varepsilon,t_0}) = \int (\psi \circ f_k)\, d\mu^k_{\varepsilon,t_0} &= \frac{1}{Z^k_{\varepsilon,t_0}} \int (\psi \circ f_k)\, (\rho_{\varepsilon,t_0} \circ f_k)\, d\nu_k \\ &= \frac{1}{Z^k_{\varepsilon,t_0}} \int \psi \rho_{\varepsilon,t_0}\, d((f_k)_\# \nu_k). \end{aligned} \tag{29.37}$$

On the one hand, Z^k_{ε,t_0} converges to 1 as $k \to \infty$; on the other,

$$\int \psi \rho_{\varepsilon,t_0}\, d((f_k)_\# \nu_k) \xrightarrow[k \to \infty]{} \int \psi \rho_{\varepsilon,t_0}\, d\nu = \int \psi\, d\mu_{\varepsilon,t_0}.$$

Plugging this information back into (29.37), we obtain

$$(f_k)_\# \mu^k_{\varepsilon,t_0} \xrightarrow[k \to \infty]{} \mu_{\varepsilon,t_0} \qquad \text{weakly.} \tag{29.38}$$

Since each $(\mathcal{X}_k, d_k, \nu_k)$ satisfies $\mathrm{CD}(K, N)$, there is a Wasserstein geodesic $(\mu^k_{\varepsilon,t})_{0 \le t \le 1}$ joining $\mu^k_{\varepsilon,0}$ to $\mu^k_{\varepsilon,1}$, such that for all $U \in \mathcal{DC}_N$ and $t \in (0, 1)$,

$$U_{\nu_k}(\mu^k_{\varepsilon,t}) \le (1-t)\, U^{\beta^{(K,N)}_{1-t}}_{\pi^k_\varepsilon,\nu_k}(\mu^k_0) \;+\; t\, U^{\beta^{(K,N)}_t}_{\check{\pi}^k_\varepsilon,\nu}(\mu^k_1), \tag{29.39}$$

where $\beta^{(K,N)}_t$ is given by (14.61) (with the distance d_k) and π^k_ε is an optimal coupling associated with $(\mu^k_{\varepsilon,t})_{0\le t\le 1}$. This means that for each $\varepsilon \in (0,1)$ and $k \in \mathbb{N}$ there is a dynamical optimal transference plan Π^k_ε such that

$$\mu^k_{\varepsilon,t} = (e_t)_\# \Pi^k_\varepsilon, \qquad \pi^k_\varepsilon = (e_0, e_1)_\# \Pi^k_\varepsilon,$$

where e_t is the evaluation at time t.

By Theorem 28.9, up to extraction of a subsequence in k, there is a dynamical optimal transference plan Π_ε on $\Gamma(\mathcal{X})$ such that, as $k \to \infty$,

$$\begin{cases} (f_k \circ)_\# \Pi^k_\varepsilon \longrightarrow \Pi_\varepsilon & \text{weakly in } P(P([0,1]\times\mathcal{X})); \\ (f_k, f_k)_\# \pi^k_\varepsilon \longrightarrow \pi_\varepsilon & \text{weakly in } P(\mathcal{X}\times\mathcal{X}); \\ \sup_{0\le t\le 1} W_2\big((f_k)_\#\mu^k_{\varepsilon,t}, \mu_{\varepsilon,t}\big) \longrightarrow 0; \end{cases}$$

where

$$\mu_{\varepsilon,t} = (e_t)_\# \Pi_\varepsilon, \qquad \pi_\varepsilon = (e_0, e_1)_\# \Pi_\varepsilon.$$

Each curve $(\mu_{\varepsilon,t})_{0\le t\le 1}$ is D-Lipschitz with $D = \operatorname{diam}(\mathcal{X})$. By Ascoli's theorem, from $\varepsilon \in (0,1)$ we may extract a subsequence (still denoted ε for simplicity) such that

$$\sup_{0\le t\le 1} W_2\big(\mu_{\varepsilon,t}, \mu_t\big) \xrightarrow[\varepsilon\to 0]{} 0, \tag{29.40}$$

where $(\mu_t)_{0\le t\le 1}$ is a Wasserstein geodesic joining μ_0 to μ_1. It remains to "pass to the limit" in inequality (29.39), letting first $k \to \infty$, then $\varepsilon \to 0$, in order to show that

$$U_\nu(\mu_t) \le (1-t)\, U^{\beta^{(K,N)}_{1-t}}_{\pi,\nu}(\mu_0) \;+\; t\, U^{\beta^{(K,N)}_t}_{\check{\pi},\nu}(\mu_1). \tag{29.41}$$

By Proposition 29.12, it is sufficient to establish (29.41) when $U \in \mathcal{DC}_N$ is nonnegative and either Lipschitz, or behaving at infinity like $r \to a\, r\log r + b\, r$; in particular we may assume that U grows at most polynomially in the sense of Theorem 29.20(iii). In the sequel, U will be such a nonlinearity, and t will be an arbitrary time in $(0,1)$. Moreover, I shall first assume that the coefficients $\beta^{(K,N)}_t$ are *continuous and bounded.*

By the joint lower semicontinuity of $(\mu,\nu) \longmapsto U_\nu(\mu)$ (Theorem 29.20(i)) and the contraction property (Theorem 29.20(ii)), we have

$$U_\nu(\mu_{\varepsilon,t}) \leq \liminf_{k\to\infty} U_{(f_k)_\#\nu_k}\big((f_k)_\#\mu^k_{\varepsilon,t}\big) \leq \liminf_{k\to\infty} U_{\nu_k}(\mu^k_{\varepsilon,t}). \tag{29.42}$$

Then by lower semicontinuity again,

$$U_\nu(\mu_t) \leq \liminf_{\varepsilon\to 0} U_\nu(\mu_{\varepsilon,t}). \tag{29.43}$$

Inequalities (29.42) and (29.43) take care of the left-hand side of (29.39):

$$U_\nu(\mu_t) \leq \liminf_{\varepsilon\to 0}\ \liminf_{k\to\infty} U_{\nu_k}(\mu^k_{\varepsilon,t}). \tag{29.44}$$

It remains to pass to the limit in the right-hand side.

Let $\beta(x,y) = \beta^{(K,N)}_{1-t}(x,y)$. Since $\beta(x,y)$ is only a function of the distance $d_k(x,y)$, since

$$\lim_{k\to\infty}\ \sup_{x,y\in\mathcal{X}_k} \big|d_k(x,y) - d(f_k(x),f_k(y))\big| = 0,$$

and since $\rho^k_{\varepsilon,0}$ and U are continuous, $\beta(x_0,x_1)\,U(\rho^k_{\varepsilon,0}(x_0)/\beta(x_0,x_1))$ is uniformly close to $\beta(f_k(x_0),f_k(x_1))\,U(\rho^k_{\varepsilon,0}(x_0)/\beta(f_k(x_0),f_k(x_1)))$ as $k\to\infty$. So

$$\lim_{k\to 0}\left|\int \beta(x_0,x_1)\,U\left(\frac{\rho^k_{\varepsilon,0}(x_0)}{\beta(x_0,x_1)}\right)\pi^k_\varepsilon(dx_1|x_0)\,\nu_k(dx_0)\right.$$
$$\left. - \int \beta(f_k(x_0),f_k(x_1))\,U\left(\frac{\rho^k_{\varepsilon,0}(x_0)}{\beta(f_k(x_0),f_k(x_1))}\right)\pi^k_\varepsilon(dx_1|x_0)\,\nu_k(dx_0)\right| = 0. \tag{29.45}$$

(Of course, in the second integral β is computed with the distance d, while in the first integral it is computed with the distance d_k.)

Let $v(r) = U(r)/r$ (this is a continuous function if $v(0) = U'(0)$); by Lemma 29.6,

$$\int \beta(f_k(x_0),f_k(x_1))\,U\left(\frac{\rho^k_{\varepsilon,0}(x_0)}{\beta(f_k(x_0),f_k(x_1))}\right)\pi^k_\varepsilon(dx_1|x_0)\,\nu_k(dx_0)$$
$$= \int v\left(\frac{\rho^k_{\varepsilon,0}(x_0)}{\beta(f_k(x_0),f_k(x_1))}\right)\pi^k_\varepsilon(dx_0\,dx_1)$$
$$= \int v\left(\frac{\rho_{\varepsilon,0}(y_0)}{Z^k_{\varepsilon,0}\,\beta(y_0,y_1)}\right) d\big[(f_k,f_k)_\#\pi^k_\varepsilon\big](y_0,y_1). \tag{29.46}$$

Since the integrand is a continuous function of (y_0, y_1) converging uniformly as $k \to \infty$, and $(f_k, f_k)_\# \pi_\varepsilon^k$ converges weakly to π_ε, we may pass to the limit in (29.46):

$$\begin{aligned}\lim_{k\to\infty} \int v\left(\frac{\rho_{\varepsilon,0}(y_0)}{Z_{\varepsilon,0}^k\,\beta(y_0,y_1)}\right) d\big[(f_k, f_k)_\#\pi_\varepsilon^k\big](y_0,y_1) \\ = \int v\left(\frac{\rho_{\varepsilon,0}(y_0)}{\beta(y_0,y_1)}\right) d\pi(y_0,y_1) \\ = \int \beta(y_0,y_1)\, U\left(\frac{\rho_{\varepsilon,0}(y_0)}{\beta(y_0,y_1)}\right) \pi_\varepsilon(dy_1|y_0)\,\nu(dy_0), \qquad (29.47)\end{aligned}$$

where the latter equality follows again by Lemma 29.6.

Now it remains to pass to the limit as $\varepsilon \to 0$. But Theorem 29.20(iii) guarantees precisely that

$$\begin{aligned}\limsup_{\varepsilon\downarrow 0} \int_{\mathcal{X}\times\mathcal{X}} \beta(x_0,x_1)\, U\left(\frac{\rho_{\varepsilon,0}(x_0)}{\beta(x_0,x_1)}\right) \pi_\varepsilon(dx_1|x_0)\,\nu(dx_0) \\ \le \int_{\mathcal{X}\times\mathcal{X}} \beta(x_0,x_1)\, U\left(\frac{\rho_0(x_0)}{\beta(x_0,x_1)}\right) \pi(dx_1|x_0)\,\nu(dx_0).\end{aligned}$$

This combined with (29.45), (29.46) and (29.47) shows that

$$\limsup_{\varepsilon\downarrow 0}\ \limsup_{k\to\infty}\ U^{\beta_t^{(K,N)}}_{\pi_\varepsilon^k,\nu_k}(\mu_{\varepsilon,0}^k) \le U^{\beta_t}_{\pi,\nu}(\mu_0). \qquad (29.48)$$

Similarly,

$$\limsup_{\varepsilon\downarrow 0}\ \limsup_{k\to\infty}\ U^{\beta_t^{(K,N)}}_{\check{\pi}_\varepsilon^k,\nu_k}(\mu_{\varepsilon,1}^k) \le U^{\beta_t}_{\check{\pi},\nu}(\mu_1). \qquad (29.49)$$

To summarize: Starting from (29.39), we can apply (29.44) to pass to the limit in the left-hand side, and (29.48)–(29.49) to pass to the limit in the right-hand side; and we recover the desired inequality (29.41).

This concludes the proof of the theorem in the case when $\beta_t^{(K,N)}$ is bounded, which is true if any one of the following conditions is satisfied: (a) $K \le 0$ and $N > 1$; (b) $K > 0$ and $N = \infty$; (c) $K > 0$, $1 < N < \infty$ and $\sup \operatorname{diam}(\mathcal{X}_k) < D_{K,N} = \pi\sqrt{(N-1)/K}$.

If $K \le 0$ and $N = 1$, we can use the inequality

$$U^{\beta_t^{(K,1)}}_{\pi,\nu}(\mu) \le U^{\beta_t^{(K,N')}}_{\pi,\nu}(\mu)$$

where $N' > 1$ (recall Remark 17.31) to deduce that for any two probability measures μ_0^k, μ_1^k on $\mathcal{X}_k$, there is a Wasserstein geodesic $(\mu_t^k)_{t\in[0,1]}$ and an associated coupling π^k such that

$$U_{\nu_k}(\mu_t^k) \le (1-t)\, U_{\pi_k,\nu_k}^{\beta_{1-t}^{(K,N')}}(\mu_0^k) + t\, U_{\check{\pi}_k,\nu_k}^{\beta_t^{(K,N')}}(\mu_1^k).$$

Then the same proof as before shows that for any two probability measures μ_0, μ_1 on $\mathcal{X}$, there is a Wasserstein geodesic $(\mu_t)_{t\in[0,1]}$ and an associated coupling π such that

$$U_\nu(\mu_t) \le (1-t)\, U_{\pi,\nu}^{\beta_{1-t}^{(K,N')}}(\mu_0) + t\, U_{\check{\pi},\nu}^{\beta_t^{(K,N')}}(\mu_1).$$

It remains to pass to the limit as $N' \downarrow 1$ to get

$$U_\nu(\mu_t) \le (1-t)\, U_{\pi,\nu}^{\beta_{1-t}^{(K,1)}}(\mu_0) + t\, U_{\check{\pi},\nu}^{\beta_t^{(K,1)}}(\mu_1).$$

If $K > 0$, $1 < N < \infty$ and $\sup \operatorname{diam}(\mathcal{X}_k) = D_{K,N}$, then we can apply a similar reasoning, introducing again the bounded coefficients $\beta_t^{(K,N')}$ for $N' > N$ and then passing to the limit as $N' \downarrow N$.

This concludes the proof of Theorem 29.24. □

Remark 29.27. What the previous proof really shows is that under certain assumptions the property of distorted displacement convexity is stable under measured Gromov–Hausdorff convergence. The usual displacement convexity is a particular case (take $\beta_t \equiv 1$).

Proof of Theorem 29.25. Let $\star_k$ (resp. $\star$) be the reference point in $\mathcal{X}_k$ (resp. $\mathcal{X}$). The same arguments as in the proof of Theorem 29.24 will work here, provided that we can localize the problem. So pick up compactly supported probability measures μ_0 and μ_1 on $\mathcal{X}$, and define μ_{ε,t_0}^k ($t_0 = 0, 1$) in exactly the same way as in the proof of Theorem 29.24. Let $R > 0$ be such that the supports of ρ_0 and ρ_1 are included in $B_{R]}(\star)$; then for k large enough and ε small enough, the supports of $\mu_{\varepsilon,0}^k$ and $\mu_{\varepsilon,1}^k$ are contained in $B_{R+1]}(\star_k)$. So a geodesic which starts from the support of $\mu_{\varepsilon,0}^k$ and ends in the support of $\mu_{\varepsilon,1}^k$ will necessarily have its image included in $B_{2(R+1)]}(\star_k)$; thus each measure $\mu_{\varepsilon,t}^k$ has its support included in $B_{2(R+1)]}(\star_k)$.

From that point on, the very same reasoning as in the proof of Theorem 29.24 can be applied, since, say, the ball $B_{2(R+2)]}(\star_k)$ in $\mathcal{X}_k$ converges in the measured Gromov–Hausdorff topology to the ball $B_{2(R+2)]}(\star)$ in $\mathcal{X}$, etc. □

An application in Riemannian geometry

In this section, by convention I shall say that a metric-measure space (M, d, ν) is a smooth Riemannian manifold if the distance d is the geodesic distance induced by a Riemannian metric g on M, and ν is a reference measure that can be written $e^{-V}\mathrm{vol}$, where vol is the volume measure on M and $V \in C^2(M)$. This definition extends in an obvious way to pointed metric-measure spaces. Theorem 29.9 guarantees that the synthetic and analytic definitions of $\mathrm{CD}(K, N)$ bounds coincide for Riemannian manifolds.

The next theorem, which is a simple consequence of our previous results, may be seen as one noticeable outcome of the theory of weak $\mathrm{CD}(K, N)$ spaces. Note that it is an external result, in the sense that its statement does not involve the definition of weak $\mathrm{CD}(K, N)$ spaces, nor any reference to optimal transport.

Theorem 29.28 (Smooth MGH limits of $\mathrm{CD}(K, N)$ manifolds are $\mathrm{CD}(K, N)$). *Let $K \in \mathbb{R}$ and $N \in [1, \infty]$. If a sequence $(M_k)_{k\in\mathbb{N}}$ of smooth* $\mathrm{CD}(K, N)$ *Riemannian manifolds converges to some smooth manifold M in the (pointed) measured Gromov–Hausdorff topology, then the limit also satisfies the* $\mathrm{CD}(K, N)$ *curvature-dimension bound.*

Proof of Theorem 29.28. The statement results at once from Theorems 29.9 and 29.25. □

Remark 29.29. All the interest of this theorem lies in the fact that the measured Gromov–Hausdorff convergence is a very weak notion of convergence, which does not imply the convergence of the Ricci tensor.

Remark 29.30. There is no analog of Theorem 29.28 for *upper* bounds on the Ricci curvature. In fact, any compact Riemannian manifold (M, g) can be approximated by a sequence $(M, g_k)_{k\in\mathbb{N}}$ with (arbitrarily large) negative Ricci curvature.

The space of $\mathrm{CD}(K, N)$ spaces

Theorem 29.24 can be summarized as follows: *The space of all compact metric-measure geodesic spaces satisfying a weak* $\mathrm{CD}(K, N)$ *bound is closed under measured Gromov–Hausdorff convergence.*

In connection with this, recall Gromov's precompactness theorem (Corollary 27.34): Given $K \in \mathbb{R}$, $N < \infty$ and $D < \infty$, the set $\mathcal{M}(K,N,D)$ of all smooth compact manifolds with dimension bounded above by N, Ricci curvature bounded below by K and diameter bounded above by D is *precompact* in the Gromov–Hausdorff topology. Then Theorem 29.24 implies that any element of the closure of $\mathcal{M}(K,N,D)$ is a compact metric-measure geodesic space satisfying CD(K,N), in the weak sense of Definition 29.8.

Even if it is smooth, the limit space might have reference measure $\nu = e^{-\Psi}\mathrm{vol}$, for some nonconstant Ψ. Such phenomena do indeed occur in examples where there is a *collapse* in the dimension; that is, when the dimension of the limit manifold is strictly less than the dimension of the manifolds in the converging sequence. The next example shows that basically any reference measure can be obtained as a limit of volume measures of higher-dimensional manifolds; it is a strong motivation to replace the class of Riemannian manifolds by the class of metric-measure spaces.

Example 29.31. Let (M,g) be a compact n-dimensional Riemannian manifold, equipped with its geodesic distance d and its volume vol; let V be any C^2 function on M, and let $\nu(dx) = e^{-V(x)}\,d\mathrm{vol}(x)$. Let S^2 stand for the usual 2-dimensional sphere, equipped with its usual metric σ. For $\varepsilon \in (0,1)$, define M_ε to be the e^{-V}-warped product of (M,g) by $\varepsilon\, S^2$: This is the $(n+2)$-dimensional manifold $M \times S^2$, equipped with the metric $g_\varepsilon(dx,ds) = g(dx) + \varepsilon^2\, e^{-V(x)}\,\sigma(ds)$. As $\varepsilon \to 0$, M_ε collapses to M; more precisely the manifold $(M_\varepsilon, g_\varepsilon)$, seen as a metric-measure space, converges in the measured Gromov–Hausdorff sense to (M,d,ν). Moreover, if $\mathrm{Ric}_{n+2,\nu} \geq K$, then M_ε has Ricci curvature bounded below by K_ε, where $K_\varepsilon \to K$.

We shall see later (Theorem 30.14) that if $(\mathcal{X},d,\nu)$ is a weak CD(K,N) space, then the reference measure ν is locally doubling on its support. More precisely, for any base point $\star$, there is a constant $D = D(K,N,R)$ such that ν is D-doubling on $B[\star,R] \cap \mathrm{Spt}\,\nu$. Combining this with Theorem 27.32, we arrive at the following **compactness theorem:**

Theorem 29.32 (Compactness of the space of weak CD(K,N) spaces). *(i) Let $K \in \mathbb{R}$, $N < \infty$, $D < \infty$, and $0 < m \leq M < \infty$. Let $\mathcal{CDD}(K,N,D,m,M)$ be the space of all compact metric-measure geodesic spaces $(\mathcal{X},d,\nu)$ satisfying the weak curvature-dimension bound*

$\mathrm{CD}(K,N)$ *of Definition 29.8, together with* $\operatorname{diam}(\mathcal{X},d) \le D$, $m \le \nu[\mathcal{X}] \le M$, *and* $\operatorname{Spt}\nu = \mathcal{X}$. *Then* $\mathcal{CDD}(K,N,D,m,M)$ *is compact in the measured Gromov–Hausdorff topology.*

(ii) Let $K \in \mathbb{R}$, $N < \infty$, $0 < m \le M < \infty$. *Let* $p\mathcal{CDD}(K,N,m,M)$ *be the space of all pointed locally compact Polish metric-measure geodesic spaces* $(\mathcal{X},d,\nu,\star)$ *satisfying the weak* $\mathrm{CD}(K,N)$ *curvature-dimension bound of Definition 29.8, together with* $m \le \nu[B_1(\mathcal{X})] \le M$, *and* $\operatorname{Spt}\nu = \mathcal{X}$. *Then* $p\mathcal{CDD}(K,N,m,M)$ *is compact in the measured Gromov–Hausdorff topology.*

Remark 29.33. It is a natural question whether smooth Riemannian manifolds, equipped with their geodesic distance and their volume measure (multiplied by a positive constant), form a dense set in, say, $\mathcal{CDD}(K,N,D,m,M)$. The answer is negative, as will be discussed in the concluding chapter.

First Appendix: Regularization in metric-measure spaces

Regularization by convolution is a fundamental tool in real analysis. It is still available, to some extent, in metric-measure spaces, as I shall now explain. Recall that a boundedly compact metric space is a metric space in which closed balls are compact; and a locally finite measure is a measure which gives finite mass to balls.

Definition 29.34 (Regularizing kernels). *Let* $(\mathcal{X},d)$ *be a boundedly compact metric space equipped with a locally finite measure* ν, *and let* $\mathcal{Y}$ *be a compact subset of* $\mathcal{X}$. *A* $(\mathcal{Y},\nu)$*-regularizing kernel is a family of nonnegative continuous symmetric functions* $(K_\varepsilon)_{\varepsilon>0}$ *on* $\mathcal{X}\times\mathcal{X}$, *such that*

$$\text{(i) } \forall x \in \mathcal{Y}, \qquad \int_{\mathcal{X}} K_\varepsilon(x,y)\,\nu(dy) = 1;$$

$$\text{(ii) } d(x,y) > \varepsilon \Longrightarrow \quad K_\varepsilon(x,y) = 0.$$

For any compact subset $\mathcal{Y}$ of $\operatorname{Spt}\nu$, there is a $(\mathcal{Y},\nu)$-regularizing kernel, which can be constructed as follows. Cover $\mathcal{Y}$ by a finite number of balls $B(x_i,\varepsilon/2)$. Introduce a continuous subordinate partition of unity $(\phi_i)_{i\in I}$: These are continuous functions satisfying $0 \le \phi_i \le 1$,

$\operatorname{Spt}(\phi_i) \subset B(x_i, \varepsilon/2)$, $\sum_i \phi_i = 1$ on $\mathcal{Y}$; only keep those functions ϕ_i such that $\operatorname{Spt} \phi_i \cap \mathcal{Y} \neq \emptyset$, so that $\int \phi_i \, d\nu > 0$. Next define

$$K_\varepsilon(x,y) := \sum_i \frac{\phi_i(x)\,\phi_i(y)}{\int \phi_i \, d\nu}. \tag{29.50}$$

For any $x \in \mathcal{Y}$, $\int K_\varepsilon(x,y)\,\nu(dy) = \sum_i \phi_i(x) = 1$. Also $\phi_i(x)\,\phi_i(y)$ can be nonzero only if x and y belong to $B(x_i, \varepsilon/2)$, which implies that $d(x,y) < \varepsilon$. So K_ε does the job.

As soon as μ is a finite measure on $\mathcal{X}$, one may define a *continuous* function $K_\varepsilon\mu$ on $\mathcal{X}$ by

$$(K_\varepsilon\mu)(x) := \int_{\mathcal{X}} K_\varepsilon(x,y)\,\mu(dy).$$

Further, if $f \in L^1(\mathcal{X}, \nu)$, define $K_\varepsilon f := K_\varepsilon(f\nu)$.

The linear operator $K_\varepsilon : \mu \to (K_\varepsilon\mu)\nu$ is mass-preserving, in the sense that for any nonnegative finite measure μ on $\mathcal{Y}$, one has $((K_\varepsilon\mu)\nu)[\mathcal{Y}] = \mu[\mathcal{Y}]$. More generally, K_ε defines a (nonstrict) contraction operator on $M(\mathcal{Y})$. Moreover, as $\varepsilon \to 0$,

- If $f \in C(\mathcal{X})$, then $K_\varepsilon f$ converges *uniformly* to f on $\mathcal{Y}$;
- If μ is a finite measure supported in $\mathcal{Y}$, then $(K_\varepsilon\mu)\nu$ converges weakly (against $C_b(\mathcal{X})$) to μ (this follows from the previous property by a duality argument);
- If $f \in L^1(\mathcal{Y})$, then $K_\varepsilon f$ converges to f in $L^1(\mathcal{Y})$ (this follows from the density of $C(\mathcal{Y})$ in $L^1(\mathcal{Y}, \nu)$, the fact that $K_\varepsilon f$ converges uniformly to f if f is continuous, and the contraction property of K_ε). There is in fact a more precise statement: For any $f \in L^1(\mathcal{Y}, \nu)$,

$$\int_{\mathcal{Y}\times\mathcal{Y}} |f(x) - f(y)|\, K_\varepsilon(x,y)\,\nu(dx)\,\nu(dy) \xrightarrow[\varepsilon \to 0]{} 0.$$

Remark 29.35. If the measure ν is (locally) doubling, then one can ask more of the kernel (K_ε), than just properties (i) and (ii) in Definition 29.34. Indeed, by Vitali's covering lemma, one can make sure that the covering $(B(x_i, \varepsilon/2))$ used in the definition of K_ε is such that the balls $B(x_i, \varepsilon/10)$ are disjoint. If (ϕ_i) is a partition of unity associated to the covering $(B(x_i, \varepsilon/2))$, necessarily ϕ_i is identically 1 on $B(x_i, \varepsilon/10)$, so $\int \phi_i \, d\nu \geq \nu[B(x_i, \varepsilon/10)] \geq C\nu[B(x_i, \varepsilon)]$, where C is a constant depending on the doubling constant of ν. Then the following uniform bound comes easily:

(iii) $K_\varepsilon(x,y) \leq \dfrac{C}{\nu[B_\varepsilon(x)]}$.

(Here C is another numerical constant, depending on the doubling constant of ν.) Assumptions (i), (ii) and (iii), together with the doubling property of ν, and classical Lebesgue density theory, guarantee that for any $f \in L^1(\mathcal{Y})$ the convergence of $K_\varepsilon f$ to f holds not only in $L^1(\mathcal{Y})$ but also *almost everywhere.*

Second Appendix: Separability of $L^1(\mathcal{X}; C(\mathcal{Y}))$

In the course of the proof of Theorem 29.20(iii), I used the density of $C(\mathcal{X} \times \mathcal{X})$ in $L^1(\mathcal{X}; C(\mathcal{X}))$. Here is a precise statement and a short proof.

Lemma 29.36 (Separability of $L^1(C)$). *Let $(\mathcal{X}, d)$ be a compact metric space equipped with a finite Borel measure μ, let $\mathcal{Y}$ be another compact metric space, and let f be a measurable function $\mathcal{X} \times \mathcal{Y} \to \mathbb{R}$, such that*

$$\begin{cases} \text{(i) } f(x,\,\cdot\,) \text{ is continuous for all } x; \\ \text{(ii) } \displaystyle\int_{\mathcal{X}} \sup_{\mathcal{Y}} |f(x,y)|\, d\mu(x) < +\infty. \end{cases}$$

Then for any $\varepsilon > 0$ there is $\Psi \in C(\mathcal{X} \times \mathcal{Y})$ such that

$$\int_{\mathcal{X}} \sup_{y \in \mathcal{Y}} \big|f(x,y) - \Psi(x,y)\big|\, d\mu(x) \leq \varepsilon.$$

Moreover, if a (possibly empty) compact subset K of $\mathcal{X}$, and a function $h \in C(K)$ are given, such that $f(x,y) = h(x)$ for all $x \in K$, then it is possible to impose that $\Psi(x,y) = h(x)$ for all $x \in K$.

Proof of Lemma 29.36. Let us first treat the case when $\mathcal{Y}$ is just a point. Then the first part of the statement of Lemma 29.36 is just the density of $C(\mathcal{X})$ in $L^1(\mathcal{X}, \mu)$, which is a classical result. To prove the second part of the statement, let $f \in L^1(\mu)$, let K be a compact subset of $\mathcal{X}$, let $h \in C(K)$ such that $f = h$ on K, and let $\varepsilon > 0$. Let $\psi \in C_c(\mathcal{X} \setminus K)$ be such that $\|\psi - f\|_{L^1(\mathcal{X}\setminus K,\mu)} \leq \varepsilon$.

Since μ and $f\,\nu$ are regular, there is an open set O_ε containing K such that $\mu[O_\varepsilon \setminus K] \leq \varepsilon/(\sup|\psi| + \sup|h|)$ and $\|f - h\|_{L^1(O_\varepsilon)} \leq \varepsilon$.

The sets O_ε and $\mathcal{X} \setminus K$ are open and cover $\mathcal{X}$, so there are continuous functions χ and η, defined on $\mathcal{X}$ and valued in $[0,1]$, such that $\chi+\eta=1$, χ is supported in O_ε and η is supported in $\mathcal{X} \setminus K$. (In particular $\chi \equiv 1$ in K.) Let

$$\Psi = h\,\chi + \psi\,\eta.$$

Then Ψ coincides with h (and therefore with f) in K, Ψ is continuous, and

$$\begin{aligned}
\|\Psi - f\|_{L^1(\mathcal{X})} &\le \|f-h\|_{L^1(O_\varepsilon)} + (\sup|h|)\,\|\chi - 1\|_{L^1(O_\varepsilon)} \\
&\qquad + (\sup|\psi|)\,\|\eta - 1\|_{L^1(\mathcal{X}\setminus K)} + \|\psi - f\|_{L^1(\mathcal{X}\setminus K)} \\
&\le \varepsilon + (\sup|h| + \sup|\psi|)\,\mu[O_\varepsilon \setminus K] + \varepsilon \\
&\le 3\,\varepsilon.
\end{aligned}$$

Since ε is arbitrarily small, this concludes the proof.

Now let us turn to the case when $\mathcal{Y}$ is any compact set. For any $\delta > 0$ there are $L = L(\delta) \in \mathbb{N}$ and $\{y_1, \dots, y_L\}$ such that the open balls $B_\delta(y_\ell)$ $(1 \le \ell \le L)$ cover $\mathcal{Y}$. Let $(\zeta_\ell)_{1\le \ell \le L}$ be a partition of unity subordinated to that covering: As in the previous Appendix, these are continuous functions satisfying $0 \le \zeta_\ell \le 1$, $\mathrm{Spt}(\zeta_\ell) \subset B_\delta(y_\ell)$, and $\sum_\ell \zeta_\ell = 1$ on $\mathcal{Y}$. Let $\eta > 0$.

For each $\ell \in \mathbb{N}$, the function $f(\,\cdot\,, y_\ell)$ is μ-integrable, so there exists $\psi_\ell \in C(\mathcal{X})$ such that

$$\int \bigl|f(x,y_\ell) - \psi_\ell(x)\bigr|\, d\mu(x) \le \eta.$$

Moreover, we may require that $\psi_\ell(x) = h(x)$ when $x \in K$.

Define Ψ by the formula

$$\Psi(x,y) := \sum_{\ell \le L} \psi_\ell(x)\,\zeta_\ell(y);$$

note that $\Psi(x,y) = h(x)$ if $x \in K$. Then

$$\begin{aligned}
f(x,y) - &\Psi(x,y) \\
&= f(x,y)\Bigl(\sum_\ell \zeta_\ell(y)\Bigr) - \sum_\ell \psi_\ell(x)\,\zeta_\ell(y) \\
&= \sum_\ell \bigl[f(x,y) - \psi_\ell(x)\bigr]\,\zeta_\ell(y) \\
&= \sum_\ell \bigl[f(x,y) - f(x,y_\ell)\bigr]\,\zeta_\ell(y) + \sum_\ell \bigl[f(x,y_\ell) - \psi_\ell(x)\bigr]\,\zeta_\ell(y).
\end{aligned}$$

Since ζ_ℓ is supported in $B_\delta(y_\ell)$,

$$\begin{aligned}
&|f(x,y) - \Psi(x,y)| \\
&\le \Big(\sup_{d(z,z')\le\delta} |f(x,z) - f(x,z')|\Big) \sum_\ell \zeta_\ell(y) \;+\; \sum_\ell |f(x,y_\ell) - \psi_\ell(x)|\,\zeta_\ell(y) \\
&\le \sup_{d(z,z')\le\delta} \big|f(x,z) - f(x,z')\big| \;+\; \sum_\ell |f(x,y_\ell) - \psi_\ell(x)|.
\end{aligned}$$

So

$$\begin{aligned}
&\int_{\mathcal{X}} \Big(\sup_y |f(x,y) - \Psi(x,y)|\Big)\,\mu(dx) \\
&\le \int_{\mathcal{X}} \Big(\sup_{d(z,z')\le\delta} |f(x,z) - f(x,z')|\Big)\,\mu(dx) \\
&\qquad\qquad + \sum_\ell \int_{\mathcal{X}} |f(x,y_\ell) - \psi_\ell(x)|\,\mu(dx) \\
&\le \int_{\mathcal{X}} m_\delta(x)\,\mu(dx) \;+\; L(\delta)\,\eta, \qquad (29.51)
\end{aligned}$$

where

$$m_\delta(x) = \sup\Big\{|f(x,z) - f(x,z')|;\; d(z,z')\le\delta\Big\}.$$

Obviously, $m_\delta(x) \le 2\sup_z |f(x,z)|$, so $m_\delta \in L^1(\mu)$, for all δ. On the other hand, $m_\delta(x)$ is nonincreasing as $\delta \downarrow 0$, and since each $f(x,\cdot)$ is continuous, actually m_δ decreases to 0 as $\delta \downarrow 0$. By monotone convergence,

$$\int m_\delta(x)\,d\mu(x) \xrightarrow[\delta\to 0]{} 0.$$

So if we choose δ small enough, then in turn η small enough, we can make sure that the right-hand side of (29.51) is as small as desired. This concludes the proof. □

Bibliographical notes

Here are some (probably too lengthy) comments about the genesis of Definition 29.8. It comes after a series of particular cases and/or variants studied by Lott and myself [577, 578] on the one hand, Sturm [762, 763] on the other. To summarize: In a first step, Lott and I [577] treated $\mathrm{CD}(K,\infty)$ and $\mathrm{CD}(0,N)$, while Sturm [762] independently treated $\mathrm{CD}(K,\infty)$. These cases can be handled with just displacement convexity. Then it took some time before Sturm [763] came up with the brilliant idea to use distorted displacement as the basis of the definition of $\mathrm{CD}(K,N)$ for $N<\infty$ and $K\neq 0$.

There are slight variations in the definitions appearing in all these works; and they are not exactly the ones appearing in this course either. I shall describe the differences in some detail below.

In the case $K=0$, for compact spaces, Definition 29.8 is exactly the definition that was used in [577]. In the case $N=\infty$, the definition in [577] was about the same as Definition 29.8, but it was based on inequality (29.2) (which is very simple in the case $K=\infty$) instead of (29.3). Sturm [762] also used a similar definition, but preferred to impose the weak displacement convexity inequality only for the Boltzmann H functional, i.e. for $U(r)=r\log r$, not for the whole class $\mathcal{DC}_\infty$. It is interesting to note that precisely for the H functional and $N=\infty$, inequalities (29.2) and (29.3) are the same, while in general the former is a priori weaker. So the definition which I have adopted here is a priori stronger than both definitions in [577] and [762].

Now for the general $\mathrm{CD}(K,N)$ criterion. Sturm's original definition [763] is quite close to Definition 29.8, with three differences. First, he does not impose the basic inequality to hold true for all members of the class $\mathcal{DC}_N$, but only for functions of the form $-r^{1-1/N'}$ with $N'\geq N$. Secondly, he does not require the displacement interpolation $(\mu_t)_{0\leq t\leq 1}$ and the coupling π to be related via some dynamical optimal transference plan. Thirdly, he imposes μ_0, μ_1 to be absolutely continuous with respect to ν, rather than just to have their support included in $\mathrm{Spt}\,\nu$. After becoming aware of Sturm's work, Lott and I [578] modified his definition, imposing the inequality to hold true for all $U\in\mathcal{DC}_N$, imposing a relation between (μ_t) and π, and allowing in addition μ_0 and μ_1 to be singular. In the present course, I decided to extend the new definition to the case $N=\infty$.

Sturm [763] proved the stability of his definition under a variant of measured Gromov–Hausdorff convergence, provided that one stays away from the limit Bonnet–Myers diameter. Then Lott and I [578] briefly sketched a proof of stability for our modified definition. Details appear here for the first time, in particular the painful[2] proof of upper semicontinuity of $U^\beta_{\pi,\nu}(\mu)$ under regularization (Theorem 29.20(iii)). It

[2] As a matter of fact, I was working on precisely this problem when my left lung collapsed, earning me a one-week holiday in hospital with unlimited amounts of pain-killers.

should be noted that Sturm manages to prove the stability of his definition without using this upper semicontinuity explicitly; but this might be due to the particular form of the functions U that he is considering, and the assumption of absolute continuity.

The treatment of noncompact spaces here is not exactly the same as in [577] or [763]. In the present set of notes I adopted a rather weak point of view in which every "noncompact" statement reduces to the compact case; in particular in Definition 29.8 I only consider compactly supported probability densities. This leads to simpler proofs, but the treatment in [577, Appendix E] is more precise in that it passes to the limit directly in the inequalities for probability measures that are not compactly supported.

Other tentative definitions have been rejected for various reasons. Let me mention four of them:

(i) Imposing the displacement convexity inequality along *all* displacement interpolations in Definition 29.8, rather than along *some* displacement interpolation. This concept is not stable under measured Gromov–Hausdorff convergence. (See the last remark in the concluding chapter.)

(ii) Replace the *integrated* displacement convexity inequalities by *pointwise* inequalities, in the style of those appearing in Chapter 14. For instance, with the same notation as in Definition 29.8, one may define

$$\mathcal{J}_t(\gamma_0) := \frac{\rho_0(x)}{\mathbb{E}\left[\rho_t(\gamma_t)|\gamma_0\right]},$$

where γ is a random geodesic with law $(\gamma_t) = \mu_t$, and ρ_t is the absolutely continuous part of μ_t with respect to ν. Then $\mathcal{J}$ is a continuous function of t, and it makes sense to require that inequality (29.1) be satisfied $\nu(dx)$-almost everywhere (as a function of t, in the sense of distributions). This notion of a weak $\mathrm{CD}(K,N)$ space makes perfect sense, and is a priori stronger than the notion discussed in this chapter. But there is no evidence that it should be stable under measured Gromov–Hausdorff convergence. Integrated convexity inequalities enjoy better stability properties. (One might hope that integrated inequalities lead to pointwise inequalities by a localization argument, as in Chapter 19; but this is not obvious at all, due to the a priori nonuniqueness of displacement interpolation in a nonsmooth context.)

(iii) Choose inequality (29.2) as the basis for the definition, instead of (29.3). In the case $K < 0$, this inequality is stable, due to the convexity of $-r^{1-1/N}$, and the a priori regularity of the speed field provided by Theorem 28.5. (This was actually my original motivation for Theorem 28.5.) In the case $K > 0$ there is no reason to expect that the inequality is stable, but then one can weaken even more the formulation of $\mathrm{CD}(K,N)$ and replace it by

$$U_\nu(\mu_t) \le (1-t)\, U_\nu(\mu_0) \;+\; t\, U_\nu(\mu_1) \\ \qquad - \frac{K_{N,U}}{2}\, \Big[\max\big(\sup \rho_0, \sup \rho_1\big)\Big]^{-1/N} W_2(\mu_0,\mu_1)^2, \quad (29.52)$$

which in turn is stable, and still equivalent to the usual $\mathrm{CD}(K,N)$ when applied to smooth manifolds. For the purpose of the present chapter, this approach would have worked fine; as far as I know, Theorem 29.28 was first proved for general K, N by this approach (in an unpublished letter of mine from September 2004). But basing the definition of the general $\mathrm{CD}(K,N)$ criterion on (29.52) has a major drawback: It seems very difficult, if not impossible, to derive any sharp geometric theorem such as Bishop–Gromov or Bonnet–Myers from this inequality. We shall see in the next chapter that these sharp inequalities do follow from Definition 29.8.

(iv) Base the definition of $\mathrm{CD}(K,N)$ on other inequalities, involving the volume growth. The main instance is the so-called **measure contraction property** (MCP). This property involves a conditional probability $P^{(t)}(x,y;dz)$ on the set $[x,y]_t$ of t-barycenters of x, y (think of $P^{(t)}$ as a measurable rule to choose barycenters of x and y). By definition, a metric-measure space $(\mathcal{X}, d)$ satisfies $\mathrm{MCP}(K,N)$ if for any Borel set $A \subset \mathcal{X}$,

$$\int \beta_t^{(K,N)}(x,y)\, P^{(t)}(x,y;A)\, \nu(dy) \le \frac{\nu[A]}{t^N};$$

and symetrically

$$\int \beta_{1-t}^{(K,N)}(x,y)\, P^{(t)}(x,y;A)\, \nu(dx) \le \frac{\nu[A]}{(1-t)^N}.$$

In the case $K=0$, these inequalities basically reduce to $\nu[[x,B]_t] \ge t^N \nu[B]$. (Recall Theorem 19.6.) This approach has two drawbacks: First, it does not extend to the case $N=\infty$; secondly, if M is a Riemannian manifold, then $\mathrm{MCP}(K,N)$ *does not* imply $\mathrm{CD}(K,N)$, unless N coincides with the true dimension of the manifold. On the other hand, $\mathrm{CD}(K,N)$ implies $\mathrm{MCP}(K,N)$, at least in a nonbranching space. All this is discussed in independent works by Sturm [763, Section 6] and Ohta [654]; see also [573, Remark 4.9]. Unlike the weak $\mathrm{CD}(K,N)$ property, the MCP property is known to be true for finite-dimensional Alexandrov spaces with curvature bounded below [654, Section 2]. An interesting field of application of the MCP property is the analysis on the Heisenberg group, which does not satisfy any $\mathrm{CD}(K,N)$ bound but still satisfies $\mathrm{MCP}(0,5)$ [496].

Bonciocat and Sturm [143] modified the definition of weak $\mathrm{CD}(K,\infty)$ spaces to allow for a fixed "resolution error" in the measurement of distances (implying of course a modification of the transport cost). This approach defines spaces which are "δ-approximately $\mathrm{CD}(K,\infty)$"; it gives the possibility to study Ricci curvature bounds outside the scope of length spaces, including even discrete spaces in the analysis. For instance, any weak $\mathrm{CD}(K,\infty)$ space is a limit of δ-approximate $\mathrm{CD}(K,\infty)$ discrete spaces, with $\delta \to 0$. The procedure is a bit similar in spirit to the construction of $\mathrm{CAT}_\delta(K)$ spaces [439], modulo of course the introduction of the formalism of entropies and Wasserstein geodesics. (So this theory provides one answer to the very last question raised by Gromov in [439].) Bonciocat and Sturm apply it to classes of planar homogeneous graphs.

Cordero-Erausquin suggested the use of the Prékopa–Leindler inequality as a possible alternative basis for a synthetic $\mathrm{CD}(K,N)$ criterion. This is motivated by the fact that the Prékopa–Leindler inequality has many geometric consequences, including Sobolev or Talagrand inequalities (see the bibliographical notes for Chapters 21 and 22). So far nobody has undertaken this program seriously and it is not known whether it includes some serious analytical difficulties.

Lott [576] noticed that (for a Riemannian manifold) at least $\mathrm{CD}(0,N)$ bounds can be formulated in terms of displacement convexity of certain functionals explicitly involving the time variable. For instance, $\mathrm{CD}(0,N)$ is equivalent to the convexity of $t \to t\,U_\nu(\mu_t) + N\,t\log t$ on $[0,1]$, along displacement interpolation, for all $U \in \mathcal{DC}_\infty$; rather than convexity of $U_\nu(\mu_t)$ for all $U \in \mathcal{DC}_N$. (Note carefully: in one formulation the dimension is taken care of by the time-dependence of the functional, while in the other one it is taken care of by the class of nonlinearities.) More general curvature-dimension bounds can probably be encoded by refined convexity estimates: for instance, $\mathrm{CD}(K,N)$ seems to be equivalent to the (displacement) convexity of $tH_\nu(\mu_t) + N\,t\log t - K(t^3/6)\,W_2(\mu_0,\mu_1)^2$.

It seems likely that this observation can be developed into a complete theory. For geometric applications, this point of view is probably less sharp than the one based on distortion coefficients, but it may be technically simpler.

A completely different approach to Ricci bounds in metric-measure spaces has been under consideration in a work by Kontsevich and Soibelman [528], in relation to Quantum Field Theory, mirror symmetry and heat kernels; see also [756]. Kontsevich pointed out to me that the class of spaces covered by this approach is probably strictly smaller than the class defined here, since it does not seem to include the normed spaces considered in Example 29.16; he also suggested that this point of view is related to the one of Ollivier, described below.

To close this list, I shall evoke the recent independent contributions by Joulin [494, 495] and Ollivier [662] who suggested defining the infimum of the Ricci curvature as the best constant K in the contraction inequality

$$W_1(P_t\delta_x, P_t\delta_y) \leq e^{-Kt}\, d(x,y),$$

where P_t is the heat semigroup (defined on probability measures); or equivalently, as the best constant K in the inequality

$$\|P_t^* f\|_{\mathrm{Lip}} \leq e^{-Kt}\, \|f\|_{\mathrm{Lip}},$$

where P_t^* is the adjoint of P_t (that is, the heat semigroup on functions). Similar inequalities have been used before in concentration theory [305, 595], and in the study of spectral gaps [231, 232] or large-time convergence [310, 458, 679, 729]. In a Riemannian setting, this definition is justified by Sturm and von Renesse [764], who showed that one recovers in this way the usual Ricci curvature bound, the key observation being that the parallel transport is close to be optimal in some sense. (The distance W_1 may be replaced by any W_r, $1 \leq r < \infty$.) This point of view is natural when the problem includes a distinguished Markov kernel; in particular, it leads to the possibility of treating discrete spaces equipped with a random walk. Ollivier [662] has demonstrated the geometric interest of this notion on an impressive list of examples and applications, most of them in discrete spaces; and he also derived geometric consequences such as concentration of measure. The dimension can probably be taken into account by using refined estimates on the rates of convergence. On the other hand, the stability of this notion requires a stronger topology than just measured Gromov–Hausdorff convergence: also the Markov process should converge. (The topology explored in [97] might be useful.) To summarize:

- there are two main classes of synthetic definitions of $\mathrm{CD}(K,N)$ bounds: those based on displacement convexity of integral functionals (as in Chapter 17) and those based on contraction properties of diffusion equations in Wasserstein distances (as in [764]), the Jordan–Kinderlehrer–Otto theorem [493] ensuring the formal equivalence of these points of view;
- the first definition was discretized by Bonciocat and Sturm [143], while the second one was discretized by Joulin [495] and Ollivier [662].

Remark 29.30 essentially means that manifolds with (strongly) negative Ricci curvature bounds are dense in Gromov–Hausdorff topology. Statements of this type are due in particular to Lokhamp; see Berger [99, Section 12.3.5] and the references there provided.

Next, here are some further comments about the ingredients in the proof of Theorem 29.24.

The definition and properties of the functional U_ν acting on singular measures (Definition 29.1, Proposition 29.19, Theorem 29.20(i)–(ii)) were worked out in detail in [577]. At least some of these properties belong to folklore, but it is not so easy to find precise references. For the particular case $U(r) = r \log r$, there is a detailed alternative proof of Theorem 29.20(i)–(ii) in [30, Lemmas 9.4.3 to 9.4.5, Corollary 9.4.6] when $\mathcal{X}$ is a separable Hilbert space, possibly infinite-dimensional; the proof of the contraction property in that reference does not rely on the Legendre representation. There is also a proof of the lower semicontinuity and the contraction property, for general functions U, in [556, Chapter 1]; the arguments there do not rely on the Legendre representation either. I personally advocate the use of the Legendre representation, as an efficient and versatile tool.

In [577], we also discussed the extension of these properties to spaces that are not necessarily compact, but only locally compact, and reference measures that are not necessarily finite, but only locally finite. Integrability conditions at infinity should be imposed on μ, as in Theorem 17.8. The discussion on the Legendre representation in this generalized setting is a bit subtle, for instance it is in general impossible to impose at the same time $\varphi \in C_c(\mathcal{X})$ and $U^*(\varphi) \in C_c(\mathcal{X})$. Here I preferred to limit the use of the Legendre representation (and the lower semicontinuity) to the compact case; but another approximation argument will be used in the next chapter to extend the displacement convexity inequalities to probability measures that are not compactly supported.

The density of $C(\mathcal{X})$ in $L^1(\mathcal{X},\mu)$, where $\mathcal{X}$ is a locally compact space and ν is a finite Borel measure, is a classical result that can be found e.g. in [714, Theorem 3.14]. It is also true that nonnegative continuous functions are dense in $L^1_+(\mathcal{X},\mu)$, or that continuous probability densities are dense in the space of probability densities, equipped with the L^1 norm. All these results can be derived from Lusin's approximation theorem [714, Theorem 2.24].

The Tietze–Urysohn extension theorem states the following: If $(\mathcal{X},d)$ is a metric space, C is a closed subset of $\mathcal{X}$, and $f : C \to \mathbb{R}$ is uniformly continuous on C, then it is possible to extend f into a continuous function on the whole of $\mathcal{X}$, with preservation of the supremum norm of f; see [318, Theorem 2.6.4].

The crucial approximation scheme based on regularization kernels was used by Lott and myself in [577]. In Appendix C of that reference, we worked out in detail the properties stated after Definition 29.34. We used this tool extensively, and also discussed regularization in noncompact spaces. Even in the framework of absolutely continuous measures, the approach based on the regularizing kernel has many advantages over Lusin's approximation theorem (it is explicit, linear, preserves convexity inequalities, etc.).

The existence of continuous partitions of unity was used in the First Appendix to construct the regularizing kernels, and in the Second Appendix about the separability of $L^1(\mathcal{X};C(\mathcal{Y}))$; the proof of this classical result can be found e.g. in [714, Theorem 2.13].

Apart from Theorem 29.28, other "external" consequences of the theory of weak $\mathrm{CD}(K,N)$ spaces are discussed in [577, Section 7.2], in the cases $K=0$ and $N=\infty$.

Lemma 29.7 is taken from a recent work of mine with Figalli [372]. It will be used later in the proof of Theorems 30.37 and 30.42.

I shall conclude with some remarks about the examples considered in this chapter.

The following generalization of Example 29.13 is proven in [30, Theorems 9.4.10 and 9.4.11]: If ν is a finite measure on $\mathbb{R}^n$ such that H_ν is displacement convex, then ν takes the form $e^{-V}\mathcal{H}^k$, where V is lower semicontinuous and $\mathcal{H}^k$ is the k-dimensional Hausdorff measure, $k=\dim(\operatorname{Spt}\nu)$. The same reference extends to infinite-dimensional separable Hilbert spaces the result according to which H_ν is displacement convex if and only if ν is log-concave.

Example 29.15 was treated in [577]: We show that the quotient of a $\mathrm{CD}(K,N)$ Riemannian manifold by a compact Lie group action is still a weak $\mathrm{CD}(K,N)$ space, if $K=0$ or $N=\infty$. (The definition of $\mathrm{CD}(K,\infty)$ space used in [577] is not exactly the same, but Theorem 30.32 in the next chapter shows that both definitions coincide in nonbranching spaces.) The same theorem is also certainly true for all values of K and N, although this was never written down explicitly. More problematic is the extension to noncompact spaces $\mathcal{X}$ or noncompact Lie groups G. The proof uses indeed an isomorphism between $P_2(\mathcal{X}/G)$ and $P_2(\mathcal{X})^G$, the set of probability measures on $\mathcal{X}$ which are invariant under the action of G; but this isomorphism might not exist in noncompact situations. Take for instance $\mathcal{X}=\mathbb{R}$, $G=\mathbb{Z}$, then $P_2(\mathcal{X})^G$ is the set of probability measures invariant by integer translation, which is empty.

Elementary background on Lie group actions, and possibly singular spaces obtained by this procedure, can be found in Burago, Burago and Ivanov [174, Chapter 3]. This topic is also treated in dozens of books on Riemannian geometry.

Example 29.16 will be considered in more detail in the final chapter. It can be considered as the precursor of a large work by Ohta [657] investigating curvature-dimension conditions in Finsler spaces.

Example 29.31 was explained to me by Lott; it is studied in [574]. This example shows that a lower bound on the Ricci curvature is not enough to ensure the continuity of the Hausdorff measure under measured Gromov–Hausdorff convergence. Such a phenomenon is necessarily linked with *collapsing* (loss of dimension), as shown by the results of continuity of the Hausdorff measure for n-dimensional Alexandrov spaces [751] or for limits of n-dimensional Riemannian manifolds with Ricci curvature bounded below [228, Theorem 5.9].

Example 29.17 was also suggested by Lott.

30

Weak Ricci curvature bounds II: Geometric and analytic properties

In the previous chapter I introduced the concept of weak curvature-dimension bound, which extends the classical notion of curvature-dimension bound from the world of smooth Riemannian manifolds to the world of metric-measure geodesic spaces; then I proved that such bounds are stable under measured Gromov–Hausdorff convergence.

Still, this notion would be of limited value if it could not be used to establish nontrivial conclusions. Fortunately, weak curvature-dimension bounds do have interesting geometric and analytic consequences. This might not be a surprise to the reader who has already browsed Part II of these notes, since there many geometric and analytic statements of Riemannian geometry were derived from optimal transport theory.

In this last chapter, I shall present the state of the art concerning properties of weak $\mathrm{CD}(K, N)$ spaces. This direction of research is growing relatively fast, so the present list might soon become outdated.

Convention: Throughout the sequel, a "weak $\mathrm{CD}(K, N)$ space" is a *locally compact, complete separable geodesic space* $(\mathcal{X}, d)$ *equipped with a locally finite Borel measure* ν*, satisfying a weak* $\mathrm{CD}(K, N)$ *condition as in Definition 29.8.*

Elementary properties

The next proposition gathers some almost immediate consequences of the definition of weak $\mathrm{CD}(K, N)$ spaces. I shall say that a subset $\mathcal{X}'$ of a geodesic space $(\mathcal{X}, d)$ is *totally convex* if any geodesic whose endpoints belong to $\mathcal{X}'$ is entirely contained in $\mathcal{X}'$.

C. Villani, *Optimal Transport*, Grundlehren der mathematischen Wissenschaften 338,

Proposition 30.1 (Elementary consequences of weak CD(K,N) bounds). *Let* $(\mathcal{X},d,\nu)$ *be a weak* CD(K,N) *space. Then:*

(i) If $\mathcal{X}'$ *is a totally convex closed subset of* $\mathcal{X}$*, then* $\mathcal{X}'$ *inherits from* $(\mathcal{X},d,\nu)$ *a natural structure of metric-measure geodesic space, and* $\mathcal{X}'$ *is also a weak* CD(K,N) *space;*

(ii) For any $\alpha > 0$*,* $(\mathcal{X},d,\alpha\nu)$ *is a weak* CD(K,N) *space;*

(iii) For any $\lambda > 0$*,* $(\mathcal{X},\lambda d,\nu)$ *is a weak* CD$(\lambda^{-2}K,N)$ *space.*

Proof of Proposition 30.1. Let $\mathcal{X}'$ be a totally convex subset of $\mathcal{X}$, equipped with the restriction of the distance d and the measure ν. Let μ_0, μ_1 be two probability measures in $P_2(\mathcal{X}')$. The notion of optimal coupling is the same whether one considers them as measures on $\mathcal{X}'$ or on $\mathcal{X}$. Also, since $\mathcal{X}'$ is totally convex, a path $[0,1] \to \mathcal{X}$ with endpoints in $\mathcal{X}'$ is a geodesic in $\mathcal{X}'$ if and only if it is a geodesic in $\mathcal{X}$. So $\mathcal{X}'$ is a geodesic space, and the representation theorem for Wasserstein geodesics (Theorem 7.22) ensures that a path $(\mu_t)_{0\le t\le 1}$ valued in $P_2(\mathcal{X}')$ is a geodesic in $P_2(\mathcal{X}')$ if and only if it is a geodesic in $P_2(\mathcal{X})$. Property (i) follows immediately.

To prove (ii), note that the replacement of ν by $\alpha\nu$ induces a multiplication of the density ρ by α^{-1}; so

$$U_{\alpha\nu}(\mu) = (U_\alpha)_\nu(\mu), \qquad U^\beta_{\pi,\alpha\nu}(\mu) = (U_\alpha)^\beta_{\pi,\nu}(\mu),$$

where $U_\alpha(r) = \alpha\, U(\alpha^{-1} r)$. But the transform $U \to U_\alpha$ leaves the class $\mathcal{DC}_N$ invariant. So the inequalities defining the CD(K,N) condition will hold just the same in $(\mathcal{X},d,\alpha\nu)$ or in $(\mathcal{X},d,\nu)$.

As for (iii), recall the definition of $\beta^{(K,N)}$:

$$\beta_t^{(K,N)}(x,y) = \left(\frac{\sin t\alpha(N,K,d(x,y))}{t\sin\alpha(N,K,d(x,y))}\right)^{N-1},$$

where

$$\alpha(N,K,d) = \sqrt{\frac{K}{N-1}}\, d(x,y).$$

Then $\alpha(N,K,d) = \alpha(N,\lambda^{-2}K,\lambda d)$, and Property (iii) follows. □

The next theorem shows that the property of being a CD(K,N) space does not involve the whole space $\mathcal{X}$, but only the support of ν:

Theorem 30.2 (Restriction of the CD(K,N) property to the support). *A metric-measure space* $(\mathcal{X},d,\nu)$ *is a weak* CD(K,N) *space if and only if* $(\operatorname{Spt}\nu,d,\nu)$ *is itself a weak* CD(K,N) *space.*

Remark 30.3. Theorem 30.2 allows us to systematically reduce to the case when $\operatorname{Spt}\nu = \mathcal{X}$ in the study of properties of weak $\mathrm{CD}(K,N)$ spaces. Then why not impose this in the definition of these spaces? The answer is that on some occasions it is useful to allow $\mathcal{X}$ to be larger than $\operatorname{Spt}\nu$, in particular for *convergence* issues. Indeed, it may very well happen that a sequence of weak $\mathrm{CD}(K,N)$ spaces $(\mathcal{X}_k, d_k, \nu_k)_{k\in\mathbb{N}}$ with $\operatorname{Spt}\nu_k = \mathcal{X}_k$ converges in the measured Gromov–Hausdorff sense to a metric-measure space $(\mathcal{X}, d, \nu)$ such that $\operatorname{Spt}\nu$ is strictly smaller than $\mathcal{X}$. This phenomenon of "reduction of support" is impossible if $N < \infty$, as shown by Theorem 29.32, but can occur in the case $N = \infty$. As a simple example, consider $\mathcal{X}_k = (\mathbb{R}^n, |\cdot|)$ (Euclidean space) equipped with the sharply peaked Gaussian probability measure $e^{-k|x|^2}\,dx/Z_k$, Z_k being a normalizing constant. As $k\to\infty$, $\mathcal{X}_k$ converges in measured Gromov–Hausdorff sense to $\mathcal{X} = (\mathbb{R}^n, |\cdot|, \delta_0)$. Each of the spaces $\mathcal{X}_k$ is a weak $\mathrm{CD}(0,\infty)$ space and satisfies $\operatorname{Spt}\nu_k = \mathcal{X}_k$, however, the limit measure is supported in just a point. To summarize: *For weak* $\mathrm{CD}(K,N)$ *spaces* $(\mathcal{X}, d, \nu)$ *with* $N < \infty$, *one probably does not lose anything by assuming* $\operatorname{Spt}\nu = \mathcal{X}$; *but in the class of weak* $\mathrm{CD}(K,\infty)$ *spaces, the stability theorem would not be true if one would not allow the support of* ν *to be strictly smaller than the whole space.*

Proof of Theorem 30.2. First assume that $(\operatorname{Spt}\nu, d, \nu)$ has the weak $\mathrm{CD}(K,N)$ property. Replacing $\operatorname{Spt}\nu$ by $\mathcal{X}$ does not enlarge the class of probability measures that can be chosen for μ_0 and μ_1 in Definition 29.8, and does not change the functionals U_ν or $U_{\pi,\nu}^{\beta_t^{(K,N)}}$ either. Because $\operatorname{Spt}\nu$ is (by assumption) a length space, geodesics in $\operatorname{Spt}\nu$ are also geodesics in $\mathcal{X}$. So geodesics in $P_2(\operatorname{Spt}\nu)$ are also geodesics in $P_2(\mathcal{X})$ (it is the converse that might be false), and then the property of $\mathcal{X}'$ being a weak $\mathrm{CD}(K,N)$ space implies that $\mathcal{X}$ is also a weak $\mathrm{CD}(K,N)$ space.

The converse implication is more subtle. Assume that $(\mathcal{X}, d, \nu)$ is a weak $\mathrm{CD}(K,N)$ space. Let μ_0, μ_1 be two compactly supported probability measures on $\mathcal{X}$ with $\operatorname{Spt}\mu_0 \subset \operatorname{Spt}\nu$, $\operatorname{Spt}\mu_1 \subset \operatorname{Spt}\nu$. For any $t_0 \in \{0,1\}$, choose a sequence of probability measures $(\mu_{k,t_0})_{k\in\mathbb{N}}$ converging to μ_{t_0}, satisfying the conclusion of Theorem 29.20(iii), such that the supports of all measures μ_{k,t_0} are included in a common compact set. By definition of the weak $\mathrm{CD}(K,N)$ property, for each $k\in\mathbb{N}$ there is a Wasserstein geodesic $(\mu_{k,t})_{0\le t\le 1}$ and an associated coupling $\pi_k \in \Pi(\mu_k, \nu_k)$ such that for all $t\in[0,1]$ and $U \in \mathcal{DC}_N$,

$$U_\nu(\mu_{k,t}) \le (1-t)\, U_{\pi_k,\nu}^{\beta_{1-t}^{(K,N)}}(\mu_{k,0}) \;+\; t\, U_{\check{\pi}_k,\nu}^{\beta_t^{(K,N)}}(\mu_{k,1}). \tag{30.1}$$

Choosing $H(r) = r \log r$, and using the monotonicity of the $\mathrm{CD}(K,N)$ condition with respect to N, we deduce

$$H_\nu(\mu_{k,t}) \le (1-t)\, H_{\pi_k,\nu}^{\beta_{1-t}^{(K,\infty)}}(\mu_{k,0}) \;+\; t\, H_{\check{\pi}_k,\nu}^{\beta_t^{(K,\infty)}}(\mu_{k,1}). \tag{30.2}$$

By an explicit calculation (as in the proof of (30.6) later in this chapter) the right-hand side is equal to

$$(1-t)\, H_\nu(\mu_{k,0}) \;+\; t\, H_\nu(\mu_{k,1}) - K\, \frac{t(1-t)}{2} \int d(x_0,x_1)^2\, \pi_k(dx_0\, dx_1),$$

and this quantity is finite since $\mu_{k,0}$, $\mu_{k,1}$ are compactly supported. Then by (30.2), $H_\nu(\mu_{k,t}) < +\infty$ for all $t \in [0,1]$ and for all $k \in \mathbb{N}$. Since $H'(\infty) = \infty$, this implies that $\mu_{k,t}$ is absolutely continuous with respect to ν, and in particular it is supported in $\mathrm{Spt}\,\nu$.

Next, by Ascoli's theorem, some subsequence of the family $(\mu_{k,t})$ converges uniformly in $C([0,1], P_2(\mathcal{X}))$ to some Wasserstein geodesic $(\mu_t)_{0\le t\le 1}$. Since $\mathrm{Spt}\,\nu$ is closed, μ_t is also supported in $\mathrm{Spt}\,\nu$, for each $t \in [0,1]$.

Let $(\gamma_t)_{0\le t\le 1}$ be a random geodesic such that $\mu_t = \mathrm{law}\,(\gamma_t)$. By the preceding argument, $\mathbb{P}\,[\gamma_t \notin \mathrm{Spt}\,\nu] = 0$ for any $t \in [0,1]$. If $(t_j)_{j\in\mathbb{N}}$ is a dense sequence of times in $[0,1]$, then $\mathbb{P}\,[\exists j;\ \gamma_{t_j} \notin \mathrm{Spt}\,\nu] = 0$. But since γ is continuous and $\mathrm{Spt}\,\nu$ closed, this can be reinforced into $\mathbb{P}\,[\exists t;\ \gamma_t \notin \mathrm{Spt}\,\nu] = 0$. So γ lies entirely in $\mathrm{Spt}\,\nu$, with probability 1. The path $(\mu_t)_{0\le t\le 1}$ is valued in $P_2(\mathrm{Spt}\,\nu)$, and it is a geodesic in the larger space $P_2(\mathcal{X})$; so it is also a geodesic in $P_2(\mathrm{Spt}\,\nu)$.

Admit for the moment that $\mathrm{Spt}\,\nu$ is a geodesic space. By Proposition 29.12, to show that $\mathrm{Spt}\,\nu$ is a weak $\mathrm{CD}(K,N)$ space it is sufficient to establish the displacement convexity property when $U \in \mathcal{DC}_N$ is Lipschitz (for $N < \infty$) or behaves like $r \log r$ at infinity (for $N = \infty$). For such a nonlinearity U, we can pass to the limit in (30.1), invoking Theorem 29.20(i) and (iv):

$$\begin{aligned}
U_\nu(\mu_t) &\le \liminf_{k\to\infty}\, U_\nu(\mu_{k,t}) \\
&\le \limsup_{k\to\infty} \left[(1-t)\, U_{\pi_k,\nu}^{\beta_{1-t}^{(K,N)}}(\mu_{k,0}) \;+\; t\, U_{\check{\pi}_k,\nu}^{\beta_t^{(K,N)}}(\mu_{k,1})\right] \\
&\le\; (1-t)\, U_{\pi,\nu}^{\beta_{1-t}^{(K,N)}}(\mu_0) \;+\; t\, U_{\check{\pi},\nu}^{\beta_t^{(K,N)}}(\mu_1),
\end{aligned} \tag{30.3}$$

where $\mu_0 = \rho_0\,\nu$, $\mu_1 = \rho_1\,\nu$, and π is an optimal coupling between μ_0 and μ_1.

It only remains to check that $\operatorname{Spt}\nu$ is indeed geodesic; this is not a priori obvious since it needs not be totally convex. Let x_0, x_1 be any two points in $\operatorname{Spt}\nu$; then for any $r > 0$, we have $\nu[B_r(x_0)] > 0$, $\nu[B_r(x_1)] > 0$; so it makes sense to define $\mu_0 = 1_{B_r(x_0)}/\nu[B_r(x_0)]$, and $\mu_1 = 1_{B_r(x_1)}/\nu[B_r(x_1)]$. The preceding reasoning shows that there is a random geodesic $\gamma^{(r)}$ which lies entirely in $\operatorname{Spt}\nu$, and whose endpoints belong to $B_r(x_0)$ and $B_r(x_1)$. By Ascoli's theorem, there is a subsequence $r_j \to 0$ such that $\gamma^{(r_j)}$ converges uniformly to some random geodesic γ, which necessarily satisfies $\gamma_0 = x_0$, $\gamma_1 = x_1$, and lies entirely in $\operatorname{Spt}\nu$. So $\operatorname{Spt}\nu$ satisfies the axioms of a geodesic space. □

Displacement convexity

The definition of weak $\mathrm{CD}(K,N)$ spaces is based upon displacement convexity inequalities, but these inequalities are only required to hold under the assumption that μ_0 and μ_1 are compactly supported. To exploit the full strength of displacement convexity inequalities, it is important to get rid of this restriction.

The next theorem shows that the functionals appearing in Definition 29.1 can be extended to measures μ that are not compactly supported, provided that the nonlinearity U belongs to some $\mathcal{DC}_N$ class, and the measure μ admits a moment of order p, where N and p are related through the behavior of ν at infinity. Recall Conventions 17.10 and 17.30 from Part II.

Theorem 30.4 (Domain of definition of U_ν and $U^\beta_{\pi,\nu}$ on noncompact spaces). *Let $(\mathcal{X}, d)$ be a boundedly compact Polish space, equipped with a locally finite measure ν, and let z be any point in $\mathcal{X}$. Let $K \in \mathbb{R}$, $N \in [1,\infty]$, and $U \in \mathcal{DC}_N$. For any measure μ on $\mathcal{X}$, introduce its Lebesgue decomposition with respect to ν:*

$$\mu = \rho\,\nu + \mu_s.$$

Let $\pi(dy|x)$ be a family of conditional probability measures on $\mathcal{X}$, indexed by $x \in \mathcal{X}$, and let $\pi(dx\,dy) = \mu(dx)\,\pi(dy|x)$ be the associated coupling with first marginal μ. Assume that

$$\int_{\mathcal{X}\times\mathcal{X}} d(x,y)^2\,\pi(dx\,dy) < +\infty; \qquad \int_{\mathcal{X}} d(z,x)^p\,\mu(dx) < +\infty,$$

where $p \geq 2$ is such that

$$\begin{cases} \displaystyle\int_{\mathcal{X}} \frac{d\nu(x)}{[1+d(z,x)]^{p(N-1)}} < +\infty & (N < \infty) \\ \exists c > 0; \quad \displaystyle\int_{\mathcal{X}} e^{-c\, d(z,x)^p}\, d\nu(x) < +\infty & (N = \infty). \end{cases} \tag{30.4}$$

Then for any $t \in [0,1]$ the following expressions make sense in $\mathbb{R} \cup \{\pm\infty\}$ and can be taken as generalized definitions of the functionals appearing in Definition 29.1:

$$U_\nu(\mu) := \int_{\mathcal{X}} U(\rho(x))\, \nu(dx) \;+\; U'(\infty)\, \mu_s[\mathcal{X}];$$

$$U_{\pi,\nu}^{\beta_t^{(K,N)}}(\mu) := \int_{\mathcal{X}\times\mathcal{X}} U\left(\frac{\rho(x)}{\beta_t^{(K,N)}(x,y)}\right) \beta_t^{(K,N)}(x,y)\, \pi(dy|x)\, \nu(dx) \\ + \; U'(\infty)\, \mu_s[\mathcal{X}],$$

where the right-hand side really should be understood as

$$\lim_{N'\downarrow N} \int_{\mathcal{X}\times\mathcal{X}} U\left(\frac{\rho(x)}{\beta_t^{(K,N')}(x,y)}\right) \beta_t^{(K,N')}(x,y)\, \pi(dy|x)\, \nu(dx) \\ + \; U'(\infty)\, \mu_s[\mathcal{X}].$$

Even if there is no such p, $U_{\pi,\nu}^{\beta}(\mu)$ still makes sense for any $\mu \in P_c(\mathcal{X})$.

Proof of Theorem 30.4. The proof is the same as for Theorem 17.28 and Application 17.29, with only two minor differences: (a) ρ is not necessarily a probability density, but still its integral is bounded above by 1; (b) there is an additional term $U'(\infty)\, \mu_s[\mathcal{X}] \in \mathbb{R} \cup \{+\infty\}$. □

The next theorem is the final goal of this section: It extends the displacement convexity inequalities of Definition 29.8 to noncompact situations.

Theorem 30.5 (Displacement convexity inequalities in weak CD(K,N) spaces). *Let $N \in [1,\infty]$, let $(\mathcal{X}, d, \nu)$ be a weak* CD(K,N) *space, and let $p \in [2,+\infty) \cup \{c\}$ satisfy condition (30.4). Let μ_0 and μ_1 be two probability measures in $P_p(\mathcal{X})$, whose supports are included in* Spt ν. *Then there exists a Wasserstein geodesic $(\mu_t)_{0\leq t\leq 1}$, and an* *associated optimal coupling π of (μ_0,μ_1) such that, for all $U \in \mathcal{DC}_N$ and for all $t \in [0,1]$,*

$$U_\nu(\mu_t) \leq (1-t)\, U_{\pi,\nu}^{\beta_{1-t}^{(K,N)}}(\mu_0) \;+\; t\, U_{\check{\pi},\nu}^{\beta_t^{(K,N)}}(\mu_1). \tag{30.5}$$

Furthermore, if $N=\infty$, one also has

$$U_\nu(\mu_t) \leq (1-t)\, U_\nu(\mu_0) + t\, U_\nu(\mu_1) - \frac{\lambda(K,U)\, t(1-t)}{2}\, W_2(\mu_0,\mu_1)^2, \tag{30.6}$$

where

$$\lambda(K,U) = \inf_{r>0} \frac{K p(r)}{r} = \begin{cases} K\, p'(0) & \text{if } K>0 \\ 0 & \text{if } K=0 \\ K\, p'(\infty) & \text{if } K<0. \end{cases} \tag{30.7}$$

These inequalities are the starting point for *all* subsequent inequalities considered in the present chapter.

The proof of Theorem 30.5 will use two auxiliary results, which generalize Theorems 29.20(i) and (iii) to noncompact situations. These are definitely not the most general results of their kind, but they will be enough to derive displacement convexity inequalities with a lot of generality. As usual, I shall denote by $M_+(\mathcal{X})$ the set of finite (nonnegative) Borel measures on $\mathcal{X}$, and by $L^1_+(\mathcal{X})$ the set of nonnegative ν-integrable measurable functions on $\mathcal{X}$. Recall from Definition 6.8 the notion of (weak) convergence in $P_2(\mathcal{X})$.

Theorem 30.6 (Lower semicontinuity of U_ν again). *Let $(\mathcal{X},d)$ be a boundedly compact Polish space, equipped with a locally finite measure ν such that $\operatorname{Spt}\nu = \mathcal{X}$. Let $U : \mathbb{R}_+ \to \mathbb{R}$ be a continuous convex function, with $U(0)=0$, $U(r) \geq -c\, r$ for some $c \in \mathbb{R}$. Then*

(i) For any $\mu \in M_+(\mathcal{X})$ and any sequence $(\mu_k)_{k\in\mathbb{N}}$ converging weakly to μ in $M_+(\mathcal{X})$,

$$U_\nu(\mu) \leq \liminf_{k\to\infty} U_\nu(\mu_k).$$

(ii) Assume further that $\mu \in P_2(\mathcal{X})$, and let $\beta(x,y)$ be a positive measurable function on $\mathcal{X}\times\mathcal{X}$, with $|\log\beta(x,y)| = O(d(x,y)^2)$. Then there is a sequence $(\mu_k)_{k\in\mathbb{N}}$ of compactly supported probability measures on $\mathcal{X}$ such that

$$\mu_k \xrightarrow[k\to\infty]{} \mu \qquad \text{in } P_2(\mathcal{X}),$$

and for any sequence of probability measures $(\pi_k)_{k\in\mathbb{N}}$ *such that the first marginal of* π_k *is* μ_k*, and the second one is* $\mu_{k,1}$*, the limits*

$$\begin{cases} \pi_k \xrightarrow[k\to\infty]{} \pi \\ \displaystyle\int d(x,y)^2\,\pi_k(dx\,dy) \xrightarrow[k\to\infty]{} \int d(x,y)^2\,\pi(dx\,dy) \\ \mu_{k,1} \xrightarrow[k\to\infty]{} \mu_1 \quad \text{in } P_2(\mathcal{X}) \end{cases}$$

imply

$$\lim_{k\to\infty} U^\beta_{\pi_k,\nu}(\mu_k) = U^\beta_{\pi,\nu}(\mu).$$

Proof of Theorem 30.6. First of all, we may reduce to the case when U is valued in $\mathbb{R}_+$, just replacing U by $r \to U(r) + c\,r$. So in the sequel U will be nonnegative.

Let us start with (i). Let z be an arbitrary base point, and let $(\chi_R)_{R>0}$ be a z-cutoff as in the Appendix (that is, a family of cutoff continuous functions that are identically equal to 1 on a ball $B_R(z)$ and identically equal to 0 outside $B_{R+1]}(z)$). For any $R > 0$, write

$$U_\nu(\chi_R\,\mu) = \int U(\chi_R\,\rho)\,d\nu + U'(\infty)\int \chi_R\,d\mu_s.$$

Since U is convex nonnegative with $U(0) = 0$, it is nondecreasing; by the monotone convergence theorem,

$$U_\nu(\chi_R\,\mu) \xrightarrow[R\to\infty]{} U_\nu(\mu).$$

In particular,

$$U_\nu(\mu) = \sup_{R>0} U_\nu(\chi_R\,\mu). \tag{30.8}$$

On the other hand, for each fixed R, we have

$$U_\nu(\chi_R\,\mu) = U_{\chi_{R+1}\nu}(\chi_R\,\mu),$$

and then we can apply Proposition 29.19(i) with the compact space $(B_{R+1]}(z), \nu)$, to get

$$U_\nu(\chi_R\,\mu) = \sup\left\{\int_{\mathcal{X}} \varphi\,\chi_R\,d\mu - \int_{\mathcal{X}} U^*(\varphi)\,\chi_{R+1}d\nu;\right.$$
$$\left.\varphi \in C_b(B_{R+1]}(z)),\ \varphi \le U'(\infty)\right\}.$$

The function $\varphi\chi_R$, extended by 0 outside of $B_{R+1]}$, defines a bounded continuous function on the whole of $\mathcal{X}$, so $\mu \longmapsto \int \varphi\,\chi_R\,d\mu$ is continuous with respect to the weak topology of convergence against bounded continuous functions. Thus $U_\nu(\chi_R\mu)$ is a lower semicontinuous function of μ. This combined with (30.8) shows that U_ν is a lower semicontinuous function of μ, which establishes (i).

Let us now turn to the proof of (ii), which is more tricky. As before I shall assume that U is nonnegative. Let $B_R = B_{R]}(z)$. Again, let $(\chi_R)_{R>0}$ be a z-cutoff as in the Appendix. For k large enough, $\int \chi_R\,d\mu \geq 1/2$. Define

$$\mu_k = \frac{\chi_k\,\mu}{\int \chi_k\,d\mu}.$$

We'll see that μ_k does the job. The proof will use truncation and regularization into continuous functions.

So let (π_k) be as in the theorem. Define

$$\mu^{(R)} = \frac{\chi_R\,\mu}{\int \chi_R\,d\mu}; \qquad Z^{(R)} = \int \chi_R\,d\mu.$$

Let $\rho^{(R)}$ be the density of the absolutely continuous part of $\mu^{(R)}$, and $\mu_s^{(R)}$ be the singular part. It is obvious that $\rho^{(R)}$ converges to ρ in $L^1(\nu)$ and $\mu_s^{(R)}[\mathcal{X}] \to \mu_s[\mathcal{X}]$ as $R \to \infty$.

Next, define

$$\begin{cases} \pi^{(R)}(dy|x) = \chi_R(y)\,\pi(dy|x) + \left(\displaystyle\int \big[1 - \chi_R(y')\big]\,\pi(dy'|x)\right)\,\delta_z; \\[2ex] \pi^{(R)}(dx\,dy) = \mu^{(R)}(dx)\,\pi^{(R)}(dy|x). \end{cases}$$

Note that $\pi^{(R)}$ is a probability measure supported in $B_R \times B_R$.

Further, define $\mu_k^{(R)}$, $Z_k^{(R)}$, $\rho_k^{(R)}$, $\mu_{k,s}^{(R)}$, $\pi_k^{(R)}(dy|x)$, $\pi_k^{(R)}(dx\,dy)$ in a similar way, just replacing μ by μ_k. The explicit formula

$$\int \Psi\,d\pi_k^{(R)} = \int \Psi(x,y)\,\chi_R(y)\,\pi_k(dx\,dy) + \int \Psi(x,z)\,\big[1 - \chi_R(y)\big]\,\pi_k(dx\,dy)$$

shows that $\pi_k^{(R)}$ converges to $\pi^{(R)}$ as $k \to \infty$, for any fixed R.

The plan is to first replace the original expressions by the expressions with the superscript (R), and then to pass to the limit as $k \to \infty$ for fixed R. For that I will distinguish two cases.

Case 1: U is Lipschitz. Then

$$\begin{aligned}
&\left|U^{\beta}_{\pi^{(R)},\nu}(\mu^{(R)}) - U^{\beta}_{\pi,\nu}(\mu)\right| \\
&\leq \left|\int U\left(\frac{\rho^{(R)}(x)}{\beta(x,y)}\right)\beta(x,y)\,\pi^{(R)}(dy|x)\,\nu(dx)\right. \\
&\qquad\qquad\left. - \int U\left(\frac{\rho(x)}{\beta(x,y)}\right)\beta(x,y)\,\pi(dy|x)\,\nu(dx)\right| \\
&\quad + U'(\infty)\left|\mu_s^{(R)}[\mathcal{X}] - \mu_s[\mathcal{X}]\right| \\
&\leq \int\left|U\left(\frac{\rho^{(R)}(x)}{\beta(x,y)}\right) - U\left(\frac{\rho(x)}{\beta(x,y)}\right)\right|\beta(x,y)\,\pi^{(R)}(dy|x)\,\nu(dx) \\
&\quad + \int U\left(\frac{\rho(x)}{\beta(x,y)}\right)\beta(x,y)\,[1-\chi_R(y)]\,\pi(dy|x)\,\nu(dx) \\
&\quad + \int U\left(\frac{\rho(x)}{\beta(x,y)}\right)\beta(x,y)\left(\int[1-\chi_R(y')]\,\pi(dy'|x)\right)\delta_{y=z}\,\nu(dx) \\
&\quad + \|U\|_{\mathrm{Lip}}\left|\mu_s^{(R)}[\mathcal{X}] - \mu_s[\mathcal{X}]\right| \\
&\leq \|U\|_{\mathrm{Lip}}\left(\int\left|\rho^{(R)}(x) - \rho(x)\right|\pi^{(R)}(dy|x)\,\nu(dx)\right. \\
&\qquad\qquad + \int\rho(x)\,[1-\chi_R(y)]\,\pi(dy|x)\,\nu(dx) \\
&\qquad\qquad + \int\rho(x)\,[1-\chi_R(y')]\,\pi(dy'|x)\,\nu(dx) \\
&\qquad\qquad \left. + \left|\mu_s^{(R)}[\mathcal{X}] - \mu_s[\mathcal{X}]\right|\right) \\
&\leq \|U\|_{\mathrm{Lip}}\left[\int|\rho^{(R)} - \rho|\,d\nu + 2\int[1-\chi_R(y)]\,\pi(dx\,dy) + \left|\mu_s^{(R)}[\mathcal{X}] - \mu_s[\mathcal{X}]\right|\right].
\end{aligned}$$

In the last expression, the second term inside brackets is bounded by

$$2\int_{d(z,y)\geq R}\pi(dx\,dy) = 2\int_{d(z,y)\geq R}\mu_1(dy) \leq \frac{2}{R^2}\int d(z,y)^2\,\mu_1(dy).$$

Conclusion: There is a constant C such that

$$\begin{aligned}
\left|U^{\beta}_{\pi^{(R)},\nu}(\mu^{(R)}) - U^{\beta}_{\pi,\nu}(\mu)\right| \leq C\Big(\|\rho^{(R)} - \rho\|_{L^1(\nu)} &+ \left|\mu_s^{(R)}[\mathcal{X}] - \mu_s[\mathcal{X}]\right| \\
&+ \frac{1}{R^2}\int d(z,y)^2\,\mu_1(dy)\Big). \qquad (30.9)
\end{aligned}$$

Similarly,[1]

$$\left|U^\beta_{\pi_k^{(R)},\nu}(\mu_k^{(R)}) - U^\beta_{\pi_k,\nu}(\mu_k)\right| \le C\Big(\|\rho_k^{(R)} - \rho_k\|_{L^1(\nu)} + \left|\mu_{k,s}^{(R)}[\mathcal{X}] - \mu_{k,s}[\mathcal{X}]\right| + \frac{1}{R^2}\int d(z,y)^2\,\mu_{k,1}(dy)\Big). \quad (30.10)$$

Note that for $k \ge R$, $\rho_k^{(R)} = \rho^{(R)}$, and $\mu_{k,s}^{(R)} = \mu_s^{(R)}$. Then in view of the definition of μ_k and the fact that $\int d(z,y)^2\,\mu_{k,1}(dy)$ is bounded, we easily deduce from (30.9) and (30.10) that

$$\begin{cases} \lim_{R\to\infty} \left|U^\beta_{\pi^{(R)},\nu}(\mu^{(R)}) - U^\beta_{\pi,\nu}(\mu)\right| = 0; \\ \lim_{R\to\infty}\ \limsup_{k\to\infty} \left|U^\beta_{\pi_k^{(R)},\nu}(\mu_k^{(R)}) - U^\beta_{\pi_k,\nu}(\mu_k)\right| = 0. \end{cases}$$

So to prove the result, it is sufficient to establish that for fixed R,

$$\lim_{k\to\infty} U^\beta_{\pi_k^{(R)},\nu}(\mu_k^{(R)}) = U^\beta_{\pi^{(R)},\nu}(\mu^{(R)}). \quad (30.11)$$

The interest of this reduction is that all probability measures $\mu_k^{(R)}$ (resp. $\pi_k^{(R)}$) are now supported in a common compact set, namely the closed ball B_{2R} (resp. $B_{2R}\times B_{2R}$). Note that $\mu_k^{(R)}$ converges to $\mu^{(R)}$.

If k is large enough, $\mu_k^{(R)} = \mu^{(R)}$, so (30.11) becomes

$$U^\beta_{\pi_k^{(R)},\nu}(\mu^{(R)}) \xrightarrow[k\to\infty]{} U^\beta_{\pi^{(R)},\nu}(\mu^{(R)}).$$

In the sequel, I shall drop the superscript (R), so the goal will be

$$U^\beta_{\pi_k,\nu}(\mu) \xrightarrow[k\to\infty]{} U^\beta_{\pi,\nu}(\mu).$$

The argument now is similar to the one used in the proof of Theorem 29.20(iii). Define

$$g(x,y) = U\left(\frac{\rho(x)}{\beta(x,y)}\right)\frac{\beta(x,y)}{\rho(x)},$$

with the convention that $g(x,y) = U'(\infty)$ when $x \in \operatorname{Spt}\mu_s$, and $g(x,y) = U'(0)$ when $\rho(x) = 0$ and $x \notin \operatorname{Spt}\mu_s$. Then

[1] Here again the notation might be confusing: $\mu_{k,s}$ stands for the singular part of μ_k, while $\mu_{k,1}$ is the second marginal of π_k.

$$U^{\beta}_{\pi_k,\nu}(\mu) = \int g(x,y)\,\mu(dx)\,\pi_k(dy|x) = \int g(x,y)\,\pi_k(dx\,dy);$$

$$U_{\pi,\nu}(\mu) = \int g(x,y)\,\pi(dx\,dy).$$

Since $g \in L^1((B_{2R},\mu); C(B_{2R}))$, by Lemma 29.36 there is a sequence $(\Psi_j)_{j\in\mathbb{N}}$ in $C(B_{2R}\times B_{2R})$ such that $\|\Psi_j - g\|_{L^1((B_{2R},\mu);C(B_{2R}))} \xrightarrow[j\to\infty]{} 0$. Then

$$\begin{aligned}\sup_{k\in\mathbb{N}} \left| U^{\beta}_{\pi_k,\nu}(\mu) - \int \Psi_j\, d\pi_k \right| &\le \int \bigl|g(x,y) - \Psi_j(x,y)\bigr|\,\pi_k(dx\,dy)\\ &\le \int \sup_y \bigl|g(x,y) - \Psi_j(x,y)\bigr|\,\mu(dx) \xrightarrow[j\to\infty]{} 0;\end{aligned} \tag{30.12}$$

$$\left| U^{\beta}_{\pi,\nu}(\mu) - \int \Psi_j\, d\pi \right| \xrightarrow[j\to\infty]{} 0; \tag{30.13}$$

and for each fixed k,

$$\int \Psi_j\, d\pi_k \longrightarrow \int \Psi_j\, d\pi. \tag{30.14}$$

The combination of (30.12), (30.13) and (30.14) closes the case.

Case 2: U is not Lipschitz. If $U^{\beta}_{\pi,\nu}(\mu) < +\infty$ then necessarily $\mu_s[\mathcal{X}] = 0$. Moreover, there exist positive constants a, c such that $U(r) \le a\,r\,\log(2+r)$ and $|U'(r)| \le c\,\log(2+r)$; in particular, there is $C > 0$ such that

$$\forall x, y \ge 0, \qquad |U(x) - U(y)| \le C\,|x-y|\,\Bigl(\log(2+x) + \log(2+y)\Bigr).$$

Possibly increasing the value of C, we deduce that

$$\begin{aligned}&\Bigl|U^{\beta}_{\pi^{(R)},\nu}(\mu^{(R)}) - U^{\beta}_{\pi,\nu}(\mu)\Bigr|\\ &\le C \int \left| \frac{\rho^{(R)}(x)}{\beta(x,y)} - \frac{\rho(x)}{\beta(x,y)} \right| \left(\log(2+\rho(x)) + \log\Bigl(2 + \frac{1}{\beta(x,y)}\Bigr) \right.\\ &\qquad\qquad \left. + \log(2+\rho^{(R)}(x)) \right) \beta(x,y)\,\pi^{(R)}(dy|x)\,\nu(dx)\\ &+ C \int \frac{\rho(x)}{\beta(x,y)} \left[\log(2+\rho(x)) + \log\Bigl(2+\frac{1}{\beta(x,y)}\Bigr) \right] \beta(x,y)\\ &\qquad\qquad [1 - \chi_R(y)]\,\pi(dy|x)\,\nu(dx)\\ &+ C \int \frac{\rho(x)}{\beta(x,y)} \left[\log(2+\rho(x)) + \log\Bigl(2+\frac{1}{\beta(x,y)}\Bigr) \right] \beta(x,y)\\ &\qquad\qquad \left(\int [1 - \chi_R(y')]\,\pi(dy'|x) \right) \delta_{y=z}\,\nu(dx).\end{aligned}$$

Using $\rho^{(R)} \le 2\rho$, $\log(1/\beta) \le C\, d(x,y)^2$ and reasoning as in the first case, we can bound the above expression by

$$
\begin{aligned}
& C \int \left|\rho^{(R)}(x) - \rho(x)\right| \log(2+\rho(x))\, \nu(dx) \\
& + C \int \left|\rho^{(R)}(x) - \rho(x)\right| \big(1 + d(x,y)\big)^2\, \pi^{(R)}(dy|x)\, \nu(dx) \\
& + C \int \rho(x) \log(2+\rho(x)) \big[1 - \chi_R(y)\big]\, \pi(dy|x)\, \nu(dx) \\
& + C \int \rho(x) \left(1 + d(x,z)^2\right) \big[1 - \chi_R(y)\big]\, \pi(dy|x)\, \nu(dx)
\end{aligned}
$$

$$
\begin{aligned}
\le\; & C \int \left|\rho^{(R)}(x) - \rho(x)\right| \log(2+\rho(x))\, \nu(dx) \\
& + C\,(1+D^2) \int \left|\rho^{(R)}(x) - \rho(x)\right| \nu(dx) \\
& + C \int_{d(x,y)\ge D} \left|\rho^{(R)}(x) - \rho(x)\right| \left(1 + d(x,y)^2\right) \big(\pi(dy|x) + \delta_z\big)\, \nu(dx) \\
& + C \log(2+M) \int \rho(x) \big[1 - \chi_R(y)\big]\, \pi(dy|x)\, \nu(dx) \\
& + C \int_{\rho(x)\ge M} \rho(x) \log(2+\rho(x))\, \pi(dy|x)\, \nu(dx) \\
& + C\,(1+D^2) \int \rho(x) \big[1 - \chi_R(y)\big]\, \pi(dy|x)\, \nu(dx) \\
& + C \int_{d(x,z)\ge D} \rho(x) \left(1 + d(x,z)^2\right) \pi(dy|x)\, \nu(dx)
\end{aligned}
$$

$$
\begin{aligned}
\le\; & C\,(1+D^2) \int \left|\rho^{(R)}(x) - \rho(x)\right| \log(2+\rho(x))\, \nu(dx) \\
& + C \int_{d(x,y)\ge D} \big[1 + d(x,y)^2\big]\, \pi(dx\,dy) \\
& + C \int_{d(x,z)\ge D} \big[1 + d(x,z)^2\big]\, \pi(dx\,dy) \\
& + C \left(\log(2+M) + (1+D^2)\right) \int \big[1 - \chi_R(y)\big]\, \pi(dx\,dy) \\
& + C \int_{\rho\ge M} \rho \log(2+\rho)\, d\nu.
\end{aligned}
$$

Since $d(x,y)^2\, 1_{d(x,y)\geq D} \leq d(x,z)^2\, 1_{d(x,z)\geq D/2} + d(y,z)^2\, 1_{d(y,z)\geq D/2}$, the above expression can in turn be bounded by

$$C\,(1{+}D^2)\int \bigl|\rho^{(R)}{-}\rho\bigr|\,\log(2+\rho)\,d\nu + C\int_{d(z,x)\geq D/2}\bigl[1+d(x,z)^2\bigr]\,\mu(dx)$$
$$+\,C\int_{d(z,y)\geq D/2}\bigl[1+d(y,z)^2\bigr]\,\mu_1(dy) + C\int_{\rho\geq M}\rho\log(2+\rho)\,d\nu$$
$$+\,C\,\frac{\bigl(\log(2+M)+D^2\bigr)}{R^2}\int_{d(z,y)\geq D/2} d(z,y)^2\,\mu_1(dy).$$

Of course this bound converges to 0 if $R\to\infty$, then $M,D\to\infty$.

Similarly, $\left|U^\beta_{\pi_k^{(R)},\nu}(\mu_k^{(R)}) - U^\beta_{\pi_k,\nu}(\mu_k)\right|$ is bounded by

$$C\,(1{+}D^2)\int \bigl|\rho^{(R)}{-}\rho\bigr|\,\log(2+\rho)\,d\nu + C\int_{d(z,x)\geq D/2}\bigl[1+d(x,z)^2\bigr]\,\mu(dx)$$
$$+\,C\int_{d(z,y)\geq D/2}\bigl[1+d(y,z)^2\bigr]\,\mu_{k,1}(dy) + C\int_{\rho\geq M}\rho\log(2+\rho)\,d\nu$$
$$+\,C\,\frac{\bigl(\log(2+M)+D^2\bigr)}{R^2}\int_{d(z,y)\geq D/2} d(z,y)^2\,\mu_{k,1}(dy).$$

By letting $k\to\infty$, then $R\to\infty$, then $M,D\to\infty$ and using the definition of convergence in $P_2(\mathcal{X})$, we conclude that

$$\limsup_{k\to\infty}\left|U^\beta_{\pi_k^{(R)},\nu}(\mu_k^{(R)}) - U^\beta_{\pi_k,\nu}(\mu_k)\right| \xrightarrow[R\to\infty]{} 0.$$

From that point on, the proof is similar to the one in the first case. (To prove that $g\in L^1(B_{2R};C(B_{2R}))$ one can use the fact that β is bounded from above and below by positive constants on $B_{2R}\times B_{2R}$, and apply the same estimates as in the proof of Theorem 29.20(ii).) □

Proof of Theorem 30.5. By an approximation theorem as in the proof of Proposition 29.12, we may restrict to the case when U is nonnegative; we may also assume that U is Lipschitz (in case $N<\infty$) or that it behaves at infinity like $a\,r\log r + b\,r$ (in case $N=\infty$). By approximating N by $N'>N$, we may also assume that the distortion coefficients $\beta_t^{(K,N)}(x,y)$ are locally bounded and $|\log\beta_t^{(K,N)}(x,y)| = O(d(x,y)^2)$.

Let $(\mu_{k,0})_{k\in\mathbb{N}}$ (resp. $(\mu_{k,1})_{k\in\mathbb{N}}$) be a sequence converging to μ_0 (resp. to μ_1) and satisfying the conclusions of Theorem 30.6(ii). For each k there is a Wasserstein geodesic $(\mu_{k,t})_{0\leq t\leq 1}$ and an associated coupling π_k of $(\mu_{k,0},\mu_{k,1})$ such that

$$U_\nu(\mu_{k,t}) \leq (1-t)\, U^{\beta^{(K,N)}_{1-t}}_{\pi_k,\nu}(\mu_{k,0}) + t\, U^{\beta^{(K,N)}_t}_{\check{\pi}_k,\nu}(\mu_{k,1}). \tag{30.15}$$

Further, let Π_k be a dynamical optimal transference plan such that $(e_t)_\#\Pi_k = \mu_{k,t}$ and $(e_0,e_1)_\#\Pi_k = \pi_k$. Since the sequence $\mu_{k,0}$ converges weakly to μ_0, its elements belong to a compact subset of $P(\mathcal{X})$; the same is true of the measures $\mu_{k,1}$. By Theorem 7.21 the families $(\mu_{k,t})_{0\leq t\leq 1}$ belong to a compact subset of $C([0,1];P(\mathcal{X}))$; and also the dynamical optimal transference plans Π_k belong to a compact subset of $P(C([0,1];\mathcal{X}))$. So up to extraction of a subsequence we may assume that Π_k converges to some Π, $(\mu_{k,t})_{0\leq t\leq 1}$ converges to some path $(\mu_t)_{0\leq t\leq 1}$ (uniformly in t), and π_k converges to some π. Since the evaluation map is continuous, it is immediate that $\pi = (e_0,e_1)_\#\Pi$ and $\mu_t = (e_t)_\#\Pi$.

By Theorem 30.6(i), $U_\nu(\mu_t) \leq \liminf_{k\to\infty} U_\nu(\mu_{k,t})$. Then, by construction (and Theorem 30.6(ii)),

$$\begin{cases} \limsup_{k\to\infty} U^{\beta^{(K,N)}_{1-t}}_{\pi_k,\nu}(\mu_{k,0}) \leq U^{\beta^{(K,N)}_{1-t}}_{\pi,\nu}(\mu_0) \\ \limsup_{k\to\infty} U^{\beta^{(K,N)}_t}_{\check{\pi}_k,\nu}(\mu_{k,1}) \leq U^{\beta^{(K,N)}_t}_{\check{\pi},\nu}(\mu_1). \end{cases}$$

The desired inequality (30.5) follows by plugging the above into (30.15).

To deduce (30.6) from (30.5), I shall use a reasoning similar to the one in the proof of Theorem 20.10. Since $N=\infty$, Proposition 17.7(ii) implies $U'(\infty)=+\infty$ (save for the trivial case where U is linear), so we may assume that μ_0 and μ_1 are absolutely continuous with respect to ν, with respective densities ρ_0 and ρ_1. The convexity of $u:\delta\longmapsto U(e^{-\delta})e^\delta$ implies

$$\begin{aligned}
&U\left(\frac{\rho_1(x_1)}{\beta_t^{(K,\infty)}(x_0,x_1)}\right)\frac{\beta_t^{(K,\infty)}(x_0,x_1)}{\rho_1(x_1)} \\
&\leq U(\rho_1(x_1))\,\frac{1}{\rho_1(x_1)} \\
&\quad + \frac{\beta_t^{(K,\infty)}(x_0,x_1)}{\rho_1(x_1)}\, p\left(\frac{\rho_1(x_1)}{\beta_t^{(K,\infty)}(x_0,x_1)}\right)\left(\log\frac{1}{\rho_1(x_1)} - \log\frac{\beta_t^{(K,\infty)}(x_0,x_1)}{\rho_1(x_1)}\right) \\
&= \frac{U(\rho_1(x_1))}{\rho_1(x_1)} - \frac{\beta_t^{(K,\infty)}(x_0,x_1)}{\rho_1(x_1)}\, p\left(\frac{\rho_1(x_1)}{\beta_t^{(K,\infty)}}\right)\frac{K\,(1-t^2)}{6}\, d(x_0,x_1)^2 \\
&\leq \frac{U(\rho_1(x_1))}{\rho_1(x_1)} - \lambda(K,U)\left(\frac{1-t^2}{6}\right) d(x_0,x_1)^2;
\end{aligned}$$

so

$$U_{\tilde{\pi},\nu}^{\beta_t^{(K,\infty)}}(\mu_1) \leq U_\nu(\mu_1) - \lambda(K,U)\left(\frac{1-t^2}{6}\right) W_2(\mu_0,\mu_1)^2. \quad (30.16)$$

Similarly,

$$U_{\pi,\nu}^{\beta_{1-t}^{(K,\infty)}}(\mu_0) \leq U_\nu(\mu_0) - \lambda(K,U)\left(\frac{1-(1-t)^2}{6}\right) W_2(\mu_0,\mu_1)^2. \quad (30.17)$$

Then (30.6) follows from (30.5), (30.16), (30.17) and the identity

$$t\left(\frac{1-t^2}{6}\right) + (1-t)\left(\frac{1-(1-t)^2}{6}\right) = \frac{t(1-t)}{2}.$$

□

Brunn–Minkowski inequality

The next theorem can be taken as the first step to control volumes in weak $\mathrm{CD}(K,N)$ spaces:

Theorem 30.7 (Brunn–Minkowski inequality in weak CD(K,N) spaces). *Let $K \in \mathbb{R}$ and $N \in [1,\infty]$. Let $(\mathcal{X},d,\nu)$ be a weak* $\mathrm{CD}(K,N)$ *space, let A_0, A_1 be two compact subsets of* $\mathrm{Spt}\,\nu$, *and let $t \in (0,1)$. Then:*

- *If $N < \infty$,*

$$\nu\big[[A_0,A_1]_t\big]^{\frac{1}{N}} \geq (1-t)\left[\inf_{(x_0,x_1)\in A_0\times A_1} \beta_{1-t}^{(K,N)}(x_0,x_1)^{\frac{1}{N}}\right]\nu[A_0]^{\frac{1}{N}} + t\left[\inf_{(x_0,x_1)\in A_0\times A_1} \beta_t^{(K,N)}(x_0,x_1)^{\frac{1}{N}}\right]\nu[A_1]^{\frac{1}{N}}. \quad (30.18)$$

- *In particular, if $N < \infty$ and $K \geq 0$, then*

$$\nu\big[[A_0,A_1]_t\big]^{\frac{1}{N}} \geq (1-t)\,\nu[A_0]^{\frac{1}{N}} + t\,\nu[A_1]^{\frac{1}{N}}. \quad (30.19)$$

- *If $N = \infty$, then*

$$\log\frac{1}{\nu\big[[A_0,A_1]_t\big]} \leq (1-t)\log\frac{1}{\nu[A_0]} + t\log\frac{1}{\nu[A_1]} - \frac{K\,t(1-t)}{2}\sup_{x_0\in A_0,\,x_1\in A_1} d(x_0,x_1)^2. \quad (30.20)$$

Proof of Theorem 30.7. The proof is the same, *mutatis mutandis*, as the proof of Theorem 18.5: Use the regularity of ν to reduce to the case when $\nu[A_0], \nu[A_1] > 0$; then define $\mu_0 = (1_{A_0}/\nu[A_0])\nu$, $\mu_1 = (1_{A_1}/\nu[A_1])\nu$, and apply the displacement convexity inequality from Theorem 30.5 with the nonlinearity $U(r) = -r^{1-1/N}$ if $N < \infty$, $U(r) = r \log r$ if $N = \infty$. □

Remark 30.8. The result fails if A_0, A_1 are not assumed to lie in the support of ν. (Take $\nu = \delta_{x_0}$, $x_1 \neq x_0$, and $A_0 = \{x_0\}$, $A_1 = \{x_1\}$.)

Here are two interesting corollaries:

Corollary 30.9 (Nonatomicity of the support). *Let $K \in \mathbb{R}$ and $N \in [1, \infty]$. If $(\mathcal{X}, d, \nu)$ is a weak* $\mathrm{CD}(K,N)$ *space, then either ν is a Dirac mass, or ν has no atom.*

Corollary 30.10 (Exhaustion by intermediate points). *Let $K \in \mathbb{R}$and $N \in [1, \infty)$. Let $(\mathcal{X}, d, \nu)$ be a weak* $\mathrm{CD}(K,N)$ *space, let A be a compact subset of* $\operatorname{Spt} \nu$, *and let $x \in A$. Then*

$$\nu\big[[x, A]_t\big] \xrightarrow[t\to 1]{} \nu[A].$$

Proof of Corollary 30.9. This corollary will be derived as a consequence of (30.20). By Theorem 30.2, we may assume without loss of generality that $\operatorname{Spt} \nu = \mathcal{X}$. Suppose that ν has an atom, i.e. some $x_0 \in \mathcal{X}$ with $\nu[\{x_0\}] > 0$; and yet $\nu \neq \delta_{x_0}$, so that $\nu[\mathcal{X} \setminus \{x_0\}] > 0$. Define $A_0 = \{x_0\}$ and let A_1 be some compact subset of $\mathcal{X} \setminus \{x_0\}$ such that $\nu[A_1] > 0$. For $t > 0$, $[A_0, A_1]_t$ does not contain x_0, but it is included in a ball that shrinks around x_0; it follows that $\nu[[A_0, A_1]_t]$ converges to 0 as $t \to 0$. So $\log(1/\nu[[A_0, A_1]_t]) \to +\infty$ as $t \to 0$; but this contradicts (30.20). □

Proof of Corollary 30.10. Define $R = \max\{d(x,a);\ a \in A\}$. Then $[x, A]_t \subset A^{(1-t)\,R]} = \{y;\ d(y, A) \leq (1-t)\,R\}$; so $\nu[[x,A]_t] \leq \nu[A^{(1-t)\,R]}]$. The limit $t \to 1$ yields

$$\limsup_{t\to 1}\ \nu\big[[x, A]_t\big] \leq \nu[A]. \tag{30.21}$$

This was the easy part, which does not need the $\mathrm{CD}(K,N)$ condition.

To prove the lower bound, apply (30.18) with $A_0 = \{x\}$, $A_1 = A$: This results in

$$\nu\big[[x, A]_t\big]^{\frac{1}{N}} \geq t\ \inf_{a\in A} \beta_t^{(K,N)}(x,a)^{\frac{1}{N}}\, \nu[A]^{\frac{1}{N}}.$$

As $t \to 1$, $\inf \beta_t^{(K,N)}(x,a)$ converges to 1, so we may pass to the lim inf and recover

$$\liminf_{t\to 1} \nu\big[[x,A]_t\big] \geq \nu[A].$$

This combined with (30.21) proves the claim. □

Bishop–Gromov inequalities

Once we know that ν has no atom, we can get much more precise information and control on the growth of the volume of balls, and in particular prove sharp Bishop–Gromov inequalities for weak $\mathrm{CD}(K,N)$ spaces with $N < \infty$:

Theorem 30.11 (Bishop–Gromov inequality in metric-measure spaces). *Let $(\mathcal{X}, d, \nu)$ be a weak $\mathrm{CD}(K,N)$ space and let $x_0 \in \operatorname{Spt} \nu$. Then, for any $r > 0$, $\nu[B[x_0,r]] = \nu[B(x_0,r)]$. Moreover,*

- *If $N < \infty$, then*

$$\frac{\nu[B_r(x_0)]}{\displaystyle\int_0^r s^{(K,N)}(t)\,dt} \quad \text{is a nonincreasing function of } r, \tag{30.22}$$

  where $s^{(K,N)}$ is defined as in Theorem 18.8.
- *If $N = \infty$, then for any $\delta > 0$ there exists a constant $C = C\big(K_-, \delta, \nu[B_\delta(x_0)], \nu[B_{2\delta}(x_0)]\big)$, such that for all $r \geq \delta$,*

$$\nu[B_r(x_0)] \leq e^{Cr}\, e^{(K_-)\frac{r^2}{2}}; \tag{30.23}$$

$$\nu[B_{r+\delta}(x_0) \setminus B_r(x_0)] \leq e^{Cr}\, e^{-K\frac{r^2}{2}} \qquad \text{if } K > 0. \tag{30.24}$$

In particular, if $K' < K$ then

$$\int e^{\frac{K'}{2} d(x_0,x)^2}\, \nu(dx) < +\infty. \tag{30.25}$$

Before providing the proof of this theorem, I shall state three immediate but important corollaries, all of them in finite dimension.

Corollary 30.12 (Measure of small balls in weak CD(K,N) spaces). *Let $(\mathcal{X}, d, \nu)$ be a weak* CD(K,N) *space and let $z \in \operatorname{Spt}\nu$. Then for any $R > 0$ there is a constant $c = c(K,N,R)$ such that if $B(x_0, r) \subset B(z, R)$ then*

$$\nu[B(x_0,r)] \geq \big(c\,\nu[B(z,R)]\big)\, r^N.$$

Corollary 30.13 (Dimension of weak CD(K,N) spaces). *If $\mathcal{X}$ is a weak* CD(K,N) *space with $K \in \mathbb{R}$ and $N \in [1,\infty)$, then the Hausdorff dimension of* $\operatorname{Spt}\nu$ *is at most N.*

Corollary 30.14 (Weak CD(K,N) spaces are locally doubling). *If $\mathcal{X}$ is a weak* CD(K,N) *space with $K \in \mathbb{R}$, $N < \infty$, $\operatorname{Spt}\nu = \mathcal{X}$, then $(\mathcal{X}, d, \nu)$ is C-doubling on each ball $B(z,R)$, with a constant C depending only on K, N and R. In particular if* $\operatorname{diam}(\mathcal{X}) \leq D$ *then $(\mathcal{X}, d, \nu)$ is C-doubling with a constant $C = C(K,N,D)$.*

Remark 30.15. Corollary 30.14, combined with the general theory of Gromov–Hausdorff convergence (as exposed in Chapter 27), implies the compactness Theorem 29.32.

Remark 30.16. It is natural to ask whether the equality $N = \dim(\mathcal{X})$ in Corollary 30.13 forces ν to be proportional to the N-dimensional Hausdorff measure.

Proofs of Corollaries 30.12 to 30.14. If $B(x_0, r) \subset B(z,R)$, then obviously $B(z,R) \subset B(x_0, 2R)$, so by (30.22),

$$\nu[B(x_0,r)] \geq \left(\frac{\int_0^r s^{(K,N)}(t)\,dt}{\int_0^{2R} s^{(K,N)}(t)\,dt}\right) \nu[B(z,R)].$$

Then Corollary 30.12 follows from the elementary observation that $(\int_0^r s^{(K,N)}(t)\,dt)/r^N$ is bounded below by a positive constant $K(R)$ if $r \leq R$.

Next, Corollary 30.12 and the definition of Hausdorff measure imply $\mathcal{H}^d[B(z,R)] = 0$ for any $d > N$, where $\mathcal{H}^d$ stands for the d-dimensional Hausdorff measure. This gives Corollary 30.13 at once.

Finally, Corollary 30.14 is a consequence of the elementary estimate

$$r \leq R \Longrightarrow \quad \frac{\int_0^{2r} s^{(K,N)}}{\int_0^r s^{(K,N)}} \leq C(R).$$

□

Proof of Theorem 30.11. The open ball (resp. closed ball) of center x_0 and radius r in the space $(\mathrm{Spt}\,\nu, d)$ coincides with $B(x_0, r) \cap \mathrm{Spt}\,\nu$ (resp. $B[x_0, r] \cap \mathrm{Spt}\,\nu$). So we may use Theorem 30.2 to reduce to the case $\mathcal{X} = \mathrm{Spt}\,\nu$.

Next, we may dismiss the case where ν is a Dirac mass as trivial; then by Corollary 30.9, we may assume that ν has no atom.

Let $x_0 \in \mathcal{X}$ and $r > 0$. The open ball $B_r(x_0)$ contains $[x_0, B_{r]}(x_0)]_t$, for all $t \in (0, 1)$. By Corollary 30.10,

$$\nu[B_r(x_0)] \geq \lim_{t \to 1} \nu\big[x_0, B_{r]}(x_0)\big]_t = \nu[B_{r]}(x_0)],$$

so $\nu[B_r(x_0)] = \nu[B_{r]}(x_0)]$.

To prove (30.22), apply the displacement convexity inequality (30.5) in the case when $U(r) = -\,r^{1-1/N}$, $\mu_0 = \delta_{x_0}$, $\mu_1 = \rho_1\,d\nu$, where ρ_1 is the normalized indicator function of the set $A_1 = B_{r+\varepsilon}(x) \setminus B_r(x)$; and $U(r) = -\,r^{1-1/N}$. Note that $U_\nu(\mu_0) = 0$, and apply the same reasoning as in the proof of Theorem 18.8.

To prove (30.23), use inequality (30.20) and a reasoning similar to the proof of Theorem 18.12. □

Uniqueness of geodesics

It is an important result in Riemannian geometry that for almost any pair of points (x, y) in a complete Riemannian manifold, x and y are linked by a unique geodesic. This statement *does not* extend to general weak $\mathrm{CD}(K, N)$ spaces, as will be discussed in the concluding chapter; however, it becomes true if the weak $\mathrm{CD}(K, N)$ criterion is supplemented with a *nonbranching* condition. Recall that a geodesic space $(\mathcal{X}, d)$ is said to be nonbranching if two distinct constant-speed geodesics cannot coincide on a nontrivial interval.

Theorem 30.17 (Unique geodesics in nonbranching CD(K, N) spaces). *Let $(\mathcal{X}, d, \nu)$ be a nonbranching weak $\mathrm{CD}(K, N)$ space with $K \in \mathbb{R}$ and $N \in [1, \infty)$. Then for $\nu \otimes \nu$-almost any $(x, y) \in \mathcal{X} \times \mathcal{X}$, there is a unique geodesic joining x to y. More precisely, for any $x \in \mathrm{Spt}\,\nu$, the set of points $y \in \mathrm{Spt}\,\nu$ which can be joined to x by several geodesics has zero measure.*

Remark 30.18. The restriction $N < \infty$ seems natural, but I don't have a counterexample for $N = \infty$.

Proof of Theorem 30.17. By Theorem 30.2 we may assume $\operatorname{Spt} \nu = \mathcal{X}$. Let $x \in \mathcal{X}$, $r > 0$, $A = B_r(x)$ and $A_t = [x, B_r(x)]_t \subset B_{tr}(x)$. For any $z \in A_t$, there is a geodesic γ joining x to some $y \in Z$, with $\gamma(t) = z$. Assume that there is another distinct geodesic $\widetilde{\gamma}$ joining x to z; up to a rescaling of time, one may assume that also $\widetilde{\gamma}$ is defined on $[0, t]$, so that $\gamma(0) = x$, $\widetilde{\gamma}(t) = z$. ($\widetilde{\gamma}$ might not be defined after time t.) Then the curve obtained by concatenation of $\widetilde{\gamma}$ on $[0, t]$ and γ on $[t, 1]$ is also a geodesic, and it is distinct from γ, which is impossible since it coincides with γ on the nontrivial interval $[t, 1]$. The conclusion is that there is one and only one geodesic joining x to z; it is obtained by reparametrizing the restriction of γ to the interval $[0, t]$.

Let $Z := \cup_{0<t<1} A_t \subset A$. The preceding reasoning shows that for any $z \in Z$ there is only one geodesic path joining x to z. The sets A_t are nondecreasing in t, so

$$\nu\Big[\bigcup_{0<t<1} A_t\Big] = \lim_{t \to 1} \nu\big[[x, A]_t\big] = \nu[A],$$

where the first equality follows from the monotone convergence theorem and the second from Corollary 30.10.

So for any $k \in \mathbb{N}$, the set Z_k of points in $B_k(x)$ which can be joined to x by several geodesics is of zero measure. The set of points in $\mathcal{X}$ which can be joined to x by several geodesics is contained in the union of all Z_k, and is therefore of zero measure too. □

Regularity of the interpolant

If M is a smooth Riemannian manifold, then the displacement interpolant μ_t between μ_0 and μ_1 is absolutely continuous with respect to the volume measure, as soon as *either* μ_0 or μ_1 is absolutely continuous (Theorem 8.5(ii)). It is not known whether this property is still true in weak $\mathrm{CD}(K, N)$ spaces; but a weakened version is still available. Moreover, some regularity properties are known to be inherited by the displacement interpolant if they are satisfied by *both* μ_0 and μ_1. This is the content of the next theorem. Just as Theorem 30.17, it uses an assumption of finite dimension.

Theorem 30.19 (Regularity of interpolants in weak CD(K,N) spaces). *Let $(\mathcal{X}, d, \nu)$ be a weak* CD(K,N) *space with $K \in \mathbb{R}$ and $N \in [1, \infty)$. Further, let μ_0, μ_1 be two probability measures in $P_2(\mathcal{X})$ with* $\operatorname{Spt} \mu_0 \subset \operatorname{Spt} \nu$, $\operatorname{Spt} \mu_1 \subset \operatorname{Spt} \nu$. *Then:*

(i) Assume that both μ_0 and μ_1 are absolutely continuous with respect to ν; if $K < 0$, further assume that they are compactly supported. Let $(\mu_t)_{0 \le t \le 1}$ be a Wasserstein geodesic satisfying the displacement convexity inequalities of Theorem 30.5. Then also μ_t is absolutely continuous, for all $t \in [0,1]$.

(ii) If either μ_0 or μ_1 is absolutely continuous, and $t_0 \in (0,1)$ is given, then one can find a Wasserstein geodesic joining μ_0 to μ_1, such that μ_{t_0} is also absolutely continuous.

(iii) If either μ_0 or μ_1 is not purely singular, then one can find a Wasserstein geodesic joining μ_0 to μ_1, such that for any $t \in [0,1]$, μ_t is not purely singular.

Theorem 30.20 (Uniform bound on the interpolant in nonnegative curvature). *Let $(\mathcal{X}, d, \nu)$ be a weak* CD($0,\infty$) *space and let $\mu_0, \mu_1 \in P^{ac}(\mathcal{X})$, with bounded respective densities ρ_0, ρ_1. Let $(\mu_t)_{0 \le t \le 1}$ be a Wasserstein geodesic satisfying the displacement convexity inequalities of Theorem 30.5. Then the density ρ_t of μ_t is bounded by* $\max(\sup \rho_0, \sup \rho_1)$.

Proof of Theorem 30.19. First assume that $K \ge 0$, or, what amounts to the same, $K = 0$. Since μ_0 and μ_1 are absolutely continuous, by the Dunford–Pettis theorem there exists $\Psi : \mathbb{R}_+ \to \mathbb{R}_+$ such that

$$\lim_{r \to \infty} \frac{\Psi(r)}{r} = +\infty, \qquad \int \Psi(\rho_0)\, d\nu < +\infty, \qquad \int \Psi(\rho_1)\, d\nu < +\infty.$$

Thanks to Proposition 17.7(i), one may assume that Ψ belongs to $\mathcal{DC}_N$. Then the convexity inequality

$$\Psi_\nu(\mu_t) \le (1-t)\, \Psi_\nu(\mu_0) + t\, \Psi_\nu(\mu_1) \ < +\infty$$

shows that μ_t is absolutely continuous. This proves (i).

Now consider the case when $K < 0$. It is not hard to prove that in the Dunford–Pettis theorem, one may impose Ψ to have polynomial growth, in the sense that $0 \le U(ar) \le C(a)[U(r) + 1]$ for any $a > 1$. Since μ_0 and μ_1 are compactly supported, the distortion coefficients $\beta(x_0, x_1)$ appearing in the right-hand side of the inequalities in Theorem 30.5 are bounded from above and below; then the integrability of $\Psi(\rho)$ also implies the finiteness of $\int \beta(x_0,x_1)\,\Psi(\rho(x_0)/\beta(x_0,x_1))\,\pi(dx_1|x_0)\,\nu(dx_0)$, and the same reasoning as before applies.

To prove (ii), let $U(r) = -Nr^{1-1/N}$; $U \in \mathcal{DC}_N$ and $U'(\infty) = 0$. Moreover, $U_\nu(\mu) < 0$ as soon as μ is absolutely continuous. Among all dynamical optimal transport plans Π with $(e_0)_\#\Pi = \mu_0$, $(e_1)_\#\Pi = \mu_1$ which satisfy the CD(K,N) displacement convexity inequality, choose one with minimal $U_\nu(\mu_{t_0})$. There exists one such dynamical optimal transport plan by compactness of the set of admissible transference plans (Lemma 4.4) and lower semicontinuity of U_ν (Theorem 30.6(i)).[2]

Assume that μ_{t_0} is not absolutely continuous, i.e. there exists a Borel set $Z \subset \mathcal{X}$ with $\nu[Z] = 0$ and $\mu_{t_0}[Z] > 0$. Let Π' be the restricted transport obtained by conditioning Π by the event "$\gamma(t_0) \in Z$". The measures $(e_t)_\#\Pi' = \mu'_t$ satisfy the following properties: (a) $\mu'_t \le \mu_t/\mu_{t_0}[Z]$, and in particular μ'_0 is absolutely continuous; (b) μ'_{t_0} is concentrated on Z. So $U_\nu(\mu'_{t_0}) = 0$, but $U_{\pi,\nu}^{\beta_{1-t_0}^{(K,N)}}(\mu'_0) < 0$, and

$$U_\nu(\mu'_{t_0}) \le (1-t_0)\,U_{\pi,\nu}^{\beta_{1-t_0}^{(K,N)}}(\mu'_0) + t_0\,U_{\check{\pi},\nu}^{\beta_{t_0}^{(K,N)}}(\mu'_1) \tag{30.26}$$

cannot hold.

On the other hand, there has to be some dynamical optimal transport plan Π'' such that $(e_0)_\#\Pi'' = \mu'_0$, $(e_1)_\#\Pi'' = \mu'_1$ and inequality (30.26) holds true with μ'_{t_0} replaced by $\mu''_{t_0} = (e_{t_0})_\#\Pi''$. In particular, $U_\nu(\mu''_{t_0}) < U_\nu(\mu'_{t_0}) = 0$, which implies that μ''_{t_0} is not purely singular.

Now consider the plan $\widehat{\Pi}$ defined by

$$\widehat{\Pi} = \mathbb{P}\left[\gamma_{t_0} \in Z\right]\Pi'' \;+\; 1_{[\gamma_{t_0}\notin Z]}\,\Pi. \tag{30.27}$$

This is still a dynamical optimal transport plan, because

$$\begin{aligned}\int d(\gamma_0,\gamma_1)^2\,\widehat{\Pi}(d\gamma)\\ &= \mathbb{P}\left[\gamma_{t_0} \in Z\right]\int d(\gamma_0,\gamma_1)^2\,\Pi''(d\gamma) + \int 1_{\gamma_{t_0}\notin Z}\,d(\gamma_0,\gamma_1)^2\,\Pi(d\gamma)\end{aligned}$$

---

[2] Here I am cheating a bit because Theorem 30.6, in the version which I have stated, assumes $U(r) \ge -cr$. To deal with this issue, one could prove a more general version of Theorem 30.6; but a simpler remedy is to introduce $\varepsilon > 0$ and choose $U(r) = -Nr\,(r+\varepsilon)^{-1/N}$ instead of $-Nr^{1-1/N}$. If μ_0 and μ_1 are compactly supported, one may keep the proof as it is and use Theorem 29.20(i) instead of Theorem 30.6(i).

$$= \mathbb{P}\,[\gamma_{t_0} \in Z] \int d(\gamma_0,\gamma_1)^2\, \Pi'(d\gamma) + \int 1_{\gamma_{t_0}\notin Z}\, d(\gamma_0,\gamma_1)^2\, \Pi(d\gamma)$$
$$= \int d(\gamma_0,\gamma_1)^2\, \Pi(d\gamma).$$

(To pass from the first to the second line I used the fact that Π' and Π'' are displacement optimal transference plans between the same two measures.)

It follows from (30.27) and the ν-negligibility of Z that

$$\widehat{\rho}_{t_0} = a\, \rho''_{t_0} + 1_{\gamma_{t_0}\notin Z}\rho_{t_0} = a\, \rho''_{t_0} + \rho_{t_0} \qquad (\nu\text{-almost surely}),$$

where $\widehat{\rho}_{t_0}$, ρ''_{t_0} and ρ_{t_0} respectively stand for the density of the absolutely continuous parts of $\widehat{\mu}_{t_0}$, μ''_{t_0} and μ_{t_0}, and $a = \mathbb{P}\,[\gamma_{t_0} \in Z] > 0$. Then from the minimality property of μ_{t_0},

$$\int U(\rho_{t_0}(x))\, d\nu(x) = U_\nu(\mu_{t_0}) \le U_\nu(\widehat{\mu}_{t_0}) = \int U\big(\rho_{t_0}(x) + a\, \rho''_{t_0}(x)\big)\, d\nu(x).$$

Since a is positive and U is strictly decreasing, this inequality is possible only if $\rho''_{t_0} = 0$ almost everywhere, but this would contradict the fact that μ''_{t_0} is not purely singular. The only possibility left out is that μ_{t_0} is absolutely continuous. This proves (ii).

Statement (iii) is based on the same principle as (ii), but now this is much simpler: Choose $U(r) = -N r^{1-1/N}$, and choose a displacement interpolation $(\mu_t)_{0\le t\le 1}$ satisfying the convexity inequality

$$U_\nu(\mu_t) \le (1-t)\, U_{\pi,\nu}^{\beta_{1-t}^{(K,N)}}(\mu_0) \;+\; t\, U_{\check{\pi},\nu}^{\beta_t^{(K,N)}}(\mu_1).$$

If, say, μ_0 is not purely singular, then the first term on the right-hand side is negative, while the second one is nonpositive. It follows that $U_\nu(\mu_t) < 0$, and therefore μ_t is not purely singular. □

Proof of Theorem 30.20. Let ρ_t be the density of μ_t, and let $U(r) = r^p$, where $p \ge 1$. Since $U \in \mathcal{DC}_\infty$, (30.5) implies

$$\begin{aligned} \|\rho_t\|_{L^p(\nu)}^p \le U_\nu(\mu_t) &\le (1-t)\, \|\rho_0\|_{L^p(\nu)}^p + t\, \|\rho_1\|_{L^p(\nu)}^p \\ &\le \max\big(\|\rho_0\|_{L^p(\nu)}, \|\rho_1\|_{L^p(\nu)}\big)^p. \end{aligned} \tag{30.28}$$

Since ρ_0 and ρ_1 belong to $L^1(\nu)$ and $L^\infty(\nu)$, by elementary interpolation they belong to all L^p spaces, so the right-hand side in (30.28) is also finite, and ρ_t belongs to L^p. Then take powers $1/p$ in both sides of (30.28) and pass to the limit as $p \to \infty$, to recover

$$\|\rho_t\|_{L^\infty(\nu)} \leq \max\Big(\|\rho_0\|_{L^\infty(\nu)}, \|\rho_1\|_{L^\infty(\nu)}\Big).$$

□

Remark 30.21. The above argument exploited the fact that in the definition of weak $\mathrm{CD}(K,N)$ spaces the displacement convexity inequality (29.11) is required to hold *for all members of* $\mathcal{DC}_N$ and *along a common Wasserstein geodesic.*

HWI and logarithmic Sobolev inequalities

There is a generalized notion of Fisher information in a metric-measure space $(\mathcal{X},d,\nu)$:

$$I_\nu(\mu) = \int \frac{|\nabla^-\rho|^2}{\rho}\,d\nu, \qquad \mu = \rho\,\nu,$$

where $|\nabla^-\rho|$ is defined by (20.2) (one may also use $|\nabla\rho|$ in place of $|\nabla^-\rho|$). With this notion, one has the following estimates:

Theorem 30.22 (HWI and log Sobolev inequalities in weak $\mathrm{CD}(K,\infty)$ spaces). *Let $K \in \mathbb{R}$ and let $(\mathcal{X},d,\nu)$ be a weak $\mathrm{CD}(K,\infty)$ space. Further, let μ_0 and μ_1 be two probability measures in $P_2(\mathcal{X})$, such that $\mu_0 = \rho_0\,\nu$ with ρ_0 Lipschitz. Then*

$$H_\nu(\mu_0) \leq H_\nu(\mu_1) + W_2(\mu_0,\mu_1)\sqrt{I_\nu(\mu_0)} - \frac{K\,W_2(\mu_0,\mu_1)^2}{2}. \tag{30.29}$$

In particular, if $\nu \in P_2(\mathcal{X})$, then, for any $\mu \in P_2(\mathcal{X})$ with Lipschitz-continuous density,

$$H_\nu(\mu) \leq W_2(\mu,\nu)\sqrt{I_\nu(\mu)} - \frac{K\,W_2(\mu,\nu)^2}{2}. \tag{30.30}$$

Consequently, if $K>0$, ν satisfies a logarithmic Sobolev inequality with constant K:

$$H_\nu \leq \frac{I_\nu}{2K}.$$

Proof of Theorem 30.22. By an easy approximation argument, it suffices to establish (30.29) when $H_\nu(\mu_0) < +\infty$, $H_\nu(\mu_1) < +\infty$. Let $(\mu_t)_{0\le t\le 1}$ be a displacement interpolation satisfying the inequalities of Theorem 30.5. By Theorem 30.19(i), or more directly by the inequality $H_\nu(\mu_t) \le (1-t)\, H_\nu(\mu_0) + t\, H_\nu(\mu_1) - K\, t(1-t)\, W_2(\mu_0,\mu_1)^2/2$, each μ_t is absolutely continuous. Then we can repeat the proof of Theorem 20.1, Corollary 20.13 and Theorem 25.1. □

Sobolev inequalities

Theorem 30.22 provides a Sobolev inequality of infinite-dimensional nature; but in weak $\mathrm{CD}(K,N)$ spaces with $N < \infty$, one also has access to finite-dimensional Sobolev inequalities. An example is the following statement.

Theorem 30.23 (Sobolev inequality in weak CD(K,N) spaces). *Let $(\mathcal{X}, d, \nu)$ be a weak $\mathrm{CD}(K,N)$ space, where $K < 0$ and $N \in [1,\infty)$. Then, for any $R > 0$ there exist constants $A = A(K,N,R)$ and $B = B(K,N,R)$ such that for any Lipschitz function u supported in a ball $B(z,R)$,*

$$\|u\|_{L^{\frac{N}{N-1}}(\nu)} \le A\, \|\nabla^- u\|_{L^1(\nu)} + B\, \|u\|_{L^1(\nu)}. \tag{30.31}$$

Proof of Theorem 30.23. Since $\|\nabla^- u\| \le \|\nabla^- |u|\|$, it is sufficient to treat the case when u is nonnegative (and nonzero). Let $\rho = u^{N/(N-1)}/Z$, where $Z = \int u^{N/(N-1)}$, so that ρ is a probability density. Let $\mu_0 = \rho\,\nu$, and $\mu_1 = (1_{B(z,R)})/\nu[B(z,R)]$. Let $(\mu_t)_{0\le t\le 1}$ be a displacement interpolation satisfying the displacement convexity inequalities of Theorem 30.5. By Theorem 30.19(i), each μ_t is absolutely continuous. Then we can repeat the proof of Theorems 20.1, 20.10 and 21.15. □

Remark 30.24. It is not known whether weak $\mathrm{CD}(K,N)$ spaces with $K > 0$ and $N < \infty$ satisfy sharp Sobolev inequalities such as (21.8).

Diameter control

Recall from Proposition 29.11 that a weak $\mathrm{CD}(K,N)$ space with $K > 0$ and $N < \infty$ satisfies the Bonnet–Myers diameter bound

$$\operatorname{diam}(\operatorname{Spt}\nu) \leq \pi\sqrt{\frac{N-1}{K}}.$$

Slightly weaker conclusions can also be obtained under a priori weaker assumptions: For instance, if $\mathcal{X}$ is at the same time a weak $\mathrm{CD}(0,N)$ space and a weak $\mathrm{CD}(K,\infty)$ space, then there is a universal constant C such that

$$\operatorname{diam}(\operatorname{Spt}\nu) \leq C\sqrt{\frac{N-1}{K}}. \tag{30.32}$$

See the bibliographical notes for more details.

Poincaré inequalities

As I already mentioned in Chapter 19, there are many kinds of Poincaré inequalities, which can be roughly divided into *global* and *local* inequalities. In a nonsmooth context, global Poincaré inequalities can be seen as a replacement for *spectral gap* estimates in a context where a Laplace operator is not necessarily defined.

If one does not care about dimension, there is a general principle (independent of optimal transport) according to which a logarithmic Sobolev inequality with constant K implies a global Poincaré inequality with the same constant; and then from Theorem 30.22 we know that a weak $\mathrm{CD}(K,\infty)$ condition does imply such a logarithmic Sobolev inequality. If one does care about the dimension, it is possible to adapt the proof of Theorem 21.20 and establish the following Poincaré inequality with the optimal constant $KN/(N-1)$.

Theorem 30.25 (Global Poincaré inequalities in weak $\mathrm{CD}(K,N)$ spaces). *Let $(\mathcal{X},d,\nu)$ be a weak $\mathrm{CD}(K,N)$ space with $K>0$ and $N\in(1,\infty]$. Then, for any Lipschitz function $f:\operatorname{Spt}\nu\to\mathbb{R}$,*

$$\int f\,d\nu = 0 \Longrightarrow \qquad \int f^2\,d\nu \leq \left(\frac{N-1}{NK}\right)\int |\nabla^- f|^2\,d\nu,$$

with the convention that $(N-1)/N=1$ if $N=\infty$.

I omit the proof of Theorem 30.25 since it is almost a copy–paste of the proof of Theorem 21.20.

Local Poincaré inequalities play a key role in the modern geometry of metric spaces, and it is natural to ask whether weak $\mathrm{CD}(K,N)$ spaces with $N<\infty$ satisfy them. (The restriction to finite N is natural because these inequalities are morally finite-dimensional, contrary to global Poincaré inequalities.) For the moment, this is only known to be true under a *nonbranching* assumption.

Theorem 30.26 (Local Poincaré inequalities in nonbranching $\mathrm{CD}(K,N)$ spaces). *Let $K\in\mathbb{R}$, $N\in[1,\infty)$, and let $(\mathcal{X},d,\nu)$ be a nonbranching weak* $\mathrm{CD}(K,N)$ *space. Let $u:\operatorname{Spt}\nu\to\mathbb{R}$ be a Lipschitz function, and let $x_0\in\operatorname{Spt}\nu$. For any $R>0$, if $r\le R$ then*

$$\fint_{B_r(x_0)} \Big|u(x)-\langle u\rangle_{B_r(x_0)}\Big|\,d\nu(x) \le P(K,N,R)\,r \fint_{B_{2r}(x_0)} |\nabla u|(x)\,d\nu(x), \tag{30.33}$$

where $\fint_B=(\nu[B])^{-1}\int_B$ stands for the averaged integral over B; $\langle u\rangle_B=\fint_B u\,d\nu$ stands for the average of the function u on B; $P(K,N,R)=2^{2N+1}\,C(K,N,R)\,D(K,N,R)$; $C(K,N,R)$, $D(K,N,R)$ are defined by (19.11) *and* (18.10) *respectively.*

In particular, if $K\ge 0$ then $P(K,N,R)=2^{2N+1}$ is admissible; so ν satisfies a uniform local Poincaré inequality. Moreover, (30.33) *still holds true if the local "norm of the gradient" $|\nabla u|$ is replaced by any upper gradient of u, that is a function g such that for any Lipschitz path $\gamma:[0,1]\to\mathcal{X}$,*

$$\big|g(\gamma(1))-g(\gamma(0))\big| \le \int_0^1 g(\gamma(t))\,|\dot\gamma(t)|\,dt.$$

Remark 30.27. It would be desirable to eliminate the nonbranching condition, since it is not always satisfied by weak $\mathrm{CD}(K,N)$ spaces, and rather unnatural in the theory of local Poincaré inequalities.

Proof of Theorem 30.26. Modulo changes of notation, the proof is the same as the proof of Theorem 19.13, once Theorem 30.17 guarantees the almost sure uniqueness of geodesics. □

Talagrand inequalities

With logarithmic Sobolev inequalities come a rich functional apparatus for treating concentration of measure. One may also get concentration from curvature bounds $\mathrm{CD}(K, \infty)$ via Talagrand inequalities. As for the links between logarithmic Sobolev and Talagrand inequalities, they also remain true, at least under mild stringent regularity assumptions on $\mathcal{X}$:

Theorem 30.28 (Talagrand inequalities and weak curvature bounds). *(i) Let $(\mathcal{X}, d, \nu)$ be a weak $\mathrm{CD}(K, \infty)$ space with $K > 0$. Then ν lies in $P_2(\mathcal{X})$ and satisfies the Talagrand inequality $T_2(K)$.*

(ii) Let $(\mathcal{X}, d, \nu)$ be a locally compact Polish geodesic space equipped with a locally doubling measure ν, satisfying a local Poincaré inequality. If ν satisfies a logarithmic Sobolev inequality for some constant $K > 0$, then ν lies in $P_2(\mathcal{X})$ and satisfies the Talagrand inequality $T_2(K)$.

(iii) Let $(\mathcal{X}, d, \nu)$ be a locally compact Polish geodesic space. If ν satisfies a Talagrand inequality $T_2(K)$ for some $K > 0$, then it also satisfies a global Poincaré inequality with constant K.

(iv) Let $(\mathcal{X}, d, \nu)$ be a locally compact Polish geodesic space equipped with a locally doubling measure ν, satisfying a local Poincaré inequality. If ν satisfies a global Poincaré inequality, then it also satisfies a modified logarithmic Sobolev inequality and a quadratic-linear transportation inequality as in Theorem 22.25.

Remark 30.29. In view of Corollary 30.14 and Theorem 30.26, the regularity assumptions required in (ii) are satisfied if $(\mathcal{X}, d, \nu)$ is a nonbranching weak $\mathrm{CD}(K', N')$ space for some $K' \in \mathbb{R}$, $N' < \infty$; note that the values of K' and N' do not play any role in the conclusion.

Proof of Theorem 30.28. Part (i) is an immediate consequence of (30.25) and (30.29) with $\mu_0 = \nu$.

The proof of (ii) and (iii) is the same as the proof of Theorem 22.17, once one has an analog of Proposition 22.16. It turns out that properties (i)–(vi) of Proposition 22.16 and Theorem 22.46 are still satisfied when the Riemannian manifold M is replaced by any metric space $\mathcal{X}$, but property (vii) might fail in general. It is still true that this property holds true *for ν-almost all x*, under the assumption that ν is locally doubling and satisfies a local Poincaré inequality. See Theorem 30.30 below for a precise statement (and the bibliographical notes for references). This is enough for the proof of Theorem 22.17 to go through. □

The next theorem was used in the proof of Theorem 30.28:

Theorem 30.30 (Hamilton–Jacobi semigroup in metric spaces). *Let $L : \mathbb{R}_+ \to \mathbb{R}_+$ be a strictly increasing, locally semiconcave, convex continuous function such that $L(0) = 0$. Let $(\mathcal{X}, d)$ be a locally compact geodesic Polish space equipped with a reference measure ν, locally doubling and satisfying a local Poincaré inequality. For any $f \in C_b(\mathcal{X})$, define the evolution $(H_t f)_{t \geq 0}$ by*

$$\begin{cases} H_0 f = f \\[2ex] (H_t f)(x) = \inf_{y \in \mathcal{X}} \left[f(y) + t\, L\left(\dfrac{d(x,y)}{t}\right)\right] \qquad (t > 0,\ x \in \mathcal{X}). \end{cases} \tag{30.34}$$

Then Properties (i)–(vi) of Theorem 22.46 remain true, up to the replacement of M by $\mathcal{X}$. Moreover, the following weakened version of (vii) holds true:

(vii') For ν-almost any $x \in \mathcal{X}$ and any $t > 0$,

$$\lim_{s \downarrow 0} \frac{(H_{t+s} f)(x) - (H_t f)(x)}{s} = -\, L^*\big(|\nabla^- H_t f|\big);$$

this conclusion extends to $t = 0$ if $\|f\|_{\mathrm{Lip}} \leq L'(\infty)$ and f is locally Lipschitz.

Remark 30.31. There are also dimensional versions of Talagrand inequalities available, for instance the analog of Theorem 22.37 holds true in weak $\mathrm{CD}(K,N)$ spaces with $K > 0$ and $N < \infty$.

Equivalence of definitions in nonbranching spaces

In the definition of weak $\mathrm{CD}(K,N)$ spaces we chose to impose the displacement convexity inequality for all $U \in \mathcal{DC}_N$, but only along some displacement interpolation. We could have chosen otherwise, for instance impose the inequality for just some particular functions U, or along all displacement interpolations. In the end our choice was dictated partly by the will to get a stable definition, partly by convenience. It turns out that in *nonbranching* metric-measure spaces, the choice really does not matter. It is equivalent:

- to require the displacement convexity inequality to hold true for any $U \in \mathcal{DC}_N$; or just for $U = U_N$, where as usual $U_N(r) = -N r^{1-1/N}$ if $1 < N < \infty$, and $U_\infty(r) = r \log r$;
- to require the inequality to hold true for compactly supported, absolutely continuous probability measures μ_0, μ_1; or for any two probability measures with suitable moment conditions;
- to require the inequality to hold true along some displacement interpolation, or along any displacement interpolation.

The next statement makes this claim precise. Note that I leave aside the case $N = 1$, which is special (for instance U_1 is not defined). I shall write $(U_N)_\nu = H_{N,\nu}$, and $(U_N)^\beta_{\pi,\nu} = H^\beta_{N,\pi,\nu}$. Recall Convention 17.10.

Theorem 30.32 (Equivalent definitions of CD(K,N) in nonbranching spaces). *Let $(\mathcal{X}, d, \nu)$ be a nonbranching locally compact Polish geodesic space equipped with a locally finite measure ν. Let $K \in \mathbb{R}$, $N \in (1, \infty]$, and let $p \in [2, +\infty) \cup \{c\}$ satisfy the assumptions of Theorem 30.4. Then the following three properties are equivalent:*

(i) $(\mathcal{X}, d, \nu)$ is a weak CD(K,N) *space, in the sense of Definition 29.8;*

(ii) For any two compactly supported continuous probability densities ρ_0 and ρ_1, there is a displacement interpolation $(\mu_t)_{0 \le t \le 1}$ joining $\mu_0 = \rho_0\, \nu$ to $\mu_1 = \rho_1\, \nu$, and an associated optimal plan π, such that for all $t \in [0,1]$,

$$H_{N,\nu}(\mu_t) \le (1-t)\, H^{\beta^{(K,N)}_{1-t}}_{N,\pi,\nu}(\mu_0) \;+\; t\, H^{\beta^{(K,N)}_t}_{N,\check{\pi},\nu}(\mu_1). \tag{30.35}$$

(iii) For any displacement interpolation $(\mu_t)_{0 \le t \le 1}$ with $\mu_0, \mu_1 \in P_p(\mathcal{X})$, for any associated transport plan π, for any $U \in \mathcal{DC}_N$ and for any $t \in [0,1]$,

$$U_\nu(\mu_t) \le (1-t)\, U^{\beta^{(K,N)}_{1-t}}_{\pi,\nu}(\mu_0) \;+\; t\, U^{\beta^{(K,N)}_t}_{\check{\pi},\nu}(\mu_1). \tag{30.36}$$

Remark 30.33. In the case $N = 1$, (30.35) does not make any sense, but the equivalence (i) $\Rightarrow$ (iii) still holds. This can be seen by working in dimension $N' > 1$ and letting $N' \downarrow 1$, as in the proof of Theorem 17.41.

Theorem 30.32 is interesting even for smooth Riemannian manifolds, since it covers singular measures, for which there is a priori no uniqueness of displacement interpolant. Its proof is based on the idea, already used in Theorem 19.4, that we may condition the optimal transport to lie in a very small ball at time t, and, by passing to the limit, retrieve a pointwise control of the density ρ_t. This will work because the nonbranching property implies the uniqueness of the displacement interpolation between intermediate times, and forbids the crossing of geodesics used in the optimal transport, as in Theorem 7.30. Apart from this simple idea, the proof is quite technical and can be skipped at first reading.

Proof of Theorem 30.32. Let us first consider the case $N < \infty$.

Clearly, (iii) $\Rightarrow$ (i) $\Rightarrow$ (ii). So it is sufficient to show (ii) $\Rightarrow$ (iii). In the sequel, I shall assume that Property (ii) is satisfied. By the same arguments as in the proof of Proposition 29.12, it is sufficient to establish (30.36) when U is nonnegative and Lipschitz continuous, and $u(r) := U(r)/r$ (extended at 0 by $u(0) = U'(0)$) is a continuous function of r. I shall fix $t \in (0,1)$ and establish Property (iii) for that t. For simplicity I shall abbreviate $\beta_t^{(K,N)}$ into just β_t.

First of all, let us establish that Property (ii) also applies if μ_0 and μ_1 are not absolutely continuous. The scheme of reasoning is the same as we already used several times. Let μ_0 and μ_1 be any two compactly supported measures. As in the proof of Theorem 29.24 we can construct probability measures $\mu_{k,0}$ and $\mu_{k,1}$, absolutely continuous with continuous densities, supported in a common compact set, such that $\mu_{k,0}$ converges to μ_0, $\mu_{k,1}$ converges to μ_1, in such a way that

$$\limsup_{k\to\infty} U_{\pi_k,\nu}^{\beta_{1-t}}(\mu_{k,0}) \le U_{\pi,\nu}^{\beta_{1-t}}(\mu_0); \qquad \limsup_{k\to\infty} U_{\check{\pi}_k,\nu}^{\beta_t}(\mu_{k,1}) \le U_{\check{\pi},\nu}^{\beta_t}(\mu_1), \tag{30.37}$$

where π_k is any optimal transference plan between $\mu_{k,0}$ and $\mu_{k,1}$ such that $\pi_k \to \pi$. Since $\mu_{k,0}$ and $\mu_{k,1}$ are absolutely continuous with continuous densities, for each k we may choose an optimal tranference plan π_k and an associated displacement interpolation $(\mu_{k,t})_{0\le t\le 1}$ such that

$$H_{N,\nu}(\mu_{k,t}) \le (1-t)\, H_{N,\pi_k,\nu}^{\beta_{1-t}}(\mu_{k,0}) \;+\; t\, H_{N,\check{\pi}_k,\nu}^{\beta_t}(\mu_{k,1}). \tag{30.38}$$

Since all the measures $\mu_{k,0}$ and $\mu_{k,1}$ are supported in a uniform compact set, Corollary 7.22 guarantees that the sequence $(\Pi_k)_{k\in\mathbb{N}}$ converges, up to extraction, to some dynamical optimal transference plan Π with $(e_0)_\#\Pi = \mu_0$ and $(e_1)_\#\Pi = \mu_1$. Then $\mu_{k,t}$ converges weakly to $\mu_t = (e_t)_\#\Pi$, and $\pi_k := (e_0,e_1)_\#\Pi_k$ converges weakly to $\pi = (e_0,e_1)_\#\Pi$. It remains to pass to the limit as $k \to \infty$ in the inequality (30.38); this is easy in view of (30.37) and Theorem 29.20(i), which imply

$$H_{N,\nu}(\mu_t) \le \liminf_{k\to\infty} H_{N,\nu}(\mu_{k,t}). \tag{30.39}$$

Next, the proofs of Theorem 30.11 and Corollary 30.14 go through, since they only use the convex function $U = U_N$; in particular the measure ν is *locally doubling* on its support.

Also the proof of Theorem 30.19(ii)-(iii) can be easily adapted in the present setting, as soon as μ_0 and μ_1 are compactly supported.

Now we can start the core of the argument. It will be decomposed into four steps.

Step 1: Assume that *μ_0 and μ_1 are compactly supported, μ_t is absolutely continuous and there exists a dynamical optimal transference plan Π joining μ_0 to μ_1, such that for any subplan $\Pi' = \widetilde{\Pi}/\widetilde{\Pi}[\Gamma]$, $0 \le \widetilde{\Pi} \le \Pi$, it happens that Π' is the unique dynamical optimal transference plan between $\mu_0' = (e_0)_\#\Pi'$ and $\mu_1' = (e_1)_\#\Pi'$.*

In particular, Π is the unique dynamical optimal transference plan between μ_0 and μ_1, and, by Corollary 7.23, $\mu_t = (e_t)_\#\Pi$ defines the unique displacement interpolation between μ_0 and μ_1. In the sequel, I shall denote by ρ_t the density of μ_t, and by ρ_0, ρ_1 the densities of the absolutely continuous parts of μ_0, μ_1 respectively. I shall also fix Borel sets S_0, S_1 such that $\nu[S_0] = \nu[S_1] = 0$, $\mu_{0,s}$ is concentrated on S_0 and $\mu_{1,s}$ is concentrated on S_1. By convention ρ_0 is defined to be $+\infty$ on S_0; similarly ρ_1 is defined to be $+\infty$ on S_1.

Then let $y \in \operatorname{Spt}\mu_t$, and let $\delta > 0$. Define

$$\mathcal{Z} = \Big\{\gamma \in \Gamma;\ \gamma_t \in B_\delta(y)\Big\},$$

and let $\Pi' = (1_{\mathcal{Z}}\Pi)/\Pi[\mathcal{Z}]$. (If γ is a random variable distributed according to Π, then Π' is the law of γ conditioned by the event "$\gamma_t \in B_\delta(y)$".) Let $\mu_t' = (e_t)_\#\Pi'$, let ρ_t' be the density of the absolutely continuous part of μ_t', and let $\pi' := (e_0, e_1)_\#\Pi'$. Since Π' is the unique dynamical optimal transference plan between μ_0' and μ_1', we can write the displacement convexity inequality

$$H_{N,\nu}(\mu_t') \le (1-t)\, H^{\beta_{1-t}}_{N,\pi,\nu}(\mu_0') \;+\; t\, H^{\beta_t}_{N,\check{\pi},\nu}(\mu_1').$$

In other words,

$$\int_{\mathcal{X}} (\rho'_t)^{1-\frac{1}{N}}\,d\nu \geq (1-t)\int_{\mathcal{X}\times\mathcal{X}} (\rho'_0(x_0))^{-\frac{1}{N}}\beta_{1-t}(x_0,x_1)^{\frac{1}{N}}\,\pi'(dx_0\,dx_1) + t\int_{\mathcal{X}\times\mathcal{X}} (\rho'_1(x_1))^{-\frac{1}{N}}\beta_t(x_0,x_1)^{\frac{1}{N}}\,\pi'(dx_0\,dx_1), \tag{30.40}$$

with the understanding that $\rho'_0(x_0) = +\infty$ when $x_0 \in S_0$, and similarly $\rho'_1(x_1) = +\infty$ when $x_1 \in S_1$.

By reasoning as in the proof of Theorem 19.4, we obtain

$$\frac{\nu[B_\delta(y)]^{\frac{1}{N}}}{\mu_t[B_\delta(y)]^{\frac{1}{N}}} \geq \mathbb{E}_{\Pi}\left[(1-t)\left(\frac{\beta_{1-t}(\gamma_0,\gamma_1)}{\rho_0(\gamma_0)}\right)^{\frac{1}{N}} + t\left(\frac{\beta_t(\gamma_0,\gamma_1)}{\rho_1(\gamma_1)}\right)^{\frac{1}{N}} \Bigm| \gamma_t \in B_\delta(y)\right].$$

If we define

$$f(\gamma) := (1-t)\left(\frac{\beta_{1-t}(\gamma_0,\gamma_1)}{\rho_0(\gamma_0)}\right)^{\frac{1}{N}} + t\left(\frac{\beta_t(\gamma_0,\gamma_1)}{\rho_1(\gamma_1)}\right)^{\frac{1}{N}},$$

then the conclusion can be rewritten

$$\frac{\nu[B_\delta(y)]^{\frac{1}{N}}}{\mu_t[B_\delta(y)]^{\frac{1}{N}}} \geq \mathbb{E}_{\Pi}\left[f(\gamma)|\gamma_t \in B_\delta(y)\right] = \frac{\mathbb{E}\, f(\gamma) 1_{[\gamma_t\in B_\delta(y)]}}{\mu_t[B_\delta(y)]}. \tag{30.41}$$

In view of the nonbranching property, Π only sees geodesics which do not cross each other; recall Theorem 7.30(iv)-(v). Let F_t be the map appearing in that theorem, defined by $F_t(\gamma_t) = \gamma$. Then (30.41) becomes

$$\frac{\nu[B_\delta(y)]^{\frac{1}{N}}}{\mu_t[B_\delta(y)]^{\frac{1}{N}}} \geq \frac{\mathbb{E}\,\left[f(F_t(\gamma_t))\,1_{[\gamma_t\in B_\delta(y)]}\right]}{\mu_t[B_\delta(y)]} = \frac{\displaystyle\int_{B_\delta(y)} f(F_t(x))\,d\mu_t(x)}{\mu_t[B_\delta(y)]}. \tag{30.42}$$

Since the measure ν is locally doubling, we can apply Lebesgue's density theorem: There is a set Z of zero ν-measure such that if $y \notin Z$, then

$$\frac{\nu[B_\delta(y)]^{\frac{1}{N}}}{\mu_t[B_\delta(y)]^{\frac{1}{N}}} \xrightarrow[\delta\to 0]{} \frac{1}{\rho_t(y)^{\frac{1}{N}}}.$$

Similarly, outside of a set of zero measure,

$$\frac{\int_{B_\delta(y)} f(F_t(x))\,d\mu_t(x)}{\mu_t[B_\delta(y)]} = \frac{\int_{B_\delta(y)} f(F_t(x))\,\rho_t(x)\,d\nu(x)}{\nu[B_\delta(y)]}\,\frac{\nu[B_\delta(y)]}{\mu_t[B_\delta(y)]}$$

$$\xrightarrow[\delta\to 0]{} \frac{f(F_t(y))\,\rho_t(y)}{\rho_t(y)},$$

and this coincides with $f(F_t(y))$ if $\rho_t(y) \neq 0$. All in all, $\mu_t(dy)$-almost surely,

$$\frac{1}{\rho_t(y)^{\frac{1}{N}}} \geq f(F_t(y)).$$

Equivalently, $\Pi(d\gamma)$-almost surely,

$$\frac{1}{\rho_t(\gamma_t)^{\frac{1}{N}}} \geq f(F_t(\gamma_t)) = f(\gamma).$$

Let us recapitulate: We have shown that $\Pi(d\gamma)$-almost surely,

$$\frac{1}{\rho_t(\gamma_t)^{\frac{1}{N}}} \geq (1-t)\left(\frac{\beta_{1-t}(\gamma_0,\gamma_1)}{\rho_0(\gamma_0)}\right)^{\frac{1}{N}} + t\left(\frac{\beta_t(\gamma_0,\gamma_1)}{\rho_1(\gamma_1)}\right)^{\frac{1}{N}}. \tag{30.43}$$

Step 2: Now we shall prove inequality (30.36) when μ_0 *and* μ_1 *are compactly supported, and* μ_t *is absolutely continuous.* So let $(\mu_s)_{0\leq s\leq 1}$ be a displacement interpolation joining μ_0 to μ_1, and let Π be a dynamical optimal transport plan with $\mu_s = (e_s)_\#\Pi$. Let $\varepsilon \in (0, 1-t)$ be given. By the nonbranching property and Theorem 7.30(iii), the restricted plan $\Pi^{0,1-\varepsilon}$ obtained as the push-forward of Π by the restriction map from $C([0,1];\mathcal{X})$ to $C([0,1-\varepsilon];\mathcal{X})$ is the only dynamical optimal transport plan between μ_0 and $\mu_{1-\varepsilon}$; and more generally, if $0 \leq \widetilde{\Pi} \leq \Pi^{0,1-\varepsilon}$ with $\widetilde{\Pi}[\Gamma] > 0$, then $\Pi' := \widetilde{\Pi}/\widetilde{\Pi}[\Gamma]$ is the only dynamical optimal transport plan between its endpoints measures. In other words, $\widetilde{\mu}_0 = \mu_0$ and $\widetilde{\mu}_1 = \mu_{1-\varepsilon}$ satisfy the assumptions used in Step 1. The only displacement interpolation between $\widetilde{\mu}_0$ and $\widetilde{\mu}_1$ is $\widetilde{\mu}_t = \mu_{(1-\varepsilon)t}$, so we can apply formula (30.43) to that path, after time-reparametrization. Writing

$$t = \left(\frac{1-t-\varepsilon}{1-\varepsilon}\right)\times 0 + \left(\frac{t}{1-\varepsilon}\right)\times(1-\varepsilon),$$

we see that, $\Pi(d\gamma)$-almost surely,

$$\frac{1}{\rho_t(\gamma_t)^{\frac{1}{N}}} \geq \left(\frac{1-t-\varepsilon}{1-\varepsilon}\right)\left(\frac{\beta_{\frac{1-t-\varepsilon}{1-\varepsilon}}(\gamma_0,\gamma_{1-\varepsilon})}{\rho_0(\gamma_0)}\right)^{\frac{1}{N}} + \left(\frac{t}{1-\varepsilon}\right)\left(\frac{\beta_{\frac{t}{1-\varepsilon}}(\gamma_0,\gamma_{1-\varepsilon})}{\rho_{1-\varepsilon}(\gamma_{1-\varepsilon})}\right)^{\frac{1}{N}}. \qquad (30.44)$$

Next, let us apply the same reasoning on the time-interval $[t,1]$ rather than $[0,1-\varepsilon]$. Write $1-\varepsilon$ as an intermediate point between t and 1:

$$1-\varepsilon = \left(\frac{\varepsilon}{1-t}\right)\times t + \left(\frac{1-t-\varepsilon}{1-t}\right)\times 1.$$

Since μ_t is absolutely continuous and $\mu_{1-\varepsilon}$ belongs to the unique displacement interpolation between μ_t and μ_1, it follows from Theorem 30.19(ii) that $\mu_{1-\varepsilon}$ is absolutely continuous too. Then (30.43) becomes, after time-reparametrization,

$$\frac{1}{\rho_{1-\varepsilon}(\gamma_{1-\varepsilon})^{\frac{1}{N}}} \geq \left(\frac{\varepsilon}{1-t}\right)\left(\frac{\beta_{\frac{\varepsilon}{1-t}}(\gamma_t,\gamma_1)}{\rho_t(\gamma_t)}\right)^{\frac{1}{N}} + \left(\frac{1-t-\varepsilon}{1-t}\right)\left(\frac{\beta_{\frac{1-t-\varepsilon}{1-t}}(\gamma_t,\gamma_1)}{\rho_1(\gamma_1)}\right)^{\frac{1}{N}}. \qquad (30.45)$$

The combination of (30.44) and (30.45) yields

$$\left(1-\left(\frac{t}{1-\varepsilon}\right)\left(\frac{\varepsilon}{1-t}\right)\beta_{\frac{t}{1-\varepsilon}}(\gamma_0,\gamma_{1-\varepsilon})^{\frac{1}{N}}\,\beta_{\frac{\varepsilon}{1-t}}(\gamma_t,\gamma_1)^{\frac{1}{N}}\right)\frac{1}{\rho_t(\gamma_t)^{\frac{1}{N}}}$$
$$\geq \left(\frac{1-t-\varepsilon}{1-\varepsilon}\right)\left(\frac{\beta_{\frac{1-t-\varepsilon}{1-\varepsilon}}(\gamma_0,\gamma_{1-\varepsilon})}{\rho_0(\gamma_0)}\right)^{\frac{1}{N}} + \left(\frac{1-t-\varepsilon}{1-t}\right)\left(\frac{t}{1-\varepsilon}\right)\left(\frac{\beta_{\frac{1-t-\varepsilon}{1-t}}(\gamma_t,\gamma_1)\,\beta_{\frac{t}{1-\varepsilon}}(\gamma_0,\gamma_{1-\varepsilon})}{\rho_1(\gamma_1)}\right)^{\frac{1}{N}}.$$

Then we can pass to the limit as $\varepsilon\to 0$ thanks to the continuity of γ and β; since $\beta_1(x,y)=1$ for all x,y, we conclude that inequality (30.43) holds true almost surely.

Now let $w(\delta)=u(\delta^{-N})=\delta^N U(\delta^{-N})$, with the convention $w(0)=U'(\infty)$. By assumption w is a convex nonincreasing function of δ. So (30.43) implies

$$\mathbb{E}\, u(\rho_t(\gamma_t)) = \mathbb{E}\, w\left(\frac{1}{\rho_t(\gamma_t)^{\frac{1}{N}}}\right) \leq (1-t)\,\mathbb{E}\, w\left(\left(\frac{\beta_{1-t}(\gamma_0,\gamma_1)}{\rho_0(\gamma_0)}\right)^{\frac{1}{N}}\right) + t\,\mathbb{E}\, w\left(\left(\frac{\beta_t(\gamma_0,\gamma_1)}{\rho_1(\gamma_1)}\right)^{\frac{1}{N}}\right). \quad (30.46)$$

The left-hand side is just $\int U(\rho_t(x))/\rho_t(x)\, d\mu_t(x) = \int U(\rho_t(x))\, d\nu(x) = U_\nu(\mu_t)$. The first term in the right-hand side is $(1-t)\, U^{\beta_{1-t}}_{\pi,\nu}(\mu_0)$, since we chose to define $\rho_0(x_0) = +\infty$ when x_0 belongs to the singular set S_0. Similarly, the second term is $t\, U^{\beta_t}_{\check{\pi},\nu}(\mu_1)$. So (30.46) reads

$$U_\nu(\mu_t) \leq (1-t)\, U^{\beta_{1-t}}_{\pi,\nu}(\mu_0) \;+\; t\, U^{\beta_t}_{\check{\pi},\nu}(\mu_1),$$

as desired.

Step 3: Now we wish to establish inequality (30.36) in the case when μ_t *is absolutely continuous*, that is, we just want to drop the assumption of compact support.

It follows from Step 2 that $(\mathcal{X}, d, \nu)$ is a weak $\mathrm{CD}(K,N)$ space, so we now have access to Theorem 30.19 even if μ_0 and μ_1 are not compactly supported; and also we can appeal to Theorem 30.5 to guarantee that Property (ii) is verified for probability measures that are not necessarily compactly supported. Then we can repeat Steps 1 and 2 without the assumption of compact support, and in the end establish inequality (30.36) for measures that are not compactly supported.

Step 4: Now we shall consider the case when μ_t is not absolutely continuous. (This is the part of the proof which has interest even in a smooth setting.) Let $(\mu_t)_s$ stand for the singular part of μ_t, and $m := (\mu_t)_s[\mathcal{X}] > 0$.

Let $E^{(a)}$ and $E^{(s)}$ be two disjoint Borel sets in $\mathcal{X}$ such that the absolutely continuous part of μ_t is concentrated on $E^{(a)}$, and the singular part of μ_t is concentrated on $E^{(s)}$. Obviously, $\Pi[\gamma_t \in E^{(s)}] = (\mu_t)_s[\mathcal{X}] = m$, and $\Pi[\gamma_t \in E^{(a)}] = 1-m$. Let us decompose Π into $\Pi = (1-m)\,\Pi^{(a)} + m\,\Pi^{(s)}$, where

$$\Pi^{(a)}(d\gamma) = \frac{1_{[\gamma_t \in E^{(a)}]}\Pi(d\gamma)}{\Pi[\gamma_t \in E^{(a)}]}, \qquad \Pi^{(s)}(d\gamma) = \frac{1_{[\gamma_t \in E^{(s)}]}\Pi(d\gamma)}{\Pi[\gamma_t \in E^{(s)}]}.$$

Further, for any $s \in [0,1]$, let

$$\mu_s^{(a)} = (e_s)_\#\Pi^{(a)}, \qquad \mu_s^{(s)} = (e_s)_\#\Pi^{(s)},$$
$$\pi^{(a)} = (e_0,e_1)_\#\Pi^{(a)}, \qquad \pi^{(s)} = (e_0,e_1)_\#\Pi^{(s)}.$$

Since it has been obtained by conditioning of a dynamical optimal transference plan, $\Pi^{(a)}$ is itself a dynamical optimal transference plan (Theorem 7.30(ii)), and by construction $\mu_t^{(a)}$ is the absolutely continuous part of μ_t, while $\mu_t^{(s)}$ is its singular part. So the result of Step 2 applies to the path $(\mu_s^{(a)})_{0\le s\le 1}$:

$$U_\nu(\mu_t^{(a)}) \le (1-t)\, U^{\beta_{1-t}}_{\pi^{(a)},\nu}(\mu_0^{(a)}) \;+\; t\, U^{\beta_t}_{\check{\pi}^{(a)},\nu}(\mu_1^{(a)}).$$

Actually, we shall not apply this inequality with the nonlinearity U, but rather with $U_m(r) = U((1-m)r)$, which lies in $\mathcal{DC}_N$ if U does. So

$$(U_m)_\nu(\mu_t^{(a)}) \le (1-t)\, (U_m)^{\beta_{1-t}}_{\pi^{(a)},\nu}(\mu_0^{(a)}) \;+\; t\, (U_m)^{\beta_t}_{\check{\pi}^{(a)},\nu}(\mu_1^{(a)}). \tag{30.47}$$

Since $\mu_t^{(s)}$ is purely singular and $\mu_t = (1-m)\,\mu_t^{(a)} + m\,\mu_t^{(s)}$, the definition of U_ν implies

$$U_\nu(\mu_t) = (U_m)_\nu(\mu_t^{(a)}) + m\, U'(\infty). \tag{30.48}$$

By Theorem 30.19(iii), $\mu_0^{(s)}$ is purely singular. In view of the identity $\mu_0 = (1-m)\,\mu_0^{(a)} + m\,\mu_0^{(s)}$,

$$U^{\beta_{1-t}}_{\pi,\nu}(\mu_0) = (U_m)^{\beta_{1-t}}_{\pi^{(a)},\nu}(\mu_0^{(a)}) \;+\; m\, U'(\infty). \tag{30.49}$$

Similarly,

$$U^{\beta_t}_{\check{\pi},\nu}(\mu_1) = (U_m)^{\beta_t}_{\check{\pi}^{(a)},\nu}(\mu_1^{(a)}) \;+\; m\, U'(\infty). \tag{30.50}$$

The combination of (30.47), (30.48), (30.49) and (30.50) implies

$$U_\nu(\mu_t) \le (1-t)\, U^{\beta_{1-t}}_{\pi,\nu}(\mu_0) \;+\; t\, U^{\beta_t}_{\check{\pi},\nu}(\mu_1).$$

This concludes the proof in the case $N < \infty$.

When $N = \infty$, at first sight things look pretty much the same; formula (30.43) should be replaced by

$$\log \frac{1}{\rho_t(\gamma_t)} \ge (1-t)\, \log \frac{1}{\rho_0(\gamma_0)} + t\, \log \frac{1}{\rho_1(\gamma_1)} + \frac{K\, t(1-t)}{2}\, d(\gamma_0,\gamma_1)^2. \tag{30.51}$$

At a technical level, there is a small simplification since it is not necessary to treat singular measures (if μ is singular and U is not linear, then according to Proposition 17.7(ii) $U'(\infty) = +\infty$, so $U_\nu(\mu) = +\infty$).

On the other hand, there is a serious complication: The proof of Step 1 breaks down since the measure ν is not a priori locally doubling, and Lebesgue's density theorem does not apply!

It seems a bit of a miracle that the method of proof can still be saved, as I shall now explain. First assume that ρ_0 and ρ_1 satisfy the same assumptions as in Step 1 above, but that in addition they are *upper semicontinuous*. As in Step 1, define

$$\begin{aligned} f(\gamma) &= (1-t)\,\log\left(\frac{\beta_{1-t}(\gamma_0,\gamma_1)}{\rho_0(\gamma_0)}\right) + t\,\log\left(\frac{\beta_t(\gamma_0,\gamma_1)}{\rho_1(\gamma_1)}\right) \\ &= (1-t)\,\log\frac{1}{\rho_0(\gamma_0)} + t\,\log\frac{1}{\rho_1(\gamma_1)} + \frac{K\,t(1-t)}{2}\,d(\gamma_0,\gamma_1)^2. \end{aligned}$$

The argument of Step 1 shows that

$$\log\left(\frac{\nu[B_\delta(y)]}{\mu_t[B_\delta(y)]}\right) \geq \frac{\int_{B_\delta(y)} f(F_t(x))\,\mu_t(dx)}{\mu_t[B_\delta(y)]};$$

in particular $\mu_t[B_\delta(y)] \leq \exp\left(-\inf_{x\in B_\delta(y)} f(F_t(x))\right)\,\nu[B_\delta(y)]$. Similarly, for any $z \in B_{\delta/2}(y)$ and $r \leq \delta/2$,

$$\mu_t[B_r(z)] \leq \exp\left(-\inf_{x\in B_\delta(y)} f(F_t(x))\right)\,\nu[B_r(z)]. \qquad (30.52)$$

The family of balls $\{B_r(z);\ z\in B_{\delta/2}(y);\ r\leq\delta/2\}$ generates the Borel σ-algebra of $B_{\delta/2}(y)$, so (30.52) holds true for any measurable set $S \subset B_{\delta/2}(y)$ instead of $B_r(z)$. Then we can pass to densities:

$$\rho_t(z) \leq \exp\left(-\inf_{x\in B_\delta(y)} f(F_t(x))\right) \qquad \text{almost surely in } B_{\delta/2}(y).$$

In particular, for almost any z,

$$\rho_t(z) \leq \sup_{x\in B_{2\delta}(z)} e^{-f(F_t(x))}. \qquad (30.53)$$

Now note that the map F_t from Theorem 7.30(v) is *continuous* on $\operatorname{Spt}\Pi$. Indeed, if a sequence $(\gamma_k)_{k\in\mathbb{N}}$ of geodesics is given in $\operatorname{Spt}\Pi$, in such a way that $\gamma_k(t)\to\gamma(t)$, by compactness there is a subsequence, still denoted γ_k, which converges uniformly to some geodesic $\widetilde{\gamma}\in\operatorname{Spt}\Pi$ and satisfying $\widetilde{\gamma}(t)=\gamma(t)$; which implies that $\widetilde{\gamma}=\gamma$. (In fact I used the same argument to prove the measurability of F_t in the case when Spt Π is not necessarily compact.) Since ρ_0 and ρ_1 are upper semicontinuous, f is lower semicontinuous; so $e^{-f(F_t)}$ is upper semicontinuous, and

$$\lim_{\delta\downarrow 0} \sup_{x\in B_\delta(z)} e^{-f(F_t(x))} \le e^{-f(F_t(z))}.$$

So we may pass to the limit as $\delta \to 0$ in (30.53) and recover $\rho_t \le e^{-f\circ F_t}$, or in other words

$$\log \frac{1}{\rho_t(\gamma_t)} \ge (1-t)\,\log\left(\frac{\beta_{1-t}(\gamma_0,\gamma_1)}{\rho_0(\gamma_0)}\right) + t\,\log\left(\frac{\beta_t(\gamma_0,\gamma_1)}{\rho_1(\gamma_1)}\right). \qquad (30.54)$$

This is the desired estimate, but under the additional assumption of upper semicontinuity of ρ_0 and ρ_1. In the general case, we still have

$$\log \frac{1}{\rho_t(\gamma_t)} \ge (1-t)\,\log\left(\frac{\beta_{1-t}(\gamma_0,\gamma_1)}{\overline{\rho}_0(\gamma_0)}\right) + t\,\log\left(\frac{\beta_t(\gamma_0,\gamma_1)}{\overline{\rho}_1(\gamma_1)}\right), \qquad (30.55)$$

where $\overline{\rho}_0$ and $\overline{\rho}_1$ are upper semicontinuous and $\rho_0 \le \overline{\rho}_0$, $\rho_1 \le \overline{\rho}_1$.

Next recall that if g is any nonnegative measurable function, there is a sequence $(g_k)_{k\in\mathbb{N}}$ of nonnegative upper semicontinuous functions such that $0 \le g_k \le g$ out of a set of zero measure, and $g_k \uparrow g$ almost surely as $k \to \infty$. Indeed, one can write g as a nondecreasing limit of simple functions $h_j = \sum_\ell \lambda_j^\ell 1_{B_j^\ell}$, where $(B_j^\ell)_{1\le \ell\le L_j}$ is a finite family of Borel sets. For each B_j^ℓ, the regularity of the measure allows one to find a nondecreasing sequence of compact sets $(K_{j,m}^\ell)_{m\in\mathbb{N}}$ included in B_j^ℓ, such that $\nu[K_{j,m}^\ell] \longrightarrow \nu[B_j^\ell]$. So $h_{j,m} = \sum \lambda_j^\ell 1_{K_{j,m}^\ell}$ converges monotonically to h_j as $m \to \infty$, up to a set of zero ν-measure. Each $h_{j,m}$ is obviously upper semicontinuous. Then choose $g_k = \max\{h_{j,j};\ j \le k\}$: this is still upper semicontinuous (the maximum is over a finite set of upper semicontinuous functions). For any ℓ and any $k \ge \ell$ we have $g_k \ge h_{\ell,k}$, which converges to g_ℓ as $k \to \infty$; so $\liminf g_k \ge g$, almost surely.

Coming back to the proof of Theorem 30.32, I shall now proceed to approximate Π. Let $(g_k)_{k\in\mathbb{N}}$ be a sequence of upper semicontinuous functions such that $0 \le g_k \le \rho_0$ and $g_k \uparrow \rho_0$ up to a ν-negligible set. Let

$$Z_k = \int g_k\, d\nu, \qquad \rho_{k,0} = \frac{g_k}{Z_k}.$$

Next disintegrate Π with respect to its marginal $(e_0)_\#\Pi$:

$$\Pi(d\gamma) = \rho(\gamma_0)\,\nu(d\gamma_0)\,\Pi(d\gamma|\gamma_0),$$

and define

$$\Pi'_k(d\gamma) = g_k(\gamma_0)\,\nu(d\gamma_0)\,\Pi(d\gamma|\gamma_0); \qquad \Pi_k = \frac{\Pi'_k}{Z_k}.$$

Then Π_k is a probability measure on geodesics, and since it has been obtained from Π by restriction, it is actually the unique dynamical optimal transference plan between the two probability measures $\mu_{k,0} = (e_0)_\#\Pi_k$ and $\mu_{k,1} = (e_1)_\#\Pi_k$. From the construction of Π_k,

$$\mu_{k,0} = \rho_{k,0}\,\nu; \quad \mu_{k,1} = \rho_{k,1}\,\nu; \quad \rho_{k,1} \le \frac{\rho_1}{Z_k}.$$

Next we repeat the process at the other end: Let $(h_{k,\ell})_{\ell\in\mathbb{N}}$ be a nonincreasing sequence of upper semicontinuous functions such that $0 \le h_{k,\ell} \le \rho_{k,1}$ and $h_{k,\ell} \uparrow \rho_{k,1}$ (up to a set of zero measure). Define

$$Z_{k,\ell} = \int h_{k,\ell}\,d\nu, \qquad \rho_{k,\ell,1} = \frac{h_{k,\ell}}{Z_{k,\ell}};$$
$$\Pi'_{k,\ell}(d\gamma) = \Pi_k(d\gamma|\gamma_1)\,h_{k,\ell}(\gamma_1)\,\nu(d\gamma_1); \qquad \Pi_{k,\ell}(d\gamma) = \frac{\Pi'_{k,\ell}(d\gamma)}{Z_{k,\ell}}.$$

Then again $\Pi_{k,\ell}$ is the unique dynamical optimal transference plan between its marginals $\mu_{k,\ell,0} = (e_0)_\#\Pi_{k,\ell}$ and $\mu_{k,\ell,1} = \rho_{k,\ell,1}\,\nu$.

If t is any time in $[0,1]$ and $\rho_{k,t}$ (resp. $\rho_{k,\ell,t}$) stands for the density of $(e_t)_\#\Pi_k$ (resp. $(e_t)_\#\Pi_{k,\ell}$) with respect to ν, then

$$Z_k\,\rho_{k,t} \uparrow \rho_t \quad \text{as } k \to \infty;$$
$$Z_{k,\ell}\,\rho_{k,\ell,t} \uparrow \rho_{k,t} \quad \text{as } \ell \to \infty.$$

Moreover, $\rho_{k,0}$ and $\rho_{k,\ell,1}$ are upper semicontinuous; in particular $\rho_{k,\ell,0} \le \rho_{k,0}/Z_{k,\ell}$, which is upper semicontinuous. Then we can apply (30.55) with the dynamical plan $\Pi_{k,\ell}$ and get

$$\log\frac{1}{\rho_{k,\ell,t}(\gamma_t)} \ge (1-t)\,\log\left(\frac{Z_{k,\ell}\,\beta_{1-t}(\gamma_0,\gamma_1)}{\rho_{0,k}(\gamma_0)}\right) + t\,\log\left(\frac{\beta_t(\gamma_0,\gamma_1)}{\rho_{k,\ell,1}(\gamma_1)}\right).$$

By letting $\ell \to \infty$ and then $k \to \infty$, we conclude that (30.54) is true, but this time without any upper semicontinuity assumption.

This concludes the proof of Step 1 in the case $N = \infty$. Then Steps 2 and 3 are done as before. □

Locality

Locality is one of the most fundamental properties that one may expect from any notion of curvature. In the setting of weak $\mathrm{CD}(K,N)$ spaces, the locality problem may be loosely formulated as follows: *If* $(\mathcal{X},d,\nu)$ *is weakly* $\mathrm{CD}(K,N)$ *in the neighborhood of any of its points, then* $(\mathcal{X},d,\nu)$ *should be a weakly* $\mathrm{CD}(K,N)$ *space.*

So far it is not known whether this "local-to-global" property holds in general. However, it is true at least in a *nonbranching space*, if $K=0$ and $N<\infty$. The validity of a more general statement may depend on the following:

Conjecture 30.34 (Local-to-global $\mathrm{CD}(K,N)$ property along a path). *Let* $\theta\in(0,1)$ *and* $\alpha\in[0,\pi]$. *Let* $f:[0,1]\to\mathbb{R}_+$ *be a measurable function such that for all* $\lambda\in[0,1]$, $t,t'\in[0,1]$, *the inequality*

$$f\big((1-\lambda)\,t+\lambda\,t'\big)\geq(1-\lambda)\left(\frac{\sin\big((1-\lambda)\,\alpha|t-t'|\big)}{(1-\lambda)\,\sin(\alpha|t-t'|)}\right)^{\theta}f(t)+\lambda\left(\frac{\sin\big(\lambda\,\alpha|t-t'|\big)}{\lambda\,\sin(\alpha|t-t'|)}\right)^{\theta}f(t') \qquad (30.56)$$

holds true as soon as $|t-t'|$ *is small enough. Then* (30.56) *automatically holds true for all* $t,t'\in[0,1]$.

The same if sin *is replaced by* sinh *and* α *is allowed to vary in* $\mathbb{R}_+$.

I really don't have much to support this conjecture, except that it would imply a really nice (to my taste) result. It might be trivially false or trivially true, but I was unable to prove or disprove it. (If it would hold true only under additional regularity assumptions such as local integrability or continuity of f, this might be fine.)

To understand the relation of (30.56) to optimal transport, take $\theta=1-1/N$, $\alpha=\sqrt{|K|/(N-1)}\,d(\gamma_0,\gamma_1)$, $f(t)=\rho_t(\gamma_t)^{-1/N}$, and write $I_t(\gamma_0,\gamma_t,\gamma_1)$ for the inequality appearing in (30.43). Then Conjecture 30.34, if true, means that this inequality is local, in the sense that if $I_t(\gamma_{t_0},\gamma_{(1-t)\,t_0+t\,t_1},\gamma_{t_1})$ holds true for $|t_0-t_1|$ small enough, then it holds true for all t_0,t_1, and in particular $t_0=0$, $t_1=1$.

There are at least two limit cases in which Conjecture 30.34 becomes true. The first one is for $\alpha=0$ and θ fixed (this corresponds to $\mathrm{CD}(0,N)$, $N=1/(1-\theta)$); the second one is the limit when $\theta\to1$,

$\alpha \to 0$ in such a way that $\alpha^2/(1-\theta)$ converges to a finite limit (this corresponds to $\mathrm{CD}(K,\infty)$, and the limit of $\alpha^2/(1-\theta)$ would be $K\,d(\gamma_0,\gamma_1)^2$). In the first case, Conjecture 30.56 reduces to the locality of the property of concavity:

$$f\big((1-\lambda)\,t+\lambda\,t'\big) \geq (1-\lambda)\,f(t)+\lambda\,f(t');$$

while in the second case, it reduces to the locality of the more general property of κ-concavity ($\kappa \in \mathbb{R}$):

$$f\big((1-\lambda)\,t+\lambda\,t'\big) \geq (1-\lambda)\,f(t)+\lambda\,f(t')+\frac{\kappa\,\lambda(1-\lambda)}{2}\,|t-t'|^2. \quad (30.57)$$

These properties do satisfy a local-to-global principle, for instance because they are equivalent to the differential inequality $f'' \leq 0$, or $f'' \leq -\kappa$, to be understood in the distributional sense.

To summarize: If $K = 0$ (resp. $N = \infty$), inequality (30.43) (resp. (30.51)) satisfies a local-to-global principle; in the other cases I don't know.

Next, I shall give a precise definition of what it means to satisfy $\mathrm{CD}(K,N)$ locally:

Definition 30.35 (Local $\mathrm{CD}(K,N)$ space). *Let $K \in \mathbb{R}$ and $N \in [1,\infty]$. A locally compact Polish geodesic space $(\mathcal{X},d)$ equipped with a locally finite measure ν is said to be a local weak* $\mathrm{CD}(K,N)$ *space if for any $x_0 \in \mathcal{X}$ there is $r>0$ such that whenever μ_0, μ_1 are two probability measures supported in $B_r(x_0) \cap \mathrm{Spt}\,\nu$, there is a displacement interpolation $(\mu_t)_{0\leq t\leq 1}$ joining μ_0 to μ_1, and an associated optimal coupling π, such that for all $t \in [0,1]$ and for all $U \in \mathcal{DC}_N$,*

$$U_\nu(\mu_t) \leq (1-t)\,U_{\pi,\nu}^{\beta_{1-t}^{(K,N)}}(\mu_0)+t\,U_{\check{\pi},\nu}^{\beta_t^{(K,N)}}(\mu_1). \qquad (30.58)$$

Remark 30.36. In the previous definition, one could also have imposed that the whole path $(\mu_t)_{0\leq t\leq 1}$ is supported in $B_r(x_0)$. Both formulations are equivalent: Indeed, if μ_0 and μ_1 are supported in $B_{r/3}(x_0)$ then all measures μ_t are supported in $B_r(x_0)$.

Now comes the main result in this section:

Theorem 30.37 (From local to global $\mathrm{CD}(K,N)$). *Let $K \in \mathbb{R}$, $N \in [1,\infty)$, and let $(\mathcal{X},d,\nu)$ be a nonbranching local weak* $\mathrm{CD}(K,N)$ *space with* $\mathrm{Spt}\,\nu = \mathcal{X}$. *If $K=0$, then $\mathcal{X}$ is also a weak* $\mathrm{CD}(K,N)$ *space. The same is true for all values of K if Conjecture 30.34 has an affirmative answer.*

Remark 30.38. If the assumption $\operatorname{Spt}\nu = \mathcal{X}$ is dropped then the result becomes trivially false. As a counterexample, take $\mathcal{X} = \mathbb{R}^3$, equipped with the Euclidean distance, and let ν be the 2-dimensional Lebesgue measure on each horizontal plane of integer altitude. (So the measure is concentrated on well-separated parallel planes.) This is a local weak $\mathrm{CD}(0,2)$ space but not a weak $\mathrm{CD}(0,2)$ space.

Remark 30.39. I don't know if the nonbranching condition can be removed in Theorem 30.37.

As in the proof of Theorem 30.32, one of the main ideas in the proof of Theorem 30.37 consists in using the nonbranching condition to translate integral conditions into pointwise density bounds along geodesic paths. Another idea consists in "cutting" dynamical optimal transference plans into small pieces, each of which is "small enough" that the local displacement convexity can be applied. The fact that we work along geodesic paths parametrized by $[0,1]$ explains that the whole locality problem is reduced to the one-dimensional "local-to-global" problem exposed in Conjecture 30.34.

Proof of Theorem 30.37. If we can treat the case $N > 1$, then the case $N = 1$ will follow by letting N go to 1 (as in the proof of Theorem 29.24). So let us assume $1 < N < \infty$. In the sequel, I shall use the shorthand $\beta_t = \beta_t^{(K,N)}$.

Let $(\mathcal{X}, d, \nu)$ be a nonbranching local weak $\mathrm{CD}(K,N)$ space. By repeating the proof of Theorem 30.32, we can show that for any $x_0 \in \mathcal{X}$ there is $r = r(x_0) > 0$ such that (30.58) holds true along *any* displacement interpolation $(\mu_t)_{0 \le t \le 1}$ which is supported in $B(x_0, r)$. Moreover, if Π is a dynamical optimal transference plan such that $(e_t)_\# \Pi = \mu_t$, and each measure μ_t is absolutely continuous with density ρ_t, then $\Pi(d\gamma)$-almost all geodesics will satisfy inequality (30.43), which I recast below:

$$\frac{1}{\rho_t(\gamma_t)^{\frac{1}{N}}} \ge (1-t) \left(\frac{\beta_{1-t}(\gamma_0,\gamma_1)}{\rho_0(\gamma_0)}\right)^{\frac{1}{N}} + t \left(\frac{\beta_t(\gamma_0,\gamma_1)}{\rho_1(\gamma_1)}\right)^{\frac{1}{N}}. \tag{30.59}$$

Let μ_0, μ_1 be any two compactly supported probability measures on $\mathcal{X}$, and let $B = B(z,R)$ be a large ball such that any geodesic going from $\operatorname{Spt}\mu_0$ to $\operatorname{Spt}\mu_1$ lies within B. Let Π be a dynamical optimal transference plan between μ_0 and μ_1. The goal is to prove that for all $U \in \mathcal{DC}_N$,

$$U_\nu(\mu_t) \le (1-t)\, U_{\pi,\nu}^{\beta_{1-t}}(\mu_0) + t\, U_{\check{\pi},\nu}^{\beta_t}(\mu_1). \tag{30.60}$$

The plan is to cut Π into very small pieces, each of which will be included in a sufficiently small ball that the local weak $\mathrm{CD}(K,N)$ criterion can be used. I shall first proceed to construct these small pieces.

Cover the closed ball $B[z,R]$ by a finite number of balls $B(x_j, r_j/3)$ with $r_j = r(x_j)$, and let $\overline{r} := \inf(r_j/3)$. For any $y \in B[z,R]$, the ball $B(y,\overline{r})$ lies inside some $B(x_j,r_j)$; so if $(\overline{\mu}_t)_{0\le t\le 1}$ is any displacement interpolation supported in some ball $B(y,\overline{r})$, $\overline{\Pi}$ is an associated dynamical optimal transference plan, and $\overline{\mu}_0$, $\overline{\mu}_1$ are absolutely continuous, then the density $\overline{\rho}_t$ of $\overline{\mu}_t$ will satisfy the inequality

$$\frac{1}{\overline{\rho}_t(\gamma_t)^{\frac{1}{N}}} \ge (1-t)\left(\frac{\beta_{1-t}(\gamma_0,\gamma_1)}{\overline{\rho}_0(\gamma_0)}\right)^{\frac{1}{N}} + t\left(\frac{\beta_t(\gamma_0,\gamma_1)}{\overline{\rho}_1(\gamma_1)}\right)^{\frac{1}{N}}, \qquad (30.61)$$

$\overline{\Pi}(d\gamma)$-almost surely. The problem now is to cut Π into many small subplans $\overline{\Pi}$ and to apply (30.61) to all these subplans.

Let $\delta \in 1/\mathbb{N}$ be small enough that $4R\,\delta \le \overline{r}/3$, and let $B(y_\ell,\delta)_{1\le\ell\le L}$ be a finite covering of $B[z,R]$ by balls of radius δ. Define $A_1 = B(y_1,\delta)$, $A_2 = B(y_2,\delta)\setminus A_1$, $A_3 = B(y_3,\delta)\setminus(A_1\cup A_2)$, etc. This provides a covering of $B(z,R)$ by *disjoint* sets $(A_\ell)_{1\le\ell\le L}$, each of which is included in a ball of radius δ. (Without loss of generality, we can assume that they are all nonempty.)

Let $m = 1/\delta \in \mathbb{N}$. We divide the set Γ of all geodesics going from $\mathrm{Spt}\,\mu_0$ to $\mathrm{Spt}\,\mu_1$ into pieces, as follows. For any finite sequence $\underline{\ell} = (\ell_0,\ell_1,\ldots,\ell_m)$, let

$$\Gamma_{\underline{\ell}} = \Big\{\gamma\in\Gamma;\quad \gamma_0\in A_{\ell_0},\ \gamma_\delta\in A_{\ell_1},\ \gamma_{2\delta}\in A_{\ell_2},\ldots,\ \gamma_{m\delta}=\gamma_1\in A_{\ell_m}\Big\}.$$

The sets $\Gamma_{\underline{\ell}}$ are disjoint. We discard the sequences $\underline{\ell}$ such that $\Pi[\Gamma_{\underline{\ell}}]=0$. Then let $Z_{\underline{\ell}} = \Pi[\Gamma_{\underline{\ell}}]$, and let

$$\Pi_{\underline{\ell}} = \frac{1_{\Gamma_{\underline{\ell}}}\,\Pi_{\underline{\ell}}}{Z_{\underline{\ell}}}$$

be the law of γ conditioned by the event $\{\gamma\in\Gamma_{\underline{\ell}}\}$. Further, let $\mu_{\underline{\ell},t} = (e_t)_\#\Pi_{\underline{\ell}}$, and $\pi_{\underline{\ell}} = (e_0,e_1)_\#\Pi_{\underline{\ell}}$.

For each $\underline{\ell}$ and $k\in\{0,\ldots,m-2\}$, we define $\Pi^k_{\underline{\ell}}$ to be the image of $\Pi_{\underline{\ell}}$ by the restriction map $[0,1]\to[k\delta,(k+2)\delta]$. Up to affine reparametrization of time, $\Pi^k_{\underline{\ell}}$ is a dynamical optimal transference plan between the measures $\mu_{\underline{\ell},k\delta}$ and $\mu_{\underline{\ell},(k+2)\delta}$ (Theorem 7.30(i)–(ii)).

Let γ be a random geodesic distributed according to the law $\Pi^k_{\underline{\ell}}$. Almost surely, $\gamma(k\delta)$ belongs to A_{ℓ_k}, which has a diameter at most $\overline{r}/3$. Moreover, the speed of γ is bounded above by $\operatorname{diam}(B[z,R]) \leq 2R$, so on the time-interval $[k\delta, (k+2)\delta]$, γ moves at most by a distance $(2\delta)(2R) \leq \overline{r}/3$. Thus γ is entirely contained in a set of diameter $2\overline{r}/3$. In particular, $(\mu^k_{\underline{\ell},t})_{k\delta\leq t\leq(k+2)\delta}$ is entirely supported in a set of diameter $\overline{r}$, and satisfies the displacement convexity inequalities which are typical of the curvature-dimension bound $\mathrm{CD}(K,N)$.

By Theorem 7.30(iii), $\mu^k_{\underline{\ell},t}$ is (up to time-reparametrization) the unique optimal dynamical transference plan between $\mu_{\underline{\ell},k\delta}$ and $\mu_{\underline{\ell},(k+2)\delta}$. So by Theorem 30.19(ii), the absolute continuity of $\mu_{\underline{\ell},k\delta}$ implies the absolute continuity of $\mu_{\underline{\ell},t}$ for all $t \in [k\delta, (k+2)\delta)$. Since $\mu_{\underline{\ell},0}$ is absolutely continuous, an immediate induction shows that $\mu_{\underline{\ell},t}$ is absolutely continuous for all times. Then we can apply (30.61) to each path $(\mu_{\underline{\ell},t})_{k\delta\leq t\leq(k+2)\delta}$; after time reparametrization, this becomes:

$$\begin{gathered}\forall k \in \{0,\ldots,m-2\},\\ \Pi_{\underline{\ell}}(d\gamma)\text{-almost surely, } \forall t \in [0,1],\ \forall (t_0,t_1) \in [k\delta,(k+2)\delta],\\ \frac{1}{\rho_{\underline{\ell},(1-t)t_0+tt_1}(\gamma_{(1-t)t_0+tt_1})^{\frac{1}{N}}} \geq (1-t)\left(\frac{\beta_{1-t}(\gamma_{t_0},\gamma_{t_1})}{\rho_{\underline{\ell},t_0}(\gamma_{t_0})}\right)^{\frac{1}{N}} + t\left(\frac{\beta_t(\gamma_{t_0},\gamma_{t_1})}{\rho_{\underline{\ell},t_1}(\gamma_{t_1})}\right)^{\frac{1}{N}}.\end{gathered} \tag{30.62}$$

So inequality (30.62) is satisfied whenever $|t_0 - t_1| \leq \delta$. Then our assumptions, and the discussion following Conjecture 30.34, imply that the same inequality is satisfied for all values of t_0 and t_1 in $[0,1]$. In particular, $\Pi_{\underline{\ell}}$-almost surely,

$$\frac{1}{\rho_{\underline{\ell},t}(\gamma_t)^{\frac{1}{N}}} \geq (1-t)\left(\frac{\beta_{1-t}(\gamma_0,\gamma_1)}{\rho_{\underline{\ell},0}(\gamma_0)}\right)^{\frac{1}{N}} + t\left(\frac{\beta_t(\gamma_0,\gamma_1)}{\rho_{\underline{\ell},1}(\gamma_1)}\right)^{\frac{1}{N}}. \tag{30.63}$$

By reasoning as in the proof of Theorem 30.32 (end of Step 2), we deduce the inequality

$$U_\nu(\mu_{\underline{\ell},t}) \leq (1-t)\, U^{\beta_{1-t}}_{\pi_{\underline{\ell}},\nu}(\mu_{\underline{\ell},0}) + t\, U^{\beta_t}_{\check{\pi}_{\underline{\ell}},\nu}(\mu_{\underline{\ell},1}). \tag{30.64}$$

Recall that $\mu_t = \sum Z_{\underline{\ell}}\, \mu_{\underline{\ell},t}$; so the issue is now to add up the various contributions coming from different values of $\underline{\ell}$.

For each $\underline{\ell}$, we apply (30.64) with U replaced by $U_{\underline{\ell}} = U(Z_{\underline{\ell}}\,\cdot)/Z_{\underline{\ell}}$. Then, with the shorthand $U_{\underline{\ell},\nu} = (U_{\underline{\ell}})_\nu$ and $U^\beta_{\underline{\ell},\pi_{\underline{\ell}},\nu} = (U_{\underline{\ell}})^\beta_{\pi_{\underline{\ell}},\nu}$, we obtain

$$U_{\underline{\ell},\nu}(\mu_{\underline{\ell},t}) \leq (1-t)\, U^{\beta_{1-t}}_{\underline{\ell},\pi_{\underline{\ell}},\nu}(\mu_{\underline{\ell},0}) + t\, U^{\beta_t}_{\underline{\ell},\check{\pi}_{\underline{\ell}},\nu}(\mu_{\underline{\ell},1}). \tag{30.65}$$

For any $t \in (0,1)$, the map $\gamma_t \to \gamma$ is injective, as a consequence of Theorem 7.30(iv)–(v), and in particular the measures $\mu_{\underline{\ell},t}$ are mutually singular as $\underline{\ell}$ varies. Then it follows from Lemma 29.7 that

$$U_\nu(\mu_t) = \sum_{\underline{\ell}} Z_{\underline{\ell}}\, U_{\ell,\nu}(\mu_{\underline{\ell},t}). \tag{30.66}$$

Since $\pi = \sum_{\underline{\ell}} Z_{\underline{\ell}}\, \pi_{\underline{\ell}}$, Lemma 29.7 also implies

$$\sum_{\underline{\ell}} Z_{\underline{\ell}}\, U^{\beta_{1-t}}_{\underline{\ell},\pi_{\underline{\ell}},\nu}(\mu_{\underline{\ell},0}) \leq U^{\beta_{1-t}}_{\pi,\nu}(\mu_0);$$

$$\sum_{\underline{\ell}} Z_{\underline{\ell}}\, U^{\beta_t}_{\underline{\ell},\check{\pi}_{\underline{\ell}},\nu}(\mu_{\underline{\ell},1}) \leq U^{\beta_t}_{\check{\pi},\nu}(\mu_1). \tag{30.67}$$

The combination of (30.65), (30.66) and (30.67) implies the desired conclusion (30.60). □

In the case $N = \infty$, Conjecture 30.34 is satisfied; however, I don't know if Theorem 30.37 can be extended to that case without additional assumptions; the problem is that $H_\nu(\mu)$ might be $+\infty$. More precisely, if either $H_\nu(\mu_{k\delta})$ or $H_\nu(\mu_{(k+2)\delta})$ is $+\infty$, then we cannot derive (30.51) between times $k\delta$ and $(k+2)\delta$, so the proof breaks down.

To get around this problem, I shall impose further assumptions ensuring that the space is "almost everywhere" finite-dimensional. Let us agree that a point x in a metric-measure space $(\mathcal{X}, d, \nu)$ is *finite-dimensional* if there is a small ball $B_r(x)$ in which the criterion for $\mathrm{CD}(K', N')$ is satisfied, where $K' \in \mathbb{R}$ and $N' < \infty$. More explicitly, it is required that for any two probability measures μ_0, μ_1 supported in $B_r(x) \cap \operatorname{Spt} \nu$, there is a displacement interpolation $(\mu_t)_{0\leq t\leq 1}$ and an associated coupling π such that for all $U \in \mathcal{DC}_N$,

$$U_\nu(\mu_t) \leq (1-t)\, U^{\beta^{(K,N)}_{1-t}}_{\pi,\nu}(\mu_0) + t\, U^{\beta^{(K,N)}_t}_{\check{\pi},\nu}(\mu_1).$$

A point will be called *infinite-dimensional* if it is not finite-dimensional.

Example 30.40. Let $\varphi : \mathbb{R} \to \mathbb{R} \cup \{+\infty\}$ be a convex function with domain (a,b), where a, b are two real numbers. (So φ takes the value $+\infty$ outside of (a,b).) Equip $\mathbb{R}$ with the usual distance and the measure $\nu(dx) = e^{-\varphi(x)}\, dx$; this gives a weak $\mathrm{CD}(0,\infty)$ space. Then the support $[a,b]$ of ν consists in finite-dimensional points, which fill up the open interval (a,b); and the two infinite-dimensional points a and b.

Example 30.41. The space $\mathcal{X}$ in Example 29.17 is "genuinely infinite-dimensional" in the sense that none of its points is finite-dimensional (such a point would have a neighborhood of finite Hausdorff dimension by Corollary 30.13).

Theorem 30.42 (From local to global CD(K,∞)). *Let* $K \in \mathbb{R}$ *and let* $(\mathcal{X}, d, \nu)$ *be a local weak* $\mathrm{CD}(K,\infty)$ *space with* $\operatorname{Spt}\nu = \mathcal{X}$*. Assume that* $\mathcal{X}$ *is nonbranching and that there is a totally convex measurable subset* $\mathcal{Y}$ *of* $\mathcal{X}$ *such that all points in* $\mathcal{Y}$ *are finite-dimensional and* $\nu[\mathcal{X}\setminus\mathcal{Y}] = 0$*. Then* $(\mathcal{X}, d, \nu)$ *is a weak* $\mathrm{CD}(K,\infty)$ *space.*

Remark 30.43. I don't know if the assumption of the existence of $\mathcal{Y}$ can be removed from the above theorem.

Proof of Theorem 30.42. Let Γ be the set of all geodesics in $\mathcal{X}$. Let K be a compact subset of $\mathcal{Y}$, and let μ_0, μ_1 be two probability measures supported in K. Let $(\mu_t)_{0\le t\le 1}$ be a displacement interpolation. The set Γ_K of geodesics $(\gamma_t)_{0\le t\le 1}$ such that $\gamma_0, \gamma_1 \in K$ is a compact subset of $\Gamma(\mathcal{X})$ (the set of all geodesics in $\mathcal{X}$). So

$$\mathcal{X}_K := \Big\{\gamma_t;\ 0\le t\le 1;\ \gamma_0\in K,\ \gamma_1\in K\Big\}$$

is a compact set too, as the image of $\Gamma_K\times[0,1]$ by the continuous map $(\gamma,t)\to\gamma_t$.

For each $x \in \mathcal{X}_K$, we may find a small ball $B_r(x)$ such that the displacement convexity inequality defining $\mathrm{CD}(K,\infty)$ is satisfied for all displacement interpolations supported in $B_r(x)$; but also the displacement convexity inequality defining $\mathrm{CD}(K',N')$, for some $K'\in\mathbb{R}$, $N'<\infty$. (Both K' and N' will depend on x.) In particular, if $(\mu'_t)_{t_1\le t\le t_2}$ is a displacement interpolation supported in $B_r(x)$, with μ'_{t_1} absolutely continuous, then also μ'_t is absolutely continuous for all $t\in(t_1,t_2)$ (the proof is the same as for Theorem 30.19(ii)). By reasoning as in the proof of Theorem 30.37, one deduces that μ_t is absolutely continuous for all $t\in[0,1]$ if μ_0 and μ_1 are absolutely continuous.

Of course this is not yet sufficient to imply the finiteness of $H_\nu(\mu_t)$, but now we shall be able to reduce to this case by approximation. More precisely, we shall construct a sequence $(\Pi^{(k)})_{k\in\mathbb{N}}$ of dynamical optimal transference plans, such that

$$\Pi^{(k)} = \frac{\widehat{\Pi}^{(k)}}{Z^{(k)}},\qquad 0\le\widehat{\Pi}^{(k)}\le\Pi;\qquad Z^{(k)}\uparrow 1;\qquad Z^{(k)}\,\Pi^{(k)}\uparrow\Pi, \tag{30.68}$$

and

$$\forall k \in \mathbb{N}, \quad \forall j \in \mathbb{N}\ (j \leq 1/\delta), \qquad \sup \rho_{j\delta}^{(k)} < +\infty, \tag{30.69}$$

where the supremum really is an essential supremum, and $\rho_t^{(k)}$ is the density of $\mu_t^{(k)} = (e_t)_\# \Pi^{(k)}$ with respect to ν.

If we can do this, then by repeating the proof of Theorem 30.37 we shall obtain

$$H_\nu(\mu_t^{(k)}) \leq (1-t)\, H_{\pi_k,\nu}^{\beta_{1-t}^{(K,\infty)}}(\mu_0^{(k)}) + t\, H_{\check{\pi}_k,\nu}^{\beta_t^{(K,\infty)}}(\mu_1^{(k)}).$$

Then by monotonicity we may pass to the limit as $k \to \infty$ (as in the proof of Theorem 17.37, say) and deduce

$$H_\nu(\mu_t) \leq (1-t)\, H_{\pi,\nu}^{\beta_{1-t}^{(K,\infty)}}(\mu_0) + t\, H_{\check{\pi},\nu}^{\beta_t^{(K,\infty)}}(\mu_1). \tag{30.70}$$

Here μ_0 and μ_1 are assumed to be supported in a compact subset K of $\mathcal{Y}$. But then, by regularity of ν, we may introduce an increasing sequence of compact sets $(K_m)_{m\in\mathbb{N}}$ such that $\cup K_m = \mathcal{Y}$, up to a ν-negligible set. Observe that $\overline{\mathcal{Y}} = \mathcal{X}$ (otherwise $\operatorname{Spt}\nu$ would be included in $\overline{\mathcal{Y}}$ and strictly smaller than $\mathcal{X}$); that $\mathcal{X}\setminus\mathcal{Y}$ has zero measure; that any μ such that $H_\nu(\mu) < +\infty$ satisfies $\mu_s = 0$, so $H_\nu(\mu) = \int_{\mathcal{X}} \rho \log \rho\, d\nu = \int_{\mathcal{Y}} \rho \log\rho\, d\nu$. This makes it possible to run again a classical scheme to approximate any $\mu \in P_c(\mathcal{X})$ with $H_\nu(\mu) < +\infty$ by a sequence $(\mu_m)_{m\in\mathbb{N}}$, such that μ_m is supported in K_m, μ_m converges weakly to μ and $H_{\pi_m,\nu}^{\beta_{1-t}^{(K,\infty)}}$ converges to $H_{\pi,\nu}^{\beta_{1-t}^{(K,\infty)}}$ if $\pi_m \to \pi$. (Choose for instance $\mu_m = \chi_m\, \mu/(\int \chi_m\, d\mu)$, where χ_m is a cutoff function satisfying $0 \leq \chi_m \leq 1$, $\chi_m = 0$ outside K_{m+1}, $\chi_m = 1$ on K_m, and argue as in the proof of Theorem 30.5.) A limit argument will then establish (30.70) for any two compactly supported probability measures μ_0, μ_1.

So it all boils down to providing an approximation sequence $\Pi^{(k)}$ satisfying (30.68), (30.69). This is done in m (simple) steps as follows.

First approximate ρ_δ by a nondecreasing sequence of bounded densities: $0 \leq h_\delta^{k_1} \leq \rho_\delta$, $k_1 \in \mathbb{N}$, where each $h_\delta^{k_1}$ is bounded and $h_\delta^{k_1} \uparrow \rho_\delta$ as $k_1 \to \infty$. Define

$$\widehat{\Pi}^{k_1}(d\gamma) = (h_\delta^{k_1}\, \nu)(d\gamma_\delta)\, \Pi(d\gamma|\gamma_\delta),$$

where $\Pi(d\gamma|\gamma_t)$ stands for the conditional probability of γ, distributed according to Π and conditioned by its value at time t. Then let

$$Z^{k_1} = \widehat{\Pi}^{k_1}[\Gamma]; \qquad \Pi^{k_1} = \frac{\widehat{\Pi}^{k_1}}{Z^{k_1}}.$$

As k_1 goes to infinity, it is clear that $Z^{k_1} \uparrow 1$ (in particular, we may assume without loss of generality that $Z^{k_1} > 0$) and $Z^{k_1}\,\Pi^{k_1} \uparrow \Pi$. Moreover, if $\rho_t^{k_1}$ stands for the density of $(e_t)_\#\Pi^{k_1}$, then $\rho_\delta^{k_1} = (Z^{k_1})^{-1} h_\delta^{k_1}$ is bounded.

Now comes the second step: For each k_1, let $(h_{2\delta}^{k_1,k_2})_{k_2\in\mathbb{N}}$ be a nondecreasing sequence of bounded functions converging almost surely to $\rho_{2\delta}^{k_1}$ as $k_2 \to \infty$. Let

$$\widehat{\Pi}^{k_1,k_2}(d\gamma) = (h_{2\delta}^{k_1,k_2}\,\nu)(d\gamma_{2\delta})\,\Pi^{k_1}(d\gamma|\gamma_{2\delta}),$$

$$Z^{k_1,k_2} = \widehat{\Pi}^{k_1,k_2}[\Gamma], \qquad \Pi^{k_1,k_2} = \frac{\widehat{\Pi}^{k_1,k_2}}{Z^{k_1,k_2}},$$

and let $\rho_t^{k_1,k_2}$ stand for the density of $(e_t)_\#\Pi^{k_1,k_2}$. Then $\rho_\delta^{k_1,k_2} \le (Z^{k_1,k_2})^{-1}\rho_\delta^{k_1} = (Z^{k_1,k_2}Z^{k_1})^{-1}h_\delta^{k_1}$ and $\rho_{2\delta}^{k_1,k_2} = (Z^{k_1,k_2})^{-1}\,h_{2\delta}^{k_1,k_2}$ are both bounded.

Then repeat the process: If $\Pi^{k_1,\ldots,k_j}$ has been constructed for any $k_1,\ldots,k_j$ in $\mathbb{N}$, introduce a nonincreasing sequence $(h_{(j+1)\delta}^{k_1,\ldots,k_{j+1}})_{k_{j+1}\in\mathbb{N}}$ converging almost surely to $\rho_{(j+1)\delta}^{k_1,\ldots,k_j}$ as $k_{j+1} \to \infty$; define

$$\begin{aligned}
\widehat{\Pi}^{k_1,\ldots,k_{j+1}}(d\gamma) &= (h_{(j+1)\delta}^{k_1,\ldots,k_{j+1}}\,\nu)(d\gamma_{(j+1)\delta})\,\Pi^{k_1,\ldots,k_{j+1}}(d\gamma|\gamma_{(j+1)\delta}),\\
Z^{k_1,\ldots,k_{j+1}} &= \widehat{\Pi}^{k_1,\ldots,k_{j+1}}[\Gamma],\\
\Pi^{k_1,\ldots,k_{j+1}} &= \widehat{\Pi}^{k_1,\ldots,k_{j+1}}/Z^{k_1,\ldots,k_{j+1}},\\
\mu_t^{k_1,\ldots,k_{j+1}} &= (e_t)_\#\Pi^{k_1,\ldots,k_{j+1}}.
\end{aligned}$$

Then for any $t \in \{\delta, 2\delta, \ldots, (j+1)\delta\}$, the density $\rho_t^{k_1,\ldots,k_{j+1}}$ of $\mu_t^{k_1,\ldots,k_{j+1}}$ is bounded.

After m operations this process has constructed $\Pi^{k_1,\ldots,k_m}$ such that all densities $\rho_{j\delta}^{k_1,\ldots,k_m}$ are bounded. The proof is completed by choosing $\Pi^{(k)} = \Pi^{k,\ldots,k}$, $Z^{(k)} = Z^k \cdot Z^{k,k} \cdot \ldots \cdot Z^{k,\ldots,k}$. □

Appendix: Localization in measure spaces

In this Appendix I recall some basic facts about the use of cutoff functions to reduce to compact sets. Again, the natural setting is that of boundedly compact metric spaces, i.e. metric spaces where closed balls are compact.

Definition 30.44 (Cutoff functions). *Let $(\mathcal{X}, d)$ be a boundedly compact metric space, and let $\star$ be an arbitrary base point. For any $R > 0$, let B_R be the closed ball $B[\star, R]$. A $\star$-cutoff is a family of non-negative continuous functions $(\chi_R)_{R>0}$ such that $1_{B_R} \le \chi_R \le 1_{B_{R+1}}$ for all R.*

More explicitly: χ_R is valued in $[0,1]$, and χ_R is identically equal to 1 on B_R, identically equal to 0 on B_{R+1}.

The existence of a $\star$-cutoff follows from Urysohn's lemma.

If μ is any finite measure on $\mathcal{X}$, then $\chi_R\,\mu$ converges to μ in total variation norm; moreover, for any $R > 0$, the truncation operator $T_R : \mu \to \chi_R\,\mu$ is a (nonstrict) contraction. As a particular case, if ν is any measure on $\mathcal{X}$, and $f \in L^1(\mathcal{X}, \nu)$, then $\chi_R\, f$ converges to f in $L^1(\nu)$.

A consequence is the density of $C_c(\mathcal{X})$ in $L^1(\mathcal{X}, \nu)$, as soon as ν is locally finite. Indeed, if f is given in $L^1(\mathcal{X}, \nu)$, first choose R such that $\|f\|_{L^1(\mathcal{X}\setminus B_R,\nu)} \le \delta$; then pick up some $g \in C(B_{R+1})$ such that $\|f - g\|_{L^1(B_{R+1},\nu)} \le \delta$. (Since B_{R+1} is compact, this can be done with Lusin's theorem.) Finally define $\widetilde{g} := g\,\chi_R$, extended by 0 outside of B_R: then $\widetilde{g}$ is a continuous function with compact support, and it is easy to check that $\|f - \widetilde{g}\|_{L^1(\mathcal{X},\nu)} \le 2\delta$.

Bibliographical notes

Most of the material in this chapter comes from papers by Lott and myself [577, 578, 579] and by Sturm [762, 763]. Some of the results are new. Prior to these references, there had been an important series of papers by Cheeger and Colding [227, 228, 229, 230], with a follow-up by Ding [303], about the structure of measured Gromov–Hausdorff limits of sequences of Riemannian manifolds satisfying a uniform $\mathrm{CD}(K, N)$ bound. Some of the results by Cheeger and Colding can be re-interpreted in the present framework, but for many other theorems this remains open: Examples are the generalized splitting theorem [227, Theorem 6.64], the theorem of mutual absolute continuity of admissible reference measures [230, Theorem 4.17], or the theorem of continuity of the volume in absence of collapsing [228, Theorem 5.9]. (A positive answer to the problem raised in Remark 30.16 would solve the latter issue.)

Theorem 30.2 is taken from work by Lott and myself [577] as well as Corollary 30.9, Theorems 30.22 and 30.23, and the first part of Theorem 30.28. Theorem 30.7, Corollary 30.10, Theorems 30.11 and 30.17 are due to Sturm [762, 763]. Part (i) of Theorem 30.19 was proven by Lott and myself in the case $K = 0$. Part (ii) follows a scheme of proof communicated to me by Sturm. In a Euclidean context, Theorem 30.20 is well-known to specialists and used in several recent works about optimal transport; I don't know who first made this observation.

The Poincaré inequalities appearing in Theorems 30.25 and 30.26 (in the case $K = 0$) are due to Lott and myself [578]. The concept of upper gradient was put forward by Heinonen and Koskela [470] and other authors; it played a key role in Cheeger's construction [226] of a differentiable structure on metric spaces satisfying a doubling condition and a local Poincaré inequality. Independently of [578], there were several simultaneous treatments of local Poincaré inequalities under weak $CD(K, N)$ conditions, by Sturm [763] on the one hand, and von Renesse [825] on the other. The proofs in all of these works have many common points, and also common features with the proof by Cheeger and Colding [230]. But the argument by Cheeger and Colding uses another inequality called the "segment inequality" [227, Theorem 2.11], which as far as I know has not been adapted to the context of metric-measure spaces. In [578] on the other hand we used the concept of "democratic condition", as in Theorem 19.13.

All these notions (possibly coupled with a doubling condition) are stable under the Gromov–Hausdorff limit: this was checked in [226, 510, 529] for the Poincaré inequality, in [230] for the segment inequality, and in [578] for the democratic condition.

Theorem 30.28(ii) is due to Lott and myself [579]; it uses Proposition 30.30 with $L(d) = d^2/2$. In this particular case (and under a nonessential compactness assumption), a complete proof of Proposition 30.30 can be found in [579]. It is also shown there that the conclusions of Proposition 22.16 all remain true if $(\mathcal{X}, d)$ is a finite-dimensional Alexandrov space with curvature bounded below; this is a *pointwise* property, as opposed to the "almost everywhere" statement appearing in Proposition 30.30(vii). As for the proof of this almost everywhere result, it is based on Cheeger's generalized Rademacher theorem [226, Theorem 10.2]. (The argument is written down in [579] only for the case $L(d) = d^2/2$ but this is no big deal.)

An even more general proof of Theorem 30.28(ii) was given by Gozlan [429].

As a consequence of the doubling and local Poincaré inequalities, a weak $\mathrm{CD}(K, N)$ space with $N < \infty$ automatically has some regularity (a differentiable structure defined almost everywhere); see again [226]. For Alexandrov spaces, such "automatic regularity" results have been obtained in [665, 677]; see [174, Chapter 10] for a survey.

Inequality (30.32) was proven by Lott and myself in [577], at a time when we did not have the general definition of weak $\mathrm{CD}(K, N)$ spaces. The argument is inspired by previous works of Otto and myself [671, Theorem 4], and Ledoux [544]. It might still have some interest, since there is no reason why $\mathrm{CD}(0, N)$ and $\mathrm{CD}(K, \infty)$ together should imply $\mathrm{CD}(K, N)$.

As discussed in [654, 656, 763], several of the inequalities established in the present chapter (and elsewhere) from the $\mathrm{CD}(K, N)$ property also follow from the measure-contraction property $\mathrm{MCP}(K, N)$, which, at least in nonbranching spaces, is weaker than $\mathrm{CD}(K, N)$. (But $\mathrm{MCP}(K, N)$ cannot be directly related to Ricci curvature bounds, as explained in the bibliographical notes of Chapter 29.) Building on previous works such as [226, 538, 539, 459, 760], Sturm [763, Section 7] discussed other consequences of $\mathrm{MCP}(K, N)$, under the additional assumption that $\lim_{r \to 0} \nu[B_r(x)]/r^N$ is bounded. These consequences include results on Dirichlet forms, Sobolev spaces, Harnack inequalities, Hölder continuity of harmonic functions, Gaussian estimates for heat kernels.

I proved Theorem 30.32 specifically for these notes, but a very close statement was also obtained shortly after and independently by Sturm [763, Proposition IV.2], at least for absolutely continuous measures. Sturm's proof is different from mine, although many common ingredients can be recognized.

The treatment of singular measures in Theorem 30.32 (Step 4 of the proof) grew out of a joint work of mine with Figalli [372]; there we proved Theorem 30.32 (more precisely, the parts which were not proven in [577]) in smooth Riemannian manifolds. The proof in [372] is slightly different from the one which I gave here; it uses Lemma 29.7. An alternative "Eulerian" approach to displacement convexity for singular measures was implemented by Daneri and Savaré [271].

In Alexandrov spaces, the locality of the notion "curvature is bounded below by κ" is called **Toponogov's theorem**; in full generality it is due to Perelman [175]. A proof can be found in [174, Theorem 10.3.1], along with bibliographical comments.

The conditional locality of $\mathrm{CD}(K,\infty)$ in nonbranching spaces (Theorem 30.42) was first proven by Sturm [762, Theorem 4.17], with a different argument than the one used here. Sturm does not make any assumption about infinite-dimensional points, but he assumes that the space of probability measures μ with $H_\nu(\mu) < +\infty$ is geodesically convex. It is clear that the proof of Theorem 30.42 can be adapted and simplified to cover this assumption. Theorem 30.37 is new as far as I know. Example 30.41 was suggested to me by Lott.

When one restricts to $\lambda = 1/2$, Conjecture 30.34 takes a simpler form, and at least seems to be true for all θ outside $(0, 1)$; but of course we are interested precisely in the range $\theta \in (0, 1)$. I once hoped to prove Conjecture 30.34 by reinterpreting it as the locality of the $\mathrm{CD}(K, N)$ property in 1-dimensional spaces, and classifying 1-dimensional local weak $\mathrm{CD}(K, N)$ spaces; but I did not manage to get things to work.

Natural geometric questions, related to the locality problem, are the stability of $\mathrm{CD}(K, N)$ under *quotient* by Lie group action and *lifting* to the universal covering. I shall briefly discuss what is known about these issues.

- About the quotient problem, there are some results. In [577, Section 5.5], Lott and I proved that the quotient of a $\mathrm{CD}(K, N)$ metric-measure space $\mathcal{X}$ by a Lie group of isometries G is itself $\mathrm{CD}(K, N)$, under the assumptions that (a) $\mathcal{X}$ and G are compact; (b) $K = 0$ or $N = \infty$; (c) any two absolutely continuous probability measures on $\mathcal{X}$ are joined by a unique displacement interpolation which is absolutely continuous for all times. The definition of $\mathrm{CD}(K, \infty)$ which was used in [577] is not exactly the same as in these notes, but Theorem 30.32 guarantees that there is no difference if $\mathcal{X}$ is nonbranching. Assumption (c) was used only to make sure that any displacement interpolation between absolutely continuous probability measures would satisfy the displacement interpolation inequalities which are characteristic of $\mathrm{CD}(0, N)$; but Theorem 30.32 ensures that this is the case in nonbranching $\mathrm{CD}(0, N)$ spaces, so the proof would go through if Assumption (c) were replaced by just a nonbranching property. Assumption (b) is probably easy to remove. On the other hand, relaxing Assumption (a) does not seem trivial at all, and would require more thinking.

• About the lifting problem, one might first think that it follows from the locality, as stated for instance in Theorems 30.37 or 30.42. But even in situations where $\mathrm{CD}(K, N)$ has been shown to be local (say $N = 0$ and $\mathcal{X}$ is nonbranching), the existence of the universal covering is not obvious. Abstract topology shows that the existence of the universal covering of $\mathcal{X}$ is equivalent to $\mathcal{X}$ being semi-locally simply connected ("délaçable" in the terminology of Bourbaki). This property is satisfied if $\mathcal{X}$ is locally contractible, that is, every point x has a neighborhood which can be contracted into x. For instance, an Alexandrov space with curvature bounded below is locally contractible because any point x has a neighborhood which is homeomorphic to the tangent cone at x; no such theory is known for weak $\mathrm{CD}(K, N)$ spaces. (All this was explained to me by Lott.)

Here are some technical notes to conclude.

An introduction to Hausdorff measure and Hausdorff dimension can be found, e.g., in Falconer's broad-audience books [337, 338].

The Dunford–Pettis theorem provides a sufficient condition for uniform equi-integrability: If a family $\mathcal{F} \subset L^1(\nu)$ is weakly sequentially compact in $L^1(\nu)$, then there exists a function $\Psi : \mathbb{R}_+ \to \mathbb{R}_+$ such that $\Psi(r)/r \to +\infty$ as $r \to \infty$ and $\sup_{f\in\mathcal{F}} \int \Psi(f)\, d\nu < +\infty$. A proof can be found, e.g., in [177, Theorem 2.12] (there the theorem is stated in $\mathbb{R}^n$ but the proof is the same in a more general space), or in my own course on integration [819, Section VII-5].

Urysohn's lemma [318, Theorem 2.6.3] states the following: If $(\mathcal{X}, d)$ is a locally compact metric space (or even just a locally compact Hausdorff space), K is a compact subset of $\mathcal{X}$ and O is an open subset of $\mathcal{X}$ with $K \subset O$, then there is $f \in C_c(\mathcal{X})$ with $1_K \le f \le 1_O$.

Analysis on metric spaces (in terms of regularity, Sobolev spaces, etc.) has undergone rapid development in the past ten years, after the pioneering works by Hajłasz and Koskela [459, 461] and others. Among dozens and dozens of papers, I shall only quote two reference books [37, 469] and a survey paper [460]; see also the bibliographical notes of Chapter 26 for more references about analysis in Alexandrov spaces. The thesis developed in the present course is that optimal transport has suddenly become an important actor in this theory.

Conclusions and open problems

In these notes I have tried to present a consistent picture of the theory of optimal transport, with a *dynamical*, *probabilistic* and *geometric* point of view, insisting on the notions of displacement interpolation, probabilistic representation, and curvature effects.

The qualitative description of optimal transport, developed in Part I, now seems to be more or less under control. Even the smoothness of the transport map in curved geometries starts to be better understood, thanks in particular to the recent works of Grégoire Loeper, Xinan Ma, Neil Trudinger and Xu-Jia Wang which were described in Chapter 12. Among issues which seem to be of interest I shall mention:

• find relevant examples of cost functions with nonnegative, or positive c-curvature (Definition 12.27), and theorems guaranteeing that the optimal transport does not approach singularities of the cost function — so that the smoothness of the transport map can be established;

• get a precise description of the singularities of the optimal transport map when the latter is not smooth;

• further analyze the displacement interpolation on singular spaces, maybe via nonsmooth generalizations of Mather's estimates (as in Open Problem 8.21).

For the applications of optimal transport to Riemannian geometry, a consistent picture is also emerging, as I have tried to show in Part II. The main regularity problems seem to be under control here, but there remain several challenging "structural" problems:

• How can one best understand the relation between plain displacement convexity and distorted displacement convexity, as described in Chapter 17? Is there an Eulerian counterpart of the latter concept? See Open Problems 17.38 and 17.39 for more precise formulations.

• Optimal transport seems to work well to establish sharp geometric inequalities when the "natural dimension of the inequality" coincides with the dimension bound; on the other hand, so far it has failed to establish for instance sharp logarithmic Sobolev or Talagrand inequalities (infinite-dimensional) under a $\mathrm{CD}(K, N)$ condition for $N < \infty$ (Open Problems 21.6 and 22.44). The sharp L^2-Sobolev inequality (21.9) has also escaped investigations based on optimal transport (Open Problems 21.11). Can one find a more precise strategy to attack such problems by a displacement convexity approach? A seemingly closely related question is whether one can mimick (maybe by changes of unknowns in the transport problem?) the changes of variables in the Γ_2 formalism, which are often at the basis of the derivation of such sharp inequalities, as in the recent papers of Jérôme Demange. To add to the confusion, the mysterious structure condition (25.10) has popped out in these works; it is natural to ask whether this condition has any interpretation in terms of optimal transport.

• Are there interesting **examples of displacement convex functionals** apart from the ones that have already been explored during the past ten years — basically all of the form $\int_M U(\rho)\,d\nu + \int_{M^k} V\,d\mu^{\otimes k}$? It is frustrating that so few examples of displacement convex functionals are known, in contrast with the enormous amount of plainly convex functionals that one can construct. Open Problem 15.11 might be related to this question.

• Is there a transport-based proof of the famous **Lévy–Gromov isoperimetric inequalities** (Open Problem 21.16), that would not involve so much "hard analysis" as the currently known arguments? Besides its intrinsic interest, such a proof could hopefully be adapted to nonsmooth spaces such as the weak $\mathrm{CD}(K,N)$ spaces studied in Part III.

• Caffarelli's **log concave perturbation theorem** (alluded to in Chapter 2) is another riddle in the picture. The Gaussian space can be seen as the infinite-dimensional version of the sphere, which is the Riemannian "reference space" with positive constant (sectional) curvature; and the space $\mathbb{R}^n$ equipped with a log concave measure is a space of nonnegative Ricci curvature. So Caffarelli's theorem can be restated as follows: *If the Euclidean space* $(\mathbb{R}^n, d_2)$ *is equipped with a probability measure* ν *that makes it a* $\mathrm{CD}(K,\infty)$ *space, then* ν *can be realized as a 1-Lipschitz push-forward of the reference Gaussian measure with curvature* K. This implies almost obviously that isoperimetric inequalities in $(\mathbb{R}^n, d_2, \nu)$ are not worse than isoperimetric inequalities in the Gaussian space; so there is a strong analogy between Caffarelli's theorem on the one hand, and the Lévy–Gromov isoperimetric inequality on the other hand. It is natural to ask whether there is a common framework for both results; this does not seem obvious at all, and I have not been able to formulate even a decent guess of what could be a geometric generalization of Caffarelli's theorem.

• Another important remark is that the geometric theory has been almost exclusively developed in the case of the optimal transport with *quadratic* cost function; the exponent $p = 2$ here is natural in the context of Riemannian geometry, but working with other exponents (or with radically different Lagrangian cost functions) might lead to new geometric territories. An illustration is provided by the recent work of Shin-ichi Ohta in Finsler geometry. A related question is Open Problem 15.12.

In Part III of these notes, I discussed the emerging theory of weak Ricci curvature lower bounds in metric-measure spaces, based on displacement convexity inequalities. The theory has grown very fast and is starting to be rather well-developed; however, some challenging issues remain to be solved before one can consider it as mature. Here are three missing pieces of the puzzle:

• A **globalization theorem** that would play the role of the Toponogov–Perelman theorem for Alexandrov spaces with a lower bound on the curvature. This result should state that a weak local $\mathrm{CD}(K,N)$ space is automatically a weak $\mathrm{CD}(K,N)$ space. Theorem 30.37 shows that this is true at least if $K=0$, $N<\infty$ and $\mathcal{X}$ is nonbranching; if Conjecture 30.34 turns out to be true, the same result will be available for all values of K.

• The **compatibility with the theory of Alexandrov spaces** (with lower curvature bounds). Alexandrov spaces have proven their flexibility and have gained a lot of popularity among geometers. Since Alexandrov bounds are weak sectional curvature bounds, they should in principle be able to control weak Ricci curvature bounds. The natural question here can be stated as follows: Let $(\mathcal{X},d)$ be a finite-dimensional Alexandrov space with dimension n and curvature bounded below by κ, and let $\mathcal{H}^n$ be the n-dimensional Hausdorff measure on $\mathcal{X}$; is $(\mathcal{X},d,\mathcal{H}^n)$ a weak $\mathrm{CD}((n-1)\kappa,n)$ space?

• A thorough discussion of the **branching** problem: Find examples of weak $\mathrm{CD}(K,N)$ spaces that are branching; that are singular but nonbranching; identify simple regularity conditions that prevent branching; etc. It is also of interest to enquire whether the nonbranching assumption can be dispensed with in Theorems 30.26 and 30.37 (recall Remarks 30.27 and 30.39).

More generally, we would like to have more information about the structure of weak $\mathrm{CD}(K,N)$ spaces, at least when N is finite. It is known from the work of Jeff Cheeger and others that metric-measures spaces in which the measure is (locally) doubling and satisfies a (local) Poincaré inequality have at least some little bit of regularity: There is a tangent space defined almost everywhere, varying in a measurable way.

In the context of Alexandrov spaces with curvature bounded below, some rather strong structure theorems have been established by Grigori Perelman and others; it is natural to ask whether similar results hold true for weak $\mathrm{CD}(K,N)$ spaces.

Another relevant problem is to check the compatibility of the $\mathrm{CD}(K,N)$ condition with the operations of quotient by Lie group actions, and lifting to the universal covering. As explained in the bibliographical notes of Chapter 30, only partial results are known in these directions.

Besides these issues, it seems important to find further *examples* of weak $\mathrm{CD}(K,N)$ spaces, apart from the ones presented in Chapter 29, mostly constructed as limits or quotients of manifolds. It was realized in a recent Oberwolfach meeting, as a consequence of discussions between Dario Cordero-Erausquin, Karl-Theodor Sturm and myself, that the Euclidean space $\mathbb{R}^n$, equipped with *any* norm $\|\cdot\|$, is a weak $\mathrm{CD}(0,n)$ space:

Theorem. *Let $\|\cdot\|$ be a norm on $\mathbb{R}^n$ (considered as a distance on $\mathbb{R}^n \times \mathbb{R}^n$), and let λ_n be the n-dimensional Lebesgue measure. Then the metric-measure space $(\mathbb{R}^n, \|\cdot\|, \lambda_n)$ is a weak* $\mathrm{CD}(0,n)$ *space in the sense of Definition 29.8.*

I did not include this theorem in the body of these notes, because it appeals to some results that have not yet been adapted to a genuinely geometric context, and which I preferred not to discuss. I shall sketch the proof at the end of this text, but first I would like to explain why this result is at the same time motivating, and a bit shocking:

(a) As pointed out to me by John Lott, if $\|\cdot\|$ is not Euclidean, then the metric-measure space $(\mathbb{R}^n, \|\cdot\|, \lambda_n)$ cannot be realized as a limit of smooth Riemannian manifolds with a uniform $\mathrm{CD}(0,N)$ bound, because it fails to satisfy the *splitting principle*. (If a nonnegatively curved space admits a line, i.e. a geodesic parametrized by $\mathbb{R}$, then the space can be "factorized" by this geodesic.) Results by Jeff Cheeger, Toby Colding and Detlef Gromoll say that the splitting principle holds for $\mathrm{CD}(0,N)$ manifolds and their measured Gromov–Hausdorff limits.

(b) If $\|\cdot\|$ is not the Euclidean norm, the resulting metric space is very singular in certain respects: It is in general not an Alexandrov space, and it can be extremely branching. For instance, if $\|\cdot\|$ is the ℓ_∞ norm, then any two distinct points are joined by an uncountable infinity of geodesics. Since $(\mathbb{R}^n, \|\cdot\|_{\ell_\infty}, \lambda_n)$ is the (pointed) limit of the nonbranching spaces $(\mathbb{R}^n, \|\cdot\|_{\ell_p}, \lambda_n)$ as $p \to \infty$, we also realize that weak $\mathrm{CD}(K,N)$ bounds do not prevent the appearance of branching in measured Gromov-Hausdorff limits, at least if $K \leq 0$.

On the other hand, the study of optimal Sobolev inequalities in $\mathbb{R}^n$ which I performed together with Bruno Nazaret and Dario Cordero-Erausquin shows that optimal Sobolev inequalities basically do not depend on the choice of the norm on $\mathbb{R}^n$. In a Riemannian context, Sobolev inequalities strongly depend on Ricci curvature bounds; so, our result suggests that it is not absurd to decide that $\mathbb{R}^n$ is a weak $\mathrm{CD}(0,n)$ space independently of the norm. Shin-ichi Ohta has developed this point of view by studying curvature-dimension conditions in certain classes of Finsler spaces.

One can also ask whether there are additional regularity conditions that might be added to the definition of weak $\mathrm{CD}(K,N)$ space, in order to enforce nonbranching, or the splitting principle, or both, and in particular rule out non-Euclidean norms.

As a side consequence of point (a) above, we realize that smooth $\mathrm{CD}(K,N)$ manifolds are *not* dense in the spaces $\mathcal{CDD}(K,N,D,m,M)$ introduced in Theorem 29.32.

The interpretation of dissipative equations as gradient flows with respect to optimal transport, and the theory reviewed in Chapters 23 to 25, also lead to fascinating issues that are relevant in smooth or nonsmooth geometry as well as in partial differential equations. For instance,

(a) Can one define a reasonably well-behaved **heat flow** on weak $\mathrm{CD}(K,N)$ spaces by taking the gradient flow for Boltzmann's H functional? The theory of gradient flows in abstract metric spaces has been pushed very far, in particular by Luigi Ambrosio, Giuseppe Savaré and collaborators; so it might not be so difficult to define an object that would play the role of a heat semigroup. But this will be of limited value unless one can prove relevant theorems about this object.

Shin-ichi Ohta, and independently Giuseppe Savaré, recently made progress in this direction by constructing gradient flows in the Wasserstein space over a finite-dimensional Alexandrov space of curvature bounded below, or over more general spaces satisfying very weak regularity assumptions expressed in terms of distances and angles. In the particular case when the energy functional is the Boltzmann entropy, this provides a natural notion of heat equation and heat semigroup.

Savaré uses a very elegant argument, based on properties of Wasserstein distances and entropy, to prove the linearity of this semigroup, and other properties as well (positivity, contraction in W_p for $1 \le p \le 2$, contraction in L^p, some regularizing effect).

This problem might be related to the possibility of defining a Laplace operator on a singular space, an issue which has been addressed in particular by Jeff Cheeger and Toby Colding, for limits of Riemannian manifolds. However, their construction is strongly based on regularity properties enjoyed by such limits, and breaks down, e.g., for $\mathbb{R}^n$ equipped with a non-Euclidean norm $\|\cdot\|$. In fact, as noted by Karl-Theodor Sturm, the gradient flow of the H functional in $P_2((\mathbb{R}^N, \|\cdot\|))$ yields a *nonlinear* evolution. The fact that this equation has the same fundamental solution (in the sense of a solution evolving from a Dirac mass) as the Euclidean one is one argument to believe that this is a natural notion of heat equation on the non-Euclidean $\mathbb{R}^N$ — and may reinforce us in the conviction that this space deserves its status of weak $\mathrm{CD}(0, N)$ space.

(b) Can one extend the theory of dissipative equations to other equations, which are of Hamiltonian, or, even more interestingly, of dissipative Hamiltonian nature? As explained in the bibliographical notes of Chapter 23, there has been some recent work in that direction by Luigi Ambrosio, Wilfrid Gangbo and others, however the situation is still far from clear.

A loosely related issue is the study of the semi-geostrophic system, which in the simplest situations can formally be written as a Hamiltonian flow, where the Hamiltonian function is the square Wasserstein distance with respect to some uniform reference measure. I think that the rigorous qualitative understanding of the semi-geostrophic system is one of the most exciting problems that I am aware of in theoretical fluid mechanics; and discussions with Mike Cullen convinced me that it is very relevant in applications to meteorology. Although the theory of the semi-geostrophic system is still full of fundamental open problems, enough has already been written on it to make the substance of a complete monograph.

On a much more theoretical level, the geometric understanding of the Wasserstein space $P_2(\mathcal{X})$, where $\mathcal{X}$ is a Riemannian manifold or just a geodesic space, has been the object of several recent studies, and still retains many mysteries. For instance, there is a neat statement according to which $P_2(\mathcal{X})$ is nonnegatively curved, in the sense of Alexandrov, if and only if $\mathcal{X}$ itself is nonnegatively curved. But there is no similar statement for nonzero lower bounds on the curvature! In fact, if x is a point of negative curvature, then the curvature of $P_2(\mathcal{X})$ seems to be unbounded in both directions ($+\infty$ and $-\infty$) in the neighborhood of δ_x. Also it is not clear what exactly is "the right" structure on, say, $P_2(\mathbb{R}^n)$; recent works on the subject have suggested differing answers. Another relevant open problem is whether there is a natural "volume" measure on $P_2(M)$. Karl-Theodor Sturm and Max-Kostja von Renesse have recently managed to construct a natural one-parameter family of "Gibbs" probability measures on $P_2(S^1)$. A multi-dimensional generalization would be of great interest.

In their book on gradient flows, Luigi Ambrosio, Nicola Gigli and Giuseppe Savaré make an intriguing observation: One may define "generalized geodesics" in $P_2(\mathbb{R}^n)$ by considering the law of $(1-t)\,X_0+t\,X_1$, where (X_0,Z) and (X_1,Z) are optimal couplings. These generalized geodesics have intriguing properties: For instance, they still satisfy the characteristic displacement interpolation inequalities; and they provide curves of "nonpositive curvature", that can be exploited for various purposes, such as error estimates for approximate gradient flow schemes. It is natural to further investigate the properties of these objects, which are reminiscent of the c-segments considered in Chapter 12.

The list above provides but a sample among the many problems that remain open in the theory of optimal transport.

As I already mentioned in the preface, another crucial issue which I did not address at all is the numerical analysis of optimal transport. This topic also has a long and complex history, with some famous schemes such as the old simplex algorithm, described for instance in Alexander Schrijver's monograph *Combinatorial Optimization: Polyhedra and Efficiency*; or the more recent auction algorithm developed by Dimitri Bertsekas. Numerical schemes based on Monge–Ampère equations have been suggested, but hardly implemented yet. Recent works by Uriel Frisch and collaborators in cosmology provide an example where one would like to efficiently solve the optimal transport problem with *huge* sets of data. To add to the variety of methods, continuous schemes based on partial differential equations have been making their way lately. All in all, this subject certainly deserves a systematic study on its own, with experiments, comparisons of algorithms, benchmark problems and so forth. By the way, the optimum matching problem is one of the topics that Donald Knuth has planned to address in his long-awaited Volume 4 of *The Art of Computer Programming.*

Needless to say, the theory might also decide to explore new horizons which I am unable to foresee.

Sketch of proof of the Theorem. First consider the case when $N = \|\cdot\|$ is a uniformly convex, smooth norm, in the sense that

$$\lambda I_n \leq \nabla^2 N^2 \leq \Lambda I_n$$

for some positive constants λ and Λ. Then the cost function $c(x,y) = N(x-y)^2$ is both strictly convex and $C^{1,1}$, i.e. uniformly semiconcave. This makes it possible to apply Theorem 10.28 (recall Example 10.35) and deduce the following theorem about the structure of optimal maps: If μ_0 and μ_1 are compactly supported and absolutely continuous, then there is a unique optimal transport, and it takes the form

$$T(x) = x - \nabla(N^2)^*(-\nabla\psi(x)), \qquad \psi \text{ a } c\text{-convex function.}$$

Since the norm is uniformly convex, geodesic lines are just straight lines; so the displacement interpolation takes the form $(T_t)_\#(\rho_0\,\lambda_n)$, where

$$T_t(x) = x - t\,\nabla(N^2)^*(-\nabla\psi(x)) \qquad 0 \leq t \leq 1.$$

Let $\theta(x) = \nabla(N^2)^*(-\nabla\psi(x))$. By [814, Remark 2.56], the Jacobian matrix $\nabla\theta$, although not symmetric, is pointwise diagonalizable, with eigenvalues bounded above by 1 (this remark goes back at least to a 1996 preprint by Otto [666, Proposition A.4]; a more general statement is in [30, Theorem 6.2.7]). It follows easily that $t \to \det(I_n - t\nabla\theta)^{1/n}$ is a concave function of t [814, Lemma 5.21], and one can reproduce the proof of displacement convexity for U_{λ_n}, as soon as $U \in \mathcal{DC}_n$ [814, Theorem 5.15 (i)].

This shows that $(\mathbb{R}^n, N, \lambda_n)$ satisfies the $\mathrm{CD}(0,n)$ displacement convexity inequalities when N is a smooth uniformly convex norm. Now if N is arbitrary, it can be approximated by a sequence $(N_k)_{k\in\mathbb{N}}$ of smooth uniformly convex norms, in such a way that $(\mathbb{R}^n, N, \lambda_n, 0)$ is the pointed measured Gromov–Hausdorff limit of $(\mathbb{R}^n, N_k, \lambda_n, 0)$ as $k \to \infty$. Then the general conclusion follows by stability of the weak $\mathrm{CD}(0,n)$ criterion (Theorem 29.24). □

Remark. In the above argument the spaces $(\mathbb{R}^n, N_k, \lambda_n)$ satisfy the property that the displacement interpolation between any two absolutely continuous, compactly supported probability measures is unique, while the limit space $(\mathbb{R}^n, N, \lambda_n)$ does not necessarily satisfy this property. For instance, if $N = \|\cdot\|_{\ell_\infty}$, there is an enormous number of displacement interpolations between two given probability measures; and most of them do not satisfy the displacement convexity inequalities that are used to define $\mathrm{CD}(0, n)$ bounds. This shows that if in Definition 29.8 one requires the inequality (29.11) to hold true for *any* Wasserstein geodesic, rather than for *some* Wasserstein geodesic, then the resulting $\mathrm{CD}(K, N)$ property is not stable under measured Gromov–Hausdorff convergence.

References

1. Abdellaoui, T., and Heinich, H. Sur la distance de deux lois dans le cas vectoriel. *C. R. Acad. Sci. Paris Sér. I Math. 319*, 4 (1994), 397–400.
2. Abdellaoui, T., and Heinich, H. Caractérisation d'une solution optimale au problème de Monge–Kantorovitch. *Bull. Soc. Math. France 127*, 3 (1999), 429–443.
3. Agrachev, A., and Lee, P. Optimal transportation under nonholonomic constraints. To appear in *Trans. Amer. Math. Soc.*
4. Agueh, M. Existence of solutions to degenerate parabolic equations via the Monge–Kantorovich theory. *Adv. Differential Equations 10*, 3 (2005), 309–360.
5. Agueh, M., Ghoussoub, N., and Kang, X. Geometric inequalities via a general comparison principle for interacting gases. *Geom. Funct. Anal. 14*, 1 (2004), 215–244.
6. Ahmad, N. The geometry of shape recognition via the Monge–Kantorovich optimal transport problem. PhD thesis, Univ. Toronto, 2004.
7. Aida, S. Uniform positivity improving property, Sobolev inequalities, and spectral gaps. *J. Funct. Anal. 158*, 1 (1998), 152–185.
8. Ajtai, M., Komlós, J., and Tusnády, G. On optimal matchings. *Combinatorica 4*, 4 (1984), 259–264.
9. Alberti, G. On the structure of singular sets of convex functions. *Calc. Var. Partial Differential Equations 2*, 1 (1994), 17–27.
10. Alberti, G. Some remarks about a notion of rearrangement. *Ann. Scuola Norm. Sup. Pisa Cl. Sci (4) 29*, 2 (2000), 457–472.
11. Alberti, G., and Ambrosio, L. A geometrical approach to monotone functions in $\mathbb{R}^n$. *Math. Z. 230*, 2 (1999), 259–316.
12. Alberti, G., Ambrosio, L., and Cannarsa, P. On the singularities of convex functions. *Manuscripta Math. 76*, 3-4 (1992), 421–435.
13. Alberti, G., and Bellettini, G. A nonlocal anisotropic model for phase transitions. I: The optimal profile problem. *Math. Ann. 310*, 3 (1998), 527–560.
14. Albeverio, S., and Cruzeiro, A. B. Global flows with invariant (Gibbs) measures for Euler and Navier–Stokes two-dimensional fluids. *Comm. Math. Phys. 129*, 3 (1990), 431–444.
15. Alesker, S., Dar, S., and Milman, V. D. A remarkable measure preserving diffeomorphism between two convex bodies in $\mathbf{R}^n$. *Geom. Dedicata 74*, 2 (1999), 201–212.

16. Alexandrov, A. D. Existence and uniqueness of a convex surface with given integral curvature. *Dokl. Akad. Nauk. SSSR 35* (1942), 131–134.
17. Alexandrov, A. D. Smoothness of a convex surface of bounded Gaussian curvature. *Dokl. Akad. Nauk. SSSR 36* (1942), 195–199.
18. Aliprantis, C. D., and Border, K. C. *Infinite dimensional analysis. A hithchiker's guide.* Third edition. Springer, Berlin, 2006.
19. Ambrosio, L. Minimizing movements. *Rend. Accad. Naz. Sci. XL Mem. Mat. Appl. (5) 19* (1995), 191–246.
20. Ambrosio, L. Lecture notes on optimal transport problems. In *Mathematical aspects of evolving interfaces (Funchal, 2000)*, vol. 1812 of *Lecture Notes in Math.* Springer, Berlin, 2003, pp. 1–52.
21. Ambrosio, L. Transport equation and Cauchy problem for BV vector fields. *Invent. Math. 158*, 2 (2004), 227–260.
22. Ambrosio, L. Steepest descent flows and applications to spaces of probability measures. In *Recent trends in partial differential equations*, vol. 409 of *Contemp. Math.*, Amer. Math. Soc., Providence, RI, 2006, pp. 1–32.
23. Ambrosio, L. Disintegrations of Gaussian measures. Personal communication, 2008.
24. Ambrosio, L., and Figalli, A. Geodesics in the space of measure-preserving maps and plans. To appear in *Arch. Ration. Mech. Anal.*
25. Ambrosio, L., and Figalli, A. On the regularity of the pressure field of Brenier's weak solutions to incompressible Euler equations. To appear in *Calc. Var. Partial Differential Equations.*
26. Ambrosio, L., Fusco, N., and Pallara, D. *Functions of bounded variation and free discontinuity problems.* Oxford Mathematical Monographs. The Clarendon Press Oxford University Press, New York, 2000.
27. Ambrosio, L., and Gangbo, W. Hamiltonian flows in $P_2(\mathbb{R}^{2d})$. Report at the MSRI Optimal Transport meeting, Berkeley, November 2005.
28. Ambrosio, L., and Gangbo, W. Hamiltonian ODE's in the Wasserstein space of probability measures. *Comm. Pure Appl. Math. 51* (2007), 18–53.
29. Ambrosio, L., and Gigli, N. Construction of the parallel transport in the Wasserstein space. To appear in *Meth. Appl. Anal.*
30. Ambrosio, L., Gigli, N., and Savaré, G. *Gradient flows in metric spaces and in the space of probability measures.* Lectures in Mathematics ETH Zürich. Birkhäuser Verlag, Basel, 2005.
31. Ambrosio, L., Kirchheim, B., and Pratelli, A. Existence of optimal transport maps for crystalline norms. *Duke Math. J. 125*, 2 (2004), 207–241.
32. Ambrosio, L., and Pratelli, A. Existence and stability results in the L^1 theory of optimal transportation. In *Optimal transportation and applications (Martina Franca, 2001)*, vol. 1813 of *Lecture Notes in Math.* Springer, Berlin, 2003, pp. 123–160.
33. Ambrosio, L., and Rigot, S. Optimal mass transportation in the Heisenberg group. *J. Funct. Anal. 208*, 2 (2004), 261–301.
34. Ambrosio, L., and Savaré, G. Personal communication, 2006.
35. Ambrosio, L., Savaré, G., and Zambotti, L. Existence and stability for Fokker–Planck equations with log-concave reference measure. To appear in *Probab. Theory Related Fields.*
36. Ambrosio, L., and Serfaty, S. A gradient flow approach to an evolution problem arising in superconductivity. To appear in *Comm. Pure Appl. Math.*

37. Ambrosio, L., and Tilli, P. *Topics on analysis in metric spaces*, vol. 25 of *Oxford Lecture Series in Mathematics and its Applications.* Oxford University Press, Oxford, 2004.
38. Andres, S., and von Renesse, M.-K. Particle approximation of the Wasserstein diffusion. Preprint, 2007.
39. Andreu, F., Caselles, V., and Mazón, J. M. The Cauchy problem for a strongly degenerate quasilinear equation. *J. Eur. Math. Soc. (JEMS) 7*, 3 (2005), 361–393.
40. Andreu, F., Caselles, V., Mazón, J., and Moll, S. Finite propagation speed for limited flux diffusion equations. *Arch. Rational Mech. Anal. 182*, 2 (2006), 269–297.
41. Ané, C., Blachère, S., Chafaï, D., Fougères, P., Gentil, I., Malrieu, F., Roberto, C., and Scheffer, G. *Sur les inégalités de Sobolev logarithmiques*, vol. 10 of *Panoramas et Synthèses.* Société Mathématique de France, 2000.
42. Appell, P. Mémoire sur les déblais et les remblais des systèmes continus ou discontinus. *Mémoires présentés par divers Savants à l'Académie des Sciences de l'Institut de France, Paris No. 29* (1887), 1–208. Available online at `gallica.bnf.fr`.
43. Arnold, A., Markowich, P., Toscani, G., and Unterreiter, A. On logarithmic Sobolev inequalities and the rate of convergence to equilibrium for Fokker–Planck type equations. *Comm. Partial Differential Equations 26*, 1–2 (2001), 43–100.
44. Arnol'd, V. I. *Mathematical methods of classical mechanics*, vol. 60 of *Graduate Texts in Mathematics.* Springer-Verlag, New York. Translated from the 1974 Russian original by K. Vogtmann and A. Weinstein. Corrected reprint of the second (1989) edition.
45. Aronson, D. G., and Bénilan, Ph. Régularité des solutions de l'équation des milieux poreux dans $\mathbf{R}^N$. *C. R. Acad. Sci. Paris Sér. A-B 288*, 2 (1979), A103–A105.
46. Aronsson, G. A mathematical model in sand mechanics: Presentation and analysis. *SIAM J. Appl. Math. 22* (1972), 437–458.
47. Aronsson, G., and Evans, L. C. An asymptotic model for compression molding. *Indiana Univ. Math. J. 51*, 1 (2002), 1–36.
48. Aronsson, G., Evans, L. C., and Wu, Y. Fast/slow diffusion and growing sandpiles. *J. Differential Equations 131*, 2 (1996), 304–335.
49. Attouch, L., and Soubeyran, A. From procedural rationality to routines: A "Worthwile to Move" approach of satisficing with not too much sacrificing. Preprint, 2005. Available online at `www.gate.cnrs.fr/seminaires/2006_2007`.
50. Aubry, P. *Monge, le savant ami de Napoléon Bonaparte, 1746–1818.* Gauthier-Villars, Paris, 1954.
51. Aubry, S. The twist map, the extended Frenkel–Kontorova model and the devil's staircase. In *Order in chaos* (Los Alamos, 1982), *Phys. D 7*, 1–3 (1983), 240–258.
52. Aubry, S., and Le Daeron, P. Y. The discrete Frenkel–Kantorova model and its extensions, I. *Phys. D. 8*, 3 (1983), 381–422.
53. Bakelman, I. J. *Convex analysis and nonlinear geometric elliptic equations.* Edited by S. D. Taliaferro. Springer-Verlag, Berlin, 1994.

54. Bakry, D. L'hypercontractivité et son utilisation en théorie des semigroupes. In *École d'été de Probabilités de Saint-Flour*, no. 1581 in Lecture Notes in Math. Springer, 1994.
55. Bakry, D., Cattiaux, P., and Guillin, A. Rate of convergence for ergodic continuous Markov processes: Lyapunov versus Poincaré. *J. Funct. Anal. 254*, 3 (2008), 727–759.
56. Bakry, D., and Émery, M. Diffusions hypercontractives. In *Sém. Proba. XIX*, no. 1123 in Lecture Notes in Math. Springer, 1985, pp. 177–206.
57. Bakry, D., and Ledoux, M. Lévy–Gromov's isoperimetric inequality for an infinite-dimensional diffusion generator. *Invent. Math. 123*, 2 (1996), 259–281.
58. Bakry, D., and Ledoux, M. Sobolev inequalities and Myers's diameter theorem for an abstract Markov generator. *Duke Math. J. 85*, 1 (1996), 253–270.
59. Bakry, D., and Ledoux, M. A logarithmic Sobolev form of the Li–Yau parabolic inequalities. *Rev. Mat. Iberoamericana 22*, 2 (2006), 683–702.
60. Bakry, D., and Qian, Z. M. Harnack inequalities on a manifold with positive or negative Ricci curvature. *Rev. Mat. Iberoamericana 15*, 1 (1999), 143–179.
61. Bakry, D., and Qian, Z. M. Some new results on eigenvectors via dimension, diameter, and Ricci curvature. *Adv. Math. 155*, 1 (2000), 98–153.
62. Ball, K., Carlen, E., and Lieb, E. H. Sharp uniform convexity and smoothness inequalities for trace norms. *Invent. Math. 115*, 3 (1994), 463–482.
63. Ball, J. M., and Mizel, V. J. Singular minimizers for regular one-dimensional problems in the calculus of variations. *Bull. Amer. Math. Soc. (N.S.) 11*, 1 (1984), 143–146.
64. Ball, J. M., and Mizel, V. J. One-dimensional variational problems whose minimizers do not satisfy the Euler-Lagrange equation. *Arch. Rational Mech. Anal. 90*, 4 (1985), 325–388.
65. Banavar, J. R., Colaiori, F., Flammini, A., Maritan, A., and Rinaldo, A. Scaling, optimality and landscape evolution. *J. Statist. Phys. 104*, 1/2 (2001), 1–48.
66. Bangert, V. Mather sets for twist maps and geodesics on tori. In *Dynamics reported*, Dynam. Report. Ser. Dynam. Systems Appl., no. 1, Wiley, Chichester, 1988, pp. 1–56.
67. Bangert, V. Minimal measures and minimizing closed normal one-currents. *Geom. Funct. Anal. 9*, 3 (1999), 413–427.
68. Barles, G. *Solutions de viscosité des équations de Hamilton–Jacobi*, vol. 17 of *Mathématiques & Applications*. Springer-Verlag, Paris, 1994.
69. Barles, G., and Souganidis, P. E. On the large time behavior of solutions of Hamilton–Jacobi equations. *SIAM J. Math. Anal. 31*, 4 (2000), 925–939.
70. Barthe, F. Log-concave and spherical models in isoperimetry. *Geom. Funct. Anal. 12*, 1 (2002), 32–55.
71. Barthe, F. Autour de l'inégalité de Brunn–Minkowski. *Ann. Fac. Sci. Toulouse Math. (6) 12*, 2 (2003), 127–178.
72. Barthe, F. L'inégalité de Brunn–Minkowski. *Images des Mathématiques 2006*, CNRS, Paris, 2006, pp. 5–10. Available online at `www.math.cnrs.fr/imagesdesmaths`.
73. Barthe, F. Handwritten notes about Talagrand's inequality for the exponential measure on the real line, 2006.
74. Barthe, F., Bakry, D., Cattiaux, P., and Guillin, A. Poincaré inequalities for logconcave probability measures: a Lyapunov function approach. To appear in *Electron. Comm. Probab.*

75. Barthe, F., Cattiaux, P., and Roberto, C. Interpolated inequalities between exponential and Gaussian, Orlicz hypercontractivity and application to isoperimetry. *Rev. Mat. Iberoamericana 22*, 3 (2006), 993–1067.
76. Barthe, F., and Kolesnikov, A. Mass transport and variants of the logarithmic Sobolev inequality. Preprint, 2008.
77. Barthe, F., and Roberto, C. Sobolev inequalities for probability measures on the real line. *Studia Math. 159*, 3 (2003), 481–497.
78. Beckner, W. A generalized Poincaré inequality for Gaussian measures. *Proc. Amer. Math. Soc. 105*, 2 (1989), 397–400.
79. Beiglböck, M., Goldstern, M., Maresch, G., and Schachermayer, W. Optimal and better transport plans. Preprint, 2008. Available online at `www.fam.tuwien.ac.at/~wschach/pubs`.
80. Bell, E. T. *Men of Mathematics.* Simon and Schuster, 1937.
81. Benachour, S., Roynette, B., Talay, D., and Vallois, P. Nonlinear self-stabilizing processes. I. Existence, invariant probability, propagation of chaos. *Stochastic Process. Appl. 75*, 2 (1998), 173–201.
82. Benachour, S., Roynette, B., and Vallois, P. Nonlinear self-stabilizing processes. II. Convergence to invariant probability. *Stochastic Process. Appl. 75*, 2 (1998), 203–224.
83. Benaïm, M., Ledoux, M., and Raimond, O. Self-interacting diffusions. *Probab. Theory Related Fields 122*, 1 (2002), 1–41.
84. Benaïm, M., and Raimond, O. Self-interacting diffusions. II. Convergence in law. *Ann. Inst. H. Poincaré Probab. Statist. 39*, 6 (2003), 1043–1055.
85. Benaïm, M., and Raimond, O. Self-interacting diffusions. III. Symmetric interactions. *Ann. Probab. 33*, 5 (2005), 1717–1759.
86. Benamou, J.-D. *Transformations conservant la mesure, mécanique des fluides incompressibles et modèle semi-géostrophique en météorologie. Mémoire présenté en vue de l'Habilitation à Diriger des Recherches.* PhD thesis, Univ. Paris-Dauphine, 1992.
87. Benamou, J.-D., and Brenier, Y. Weak existence for the semigeostrophic equations formulated as a coupled Monge–Ampère/transport problem. *SIAM J. Appl. Math. 58*, 5 (1998), 1450–1461.
88. Benamou, J.-D., and Brenier, Y. A numerical method for the optimal time-continuous mass transport problem and related problems. In *Monge Ampère equation: applications to geometry and optimization* (Deerfield Beach, FL, 1997). Amer. Math. Soc., Providence, RI, 1999, pp. 1–11.
89. Benamou, J.-D., and Brenier, Y. A computational fluid mechanics solution to the Monge–Kantorovich mass transfer problem. *Numer. Math. 84*, 3 (2000), 375–393.
90. Benamou, J.-D., and Brenier, Y. Mixed L^2-Wasserstein optimal mapping between prescribed density functions. *J. Optim. Theory Appl. 111*, 2 (2001), 255–271.
91. Benamou, J.-D., Brenier, Y., and Guittet, K. The Monge–Kantorovitch mass transfer and its computational fluid mechanics formulation. *ICFD Conference on Numerical Methods for Fluid Dynamics* (Oxford, 2001). *Internat. J. Numer. Methods Fluids 40*, 1–2 (2002), 21–30.
92. Benedetto, D., Caglioti, E., Carrillo, J. A., and Pulvirenti, M. A non-Maxwellian steady distribution for one-dimensional granular media. *J. Statist. Phys. 91*, 5–6 (1998), 979–990.

93. Benedetto, D., Caglioti, E., and Pulvirenti, M. A one-dimensional Boltzmann equation with inelastic collisions. *Rend. Sem. Mat. Fis. Milano 67* (1997), 169–179 (2000).
94. Benfatto, G., Picco, P., and Pulvirenti, M. On the invariant measures for the two-dimensional Euler flow. *J. Statist. Phys. 46*, 3–4 (1987), 729–742.
95. Bénilan, Ph. Solutions intégrales d'équations d'évolution dans un espace de Banach. *C. R. Acad. Sci. Paris Sér. A–B 274* (1972), A47–A50.
96. Bennett, G. Probability inequalities for the sum of independent random variables. *J. Amer. Statist. Assoc. 57*, 297 (1962), 33–45.
97. Bérard, P., Besson, G., and Gallot, S. Embedding Riemannian manifolds by their heat kernel. *Geom. Funct. Anal. 4*, 4 (1994), 373–398.
98. Berbee, H. C. P. *Random walks with stationary increments and renewal theory.* Mathematical Centre Tracts, vol. 112. Mathematisch Centrum, Amsterdam, 1979.
99. Berger, M. *A panoramic view of Riemannian geometry.* Springer-Verlag, 2003.
100. Berger, M., Gauduchon, P., and Mazet, E. *Le spectre d'une variété riemannienne.* Lecture Notes in Math., vol. 194, Springer-Verlag, Berlin, 1971.
101. Berkes, I., and Philipp, W. An almost sure invariance principle for the empirical distribution function of mixing random variables. *Z. Wahrscheinlichkeitstheorie und Verw. Gebiete 41*, 2 (1977/78), 115–137.
102. Bernard, P. Smooth critical sub-solutions of the Hamilton–Jacobi equation. *Math. Res. Lett. 14*, 3 (2007), 503–511.
103. Bernard, P. Existence of $C^{1,1}$ critical sub-solutions of the Hamilton–Jacobi equation on compact manifolds. To appear in *Ann. Sci. École Norm. Sup.*
104. Bernard, P., and Buffoni, B. The Monge problem for supercritical Mañé potentials on compact manifolds. *Adv. Math. 207*, 2 (2006), 691–706.
105. Bernard, P., and Buffoni, B. Optimal mass transportation and Mather theory. *J. Eur. Math. Soc. (JEMS) 9*, 1 (2007), 85–121.
106. Bernard, P., and Buffoni, B. Weak KAM pairs and Monge–Kantorovich duality. To appear in *Adv. Studies Pure Math.* Available online at `www.ceremade.dauphine.fr/~pbernard/publi.html`.
107. Bernard, P., and Roquejoffre, J.-M. Convergence to time-periodic solutions in time-periodic Hamilton–Jacobi equations on the circle. *Comm. Partial Differential Equations 29*, 3–4 (2004), 457–469.
108. Bernot, M. *Transport optimal et irrigation.* PhD thesis, ENS Cachan, 2005.
109. Bernot, M., Caselles, V., and Morel, J.-M. Traffic plans. *Publ. Mat. 49*, 2 (2005), 417–451.
110. Bernot, M., Caselles, V., and Morel, J.-M. Are there infinite irrigation trees? *J. Math. Fluid Mech. 8*, 3 (2006), 311–332.
111. Bernot, M., Caselles, V., and Morel, J.-M. The structure of branched transportation networks. To appear in *Calc. Var. Partial Differential Equations.*
112. Bernot, M., Caselles, V., and Morel, J.-M. The models and theory of optimal transportation networks. To appear in *Lect. Notes in Math.*
113. Bernot, M., and Figalli, A. Synchronized traffic plans and stability of optima. To appear in *ESAIM Control Optim. Calc. Var.*
114. Bertrand, J. Existence and uniqueness of optimal maps on Alexandrov spaces. Preprint, 2007.

115. Bertsekas, D. *Network optimization: Continuous and discrete models.* Athena Scientific, 1998. Referenced online at www.athenasc.com/netbook.html.
116. Bestvina, M. Degenerations of the hyperbolic space. *Duke Math. J. 56*, 1 (1988), 143–161.
117. Bialy, M. L., and Polterovich, L. V. Geodesic flows on the two-dimensional torus and phase transitions "commensurability-noncommensurability". *Funct. Anal. Appl. 20* (1986), 260–266.
118. Biane, P., and Voiculescu, D. A free probability analogue of the Wasserstein metric on the trace-state space. *Geom. Funct. Anal. 11*, 6 (2001), 1125–1138.
119. Biler, P., Dolbeault, J., and Esteban, M. J. Intermediate asymptotics in L^1 for general nonlinear diffusion equations. *Appl. Math. Lett. 15*, 1 (2002), 101–107.
120. Billingsley, P. *Convergence of probability measures*, second ed. John Wiley & Sons Inc., New York, 1999. A Wiley–Interscience Publication.
121. Blanchet, A., Calvez, V., and Carrillo, J. A. Convergence of the mass-transport steepest descent scheme for the sub-critical Patlak–Keller–Segel model. Preprint, 2007.
122. Blower, G. The Gaussian isoperimetric inequality and transportation. *Positivity 7*, 3 (2003), 203–224.
123. Blower, G. Displacement convexity for the generalized orthogonal ensemble. *J. Statist. Phys. 116*, 5–6 (2004), 1359–1387.
124. Blower, G., and Bolley, F. Concentration inequalities on product spaces with applications to Markov processes. *Studia Math. 175*, 1 (2006), 47–72. Archived online at arxiv.org/abs/math.PR/0505536.
125. Bobkov, S. G. A functional form of the isoperimetric inequality for the Gaussian measure. *J. Funct. Anal. 135*, 1 (1996), 39–49.
126. Bobkov, S. G. Isoperimetric and analytical inequalities for log-concave probability measures. *Ann. Probab. 27*, 4 (1999), 1903–1921.
127. Bobkov, S. G., Gentil, I., and Ledoux, M. Hypercontractivity of Hamilton–Jacobi equations. *J. Math. Pures Appl. 80*, 7 (2001), 669–696.
128. Bobkov, S. G., and Götze, F. Exponential integrability and transportation cost related to logarithmic Sobolev inequalities. *J. Funct. Anal. 163*, 1 (1999), 1–28.
129. Bobkov, S. G., and Houdré, Ch. Some connections between isoperimetric and Sobolev-type inequalities. *Mem. Amer. Math. Soc. 129*, 616 (1997).
130. Bobkov, S. G., and Ledoux, M. Poincaré's inequalities and Talagrand's concentration phenomenon for the exponential distribution. *Probab. Theory Related Fields 107*, 3 (1997), 383–400.
131. Bobkov, S. G., and Ledoux, M. From Brunn–Minkowski to Brascamp–Lieb and to logarithmic Sobolev inequalities. *Geom. Funct. Anal. 10*, 5 (2000), 1028–1052.
132. Bobkov, S. G., and Ledoux, M. From Brunn–Minkowski to sharp Sobolev inequalities. To appear in *Ann. Mat. Pura Appl.* Available online at www.lsp.ups-tlse.fr/Ledoux.
133. Bobylev, A., and Toscani, G. On the generalization of the Boltzmann H-theorem for a spatially homogeneous Maxwell gas. *J. Math. Phys. 33*, 7 (1992), 2578–2586.

134. Bogachev, V. I., Kolesnikov, A. V., and Medvedev, K. V. On triangular transformations of measures. *Dokl. Ross. Akad. Nauk 396* (2004), 438–442. Translation in *Dokl. Math. 69* (2004).
135. Bogachev, V. I., Kolesnikov, A. V., and Medvedev, K. V. Triangular transformations of measures. *Mat. Sb. 196*, 3 (2005), 3–30. Translation in *Sb. Math. 196*, 3-4 (2005), 309–335.
136. Bolley, F. Separability and completeness for the Wasserstein distance. To appear in *Seminar on Probability*, Lect. Notes in Math.
137. Bolley, F., Brenier, Y., and Loeper, G. Contractive metrics for scalar conservation laws. *J. Hyperbolic Differ. Equ. 2*, 1 (2005), 91–107.
138. Bolley, F., and Carrillo, J. A. Tanaka theorem for inelastic Maxwell models. *Comm. Math. Phys. 276*, 2 (2007), 287–314.
139. Bolley, F., Guillin, A., and Villani, C. Quantitative concentration inequalities for empirical measures on non-compact spaces. *Prob. Theory Related Fields 137*, 3–4 (2007), 541–593.
140. Bolley, F., and Villani, C. Weighted Csiszár–Kullback–Pinsker inequalities and applications to transportation inequalities. *Ann. Fac. Sci. Toulouse Math. (6) 14*, 3 (2005), 331–352.
141. Boltzmann, L. *Lectures on gas theory.* University of California Press, Berkeley, 1964. Translated by Stephen G. Brush. Reprint of the 1896–1898 Edition by Dover Publications, 1995.
142. Bonami, A. Étude des coefficients de Fourier des fonctions de $L^p(G)$. *Ann. Inst. Fourier (Grenoble) 20*, fasc. 2 (1970), 335–402 (1971).
143. Bonciocat, A.-I., and Sturm, K.-T. Mass transportation and rough curvature bounds for discrete spaces. Preprint, 2007.
144. Borell, C. Convex set functions in d-space. *Period. Math. Hungar. 6*, 2 (1975), 111–136.
145. Borell, C. The Brunn–Minkowski inequality in Gauss space. *Invent. Math. 30*, 2 (1975), 207–216.
146. Borovkov, A. A., and Utev, S. A. An inequality and a characterization of the normal distribution connected with it. *Teor. Veroyatnost. i Primenen. 28*, 2 (1983), 209–218.
147. Bouchitté, G., and Buttazzo, G. Characterization of optimal shapes and masses through Monge–Kantorovich equation. *J. Eur. Math. Soc. (JEMS) 3*, 2 (2001), 139–168.
148. Bouchitté, G., Buttazzo, G., and Seppecher, P. Shape optimization solutions via Monge–Kantorovich equation. *C. R. Acad. Sci. Paris Sér. I Math. 324*, 10 (1997), 1185–1191.
149. Bouchitté, G., Champion, Th., and Jiemenez, Ch. Completion of the space of measures in the Kantorovich norm. *Riv. Mat. Univ. Parma 7*, 4 (2005), 127–139.
150. Bouchitté, G., Jimenez, Ch., and Rajesh. M. Asymptotique d'un problème de positionnement optimal. *C. R. Acad. Sci. Paris, Sér. I, 335* (2002), 835–858.
151. Bouchitté, G., Jimenez, Ch., and Rajesh, M. A new L^∞ estimate in optimal mass transport. *Proc. Amer. Math. Soc. 135*, 11 (2007), 3525–3535.
152. Brancolini, A., Buttazzo, G., and Santambrogio, F. Path functionals over Wasserstein spaces. *J. Eur. Math. Soc. (JEMS) 8*, 3 (2006), 415–434.
153. Brascamp, H. J., and Lieb, E. H. On extensions of the Brunn–Minkowski and Prékopa–Leindler theorems, including inequalities for log concave func-

tions, and with an application to the diffusion equation. *J. Functional Analysis 22*, 4 (1976), 366–389.
154. Brenier, Y. Décomposition polaire et réarrangement monotone des champs de vecteurs. *C. R. Acad. Sci. Paris Sér. I Math. 305*, 19 (1987), 805–808.
155. Brenier, Y. The least action principle and the related concept of generalized flows for incompressible perfect fluids. *J. Amer. Math. Soc. 2*, 2 (1989), 225–255.
156. Brenier, Y. Polar factorization and monotone rearrangement of vector-valued functions. *Comm. Pure Appl. Math. 44*, 4 (1991), 375–417.
157. Brenier, Y. The dual least action problem for an ideal, incompressible fluid. *Arch. Rational Mech. Anal. 122*, 4 (1993), 323–351.
158. Brenier, Y. A homogenized model for vortex sheets. *Arch. Rational Mech. Anal. 138*, 4 (1997), 319–353.
159. Brenier, Y. Minimal geodesics on groups of volume-preserving maps and generalized solutions of the Euler equations. *Comm. Pure Appl. Math. 52*, 4 (1999), 411–452.
160. Brenier, Y. Derivation of the Euler equations from a caricature of Coulomb interaction. *Comm. Math. Phys. 212*, 1 (2000), 93–104.
161. Brenier, Y. A Monge–Kantorovich approach to the Maxwell equations. In: *Hyperbolic problems: theory, numerics, applications, vol. I, II* (Magdeburg, 2000), 2001, pp. 179–186.
162. Brenier, Y. Volume preserving maps, Euler equations and Coulomb interaction. *XIIIth International Congress on Mathematical Physics* (London, 2000), Int. Press, Boston, MA, 2001, pp. 303–309.
163. Brenier, Y. Some geometric PDEs related to hydrodynamics and electrodynamics. *Proceedings of the International Congress of Mathematicians, Vol. III* (Beijing, 2002), Higher Ed. Press, Beijing, 2002, pp. 761–772.
164. Brenier, Y. Extended Monge–Kantorovich theory. In *Optimal transportation and applications (Martina Franca, 2001)*, vol. 1813 of *Lecture Notes in Math.* Springer, Berlin, 2003, pp. 91–121.
165. Brenier, Y. A note on deformations of 2D fluid motions using 3D Born–Infeld equations. *Monatsh. Math. 142*, 1-2 (2004), 113–122.
166. Brenier, Y. Extension of the Monge–Kantorovich theory to classical electrodynamics. In *Recent advances in the theory and applications of mass transport*, vol. 353 of *Contemp. Math.*, Amer. Math. Soc., Providence, RI, 2004, pp.19–41.
167. Brenier, Y. Personal communication, 2006.
168. Brenier, Y., Frisch, U., Hénon, M., Loeper, G., Matarrese, S., Mohayaee, R., and Sobolevskiĭ, A. Reconstruction of the early Universe as a convex optimization problem. *Mon. Not. R. Astron. Soc. 346* (2003), 501–524.
169. Brenier, Y., and Grenier, E. Sticky particles and scalar conservation laws. *SIAM J. Numer. Anal. 35*, 6 (1998), 2317–2328.
170. Brenier, Y., and Loeper, G. A geometric approximation to the Euler equation: The Vlasov–Monge–Ampère equation. *Geom. Funct. Anal. 14*, 6 (2004), 1182–1218.
171. Brézis, H. *Opérateurs maximaux monotones et semi-groupes de contractions dans les espaces de Hilbert.* North-Holland Publishing Co., Amsterdam, 1973. North-Holland Mathematics Studies, No. 5.
172. Brézis, H. *Analyse fonctionnelle. Théorie et applications.* Masson, Paris, 1983.

173. Brézis, H., and Lieb, E. H. Sobolev inequalities with a remainder term. *J. Funct. Anal. 62* (1985), 73–86.
174. Burago, D., Burago, Y., and Ivanov, S. *A course in metric geometry*, vol. 33 of *Graduate Studies in Mathematics*. American Mathematical Society, Providence, RI, 2001. Errata list available online at `www.pdmi.ras.ru/staff/burago.html`.
175. Burago, Y., Gromov, M., and Perel′man, G. A. D. Aleksandrov spaces with curvatures bounded below. *Uspekhi Mat. Nauk 47*, 2(284) (1992), 3–51, 222.
176. Burago, Y. D., and Zalgaller, V. A. *Geometric inequalities*. Springer-Verlag, Berlin, 1988. Translated from the Russian by A. B. Sosinskiĭ, Springer Series in Soviet Mathematics.
177. Buttazzo, G., Giaquinta, M., and Hildebrandt, S. *One-dimensional variational problems. An introduction*, vol. 15 of *Oxford Lecture Series in Mathematics and its Applications*. The Clarendon Press Oxford University Press, New York, 1998.
178. Buttazzo, G., Pratelli, A., and Stepanov, E. Optimal pricing policies for public transportation networks. *SIAM J. Optim. 16*, 3 (2006), 826–853.
179. Buttazzo, G., Pratelli, A., Solimini, S., and Stepanov, E. *Optimal urban networks via mass transportation*. Preprint, 2007. Available online via `cvgmt.sns.it`.
180. Buttazzo, G., and Santambrogio, F. A model for the optimal planning of an urban area. *SIAM J. Math. Anal. 37*, 2 (2005), 514–530.
181. Cabré, X. Nondivergent elliptic equations on manifolds with nonnegative curvature. *Comm. Pure Appl. Math. 50*, 7 (1997), 623–665.
182. Cabré, X. The isoperimetric inequality via the ABP method. Preprint, 2005.
183. Cáceres, M., and Toscani, G. Kinetic approach to long time behavior of linearized fast diffusion equations. *J. Stat. Phys. 128*, 4 (2007), 883–925.
184. Caffarelli, L. A. Interior $W^{2,p}$ estimates for solutions of the Monge–Ampère equation. *Ann. of Math. (2) 131*, 1 (1990), 135–150.
185. Caffarelli, L. A. The regularity of mappings with a convex potential. *J. Amer. Math. Soc. 5*, 1 (1992), 99–104.
186. Caffarelli, L. A. Boundary regularity of maps with convex potentials. *Comm. Pure Appl. Math. 45*, 9 (1992), 1141–1151.
187. Caffarelli, L. A. Boundary regularity of maps with convex potentials. II. *Ann. of Math. (2) 144*, 3 (1996), 453–496.
188. Caffarelli, L. A. Monotonicity properties of optimal transportation and the FKG and related inequalities. *Comm. Math. Phys. 214*, 3 (2000), 547–563. Erratum in *Comm. Math. Phys. 225* (2002), 2, 449–450.
189. Caffarelli, L. A., and Cabré, X. *Fully nonlinear elliptic equations*, vol. 43 of *American Mathematical Society Colloquium Publications*. American Mathematical Society, Providence, RI, 1995.
190. Caffarelli, L. A., Feldman, M., and McCann, R. J. Constructing optimal maps for Monge's transport problem as a limit of strictly convex costs. *J. Amer. Math. Soc. 15*, 1 (2002), 1–26.
191. Caffarelli, L. A., Gutiérrez, C., and Huang, Q. On the regularity of reflector antennas. *Ann. of Math. (2)*, to appear.
192. Caffarelli, L. A., and McCann, R. J. Free boundaries in optimal transport and Monge–Ampère obstacle problems. To appear in *Ann. of Math. (2)*.

193. Caglioti, E., Pulvirenti, M., and Rousset, F. 2-D constrained Navier–Stokes equation and intermediate asymptotics. To appear in *Physica A.*
194. Caglioti, E., Pulvirenti, M., and Rousset, F. Quasi-stationary states for the 2-D Navier–Stokes equation. Preprint, 2007.
195. Caglioti, E., and Villani, C. Homogeneous cooling states are not always good approximations to granular flows. *Arch. Ration. Mech. Anal. 163*, 4 (2002), 329–343.
196. Calvez, V. *Modèles et analyses mathématiques pour les mouvements collectifs de cellules.* PhD thesis, Univ. Paris VI, 2007.
197. Calvez, V. Personal communication.
198. Calvez, V., and Carrillo, J. A. Work in progress.
199. Cannarsa, P., and Sinestrari, C. *Semiconcave functions, Hamilton–Jacobi equations, and optimal control.* Progress in Nonlinear Differential Equations and their Applications, 58. Birkhäuser Boston Inc., Boston, MA, 2004.
200. Carfora, M. Fokker–Planck dynamics and entropies for the normalized Ricci flow. Preprint, 2006. Archived online at `arxiv.org/abs/math.DG/0507309`.
201. Carlen, E. Behind the wave function: stochastic mechanics today. In *Proceedings of the Joint Concordia Sherbrooke Seminar Series on Functional Integration Methods in Stochastic Quantum Mechanics* (Sherbrooke, PQ and Montreal, PQ, 1987) (1991), vol. 25, pp. 141–156.
202. Carlen, E. A., Carvalho, M. C., Esposito, R., Lebowitz, J. L., and Marra, R. Displacement convexity and minimal fronts at phase boundaries. Preprint, 2007.
203. Carlen, E. A., and Gangbo, W. Constrained steepest descent in the 2-Wasserstein metric. *Ann. of Math. (2) 157*, 3 (2003), 807–846.
204. Carlen, E. A., and Gangbo, W. Solution of a model Boltzmann equation via steepest descent in the 2-Wasserstein metric. *Arch. Ration. Mech. Anal. 172*, 1 (2004), 21–64.
205. Carlen, E. A., and Soffer, A. Entropy production by block variable summation and central limit theorems. *Comm. Math. Phys. 140* (1991), 339–371.
206. Carlier, G., and Jimenez, Ch. On Monge's problem for Bregman-like cost functions. To appear in *J. Convex Anal.*
207. Carlier, G., Jimenez, Ch., and Santambrogio, S. Optimal transportation with traffic congestion and Wardrop equilibria. Preprint, 2006.
208. Carrillo, J. A., Di Francesco, M., and Lattanzio, C. Contractivity of Wasserstein metrics and asymptotic profiles for scalar conservation laws. *J. Differential Equations 231*, 2 (2006), 425–458.
209. Carrillo, J. A., Di Francesco, M., and Lattanzio, C. Contractivity and asymptotics in Wasserstein metrics for viscous nonlinear scalar conservation laws. *Boll. Unione Mat. Ital. Ser. B Artic. Ric. Mat. 10*, 2 (2007), 277–292.
210. Carrillo, J. A., Di Francesco, M., and Toscani, G. Intermediate asymptotics beyond homogeneity and self-similarity: long time behavior for $u_t = \Delta\phi(u)$. *Arch. Ration. Mech. Anal. 180*, 1 (2006), 127–149.
211. Carrillo, J. A., Di Francesco, M., and Toscani, G. Strict contractivity of the 2-Wasserstein distance for the porous medium equation by mass centering. *Proc. Amer. Math. Soc. 135*, 2 (2007), 353–363.
212. Carrillo, J. A., Gualdani, M. P., and Toscani, G. Finite speed of propagation in porous media by mass transportation methods. *C. R. Math. Acad. Sci. Paris 338*, 10 (2004), 815–818.

213. Carrillo, J. A., McCann, R. J., and Villani, C. Kinetic equilibration rates for granular media and related equations: entropy dissipation and mass transportation estimates. *Rev. Mat. Iberoamericana 19*, 3 (2003), 971–1018.
214. Carrillo, J. A., McCann, R. J., and Villani, C. Contractions in the 2-Wasserstein length space and thermalization of granular media. *Arch. Ration. Mech. Anal. 179*, 2 (2006), 217–263.
215. Carrillo, J. A., and Toscani, G. Asymptotic L^1-decay of solutions of the porous medium equation to self-similarity. *Indiana Univ. Math. J. 49*, 1 (2000), 113–142.
216. Carrillo, J. A., and Toscani, G. Contractive probability metrics and asymptotic behavior of dissipative kinetic equations. Proceedings of the 2006 Porto Ercole Summer School. *Riv. Mat. Univ. Parma 6* (2007), 75–198.
217. Carrillo, J. A., and Vázquez, J. L. Fine asymptotics for fast diffusion equations. *Comm. Partial Differential Equations 28*, 5–6 (2003), 1023–1056.
218. Carrillo, J. A., and Vázquez, J. L. Asymptotic complexity in filtration equations. *J. Evol. Equ. 7*, 3 (2007), 471–495.
219. Cattiaux, P., and Guillin, A. On quadratic transportation cost inequalities. *J. Math. Pures Appl. (9) 86*, 4 (2006), 341–361.
220. Cattiaux, P., and Guillin, A. Trends to equilibrium in total variation distance. To appear in *Ann. Inst. H. Poincaré Probab. Statist.*
221. Cattiaux, P., Guillin, A., and Malrieu, F. Probabilistic approach for granular media equations in the nonuniformly convex case. *Probab. Theory Related Fields 140*, 1–2 (2008), 19–40.
222. Champion, Th., De Pascale, L., and Juutinen, P. The ∞-Wasserstein distance: local solutions and existence of optimal transport maps. Preprint, 2007.
223. Chavel, I. *Riemannian geometry — a modern introduction*, vol. 108 of *Cambridge Tracts in Mathematics.* Cambridge University Press, Cambridge, 1993.
224. Chazal, F., Cohen-Steiner, D., and Mérigot, Q. Stability of boundary measures. *INRIA Report, 2007.* Available online at `www-sop.inria.fr/geometrica/team/Quentin.Merigot/`
225. Cheeger, J. A lower bound for the smallest eigenvalue of the Laplacian. In *Problems in analysis* (Papers dedicated to Salomon Bochner, 1969), pp. 195–199. Princeton Univ. Press, Princeton, N. J., 1970.
226. Cheeger, J. Differentiability of Lipschitz functions on metric measure spaces. *Geom. Funct. Anal. 9*, 3 (1999), 428–517.
227. Cheeger, J., and Colding, T. H. Lower bounds on Ricci curvature and the almost rigidity of warped products. *Ann. of Math. (2) 144*, 1 (1996), 189–237.
228. Cheeger, J., and Colding, T. H. On the structure of spaces with Ricci curvature bounded below. I. *J. Differential Geom. 46*, 3 (1997), 406–480.
229. Cheeger, J., and Colding, T. H. On the structure of spaces with Ricci curvature bounded below. II. *J. Differential Geom. 54*, 1 (2000), 13–35.
230. Cheeger, J., and Colding, T. H. On the structure of spaces with Ricci curvature bounded below. III. *J. Differential Geom. 54*, 1 (2000), 37–74.
231. Chen, M.-F. Trilogy of couplings and general formulas for lower bound of spectral gap. *Probability towards 2000* (New York, 1995), 123–136, *Lecture Notes in Statist. 128*, Springer, New York, 1998.
232. Chen, M.-F., and Wang, F.-Y. Application of coupling method to the first eigenvalue on manifold. *Science in China (A) 37*, 1 (1994), 1–14.

233. Christensen, J. P. R. Measure theoretic zero sets in infinite dimensional spaces and applications to differentiability of Lipschitz mappings. *Publ. Dép. Math. (Lyon) 10*, 2 (1973), 29–39. Actes du Deuxième Colloque d'Analyse Fonctionnelle de Bordeaux (Univ. Bordeaux, 1973), I, pp. 29–39.
234. Cianchi, A., Fusco, N., Maggi, F., and Pratelli, A. The sharp Sobolev inequality in quantitative form. Preprint, 2007.
235. Clarke, F. H. *Methods of dynamic and nonsmooth optimization*, vol. 57 of *CBMS-NSF Regional Conference Series in Applied Mathematics*. Society for Industrial and Applied Mathematics (SIAM), Philadelphia, PA, 1989.
236. Clarke, F. H., and Vinter, R. B. Regularity properties of solutions in the basic problem of calculus of variations. *Trans. Amer. Math. Soc. 289* (1985), 73–98.
237. Clement, Ph., and Desch, W. An elementary proof of the triangle inequality for the Wasserstein metric. *Proc. Amer. Math. Soc. 136*, 1 (2008), 333–339.
238. Contreras, G., and Iturriaga, R. *Global minimizers of autonomous Lagrangians.* 22° Colóquio Brasileiro de Matemática. Instituto de Matemática Pura e Aplicada (IMPA), Rio de Janeiro, 1999.
239. Contreras, G., Iturriaga, R., Paternain, G. P., and Paternain, M. The Palais–Smale condition and Mañé's critical values. *Ann. Henri Poincaré 1*, 4 (2000), 655–684.
240. Cordero-Erausquin, D. Sur le transport de mesures périodiques. *C. R. Acad. Sci. Paris Sér. I Math. 329*, 3 (1999), 199–202.
241. Cordero-Erausquin, D. Inégalité de Prékopa–Leindler sur la sphère. *C. R. Acad. Sci. Paris Sér. I Math. 329*, 9 (1999), 789–792.
242. Cordero-Erausquin, D. Some applications of mass transport to Gaussian-type inequalities. *Arch. Ration. Mech. Anal. 161*, 3 (2002), 257–269.
243. Cordero-Erausquin, D. Non-smooth differential properties of optimal transport. In *Recent advances in the theory and applications of mass transport*, vol. 353 of *Contemp. Math.*, Amer. Math. Soc., Providence, 2004, pp. 61–71.
244. Cordero-Erausquin, D. Quelques exemples d'application du transport de mesure en géométrie euclidienne et riemannienne. In *Séminaire de Théorie Spectrale et Géométrie. Vol. 22, Année 2003–2004*, pp. 125–152.
245. Cordero-Erausquin, D., Gangbo, W., and Houdré, Ch. Inequalities for generalized entropy and optimal transportation. In *Recent advances in the theory and applications of mass transport*, vol. 353 of *Contemp. Math.*, Amer. Math. Soc., Providence, RI, 2004, pp. 73–94.
246. Cordero-Erausquin, D., McCann, R. J., and Schmuckenschläger, M. A Riemannian interpolation inequality à la Borell, Brascamp and Lieb. *Invent. Math. 146*, 2 (2001), 219–257.
247. Cordero-Erausquin, D., McCann, R. J., and Schmuckenschläger, M. Prékopa–Leindler type inequalities on Riemannian manifolds, Jacobi fields, and optimal transport. *Ann. Fac. Sci. Toulouse Math. (6) 15*, 4 (2006), 613–635.
248. Cordero-Erausquin, D., Nazaret, B., and Villani, C. A mass-transportation approach to sharp Sobolev and Gagliardo–Nirenberg inequalities. *Adv. Math. 182*, 2 (2004), 307–332.
249. Coulhon, Th., and Saloff-Coste, L. Isopérimétrie pour les groupes et les variétés. *Rev. Mat. Iberoamericana 9*, 2 (1993), 293–314.
250. Coulhon, Th., and Saloff-Coste, L. Variétés riemanniennes isométriques à l'infini. *Rev. Mat. Iberoamericana 11*, 3 (1995), 687–726.

251. Courtault, J.-M., Kabanov, Y., Bru, B., Crépel, P., Lebon, I., and Le Marchand, A. Louis Bachelier on the centenary of "Théorie de la spéculation." *Math. Finance 10*, 3 (2000), 341–353.
252. Cover, T. M., and Thomas, J. A. *Elements of information theory.* John Wiley & Sons Inc., New York, 1991. A Wiley–Interscience Publication.
253. Csiszár, I. Information-type measures of difference of probability distributions and indirect observations. *Stud. Sci. Math. Hung. 2* (1967), 299–318.
254. Cuesta-Albertos, J. A., and Matrán, C. Notes on the Wasserstein metric in Hilbert spaces. *Ann. Probab. 17*, 3 (1989), 1264–1276.
255. Cuesta-Albertos, J. A., and Matrán, C. Stochastic convergence through Skorohod representation theorems and Wasserstein distances. *Rend. Circ. Mat. Palermo (2) Suppl. No. 35* (1994), 89–113.
256. Cuesta-Albertos, J. A., Matrán, C., Rachev, S. T., and Rüschendorf, L. Mass transportation problems in probability theory. *Math. Sci. 21*, 1 (1996), 34–72.
257. Cuesta-Albertos, J. A., Matrán, C., and Rodríguez-Rodríguez, J. Approximation to probabilities through uniform laws on convex sets. *J. Theoret. Probab. 16*, 2 (2003), 363–376.
258. Cuesta-Albertos, J. A., Matrán, C., and Tuero-Díaz, A. On lower bounds for the L^2-Wasserstein metric in a Hilbert space. J. Theoret. Probab. 9, 2 (1996), 263–283.
259. Cuesta-Albertos, J. A., Matrán, C., and Tuero-Díaz, A. On the monotonicity of optimal transportation plans. *J. Math. Anal. Appl. 215* (1997), 1, 86–94.
260. Cuesta-Albertos, J. A., Matrán, C., and Tuero-Díaz, A. Optimal transportation plans and convergence in distribution. J. Multivariate Anal. 60, 1 (1997), 72–83.
261. Cuesta-Albertos, J. A., and Tuero-Díaz, A. A characterization for the solution of the Monge–Kantorovich mass transference problem. *Statist. Probab. Lett. 16*, 2 (1993), 147–152.
262. Cullen, M. J. P. *A mathematical theory of large-scale atmosphere/ocean flow.* World Scientific, 2006.
263. Cullen, M. J. P., and Douglas, R. J. Applications of the Monge–Ampère equation and Monge transport problem to meteorology and oceanography. In *Monge Ampère equation: applications to geometry and optimization* (Deerfield Beach, FL, 1997). Amer. Math. Soc., Providence, RI, 1999, pp. 33–53.
264. Cullen, M. J. P., Douglas, R. J., Roulstone, I., and Sewell, M. J. Generalised semigeostrophic theory on a sphere. *J. Fluid Mech. 531* (2005), 123–157.
265. Cullen, M., and Feldman, M. Lagrangian solutions of semigeostrophic equations in physical space. *SIAM J. Math. Anal. 37*, 5 (2006), 1371–1395.
266. Cullen, M. J. P., and Gangbo, W. A variational approach for the 2-dimensional semigeostrophic shallow water equations. *Arch. Ration. Mech. Anal. 156*, 3 (2001), 241–273.
267. Cullen, M. J. P., Gangbo, W., and Pisante, G. The semigeostrophic equations discretized in reference and dual variables. To appear in *Arch. Ration. Mech. Anal.*
268. Cullen, M. J. P., and Maroofi, H. The fully compressible semi-geostrophic system from meteorology. *Arch. Ration. Mech. Anal. 167*, 4 (2003), 309–336.

269. Cullen, M. J. P., and Purser, R. J. Properties of the Lagrangian semi-geostrophic equations. *J. Atmospheric Sci. 46*, 17 (1989), 2684–2697.
270. Dacorogna, B., and Moser, J. On a partial differential equation involving the Jacobian determinant. *Ann. Inst. H. Poincaré Anal. Non Linéaire 7*, 1 (1990), 1–26.
271. Daneri, S., and Savaré, G. Eulerian calculus for the displacement convexity in the Wasserstein distance. Preprint, 2008.
272. Davies, E. B. *Heat kernels and spectral theory*, vol. 92 of *Cambridge Tracts in Mathematics.* Cambridge University Press, Cambridge, 1989.
273. de Acosta, A. Invariance principles in probability for triangular arrays of B-valued random vectors and some applications. *Ann. Probab. 10*, 2 (1982), 346–373.
274. De Giorgi, E. Sulla differenziabilità e l'analiticità delle estremali degli integrali multipli regolari. *Mem. Accad. Sci. Torino. Cl. Sci. Fis. Mat. Nat. (3) 3* (1957), 25–43.
275. De Giorgi, E. Movimenti minimizzanti. In *Aspetti e Problemi della Matematica Oggi* (Lecce, 1992).
276. De Giorgi, E., Marino, A., and Tosques, M. Problems of evolution in metric spaces and maximal decreasing curve. *Atti Accad. Naz. Lincei Rend. Cl. Sci. Fis. Mat. Natur. (8) 68*, 3 (1980), 180–187.
277. De Launay, L. *Un grand français – Monge, fondateur de l'École Polytechnique.* Éd. Pierre Roger, Paris, 1933. Available online at www.annales.org/archives/x/monge0.html.
278. De Pascale, L., Gelli, M. S., and Granieri, L. Minimal measures, one-dimensional currents and the Monge-Kantorovich problem. *Calc. Var. 27*, 1 (2006), 1–23.
279. De Pascale, L., Evans, L. C., and Pratelli, A. Integral estimates for transport densities. *Bull. London Math. Soc. 36*, 3 (2004), 383–395.
280. De Pascale, L., and Pratelli, A. Regularity properties for Monge transport density and for solutions of some shape optimization problem. *Calc. Var. Partial Differential Equations 14*, 3 (2002), 249–274.
281. De Pascale, L., and Pratelli, A. Sharp summability for Monge transport density via interpolation. *ESAIM Control Optim. Calc. Var. 10*, 4 (2004), 549–552.
282. del Barrio, E., Cuesta-Albertos, J. A., Matrán, C., and Rodríguez-Rodríguez, J. M. Tests of goodness of fit based on the L_2-Wasserstein distance. *Ann. Statist. 27*, 4 (1999), 1230–1239.
283. Del Pino, M., and Dolbeault, J. Best constants for Gagliardo–Nirenberg inequalities and applications to nonlinear diffusions. *J. Math. Pures Appl. (9) 81*, 9 (2002), 847–875.
284. Delanoë, Ph. Classical solvability in dimension two of the second boundary-value problem associated with the Monge-Ampère operator. *Ann. Inst. H. Poincaré Anal. Non Linéaire 8*, 5 (1991), 443–457.
285. Delanoë, Ph. Gradient rearrangement for diffeomorphisms of a compact manifold. *Differential Geom. Appl. 20*, 2 (2004), 145–165.
286. Delanoë, Ph., and Loeper, G. Gradient estimates for potentials of invertible gradient-mappings on the sphere. *Calc. Var. Partial Differential Equations 26*, 3 (2006), 297–311.
287. Delanoë, Ph., and Ge, Y. Regularity of optimal transportation maps on compact, locally nearly spherical, manifolds. Work in progress.

288. DELLACHERIE, C. Ensembles analytiques. Théorèmes de séparation et applications. In *Séminaire de Probabilités, IX* (Seconde Partie, Univ. Strasbourg, années universitaires 1973/1974 et 1974/1975). Springer, Berlin, 1975, pp. 336–372. Lecture Notes in Math., vol. 465.
289. DELON, J. Personal communication, 2005.
290. DEMANGE, J. From porous media equations to generalized Sobolev inequalities on a Riemannian manifold. Preprint, 2004.
291. DEMANGE, J. Improved Sobolev inequalities under positive curvature. Preprint, 2004.
292. DEMANGE, J. *Des équations à diffusion rapide aux inégalités de Sobolev sur les modèles de la géométrie.* PhD thesis, Univ. Paul Sabatier (Toulouse), 2005.
293. DEMANGE, J. Porous media equation and Sobolev inequalities under negative curvature. *Bull. Sci. Math. 129*, 10 (2005), 804–830.
294. DEMBO, A. Information inequalities and concentration of measure. *Ann. Probab. 25*, 2 (1997), 927–939.
295. DEMBO, A., COVER, T. M., AND THOMAS, J. A. Information theoretic inequalities. *IEEE Trans. Inform. Theory 37*, 6 (1991), 1501–1518.
296. DEMBO, A., AND ZEITOUNI, O. *Large deviations techniques and applications*, second ed. Springer-Verlag, New York, 1998.
297. DENZLER, J. Mather sets for plane Hamiltonian systems. *Z. Angew. Math. Phys. 38*, 6 (1987), 791–812.
298. DENZLER, J., AND MCCANN, R. J. Phase transitions and symmetry breaking in singular diffusion. *Proc. Natl. Acad. Sci. USA 100*, 12 (2003), 6922–6925.
299. DENZLER, J., AND MCCANN, R. J. Fast diffusion to self-similarity: complete spectrum, long-time asymptotics, and numerology. *Arch. Ration. Mech. Anal. 175*, 3 (2005), 301–342.
300. DESVILLETTES, L., AND VILLANI, C. On the spatially homogeneous Landau equation for hard potentials. II. H-theorem and applications. *Comm. Partial Differential Equations 25*, 1–2 (2000), 261–298.
301. DESVILLETTES, L., AND VILLANI, C. On the trend to global equilibrium in spatially inhomogeneous entropy-dissipating systems: the linear Fokker–Planck equation. *Comm. Pure Appl. Math. 54*, 1 (2001), 1–42.
302. DEUSCHEL, J.-D., AND STROOCK, D. W. *Large deviations*, vol. 137 of *Pure and Applied Mathematics.* Academic Press Inc., Boston, MA, 1989.
303. DING, Y. Heat kernels and Green's functions on limit spaces. *Comm. Anal. Geom. 10*, 3 (2002), 475–514.
304. DIPERNA, R. J., AND LIONS, P.-L. Ordinary differential equations, transport theory and Sobolev spaces. *Invent. Math. 98* (1989), 511–547.
305. DJELLOUT, H., GUILLIN, A., AND WU, L. Transportation cost-information inequalities and applications to random dynamical systems and diffusions. *Ann. Probab. 32*, 3B (2004), 2702–2732.
306. DO CARMO, M. P. *Riemannian geometry.* Mathematics: Theory & Applications. Birkhäuser Boston Inc., Boston, MA, 1992. Translated from the second Portuguese edition by Francis Flaherty.
307. DOBRIĆ, V., AND YUKICH, J. E. Asymptotics for transportation cost in high dimensions. *J. Theoret. Probab. 8*, 1 (1995), 97–118.
308. DOBRUŠIN, R. L. Prescribing a system of random variables by conditional distributions. *Teor. Verojatnost. i Primenen. 15* (1970), 469–497. English translation in *Theor. Prob. Appl. 15* (1970), 458–486.

309. DOBRUŠIN, R. L. Vlasov equations. *Funktsional. Anal. i Prilozhen. 13*, 2 (1979), 48–58, 96.
310. DOBRUŠIN, R. L. Perturbation methods of the theory of Gibbsian fields. In *Lectures on probability theory and statistics* (Saint-Flour, 1994). Springer, Berlin, 1996, pp. 1–66.
311. DOBRUŠIN, R. L., AND SHLOSMAN, S. B. Constructive criterion for the uniqueness of Gibbs field. In *Statistical physics and dynamical systems* (Köszeg, 1984), Progr. Phys., 10, Birkhäuser Boston, 1985, pp. 347–370.
312. DOLBEAULT, J., NAZARET, B., AND SAVARÉ, G. Work in progress.
313. DONSKER, M., AND VARADHAN, S. R. S. Asymptotic evaluation of certain Markov process expectations for large time, I. *Comm. Pure Appl. Math. 28* (1975), 1–47.
314. DUDLEY, R. M. The speed of mean Glivenko–Cantelli convergence. *Ann. Math. Statist. 40* (1968), 40–50.
315. DUDLEY, R. M. Speeds of metric probability convergence. *Z. Wahrscheinlichkeitstheorie verw. Geb. 22* (1972), 323–332.
316. DUDLEY, R. M. *Probabilities and metrics: Convergence of laws on metric spaces, with a view to statistical testing.* Aarhus Universitet, 1976.
317. DUDLEY, R. M. *Uniform central limit theorems*, vol. 63 of *Cambridge Studies in Advanced Mathematics.* Cambridge University Press, Cambridge, 1999.
318. DUDLEY, R. M. *Real analysis and probability*, vol. 74 of *Cambridge Studies in Advanced Mathematics.* Cambridge University Press, Cambridge, 2002. Revised reprint of the 1989 original.
319. DUPIN, C. Applications de la géométrie et de la mécanique. Re-edition by Bachelier, Paris (1922).
320. E, W., RYKOV, YU. G., AND SINAI, YA. G. Generalized variational principles, global weak solutions and behavior with random initial data for systems of conservation laws arising in adhesion particle dynamics. *Comm. Math. Phys. 177*, 2 (1996), 349–380.
321. ECKMANN, J.-P., AND HAIRER, M. Uniqueness of the invariant measure for a stochastic PDE driven by degenerate noise. *Comm. Math. Phys. 219*, 3 (2001), 523–565.
322. EINSTEIN, A. Mouvement des particules en suspension dans un fluide au repos, comme conséquence de la théorie cinétique moléculaire de la chaleur. *Annalen der Physik 17* (1905), 549–560. Trad. française.
323. EKELAND, I. An optimal matching problem. *ESAIM Control Optim. Calc. Var. 11*, 1 (2005), 57–71.
324. EKELAND, I., AND TÉMAM, R. *Convex analysis and variational problems*, English ed., vol. 28 of *Classics in Applied Mathematics.* Society for Industrial and Applied Mathematics (SIAM), Philadelphia, PA, 1999. Translated from the French.
325. ELIASSEN, A. The quasi-static equations of motion. *Geofys. Publ. 17*, 3 (1948).
326. EVANS, L. C. Classical solutions of fully nonlinear convex, second order elliptic equations. *Comm. Pure Appl. Math. 25* (1982), 333–363.
327. EVANS, L. C. *Partial differential equations.* American Mathematical Society, Providence, RI, 1998.
328. EVANS, L. C. Partial differential equations and Monge–Kantorovich mass transfer. In *Current developments in mathematics, 1997* (Cambridge, MA). Int. Press, Boston, MA, 1999, pp. 65–126.

329. Evans, L. C., Feldman, M., and Gariepy, R. F. Fast/slow diffusion and collapsing sandpiles. *J. Differential Equations 137*, 1 (1997), 166–209.
330. Evans, L. C. and Gangbo, W. Differential equations methods for the Monge–Kantorovich mass transfer problem. *Mem. Amer. Math. Soc. 137*, no. 653, 1999.
331. Evans, L. C., and Gariepy, R. *Measure theory and fine properties of functions.* CRC Press, Boca Raton, FL, 1992.
332. Evans, L. C., and Gomes, D. A. Effective Hamiltonians and averaging for Hamiltonian dynamics. I. *Arch. Ration. Mech. Anal. 157*, 1 (2001), 1–33.
333. Evans, L. C., and Gomes, D. A. Effective Hamiltonians and averaging for Hamiltonian dynamics. II. *Arch. Ration. Mech. Anal. 161*, 4 (2002), 271–305.
334. Evans, L. C., and Gomes, D. A. Linear programming interpretations of Mather's variational principle. *ESAIM Control Optim. Calc. Var. 8* (2002), 693–702.
335. Evans, L. C., and Ishii, H. A PDE approach to some asymptotic problems concerning random differential equations with small noise intensities. *Ann. Inst. H. Poincaré Anal. Non Linéaire 2*, 1 (1985), 1–20.
336. Evans, L. C., and Rezakhanlou, F. A stochastic model for growing sandpiles and its continuum limit. *Comm. Math. Phys. 197*, 2 (1998), 325–345.
337. Falconer, K. J. *The geometry of fractal sets*, vol. 85 of *Cambridge Tracts in Mathematics.* Cambridge University Press, Cambridge, 1986.
338. Falconer, K. J. *Fractal geometry*, second ed. John Wiley & Sons Inc., Hoboken, NJ, 2003. Mathematical foundations and applications.
339. Fang, S., and Shao, J. Optimal transportation maps for Monge–Kantorovich problem on loop groups. Preprint Univ. Bourgogne No. 470, 2006.
340. Fang, S., and Shao, J. Distance riemannienne, théorème de Rademacher et inégalité de transport sur le groupe des lacets. *C. R. Math. Acad. Sci. Paris 341*, 7 (2005), 445–450.
341. Fang, S., and Shao, J. Transportation cost inequalities on path and loop groups. *J. Funct. Anal. 218*, 2 (2005), 293–317.
342. Fang, S., Shao, J., and Sturm, K.-T. Wasserstein space over the Wiener space. Preprint, 2007.
343. Faris, W., Ed. *Diffusion, Quantum Theory, and Radically Elementary Mathematics*, vol. 47 of *Mathematical Notes.* Princeton University Press, 2006.
344. Fathi, A. Solutions KAM faibles conjuguées et barrières de Peierls. *C. R. Acad. Sci. Paris Sér. I Math. 325*, 6 (1997), 649–652.
345. Fathi, A. Théorème KAM faible et théorie de Mather sur les systèmes lagrangiens. *C. R. Acad. Sci. Paris Sér. I Math. 324*, 9 (1997), 1043–1046.
346. Fathi, A. Sur la convergence du semi-groupe de Lax–Oleinik. *C. R. Acad. Sci. Paris Sér. I Math. 327*, 3 (1998), 267–270.
347. Fathi, A. *Weak KAM theorem in Lagrangian dynamics.* To be published by *Cambridge University Press.*
348. Fathi, A., and Figalli, A. Optimal transportation on non-compact manifolds. To appear in *Israel J. Math.*
349. Fathi, A., and Mather, J. N. Failure of convergence of the Lax–Oleinik semigroup in the time-periodic case. *Bull. Soc. Math. France 128*, 3 (2000), 473–483.
350. Fathi, A., and Siconolfi, A. Existence of C^1 critical subsolutions of the Hamilton–Jacobi equation. *Invent. Math. 155*, 2 (2004), 363–388.

351. FEDERBUSH, P. Partially alternate derivation of a result by Nelson. *J. Math. Phys. 10*, 1 (1969), 50–52.
352. FEDERER, H. *Geometric measure theory.* Die Grundlehren der mathematischen Wissenschaften, Band 153. Springer-Verlag New York Inc., New York, 1969.
353. FELDMAN, M. Variational evolution problems and nonlocal geometric motion. *Arch. Rational Mech. Anal. 146*, 3 (1999), 221–274.
354. FELDMAN, M., AND MCCANN, R. J. Uniqueness and transport density in Monge's mass transportation problem. *Calc. Var. Partial Differential Equations 15*, 1 (2002), 81–113.
355. FENG, J., AND KURTZ, T. G. *Large deviations for stochastic processes*, vol. 131 of *Mathematical Surveys and Monographs.* American Mathematical Society, Providence, RI, 2006.
356. FERNIQUE, X. Sur le théorème de Kantorovitch–Rubinstein dans les espaces polonais. In *Seminar on Probability, XV* (Univ. Strasbourg, 1979/1980), vol. 850 of *Lecture Notes in Math.* Springer, Berlin, 1981, pp. 6–10.
357. FEYEL, D. Une démonstration simple du théorème de Kantorovich. Personal communication, 2003.
358. FEYEL, D., AND ÜSTÜNEL, A. S. Measure transport on Wiener space and the Girsanov theorem. *C. R. Math. Acad. Sci. Paris 334*, 11 (2002), 1025–1028.
359. FEYEL, D., AND ÜSTÜNEL, A. S. Monge–Kantorovitch measure transportation and Monge–Ampère equation on Wiener space. *Probab. Theory Related Fields 128*, 3 (2004), 347–385.
360. FEYEL, D., AND ÜSTÜNEL, A. S. Monge–Kantorovitch measure transportation, Monge–Ampère equation and the Itô calculus. In *Stochastic analysis and related topics in Kyoto*, vol. 41 of *Adv. Stud. Pure Math.* Math. Soc. Japan, Tokyo, 2004, pp. 49–74.
361. FEYEL, D., AND ÜSTÜNEL, A. S. The strong solution of the Monge–Ampère equation on the Wiener space for log-concave densities. *C. R. Math. Acad. Sci. Paris 339*, 1 (2004), 49–53.
362. FEYEL, D., AND ÜSTÜNEL, A. S. Solution of the Monge–Ampère equation on Wiener space for general log-concave measures. *J. Funct. Anal. 232*, 1 (2006), 29–55.
363. FIGALLI, A. Existence, uniqueness and regularity of optimal transport maps. *SIAM J. Math. Anal. 39*, 1 (2007), 126–137.
364. FIGALLI, A. The Monge problem on noncompact manifolds. *Rend. Sem. Mat. Univ. Padova 117* (2007), 147–166.
365. FIGALLI, A. The optimal partial transfer problem. Preprint, 2008.
366. FIGALLI, A., AND JUILLET, N. Absolute continuity of Wasserstein geodesics in the Heisenberg group. Preprint, 2007.
367. FIGALLI, A., KIM, Y.-H., AND MCCANN, R. Work in progress.
368. FIGALLI, A., AND LOEPER, G. C^1 regularity in two dimensions for potentials of the optimal transport problem. Preprint, 2007.
369. FIGALLI, A., MAGGI, F., AND PRATELLI, A. A mass transportation approach to quantitative isoperimetric inequalities. Preprint, 2007.
370. FIGALLI, A., AND RIFFORD, L. Mass transportation on sub-Riemannian manifolds. Preprint, 2008.
371. FIGALLI, A., AND RIFFORD, L. Continuity of optimal transport maps on small deformations of S^2. Preprint, 2008.
372. FIGALLI, A., AND VILLANI, C. Strong displacement convexity on Riemannian manifolds. *Math. Z. 257*, 2 (2007), 251–259.

373. Fisher, R. A. Theory of statistical estimation. *Math. Proc. Cambridge Philos. Soc. 22* (1925), 700–725.
374. Fontbona, J., Guérin, H., and Méléard, S. Measurability of optimal transportation and convergence rate for Landau type interacting particle systems. Preprint, 2007.
375. Fortuin, C. M., Kasteleyn, P. W., and Ginibre, J. Correlation inequalities on some partially ordered sets. *Comm. Math. Phys. 22* (1971), 89–103.
376. Forzani, L., and Maldonado, D. Properties of the solutions to the Monge–Ampère equation. *Nonlinear Anal. 57*, 5–6 (2004), 815–829.
377. Fournier, N. Uniqueness for a class of spatially homogeneous Boltzmann equations without angular cutoff. *J. Statist. Phys. 125*, 4 (2006), 927–946.
378. Fournier, N., and Guérin, H. On the uniqueness for the spatially homogeneous Boltzmann equation with a strong angular singularity. To appear in *J. Stat. Phys.*
379. Fournier, N., and Mouhot, C. On the well-posedness of the spatially homogeneous Boltzmann equation with a moderate angular singularity. Preprint, 2007.
380. Fragalà, I., Gelli, M. S., and Pratelli, A. Continuity of an optimal transport in Monge problem. *J. Math. Pures Appl. (9) 84*, 9 (2005), 1261–1294.
381. Fréchet, M. Sur la distance de deux lois de probabilité. *C. R. Acad. Sci. Paris 244* (1957), 689–692.
382. Frisch, U., Matarrese, S., Mohayaee, R., and Sobolevskĭ, A. A reconstruction of the initial conditions of the Universe by optimal mass transportation. *Nature 417* (2002), 260–262.
383. Fujita, Y., Ishii, H., and Loreti, P. Asymptotic solutions of viscous Hamilton–Jacobi equations with Ornstein–Uhlenbeck operator. *Comm. Partial Differential Equations 31*, 4–6 (2006), 827–848.
384. Fujita, Y., Ishii, H., and Loreti, P. Asymptotic solutions for large time of Hamilton–Jacobi equations in Euclidean n space. To appear in *Ann. Inst. H. Poincaré Anal. Non Linéaire.*
385. Fujita, Y., Ishii, H., and Loreti, P. Asymptotic solutions of Hamilton–Jacobi equations in Euclidean n space. *Indiana Univ. Math. J. 55*, 5 (2006), 1671–1700.
386. Fukaya, K. Collapsing of Riemannian manifolds and eigenvalues of Laplace operator. *Invent. Math. 87*, 3 (1987), 517–547.
387. Fukaya, K. Hausdorff convergence of Riemannian manifolds and its applications. In *Recent topics in differential and analytic geometry*, vol. 18 of *Adv. Stud. Pure Math.* Academic Press, Boston, MA, 1990, pp. 143–238.
388. Fusco, N., Maggi, F., and Pratelli, A. The sharp quantitative Sobolev inequality for functions of bounded variation. *J. Funct. Anal. 244*, 1 (2007), 315–341.
389. Fusco, N., Maggi, F., and Pratelli, A. The sharp quantitative isoperimetric inequality. To appear in *Ann. of Math.*
390. Gaffke, N., and Rüschendorf, L. On a class of extremal problems in statistics. *Math. Operationsforsch. Statist. Ser. Optim. 12*, 1 (1981), 123–135.
391. Galichon, A., and Henry, M. Testing non-identifying restrictions and confidence regions for partially identified parameters. Preprint, 2007.

392. Gallay, T., and Wayne, C. E. Global stability of vortex solutions of the two-dimensional Navier–Stokes equation. *Comm. Math. Phys. 255*, 1 (2005), 97–129.
393. Gallot, S. Isoperimetric inequalities based on integral norms of Ricci curvature. *Astérisque*, 157–158 (1988), 191–216.
394. Gallot, S., Hulin, D., and Lafontaine, J. *Riemannian geometry*, second ed. Universitext. Springer-Verlag, Berlin, 1990.
395. Gangbo, W. An elementary proof of the polar factorization of vector-valued functions. *Arch. Rational Mech. Anal. 128*, 4 (1994), 381–399.
396. Gangbo, W. The Monge mass transfer problem and its applications. In *Monge Ampère equation: applications to geometry and optimization* (Deerfield Beach, FL, 1997), vol. 226 of *Contemp. Math.*, Amer. Math. Soc., Providence, RI, 1999, pp. 79–104.
397. Gangbo, W. Review on the book *Gradient flows in metric spaces and in the space of probability measures* by Ambrosio, Gigli and Savaré, 2006. Available online at www.math.gatech.edu/~gangbo.
398. Gangbo, W., and McCann, R. J. Optimal maps in Monge's mass transport problem. *C. R. Acad. Sci. Paris Sér. I Math. 321*, 12 (1995), 1653–1658.
399. Gangbo, W., and McCann, R. J. The geometry of optimal transportation. *Acta Math. 177*, 2 (1996), 113–161.
400. Gangbo, W., and McCann, R. J. Shape recognition via Wasserstein distance. *Quart. Appl. Math. 58*, 4 (2000), 705–737.
401. Gangbo, W., Nguyen, T., and Tudorascu, A. Euler–Poisson systems as action-minimizing paths in the Wasserstein space. Preprint, 2006. Available online at www.math.gatech.edu/~gangbo/publications.
402. Gangbo, W., and Oliker, V. I. Existence of optimal maps in the reflector-type problems. *ESAIM Control Optim. Calc. Var. 13*, 1 (2007), 93–106.
403. Gangbo, W., and Święch, A. Optimal maps for the multidimensional Monge–Kantorovich problem. *Comm. Pure Appl. Math. 51*, 1 (1998), 23–45.
404. Gangbo, W., and Westdickenberg, M. Optimal transport for the system of isentropic Euler equations. Work in progress.
405. Gao, F., and Wu, L. Transportation-Information inequalities for Gibbs measures. Preprint, 2007.
406. Gardner, R. The Brunn–Minkowski inequality. *Bull. Amer. Math. Soc. (N.S.) 39*, 3 (2002), 355–405.
407. Gelbrich, M. On a formula for the L^2 Wasserstein metric between measures on Euclidean and Hilbert spaces. *Math. Nachr. 147* (1990), 185–203.
408. Gentil, I. *Inégalités de Sobolev logarithmiques et hypercontractivité en mécanique statistique et en E.D.P.* PhD thesis, Univ. Paul-Sabatier (Toulouse), 2001. Available online at www.ceremade.dauphine.fr/~gentil/maths.html.
409. Gentil, I. Ultracontractive bounds on Hamilton–Jacobi solutions. *Bull. Sci. Math. 126*, 6 (2002), 507–524.
410. Gentil, I., Guillin, A., and Miclo, L. Modified logarithmic Sobolev inequalities and transportation inequalities. *Probab. Theory Related Fields 133*, 3 (2005), 409–436.
411. Gentil, I., Guillin, A., and Miclo, L. Modified logarithmic Sobolev inequalities in null curvature. *Rev. Mat. Iberoamericana 23*, 1 (2007), 237–260.
412. Gentil, I., and Malrieu, F. Équations de Hamilton–Jacobi et inégalités entropiques généralisées. *C. R. Acad. Sci. Paris 335* (2002), 437–440.

413. Giacomin, G. On stochastic domination in the Brascamp–Lieb framework. *Math. Proc. Cambridge Philos. Soc. 134*, 3 (2003), 507–514.
414. Gianazza, U., Savaré, G., and Toscani, G. The Wasserstein gradient flow of the Fisher information and the quantum drift-diffusion equation. To appear in *Arch. Ration. Mech. Anal.*
415. Gigli, N. On the geometry of the space of probability measures endowed with the Kantorovich–Rubinstein distance. PhD Thesis, SNS Pisa, 2008.
416. Gilbarg, D., and Trudinger, N. S. *Elliptic partial differential equations of second order.* Classics in Mathematics. Springer-Verlag, Berlin, 2001. Reprint of the 1998 edition.
417. Gini, C. Sulla misura della concentrazione e della variabilità del caratteri. *Atti del R. Instituto Veneto 73* (1914), 1913–1914.
418. Gini, C. Measurement of inequality and incomes. *The Economic Journal 31*, 124–126.
419. Glimm, T., and Oliker, V. I. Optical design of single reflector systems and the Monge–Kantorovich mass transfer problem. *J. of Math. Sciences 117*, 3 (2003), 4096–4108.
420. Glimm, T., and Oliker, V. I. Optical design of two-reflector systems, the Monge–Kantorovich mass transfer problem and Fermat's principle. *Indiana Univ. Math. J. 53*, 5 (2004), 1255–1277.
421. Goldman, D., and McCann, R. J. Chaotic response of the 2D semi-geostrophic and 3D quasi-geostrophic equations to gentle periodic forcing. Preprint, 2007. Available online at `www.math.toronto.edu/mccann`.
422. Gomes, D. A. A stochastic analogue of Aubry–Mather theory. *Nonlinearity 15*, 3 (2002), 581–603.
423. Gomes, D. A., and Oberman, A. M. Computing the effective Hamiltonian using a variational approach. *SIAM J. Control Optim. 43*, 3 (2004), 792–812.
424. Goudon, T., Junca, S., and Toscani, G. Fourier-based distances and Berry–Esseen like inequalities for smooth densities. *Monatsh. Math. 135*, 2 (2002), 115–136.
425. Gozlan, N. Characterization of Talagrand's like transportation inequalities on the real line. *J. Funct. Anal. 250*, 2 (2007), 400–425.
426. Gozlan, N. Un critère général pour les inégalités de transport sur $\mathbb{R}$ associées à une fonction de coût convexe. Preprint, 2006.
427. Gozlan, N. *Principe conditionnel de Gibbs pour des contraintes fines approchées et Inégalités de Transport.* PhD thesis, Univ. Paris X–Nanterre, 2005.
428. Gozlan, N. Integral criteria for transportation-cost inequalities. *Electron. Comm. Probab. 11* (2006), 64–77.
429. Gozlan, N. A characterization of dimension free concentration in terms of transportation inequalities. Preprint, 2008.
430. Gozlan, N., and Léonard, Ch. A large deviation approach to some transportation cost inequalities. *Probab. Theory Related Fields 139*, 1–2 (2007), 235–283.
431. Granieri, L. On action minimizing measures for the Monge–Kantorovich problem. To appear in *NoDEA Nonlinear Differential Equations Appl.*
432. Greene, R. E., and Shiohama, K. Diffeomorphisms and volume-preserving embeddings of noncompact manifolds. *Trans. Amer. Math. Soc. 255* (1979), 403–414.
433. Grigor′yan, A. A. The heat equation on noncompact Riemannian manifolds. *Mat. Sb. 182*, 1 (1991), 55–87.

434. Grigor′yan, A. A. Analytic and geometric background of recurrence and non-explosion of the Brownian motion on Riemannian manifolds. *Bull. Amer. Math. Soc. (N.S.) 36*, 2 (1999), 135–249.
435. Gromov, M. Paul Lévy's isoperimetric inequality. Preprint IHES, 1980.
436. Gromov, M. Groups of polynomial growth and expanding maps. *Inst. Hautes Études Sci. Publ. Math.*, 53 (1981), 53–73.
437. Gromov, M. Sign and geometric meaning of curvature. *Rend. Sem. Mat. Fis. Milano 61* (1991), 9–123 (1994).
438. Gromov, M. *Metric structures for Riemannian and non-Riemannian spaces*, vol. 152 of *Progress in Mathematics*. Birkhäuser Boston Inc., Boston, MA, 1999. Based on the 1981 French original. With appendices by M. Katz, P. Pansu and S. Semmes. Translated from the French by S. M. Bates.
439. Gromov, M. Mesoscopic curvature and hyperbolicity. In *Global differential geometry: the mathematical legacy of Alfred Gray* (Bilbao 2000), vol. 228 of *Contemp. Math.*, Amer. Math. Soc., Providence, RI, 2001, pp. 58–69.
440. Gromov, M., and Milman, V. D. A topological application of the isoperimetric inequality. *Amer. J. Math. 105*, 4 (1983), 843–854.
441. Gross, L. Logarithmic Sobolev inequalities. *Amer. J. Math. 97* (1975), 1061–1083.
442. Gross, L. Logarithmic Sobolev inequalities and contractivity properties of semigroups. In *Dirichlet forms* (Varenna, 1992). Springer, Berlin, 1993, pp. 54–88. Lecture Notes in Math., 1563.
443. Grove, K., and Petersen, P. Manifolds near the boundary of existence. *J. Differential Geom. 33*, 2 (1991), 379–394.
444. Grunewald, N., Otto, F., Villani, C., and Westdickenberg, M. A two-scale approach to logarithmic Sobolev inequalities and the hydrodynamic limit. To appear in *Ann. Inst. H. Poincaré Probab. Statist.*
445. Grunewald, N., Otto, F., Villani, C., and Westdickenberg, M. Work in progress.
446. Guan, P. Monge–Ampère equations and related topics. Lecture notes. Available online at `www.math.mcgill.ca/guan/notes.html`, 1998.
447. Guillin, A. Personal communication.
448. Guillin, A., Léonard, C., Wu, L., and Yao, N. Transportation cost vs. Fisher–Donsker–Varadhan information. Preprint, 2007. Available online at `www.cmi.univ-mrs.fr/guillin/index3.html`
449. Guionnet, A. First order asymptotics of matrix integrals; a rigorous approach towards the understanding of matrix models. *Comm. Math. Phys. 244*, 3 (2004), 527–569.
450. Guittet, K. *Contributions à la résolution numérique de problèmes de transport optimal de masse.* PhD thesis, Univ. Paris VI – Pierre et Marie Curie (France), 2003.
451. Guittet, K. On the time-continuous mass transport problem and its approximation by augmented Lagrangian techniques. *SIAM J. Numer. Anal. 41*, 1 (2003), 382–399.
452. Gutiérrez, C. E. *The Monge–Ampère equation*, vol. 44 of *Progress in Nonlinear Differential Equations and their Applications.* Birkhäuser Boston Inc., Boston, MA, 2001.
453. Gutiérrez, C. E., and Huang, Q. The refractor problem in reshaping light beams. Preprint, 2007.

454. HACKENBROCH, W., AND THALMAIER, A. *Stochastische Analysis. Eine Einführung in die Theorie der stetigen Semimartingale.* Mathematische Leitfäden. B. G. Teubner, Stuttgart, 1994.
455. HAIRER, M. Exponential mixing properties of stochastic PDEs through asymptotic coupling. *Probab. Theory Related Fields 124*, 3 (2002), 345–380.
456. HAIRER, M., AND MATTINGLY, J. C. Ergodic properties of highly degenerate 2D stochastic Navier–Stokes equations. *C. R. Math. Acad. Sci. Paris 339*, 12 (2004), 879–882.
457. HAIRER, M., AND MATTINGLY, J. C. Ergodicity of the 2D Navier–Stokes equations with degenerate stochastic forcing. *Ann. of Maths (2) 164*, 3 (2006), 993–1032.
458. HAIRER, M., AND MATTINGLY, J. C. Spectral gaps in Wasserstein distances and the 2D Navier–Stokes equation. To appear in *Ann. Probab.*
459. HAJŁASZ, P. Sobolev spaces on an arbitrary metric space. *Potential Anal. 5*, 4 (1996), 403–415.
460. HAJŁASZ, P. Sobolev spaces on metric-measure spaces. In *Heat kernels and analysis on manifolds, graphs, and metric spaces* (Paris, 2002), vol. 338 of *Contemp. Math.*, Amer. Math. Soc., Providence, RI, 2003, pp. 173–218.
461. HAJŁASZ, P., AND KOSKELA, P. Sobolev meets Poincaré. *C. R. Acad. Sci. Paris Sér. I Math. 320*, 10 (1995), 1211–1215.
462. HAKER, S., AND TANNENBAUM, A. On the Monge–Kantorovich problem and image warping. In *Mathematical methods in computer vision*, vol. 133 of *IMA Vol. Math. Appl.* Springer, New York, 2003, pp. 65–85.
463. HAKER, S., ZHU, L., TANNENBAUM, A., AND ANGENENT, S. Optimal mass transport for registration and warping. *International Journal on Computer Vision 60*, 3 (2004), 225–240.
464. HANIN, L. G. An extension of the Kantorovich norm. In *Monge Ampère equation: applications to geometry and optimization* (Deerfield Beach, FL, 1997). Amer. Math. Soc., Providence, RI, 1999, pp. 113–130.
465. HARGÉ, G. A convex/log-concave correlation inequality for Gaussian measure and an application to abstract Wiener spaces. *Probab. Theory Related Fields 130*, 3 (2004), 415–440.
466. HAURAY, M., AND JABIN, P.-E. N-particles approximation of the Vlasov equations with singular potential. *Arch. Ration. Mech. Anal. 183*, 3 (2007), 489–524.
467. HEDLUND, G. Geodesics on a two-dimensional Riemannian manifold with periodic coefficients. *Ann. of Math. (2) 33* (1932), 719–739.
468. HEINICH, H., AND LOOTGIETER, J.-C. Convergence des fonctions monotones. *C. R. Acad. Sci. Paris Sér. I Math. 322*, 9 (1996), 869–874.
469. HEINONEN, J. *Lectures on analysis on metric spaces.* Universitext. Springer-Verlag, New York, 2001.
470. HEINONEN, J., AND KOSKELA, P. Quasiconformal maps in metric spaces with controlled geometry. *Acta Math. 181*, 1 (1998), 1–61.
471. HEINZ, E. Über die Differentialungleichung $0 < \alpha \leq rt - s^2 \leq \beta < \infty$. *Math. Z. 72* (1959/1960), 107–126.
472. HÉRAU, F., AND NIER, F. Isotropic hypoellipticity and trend to equilibrium for the Fokker–Planck equation with a high-degree potential. *Arch. Ration. Mech. Anal. 171*, 2 (2004), 151–218.

473. Hiai, F., Ohya, M., and Tsukada, M. Sufficiency and relative entropy in *-algebras with applications in quantum systems. *Pacific J. Math. 107*, 1 (1983), 117–140.
474. Hiai, F., Petz, D., and Ueda, Y. Free transportation cost inequalities via random matrix approximation. *Probab. Theory Related Fields 130*, 2 (2004), 199–221.
475. Hiai, F., and Ueda, Y. Free transportation cost inequalities for noncommutative multi-variables. *Infin. Dimens. Anal. Quantum Probab. Relat. Top. 9*, 3 (2006), 391–412.
476. Hohloch, S. *Optimale Massebewegung im Monge–Kantorovich-Transportproblem.* Diploma thesis, Freiburg Univ., 2002.
477. Holley, R. Remarks on the FKG inequalities. *Comm. Math. Phys. 36* (1974), 227–231.
478. Holley, R., and Stroock, D. W. Logarithmic Sobolev inequalities and stochastic Ising models. *J. Statist. Phys. 46*, 5–6 (1987), 1159–1194.
479. Horowitz, J., and Karandikar, R. L. Mean rates of convergence of empirical measures in the Wasserstein metric. *J. Comput. Appl. Math. 55*, 3 (1994), 261–273. (See also the reviewer's note on MathSciNet.)
480. Hoskins, B. J. Atmospheric frontogenesis models: some solutions. *Q.J.R. Met. Soc. 97* (1971), 139–153.
481. Hoskins, B. J. The geostrophic momentum approximation and the semi-geostrophic equations. *J. Atmosph. Sciences 32* (1975), 233–242.
482. Hoskins, B. J. The mathematical theory of frontogenesis. *Ann. Rev. of Fluid Mech. 14* (1982), 131–151.
483. Hsu, E. P. *Stochastic analysis on manifolds*, vol. 38 of *Graduate Studies in Mathematics*. American Mathematical Society, Providence, RI, 2002.
484. Hsu, E. P., and Sturm, K.-T. Maximal coupling of Euclidean Brownian motions. FB-Preprint No. 85, Bonn. Available online at `www-wt.iam.uni-bonn.de/~sturm/de/index.html`.
485. Huang, C., and Jordan, R. Variational formulations for Vlasov–Poisson–Fokker–Planck systems. *Math. Methods Appl. Sci. 23*, 9 (2000), 803–843.
486. Ichihara, K. Curvature, geodesics and the Brownian motion on a Riemannian manifold. I. Recurrence properties, II. Explosion properties. *Nagoya Math. J. 87* (1982), 101–114; 115–125.
487. Ishii, H. Asymptotic solutions for large time of Hamilton–Jacobi equations. *International Congress of Mathematicians, Vol. III*, Eur. Math. Soc., Zürich, 2006, pp. 213–227. Short presentation available online at `www.edu.waseda.ac.jp/~ishii`.
488. Ishii, H. Unpublished lecture notes on the weak KAM theorem, 2004. Available online at `www.edu.waseda.ac.jp/~ishii`.
489. Itoh, J.-i., and Tanaka, M. The Lipschitz continuity of the distance function to the cut locus. *Trans. Amer. Math. Soc. 353*, 1 (2001), 21–40.
490. Jian, H.-Y., and Wang, X.-J. Continuity estimates for the Monge–Ampère equation. Preprint, 2006.
491. Jimenez, Ch. Dynamic formulation of optimal transport problems. To appear in *J. Convex Anal.*
492. Jones, P., Maggioni, M., and Schul, R. Universal local parametrizations via heat kernels and eigenfunctions of the Laplacian. Preprint, 2008.
493. Jordan, R., Kinderlehrer, D., and Otto, F. The variational formulation of the Fokker–Planck equation. *SIAM J. Math. Anal. 29*, 1 (1998), 1–17.

494. JOULIN, A. Poisson-type deviation inequalities for curved continuous time Markov chains. *Bernoulli 13*, 3 (2007), 782–798.
495. JOULIN, A. A new Poisson-type deviation inequality for Markov jump process with positive Wasserstein curvature. Preprint, 2007.
496. JUILLET, N. Geometric inequalities and generalized Ricci bounds in the Heisenberg group. Preprint, 2006.
497. JÜNGEL, A., AND MATTHES, D. The Derrida–Lebowitz–Speer–Spohn equation: Existence, nonuniqueness, and decay rates of the solutions. Preprint, 2006. Available online at `www.juengel.de.vu`.
498. KALASHNIKOV, V. V., AND RACHEV, S. T. *Mathematical methods for construction of queueing models. The Wadsworth & Brooks/Cole Operations Research Series*, Pacific Grove, CA, 1990.
499. KANTOROVICH, L. V. *Mathematical methods in the organization and planning of production.* Leningrad Univ., 1939. English translation in *Management Science 6*, 4 (1960), 363-422.
500. KANTOROVICH, L. V. On an effective method of solving certain classes of extremal problems. *Dokl. Akad. Nauk. USSR 28* (1940), 212–215.
501. KANTOROVICH, L. V. On the translocation of masses. *Dokl. Akad. Nauk. USSR 37* (1942), 199–201. English translation in *J. Math. Sci. 133*, 4 (2006), 1381–1382.
502. KANTOROVICH, L. V. On a problem of Monge. *Uspekhi Mat. Nauk. 3* (1948), 225–226. English translation in *J. Math. Sci. 133*, 4 (2006), 1383.
503. KANTOROVICH, L. V. *The best use of economic resources.* Oxford Pergamon Press, 1965.
504. KANTOROVICH, L. V. Autobiography. In *Nobel Lectures, Economics 1969–1980*, A. Lindbeck, Ed., Les Prix Nobel/Nobel Lectures. World Scientific Publishing Co., Singapore, 1992. Available online at `nobelprize.org/economics/laureates/1975/kantorovich-autobio.html`.
505. KANTOROVICH, L. V., AND GAVURIN, M. L. Application of mathematical methods to problems of analysis of freight flows. In *Problems of raising the efficiency of transport performance* (1949), Moscow-Leningrad Univ., pp. 110–138.
506. KANTOROVICH, L. V., AND RUBINSHTEIN, G. S. On a space of totally additive functions. *Vestn. Leningrad. Univ. 13*, 7 (1958), 52–59.
507. KARAKHANYAN, A., AND WANG, X.-J. The reflector design problem. Preprint, 2007.
508. KASUE, A., AND KUMURA, H. Spectral convergence of Riemannian manifolds. *Tohoku Math. J. (2) 46*, 2 (1994), 147–179.
509. KASUE, A., AND KUMURA, H. Spectral convergence of Riemannian manifolds. II. *Tohoku Math. J. (2) 48*, 1 (1996), 71–120.
510. KEITH, S. Modulus and the Poincaré inequality on metric measure spaces. *Math. Z. 245*, 2 (2003), 255–292.
511. KEITH, S. A differentiable structure for metric measure spaces. *Adv. Math. 183*, 2 (2004), 271–315.
512. KELLERER, H. G. Duality theorems for marginal problems. *Z. Wahrsch. Verw. Gebiete 67*, 4 (1984), 399–432.
513. KHESIN, B., AND LEE, P. Poisson geometry and first integrals of geostrophic equations. Preprint, 2008.
514. KHESIN, B., AND LEE, P. A nonholomic Moser theorem and optimal mass transport. Preprint, 2008.

515. Khesin, B., and Misiołek, G. Shock waves for the Burgers equation and curvatures of diffeomorphism groups. Shock waves for the Burgers equation and curvatures of diffeomorphism groups. *Proc. Steklov Inst. Math. 250* (2007), 1–9.
516. Kim, S. Harnack inequality for nondivergent elliptic operators on Riemannian manifolds. *Pacific J. Math. 213*, 2 (2004), 281–293.
517. Kim, Y. J., and McCann, R. J. Sharp decay rates for the fastest conservative diffusions. *C. R. Math. Acad. Sci. Paris 341*, 3 (2005), 157–162.
518. Kim, Y.-H. Counterexamples to continuity of optimal transportation on positively curved Riemannian manifolds. Preprint, 2007.
519. Kim, Y.-H., and McCann, R. J. On the cost-subdifferentials of cost-convex functions. Preprint, 2007. Archived online at `arxiv.org/abs/0706.1266`.
520. Kim, Y.-H., and McCann, R. J. Continuity, curvature, and the general covariance of optimal transportation. Preprint, 2007.
521. Kleptsyn, V., and Kurtzmann, A. Ergodicity of self-attracting Brownian motion. Preprint, 2008.
522. Kloeckner, B. A geometric study of the Wasserstein space of the line. Preprint, 2008.
523. Knothe, H. Contributions to the theory of convex bodies. *Michigan Math. J. 4* (1957), 39–52.
524. Knott, M., and Smith, C. S. On the optimal mapping of distributions. *J. Optim. Theory Appl. 43*, 1 (1984), 39–49.
525. Knott, M., and Smith, C. S. On a generalization of cyclic monotonicity and distances among random vectors. *Linear Algebra Appl. 199* (1994), 363–371.
526. Kolesnikov, A. V. Convexity inequalities and optimal transport of infinite-dimensional measures. *J. Math. Pures Appl. (9) 83*, 11 (2004), 1373–1404.
527. Kontorovich, L. A linear programming inequality with applications to concentration of measures. Preprint, 2006. Archived online at `arxiv.org/abs/math.FA/0610712`.
528. Kontsevich, M., and Soibelman, Y. Homological mirror symmetry and torus fibrations. In *Symplectic geometry and mirror symmetry* (Seoul, 2000). World Sci. Publ., River Edge, NJ, 2001, pp. 203–263.
529. Koskela, P. Upper gradients and Poincaré inequality on metric measure spaces. In *Lecture notes on Analysis in metric spaces* (Trento, 1999), *Appunti Corsi Tenuti Docenti Sc.*, Scuola Norm. Sup., Pisa, 2000, pp. 55–69.
530. Krylov, N. V. Boundedly nonhomogeneous elliptic and parabolic equations. *Izv. Akad. Nauk SSSR, Ser. Mat. 46*, 3 (1982), 487–523. English translation in *Math. USSR Izv. 20*, 3 (1983), 459–492.
531. Krylov, N. V. Fully nonlinear second order elliptic equations: recent development. *Ann. Scuola Norm. Sup. Pisa Cl. Sci. 25*, 3–4 (1998), 569–595.
532. Krylov, N. V., and Safonov, M. V. An estimate of the probability that a diffusion process hits a set of positive measure. *Dokl. Acad. Nauk SSSR 245*, 1 (1979), 18–20.
533. Kuksin, S., Piatnitski, A., and Shirikyan, A. A coupling approach to randomly forced nonlinear PDE's, II. *Comm. Math. Phys. 230*, 1 (2002), 81–85.
534. Kullback, S. A lower bound for discrimination information in terms of variation. *IEEE Trans. Inform. Theory 4* (1967), 126–127.
535. Kurtzmann, A. The ODE method for some self-interacting diffusions on $\mathbb{R}^d$. Preprint, 2008.

536. KUWAE, K., MACHIGASHIRA, Y., AND SHIOYA, T. Beginning of analysis on Alexandrov spaces. In *Geometry and topology: Aarhus (1998)*, vol. 258 of *Contemp. Math.*, Amer. Math. Soc., Providence, RI, 2000, pp. 275–284.
537. KUWAE, K., MACHIGASHIRA, Y., AND SHIOYA, T. Sobolev spaces, Laplacian, and heat kernel on Alexandrov spaces. *Math. Z. 238*, 2 (2001), 269–316.
538. KUWAE, K., AND SHIOYA, T. On generalized measure contraction property and energy functionals over Lipschitz maps. *Potential Anal. 15*, 1–2 (2001), 105–121.
539. KUWAE, K., AND SHIOYA, T. Sobolev and Dirichlet spaces over maps between metric spaces. *J. Reine Angew. Math. 555* (2003), 39–75.
540. LATAŁA, R., AND OLESZKIEWICZ, K. Between Sobolev and Poincaré. In *Geometric aspects of functional analysis*, vol. 1745 of *Lecture Notes in Math.* Springer, Berlin, 2000, pp. 147–168.
541. LEDOUX, M. On an integral criterion for hypercontractivity of diffusion semigroups and extremal functions. *J. Funct. Anal. 105*, 2 (1992), 444–465.
542. LEDOUX, M. Inégalités isopérimétriques en analyse et probabilités. *Astérisque*, 216 (1993), Exp. No. 773, 5, 343–375. Séminaire Bourbaki, vol. 1992/93.
543. LEDOUX, M. Isoperimetry and Gaussian analysis. In *Lectures on probability theory and statistics* (Saint-Flour, 1994). Springer, Berlin, 1996, pp. 165–294.
544. LEDOUX, M. Concentration of measure and logarithmic Sobolev inequalities. In *Séminaire de Probabilités, XXXIII.* Springer, Berlin, 1999, pp. 120–216.
545. LEDOUX, M. The geometry of Markov diffusion generators. *Ann. Fac. Sci. Toulouse Math. (6) 9*, 2 (2000), 305–366.
546. LEDOUX, M. *The concentration of measure phenomenon.* American Mathematical Society, Providence, RI, 2001.
547. LEDOUX, M. Measure concentration, transportation concentration, and functional inequalities. Lecture Notes for the Instructional Conference on Combinatorial Aspects of Mathematical Analysis (Edinburgh, March 2002) and the Summer School on Singular Phenomena and Scaling in Mathematical Models (Bonn, June 2003), 2003. Available online at `www.lsp.ups-tlse.fr/Ledoux`.
548. LEINDLER, L. On a certain converse of Hölder's inequality. In *Linear operators and approximation* (Proc. Conf., Oberwolfach, 1971). Birkhäuser, Basel, 1972, pp. 182–184. Internat. Ser. Numer. Math., vol. 20.
549. LELIÈVRE, T., ROUSSET, M., AND STOLTZ, G. Long-time convergence of an adaptive biasing force method. Preprint, 2007.
550. LÉONARD, CH. A saddle-point approach to the Monge–Kantorovich optimal transport problem. Preprint, 2008.
551. LI, Y., AND NIRENBERG, L. The distance function to the boundary, Finsler geometry, and the singular set of viscosity solutions of some Hamilton–Jacobi equations. *Comm. Pure Appl. Math. 58* (2004), 85–146.
552. LI, P., AND YAU, S.-T. On the parabolic kernel of the Schrödinger operator. *Acta Math. 156*, 3–4 (1986), 153–201.
553. LICHNEROWICZ, A. *Géométrie des groupes de transformations.* Travaux et recherches mathématiques, Vol. III, Dunod, Paris, 1958.
554. LIEB, E. H., AND LOSS, M. *Analysis*, second ed. American Mathematical Society, Providence, RI, 2001.
555. LIEBERMANN, G. M., AND TRUDINGER, N. S. Nonlinear oblique boundary value problems for nonlinear elliptic equations. *Trans. Amer. Math. Soc. 295*, 2 (1986), 509–546.

556. Liese, F., and Vajda, I. *Convex statistical distances*, vol. 95 of *Teubner-Texte zur Mathematik*. BSB B. G. Teubner Verlagsgesellschaft, Leipzig, 1987. With German, French and Russian summaries.
557. Lions, J.-L. *Quelques méthodes de résolution des problèmes aux limites non linéaires*. Dunod, 1969.
558. Lions, P.-L. *Generalized solutions of Hamilton–Jacobi equations*. Pitman (Advanced Publishing Program), Boston, Mass., 1982.
559. Lions, P.-L. Personal communication, 2003.
560. Lions, P.-L., and Trudinger, N. S. Linear oblique derivative problems for the uniformly elliptic Hamilton–Jacobi–Bellman equation. *Math. Z. 191*, 1 (1986), 1–15.
561. Lions, P.-L., Trudinger, N. S., and Urbas, J. The Neumann problem for equations of Monge–Ampère type. *Comm. Pure Appl. Math. 39*, 4 (1986), 539–563.
562. Lions, P.-L., Trudinger, N. S., and Urbas, J. The Neumann problem for equations of Monge–Ampère type. Proceedings of the Centre for Mathematical Analysis, Australian National University, 10, Canberra, 1986, pp. 135–140.
563. Lions, P.-L., and Lasry, J.-M. Régularité optimale de racines carrées. *C. R. Acad. Sci. Paris Sér. I Math. 343*, 10 (2006), 679–684.
564. Lions, P.-L., Papanicolaou, G., and Varadhan, S. R. S. Homogeneization of Hamilton–Jacobi equations. Unpublished preprint, 1987.
565. Lisini, S. Characterization of absolutely continuous curves in Wasserstein spaces. *Calc. Var. Partial Differential Equations 28*, 1 (2007), 85–120.
566. Liu, J. Hölder regularity of optimal mapping in optimal transportation. To appear in *Calc. Var. Partial Differential Equations*.
567. Quasi-neutral limit of the Euler–Poisson and Euler–Monge–Ampère systems. *Comm. Partial Differential Equations 30*, 7-9 (2005), 1141–1167.
568. Loeper, G. The reconstruction problem for the Euler-Poisson system in cosmology. *Arch. Ration. Mech. Anal. 179*, 2 (2006), 153–216.
569. Loeper, G. A fully nonlinear version of Euler incompressible equations: the Semi-Geostrophic system. *SIAM J. Math. Anal. 38*, 3 (2006), 795–823.
570. Loeper, G. On the regularity of maps solutions of optimal transportation problems. To appear in *Acta Math.*
571. Loeper, G. On the regularity of maps solutions of optimal transportation problems II. Work in progress, 2007.
572. Loeper, G., and Villani, C. Regularity of optimal transport in curved geometry: the nonfocal case. Preprint, 2007.
573. Lott, J. Optimal transport and Ricci curvature for metric-measure spaces. To appear in *Surveys in Differential Geometry*.
574. Lott, J. Some geometric properties of the Bakry–Émery–Ricci tensor. *Comment. Math. Helv. 78*, 4 (2003), 865–883.
575. Lott, J. Some geometric calculations on Wasserstein space. To appear in *Comm. Math. Phys.* Available online at `www.math.lsa.umich.edu/~lott`.
576. Lott, J. Optimal transport and Perelman's reduced volume. Preprint, 2008.
577. Lott, J., and Villani, C. Ricci curvature for metric-measure spaces via optimal transport. To appear in *Ann. of Math. (2)* Available online at `www.umpa.ens-lyon.fr/~cvillani`.
578. Lott, J., and Villani, C. Weak curvature bounds and functional inequalities. *J. Funct. Anal. 245*, 1 (2007), 311–333.

579. Lott, J., and Villani, C. Hamilton–Jacobi semigroup on length spaces and applications. *J. Math. Pures Appl. (9) 88*, 3 (2007), 219–229.
580. Lu, P., Ni, L., Vázquez, J.-L., and Villani, C. Local Aronson–Bénilan estimates and entropy formulae for porous medium and fast diffusion equations on manifolds. Preprint, 2008.
581. Lusternik, L. A. Die Brunn–Minkowskische Ungleichung für beliebige messbare Mengen. *Dokl. Acad. Sci. URSS*, 8 (1935), 55–58.
582. Lutwak, E., Yang, D., and Zhang, G. Optimal Sobolev norms and the L^p Minkowski problem. *Int. Math. Res. Not.* (2006), Art. ID 62987, 21.
583. Lytchak, A. Differentiation in metric spaces. *Algebra i Analiz 16*, 6 (2004), 128–161.
584. Lytchak, A. Open map theorem for metric spaces. *Algebra i Analiz 17*, 3 (2005), 139–159.
585. Ma, X.-N., Trudinger, N. S., and Wang, X.-J. Regularity of potential functions of the optimal transportation problem. *Arch. Ration. Mech. Anal. 177*, 2 (2005), 151–183.
586. Maggi, F. Some methods for studying stability in isoperimetric type problems. To appear in *Bull. Amer. Math. Soc.*
587. Maggi, F., and Villani, C. Balls have the worst best Sobolev inequalities. *J. Geom. Anal. 15*, 1 (2005), 83–121.
588. Maggi, F., and Villani, C. Balls have the worst best Sobolev inequalities. Part II: Variants and extensions. *Calc. Var. Partial Differential Equations 31*, 1 (2008), 47–74.
589. Mallows, C. L. A note on asymptotic joint normality. *Ann. Math. Statist. 43* (1972), 508–515.
590. Malrieu, F. Logarithmic Sobolev inequalities for some nonlinear PDE's. *Stochastic Process. Appl. 95*, 1 (2001), 109–132.
591. Malrieu, F. Convergence to equilibrium for granular media equations and their Euler schemes. *Ann. Appl. Probab. 13*, 2 (2003), 540–560.
592. Mañé, R. Lagrangian flows: the dynamics of globally minimizing orbits. In *International Conference on Dynamical Systems* (Montevideo, 1995), vol. 362 of *Pitman Res. Notes Math. Ser.* Longman, Harlow, 1996, pp. 120–131.
593. Mañé, R. Lagrangian flows: the dynamics of globally minimizing orbits. *Bol. Soc. Brasil. Mat. (N.S.) 28*, 2 (1997), 141–153.
594. Maroofi, H. *Applications of the Monge–Kantorovich theory.* PhD thesis, Georgia Tech, 2002.
595. Marton, K. A measure concentration inequality for contracting Markov chains. *Geom. Funct. Anal. 6* (1996), 556–571.
596. Marton, K. Measure concentration for Euclidean distance in the case of dependent random variables. *Ann. Probab. 32*, 3B (2004), 2526–2544.
597. Massart, P. Concentration inequalities and model selection. Lecture Notes from the 2003 Saint-Flour Probability Summer School. To appear in the Springer book series *Lecture Notes in Math.* Available online at `www.math.u-psud.fr/~massart`.
598. Mather, J. N. Existence of quasiperiodic orbits for twist homeomorphisms of the annulus. *Topology 21*, 4 (1982), 457–467.
599. Mather, J. N. More Denjoy minimal sets for area preserving diffeomorphisms. *Comment. Math. Helv. 60* (1985), 508–557.
600. Mather, J. N. Minimal measures. *Comment. Math. Helv. 64*, 3 (1989), 375–394.

601. Mather, J. N. Action minimizing invariant measures for positive definite Lagrangian systems. *Math. Z. 207*, 2 (1991), 169–207.
602. Mather, J. N. Variational construction of orbits of twist diffeomorphisms. *J. Amer. Math. Soc. 4*, 2 (1991), 207–263.
603. Mather, J. N. Variational construction of connecting orbits. *Ann. Inst. Fourier (Grenoble) 43*, 5 (1993), 1349–1386.
604. Mather, J. N., and Forni, G. Action minimizing orbits in Hamiltonian systems. In *Transition to chaos in classical and quantum mechanics*, vol. 1589 of *Lecture Notes in Math.*, Springer-Verlag, Berlin, 1994, pp. 92–186.
605. Mattingly, J. C. Ergodicity of stochastic partial differential equations. Lecture Notes for the 2007 Saint-Flour Summer School.
606. Mattingly, J. C., Stuart, A. M., and Higham, D. J. Ergodicity for SDEs and approximations: locally Lipschitz vector fields and degenerate noise. *Stochastic Process. Appl. 101*, 2 (2002), 185–232.
607. Maurey, B. Some deviation inequalities. *Geom. Funct. Anal. 1*, 2 (1991), 188–197.
608. Maurey, B. Inégalité de Brunn–Minkowski–Lusternik, et autres inégalités géométriques et fonctionnelles. *Astérisque*, 299 (2005), Exp. No. 928, vii, 95–113. Séminaire Bourbaki, Vol. 2003/2004.
609. Maz'ja, V. G. The negative spectrum of the higher-dimensional Schrödinger operator. *Dokl. Akad. Nauk SSSR 144* (1962), 721–722.
610. Maz'ja, V. G. On the solvability of the Neumann problem. *Dokl. Akad. Nauk SSSR 147* (1962), 294–296.
611. Maz'ja, V. G. *Sobolev spaces*. Springer Series in Soviet Mathematics. Springer-Verlag, Berlin, 1985. Translated from the Russian by T. O. Shaposhnikova.
612. McCann, R. J. *A convexity theory for interacting gases and equilibrium crystals*. PhD thesis, Princeton Univ., 1994.
613. McCann, R. J. Existence and uniqueness of monotone measure-preserving maps. *Duke Math. J. 80*, 2 (1995), 309–323.
614. McCann, R. J. A convexity principle for interacting gases. *Adv. Math. 128*, 1 (1997), 153–179.
615. McCann, R. J. Exact solutions to the transportation problem on the line. *R. Soc. Lond. Proc. Ser. A Math. Phys. Eng. Sci. 455*, 1984 (1999), 1341–1380.
616. McCann, R. J. Polar factorization of maps on Riemannian manifolds. *Geom. Funct. Anal. 11*, 3 (2001), 589–608.
617. McCann, R. J. Stable rotating binary stars and fluid in a tube. *Houston J. Math. 32*, 2 (2006), 603–631.
618. McCann, R. J., and Puel, M. Constructing a relativistic heat flow by transport time steps. Preprint, 2007. Available at `www.math.toronto.edu/mccann`.
619. McCann, R. J., and Slepčev, D. Second-order asymptotics for the fast-diffusion equation. *Int. Math. Res. Not.* (2006), Art. ID 24947, 22.
620. McCann, R. J., and Topping, P. Ricci flow, entropy and optimal transportation. Preprint, 2007. Available online at `www.math.toronto.edu/mccann`.
621. McDiarmid, C. On the method of bounded differences. In *Surveys in combinatorics, 1989* (Norwich, 1989), vol. 141 of *London Math. Soc. Lecture Note Ser.*, Cambridge University Press, Cambridge, 1989, pp. 148–188.
622. McNamara, S., and Young, W. R. Kinetics of a one-dimensional granular medium. *Phys. Fluids A 5*, 1 (1993), 34–45.

623. MECKES, M. Some remarks on transportation cost and related inequalities. In *Geometric aspects of functional analysis*, Lect. Notes in Math., vol. 1910, Springer-Verlag, Berlin, 2007, pp. 237–244.
624. MÉLÉARD, S. Probabilistic interpretation and approximations of some Boltzmann equations. In *Stochastic models* (Guanajuato, 1998). Soc. Mat. Mexicana, México, 1998, pp. 1–64.
625. MELLERAY, J., PETROV, F., AND VERSHIK, A. M. Linearly rigid metric spaces. *C. R. Math. Acad. Sci. Paris 344*, 4 (2007), 235–240.
626. MENGUY, X. Examples of nonpolar limit spaces. *Amer. J. Math. 122*, 5 (2000), 927–937.
627. MIHALAS, D., AND MIHALAS, B. *Foundations of Radiation Hydrodynamics.* Oxford University Press, 1984.
628. MIKAMI, T. Monge's problem with a quadratic cost by the zero-noise limit of h-path processes. *Probab. Theory Related Fields 129*, 2 (2004), 245–260.
629. MIKAMI, T. A simple proof of duality theorem for Monge–Kantorovich problem. *Kodai Math. J. 29*, 1 (2006), 1–4.
630. MIKAMI, T., AND THIEULLEN, M. Optimal transportation problem by stochastic optimal control. Preprint, 2005. Available online at `eprints.math.sci.hokudai.ac.jp`.
631. MIKAMI, T., AND THIEULLEN, M. Duality theorem for the stochastic optimal control problem. *Stochastic Process. Appl. 116*, 12 (2006), 1815–1835.
632. MILAKIS, E. On the regularity of optimal sets in mass transfer problems. *Comm. Partial Differential Equations 31*, 4–6 (2006), 817–826.
633. MILMAN, E. On the role of convexity in isoperimetry, spectral gap and concentration. Preprint, 2008.
634. MILMAN, V. D. A new proof of A. Dvoretzky's theorem on cross-sections of convex bodies. *Funkcional. Anal. i Priložen. 5*, 4 (1971), 28–37.
635. MILMAN, V. D., AND SCHECHTMAN, G. *Asymptotic theory of finite-dimensional normed spaces.* With an appendix by M. Gromov. Springer-Verlag, Berlin, 1986.
636. MONGE, G. Mémoire sur la théorie des déblais et des remblais. In *Histoire de l'Académie Royale des Sciences de Paris* (1781), pp. 666–704.
637. MORSE, H. M. A fundamental class of geodesics on any closed surface of genus greater than one. *Trans. Amer. Math. Soc. 26* (1924), 25–60.
638. MOSER, J. On Harnack's theorem for elliptic differential equations. *Comm. Pure Appl. Math. 14* (1961), 577–591.
639. MOSER, J. A Harnack inequality for parabolic differential equations. *Comm. Pure Appl. Math. 17* (1964), 101–134. Erratum in *Comm. Pure Appl. Math. 20* (1967), 231–236.
640. MOSER, J. On the volume elements on a manifold. *Trans. Amer. Math. Soc. 120* (1965), 286–294.
641. MOSER, J. Monotone twist mappings and the calculus of variations. *Ergodic Theory Dynam. Systems 6*, 3 (1986), 401–413.
642. MOUTSINGA, O. *Approche probabiliste des Particules Collantes et Système de Gaz Sans Pression.* PhD thesis, Univ. Lille I (France), 2003.
643. MUCKENHOUPT, B. Hardy's inequality with weights. *Studia Math. 44* (1972), 31–38.
644. MURATA, H., AND TANAKA, H. An inequality for certain functional of multidimensional probability distributions. *Hiroshima Math. J. 4* (1974), 75–81.

645. Namah, G., and Roquejoffre, J.-M. Remarks on the long time behaviour of the solutions of Hamilton–Jacobi equations. *Comm. Partial Differential Equations 24*, 5–6 (1999), 883–893.
646. Nash, J. Continuity of solutions of parabolic and elliptic equations. *Amer. J. Math. 80* (1958), 931–954.
647. Nelson, E. Derivation of the Schrödinger equation from Newtonian mechanics. *Phys. Rev. 150* (1966), 1079–1085.
648. Nelson, E. *Dynamical theories of Brownian motion.* Princeton University Press, Princeton, N.J., 1967. 2001 re-edition by J. Suzuki. Available online at `www.math.princeton.edu/~nelson/books.html`.
649. Nelson, E. The free Markoff field. *J. Functional Analysis 12* (1973), 211–227.
650. Nelson, E. Critical diffusions. In *Séminaire de probabilités, XIX, 1983/84*, vol. 1123 of *Lecture Notes in Math.* Springer, Berlin, 1985, pp. 1–11.
651. Nelson, E. *Quantum fluctuations.* Princeton Series in Physics. Princeton University Press, Princeton, NJ, 1985.
652. Nelson, E. Stochastic mechanics and random fields. In *École d'Été de Probabilités de Saint-Flour XV–XVII, 1985–87*, vol. 1362 of *Lecture Notes in Math.*, Springer, Berlin, 1988, pp. 427–450.
653. Neunzert, H. An introduction to the nonlinear Boltzmann–Vlasov equation. In *Kinetic theories and the Boltzmann equation*, C. Cercignani, Ed., vol. 1048 of *Lecture Notes in Math.*, Springer, Berlin, Heidelberg, 1984, pp. 60–110.
654. Ohta, S.-i. On the measure contraction property of metric measure spaces. *Comment. Math. Helv. 82*, 4 (2007), 805–828.
655. Ohta, S.-i. Gradient flows on Wasserstein spaces over compact Alexandrov spaces. To appear in *Amer. J. Math.*
656. Ohta, S.-i. Products, cones, and suspensions of spaces with the measure contraction property. *J. Lond. Math. Soc. (2) 76*, 1 (2007), 225–236.
657. Ohta, S.-i. Finsler interpolation inequalities. Preprint, 2008. Available online at `www.math.kyoto-u.ac.jp/~sohta`.
658. Øksendal, B. *Stochastic differential equations. An introduction with applications*, sixth ed. Universitext. Springer-Verlag, Berlin, 2003.
659. Oliker, V. Embedding $\mathbb{S}^n$ into $\mathbb{R}^{n+1}$ with given integral Gauss curvature and optimal mass transport on $\mathbb{S}^n$. *Adv. Math. 213*, 2 (2007), 600–620.
660. Oliker, V. Variational solutions of some problems in convexity via Monge–Kantorovich optimal mass transport theory. Conference in Oberwolfach, July 2006 (personal communication).
661. Oliker, V., and Prussner, L. D. A new technique for synthesis of offset dual reflector systems. In *10th Annual Review of Progress in Applied Computational Electromagnetics*, 1994, pp. 45–52.
662. Ollivier, Y. Ricci curvature of Markov chains on metric spaces. Preprint, 2007. Available online at `www.umpa.ens-lyon.fr/~yollivie/publs.html`.
663. Ollivier, Y., and Pansu, P. Courbure de Ricci et concentration de la mesure. Working seminar notes. Available online at `www.umpa.ens-lyon.fr/~yollivie`.
664. Osserman, R. The isoperimetric inequality. *Bull. Amer. Math. Soc. 84*, 6 (1978), 1182–1238.
665. Otsu, Y., and Shioya, T. The Riemannian structure of Alexandrov spaces. *J. Differential Geom. 39*, 3 (1994), 629–658.
666. Otto, F. Double degenerate diffusion equations as steepest descent. Preprint Univ. Bonn, 1996.

667. Otto, F. Dynamics of labyrinthine pattern formation in magnetic fluids: a mean-field theory. *Arch. Rational Mech. Anal. 141*, 1 (1998), 63–103.
668. Otto, F. Lubrication approximation with prescribed nonzero contact angle. *Comm. Partial Differential Equations 23*, 11–12 (1998), 2077–2164.
669. Otto, F. The geometry of dissipative evolution equations: the porous medium equation. *Comm. Partial Differential Equations 26*, 1–2 (2001), 101–174.
670. Otto, F., and Tzavaras, A. E. Continuity of velocity gradients in suspensions of rod-like molecules. Preprint, 2004. Available online at `www.math.umd.edu/ ~tzavaras/listsu.html`.
671. Otto, F., and Villani, C. Generalization of an inequality by Talagrand and links with the logarithmic Sobolev inequality. *J. Funct. Anal. 173*, 2 (2000), 361–400.
672. Otto, F., and Villani, C. Comment on: "Hypercontractivity of Hamilton–Jacobi equations", by S. G. Bobkov, I. Gentil and M. Ledoux. *J. Math. Pures Appl. (9) 80*, 7 (2001), 697–700.
673. Otto, F., and Westdickenberg, M. Eulerian calculus for the contraction in the Wasserstein distance. *SIAM J. Math. Anal. 37*, 4 (2005), 1227–1255.
674. Oxtoby, J. C. Ergodic sets. *Bull. Amer. Math. Soc. 58* (1952), 116–136.
675. Paulin, F. Topologie de Gromov équivariante, structures hyperboliques et arbres réels. *Invent. Math. 94*, 1 (1988), 53–80.
676. Perel′man, G. Alexandrov's spaces with curvatures bounded from below. Unpublished preprint, available online at `www.math.psu.edu/petrunin/papers/papers.html`.
677. Perel′man, G. DC structure on Alexandrov spaces. Unpublished preprint, preliminary version available online at `www.math.psu.edu/petrunin/papers/papers.html`.
678. Perel′man, G., and Petrunin, A. Quasigeodesics and gradient curves in Alexandrov spaces. Unpublished preprint, available online at `www.math.psu.edu/petrunin/papers/papers.html`.
679. Peres, Y. Mixing for Markov chains and spin systems. Unpublished notes, available online at `www.stat.berkeley.edu/~peres/ubc.pdf`.
680. Petersen, P. *Riemannian geometry*, vol. 171 of *Graduate Texts in Mathematics.* Springer-Verlag, New York, 1998.
681. Petrunin, A. Harmonic functions on Alexandrov spaces. Unpublished preprint, available online at `www.math.psu.edu/petrunin/papers/papers.html`.
682. Pinsker, M. S. *Information and information stability of random variables and processes.* Holden-Day, San Francisco, 1964.
683. Plakhov, A. Yu. Exact solutions of the one-dimensional Monge–Kantorovich problem. *Mat. Sb. 195*, 9 (2004), 57–74; English translation in *Sb. Math. 195*, 9–10 (2004), 1291–1307.
684. Pogorelov, A. V. *Monge-Ampère equations of elliptic type.* Groningen, Noordhoff, 1964.
685. Pogorelov, A. V. The Dirichlet problem for the multidimensional analogue of the Monge–Ampère equation. *Dokl. Akad. Nauk SSSR 201* (1971), 790–793. English translation in *Soviet Math. Dokl. 12* (1971), 1727–1731.
686. Pratelli, A. How to show that some rays are maximal transport rays in Monge problem. *Rend. Sem. Mat. Univ. Padova 113* (2005), 179–201.

687. PRATELLI, A. On the equality between Monge's infimum and Kantorovich's minimum in optimal mass transportation. *Ann. Inst. H. Poincaré Probab. Statist. 43*, 1 (2007), 1–13.
688. PRATELLI, A. On the sufficiency of c-cyclical monotonicity for optimality of transport plans. To appear in *Math. Z.*
689. PRATELLI, A. Equivalence between some definitions for the optimal mass transport problem and for the transport density on manifolds. *Ann. Mat. Pura Appl. (4) 184*, 2 (2005), 215–238.
690. PRÉKOPA, A. On logarithmic concave measures and functions. *Acta Sci. Math. (Szeged) 34* (1973), 335–343.
691. PRIGOZHIN, L. Variational model of sandpile growth. *European J. Appl. Math. 7*, 3 (1996), 225–235.
692. PULVIRENTI, A., AND TOSCANI, G. The theory of the nonlinear Boltzmann equation for Maxwell molecules in Fourier representation. *Ann. Mat. pura ed appl. 171*, 4 (1996), 181–204.
693. PULVIRENTI, M. On invariant measures for the 2-D Euler flow. In *Mathematical aspects of vortex dynamics* (Leesburg, VA, 1988). SIAM, Philadelphia, PA, 1989, pp. 88–96.
694. RACHEV, S. The Monge–Kantorovich mass transference problem and its stochastic applications. *Theory Probab. Appl. 29* (1984), 647–676.
695. RACHEV, S. T. *Probability metrics and the stability of stochastic models.* John Wiley & Sons Ltd., Chichester, 1991.
696. RACHEV, S. T., AND RÜSCHENDORF, L. *Mass Transportation Problems. Vol. I: Theory, Vol. II: Applications.* Probability and its applications. Springer-Verlag, New York, 1998.
697. RADEMACHER, H. Über partielle und totale differenzierbarkeit von Funktionen mehrerer Variabeln und über die Transformation der Doppelintegrale. *Math. Ann. 79*, 4 (1919), 340–359.
698. RAMACHANDRAN, D., AND RÜSCHENDORF, L. A general duality theorem for marginal problems. *Probab. Theory Related Fields 101*, 3 (1995), 311–319.
699. RAMACHANDRAN, D., AND RÜSCHENDORF, L. Duality and perfect probability spaces. *Proc. Amer. Math. Soc. 124*, 7 (1996), 2223–2228.
700. REINHARD, E., ASHIKHMIN, M., GOOCH, B., AND SHIRLEY, P. Color transfer between images. *IEEE Computer Graphics and Applications 21*, 5 (2001), 34–41.
701. REY-BELLET, L., AND THOMAS, L. E. Exponential convergence to non-equilibrium stationary states in classical statistical mechanics. *Comm. Math. Phys. 225*, 2 (2002), 305–329.
702. RIGOT, S. Mass transportation in groups of type H. *Commun. Contemp. Math. 7*, 4 (2005), 509–537.
703. RIO, E. Inégalités de Hoeffding pour les fonctions lipschitziennes de suites dépendantes. *C. R. Acad. Sci. Paris Sér. I Math. 330*, 10 (2000), 905–908.
704. RIO, E. Upper bounds for minimal distances in the central limit theorem. Preprint, 2007.
705. ROCKAFELLAR, R. T. *Convex analysis.* Princeton University Press, Princeton, NJ, 1997. Reprint of the 1970 original, Princeton Paperbacks.
706. RODRIGUEZ-ITURBE, I., AND RINALDO, A. *Fractal River Basins.* Cambridge University Press, 1997.

707. Roquejoffre, J.-M. Convergence to steady states or periodic solutions in a class of Hamilton–Jacobi equations. *J. Math. Pures Appl. (9) 80*, 1 (2001), 85–104.
708. Rosenau, P. Tempered diffusion: A transport process with propagating front and initial delay. *Phys. Rev. A 46* (1992), 7371–7374.
709. Rosenblatt, M. Remarks on a multivariate transformation. *Ann. Math. Statistics 23* (1952), 470–472.
710. Rothaus, O. S. Diffusion on compact Riemannian manifolds and logarithmic Sobolev inequalities. *J. Funct. Anal. 42* (1981), 102–109.
711. Royer, G. *Une initiation aux inégalités de Sobolev logarithmiques*, vol. 5 of *Cours Spécialisés*. Société Mathématique de France, Paris, 1999.
712. Rubinstein, J., and Wolansky, G. Intensity control with a free-form lens. *J. Opt. Soc. Am. A 24*, 2 (2007), 463–469.
713. Rubner, Y., Tomasi, C., and Guibas, L. J. The Earth Mover's distance as a metric for image retrieval. *International Journal of Computer Vision 40*, 2 (2000), 99–121.
714. Rudin, W. *Real and complex analysis*, third ed. McGraw-Hill Book Co., New York, 1987.
715. Rüschendorf, L. Sharpness of Fréchet-bounds. *Z. Wahrsch. Verw. Gebiete 57*, 2 (1981), 293–302.
716. Rüschendorf, L. The Wasserstein distance and approximation theorems. *Z. Wahrsch. Verw. Gebiete 70* (1985), 117–129.
717. Rüschendorf, L. Fréchet-bounds and their applications. In *Advances in probability distributions with given marginals* (Rome, 1990), vol. 67 of *Math. Appl.* Kluwer Acad. Publ., Dordrecht, 1991, pp. 151–187.
718. Rüschendorf, L. Optimal solutions of multivariate coupling problems. *Appl. Math. (Warsaw) 23*, 3 (1995), 325–338.
719. Rüschendorf, L. On c-optimal random variables. *Statist. Probab. Lett. 27*, 3 (1996), 267–270.
720. Rüschendorf, L. Wasserstein-metric. Historical note. Personal communication, 2006.
721. Rüschendorf, L. Monge–Kantorovich transportation problem and optimal couplings. Notes based on a talk at the MSRI meeting on Optimal Transport (Berkeley, November 2005), 2006.
722. Rüschendorf, L., and Rachev, S. T. A characterization of random variables with minimum L^2-distance. *J. Multivariate Anal. 32*, 1 (1990), 48–54. Corrigendum in *J. Multivariate Anal. 34*, 1 (1990), p. 156.
723. Rüschendorf, L., and Uckelmann, L. On optimal multivariate couplings. In *Distributions with given marginals and moment problems* (Prague, 1996). Kluwer Acad. Publ., Dordrecht, 1997, pp. 261–273.
724. Rüschendorf, L., and Uckelmann, L. Numerical and analytical results for the transportation problem of Monge–Kantorovich. *Metrika 51*, 3 (2000), 245–258.
725. Rüschendorf, L., and Uckelmann, L. On the n-coupling problem. *J. Multivariate Anal. 81*, 2 (2002), 242–258.
726. Saloff-Coste, L. A note on Poincaré, Sobolev, and Harnack inequalities. *Internat. Math. Res. Notices*, 2 (1992), 27–38.
727. Saloff-Coste, L. Convergence to equilibrium and logarithmic Sobolev constant on manifolds with Ricci curvature bounded below. *Colloquium Math. 67* (1994), 109–121.

728. Saloff-Coste, L. *Aspects of Sobolev-type inequalities*, vol. 289 of *London Mathematical Society Lecture Note Series*. Cambridge University Press, Cambridge, 2002.
729. Sammer, M. D. Aspects of mass transportation in discrete concentration inequalities. PhD thesis, Georgia Institute of Technology, 2005. Available online at `etd.gatech.edu/theses/etd-04112005-163457/unrestricted/sammer_marcus_d_200505_phd.pdf`
730. Samson, P.-M. Concentration of measure inequalities for Markov chains and Φ-mixing processes. *Ann. Probab. 28*, 1 (2000), 416–461.
731. Santambrogio, F. Transport and concentration problems with interaction effects. *J. Global Optim. 38*, 1 (2007), 129–141.
732. Santambrogio, F. Optimal channel networks, landscape function and branched transport. *Interfaces and Free Boundaries 9* (2007), 149–169.
733. Santambrogio, F. *Variational problems in transport theory with mass concentration.* PhD Thesis, Scuola Normale di Pisa, 2006.
734. Santambrogio, F., and Tilli, P. Blow-up of optimal sets in the irrigation problems. *J. Geom. Anal. 15*, 2 (2005), 343–362.
735. Savaré, G. Gradient flows and diffusion semigroups in metric spaces under lower curvature bounds. *C. R. Acad. Sci. Paris Sér. I Math. 345*, 3 (2007), 151–154.
736. Savaré, G. Work in progress.
737. Savin, O. A free boundary problem with optimal transportation. *Comm. Pure Appl. Math. 57*, 1 (2004), 126–140.
738. Schachermayer, W., and Teichmann, J. Characterization of optimal transport plans for the Monge–Kantorovich problem. To appear in *Proc. Amer. Math. Soc.* Available online at `www.fam.tuwien.ac.at/~wschach/pubs`.
739. Schachermayer, W., and Teichmann, J. How close are the Option Pricing Formulas of Bachelier and Black–Merton–Scholes? To appear in *Math. Finance.*
740. Scheffer, G. Isopérimétrie fonctionnelle dimensionnelle en courbure positive. *C. R. Acad. Sci. Paris Sér. I Math. 331*, 3 (2000), 251–254.
741. Schneider, R. *Convex bodies: the Brunn–Minkowski theory.* Cambridge University Press, Cambridge, 1993.
742. Schochet, S. The point-vortex method for periodic weak solutions of the 2-D Euler equations. *Comm. Pure Appl. Math. 49*, 9 (1996), 911–965.
743. Schrijver, A. *Combinatorial optimization: polyhedra and efficiency*, vol. 24 of *Algorithms and combinatorics.* Springer-Verlag, 2003.
744. Schulz, F. *Regularity theory for quasilinear elliptic systems and Monge–Ampère equations in two dimensions.* Lect. Notes in Math., vol. 1445, Springer-Verlag, Berlin, 1990.
745. Sever, M. An existence theorem in the large for zero-pressure gas dynamics. *Differential Integral Equations 14*, 9 (2001), 1077–1092.
746. Shannon, C. E. A mathematical theory of communication. *Bell System Tech. J. 27* (1948), 379–423, 623–656.
747. Shannon, C. E., and Weaver, W. *The mathematical theory of communication.* The University of Illinois Press, Urbana, Ill., 1949.
748. Shao, J. *Inégalité du coût de transport et le problème de Monge–Kantorovich sur le groupe des lacets et l'espace des chemins.* PhD thesis, Univ. Bourgogne, 2006.
749. Shao, J. Hamilton–Jacobi semigroups in infinite dimensional spaces. *Bull. Sci. Math. 130*, 8 (2006), 720–738.

750. SHENG, W., TRUDINGER, N. S., AND WANG, X.-J. The Yamabe problem for higher order curvatures. Preprint, archived online at `arxiv.org/abs/math/0505463`.
751. SHIOYA, T. Mass of rays in Alexandrov spaces of nonnegative curvature. *Comment. Math. Helv. 69*, 2 (1994), 208–228.
752. SIBURG, K. F. *The principle of least action in geometry and dynamics*, vol. 1844 of *Lecture Notes in Mathematics*. Springer-Verlag, Berlin, 2004.
753. SIMON, L. *Lectures on geometric measure theory.* Proceedings of the Centre for Mathematical Analysis, Australian National University, 3, Canberra, 1983.
754. SMITH, C., AND KNOTT, M. On Hoeffding–Fréchet bounds and cyclic monotone relations. *J. Multivariate Anal. 40*, 2 (1992), 328–334.
755. SOBOLEVSKIĬ, A., AND FRISCH, U. Application of optimal transportation theory to the reconstruction of the early Universe. *Zap. Nauchn. Sem. S.-Peterburg. Otdel. Mat. Inst. Steklov. (POMI) 312*, 11 (2004), 303–309, 317. English translation in *J. Math. Sci. (N. Y.) 133*, 4 (2006), 1539–1542.
756. SOIBELMAN, Y. Notes on noncommutative Riemannian geometry. Personal communication, 2006.
757. SPOHN, H. *Large scale dynamics of interacting particles.* Texts and Monographs in Physics. Springer-Verlag, Berlin, 1991.
758. STAM, A. Some inequalities satisfied by the quantities of information of Fisher and Shannon. *Inform. Control 2* (1959), 101–112.
759. STROOCK, D. W. *An introduction to the analysis of paths on a Riemannian manifold*, vol. 74 of *Mathematical Surveys and Monographs.* American Mathematical Society, Providence, RI, 2000.
760. STURM, K.-T. Diffusion processes and heat kernels on metric spaces. *Ann. Probab. 26*, 1 (1998), 1–55.
761. STURM, K.-T. Convex functionals of probability measures and nonlinear diffusions on manifolds. *J. Math. Pures Appl. (9) 84*, 2 (2005), 149–168.
762. STURM, K.-T. On the geometry of metric measure spaces. I. *Acta Math. 196*, 1 (2006), 65–131.
763. STURM, K.-T. On the geometry of metric measure spaces. II. *Acta Math. 196*, 1 (2006), 133–177.
764. STURM, K.-T., AND VON RENESSE, M.-K. Transport inequalities, gradient estimates, entropy and Ricci curvature. *Comm. Pure Appl. Math. 58*, 7 (2005), 923–940.
765. SUDAKOV, V. N. Geometric problems in the theory of infinite-dimensional probability distributions. *Proc. Steklov Inst. Math. 141* (1979), 1–178.
766. SUDAKOV, V. N. AND CIREL′SON, B. S. Extremal properties of half-spaces for spherically invariant measures. *Zap. Naučn. Sem. Leningrad. Otdel. Mat. Inst. Steklov. (LOMI) 41* (1974), 14–24. English translation in *J. Soviet Math. 9* (1978), 9–18.
767. SZNITMAN, A.-S. Equations de type de Boltzmann, spatialement homogènes. *Z. Wahrsch. Verw. Gebiete 66* (1984), 559–562.
768. SZNITMAN, A.-S. Topics in propagation of chaos. In *École d'Été de Probabilités de Saint-Flour XIX—1989.* Springer, Berlin, 1991, pp. 165–251.
769. SZULGA, A. On minimal metrics in the space of random variables. *Teor. Veroyatnost. i Primenen. 27*, 2 (1982), 401–405.
770. TALAGRAND, M. A new isoperimetric inequality for product measure and the tails of sums of independent random variables. *Geom. Funct. Anal. 1*, 2 (1991), 211–223.

771. Talagrand, M. Matching random samples in many dimensions. *Ann. Appl. Probab. 2*, 4 (1992), 846–856.
772. Talagrand, M. Concentration of measure and isoperimetric inequalities in product spaces. *Inst. Hautes Études Sci. Publ. Math.*, 81 (1995), 73–205.
773. Talagrand, M. New concentration inequalities in product spaces. *Invent. Math. 126*, 3 (1996), 505–563.
774. Talagrand, M. Transportation cost for Gaussian and other product measures. *Geom. Funct. Anal. 6*, 3 (1996), 587–600.
775. Talagrand, M. *Spin Glasses: a challenge for mathematicians. Cavity and mean field models.* Vol. 46 of *Ergebnisse der Mathematik und ihrer Grenzgebiete.* Springer-Verlag, Berlin, 2003.
776. Tanaka, H. An inequality for a functional of probability distributions and its application to Kac's one-dimensional model of a Maxwellian gas. *Z. Wahrscheinlichkeitstheorie und Verw. Gebiete 27* (1973), 47–52.
777. Tanaka, H. Probabilistic treatment of the Boltzmann equation of Maxwellian molecules. *Z. Wahrsch. Verw. Gebiete 46*, 1 (1978/79), 67–105.
778. Taton, R. *Gaspard Monge.* Birkhäuser, Basel, 1951.
779. Taton, R., Ed. *L'oeuvre scientifique de Gaspard Monge.* P.U.F., Paris, 1951.
780. Thirring, W. *Classical mathematical physics*, third ed. Springer-Verlag, New York, 1997. Dynamical systems and field theories, Translated from the German by Evans M. Harrell, II.
781. Thorisson, H. *Coupling, stationarity, and regeneration.* Probability and its Applications. Springer-Verlag, New York, 2000.
782. Topping, P. $\mathcal{L}$-optimal transportation for Ricci flow. Preprint, available online at `www.warwick.ac.uk/~maseq`.
783. Toscani, G. New a priori estimates for the spatially homogeneous Boltzmann equation. *Cont. Mech. Thermodyn. 4* (1992), 81–93.
784. Toscani, G. Sur l'inégalité logarithmique de Sobolev. *C.R. Acad. Sci. Paris, Série I, 324* (1997), 689–694.
785. Toscani, G. Entropy production and the rate of convergence to equilibrium for the Fokker–Planck equation. *Quart. Appl. Math. 57*, 3 (1999), 521–541.
786. Trudinger, N. S. *Lectures on nonlinear elliptic equations of second order.* Notes by N. Ishimura. Lectures in Mathematical Sciences 9, Univ. of Tokyo, 1995, pp. 1–52.
787. Trudinger, N. S. Isoperimetric inequalities for quermassintegrals. *Ann. Inst. H. Poincaré Anal. Non Linéaire 11*, 4 (1994), 411–425.
788. Trudinger, N. S. Recent developments in elliptic partial differential equations of Monge–Ampère type. *International Congress of Mathematicians, Vol. III*, Eur. Math. Soc., Zürich, 2006, pp. 291–301.
789. Trudinger, N. S., and Wang, X.-J. Hessian measures. I. *Topol. Methods Nonlinear Anal. 10*, 2 (1997), 225–239.
790. Trudinger, N. S., and Wang, X.-J. Hessian measures. II. *Ann. of Math. (2) 150*, 2 (1999), 579–604.
791. Trudinger, N. S., and Wang, X.-J. On the Monge mass transfer problem. *Calc. Var. Partial Differential Equations 13* (2001), 19–31.
792. Trudinger, N. S., and Wang, X.-J. Hessian measures. III. *J. Funct. Anal. 193*, 1 (2002), 1–23.
793. Trudinger, N. S., and Wang, X.-J. On strict convexity and continuous differentiability of potential functions in optimal transportation. To appear in *Arch. Ration. Mech. Anal.*

794. TRUDINGER, N. S., AND WANG, X.-J. On the second boundary value problem for Monge–Ampère type equations and optimal transportation. Preprint, 2006. Archived online at `arxiv.org/abs/math.AP/0601086`.
795. TUERO-DÍAZ, A. On the stochastic convergence of representations based on Wasserstein metrics. *Ann. Probab. 21*, 1 (1993), 72–85.
796. UCKELMANN, L. Optimal couplings between one-dimensional distributions. *Distributions with given marginals and moment problems* (Prague, 1996), Kluwer Acad. Publ., Dordrecht, 1997, pp. 275–281.
797. UNTERREITER, A., ARNOLD, A., MARKOWICH, P., AND TOSCANI, G. On generalized Csiszár–Kullback inequalities. *Monatsh. Math. 131*, 3 (2000), 235–253.
798. URBAS, J. On the second boundary value problem for equations of Monge–Ampère type. *J. Reine Angew. Math. 487* (1997), 115–124.
799. URBAS, J. Mass transfer problems. Lecture notes from a course given in Univ. Bonn, 1997–1998.
800. URBAS, J. The second boundary value problem for a class of Hessian equations. *Comm. Partial Differential Equations 26*, 5–6 (2001), 859–882.
801. VALDIMARSSON, S. I. On the Hessian of the optimal transport potential. Preprint, 2006.
802. VARADHAN, S. R. S. *Mathematical statistics.* Courant Institute of Mathematical Sciences New York University, New York, 1974. Lectures given during the academic year 1973–1974, Notes by Michael Levandowsky and Norman Rubin.
803. VASERSHTEIN, L. N. Markov processes over denumerable products of spaces describing large system of automata. *Problemy Peredači Informacii 5*, 3 (1969), 64–72.
804. VÁZQUEZ, J. L. An introduction to the mathematical theory of the porous medium equation. In *Shape optimization and free boundaries* (Montreal, PQ, 1990). Kluwer Acad. Publ., Dordrecht, 1992, pp. 347–389.
805. VÁZQUEZ, J. L. *The porous medium equation. Mathematical theory.* Oxford Mathematical Monographs. The Clarendon Press Oxford University Press, New York, 2006.
806. VÁZQUEZ, J. L. *Smoothing and decay estimates for nonlinear parabolic equations of porous medium type*, vol. 33 of *Oxford Lecture Series in Mathematics and its Applications.* Oxford University Press, 2006.
807. VERSHIK, A. M. Some remarks on the infinite-dimensional problems of linear programming. *Russian Math. Surveys 25*, 5 (1970), 117–124.
808. VERSHIK, A. M. L. V. Kantorovich and linear programming. Historical note (2001, updated in 2007). Archived at `www.arxiv.org/abs/0707.0491`.
809. VERSHIK, A. M. The Kantorovich metric: the initial history and little-known applications. *Zap. Nauchn. Sem. S.-Peterburg. Otdel. Mat. Inst. Steklov. (POMI) 312*, Teor. Predst. Din. Sist. Komb. i Algoritm. Metody. 11 (2004), 69–85, 311.
810. VERSHIK, A. M., Ed. *J. Math. Sci. 133*, 4 (2006). Special issue dedicated to L. V. Kantorovich. Springer, New York, 2006. Translated from the Russian: *Zapsiki Nauchn. seminarov POMI*, vol. 312: "Theory of representation of Dynamical Systems. Special Issue". Saint-Petersburg, 2004.
811. VILLANI, C. Remarks about negative pressure problems. Unpublished notes, 2002.
812. VILLANI, C. A review of mathematical topics in collisional kinetic theory. In *Handbook of mathematical fluid dynamics, Vol. I.* North-Holland, Amsterdam, 2002, pp. 71–305.

813. Villani, C. Optimal transportation, dissipative PDE's and functional inequalities. In *Optimal transportation and applications* (Martina Franca, 2001), vol. 1813 of *Lecture Notes in Mathematics*, Springer, Berlin, 2003, pp. 53–89.
814. Villani, C. *Topics in optimal transportation*, vol. 58 of *Graduate Studies in Mathematics*. American Mathematical Society, Providence, RI, 2003.
815. Villani, C. Trend to equilibrium for dissipative equations, functional inequalities and mass transportation. In *Recent advances in the theory and applications of mass transport*, vol. 353 of *Contemp. Math.*, Amer. Math. Soc., Providence, RI, 2004, pp. 95–109.
816. Villani, C. Hypocoercive diffusion operators. *International Congress of Mathematicians, Vol. III*, Eur. Math. Soc., Zürich, 2006, pp. 473–498.
817. Villani, C. Entropy dissipation and convergence to equilibrium. Notes from a series of lectures in Institut Henri Poincaré, Paris (2001). Updated in 2004. Lect. Notes in Math., vol. 1916, Springer-Verlag, Berlin, 2008, pp. 1–70.
818. Villani, C. Hypocoercivity. Preprint, 2006. To appear as a *Memoir of the AMS*.
819. Villani, C. Intégration et analyse de Fourier (in french). Incomplete lecture notes, available online at `www.umpa.ens-lyon.fr/~cvillani`.
820. Villani, C. Mathematics of granular materials. *J. Statist. Phys. 124*, 2–4 (2006), 781–822.
821. Villani, C. Transport optimal et courbure de Ricci. To appear in *Sémin. Théor. Spectr. Géom.* Available online at `www.umpa.ens-lyon.fr/~cvillani`.
822. Villani, C. Weak stability of a 4th-order curvature condition arising in optimal transport theory. Preprint, 2007.
823. Vinter, R. B. *Optimal control. Systems & Control: Foundations & Applications*, Birkhäuser, Boston, 2000.
824. von Nessi, G. T. On regularity for potentials of optimal transportation problems on spheres and related Hessian equations. PhD Thesis, Australian National University, 2008.
825. von Renesse, M.-K. On local Poincaré via transportation. To appear in *Math. Z.* Archived online at `math.MG/0505588`, 2005.
826. von Renesse, M.-K. An optimal transport view on Schrödinger's equation. Preprint, 2008.
827. von Renesse, M.-K., and Sturm, K.-T. Entropic measure and Wasserstein diffusion. To appear in *Ann. Probab.*
828. Wang, F.-Y. Logarithmic Soblev inequalities on non-compact Riemannian manifolds. *Probab. Theory Relat. Fields 109* (1997), 417–424.
829. Wang, F.-Y. Logarithmic Sobolev inequalities: conditions and counterexamples. *J. Operator Theory 46*, 1 (2001), 183–197.
830. Wang, F.-Y. Transportation cost inequalities on path spaces over Riemannian manifolds. *Illinois J. Math. 46*, 4 (2002), 1197–1206.
831. Wang, F.-Y. Probability distance inequalities on Riemannian manifolds and path spaces. *J. Funct. Anal. 206*, 1 (2004), 167–190.
832. Wang, X.-J. Some counterexamples to the regularity of Monge–Ampère equations. *Proc. Amer. Math. Soc. 123*, 3 (1995), 841–845.
833. Wang, X.-J. On the design of a reflector antenna. *Inverse Problems 12*, 3 (1996), 351–375.
834. Wang, X.-J. On the design of a reflector antenna. II. *Calc. Var. Partial Differential Equations 20*, 3 (2004), 329–341.

835. WANG, X.-J. Schauder estimates for elliptic and parabolic equations. *Chin. Ann. Math. 27B*, 6 (2006), 637–642.
836. WERNER, R. F. The uncertainty relation for joint measurement of position and momentum. *Quantum Information and Computing (QIC) 4*, 6–7 (2004), 546–562. Archived online at arxiv.org/abs/quant-ph/0405184.
837. WOLANSKY, G. On time reversible description of the process of coagulation and fragmentation. To appear in *Arch. Ration. Mech. Anal.*
838. WOLANSKY, G. Extended least action principle for steady flows under a prescribed flux. To appear in *Calc. Var. Partial Differential Equations.*
839. WOLANSKY, G. Minimizers of Dirichlet functionals on the n-torus and the weak KAM theory. Preprint, 2007. Available online at www.math.technion.ac.il/~gershonw.
840. WOLFSON, J. G. Minimal Lagrangian diffeomorphisms and the Monge–Ampère equation. *J. Differential Geom. 46*, 2 (1997), 335–373.
841. WU, L. Poincaré and transportation inequalities for Gibbs measures under the Dobrushin uniqueness condition. *Ann. Probab. 34*, 5 (2006), 1960–1989.
842. WU, L. A simple inequality for probability measures and applications. Preprint, 2006.
843. WU, L., AND ZHANG, Z. Talagrand's T_2-transportation inequality w.r.t. a uniform metric for diffusions. *Acta Math. Appl. Sin. Engl. Ser. 20*, 3 (2004), 357–364.
844. WU, L., AND ZHANG, Z. Talagrand's T_2-transportation inequality and log-Sobolev inequality for dissipative SPDEs and applications to reaction-diffusion equations. *Chinese Ann. Math. Ser. B 27*, 3 (2006), 243–262.
845. YUKICH, J. E. Optimal matching and empirical measures. *Proc. Amer. Math. Soc. 107*, 4 (1989), 1051–1059.
846. ZHU, S. The comparison geometry of Ricci curvature. In *Comparison geometry* (Berkeley, CA, 1993–94), vol. 30 of *Math. Sci. Res. Inst. Publ.*, Cambridge University Press, Cambridge, 1997, pp. 221–262.

List of short statements

List of figures

Index

Some notable cost functions

In the following table I have compiled a few remarkable cost functions which have appeared in various applications. The list is by no means exhaustive and the suggested references should be considered just as entry points to the corresponding literature.

Cost	Setting	Use	Where quoted
$\|x-y\|$	$\mathbb{R}^2$ or $\mathbb{R}^3$	Monge's original cost function shape optimization, sandpile growth, compression molding	[20, 636] [328]
$d(x,y)$	Polish space	Kantorovich's cost function definition of Kantorovich distance/norm	[501] Chapter 6
$1_{x \neq y}$	Polish space	representation of total variation	[814, Appendix 1.4]
$1_{d(x,y) \geq \varepsilon}$	Polish space	Strassen's duality theorem	[814, Appendix 1.4]
$d(x,y)^p$	Polish space	p-Wasserstein distances	Chapter 6
$\|x-y\|^2$	$\mathbb{R}^n$	Tanaka's study of Boltzmann equation Brenier's study of incompressible fluids most useful for geometric applications in $\mathbb{R}^n$ diffusion equations of Fokker–Planck type semi-geostrophic equations	[777] [814, Section 7.5] [24, 154, 156, 159] [814, Section 3.2] [814, Chapter 6] [493, 669, 671] [269]
$-\log(1-x\cdot y)$	S^2	Far-field reflector antenna design	[419, 834]
$-\log(1-\epsilon(x)\cdot y)$	$x \in$ surface, $y \in S^2$	Near-field reflector antenna design	[660]
$\log\langle x,y\rangle_+$	S^n	prescribed integral curvature problem	[659]
$-\log\|x-y\|$	$\mathbb{R}^n$	(flat) conformal geometry	[520, 788]
$1-\sqrt{1-\|x-y\|^2}$	$\mathbb{R}^3$ $\mathbb{R}^2$	relativistic theory Rubinstein–Wolansky's design of lens	[164] [712]
$\sqrt{1+\|x-y\|^2}$	$\mathbb{R}^3$	relativistic heat equation	[39, 40, 164]
$\dfrac{(x_1-y_1)^2+(x_2-y_2)^2}{f(x_3-y_3)}$	$\mathbb{R}^3$	semi-geostrophic equations	[268]
$\mathrm{erf}(\alpha\|x-y\|)$	$\mathbb{R}$	Hsu–Sturm's maximal coupling of Brownian paths	[484]
$\|x-y\|^\beta, 0<\beta<1$	$\mathbb{R}$ or $\mathbb{R}^2$	modeling in economy	[399]

Cost	Setting	Use	Where quoted
$d(x,y)^2$	Riemannian manifold	Riemannian geometry	[246, 616] and Part II
$\sum_i \min(d(x_i,y_i), d(x_i,y_i)^2)$	product metric space	Talagrand's study of exponential concentration	[774] and Chapter 22
$\inf \int L(x,v,t)\,dt$	Riemannian manifold	Mather's theory of Lagrangian mechanics	[104, 105, 601] and Chapter 8
$\inf \int \Big(\frac{\lvert v\rvert^2}{2} - p(t,x)\Big)\,dt$	subset of $\mathbb{R}^n$	incompressible Euler equation	[24, 25, 155]
$\frac{1}{2}\inf \int (\lvert v\rvert^2 + \mathrm{Scal}(t,x))\,dt$ and variants ($\mathcal{L}_0$, $\mathcal{L}_\pm$)	Riemannian manifold	Study of Ricci flow	[576, 782]
$d(x,y)^2$	geodesic space	Lott–Sturm–Villani's nonsmooth Ricci curvature bounds	[577, 762, 763] and Part III

Grundlehren der mathematischen Wissenschaften

A Series of Comprehensive Studies in Mathematics

A Selection

247. Suzuki: Group Theory I
248. Suzuki: Group Theory II
249. Chung: Lectures from Markov Processes to Brownian Motion
250. Arnold: Geometrical Methods in the Theory of Ordinary Differential Equations
251. Chow/Hale: Methods of Bifurcation Theory
252. Aubin: Nonlinear Analysis on Manifolds. Monge-Ampère Equations
253. Dwork: Lectures on ρ-adic Differential Equations
254. Freitag: Siegelsche Modulfunktionen
255. Lang: Complex Multiplication
256. Hörmander: The Analysis of Linear Partial Differential Operators I
257. Hörmander: The Analysis of Linear Partial Differential Operators II
258. Smoller: Shock Waves and Reaction-Diffusion Equations
259. Duren: Univalent Functions
260. Freidlin/Wentzell: Random Perturbations of Dynamical Systems
261. Bosch/Güntzer/Remmert: Non Archimedian Analysis – A System Approach to Rigid Analytic Geometry
262. Doob: Classical Potential Theory and Its Probabilistic Counterpart
263. Krasnosel'skiĭ/Zabreĭko: Geometrical Methods of Nonlinear Analysis
264. Aubin/Cellina: Differential Inclusions
265. Grauert/Remmert: Coherent Analytic Sheaves
266. de Rham: Differentiable Manifolds
267. Arbarello/Cornalba/Griffiths/Harris: Geometry of Algebraic Curves, Vol. I
268. Arbarello/Cornalba/Griffiths/Harris: Geometry of Algebraic Curves, Vol. II
269. Schapira: Microdifferential Systems in the Complex Domain
270. Scharlau: Quadratic and Hermitian Forms
271. Ellis: Entropy, Large Deviations, and Statistical Mechanics
272. Elliott: Arithmetic Functions and Integer Products
273. Nikol'skiĭ: Treatise on the shift Operator
274. Hörmander: The Analysis of Linear Partial Differential Operators III
275. Hörmander: The Analysis of Linear Partial Differential Operators IV
276. Liggett: Interacting Particle Systems
277. Fulton/Lang: Riemann-Roch Algebra
278. Barr/Wells: Toposes, Triples and Theories
279. Bishop/Bridges: Constructive Analysis
280. Neukirch: Class Field Theory
281. Chandrasekharan: Elliptic Functions
282. Lelong/Gruman: Entire Functions of Several Complex Variables
283. Kodaira: Complex Manifolds and Deformation of Complex Structures
284. Finn: Equilibrium Capillary Surfaces
285. Burago/Zalgaller: Geometric Inequalities
286. Andrianaov: Quadratic Forms and Hecke Operators
287. Maskit: Kleinian Groups
288. Jacod/Shiryaev: Limit Theorems for Stochastic Processes
289. Manin: Gauge Field Theory and Complex Geometry

290. Conway/Sloane: Sphere Packings, Lattices and Groups
291. Hahn/O'Meara: The Classical Groups and K-Theory
292. Kashiwara/Schapira: Sheaves on Manifolds
293. Revuz/Yor: Continuous Martingales and Brownian Motion
294. Knus: Quadratic and Hermitian Forms over Rings
295. Dierkes/Hildebrandt/Küster/Wohlrab: Minimal Surfaces I
296. Dierkes/Hildebrandt/Küster/Wohlrab: Minimal Surfaces II
297. Pastur/Figotin: Spectra of Random and Almost-Periodic Operators
298. Berline/Getzler/Vergne: Heat Kernels and Dirac Operators
299. Pommerenke: Boundary Behaviour of Conformal Maps
300. Orlik/Terao: Arrangements of Hyperplanes
301. Loday: Cyclic Homology
302. Lange/Birkenhake: Complex Abelian Varieties
303. DeVore/Lorentz: Constructive Approximation
304. Lorentz/v. Golitschek/Makovoz: Construcitve Approximation. Advanced Problems
305. Hiriart-Urruty/Lemaréchal: Convex Analysis and Minimization Algorithms I. Fundamentals
306. Hiriart-Urruty/Lemaréchal: Convex Analysis and Minimization Algorithms II. Advanced Theory and Bundle Methods
307. Schwarz: Quantum Field Theory and Topology
308. Schwarz: Topology for Physicists
309. Adem/Milgram: Cohomology of Finite Groups
310. Giaquinta/Hildebrandt: Calculus of Variations I: The Lagrangian Formalism
311. Giaquinta/Hildebrandt: Calculus of Variations II: The Hamiltonian Formalism
312. Chung/Zhao: From Brownian Motion to Schrödinger's Equation
313. Malliavin: Stochastic Analysis
314. Adams/Hedberg: Function spaces and Potential Theory
315. Bürgisser/Clausen/Shokrollahi: Algebraic Complexity Theory
316. Saff/Totik: Logarithmic Potentials with External Fields
317. Rockafellar/Wets: Variational Analysis
318. Kobayashi: Hyperbolic Complex Spaces
319. Bridson/Haefliger: Metric Spaces of Non-Positive Curvature
320. Kipnis/Landim: Scaling Limits of Interacting Particle Systems
321. Grimmett: Percolation
322. Neukirch: Algebraic Number Theory
323. Neukirch/Schmidt/Wingberg: Cohomology of Number Fields
324. Liggett: Stochastic Interacting Systems: Contact, Voter and Exclusion Processes
325. Dafermos: Hyperbolic Conservation Laws in Continuum Physics
326. Waldschmidt: Diophantine Approximation on Linear Algebraic Groups
327. Martinet: Perfect Lattices in Euclidean Spaces
328. Van der Put/Singer: Galois Theory of Linear Differential Equations
329. Korevaar: Tauberian Theory. A Century of Developments
330. Mordukhovich: Variational Analysis and Generalized Differentiation I: Basic Theory
331. Mordukhovich: Variational Analysis and Generalized Differentiation II: Applications
332. Kashiwara/Schapira: Categories and Sheaves. An Introduction to Ind-Objects and Derived Categories
333. Grimmett: The Random-Cluster Model
334. Sernesi: Deformations of Algebraic Schemes
335. Bushnell/Henniart: The Local Langlands Conjecture for GL(2)
336. Gruber: Convex and Discrete Geometry
337. Maz'ya/Shaposhnikova: Theory of Sobolev Multipliers. With Applications to Differential and Integral Operators
338. Villani: Optimal Transport: Old and New

www.ingramcontent.com/pod-product-compliance
Lightning Source LLC
Chambersburg PA
CBHW051257050925
31950CB00075B/147
* 9 7 8 3 5 4 0 7 1 0 4 9 3 *